Inverse Function
A function g is the inverse of a function f if
$$f(g(x)) = x \text{ and } g(f(x)) = x.$$

Even and Odd Functions
Even: $f(-x) = f(x)$;   Symmetry: y-axis
Odd: $f(-x) = -f(x)$;   Symmetry: origin

Shifted Graphs of a Function ($c > 0$)
Original graph: $y = f(x)$
  $y = f(x) + c$, shifted up c units
  $y = f(x) - c$, shifted down c units
  $y = f(x + c)$, shifted to left c units
  $y = f(x - c)$, shifted to right c units

Reflected Graphs
Original graph: $y = f(x)$
  $y = -f(x)$, reflected in x-axis
  $y = f(-x)$, reflected in y-axis

Factor Theorem
A number c is a zero of a polynomial function $f(x)$ if and only if $x - c$ is a factor of $f(x)$.

Fundamental Theorem of Algebra
A polynomial function $f(x)$ of degree $n > 0$ has exactly n zeros, where a zero of multiplicity m is counted m times.

Exponential and Logarithmic Functions
Exponential: $f(x) = b^x$, $b > 0$, $b \neq 1$
Logarithmic: $f(x) = \log_b x$, $x > 0$

The Number e
$e = 2.718281828459\ldots$

Properties of Logarithms
$y = \log_b x$ if and only if $b^y = x$
$\log_b b = 1$ $\log_b 1 = 0$
$b^{\log_b x} = x$ $\log_e x = \ln x$

Changing from Base b to Base e
$$\log_b x = \frac{\ln x}{\ln b}$$

Laws of Logarithms ($M > 0$, $N > 0$)
$\log_b MN = \log_b M + \log_b N$
$\log_b \dfrac{M}{N} = \log_b M - \log_b N$
$\log_b N^c = c \log_b N$, c a real number

Conic Sections
Parabola ...
Ellipse ... frequency
axis e... inor-

$$\frac{x^2}{a^2} + \frac{y^2}{b^2} = 1 \text{ where } b^2 = a^2 - c^2$$

Hyperbola with foci $(\pm c, 0)$ and vertices $(\pm a, 0)$:
$$\frac{x^2}{a^2} - \frac{y^2}{b^2} = 1 \text{ where } b^2 = c^2 - a^2$$

Trigonometric Functions

of Acute Angles
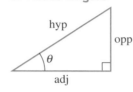
$\sin \theta = \dfrac{\text{opp}}{\text{hyp}}$ $\csc \theta = \dfrac{\text{hyp}}{\text{opp}}$
$\cos \theta = \dfrac{\text{adj}}{\text{hyp}}$ $\sec \theta = \dfrac{\text{hyp}}{\text{adj}}$
$\tan \theta = \dfrac{\text{opp}}{\text{adj}}$ $\cot \theta = \dfrac{\text{adj}}{\text{opp}}$

of General Angles
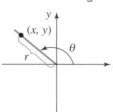
$\sin \theta = \dfrac{y}{r}$ $\csc \theta = \dfrac{r}{y}$
$\cos \theta = \dfrac{x}{r}$ $\sec \theta = \dfrac{r}{x}$
$\tan \theta = \dfrac{y}{x}$ $\cot \theta = \dfrac{x}{y}$

of Real Numbers

$\sin t = y$ $\csc t = \dfrac{1}{y}$
$\cos t = x$ $\sec t = \dfrac{1}{x}$
$\tan t = \dfrac{y}{x}$ $\cot t = \dfrac{x}{y}$

Values of Sine and Cosine for Special Angles

Algebra and Trigonometry
Third Edition

THE JONES & BARTLETT LEARNING SERIES IN MATHEMATICS

Geometry

Geometry with an Introduction to Cosmic Topology
Hitchman (978-0-7637-5457-0) © 2009

Euclidean and Transformational Geometry: A Deductive Inquiry
Libeskind (978-0-7637-4366-6) © 2008

A Gateway to Modern Geometry: The Poincaré Half-Plane, Second Edition
Stahl (978-0-7637-5381-8) © 2008

Understanding Modern Mathematics
Stahl (978-0-7637-3401-5) © 2007

Lebesgue Integration on Euclidean Space, Revised Edition
Jones (978-0-7637-1708-7) © 2001

Precalculus

Essentials of Precalculus with Calculus Previews, Fifth Edition
Zill/Dewar (978-1-4496-1497-3) © 2012

Algebra and Trigonometry, Third Edition
Zill/Dewar (978-0-7637-5461-7) © 2012

College Algebra, Third Edition
Zill/Dewar (978-1-4496-0602-2) © 2012

Trigonometry, Third Edition
Zill/Dewar (978-1-4496-0604-6) © 2012

Precalculus: A Functional Approach to Graphing and Problem Solving, Sixth Edition
Smith (978-0-7637-5177-7) © 2012

Precalculus with Calculus Previews (Expanded Volume), Fourth Edition
Zill/Dewar (978-0-7637-6631-3) © 2010

Calculus

Single Variable Calculus: Early Transcendentals, Fourth Edition
Zill/Wright (978-0-7637-4965-1) © 2011

Multivariable Calculus, Fourth Edition
Zill/Wright (978-0-7637-4966-8) © 2011

Calculus: Early Transcendentals, Fourth Edition
Zill/Wright (978-0-7637-5995-7) © 2011

Multivariable Calculus
Damiano/Freije (978-0-7637-8247-4) © 2011

Calculus: The Language of Change
Cohen/Henle (978-0-7637-2947-9) © 2005

Applied Calculus for Scientists and Engineers
Blume (978-0-7637-2877-9) © 2005

Calculus: Labs for Mathematica
O'Connor (978-0-7637-3425-1) © 2005

Calculus: Labs for MATLAB®
O'Connor (978-0-7637-3426-8) © 2005

Linear Algebra

Linear Algebra: Theory and Applications, Second Edition
Cheney/Kincaid (978-1-4496-1352-5) © 2012

Linear Algebra with Applications, Seventh Edition
Williams (978-0-7637-8248-1) © 2011

Linear Algebra with Applications, Alternate Seventh Edition
Williams (978-0-7637-8249-8) © 2011

Advanced Engineering Mathematics

A Journey into Partial Differential Equations
Bray (978-0-7637-7256-7) © 2012

Advanced Engineering Mathematics, Fourth Edition
Zill/Wright (978-0-7637-7966-5) © 2011

An Elementary Course in Partial Differential Equations, Second Edition
Amaranath (978-0-7637-6244-5) © 2009

Complex Analysis

Complex Analysis for Mathematics and Engineering, Sixth Edition
Mathews/Howell (978-1-4496-0445-5) © 2012

A First Course in Complex Analysis with Applications, Second Edition
Zill/Shanahan (978-0-7637-5772-4) © 2009

Classical Complex Analysis
Hahn (978-0-8672-0494-0) © 1996

Real Analysis

Elements of Real Analysis
Denlinger (978-0-7637-7947-4) © 2011

An Introduction to Analysis, Second Edition
Bilodeau/Thie/Keough (978-0-7637-7492-9) © 2010

Basic Real Analysis
Howland (978-0-7637-7318-2) © 2010

Closer and Closer: Introducing Real Analysis
Schumacher (978-0-7637-3593-7) © 2008

The Way of Analysis, Revised Edition
Strichartz (978-0-7637-1497-0) © 2000

Topology

Foundations of Topology, Second Edition
Patty (978-0-7637-4234-8) © 2009

Discrete Mathematics and Logic

Essentials of Discrete Mathematics, Second Edition
Hunter (978-1-4496-0442-4) © 2012

Discrete Structures, Logic, and Computability, Third Edition
Hein (978-0-7637-7206-2) © 2010

Logic, Sets, and Recursion, Second Edition
Causey (978-0-7637-3784-9) © 2006

Numerical Methods

Numerical Mathematics
Grasselli/Pelinovsky (978-0-7637-3767-2) © 2008

Exploring Numerical Methods: An Introduction to Scientific Computing Using MATLAB®
Linz (978-0-7637-1499-4) © 2003

Advanced Mathematics

Mathematical Modeling with Excel®
Albright (978-0-7637-6566-8) © 2010

Clinical Statistics: Introducing Clinical Trials, Survival Analysis, and Longitudinal Data Analysis
Korosteleva (978-0-7637-5850-9) © 2009

Harmonic Analysis: A Gentle Introduction
DeVito (978-0-7637-3893-8) © 2007

Beginning Number Theory, Second Edition
Robbins (978-0-7637-3768-9) © 2006

A Gateway to Higher Mathematics
Goodfriend (978-0-7637-2733-8) © 2006

For more information on this series and its titles, please visit us online at http://www.jblearning.com. Qualified instructors, contact your Publisher's Representative at 1-800-832-0034 or info@jblearning.com to request review copies for course consideration.

THE JONES & BARTLETT LEARNING INTERNATIONAL SERIES IN MATHEMATICS

Linear Algebra: Theory and Applications, Second Edition, International Version
Cheney/Kincaid (978-1-4496-2731-7) © 2012

Multivariable Calculus
Damiano/Freije (978-0-7637-8247-4) © 2012

Complex Analysis for Mathematics and Engineering, Sixth Edition, International Version
Mathews/Howell (978-1-4496-2870-3) © 2012

A Journey into Partial Differential Equations
Bray (978-0-7637-7256-7) © 2012

Functions of Mathematics in the Liberal Arts
Johnson (978-0-7637-8116-3) © 2013

Advanced Engineering Mathematics, Fourth Edition, International Version
Zill/Wright (978-0-7637-7994-8) © 2011

Calculus: Early Transcendentals, Fourth Edition, International Version
Zill/Wright (978-0-7637-8652-6) © 2011

Real Analysis
Denlinger (979-0-7637-7947-4) © 2011

Mathematical Modeling for the Scientific Method
Pravica/Spurr (978-0-7637-7946-7) © 2011

Mathematical Modeling with Excel®
Albright (978-0-7637-6566-8) © 2010

An Introduction to Analysis, Second Edition
Bilodeau/Thie/Keough (978-0-7637-7492-9) © 2010

Basic Real Analysis
Howland (978-0-7637-7318-2) © 2010

For more information on this series and its titles, please visit us online at http://www.jblearning.com. Qualified instructors, contact your Publisher's Representative at 1-800-832-0034 or info@jblearning.com to request review copies for course consideration.

Algebra and Trigonometry
Third Edition

Dennis G. Zill
Loyola Marymount University

Jacqueline M. Dewar
Loyola Marymount University

JONES & BARTLETT
LEARNING

World Headquarters
Jones & Bartlett Learning
40 Tall Pine Drive
Sudbury, MA 01776
978-443-5000
info@jblearning.com
www.jblearning.com

Jones & Bartlett Learning Canada
6339 Ormindale Way
Mississauga, Ontario L5V 1J2
Canada

Jones & Bartlett Learning International
Barb House, Barb Mews
London W6 7PA
United Kingdom

Jones & Bartlett Learning books and products are available through most bookstores and online booksellers. To contact Jones & Bartlett Learning directly, call 800-832-0034, fax 978-443-8000, or visit our website, www.jblearning.com.

Substantial discounts on bulk quantities of Jones & Bartlett Learning publications are available to corporations, professional associations, and other qualified organizations. For details and specific discount information, contact the special sales department at Jones & Bartlett Learning via the above contact information or send an email to specialsales@jblearning.com.

Copyright © 2012 by Jones & Bartlett Learning, LLC

All rights reserved. No part of the material protected by this copyright may be reproduced or utilized in any form, electronic or mechanical, including photocopying, recording, or by any information storage and retrieval system, without written permission from the copyright owner.

Production Credits
Chief Executive Officer: Ty Field
President: James Homer
SVP, Chief Operating Officer: Don Jones, Jr.
SVP, Chief Technology Officer: Dean Fossella
SVP, Chief Marketing Officer: Alison M. Pendergast
SVP, Chief Financial Officer: Ruth Siporin
Publisher, Higher Education: Cathleen Sether
Senior Acquisitions Editor: Timothy Anderson
Associate Editor: Melissa Potter
Production Director: Amy Rose
Production Assistant: Sara Fowles
Senior Marketing Manager: Andrea DeFronzo
Associate Photo Researcher: Carolyn Arcabascio
V.P., Manufacturing and Inventory Control: Therese Connell
Composition: Aptara, Inc.
Cover Design: Kristin E. Parker
Cover Image: © William James Warren/Science Faction/Corbis
Printing and Binding: Courier Kendallville
Cover Printing: Courier Kendallville

Library of Congress Cataloging-in-Publication Data
Zill, Dennis G., 1940–
　Algebra and trigonometry / Dennis Zill, Jacqueline Dewar. — 3rd ed.
　　p. cm.
　Includes bibliographical references and index.
　ISBN-13: 978-0-7637-5461-7 (casebd.)
　ISBN-10: 0-7637-5461-7 (casebd.)
　1. Algebra—Textbooks. 2. Trigonometry—Textbooks. I. Dewar, Jacqueline M. II. Title.
　QA154.3.Z55 2011
　512'.13—dc22　　　　　　　　　　　　　　　　　2010030089

6048

Printed in the United States of America
14 13 12 11 10　10 9 8 7 6 5 4 3 2 1

Contents

Preface xiii

R — Review of Basic Algebra   1

- **R.1** The Real Number System   2
- **R.2** The Real Number Line   11
- **R.3** Integer Exponents   17
- **R.4** Radicals   24
- **R.5** Rational Exponents   31
- **R.6** Polynomials and Special Products   36
- **R.7** Factoring Polynomials   45
- **R.8** Rational Expressions   51
  Chapter R Review Exercises   60

1 — Equations and Inequalities   65

- **1.1** Equations   66
- **1.2** Building an Equation from Words   72
- **1.3** Quadratic Equations   81
- **1.4** Complex Numbers   93
- **1.5** Linear Inequalities   99
- **1.6** Absolute-Value Equations and Inequalities   105
- **1.7** Polynomial and Rational Inequalities   110
  Chapter 1 Review Exercises   117

vii

2 Rectangular Coordinate System and Graphs 123

- **2.1** The Rectangular Coordinate System 124
- **2.2** Circles and Graphs 130
- **2.3** Equations of Lines 139
- **2.4** Variation 146
- Chapter 2 Review Exercises 151

3 Functions and Graphs 155

- **3.1** Functions and Graphs 156
- **3.2** Symmetry and Transformations 165
- **3.3** Linear and Quadratic Functions 175
- **3.4** Piecewise-Defined Functions 185
- **3.5** Combining Functions 192
- **3.6** Inverse Functions 201
- **3.7** Building a Function from Words 209
- **3.8** Least Squares Line 216
- Chapter 3 Review Exercises 220

4 Polynomial and Rational Functions 225

- **4.1** Polynomial Functions 226
- **4.2** Division of Polynomial Functions 235
- **4.3** Zeros and Factors of Polynomial Functions 242
- **4.4** Real Zeros of Polynomial Functions 249
- **4.5** Approximating Real Zeros 257
- **4.6** Rational Functions 261
- Chapter 4 Review Exercises 274

5 Exponential and Logarithmic Functions 279

- **5.1** Exponential Functions 280
- **5.2** Logarithmic Functions 287
- **5.3** Exponential and Logarithmic Equations 294

5.4 Exponential and Logarithmic Models 300
5.5 Hyperbolic Functions 312
Chapter 5 Review Exercises 315

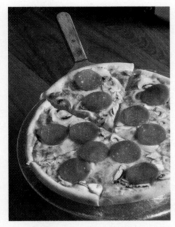

6 Right Triangle Trigonometry 319

6.1 Angles and Their Measure 320
6.2 Right Triangle Trigonometry 328
6.3 Trigonometric Functions of Special Angles 335
6.4 Trigonometric Functions of General Angles 339
Chapter 6 Review Exercises 350

7 Unit Circle Trigonometry 353

7.1 The Circular Functions 354
7.2 Graphs of Sine and Cosine Functions 361
7.3 Graphs of Other Trigonometric Functions 372
7.4 Special Identities 379
7.5 Inverse Trigonometric Functions 389
7.6 Trigonometric Equations 398
Chapter 7 Review Exercises 405

8 Applications of Trigonometry 409

8.1 Solving Right Triangles 410
8.2 Applications of Right Triangles 412
8.3 Law of Sines 418
8.4 Law of Cosines 422
8.5 Simple Harmonic Motion 427
8.6 Trigonometric Form of Complex Numbers 431
8.7 Powers and Roots of Complex Numbers 436
Chapter 8 Review Exercises 442

9 Topics in Analytic Geometry 447

- **9.1** The Parabola 448
- **9.2** The Ellipse 454
- **9.3** The Hyperbola 461
- **9.4** Rotation of Axes 470
- **9.5** Parametric Equations 476
- Chapter 9 Review Exercises 484

10 Polar Coordinates 487

- **10.1** Polar Coordinates 488
- **10.2** Graphs of Polar Equations 492
- **10.3** Conic Sections in Polar Coordinates 502
- **10.4** Vectors in the Plane 508
- **10.5** Dot Product 516
- Chapter 10 Review Exercises 523

11 Systems of Equations and Inequalities 527

- **11.1** Systems of Linear Equations 528
- **11.2** Systems of Nonlinear Equations 537
- **11.3** Partial Fractions 543
- **11.4** Systems of Inequalities 548
- **11.5** Linear Programming 554
- Chapter 11 Review Exercises 561

12 Matrices and Determinants 565

- **12.1** Introduction to Matrices 566
- **12.2** Algebra of Matrices 570
- **12.3** Determinants 581
- **12.4** Inverse of a Matrix 590
- **12.5** Linear Systems: Augmented Matrices 599

12.6 Linear Systems: Matrix Inverses 608

12.7 Linear Systems: Determinants 614

12.8 Cryptography 618

Chapter 12 Review Exercises 622

13 Sequences, Series, and Probability 627

13.1 Sequences 628

13.2 Series 635

13.3 Convergence of Sequences and Series 642

13.4 Mathematical Induction 651

13.5 The Binomial Theorem 655

13.6 Principles of Counting 661

13.7 Introduction to Probability 669

Chapter 13 Review Exercises 677

Final Examination 682

Answers to Selected Odd-Numbered Problems ANS-1

Index I-1

Credits C-1

Learn more about the complete Zill and Dewar series today!
Visit http://go.jblearning.com/precalculus/ for details.

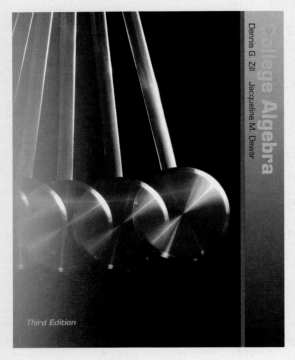

College Algebra
Third Edition
Dennis G. Zill
Jacqueline M. Dewar
ISBN-13: 978-1-4496-0602-2
Hardcover · 568 Pages · © 2012

Brief Table of Contents
Chapter 1 Review of Basic Algebra
Chapter 2 Equations and Inequalities
Chapter 3 Rectangular Coordinate System and Graphs
Chapter 4 Functions and Graphs
Chapter 5 Polynomial and Rational Functions
Chapter 6 Exponential and Logarithmic Functions
Chapter 7 Topics in Analytic Geometry
Chapter 8 Systems of Equations and Inequalities
Chapter 9 Matrices and Determinants
Chapter 10 Sequences, Series, and Probability

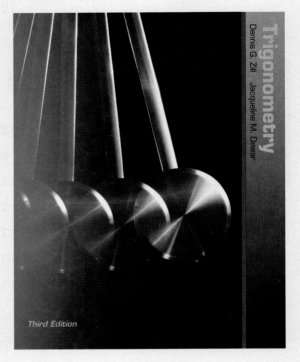

Trigonometry
Third Edition
Dennis G. Zill
Jacqueline M. Dewar
ISBN-13: 978-1-4496-0604-6
Hardcover · 416 pages · © 2012

Brief Table of Contents
Chapter 1 Prerequisites for Trigonometry
Chapter 2 Right Triangle Trigonometry
Chapter 3 Unit Circle Trigonometry
Chapter 4 Applications of Trigonometry
Chapter 5 Complex Numbers and Vectors
Chapter 6 Exponential and Logarithmic Functions
Chapter 7 Topics in Analytic Geometry
Chapter 8 Polar Coordinates

For questions, or to learn more about our complete student and instructor ancillary package, contact your Account Specialist at 1–800–832–0034.

Preface

To the Instructor

☐ **Philosophy** This text reflects our philosophy that a mathematics text at the beginning college level should be readable, straightforward, and loaded with motivation. But ultimately, students can learn mathematics only by doing mathematics. Therefore, throughout this text we have placed a strong emphasis on problem solving as a means of understanding. The examples are designed to motivate, instruct, and guide students. The exercises then give the students an opportunity to test their comprehension, challenge their understanding, and apply their knowledge to real-world situations.

☐ **Audience and Flexibility** We intend this text to provide a treatment of algebra, graphs, functions, logarithms, trigonometry, systems of equations and inequalities, matrices, analytic geometry, polar coordinates, sequences, and probability that is accessible to a college student with two years of high-school mathematics. We have provided sufficient material here for a standard one-semester or two-quarter or even a slower-paced full-year course. This wealth of topics allows the instructor to choose those best suited to the objectives of his/her courses and the backgrounds and abilities of the students. The text can serve as a prerequisite for finite mathematics, statistics, or discrete mathematics. It can also be an introductory course in college mathematics for the liberal arts or business student who plans no further study of mathematics or as a beginning course in a sequence that provides the prerequisites for calculus.

Features in the Text

☐ **Examples** It has been our experience that examples and exercises are the primary learning sources in a mathematics text. We have found that students rely on examples, not theorems and proofs. Therefore we have included numerous examples to illustrate both the theoretical concepts and the computational techniques covered in the text.

☐ **Exercises** As mentioned, we feel that students can learn only by doing. Therefore, in order to promote active participation in problem solving, the exercises are extensive and varied. The exercise sets include an abundance of drill problems, true/false questions, fill-in-the-blank questions, applications, challenging problems, graphing problems, problems that require interpretation of graphs, and discussion problems. This variety of examples gives students the opportunity to solidify their understanding of basic concepts, see practical uses for abstract mathematical ideas, and test their ingenuity. For this third edition we have reorganized and expanded almost all the exercise sets.

☐ **Motivation** While a number of proofs are included, we have typically motivated concepts in an intuitive or geometric manner. In addition, wherever possible we have used figures to illustrate an idea or aid in a solution.

☐ **Emphasis on Functions** Since functions are an essential concept in this course and in mathematics as a whole, we have increased the emphasis on functions and function notation in this edition.

☐ **Emphasis on Graphing** There is a great emphasis on graphing equations and functions. We have stressed symmetry, use of shifted graphs, reflections, intercepts, and interpretation of graphs throughout the text.

New to the Third Edition

☐ **Applications** In this revision we continue to provide applications culled from journals, newspapers, and scientific texts. These "real-life" problems show students the power and usefulness of the mathematics they learn in this course. The applications in this revision span a wide variety of disciplines including astronomy, biology, business, chemistry, ecology, engineering, geology, medicine, meteorology, optics, and physics.

☐ **Annotation Arrows** In the examples we have added many blue-colored annotation arrows within the examples and in the margin to guide the students through the various steps of the solution and to show them how concepts and properties given in theorems and definitions are used in solving a problem. Red-colored annotation arrows in the margin indicate a *Note of Caution*. These cautionary annotations indicate places in the exposition where the student should proceed slowly or even reread the text to avoid common pitfalls and misinterpretations of the material.

☐ **Chapter Openers** Each chapter now opens with its own table of contents. In addition we have provided a motivational discussion of the material and a brief historical account of one or more individuals who had influence on the development of the mathematics in the chapter.

☐ **Chapter R** The first chapter of the text has been renamed *Review of Basic Algebra* and relabeled as Chapter R because we feel that this material is fundamentally topics covered in a high-school algebra course. As such we feel that the instructor can cover this chapter at a review pace, use it as an informal review assignment to the students, or skip it entirely.

☐ **Notes from the Classroom** Selected sections in the text conclude with informal remarks called *Notes from the Classroom*. These remarks are aimed directly at the student and address a wide variety of student/textbook/classroom issues such as alternative terminology, common errors, reinforcement of important concepts, what material is or is not recommended for memorization, solution procedures, use and misuse of calculators, advice on the importance of neatness and organization, misinterpretations, and an occasional word of encouragement.

☐ **Key Concepts** Each chapter concludes with a list of the topics that we feel were most important in the chapter. The students can use this as a checklist in reviewing the material for quizzes and examinations.

☐ **Chapter Review Exercises** To aid the instructor in choosing topics for review or emphasis, we have reorganized each *Chapter Review Exercises* into three distinct parts: Part A are true/false questions, Part B are fill in the blank questions, and Part C consists of traditional problems that review the important topics and concepts covered in the chapter.

☐ **Figures** A word about the numbering of figures, definitions, theorems, and tables is in order. Because of the great number of figures in this text we were motivated to change to a double-decimal numeration system. For example, the interpretation of "Figure 1.2.3" is

$$\underset{\downarrow}{Chapter} \underset{\downarrow}{Section}$$
$$1 \cdot 2 \cdot 3 \leftarrow \text{Third figure in the section}$$

We feel that this type of numeration will make it easier to find figures, definitions, and theorems when they are referred to in later sections or chapters. In addition, to better link a figure with the text, the first textual reference to each figure is done in the same font style and color as the figure number; for example, FIGURE 1.2.3. Also, in this revision all the figures now have brief explanatory captions.

☐ **New Topics** In the bulleted list that follows we indicate some of the changes made in the subject matter.

- Almost all exercise sets now contain problems called *For Discussion*. We hope that instructors will utilize these problems, which are primarily conceptual in nature, and their expertise to engage in a classroom exchange of ideas with the students on how these problems can be solved. These problems could also be the basis for assigned writing projects. To encourage original thought we purposely have not included answers to these problems.
- We have improved the discussion of the inverse functions (Section 3.6) by providing more motivation and clarity with several additional figures.
- Section 3.7, *Building a Function from Words*, is new to Chapter 3.
- Section 3.8, *Least Squares Line*, is also new to Chapter 3. In Section 3.8 we compute the least squares line in the usual algebraic manner. The least squares line concept is covered again from the viewpoint of using an inverse matrix in Section 12.6.
- The chapter on exponential and logarithmic functions has been completely rewritten. Many new mathematical models involving these functions are considered in Section 5.4. The hyperbolic functions are now introduced in the new Section 5.5.

- The section on verification of useless trigonometric identities covered in Section 7.4 of the previous edition has been deleted. We feel that the learning value of such a section is questionable, especially when there are far more important topics that could be covered in greater depth. In this edition, Section 7.4 is devoted to the important Pythagorean identities, the sum and difference formulas, double-angle formulas, and half-angle formulas.
- A new section (Section 8.5), *Simple Harmonic Motion*, has been added to Chapter 8.
- Polar coordinates are now covered in a separate chapter (Chapter 10). The discussion of vectors in the plane have been moved to this chapter.
- Because of its simplicity we have added discussions on rotation of polar graphs to Chapter 10 (Sections 10.1–10.3).
- A new section (Section 10.5), *The Dot Product*, has been added to Chapter 10.
- In Section 12.5, *Linear Systems: Augmented Matrices*, we show how to use elementary row operations on an augmented matrix to balance chemical equations.
- In Section 12.6, *Linear Systems: Matrix Inverses*, we revisit the notion of the least squares line $y = mx + b$. In this section we compute the coefficients m and b using matrix methods.
- A new section (Section 12.8), *Cryptography*, has been added to Chapter 12. This brief section introduces the notions of encoding and decoding messages using matrices. We feel that the students will find this material interesting and perhaps will motivate them to seek further information about this important application of matrices.
- A new section (Section 13.3), *Convergence of Sequences and Series*, has been added to Chapter 13. The discussion of the notion of convergence of a sequence or an infinite series is kept at an intuitive level.
- The section on permutations and combinations in the last edition has been rewritten and is now entitled *Principles of Counting* (Section 13.6).

Supplements

For the Instructor

The following materials are available online, at

http://www.jblearning.com/catalog/9780763754617/

- *Complete Solutions Manual* (*CSM*) prepared by Warren S. Wright and Carol D. Wright.
- *Computerized Testing System* for both Windows® and Mac OS® operating systems. This system allows instructors to create customized tests and quizzes. The questions and answers are sorted by chapter and can be easily installed on a computer. Publisher-supplied .rtf files can also be uploaded to the instructor's Learning Management System.
- *PowerPoint® slides* that feature all labeled figures as they appear in the text. This useful tool allows instructors to easily display and discuss figures and problems found within the textbook.

- *WebAssign*™, developed by instructors for instructors, is a premier independent online teaching and learning environment, guiding several million students through their academic careers since 1997. With WebAssign, instructors can create and distribute algorithmic assignments using questions specific to this textbook. Instructors can also grade, record, and analyze student responses and performance instantly; offer more practice exercises, quizzes, and homework; and upload additional resources to share and communicate with their students seamlessly such as the PowerPoint slides and the test items supplied by Jones & Bartlett Learning.
- *eBook format.* As an added convenience this complete textbook is now available in eBook format for purchase by the student through WebAssign.
- *CourseSmart* is a new way for instructors and students to access this textbook in digital format, anytime from anywhere. Jones & Bartlett Learning has partnered with CourseSmart to make available many of our leading mathematics textbooks in the CourseSmart eTextbook store.

    For more information on CourseSmart Editions please visit http://www.jblearning.com/elearning/econtent/coursesmart/.

Please contact your Jones & Bartlett Learning Account Specialist for information on, access to, and online demonstrations of the supplements and services described above.

For the Student

- *Student Resource Manual (SRM)* prepared by Warren S. Wright and Carol D. Wright. This manual continues to be popular with students using any one of the Zill series of mathematics textbooks. A complete description of the content specific to this text can be found in the preface. Available in both print and online formats, this student manual can be purchased separately or ordered bundled with the textbook at substantial savings.
- *Student Companion Website* is available at www.jblearning.com/catalog/9780763754617/. This online tutorial learning center can be accessed at any time during the term. The resources are tied directly to the text and include: Practice Quizzes, an Online Glossary of Key Terms, and Animated Flashcards.
- *Graphing Calculator Manual* by Jeffery M. Gervasi, EdD of Porterville College, may be ordered through the bookstore or online at http://www.jblearning.com/mathematics/precalculus/.
- *WebAssign Access card* can be bundled with this text or purchased separately by the student online at http://www.webassign.net.
- *eBook with course access card* can also be purchased separately by the student online at http://www.webassign.net/.
- *CourseSmart* is a new way for students to access college textbooks in digital format, anytime from anywhere. Jones & Bartlett Learning has partnered with CourseSmart to make this textbook available in the CourseSmart eTextbook store.

    For students, this CourseSmart Edition has many features designed to make studying more efficient such as highlighting, online search, note-taking, and print capabilities.

    For more information on purchasing CourseSmart Editions please visit http://www.jblearning.com/elearning/econtent/coursesmart/.

Acknowledgments

It was also our good fortune to have the following individuals who either read all (or part) of the subsequent editions or participated in a detailed survey. Their criticisms and many fine suggestions are gratefully acknowledged:

Wayne Andrepont, *University of Southwestern Louisiana*
Nancy Angle, *Colorado School of Mines*
James E. Arnold, *University of Wisconsin—Milwaukee*
Judith Baxter, *University of Illinois—Chicago Circle*
Margaret Blumberg, *Southeastern Louisiana University*
Robert A. Chaffer, *Central Michigan University*
Daniel Drucker, *Wayne State University*
Chris Ennis, *Carleton College*
Jeffrey M. Gervasi, *Porterville College*
E. John Hornsby, *University of New Orleans*
Don Johnson, *New Mexico State University*
Jimmie Lawson, *Louisiana State University*
Gerald Ludden, *Michigan State University*
Stanley M. Lukawecki, *Clemson University*
Richard Marshall, *Eastern Michigan University*
Glenn Mattingly, *Sam Houston State University*
Michael Mays, *West Virginia University*
Phillip R. Montgomery, *University of Kansas*
Bruce Reed, *Virginia Polytechnic Institute and State University*
Jean Rubin, *Purdue University*
Helen Salzberg, *Rhode Island College*
George L. Szoke, *University of Akron*
Darrell Turnbridge, *Kent State University*
Carol Achs, *Mesa Community College*
Joseph Altinger, *Youngstown State University*
Phillip Barker, *University of Missouri—Kansas City*
Wayne Britt, *Louisiana State University*
Kwang Chul Ha, *Illinois State University*
Duane Deal, *Ball State University*
Richard Friedlander, *University of Missouri—St. Louis*
August Garver, *University of Missouri—Rolla*
Irving Katz, *George Washington University*
Janice Kilpatrick, *University of Toledo*
Barbara Meininger, *University of Oregon*
Eldon Miller, *University of Mississippi*
Judith Rollstin, *University of New Mexico*
Monty J. Strauss, *Texas Tech University*
Faye Thames, *Lamar University*
Waldemar Weber, *Bowling Green State University*

We would like to take this opportunity to express our appreciation to Barry A. Cipra for supplying many of the applied problems that appear in the exercise sets and to our colleague Warren S. Wright at Loyola Marymount University for giving us permission to use his material from an earlier edition, for producing the excellent instructor and student manuals, and for his careful proofreading of the first-round page proofs of this edition.

Our warm gratitude goes out to all the good people at Jones & Bartlett Learning who worked on this text. Because of their number, they perforce will remain nameless. But we do want to single out for special thanks Timothy Anderson, senior acquisitions editor, and Amy Rose, production director, for their hard work, cooperation, and patience in making this third edition a reality.

Lastly, all the mistakes in the text are ours. If you run across any of these errors we would greatly appreciate it if you would call them to our attention through our editor at:

tanderson@jblearning.com

Dennis G. Zill

Jacqueline M. Dewar

Review of Basic Algebra

In This Chapter

R.1 The Real Number System
R.2 The Real Number Line
R.3 Integer Exponents
R.4 Radicals
R.5 Rational Exponents
R.6 Polynomials and Special Products
R.7 Factoring Polynomials
R.8 Rational Expressions
   Chapter R Review Exercises

If you wish, you can wait until Chapter 13 to find the value of this fraction.

A Bit of History Most students do not realize that much of the algebraic notation used in algebra textbooks is less than 400 years old.

The greatest French mathematician of the sixteenth century was **François Viète** (1540–1603), a lawyer and a member of Parliament, who devoted most of his leisure time to mathematics. He wrote a number of works on algebra, geometry, and trigonometry, most of which were printed and distributed at his own expense. Viète's most famous work, *In Artem*, significantly advanced algebraic notation. Before Viète's work, it was common practice to use different symbols to represent various powers such as x, x^2, x^3, and so on. Viète, writing in Latin, used the same letter properly qualified for these powers: x, *x quadratum*, *x cubum*, and so on. In addition, Viète expanded the use of letters of the alphabet to denote not only variables but also constant coefficients. Viète's new notation clarified the operations that went into building up a complex series of terms.

Chapter R provides a review of fundamental concepts such as set theory, the real number system, and algebraic notation. This material forms the foundation for the remainder of the text and for any further study of mathematics.

R.1 The Real Number System

Introduction The theory of sets enables us to describe in a very precise way collections of numbers that share a common property. This can be very useful in stating the solutions to certain types of problems. You are no doubt already familiar with most of the following concepts from basic set theory. In this review section we focus on the set of real numbers.

Set Terminology A **set** is a collection of distinct objects. An object in a set is called an **element** of the set. We usually designate a set by a capital letter, such as A or B, and an element of the set by a lowercase letter, such as x. To indicate that x is an element of a set A, we write $x \in A$.

A set can be specified in two ways: by **listing** the elements in the set or by **stating a property** that determines the elements in the set. In each case, braces { } are used. For example, the set consisting of the numbers 5, 10, and 15 can be denoted by either

$$\{5, 10, 15\} \quad \text{or} \quad \{x \mid x = 5n, n = 1, 2, 3\}. \tag{1}$$

The first notation in (1), where the elements of the set are listed, is called the **roster method**. The second notation in (1) is called **set-builder notation**, and in this case, is read "the set of all numbers x such that $x = 5n$, where $n = 1, 2, 3$."

If every element of a set B is also an element of set A, we say that B is a **subset** of A and write

$$B \subset A.$$

It follows that every set is a subset of itself.

A set containing no elements is said to be **empty** and is denoted by the symbol $\varnothing$.

The **union** of two sets A and B is the set of elements belonging to at least one of the sets A or B. In set notation, we write

$$A \cup B = \{x \mid x \in A \quad \text{or} \quad x \in B\}.$$

> In this text, the word *or* should be interpreted to mean that at least one of the properties is true. This allows the possibility that both are true. Thus for the union, if $x \in A \cup B$, then x could be in A and in B.

The **intersection** of two sets A and B is the set of elements common to both sets A and B and is written as

$$A \cap B = \{x \mid x \in A \quad \text{and} \quad x \in B\}.$$

If A and B have no common elements, that is, if $A \cap B = \varnothing$, then the sets are said to be **disjoint**.

EXAMPLE 1   Union and Intersection

If $A = \{1, 2, 3, 4, 5\}$, $B = \{1, 3, 5\}$, $C = \{2, 4, 6\}$, then we have $B \subset A$ because the numbers 1, 3, and 5 are elements in A. Also,

$$A \cup C = \{1, 2, 3, 4, 5, 6\},$$
$$A \cap C = \{2, 4\},$$

and $\quad B \cap C = \varnothing.$ ← Sets B and C have no common elements.

Numbers Recall that the set of **natural numbers**, or **positive integers**, consists of

$$N = \{1, 2, 3, 4, \ldots\}.$$

CHAPTER R REVIEW OF BASIC ALGEBRA

The set N is a subset of the set of **integers**:

$$Z = \{\ldots, -3, -2, -1, 0, 1, 2, 3, \ldots\}.$$

The three dots ($\ldots$) in the sets N and Z are called an **ellipsis** and indicate that the elements follow indefinitely the same pattern as that set by the elements given. The set Z includes both the positive and the negative integers and the number 0, which is neither positive nor negative. The set Z of integers is, in turn, a subset of the set of **rational numbers**:

$$Q = \left\{\frac{p}{q} \,\middle|\, p \text{ and } q \text{ are integers}, q \neq 0\right\}.$$

The set Q consists of all numbers that are quotients of two integers, provided that the denominator is nonzero; for example,

$$\frac{-1}{2}, \quad \frac{17}{5}, \quad \frac{10}{-2} = -5, \quad \frac{22}{7}, \quad \frac{36}{4} = 9, \quad \frac{0}{8} = 0.$$

The quotient p/q is said to be undefined whenever $q = 0$. For example, 8/0 and 0/0 are undefined. ◂ Note of Caution

The set of rational numbers is not sufficient to solve certain elementary algebraic and geometric problems. For example, there is no rational number p/q for which

$$\left(\frac{p}{q}\right)^2 = 2.$$

See Problem 69 in Exercises R.1. Thus we cannot use rational numbers to describe the length of a diagonal of a unit square. See **FIGURE R.1.1**. By the Pythagorean theorem, we know that the length of the diagonal d must satisfy

$$d^2 = (1)^2 + (1)^2 = 2.$$

FIGURE R.1.1 Unit square

We write $d = \sqrt{2}$ and call d "the square root of 2." As we have just indicated, $\sqrt{2}$ is not a rational number. It belongs instead to the set of **irrational numbers**, that is, the set of numbers that cannot be expressed as a quotient of two integers. Other examples of irrational numbers are π, $-\sqrt{3}$, $\frac{1}{\sqrt{7}}$ and $\frac{\sqrt{5}}{4}$.

If we denote the set of irrational numbers by the symbol H, then the set of **real numbers** R can be written as the union of two disjoint sets:

$$R = Q \cup H.$$

We also note that the set of real numbers R can be written as the union of three disjoint sets: $R = R^- \cup \{0\} \cup R^+$, where R^- is the set of **negative** real numbers and R^+ is the set of **positive** real numbers. Elements of the set $\{0\} \cup R^+$ are called **nonnegative** real numbers.

The chart in **FIGURE R.1.2** summarizes the relationship between some principal sets of real numbers.

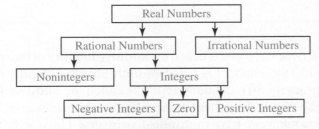

FIGURE R.1.2 Real numbers are either rational or irrational

☐ **Decimals** Every real number can be written in **decimal form**. For example,

$$\frac{1}{4} = 0.25, \qquad \frac{25}{7} = 3.571428571428\ldots,$$

$$\frac{7}{3} = 2.3333\ldots, \qquad \pi = 3.14159265\ldots.$$

Numbers such as 0.25 and 1.6 are said to be **terminating decimals**, whereas numbers such as

$$1.3\overset{\text{repeats}}{\overbrace{2323}}2\ldots \quad \text{and} \quad 3.5\overset{\text{repeats}}{\overbrace{71428}}571428\ldots \qquad (2)$$

are called **repeating decimals**. A repeating decimal, such as $1.323232\ldots$, is often written as $1.\overline{32}$, where the bar indicates the number or numbers that repeat. It can be shown that every rational number has either a repeating or a terminating decimal representation. Conversely, every repeating or terminating decimal is a rational number. Thus the two numbers in (2) are rational. It is also a basic fact that every decimal number is a real number. It follows, then, that the set of irrational numbers consists of all decimals that neither terminate nor repeat. Thus π and $\sqrt{2}$ have nonrepeating and nonterminating decimal representations.

The number π is a nonterminating and nonrepeating decimal

☐ **The Real Number System** The set R of real numbers together with the operations of addition and multiplication is called the **real number system**. The basic rules of algebra for this system enable us to express mathematical facts in simple, concise forms and to solve equations to find answers to mathematical questions. The **basic properties** of the real number system with respect to the operations of *addition* (denoted by $+$) and *multiplication* (denoted by $\cdot$ or $\times$) are given next. In each list the letters a, b, and c denote real numbers.

BASIC PROPERTIES OF REAL NUMBERS

Addition	Multiplication
1. Closure Properties:	
(i) $a + b$ is a real number	(ii) $a \cdot b$ is a real number
2. Commutive Properties:	
(i) $a + b = b + a$	(ii) $a \cdot b = b \cdot a$
3. Associative Properties:	
(i) $a + (b + c) = (a + b) + c$	(ii) $a \cdot (b \cdot c) = (a \cdot b) \cdot c$
4. Identity Properties:	
(i) $a + 0 = 0 + a = a$	(ii) $a \cdot 1 = 1 \cdot a = a$
5. Inverse Properties:	
(i) $a + (-a) = (-a) + a = 0$	(ii) $a \cdot \dfrac{1}{a} = \dfrac{1}{a} \cdot a = 1$

In property 4(*i*) the number 0 is called the **additive identity** of the real number system; in property 4(*ii*) the number 1 is called the **multiplicative identity** of the real number system. In property 5(*i*), the number $-a$ is called the **additive inverse**, or the **negative**, of the number a. Every real number has an additive inverse, but in property 5(*ii*) every *nonzero* number a has a **multiplicative inverse** $1/a$, $a \neq 0$. The multiplicative inverse of the nonzero number a is also called the **reciprocal** of a.

CHAPTER R REVIEW OF BASIC ALGEBRA

EXAMPLE 2   Inverses

(a) The additive inverse of 10 is -10.
(b) The additive inverse of $-\frac{1}{2}$ is $\frac{1}{2}$.
(c) The multiplicative inverse, or reciprocal, of 7 is $\frac{1}{7}$.
(d) The multiplicative inverse, or reciprocal, of $\frac{2}{3}$ is $\dfrac{1}{\frac{2}{3}} = \frac{3}{2}$.   ≡

The **distributive property** of real numbers combines the two operations of addition and multiplication. As a matter of course, the product $a \cdot b$ of two real numbers a and b is usually written without the multiplication dot, that is, as ab.

BASIC PROPERTIES OF REAL NUMBERS (CONTINUED)

6. Distributive Properties:
    (i) $a(b + c) = ab + ac$ (ii) $(a + b)c = ac + bc$

The distributive property can be extended to include more than two numbers in the sum. For example,

$$a(b + c + d) = ab + ac + ad$$

and

$$(a + b + c + d)e = ae + be + ce + de.$$

EXAMPLE 3   Recognizing the Properties

State one basic algebraic property of the real number system to justify each of the following statements, where x, y, and z represent real numbers.

(a) $(6 + 8)y = y(6 + 8)$ (b) $(3 + 5) + 2 = 3 + (5 + 2)$
(c) $(x + 3)y + 2 = (xy + 3y) + 2$ (d) $(x + y) \cdot 1 = x + y$
(e) $(x + 2) + [-(x + 2)] = 0$ (f) $(y + z)\dfrac{1}{y + z} = 1$,   if $y + z \neq 0$

Solution
(a) Commutative property of multiplication ← property 2(ii)
(b) Associative property of addition ← property 3(i)
(c) Distributive property ← property 6(ii)
(d) Identity property of multiplication ← Property 4(ii)
(e) Inverse property of addition ← Property 5(i)
(f) Inverse property of multiplication ← property 5(ii)   ≡

It is possible to define the operations of **subtraction** and **division** in terms of addition and multiplication, respectively.

DEFINITION R.1.1   Difference and Quotient

For real numbers a and b the **difference**, $a - b$, is defined as

$$a - b = a + (-b).$$

If $b \neq 0$, then the **quotient**, $a \div b$, is defined as

$$a \div b = a \cdot \left(\frac{1}{b}\right) = \frac{a}{b}.$$

In the quotient a/b, a is called the **numerator** and b is called the **denominator**. Frequently, the quotient of two real numbers a/b is called a **fraction**. Note that $a \div b$ or a/b is not defined when $b = 0$. Thus $a/0$ is not defined for any real number a. As the following example shows, not all the properties that hold for addition and multiplication are valid for subtraction and division.

EXAMPLE 4   Subtraction is Not Associative

Since $1 - (2 - 3) = 2$ and $(1 - 2) - 3 = -4$, we see that
$$1 - (2 - 3) \neq (1 - 2) - 3.$$
Thus the operation of subtraction is not associative.

Many additional properties of the real numbers can be derived from the basic properties. The following properties will also be used throughout this text.

ADDITIONAL PROPERTIES

7. **Equality Properties:**
   (*i*) If $a = b$, then $a + c = b + c$ for any real number c.
   (*ii*) If $a = b$, then $ac = bc$ for any real number c.
8. **Zero Factor Properties:**
   (*i*) $a \cdot 0 = 0 \cdot a = 0$
   (*ii*) If $a \cdot b = 0$, then either $a = 0$ or $b = 0$, or both.
9. **Cancellation Properties:**
   (*i*) If $ac = bc$, and $c \neq 0$, then $a = b$.
   (*ii*) $\dfrac{ac}{bc} = \dfrac{a}{b}$, provided $c \neq 0$ and $b \neq 0$.

As we will see in Chapter 1, property 8(*ii*) is extremely important in solving certain kinds of equations. For example, if $x(x + 1) = 0$, then we can conclude that either $x = 0$ or $x + 1 = 0$.

EXAMPLE 5   Cancellation

(a) If $2x = 2y$, then $x = y$. ← by property 9(*i*)

(b) $\dfrac{36}{27} = \dfrac{4 \cdot \cancel{9}}{3 \cdot \cancel{9}} = \dfrac{4}{3}$ ← by property 9(*ii*)

> The red slanted bars through a symbol indicates that the symbol is being cancelled.

ADDITIONAL PROPERTIES (CONTINUED)

10. **Properties of Subtraction and Negatives:**
    (*i*) $-(-a) = a$
    (*ii*) $-(ab) = (-a)(b) = a(-b)$
    (*iii*) $-a = (-1)a$
    (*iv*) $(-a)(-b) = ab$

CHAPTER R REVIEW OF BASIC ALGEBRA

EXAMPLE 6 Simplify

Simplify $-(4 + x - y)$.

Solution In view of property 10(*iii*) we can write
$$-(4 + x - y) = (-1)(4 + x - y).$$

Then by the distributive law, property 6(*i*),
$$\begin{aligned}-(4 + x - y) &= (-1)(4 + x - y) \\ &= (-1)4 + (-1)x + (-1)(-y) \quad \leftarrow \text{by properties } 10(iii) \text{ and } 10(iv) \\ &= -4 - x + y.\end{aligned}$$

≡

You should already be familiar with the following list of properties of fractions a/b and c/d, where $b \neq 0$ and $d \neq 0$.

ADDITIONAL PROPERTIES (CONTINUED)

11. **Equivalent Fractions:**
$$\frac{a}{b} = \frac{c}{d} \text{ if and only if } ad = bc$$

12. **Rule of Signs:**
$$-\frac{a}{b} = \frac{-a}{b} = \frac{a}{-b}$$

13. **Addition or Subtraction with Like Denominators:**
$$\frac{a}{b} \pm \frac{c}{b} = \frac{a \pm c}{b}$$

14. **Multiplication:**
$$\frac{a}{b} \cdot \frac{c}{d} = \frac{ac}{bd}$$

15. **Division:**
$$\frac{a}{b} \div \frac{c}{d} = \frac{a/b}{c/d} = \frac{a}{b} \cdot \frac{d}{c} = \frac{ad}{bc}, \quad c \neq 0$$

EXAMPLE 7 Example 2(a) Revisited

The multiplicative inverse, or reciprocal, of $\tfrac{2}{3}$ is $\overbrace{\dfrac{1}{\tfrac{2}{3}}}^{\text{by property 15}} = 1 \cdot \tfrac{3}{2} = \tfrac{3}{2}$ since

$$\overbrace{\frac{2}{3} \cdot \frac{3}{2}}^{\text{by property 14}} = \frac{6}{6} = 1.$$

≡

> **ADDITIONAL PROPERTIES (CONTINUED)**
>
> **16. Division of and by Zero**
>
> (i) $0 \div b = \dfrac{0}{b} = 0, \quad b \neq 0$
>
> (ii) $a \div 0 = \dfrac{a}{0}$ is undefined, $a \neq 0$
>
> (iii) $0 \div 0 = \dfrac{0}{0}$ is undefined

EXAMPLE 8 — Products and Quotients

Evaluate each of the following expressions.

(a) $(-x)(-y)$

(b) $\dfrac{-(-a)}{-b}$

(c) $\dfrac{2(u+v)}{2v}$

(d) $\dfrac{y}{\frac{1}{4} + \frac{3}{5}}$

(e) $z \cdot \dfrac{0}{5}$

(f) $\dfrac{w}{2 - (5-3)}$

Solution

(a) $(-x)(-y) = xy \leftarrow$ by property 10(iv)

(b) $\dfrac{-(-a)}{-b} = \dfrac{a}{-b} = -\dfrac{a}{b} \leftarrow$ by properties 10(i) and 12

(c) $\dfrac{2(u+v)}{2v} = \dfrac{u+v}{v} \leftarrow$ by property 9(ii)

(d) To evaluate $y/(1/4 + 3/5)$, we first evaluate the denominator:

$$\dfrac{1}{4} + \dfrac{3}{5} = \dfrac{(1)(5) + (4)(3)}{(4)(5)} = \dfrac{17}{20}. \leftarrow \text{common denominator}$$

Then we have

$$\dfrac{y}{\frac{1}{4} + \frac{3}{5}} = \dfrac{y}{\frac{17}{20}} = \dfrac{y}{1} \cdot \dfrac{20}{17} = \dfrac{20y}{17}. \leftarrow \text{by property 15}$$

(e) $z \cdot \dfrac{0}{5} = z \cdot 0 = 0 \leftarrow$ by property 8(i)

(f) The expression $w/[2 - (5-3)]$ is undefined, because its denominator is zero; that is, $2 - (5-3) = 2 - 2 = 0$. See property 16(ii).

NOTES FROM THE CLASSROOM

In the solution of part (c) of Example 8, a common mistake (seen too often on students' homework and tests) is to cancel the v's in the numerator and denominator:

$$\dfrac{u + \cancel{v}}{\cancel{v}} = u. \leftarrow \text{INCORRECT}$$

No further cancellation can be performed in the simplification of $\dfrac{2(u+v)}{2v}$, since v is *not a multiplicative factor* of both the numerator and the denominator as required by the cancellation property 9(ii).

CHAPTER R REVIEW OF BASIC ALGEBRA

R.1 Exercises

Answers to selected odd-numbered problems begin on page ANS-1.

In Problems 1–8, find the indicated set if $A = \{1, 4, 6, 8, 10, 15\}$, $B = \{3, 9, 11, 12, 14\}$, and $C = \{1, 2, 5, 7, 8, 13, 14\}$.

1. $A \cup B$
2. $A \cup C$
3. $B \cup C$
4. $A \cap B$
5. $A \cap C$
6. $B \cap C$
7. $(A \cap B) \cup B$
8. $A \cup (B \cup C)$

In Problems 9–12, list the elements of the given set.

9. $\{r \mid r = p/q, p = 1, 2, q = -1, 1\}$
10. $\{t \mid t = 4 + z, z = -1, -3, -5\}$
11. $\{x \mid x = 2y, y = \frac{1}{3}, \frac{2}{3}\}$
12. $\{y \mid y - 5 = 2\}$

In Problems 13–16, use set notation to express the given set.

13. The set of negative integers greater than -3
14. The set of real numbers whose square is 9
15. The set of even integers
16. The set of odd integers

In Problems 17–32, state one of the basic properties of the real number system (properties 1–6) to justify each of the given statements.

17. $(2 + 3) + 5 = 2 + (3 + 5)$
18. $[(1)(2)](3) = [(2)(1)](3)$
19. $(x + y) + 3 = (y + x) + 3$
20. $(a + 2) + \pi = \pi + (a + 2)$
21. $[(-2)(\frac{1}{2})]z = -2[(\frac{1}{2})(z)]$
22. $(1 + 2)(-3) = 1(-3) + 2(-3)$
23. $1 \cdot (\sqrt{2}) = \sqrt{2}$
24. $(3 + 4)(5 + 2) = (3 + 4)5 + (3 + 4)2$
25. $(\frac{1}{5}) \cdot 5 = 1$
26. $\frac{1}{4} + (-\frac{1}{4}) = 0$
27. $x(y + 0) + z = xy + z$
28. $\{3 + [(-5)(1)]\} + 4 = \{3 + (-5)\} + 4$
29. $[(w + 3)2]z = [2(w + 3)]z$
30. $(-13 + z)(2) + 7 = [z + (-13)](2) + 7$
31. $(a - b) + [-(a - b)] = 0$
32. $(x - y)\left(\dfrac{1}{x - y}\right) = 1, x \neq y$

In Problems 33–44, state one of the additional properties of the real number system (properties 7–16) to justify each of the given statements.

33. $(-5)(-x) = 5x$
34. $-(-17) = 17$
35. If $x + 3 = y + 3$, then $x = y$.
36. If $y + z = 5 + z$, then $y = 5$.
37. If $(x + 2)(3) = 4(3)$, then $x + 2 = 4$.
38. If $z^2 = 0$, then $z = 0$.
39. If $(x + 1)(x - 2) = 0$, then $x + 1 = 0$ or $x - 2 = 0$.
40. $(a + b + c) \cdot 0 = 0$
41. $\dfrac{0}{a^2 + 1} = 0$
42. $\dfrac{2(x^2 + 1)}{x^2 + 1} = 2$
43. $\dfrac{x + y}{2} = \dfrac{x}{2} + \dfrac{y}{2}$
44. $\dfrac{-x}{y^2 + 9} = -\dfrac{x}{y^2 + 9}$

In Problems 45–50, simplify the given expression.

45. $-(-a)[2 - 3]$
46. $\dfrac{-(-b)}{-bc}$
47. $\dfrac{4(3 + c)}{4c}$
48. $[(4)(\frac{1}{2})(-\frac{1}{2})](-z) + z$
49. $\dfrac{(14)(0)(x)}{\sqrt{2} - \sqrt{3}}$
50. $(\pi - \pi)(x + y - 3)$

Miscellaneous Applications

51. Old Math The Rhind papyrus (c. 1650 BCE), purchased by the Scottish Egyptologist Alexander Henry Rhind in 1858, is considered one of the best examples of Egyptian mathematics. In it the Egyptians used $\left(\frac{16}{9}\right)^2$ as the value of π.
  (a) Is this approximation greater than or less than π?
  (b) Show that the error in using this approximation is less than 1% of π.

52. Bible Study Using the fact that the circumference of a circle is π times the diameter, determine what value of π is implied from the following biblical quotation. "Also he made the molten sea of ten cubits from brim to brim, round in compass, and the height thereof was five cubits; and a line of thirty cubits did compass it round about." (This is taken from II Chronicles 4:2 and I Kings 7:23, which date from the tenth century BCE)

The Rhind papyrus

For Discussion

For Problems 53–68, answer true or false.

53. $\frac{1}{3}$ is an element of Z. ___
54. $-\frac{1}{2}$ is an element of Q. ___
55. $\sqrt{3}$ is an element of R. ___
56. $\sqrt{2}$ is a rational number. ___
57. 0.1333... is an irrational number. ___
58. 1.5 is a rational number. ___
59. 0.121212... is a rational number. ___
60. $\frac{8}{0}$ is an element of Q. ___
61. -4 is an element of Z, but -4 is not an element of N. ___
62. π is an element of R, but π is not an element of Q. ___
63. Every irrational number is a real number. ___
64. Every integer is a rational number. ___
65. Every decimal number is a real number. ___
66. The intersection of the set of rational numbers and the set of irrational numbers is the empty set. ___
67. If $c \neq 0$, then $(a + b) \div c = (a \div c) + (b \div c)$. ___
68. If $a \neq 0$, $b \neq 0$, and $a + b \neq 0$, then $c \div (a + b) = (c \div a) + (c \div b)$. ___
69. Show that $\sqrt{2}$ cannot be written as a quotient of integers. [*Hint*: Assume that there is a fraction p/q, reduced to lowest terms, such that $(p/q)^2 = 2$. This simplifies to $p^2 = 2q^2$, which implies that p^2, hence p is an even integer, say $p = 2r$. Make this substitution and consider $(2r/q)^2 = 2$. You should arrive at a contradiction to the fact that p/q was reduced to lowest terms.]
70. Discuss: The sum of an irrational number and a rational number must be irrational. [*Hint*: If the sum of the two numbers were rational, then it could be written as a quotient of integers p/q. Why does this lead to a contradiction?]
71. Discuss: Is the sum of two irrational numbers necessarily irrational?
72. Discuss: Is the product of two irrational numbers necessarily irrational?
73. Discuss: Is the quotient of two irrational numbers necessarily irrational?
74. In general, $a + (-b) \neq b + (-a)$. What does this say about the operation of subtraction?
75. Some secret codes work by shifting letters of the alphabet. **FIGURE R.1.3** shows a shift of 2. Each letter in a message can be represented by the digits in a decimal number. For example, the decimal number 0.12121212... codes the message STUDY MATH into TVVFZ OBVI. If using 9/37 produces the coded message RCWJEJQVDU PLXIV, what was the original message?
76. Suppose that the sets A and B have a finite number of elements. Let $n(A)$ and $n(A)$ denote the number of elements in sets A and B, respectively. Discuss why the formula
$$n(A \cup B) = n(A) + n(B) - n(A \cap B)$$
gives the number of elements in the union $A \cup B$.

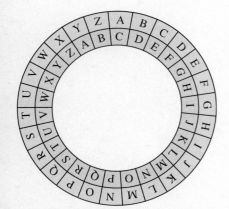

FIGURE R.1.3 Code wheel in Problem 75

R.2 The Real Number Line

≡ **Introduction** For any two distinct real numbers a and b, there is always a third real number between them; for example, their average $(a + b)/2$ is midway between them. Similarly, for any two distinct points A and B on a straight line, there is always a third point between them; for example, the midpoint M of the line segment AB. There are many such similarities between the set R of real numbers and the set of points on a straight line that suggest using a line to "picture" the set of real numbers $R = R^- \cup \{0\} \cup R^+$. This can be done as follows.

FIGURE R.2.1 Number line

□ **Real Number Line** Given any straight line, we choose a point O on the line to represent the number 0. This particular point is called the **origin**. If we now select a line segment of unit length as shown in FIGURE R.2.1, each positive real number x can be represented by the point at a distance x to the *right* of the origin. Similarly, each negative real number $-x$ can be represented by the point at a distance x to the *left* of the origin. This association results in a one-to-one correspondence between the set R of real numbers and the set of points on a straight line, called the **real number line**. For any given point P on the number line, the number p, which corresponds to this point, is called the **coordinate** of P. Thus the set R^- of negative real numbers consists of the coordinates of points to the left of the origin, the set R^+ of negative real numbers consists of the coordinates of points to the right of the origin; the number 0 is the coordinate of the origin O. See FIGURE R.2.2.

FIGURE R.2.2 Positive and negative directions on the number line

In general, we will not distinguish between a point on the number line and its coordinate. Thus, for example, we will sometimes refer to the point on the real number line with coordinate 5 as "the point 5."

□ **Less Than and Greater Than** Two real numbers a and b, $a \neq b$, can be compared by the order relation **less than**. We have the following definition.

DEFINITION R.2.1 Less Than

The real number a is said to be **less than** b, written $a < b$, if and only if the difference $b - a$ is positive.

If a is less than b, then equivalently we can say that b is **greater than** a and write $b > a$. For example, $-7 < 5$, because $5 - (-7) = 12$ is positive. Alternatively, we can write $5 > -7$.

EXAMPLE 1   An Inequality

Using the order relation greater than, compare the real numbers π and $\frac{22}{7}$.

Solution From $\pi = 3.1415\ldots$ and $\frac{22}{7} = 3.1428\ldots$, we find that

$$\frac{22}{7} - \pi = (3.1428\ldots) - (3.1415\ldots) = 0.001\ldots.$$

Since this difference is positive, we conclude that $\frac{22}{7} > \pi$.   ≡

□ **Inequalities** The number line is useful in demonstrating order relations between two real numbers a and b. As shown in FIGURE R.2.3, we say that the number a is **less**

FIGURE R.2.3 The number a is to the left of the number b

than the number b, and write $a < b$, whenever the number a lies to the left of the number b on the number line. Equivalently, because the number b lies to the right of a on the number line we say that b is **greater than** a and write $b > a$. For example, $4 < 9$ is the same as $9 > 4$. We also use the notation $a \leq b$ if the number a is **less than or equal to** the number b. Similarly, $b \geq a$ means b is **greater than or equal to** a. For example, $2 \leq 5$ since $2 < 5$. Also, $4 \geq 4$ because $4 = 4$.

For any two real numbers a and b, exactly *one* of the following is true:

$$a < b, \quad a = b, \quad \text{or} \quad a > b. \tag{1}$$

The property given in (1) is called the **trichotomy law**.

☐ **Terminology** The symbols $<, >, \leq$, and $\geq$ are called **inequality symbols** and expressions such as $a < b$ or $b \geq a$ are called **inequalities**. An inequality $a < b$ is often called a **strict inequality**, whereas an inequality such as $b \geq a$ is called a **nonstrict inequality**. The inequality $a > 0$ means the number a lies to the right of the number 0 on the number line and so a is **positive**. We signify that a number a is **negative** by the inequality $a < 0$. Because the inequality $a \geq 0$ means a is either greater than 0 (positive) or equal to 0 (which is neither positive nor negative), we say that a is **nonnegative**. Similarly, if $a \leq 0$, we say that a is **nonpositive**.

Inequalities also have the following transitivity property.

THEOREM R.2.1 Transitivity Property

If $a < b$ and $b < c$, then $a < c$.

For example, if $x < 12$ and $12 < y$, we conclude from the transitivity property that $x < y$. Theorem R.2.1 can easily be visualized on the number line by placing a anywhere on the line, b to the right of a, and the number c to the right of b.

☐ **Absolute Value** We can also use the real number line to picture distance. As shown in FIGURE R.2.4, the distance from the point 3 to the origin is 3 units, and the distance from the point -3 to the origin is 3, or $-(-3)$, units. It follows from our discussion of the number line that, in general, the distance from any number to the origin is the "unsigned value" of that number.

FIGURE R.2.4 Distance on the number line

FIGURE R.2.5 The distance from 0 to x is x; the distance from 0 to y is $-y$

More precisely, as shown in FIGURE R.2.5, for any positive real number x, the distance from the point x to the origin is x, but for any *negative* number y, the distance from the point y to the origin is $-y$. Of course, for $x = 0$, the distance to the origin is 0. The concept of the distance from a point on the number line to the origin is described by the notion of the **absolute value** of a real number.

DEFINITION R.2.2 Absolute Value

For any real number a, the absolute value of a, denoted by $|a|$, is

$$|a| = \begin{cases} a, & \text{if } a \geq 0, \\ -a, & \text{if } a < 0. \end{cases} \tag{2}$$

EXAMPLE 2   Absolute Values

Since 3 and $\sqrt{2}$ are positive numbers,
$$|3| = 3 \quad \text{and} \quad |\sqrt{2}| = \sqrt{2}.$$
But since -3 and $-\sqrt{2}$ are negative numbers; that is, $-3 < 0$ and $-\sqrt{2} < 0$, we see from (2) that
$$|-3| = -(-3) = 3 \quad \text{and} \quad |-\sqrt{2}| = -(-\sqrt{2}) = \sqrt{2}.$$

EXAMPLE 3   Absolute Values

(a) $|2 - 2| = |0| = 0 \leftarrow$ from (2), $0 \geq 0$
(b) $|2 - 6| = |-4| = -(-4) = 4 \leftarrow$ from (2), $-4 < 0$
(c) $|2| - |-5| = 2 - [-(-5)] = 2 - 5 = -3 \leftarrow$ from (2), $-5 < 0$

EXAMPLE 4   Absolute Value

Find $|\sqrt{2} - 3|$.

Solution To find $|\sqrt{2} - 3|$, we must determine first whether the number $\sqrt{2} - 3$ is positive or negative. Since $\sqrt{2} \approx 1.4$, we see that $\sqrt{2} - 3$ is a negative number. Thus,
$$|\sqrt{2} - 3| = -(\sqrt{2} - 3) = -\sqrt{2} + 3 \quad \leftarrow \text{The distributive law is used here.}$$
$$= 3 - \sqrt{2}.$$

It is a common mistake to think that $-y$ represents a negative number because the symbol y is preceded by a minus sign. We emphasize that if y represents a negative number, then the negative of y, that is, $-y$, is a positive number. Hence, if y is *negative*, then $|y| = -y$.

◀ Note of Caution

EXAMPLE 5   Value of an Absolute Value Expression

Find $|x - 6|$ if (a) $x > 6$, (b) $x = 6$, and (c) $x < 6$.

Solution

(a) If $x > 6$, then $x - 6$ is positive. Then from the definition of absolute value in (2), we conclude that $|x - 6| = x - 6$.
(b) If $x = 6$, then $x - 6 = 0$; hence $|x - 6| = |0| = 0$.
(c) If $x < 6$, then $x - 6$ is negative and we have that $|x - 6| = -(x - 6) = 6 - x$.

For any real number x and its negative, $-x$, the distance to the origin is the same. That is, $|x| = |-x|$. This is one of several special properties of the absolute value, which we list in the following theorem.

THEOREM R.2.2   Properties of Absolute Value

Suppose x and y are real numbers. Then
  (i) $|x| \geq 0$
  (ii) $|x| = 0$ if and only if $x = 0$
  (iii) $|x| = |-x|$
  (iv) $|xy| = |x||y|$
  (v) $\left|\dfrac{x}{y}\right| = \dfrac{|x|}{|y|}, y \neq 0$
  (vi) $|x + y| \leq |x| + |y|$

Restating these properties in words is one way of increasing your understanding of them. For example, property (*i*) states that the absolute value of a quantity is always nonnegative. Property (*iv*) says that the absolute value of a product equals the product of the absolute values of the two factors. Part (*vi*) of Theorem R.2.2 is an important property and is called the **triangle inequality**.

☐ **Distance Between Points** The concept of absolute value not only describes the distance from a point to the origin. It is also useful in finding the distance between two points on the number line. Since we want to describe distance as a positive quantity, we subtract one coordinate from the other and then take the absolute value of the difference. See FIGURE R.2.6.

FIGURE R.2.6 Distance on the number line

DEFINITION R.2.3   Distance on the Number Line

If a and b are two points on the real number line, then the **distance** from a to b is given by

$$d(a, b) = |b - a|. \quad (3)$$

EXAMPLE 6   **Distances**

(a) The distance from -5 to 2 is
$$d(-5, 2) = |2 - (-5)| = |7| = 7.$$

(b) The distance from 3 to $\sqrt{2}$ is
$$d(3, \sqrt{2}) = |\sqrt{2} - 3| = 3 - \sqrt{2}. \quad \leftarrow \text{see Example 4}$$

We see that the distance from a to b is the same as the distance from b to a, since by (*iii*) of Theorem R.2.2,

$$d(a, b) = |b - a| = |-(b - a)| = |a - b| = d(b, a). \quad \leftarrow \begin{smallmatrix}b - a \text{ plays the part of } x \\ \text{in } (iii) \text{ of Theorem R.2.2.}\end{smallmatrix}$$

Thus, $d(a, b) = d(b, a)$.

☐ **Coordinate of the Midpoint** Definition R.2.3 can be used to find an expression for the **midpoint** of a line segment. The midpoint m of the line segment joining a and b is the average of the two endpoints:

$$m = \frac{a + b}{2}. \quad (4)$$

FIGURE R.2.7 Distance from a to m equals the distance from m to b

See FIGURE R.2.7.

EXAMPLE 7   **Midpoint**

From (4), the midpoint of the line segment joining the points 5 and -2 is
$$\frac{5 + (-2)}{2} = \frac{3}{2}.$$

FIGURE R.2.8 Midpoint in Example 7

See FIGURE R.2.8.

EXAMPLE 8 Given the Midpoint

The line segment joining a to b has midpoint $m = 4$. If the distance from a to b is 7, find a and b.

Solution As we can see in FIGURE R.2.9, since m is the midpoint,
$$l = d(a, m) = d(m, b).$$
Thus, $2l = 7$ or $l = \frac{7}{2}$. We now have $a = 4 - \frac{7}{2} = \frac{1}{2}$ and $b = 4 + \frac{7}{2} = \frac{15}{2}$. ≡

FIGURE R.2.9 Distances are equal in Example 8

R.2 Exercises
Answers to selected odd-numbered problems begin on page ANS-1.

In Problems 1 and 2, construct a number line and locate the given points on it.

1. $0, -\frac{1}{2}, 1, -1, 2, -2, \frac{4}{3}, 2.5$
2. $0, 1, -1, \sqrt{2}, -3, -\sqrt{2} + 1$

In Problems 3–10, write the statement as an inequality.

3. x is positive
4. y is negative
5. $x + y$ is nonnegative
6. a is less than -3
7. b is greater than or equal to 100
8. $c - 1$ is less than or equal to 5
9. $|t - 1|$ is less than 50
10. $|s + 4|$ is greater than or equal to 7

In Problems 11–16, compare the pair of numbers using the order relation "less than."

11. $15, -3$
12. $-9, 0$
13. $\frac{4}{3}, 1.33$
14. $-\frac{7}{15}, -\frac{5}{11}$
15. $\pi, 3.14$
16. $1.732, \sqrt{3}$

In Problems 17–22, compare the pair of numbers using the order relation "greater than or equal to."

17. $-2, -7$
18. $-\frac{1}{7}, -0.143$
19. $2.5, \frac{5}{2}$
20. $0.333, \frac{1}{3}$
21. $\frac{423}{157}, 2.6$
22. $\sqrt{2}, 1.414$

In Problems 23–44, find the absolute value.

23. $|7|$
24. $|-7|$
25. $|22|$
26. $\left|\frac{22}{7}\right|$
27. $\left|\frac{-22}{7}\right|$
28. $|\sqrt{5}|$
29. $|-\sqrt{5}|$
30. $|0.13|$
31. $|\pi - 4|$
32. $|2 - 6|$
33. $|6 - 2|$
34. $||2| - |-6||$
35. $|-6| - |-2|$
36. $|\sqrt{5} - 3|$
37. $|3 - \sqrt{5}|$
38. $|8 - \sqrt{7}|$
39. $|\sqrt{7} - 8|$
40. $|-(\sqrt{7} - 8)|$
41. $|\sqrt{5} - 2.3|$
42. $\left|\frac{\pi}{2} - 1.57\right|$
43. $|6.28 - 2\pi|$
44. $|\sqrt{7} - 4.123|$

In Problems 45–56, write the expression without using absolute value symbols.

45. $|h|$, if h is negative
46. $|-h|$, if h is negative
47. $|x - 2|$, if $x < 2$
48. $|x - 2|$, if $x = 2$
49. $|x - 2|$, if $x > 2$
50. $|5 - x|$, if $x < 5$
51. $|5 - x|$, if $x = 5$
52. $|5 - x|$, if $x > 5$
53. $|x - y| - |y - x|$
54. $\dfrac{|x - y|}{|y - x|}, x \neq y$
55. $\dfrac{|h|}{h}, h < 0$
56. $\dfrac{z}{|-z|}, z > 0$

In Problems 57–64, **(a)** find the distance between the given points and **(b)** find the coordinate of the midpoint of the line segment joining the given points.

57. 7, 3
58. 2, 5
59. 0.6, 0.8
60. −100, 255
61. −5, −8
62. 6, −4.5
63. $\frac{3}{2}, -\frac{3}{2}$
64. $-\frac{1}{4}, \frac{7}{4}$

In Problems 65–72, m is the midpoint of the line segment joining a (the left endpoint) and b (the right endpoint). Use the given conditions to find the indicated values.

65. $m = 5, d(a, m) = 3; a, b$
66. $m = -1, d(m, b) = 2; a, b$
67. $m = 2, d(a, b) = 7; a, b$
68. $m = \sqrt{2}, d(a, b) = 1; a, b$
69. $a = 4, d(a, m) = \pi; b, m$
70. $a = 10, d(b, m) = 5; b, m$
71. $b = -3, d(a, b) = \sqrt{2}; a, m$
72. $b = -\frac{3}{2}, d(a, b) = \frac{1}{2}; a, m$

In Problems 73–80, determine which statement from the trichotomy law ($a < b, a = b,$ or $a > b$) is true for the given pair of numbers a, b.

73. (10)(10), 100
74. $\sqrt{3} - 3, 0$
75. $\pi, 3.14$
76. $|-15|, 15$
77. $\frac{7}{11}, 0.\overline{63}$
78. $\frac{2}{9}, 0.2$
79. $\sqrt{2}, 1.4$
80. $-\sqrt{2}, -1.4$

Miscellaneous Applications

81. How Far? Greg, Tricia, Ethan, and Natalie live on Real Street. Tricia lives a mile from Greg, and Ethan lives one-half mile from Tricia. Natalie lives halfway between Ethan and Tricia. How far does Natalie live from Greg? [*Hint*: There are two solutions.]

82. Shipping Distance A company that owned one manufacturing plant next to a river bought two additional manufacturing plants, one x miles upstream and the other y miles downstream. Now the company wants to build a processing plant located so that the total shipping distance from the processing plant should be built at the same location as the original manufacturing plant. [*Hint*: Think of the plants as being located at 0, x, and $-y$ on a number line.] See **FIGURE R.2.10**. Using absolute values, find an expression for the total shipping distance if the processing plant is located at point d.

FIGURE R.2.10 Plants in Problem 82

For Discussion

For Problems 83–90, answer true or false for any real number a.

83. $\dfrac{|a \cdot a|}{|a|} = |a|, a \neq 0$ ____
84. $|a| > -1$ ____
85. $-|a| \leq |a|$ ____
86. $-a \leq a$ ____
87. $a \leq |a|$ ____
88. $-|a| \leq a$ ____
89. If $x < a$ and $a < z$, then $x < z$ ____
90. $|a + 1| \leq |a| + 1$ ____
91. For what values of x is it true that $x \leq |x|$?
92. For what values of x is it true that $x = |x|$?

93. Use Definition R.2.2 to prove that $|xy| = |x||y|$ for any real numbers x and y.
94. Use Definition R.2.2 to prove that $|x/y| = |x|/|y|$ for any real number x and any nonzero real number y.
95. Under what conditions does equality hold in the triangle inequality; that is, when is it true that $|a + b| = |a| + |b|$?
96. Use the triangle inequality to prove $|a - b| \le |a| + |b|$.
97. Use the triangle inequality to prove $|a - b| \ge |a| - |b|$. [*Hint*: $a = (a - b) + b$.]
98. Prove the midpoint formula (4).

R.3 Integer Exponents

≡ **Introduction** We find it more convenient to write a repeated sum such as $x + x + x + x$ as $4x$. Similarly, we can write a repeated product such as $x \cdot x \cdot x$ more efficiently using **exponents**. In this section we review the laws of integer exponents.

We begin with a definition of "x to the nth power."

DEFINITION R.3.1 Positive Integer Power of x

For any real number x and any positive integer n, the symbol x^n represents the product of n factors of x. That is,

$$x^n = \overbrace{x \cdot x \cdots x}^{n \text{ factors of } x}. \tag{1}$$

For example, $x \cdot x \cdot x = x^3$. In the case when $n = 1$ we have $x^1 = x$.

In the expression x^n, n is called the **exponent** or **power** of x, and x is called the **base**.

EXAMPLE 1 Using (1)

(a) $5^2 = 5 \cdot 5 = 25$ ← by (1) of Definition R.3.1 with $x = 5$
(b) $y^3 = y \cdot y \cdot y$ ← by (1) of Definition R.3.1 with x replaced by y
(c) $(2x)^3 = 2x \cdot 2x \cdot 2x = 8x^3$ ← by (1) of Definition R.3.1 with x replaced by $2x$
(d) $(-3)^4 = (-3) \cdot (-3) \cdot (-3) \cdot (-3) = 81$ ← by (1) of Definition R.3.1 with $x = -3$

≡

The exponential statement in part (a) of Example 1 is read "5 squared," whereas in part (b) we say "y cubed."

Negative powers of x are defined next.

DEFINITION R.3.2 Negative Integer Power of x

For any nonzero real number x and any positive integer n, the symbol x^{-n} represents the reciprocal of the product of n factors of x. That is,

$$x^{-n} = \frac{1}{x^n}, \quad x \ne 0. \tag{2}$$

EXAMPLE 2   Using (2)

(a) $2^{-3} = \dfrac{1}{2^3} = \dfrac{1}{2 \cdot 2 \cdot 2} = \dfrac{1}{8}$ ← by (2) with $x = 2$

(b) $\left(-\dfrac{1}{10}\right)^{-4} = \dfrac{1}{\left(-\frac{1}{10}\right)^4}$ ← by (2) with $x = -\frac{1}{10}$

$= \dfrac{1}{\left(-\frac{1}{10}\right) \cdot \left(-\frac{1}{10}\right) \cdot \left(-\frac{1}{10}\right) \cdot \left(-\frac{1}{10}\right)} = \dfrac{1}{\frac{1}{10{,}000}} = 10{,}000.$ ≡

Finally, for any nonzero base x, we define

$$x^0 = 1. \qquad (3)$$

Thus,

$$2^0 = 1 \quad \text{and} \quad (\sqrt{2} + \sqrt{3})^0 = 1.$$

See Problem 93 in Exercises R.3 for a rationale for the special definition (3). Note that 0^0 is not defined.

Note of caution ▶

☐ **Laws of Exponents** Several rules for combining powers, called the **laws of exponents**, have been established. As an example, consider the product $3^2 \cdot 3^4$. By counting factors we see that

$$3^2 \cdot 3^4 = \overbrace{(3 \cdot 3)}^{\text{2 factors}} \overbrace{(3 \cdot 3 \cdot 3 \cdot 3)}^{\text{4 factors}} = \overbrace{3 \cdot 3 \cdot 3 \cdot 3 \cdot 3 \cdot 3}^{\text{6 factors}} = 3^6,$$

that is, $3^2 \cdot 3^4 = 3^{2+4}.$

In general, if x is any number and m and n are positive integers, then

$$x^m x^n = \overbrace{x \cdot x \cdots x}^{m \text{ factors}} \cdot \overbrace{x \cdot x \cdots x}^{n \text{ factors}} = \overbrace{x \cdot x \cdots x}^{m+n \text{ factors}} = x^{m+n}.$$

When both m and n are negative, the factors are counted in the same way, although they are in the denominator of the resulting fraction. If $m \geq 0$ and n is negative, we let $n = -q$, where $q > 0$. Then

$$x^m x^n = x^m x^{-q} = \dfrac{x^m}{x^q} = \dfrac{\overbrace{x \cdot x \cdots x}^{m \text{ factors}}}{\underbrace{x \cdot x \cdots x}_{q \text{ factors}}}.$$

After all possible factors have been cancelled, either $m - q$ factors remain in the numerator or $q - m$ factors remain in the denominator. In the first case,

$$x^m x^n = x^m x^{-q} = x^{m-q} = x^{m+n};$$

and in the second case,

$$x^m x^n = x^m x^{-q} = \dfrac{1}{x^{q-m}} = x^{-(q-m)} = x^{m-q} = x^{m+n}.$$

By a similar argument, $x^m x^n = x^{m+n}$ can be verified if m is negative and $n \geq 0$.

This and several other formulas involving exponents are listed next.

THEOREM R.3.1 **Laws of Integer Exponents**

Let x and y be real numbers and m and n be integers. Then,

(i) $x^m x^n = x^{m+n}$ (ii) $(x^m)^n = x^{mn}$ (iii) $(xy)^n = x^n y^n$

(iv) $\left(\dfrac{x}{y}\right)^n = \dfrac{x^n}{y^n}$ (v) $\dfrac{x^m}{x^n} = x^{m-n}$

provided each expression represents a real number.

In the statement of these laws, whenever x or y occurs in a denominator or with a negative exponent, x or y must be nonzero. Also, (iii) of Theorem R.3.1 extends to more than two variables; for example

$$(xyzw)^n = x^n y^n z^n w^n.$$

In the following examples, we illustrate each of the laws of exponents.

EXAMPLE 3   **Using the Laws of Exponents**

(a) $a^5 a^4 = a^{5+4} = a^9$ ← by (i) of Theorem R.3.1

(b) $(b^3)^{-2} = b^{3(-2)} = b^{-6} = \dfrac{1}{b^6}$ ← by (ii) of Theorem R.3.1 with x replaced by b

(c) $(3x)^4 = 3^4 x^4 = 81 x^4$ ← by (iii) of Theorem R.3.1 with x replaced by 3 and y replaced by x

(d) $\left(\dfrac{y}{4}\right)^{-5} = \dfrac{y^{-5}}{4^{-5}} = \dfrac{\frac{1}{y^5}}{\frac{1}{4^5}} = \dfrac{4^5}{y^5} = \dfrac{1024}{y^5}$ ← by (iv) of Theorem R.3.1 with x replaced by y and y replaced by 4

(e) $\dfrac{a^{-5}}{a^{-3}} = a^{-5-(-3)} = a^{-2} = \dfrac{1}{a^2}$ ← by (v) of Theorem R.3.1 with x replaced by a

                                                                                                             ≡

The laws of exponents are useful in simplifying algebraic expressions, as we see in the following example.

EXAMPLE 4   **Using the Laws of Exponents**

Simplify $\dfrac{(-6xy^2)^3}{x^2 y^5}$.

Solution By the laws of exponents, we have

$$\dfrac{(-6xy^2)^3}{x^2 y^5} = \dfrac{(-6)^3 x^3 (y^2)^3}{x^2 y^5} \quad \leftarrow \text{by (iii) of Theorem R.3.1}$$

$$= -\dfrac{216 x^3 y^6}{x^2 y^5} \quad \leftarrow \text{by (i) and (ii) of Theorem R.3.1}$$

$$= -216 x^{3-2} y^{6-5} \quad \leftarrow \text{by (v) of Theorem R.3.1}$$

$$= -216 xy.$$

☐ **Scientific Notation** Integer exponents are frequently used to write very large or very small numbers in a convenient way. Any positive real number can be written in the form

$$a \times 10^n,$$

where $1 \leq a < 10$ and n is an integer. We say that a number written in this form is in **scientific notation**. For example,

$$1{,}000{,}000 = 1 \times 10^6 = 10^6 \quad \text{and} \quad 0.0000000537 = 5.37 \times 10^{-8}.$$

Scientific notation is most useful in chemistry and physics, where numbers such as

$$92{,}900{,}000 = 9.29 \times 10^7 \quad \text{and} \quad 0.000000000251 = 2.51 \times 10^{-10}$$

frequently occur. These numbers are the average distance from the Earth to the Sun in miles and the average lifetime of a lambda particle in seconds, respectively. Numbers such as these are certainly easier to write and to remember when given in scientific notation and are easily simplified. This is illustrated in the following example.

EXAMPLE 5   **Using Scientific Notation**

Find the value of

$$\frac{(4000)^3(1{,}000{,}000)}{(20{,}000{,}000)^5}.$$

Solution We write the numbers in scientific notation and then use the laws of exponents to simplify the expression:

$$\begin{aligned}
\frac{(4000)^3(1{,}000{,}000)}{(20{,}000{,}000)^5} &= \frac{(4 \times 10^3)^3(1 \times 10^6)}{(2 \times 10^7)^5} \\
&= \frac{(4)^3(10^3)^3 10^6}{2^5(10^7)^5} \\
&= \frac{64(10^9)(10^6)}{32(10^{35})} \\
&= 2 \times 10^{-20} = 0.00000000000000000002.
\end{aligned}$$

Most calculators automatically convert a number to scientific notation whenever it is too large or too small to be displayed in decimal form. For example, the number 1.234×10^{15} requires 16 digits for its decimal form, but since few calculators can display more than 10 digits, the multiplication sign and the base 10 are not shown. Therefore the number 1.234×10^{15} may appear as $\boxed{1.234 \quad 15}$. On many calculators it is possible to use scientific notation when entering a number. Consult your calculator manual for details.

☐ **Significant Digits** Most real-world applications of mathematics involve measurements that are subject to error and therefore are considered to be approximations. We can describe the accuracy of an approximation by stating how many **significant digits** it has.

Suppose that the result of a measurement is expressed in scientific notation

$$x = a \times 10^n, \quad \text{where } 1 \leq a < 10,$$

and the digits in a are known to be accurate (except possibly for the last digit, which may be off by one if the number was rounded.) If a contains k decimal places (that is, k digits to the right of the decimal point), then x is said to have $k + 1$ significant digits. According to this convention, 2.0285×10^{23} has five significant digits and 9.30×10^{-20} has three significant digits.

EXAMPLE 6   Length of a Light Year

A light year is the distance traveled by light in one Earth year (365.25 days). The speed of light is 3.00×10^5 kilometers per second (accurate to three significant digits). Find the length of one light year in kilometers. In miles.

Solution To determine the length of one light year in kilometers, we multiply the speed of light in kilometers per second by the number of seconds in an Earth year. We first set up the conversion of one Earth year to seconds:

$$1 \text{ Earth year} \approx 365.25 \text{ days} \times 24\frac{\text{hours}}{\text{day}} \times 60\frac{\text{minutes}}{\text{hour}} \times 60\frac{\text{seconds}}{\text{minute}}.$$

Then the length of one light year in kilometers is given by

$$3.00 \times 10^5 \times 365.25 \times 24 \times 60 \times 60 \approx 9.47 \times 10^{12} \text{ km}.$$

Now 1 km = 6.21×10^{-1} mi and so the length of a light year in miles is

$$3.00 \times 10^5 \times 6.21 \times 10^{-1} \times 365.25 \times 24 \times 60 \times 60 \approx 5.88 \times 10^{12} \text{ mi}. \equiv$$

NOTES FROM THE CLASSROOM

You should get in the habit of taking a little extra time in reading a mathematical expression containing powers of x. For example, the distinction between the quantities $5x^3$ and $(5x)^3$ is often overlooked in the rush to complete homework or a test. The parentheses indicate that the exponent 3 applies to $5x$ and not just to x. In other words,

$$5x^3 = 5 \cdot x \cdot x \cdot x$$

whereas $(5x^3) = 5x \cdot 5x \cdot 5x = 125x^3.$

Similarly, $-3^4 = -(3 \cdot 3 \cdot 3 \cdot 3) = -81$

whereas $(-3)^4 = (-3)(-3)(-3)(-3) = 81.$

R.3 Exercises
Answers to selected odd-numbered problems begin on page ANS-1.

Assume in Problems 1–86 that all variables are nonzero.

In Problems 1–4, write the expression with positive exponents.

1. $\dfrac{1}{8 \cdot 8 \cdot 8}$
2. $3 \cdot 3 \cdot 3$
3. $2y \cdot 2y \cdot 2y \cdot 2y$
4. $\dfrac{1}{z} \cdot \dfrac{1}{z}$

In Problems 5–8, write the expression with negative exponents.

5. $\dfrac{1}{4^5}$
6. $\dfrac{x^2}{y^2}$
7. $\dfrac{1}{x^3}$
8. $\left(\dfrac{1}{z}\right)^2$

In Problems 9–14, find the indicated numbers.

9. (a) 3^4 (b) 3^{-4} (c) -3^4
10. (a) $\left(\frac{1}{3}\right)^3$ (b) $\left(-\frac{1}{3}\right)^{-3}$ (c) $\left(\frac{1}{3}\right)^{-3}$
11. (a) $(-7)^2$ (b) $(-7)^{-2}$ (c) $-(7)^{-2}$
12. (a) $\left(-\frac{2}{3}\right)^5$ (b) $\left(-\frac{2}{3}\right)^{-5}$ (c) $-\left(-\frac{2}{3}\right)^5$
13. (a) $(5)^0$ (b) $(-5)^0$ (c) -5^0
14. (a) $(-1)^{-1}$ (b) $(1)^{-1}$ (c) $-(-1)^{-1}$

In Problems 15–20, evaluate the expression.

15. $2^{-1} - 2^1$
16. $\dfrac{2^{-2}}{3^{-3}}$
17. $\dfrac{2^{-1} - 3^{-1}}{2^{-1} + 3^{-1}}$
18. $\dfrac{(-1)^5 - 2^6}{(-1)^{-1}}$
19. $\dfrac{0^1}{1^0}$
20. $\dfrac{(1-1)^0}{1^0}$

In Problems 21–26, find the value of the expression if $a = 2$, $b = -3$, and $c = -1$.

21. $-2ab + c^2$
22. $ab^2 - c^3$
23. $ab^2 + bc^2 + ca^2$
24. $a^{-1}b^{-1}c^{-1}$
25. $ab^{-1} + ca^{-1}$
26. $a^{-1} + b^{-1} + c^{-1}$

In Problems 27–50, simplify and eliminate any negative exponents.

27. $x^6 x^{-2}$
28. $2^{10} 2^{12}$
29. $(7x^4)(-3x^2)$
30. $(-5x^2y^3)(3xy^{-2})$
31. $\dfrac{2^8}{2^3}$
32. $\dfrac{3^4}{3^{-2}}$
33. $\dfrac{10^{-7}}{10^4}$
34. $\dfrac{35y^8x^5}{-21y^{-1}x^9}$
35. $(5x)^2$
36. $(-4x)^3$
37. $(5^2)^3$
38. $(x^4)^{-5}$
39. $(4x^2y^{-1})^3$
40. $(3x^2y^4)^{-2}$
41. $x^2 x^3 x^{-4}$
42. $\dfrac{-x^5(y^2)^3}{(xy)^2}$
43. $\dfrac{(7a^2b^3)^2}{a^3b^5}$
44. $\dfrac{(-4x^5y^{-2})^3}{x^7y^{-3}}$
45. $(-3xy^5)^2(x^3y)^{-1}$
46. $\left(\dfrac{a^4b^{-5}}{b^2}\right)^{-1}$
47. $\left(\dfrac{a^3b^3}{b^{-2}}\right)^2$
48. $(-x^2y^4)^3(x^3y^{-1})^2$
49. $\dfrac{-xy^2z^3}{(xy^2z^3)^{-1}}$
50. $\dfrac{(3abc)^3}{(2a^{-1}b^{-2}c)^2}$

In Problems 51–56, determine whether the given number is positive or negative.

51. $(-4)^{-3}(2^{-4})$
52. $(-1)^{-1}(-1)^0(-1)$
53. $[10^{-5}(-10)^5(-10)^{-5}]^2$
54. $[(-1)^{-2}]^{-3}$
55. $[-10 - 10]^{-10+10}$
56. $[\pi^2\pi^3\pi^{-4}]^{-1}$

In Problems 57–62, write a formula for the given quantity using exponents.

57. The area A of a square is the square of the length s of a side.
58. The volume V of a cube is the cube of the length s of a side.
59. The area A of a circle is π times the square of the radius r.
60. The volume V of a sphere is $\frac{4}{3}\pi$ times the cube of the radius r.
61. The volume V of a right circular cylinder is π times the square of the radius r times the height h.
62. The area A of an equilateral triangle is $\sqrt{3}/4$ times the square of the length s of a side.

In Problems 63–66, write the given numbers in scientific notation.

63. (a) 1050000   (b) 0.0000105
64. (a) 341000000   (b) 0.00341
65. (a) 1200000000   (b) 0.000000000120
66. (a) 825600   (b) 0.0008256

In Problems 67–70, write the given numbers in decimal form.

67. (a) 3.25×10^7 (b) 3.25×10^{-5}
68. (a) 4.02×10^{10} (b) 4.02×10^{-4}
69. (a) 9.87×10^{-17} (b) 9.87×10^{12}
70. (a) 1.423×10^5 (b) 1.423×10^{-4}

In Problems 71–76, use a calculator to perform the computation. Write your answer in scientific notation using five significant digits

71. $(0.90324)(0.0005432)$ **72.** $\dfrac{0.2315}{(5480)^2}$

73. $\dfrac{0.143}{15000}$ **74.** $\dfrac{4033}{0.00000021}$

75. $(2.75 \times 10^3)(3.0 \times 10^{10})$ **76.** $\dfrac{8.25 \times 10^{-12}}{3.01 \times 10^{12}}$

In Problems 77–80, find the value of the given expression without the aid of a calculator. Write your answer **(a)** in decimal form and **(b)** in scientific notation.

77. $(3000)^2(200{,}000)^3(0.0000000001)$ **78.** $[(1{,}000{,}000)^{-1}(0.00001)]^{-1}$

79. $\dfrac{(80{,}000)^2}{(2{,}000{,}000)(0.0001)^4}$ **80.** $\dfrac{(21{,}000)(0.00005)^3}{3{,}000{,}000}$

Miscellaneous Applications

81. Population The estimated population of China for 2009 was 1,335,000,000. Write this number in scientific notation.

82. Population If the average annual growth rate for the population of China is 1.4%, use the information given in Problem 81 to compute the population of China **(a)** in 2010 and **(b)** in 2020. Write your answers in scientific notation.

83. GDP The gross domestic product is a basic measure of a country's overall economic output. In October 2009 it was predicted that the GDP for the United States would be 14.261 trillion dollars. Write this number **(a)** in decimal form and **(b)** in scientific notation.

84. Of Things to Come Future computers may be photonic (that is, operating on light signals rather than electronic. The speed of light (3×10^{10} cm/s) will be a limiting factor on the size and speed of such computers. Suppose that a signal must get from one component of a photonic computer to another within 1 nanosecond (1×10^{-9} s). What is the maximum possible distance between these two components? [*Hint*: How far does light travel in 1 nanosecond?] State your answer **(a)** in centimeters and **(b)** in inches (1 in. $\approx$ 2.5 cm).

85. Galactic Distance The distance to the Andromeda galaxy (Messier 31), located in the direction of the constellation Andromeda, is 2,500,000 light years from our own Milky Way galaxy. As we have seen in Example 6, a light year is a measure of distance. If a light year is (approximately) 6 million million miles, write the approximate distance (in miles) to the Andromeda galaxy in scientific notation.

Andromeda galaxy

86. **Average Speed** Pioneer 10, a deep-space probe, took 21 months to travel from Mars to Jupiter. If the distance from Mars to Jupiter is 998 million kilometers, find the average speed of Pioneer 10 in kilometers per hour. (Assume that there are 30.4 days in a month.)

For Discussion

In Problems 87–92, answer true or false.

87. $0^0 = 0$ ___

88. $\left(-\dfrac{1}{x}\right)^{-1} = x, x \neq 0$ ___

89. If n is even, $x^n \geq 0$ for all real numbers x. ___
90. $x^{-n} \leq 0$, for all real numbers x ___
91. $(x + y)^2 = x^2 + y^2$ ___
92. $(x^n)^{-1} = x^{-n}$, for $x \neq 0$ ___
93. By (v) of the laws of exponents, if $x \neq 0$, then what does x^n/x^n equal? However, what does any nonzero number divided by itself equal? Use the answer to these two questions to explain the rationale behind the definition $x^0 = 1$, for any nonzero base x.

R.4 Radicals

≡ **Introduction** Many problems in science, business, or engineering lead to statements such as $s^2 = 25$ or $x^3 = 64$. The numbers that satisfy these simple exponential equations are called **roots**. In particular, a number s that satisfies $s^2 = 25$ is said to be a **square root** of 25 and a number x satisfying $x^3 = 64$ is a **cube root** of 64.

There are two real number square roots of the number 25 because

$$(-5)^2 = 25 \quad \text{and} \quad 5^2 = 25.$$

By convention the symbol $\sqrt{}$ denotes the principal square root, which is a *nonnegative* real number. Thus, $\sqrt{25} = 5$.

In this section we review the definition and the properties of the **principal nth roots** of a real number x, where n is a positive integer.

DEFINITION R.4.1 Principal nth Root

Suppose x is a real number and $n \geq 2$ is a positive integer.

(i) If $x > 0$, then the **principal nth root** $\sqrt[n]{x}$ is the *positive* number r such that $x = r^n$.

(ii) If $x < 0$ and n is an odd positive integer, then the **principal nth root** $\sqrt[n]{x}$ is a *negative* number r such that $x = r^n$.

(iii) If $x < 0$ and n is an even positive integer, then $\sqrt[n]{x}$ is not a real number.

(iv) If $x = 0$, then $\sqrt[n]{x} = 0$.

To summarize (*i*), (*ii*), and (*iv*) of Definition R.4.1, the statement

$$\sqrt[n]{x} = r \quad \text{means} \quad x = r^n.$$

CHAPTER R REVIEW OF BASIC ALGEBRA

☐ **Terminology** The symbol $\sqrt[n]{x}$ that denotes the principal nth root of x is called a **radical**, the integer n is the **index** of the radical, and the real number x is called the **radicand**. If the index n is 2, it is usually omitted from the radical; that is, $\sqrt[2]{25}$ is written $\sqrt{25}$. When $n = 2$ we say $\sqrt{x}$ is the **square root** of x and when $n = 3$ we say $\sqrt[3]{x}$ is the **cube root** of x. If the index n is an *odd positive integer*, it can be shown that for any real number x, there is exactly *one* real nth root of x. For example,

$$\sqrt[3]{125} = 5 \quad \text{and} \quad \sqrt[5]{-32} = -2.$$

If the index n is an *even positive integer* and x is positive, then there are two real nth roots of x. The symbol $\sqrt[n]{x}$, however, is reserved for the positive (principal) nth root; we denote the negative nth root by $-\sqrt[n]{x}$. Thus, for example,

◀ Reread the first two sentences.

$$\sqrt{4} = 2 \quad \text{and} \quad -\sqrt{4} = -2,$$
$$\sqrt[4]{\frac{1}{81}} = \frac{1}{3} \quad \text{and} \quad -\sqrt[4]{\frac{1}{81}} = -\frac{1}{3}.$$

If n is even and x is negative, there is no real nth root of x.*

EXAMPLE 1 Roots

Find **(a)** $\sqrt{100}$ **(b)** $\sqrt[3]{-64}$ **(c)** $\sqrt[4]{\frac{16}{81}}$.

Solution There will be a single answer in each case since we are finding, in turn, the principal square root, the principal cube root, and the principal fourth root.

(a) $\sqrt{100} = 10$, since $10^2 = 100$. ← by (*i*) of Definition R.4.1
(b) $\sqrt[3]{-64} = -4$, since $(-4)^3 = -64$. ← by (*ii*) of Definition R.4.1
(c) $\sqrt[4]{\frac{16}{81}} = \frac{2}{3}$, since $\left(\frac{2}{3}\right)^4 = \frac{16}{81}$. ← by (*i*) of Definition R.4.1

≡

☐ **The Laws of Radicals** The following properties can often be used to simplify expressions involving radicals.

THEOREM R.4.1 Laws of Radicals

Let m and n be positive integers and x and y be real numbers. Then,

(*i*) $\left(\sqrt[n]{x}\right)^n = x$ (*ii*) $\sqrt[n]{x^n} = \begin{cases} x, & \text{if } n \text{ is odd} \\ |x| & \text{if } n \text{ is even} \end{cases}$

(*iii*) $\sqrt[n]{xy} = \sqrt[n]{x}\,\sqrt[n]{y}$ (*iv*) $\sqrt[n]{\dfrac{x}{y}} = \dfrac{\sqrt[n]{x}}{\sqrt[n]{y}}$

(*v*) $\sqrt[m]{\sqrt[n]{x}} = \sqrt[mn]{x}$

provided each radical represents a real number.

PARTIAL PROOF: The laws of radicals (*iii*)–(*v*) can be verified using the laws of exponents discussed in Section R.3. For example, to prove (*iii*) we let

$$\sqrt[n]{x} = a \quad \text{and} \quad \sqrt[n]{y} = b. \tag{1}$$

---

*An even root of a negative number, for example, $\sqrt{-5}$, is called a complex number. Complex numbers are discussed in Section 1.4.

Then, by definition
$$x = a^n \quad \text{and} \quad y = b^n.$$
Thus,
$$xy = a^n b^n = (ab)^n,$$
which can be written in radical form as
$$\sqrt[n]{xy} = ab. \tag{2}$$
Combining (1) and (2), we obtain $\sqrt[n]{xy} = ab = \sqrt[n]{x}\sqrt[n]{y}$. ≡

Each of the foregoing laws is illustrated in the next example. Probably the most familiar property of radicals, the square root of a product is the product of the square roots,
$$\sqrt{xy} = \sqrt{x}\sqrt{y} \tag{3}$$
for $x \geq 0$, $y \geq 0$, is just a special case of (*iii*) of Theorem R.4.1 when $n = 2$. For example, $\sqrt{40} = \sqrt{4 \cdot 10} = \sqrt{4}\sqrt{10} = 2\sqrt{10}$.

EXAMPLE 2 Simplifying Using the Laws of Radicals

Simplify each of the following expressions.

(a) $\sqrt[3]{\sqrt[7]{x^{21}}}$ (b) $\sqrt{\dfrac{x^2}{25}}$ (c) $\sqrt[3]{27y^6}$ (d) $\sqrt[3]{81a^3b^5c^6}$ (e) $(\sqrt[5]{r}\sqrt[5]{s})^5$

Solution In each case we use one or more laws of radicals.

(a) $\sqrt[3]{\sqrt[7]{x^{21}}} = \sqrt[21]{x^{21}} = x \;\leftarrow$ by (*i*) of Theorem R.4.1

(b) $\sqrt{\dfrac{x^2}{25}} = \dfrac{\sqrt{x^2}}{\sqrt{25}} = \dfrac{|x|}{5} \;\leftarrow$ by (*ii*) and (*iv*) of Theorem R.4.1

(c) $\sqrt[3]{27y^6} = \sqrt[3]{27}\sqrt[3]{y^6} = \sqrt[3]{27}\sqrt[3]{(y^2)^3} = 3y^2 \;\leftarrow$ by (*ii*) and (*iii*) of Theorem R.4.1

(d) $\sqrt[3]{81a^3b^5c^6} = \sqrt[3]{(27a^3b^3c^6)3b^2} = \sqrt[3]{(3abc^2)^3 3b^2} \;\leftarrow$ factoring in the radicand
$= \sqrt[3]{(3abc^2)^3}\sqrt[3]{3b^2} = 3abc^2\sqrt[3]{3b^2} \;\leftarrow$ by (*ii*) and (*iii*) of Theorem R.4.1

(e) $(\sqrt[5]{r}\sqrt[5]{s})^5 = (\sqrt[5]{rs})^5 = rs \;\leftarrow$ by (*i*) and (*iii*) of Theorem R.4.1 ≡

EXAMPLE 3 Using the Laws of Radicals

Simplify each of the following expressions.

(a) $\sqrt[3]{2x^2y^3}\sqrt[3]{4xz^3}$ (b) $\dfrac{\sqrt[4]{32a^{10}b^{16}}}{\sqrt[4]{2a^2}}$

Solution In parts (a) and (b), supply a reason on the colored line above the correspondingly colored equality sign.

(a) $\sqrt[3]{2x^2y^3}\sqrt[3]{4xz^3} = \sqrt[3]{8x^3y^3z^3} = \sqrt[3]{(2xyz)^3} = 2xyz$

(b) $\dfrac{\sqrt[4]{32a^{10}b^{16}}}{\sqrt[4]{2a^2}} = \sqrt[4]{16a^8b^{16}} = \sqrt[4]{(2a^2b^4)^4} = 2a^2b^4$ ≡

As we have just seen in Example 2, the laws of radicals in Theorem R.4.1 allow us to simplify products and quotients of radicals with the same index. We can frequently simplify sums and differences of radicals with the same index by using the distributive laws, as shown in the following example.

EXAMPLE 4   Simplifying

Simplify each of the following expressions.
(a) $\sqrt{10} - \sqrt{40x^4} + \sqrt{90x^4y^8}$ (b) $\sqrt[3]{8x^4} + \sqrt[3]{x^4y^3}$

Solution   We again use the laws of radicals given in Theorem R.4.1.

(a) $\sqrt{10} - \sqrt{40x^4} + \sqrt{90x^4y^8} = \sqrt{10} - 2x^2\sqrt{10} + 3x^2y^4\sqrt{10}$ ← $\sqrt{10}$ is a common factor
$= \sqrt{10}(1 - 2x^2 + 3x^2y^4)$

(b) $\sqrt[3]{8x^4} + \sqrt[3]{x^4y^3} = 2x\sqrt[3]{x} + xy\sqrt[3]{x}$ ← $x\sqrt[3]{x}$ is a common factor
$= x\sqrt[3]{x}(2 + y)$ ≡

□ **Rationalizing**   When we remove radicals from the numerator or the denominator of a fraction, we say that we are **rationalizing**. In algebra, we usually rationalize the denominator, but in calculus it is sometimes important to rationalize the numerator. The procedure of rationalizing involves multiplying the fraction by 1 written in a special way. For example,

$$\frac{1}{\sqrt{5}} = \frac{1}{\sqrt{5}} \cdot \overbrace{\frac{\sqrt{5}}{\sqrt{5}}}^{\text{fraction equals 1}} = \frac{\sqrt{5}}{5}.$$ ≡

EXAMPLE 5   Rationalizing a Denominator

Rationalize the denominator of each of the following expressions.

(a) $\dfrac{\sqrt{3}}{\sqrt{2}}$ (b) $\dfrac{1}{\sqrt[3]{2}}$

Solution

(a) $\dfrac{\sqrt{3}}{\sqrt{2}} = \dfrac{\sqrt{3}}{\sqrt{2}} \cdot \overbrace{\dfrac{\sqrt{2}}{\sqrt{2}}}^{1} = \dfrac{\sqrt{3\cdot 2}}{(\sqrt{2})^2} = \dfrac{\sqrt{6}}{2}$ ← we used (3) in the numerator

(b) Since $\sqrt[3]{2} \cdot (\sqrt[3]{2})^2 = \sqrt[3]{2} \cdot \sqrt[3]{2} \cdot \sqrt[3]{2} = 2$, we multiply the numerator and the denominator by $(\sqrt[3]{2})^2$:

$$\frac{1}{\sqrt[3]{2}} = \frac{1}{\sqrt[3]{2}} \cdot \frac{(\sqrt[3]{2})^2}{(\sqrt[3]{2})^2} = \frac{(\sqrt[3]{2})^2}{(\sqrt[3]{2})^3} = \frac{(\sqrt[3]{2})^2}{2}$$

$$= \frac{\sqrt[3]{2} \cdot \sqrt[3]{2}}{2} = \frac{\sqrt[3]{2 \cdot 2}}{2} = \frac{\sqrt[3]{4}}{2}.$$ ≡

□ **Using a Conjugate Factor**   If a fraction contains an expression such as $\sqrt{x} + \sqrt{y}$, we make use of the fact that the product of $\sqrt{x} + \sqrt{y}$ and its *conjugate* $\sqrt{x} + \sqrt{y}$ contains no radicals:

$$(\sqrt{x} + \sqrt{y})(\sqrt{x} - \sqrt{y}) = \sqrt{x}(\sqrt{x} - \sqrt{y}) + \sqrt{y}(\sqrt{x} - \sqrt{y})$$
$$= \sqrt{x}\sqrt{x} - \sqrt{x}\sqrt{y} + \sqrt{y}\sqrt{x} - \sqrt{y}\sqrt{y}$$
$$= (\sqrt{x})^2 - \sqrt{xy} + \sqrt{xy} - (\sqrt{y})^2$$
$$= x - y.$$

The conjugates of the expressions $x + \sqrt{y}$, $\sqrt{x} + y$, $x - \sqrt{y}$, $\sqrt{x} - y$ are, respectively, $x - \sqrt{y}$, $\sqrt{x} - y$, $x + \sqrt{y}$, and $\sqrt{x} + y$. You should verify that the product of each of these expressions with its conjugate contains no radicals.

EXAMPLE 6   Rationalizing a Denominator

Rationalize the denominator of the expression.
$$\frac{1}{\sqrt{x} + 3}$$

Solution The conjugate of the denominator is $\sqrt{x} - 3$. To eliminate radicals from the denominator, we multiply the given expression by
$$\frac{\sqrt{x} - 3}{\sqrt{x} - 3}.$$

Thus,
$$\frac{1}{\sqrt{x} + 3} = \frac{1}{\sqrt{x} + 3} \cdot \overbrace{\frac{\sqrt{x} - 3}{\sqrt{x} - 3}}^{1} = \frac{\sqrt{x} - 3}{x - 3}.$$

EXAMPLE 7   Rationalizing a Numerator

Eliminate the radicals in the numerator of
$$\frac{\sqrt{x + h} - \sqrt{x}}{h}.$$

Solution Since the conjugate of the numerator is $\sqrt{x + h} + \sqrt{x}$, we proceed as follows:

$$\frac{\sqrt{x + h} - \sqrt{x}}{h} \cdot \overbrace{\frac{\sqrt{x + h} + \sqrt{x}}{\sqrt{x + h} + \sqrt{x}}}^{1} = \frac{(x + h) - x}{h(\sqrt{x + h} + \sqrt{x})}$$
$$= \frac{\not{h}}{\not{h}(\sqrt{x + h} + \sqrt{x})} \quad \leftarrow \text{cancel the } h\text{'s}$$
$$= \frac{1}{\sqrt{x + h} + \sqrt{x}}.$$

The rationalization of a numerator, as illustrated in Example 7, occurs often in calculus.

EXAMPLE 8   Artificial Gravity

An artificial gravity can be created in a space station (or interplanetary spaceship) by rotating the station like a giant centrifuge. The rotation will produce a force against the astronauts on board that cannot be distinguished from gravity. The rate of rotation N, measured in rotations per second, required to produce an acceleration of a m/s² at a point r meters from the center of rotation is given by
$$N = \frac{1}{2\pi}\sqrt{\frac{a}{r}}.$$

If the radius of the station is 150 m, calculate the rotation rate necessary to produce the equivalent of Earth's gravity.

Early design of a space station with artificial gravity

Solution The acceleration due to gravity on Earth is 9.8 m/s². Therefore, we identify $a = 9.8$ and $r = 150$ and obtain

$$N = \frac{1}{2\pi}\sqrt{\frac{9.8}{150}}.$$

Using the square root $\boxed{\sqrt{}}$ and $\boxed{\pi}$ keys on a calculator, we find that $N \approx 0.04$. Therefore, approximately 0.04 rotations per second (or, equivalently, 2.4 rotations per minute) are required to produce the equivalent of Earth's gravity. ≡

NOTES FROM THE CLASSROOM

In this note we discuss some common mistakes using radicals and the laws of radicals.

(*i*) It is a common mistake to simplify $\sqrt{x^2}$ as x. This is valid only for *nonnegative x*. For example, if $x = -3$, we see that

$$\sqrt{(-3)^2} = \sqrt{9} = 3 \neq -3.$$

The correct result is given by (*ii*) of the laws of radicals:

$$\sqrt{(-3)^2} = |-3| = 3.$$

(*ii*) In part (b) of Example 5, it would be incorrect to try to rationalize $1/\sqrt[3]{2}$ by multiplying the numerator and the denominator by $\sqrt[3]{2}$:

$$\frac{1}{\sqrt[3]{2}} \cdot \frac{\sqrt[3]{2}}{\sqrt[3]{2}} = \frac{\sqrt[3]{2}}{(\sqrt[3]{2})^2} \neq \frac{\sqrt[3]{2}}{2}.$$

R.4 Exercises
Answers to selected odd-numbered problems begin on page ANS-1.

Assume in Problems 1–62 that all variables are positive.

In Problems 1–32, evaluate the radical(s).

1. $\sqrt[3]{-125}$
2. $\sqrt[4]{\frac{1}{4}}\sqrt[4]{\frac{1}{4}}$
3. $\sqrt[5]{100{,}000}$
4. $\sqrt[3]{16}$
5. $\sqrt[4]{0.0001}$
6. $\sqrt[5]{32}$
7. $\sqrt[3]{-64/27}$
8. $\sqrt[3]{-1000/8}$
9. $\sqrt{\dfrac{1}{x^2 y^4}}$
10. $\sqrt{\dfrac{10a^2}{bc^4}}$
11. $\sqrt[3]{\dfrac{x^3 y^6}{z^9}}$
12. $\sqrt[4]{\dfrac{x^4 y^{16}}{16 z^{18}}}$
13. $\sqrt{0.25 x^4}\sqrt{z^4}$
14. $\sqrt{8x^2 y z^2}\sqrt{yzw}\sqrt{2zw^3}$
15. $\sqrt[3]{4ab^3}\sqrt[3]{16a^2}$
16. $\sqrt[4]{16x^5}\sqrt[4]{2x^3 y^4}$
17. $\dfrac{\sqrt{3}}{\sqrt{27}}$
18. $\dfrac{\sqrt{125}}{\sqrt{5}}$
19. $\dfrac{\sqrt{7ab^2}}{\sqrt{49a}\sqrt{7b^4}}$
20. $\dfrac{\sqrt[4]{4xy}\sqrt[3]{2xy^2}}{\sqrt[3]{x^2 z^3}}$

21. $\sqrt{\sqrt{0.0016}}$
22. $\sqrt{2\sqrt{4}}$
23. $\sqrt[3]{\sqrt{a^6 b^{12}}}$
24. $\sqrt{x^3 \sqrt{(x^2 y)^2}}$
25. $\left(-\sqrt{xyz^5}\right)^2$
26. $\sqrt{(-2x^3 y)^2}$
27. $\sqrt{(-abc)^2}$
28. $\left(-\sqrt[3]{\dfrac{-27x}{xy^3}}\right)^3$
29. $\sqrt[4]{(-4r^2 s^6)^2}$
30. $\sqrt[3]{-(p^{-1} q^2)^3}$
31. $\sqrt{\dfrac{-16x^2}{-8x^{-2}}}$
32. $\sqrt[3]{\dfrac{(-2x)^3}{-z^6}}$

In Problems 33–44, rationalize the denominator of the expression.

33. $\dfrac{1}{\sqrt{27}}$
34. $\dfrac{\sqrt{3}}{\sqrt{2}}$
35. $\dfrac{1}{\sqrt{x+1}}$
36. $\dfrac{\sqrt{a}}{1+\sqrt{a}}$
37. $\dfrac{\sqrt{2}-\sqrt{5}}{\sqrt{2}+\sqrt{5}}$
38. $\dfrac{\sqrt{3}-\sqrt{7}}{\sqrt{3}+\sqrt{7}}$
39. $\dfrac{\sqrt{x}+\sqrt{y}}{\sqrt{x}-\sqrt{y}}$
40. $\dfrac{1}{\sqrt{a}-\sqrt{b}}$
41. $\dfrac{2}{\sqrt[3]{4}}$
42. $\dfrac{1}{\sqrt[3]{xy}}$
43. $\dfrac{4}{\sqrt[3]{x-1}}$
44. $\dfrac{1}{\sqrt[4]{2x}}$

In Problems 45–48, rationalize the numerator of the expression.

45. $\dfrac{\sqrt{2(x+h)}-\sqrt{2x}}{h}$
46. $\dfrac{\sqrt{(x+h)^2+1}-\sqrt{x^2+1}}{h}$
47. $\dfrac{\sqrt{x+h+1}-\sqrt{x+1}}{h}$
48. $\dfrac{\dfrac{1}{\sqrt{x+h}}-\dfrac{1}{\sqrt{x}}}{h}$ [*Hint*: First combine terms in the numerator.]

In Problems 49–56, combine the radicals and simplify.

49. $4\sqrt{x}+3\sqrt{x}-2\sqrt{x}$
50. $\sqrt{2}-\sqrt{6}+\sqrt{8}$
51. $4\sqrt[3]{2}-\sqrt[3]{16}$
52. $\sqrt[3]{xy}+3\sqrt[3]{x}-\sqrt[3]{xz^3}$
53. $3\sqrt{8x^3}-\sqrt{18xy^2}+\sqrt{32x^5}$
54. $\sqrt[3]{x^4 yz}-\sqrt[3]{xy^4 z}+\sqrt[3]{xyz^4}$
55. $\sqrt{\dfrac{a}{b}}-\sqrt{\dfrac{a^3}{b}}$
56. $\sqrt[3]{\dfrac{x}{y}}-\sqrt[3]{\dfrac{x}{y^2}}-\sqrt[3]{\dfrac{xy}{y^2}}$

In Problems 57–60, write a formula for the given quantity using radical notation.

57. The length s of the side of a square is the square root of the area A.
58. The length s of the side of a cube is the cube root of the volume V.
59. The length c of the hypotenuse of a right triangle is equal to the square root of the sum of the squares of the lengths a and b of the other two sides.
60. The velocity v of a satellite in a circular orbit around the Earth is equal to the square root of the product of the radius r of the orbit and the acceleration due to gravity g_r at the orbit.

Miscellaneous Applications

61. **Earth Satellite** If a satellite revolves around the Earth in a circular orbit of radius $r = 6.70 \times 10^6$ m, find its velocity v if $v = R\sqrt{g/r}$, where R is the radius of the Earth and g is the acceleration due to gravity at the Earth's surface. Use the values $R = 6.40 \times 10^6$ m and $g = 9.80$ m/s^2.

62. Relativity According to Einstein's theory of relativity, the mass m of an object moving at velocity v is given by

$$m = \frac{m_0}{\sqrt{1 - v^2/c^2}},$$

where m_0 is the mass of the object at rest and c is the speed of light. Find the mass of an electron traveling with velocity $0.6c$ if its rest mass is 9.1×10^{-31} kg.

For Discussion

In Problems 63–70, answer true or false.

63. $\sqrt{a+b} = \sqrt{a} + \sqrt{b}$, for $a, b \geq 0$ ___
64. $\sqrt{ab} = \sqrt{a}\sqrt{b}$, for $a, b \geq 0$ ___
65. $\sqrt{a^2} = a$, for any real number a ___
66. $(\sqrt{a})^2 = a$, for any real number a ___
67. If n is odd, $\sqrt[n]{x}$ is defined for any real number x. ___
68. If n is even, $\sqrt[n]{x}$ is defined for any real number x. ___
69. $\sqrt[4]{x^2} = \sqrt{x}$, for any real number x ___
70. $\sqrt{a^2/b^2} = |a/b|$, for any real numbers a and $b \neq 0$ ___

R.5 Rational Exponents

≡ **Introduction** The concept of the nth root of a number enables us to extend the definition of x^n from integer exponents to **rational exponents**. And, as we will see, it is often easier to work with rational exponents than with radicals.

DEFINITION R.5.1 Rational Power of x

Suppose x is a real number and $n \geq 2$ is a positive integer.

(*i*) If $\sqrt[n]{x}$ exists, then

$$x^{1/n} = \sqrt[n]{x}.$$

(*ii*) If $\sqrt[n]{x}$ exists and m is any integer such that m/n is in lowest terms, then

$$x^{m/n} = \sqrt[n]{x^m} = (\sqrt[n]{x})^m.$$

In part (*i*) of Definition R.5.1, $x^{1/n}$ is simply another way of designating the principal nth root of x. In part (*ii*) of Definition R.5.1, keep in mind that n is a positive integer greater than or equal to 2 and m can be any integer (positive, zero, or negative) and that the rational number m/n is reduced to lowest terms. Finally, $x^{m/n}$ can be computed by using either $\sqrt[n]{x^m}$ or $(\sqrt[n]{x})^m$ but as a matter of practicality, it is *usually* easier to take the nth root of the number x first and then raise it to the mth power; in other words, use $(\sqrt[n]{x})^m$.

◂ *Lowest terms* means m and n have no common integer factors.

EXAMPLE 1   Using Definition R.5.1(*i*)

We evaluate each of the following rational powers using part (*i*) of Definition R.5.1.

(a) $(25)^{1/2} = \sqrt{25} = 5$ ← principal square root
(b) $(64)^{1/3} = \sqrt[3]{64} = 4$ ← principal cube root

≡

EXAMPLE 2 — Using Definition R.5.1(*ii*)

We evaluate each of the following rational powers using part (*ii*) of Definition R.5.1.

(a) $(0.09)^{5/2} = [(0.09)^{1/2}]^5 = (\sqrt{0.09})^5 \;\leftarrow m=5, n=2$
$\phantom{(0.09)^{5/2}} = (0.3)^5 = 0.00243 \;\leftarrow$ principal square root of 0.09 is 0.3

(b) $(-27)^{-4/3} = [(-27)^{1/3}]^{-4} = [\sqrt[3]{-27}]^{-4} \;\leftarrow m=-4, n=3$
$\phantom{(-27)^{-4/3}} = (-3)^{-4} = \dfrac{1}{(-3)^4} = \dfrac{1}{81} \;\leftarrow$ principal cube root of -27 is -3, and (2) of Definition R.3.2

EXAMPLE 3 — A Comparison

Although part (*ii*) of Definition R.5.1 stipulates the equality

$$(125)^{2/3} = [(125)^{1/3}]^2 = [(125)^2]^{1/3},$$

the computation

$$[(125)^{1/3}]^2 = [\sqrt[3]{125}]^2 = 5^2 = 25 \;\leftarrow \text{using } (\sqrt[n]{x})^m$$

can be done mentally, whereas

$$[(125)^2]^{1/3} = [15{,}625]^{1/3} = \sqrt[3]{15{,}625} = 25 \;\leftarrow \text{using } \sqrt[n]{x^m}$$

may require the use of a calculator.

The following example illustrates a case in which $x^{m/n}$, $(x^m)^{1/n}$, and $(x^{1/n})^m$ are not equivalent. This example illustrates why m/n must be in lowest terms in part (*ii*) of Definition R.5.1.

EXAMPLE 4 — Comparison of Three Results

Compare **(a)** $x^{m/n}$, **(b)** $(x^m)^{1/n}$, and **(c)** $(x^{1/n})^m$ for $x = -9$, $m = 2$, and $n = 2$.

Solution Substituting $x = -9$, $m = 2$, and $n = 2$, we find:

(a) $x^{m/n} = (-9)^{2/2} = (-9)^1 = -9$
(b) $(x^m)^{1/n} = [(-9)^2]^{1/2} = 81^{1/2} = 9$
(c) $(x^{1/n})^m = [(-9)^{1/2}]^2 = (\sqrt{-9})^2$, which is not a real number, since it involves the square root of a negative number.

▶ $\sqrt{-9}$ is an example of a complex number. Complex numbers will be studied in Section 1.4.

☐ **Laws of Exponents** The laws of exponents given for integer exponents in Theorem R.3.1 of Section R.3 also hold true for rational exponents.

THEOREM R.5.1 Laws of Rational Exponents

Let x and y be real numbers and r and s be rational numbers. Then,

(i) $x^r x^s = x^{r+s}$ (ii) $(x^r)^s = x^{rs} = (x^s)^r$ (iii) $(xy)^r = x^r y^r$

(iv) $\left(\dfrac{x}{y}\right)^r = \dfrac{x^r}{y^r}$ (v) $\dfrac{x^r}{x^s} = x^{r-s}$

provided each expression represents a real number.

As shown in the following examples, these laws enables us to simplify algebraic expressions. For the remainder of this section we will assume that all variable bases x, y, a, b, and so on represent positive numbers so that all rational powers are defined.

EXAMPLE 5 Using the Laws of Exponents

(a) $(3x^{1/2})(2x^{1/5}) = 3(2)x^{1/2}x^{1/5} = 6x^{1/2+1/5}$ ← by (i) of Theorem R.5.1
$= 6x^{(5+2)/10} = 6x^{7/10}$

(b) $(a^2 b^{-8})^{1/4} = (a^2)^{1/4}(b^{-8})^{1/4} = a^{2/4} b^{-8/4}$ ← by (ii) and (iii) of Theorem R.5.1
$= a^{1/2} b^{-2} = \dfrac{a^{1/2}}{b^2}$

(c) $\dfrac{x^{2/3} y^{1/2}}{x^{1/4} y^{3/2}} = x^{2/3-1/4} y^{1/2-3/2} = x^{(8-3)/12} y^{-1} = \dfrac{x^{5/12}}{y}$ ← by (v) of Theorem R.5.1

(d) $\left(\dfrac{3x^{3/4}}{y^{1/3}}\right)^3 = \dfrac{(3x^{3/4})^3}{(y^{1/3})^3} = \dfrac{3^3(x^{3/4})^3}{(y^{1/3})^3} = \dfrac{27 x^{9/4}}{y}$ ← by (ii), (iii), and (iv) of Theorem R.5.1 ≡

EXAMPLE 6 Simplifying

Simplify $\left(\dfrac{5r^{3/4}}{s^{1/3}}\right)^2 \left(\dfrac{2r^{-3/2}}{s^{1/2}}\right)$.

Solution You should supply the parts of the laws of rational exponents (Theorem R.5.1) used in the following simplification:

$$\left(\dfrac{5r^{3/4}}{s^{1/3}}\right)^2 \left(\dfrac{2r^{-3/2}}{s^{1/2}}\right) = \left(\dfrac{25 r^{3/2}}{s^{2/3}}\right)\left(\dfrac{2r^{-3/2}}{s^{1/2}}\right) = \dfrac{50 r^0}{s^{7/6}} = \dfrac{50}{s^{7/6}}.$$ ≡

As we will see in the next two examples, certain radical expressions can be simplified more easily if they are rewritten using rational exponents.

EXAMPLE 7 Write as One Radical

Write $\sqrt{x \sqrt[4]{x}}$ as a single radical.

Solution We rewrite $\sqrt{x \sqrt[4]{x}}$ using rational exponents and then simplify using the laws of rational exponents:

$$\sqrt{x \sqrt[4]{x}} = (x \sqrt[4]{x})^{1/2} = \overbrace{(x \cdot x^{1/4})^{1/2}}^{\text{use } (i) \text{ of Theorem R.5.1}} = (x^{5/4})^{1/2} = x^{5/8}.$$

From (ii) of Definition R.5.1, we can write $x^{5/8}$ as $\sqrt[8]{x^5}$. ≡

EXAMPLE 8 Write as One Radical

Write $\sqrt[3]{16}/\sqrt{2}$ as a single radical.

Solution We rewrite the expression using rational exponents:

$$\dfrac{\sqrt[3]{16}}{\sqrt{2}} = \dfrac{16^{1/3}}{2^{1/2}}.$$

Next we must find a common base so that we can use the properties of rational exponents to simplify the expression. Since $16 = 2^4$, we have

$$\dfrac{16^{1/3}}{2^{1/2}} = \dfrac{(2^4)^{1/3}}{2^{1/2}} = \dfrac{2^{4/3}}{2^{1/2}} = 2^{4/3-1/2} = 2^{5/6}.$$

From (ii) of Definition R.5.1, the last term $2^{5/6}$ is the same as $\sqrt[6]{2^5} = \sqrt[6]{32}$. ≡

EXAMPLE 9   Simplifying

Simplify $(8{,}000{,}000)^{2/3} \sqrt[4]{0.0001 r^8 t^{12}}$.

Solution We write the numbers in scientific notation and use the laws of rational exponents:

$$\begin{aligned}
(8{,}000{,}000)^{2/3} \sqrt[4]{0.0001 r^8 t^{12}} &= (8 \times 10^6)^{2/3}(1 \times 10^{-4} \cdot r^8 t^{12})^{1/4} \\
&= 8^{2/3}(10^6)^{2/3}(10^{-4})^{1/4}(r^8)^{1/4}(t^{12})^{1/4} \\
&= (\sqrt[3]{8})^2 (10^4)(10^{-1}) \, r^2 t^3 \\
&= (4 \times 10^3) r^2 t^3 \\
&= 4000 r^2 t^3.
\end{aligned}$$

EXAMPLE 10   Inflation

Suppose that a piece of property cost p dollars n years ago. If it is now worth q dollars, then the average annual inflation rate for the property r is given by

$$r = \left(\frac{q}{p}\right)^{1/n} - 1.$$

Find the average annual inflation rate for a home now worth $500{,}000$ if it was purchased 12 years ago for $80{,}000$.

Solution We first identify $p = 80{,}000$, $q = 500{,}000$, and $n = 12$. Substituting, we then obtain

$$r = \left(\frac{500{,}000}{80{,}000}\right)^{1/12} - 1 = (6.25)^{1/12} - 1.$$

Using the $\boxed{y^x}$ key on a calculator with $y = 6.25$ and $x = \frac{1}{12}$, we find $r \approx 0.165$. Therefore, the average annual inflation rate for this property has been 16.5%.

In Section 5.1 we shall indicate how expressions with irrational exponents such as $x^{\sqrt{2}}$ or x^{π} can be defined. *The laws of exponents also hold for irrational exponents.*

R.5   Exercises
Answers to selected odd-numbered problems begin on page ANS-2.

Assume throughout this exercise set that all variables are positive.

In Problems 1–8, rewrite the expression using rational exponents.

1. $\sqrt[3]{ab}$
2. $\sqrt[5]{7x}$
3. $\dfrac{1}{(\sqrt[3]{x})^4}$
4. $\dfrac{1}{(\sqrt[4]{a})^3}$
5. $\sqrt[7]{x+y}$
6. $\sqrt[3]{a^2 + b^2}$
7. $\sqrt{x + \sqrt{x}}$
8. $\sqrt{x^2 - y^2}$

In Problems 9–16, rewrite the expression using radical notation.

9. $a^{2/3}$
10. $2a^{1/3}$
11. $(3a)^{2/3}$
12. $3a^{2/3}$
13. $3 + a^{2/3}$
14. $(3 + a)^{2/3}$
15. $\dfrac{3}{a^{2/3}}$
16. $(3a)^{-3/2}$

In Problems 17–22, find the indicated numbers.

17. (a) $(49)^{1/2}$ (b) $(49)^{-1/2}$
18. (a) $(-8)^{1/3}$ (b) $(-8)^{1/3}$
19. (a) $(0.04)^{7/2}$ (b) $(0.04)^{-7/2}$
20. (a) $\left(\frac{1}{64}\right)^{2/3}$ (b) $\left(\frac{1}{64}\right)^{-2/3}$
21. (a) $(27)^{7/3}$ (b) $(-27)^{-7/3}$
22. (a) $\left(\frac{81}{16}\right)^{3/4}$ (b) $\left(\frac{81}{16}\right)^{-3/4}$

In Problems 23–48, simplify and eliminate any negative exponents.

23. $(4x^{1/2})(3x^{1/3})$
24. $(3w^{3/2})(7w^{5/2})$
25. $a^{3/2}(4a^{2/3})$
26. $(-5x^3)x^{5/3}$
27. $x^{1/2}x^{1/4}x^{1/8}$
28. $(2a^{1/2})(2a^{1/3})(2a^{1/6})$
29. $(a^2b^4)^{1/4}$
30. $(100x^4)^{-3/2}$
31. $(25x^{1/3}y)^{3/2}$
32. $(4x^4y^{-6})^{1/2}$
33. $\dfrac{cd^{1/3}}{c^{1/3}d}$
34. $\dfrac{4x^{1/2}}{(8x)^{1/3}}$
35. $\left(\dfrac{2x^{1/2}}{z^{-1/6}y^{2/3}}\right)^6$
36. $\left(\dfrac{-y^{1/2}}{y^{-1/2}}\right)^{-1}$
37. $((-27a^3b^{-6})^{1/3})^2$
38. $a^{1/3}(a^{2/3}(ab)^{5/3}b^{-1/3})^{-1/2}$
39. $(x^{1/2})(x^{-1/2})^2x^{1/2}$
40. $y^{1/4}(y^{1/4}(y^{1/2})^{-4})^{-1/2}$
41. $(5x^{2/3}(x^{4/3})^{1/4})^3$
42. $(2z^{1/2}(2z^{1/2})^{-1/2})^{1/2}$
43. $\dfrac{(a^{-1/3}b^{2/9}c^{1/6})^9}{(a^{1/6}b^{-2/3})^6}$
44. $\left(\dfrac{x^{1/5}y^{3/10}}{x^{-2/5}y^{1/2}}\right)^{-10}$
45. $\dfrac{(p^{1/3}q^{1/2})^{-1}}{(p^{-1}q^{-2})^{1/2}}$
46. $\left(\dfrac{2x^{1/2}}{x^{-3/2}}\right)\left(\dfrac{1}{4x}\right)^{-1/2}$
47. $\left(\dfrac{r^2s^{-4}t^6}{r^{-4}s^2t^6}\right)^{1/6}$
48. $\left(\dfrac{-x^{1/2}y^{1/4}}{8x^2y^4}\right)^{1/3}$

In Problems 49–56, rewrite the expression as a single radical.

49. $\sqrt{5}\sqrt[3]{2}$
50. $\sqrt[3]{4}\sqrt{2}$
51. $\dfrac{\sqrt[3]{16}}{\sqrt[6]{4}}$
52. $\dfrac{\sqrt[3]{81}}{\sqrt[3]{3}}$
53. $\sqrt{x}\sqrt{x}$
54. $\sqrt{x}\sqrt[3]{x}$
55. $\dfrac{\sqrt{a}}{\sqrt[8]{a}}$
56. $\dfrac{\sqrt[3]{y^2}\sqrt{y}}{\sqrt[4]{y}}$

In Problems 57–60, use scientific notation to simplify the expression.

57. $\sqrt{0.000004}(8000)^{2/3}(100{,}000)^{4/5}$
58. $\dfrac{\sqrt{40{,}000}\,\sqrt[3]{8000}}{(0.000004)^{3/2}}$
59. $\left(\sqrt[3]{\sqrt{0.000001x^6y^{12}}}\right)^5$
60. $\sqrt[4]{(160{,}000a^{8/3})^3}$

Miscellaneous Applications

61. **Pendulum Motion** For a simple pendulum the period of time required for one complete oscillation is approximately $T \approx 2\pi(L/g)^{1/2}$, where L is the length of the pendulum string and g is the gravitational constant. Use a calculator to approximate the period of a pendulum with a 10-in. string if the value of g is 32 ft/s². [*Hint*: Use consistent units.]

62. **Sphere** The radius r of a sphere with volume V is given by $r = (3V/4\pi)^{1/3}$. Use a calculator to find the radius of a sphere of volume 100 cm³.
63. **Speed of Sound** The speed of sound v measured in feet per second through air of temperature t degrees Celsius is given by
$$v = \frac{1087(273 + t)^{1/2}}{16.52}.$$
Use a calculator to find the speed of sound through air when the temperature is 20°C.
64. **Flowing Water** A fast-running stream can carry along larger particles than a slow-moving one. Laboratory studies have shown that the critical velocity v_t of the water needed to start a particle in a stream bed moving is given by the formula
$$v_t = 0.152\, d^{4/9}(G - 1)^{1/2},$$
where v_t is measured in meters per second, d is the diameter of the particle in millimeters, and G is the specific gravity of the particle. Find the critical velocity needed to begin moving a grain of feldspar that has a specific gravity of 2.56 and a diameter of 3 mm.

For Discussion

In Problems 65–74, answer true or false.

65. $(z^2 + 25)^{1/2} = z + 5$ _____
66. $36x^{1/2} = 6\sqrt{x}$ _____
67. $((-4)^2)^{1/2} = 4$ _____
68. $[(-4)^{1/2}]^2 = -4$ _____
69. $((-1)^{-1})^{-1} = -1$ _____
70. $(-1)^{-1}(-1)^{-1} = 1$ _____
71. $x^{-3/2} = \dfrac{1}{x^{2/3}}$ _____
72. $x^{2/3} y^{-2/3} = 1$ _____
73. $(b^{4/3})^{3/4} = b$ _____
74. $\dfrac{a^{3/2}}{a^{-3/2}} = a^2$ _____

R.6 Polynomials and Special Products

≡ Introduction We have already found it convenient to use letters such as x or y to represent numbers. Such a symbol is called a **variable**. An **algebraic expression** is the result of performing a finite number of additions, subtractions, multiplications, divisions, or roots on a collection of variables and real numbers. The following are examples of algebraic expressions:

$$x^3 - 2x^2 + \sqrt{x} - \pi, \quad \frac{4xy - x}{x + y}, \quad \text{and} \quad \sqrt[3]{\frac{7y - 3}{x^5 y^{-2} + z}}.$$

Sometimes an algebraic expression represents a real number only for certain values of a variable. In considering the expression $\sqrt{x}$, we find that we must have $x \geq 0$ in order for $\sqrt{x}$ to represent a real number. When we work with algebraic expressions, we assume that the variables are restricted so that the expression represents a real number. The set of permissible values for the variable is called the **domain of the variable**. Thus the domain of the variable in $\sqrt{x}$ is the set of all nonnegative real numbers $\{x \mid x \geq 0\}$, and for $3/(x + 1)$ the domain is the set of all real numbers except $x = -1$, that is, $\{x \mid x \neq 1\}$.

If specific numbers are substituted for the variables in an algebraic expression, the resulting real number is called the **value** of the expression. For example, the value of $x^2 + 2y$ when $x = 1$ and $y = 2$ is $(1)^2 + 2(2) = 5$.

☐ **Polynomials** Certain algebraic expressions have special names. A **monomial** in one variable is any algebraic expression of the form

$$ax^n,$$

where a is a real number, x is a variable, and n is a nonnegative integer. The number a is called the **coefficient** of the monomial and n is called the **degree**. For example, $17x^5$ is a monomial of degree 5 with coefficient 17 and the constant -5 is a monomial of degree 0. The sum of two monomials is called a **binomial**. The sum of three monomials is called a **trinomial**. For example,

$$3x - 2 \quad \text{and} \quad x^3 + 6x$$

are binomials, whereas

$$4x^2 - 2x - 1 \quad \text{and} \quad 8x^4 + x^2 - 4x$$

are trinomials.

A **polynomial** is any finite sum of monomials. More formally, we have the following definition.

DEFINITION R.6.1 Polynomial

A **polynomial of degree n in the variable x** is any algebraic expression of the form

$$a_n x^n + a_{n-1} x^{n-1} + \cdots + a_2 x^2 + a_1 x + a_0, \quad a_n \neq 0, \qquad (1)$$

where n is a nonnegative integer and a_i, $i = 0, 1, \ldots, n$, are real numbers.

The expression (1) is called the **standard form** of a polynomial; that is, the polynomial is written in descending powers of x. Of course, not all powers need be present in a polynomial; some of the coefficients a_i, $i = 0, 1, \ldots, n$ could be 0.

Since a polynomial in x represents a real number for any real number x, the domain of a polynomial is the set of all real numbers R. The monomials $a_i x^i$ in the polynomial are called **terms** of the polynomial, and the coefficient a_n of the highest power of x is called the **leading coefficient**. For example, $6x^5 - 7x^3 + 3x^2 - 1$ is a polynomial of degree 5 with leading coefficient 6. The terms of this polynomial are $6x^5$, $-7x^3$, $3x^2$, and -1. The number a_0 is called the **constant term** of the polynomial. It may be 0, as in the polynomial $6x^2 - x$. If all the coefficients of a polynomial are zero, then the polynomial is called the **zero polynomial** and is denoted by 0.

Polynomials can be classified by their degrees, although the zero polynomial is not assigned a degree. Special names are used to describe the lower-degree polynomials, as listed in the following table.

Polynomial	Degree	Standard Form	Example
Constant	0	$a_0 \; (a_0 \neq 0)$	5
Linear	1	$a_1 x + a_0 \; (a_1 \neq 0)$	$3x - 5$
Quadratic	2	$a_2 x^2 + a_1 x + a_0 \; (a_2 \neq 0)$	$-\frac{1}{2}x^2 + x - 2$
Cubic	3	$a_3 x^3 + a_2 x^2 + a_1 x + a_0 \; (a_3 \neq 0)$	$x^3 - 6x + \sqrt{3}$
nth degree	n	$a_n x^n + a_{n-1} x^{n-1} + \cdots + a_1 x + a_0 \; (a_n \neq 0)$	$x^n - 1$

In each term in a polynomial, the exponent of the variable must be a nonnegative integer. For example,

$$\underset{\substack{\uparrow \\ \text{negative integer}}}{x^{-1}} + x - 1 \quad \text{and} \quad x^2 - 2\underset{\substack{\uparrow \\ \text{not an integer}}}{x^{1/2}} + 6$$

are *not* polynomials. However,

$$\tfrac{1}{3}x^2 + 4 \quad \text{and} \quad 0.5x^3 + \sqrt{6}\,x^2 - \pi x + 9$$

are polynomials, since the coefficients can be any real numbers.

EXAMPLE 1   Recognition of a Polynomial

Determine which of the following algebraic expressions are polynomials. If the expression is a polynomial, give its degree and its leading coefficient.

(a) $x^2 + \sqrt{x} - 1$
(b) $\sqrt{2} - x + 3x^2 - 17x^8$
(c) $7x^5 - x^2 + \tfrac{1}{2}x + x^{-2}$
(d) $x^4 - x^2$

Solution Since the variable in each term must be raised to a nonnegative integer power, (a) and (c) are not polynomials. The polynomials in (b) and (d) are of degree 8 and degree 4, respectively. Writing (b) in the standard form (1), $-17x^8 + 3x^2 - x + \sqrt{2}$, we see that the leading coefficient is -17. Since (d) is already written in standard form, the leading coefficient is 1. ≡

☐ **Algebra of Polynomials** Since each symbol in a polynomial represents a real number, we can use the properties of the real number system discussed in Section R.1 to add, subtract, and multiply polynomials. In other words, the sum, difference, and product of two polynomials is a polynomial.

EXAMPLE 2   Sum of Two Polynomials

Find the sum of the polynomials $x^4 - 3x^2 + 7x - 8$ and $2x^4 + x^2 + 3x$.

Solution Rearranging terms and using the distributive properties, we have

$$(x^4 - 3x^2 + 7x - 8) + (2x^4 + x^2 + 3x)$$
$$= x^4 + 2x^4 - 3x^2 + x^2 + 7x + 3x - 8$$
$$= (1 + 2)x^4 + (-3 + 1)x^2 + (7 + 3)x - 8$$
$$= 3x^4 - 2x^2 + 10x - 8.$$

Example 2 indicates that we can add two polynomials in x by adding the coefficients of like powers. Some students find it easier to add polynomials by lining up terms with like powers of x in a vertical format, as shown below:

$$\begin{array}{r} x^4 - 3x^2 + 7x - 8 \\ 2x^4 + x^2 + 3x \\ \hline 3x^4 - 2x^2 + 10x - 8. \end{array}$$

The choice of which format to use is simply a matter of personal preference. Generally, the vertical format requires more space, so after this section of the text we will use the horizontal format.

As the next example shows, subtraction of polynomials is performed in a manner similar to addition.

EXAMPLE 3 Difference of Two Polynomials

Subtract $2x^3 - 3x - 4$ from $x^3 + 5x^2 - 10x + 6$.

Solution Subtracting terms with like powers of x, we have

$$\begin{array}{r} x^3 + 5x^2 - 10x + 6 \\ -(2x^3 - 3x - 4) \\ \hline -x^3 + 5x^2 - 7x + 10. \end{array}$$

To perform this subtraction using a horizontal format, we proceed as follows:

$(x^3 + 5x^2 - 10x + 6) - (2x^3 - 3x - 4)$ ← use distributive law here
$= x^3 + 5x^2 - 10x + 6 - 2x^3 + 3x + 4$
$= (x^3 - 2x^3) + 5x^2 + (-10x + 3x) + (6 + 4)$ ← grouping like terms
$= -x^3 + 5x^2 - 7x + 10.$ ≡

In order to find the **product** of two polynomials, we use the distributive properties and the laws of exponents, as the following example shows.

EXAMPLE 4 Product of Two Polynomials

Multiply $x^3 + 3x - 1$ and $2x^2 - 4x + 5$.

Solution We begin by using the distributive law several times:

$(x^3 + 3x - 1)(2x^2 - 4x + 5)$
$= (x^3 + 3x - 1)(2x^2) + (x^3 + 3x - 1)(-4x) + (x^3 + 3x - 1)(5)$
$= (2x^5 + 6x^3 - 2x^2) + (-4x^4 - 12x^2 + 4x) + (5x^3 + 15x - 5).$

Combining like terms, we find the product to be

$(x^3 + 3x - 1)(2x^2 - 4x + 5)$
$= 2x^5 - 4x^4 + (6x^3 + 5x^3) + (-2x^2 - 12x^2) + (4x + 15x) - 5$
$= 2x^5 - 4x^4 + 11x^3 - 14x^2 + 19x - 5.$ ≡

As in Example 4, when multiplying two polynomials, we must multiply each term of the first polynomial by each term of the second. A vertical format can be used (provided we keep like terms aligned), as follows:

$$\begin{array}{r} x^3 + 3x - 1 \\ 2x^2 - 4x + 5 \\ \hline 5x^3 + 15x - 5 \\ -4x^4 - 12x^2 + 4x \\ 2x^5 + 6x^3 - 2x^2 \\ \hline 2x^5 - 4x^4 + 11x^3 - 14x^2 + 19x - 5. \end{array}$$

← $5(x^3 + 3x - 1)$
← $-4x(x^3 + 3x - 1)$
← $2x^2(x^3 + 3x - 1)$

☐ **Special Products** Certain products of binomials occur so frequently that you should learn to recognize them. We begin with the product of two binomials $ax + b$ and $cx + d$:

$$(ax + b)(cx + d) = acx^2 + (ad + bc)x + bd. \tag{2}$$

Some polynomials can be expressed as a positive integer power of a binomial. The **square** and **cube** of a binomial $x + a$ are, in turn,

$$(x + a)^2 = x^2 + 2ax + a^2 \tag{3}$$

and

$$(x + a)^3 = x^3 + 3ax^2 + 3a^2x + a^3. \tag{4}$$

R.6 Polynomials and Special Products

It is readily shown that the product of a binomial $x + a$ and its conjugate $x - a$ results in the **difference of two squares**:

$$(x + a)(x - a) = x^2 - a^2. \qquad (5)$$

A classic mnemonic for carrying out the multiplication in (2) is the so-called **FOIL** method. The idea is outlined schematically in FIGURE R.6.1; the letters F, O, I, and L, are, respectively, the first letter in the words **f**irst, **o**uter, **i**nner, and **l**ast.

$$(ax + b)(cx + d) = \overbrace{ax \cdot cx}^{F} + \overbrace{ax \cdot d}^{O} + \overbrace{b \cdot cx}^{I} + \overbrace{b \cdot d}^{L}$$
$$= acx^2 + [adx + bcx] + bd$$
$$= acx^2 + (ad + bc)x + bd$$

FIGURE R.6.1 The FOIL method of multiplying two binomials

EXAMPLE 5 Using the FOIL Method

Find the product $\left(\frac{2}{3}x - 2\right)\left(x - \frac{1}{3}\right)$.

Solution We identify $a = \frac{2}{3}$, $b = -2$, $c = 1$, and $d = -\frac{1}{3}$. Using the FOIL method we get

$$\left(\tfrac{2}{3}x - 2\right)\left(x - \tfrac{1}{3}\right) = \overbrace{\tfrac{2}{3}(1)x^2}^{\text{first}} + \overbrace{\left[\left(\tfrac{2}{3}\right)\left(-\tfrac{1}{3}\right)x\right.}^{\text{outer}} + \overbrace{\left.(-2)(1)x\right]}^{\text{inner}} + \overbrace{(-2)\left(-\tfrac{1}{3}\right)}^{\text{last}}$$
$$= \tfrac{2}{3}x^2 + \left(-\tfrac{2}{9} - 2\right)x + \tfrac{2}{3}$$
$$= \tfrac{2}{3}x^2 - \tfrac{20}{9}x + \tfrac{2}{3}. \qquad \equiv$$

At first glance some products may not appear to be in the form (2) when, in fact, they are. With practice, however, you will become adept at recognizing them.

EXAMPLE 6 Using the FOIL Method

Find the product $(5x^2 + 3)(4x^2 - 6)$.

Solution In (2) we simply replace ax by $5x^2$ and cx by $4x^2$:

$$(5x^2 + 3)(4x^2 - 6) = \overbrace{5(4)(x^2)^2}^{\text{first}} + \overbrace{[5(-6)x^2}^{\text{outer}} + \overbrace{3(4)x^2]}^{\text{inner}} + \overbrace{3(-6)}^{\text{last}}$$
$$= 20x^4 + (-30 + 12)x^2 - 18$$
$$= 20x^4 - 18x^2 - 18. \qquad \equiv$$

So too in each of the special products (3), (4), and (5) bear in mind that the symbols x and a can be replaced by another variable, a number, or a more complicated expression.

EXAMPLE 7 Squares

Find each of the following products.

(a) $(3x + 7)^2$ (b) $(5x - 4)^2$

Solution (a) From (3) with x replaced by $3x$ and a replaced by 7, we have
$$(3x + 7)^2 = (3x)^2 + 2(3x)(7) + (7)^2$$
$$= 9x^2 + 42x + 49.$$

(b) From (3) with x replaced by $5x$ and a replaced by -4 we have
$$(5x - 4)^2 = (5x)^2 + 2(5x)(-4) + (-4)^2$$
$$= 25x^2 - 40x + 16.$$

EXAMPLE 8   Cubes

Find each of the following products.

(a) $\left(\tfrac{1}{2}x + 2\right)^3$ (b) $\left(4x - \dfrac{1}{x^2}\right)^3$

Solution (a) With x replaced by $\tfrac{1}{2}x$, a replaced by 2, and using the laws of exponents, (4) yields
$$\left(\tfrac{1}{2}x + 2\right)^3 = \left(\tfrac{1}{2}x\right)^3 + 3\left(\tfrac{1}{2}x\right)^2(2) + 3\left(\tfrac{1}{2}x\right)(2) + (2)^3$$
$$= \tfrac{1}{8}x^3 + \tfrac{3}{2}x^2 + 3x + 8.$$

(b) Before proceeding, we note that the answer we obtain will not be a polynomial because the two-term expression $4x - 1/x^2$ is, strictly speaking, not a binomial. Nevertheless, (4) can be used with the symbol x replaced by $4x$ and the symbol a replaced by $-1/x^2$:

$$\left(4x - \dfrac{1}{x^2}\right)^3 = (4x)^3 + 3(4x)^2\left(-\dfrac{1}{x^2}\right) + 3(4x)\left(-\dfrac{1}{x^2}\right)^2 + \left(-\dfrac{1}{x^2}\right)^3$$
$$= 64x^3 - 48 + \dfrac{12}{x^3} - \dfrac{1}{x^6}.$$

EXAMPLE 9   Difference of Two Squares

Find the product $(6y + \sqrt{2})(6y - \sqrt{2})$.

Solution With x replaced by $6y$ and a replaced by $\sqrt{2}$, the product formula (5) gives
$$(6y + \sqrt{2})(6y - \sqrt{2}) = (6y)^2 - (\sqrt{2})^2 = 36y^2 - 2.$$

☐ **Polynomials in Two Variables** So far we have primarily considered polynomials in one variable. We may have polynomials in x or in other variables, such as $2y^2 - y = 5$ or $\sqrt{2}z^3 - 17$, or we may have polynomials in two or more variables. A **polynomial in two variables x and y** is a sum of monomials (or terms) of the form $ax^n y^m$, where a is a real number, x and y are variables, and n and m are nonnegative integers. Examples are

$$5x - 2y, \quad x^2 + xy - y^2, \quad \text{and} \quad 8x^3y + xy^2 - x + \tfrac{7}{2}.$$

Similarly, a **polynomial in three variables x, y, and z** is a sum of monomials of the form $ax^n y^m z^k$, where n, m, and k are nonnegative integers. For example, $xy^2z^3 - 2xy + z - 1$ is a polynomial in three variables. Polynomials of four or more variables are defined in a similar manner. For example, $xy + 5y - 3yz^3 + 6xy^2z^3w^4$ is a polynomial in four variables.

We add, subtract, and multiply polynomials of several variables using the properties of real numbers just as we did for polynomials in one variable.

EXAMPLE 10   Sum of Two Polynomials in x and y

Find the sum of $xy^3 + x^3y - 3$ and $x^3 - y^3 + 3xy^3 - x^3y$.

Solution   We simply add like terms that are indicated by the same color:

$$(xy^3 + x^3y - 3) + (x^3 - y^3 + 3xy^3 - x^3y)$$
$$= x^3 - y^3 + 4xy^3 + 0x^3y - 3$$
$$= x^3 - y^3 + 4xy^3 - 3.$$

EXAMPLE 11   Product of Two Polynomials in x and y

Multiply $x + y$ and $x^2 - xy + y^2$.

Solution   As in Example 4 we use the distributive law several times and then combine like terms:

$$(x + y)(x^2 - xy + y^2) = x(x^2 - xy + y^2) + y(x^2 - xy + y^2)$$
$$= x^3 - x^2y + xy^2 + x^2y - xy^2 + y^3$$
$$= x^3 + y^3.$$

In Example 11 we have verified one of our last two special products formulas. The **difference of two cubes** is

$$(x - a)(x^2 + ax + a^3) = x^3 - a^3, \tag{6}$$

whereas the **sum of two cubes** is

$$(x + a)(x^2 - ax + a^3) = x^3 + a^3. \tag{7}$$

Formulas (6) and (7) are probably more important in factoring polynomials than they are as formulas to remember for carrying out a multiplication.

Division by a monomial uses the properties of fractions and the laws of exponents, as shown in Example 12. Division of two polynomials is more complicated and is discussed in Chapter 4.

EXAMPLE 12   Dividing Two Polynomials

Divide $15xy^3 + 25x^2y^2 - 5xy^2$ by $5xy^2$.

Solution   We use the common denominator property

$$\frac{a}{d} + \frac{b}{d} + \frac{c}{d} = \frac{a + b + c}{d}$$

reading it right to left. With the identification $d = 5xy^2$ and the laws of exponents we get

$$\frac{15xy^3 + 25x^2y^2 - 5xy^2}{5xy^2} = \frac{15xy^3}{5xy^2} + \frac{25x^2y^2}{5xy^2} - \frac{5xy^2}{5xy^2}$$
$$= 3y + 5x - 1.$$

EXAMPLE 13 Using a Product Formula Twice

Find the product $(2x + y)(2x - y)(4x^2 + y^2)$.

Solution We use (5) twice in succession:

$$(2x + y)(2x - y)(4x^2 + y^2) = \overbrace{[(2x + y)(2x - y)]}^{\text{use (5) with } x \text{ replaced by } 2x \text{ and } a \text{ replaced by } y}(4x^2 + y^2)$$

$$= \overbrace{(4x^2 - y^2)}^{\text{use (5) again with } x \text{ replaced by } 4x^2 \text{ and } a \text{ replaced by } y^2}(4x^2 + y^2)$$

$$= 16x^4 - y^4.\quad\equiv$$

The more familiar you are with these special products (2)–(5) the easier it will be to understand factoring, which is reviewed in the next section.

NOTES FROM THE CLASSROOM

A very common mistake when subtracting polynomials in the horizontal format is to fail to apply the distributive property. The sign of each term of the polynomial being subtracted must be changed:

$$-(2x^3 - 3x - 4) = (-1)(2x^3 - 3x - 4)$$
$$= (-1)(2x^3) + (-1)(-3x) + (-1)(-4)$$
$$= -2x^3 + 3x + 4 \neq -2x^3 - 3x - 4.$$

R.6 Exercises
Answers to selected odd-numbered problems begin on page ANS-2.

In Problems 1–8, find the value of the polynomial for **(a)** $x = -3$, **(b)** $x = \frac{1}{2}$, and **(c)** $x = 0$.

1. $x^2 - 5x + 6$
2. $\sqrt{2}x^2 + 3x - 4\sqrt{2}$
3. $x - 3x^2 + 6x^3$
4. $x^4 - x^3 + x^2 - x + 1$
5. $\frac{1}{2}x - 1$
6. $(x - 1)^2 + (x - 1)$
7. $0.1x^2 - 0.5x + 0.2$
8. $(2x + 1)^3$

In Problems 9–16, determine whether the algebraic expression is a polynomial in one variable. If it is, give its degree and its leading coefficient.

9. $\sqrt{3} + 8x$
10. $0.5x^{10} - 1.7x^3 + 3.4x - 7.2$
11. $y^3 - y^2 + y^{1/3} - 7$
12. $t^4 - t^3 + t^{-1} - 1$
13. $7x^{100} - 4x^{99} + 26x^{101} - 5$
14. $3 + 2x - \sqrt{7}x^3 + \frac{1}{2}x^{-10}$
15. $\sqrt{r} - 4$
16. $z^2(5z^3 - 4z + 18)$

In Problems 17–30, perform the indicated operation and express the result as a polynomial in standard form.

17. $(3x^5 - 5x^2 + 4x - 7) + (x^3 - 3x^2 + 2x + 1)$
18. $(4x^{10} - 7x^5 + 1) + (3x^5 + 2x^2 - 7x + 1)$

19. $(y^3 - 3y^2 + 7y - 8) + (5y^3 + 4y^2 - 9y + 1)$
20. $(\sqrt{2}z^5 - 6z^3 + 17z + \sqrt[3]{6}) + (z^4 + 16z^3 - 5z + \sqrt{6})$
21. $(x^2 + 2x - 1) - (3x^4 - 4x^2 + 2x)$
22. $(3y^4 - 2y^2 + 8y - 16) - (6y^4 + 5y^2 + 10y - 11)$
23. $(3x^7 - 7x^6 + x^5 - 14) - (x^4 - 2x^2 + 8x)$
24. $(4s^{10} - 5s^5 + \frac{9}{2}) - (s^{10} + \frac{1}{2}s^5 - s + \frac{1}{2})$
25. $3(t^3 - 4t^2 + 6t - 3) + 5(-t^3 + 2t^2 - 9t + 11)$
26. $6(2x^4 - 5x^3 - 10x^2 + 4x - 8) - 4(-5x^4 + 7x^3 + 9x^2 - 3x - 13)$
27. $(2v + 4)(v^2 - 6v)$
28. $(w^2 - w + 1)(w^4 - w^2)$
29. $(y^2 + 2y - 4)(y^2 - y + 5)$
30. $(z^3 + 4z - 3)(2z^3 - 7z + 1)$

In Problems 31–38, perform the indicated operations and simplify.

31. $(8a^4 + 7a^2b^2 + 6b^4) + (7a^4 - a^3b + a^2b^2 - 8ab^3 + 5b^4)$
32. $(\sqrt{2}xy^3 - \sqrt{3}y^2) - (x^3 + y^3 - \sqrt{2}xy^3 + 6\sqrt{3}y^2 - \sqrt{5})$
33. $(2a - b)(3a^2 - ab + b^2)$
34. $(x^2 - xy + y)(5x - 3y^2)$
35. $\dfrac{5s^2(2rs - 8rs^2)}{2rs^3}$
36. $\dfrac{7p^3q^3 - 4p^2q^5}{p^2q^3}$
37. $\dfrac{4x^2y^2 - (2x^2y)^2 + 8x^8y^3}{4x^2y^2}$
38. $\dfrac{3a^2b^2c^2 - 2ab^2c + \sqrt{5}abc}{abc}$

In Problems 39–80, find the given product.

39. $(x - 1)(x + 2)$
40. $(4x - 5)(x + 3)$
41. $(2r^3 + 1)(r^3 - 7)$
42. $(v^2 + 3)(v^2 - 5)$
43. $(5t - 7)(2t + 8)$
44. $(3z - 5)(7z + 1)$
45. $(4\sqrt{x} + 1)(6\sqrt{x} - 2)$
46. $(2\sqrt{x} - 3)(5\sqrt{x} + 8)$
47. $(0.3x + 0.7)(10x + 2.1)$
48. $(1.2x + 0.4)(2x - 1.3)$
49. $(\frac{1}{2}x - \frac{1}{3})(2x + \frac{1}{4})$
50. $(\frac{2}{5}x + 5)(\frac{1}{5}x + 1)$
51. $(1 + 5b)^2$
52. $(2c - 4)^2$
53. $(5x + 2)(10x + 4)$
54. $(-3x^2 + 9)(x^2 - 3)$
55. $(2 + \sqrt{3x})(2 - \sqrt{3x})$
56. $[4(x + 1) + 3][4(x + 1) - 3]$
57. $(y^{-1} - 2x)(y^{-1} + 2x)$
58. $(2z^2 + z)(2z^2 - z)$
59. $(2x - 3)^3$
60. $(x + 5)^3$
61. $(x^2y^3 + 2)^3$
62. $\left(\dfrac{x^2 - 1}{x^2}\right)^3$
63. $(x + y)(x^2 + 2xy + y^2)$
64. $(2a^2 - 1)(4a^4 - 4a^2 + 1)$
65. $(a - 3)(a^2 + 3a + 9)$
66. $(2 - y)(4 + 2y + y^2)$
67. $(9 + y)(81 - 9y + y^2)$
68. $(x + z^2)(x^2 - xz^2 + z^4)$
69. $(5x - y)(5x + y)(25x^2 + y^2)$
70. $(2 - x + y)(2 - x - y)$
71. $(x + y + 1)^2$
72. $(x + x^2 + x^3)^2$
73. $(x + y + 1)^3$
74. $(x + x^2 + x^3)^3$
75. $(x^{2/3} - x^{1/3})(x^{2/3} + x^{1/3})$
76. $(\sqrt{x} - y + 1)(\sqrt{x} + y - 1)$
77. $\left(\dfrac{1}{y^2} - \dfrac{1}{x^2}\right)\left(\dfrac{1}{y^4} + \dfrac{1}{y^2x^2} + \dfrac{1}{x^4}\right)$
78. $(\sqrt{x} + \sqrt{y})(x - \sqrt{xy} + y)$
79. $(x^5 - x^2)(x - 1)$
80. $(2x^{1/2} - x)(x + 5)$

81. Write polynomials in standard form for **(a)** the volume and **(b)** the surface area of the solid object shown in FIGURE R.6.2.
82. Write a polynomial in the variables r and s for the area of the region (a rectangle with semicircular ends) shown in FIGURE R.6.3.

FIGURE R.6.2 Solid object in Problem 81

FIGURE R.6.3 Region in Problem 82

For Discussion

In Problems 83–88, answer true or false.

83. $(t + 1)^2 = t^2 + 1$ _____
84. The degree of the polynomial $x^4 - 3x^2 + x^5$ is 4. _____
85. The leading coefficient of $2y^3 - y^8 + 4$ is -1. _____
86. The expression $3r^{14} - \sqrt{2}r + \pi$ is a polynomial in the variable r. _____
87. $4t^3 + 3t - (2t^3 + t + 7) = 2t^3 + 4t + 7$ _____
88. The value of $z^4 - 3z + 1$ when $z = \sqrt{2}$ is $5 - 3\sqrt{2}$. _____

In Problems 89 and 90, the polynomials are of a single variable x.

89. If a polynomial of degree 2 and a polynomial of degree 3 are added, what is the degree of the resulting polynomial? What is the degree of their product?
90. For two polynomials of degree n, what can be said about the degree of their sum? Their product? Their difference?

R.7 Factoring Polynomials

≡ **Introduction** In the preceding section we multiplied polynomials. Now we reverse the procedure and try to write a polynomial as a product of other polynomials. This process is called **factoring**, and each polynomial in the product is called a **factor** of the original polynomial. For example, $3x^2$ and $x^2 + 2$ are factors of $3x^4 + 6x^2$ because

$$3x^4 + 6x^2 = 3x^2(x^2 + 2).$$

Generally, we seek polynomial factors of degree 1 or higher.

By factoring, we can sometimes replace a complicated expression by a product of linear factors. An example is

$$5x^3 + 6x^2 - 29x - 6 = (5x + 1)(x - 2)(x + 3).$$

Thus factoring can be very useful in simplifying expressions. As we will see in Chapter 1, it is particularly useful in solving equations. We will study factoring of polynomials in greater depth in Section 4.3.

In general, the first step in factoring any algebraic expression is to determine whether the terms have a common factor.

EXAMPLE 1   Factoring

Factor $6x^4y^4 - 4x^2y^2 + 10\sqrt{2}xy^3 - 2xy^2$.

Solution Since $2xy^2$ is a common factor of the terms, we have

$$6x^4y^4 - 4x^2y^2 + 10\sqrt{2}xy^3 - 2xy^2$$
$$= 2xy^2(3x^3y^2) - 2xy^2(2x) + 2xy^2(5\sqrt{2}y) - 2xy^2(1)$$
$$= 2xy^2(3x^3y^2 - 2x + 5\sqrt{2}y - 1).$$

≡

When the terms of an expression do not have a common factor, it may still be possible to factor by **grouping** the terms in an appropriate manner.

EXAMPLE 2 — Grouping

Factor $x^2 + 2xy - x - 2y$.

Solution Grouping the first two terms and the last two terms gives
$$x^2 + 2xy - x - 2y = (x^2 + 2xy) + (-x - 2y)$$
$$= x(x + 2y) + (-1)(x + 2y).$$

We observe the common factor $x + 2y$ and complete the factorization as
$$x^2 + 2xy - x - 2y = (x - 1)(x + 2y).$$

☐ **Factoring Quadratic Polynomials** It is *sometimes* possible to factor the quadratic polynomial $ax^2 + bx + c$, where a, b, and c are integers, as
$$(Ax + B)(Cx + D),$$
where the symbols A, B, C, and D also represent integers.

Initially, to simplify our discussion we assume that the quadratic polynomial has as its leading coefficient $a = 1$. If $x^2 + bx + c$ has a factorization using integer coefficients, then it will be of the form
$$(x + B)(x + D),$$
where B and D are integers. Carrying out the multiplication and comparing coefficients,
$$(x + B)(x + D) = x^2 + (B + D)x + BD = x^2 + bx + c,$$
with $B + D = b$ and $BD = c$,

we see that
$$B + D = b \quad \text{and} \quad BD = c.$$

Thus to factor $x^2 + bx + c$ with integer coefficients, we list all possible factorizations of c as a product of two integers B and D. We then check which, if any, of the sums $B + D$ equals b.

EXAMPLE 3 — Factoring a Polynomial

Factor $x^2 - 9x + 18$.

Solution With $b = -9$ and $c = 18$, we look for integers B and D such that
$$B + D = -9 \quad \text{and} \quad BD = 18.$$
We can write 18 as a product BD in the following ways:
$$1(18), \quad 2(9), \quad 3(6), \quad (-1)(-18), \quad (-2)(-9), \quad (-3)(-6).$$
Since -9 is the sum $B + D$ when $B = -3$ and $D = -6$, the factorization is
$$x^2 - 9x + 18 = (x - 3)(x - 6).$$

Note that it is always possible to check a factorization by multiplying the factors.

EXAMPLE 4 — Factoring a Polynomial

Factor $x^2 + 3x - 1$.

Solution The number -1 can be written as a product of two integers BD in only one way, namely, $(-1)(1)$. With $B = -1$ and $D = 1$ we conclude from
$$B + D = -1 + 1 \neq 3$$
that $x^2 + 3x - 1$ cannot be factored using integer coefficients.

It is more complicated to factor the general quadratic polynomial $ax^2 + bx + c$, with $a \neq 1$, since we must consider factors of a as well as of c. Finding the product and comparing coefficients

$$(Ax + B)(Cx + D) = ACx^2 + (AD + BC)x + BD = ax^2 + bx + c,$$

with $AC = a$, $BD = c$, $AD + BC = b$

we see that $ax^2 + bx + c$ factors as $(Ax + b)(Cx + D)$ if we can find integers A, B, and C that satisfy

$$AC = a, \quad AD + BC = b, \quad BD = c.$$

EXAMPLE 5 Factoring a Polynomial

Factor $2x^2 + 11x - 6$.

Solution The factors will be

$$(2x + \underline{})(1x + \underline{}),$$

where the blanks are to be filled with a pair of integers B and D whose product BD equals -6. Possible pairs are:

$$1 \text{ and } -6, \quad -1 \text{ and } 6, \quad 3 \text{ and } -2, \quad -3 \text{ and } 2.$$

Now we must check to see if one of the pairs gives 11 as the value of $AD + BC$ (the coefficient of the middle term), where $A = 2$ and $C = 1$. We find

$$2(6) + 1(-1) = 11;$$

therefore,

$$2x^2 + 11x - 6 = (2x - 1)(x + 6). \qquad \equiv$$

This general method can be applied to polynomials in two variables x and y of the form

$$ax^2 + bxy + cy^2,$$

where a, b, and c are integers.

EXAMPLE 6 Factoring

Factor $15x^2 + 17xy + 4y^2$.

Solution The factors could have the form

$$(5x + \underline{}y)(3x + \underline{}y) \quad \text{or} \quad (15x + \underline{}y)(1x + \underline{}y) \qquad (1)$$

There is no need to consider the cases

$$(-5x + \underline{}y)(-3x + \underline{}y) \quad \text{and} \quad (-15x + \underline{}y)(-x + \underline{}y).$$

(Why?) The blanks in (1) must be filled with a pair of integers whose product is 4. Possible pairs are

$$1 \text{ and } 4, \quad -1 \text{ and } -4, \quad 2 \text{ and } 2, \quad -2 \text{ and } -2.$$

We check each pair with the possible forms in (1) to see which combination, if any, gives a coefficient of 17 for the middle term. We find

$$15x^2 + 17xy + 4y^2 = (5x + 4y)(3x + y). \qquad \equiv$$

EXAMPLE 7 Factoring a Polynomial

Factor $2t^4 + 11t^2 + 12$.

Solution If we let $x = t^2$, then we can regard this expression as a quadratic polynomial in the variable x,
$$2x^2 + 11x + 12.$$
We then factor this quadratic polynomial. The factors will have the form
$$(x + \underline{})(2x + \underline{}), \tag{2}$$
where the blanks are to be filled with a pair of integers whose product is 12. Possible pairs are

$$1 \text{ and } 12, \quad -1 \text{ and } -12, \quad 2 \text{ and } 6, \quad -2 \text{ and } -6, \quad 3 \text{ and } 4, \quad -3 \text{ and } -4.$$

We check each pair with (2) to see which combination, if any, gives a coefficient of 11 for the middle term. We find
$$2x^2 + 11x + 12 = (x + 4)(2x + 3).$$
Substituting t^2 for x gives us the desired factorization
$$2t^4 + 11t^2 + 12 = (t^2 + 4)(2t^2 + 3). \qquad \equiv$$

In the preceding example you should verify that neither $t^2 + 4$ nor $2t^2 + 3$ will factor using integer coefficients, or for that matter, using real numbers.

☐ **Factorization Formulas** By reversing the special product formulas from Section R.6, we have the following important **factorization formulas**. These formulas are simply (3), (5), (6), and (7) of Section R.6 written in reverse manner.

Perfect square:	$x^2 + 2ax + a^2 = (x + a)^2$	(3)
Difference of two squares:	$x^2 - a^2 = (x + a)(x - a).$	(4)
Difference of two cubes:	$x^3 - a^3 = (x - a)(x^2 + ax + a^2)$	(5)
Sum of two cubes:	$x^3 + a^3 = (x + a)(x^2 - ax + a^2).$	(6)

As noted in Section R.6 the symbols x and a can be replaced by another variable, a number, or a more complicated expression.

EXAMPLE 8 Perfect Square

Factor $y^2 - 6y + 9$.

Solution With the symbol y playing the part of x and with $a = -3$ we see from (3) that
$$y^2 - 6y + 9 = y^2 + 2(-3)y + (-3)^2 = (y - 3)^2. \qquad \equiv$$

EXAMPLE 9 Difference of Two Squares

Factor $16x^4y^2 - 25$.

Solution By rewriting the expression as
$$16x^4y^2 - 25 = (4xy^2)^2 - 5^2$$
we recognize the difference of two squares. Thus from formula (4) with the symbol x replaced by $4x^2y$ and $a = 5$, we have
$$16x^4y^2 - 25 = (4x^2y)^2 - (5)^2$$
$$= (4x^2y - 5)(4x^2y + 5). \qquad \equiv$$

EXAMPLE 10 Sum of Two Cubes

Factor $8a^3 + 27b^6$.

Solution Because we can write the given expression as the sum of two cubes,

$$8a^3 + 27b^6 = (2a)^3 + (3b^2)^3,$$

it can be factored using formula (6). If we replace x by $2a$ and a by $3b^2$, then it follows from formula (6) that

$$\begin{aligned} 8a^3 + 27b^6 &= (2a)^3 + (3b^2)^3 \\ &= (2a + 3b^2)[(2a)^2 - (2a)(3b^2) + (3b^2)^2] \\ &= (2a + 3b^2)(4a^2 - 6ab^2 + 9b^4) \end{aligned}$$
≡

Observe that formulas (4)–(6) indicate that the difference of two squares and the sum and difference of two cubes *always* factor, provided that we do not restrict the coefficients to integers. For example, using formula (4) to factor $x^2 - 5$, we identify $a = \sqrt{5}$ so that

$$\begin{aligned} x^2 - 5 &= x^2 - (\sqrt{5})^2 \\ &= (x - \sqrt{5})(x + \sqrt{5}). \end{aligned}$$

We now consider an example in which a first factorization yields expressions that can be factored again. In general, we require that an expression be **factored completely**, that is, factored until none of the factors can themselves be factored into polynomials of degree 1 or higher with integer coefficients.

EXAMPLE 11 Two Different Methods

Factor completely $x^6 - y^6$.

Solution We can view the expression $x^6 - y^6$ in two ways: as a difference of two squares or as a difference of two cubes. Using the difference of two cubes, formula (5), we write

$$\begin{aligned} x^6 - y^6 &= (x^2)^3 - (y^2)^3 \\ &= (x^2 - y^2)(x^4 + x^2y^2 + y^4) \\ &= (x - y)(x + y)(x^4 + x^2y^2 + y^4). \end{aligned} \quad (7)$$

From this foregoing result we might conclude that the factorization is complete. However, treating the expression $x^6 - y^6$ as a difference of two squares is more revealing, since

$$\begin{aligned} x^6 - y^6 &= (x^3)^2 - (y^3)^2 \\ &= (x^3 - y^3)(x^3 + y^3) \quad \leftarrow \text{difference of two cubes and the sum of two cubes} \\ &= (x - y)(x^2 + xy + y^2)(x + y)(x^2 - xy + y^2) \\ &= (x - y)(x + y)(x^2 + xy + y^2)(x^2 - xy + y^2). \end{aligned} \quad (8)$$
≡

In Example 11, if we compare the results in the last lines of the factorizations in (7) and (8), we discover the additional factorization:

$$x^4 + x^2y^2 + y^4 = (x^2 + xy + y^2)(x^2 - xy + y^2).$$

We leave it to you to check that neither of the expressions on the right-hand side of the equality will factor further.

R.7 Exercises

Answers to selected odd-numbered problems begin on page ANS-2.

In Problems 1–10, factor the polynomial by finding a common factor or by grouping.

1. $12x^3 + 2x^2 + 6x$
2. $6x^3y^4 - 3\sqrt{3}x^2y^2 - 3x^2y + 3xy$
3. $2y^2 - yz + 6y - 3z$
4. $6x^5y^5 + \sqrt{2}x^2y^3 + 14xy^3$
5. $15at + 3bt + 5as + bs$
6. $3a^2b^3 - 3\sqrt{2}a^4b^2 + 9a^2b$
7. $xyz^3 - xy^3z + x^3yz$
8. $x^3 + 2x + x^2 + 2$
9. $2p^3 - p^2 + 2p - 1$
10. $2uv - 5wz + 2uz - 5wv$

In Problems 11–22, use the factorization formulas (3)–(6) to factor the given polynomial.

11. $36x^2 - 25$
12. $a^2 - 4b^2$
13. $4x^2y^2 - 1$
14. $49x^2 - 64y^2$
15. $x^4 - y^4$
16. $x^6 + y^6$
17. $x^8 - y^8$
18. $a^3 - 64b^3$
19. $8x^3y^3 + 27$
20. $y^3 + 125$
21. $y^6 - 1$
22. $1 - x^3$

In Problems 23–42, use techniques for factoring quadratic polynomials to factor the given polynomial, if possible.

23. $x^2 - 5x + 6$
24. $x^2 - 10x + 24$
25. $y^2 + 7y + 10$
26. $y^4 + 10y^2 + 21$
27. $x^4 - 3x^2 - 4$
28. $x^2 + 4x - 12$
29. $r^2 + 2r + 1$
30. $s^2 + 5s - 14$
31. $x^2 - xy - 2y^2$
32. $x^2 - 4xy + 3y^2$
33. $x^2 + 10x + 25$
34. $4x^2 + 12x + 9$
35. $s^2 - 8st + 16t^2$
36. $9m^2 - 6mv + v^2$
37. $2p^2 + 7p + 5$
38. $8q^2 + 2q - 3$
39. $6a^4 + 13a^2 - 15$
40. $10b^4 - 23b^2 + 12$
41. $2x^2 - 7xy + 3y^2$
42. $-3x^2 - 5xy + 12y^2$

In Problems 43–60, use any method to factor the given polynomial.

43. $(x^2 + 1)^3 + (y^2 - 1)^3$
44. $(4 - x^2)^3 - (4 - y^2)^3$
45. $x(x - y) + y(y - x)$
46. $x(x - y) - y(y - x)$
47. $(1 - x^2)^3 - (1 - y^2)^3$
48. $(x^2 - 4)^3 + (4 - y^2)^3$
49. $1 - 256v^8$
50. $s^8 - 6561$
51. $x^6 + 7x^3 - 8$
52. $z^{10} - 5z^5 - 6$
53. $r^3s^3 - 8t^3$
54. $25c^2d^2 - x^2y^2$
55. $a^3 + a^2b - b^3 - ab^2$
56. $p^3 - pq^2 + p^2q - q^3$
57. $4z^2 + 7zy - 2y^2$
58. $36x^2 + 12xy + y^2$
59. $16a^2 - 24ab + 9b^2$
60. $4m^2 + 2mm - 12n^2$

In Problems 61–70, use the factorization formulas (3) and (4) to factor the expression into linear factors. [*Hint:* Some coefficients will not be integers.]

61. $x^2 - 3$
62. $2r^2 - 1$
63. $5y^2 - 1$
64. $\frac{1}{4}a^2 - b^2$
65. $a^2 + a + \frac{1}{4}$
66. $t^2 - \frac{2}{5}t + \frac{1}{25}$
67. $a^2 - 2b^2$
68. $3u^2 - 4v^2$
69. $24 - x^2$
70. $x^2 - 2\sqrt{2}xy + 2y^2$

CHAPTER R REVIEW OF BASIC ALGEBRA

For Discussion

In Problems 71–74, answer true or false.

71. $x^2 + y^2 = (x + y)(x + y)$ ___
72. $a^3 + b^3 = (a + b)^3$ ___
73. $(r - 1)(r - 1) = r^2 + 1$ ___
74. $r^3 - s^3 = (r - s)(r^2 + rs + s^2)$ ___

In Problems 75–77, several of the factorization formulas are discussed in geometric terms.

In Book II of Euclid's *Elements* (c. 300 BCE), algebraic problems are discussed and solved in geometric terms, because the Greeks lacked algebraic notation. For example, the product of two positive numbers a and b is represented as the area of a rectangle whose sides have lengths a and b, respectively.

75. Explain how FIGURE R.7.1 justifies the factorization formula $a^2 + 2ab + b^2 = (a + b)^2$ for positive numbers a and b.
76. Explain how FIGURE R.7.2 justifies the factorization formula $a^2 - b^2 = (a - b)(a + b)$, where $a > b > 0$.

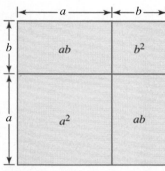

FIGURE R.7.1 Rectangles in Problem 75

FIGURE R.7.2 Rectangles in Problem 76

77. FIGURE R.7.3 suggests that the factorization formula for the difference of two cubes, $a^3 - b^3 = (a - b)(a^2 + ab + b^2)$ for $a > b > 0$, can be justified geometrically. Complete the proof. [*Hint*: Label the four boxes inside the cube and compute the volume of each.]

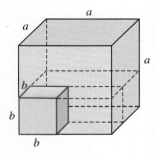

FIGURE R.7.3 Cube in Problem 77

R.8 Rational Expressions

☰ **Introduction** When one polynomial is divided by another, the result need not be a polynomial. A quotient of two polynomials is called a **rational expression**. For example,

$$\frac{2x^2 + 5}{x + 1} \quad \text{and} \quad \frac{3}{2x^3 - x + 8}$$

are rational expressions. The **domain** of the variable in a rational expression consists of all real numbers for which the value of the denominator is nonzero. For example, in $(2x^2 + 5)/(x + 1)$ the domain of the variable is $\{x \mid x \neq -1\}$.

In solving problems, we often must combine rational expressions and then simplify the results. Since a rational expression represents a real number, we can apply the properties of the real number system to combine and simplify rational expressions. The properties of fractions from Section R.1 are particularly helpful. We repeat the most frequently used ones here for convenience.

FREQUENTLY USED PROPERTIES OF REAL NUMBERS

For any real numbers a, b, c, and d:

(i) Cancellation
$$\frac{ac}{bc} = \frac{a}{b}, \quad c \neq 0$$

(ii) Addition or Subtraction
$$\frac{a}{b} \pm \frac{c}{b} = \frac{a \pm c}{b}$$

(iii) Multiplication
$$\frac{a}{b} \cdot \frac{c}{d} = \frac{ac}{bd}$$

(iv) Division
$$\frac{a}{b} \div \frac{c}{d} = \frac{a}{b} \cdot \frac{d}{c} = \frac{ad}{bc}$$

provided that each denominator is nonzero.

EXAMPLE 1 Simplifying

Simplify the rational expression $\dfrac{2x^2 - x - 1}{x^2 - 1}$.

Solution We factor the numerator and the denominator and cancel common factors using the cancellation property (i):

$$\frac{2x^2 - x - 1}{x^2 - 1} = \frac{(2x+1)(x-1)}{(x+1)(x-1)} = \frac{2x+1}{x+1}. \qquad \equiv$$

Note that in Example 1 the cancellation of the common factor $x - 1$ is valid only for those values of x such that $x - 1$ is nonzero; that is, for $x \neq 1$. However, since the expression $(2x^2 - x - 1)/(x^2 - 1)$ is not defined for $x = 1$, our simplification is valid for all real numbers in the domain of the variable x in the *original* expression. We emphasize that the equality

$$\frac{2x^2 - x - 1}{x^2 - 1} = \frac{2x+1}{x+1}$$

is not valid for $x = 1$, even though the right-hand side, $(2x + 1)/(x + 1)$, is defined for $x = 1$. Considerations of this sort will be important in the next chapter when we solve equations involving rational expressions.

Note ▶ For the remainder of this chapter we will assume without further comment that variables are restricted to values for which all denominators in an equation are nonzero.

EXAMPLE 2 Simplifying

Simplify the rational expression $\dfrac{4x^2 + 11x - 3}{2 - 5x - 12x^2}$.

Solution

$$\begin{aligned}
\frac{4x^2 + 11x - 3}{2 - 5x - 12x^2} &= \frac{(4x - 1)(x + 3)}{(1 - 4x)(2 + 3x)} \\
&= \frac{\cancel{(4x - 1)}(x + 3)}{-\cancel{(4x - 1)}(2 + 3x)} \quad \leftarrow \text{by cancellation property } (i) \\
&= -\frac{x + 3}{2 + 3x}.
\end{aligned}$$

□ **Least Common Denominator** In order to add or subtract rational expressions, we proceed just as we do when adding or subtracting fractions. We first find a common denominator and then apply property (*ii*). Although any common denominator will do, less work is involved if we use the **least common denominator** (**LCD**). This is found by factoring each denominator completely and forming a product of the distinct factors, using each factor with the highest exponent with which it occurs in any single denominator.

EXAMPLE 3 Least Common Denominator

Find the LCD of

$$\frac{1}{x^4 - x^2}, \quad \frac{x + 2}{x^2 + 2x + 1}, \quad \text{and} \quad \frac{1}{x}.$$

Solution Factoring the denominators in the rational expressions, we obtain

$$\frac{1}{x^2(x - 1)(x + 1)}, \quad \frac{x + 2}{(x + 1)^2}, \quad \text{and} \quad \frac{1}{x}. \quad \leftarrow \text{See (3) and (5) in Section R.6.}$$

The distinct factors of the denominators are x, $x - 1$, and $x + 1$. We use each factor with the highest exponent with which it occurs in any single denominator. Thus the LCD is $x^2(x - 1)(x + 1)^2$.

EXAMPLE 4 Combining Terms

Combine

$$\frac{x}{x^2 - 4} + \frac{1}{x^2 + 4x + 4}$$

and simplify the resulting rational expression.

Solution In factored form, the denominators are $(x - 2)(x + 2)$ and $(x + 2)^2$. Thus the LCD is $(x - 2)(x + 2)^2$. We use property (*i*) in reverse to rewrite each of the rational expressions with the LCD as denominator:

first term $\rightarrow \dfrac{x}{x^2 - 4} = \dfrac{x}{(x - 2)(x + 2)} = \dfrac{x(x + 2)}{(x - 2)(x + 2)(x + 2)} = \dfrac{x(x + 2)}{(x - 2)(x + 2)^2}$

second term $\rightarrow \dfrac{1}{x^2 + 4x + 4} = \dfrac{1}{(x + 2)^2} = \dfrac{1 \cdot (x - 2)}{(x + 2)^2(x - 2)} = \dfrac{x - 2}{(x - 2)(x + 2)^2}.$

Then using property (*ii*) we add and simplify:

$$\frac{x}{x^2-4} = \frac{1}{x^2+4x+4} = \frac{x(x+2)}{(x-2)(x+2)^2} + \frac{x-2}{(x-2)(x+2)^2}$$
$$= \frac{x(x+2)+x-2}{(x-2)(x+2)^2}$$
$$= \frac{x^2+2x+x-2}{(x-2)(x+2)^2}$$
$$= \frac{x^2+3x-2}{(x-2)(x+2)^2}.$$

≡

To multiply or divide rational expressions, we apply property (*iii*) or property (*iv*) and then simplify.

EXAMPLE 5   Combining Terms

Combine

$$\frac{x}{5x^2+21x+4} \cdot \frac{25x^2+10x+1}{3x^2+x}$$

and simplify the resulting rational expression.

Solution We begin by using property (*iii*):

$$\frac{x}{5x^2+21x+4} \cdot \frac{25x^2+10x+1}{3x^2+x} = \frac{x(25x^2+10x+1)}{(5x^2+21x+4)(3x^2+x)}$$
$$= \frac{\cancel{x}(5x+1)\cancel{(5x+1)}}{\cancel{(5x+1)}(x+4)\cancel{x}(3x+1)} \quad \leftarrow \text{factoring numerator and denominator and cancellation}$$
$$= \frac{5x+1}{(x+4)(3x+1)}.$$

≡

EXAMPLE 6   Combining Terms

Combine

$$\frac{2x^2+9x+10}{x^2+4x+3} \div \frac{2x+5}{x+3}$$

and simplify the resulting rational expression.

Solution We begin by writing the given expression as a product:

$$\frac{2x^2+9x+10}{x^2+4x+3} \div \frac{2x+5}{x+3} = \frac{2x^2+9x+10}{x^2+4x+3} \cdot \frac{x+3}{2x+5} \quad \leftarrow \text{by property (\textit{iv})}$$
$$= \frac{(2x^2+9x+10)(x+3)}{(x^2+4x+3)(2x+5)} \quad \leftarrow \text{by property (\textit{iii})}$$
$$= \frac{\cancel{(2x+5)}(x+2)\cancel{(x+3)}}{\cancel{(x+3)}(x+1)\cancel{(2x+5)}} \quad \leftarrow \text{factoring numerator and denominator and cancellation}$$
$$= \frac{x+2}{x+1}.$$

≡

☐ **Fractional Expressions** A quotient of two algebraic expressions that are not polynomials, such as $(\sqrt{x} - 1)/(\sqrt[3]{x} + 1)$, is called a **fractional expression**. The techniques used to simplify fractional expressions are similar to those used for rational expressions.

EXAMPLE 7   Simplifying

Simplify
$$\frac{\dfrac{1}{x} - \dfrac{x}{x+1}}{1 + \dfrac{1}{x}}.$$

Solution First we obtain single rational expressions for the numerator,
$$\frac{1}{x} - \frac{x}{x+1} = \frac{1(x+1)}{x(x+1)} - \frac{x \cdot x}{(x+1)x} = \frac{x+1-x^2}{x(x+1)} = \frac{-x^2+x+1}{x(x+1)}$$

and the denominator,
$$1 + \frac{1}{x} = \frac{x}{x} + \frac{1}{x} = \frac{x+1}{x}.$$

Hence the given expression is the same as
$$\frac{\dfrac{1}{x} - \dfrac{x}{x+1}}{1 + \dfrac{1}{x}} = \frac{\dfrac{-x^2+x+1}{x(x+1)}}{\dfrac{x+1}{x}}.$$

Next we apply property (*iv*) to this quotient to obtain
$$\frac{\dfrac{-x^2+x+1}{x(x+1)}}{\dfrac{x+1}{x}} = \frac{-x^2+x+1}{\cancel{x}(x+1)} \cdot \frac{\cancel{x}}{x+1} = \frac{-x^2+x+1}{(x+1)^2}. \quad \equiv$$

An alternative method of simplifying a complex fraction is to multiply both the numerator and the denominator by the LCD of all the fractions that occur in the complex fraction. Using this approach here, we multiply the numerator and the denominator by $x(x+1)$ and simplify as follows:

$$\frac{\dfrac{1}{x} - \dfrac{x}{x+1}}{1 + \dfrac{1}{x}} = \frac{\left(\dfrac{1}{x} - \dfrac{x}{x+1}\right) \cdot x(x+1)}{\left(1 + \dfrac{1}{x}\right) \cdot x(x+1)}$$
$$= \frac{(x+1) - x^2}{x(x+1) + (x+1)}$$
$$= \frac{-x^2+x+1}{(x+1)(x+1)} = \frac{-x^2+x+1}{(x+1)^2}.$$

The techniques discussed in this section can often be applied to expressions containing negative exponents, as we see in the following example.

EXAMPLE 8 — Simplifying

Simplify $(a^{-1} + b^{-1})^{-1}$.

Solution We first replace all negative exponents by the equivalent quotients and then use the properties of fractions to simplify the resulting algebraic expression:

$$(a^{-1} + b^{-1})^{-1} = \frac{1}{a^{-1} + b^{-1}} \qquad \leftarrow \text{reciprocal of } a^{-1} + b^{-1}$$

$$= \frac{1}{\frac{1}{a} + \frac{1}{b}} = \frac{1}{\frac{b + a}{ab}} \qquad \leftarrow \text{reciprocal of } a \text{ and } b \text{ and LCD}$$

$$= \frac{ab}{b + a}. \qquad \leftarrow \text{property } (iv)$$

EXAMPLE 9 — Combining Terms

Combine

$$\frac{x}{\sqrt{y}} + \frac{y}{\sqrt{x}}$$

and simplify the resulting fractional expression.

Solution First we find a common denominator and then add:

$$\frac{x}{\sqrt{y}} + \frac{y}{\sqrt{x}} = \frac{x\sqrt{x}}{\sqrt{y}\sqrt{x}} + \frac{y\sqrt{y}}{\sqrt{x}\sqrt{y}} = \frac{x\sqrt{x} + y\sqrt{y}}{\sqrt{x}\sqrt{y}}.$$

If we desire to rationalize the denominator, the final result would be

$$\frac{x\sqrt{x} + y\sqrt{y}}{\sqrt{x}\sqrt{y}} \cdot \frac{\sqrt{x}\sqrt{y}}{\sqrt{x}\sqrt{y}} = \frac{x^2\sqrt{y} + y^2\sqrt{x}}{xy}.$$

Examples 10 and 11 illustrate how to simplify certain types of fractional expressions that occur in calculus.

EXAMPLE 10 — Simplifying

Simplify $\dfrac{\dfrac{1}{x + h} - \dfrac{1}{x}}{h}$.

Solution We begin by combining the terms in the numerator:

$$\frac{\frac{1}{x + h} - \frac{1}{x}}{h} = \frac{\frac{x - (x + h)}{(x + h)x}}{h} = \frac{\frac{-h}{x(x + h)}}{h}.$$

Then by properties (i) and (iv),

$$\frac{\frac{1}{x + h} - \frac{1}{x}}{h} = \frac{-\not{h}}{x(x + h)\not{h}} = \frac{-1}{x(x + h)}.$$

EXAMPLE 11 — Simplifying

Combine

$$(2x)(x - 1)^{1/2} + \left(\tfrac{1}{2}\right)(x - 1)^{-1/2}(x^2)$$

and simplify the resulting fractional expression.

CHAPTER R REVIEW OF BASIC ALGEBRA

Solution In the second term we use $(x-1)^{-1/2} = 1/(x-1)^{1/2}$ and then use $2(x-1)^{1/2}$ as the LCD:

$$(2x)(x-1)^{1/2} + \left(\tfrac{1}{2}\right)(x-1)^{-1/2}(x^2) = (2x)(x-1)^{1/2} + \frac{x^2}{2(x-1)^{1/2}}$$

$$= \frac{(2)(2x)(x-1) + x^2}{2(x-1)^{1/2}}$$

$$= \frac{4x^2 - 4x + x^2}{2(x-1)^{1/2}}$$

$$= \frac{5x^2 - 4x}{2(x-1)^{1/2}} \qquad \equiv$$

R.8 Exercises
Answers to selected odd-numbered problems begin on page ANS-2.

In Problems 1–8, simplify the rational expression.

1. $\dfrac{x^2 + 3x + 2}{x^2 + 6x + 8}$

2. $\dfrac{v^4 + 4v^2 + 4}{4 - v^4}$

3. $\dfrac{z^2 - 9}{z^3 + 27}$

4. $\dfrac{x^2 - 2xy - 3y^2}{x^2 - 4xy + 3y^2}$

5. $\dfrac{3x^2 - 7x - 20}{2x^2 - 5x - 12}$

6. $\dfrac{4y^2 + 20y + 25}{2y^3 + 3y^2 - 5y}$

7. $\dfrac{w^3 - 9w}{w^3 - 6w^2 + 9w}$

8. $\dfrac{a^2b + ab^2}{a^2 - b^2}$

In Problems 9–16, find the least common denominator (LCD) of the rational expressions.

9. $\dfrac{1}{x^2 + x - 2}, \dfrac{4}{x + 2}$

10. $\dfrac{5}{v^2 + 2v + 1}, \dfrac{v}{v^2 - 3v - 4}$

11. $\dfrac{10}{b^3 + b^2 - 6b}, \dfrac{1}{b^3 - 6b^2}, \dfrac{b}{b - 2}$

12. $\dfrac{1}{x^2 - 10x + 25}, \dfrac{x}{x^2 - 25}, \dfrac{1}{x^2 + 10x + 25}$

13. $\dfrac{1}{c^2 + c}, \dfrac{c}{c^2 + 2c + 1}, \dfrac{1}{c^2 - 1}$

14. $\dfrac{p}{p + r}, \dfrac{r}{p^2 + 2pr + r^2}, \dfrac{1}{p^3 + r^3}$

15. $\dfrac{1}{x^3 - x^2}, \dfrac{x}{x^2 - 1}, \dfrac{1}{x^3 + 2x^2 + x}$

16. $\dfrac{y + 5}{3y^3 - 14y^2 - 5y}, \dfrac{1}{y}, \dfrac{y + 5}{y^2 - 5y}$

In Problems 17–42, combine terms and simplify the rational expression.

17. $\dfrac{4x}{4x + 5} + \dfrac{5}{4x + 5}$

18. $\dfrac{3}{s - 2} + \dfrac{4}{2 - s}$

19. $\dfrac{7z}{7z - 1} - \dfrac{1}{1 - 7z}$

20. $\dfrac{3}{a - 2} - \dfrac{6}{a^2 + 4}$

21. $\dfrac{2x}{x + 1} + \dfrac{5}{x^2 - 1}$

22. $\dfrac{b}{2b + 1} - \dfrac{2b}{b - 2}$

23. $\dfrac{y}{y - x} - \dfrac{x}{y + x}$

24. $\dfrac{x}{x - y} + \dfrac{x}{y - x}$

25. $\dfrac{2}{r^2 - r - 12} + \dfrac{r}{r + 3}$

26. $\dfrac{1}{w + 3} + \dfrac{w}{w + 1} + \dfrac{w^2 + 1}{w^2 + 4w + 3}$

27. $\dfrac{x}{2x^2 + 3x - 2} - \dfrac{1}{2x - 1} - \dfrac{4}{x + 2}$

28. $\dfrac{z}{2z + 3} - \dfrac{3}{4z^2 - 3z - 1} + \dfrac{4z + 1}{2z^2 + z - 3}$

29. $\dfrac{t - 4}{t + 3} \cdot \dfrac{t + 5}{t - 2}$

30. $\dfrac{x^2 + x}{x^2 - 1} \cdot \dfrac{x + 1}{x^2}$

31. $(x^2 - 2x + 1) \cdot \dfrac{x + 1}{x^3 - 1}$

32. $\dfrac{2p + 8}{p - 1} \cdot \dfrac{p + 4}{2p}$

33. $\dfrac{6x + 5}{3x + 3} \cdot \dfrac{x + 1}{6x^2 - 7x - 10}$

34. $\dfrac{1 + x}{2 + x} \cdot \dfrac{x^2 + x - 12}{3 + 2x - x^2}$

35. $\dfrac{u + 1}{u + 2} \div \dfrac{u + 1}{u + 7}$

36. $\dfrac{3w + 1}{w - 4} \div \dfrac{2w + 1}{w}$

37. $\dfrac{x}{x + 4} \div \dfrac{x + 5}{x}$

38. $\dfrac{x - 3}{x + 1} \div \dfrac{x + 1}{2x + 1}$

39. $\dfrac{q^2 - 1}{q^2 + 2q - 3} \div \dfrac{q - 4}{q + 3}$

40. $\dfrac{x^2 - 3x + 2}{x^2 - 7x + 12} \div \dfrac{x - 2}{x - 3}$

41. $\dfrac{s^2 - 5s + 6}{s^2 + 7s + 10} \div \dfrac{2 - s}{s + 2}$

42. $\dfrac{x}{x + y} \div \dfrac{y}{x + y}$

In Problems 43–64, simplify the given fractional expression.

43. $\dfrac{\dfrac{1}{x^2} - x}{\dfrac{1}{x^2} + x}$

44. $\dfrac{\dfrac{1}{s} + \dfrac{1}{t}}{\dfrac{1}{s} - \dfrac{1}{t}}$

45. $\dfrac{z + \dfrac{1}{2}}{2 + \dfrac{1}{z}}$

46. $\dfrac{\dfrac{1 + r}{r} + \dfrac{r}{1 - r}}{\dfrac{1 - r}{r} + \dfrac{r}{1 + r}}$

47. $\dfrac{x^2 + xy + y^2}{\dfrac{x^2}{y} - \dfrac{y^2}{x}}$

48. $\dfrac{\dfrac{a}{a - 1} - \dfrac{a + 1}{a}}{1 - \dfrac{a}{a - 1}}$

49. $\dfrac{\dfrac{1}{(x + h)^2} - \dfrac{1}{x^2}}{h}$

50. $\dfrac{\dfrac{1}{2x + 2h + 1} - \dfrac{1}{2x + 1}}{h}$

51. $(a^{-2} - b^{-2})^{-1}$

52. $\dfrac{a + b}{a^{-1} + b^{-1}}$

53. $\dfrac{u^{-2} - v^{-2}}{u^2 v^2}$

54. $\dfrac{u^{-2} + v^{-2}}{u^2 + v^2}$

55. $\dfrac{1}{\sqrt{u}} + \dfrac{1}{\sqrt{w}}$

56. $\dfrac{v}{\sqrt{z}} - \dfrac{z}{\sqrt{v}}$

57. $\dfrac{1 + \dfrac{1}{\sqrt{x}}}{1 + \dfrac{1}{\sqrt{y}}}$

58. $\dfrac{\dfrac{1}{a} + \dfrac{1}{\sqrt{a}}}{\dfrac{1}{b} - \dfrac{1}{\sqrt{b}}}$

59. $\dfrac{\dfrac{1}{(x + h)^3} - \dfrac{1}{x^3}}{h}$

60. $\dfrac{\dfrac{2}{3x + 3h} - \dfrac{2}{3x}}{h}$

61. $\dfrac{\dfrac{5}{2x+2h-1}-\dfrac{5}{2x-1}}{h}$

62. $(x^2-1)(\tfrac{1}{3})(x+1)^{-2/3}+(x+1)^{1/3}(2x)$

63. $\dfrac{(x^2+1)(\tfrac{1}{2})(x^{-1/2})-(x^{1/2})(2x)}{(x^2+1)^2}$

64. $\dfrac{(x^2+8)^{1/5}(5)-(5x)(\tfrac{1}{5})(x^2+8)^{-4/5}(2x)}{[(x^2+8)^{1/5}]^2}$

Miscellaneous Applications

65. **Resistance in a Circuit** If three resistors in an electrical circuit with resistances R_1, R_2, and R_3 ohms, respectively, are connected in parallel, then the resistance (in ohms) of the combination is given by

$$\dfrac{1}{\dfrac{1}{R_1}+\dfrac{1}{R_2}+\dfrac{1}{R_3}}.$$

Simplify this fractional expression.

66. **Optics** In the field of optics, if p is the distance from the object to the lens and q is the distance from the image to the lens, then the focal length of the lens is given by

$$\dfrac{1}{\dfrac{1}{p}+\dfrac{1}{q}}.$$

Simplify this fractional expression.

CONCEPTS REVIEW *You should be able to give the meaning of each of the following concepts.*

Set:
  subset
  union
  intersection
  disjoint
Real numbers:
  natural number
  integer
  rational number
  irrational number
  negative number
  nonnegative number
  positive number
Multiplicative identity
Reciprocal
Distributive property
Real number line:
  origin
  coordinate

Algebraic expression
Order relations:
  less than
  less than or equal to
  greater than
  greater than or equal to
Absolute value
Triangle inequality
Distance on number line
Midpoint of a line segment
Laws of exponents
Exponent:
  base
  scientific notation
Radical:
  square root
  cube root
  nth root
Rationalizing the denominator

Polynomial in one variable:
  monomial
  binomial
  trinomial
  coefficient
  degree
  term
  leading coefficient
  constant term
Factor completely
Rational expression:
  domain
  numerator
  denominator
Polynomial in two variables
Least common denominator

CHAPTER R Review Exercises
Answers to selected odd-numbered problems begin on page ANS-3.

A. True/False

In Problems 1–26, answer true or false.

1. -3.3 is greater than -3. ___
2. Every real number has a reciprocal. ___
3. $0/0$ is a real number. ___
4. π is a rational number. ___
5. Every real number can be written as a quotient of two integers. ___
6. No irrational number can be written as a fraction. ___
7. $\sqrt{(-10)^2} = -10$ ___
8. $\sqrt{100} = \pm 10$ ___
9. For $p > 0$, $\dfrac{p^{1/2}}{p^{-1/2}} = p$. ___
10. If $x^{1/n} = r$, then $r^n = x$. ___
11. The LCD of $\dfrac{1}{(r+2)^2}$ and $\dfrac{1}{(r+3)^3(r+2)}$ is $(r+3)^3(r+2)^3$. ___
12. For $a > 0$, $m \geq 2$ and $n \geq 2$ positive integers, $\sqrt[n]{a^m} = (\sqrt[n]{a})^m$. ___
13. For all t, $\dfrac{t-1}{1-t} = -1$. ___
14. $(u^{-2} + v^{-2})^{-1} = u^2 + v^2$ ___
15. $\dfrac{x+y}{x} = y$ ___
16. $|-6x| = 6|x|$ ___
17. If a and b are real numbers such that $a < b$, then $a^2 < b^2$. ___
18. Every real number x possesses a multiplicative inverse. ___
19. The algebraic expression $6x^{-2} + \sqrt{2}x$ is not a polynomial. ___
20. The cube root of a negative number is undefined. ___
21. $((x^{-1})^{-1})^{-1} = \dfrac{1}{x}$, $x \neq 0$ ___
22. $(a + b + c)(a + b - c) = (a + b)^2 - c^2$ ___
23. $\dfrac{2+3}{4+5} = \dfrac{2}{4} + \dfrac{3}{5}$ ___
24. $(-1)(-a + b - c) = a + b - c$ ___
25. The sum of two rational numbers is rational. ___
26. The sum of two irrational numbers is irrational. ___

B. Fill in the Blanks

In Problems 1–22, fill in the blanks.

1. The first three nonnegative integers are _____.
2. $-10(x - y) = -10x + 10y$ is an example of the _____ law.
3. The quotient C/d of a circle's circumference C and its diameter d is a _____ (rational or irrational) number.
4. On the number line, the _____ of the segment joining -1 and 5 is 2.
5. Geometrically, $a < b$ means that the point corresponding to a on the number line lies to the _____ of the point corresponding to b.

6. If x is a negative, then $|x| = $ _____.
7. The _____ of $x \neq 0$ can be written as $1/x$ or as x^{-1}.
8. For $x \neq 0$, $x^0 = $ _____.
9. Using rational exponents, $\sqrt{x}\sqrt{x} = x$ _____.
10. The domain of the variable x in $(3x + 1)/(x^2 - 1)$ is _____.
11. The expression $3x^4 - x^2 + 5x$ is a _____ of degree _____ with leading coefficient _____ and constant term _____.
12. In simplifying $x(x + 2)/((x - 2)(x + 2))$ to $x/(x - 2)$, we used the _____ property.
13. The distance from a to b is given by _____.
14. On the number line, the absolute value of a number measures its distance to the _____.
15. The number 4.2×10^{-5} is written in _____.
16. The real numbers a and b for which $\sqrt{a^2 + b^2} = a + b$ are _____.
17. The sets $\{1, 3, 5\}$ and $\{2, 4\}$ with no common elements are said to be _____.
18. On the number line, if the distance between x and 7 is 3, then x is _____.
19. In the expression x^3, x is called the _____ and 3 is called the _____.
20. The expression $\sqrt[4]{x^2 + y^2}$ is called a _____ of index _____.
21. For real numbers x and y, $xy = yx$ is an example of the _____ law of multiplication.
22. If $a < b$, then _____ is positive.

C. Review Exercises

In Problems 1–6, find the indicated set if $A = \{1, 3, 5, 7, 9\}$, $B = \{1, 2, 3, 4, 5\}$, and $C = \{2, 4, 6, 8\}$.

1. $A \cup B$
2. $A \cap B$
3. $(A \cup B) \cap C$
4. $(A \cap C) \cup B$
5. $(A \cup B) \cup C$
6. $(A \cap B) \cap C$

In Problems 7 and 8, write the given statement as an inequality.

7. $x - y$ is greater than or equal to 10.
8. z is nonnegative.

In Problems 9–12, insert the appropriate symbol: $<$, $>$, or $=$.

9. -1.4, $-\sqrt{2}$
10. 0.50, $\frac{1}{2}$
11. $\frac{2}{3}$, 0.67
12. -0.9, -0.8

In Problems 13–18, find the indicated absolute value.

13. $|\sqrt{8} - 3|$
14. $|-(\sqrt{15} - 4)|$
15. $|x^2 + 5|$
16. $\dfrac{|x|}{|-x|}$, $x \neq 0$
17. $|t + 5|$, if $t < -5$
18. $|r - s|$, if $r > s$

In Problems 19 and 20, find (a) the distance between the given points and (b) the coordinate of the midpoint of the line segment joining the given points.

19. -3.5, 5.8
20. $\sqrt{2}$, $-\sqrt{2}$

In Problems 21–38, eliminate negative and zero exponents and simplify. Assume that all variables are positive.

21. $(3uv^2)(6u^2v^3)^2$

22. $\dfrac{4a^3b^2}{16ab^3}$

23. $\dfrac{(2x^{-4}y^2)^{-1}}{x^0y^{-1}}$

24. $\dfrac{2x^5y^{-3}z^2}{6x^3y^{-3}z^{-5}}$

25. $\left(\dfrac{-8c^3d^6}{c^{-9}d^{12}}\right)^{2/3}$

26. $\dfrac{s^{-1}t^{-1}}{s^{-1}+t^{-1}}$

27. $\dfrac{x^{1/3}y^{-2/3}}{x^{4/3}y^{-7/9}}$

28. $((81w^2z^{-1/2})^{-1})^{1/4}$

29. $\dfrac{\sqrt[3]{125}}{25^{-1/2}}$

30. $\sqrt{\dfrac{ab^2c^4}{a^2}}$

31. $\sqrt{\sqrt[3]{(x^3y^9)^2}}$

32. $\sqrt{125xy}\sqrt{5yz}\sqrt{xz}$

33. $\sqrt[3]{-(p^{-2}q^3)^3}$

34. $\sqrt{(x^2+y^2)^2}$

35. $4\sqrt{xy}-\sqrt{\sqrt{x^2y^2}}+\sqrt{2xy}$

36. $\dfrac{\sqrt[3]{ab^3}-\sqrt[3]{b^4}}{b}$

37. $\sqrt[3]{x\sqrt{x}}$

38. $\dfrac{\sqrt{x^3}}{\sqrt[4]{x}\sqrt[4]{x}}$

In Problems 39 and 40, write the given number in scientific notation.

39. 0.0000007023

40. 158,000,000,000

In Problems 41 and 42, use scientific notation to evaluate the given expression.

41. $\dfrac{(16,000)(5,000,000)^2}{0.00008}$

42. $\sqrt{\dfrac{(0.0001)(480,000)}{0.03}}$

43. In 2009 it is estimated that American taxpayers spent $52.67 billion on the war on drugs. Write this number **(a)** in decimal form and **(b)** in scientific notation.

44. One nanosecond is 0.000000001 second.
    (a) Write 0.000000001 second in scientific notation.
    (b) One second equals how many nanoseconds?

In Problems 45–52, perform the indicated operations and simplify.

45. $(4x^3-3x^2+6x-2)-(x^2-3x+4)$

46. $(3x^4-\sqrt{2}x^2)+x(\sqrt{2}x+5)$

47. $(a+1)(a-2)(a+3)$

48. $\dfrac{c^2d^2-3cd^3+5c^3d^2}{cd^2}$

49. $(3z^4-2z)^2$

50. $(x^2+2y)^3$

51. $(3x^2+5y)(3x^2-5y)$

52. $(u-v)(u^2+uv+v^2)$

In Problems 53–60, factor the given polynomials using integer coefficients.

53. $12x^2-19x-18$

54. $16a^4-81b^4$

55. $2xy+3y-6x-9$

56. $4w^2+40wz+100z^2$

57. $8x^3+125y^6$

58. $2x^3+3x^2-18x-27$

59. $4t^4-4t^2s+s^2$

60. $125+75uv+15u^2v^2+u^3v^3$

In Problems 61–72, perform the indicated operations and simplify.

61. $\dfrac{1}{x-2} + \dfrac{2}{x+2} - \dfrac{1}{x^2-4}$

62. $\dfrac{x^2-1}{x} \div \dfrac{x^3-1}{x^2}$

63. $\left(\dfrac{1}{x} + \dfrac{1}{y}\right)\left(\dfrac{1}{x+y}\right)$

64. $(u^{-2} - v^{-2})(v-u)^{-1}$

65. $\dfrac{x + \dfrac{1}{x^2}}{x^2 + \dfrac{1}{x}}$

66. $\dfrac{\sqrt{c} + \dfrac{1}{d}}{d + \dfrac{1}{\sqrt{c}}}$

67. $\dfrac{\dfrac{r}{s} + 2}{\dfrac{s}{r} + 2}$

68. $\dfrac{1 + t^{-3}}{1 - t^{-3}}$

69. $\dfrac{\dfrac{4}{(x+h)^3} - \dfrac{4}{x^3}}{h}$

70. $\dfrac{\dfrac{1}{2(3+h)^2} - \dfrac{1}{2(3)^2}}{h}$

71. $(8x)\left(\tfrac{1}{4}\right)(2x+1)^{-3/4}(2) + (2x+1)^{1/4}(8)$

72. $\dfrac{(x+1)^{5/2}(3x^2) - (x^3)\left(\tfrac{5}{2}\right)(x+1)^{3/2}}{[(x+1)^{5/2}]^2}$

In Problems 73 and 74, rationalize the denominator and simplify.

73. $\dfrac{2}{\sqrt{s} + \sqrt{t}}$

74. $\dfrac{4}{\sqrt[5]{8}}$

In Problems 75 and 76, rationalize the numerator and simplify.

75. $\dfrac{\sqrt{2x+2h+3} - \sqrt{2x+3}}{h}$

76. $\dfrac{\sqrt{(x+h)^2 - (x+h)} - \sqrt{x^2-x}}{h}$

1 Equations and Inequalities

In This Chapter

1.1 Equations

1.2 Building an Equation from Words

1.3 Quadratic Equations

1.4 Complex Numbers

1.5 Linear Inequalities

1.6 Absolute-Value Equations and Inequalities

1.7 Polynomial and Rational Inequalities

Chapter 1 Review Exercises

The 1911 Oldsmobile Limited Touring Car plays a part in Problem 82 of the Chapter 1 Review Exercises.

A Bit of History Little is known of the personal life of the Greek mathematician **Diophantus**, who lived in Alexandria, Egypt, in the third century CE. His work, however, was of tremendous importance to the development of algebra and greatly influenced seventh-century European mathematicians. He wrote several treatises, the most famous of which are the 13 books of his *Arithmetica*. This series of texts deals primarily with special types of equations now called *diophantine equations*. Legend has it that the following epitaph marked Diophantus' grave:

Diophantus passed one-sixth of his life in childhood, one-twelfth in youth, and one-seventh as a bachelor. Five years after his marriage was born a son who died four years before his father at one-half his father's (final) age.

If the symbol x represents Diophantus' age at death, then the information above can be interpreted as the equation

$$\frac{x}{6} + \frac{x}{12} + \frac{x}{7} + 5 + \frac{x}{2} + 4 = x.$$

In this chapter we will examine techniques for solving various types of equations (including the one above) and inequalities. We will also discuss how to apply these methods to obtain solutions to practical problems.

1.1 Equations

≡ Introduction An **equation** is a statement that two expressions are equal, whereas an **inequality** states that one expression is less than another. A wide variety of real-world problems can be expressed as either equations or inequalities. We begin this section with some terminology describing equations and their solutions.

☐ Terminology When two expressions, and at least one of the expressions contains a variable, are set equal to each other we say that the mathematical statement is an **equation in one variable**. For example,

$$\sqrt{x-1} = 2, \quad x^2 - 1 = (x+1)(x-1), \quad \text{and} \quad |x+1| = 5$$

are equations in the variable x. A **solution**, or **root**, of an equation is a number that, when substituted into the equation, makes it a true statement. A number is said to **satisfy an equation** if it is a solution of the equation. To **solve an equation** means to find all its solutions.

EXAMPLE 1   Verifying a Solution

The number 2 is a solution of the equation $3x - 2 = x + 2$, because when substituted into the equation we get the true statement:

$$\underbrace{3(2) - 2}_{\text{left-hand side is } 6 - 2 = 4} = \underbrace{2 + 2}_{\text{right-hand side is } 2 + 2 = 4}$$

As we will see later, there are no other values of x that satisfy this equation. ≡

See Section R.6 for a review of the concept *domain of a variable*. ▶

An equation is called an **identity** if it is satisfied by all numbers in the domain of the variable. If there is at least one number in the domain of the variable for which the equation is *not* satisfied, then it is said to be a **conditional equation**.

EXAMPLE 2   An Identity and a Conditional Equation

(a) The equation

$$\frac{x^2 - 1}{x - 1} = x + 1$$

is satisfied by the set of all real numbers except $x = 1$. Because the number 1 is not in the domain of the variable, the equation is an identity.

(b) The number 3 is in the domain of the variable in the equation $4x - 1 = 2$, but it does not satisfy this equation since $4(3) - 1 \neq 2$. Thus, $4x - 1 = 2$ is a conditional equation. ≡

The set of all solutions of an equation is called its **solution set**. In Example 1 the solution set of $3x - 2 = x + 2$ is written $\{2\}$. You are encouraged to verify that the solution set of the equation $|x + 1| = 5$ is $\{-6, 4\}$.

☐ **Equivalent Equations** We say that two equations are **equivalent** if they have the same solutions, that is, their solution sets are exactly the same. For example,

$$2x - 1 = 0, \quad 2x = 1, \quad \text{and} \quad x = \tfrac{1}{2}$$

are equivalent equations. Generally, we solve an equation by finding an equivalent equation with solutions that are determined easily. The following operations yield equivalent equations.

THEOREM 1.1.1 Operations That Yield Equivalent Equations

(*i*) Add to or subtract from each side of an equation the same expression representing a real number.
(*ii*) Multiply or divide each side of an equation by the same expression representing a nonzero real number.

EXAMPLE 3   **A Simple Equation**

Solve $3x - 18 = 0$.

Solution We obtain the following list of equivalent equations:

$$\begin{aligned}
3x - 18 &= 0 \\
3x - 18 + 18 &= 0 + 18 \quad &\leftarrow \text{by (}i\text{) of Theorem 1.1.1} \\
3x &= 18 \\
\tfrac{1}{3}(3x) &= \tfrac{1}{3}(18) \quad &\leftarrow \text{by (}ii\text{) of Theorem 1.1.1} \\
x &= 6.
\end{aligned}$$

The solution set of the equation is $\{6\}$.   ≡

Since it is not unusual to make arithmetic or algebraic errors when solving an equation, it is always good practice to verify each solution by substituting it into the original equation. To verify the solution in Example 3, we substitute 6 for x in $3x - 18 = 0$: ◀ Note of Caution

$$\begin{aligned}
3(6) - 18 &\stackrel{?}{=} 0 \\
18 - 18 &\stackrel{?}{=} 0 \\
0 &= 0.
\end{aligned}$$

☐ **Linear Equations** An equation of the form

$$ax + b = 0, \quad a \neq 0, \tag{1}$$

where b is a real number, is called a **linear equation**. The equation in Example 3 is a linear equation. To solve (1), we proceed in a manner similar to Example 3:

$$\begin{aligned}
ax + b &= 0 \\
ax + b - b &= 0 - b \quad &\leftarrow \text{by (}i\text{) of Theorem 1.1.1} \\
ax &= -b \\
\frac{1}{a}(ax) &= \frac{1}{a}(-b) \quad &\leftarrow \text{by (}ii\text{) of Theorem 1.1.1} \\
x &= -\frac{b}{a}.
\end{aligned}$$

Thus the linear equation $ax + b = 0$, $a \neq 0$, has exactly one solution, $\{-b/a\}$.

EXAMPLE 4   A Linear Equation

Solve $2x - 7 = 5x + 6$.

Solution You should supply reasons why the following equations are equivalent.

$$2x - 7 = 5x + 6$$
$$2x - 7 - 5x = 5x + 6 - 5x$$
$$-3x - 7 = 6$$
$$-3x - 7 + 7 = 6 + 7$$
$$-3x = 13$$
$$-\tfrac{1}{3}(-3x) = -\tfrac{1}{3}(13)$$
$$x = -\tfrac{13}{3}.$$

Thus the solution set of the original equation is $\{-\tfrac{13}{3}\}$.

☐ **Extraneous Solutions** When both sides of an equation are multiplied by an expression containing a variable, the resulting equation may *not* be equivalent to the original, since we have excluded multiplication by 0 in operation (*ii*) of Theorem 1.1.1. For example, multiplication of the equation $2x = 4$ by x yields $2x^2 = 4x$. The two equations are not equivalent, since obviously 0 is a solution of the latter equation but is not a solution of the former equation. We say that 0 is an **extraneous solution** of the original equation.

EXAMPLE 5   Multiplying an Equation by a Variable

Solve

$$2 - \frac{1}{z+1} = \frac{z}{z+1}. \tag{2}$$

Solution Multiplying both sides of the given equation by $z + 1$ yields a linear equation:

$$(z + 1)\left(2 - \frac{1}{z+1}\right) = (z+1) \cdot \frac{z}{z+1} \quad \leftarrow \text{cancel on the right}$$
$$(z + 1) \cdot 2 - (z+1)\frac{1}{z+1} = z \quad \leftarrow \text{on the left: distributive law and cancellation}$$
$$2z + 2 - 1 = z \quad \leftarrow \text{on the left: distributive law again}$$
$$z = -1.$$

Because we have multiplied by an expression containing a variable, we must check $z = -1$ by substituting it into the original equation (2). We obtain

$$2 + \frac{1}{-1+1} = \frac{-1}{-1+1} \quad \text{or} \quad 2 + \frac{1}{0} = \frac{-1}{0}.$$

Since division by 0 is not defined, $z = -1$ is not a solution of the original equation. Thus, -1 is an extraneous solution, and so we conclude that equation (2) has no solutions. That is, the solution set is the empty set $\varnothing$.

Note of Caution ▶   As Example 5 shows, it is essential to check a "solution" that has been obtained as a result of multiplying both sides of an equation by an expression that may be 0 for some values of the variable.

EXAMPLE 6   Multiplying an Equation by a Variable

Solve

$$\frac{1}{x} + \frac{1}{x-4} = \frac{2}{x^2 - 4x}. \tag{3}$$

Solution To clear the denominators in (3), we multiply both sides of the equation by the LCD $x(x - 4)$ of the fractions in the equation

$$x(x - 4)\left[\frac{1}{x} + \frac{1}{x - 4}\right] = x(x - 4)\left[\frac{2}{x^2 - 4x}\right]$$

$$\cancel{x}(x - 4) \cdot \frac{1}{\cancel{x}} + x\cancel{(x - 4)} \cdot \frac{1}{\cancel{x - 4}} = x\cancel{(x - 4)} \cdot \frac{2}{\cancel{x(x - 4)}} \quad \leftarrow \begin{cases} \text{distributive law} \\ \text{and cancellation} \end{cases}$$

$$(x - 4) + x = 2$$
$$2x = 6$$
$$x = 3.$$

Substituting $x = 3$ into (3), we find that this value satisfies the original equation:

$$\frac{1}{3} + \frac{1}{3 - 4} \stackrel{?}{=} \frac{2}{3^2 - 4 \cdot 3}$$

$$-\frac{2}{3} = -\frac{2}{3}$$

and so the solution set is $\{3\}$. ≡

☐ **Solving for a Variable** In other courses, especially physics, you will often encounter equations that contain several variables. It is often necessary to solve for a particular variable in terms of the remaining variables. The next example illustrates the idea.

EXAMPLE 7   **Solving for Another Variable**

The total resistance R of an electric circuit containing two resistors, of resistance R_1 and R_2, connected in parallel is given by

$$\frac{1}{R} = \frac{1}{R_1} + \frac{1}{R_2}.$$

See **FIGURE 1.1.1**. Solve for R_2 in terms of R and R_1.

FIGURE 1.1.1 Electric circuit in Example 7

Solution First we clear the equation of fractions by multiplying both sides by the quantity RR_1R_2, which is the least common denominator of the fractions in the equation

$$RR_1R_2 \cdot \left(\frac{1}{R}\right) = RR_1R_2 \cdot \left(\frac{1}{R_1} + \frac{1}{R_2}\right)$$

$$\cancel{R}R_1R_2 \cdot \frac{1}{\cancel{R}} = R\cancel{R_1}R_2 \cdot \frac{1}{\cancel{R_1}} + RR_1\cancel{R_2} \cdot \frac{1}{\cancel{R_2}} \quad \leftarrow \begin{cases} \text{distributive law} \\ \text{and cancellation} \end{cases}$$

$$R_1R_2 = RR_2 + RR_1.$$

To obtain an equivalent equation with all terms containing R_2 on the left-hand side, we subtract RR_2 from both sides:

$$R_1R_2 - RR_2 = RR_1.$$

Since R_2 is a common factor of each term on the left-hand side, we can write

$$R_2(R_1 - R) = RR_1.$$

Dividing both sides by $R_1 - R$ gives us the desired result

$$R_2 = \frac{RR_1}{R_1 - R}.$$

≡

1.1 Exercises
Answers to selected odd-numbered problems begin on page ANS-3.

In Problems 1–6, determine whether the given pairs of equations are equivalent.

1. $x = 8$; $x - 8 = 0$
2. $x^2 = x$; $x = 1$
3. $4y - (y - 1) = 2$; $3y = 1$
4. $-2z - 4 = 6z + 10$; $-4z = 7$
5. $t + 1 = 1$; $\dfrac{t+1}{t} = \dfrac{1}{t}$
6. $x^2 = (x + 1)^2$; $2x + 1 = 0$

In Problems 7–48, solve the given equation.

7. $2x + 14 = 0$
8. $3x - 5 = 0$
9. $-5w + 1 = 2$
10. $7z + 8 = -6$
11. $7(y + 1) - 2 = 5(y + 1) + 2$
12. $3y - 2 = y + 6$
13. $x - (2 - x) = 3(x + 1) + x$
14. $[2x - 2(x - 1)]5 = 4 - x$
15. $\frac{1}{2}x - \frac{1}{4} = 0$
16. $\frac{2}{5}x + \frac{1}{5} = -1$
17. $-5t + 3 = 4(t - 6)$
18. $\frac{1}{3}(t - 2) + \frac{2}{3}t = 2t + \frac{4}{3}$
19. $\frac{1}{2}(u - 3) = 2u - \frac{3}{2}$
20. $\frac{1}{4}s + \frac{1}{2} = \frac{1}{2} - \frac{3}{4}s$
21. $0.2x + 1.2 = 0.5$
22. $2.1x - 3 = 0.5x + 0.2$
23. $-3.6z + 1.3 = 0.2(z - 3)$
24. $4.5x - 1.5x = 0.3(2 - x)$
25. $\sqrt{2}x - \dfrac{1}{\sqrt{2}} = \sqrt{8}x$
26. $\dfrac{-2x}{\sqrt{3}} + 1 = 2\sqrt{3} - \sqrt{3}x$
27. $p^2 + 6p - 1 = p^2 - p + 6$
28. $r^2 + 5 = -(10r - r^2)$
29. $(2t - 1)^2 = 4t^2 + 1$
30. $(w - 1)(w + 1) = w(w - 4)$
31. $(x - 1)^3 = x^2(x - 3) + x$
32. $(x + 3)^2 + (x + 2)^3 = x^3 + 7x^2 + 9$
33. $2 + \dfrac{1}{x} = 3 + \dfrac{2}{x}$
34. $\dfrac{2}{t} - 1 = 5 - \dfrac{1}{t}$
35. $\dfrac{1}{s - 1} = \dfrac{2}{s + 1}$
36. $\dfrac{1}{x - 1} + \dfrac{3}{4 - x} = 0$
37. $\dfrac{1}{y - 2} = \dfrac{2y + 1}{y^2 - 4}$
38. $\dfrac{x}{x - 5} = 2 + \dfrac{5}{x - 5}$
39. $\dfrac{2x}{x - 2} - 2 = \dfrac{-4}{2 - x}$
40. $\dfrac{3 - z}{z - 2} = \dfrac{z}{z + 2} - 2$
41. $\dfrac{3}{x + 5} - \dfrac{1}{x - 2} = \dfrac{7}{x^2 + 3x - 10}$
42. $\dfrac{3}{x + 1} + \dfrac{4}{x^2 - 1} = \dfrac{2}{x - 1}$
43. $\dfrac{q}{q - 3} - \dfrac{6}{q^2 - 2q - 3} = 1$
44. $\dfrac{6}{3w + 9} - \dfrac{4}{2w + 6} = 0$
45. $\dfrac{3x}{x - 2} = \dfrac{6}{x - 2} + 1$
46. $\dfrac{x^2 + 3}{x - 3} - \dfrac{x + 6}{3 - x} = 1$
47. $\dfrac{4}{\sqrt{x}} + \sqrt{x} = \dfrac{12}{\sqrt{x}} - \sqrt{x}$
48. $\dfrac{1}{2\sqrt{x}} - \dfrac{2}{\sqrt{x}} = \dfrac{5}{\sqrt{x}}$

49. Find a so that the solution of $3x + 3a = 6x - a$ is 4.
50. Find d so that the equation
$$\dfrac{2x - 1}{x + 2} - \dfrac{x + d}{x + 2} = 0$$
has no solutions.
51. Find c so that $3(y - c) = 3y + 7$ is an identity.
52. Find a so that $(x - 1)(x + a) = x^2 - 2x - a$ is an identity.
53. Find a so that $5 - z = 1$ and $3z + 2a = 10$ are equivalent equations.
54. Find the relationship between a and b if $ax + b = 0$ has the solution $x = 5$.

In Problems 55–66, solve for the indicated variable in terms of the remaining variables.

55. Circumference of a circle
$C = 2\pi r$, for r
56. Perimeter of a rectangle
$P = 2w + 2l$, for l
57. Simple interest
$I = Prt$, for t
58. Lateral surface area of a cylinder
$S = 2\pi rh$, for h
59. Amount accrued by simple interest
$A = P + Prt$, for P
60. Volume of a rectangular parallelepiped
$V = lwh$, for h
61. nth term of an arithmetic sequence
$a_n = a + (n - 1)d$, for n
62. Sum of a geometric series
$S = \dfrac{a}{1 - r}$, for r
63. Newton's universal gravitation law
$F = g\dfrac{m_1 m_2}{r^2}$, for m_1
64. Free-falling body
$s = \tfrac{1}{2}gt^2 + v_0 t$, for v_0
65. Resistance in a parallel circuit
$R = \dfrac{R_1 R_2}{R_1 + R_2}$, for R_2
66. Surface area of a cylinder
$A = 2\pi r(r + h)$, for h

Miscellaneous Applications

67. **Temperature** The relationship between temperature measured in degrees Celsius (T_C) and in degrees Fahrenheit (T_F) is given by $T_C = \tfrac{5}{9}(T_F - 32)$.
    (a) Solve the last equation for T_F.
    (b) Use the result in part (a) to convert the Celsius temperatures $-5°C$, $0°C$, $16°C$, $35°C$, and $100°C$ to degrees Fahrenheit.
68. **Tennis Anyone?** The velocity v in ft/s of a tennis ball t seconds after it has been thrown upward with an initial velocity of 8 ft/s is given by $v = -32t + 8$. How many seconds have elapsed when (a) $v = 4$ ft/s and (b) $v = 0$ ft/s?
69. **Diophantus' Age** In the chapter opening, we saw that the equation

$$\frac{x}{6} + \frac{x}{12} + \frac{x}{7} + 5 + \frac{x}{2} + 4 = x$$

described the age x of Diophantus when he died. Determine how long Diophantus lived.
70. **Heart Rate** As reported in *Science and Sport* by Thomas Vaughan (Boston: Little, Brown & Co., 1970), a series of 4200 readings taken on 136 world-class athletes resulted in the formula for the maximum heart rate r_{max} in heartbeats per minute during exercise:

$$r_{max} = 0.981 r_5 + 5.948,$$

where r_5 is the heart rate taken within 5 seconds after ceasing to exercise.
    (a) If a walking champion has a maximum heart rate of 215, find the heart rate immediately after exercising.
    (b) If a world-class cyclist has a maximum heart rate of 180, find the heart rate immediately after exercising.
71. **Ideal Gas** For an ideal gas at low pressure, the volume V at T degrees Celsius is given by

$$V = V_0\left(1 + \frac{T}{273.15}\right),$$

where V_0 is the volume at $0°C$. At what temperature does $V = \tfrac{3}{4}V_0$ for an ideal gas at low pressure?

72. Snow Cover Empirical studies of snowfall on the island of Great Britain found that the number of days D in a year on which the ground is covered with snow increases linearly with elevation:

$$D = 0.155H + 11,$$

where H is the elevation measured in meters.
(a) According to this formula, how many days of snow cover are there at sea level?
(b) At what elevation does this formula predict year-round (365-day) snow cover?

For Discussion

73. Point out the error in the following reasoning:

$$x = -1$$
$$x^2 + x = x^2 - 1$$
$$x(x + 1) = (x - 1)(x + 1)$$
$$x = x - 1$$
$$0 = -1.$$

74. Consider the following sequence of equations:

$$x^2 - 1 = x^2 + 4x - 5$$
$$(x + 1)(x - 1) = (x + 5)(x - 1)$$
$$x + 1 = x + 5$$
$$1 = 5.$$

(a) What is the solution to the first equation in the sequence?
(b) Find an equation in the sequence that is not equivalent to the equation that precedes it.

1.2 Building an Equation from Words

≡ Introduction Algebra is useful in solving many practical problems involving rates, mixtures, money, and so on. Because these problems are often stated in words, the basic idea is to translate the words and build or construct an appropriate algebraic equation. Since there is no single procedure for making this translation, work, practice, and patience are required to become proficient in solving problems of this sort. The following suggestions may be helpful.

GUIDELINES FOR BUILDING AN EQUATION

(*i*) Read the problem very carefully.
(*ii*) Reread the problem and identify an unknown quantity to be found.
(*iii*) When possible, draw a figure.
(*iv*) Let a variable, say x, represent the unknown quantity. Write the definition of this variable on your paper.
(*v*) If possible, represent any other quantities in the problem in terms of x.
(*vi*) Write an equation that accurately expresses the relationship described in the problem.
(*vii*) Solve the equation.
(*viii*) Check that your answer agrees with all the conditions stated in the problem.

☐ **Age Problems** As our first example, consider the following age problem.

EXAMPLE 1 An Age Problem

Two years ago John was five times as old as Bill. Now he is 8 years older than Bill. Find John's present age.

Solution The unknown quantity to be determined is John's present age, so we let

$$x = \text{John's present age.}$$

Then we can represent the other quantities in the problem in terms of x:

$$x - 8 = \text{Bill's present age,}$$
$$x - 2 = \text{John's age 2 years ago,}$$
$$(x - 8) - 2 = x - 10 = \text{Bill's age 2 years ago.}$$

You may find it helpful to list this information in tabular form, as shown below.

	Age now	Age 2 years ago
John	x	$x - 2$
Bill	$x - 8$	$x - 10$

An equation that expresses the relationship of their ages 2 years ago is

$$x - 2 = 5(x - 10).$$

We solve this equation:

$$x - 2 = 5x - 50$$
$$48 = 4x$$
$$x = 12.$$

Thus John's present age is 12.

Check: If John is now 12, Bill must be 4. Two years ago John was 10 and Bill was 2. Since $10 = 5(2)$, the answer checks. ≡

☐ **Investment Problems** Many **investment problems** utilize the simple-interest formula

$$I = Prt, \tag{1}$$

where I is the amount of interest earned on a sum of money P (called the principal) invested at a simple interest rate of r percent for t years. As the following example shows, it can be helpful to organize the data in tabular form.

EXAMPLE 2 Simple Interest

A businesswoman plans to invest a total of $30,000. Part of it will be put into a certificate of deposit (CD) paying 3% simple interest and the remainder into an investment fund yielding 5.5% simple interest. How much should she invest in each in order to obtain a 4% return on her money after 1 year?

Solution In (1) we identify $r = 0.03$ and $t = 1$. Now if x represents the amount (in dollars) invested in the CD, then

$$30{,}000 - x = \text{the amount in dollars put into the investment fund.}$$

The following table summarizes the given information.

	Principal P	Interest rate r	Time t	Interest earned $I = Prt$
Certificate of deposit	x	0.03	1 year	$x(0.03)(1) = 0.03x$
Investment fund	$30{,}000 - x$	0.055	1 year	$(30{,}000 - x)(0.055)(1)$ $= 0.055(30{,}000 - x)$
Equivalent investment	30,000	0.04	1 year	$30{,}000(0.04)(1) = 1200$

Since the combined interest from the CD and the investment fund is to equal that of an equivalent total investment made at 4% simple interest, we have

$$0.03x + (0.055)(30{,}000 - x) = 1200.$$

We begin our solution of the foregoing equation by multiplying it by 100:

$$3x + (5.5)(30{,}000 - x) = 100(1200)$$
$$3x + 165{,}000 - 5.5x = 120{,}000$$
$$-2.5x = -45{,}000$$
$$x = 18{,}000.$$

Thus, $18,000 should be invested in the CD, and $30,000 − $18,000 = $12,000 should be put into the investment fund.

Check: The sum of $18,000 and $12,000 is $30,000. The interest earned on the CD is ($18,000)(0.03)(1) = $540. The interest earned on the investment fund is ($12,000)(0.055)(1) = $660. If the $30,000 were invested at 4%, the interest earned would be ($30,000)(0.04)(1) = $1200. Since $540 + $660 = $1200, the answer checks. ≡

☐ **Rate Problems** If an object moves at a constant rate r, then the distance d it travels in t units of time is given by distance = rate × time, or in symbols:

$$d = rt. \qquad (2)$$

Other forms of (2) that may be useful in solving certain rate problems are

$$r = \frac{d}{t} \quad \text{and} \quad t = \frac{d}{r}. \qquad (3)$$

Usually the most difficult part of solving a distance problem is determining what relationship to express as an equation. It can be helpful to consider the following questions:

- Are there two distances (or times or rates) that are equal?
- Is the sum of two distances (or times or rates) a constant?
- Is the difference of two distances (or times or rates) a constant?

The second equation in (3) is used in the next problem.

EXAMPLE 3   **A Rate Problem**

It takes a motorist 1 hour and 30 minutes longer to travel between two cities at night than it does during the day. At night she averages 40 mi/h while during the day she can average 55 mi/h. Find the distance between the two cities.

Solution Let the symbol d represent the distance between the two cities. The following table displays the distance, rate, and time for each trip.

	Distance	Rate	Time
Night	d	40	$\dfrac{d}{40}$
Day	d	55	$\dfrac{d}{55}$

Since it takes 1.5 h longer to travel the distance between the two cities at night, we have

$$\frac{d}{40} - \frac{d}{55} = 1.5.$$

We multiply both sides of this equation by $(40)(55) = 2200$ and solve:

$$55d - 40d = 3300$$
$$15d = 3300$$
$$d = 220.$$

The distance between the two cities is 220 mi.

Check: Her time at night is $220/40 = 5.5$ h, and her time during the day is $220/55 = 4$ h. Since $5.5 - 4 = 1.5$, the answer checks. ≡

☐ **Mixture Problems** occur frequently in chemistry, pharmacology, manufacturing, and everyday situations. When solving mixture problems, we focus on the amount of one element in each of the different mixtures. Again, it can be helpful to organize the information in tabular form, as we see in the following example.

EXAMPLE 4   A Mixture Problem

Find how many liters of pure alcohol must be added to a 15-liter solution containing 20% alcohol so that the resulting mixture is 30% alcohol.

Solution If x represents the amount of pure alcohol added, then

$$15 + x = \text{the amount in liters in the new solution.}$$

The following table summarizes the given information.

	Liters of solution	Concentration of alcohol	Liters of alcohol
Original solution	15	0.20	0.20(15)
Pure alcohol	x	1.00	1.00x
Resulting mixture	$15 + x$	0.30	$0.30(15 + x)$

Since the amount of alcohol in the original solution plus the amount of pure alcohol added equals the amount of alcohol in the resulting mixture, we have

$$0.20(15) + 1.00x = 0.30(15 + x)$$
$$3 + x = 4.5 + 0.3x$$
$$0.7x = 1.5$$
$$x = \tfrac{15}{7}.$$

The amount of pure alcohol added is $\tfrac{15}{7}$ L.

Check: If $\tfrac{15}{7}$ L of alcohol is added, the new solution totaling $15 + \tfrac{15}{7} = \tfrac{120}{7}$ L contains $(0.20)(15) + \tfrac{15}{7} = \tfrac{36}{7}$ L of alcohol. Since $\tfrac{36}{7}/\tfrac{120}{7} = 0.30$, the new solution is 30% alcohol and the answer checks. ≡

☐ **Work Problems** Several people (or machines) doing the same job, each working at a constant rate, can complete the job faster than any one of them working alone. Therefore, when solving **work problems**, we use the following basic principle:

- *If one individual can do the entire job in T units of time, then in x units of time, x/T of the job is completed.* (4)

For example, if one person can do an entire job in 5 h, then in 3 h, $\frac{3}{5}$ of the job can be done.

EXAMPLE 5 A Work Problem

Working alone, pump A can fill a tank in 2 h and pump B can fill the same tank in 3 h. Determine how fast the pumps can fill the tank working together.

Solution If we let x be the number of hours required for both pumps working together to fill the tank, then

$$\frac{x}{2} = \text{the fractional part of the entire job completed in } x \text{ hours by pump } A$$

and

$$\frac{x}{3} = \text{the fractional part of the entire job completed in } x \text{ hours by pump } B.$$

This information is summarized in the following table.

	Time (in hours) to complete the total job	Fraction of the job completed in x hours
Pump A	2	$\frac{x}{2}$
Pump B	3	$\frac{x}{3}$
Together	x	1

The sum of the fractional parts done by each pump in x hours is 1, since the two pumps working together complete the entire job in x hours. Thus we have

$$\frac{x}{2} + \frac{x}{3} = 1.$$

We begin by multiplying the equation by the least common denominator of the fractions in the equation. Then solving for x, we find

$$3x + 2x = 6$$
$$5x = 6$$
$$x = \tfrac{6}{5}.$$

Together it takes the pumps $\frac{6}{5} = 1.2$ h (or 1 h and 12 min.) to fill the tank.

Check: In $\frac{6}{5}$ h pump A fills $\frac{6}{5}/2 = \frac{3}{5}$ of the tank, and pump B fills $\frac{6}{5}/3 = \frac{2}{5}$ of the tank. Because $\frac{3}{5} + \frac{2}{5} = 1$, the solution checks. ≡

☐ **Miscellaneous Problems** In addition to the age, investment, rate, mixture, and work problems that we have just considered, there is a wide variety of word problems. We end this section with two additional examples.

EXAMPLE 6   A Fence Problem

A rectangular field that is 20 m longer than it is wide is enclosed with exactly 100 m of fencing. What are the dimensions of the field?

Solution The geometric description of this problem compels us to draw a diagram. See **FIGURE 1.2.1**. If we let w be the width of the field in meters, then

$$w + 20 = \text{the length of the field in meters.}$$

Since the perimeter of the field is 100 m, we have

$$100 = \overbrace{w + w}^{\text{two widths}} + \overbrace{(w + 20) + (w + 20)}^{\text{two lengths}}$$

or

$$100 = 2w + 2(w + 20).$$

FIGURE 1.2.1 Field in Example 6

Solving for w, we find

$$100 = 4w + 40$$
$$60 = 4w$$
$$15 = w.$$

Thus the width is $w = 15$ m and the length is $w + 20 = 35$ m.

Check: The length is 20 m longer than the width since $35 - 15 = 20$, and the amount of fencing required is $2(35) + 2(15) = 70 + 30 = 100$. Thus the answers check. ≡

EXAMPLE 7   Improving an Exam Grade Average

A student scores 75 and 82 on his first two exams. What score on the next exam will raise the student's average to 85?

Solution We begin by letting x denote the score on the future third exam. Then the average of the three exam scores is

$$\frac{75 + 82 + x}{3}.$$

Since this average must equal 85, we want

$$\frac{75 + 82 + x}{3} = 85.$$

We multiply each side of the last equation by 3 and solve for x:

$$75 + 82 + x = 3(85)$$
$$157 + x = 255$$
$$x = 98.$$

Therefore, a score of 98 on the third exam will raise the student's average score to 85.

Check: If the three exam scores are 75, 82, and 98, the student's average score will be

$$\frac{75 + 82 + 98}{3} = 85.$$

Hence the answer checks. ≡

NOTES FROM THE CLASSROOM

As you begin working the problems in Exercises 1.2, remember to follow the suggestions given on page 72. It may help to read and work through the examples in this section several times and supply any missing steps in the solutions. Examine how we followed the suggestions. Finally, and above all, don't get discouraged!

1.2 Exercises
Answers to selected odd-numbered problems begin on page ANS-3.

In Problems 1–46, build and solve an equation from the given words.

Number Problems

1. Find two integers whose sum is 50 and whose difference is 26.
2. The quotient of two numbers is 4. If one number is 39 less than the other, find the two numbers.
3. Find three consecutive integers whose sum is 48.
4. The difference of the squares of two consecutive even integers is 92. Find the two numbers.

Age Problems

5. In 5 years Bryan will be three times as old as he was 7 years ago. How old is Bryan?
6. The plumbing firm Papik and Son advertises a total of "30 years of experience" in plumbing. If the father has 16 years more plumbing experience than his son, how long has each been a plumber?

Investment Problems

7. A couple has $40,000 to invest. If they invest $16,000 at 12% and $14,000 at 8%, at what rate should they invest the remainder in order to have a yearly income of $4000 from their investments?
8. Mr. Janette has three investments from which he receives an annual income of $2780. One investment of $7000 is at an annual rate of 8%. Another investment of $10,000 is at an annual rate of 9%. What is the annual rate of interest he receives on the third investment of $12,000?
9. Ms. Beecham invested part of $10,000 in a savings certificate at 7% simple interest. She invested the remainder in a bond fund yielding 12%. If she received a total of $900 in interest for the first year, how much money was invested in the bond fund?
10. The Wilsons have $30,000 invested at 12% and another sum invested at 8.5%. If the yearly income on the total amount invested is equivalent to a rate of 10% on the total, how much do they have invested at 8.5%?

Rate Problems

11. A car travels from point *A* to point *B* at an average speed of 55 mi/h and returns at an average of 50 mi/h. If the round trip takes 7 h, find the distance between *A* and *B*.

12. A jet airplane flies with the wind between Los Angeles and Chicago in 3.5 h and flies against the wind from Chicago to Los Angeles in 4 h. If the speed of the plane in still air is 600 mi/h, find the speed of the wind. What is the distance between Los Angeles and Chicago?
13. A woman can walk to work at a rate of 3 mi/h, or she can bicycle to work at a rate of 12 mi/h. If it takes her 1 h longer to walk than to ride, find the time it takes her to walk to work.
14. A boy leaves point P on a bicycle and rides at a speed of 15 km/h. Thirty minutes later another boy leaves point P on a bicycle and rides at a different speed. He overtakes the first rider $2\frac{1}{2}$ h later. Find the speed of the second rider.
15. A man drove a car 280 km and then rode a bicycle an additional 50 km. If the total time for the trip was 12 h and the speed on the bicycle was $\frac{1}{4}$ of the speed driving, find each rate.
16. An instrument capsule was carried into the atmosphere by a rocket. The capsule landed 72 min later, after making a controlled descent with an average vertical speed of 420 km/h. If the rocket had an average vertical speed of 1010 km/h from liftoff to release of the capsule, at what height was the capsule released?

Mixture Problems

17. An automobile radiator contains 10 qt of a mixture of water and antifreeze that is 20% antifreeze. How much of this mixture must be drained out and replaced with pure antifreeze in order to obtain a 50% mixture in the radiator?
18. A lawn mower uses a fuel mixture of 23 parts gasoline to 1 part oil. How much gasoline should be added to a liter of a mixture that is 5 parts gasoline and 1 part oil in order to obtain the right mixture?
19. A certain brand of potting soil contains 10% peat and another brand contains 30% peat. How much of each soil should be mixed in order to produce 2 ft^3 (cubic feet) of potting soil that is 25% peat?
20. The manager of a service station bought a total of 15,000 gallons of regular and premium gasoline for $37,000. If the wholesale price was $2.40 per gallon for the regular gas and $2.60 per gallon for the premium gas, determine how many gallons of each type of gasoline were purchased.
21. A grocer sells one grade of ground beef at $3.95 per pound and another grade at $4.20 per pound. He wants to mix the two grades in order to obtain a mixture that sells for $4.15 per pound. What percentage of each grade should be used?

Work Problems

22. If Meagan can complete a task in 50 min working alone and Colleen can do it alone in 25 min, how long will it take them working together?
23. If Karen can pick a raspberry patch in 6 h and Stan can pick it in 8 h, find how fast they can harvest the patch together.
24. Using two hoses of different diameters, a homeowner can fill a hot tub in 40 min. If one hose alone can fill the tub in 90 min, find how fast the other hose would fill the tub by itself.
25. Margot can clean her room alone in 50 min. If Jeremy helps her, it takes 30 min. How long would it take Jeremy alone to clean the room?
26. An outlet pipe can empty a tank in 4 h. The pipe was opened for 1.5 h and then closed. At that time a second outlet pipe was opened and it took 2 h to finish emptying the tank. How long would it have taken the second pipe alone to empty the tank?

Dimension Problems

27. The perimeter of a rectangle is 50 cm and the width is $\frac{2}{3}$ of the length. Find the dimensions of the rectangle.
28. The area of a trapezoid is 250 ft^2 and the height is 10 ft. What is the length of the long base if the short base is 20 ft?
29. The longest side of a triangle is 2 cm longer than the shortest side. The third side is 5 cm less than twice the length of the shortest side. If the perimeter is 21 cm, how long is each side?
30. A farmer wishes to enclose a rectangular field and to divide it into three equal parts with fencing. See FIGURE 1.2.2. If the length of the field is three times the width and 1000 m of fencing is required, what are the dimensions of the field?

FIGURE 1.2.2 Fenced field in Problem 30

31. The area of a circle is 80π cm^2 less than the area of a circle whose radius is 4 cm larger. Find the radius of the smaller circle.

Miscellaneous Problems

32. **Dose of a Medicine** Friend's rule for converting an adult dose of a drug to a child's dose assumes a relationship between age and dosage and is used for children under 2 years of age:

$$\frac{\text{age in months}}{150} \times \text{adult dose} = \text{child's dose}.$$

At what age is the adult dosage 10 times that of the child's dosage?

33. **Joshua Battles for a Passing Grade** Joshua has taken one exam. If he must score 99 on a second exam in order to have an average of 73 for both exams, what did he score on the first exam?
34. **Trying for a B** Going into the final exam a student has test scores of 72 and 86. If the final exam counts as one half of the final grade, what exam score must the student make in order to finish the course with an average of 80?
35. **Class Politics** Judy defeated John in a close election for senior class president in which 211 votes were cast. If five students had voted for John instead of Judy, then John would have won by one vote. How many students voted for Judy?
36. **How Many?** At Cayley Avenue School, 40 more than half the students are boys. If the number of girls at the school is two less than half the number of boys, how many students attend the school?
37. **A Coin Problem** Kurt has four more dimes than nickels. If the total value of these coins is $2.35, find how many nickels and dimes Kurt has.
38. **Another Coin Problem** Heidi has $4.65 in nickels, dimes, and quarters. She has four more quarters than dimes and five more nickels than quarters. How many coins of each type does Heidi have?
39. **Playing with Numbers** The units digit of a 2-digit number is five more than the 10's digit. If the original number is divided by the number with its digits reversed, the result is $\frac{3}{8}$. Find the original number.

40. **More Playing with Numbers** The denominator of a fraction is two more than the numerator. If both the numerator and the denominator are increased by one, the resulting fraction equals $\frac{2}{3}$. Find the original number.

41. **Bribery?** Mr. Chaney and his son Ryan agreed that Mr. Chaney would give Ryan $5 for each word problem Ryan solved correctly but Ryan would pay his father $2 for each incorrect solution. After Ryan had completed 70 problems, neither owed the other any money. How many word problems had Ryan solved correctly?

42. **Got a Raise** A worker gets a 6% raise, which is $480. What was the old salary? What is the new salary?

43. **Pay for Itself?** A local gas company sells an insulating blanket for a hot water heater for $20. It claims that the blanket will reduce fuel costs by 10%. If the average monthly fuel cost for heating water is $20, how soon will the insulation "pay for itself"?

44. **Paying Taxes** A restaurant manager provides a no-host bar at a banquet with drinks priced at round dollar amounts to simplify the transactions. The 5% sales tax is included in the rounded prices. At the end of the banquet, the manager finds exactly $200 in the register. He is smart enough to know that $10 is too much to take out for the tax, but he is unable to figure out the proper amount of tax to pay.
    (a) Explain why $10 is too much tax.
    (b) Find the correct amount of sales tax (to the nearest penny).

45. **Clerk in Need of More Training** During 25%-off sales one store clerk always computes the discount first and then adds on the 6% sales tax. Another clerk in the same store always adds the sales tax first and then applies the discount.
    (a) Does it make a difference?
    (b) Can you show that this is always the case for any discount $d\%$ and any sales tax $t\%$?

46. **Competing Plants** Two industrial plants that manufacture identical engine components are located 100 mi apart on the Watchacallit River. See **FIGURE 1.2.3**. Both plants sell components at the same price, $150. However, because one plant is upstream, its shipping costs are lower for customers located between the two plants: 30 cents per mile per component rather than 75 cents per mile.
    (a) Assuming that a customer will buy from the plant offering the lowest total cost, how far up the river will the downstream plant have customers?
    (b) How does your answer to part (a) change if both plants raise the price of their components from $150 to $160?
    (c) How does your answer to part (a) change if shipping costs double?
    (d) What selling price would the downstream plant have to offer in order to capture half of the territory between the two industrial plants?

FIGURE 1.2.3 Industrial plants in Problem 46

1.3 Quadratic Equations

≡ **Introduction** In Section 1.1 we saw that a **linear equation** is an equation that can be put into the standard form $ax + b = 0, a \neq 0$. The equation $ax + b = 0, a \neq 0$, is a special type of polynomial equation. A **polynomial equation of degree n** is an equation of the form

◀ Review the definition of a polynomial in Section R.6.

$$a_n x^n + a_{n-1} x^{n-1} + \cdots + a_2 x^2 + a_1 x + a_0 = 0, \quad a_n \neq 0, \quad (1)$$

where n is a nonnegative integer and a_i, $i = 0, 1, \ldots, n$, are real numbers. A linear equation corresponds to the degree $n = 1$ in equation (1). Except for different symbols, $ax + b$ is the same as $a_1 x + a_0$.

A solution of a polynomial equation is also called a **root** of the equation. For example, we know that $-b/a$ is the only root of the linear or first-degree polynomial equation $ax + b = 0$.

In this section we examine polynomial equations of degree 2 or **quadratic equations**. A quadratic equation is a polynomial equation that can be written in the **standard form**:

$$ax^2 + bx + c = 0, \quad a \neq 0. \tag{2}$$

Polynomial equations of higher degrees will be studied in Chapter 4.

Many problems about moving objects involve quadratic equations. For example, if a water balloon is thrown with an initial velocity of 48 ft/s, straight down from a dormitory window 64 ft above the ground, then the height s (in feet) above the ground after t seconds can be shown to be

$$s = -16t^2 - 48t + 64.$$

When the balloon hits the ground, its height s equals 0, so we can determine the elapsed time by solving

$$-16t^2 - 48t + 64 = 0.$$

By dividing by -16 the foregoing equation is equivalent to

$$t^2 + 3t - 4 = 0. \tag{3}$$

☐ **Method of Factoring** As we will see, equation (3) is easily solved by the **method of factoring**. This method is based on the **zero factor property**, which was introduced in Section R.1. Recall, if a and b represent real numbers and $a \cdot b = 0$, then either $a = 0$ or $b = 0$. The technique is illustrated in the following example.

See page 6 of Section R.7.
See property 8(ii) in Section R.1.

EXAMPLE 1   Solution by Factoring

Solve $2x^2 + 5x - 3 = 0$.

Solution The equation is already in standard form. By factoring the left-hand side of the given equation, we obtain the equivalent equation

$$(x + 3)(2x - 1) = 0.$$

If we apply the zero factor property, we conclude that either

$$x + 3 = 0 \quad \text{or} \quad 2x - 1 = 0.$$

The solutions of these linear equations are $x = -3$ and $x = \frac{1}{2}$, respectively. The fact that these are roots of the given equation can be verified by substitution. The solution set is $\{-3, \frac{1}{2}\}$. ≡

We can now readily solve equation (3) to find the time it takes the water balloon to strike the ground. We first write $t^2 + 3t - 4 = 0$ in the factored form:

$$(t + 4)(t - 1) = 0.$$

From the zero factor property we find that we must solve $t + 4 = 0$ and $t - 1 = 0$. The solutions of these equations are $t = -4$ and $t = 1$. By substitution we can verify that both $t = -4$ and $t = 1$ satisfy the original quadratic equation $-16t^2 - 48t + 64 = 0$. Because $t = 1$ second is the only positive solution, it is the only meaningful answer to the physical problem.

As we can see from the example of the water balloon, not all the solutions of an equation will necessarily fit the required conditions of the problem.

EXAMPLE 2 Solution by Factoring

Solve $12x^2 + 15x = 18$.

Solution Since we plan to try the method of factoring, we must begin by writing the equation in the standard form $ax^2 + bx + c = 0$:

$$12x^2 + 15x - 18 = 0.$$

Eliminating the common factor 3 by division simplifies the equation

$$3(4x^2 + 5x - 6) = 0$$

to

$$4x^2 + 5x - 6 = 0.$$

Factoring then gives

$$(4x - 3)(x + 2) = 0.$$

Using the zero factor property, we set each factor equal to zero to obtain $4x - 3 = 0$ and $x + 2 = 0$. Solving each of these equations gives $x = \frac{3}{4}$ and $x = -2$. The roots of the quadratic equation are $\frac{3}{4}$ and -2; the solution set is $\{-2, \frac{3}{4}\}$. ≡

EXAMPLE 3 Solution by Factoring

Solve $4x^2 + 4x + 1 = 0$.

Solution The left-hand side of the equation readily factors as

$$(2x + 1)(2x + 1) = 0$$

and so $x = -\frac{1}{2}$ or $x = -\frac{1}{2}$. The solution set is $\{-\frac{1}{2}\}$. ≡

In Example 3, we say that $x = -\frac{1}{2}$ is a **repeated root** or a **root of multiplicity 2**. When counting roots we count such roots twice.

☐ **Square Root Method** If a quadratic equation has the special form

$$x^2 = d, \quad d \geq 0, \tag{4}$$

we can solve it by factoring:

$$x^2 - d = 0 \quad \leftarrow \text{difference of two squares} \\ \text{See Section R.7.}$$
$$(x - \sqrt{d})(x + \sqrt{d}) = 0,$$

which yields $x = \sqrt{d}$ or $x = -\sqrt{d}$. An alternative approach to solving Equation (4) is to take the square root of both sides of the equation. This is summarized as the **square root method**:

- If $x^2 = d$, $d \geq 0$, then $x = \pm\sqrt{d}$.

EXAMPLE 4 Solution by Square Root Method

Use the square root method to solve **(a)** $2x^2 = 6$ **(b)** $(y - 3)^2 = 5$.

Solution **(a)** We multiply both sides of $2x^2 = 6$ by $\frac{1}{2}$ to obtain the special form (4):

$$\tfrac{1}{2}(2x^2) = \tfrac{1}{2}(6)$$
$$x^2 = 3$$
$$x = \pm\sqrt{3}.$$

From the last line we see that the solution set is $\{-\sqrt{3}, \sqrt{3}\}$.

(b) We note that for $x = y - 3$ and $d = 5$, the equation $(y - 3)^2 = 5$ has the special form (4). Therefore, we take the square root of both sides of the equation:

$$(y - 3)^2 = 5$$
$$y - 3 = \pm\sqrt{5}.$$

This yields two linear equations $y - 3 = -\sqrt{5}$ and $y - 3 = \sqrt{5}$. Solving each of these, we find $y = 3 - \sqrt{5}$ and $y = 3 + \sqrt{5}$, respectively. Therefore, the solution set is $\{3 - \sqrt{5}, 3 + \sqrt{5}\}$. ≡

☐ **Completing the Square** When a quadratic expression cannot be factored easily and the equation does not have the special form (4), we can find the roots by **completing the square**. This technique is applied to quadratic expressions of the form $x^2 + Bx + C$; that is, the quadratic expression must have 1 as its leading coefficient. We rewrite the equation

$$x^2 + Bx + C = 0 \tag{5}$$

so that only the terms containing the variable x are on the left-hand side of the equation:

$$x^2 + Bx = -C.$$

By adding $(B/2)^2$ to both sides of the last equation,

$$x^2 + Bx + \left(\frac{B}{2}\right)^2 = -C + \left(\frac{B}{2}\right)^2,$$

the left-hand side of the resulting equation is a perfect square:

$$\left(x + \frac{B}{2}\right)^2 = \left(\frac{B}{2}\right)^2 - C.$$

It is now easy to solve for x using the square root method. This procedure is illustrated in the following example.

EXAMPLE 5   **Solution by Completing the Square**

Solve $2x^2 + 2x - 1 = 0$ by completing the square.

Solution We begin by dividing both sides of the equation by the leading coefficient 2 to obtain the form (5):

$$x^2 + x - \frac{1}{2} = 0.$$

Now we write this equation as

$$x^2 + x = \frac{1}{2}$$

and add the square of half the coefficient of x (in this case it is 1) to both sides of the equation:

$$x^2 + x + \left(\frac{1}{2}\right)^2 = \frac{1}{2} + \left(\frac{1}{2}\right)^2.$$

Thus we have

$$\left(x + \frac{1}{2}\right)^2 = \frac{3}{4}.$$

Taking the square root of both sides of the equation gives

$$x + \frac{1}{2} = \pm \frac{\sqrt{3}}{2} \quad \text{or} \quad x = -\frac{1}{2} \pm \frac{1}{2}\sqrt{3}.$$

The two solutions, or roots, then are $-\frac{1}{2} - \frac{1}{2}\sqrt{3}$ and $-\frac{1}{2} + \frac{1}{2}\sqrt{3}$, respectively. The solution set of the equation is $\{-\frac{1}{2} - \frac{1}{2}\sqrt{3}, -\frac{1}{2} + \frac{1}{2}\sqrt{3}\}$. ≡

☐ **The Quadratic Formula** The technique of completing the square on a quadratic expression is very useful in other situations. We will encounter it again in Chapters 2, 3, and 9. For now its value is to help us derive a formula that expresses the roots of $ax^2 + bx + c = 0$, $a \neq 0$, in terms of the coefficients a, b, and c. First we rewrite the equation so that it has leading coefficient 1:

$$x^2 + \frac{b}{a}x + \frac{c}{a} = 0.$$

Then we complete the square and solve for x:

$$x^2 + \frac{b}{a}x = -\frac{c}{a}$$

$$x^2 + \frac{b}{a}x + \left(\frac{b}{2a}\right)^2 = -\frac{c}{a} + \left(\frac{b}{2a}\right)^2$$

$$\left(x + \frac{b}{2a}\right)^2 = \frac{b^2 - 4ac}{4a^2} \quad \leftarrow \text{now use the square root method}$$

$$x + \frac{b}{2a} = \pm\sqrt{\frac{b^2 - 4ac}{4a^2}}$$

$$x + \frac{b}{2a} = \frac{\pm\sqrt{b^2 - 4ac}}{\sqrt{4a^2}}. \quad \leftarrow \text{square root of a quotient is the quotient of the square roots}$$

If $a > 0$, then $\sqrt{4a^2} = |2a| = 2a$, and we have

$$x = \frac{-b \pm \sqrt{b^2 - 4ac}}{2a}. \tag{6}$$

This result is known as the **quadratic formula**. If $a < 0$, then $\sqrt{4a^2} = |2a| = -2a$ and, after simplification, we see that the result in (6) is still valid.

THEOREM 1.3.1 Roots of a Quadratic Equation

If $a \neq 0$, then the roots x_1 and x_2 of $ax^2 + bx + c = 0$ are given by

$$x_1 = \frac{-b - \sqrt{b^2 - 4ac}}{2a} \quad \text{and} \quad x_2 = \frac{-b + \sqrt{b^2 - 4ac}}{2a}. \tag{7}$$

☐ **The Discriminant** The nature of the roots x_1 and x_2 of a quadratic equation $ax^2 + bx + c = 0$ is determined by the radicand $b^2 - 4ac$ in the quadratic formula (6). The quantity $b^2 - 4ac$ is called the **discriminant** of the quadratic equation. The discriminant must be positive, zero, or negative. These three possible cases are summarized next.

Discriminant	Roots
(i) $b^2 - 4ac > 0$	Two real distinct roots
(ii) $b^2 - 4ac = 0$	Real but equal roots
(iii) $b^2 - 4ac < 0$	No real roots

EXAMPLE 6 Solution by the Quadratic Formula

Solve $3x^2 - 2x - 4 = 0$.

Solution Here we identify $a = 3$, $b = -2$, and $c = -4$. The positive discriminant $b^2 - 4ac = 13$ implies that the given equation has two real distinct roots. From the quadratic formula (6) we find

$$x = \frac{-(-2) \pm \sqrt{(-2)^2 - 4(3)(-4)}}{2(3)}$$

$$= \frac{2 \pm \sqrt{52}}{6} = \frac{2 \pm 2\sqrt{13}}{6}$$

$$= \frac{1 \pm \sqrt{13}}{3}.$$

Therefore, the roots are $\frac{1}{3} - \frac{1}{3}\sqrt{13}$ and $\frac{1}{3} + \frac{1}{3}\sqrt{13}$. ≡

EXAMPLE 7 Repeated Roots

Solve $9x^2 + 16 = 24x$.

Solution To use the quadratic formula, we must first write the equation in the form $9x^2 - 24x + 16 = 0$. The quadratic formula

$$x = \frac{-(-24) \pm \sqrt{(-24)^2 - 4(9)(16)}}{2(9)}$$

$$= \frac{24 \pm \sqrt{576 - 576}}{18} \leftarrow \text{radicand is 0}$$

$$= \frac{24}{18} = \frac{4}{3}$$

shows that $\frac{4}{3}$ is a repeated root or a root of multiplicity 2. ≡

EXAMPLE 8 No Real Roots

Solve $3x^2 - x + 2 = 0$.

Solution Since the discriminant

$$b^2 - 4ac = (-1)^2 - 4(3)(2) = -23$$

is negative, we conclude that the given equation has no real roots. ≡

▸ Occasionally we even run into some non-polynomial equations that can be solved by the quadratic formula.

☐ **Quadratic Forms** Certain polynomial equations of degree greater than 2 can be solved using the quadratic formula. This demands that we recognize that the equation can be put into the standard quadratic form $at^2 + bt + c = 0$, where the symbol t represents some positive integer power of x. The next example illustrates this idea.

EXAMPLE 9   A Fourth-Degree Polynomial Equation

Solve $x^4 - 2x^2 - 2 = 0$.

Solution This polynomial equation can be considered as a quadratic equation in the variable x^2, that is,
$$(x^2)^2 - 2(x^2) - 2 = 0.$$

◀ If we let $t = x^2$, then the equation has the quadratic form $t^2 - 2t - 2 = 0$.

Using the quadratic formula to solve for the symbol x^2 we have
$$x^2 = \frac{2 \pm \sqrt{12}}{2} = 1 \pm \sqrt{3}.$$

Thus, either
$$x^2 = 1 - \sqrt{3} \quad \text{or} \quad x^2 = 1 + \sqrt{3}.$$

Now, the quadratic equation $x^2 = 1 - \sqrt{3}$ has no real roots since $1 - \sqrt{3} < 0$. But from $x^2 = 1 + \sqrt{3}$ we obtain two real roots, $-\sqrt{1 + \sqrt{3}}$ and $\sqrt{1 + \sqrt{3}}$ of the original equation.

◀ The square of any real number is nonnegative.

We conclude this section with several applications that involve quadratic equations.

EXAMPLE 10   A Rectangle Problem

The area of a rectangle is 138 in². If the length is 5 in. more than 3 times the width, find the dimensions of the rectangle.

Solution We begin by drawing and labeling a rectangle, as shown in FIGURE 1.3.1. We let
$$w = \text{the width of the rectangle in inches.}$$

Then $3w + 5 =$ the length of the rectangle in centimeters

and $w(3w + 5) = 138.$

In order to use the quadratic formula we rewrite this equation in standard form:
$$3w^2 + 5w - 138 = 0.$$

From the quadratic formula, we find that either $w = -\frac{23}{3}$ or $w = 6$. Since the width of a rectangle must be positive, we discard the solution $w = -\frac{23}{3}$. Therefore, we take $w = 6$. Then the length is $3(6) + 5 = 23$, and the dimensions of the rectangle are 6 in. by 23 in.

Check: Since $23 = 3(6) + 5$ and $6(23) = 138$, the answer checks.

FIGURE 1.3.1 Rectangle in Example 10

□ **Pythagorean Theorem** The **Pythagorean theorem** is one of the most widely used theorems from geometry. Many of its applications involve quadratic equations. Although it is named after the Greek mathematician **Pythagoras**, (c. 540 BCE), the result was known prior to his time. The theorem states that in a right triangle, the square of the length of the hypotenuse equals the sum of the squares of the lengths of the other two sides. For a right triangle as shown in FIGURE 1.3.2, we have the following formula:
$$a^2 + b^2 = c^2.$$

There are a wide variety of algebraic and geometric proofs of this theorem. See Problems 91 and 92 in Exercises 1.3.

Pythagoras

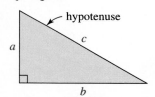

FIGURE 1.3.2 Right triangle

playground
P

x

500

R 700 − *x* *L*
refreshments parking lot

FIGURE 1.3.3 Sidewalks and diagonal shortcut in Example 11

EXAMPLE 11 A Sidewalk Problem

In a park two sidewalks at right angles join the playground P, the refreshment stand R, and the parking lot L, as shown in **FIGURE 1.3.3**. The total length of the sidewalks is 700 m. By walking diagonally across the grass (dashed red line) directly from the parking lot L to the playground P, children can shorten the distance by 200 m. What are the lengths of the sidewalks?

Solution If we let

$$x = \text{the length of the sidewalk from } P \text{ to } R,$$

then $\quad 700 - x = $ the length of the sidewalk from R to L.

Since the distance from P to L is 200 m less than the total length of the two sidewalks, we have

$$700 - 200 = 500 = \text{the distance from } P \text{ to } L.$$

From the Pythagorean theorem, we obtain the following relationship:

$$x^2 + (700 - x)^2 = (500)^2.$$

We rewrite this equation and solve by factoring:

$$2x^2 - 1400x + 240{,}000 = 0$$
$$x^2 - 700x + 120{,}000 = 0$$
$$(x - 400)(x - 300) = 0.$$

From the last form of the equation we see immediately that $x = 400$ or $x = 300$. Referring to Figure 1.3.3, if we use $x = 400$, we find that the length of the sidewalk from the playground to the refreshment stand is 400 m and the length of the sidewalk from the refreshment stand to the parking lot is $700 - 400 = 300$ m. From $x = 300$, we find that these distances are reversed. Thus there are two possible solutions to this problem.

Check: The solution checks because

$$700 = 300 + 400 \quad \text{and} \quad (500)^2 = (300)^2 + (400)^2. \quad \equiv$$

EXAMPLE 12 Bottles of Wine

A wine broker spent $800 for some bottles of reserve California cabernet sauvignon wine. If each bottle had cost $4 more, the broker would have obtained 10 fewer bottles for the $800. How many bottles were purchased?

Solution The solution of this problem is based on the following relationship:

$$(\text{cost per bottle})(\text{number of bottles}) = 800. \tag{8}$$

For the actual purchase, if we let

$$x = \text{the number of bottles purchased},$$

then $\quad \dfrac{800}{x} = $ the cost per bottle.

At the more expensive price,

$$x - 10 = \text{the number of bottles purchased}$$

and

$$\frac{800}{x} + 4 = \text{the cost per bottle.}$$

Using this information in the relationship (8), we obtain the equation

$$\left(\frac{800}{x} + 4\right)(x - 10) = 800,$$

which we solve for x as follows:

$$(800 + 4x)(x - 10) = 800x$$
$$4x^2 - 40x - 8000 = 0$$
$$x^2 - 10x - 2000 = 0.$$

The quadratic formula gives

$$x = \frac{-(-10) \pm \sqrt{8100}}{2} = \frac{10 \pm 90}{2}$$

and so $x = 50$ or $x = -40$. Since we must have a positive number of bottles purchased, we find that 50 bottles of wine were purchased.

Check: If 50 bottles were purchased for $800, the cost per bottle was $800/50 = $16. If each bottle cost $4 more, then the price per bottle would have been $20. At this more expensive price, only $800/20 = 40$ bottles could have been purchased for $800. Since $50 - 10 = 40$, the answer checks. ≡

NOTES FROM THE CLASSROOM

(*i*) When solving a quadratic equation by factoring, you must set the quadratic expression equal to zero. In Example 2, it serves no purpose to factor

$$4x^2 + 5x = 6 \quad \text{as} \quad x(4x + 5) = 6.$$

Because the right-hand side is 6 (not 0), we cannot conclude anything about the factors x and $4x + 5$.

(*ii*) When $b^2 - 4ac = 0$, a quadratic equation $ax^2 + bx + c = 0$ has a repeated root. This means that the left-hand side of such quadratic equations is a perfect square of a binomial. In Example 3, the discriminant of $4x^2 + 4x + 1 = 0$ is $b^2 - 4ac = 4^2 - 4 \cdot 4 \cdot 1 = 0$. We also saw that the left-hand side of the equation had the equivalent form $(2x + 1)^2$. In Example 7, we leave it to you to express the left-hand side of $9x^2 - 24x + 16 = 0$ as a perfect square.

(*iii*) On page 86 we saw that a quadratic equation $ax^2 + bx + c = 0$ does not have real roots when the discriminant of the equation was negative, that is, when $b^2 - 4ac < 0$. Do not interpret "no real roots" as meaning "no roots." If we count a repeated root, or a root of multiplicity 2, as two roots, then *a quadratic equation always has two roots*, either two real roots or two nonreal roots. Nonreal numbers are also called **complex numbers** and will be the subject of the discussion in the section that follows.

1.3 Exercises

Answers to selected odd-numbered problems begin on page ANS-3.

In Problems 1–16, solve the equation by factoring.

1. $x^2 - 16 = 0$
2. $y^2 - 17y + 16 = 0$
3. $2x^2 + x - 1 = 0$
4. $8t^2 - 22t + 15 = 0$
5. $1 + 4x + 4x^2 = 0$
6. $4 + 5z - 6z^2 = 0$
7. $u^2 + 12 = 7u$
8. $v^2 + 5v = -4$
9. $25y^2 + 15y = -2$
10. $2a^2 = a + 1$
11. $16b^2 - 1 = 0$
12. $25 - c^2 = 0$
13. $x^3 - 9x = 0$
14. $16p^4 - p^2 = 0$
15. $4q^5 - 25q^3 = 0$
16. $x^4 - 18x^2 + 32 = 0$

In Problems 17–22, solve using the square root method.

17. $x^2 = 17$
18. $2y^2 = 100$
19. $(v + 5)^2 = 5$
20. $5(w - 1)^2 = 4$
21. $3(t + 1)^2 = 9$
22. $4(s - 3)^2 = 5$

In Problems 23–26, solve for x. Assume a, b, c, and d represent positive real numbers.

23. $x^2 - b^2 = 0$
24. $x^2 + 2dx + d^2 = 0$
25. $(x - a)^2 = b^2$
26. $(x + c)^2 = d^2$

In Problems 27–34, solve by completing the square.

27. $u^2 + 2u - 1 = 0$
28. $v^2 + 3v - 2 = 0$
29. $2k^2 + 5k + 3 = 0$
30. $4b^2 - 4b - 35 = 0$
31. $10x^2 - 20x + 1 = 0$
32. $36 - 16w - w^2 = 0$
33. $9t^2 = 36t - 1$
34. $r = 4r^2 - 1$

In Problems 35–46, solve using the quadratic formula.

35. $3x^2 - 7x + 2 = 0$
36. $4x^2 - 12x + 9 = 0$
37. $9z^2 + 30z + 25 = 0$
38. $1 + 2w - 6w^2 = 0$
39. $2 + 5r - 10r^2 = 0$
40. $8t = -(16t^2 + 1)$
41. $3s - 2s^2 = \frac{3}{2}$
42. $\frac{1}{2}x^2 + x = 5$
43. $2c(c - 1) = 1$
44. $4x^2 = 2(x + 1)$
45. $x^4 - 6x^2 + 7 = 0$
46. $y^4 - 2y^2 = 4$

In Exercises 47–56, solve the equation using any method.

47. $3s^2 - 13s + 4 = 0$
48. $4x^2 + 8x + 4 = 0$
49. $s^2 - 4s - 4 = 0$
50. $2.4 + 1.0y + 0.1y^2 = 0$
51. $8t^2 + 10t + 5 = 0$
52. $r^2 + 2r = 35$
53. $24t^3 - 3t = 0$
54. $9u^2 + 25 = 30u$
55. $4p^2 = 60$
56. $5(c + 1)^2 = 25$

In Problems 57–62, the given formula occurs frequently in applications. Solve for the indicated variables in terms of the remaining variables. Assume that all variables represent positive real numbers.

57. Volume of a cylinder
$V = \pi r^2 h$, for r

58. Area of a circle
$A = \pi r^2$, for r

59. Surface area of a cylinder
$A = 2\pi r(r + h)$, for r

60. Equation of an ellipse
$\dfrac{x^2}{a^2} + \dfrac{y^2}{b^2} = 1$, for y

CHAPTER 1 EQUATIONS AND INEQUALITIES

61. Free-falling body

$$s = \tfrac{1}{2}gt^2 + v_0 t, \quad \text{for } t$$

62. Newton's universal gravitation law

$$F = g\frac{m_1 m_2}{r^2}, \quad \text{for } r$$

63. Determine all values of d so that $x^2 + (d + 6)x + 8d = 0$ has two equal roots.
64. Determine all values of d so that $3dx^2 - 4dx + d + 1 = 0$ has two equal roots.
65. Determine the other root of $(k - 2)x^2 - x - 4k = 0$, given that one root is -3.
66. If x_1 and x_2 are two real roots of the quadratic equation $ax^2 + bx + c = 0$, then show that $x_1 + x_2 = -b/a$ and $x_1 \cdot x_2 = c/a$.

Miscellaneous Applications

67. **Playing with Numbers** The sum of two numbers is 22, and the sum of their squares is 274. Find the numbers.
68. **Playing with Numbers** The product of two numbers is 1 more than 3 times their sum. Find the numbers if their difference is 9.
69. **Area of a Triangle** The base of a triangle is 3 cm longer than the altitude. If the area of the triangle is 119 cm², find the base and the altitude.
70. **How Long?** On a 35-km hike one boy walks $\tfrac{1}{2}$ km/h faster than another boy. If he makes the trip in 1 h and 40 min less time than the other boy, find how long it takes each boy to make the hike.
71. **Planting a Garden** Barbara has planned a rectangular vegetable garden with a perimeter of 76 m and an area of 360 m². Find the dimensions of her garden.
72. **Distance** A baseball diamond is a square that is 90 ft on a side. Find the distance from third base to first base.
73. **Area** If a square playing field has a diagonal length of 100 ft, find the area of the field.
74. **Garden Trim** A flower garden is in the shape of an isosceles right triangle with a hypotenuse of 50 ft. How many feet of redwood bender board are required to edge it?
75. **Length of Sides** Suppose that the hypotenuse of a right triangle is 10 cm longer than one of the sides and that side is 10 cm longer than the other side. Find the lengths of the three sides of this right triangle.
76. **How Far?** A 17-ft ladder is positioned against the side of a house so that its base is 8 ft from the house. If it slips so that the base of the ladder is 10 ft from the house, how far does the top of the ladder slide down the side of the house?
77. **Distance** Two motorboats leave a dock at the same time. One travels north at a speed of 18 mi/h and the other travels west at 24 mi/h. Find the distance between them after 3 h.
78. **Speed** A motorcyclist travels at a constant speed for 60 mi. If he had gone 10 mi/h faster, he would have shortened his traveling time by 1 h. Find the speed of the motorcyclist.
79. **How Fast?** James took 1 h more than John to drive a 432-mi trip at an average speed of 6 mi/h less than John. How fast did they each drive?
80. **How Many?** A group of women plan to share equally in the $14,000 cost of a boat. At the last minute three of the women drop out, which raises the share for each of the remaining women by $1500. How many women were in the original group?
81. **How Many?** Mr. Arthur buys some stock for $720. If he had bought the stock the day before when the price per share was $15 less, he could have purchased four additional shares. How many shares did Mr. Arthur buy?
82. **Find the Dimensions** A rectangular garden is surrounded by a gravel path that is 2 ft wide. The area covered by the garden is 80 ft², and the area covered by the sidewalk is 108 ft². Find the dimensions of the garden.

83. **Width** A 50-m by 24-m rectangular grassy area has a sidewalk surrounding it. If the area covered by the sidewalk is 480 m², what is its width?

84. **Building an Open Container** A container is made from a square piece of tin by cutting a 3 in. square piece from each corner and bending up the sides. See **FIGURE 1.3.4**. If the container is to have a volume of 48 in³, find the length of a side of the original piece of tin.

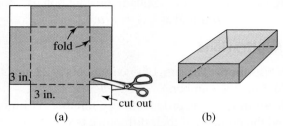

FIGURE 1.3.4 Container in Problem 84

85. **Building an Open Box** Maria has a piece of cardboard that is twice as long as it is wide. If she cuts out 2-in. squares from the corner and bends up the sides to form a box with no top, she will have a box with a volume of 140 in³. Find the dimensions of the original piece of cardboard.

86. **Length** A wire that is 32 cm long is cut into two pieces, and each piece is bent to form a square. The total area enclosed is 34 cm². Find the length of each piece of wire.

87. **How Far?** If an object is projected upward from the ground at an angle of 45° with an initial velocity of v_0 meters per second, then the height y in meters above the ground at a horizontal distance of x meters from the point of projection is given by the formula

$$y = x - \frac{9.8}{v_0^2}x^2.$$

See **FIGURE 1.3.5**. If a projectile is launched at a 45° angle with an initial velocity of 12 m/s, how far from the projection point will it land?

FIGURE 1.3.5 Projectile in Problem 87

88. **How Far?** If a fountain spouts water at an angle of 45° with a velocity of 7 m/s, how far from the spout will the water splash into the pool? See **FIGURE 1.3.6** and Problem 87.

FIGURE 1.3.6 Fountain in Problem 88

For Discussion

In Problems 89 and 90, use the notion of a quadratic equation to solve the given equation. [*Hint*: Review Section R.5.]

89. $x^{2/3} + 4x^{1/3} - 5 = 0$

90. $r^{1/2} + 30r^{-1/2} - 11 = 0$

91. One of the shortest proofs of the Pythagorean theorem was given by the Hindu scholar **Bhaskara** (c. 1150 CE). He offered the diagram shown in FIGURE 1.3.7 without the labeling to assist the reader. His only "explanation" was the word "Behold!" Assume that a square of side c can be divided into four congruent right triangles and a square of length $b - a$ as shown. Prove that $a^2 + b^2 = c^2$.

92. Assuming that a square of side $a + b$ can be divided in two ways, as in FIGURE 1.3.8, prove the Pythagorean theorem.

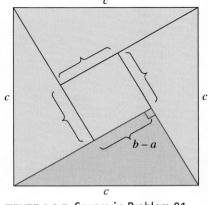

FIGURE 1.3.7 Square in Problem 91

 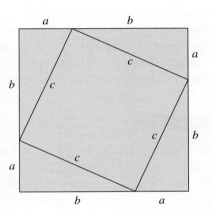

FIGURE 1.3.8 Squares in Problem 92

1.4 Complex Numbers

≡ **Introduction** In the preceding section we saw that some quadratic equations have no real solutions. For example, $x^2 + 1 = 0$ has no real roots because there is no real number x such that $x^2 = -1$. In this section we study the set of **complex numbers**, which contains solutions to equations such as $x^2 + 1 = 0$. The set of complex numbers C contains the set of real numbers R as well as numbers whose squares are negative.

To obtain the complex numbers C, we begin by defining the **imaginary unit**, denoted by the letter i, as the number that satisfies

$$i^2 = -1.$$

It is common practice to write

$$i = \sqrt{-1}.$$

With i we are able to define the principal square root of a negative number, as follows. If c is a positive real number, then the **principal square root of** $-c$, denoted $\sqrt{-c}$, is defined by

$$\sqrt{-c} = \sqrt{(-1)c} = \sqrt{-1}\sqrt{c} = i\sqrt{c} = \sqrt{c}i. \tag{1}$$

EXAMPLE 1 Principal Square Roots

Find the principal square root of (a) $\sqrt{-4}$ and (b) $\sqrt{-5}$.

Solution From (1),

(a) $\sqrt{-4} = \sqrt{(-1)(4)} = \sqrt{-1}\sqrt{4} = i(2) = 2i$
(b) $\sqrt{-5} = \sqrt{(-1)(5)} = \sqrt{-1}\sqrt{5} = i\sqrt{5} = \sqrt{5}i$. ≡

☐ **Terminology** The complex number system contains the imaginary unit i, all real numbers, products such as bi, b real, and sums such as $a + bi$, where a and b are real numbers. In particular, a **complex number** is defined to be any expression of the form

$$z = a + bi, \qquad (2)$$

where a and b are real numbers and $i^2 = -1$. The form given in (2) is called the **standard form** of a complex number. The numbers a and b are called the **real** and **imaginary parts** of z, respectively. A complex number of the form $0 + bi$ is said to be a **pure imaginary number**. Note that by choosing $b = 0$ in (2), we obtain a **real number**. Thus the set R of real numbers is a subset of the set C of complex numbers.

▶ Be careful here, the imaginary part of $a + bi$, is *not bi*; it is the real number b.

EXAMPLE 2 Real and Imaginary Parts

(a) The complex number $z = 4 + (-5)i$ is written as $z = 4 - 5i$. The real part of z is 4 and its imaginary part is -5.
(b) $z = 10i$ is a pure imaginary number.
(c) $z = 6 + 0i = 6$ is a real number. ≡

EXAMPLE 3 Writing in the Standard Form $a + bi$

Express each of the following in the standard form $a + bi$.

(a) $-3 + \sqrt{-7}$ \qquad (b) $2 - \sqrt{-25}$

Solution Using $\sqrt{-c} = \sqrt{c}i$, $c > 0$, we can write

(a) $-3 + \sqrt{-7} = -3 + i\sqrt{7} = -3 + \sqrt{7}i$,
(b) $2 - \sqrt{-25} = 2 - i\sqrt{25} = 2 - 5i$. ≡

In order to solve certain equations involving complex numbers, it is necessary to specify when two complex numbers are equal.

DEFINITION 1.4.1 Equality of Complex Numbers

Two complex numbers are equal if and only if their real parts are equal and imaginary parts are equal. That is, if $z_1 = a + bi$ and $z_2 = c + di$,

$$z_1 = z_2 \text{ if and only if } a = c \text{ and } b = d.$$

EXAMPLE 4 A Simple Equation

Solve for x and y:

$$(2x + 1) + (-2y + 3)i = 2 - 4i.$$

CHAPTER 1 EQUATIONS AND INEQUALITIES

Solution By Definition 1.4.1 we must have
$$2x + 1 = 2 \quad \text{and} \quad -2y + 3 = -4.$$
Solving each equation yields $x = \frac{1}{2}$ and $y = \frac{7}{2}$. ≡

Addition and multiplication for complex numbers are defined as follows.

DEFINITION 1.4.2 Sum, Difference, and Product

If $z_1 = a + bi$ and $z_2 = c + di$, then

(i) their **sum** is given by $\quad z_1 + z_2 = (a + c) + (b + d)i,$
(ii) their **difference** is given by $\quad z_1 - z_2 = (a - c) + (b - d)i,$
(iii) and their **product** is given by $\quad z_1 z_2 = (ac - bd) + (bc + ad)i.$

☐ **Properties of Complex Numbers** In Section R.1 we stated the basic properties for the real number system. Using the definition of addition and multiplication of complex numbers, it can be shown that these basic properties also apply to the complex number system. In particular, the associative, commutative, and distributive laws hold for complex numbers. We further observe that in Definition 1.4.2(*i*):

- *The **sum** of two complex numbers is obtained by adding their corresponding real and imaginary parts.*

Similarly, Definition 1.4.2(*ii*) shows that

- *The **difference** of two complex numbers is obtained by subtracting their corresponding real and imaginary parts.*

Also, rather than memorizing (*iii*) of Definition 1.4.2:

- *The **product** of two complex numbers can be obtained by using the associative, commutative, and distributive laws and the fact that $i^2 = -1$.*

Applying this approach, we find that

$$\begin{aligned}(a + bi)(c + di) &= (a + bi)c + (a + bi)di && \leftarrow \text{distributive law} \\ &= ac + (bc)i + (ad)i + (bd)i^2 && \leftarrow \text{distributive law} \\ &= ac + (bc)i + (ad)i + (bd)(-1) \\ &= ac + (bd)(-1) + (bc)i + (ad)i && \leftarrow \text{factor out } i \\ &= (ac - bd) + (bc + ad)i.\end{aligned}$$

This is the same result as the product given by Definition 1.4.2(*iii*). These techniques are illustrated in the following example.

EXAMPLE 5   Sum, Difference, and Product

If $z_1 = 5 - 6i$ and $z_2 = 2 + 4i$, find **(a)** $z_1 + z_2$, **(b)** $z_1 - z_2$, and **(c)** $z_1 z_2$.

Solution (a) The colors in the diagram below show how to add z_1 and z_2:

$$z_1 + z_2 = (5 - 6i) + (2 + 4i) = (5 + 2) + (-6 + 4)i = 7 - 2i.$$

(add the real parts / add the imaginary parts)

(b) Analogous to part (a) we now subtract the real and imaginary parts:

$$z_1 - z_2 = (5 - 6i) - (2 + 4i) = (5 - 2) + (-6 - 4)i = 3 - 10i.$$

(c) Using the distributive law, we write the product $(5 - 6i)(2 + 4i)$ as

$$(5 - 6i)(2 + 4i) = (5 - 6i)2 + (5 - 6i)4i \quad \leftarrow \text{distributive law}$$
$$= 10 - 12i + 20i - 24i^2 \quad \leftarrow \begin{array}{l}\text{factor } i \text{ from the two middle} \\ \text{terms and replace } i^2 \text{ by } -1\end{array}$$
$$= 10 - 24(-1) + (-12 + 20)i$$
$$= 34 + 8i. \quad \equiv$$

Note of Caution ▶

Not all the properties of the real number system hold for complex numbers. In particular, the property of radicals $\sqrt{a}\sqrt{b} = \sqrt{ab}$ is *not* true when both a and b are negative. To see this, consider that

$$\sqrt{-1}\sqrt{-1} = ii = i^2 = -1 \quad \text{whereas} \quad \sqrt{(-1)(-1)} = \sqrt{1} = 1.$$

Thus, $\sqrt{-1}\sqrt{-1} \neq \sqrt{(-1)(-1)}$. However, if *only one of* a or b is negative, then we do have $\sqrt{a}\sqrt{b} = \sqrt{ab}$.

In the set C of complex numbers, the **additive identity** is the number $0 = 0 + 0i$, and the **multiplicative identity** is the number $1 = 1 + 0i$. The number $-z = -a - bi$ is called the **additive inverse of** $z = a + bi$ because

$$z + (-z) = z - z = (a - a) + (b - b)i = 0 + 0i = 0.$$

In order to obtain the **multiplicative inverse** of a nonzero complex number $z = a + bi$, we introduce the concept of the **conjugate** of a complex number.

DEFINITION 1.4.3 Conjugate

If $z = a + bi$ is a complex number, then the number $\bar{z} = a - bi$ is called the **conjugate** of z.

In other words, the conjugate of a complex number $z = a + bi$ is the complex number obtained by changing the sign of its imaginary part. For example, the conjugate of $8 + 13i$ is $8 - 13i$, and the conjugate of $-5 - 2i$ is $-5 + 2i$.

The following computations show that both the sum and the product of a complex number z and its conjugate $\bar{z}$ are *real* numbers:

$$z + \bar{z} = (a + bi) + (a - bi) = 2a \tag{3}$$
$$z\bar{z} = (a + bi)(a - bi) = a^2 - b^2i^2 = a^2 + b^2. \tag{4}$$

The latter property makes conjugates very useful in finding the multiplicative inverse $1/z$, $z \neq 0$, and in dividing two complex numbers.

We summarize the procedure.

- To **divide** a complex number z_1 by a complex number z_2, multiply the numerator and denominator of z_1/z_2 by the conjugate of the denominator z_2. That is,

$$\frac{z_1}{z_2} = \frac{z_1}{z_2} \cdot \frac{\bar{z}_2}{\bar{z}_2} = \frac{z_1\bar{z}_2}{z_2\bar{z}_2}$$

and then use the fact that the product $z_2\bar{z}_2$ is the sum of the squares of the real and imaginary parts of z_2.

EXAMPLE 6 Division

For $z_1 = 3 - 2i$ and $z_2 = 4 + 5i$, express the given complex number in the form $a + bi$.

(a) $\dfrac{1}{z_1}$ (b) $\dfrac{z_1}{z_2}$

Solution In each case, we multiply both the numerator and the denominator by the conjugate of the denominator and simplify.

(a) $\dfrac{1}{z_1} = \dfrac{1}{3-2i} = \dfrac{1}{3-2i} \cdot \dfrac{\overset{\text{conjugate of } z_1}{\downarrow}}{3+2i}{3+2i} = \dfrac{3+2i}{3^2+(-2)^2} = \overset{\text{standard form } a+bi}{\overbrace{\dfrac{3}{13} + \dfrac{2}{13}i}}$

 $\uparrow$ from (4)

(b) $\dfrac{z_1}{z_2} = \dfrac{3-2i}{4+5i} = \dfrac{3-2i}{4+5i} \cdot \dfrac{4-5i}{4-5i} = \dfrac{12-8i-15i+10i^2}{4^2+5^2}$

$= \dfrac{2-23i}{41} = \dfrac{2}{41} - \dfrac{23}{41}i \;\;\leftarrow \text{standard form } a+bi$ ≡

From the definition of addition and subtraction of two complex numbers, it is readily shown that the conjugate of a sum and difference of two complex numbers is the sum and difference of the conjugates. This property, along with three other properties of the conjugate are summarized as a theorem.

THEOREM 1.4.1   Properties of the Conjugate

Let z_1 and z_2 be any two complex numbers. Then

(i) $\overline{z_1 \pm z_2} = \bar{z}_1 \pm \bar{z}_2$,   (ii) $\overline{z_1 z_2} = \bar{z}_1 \bar{z}_2$,

(iii) $\overline{\left(\dfrac{z_1}{z_2}\right)} = \dfrac{\bar{z}_1}{\bar{z}_2}$,   (iv) $\bar{\bar{z}} = z$.

Of course, the conjugate of any finite sum (product) of complex numbers is the sum (product) of the conjugates.

☐ **Quadratic Equations** Complex numbers make it possible to solve quadratic equations $ax^2 + bx + c = 0$ when the discriminant $b^2 - 4ac$ is negative. We now see that the solutions from the quadratic formula

$$x_1 = \dfrac{-b - \sqrt{b^2-4ac}}{2a} \quad \text{and} \quad x_2 = \dfrac{-b + \sqrt{b^2-4ac}}{2a} \tag{5}$$

represent complex numbers. Note that in fact the solutions are conjugates of each other. As the next example shows these solutions can be written in the form $z = a + bi$.

EXAMPLE 7   **Complex Solutions**

Solve $x^2 - 8x + 25 = 0$.

Solution From the quadratic formula, we obtain

$$x = \dfrac{-(-8) \pm \sqrt{(-8)^2 - 4(1)(25)}}{2} = \dfrac{8 \pm \sqrt{-36}}{2}.$$

Using $\sqrt{-36} = 6i$ we obtain

$$x = \dfrac{8 \pm 6i}{2} = 4 \pm 3i.$$

Thus, the solution set of the equation is $\{4 - 3i, 4 + 3i\}$.   ≡

☐ **Conjugate Solutions** We can now obtain solutions to any quadratic equation. In particular, if the coefficients in $ax^2 + bx + c = 0$ are real numbers and the discriminant is negative, we see from (5) that the roots appear as conjugate pairs. Observe in Example 7 that if $x_1 = 4 - 3i$ and $x_2 = 4 + 3i$, then $\bar{x}_2 = x_1$. Moreover, it is easily seen that $\bar{x}_1 = x_2$.

1.4 Exercises
Answers to selected odd-numbered problems begin on page ANS-3.

In Problems 1–10, find the indicated power of i.

1. i^3
2. i^4
3. i^5
4. i^6
5. i^7
6. i^8
7. i^{-1}
8. i^{-2}
9. i^{-3}
10. i^{-6}

In Problems 11–56, perform the indicated operation. Write the answer in standard form $a + bi$.

11. $\sqrt{-100}$
12. $-\sqrt{-8}$
13. $-3 - \sqrt{-3}$
14. $\sqrt{-5} - \sqrt{-125} + 5$
15. $(3 + i) - (4 - 3i)$
16. $(5 + 6i) + (-7 + 2i)$
17. $2(4 - 5i) + 3(-2 - i)$
18. $-2(6 + 4i) + 5(4 - 8i)$
19. $i(-10 + 9i) - 5i$
20. $i(4 + 13i) - i(1 - 9i)$
21. $3i(1 + i) - 4(2 - i)$
22. $i + i(1 - 2i) + i(4 + 3i)$
23. $(3 - 2i)(1 - i)$
24. $(4 + 6i)(-3 + 4i)$
25. $(7 + 14i)(2 + i)$
26. $(-5 - \sqrt{3}i)(2 - \sqrt{3}i)$
27. $(4 + 5i) - (2 - i)(1 + i)$
28. $(-3 + 6i) + (2 + 4i)(-3 + 2i)$
29. $i(1 - 2i)(2 + 5i)$
30. $i(\sqrt{2} - i)(1 - \sqrt{2}i)$
31. $(1 + i)(1 + 2i)(1 + 3i)$
32. $(2 + i)(2 - i)(4 - 2i)$
33. $(1 - i)[2(2 - i) - 5(1 + 3i)]$
34. $(4 + i)[i(1 + 3i) - 2(-5 + 3i)]$
35. $(4 + i)^2$
36. $(3 - 5i)^2$
37. $(1 - i)^2(1 + i)^2$
38. $(2 + i)^2(3 + 2i)^2$
39. $\dfrac{1}{4 - 3i}$
40. $\dfrac{5}{3 + i}$
41. $\dfrac{4}{5 + 4i}$
42. $\dfrac{1}{-1 + 2i}$
43. $\dfrac{i}{1 + i}$
44. $\dfrac{i}{4 - i}$
45. $\dfrac{4 + 6i}{i}$
46. $\dfrac{3 - 5i}{i}$
47. $\dfrac{1 + i}{1 - i}$
48. $\dfrac{2 - 3i}{1 + 2i}$
49. $\dfrac{4 + 2i}{2 - 7i}$
50. $\dfrac{\frac{1}{2} - \frac{7}{2}i}{4 + 2i}$
51. $i\left(\dfrac{10 - i}{1 + i}\right)$
52. $i\left(\dfrac{1 - 2\sqrt{3}i}{1 + \sqrt{3}i}\right)$
53. $(1 + i)\dfrac{2i}{1 - 5i}$
54. $(5 - 3i)\dfrac{1 - i}{2 - i}$
55. $4 - 9i + \dfrac{25i}{2 + i}$
56. $i\left(-6 + \dfrac{11}{5}i\right) + \dfrac{2 + i}{2 - i}$

In Problems 57–64, use Definition 1.4.1 to solve for x and y.

57. $2(x + yi) = i(3 - 4i)$
58. $(x + yi) + 4(1 - i) = 5 - 7i$
59. $i(x + yi) = (1 - 6i)(2 + 3i)$
60. $10 + 6yi = 5x + 24i$
61. $(1 + i)(x - yi) = i(14 + 7i) - (2 + 13i)$
62. $i^2(1 - i)(1 + i) = 3x + yi + i(y + xi)$
63. $x + yi = \dfrac{i^3}{2 - i}$
64. $25 - 49i = x^2 - y^2i$

CHAPTER 1 EQUATIONS AND INEQUALITIES

In Problems 65–76, solve the given equation.

65. $x^2 + 9 = 0$
66. $x^2 + 8 = 0$
67. $2x^2 = -5$
68. $3x^2 = -1$
69. $2x^2 - x + 1 = 0$
70. $x^2 - 2x + 10 = 0$
71. $x^2 + 8x + 52 = 0$
72. $3x^2 + 2x + 5 = 0$
73. $4x^2 - x + 2 = 0$
74. $x^2 + x + 2 = 0$
75. $x^4 + 3x^2 + 2 = 0$
76. $2x^4 + 9x^2 + 4 = 0$

77. The two square roots of the complex number i are the two numbers z_1 and z_2 that are solutions of the equation $z^2 = i$. Let $z = x + iy$ and find z^2. Then use Definition 1.4.1 to find z_1 and z_2.
78. Proceed as in Problem 77 to find two numbers z_1 and z_2 that satisfy the equation $z^2 = -3 + 4i$.

For Discussion

In Problems 79–82, prove the given properties involving the conjugates of $z_1 = a + bi$ and $z_2 = c + di$.

79. $\bar{z}_1 = z_1$ if and only if z_1 is a real number.
80. $\overline{z_1 + z_2} = \bar{z}_1 + \bar{z}_2$
81. $\overline{z_1 \cdot z_2} = \bar{z}_1 \cdot \bar{z}_2$
82. $\overline{z_1^2} = (\bar{z}_1)^2$

1.5 Linear Inequalities

≡ **Introduction** In Section R.2 we defined the order relations "less than," "less than or equal," "greater than," and "greater than or equal," and we saw how to interpret these relations on the real number line. In this section we are interested in solving various kinds of inequalities containing a variable x. If a real number is substituted for the variable x in an inequality such as

$$3x - 7 > 4, \qquad (1)$$

and if the result is a true statement, then that number is said to be a **solution** of the inequality. For example, 5 is a solution of (1) because if x is replaced by 5, then the resulting inequality $3(5) - 7 > 5$ simplifies to the true statement $8 > 5$. The word **solve** means that we are to find the set of *all* solutions of an inequality such as (1). This set is called the **solution set** of the inequality. Two inequalities are said to be **equivalent** if they have exactly the same solution set. The representation of the solution set on the number line is the **graph** of the inequality.

We solve an inequality by finding an equivalent inequality with obvious solutions. The following list summarizes three operations that yield equivalent inequalities.

THEOREM 1.5.1 Operations That Yield Equivalent Inequalities

Suppose a, b, are real numbers and c is a nonzero real number. Then the inequality $a < b$ is equivalent to:

(i) $a + c < b + c$,
(ii) $a \cdot c < b \cdot c$, for $c > 0$,
(iii) $a \cdot c > b \cdot c$, for $c < 0$.

PROOF OF (i): To prove part (*i*), we begin with the assumption that $a < b$. Then from Definition R.2.1 it follows that $b - a$ is positive. If we add $c - c = 0$ to a positive number, the sum is positive. Therefore,

$$b - a = b - a + (c - c)$$
$$= b + c - a - c$$
$$= (b + c) - (a + c)$$

is a positive number. Thus we have $a + c < b + c$.

Operations (*i*)–(*iii*) of Theorem 1.5.1 also hold with $>$ in place of $<$ and $<$ in place of $>$. In addition, (*i*)–(*iii*) can be stated for the order relations $\leq$ and $\geq$. We leave the verification of (*ii*) and (*iii*) as exercises. See Problems 55 and 56 in Exercises 1.5.

Note of Caution ▶ Property (*iii*) of Theorem 1.5.1 is frequently forgotten when solving inequalities. In words, property (*iii*) states that

- *If an inequality is multiplied by a negative number, then the direction of the resulting inequality is reversed.*

For example, if we multiply the inequality $-2 < 5$ by -3, then the *less than* symbol is changed to a *greater than* symbol:

$$-2(-3) > 5(-3) \quad \text{or} \quad 6 > -15.$$

☐ **Solving Linear Inequalities** Any inequality that can be written in one of the forms

$$ax + b < 0, \quad ax + b > 0, \tag{2}$$

$$ax + b \leq 0, \quad ax + b \geq 0, \tag{3}$$

where a and b are real numbers, is said to be a **linear inequality** in the variable x. The inequality in (1) is an example of a linear inequality since by part (*i*) of Theorem 1.5.1 we can add -4 to both sides to obtain

$$3x - 7 + (-4) > 4 + (-4)$$

or $3x - 11 > 0$, which matches the second form in (2).

In the examples that follow we use operations (*i*)–(*iii*) in Theorem 1.5.1 to solve linear inequalities.

EXAMPLE 1 Solving a Linear Inequality

Solve $8x + 4 < 16 + 5x$.

Solution We obtain equivalent inequalities using the operations in Theorem 1.5.1:

$$8x + 4 < 16 + 5x$$
$$8x + 4 - 4 < 16 + 5x - 4 \quad \leftarrow \text{by (i) of Theorem 1.5.1}$$
$$8x < 12 + 5x$$
$$8x - 5x < 12 + 5x - 5x \quad \leftarrow \text{by (i) of Theorem 1.5.1}$$
$$3x < 12$$
$$\left(\tfrac{1}{3}\right)3x < \left(\tfrac{1}{3}\right)12 \quad \leftarrow \text{by (ii) of Theorem 1.5.1}$$
$$x < 4.$$

Using set-builder notation, the solution set of the given inequality is

$$\{x \mid x \text{ is real and } x < 4\}.$$

EXAMPLE 2 Solving a Linear Inequality

Solve $\frac{1}{2} - 3x \leq \frac{5}{2}$.

Solution The following inequalities are equivalent. (You should be able to supply a reason for each step.)

$$\frac{1}{2} - 3x \leq \frac{5}{2}$$
$$-\frac{1}{2} + \frac{1}{2} - 3x \leq -\frac{1}{2} + \frac{5}{2}$$
$$-3x \leq \frac{4}{2}$$
$$-3x \leq 2$$
$$\left(-\frac{1}{3}\right)(-3x) \geq \left(-\frac{1}{3}\right)2$$
$$x \geq -\frac{2}{3}.$$

Thus, the solution set of the given inequality is $\{x \mid x \text{ is real and } x \geq -\frac{2}{3}\}$. ≡

☐ **Interval Notation** The solution set in Example 1 is graphed on the number line in **FIGURE 1.5.1(a)** as a colored arrow over the line pointing to the left. In the figure, the right parenthesis at 4 indicates that number 4 is *not* included in the solution set. Because the solution set extends indefinitely to the left, or to the negative direction, the inequality $x < 4$ can also be written as $-\infty < x < 4$, where ∞ is the **infinity symbol**. In other words, the solution set of the inequality $x < 4$ is

$$\{x \mid x \text{ real and } x < 4\} = \{x \mid -\infty < x < 4\}.$$

Using **interval notation** this set of real numbers is written $(-\infty, 4)$ and is an example of an **unbounded interval**. The graph of the solution set in Example 2,

$$\{x \mid x \text{ is real and } x \geq -\frac{2}{3}\} = \{x \mid -\frac{2}{3} \leq x < \infty\}$$

is shown in Figure 1.5.1(b), where the left bracket at $-\frac{2}{3}$ indicates that $-\frac{2}{3}$ is included in the solution set. In interval notation, this set is the unbounded interval $[-\frac{2}{3}, \infty)$. Table 1.5.1 that follows summarizes various inequalities and their solution sets, as well as interval notations, names, and graphs. In each of the first four entries of the table, the numbers a and b are called the **endpoints** of the interval. As a set, the **open interval**

$$(a, b) = \{x \mid a < x < b\},$$

does not include either endpoint, whereas the **closed interval**

$$[a, b] = \{x \mid a \leq x \leq b\},$$

includes both endpoints. Note too, that the graph of the last interval in Table 1.5.1, extending indefinitely both to the left and to the right, is the entire real number line. The interval notation $(-\infty, \infty)$ is generally used to represent the set R of real numbers.

As you peruse Table 1.5.1, keep firmly in mind that the **infinity symbols** $-\infty$ ("minus infinity") and ∞ ("infinity") do not represent real numbers and should *never* be manipulated arithmetically like a number. The infinity symbols are merely notational devices; $-\infty$ and ∞ are used to indicate unboundedness in the negative direction and in the positive direction, respectively. Thus when using interval notation, the symbols $-\infty$ and ∞ can never appear next to a square bracket; that is, an expression such as

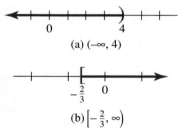

FIGURE 1.5.1 Solution sets in Examples 1 and 2 in interval notation

1.5 Linear Inequalities

TABLE 1.5.1 Inequalities and Intervals

Inequality	Solution Set	Interval Notation	Name	Graph
$a < x < b$	$\{x \mid a < x < b\}$	(a, b)	Open interval	
$a \leq x \leq b$	$\{x \mid a \leq x \leq b\}$	$[a, b]$	Closed interval	
$a < x \leq b$	$\{x \mid a < x \leq b\}$	$(a, b]$	Half-open interval	
$a \leq x < b$	$\{x \mid a \leq x < b\}$	$[a, b)$	Half-open interval	
$a < x$	$\{x \mid a < x < \infty\}$	(a, ∞)		
$x < b$	$\{x \mid -\infty < x < b\}$	$(-\infty, b)$	Unbounded intervals	
$x \leq b$	$\{x \mid -\infty < x \leq b\}$	$(-\infty, b]$		
$a \leq x$	$\{x \mid a \leq x < \infty\}$	$[a, \infty)$		
$-\infty < x < \infty$	$\{x \mid -\infty < x < \infty\}$	$(-\infty, \infty)$		

Reread the last two sentences. ▶

$(2, \infty]$ is meaningless. Also, observe that the last entry in Table 1.5.1 states that the unbounded interval $(-\infty, \infty)$ is the entire real number line. When referring to the entire set of real numbers, it is common practice to use the symbols R and $(-\infty, \infty)$ interchangeably.

☐ **Simultaneous Inequalities** An inequality of the form

$$a < x < b$$

is sometimes referred to as a **simultaneous inequality** because the number x is *between* the numbers a and b; in other words, $x > a$ *and* simultaneously $x < b$. For example, the set of real numbers that satisfy $2 < x < 5$ is the intersection of the intervals $(2, \infty)$ and $(-\infty, 5)$ defined, respectively, by the inequalities $2 < x$ and $x < 5$. Recall from Section R.1 that the **intersection** of two sets A and B, written $A \cap B$, is the set of elements that are in A *and* in B, in other words, the elements that are *common* to both sets. As illustrated in **FIGURE 1.5.2**, the solution set of the inequality $2 < x < 5$, is the intersection of sets $(2, \infty)$ and $(-\infty, 5)$. The set $(2, \infty) \cap (-\infty, 5) = (2, 5)$ can be visualized by overlapping the red arrows in the figure.

FIGURE 1.5.2 The numbers in $(2, 5)$ are the numbers common to both $(2, \infty)$ and $(-\infty, 5)$

☐ **Solving Simultaneous Inequalities** As the following examples show, we can usually solve a simultaneous inequality by isolating the variable in the middle. Operations (*i*)–(*iii*) of Theorem 1.5.1 are applied to both parts of the inequality at the same time.

EXAMPLE 3 Solving a Simultaneous Inequality

Solve $-7 \leq 2x + 1 < 19$. Give the solutions in interval notation and sketch the graph.

Solution We obtain equivalent inequalities as follows:
$$-7 \leq 2x + 1 < 19$$
$$-7 - 1 \leq 2x + 1 - 1 < 19 - 1 \quad \leftarrow \text{by } (i) \text{ of Theorem 1.5.1}$$
$$-8 \leq 2x < 18$$
$$\tfrac{1}{2}(-8) \leq \tfrac{1}{2}(2x) < \tfrac{1}{2}(18) \quad \leftarrow \text{by } (ii) \text{ of Theorem 1.5.1}$$
$$-4 \leq x < 9.$$

Thus the solutions of the given inequality are all the numbers in the interval $[-4, 9)$. The graph of the interval is shown in **FIGURE 1.5.3**. ≡

FIGURE 1.5.3 Solution set in Example 3

◀ Note of Caution

It is customary to write simultaneous inequalities with the smaller number on the left. For example, although $5 > x > 2$ is technically correct, it should be rewritten as $2 < x < 5$ to reflect the order on the number line. The following example illustrates this point.

EXAMPLE 4 Solving a Simultaneous Inequality

Solve $-1 < 1 - 2x < 3$. Give the solution set in interval notation and sketch its graph.

Solution You should supply reasons why the following inequalities are equivalent:
$$-1 < 1 - 2x < 3$$
$$-1 - 1 < -1 + 1 - 2x < -1 + 3$$
$$-2 < -2x < 2.$$

We isolate the variable x in the middle of the last simultaneous inequality by multiplying by $-\tfrac{1}{2}$. Since multiplication by a negative number reverses the direction of an inequality we have,

inequality symbols reversed
$$\left(-\tfrac{1}{2}\right)(-2) > \left(-\tfrac{1}{2}\right)(-2x) > \left(-\tfrac{1}{2}\right)2$$
$$1 > x > -1.$$

To express this solution in interval notation, we first rewrite it with the negative number on the left; that is, $-1 < x < 1$. Thus the solution set can be expressed as the open interval $(-1, 1)$, which is sketched in **FIGURE 1.5.4**. ≡

FIGURE 1.5.4 Solution set in Example 4

Example 5 illustrates an application involving inequalities.

EXAMPLE 5 Yearly Number of Sales

Mrs. Johnson is paid $15,000 a year plus a commission of 8% on her sales. What yearly sales would correspond to an annual income between $23,000 and $27,000?

Solution If we let x represent the amount in dollars of Mrs. Johnson's yearly sales, then $15,000 + 0.08x$ equals her annual income in dollars. Thus we want to find x such that
$$23,000 \leq 15,000 + 0.8x \leq 27,000.$$

We solve this inequality as follows:
$$8000 \leq 0.08x \leq 12,000$$
$$\left(\tfrac{100}{8}\right)8000 \leq \left(\tfrac{100}{8}\right)(0.08)x \leq \left(\tfrac{100}{8}\right)12,000$$
$$100,000 \leq x \leq 150,000.$$

Thus Mrs. Johnson's yearly sales must lie between $100,000 and $150,000 in order for her annual income to be between $23,000 and $27,000. ≡

1.5 Exercises
Answers to selected odd-numbered problems begin on page ANS-4.

In Problems 1–8, write the given inequality using interval notation and then graph the interval.

1. $x < 0$
2. $0 < x < 5$
3. $x \geq 5$
4. $-1 \leq x$
5. $8 < x \leq 10$
6. $-5 < x \leq -3$
7. $-2 \leq x \leq 4$
8. $x > -7$

In Problems 9–14, write the given interval as an inequality.

9. $[-7, 9]$
10. $[1, 15)$
11. $(-\infty, 2)$
12. $[-5, \infty)$
13. $(4, 20]$
14. $(-\frac{1}{2}, 10)$

In Problems 15–18, solve the given inequality and indicate where operations (*i*)–(*iii*) of Theorem 1.5.1 are used.

15. $4x + 4 \geq x$
16. $-x + 5 < 4x - 10$
17. $0 < 2(4 - x) < 6$
18. $-3 \leq \dfrac{4 - x}{4} < 7$

In Problems 19–34, solve the given linear inequality. Write the solution set using interval notation. Graph the solution set.

19. $x + 3 > -2$
20. $3x - 9 < 6$
21. $\frac{3}{2}x + 4 \leq 10$
22. $5 - \frac{5}{4}x \geq -4$
23. $\frac{3}{2} - x > x$
24. $-(1 - x) \geq 2x - 1$
25. $2 + x \geq 3(x - 1)$
26. $-7x + 3 \leq 4 - x$
27. $-\frac{20}{3} < \frac{2}{3}x < 4$
28. $-3 \leq -x < 2$
29. $-7 < x - 2 < 1$
30. $3 < x + 4 \leq 10$
31. $7 < 3 - \frac{1}{2}x \leq 8$
32. $100 + x \leq 41 - 6x \leq 121 + x$
33. $-1 \leq \dfrac{x - 4}{4} < \frac{1}{2}$
34. $2 \leq \dfrac{4x + 2}{-3} \leq 10$

In Problems 35–38, write the expression $|x - 2| + |x - 5|$ without the absolute-value symbols if x is a number in the given interval.

35. $(-\infty, 1)$
36. $(7, \infty)$
37. $(3, 4]$
38. $[2, 5]$

In Problems 39–42, write the expression $|x + 1| - |x - 3|$ without the absolute-value symbols if x is a number in the given interval.

39. $[-1, 3)$
40. $(0, 1)$
41. (π, ∞)
42. $(-\infty, -5)$

Miscellaneous Applications

43. **Playing with Numbers** If 7 times a number is decreased by 5, the result is less than 47. What can be determined about the number?
44. **Trying for a B** James has two test scores of 71 and 82 out of 100. What must he score on a third test in order to have an average of 80 or more?

45. Such a Deal A 50¢ rebate is available on a 4-oz jar of instant coffee, and the 2.5-oz jar sells for $3.00. At what price would the larger jar be more economical?

46. Fever Generally a person is considered to have a fever if he or she has an oral temperature greater than 98.6°F. What temperatures on the Celsius scale indicate a fever? [*Hint*: Recall from Problem 67 in Exercises 1.1 that $T_F = \frac{9}{5}T_C + 32$, where T_C is degrees Celsius and T_F is degrees Fahrenheit.]

47. Taken for a Ride A taxi charges 90¢ for the first quarter mile and 30¢ for each additional quarter mile. How far in quarter miles can a person ride and owe between $3 and $6?

48. At Least It's a Job A successful salesperson normally spends between 5 and 15 hours of a 40-h week working in the office. How much time does the salesperson have left to contact customers outside the office?

For Discussion

In Problems 49–54, answer true or false.

49. If $a < b$, then $a - 16 < b - 16$. _____

50. If $a < b$, then $-a < -b$. _____

51. If $0 < a$, then $a < a + a$. _____

52. If $a < 0$, then $a + a < a$. _____

53. If $1 < a$, then $\dfrac{1}{a} < 1$. _____

54. If $a < 0$, then $\dfrac{a}{-a} < 0$. _____

55. Prove operation (*ii*) of Theorem 1.5.1 using the fact that the product of two positive numbers is positive.

56. Prove operation (*iii*) of Theorem 1.5.1 using the fact that the product of a positive number and a negative number is negative.

57. If $0 < a < b$, show that $1/b < 1/a$. Is the restriction that a is positive necessary? Explain.

58. (a) If $0 < a < b$, show that $a^2 < b^2$.
(b) What is the relationship between a^2 and b^2 if $a < b < 0$? If $b < a < 0$?

1.6 Absolute-Value Equations and Inequalities

≡ **Introduction** We have seen previously that the **absolute value** of a real number x is a nonnegative quantity defined as

$$|x| = \begin{cases} -x, & x < 0 \\ x, & x \geq 0. \end{cases} \qquad (1)$$

◀ A review of Section R.2 is highly recommended.

In this section we focus on two things: solving *equations* and *inequalities* that involve an absolute value.

☐ **Absolute-Value Equations** From (1) we see immediately that $|6| = 6$ since $6 > 0$ and $|-6| = -(-6) = 6$ because $-6 < 0$. This simple example suggests that the solution of the equation $|x| = 6$ has two solutions, namely, $x = 6$ and $x = -6$. The first theorem succinctly summarizes how to solve an absolute-value equation.

THEOREM 1.6.1 Absolute-Value Equation

If a denotes a positive real number, then

$$|x| = a \quad \text{if and only if} \quad x = -a \quad \text{or} \quad x = a. \tag{2}$$

In (2), keep in mind that the symbol x is a placeholder for any quantity.

EXAMPLE 1   Absolute-Value Equations

Solve **(a)** $|-2x| = 9$ **(b)** $|5x - 3| = 8$.

Solution
(a) We use (2) with the symbol x replaced by $-2x$ and the identification $a = 9$:

$$-2x = -9 \quad \text{or} \quad -2x = 9.$$

We solve each of these equations. From $-2x = -9$, we obtain $x = \frac{-9}{-2} = \frac{9}{2}$. Then from $-2x = 9$, we get $x = \frac{9}{-2} = -\frac{9}{2}$. The solution set is $\{\frac{9}{2}, -\frac{9}{2}\}$.

(b) Now from (2) with x replaced by $5x - 3$ and $a = 8$:

$$5x - 3 = -8 \quad \text{or} \quad 5x - 3 = 8.$$

Solving each of these linear equations gives, in turn, $x = -1$ and $x = \frac{11}{5}$. The solution set of the original equation is $\{-1, \frac{11}{5}\}$.   ≡

(a) The distance between x and 0 is *less* than a

(b) The distance between x and 0 is *greater* than a

FIGURE 1.6.1 Graphical interpretation of $|x| < a$ and $|x| > a$

☐ **Absolute-Value Inequalities** Many important applications of inequalities also involve absolute values. Recall from Section R.2 that $|x|$ represents the distance along the number line from x to the origin. Thus, the inequality $|x| < a$, $(a > 0)$, indicates that the distance from x to the origin is less than a. We can see in **FIGURE 1.6.1(a)** that this is the set of real numbers x such that $-a < x < a$. On the other hand, $|x| > a$ means that the distance from x to the origin is greater than a. Therefore, as Figure 1.6.1(b) shows, either $x < -a$ or $x > a$. These geometric observations suggest the following interpretation of two kinds of absolute-value inequalities.

THEOREM 1.6.2 Absolute-Value Inequalities

(i) $|x| < a$ if and only if $-a < x < a$.
(ii) $|x| > a$ if and only if $x < -a$ or $x > a$.

Parts (*i*) and (*ii*) of Theorem 1.6.2 also hold with $\leq$ in place of $<$ and $\geq$ in place of $>$.

EXAMPLE 2   Two Absolute-Value Inequalities

Solve the inequalities **(a)** $|x| < 1$ **(b)** $|x| \geq 5$.

Solution **(a)** From (*i*) of Theorem 1.6.2 the inequality $|x| < 1$ is equivalent to the simultaneous inequality $-1 < x < 1$. Thus the solution set of $|x| < 1$ is the interval $(-1, 1)$.

(b) From (*ii*) of Theorem 1.6.2, the inequality $|x| \geq 5$ is equivalent to the pair of inequalities: $x \leq -5$ or $x \geq 5$. Thus $|x| \geq 5$ is satisfied for numbers in either of the intervals $(-\infty, -5]$ or $[5, \infty)$.

Because the solution set in part (b) of Example 2 consists of two nonintersecting, or disjoint, intervals it cannot be expressed as a single interval. The best we can do is to write the solution set as the union of the two intervals. Recall from Section R.1, the **union** of two sets A and B, written $A \cup B$, is the set of elements that are in either A or in B. Thus the solution set in Example 2(b) can be written $(-\infty, -5] \cup [5, \infty)$.

EXAMPLE 3 An Absolute-Value Inequality

Solve $|3x - 7| < 1$ and graph the solution set.

Solution If we replace x by $3x - 7$ and a by the number 1, then property (*i*) of Theorem 1.6.2 yields the equivalent simultaneous inequality

$$-1 < 3x - 7 < 1.$$

To solve we beginning by adding 7 across the inequalities:

$$-1 + 7 < 3x - 7 + 7 < 1 + 7$$
$$6 < 3x < 8.$$

Multiplying the last inequality by $\frac{1}{3}$ then gives

$$(\tfrac{1}{3})6 < (\tfrac{1}{3})3x < (\tfrac{1}{3})8$$
$$2 < x < \tfrac{8}{3}.$$

The solution set is the open interval $(2, \tfrac{8}{3})$ shown in **FIGURE 1.6.2**.

FIGURE 1.6.2 Solution set in Example 3

EXAMPLE 4 An Absolute-Value Inequality

Solve $|3x - 4| \leq 0$.

Solution Since the absolute value of an expression is never negative, the only numbers satisfying the given inequality are those for which

$$|3x - 4| = 0 \quad \text{or} \quad 3x - 4 = 0.$$

Thus the solution set consists of the single number $\tfrac{4}{3}$.

An inequality such as $|x - b| < a$ can also be interpreted in terms of distance along the number line. Since $|x - b|$ is the distance between x and b, an inequality such as $|x - b| < a$ is satisfied by all real numbers x whose distance between x and b is less than a. This interval is shown in **FIGURE 1.6.3**. Note when $b = 0$ we get property (*i*) above. Similarly, the set of numbers satisfying $|x - b| > a$ are the numbers x whose distance between x and b is greater than a.

FIGURE 1.6.3 The distance between x and b is less than a

EXAMPLE 5 An Absolute-Value Inequality

Solve $|4 - \tfrac{1}{2}x| \geq 7$ and graph the solution set.

Solution By replacing x by $4 - \tfrac{1}{2}x$, $a = 7$, and $>$ replaced by $\geq$, we see from (*ii*) of Theorem 1.6.2 that

$$|4 - \tfrac{1}{2}x| \geq 7$$

is equivalent to the two inequalities

$$4 - \tfrac{1}{2}x \leq -7 \quad \text{or} \quad 4 - \tfrac{1}{2}x \geq 7.$$

1.6 Absolute-Value Equations and Inequalities

We solve each of these inequalities separately. First we have

$$4 - \tfrac{1}{2}x \leq -7$$
$$-\tfrac{1}{2}x \leq -11$$
$$(-2)\left(-\tfrac{1}{2}\right)x \geq (-2)(-11) \quad \leftarrow \text{multiplication by } -2 \text{ reverses the direction of the inequality}$$
$$x \geq 22.$$

In interval notation this is $[22, \infty)$. Then we solve

$$4 - \tfrac{1}{2}x \geq 7$$
$$-\tfrac{1}{2}x \geq 3$$
$$x \leq -6.$$

In interval notation this is written $(-\infty, -6]$. Since any real number that satisfies either $4 - \tfrac{1}{2}x \leq -7$ or $4 - \tfrac{1}{2}x \geq 7$ also satisfies $|4 - \tfrac{1}{2}x| \geq 7$, the solution set is the union of the two disjoint intervals: $(-\infty, -6] \cup [22, \infty)$. The graph of these solutions is shown in **FIGURE 1.6.4**.

FIGURE 1.6.4 Solution set in Example 5 is the union of these two disjoint sets

Notice in Figure 1.6.1(a) that the number 0 is the midpoint of the solution interval for $|x| < a$ and in Figure 1.6.3 that the number b is the midpoint of the solution interval for the inequality $|x - b| < a$. With this in mind, work through the next example.

EXAMPLE 6 Constructing an Absolute-Value Inequality

Find an inequality of the form $|x - b| < a$ for which the open interval $(4, 8)$ is its solution set.

Solution The midpoint of the interval $(4, 8)$ is $m = \dfrac{4 + 8}{2} = 6$. The distance between the midpoint m and one of the endpoints of the interval is $d(m, 8) = |8 - 6| = 2$. Therefore, with $b = 6$ and $a = 2$, the required inequality is $|x - 6| < 2$.

1.6 Exercises Answers to selected odd-numbered problems begin on page ANS-4.

In Problems 1–10, solve the given equation.

1. $|4x - 1| = 2$
2. $|5v - 4| = 7$
3. $|\tfrac{1}{4} - \tfrac{3}{2}y| = 1$
4. $|2 - 16t| = 0$
5. $\left|\dfrac{x}{x - 1}\right| = 2$
6. $\left|\dfrac{x + 1}{x - 2}\right| = 4$
7. $|x^2 - 8| = 1$
8. $|x^2 + 3x| = 4$
9. $|x^2 - 2x| = 1$
10. $|x^2 + 5x - 3| = 3$

In Problems 11–22, solve the given inequality. Write the solution set using interval notation. Graph the solution set.

11. $|-5x| < 4$
12. $|3x| > 18$
13. $|3 + x| > 7$
14. $|x - 4| \leq 9$
15. $|2x - 7| \leq 1$
16. $|5 - \tfrac{1}{3}x| < \tfrac{1}{2}$

17. $|x + \sqrt{2}| \geq 1$
18. $|6x + 4| > 4$
19. $\left|\dfrac{3x - 1}{-4}\right| < 2$
20. $\left|\dfrac{2 - 5x}{3}\right| \geq 5$
21. $|x - 5| < 0.01$
22. $|x - (-2)| < 0.001$

In Problems 23–26, proceed as in Example 6 and find an inequality $|x - b| < a$ or $|x - b| > a$ for which the given interval is its solution set.

23. $(-3, 11)$
24. $(1, 2)$
25. $(-\infty, 1) \cup (9, \infty)$
26. $(-\infty, -3) \cup (13, \infty)$

In Problems 27 and 28, find an absolute-value inequality whose solution is the set of real numbers x satisfying the given condition. Express each set using interval notation.

27. Greater than or equal to 2 units from -3
28. Less than $\frac{1}{2}$ unit from 3.5

Miscellaneous Applications

29. **Comparing Ages** Bill's and Mary's ages, A_B and A_M, respectively, differ by at most 3 years. Write this fact as an inequality using absolute-value symbols.
30. **Survival** Your score on the first exam is 72%. The midterm grade is the average of the first exam score with the midterm exam score. If the B range is from 80 to 89%, what scores can you obtain on the midterm exam so that your midsemester grade is B?
31. **Weight of Coffee** The weight w of the coffee in cans filled by a food processing company satisfies

$$\left|\dfrac{w - 12}{0.05}\right| \leq 1,$$

where w is measured in ounces. Determine the interval in which w lies.
32. **Weight of Cans** A grocery scale is designed to be accurate to within 0.25 oz. If two identical cans of soup placed on the scale have a combined weight of 33.15 oz, what are the largest and smallest possible weights of one of the cans?
33. **The Right Part** A precision part for a small motor is specified to have a diameter of 0.623 cm. In order for the part to fit correctly, its diameter must be within 0.005 cm of the specified diameter. Write an inequality involving an absolute value that has as its solution all possible diameters of parts that will fit. Solve the inequality to determine those diameters.
34. **Water Usage** The estimated daily requirement for water by a certain city is given by

$$|N - 3{,}725{,}000| < 100{,}000,$$

where N is the number of gallons of water used each day. Find the largest and the smallest daily water requirements.

For Discussion

In Problems 35 and 36, discuss how you might solve the given inequality and equation. Carry out your ideas.

35. $\left|\dfrac{x + 5}{x - 2}\right| \leq 3$
36. $|5 - x| = |1 - 3x|$

37. The distance between the number x and 5 is $|x - 5|$.
  (a) In words, describe the graphical interpretation of the inequalities $0 < |x - 5|$ and $0 < |x - 5| < 3$.
  (b) Solve each inequality in part (a) and write each solution set using interval notation.

38. Interpret $|x - 3|$ as the distance between the number x and 3. Sketch on the number line the set of real numbers that satisfy $2 < |x - 3| < 5$.

39. Solve the simultaneous inequality $2 < |x - 3| < 5$ by first solving $|x - 3| < 5$ and then $2 < |x - 3|$. Take the intersection of the two solution sets and compare with your sketch in Problem 38.

40. Suppose a and b are real numbers and that the point a lies to the left of point b on the real number line. Without actually solving the inequality,

$$\left|x - \frac{a + b}{2}\right| < \frac{b - a}{2},$$

discuss its solution set.

1.7 Polynomial and Rational Inequalities

☰ **Introduction** In Sections 1.5 and 1.6 we solved linear inequalities, that is, inequalities containing a single variable x that can be put into the forms $ax + b > 0$, $ax + b \leq 0$, and so on. Because $ax + b$ is a linear polynomial, linear inequalities are just a special case of a larger class of inequalities that involve polynomials. If $P(x)$ denotes a polynomial of arbitrary degree, inequalities that can be put into the form

$$P(x) > 0, \quad P(x) < 0, \quad P(x) \geq 0, \quad \text{and} \quad P(x) \leq 0 \tag{1}$$

are called **polynomial inequalities**. Inequalities involving the quotient of two polynomials $P(x)$ and $Q(x)$,

$$\frac{P(x)}{Q(x)} > 0, \quad \frac{P(x)}{Q(x)} < 0, \quad \frac{P(x)}{Q(x)} \geq 0, \quad \text{and} \quad \frac{P(x)}{Q(x)} \leq 0, \tag{2}$$

are called **rational inequalities**. We will assume for a rational inequality that the polynomials $P(x)$ and $Q(x)$ have no common factors. For example,

$$x^2 - 2x - 15 \geq 0 \quad \text{and} \quad \frac{x(x - 1)}{x + 2} \leq 0$$

are a polynomial inequality and a rational inequality, respectively.

In this section we consider a method for solving polynomial and rational inequalities.

☐ **Polynomial Inequalities** In the next three examples we illustrate the **sign-chart method** for solving polynomial inequalities. The **sign properties of a product** of real numbers given next is fundamental to constructing a sign chart of a polynomial inequality:

- *The product of two real numbers is positive (negative) if and only if the numbers have the same (opposite) signs.* (3)

That is, if the signs of the numbers are $(+)(+)$ or $(-)(-)$, then their product is positive, and if they have different signs $(+)(-)$ or $(-)(+)$, then their product is negative.

Here are some of the basic steps of the sign-chart method that is illustrated in the first example. We use the fact that a polynomial $P(x)$ can change sign only at a number c for which $P(c) = 0$. A number c for which $P(c) = 0$ is called a **zero** of the polynomial.

> ### GUIDELINES FOR SOLVING POLYNOMIAL INEQUALITIES
>
> (*i*) Use the properties of inequalities to recast the given inequality into a form where all variables and nonzero constants are on the same side of the inequality symbol and the number 0 is on the other side. That is, put the inequality into one of the forms given in (1).
> (*ii*) Then, if possible, factor the polynomial $P(x)$ into linear factors $ax + b$.
> (*iii*) Mark the number line at the real zeros of $P(x)$. These numbers divide the number line into intervals.
> (*iv*) In each of these intervals, determine the sign of each factor and then determine the sign of the product using the sign properties of a product (3).

EXAMPLE 1   Solving a Polynomial Inequality

Solve $x^2 \geq -2x + 15$.

Solution We begin by rewriting the inequality with all terms to the left of the inequality symbol and 0 to the right. By the properties of inequalities,

$$x^2 \geq -2x + 15 \quad \text{is equivalent to} \quad x^2 + 2x - 15 \geq 0.$$

Factoring, the last expression is the same as $(x + 5)(x - 3) \geq 0$.

Then we indicate on the number line where each factor is 0, in this case, $x = -5$ and $x = 3$. As shown in **FIGURE 1.7.1**, this divides the number line into three disjoint, or nonintersecting, intervals $(-\infty, -5)$, $(-5, 3)$, and $(3, \infty)$. Note too, that since the given inequality requires the product to be nonnegative, that is, "greater than or *equal to* 0," the numbers -5 and 3 are two solutions. Next, we must determine the signs of the factors $x + 5$ and $x - 3$ on each of the three intervals. We are looking for those intervals on which the two factors are either both positive or both negative, for then their product will be positive. Since the linear factors $x + 5$ and $x - 3$ cannot change signs within these intervals, it suffices to obtain the sign of each factor at just *one* test number chosen from inside each interval. For example, on the interval $(-\infty, -5)$, if we use $x = -10$ as a test number, then

FIGURE 1.7.1 Three disjoint intervals in Example 1

◀ See (3) about the sign properties of products.

Interval	$(-\infty, -5)$	
Sign of $x + 5$	−	← at $x = -10$, $x + 5 = -10 + 5 < 0$
Sign of $x - 3$	−	← at $x = -10$, $x - 3 = -10 - 3 < 0$
Sign of $(x + 5)(x - 3)$	+	← $(-)(-)$ is $(+)$

1.7 Polynomial and Rational Inequalities

Continuing in this manner for the remaining two intervals we get the sign chart in FIGURE 1.7.2. As can be seen from the third line of this figure, the product $(x + 5)(x - 3)$ is nonnegative on either of the unbounded disjoint intervals $(-\infty, -5]$ or $[3, \infty)$. Thus the solution set is $(-\infty, -5] \cup [3, \infty)$.

See Example 2(b) in Section 1.6.

FIGURE 1.7.2 Sign chart for Example 1

EXAMPLE 2 Solving a Polynomial Inequality

Solve $x < 10 - 3x^2$.

Solution We begin by rewriting the inequality with all nonzero terms on the same side: $3x^2 + x - 10 < 0$. Factoring this quadratic polynomial gives us

$$(3x - 5)(x + 2) < 0.$$

From the sign chart in FIGURE 1.7.3 we see (in red) that the foregoing product is negative for the numbers in the open interval $\left(-2, \frac{5}{3}\right)$. Because of the strict inequality "less than" symbol, the endpoints of the interval numbers are not included in the solution set.

FIGURE 1.7.3 Sign chart for Example 2

EXAMPLE 3 Solving a Polynomial Inequality

Solve $(x - 4)^2(x + 8)^3 > 0$.

Solution Since the given inequality already has the form appropriate for the sign-chart method (a factored expression to the left of the inequality symbol and 0 to the right), we begin by finding the numbers where each factor is 0, in this case, $x = 4$ and $x = -8$. We place these numbers on the number line and determine three intervals. Then in each interval we consider the signs of the powers of each linear factor. Because of the even power, we see that $(x - 4)^2$ is never negative. But because of the odd power, $(x + 8)^3$ has the same sign as the factor $x + 8$. Observe that the numbers $x = 4$ and $x = -8$ are not solutions of the inequality because of the strict inequality "greater than" symbol. Therefore, as we see (in red) in FIGURE 1.7.4, the product of factors $(x - 4)^2(x + 8)^3$ is nonnegative for the numbers in the set $(-8, 4) \cup (4, \infty)$.

FIGURE 1.7.4 Sign chart for Example 3

☐ **Rational Inequalities** We turn now to rational inequalities of the type shown in (2). Rational inequalities can be solved using the procedure above except we put the zeros of both the numerator $P(x)$ and denominator $Q(x)$ on the number line and use the **sign properties of a quotient**:

- *The quotient of two real numbers is positive (negative) if and only if the numbers have the same (opposite) signs.* (4)

EXAMPLE 4 Solving a Rational Inequality

Solve $\dfrac{x+1}{x+3} \le -1$.

Solution In order to utilize sign properties of a quotient (4), we must have all nonzero terms on the same side of the inequality (just as we did for polynomial inequalities). Thus we add 1 to both sides of the inequality and then combine terms to obtain an equivalent rational inequality:

$$\frac{x+1}{x+3} + 1 \le 0$$

$$\frac{x+1}{x+3} + \overbrace{\frac{x+3}{x+3}}^{1} \le 0 \quad \leftarrow \text{common denominator}$$

$$\frac{2x+4}{x+3} \le 0.$$

Using the fact that $2x + 4 = 0$ when $x = -2$ and $x + 3 = 0$ when $x = -3$, we prepare the sign chart shown in **FIGURE 1.7.5**. From this chart we see that the number -3 is not a solution since $(2x+4)/(x+3)$ is undefined for $x = -3$, but the number -2 is included in the solution set because $(2x+4)/(x+3)$ is zero for $x = -2$. The solution set is the interval $(-3, -2]$.

```
2x + 4         - -       - -          0 + +
x + 3          - -    0   + +         + + +
2x + 4
-----          + +  undefined  - -    0 + +
x + 3
                     (              ]
                    -3             -2
```

FIGURE 1.7.5 Sign chart for Example 4

EXAMPLE 5 Solving a Rational Inequality

Solve $x \le 3 - \dfrac{6}{x+2}$.

Solution We begin by rewriting the inequality with all variables and nonzero constants to the left and 0 to the right of the inequality sign,

$$x - 3 + \frac{6}{x+2} \le 0.$$

Next we put the terms over a common denominator,

$$\frac{(x-3)(x+2) + 6}{x+2} \le 0 \quad \text{and simplify to} \quad \frac{x(x-1)}{x+2} \le 0. \quad (5)$$

◀ One thing we *don't do* is clear the denominator by multiplying the inequality by $x + 2$. See Problem 57 in Exercises 1.7.

Now the numbers that make the three linear factors in the last expression equal to 0 are $-2, 0,$ and 1. On the number line these three numbers determine four intervals. Because of the "less than or *equal to* 0" we see that 0 and 1 are members of the solution set. But -2 is excluded from the solution set since substituting this value into the fractional expression results in a zero denominator (making the fraction undefined). As we can see from the sign chart in **FIGURE 1.7.6**, the solution set is $(-\infty, -2) \cup [0, 1]$.

```
x              - -      -    - -       0 + + +   + +
x - 1          - -      -    - -       - - - 0   + +
x + 2          - -      0    + +       + + + +   + +
x(x - 1)/(x + 2)  - -  undefined + +   0 - - 0   + +
```

FIGURE 1.7.6 Sign chart for Example 5

NOTES FROM THE CLASSROOM

(*i*) Terminology used in mathematics often varies from teacher to teacher and from textbook to textbook. For example, inequalities using the symbols $<$ or $>$ are sometimes called *strict* inequalities, whereas inequalities using $\leq$ or $\geq$ are called *nonstrict*. See Section R.2. As another example, the *positive integers* 1, 2, 3, ... are often referred to as the *natural numbers*.

(*ii*) Suppose the solution set of an inequality consists of the numbers such that $x < -1$ *or* $x > 3$. An answer seen very often on homework, quizzes, and tests is $3 < x < -1$. This is a misunderstanding of the notion of *simultaneity*. The statement $3 < x < -1$ means that $x > 3$ *and* at the same time $x < -1$. If you sketch this on the number line, you will see that it is impossible for the same x to satisfy both inequalities. The best we can do in rewriting "$x < -1$ or $x > 3$" is to use the union of intervals $(-\infty, -1) \cup (3, \infty)$.

(*iii*) Here is another frequent error. The notation $a < x > b$ is meaningless. If, say, we have $x > -2$ and $x > 6$, then only the numbers $x > 6$ satisfy *both* conditions.

(*iv*) In the classroom we frequently hear the response "positive" when in reality the student means "nonnegative." Question: x under the square root sign $\sqrt{x}$ must be positive, right? Raise your hand. Invariably, lots of hands go up. Correct answer: x must be nonnegative, that is, $x \geq 0$. Don't forget that $\sqrt{0} = 0$.

1.7 Exercises Answers to selected odd-numbered problems begin on page ANS-4.

In Problems 1–40, solve the given inequality and express the solution using interval notation.

1. $x^2 + 2x - 15 > 0$
2. $3x^2 - x - 2 \leq 0$
3. $x^2 - 8x + 12 < 0$
4. $6x^2 + 14x + 4 \geq 0$

5. $x^2 - 5x \geq 0$
6. $x^2 - 4x + 4 \geq 0$
7. $3x^2 - 27 < 0$
8. $4x^2 + 7x \leq 0$
9. $4x^2 - 4x + 1 < 0$
10. $12x^2 > 27x + 27$
11. $x^2 - 16 < 0$
12. $x^2 - 5 > 0$
13. $x^2 - 12 \leq 0$
14. $9x > 2x^2 - 18$
15. $x^2 + 6x \leq -9$
16. $9x^2 + 30x > -25$
17. $\dfrac{x - 3}{x + 2} < 0$
18. $\dfrac{x + 5}{x} \geq 0$
19. $\dfrac{2x + 6}{x - 3} \leq 0$
20. $\dfrac{3x - 1}{x + 2} > 0$
21. $\dfrac{x + 1}{x - 1} + 2 > 0$
22. $\dfrac{x - 2}{x + 3} \leq 1$
23. $\dfrac{2x - 3}{5x + 2} \geq -2$
24. $\dfrac{3x - 1}{2x - 1} < -4$
25. $\dfrac{5}{x + 8} < 0$
26. $\dfrac{10}{2x + 5} \geq 0$
27. $\dfrac{1}{x^2 + 9} < 0$
28. $\dfrac{1}{x^2 - 1} < 0$
29. $\dfrac{x(x - 1)}{x + 5} \geq 0$
30. $\dfrac{(1 + x)(1 - x)}{x} \leq 0$
31. $\dfrac{x^2 - 2x + 3}{x + 1} \leq 1$
32. $\dfrac{x}{x^2 - 1} > 0$
33. $-(x + 1)(x + 2)(x + 3) < 0$
34. $-2(x - 1)(x + \tfrac{1}{2})(x - 3) \leq 0$
35. $(x^2 - 1)(x^2 - 4) \leq 0$
36. $(x - 1)^2(x + 3)(x + 5) > 0$
37. $(x - \tfrac{1}{3})^2(x + 5)^3 < 0$
38. $x^2(x - 2)(x - 3)^5 \geq 0$
39. $\dfrac{(x + 3)^2(x + 4)(x - 5)^3}{x^2 - x - 20} > 0$
40. $\dfrac{9x^2 - 6x + 1}{x^3 - x^2} \leq 0$

In Problems 41–44, solve the given inequality and express the solution using interval notation. You may need to use the quadratic formula to factor the quadratic expression.

41. $x^2 - x - 1 > 0$
42. $6x^2 < 3x + 5$
43. $\dfrac{5x - 2}{x^2 + 1} \leq 1$
44. $\dfrac{x^2}{x - 1} \geq -1$

In Problems 45 and 46, solve the given inequalities and express each solution using interval notation.

45. (a) $x^2 < x$ (b) $x^2 > x$
46. (a) $1/x < x$ (b) $x < 1/x$
47. If $x^2 \leq 1$, is it necessarily true that $x \leq 1$? Explain.
48. If $x^2 \geq 4$, is it necessarily true that $x \geq 2$? Explain.
49. If 7 times the square of a positive number is decreased by 3, the result is greater than 60. What can be determined about the number?
50. The sides of a square are extended to form a rectangle. As shown in **FIGURE 1.7.7**, one side is extended 2 cm and the other side is extended 5 cm. If the area of the resulting rectangle is less than 130 cm^2, what are the possible lengths of a side of the original square?

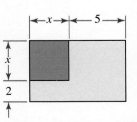

FIGURE 1.7.7 Rectangle in Problem 50

FIGURE 1.7.8 Octagon in Problem 51

51. A **polygon** is a closed figure made by joining line segments. For example, a *triangle* is a three-sided polygon. Shown in **FIGURE 1.7.8** is an eight-sided polygon called an *octagon*. A *diagonal* of a polygon is defined to be a line segment that joins any two nonadjacent vertices. The number of diagonals d in a polygon with n sides is given by $d = \frac{1}{2}(n-1)n - n$. For what polygons will the number of diagonals exceed 35?

52. The total number of dots, t, in a triangular array with n rows is given by

$$t = \frac{n(n+1)}{2}.$$

See **FIGURE 1.7.9**. How many rows can the array have if the total number of dots is to be less than 5050?

FIGURE 1.7.9 Triangular arrays of dots in Problem 52

Miscellaneous Applications

53. **Flower Garden** A rectangular flower bed is to be twice as long as it is wide. If the area enclosed must be greater than 98 m², what can you conclude about the width of the flower bed?

54. **Intensity of Light** The intensity I in lumens of a certain light source at a point r centimeters from the source is given by $I = 625/r^2$. At what distances from the light source will the intensity be less than 25 lumens?

55. **Parallel Resistors** A 5-ohm resistor and a variable resistor are placed in parallel. The resulting resistance R_T is given by $R_T = 5R/(5 + R)$. Determine the values of the variable resistor R for which the resulting resistance R_T will be greater than 2 ohms.

56. **What Goes Up...** With the aid of calculus it is easy to show that the height s of a projectile launched straight upward from an initial height s_0 with an initial velocity v_0 is given by $s = -\frac{1}{2}gt^2 + v_0 t + s_0$, where t is in seconds and $g = 32$ ft/s². If a toy rocket is shot straight upward from ground level, then $s_0 = 0$. If its initial velocity is 72 ft/s, during what time interval will the rocket be more than 80 ft above the ground?

For Discussion

57. In Example 5, explain why one should not multiply the last expression in (2) by $x + 2$.

CONCEPTS REVIEW
You should be able to give the meaning of each of the following concepts.

Equations:
   root
   identity
   conditional equation
   equivalent equations
   linear equation
   quadratic equation
   polynomial equation
   absolute-value equation
Solution:
   solution set
   extraneous solution
   check a solution
Completing the square
Quadratic formula:
   discriminant

Pythagorean theorem
Complex numbers:
   standard form
   real part
   imaginary part
   conjugate
   sum of two
   difference of two
   product of two
   quotient of two
   multiplicative inverse
Pure imaginary number
Equality of two complex numbers
Inequalities:
   equivalent inequalities
   simultaneous inequality

absolute-value inequality
linear inequality
polynomial inequality
rational inequality
Interval notation:
   open interval
   closed interval
   half-open interval
   open half-time
   closed half-time
Sign property of products
Sign chart
Zero of a polynomial

CHAPTER 1 Review Exercises
Answers to selected odd-numbered problems begin on page ANS-5.

A. True/False

In Problems 1–10, answer true or false.

1. $\sqrt{-1} \cdot \sqrt{-1} = \sqrt{(-1)(-1)} = \sqrt{1} = 1$ ___
2. $\dfrac{1}{\sqrt{-50}} = -\dfrac{\sqrt{2}}{10} i$ ___
3. $|-3t + 6| = 3|t - 2|$ ___
4. If $-3x \geq 6$, then $x \geq -2$. ___
5. The number 3.5 is in the solution set of $\left|\dfrac{2 - 4x}{5}\right| > 2.5$. ___
6. The solution set of $|4x - 6| \geq -1$ is $(-\infty, \infty)$. ___
7. For any real number x, $|-x^2 - 25| = x^2 + 25$. ___
8. If $\bar{z} = z$, then z must be a real number. ___
9. The inequality $\dfrac{100}{x^2 + 64} \leq 0$ has no solution. ___
10. If x is a real number, then $|x| = |-x|$. ___

B. Fill in the Blanks

In Problems 1–10, fill in the blanks.

1. An inequality with $(-\infty, 9]$ as its solution set is _____.
2. The solution set of the inequality $-3 < x \leq 8$ as an interval is _____.
3. The imaginary part of the complex number $4 - 6i$ is _____.
4. $(1 - \sqrt{-5})(-3 + \sqrt{-2}) =$ _____.

5. If $|15 - 2x| = x$, then $x =$ _____.
6. The solution set for the equation $|x| = |x + 1|$ is _____.
7. The set of real numbers x whose distance between x and $\sqrt{2}$ is greater than 3 is defined by the absolute-value inequality _____.
8. If x is in the interval $(4, 8)$, then $|x - 4| + |x - 8| =$ _____.
9. If $a < 0$, then $|-a| =$ _____.
10. If $y \geq 0$ and $x^2 - 2x + 1 = y$, then $x =$ _____.

C. Review Exercises

In Problems 1–26, solve the given equation.

1. $\dfrac{2}{3}x + \dfrac{4}{3} = x - \dfrac{1}{3}$
2. $4(1 - x) = x - 3(x + 1)$
3. $4 - \dfrac{1}{t} = 3 + \dfrac{3}{t}$
4. $\dfrac{4}{r + 1} - 5 = 4 - \dfrac{3}{r + 1}$
5. $\dfrac{1}{\sqrt{x}} + \sqrt{x} = \dfrac{5}{\sqrt{x}} - 2\sqrt{x}$
6. $\dfrac{2}{x^2 - 1} + \dfrac{1}{x - 1} = \dfrac{3}{x + 1}$
7. $(1 - y)(y - 1) = y^2$
8. $x(2x - 1) = 3$
9. $4x^2 + 10x - 24 = 0$
10. $3x^2 + x - 10 = 0$
11. $16x^2 + 9 = 24x$
12. $x^2 - 17 = 0$
13. $2x^2 - 6x - 3 = 0$
14. $4x^2 + 20x + 25 = 0$
15. $2x^2 + 100 = 0$
16. $x^2 + 2x + 4 = 0$
17. $x^3 - 8 = 0$
18. $x^3 + 8 = 0$
19. $x^4 + 4x^2 - 8 = 0$
20. $4x^4 - 4x^2 + 1 = 0$
21. $x^{1/4} - 2x^{1/2} + 1 = 0$
22. $8x^{2/3} - 9x^{1/3} + 1 = 0$
23. $\sqrt[3]{x^2 - 17} = 4$
24. $3 + \sqrt{3x + 1} = x$
25. $\sqrt{x + 2} - \sqrt{x - 3} = \sqrt{x - 6}$
26. $x + 3 - 28x^{-1} = 0$

In Problems 27–40, solve the given inequality. Write the solution set using interval notation.

27. $2x - 5 \geq 6x + 7$
28. $\tfrac{1}{4}x - 3 < \tfrac{1}{2}x + 1$
29. $-4 < x - 8 < 4$
30. $7 \leq 3 - 2x < 11$
31. $|x| > 10$
32. $|-6x| \leq 42$
33. $|3x - 4| < 5$
34. $|5 - 2x| \geq 7$
35. $3x \geq 2x^2 - 5$
36. $x^2 > 6x - 9$
37. $x^3 > x$
38. $(x^2 - x)(x^2 + x) \leq 0$
39. $\dfrac{1}{x} + x > 2$
40. $\dfrac{2x - 6}{x - 1} \geq 1$

In Problems 41–44, describe the given interval on the real number line using (**a**) an inequality, (**b**) interval notation.

41.
FIGURE 1.R.1 Graph for Problem 41

42.
FIGURE 1.R.2 Graph for Problem 42

43.
FIGURE 1.R.3 Graph for Problem 43

44.
FIGURE 1.R.4 Graph for Problem 44

In Problems 45–50, solve for the indicated variable in terms of the remaining variables. Assume that all variables represent positive real numbers.

45. Surface area of a rectangular parallelepiped
$$A = 2(ab + bc + ac), \quad \text{for } b$$

46. Temperature conversion
$$T_C = \tfrac{5}{9}(T_F - 32), \quad \text{for } T_F$$

47. Lens equation
$$\frac{1}{p} + \frac{1}{q} = \frac{1}{f}, \quad \text{for } f$$

48. Volume of a sphere
$$V = \tfrac{4}{3}\pi r^3, \quad \text{for } r$$

49. Equation of a hyperbola
$$\frac{x^2}{a^2} - \frac{y^2}{b^2} = 1, \quad \text{for } x$$

50. Projectile motion
$$y = x - \frac{9.8}{v_0^2} x^2, \quad \text{for } x$$

In Problems 51–58, perform the indicated operation and write the answer in the standard form $a + bi$.

51. $(6 - 5i) + (4 + 3i)$
52. $(8 + 2i) - (5 - i)$
53. $(3 + 2i)(4 - 5i)$
54. $(3 + 5i)^2$
55. $\dfrac{1}{4 - 2i}$
56. $\dfrac{i}{5 + i}$
57. $\dfrac{2 - 5i}{3 + 4i}$
58. $\dfrac{15 - 7i}{7i}$

In Problems 59–62, solve for x and y.

59. $(3x - yi)i = 4(1 + yi)$
60. $(1 + i)^2 = (x - yi)i$
61. $\dfrac{1}{i} = (2 - 3i) + (x + yi)$
62. $i^2 = -(y + xi)$

63. Playing with Numbers If the sum of two numbers is 33 and their quotient is $\tfrac{5}{6}$, find the two numbers.

64. Wind Speed In still air two airplanes fly at a speed of 180 mi/h. One plane leaves Los Angeles and flies with the wind toward Phoenix. The second plane leaves Phoenix at the same time and flies against the wind toward Los Angeles. If the distance between the cities is 400 mi and the planes pass each other 250 mi from Los Angeles, find the speed of the wind.

65. Trying for a B Again On four tests of equal weight a student has an average score of 76. If the final exam is worth twice as much as each of the four tests, find the score on the final exam necessary for the student to have an overall average of 80.

66. How Fast? Two cars travel 40 mi. One car travels 5 mi/h faster than the other car and makes the trip in 16 min less time. Find the rates of the two cars.

67. How Fast? The distance from Minneapolis to Des Moines is approximately 250 mi: 100 mi in Minnesota and 150 mi in Iowa. At one time Iowa had raised its interstate speed limit to 65 mi/h while Minnesota's speed limit was still 55 mi/h. Under these circumstances, suppose a woman wants to drive from Minneapolis to Des Moines in 4 h. If she plans to drive 65 mi/h in Iowa, how fast must she go in Minnesota?

68. **Slow Typist** Ethan can type 100 addresses in 5 h. He begins and works alone for 2 h. Then Sean starts to help on another computer and they finish the task in another 90 min. How long would it take Sean to type all 100 addresses alone?
69. **Width of a Sidewalk** A sidewalk is to be constructed around a circular mall. The diameter of the mall is 20 m. If the area of the sidewalk is 44π m², determine the width of the sidewalk.
70. **Playing with Numbers** If the digits of a two-digit number are reversed, the ratio of the original number to the new number is equal to $\frac{5}{6}$. What is the original number?
71. **Speed of Current** Fran's boat cruises at a rate of 20 mi/h in still water. If she travels the same distance in 3 h against the current that she could travel in 2 h with the current, what is the speed of the current?
72. **Dimensions** The floor of a house is a rectangle that is 5 m longer than twice its width. An addition is planned that will increase the total area of the house to 135 m². If the addition increases the width by 4 m, find the original dimensions of the house.
73. **Making the Cut** A border of uniform width is cut from a rectangular piece of cloth. The resulting piece of cloth is 20 in. by 30 in. See FIGURE 1.R.5. If the original area was twice the present area, find the width of the border that was cut off.
74. **Flat Panel TV** The "size" of a rectangular television screen is measured along the diagonal. If the length of the screen is 24 in. longer than the width and the size of the screen is given as 64 in., what are the dimensions of the screen?

FIGURE 1.R.5 Cloth in Problem 73

75. **What Goes Up...** A toy rocket is shot straight upward from the ground with an initial velocity of 72 ft/s. Its height s in feet above the ground after t seconds is given by $s = -16t^2 + 72t$. During what interval of time will the rocket be more than 80 ft above the ground?
76. **Mixture** A nurse has 2 L of a 3% boric acid solution. How much of a 10% solution must be added in order to have a 4% solution?
77. **Bond Cost** Mr. Diamond bought two bonds for a total of $30,000. One bond pays 6% interest and the other pays 8% interest. The annual interest from the 8% bond exceeds the annual interest from the 6% bond by $1000. Find the cost of each bond.
78. **Planning Retirement** Teri had worked at an aerospace firm for 21 years when Maria started at the firm. If Teri retires when she has five times as much seniority as Maria, how long will each have worked at the aerospace firm?
79. **Music in the Night** Mrs. Applebee bought a number of music boxes for her store for a total of $400. If each music box had cost $4 more, she would have obtained five fewer music boxes for the same amount of money. How many did she buy?
80. **Children's Dosage** Young's rule for converting an adult dosage of a drug to a child's dosage assumes a relationship between age and dosage and is used most frequently for children between the ages of three and twelve:

$$\frac{\text{age of child}}{\text{age of child} + 12} \times \text{adult dosage} = \text{child's dosage}.$$

At what age is the adult's dosage four times the child's dosage?
81. **Avalanche Prevention** Engineers concerned with avalanche hazards measure the hardness of a snowpack by hammering a specially designed metal tube into the snow and seeing how far it penetrates. They assign the snowpack a ram number R given by the formula

$$R = H + T + nfH/p,$$

where H is the mass of the hammer, T is the mass of the tube, n is the number of blows of the hammer, f is the distance the hammer falls for each blow, and p is the total penetration after n blows. See **FIGURE 1.R.6**. Typically, T and H are each 1 kg and f is about 50 cm. If the ram number of a snowpack is 150, how far will the tube penetrate after one blow?

FIGURE 1.R.6 Snowpack ram in Problem 81

FIGURE 1.R.7 Piston in Problem 82

82. **Classic Cars** The size of an automobile engine is measured by the volume of air that is forced upward or displaced by the motion of the pistons. This volume is given by the formula
$$V = N\pi S(B/2)^2,$$
where N is the number of pistons, S is the vertical distance or stroke that each piston moves, and B is the diameter or bore of a piston. See **FIGURE 1.R.7**.
   (a) Find the size in cubic inches of a V-8 (8-cylinder engine) with a 4-in. bore and a 3-in. stroke.
   (b) The 1909 6-cylinder Thomas Flyer had a 5.5-in. bore and stroke. Find its size. (*Note*: Early engines were large because of their low efficiency.)
   (c) The 1911 Oldsmobile Limited displaced 706.9 in³ with a 5-in. bore and a 6-in. stroke. How many cylinders did it have?
   (d) Hot-rodders will "soup up" a car by increasing the size of the bore and the length of the stroke. If the stroke is increased by $\frac{1}{4}$ in. and the bore is increased by $\frac{1}{8}$ in. in the engine described in part (a), how much displacement is gained?

1911 Oldsmobile in Problem 82

83. **Air Temperature** Vincent's formula for human skin temperature P_v in degrees Celsius is
$$P_v = 30.1 + 0.2t - (4.12 - 0.13t)v,$$
where t is the air temperature in degrees Celsius and v is the wind velocity in meters per second.
   (a) At which temperature t in still air ($v = 0$) is skin temperature less than blood temperature (37°C)?
   (b) Is there a wind velocity at which both air temperature t and skin temperature P_v equal blood temperature (37° C)? Explain your answer.
   (c) According to this formula, at what temperature t does wind velocity cause skin temperature to increase? [*Hint*: Wind velocity will increase skin temperature when the coefficient of v is positive.]

84. **Small Parking Lot** A stadium architect has designed a 20,000-car parking lot with four 4-lane exits. Under ideal conditions it is assumed that cars will exit smoothly at 10 mi/h, using all 16 lanes, with a spacing of 10 ft between cars.
   (a) If the average car is 15 ft long, how long will it take for the parking lot to empty? [*Hint*: Convert 10 mi/h to feet per minute.]
   (b) Derive a general formula that expresses the time T in minutes that it takes C cars to exit a parking lot, using N exit lanes, with a spacing of s feet between cars, all moving at v miles per hour, if the average car is L feet in length. Include the fact that converts miles per hour to feet per minute.
   (c) Solve the formula derived in part (b) for N.
   (d) Find the number of exit lanes required for 10,000 cars, each 15 ft long, to exit in no more than 30 min at 10 mi/h with 10 ft between each car.

85. **Violent Storms** A few years ago, meteorologists used the equation $D^3 = 216T^2$ as a mathematical model to describe the size and intensity of four types of violent storms—tornados, thunderstorms, hurricanes, and cyclones—where D is the diameter of the storm in miles and T is the number of hours the storm travels before dissipating. (*Source*: NCTM Student Math Notes, January 1987.)
   (a) Solve for T.
   (b) In the United States about 150 tornadoes occur each year. If the diameter of a tornado is 2 mi, use the given mathematical model to determine the number of hours it could be expected to last.

Tornado in Problem 85

86. **Plate Tectonics** Geological studies of the Earth's crust indicate that the fresh lithosphere (molten rock) that is forced up at oceanic ridges, sinks as it cools and moves away from the ridge. In plate tectonics the rate at which the lithosphere sinks is predicted by the mathematical model

$$Z = C\sqrt{T}, \quad 0 \le T \le 100,$$

Where Z is the depth in meters that the lithosphere has sunk, T is the age of the lithosphere in millions of years, and C is a constant. (The value $C = 300$ fits the data relatively well.) See **FIGURE 1.R.8**.
   (a) How far has the lithosphere sunk after 100 million years?
   (b) Solve the given equation for T. How old is the lithosphere that has sunk 500 m?

FIGURE 1.R.8 Oceanic ridge in Problem 86

2 Rectangular Coordinate System and Graphs

In This Chapter

2.1 The Rectangular Coordinate System
2.2 Circles and Graphs
2.3 Equations of Lines
2.4 Variation
    Chapter 2 Review Exercises

In Section 2.3 we will see that parallel lines have the same slope.

A Bit of History Every student of mathematics pays the French mathematician **René Descartes** (1596–1650) homage whenever he or she sketches a graph. Descartes is considered the inventor of analytic geometry, which is the melding of algebra and geometry—at the time thought to be completely unrelated fields of mathematics. In analytic geometry an equation involving two variables could be interpreted as a graph in a two-dimensional coordinate system embedded in a plane. The rectangular or Cartesian coordinate system is named in his honor. The basic tenets of analytic geometry were set forth in *La Géométrie,* published in 1637. The invention of the Cartesian plane and rectangular coordinates contributed significantly to the subsequent development of calculus by its co-inventors **Isaac Newton** (1643–1727) and **Gottfried Wilhelm Leibniz** (1646–1716).

René Descartes was also a scientist and wrote on optics, astronomy, and meteorology. But beyond his contributions to mathematics and science, Descartes is also remembered for his impact on philosophy. Indeed, he is often called the father of modern philosophy and his book *Meditations on First Philosophy* continues to be required reading to this day at some universities. His famous phrase *cogito ergo sum* (I think, therefore I am) appears in his *Discourse on the Method* and *Principles of Philosophy*. Although he claimed to be a fervent Catholic, the Church was suspicious of Descartes' philosophy and writings on the soul, and placed all his works on the *Index of Prohibited Books* in 1693.

2.1 The Rectangular Coordinate System

≡ **Introduction** In Section R.2 we saw that each real number can be associated with exactly one point on the number, or coordinate, line. We now examine a correspondence between points in a plane and ordered pairs of real numbers.

☐ **The Coordinate Plane** A **rectangular coordinate system** is formed by two perpendicular number lines that intersect at the point corresponding to the number 0 on each line. This point of intersection is called the **origin** and is denoted by the symbol O. The horizontal and vertical number lines are called the **x-axis** and the **y-axis**, respectively. These axes divide the plane into four regions, called **quadrants**, which are numbered as shown in **FIGURE 2.1.1(a)**. As we can see in Figure 2.1.1(b), the scales on the x- and y-axes need not be the same. Throughout this text, if tick marks are *not* labeled on the coordinates axes, as in Figure 2.1.1(a), then you may assume that one tick corresponds to one unit. A plane containing a rectangular coordinate system is called an **xy-plane**, a **coordinate plane**, or simply **2-space**.

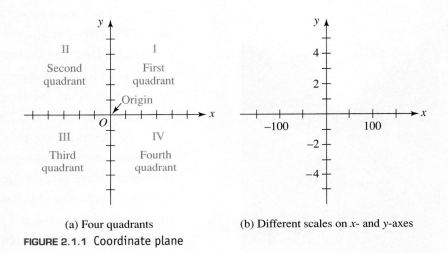

(a) Four quadrants   (b) Different scales on x- and y-axes

FIGURE 2.1.1 Coordinate plane

The rectangular coordinate system and the coordinate plane are also called the **Cartesian coordinate system** and the **Cartesian plane** after the famous French mathematician and philosopher **René Descartes** (1596–1650).

☐ **Coordinates of a Point** Let P represent a point in the coordinate plane. We associate an ordered pair of real numbers with P by drawing a vertical line from P to the x-axis and a horizontal line from P to the y-axis. If the vertical line intersects the x-axis at the number a and the horizontal line intersects the y-axis at the number b, we associate the ordered pair of real numbers (a, b) with the point. Conversely, to each ordered pair (a, b) of real numbers, there corresponds a point P in the plane. This point lies at the intersection of the vertical line through a on the x-axis and the horizontal line passing through b on the y-axis. Hereafter, we will refer to an ordered pair as a **point** and denote it by either $P(a, b)$ or (a, b).* The number a is the **x-coordinate** of the point and the number b is the **y-coordinate** of the point and we say that P has **coordinates** (a, b). For example, the coordinates of the origin are $(0, 0)$. See **FIGURE 2.1.2**.

---

*This is the same notation used to denote an open interval. It should be clear from the context of the discussion whether we are considering a point (a, b) or an open interval (a, b).

FIGURE 2.1.2 Point with coordinates (a, b)

FIGURE 2.1.3 Algebraic signs of coordinates in the four quadrants

The algebraic signs of the x-coordinate and the y-coordinate of any point (x, y) in each of the four quadrants are indicated in **FIGURE 2.1.3**. Points on either of the two axes are not considered to be in any quadrant. Since a point on the x-axis has the form $(x, 0)$, an equation that describes the x-axis is $y = 0$. Similarly, a point on the y-axis has the form $(0, y)$ and so an equation of the y-axis is $x = 0$. When we locate a point in the coordinate plane corresponding to an ordered pair of numbers and represent it using a solid dot, we say that we **plot** or **graph** the point.

EXAMPLE 1 Plotting Points

Plot the points $A(1, 2)$, $B(-4, 3)$, $C\left(-\frac{3}{2}, -2\right)$, $D(0, 4)$, and $E(3.5, 0)$. Specify the quadrant in which each point lies.

Solution The five points are plotted in the coordinate plane in **FIGURE 2.1.4**. Point A lies in the first quadrant (quadrant I), B in the second quadrant (quadrant II), and C is in the third quadrant (quadrant III). Points D and E, which lie on the y- and the x-axes, respectively, are not in any quadrant.

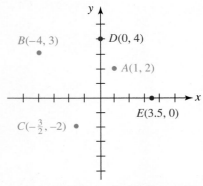

FIGURE 2.1.4 Five points in Example 1

EXAMPLE 2 Plotting Points

Sketch the set of points (x, y) in the xy-plane that satisfy both $0 \leq x \leq 2$ and $|y| = 1$.

Solution First, recall that the absolute-value equation $|y| = 1$ implies that $y = -1$ or $y = 1$. Thus the points that satisfy the given conditions are the points whose coordinates (x, y) *simultaneously* satisfy the conditions: each x-coordinate is a number in the closed interval $[0, 2]$ and each y-coordinate is either $y = -1$ or $y = 1$. For example, $(1, 1)$, $\left(\frac{1}{2}, -1\right)$, and $(2, -1)$ are a few of the points that satisfy the two conditions. Graphically, the set of all points satisfying the two conditions are points on the two parallel line segments shown in **FIGURE 2.1.5**.

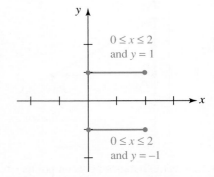

FIGURE 2.1.5 Set of points in Example 2

EXAMPLE 3 Regions Defined by Inequalities

Sketch the set of points (x, y) in the xy-plane that satisfy each of the following conditions.

(a) $xy < 0$ (b) $|y| \geq 2$

Solution (a) From (3) of the sign properties of products in Section 1.7, we know that a product of two real numbers x and y is negative when one of the numbers is positive and the other is negative. Thus, $xy < 0$ when $x > 0$ and $y < 0$ *or* when $x < 0$ and $y > 0$.

2.1 The Rectangular Coordinate System

We see from Figure 2.1.3 that $xy < 0$ for all points (x, y) in the second and fourth quadrants. Hence we can represent the set of points for which $xy < 0$ by the shaded regions in **FIGURE 2.1.6**. The coordinate axes are shown as dashed lines to indicate that the points on these axes are not included in the solution set.

(b) In Section 1.6 we saw that $|y| \geq 2$ means that either $y \geq 2$ or $y \leq -2$. Since x is not restricted in any way it can be any real number, and so the points (x, y) for which

$$y \geq 2 \text{ and } -\infty < x < \infty \quad \text{or} \quad y \leq -2 \text{ and } -\infty < x < \infty$$

can be represented by the two shaded regions in **FIGURE 2.1.7**. We use solid lines to represent the boundaries $y = -2$ and $y = 2$ of the region to indicate that the points on these boundaries are included in the solution set.

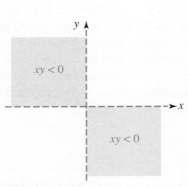

FIGURE 2.1.6 Region in the *xy*-plane satisfying condition in (a) of Example 3

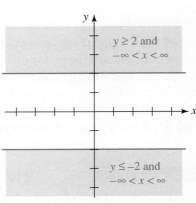

FIGURE 2.1.7 Region in the *xy*-plane satisfying condition in (b) of Example 3

☐ **Distance Formula** Suppose $P_1(x_1, y_1)$ and $P_2(x_2, y_2)$ are two distinct points in the *xy*-plane that are not on a vertical line or on a horizontal line. As a consequence, P_1, P_2, and $P_3(x_1, y_2)$ are vertices of a right triangle as shown in **FIGURE 2.1.8**. The length of the side P_3P_2 is $|x_2 - x_1|$, and the length of the side P_1P_3 is $|y_2 - y_1|$. If we denote the length of P_1P_2 by d, then

$$d^2 = |x_2 - x_1|^2 + |y_2 - y_1|^2 \tag{1}$$

by the Pythagorean theorem. Since the square of any real number is equal to the square of its absolute values, we can replace the absolute-value signs in (1) with parentheses. The distance formula given next follows immediately from (1).

FIGURE 2.1.8 Distance between points P_1 and P_2

THEOREM 2.1.1 Distance Formula

The distance between any two points $P_1(x_1, y_1)$ and $P_2(x_2, y_2)$ in the *xy*-plane is given by

$$d(P_1, P_2) = \sqrt{(x_2 - x_1)^2 + (y_2 - y_1)^2}. \tag{2}$$

Although we derived this equation for two points not on a vertical or horizontal line, (2) holds in these cases as well. Also, because $(x_2 - x_1)^2 = (x_1 - x_2)^2$, it makes no difference which point is used first in the distance formula, that is, $d(P_1, P_2) = d(P_2, P_1)$.

EXAMPLE 4　Distance Between Two Points

Find the distance between the points $A(8, -5)$ and $B(3, 7)$.

Solution From (2) with A and B playing the parts of P_1 and P_2:

$$d(A, B) = \sqrt{(3-8)^2 + (7-(-5))^2}$$
$$= \sqrt{(-5)^2 + (12)^2} = \sqrt{169} = 13.$$

The distance d is illustrated in FIGURE 2.1.9. ≡

FIGURE 2.1.9 Distance between two points in Example 4

EXAMPLE 5　Three Points Form a Triangle

Determine whether the points $P_1(7, 1)$, $P_2(-4, -1)$, and $P_3(4, 5)$ are the vertices of a right triangle.

Solution From plane geometry we know that a triangle is a right triangle if and only if the sum of the squares of the lengths of two of its sides is equal to the square of the length of the remaining side. Now, from the distance formula (2), we have

$$d(P_1, P_2) = \sqrt{(-4-7)^2 + (-1-1)^2}$$
$$= \sqrt{121 + 4} = \sqrt{125},$$
$$d(P_2, P_3) = \sqrt{(4-(-4))^2 + (5-(-1))^2}$$
$$= \sqrt{64 + 36} = \sqrt{100} = 10,$$
$$d(P_3, P_1) = \sqrt{(7-4)^2 + (1-5)^2}$$
$$= \sqrt{9 + 16} = \sqrt{25} = 5.$$

Since $\quad [d(P_3, P_1)]^2 + [d(P_2, P_3)]^2 = 25 + 100 = 125 = [d(P_1, P_2)]^2,$

we conclude that P_1, P_2, and P_3 are the vertices of a right triangle with the right angle at P_3. See FIGURE 2.1.10. ≡

FIGURE 2.1.10 Triangle in Example 5

☐ **Midpoint Formula** In Section R.2 we saw that the midpoint of a line segment between two numbers a and b on the number line is the average, $(a + b)/2$. In the xy-plane, each coordinate of the midpoint M of a line segment joining two points $P_1(x_1, y_1)$ and $P_2(x_2, y_2)$, as shown in FIGURE 2.1.11, is the average of the corresponding coordinates of the endpoints of the intervals $[x_1, x_2]$ and $[y_1, y_2]$.

To prove this, we note in Figure 2.1.11 that triangles P_1CM and MDP_2 are congruent since corresponding angles are equal and $d(P_1, M) = d(M, P_2)$. Hence, $d(P_1, C) = d(M, D)$ or $y - y_1 = y_2 - y$. Solving the last equation for y gives $y = \dfrac{y_1 + y_2}{2}$. Similarly, $d(C, M) = d(D, P_2)$ so that $x - x_1 = x_2 - x$ and therefore $x = \dfrac{x_1 + x_2}{2}$. We summarize the result.

FIGURE 2.1.11 M is the midpoint of the line segment joining P_1 and P_2

THEOREM 2.1.2　Midpoint Formula

The coordinates of the **midpoint** of the line segment joining the points $P_1(x_1, y_1)$ and $P_2(x_2, y_2)$ in the xy-plane are given by

$$\left(\frac{x_1 + x_2}{2}, \frac{y_1 + y_2}{2} \right). \qquad (3)$$

2.1 The Rectangular Coordinate System

FIGURE 2.1.12 Midpoint of line segment in Example 6

EXAMPLE 6 Midpoint of a Line Segment

Find the coordinates of the midpoint of the line segment joining $A(-2, 5)$ and $B(4, 1)$.

Solution From formula (3) the coordinates of the midpoint of the line segment joining the points A and B are given by

$$\left(\frac{-2+4}{2}, \frac{5+1}{2}\right) \quad \text{or} \quad (1, 3).$$

This point is indicated in red in FIGURE 2.1.12.

2.1 Exercises
Answers to selected odd-numbered problems begin on page ANS-5.

In Problems 1–4, plot the given points.

1. $(2, 3), (4, 5), (0, 2), (-1, -3)$
2. $(1, 4), (-3, 0), (-4, 2), (-1, -1)$
3. $\left(-\frac{1}{2}, -2\right), (0, 0), \left(-1, \frac{4}{3}\right), (3, 3)$
4. $(0, 0.8), (-2, 0), (1.2, -1.2), (-2, 2)$

In Problems 5–16, determine the quadrant in which the given point lies if (a, b) is in quadrant I.

5. $(-a, b)$
6. $(a, -b)$
7. $(-a, -b)$
8. (b, a)
9. $(-b, a)$
10. $(-b, -a)$
11. (a, a)
12. $(b, -b)$
13. $(-a, -a)$
14. $(-a, a)$
15. $(b, -a)$
16. $(-b, b)$

17. Plot the points given in Problems 5–16 if (a, b) is the point shown in FIGURE 2.1.13.
18. Give the coordinates of the points shown in FIGURE 2.1.14.

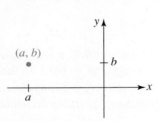

FIGURE 2.1.13 Point (a, b) in Problem 17

FIGURE 2.1.14 Points in Problem 18

19. The points $(-2, 0), (-2, 6)$, and $(3, 0)$ are vertices of a rectangle. Find the fourth vertex.
20. Describe the set of all points (x, x) in the coordinate plane. The set of all points $(x, -x)$.

In Problems 21–26, sketch the set of points (x, y) in the xy-plane that satisfy the given conditions.

21. $xy = 0$
22. $xy > 0$
23. $|x| \leq 1$ and $|y| \leq 2$
24. $x \leq 2$ and $y \geq -1$
25. $|x| > 4$
26. $|y| \leq 1$

In Problems 27–32, find the distance between the given points.

27. $A(1, 2), B(-3, 4)$
28. $A(-1, 3), B(5, 0)$
29. $A(2, 4), B(-4, -4)$
30. $A(-12, -3), B(-5, -7)$
31. $A\left(-\frac{3}{2}, 1\right), B\left(\frac{5}{2}, -2\right)$
32. $A\left(-\frac{5}{3}, 4\right), B\left(-\frac{2}{3}, -1\right)$

In Problems 33–36, determine whether the points A, B, and C are vertices of a right triangle.

33. $A(8, 1)$, $B(-3, -1)$, $C(10, 5)$
34. $A(-2, -1)$, $B(8, 2)$, $C(1, -11)$
35. $A(2, 8)$, $B(0, -3)$, $C(6, 5)$
36. $A(4, 0)$, $B(1, 1)$, $C(2, 3)$

37. Determine whether the points $A(0, 0)$, $B(3, 4)$, and $C(7, 7)$ are vertices of an isosceles triangle.
38. Find all points on the y-axis that are 5 units from the point $(4, 4)$.
39. Consider the line segment joining $A(-1, 2)$ and $B(3, 4)$.
  (a) Find an equation that expresses the fact that a point $P(x, y)$ is equidistant from A and from B.
  (b) Describe geometrically the set of points described by the equation in part (a).
40. Use the distance formula to determine whether the points $A(-1, -5)$, $B(2, 4)$, and $C(4, 10)$ lie on a straight line.
41. Find all points with x-coordinate 6 such that the distance from each point to $(-1, 2)$ is $\sqrt{85}$.
42. Which point, $(1/\sqrt{2}, 1/\sqrt{2})$ or $(0.25, 0.97)$, is closer to the origin?

In Problems 43–48, find the midpoint of the line segment joining the points A and B.

43. $A(4, 1)$, $B(-2, 4)$
44. $A\left(\frac{2}{3}, 1\right)$, $B\left(\frac{7}{3}, -3\right)$
45. $A(-1, 0)$, $B(-8, 5)$
46. $A\left(\frac{1}{2}, -\frac{3}{2}\right)$, $B\left(-\frac{5}{2}, 1\right)$
47. $A(2a, 3b)$, $B(4a, -6b)$
48. $A(x, x)$, $B(-x, x + 2)$

In Problems 49–52, find the point B if M is the midpoint of the line segment joining points A and B.

49. $A(-2, 1)$, $M\left(\frac{3}{2}, 0\right)$
50. $A\left(4, \frac{1}{2}\right)$, $M\left(7, -\frac{5}{2}\right)$
51. $A(5, 8)$, $M(-1, -1)$
52. $A(-10, 2)$, $M(5, 1)$

53. Find the distance from the midpoint of the line segment joining $A(-1, 3)$ and $B(3, 5)$ to the midpoint of the line segment joining $C(4, 6)$ and $D(-2, -10)$.
54. Find all points on the x-axis that are 3 units from the midpoint of the line segment joining $(5, 2)$ and $(-5, -6)$.
55. The x-axis is the perpendicular bisector of the line segment through $A(2, 5)$ and $B(x, y)$. Find x and y.
56. Consider the line segment joining the points $A(0, 0)$ and $B(6, 0)$. Find a point $C(x, y)$ in the first quadrant such that A, B, and C are vertices of an equilateral triangle.
57. Find points $P_1(x_1, y_1)$, $P_2(x_2, y_2)$, and $P_3(x_3, y_3)$ on the line segment joining $A(3, 6)$ and $B(5, 8)$ that divide the line segment into four equal parts.

Miscellaneous Applications

58. Going to Chicago Kansas City and Chicago are not directly connected by an interstate highway, but each city is connected to St. Louis and Des Moines. See FIGURE 2.1.15. Des Moines is approximately 40 mi east and 180 mi north of Kansas City, St. Louis is approximately 230 mi east and 40 mi south of Kansas City, and Chicago is approximately 360 mi east and 200 mi north of Kansas City. Assume that this part of the Midwest is a flat plane and that the connecting highways are straight lines. Which route from Kansas City to Chicago, through St. Louis or through Des Moines, is shorter?

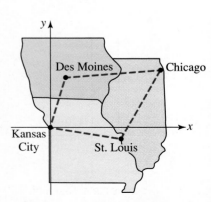

FIGURE 2.1.15 Map for Problem 58

For Discussion

59. The points $A(1, 0)$, $B(5, 0)$, $C(4, 6)$, and $D(8, 6)$ are vertices of a parallelogram. Discuss: How can it be shown that the diagonals of the parallelogram bisect each other? Carry out your ideas.

60. The points $A(0, 0)$, $B(a, 0)$, and $C(a, b)$ are vertices of a right triangle. Discuss: How can it be shown that the midpoint of the hypotenuse is equidistant from the vertices? Carry out your ideas.

2.2 Circles and Graphs

☰ Introduction In Chapter 1 we studied equations as an equality of two algebraic quantities involving one variable. Our goal then was to find the solution set of the equation. In this and subsequent sections that follow we study **equations in two variables**, say x and y. Such an equation is simply a mathematical statement that asserts two quantities involving these variables are equal. In the fields of the physical sciences, engineering, and business, equations in two (or more) variables are a means of communication. For example, if a physicist wants to tell someone how far a rock dropped from a great height travels in a certain time t, he or she will write $s = 16t^2$. A mathematician will look at $s = 16t^2$ and immediately classify it as a certain *type* of equation. The classification of an equation carries with it information about properties shared by all equations of that kind. The remainder of this text is devoted to examining different kinds of equations involving two or more variables and studying their properties. Here is a sample of some of the equations in two variables that you will see:

$$x = 1, \quad x^2 + y^2 = 1, \quad y = x^2, \quad y = \sqrt{x}, \quad y = 5x - 1, \quad y = x^3 - 3x,$$
$$y = 2^x, \quad y = \ln x, \quad y^2 = x - 1, \quad \frac{x^2}{4} + \frac{y^2}{9} = 1, \quad \frac{1}{2}x^2 - y^2 = 1. \tag{1}$$

☐ Terminology A **solution** of an equation in two variables x and y is an ordered pair of numbers (a, b) that yields a true statement when $x = a$ and $y = b$ are substituted into the equation. For example, $(-2, 4)$ is a solution of the equation $y = x^2$ because

$$\overset{y=4\downarrow \quad \downarrow x=-2}{4 = (-2)^2}$$

is a true statement. We also say that the coordinates $(-2, 4)$ **satisfy** the equation. As in Chapter 1, the set of all solutions of an equation is called its **solution set**. Two equations are said to be **equivalent** if they have the same solution set. For example, we will see in Example 4 of this section that the equation $x^2 + y^2 + 10x - 2y + 17 = 0$ is equivalent to $(x + 5)^2 + (y - 1)^2 = 3^2$.

In the list given in (1), you might object that the first equation $x = 1$ does not involve two variables. It is a matter of interpretation! Because there is no explicit y dependence in the equation, the solution set of $x = 1$ can be interpreted to mean the set

$$\{(x, y) \,|\, x = 1, \text{ where } y \text{ is any real number}\}.$$

The solutions of $x = 1$ are then ordered pairs $(1, y)$, where you are free to choose y arbitrarily so long as it is a real number. For example, $(1, 0)$ and $(1, 3)$ are solutions of the equation $x = 1$. The **graph** of an equation is the visual representation in the rectangular coordinate system of the set of points whose coordinates (a, b) satisfy the equation. The graph of $x = 1$ is the vertical line shown in **FIGURE 2.2.1**.

FIGURE 2.2.1 Graph of equation $x = 1$

☐ **Circles** The distance formula discussed in the preceding section can be used to define a set of points in the coordinate plane. One such important set is defined as follows.

DEFINITION 2.2.1 Circle

A **circle** is the set of all points $P(x, y)$ in the xy-plane that are a given distance r, called the **radius**, from a given fixed point C, called the **center**.

If the center has coordinates $C(h, k)$, then from the preceding definition a point $P(x, y)$ lies on a circle of radius r if and only if

$$d(P,C) = r \quad \text{or} \quad \sqrt{(x-h)^2 + (y-k)^2} = r.$$

Since $(x - h)^2 + (y - k)^2$ is always nonnegative, we obtain an equivalent equation when both sides are squared. We conclude that a circle of radius r and center $C(h, k)$ has the equation

$$(x - h)^2 + (y - k)^2 = r^2. \tag{2}$$

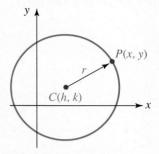

FIGURE 2.2.2 Circle with radius r and center (h, k)

In **FIGURE 2.2.2** we have sketched a typical graph of an equation of the form given in (2). Equation (2) is called the **standard form** of the equation of a circle. We note that the symbols h and k in (2) represent real numbers and as such can be positive, zero, or negative. When $h = 0$, $k = 0$, we see that the standard form of the equation of a circle with center at the origin is

$$x^2 + y^2 = r^2. \tag{3}$$

See **FIGURE 2.2.3**. When $r = 1$, we say that (2) or (3) is an equation of a **unit circle**. For example, $x^2 + y^2 = 1$ is an equation of a unit circle centered at the origin.

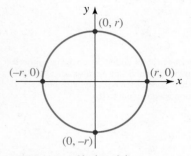

FIGURE 2.2.3 Circle with radius r and center $(0, 0)$

EXAMPLE 1   Center and Radius

Find the center and radius of the circle whose equation is

$$(x - 8)^2 + (y + 2)^2 = 49. \tag{4}$$

Solution To obtain the standard form of the equation, we rewrite (4) as

$$(x - 8)^2 + (y - (-2))^2 = 7^2.$$

Comparing this last form with (2) we identify $h = 8$, $k = -2$, and $r = 7$. Thus the circle is centered at $(8, -2)$ and has radius 7. ≡

EXAMPLE 2   Equation of a Circle

Find an equation of the circle with center $C(-5, 4)$ with radius $\sqrt{2}$.

Solution Substituting $h = -5$, $k = 4$, and $r = \sqrt{2}$ in (2) we obtain

$$(x - (-5))^2 + (y - 4)^2 = (\sqrt{2})^2 \quad \text{or} \quad (x + 5)^2 + (y - 4)^2 = 2. \quad ≡$$

EXAMPLE 3   Equation of a Circle

Find an equation of the circle with center $C(4, 3)$ and passing through $P(1, 4)$.

Solution With $h = 4$ and $k = 3$, we have from (2)

$$(x - 4)^2 + (y - 3)^2 = r^2. \tag{5}$$

2.2 Circles and Graphs

FIGURE 2.2.4 Circle in Example 3

Since the point $P(1, 4)$ lies on the circle as shown in FIGURE 2.2.4, its coordinates must satisfy equation (5). That is,

$$(1 - 4)^2 + (4 - 3)^2 = r^2 \quad \text{or} \quad 10 = r^2.$$

Thus the required equation in standard form is

$$(x - 4)^2 + (y - 3)^2 = 10.$$

☐ **Completing the Square** If the terms $(x - h)^2$ and $(y - k)^2$ are expanded and the like terms grouped together, an equation of a circle in standard form can be written as

$$x^2 + y^2 + ax + by + c = 0. \tag{6}$$

Of course in this last form the center and radius are not apparent. To reverse the process, in other words, to go from (6) to the standard form (2), we must **complete the square** in both x and y. Recall from Section 1.3 that adding $(a/2)^2$ to a quadratic expression such as $x^2 + ax$ yields $x^2 + ax + (a/2)^2$, which is the perfect square $(x + a/2)^2$. By rearranging the terms in (6),

$$(x^2 + ax \quad) + (y^2 + by \quad) = -c,$$

and then adding $(a/2)^2$ and $(b/2)^2$ to *both* sides of the last equation

$$\left(x^2 + ax + \left(\frac{a}{2}\right)^2\right) + \left(y^2 + by + \left(\frac{b}{2}\right)^2\right) = \left(\frac{a}{2}\right)^2 + \left(\frac{b}{2}\right)^2 - c,$$

The terms in color added inside the parentheses on the left-hand side are also added to the right-hand side of the equality. This new equation is equivalent to (6).

we obtain the standard form of the equation of a circle:

$$\left(x + \frac{a}{2}\right)^2 + \left(y + \frac{b}{2}\right)^2 = \frac{1}{4}(a^2 + b^2 - 4c).$$

You should *not* memorize the last equation; we strongly recommend that you work through the process of completing the square each time.

EXAMPLE 4 Completing the Square

Find the center and radius of the circle whose equation is

$$x^2 + y^2 + 10x - 2y + 17 = 0. \tag{7}$$

Solution To find the center and radius we rewrite equation (7) in the standard form (2). First, we rearrange the terms,

$$(x^2 + 10x \quad) + (y^2 - 2y \quad) = -17.$$

Then, we complete the square in x and y by adding, in turn, $(10/2)^2$ in the first set of parentheses and $(-2/2)^2$ in the second set of parentheses. Proceed carefully here because we must add these numbers to both sides of the equation:

$$[x^2 + 10x + \left(\tfrac{10}{2}\right)^2] + [y^2 - 2y + \left(\tfrac{-2}{2}\right)^2] = -17 + \left(\tfrac{10}{2}\right)^2 + \left(\tfrac{-2}{2}\right)^2$$
$$(x^2 + 10x + 25) + (y^2 - 2y + 1) = 9$$
$$(x + 5)^2 + (y - 1)^2 = 3^2.$$

From the last equation we see that the circle is centered at $(-5, 1)$ and has radius 3. See FIGURE 2.2.5.

FIGURE 2.2.5 Circle in Example 4

It is possible that an expression for which we must complete the square has a leading coefficient other than 1. For example,

$$\overset{\text{Note}}{3}x^2 + \overset{\text{Note}}{3}y^2 - 18x + 6y + 2 = 0$$

is an equation of a circle. As in Example 4 we start by rearranging the equation:

$$(3x^2 - 18x \quad) + (3y^2 + 6y \quad) = -2.$$

Now, however, we must do one extra step before attempting completion of the square; that is, we must divide both sides of the equation by 3 so that the coefficients of x^2 and y^2 are each 1:

$$(x^2 - 6x \quad) + (y^2 + 2y \quad) = -\frac{2}{3}.$$

At this point we can now add the appropriate numbers within each set of parentheses *and* to the right-hand side of the equality. You should verify that the resulting standard form is $(x - 3)^2 + (y + 1)^2 = \frac{28}{3}$.

☐ **Semicircles** If we solve (3) for y we get $y^2 = r^2 - x^2$ or $y = \pm\sqrt{r^2 - x^2}$. This last expression is equivalent to the two equations $y = \sqrt{r^2 - x^2}$ and $y = -\sqrt{r^2 - x^2}$. In like manner if we solve (3) for x we obtain $x = \sqrt{r^2 - y^2}$ and $x = -\sqrt{r^2 - y^2}$. By convention, the symbol $\sqrt{}$ denotes a nonnegative quantity, thus the y-values defined by an equation such as $y = \sqrt{r^2 - x^2}$ are nonnegative. The graphs of the four equations highlighted in color are, in turn, the upper half, lower half, right half, and the left half of the circle shown in Figure 2.2.3. Each graph in FIGURE 2.2.6 is called a **semicircle**.

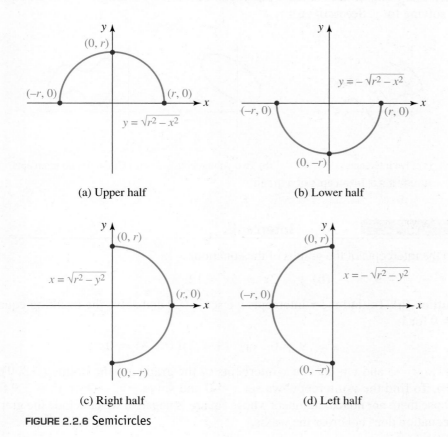

FIGURE 2.2.6 Semicircles

☐ **Inequalities** One last point about circles. On occasion we encounter problems where we must sketch the set of points in the *xy*-plane whose coordinates satisfy inequalities such as $x^2 + y^2 < r^2$ or $x^2 + y^2 \geq r^2$. The equation $x^2 + y^2 = r^2$ describes the set of points (x, y) whose distance to the origin $(0, 0)$ is exactly r. Therefore, the inequality $x^2 + y^2 < r^2$ describes the set of points (x, y) whose distance to the origin is less than r. In other words, the points (x, y) whose coordinates satisfy the inequality $x^2 + y^2 < r^2$ are in the *interior* of the circle. Similarly, the points (x, y) whose coordinates satisfy $x^2 + y^2 \geq r^2$ lie either *on* the circle or are *exterior* to it.

☐ **Graphs** It is difficult to read a newspaper, read a science or business text, surf the Internet, or even watch the news on TV without seeing graphical representations of data. It may even be impossible to get past the first page in a mathematics text without seeing some kind of graph. So many diverse quantities are connected by means of equations and so many questions about the behavior of the quantities linked by the equation can be answered by means of a graph, that the ability to graph equations quickly and accurately—like the ability to do algebra quickly and accurately—is high on the list of skills essential to your success in a course in calculus. For the rest of this section we will talk about graphs in general, and more specifically, about two important aspects of graphs of equations.

☐ **Intercepts** Locating the points at which the graph of an equation crosses the coordinates axes can be helpful when sketching a graph by hand. The ***x*-intercepts** of a graph of an equation are the points at which the graph crosses the *x*-axis. Since every point on the *x*-axis has *y*-coordinate 0, the *x*-coordinates of these points (if there are any) can be found from the given equation by setting $y = 0$ and solving for *x*. In turn, the ***y*-intercepts** of the graph of an equation are the points at which its graph crosses the *y*-axis. The *y*-coordinates of these points can be found by setting $x = 0$ in the equation and solving for *y*. See **FIGURE 2.2.7**.

(a) Five intercepts (b) Two *y*-intercepts (c) Graph has no intercepts

FIGURE 2.2.7 Intercepts of a graph

EXAMPLE 5 Intercepts

Find the intercepts of the graphs of the equations

(a) $x^2 - y^2 = 9$ **(b)** $y = 2x^2 + 5x - 12$.

Solution (a) To find the *x*-intercepts we set $y = 0$ and solve the resulting equation $x^2 = 9$ for *x*:

$$x^2 - 9 = 0 \quad \text{or} \quad (x + 3)(x - 3) = 0$$

gives $x = -3$ and $x = 3$. The *x*-intercepts of the graph are the points $(-3, 0)$ and $(3, 0)$. To find the *y*-intercepts we set $x = 0$ and solve $-y^2 = 9$ or $y^2 = -9$ for *y*. Because there are no real numbers whose square is negative we conclude the graph of the equation does not cross the *y*-axis.

(b) Setting $y = 0$ yields $2x^2 + 5x - 12 = 0$. This is a quadratic equation and can be solved either by factoring or by the quadratic formula. Factoring gives

$$(x + 4)(2x - 3) = 0$$

and so $x = -4$ and $x = \frac{3}{2}$. The x-intercepts of the graph are the points $(-4, 0)$ and $(\frac{3}{2}, 0)$. Now, setting $x = 0$ in the equation $y = 2x^2 + 5x - 12$ immediately gives $y = -12$. The y-intercept of the graph is the point $(0, -12)$.

EXAMPLE 6   Example 4 Revisited

Let's return to the circle in Example 4 and determine its intercepts from equation (7). Setting $y = 0$ in $x^2 + y^2 + 10x - 2y + 17 = 0$ and using the quadratic formula to solve $x^2 + 10x + 17 = 0$ shows the x-intercepts of this circle are $(-5 - 2\sqrt{2}, 0)$ and $(-5 + 2\sqrt{2}, 0)$. If we let $x = 0$, then the quadratic formula shows that the roots of the equation $y^2 - 2y + 17 = 0$ are complex numbers. As seen in Figure 2.2.5, the circle does not cross the y-axis.

☐ **Symmetry** A graph can also possess symmetry. You may already know that the graph of the equation $y = x^2$ is called a *parabola*. **FIGURE 2.2.8** shows that the graph of $y = x^2$ is symmetric with respect to the y-axis since the portion of the graph that lies in the second quadrant is the *mirror image* or *reflection* of that portion of the graph in the first quadrant. In general, a graph is **symmetric with respect to the y-axis** if whenever (x, y) is a point on the graph, $(-x, y)$ is also a point on the graph. Note in Figure 2.2.8 that the points $(1, 1)$ and $(2, 4)$ are on the graph. Because the graph possesses y-axis symmetry, the points $(-1, 1)$ and $(-2, 4)$ must also be on the graph. A graph is said to be **symmetric with respect to the x-axis** if whenever (x, y) is a point on the graph, $(x, -y)$ is also a point on the graph. Finally, a graph is **symmetric with respect to the origin** if whenever (x, y) is on the graph, $(-x, -y)$ is also a point on the graph. **FIGURE 2.2.9** illustrates these three types of symmetries.

FIGURE 2.2.8 Graph with y-axis symmetry

FIGURE 2.2.9 Symmetries of a graph

Observe that the graph of the circle given in Figure 2.2.3 possesses all three of these symmetries.

As a practical matter we would like to know whether a graph possesses any symmetry in advance of plotting its graph. This can be done by applying the following tests to the equation that defines the graph.

2.2 Circles and Graphs

THEOREM 2.2.1 Tests for Symmetry

The graph of an equation is symmetric with respect to the

(*i*) **y-axis** if replacing x by $-x$ results in an equivalent equation;
(*ii*) **x-axis** if replacing y by $-y$ results in an equivalent equation;
(*iii*) **origin** if replacing x and y by $-x$ and $-y$ results in an equivalent equation.

The advantage of using symmetry in graphing should be apparent: If say, the graph of an equation is symmetric with respect to the x-axis, then we need only produce the graph for $y \geq 0$ since points on the graph for $y < 0$ are obtained by taking the mirror images, through the x-axis, of the points in the first and second quadrants.

EXAMPLE 7 Test for Symmetry

By replacing x by $-x$ in the equation $y = x^2$ and using $(-x)^2 = x^2$, we see that

$$y = (-x)^2 \text{ is equivalent to } y = x^2.$$

This proves what is apparent in Figure 2.2.8; that is, the graph of $y = x^2$ is symmetric with respect to the y-axis. ≡

EXAMPLE 8 Intercepts and Symmetry

Determine the intercepts and any symmetry for the graph of

$$x + y^2 = 10. \tag{8}$$

Solution

Intercepts: Setting $y = 0$ in equation (8) immediately gives $x = 10$. The graph of the equation has a single x-intercept, $(10, 0)$. When $x = 0$, we get $y^2 = 10$, which implies that $y = -\sqrt{10}$ or $y = \sqrt{10}$. Thus there are two y-intercepts, $(0, -\sqrt{10})$ and $(0, \sqrt{10})$.

Symmetry: If we replace x by $-x$ in the equation $x + y^2 = 10$, we get $-x + y^2 = 10$. This is not equivalent to equation (8). You should also verify that replacing x and y by $-x$ and $-y$ in (8) does not yield an equivalent equation. However, if we replace y by $-y$, we find that

$$x + (-y)^2 = 10 \text{ is equivalent to } x + y^2 = 10.$$

Thus, the graph of the equation is symmetric with respect to the x-axis.

Graph: In the graph of the equation given in **FIGURE 2.2.10** the intercepts are indicated and the x-axis symmetry should be apparent. ≡

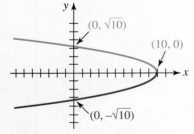

FIGURE 2.2.10 Graph of equation in Example 8

2.2 Exercises
Answers to selected odd-numbered problems begin on page ANS-5.

In Problems 1–6, find the center and the radius of the given circle. Sketch its graph.

1. $x^2 + y^2 = 5$
2. $x^2 + y^2 = 9$
3. $x^2 + (y - 3)^2 = 49$
4. $(x + 2)^2 + y^2 = 36$
5. $\left(x - \frac{1}{2}\right)^2 + \left(y - \frac{3}{2}\right)^2 = 1$
6. $(x + 3)^2 + (y - 5)^2 = 25$

In Problems 7–14, complete the square in x and y to find the center and the radius of the given circle.

7. $x^2 + y^2 + 8y = 0$
8. $x^2 + y^2 - 6x = 0$
9. $x^2 + y^2 + 2x - 4y - 4 = 0$
10. $x^2 + y^2 - 18x - 6y - 10 = 0$
11. $x^2 + y^2 - 20x + 16y + 128 = 0$
12. $x^2 + y^2 + 3x - 16y + 63 = 0$
13. $2x^2 + 2y^2 + 4x + 16y + 1 = 0$
14. $\frac{1}{2}x^2 + \frac{1}{2}y^2 + \frac{5}{2}x + 10y + 5 = 0$

In Problems 15–24, find an equation of the circle that satisfies the given conditions.

15. center $(0, 0)$, radius 1
16. center $(1, -3)$, radius 5
17. center $(0, 2)$, radius $\sqrt{2}$
18. center $(-9, -4)$, radius $\frac{3}{2}$
19. endpoints of a diameter at $(-1, 4)$ and $(3, 8)$
20. endpoints of a diameter at $(4, 2)$ and $(-3, 5)$
21. center $(0, 0)$, graph passes through $(-1, -2)$
22. center $(4, -5)$, graph passes through $(7, -3)$
23. center $(5, 6)$, graph tangent to the x-axis
24. center $(-4, 3)$, graph tangent to the y-axis

In Problems 25–28, sketch the semicircle defined by the given equation.

25. $y = \sqrt{4 - x^2}$
26. $x = 1 - \sqrt{1 - y^2}$
27. $x = \sqrt{1 - (y - 1)^2}$
28. $y = -\sqrt{9 - (x - 3)^2}$

29. Find an equation for the upper half of the circle $x^2 + (y - 3)^2 = 4$. The right half of the circle.
30. Find an equation for the lower half of the circle $(x - 5)^2 + (y - 1)^2 = 9$. The left half of the circle.

In Problems 31–34, sketch the set of points in the xy-plane whose coordinates satisfy the given inequality.

31. $x^2 + y^2 \geq 9$
32. $(x - 1)^2 + (y + 5)^2 \leq 25$
33. $1 \leq x^2 + y^2 \leq 4$
34. $x^2 + y^2 > 2y$

In Problems 35 and 36, find the x- and y-intercepts of the given circle.

35. The circle with center $(3, -6)$ and radius 7
36. The circle $x^2 + y^2 + 5x - 6y = 0$

In Problems 37–62, find any intercepts of the graph of the given equation. Determine whether the graph of the equation possesses symmetry with respect to the x-axis, y-axis, or origin. Do not graph.

37. $y = -3x$
38. $y - 2x = 0$
39. $-x + 2y = 1$
40. $2x + 3y = 6$
41. $x = y^2$
42. $y = x^3$
43. $y = x^2 - 4$
44. $x = 2y^2 - 4$
45. $y = x^2 - 2x - 2$
46. $y^2 = 16(x + 4)$
47. $y = x(x^2 - 3)$
48. $y = (x - 2)^2(x + 2)^2$
49. $x = -\sqrt{y^2 - 16}$
50. $y^3 - 4x^2 + 8 = 0$
51. $4y^2 - x^2 = 36$
52. $\dfrac{x^2}{25} + \dfrac{y^2}{9} = 1$
53. $y = \dfrac{x^2 - 7}{x^3}$
54. $y = \dfrac{x^2 - 10}{x^2 + 10}$

55. $y = \dfrac{x^2 - x - 20}{x + 6}$
56. $y = \dfrac{(x + 2)(x - 8)}{x + 1}$
57. $y = \sqrt{x} - 3$
58. $y = 2 - \sqrt{x + 5}$
59. $y = |x - 9|$
60. $x = |y| - 4$
61. $|x| + |y| = 4$
62. $x + 3 = |y - 5|$

In Problems 63–66, state all the symmetries of the given graph.

63.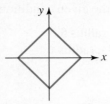

FIGURE 2.2.11 Graph for Problem 63

64.

FIGURE 2.2.12 Graph for Problem 64

65.

FIGURE 2.2.13 Graph for Problem 65

66.

FIGURE 2.2.14 Graph for Problem 66

In Problems 67–72, use symmetry to complete the given graph.

67. The graph is symmetric with respect to the y-axis.

FIGURE 2.2.15 Graph for Problem 67

68. The graph is symmetric with respect to the x-axis.

FIGURE 2.2.16 Graph for Problem 68

69. The graph is symmetric with respect to the origin.

FIGURE 2.2.17 Graph for Problem 69

70. The graph is symmetric with respect to the y-axis.

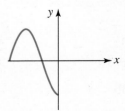

FIGURE 2.2.18 Graph for Problem 70

71. The graph is symmetric with respect to the x- and y-axes.

FIGURE 2.2.19 Graph for Problem 71

72. The graph is symmetric with respect to the origin.

FIGURE 2.2.20 Graph for Problem 72

For Discussion

73. Determine whether the following statement is true or false. Defend your answer.

If a graph has two of the three symmetries defined on page 135, then the graph must necessarily possess the third symmetry.

74. (a) The radius of the circle in FIGURE 2.2.21(a) is r. What is its equation in standard form?

(b) The center of the circle in Figure 2.2.21(b) is (h, k). What is its equation in standard form?

75. Discuss whether the following statement is true or false.

Every equation of the form $x^2 + y^2 + ax + by + c = 0$ is a circle.

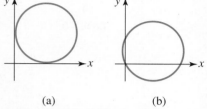

(a)   (b)

FIGURE 2.2.21 Circles in Problem 74

2.3 Equations of Lines

≡ **Introduction** Any pair of distinct points in the xy-plane determines a unique straight line. Our goal in this section is to find equations of lines. Fundamental to finding equations of lines is the concept of slope of a line.

□ **Slope** If $P_1(x_1, y_1)$ and $P_2(x_2, y_2)$ are two points such that $x_1 \neq x_2$, then the number

$$m = \frac{y_2 - y_1}{x_2 - x_1} \tag{1}$$

is called the **slope** of the line determined by these two points. It is customary to call $y_2 - y_1$ the **change in y** or **rise** of the line; $x_2 - x_1$ is the **change in x** or the **run** of the line. Therefore, the slope (1) of a line is

$$m = \frac{\text{rise}}{\text{run}}. \tag{2}$$

See FIGURE 2.3.1(a). Any pair of distinct points on a line will determine the same slope. To see why this is so, consider the two similar right triangles in Figure 2.3.1(b). Since we know that the ratios of corresponding sides in similar triangles are equal we have

$$\frac{y_2 - y_1}{x_2 - x_1} = \frac{y_4 - y_3}{x_4 - x_3}.$$

Hence the slope of a line is independent of the choice of points on the line.

(a) Rise and run

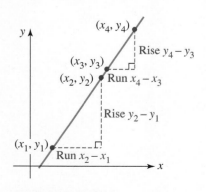

(b) Similar triangles

FIGURE 2.3.1 Slope of a line

In **FIGURE 2.3.2** we compare the graphs of lines with positive, negative, zero, and undefined slopes. In Figure 2.3.2(a) we see, reading the graph left to right, that a line with positive slope ($m > 0$) rises as x increases. Figure 2.3.2(b) shows that a line with negative slope ($m < 0$) falls as x increases. If $P_1(x_1, y_1)$ and $P_2(x_2, y_2)$ are points on a horizontal line, then $y_1 = y_2$ and so its rise is $y_2 - y_1 = 0$. Hence from (1) the slope is zero ($m = 0$). See Figure 2.3.2(c). If $P_1(x_1, y_1)$ and $P_2(x_2, y_2)$ are points on a vertical line, then $x_1 = x_2$ and so its run is $x_2 - x_1 = 0$. In this case we say that the slope of the line is **undefined** or that the line has no slope. See Figure 2.3.2(d).

FIGURE 2.3.2 Lines with slope (a)–(c); line with no slope (d)

In general, since

$$\frac{y_2 - y_1}{x_2 - x_1} = \frac{-(y_1 - y_2)}{-(x_1 - x_2)} = \frac{y_1 - y_2}{x_1 - x_2},$$

it does not matter which of the two points is called $P_1(x_1, y_1)$ and which is called $P_2(x_2, y_2)$ in (1).

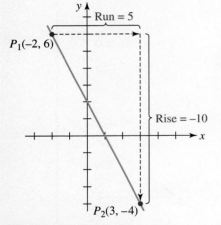

FIGURE 2.3.3 Line in Example 1

EXAMPLE 1 Slope and Graph

Find the slope of the line through the points $(-2, 6)$ and $(3, -4)$. Graph the line.

Solution Let $(-2, 6)$ be the point $P_1(x_1, y_1)$ and $(3, -4)$ be the point $P_2(x_2, y_2)$. The slope of the straight line through these points is

$$m = \frac{y_2 - y_1}{x_2 - x_1} = \frac{-4 - 6}{3 - (-2)}$$
$$= \frac{-10}{5} = -2.$$

Thus the slope is -2, and the line through P_1 and P_2 is shown in **FIGURE 2.3.3**. ≡

☐ **Point-Slope Equation** We are now in a position to find an equation of a line L. To begin, suppose L has slope m and that $P_1(x_1, y_1)$ is on the line. If $P(x, y)$ represents any other point on L, then (1) gives

$$m = \frac{y - y_1}{x - x_1}.$$

Multiplying both sides of the last equality by $x - x_1$ gives an important equation.

THEOREM 2.3.1 Point-Slope Equation

The **point-slope equation** of the line through $P_1(x_1, y_1)$ with slope m is

$$y - y_1 = m(x - x_1). \tag{3}$$

EXAMPLE 2   Point-Slope Equation

Find an equation of the line with slope 6 and passing through $\left(-\tfrac{1}{2}, 2\right)$.

Solution Letting $m = 6$, $x_1 = -\tfrac{1}{2}$, and $y_1 = 2$ we obtain from (3)
$$y - 2 = 6\left[x - \left(-\tfrac{1}{2}\right)\right].$$
Simplifying gives
$$y - 2 = 6\left(x + \tfrac{1}{2}\right) \quad \text{or} \quad y = 6x + 5.$$

EXAMPLE 3   Point-Slope Equation

Find an equation of the line passing through the points $(4, 3)$ and $(-2, 5)$.

Solution First we compute the slope of the line through the points. From (1),
$$m = \frac{5 - 3}{-2 - 4} = \frac{2}{-6} = -\frac{1}{3}.$$

The point-slope equation (3) then gives

$$y - 3 = -\frac{1}{3}(x - 4) \quad \text{or} \quad y = -\frac{1}{3}x + \frac{13}{3}.$$

(the distributive law)

◀ The distributive law
$$a(b + c) = ab + ac$$
is the source of many errors on students' papers. A common error goes something like this:
$$-(2x - 3) = -2x - 3.$$
The correct result is:
$$\begin{aligned}-(2x - 3) &= (-1)(2x - 3) \\ &= (-1)2x - (-1)3 \\ &= -2x + 3.\end{aligned}$$

□ **Slope-Intercept Equation** Any line with slope (that is, any line that is not vertical) must cross the y-axis. If this y-intercept is $(0, b)$, then with $x_1 = 0$, $y_1 = b$, the point-slope form (3) gives $y - b = m(x - 0)$. The last equation simplifies to the next result.

THEOREM 2.3.2   Slope-Intercept Equation

The **slope-intercept equation** of the line with slope m and y-intercept $(0, b)$ is
$$y = mx + b. \qquad (4)$$

When $b = 0$ in (4), the equation $y = mx$ represents a family of lines that pass through the origin $(0, 0)$. In **FIGURE 2.3.4** we have drawn a few of the members of that family.

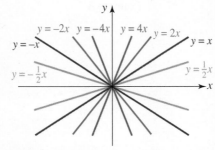

FIGURE 2.3.4 Lines through the origin are $y = mx$

EXAMPLE 4   Example 3 Revisited

We can also use the slope-intercept form (4) to obtain an equation of the line through the two points in Example 3. As in that example, we start by finding the slope $m = -\tfrac{1}{3}$. The equation of the line is then $y = -\tfrac{1}{3}x + b$. By substituting the coordinates of either point $(4, 3)$ or $(-2, 5)$ into the last equation enables us to determine b. If we use $x = 4$ and $y = 3$, then $3 = -\tfrac{1}{3} \cdot 4 + b$ and so $b = 3 + \tfrac{4}{3} = \tfrac{13}{3}$. The equation of the line is $y = -\tfrac{1}{3}x + \tfrac{13}{3}$.

□ **Horizontal and Vertical Lines** We saw in Figure 2.3.2(c) that a horizontal line has slope $m = 0$. An equation of a horizontal line passing through a point (a, b) can be obtained from (3), that is, $y - b = 0(x - a)$ or $y = b$.

2.3 Equations of Lines

THEOREM 2.3.3 Equation of Horizontal Line

The **equation of a horizontal line** with y-intercept $(0, b)$ is

$$y = b. \tag{5}$$

A vertical line through (a, b) has undefined slope and all points on the line have the same x-coordinate. The **equation of a vertical line** is then

THEOREM 2.3.4 Equation of Vertical Line

The **equation of a vertical line** with x-intercept $(a, 0)$ is

$$x = a. \tag{6}$$

EXAMPLE 5   **Vertical and Horizontal Lines**

Find equations for the vertical and horizontal lines through $(3, -1)$. Graph these lines.

Solution Any point on the vertical line through $(3, -1)$ has x-coordinate 3. The equation of this line is then $x = 3$. Similarly, any point on the horizontal line through $(3, -1)$ has y-coordinate -1. The equation of this line is $y = -1$. Both lines are graphed in **FIGURE 2.3.5**.

FIGURE 2.3.5 Horizontal and vertical lines in Example 5

☐ **Linear Equation** The equations (3), (4), (5), and (6) are special cases of the **general linear equation** in two variables x and y

$$ax + by + c = 0, \tag{7}$$

where a and b are real constants and not both zero. The characteristic that gives (7) its name *linear* is that the variables x and y appear only to the first power. Observe that

$$a = 0, b \neq 0, \text{ gives } y = -\frac{c}{b}, \quad \leftarrow \text{horizontal line}$$

$$a \neq 0, b = 0, \text{ gives } x = -\frac{c}{a}, \quad \leftarrow \text{vertical line}$$

$$a \neq 0, b \neq 0, \text{ gives } y = -\frac{a}{b}x - \frac{c}{b}. \leftarrow \text{line with nonzero slope}$$

EXAMPLE 6   **Slope and y-intercept**

Find the slope and the y-intercept of the line $3x - 7y + 5 = 0$.

Solution We solve the linear equation for y:

$$3x - 7y + 5 = 0$$
$$7y = 3x + 5$$
$$y = \frac{3}{7}x + \frac{5}{7}.$$

Comparing the last equation with (4) we see that the slope of the line is $m = \frac{3}{7}$ and the y-intercept is $\left(0, \frac{5}{7}\right)$.

If the x- and y-intercepts are distinct, the graph of the line can be drawn through the corresponding points on the x- and y-axes.

EXAMPLE 7   Graph of a Linear Equation

Graph the linear equation $3x - 2y + 8 = 0$.

Solution There is no need to rewrite the linear equation in the form $y = mx + b$. We simply find the intercepts.

y-intercept: Setting $x = 0$ gives $-2y + 8 = 0$ or $y = 4$. The *y*-intercept is $(0, 4)$.
x-intercept: Setting $y = 0$ gives $3x + 8 = 0$ or $x = -\frac{8}{3}$. The *x*-intercept is $\left(-\frac{8}{3}, 0\right)$.

As shown in **FIGURE 2.3.6**, the line is drawn through the two intercepts $(0, 4)$ and $\left(-\frac{8}{3}, 0\right)$. ≡

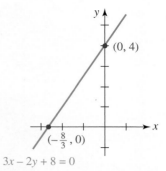

FIGURE 2.3.6 Line in Example 7

☐ **Parallel and Perpendicular Lines** Suppose L_1 and L_2 are two distinct lines with slope. This assumption means that both L_1 and L_2 are nonvertical lines. Then necessarily L_1 and L_2 are either parallel or they intersect. If the lines intersect at a right angle they are said to be perpendicular. We can determine whether two lines are parallel or are perpendicular by examining their slopes.

parallel lines

THEOREM 2.3.5   Parallel and Perpendicular Lines

If L_1 and L_2 are lines with slopes m_1 and m_2, respectively, then

   (*i*) L_1 is **parallel** to L_2 if and only if $m_1 = m_2$, and
   (*ii*) L_1 is **perpendicular** to L_2 if and only if $m_1 m_2 = -1$.

There are several ways of proving the two parts of Theorem 2.3.5. The proof of part (*i*) can be obtained using similar right triangles, as in **FIGURE 2.3.7**, and the fact that the ratios of corresponding sides in such triangles are equal. We leave the justification of part (*ii*) as an exercise. See Problems 49 and 50 in Exercises 2.3. Note that the condition $m_1 m_2 = -1$ implies that $m_2 = -1/m_1$, that is, the slopes are negative reciprocals of each other. A horizontal line $y = b$ and a vertical line $x = a$ are perpendicular, but the latter is a line with no slope.

FIGURE 2.3.7 Parallel lines

EXAMPLE 8   Parallel Lines

The linear equations $3x + y = 2$ and $6x + 2y = 15$ can be rewritten in the slope-intercept forms

$$y = -3x + 2 \quad \text{and} \quad y = -3x + \tfrac{15}{2},$$

respectively. As noted in color in the preceding line the slope of each line is -3. Therefore the lines are parallel. The graphs of these equations are shown in **FIGURE 2.3.8**. ≡

FIGURE 2.3.8 Parallel lines in Example 8

EXAMPLE 9   Perpendicular Lines

Find an equation of the line through $(0, -3)$ that is perpendicular to the graph of $4x - 3y + 6 = 0$.

Solution We express the given linear equation in slope-intercept form:

$$4x - 3y + 6 = 0 \quad \text{implies} \quad 3y = 4x + 6.$$

Dividing by 3 gives $y = \tfrac{4}{3}x + 2$. This line, whose graph is given in blue in **FIGURE 2.3.9**, has slope $\tfrac{4}{3}$. The slope of any line perpendicular to it is the negative reciprocal of $\tfrac{4}{3}$, namely, $-\tfrac{3}{4}$. Since $(0, -3)$ is the *y*-intercept of the required line, it follows from (4) that its equation is $y = -\tfrac{3}{4}x - 3$. The graph of the last equation is the red line in Figure 2.3.9. ≡

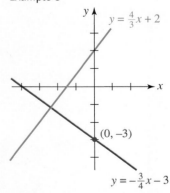

FIGURE 2.3.9 Perpendicular lines in Example 9

2.3 Equations of Lines

2.3 Exercises
Answers to selected odd-numbered problems begin on page ANS-6.

In Problems 1–6, find the slope of the line through the given points. Graph the line through the points.

1. $(3, -7), (1, 0)$
2. $(-4, -1), (1, -1)$
3. $(5, 2), (4, -3)$
4. $(1, 4), (6, -2)$
5. $(-1, 2), (3, -2)$
6. $\left(8, -\tfrac{1}{2}\right), \left(2, \tfrac{5}{2}\right)$

In Problems 7 and 8, use the graph of the given line to estimate its slope.

7.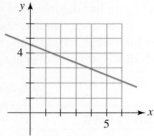

FIGURE 2.3.10 Graph for Problem 7

8.

FIGURE 2.3.11 Graph for Problem 8

In Problems 9–16, find the slope and the x- and y-intercepts of the given line. Graph the line.

9. $3x - 4y + 12 = 0$
10. $\tfrac{1}{2}x - 3y = 3$
11. $2x - 3y = 9$
12. $-4x - 2y + 6 = 0$
13. $2x + 5y - 8 = 0$
14. $\dfrac{y}{2} - \dfrac{x}{10} - 1 = 0$
15. $y + \tfrac{2}{3}x = 1$
16. $y = 2x + 6$

In Problems 17–22, find an equation of the line through $(1, 2)$ with the indicated slope.

17. $\tfrac{2}{3}$
18. $\tfrac{1}{10}$
19. 0
20. -2
21. -1
22. undefined

In Problems 23–36, find an equation of the line that satisfies the given conditions.

23. Through $(2, 3)$ and $(6, -5)$
24. Through $(5, -6)$ and $(4, 0)$
25. Through $(8, 1)$ and $(-3, 1)$
26. Through $(2, 2)$ and $(-2, -2)$
27. Through $(-2, 0)$ and $(-2, 6)$
28. Through $(0, 0)$ and (a, b)
29. Through $(-2, 4)$ parallel to $3x + y - 5 = 0$
30. Through $(1, -3)$ parallel to $2x - 5y + 4 = 0$
31. Through $(5, -7)$ parallel to the y-axis
32. Through the origin parallel to the line through $(1, 0)$ and $(-2, 6)$
33. Through $(2, 3)$ perpendicular to $x - 4y + 1 = 0$
34. Through $(0, -2)$ perpendicular to $3x + 4y + 5 = 0$
35. Through $(-5, -4)$ perpendicular to the line through $(1, 1)$ and $(3, 11)$
36. Through the origin perpendicular to every line with slope 2

In Problems 37–40, determine which of the given lines are parallel to each other and which are perpendicular to each other.

37. (a) $3x - 5y + 9 = 0$ (b) $5x = -3y$ (c) $-3x + 5y = 2$
    (d) $3x + 5y + 4 = 0$ (e) $-5x - 3y + 8 = 0$ (f) $5x - 3y - 2 = 0$

38. (a) $2x + 4y + 3 = 0$ (b) $2x - y = 2$ (c) $x + 9 = 0$
    (d) $x = 4$ (e) $y - 6 = 0$ (f) $-x - 2y + 6 = 0$
39. (a) $3x - y - 1 = 0$ (b) $x - 3y + 9 = 0$ (c) $3x + y = 0$
    (d) $x + 3y = 1$ (e) $6x - 3y + 10 = 0$ (f) $x + 2y = -8$
40. (a) $y + 5 = 0$ (b) $x = 7$ (c) $4x + 6y = 3$
    (d) $12x - 9y + 7 = 0$ (e) $2x - 3y - 2 = 0$ (f) $3x + 4y - 11 = 0$

41. Find an equation of the line L shown in **FIGURE 2.3.12** if an equation of the blue curve is $y = x^2 + 1$.
42. A **tangent to a circle** is defined to be a straight line that touches the circle at only one point P. Find an equation of the tangent line L shown in **FIGURE 2.3.13**.

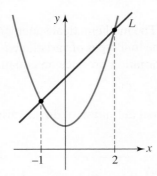

FIGURE 2.3.12 Graphs in Problem 41

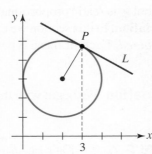

FIGURE 2.3.13 Circle and tangent line in Problem 42

For Discussion

43. How would you find an equation of the line that is the perpendicular bisector of the line segment through $(\frac{1}{2}, 10)$ and $(\frac{3}{2}, 4)$?
44. Using only the concepts of this section, how would you prove or disprove that the triangle with vertices $(2, 3)$, $(-1, -3)$, and $(4, 2)$ is a right triangle?
45. Using only the concepts of this section, how would you prove or disprove that the quadrilateral with vertices $(0, 4)$, $(-1, 3)$, $(-2, 8)$, and $(-3, 7)$ is a parallelogram?
46. If C is an arbitrary real constant, an equation such as $2x - 3y = C$ is said to define a **family of lines**. Choose four different values of C and plot the corresponding lines on the same coordinate axes. What is true about the lines that are members of this family?
47. Find the equations of the lines through $(0, 4)$ that are tangent to the circle $x^2 + y^2 = 4$.
48. For the line $ax + by + c = 0$, what can be said about a, b, and c if
    (a) the line passes through the origin,
    (b) the slope of the line is 0,
    (c) the slope of the line is undefined?

In Problems 49 and 50, to prove part (*ii*) of Theorem 2.3.5 you have to prove two things, the *only if* part (Problem 49) and then the *if* part (Problem 50) of the theorem.

49. In **FIGURE 2.3.14**, without loss of generality, we have assumed that two perpendicular lines $y = m_1 x$, $m_1 > 0$, and $y = m_2 x$, $m_2 < 0$, intersect at the origin. Use the information in the figure to prove the *only if* part:

    If L_1 and L_2 are perpendicular lines with slopes m_1 and m_2, then $m_1 m_2 = -1$.

50. Reverse your argument in Problem 49 to prove the *if* part:

    If L_1 and L_2 are lines with slopes m_1 and m_2 such that $m_1 m_2 = -1$, then L_1 and L_2 are perpendicular.

FIGURE 2.3.14 Lines through origin in Problems 49 and 50

2.4 Variation

≡ **Introduction** In many disciplines, a mathematical description by means of an equation, or **mathematical model**, of a real-life problem can be constructed using the notion of proportionality. For example, in one model of a growing population (say, bacteria) it is assumed that the rate of growth at time t is directly proportional to the population at that time. If we let R represent the rate of growth, P the population, then the preceding sentence translates into

$$R \propto P, \tag{1}$$

where the symbol $\propto$ is read "proportional to." The mathematical statement in (1) is an example of **variation**. In this section we examine four types of variation: *direct*, *inverse*, *joint*, and *combined*. Each of these types of variation produce an equation in two or more variables.

☐ **Direct Variation** We begin with the formal definition of direct variation.

DEFINITION 2.4.1 Direct Variation

A quantity y **varies directly**, or is **directly proportional to**, a quantity x if there exists a nonzero number k such that

$$y = kx. \tag{2}$$

In (2) we say that the number k is the **constant of proportionality**. Comparing (2) with Figure 2.3.4 we know that the graph of any equation of the form given in (2) is a line through the origin with slope k. **FIGURE 2.4.1** illustrates the graph of (2) in the case of when $k > 0$ and $x \geq 0$.

Of course, symbols other than x and y are often used in (2). In the study of springs in physics, the force F required to keep a spring stretched x units beyond its natural, or unstretched, length is assumed to be directly proportional to the elongation x, that is,

$$F = kx. \tag{3}$$

See **FIGURE 2.4.2**. The result in (3) is called **Hooke's law** after the irascible English physicist **Robert Hooke** (1635–1703).

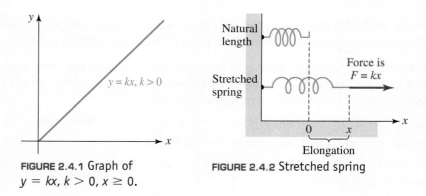

FIGURE 2.4.1 Graph of $y = kx, k > 0, x \geq 0$.

FIGURE 2.4.2 Stretched spring

EXAMPLE 1 Hooke's Law

A spring whose natural length is $\frac{1}{4}$ ft is stretched 1 in. by a force of 30 lb. How much force is necessary to stretch the spring to a length of 1 ft?

Solution The elongation of 1 in. is equivalent to $\frac{1}{12}$ ft. Hence by (2) we have

$$30 = k\left(\tfrac{1}{12}\right) \quad \text{or} \quad k = 360 \text{ lb/ft}.$$

Therefore, $F = 360x$. When the spring is stretched to a length of 1 ft, its elongation is $1 - \frac{1}{4} = \frac{3}{4}$ ft. The force necessary to stretch the spring to a length of 1 ft is

$$F = 360 \cdot \tfrac{3}{4} = 270 \text{ lb}. \qquad \equiv$$

A quantity can also be proportional to a power of another quantity. In general, we say that y **varies directly**, as the nth power of x, or is **directly proportional to** x^n, if there exists a constant k such that

$$y = kx^n, \quad n > 0. \tag{4}$$

The power n in (4) need not be an integer.

EXAMPLE 2 Direct Variation

(a) The circumference C of a circle is directly proportional to its radius r. If k is the constant of proportionality, then by (2) we can write $C = kr$.

(b) The area A of a circle is directly proportional to the square of its radius r. If k is the constant of proportionality, then by (4), $A = kr^2$.

(c) The volume V of a sphere is directly proportional to the cube of its radius r. If k is the constant of proportionality, then by (4), $V = kr^3$. $\qquad \equiv$

From geometry we know in part (a) of Example 2 that $k = 2\pi$, in part (b), $k = \pi$, and in part (c), $k = 4\pi/3$.

EXAMPLE 3 Direct Variation

Suppose that y is directly proportional to x^3. If $y = 4$ when $x = 2$, what is the value of y when $x = 4$?

Solution From (3) we can write $y = kx^3$. By substitution of $y = 4$ and $x = 2$ into this equation, we obtain the constant of proportionality k, since $4 = k \cdot 8$ implies that $k = \frac{1}{2}$. Thus, $y = \frac{1}{2}x^3$. Finally, when $x = 4$, we have $y = \frac{1}{2} \cdot 4^3$, or $y = 32$. $\qquad \equiv$

☐ **Inverse Variation** We say that a quantity y **varies inversely** with x if it is proportional to the reciprocal of x. The formal definition of this concept follows next.

DEFINITION 2.4.2 Inverse Variation

A quantity y **varies inversely**, or is **inversely proportional to**, a quantity x if there exists a nonzero number k such that

$$y = \frac{k}{x}. \tag{4}$$

Note in (4) if we let one of the quantities, say x, increase in magnitude, then correspondingly the quantity y decreases in magnitude. Alternatively, if the value of x is

FIGURE 2.4.3 Graph of $y = k/x$, $k > 0$, $x \geq 0$

small in magnitude, then the value of y is large in magnitude. This can be seen clearly in the graph given in FIGURE 2.4.3 for $x > 0$.

An alternative form of (4) is $xy = k$. In the study of gases, **Boyle's law** stipulates that the product of an ideal gas's pressure P and the volume V occupied by that gas satisfies $PV = k$. In other words, P is inversely proportional to V. If the volume V of a container containing an ideal gas is decreased, necessarily the pressure exerted by the gas on the interior walls of the container increases.

In general, we say that y **varies inversely**, or is **inversely proportional to**, the nth power of x if there exists a constant k such that

$$y = \frac{k}{x^n} = kx^{-n}, \quad n > 0.$$

☐ **Joint and Combined Variation** A variable may be directly proportional to the products of powers of several variables. If the variable z is given by

$$z = kx^m y^n, \quad m > 0, \quad n > 0, \tag{5}$$

we say that z **varies jointly** as the mth power of x and the nth power of y, or that z is **jointly proportional to** x^m and y^n. The concept of joint variation expressed in (5) can, of course, be extended to products of powers of more than two variables. Furthermore, a quantity may be directly proportional to several variables and inversely proportional to other variables. This type of variation is called **combined variation**.

EXAMPLE 4 Joint Variation

Consider the right circular cylinder and the right circular cone shown in FIGURE 2.4.4. The volume V of each is jointly proportional to the square of its radius r and its height h. That is

$$V_{\text{cylinder}} = k_1 r^2 h \quad \text{and} \quad V_{\text{cone}} = k_2 r^2 h.$$

It turns out that the constants of proportionality are $k_1 = \pi$ and $k_2 = \pi/3$. Thus the volumes are

$$V_{\text{cylinder}} = \pi r^2 h \quad \text{and} \quad V_{\text{cone}} = \frac{\pi}{3} r^2 h. \qquad \equiv$$

(a) Right circular cylinder (b) Right circular cone

FIGURE 2.4.4 Cone and Cylinder in Example 4

EXAMPLE 5 Joint Variation

The hydrodynamic resistance D to a boat moving through water is jointly proportional to the density ρ of the water, the area A of the wet portion of the boat's hull, and the square of the boat's velocity v. That is

$$D = k\rho A v^2, \tag{6}$$

where k is the constant of proportionality. See FIGURE 2.4.5. $\quad \equiv$

The same relation (6) can sometimes be used to determine the *drag force* acting on an object moving through the air.

FIGURE 2.4.5 Boat in Example 5

EXAMPLE 6 Combined Variation

Newton's Law of Universal Gravitation is a good example of combined variation:

- *Every point mass in the universe attracts every other point mass by a force that is **directly** proportional to the product of the two masses and **inversely** proportional to the square of the distance between the point masses.*

If, as shown in FIGURE 2.4.6, we denote the masses by m_1 and m_2, the distance between the masses by r, the square of the distance by r^2, the common magnitude of the force vectors $\mathbf{F}_1$ and $\mathbf{F}_2$ as F, and k the constant of proportionality, then the formulaic interpretation of the foregoing paragraph is

$$F = k\frac{m_1 m_2}{r^2}.$$

FIGURE 2.4.6 Gravitational force in Example 6

The constant of proportionality k in Example 6 is usually denoted by the symbol G and is called the **universal gravitational constant**.

2.4 Exercises
Answers to selected odd-numbered problems begin on page ANS-6.

1. Suppose that y varies directly as the square of x. If $y = 3$ when $x = 1$, what is the value of y when $x = 2$?
2. Suppose that y is directly proportional to the square root of x. If $y = 4$ when $x = 16$, what is the value of y when $x = 25$?
3. Suppose that w is inversely proportional to the cube root of t. If $w = 2$ when $t = 27$, what is the value of w when $t = 8$?
4. Suppose that s varies inversely as the square of r. If a value of r is tripled, what is the effect on s?
5. (a) Suppose a 10-lb force stretches a spring 3 in. beyond its natural length. Find a formula for the force F required to stretch the spring x ft beyond its natural length.
   (b) Determine the elongation of the spring produced by a 50-lb force.
6. A spring whose natural length of 1 ft is stretched $\frac{3}{4}$ ft by a force of 100 lb. How much force is necessary to stretch the spring to a length of 2.5 ft?

Miscellaneous Applications

7. **Falling Stone** The distance s that a stone travels when dropped from a very tall building is proportional to the square of the time t in flight. If the stone falls 64 feet in 2 seconds, find a formula that relates s and t. How far does the stone fall in 5 s? How far does the stone fall between 2 s and 3 s?
8. **Another Falling Stone** The velocity v of a stone dropped from a very tall building varies directly as the time t in flight. Find a formula relating v and t if the velocity of the stone at the end of 1 second is 32 ft/s. If the stone is dropped from the top of a building that is 144 ft tall, what is its velocity when it hits the ground? [*Hint*: Use Problem 7.]
9. **Pendulum Motion** The period T of a plane pendulum varies directly as the square root of its length L. How much should the length L be changed in order to double the period of the pendulum?
10. **Weight** The weight w of a person varies directly as the cube of the person's length l. At age 13 a person 60 in. tall weighs 120 lb. What is the person's weight at age 16 when the person is 72 in. tall?
11. **Animal Surface Area** The surface area S (in square meters) of an animal is directly proportional to the two-thirds power of its weight w measured in kg. For humans the constant of proportionality is taken to be $k = 0.11$. Find the surface area of a person whose weight is 81 kg.

FIGURE 2.4.7 Parallel wires in Problem 13

12. **Kepler's Third Law** According to Kepler's third law of planetary motion, the square of the period P of a planet (that is, the time it takes for a planet to revolve around the Sun) is proportional to the cube of its mean distance s from the sun. The period of the Earth is 365 days and its mean distance from the Sun is 92,900,000 mi. Determine the period of Mars given that its mean distance from the Sun is 142,000,000 mi.

13. **Magnetic Force** Suppose that electrical currents I_1 and I_2 flow in long parallel wires as shown in **FIGURE 2.4.7**. The force F_L per unit length exerted on a wire due to the magnetic field around the other wire is jointly proportional to the currents I_1 and I_2 and inversely proportional to the distance r between the wires. Express this combined variation as a formula. If the distance r is halved, then what is the effect on the force F_L?

14. **Energy** The kinetic energy K of a moving body varies jointly as the product of its mass m and the square of its velocity v. If the constant of proportionality is $\frac{1}{2}$, find the kinetic energy of a neutron of mass 1.7×10^{-27} kg moving at a constant rate of 3.5×10^4 m/s.

15. **Got Gas?** According to the general gas law, the pressure P of a quantity of gas is directly proportional to the absolute temperature T of the gas and inversely proportional to its volume V. Express this combined variation as a formula. A large balloon contains 500 ft³ of a gas at ground level, where the pressure is 14.7 lb/in² and the absolute temperature is 293 K (or 20°C). What is the volume occupied by this gas at an altitude of 10 mi, where the pressure is 1.5 lb/in² and the absolute temperature is 218 K (or −55°C)?

16. **Stress and Strain** In the study of elastic bodies, stress is directly proportional to strain. For a wire of length L and cross-sectional area A that is stretched an amount e by an applied force F, stress is defined to be F/A and strain is given by e/L. Find a formula that expresses e in terms of the other variables.

17. **Speed of Sound** The speed of sound in air varies with temperature according to the equation $v = 33{,}145\sqrt{T/273}$, where v is the speed of sound in centimeters per second and T is the temperature of the air in kelvin units (273 K = 0° Celsius). On which day does the sound of exploding fireworks travel faster: July 4th ($T = 310$ K) or January 1st ($T = 270$ K)? How much faster?

Fireworks in Problem 17

18. **Animal Life Span** Empirical studies indicate that the life span of a mammal in captivity is related to body size by the formula $L = (11.8)\, M^{0.20}$, where L is life span in years and M is body mass in kilograms.
    (a) What does this function predict for the life span of a 4000-kg elephant in a zoo?
    (b) What does this function predict for the life span of an 80-kg man confined to a prison?

19. **Temperature** The temperature of a Pyrex glass rod is raised from a temperature t_1 to a final temperature t_2. The thermal expansion e of the rod is jointly proportional to its length L and the rise in the temperature. When a rod of length 10 cm is heated from 20°C to 420°C, its thermal expansion is 0.012 cm. What is the thermal expansion of the same rod when it is heated from 20°C to 550°C?

20. **Pitch of a Bell** A rule of thumb has it that the pitch P of a bell is inversely proportional to the cube root of its weight w. A bell weighing 800 lb has a pitch of 512 cycles per second. How heavy would a similar bell have to be in order to produce a pitch of 256 cycles per second (middle C)?

Pitch of a bell depends on its weight

CONCEPTS REVIEW
You should be able to give the meaning of each of the following concepts.

Cartesian (rectangular) coordinate system
Coordinate axes:
   x-axis
   y-axis
Coordinates of a point
Quadrants
Point:
   coordinates
Distance formula
Midpoint formula
Circle:
   standard form
   completing the square
Semicircle

Intercepts of a graph:
   x-intercept
   y-intercept
Symmetry of a graph:
   x-axis
   y-axis
   origin
Slope of a line:
   positive
   negative
   undefined
Equations of lines:
   point-slope form
   slope-intercept form
Vertical line

Horizontal line
Parallel lines
Perpendicular lines
Variation:
   direct
   inverse
   joint
   combined
   constant of proportionality

CHAPTER 2 Review Exercises
Answers to selected odd-numbered problems begin on page ANS-6.

A. True/False

In Problems 1–22, answer true or false.

1. The point $(5, 0)$ is in quadrant I. ___
2. The point $(-3, 7)$ is in quadrant III. ___
3. The points $(0, 3)$, $(2, 2)$, and $(6, 0)$ are collinear. ___
4. Two lines with positive slopes cannot be perpendicular. ___
5. The equation of a vertical line through $(2, -5)$ is $x = 2$. ___
6. If A, B, and C are points in the Cartesian plane, then it is always true that $d(A, B) + d(B, C) > d(A, C)$. ___
7. The lines $2x + 3y = 5$ and $-2x + 3y = 1$ are perpendicular. ___
8. The circle $(x + 1)^2 + (y - 1)^2 = 1$ is tangent to both the x- and y-axes. ___
9. The graph of the equation $y = x + x^3$ is symmetric with respect to the origin. ___
10. The center of the circle $x^2 + 4x + y^2 + 10y = 0$ is $(-2, -5)$. ___
11. The circle $x^2 + 4x + y^2 + 10y = 0$ passes through the origin. ___
12. If a line has undefined slope, then it must be vertical. ___
13. The circle $(x - 3)^2 + (y + 5)^2 = 2$ has no intercepts. ___
14. If $\left(-\frac{1}{2}, \frac{3}{2}\right)$ is on a line with slope 1, then $\left(\frac{1}{2}, -\frac{3}{2}\right)$ is also on the line. ___
15. The lines $y = 2x - 5$ and $y = 2x$ are parallel. ___
16. If y is inversely proportional to x, then y decreases as x increases. ___
17. The line through the points $(-1, 2)$ and $(4, 2)$ is horizontal. ___
18. Graphs of lines of the form $y = mx$, $m > 0$, cannot contain a point with a negative x-coordinate and a positive y-coordinate. ___
19. If the graph of an equation contains the point $(2, 3)$ and is symmetric with respect to the x-axis, then the graph also contains the point $(2, -3)$. ___
20. The graph of the equation $|x| = |y|$ is symmetric with respect to the x-axis, the y-axis, and the origin. ___

21. There is no point on the circle $x^2 + y^2 - 10x + 22 = 0$ with x-coordinate 2. ___
22. The radius r of the circle centered at the origin containing the point $(1, -2)$ is 5. ___

B. Fill in the Blanks

In Problems 1–20, fill in the blanks.

1. The lines $2x - 5y = 1$ and $kx + 3y + 3 = 0$ are parallel if $k =$ _____.
2. An equation of a line through $(1, 2)$ that is perpendicular to $y = 3x - 5$ is _____.
3. The slope and the x- and y-intercepts of the line $-4x + 3y - 48 = 0$ are _____.
4. The distance between the points $(5, 1)$ and $(-1, 9)$ is _____.
5. The slope of the line $4y = 6x + 3$ is $m =$ _____.
6. The lines $2x - 5y = 1$ and $kx + 3y + 3 = 0$ are perpendicular if $k =$ _____.
7. Two points on the circle $x^2 + y^2 = 25$ with the same y-coordinate -3 are _____.
8. The graph of $y = -6$ is a _____.
9. The center and the radius of the circle $(x - 2)^2 + (y + 7)^2 = 8$ are _____.
10. The point $(1, 5)$ is on a graph. Give the coordinates of another point on the graph if the graph is
    (a) symmetric with respect to the x-axis. _____
    (b) symmetric with respect to the y-axis. _____
    (c) symmetric with respect to the origin. _____
11. If $(-2, 6)$ is the midpoint of the line segment from $P_1(x_1, 3)$ to $P_2(8, y_2)$, then $x_1 =$ _____ and $y_2 =$ _____.
12. The midpoint of the line segment from $P_1(2, -5)$ to $P_2(8, -9)$ is _____.
13. The quadrants of the xy-plane in which the quotient x/y is negative are _____.
14. A line with x-intercept $(-4, 0)$ and y-intercept $(0, 32)$ has slope _____.
15. An equation of a line perpendicular to $y = 3$ and contains the point $(-2, 7)$ is _____.
16. If the point $(a, a + \sqrt{3})$ is on the graph of $y = 2x$, then $a =$ _____.
17. The graph of $y = -\sqrt{100 - x^2}$ is a _____.
18. The equation _____ is an example of a circle with center and both x-intercepts on the negative x-axis.
19. The distance from the midpoint of the line segment joining the points $(4, -6)$ and $(-2, 0)$ to the origin is _____.
20. If p varies inversely as the cube of q and $p = 9$ when $q = -1$, then $p =$ _____ when $q = 3$.

C. Review Exercises

1. Determine whether the points $A(1, 1)$, $B(3, 3)$, and $C(5, 1)$ are vertices of a right triangle.
2. Find an equation of a circle with the points $(3, 4)$ and $(5, 6)$ as the endpoints of a diameter.
3. Find an equation of the line through the origin perpendicular to the line through $(1, 1)$ and $(2, -2)$.
4. Find an equation of the line through $(2, 4)$ parallel to the line through $(-1, -1)$ and $(4, -3)$.

5. Consider the line segment joining $(-1, 6)$ and $(1, 10)$ and the line segment joining $(7, 3)$ and $(-3, -2)$. Find an equation of the line containing the midpoints of these two line segments.
6. Find an equation of the line that passes through $(3, -8)$ and is parallel to the line $2x - y = -7$.
7. Find two distinct points, other than the intercepts, on the line $2x + 5y = 12$.
8. The y-coordinate of a point is 2. Find the x-coordinate of the point if the distance from the point to $(1, 3)$ is $\sqrt{26}$.
9. Find an equation of the circle with center at the origin if the length of its diameter is 8.
10. Find an equation of the circle that has center $(1, 1)$ and passes through the point $(5, 2)$.
11. Find equations of the circles that pass through the points $(1, 3)$ and $(-1, -3)$ and have radius 10.
12. The point $(-3, b)$ is on the graph of $y + 2x + 10 = 0$. Find b.
13. Three vertices of a rectangle are $(3, 5)$, $(-3, 7)$, and $(-6, -2)$. Find the fourth vertex.
14. Find the point of intersection of the diagonals of the rectangle in Problem 13.

In Problems 15 and 16, solve for x.

15. $P_1(x, 2)$, $P_2(1, 1)$, $d(P_1, P_2) = \sqrt{10}$
16. $P_1(x, 0)$, $P_2(-4, 3x)$, $d(P_1, P_2) = 4$
17. Find an equation that relates x and y if it is known that the distance (x, y) to $(0, 1)$ is the same as the distance from (x, y) to $(x, -1)$.
18. Show that the point $(-1, 5)$ is on the perpendicular bisector of the line segment from $P_1(1, 1)$ to $P_2(3, 7)$.
19. If M is the midpoint of the line segment from $P_1(2, 3)$ to $P_2(6, -9)$, find the midpoint of the line segment from P_1 to M and the midpoint of the line segment from M to P_2.
20. **FIGURE 2.R.1** shows the midpoints of the sides of a triangle. Determine the vertices of the triangle.
21. A tangent line to a circle at a point P on the circle is a line through P that is perpendicular to the line through P and the center of the circle. Find an equation of the tangent line L indicated in **FIGURE 2.R.2**.

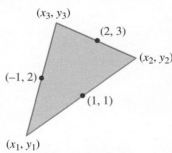

FIGURE 2.R.1 Triangle in Problem 20

FIGURE 2.R.2 Graph for Problem 21

22. Which of the following equations best describes the circle given in **FIGURE 2.R.3**? The symbols a, b, c, d, and e stand for nonzero constants.
    (a) $ax^2 + by^2 + cx + dy + e = 0$
    (b) $ax^2 + ay^2 + cx + dy + e = 0$
    (c) $ax^2 + ay^2 + cx + dy = 0$
    (d) $ax^2 + ay^2 + c = 0$
    (e) $ax^2 + ay^2 + cx + e = 0$

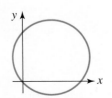

FIGURE 2.R.3 Graph for Problem 22

In Problems 23–30, match the given equation with the appropriate graph given in FIGURE 2.R.4.

(a), (b), (c), (d), (e), (f), (g), (h)

FIGURE 2.R.4 Graphs for Problems 23–30

23. $x + y - 1 = 0$
24. $x + y = 0$
25. $x - 1 = 0$
26. $y - 1 = 0$
27. $10x + y - 10 = 0$
28. $-10x + y + 10 = 0$
29. $x + 10y - 10 = 0$
30. $-x + 10y - 10 = 0$

3 Functions and Graphs

In This Chapter

3.1 Functions and Graphs
3.2 Symmetry and Transformations
3.3 Linear and Quadratic Functions
3.4 Piecewise-Defined Functions
3.5 Combining Functions
3.6 Inverse Functions
3.7 Building a Function from Words
3.8 Least Squares Line

Chapter 3 Review Exercises

The correspondence between the students in a class and the set of desks filled by these students is an example of a function.

A Bit of History If the question "What is the most important mathematical concept?" were posed to a group of mathematicians, mathematics teachers, and scientists, certainly the term *function* would appear near or even at the top of the list of their responses. In Chapters 3 and 4, we will focus primarily on the definition and the graphical interpretation of a function.

The word *function* was probably introduced by the German mathematician and "co-inventor" of calculus, **Gottfried Wilhelm Leibniz** (1646–1716), in the late seventeenth century and stems from the Latin word *functo*, meaning to act or perform. In the seventeenth and eighteenth centuries, mathematicians had only the most intuitive notion of a function. To many of them, a functional relationship between two variables was given by some smooth curve or by an equation involving the two variables. Although formulas and equations play an important role in the study of functions, we will see in Section 3.1 that the "modern" interpretation of a function (dating from the middle of the nineteenth century) is that of a special type of correspondence between the elements of two sets.

3.1 Functions and Graphs

≡ **Introduction** Using the objects and the persons around us, it is easy to make up a rule of correspondence that associates, or pairs, the members, or elements, of one set with the members of another set. For example, to each social security number there is a person, to each car registered in the state of California there is a license plate number, to each book there corresponds at least one author, to each state there is a governor, and so on. A natural correspondence occurs between a set of 20 students and a set of, say, 25 desks in a classroom when each student selects and sits in a different desk. In mathematics we are interested in a special type of correspondence, a *single-valued correspondence*, called a function.

DEFINITION 3.1.1 Function

A **function** from a set X to a set Y is a rule of correspondence that assigns to each element x in X exactly one element y in Y.

In the student/desk correspondence above suppose the set of 20 students is the set X and the set of 25 desks is the set Y. This correspondence is a function from the set X to the set Y provided no student sits in two desks at the same time.

☐ **Terminology** A function is usually denoted by a letter such as f, g, or h. We can then represent a function f from a set X to a set Y by the notation $f: X \to Y$. The set X is called the **domain** of f. The set of corresponding elements y in the set Y is called the **range** of the function. For our student/desk function, the set of students is the domain and the set of 20 desks actually occupied by the students constitutes the range. Notice that the range of f need not be the entire set Y. The unique element y in the range that corresponds to a selected element x in the domain X is called the **value** of the function at x, or the **image** of x, and is written $f(x)$. The latter symbol is read "f of x" or "f at x," and we write $y = f(x)$. See FIGURE 3.1.1. In many texts, x is also called the **input** of the function f and the value $f(x)$ is called the **output** of f. Since the value of y depends on the choice of x, y is called the **dependent variable**; x is called the **independent variable**. Unless otherwise stated, we will assume hereafter that the sets X and Y consist of real numbers.

The set Y is not necessarily the range

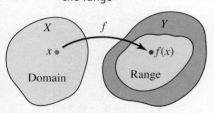

FIGURE 3.1.1 Domain and range of a function f

EXAMPLE 1 The Squaring Function

The rule for squaring a real number is given by the equation $y = x^2$ or $f(x) = x^2$. The values of f at $x = -5$ and $x = \sqrt{7}$ are obtained by replacing x, in turn, by the numbers -5 and $\sqrt{7}$:

$$f(-5) = (-5)^2 = 25 \quad \text{and} \quad f(\sqrt{7}) = (\sqrt{7})^2 = 7.$$ ≡

Occasionally for emphasis we will write a function using parentheses in place of the symbol x. For example, we can write the squaring function $f(x) = x^2$ as

$$f(\ ) = (\ )^2. \tag{1}$$

This illustrates the fact that x is a *placeholder* for any number in the domain of the function $y = f(x)$. Thus, if we wish to evaluate (1) at, say, $3 + h$, where h represents a real number, we put $3 + h$ into the parentheses and carry out the appropriate algebra:

$$f(3 + h) = (3 + h)^2 = 9 + 6h + h^2.$$

See (3) of Section R.6. ▶

If a function f is defined by means of a formula or an equation, then typically the domain of $y = f(x)$ is not expressly stated. We will see that we can usually deduce the domain of $y = f(x)$ either from the structure of the equation or from the context of the problem.

EXAMPLE 2 Domain and Range

In Example 1, since any real number x can be squared and the result x^2 is another real number, $f(x) = x^2$ is a function from R to R, that is, $f: R \to R$. In other words, the domain of f is the set R of real numbers. Using interval notation, we also write the domain as $(-\infty, \infty)$. The range of f is the set of nonnegative real numbers or $[0, \infty)$; this follows from the fact that $x^2 \geq 0$ for every real number x. ≡

☐ **Domain of a Function** As mentioned earlier, the domain of a function $y = f(x)$ that is defined by a formula is usually not specified. Unless stated or implied to the contrary, it is understood that

> *The domain of a function f is the largest subset of the set of real numbers for which $f(x)$ is a real number.*

This set is sometimes referred to as the **implicit domain** of the function. For example, we cannot compute $f(0)$ for the reciprocal function $f(x) = 1/x$ since $1/0$ is not a real number. In this case we say that f is **undefined** at $x = 0$. Since every *nonzero* real number has a reciprocal, the domain of $f(x) = 1/x$ is the set of real numbers except 0. By the same reasoning, the function $g(x) = 1/(x^2 - 4)$ is not defined at either $x = -2$ or $x = 2$, and so its domain is the set of real numbers with the numbers -2 and 2 excluded. The square root function $h(x) = \sqrt{x}$ is not defined at $x = -1$ because $\sqrt{-1}$ is not a real number. In order for $h(x) = \sqrt{x}$ to be defined in the real number system we must require the **radicand**, in this case simply x, to be nonnegative. From the inequality $x \geq 0$ we see that the domain of the function h is the interval $[0, \infty)$.

EXAMPLE 3 Domain and Range

Determine the domain and range of $f(x) = 4 + \sqrt{x - 3}$.

Solution The radicand $x - 3$ must be nonnegative. By solving the inequality $x - 3 \geq 0$ we get $x \geq 3$, and so the domain of f is $[3, \infty)$. Now, since the symbol $\sqrt{}$ denotes the principal square root of a number, $\sqrt{x - 3} \geq 0$ for $x \geq 3$ and consequently $4 + \sqrt{x - 3} \geq 4$. The smallest value of $f(x)$ occurs at $x = 3$ and is $f(3) = 4 + \sqrt{0} = 4$. Moreover, because $x - 3$ and $\sqrt{x - 3}$ increase as x takes on increasingly larger values, we conclude that $y \geq 4$. Consequently the range of f is $[4, \infty)$. ≡

◀ See Section R.4.

EXAMPLE 4 Domain of f

Determine the domain of $f(x) = \sqrt{x^2 + 2x - 15}$.

Solution As in Example 3, the expression under the radical symbol—the radicand—must be nonnegative; that is, the domain of f is the set of real numbers x for which $x^2 + 2x - 15 \geq 0$ or $(x - 3)(x + 5) \geq 0$. We have already solved the last inequality by means of a sign chart in Example 1 of Section 1.7. The solution set of the inequality $(-\infty, -5] \cup [3, \infty)$ is the domain of f. ≡

EXAMPLE 5 Domains of Two Functions

Determine the domain of **(a)** $g(x) = \dfrac{1}{\sqrt{x^2 + 2x - 15}}$ and **(b)** $h(x) = \dfrac{5x}{x^2 - 3x - 4}$.

Solution A function that is given by a fractional expression is not defined at the x-values for which its denominator is equal to 0.

(a) The expression under the radical is the same as in Example 4. Since $x^2 + 2x - 15$ is in the denominator we must have $x^2 + 2x - 15 \neq 0$. This excludes $x = -5$ and $x = 3$. In addition, since $x^2 + 2x - 15$ appears under a radical, we must have $x^2 + 2x - 15 > 0$ for all other values of x. Thus the domain of the function g is the union of two open intervals $(-\infty, -5) \cup (3, \infty)$.

(b) Since the denominator of $h(x)$ factors, $x^2 - 3x - 4 = (x + 1)(x - 4)$, we see that $(x + 1)(x - 4) = 0$ for $x = -1$ and $x = 4$. In contrast to the function in part (a), these are the *only* numbers for which h is not defined. Hence, the domain of the function h is the set of real numbers with $x = -1$ and $x = 4$ excluded. ≡

Using interval notation, the domain of the function h in part (b) of Example 5 can be written as

$$(-\infty, -1) \cup (-1, 4) \cup (4, \infty).$$

As an alternative to this ungainly union of disjoint intervals, this domain can also be written using set-builder notation as $\{x \mid x \text{ real}, x \neq -1 \text{ and } x \neq 4\}$.

☐ **Graphs** A function is often used to describe phenomena in fields such as science, engineering, and business. In order to interpret and utilize data, it is useful to display this data in the form of a graph. The graph of a function f is the graph of the set of ordered pairs $(x, f(x))$, where x is in the domain of f. In the xy-plane an ordered pair $(x, f(x))$ is a point, so that the graph of a function is a set of points. If a function is defined by an equation $y = f(x)$, then the graph of f is the graph of the equation. To obtain points on the graph of an equation $y = f(x)$, we judiciously choose numbers $x_1, x_2, x_3, \ldots$ in its domain, compute $f(x_1), f(x_2), f(x_3), \ldots$, plot the corresponding points $(x_1, f(x_1)), (x_2, f(x_2)), (x_3, f(x_3)), \ldots$ and then connect these points with a curve. See **FIGURE 3.1.2**. Keep in mind that

- a value of x is a directed distance from the y-axis, and
- a function value $f(x)$ is a directed distance from the x-axis.

FIGURE 3.1.2 Points on the graph of an equation $y = f(x)$

☐ **End Behavior** A word about the figures in this text is in order. With a few exceptions, it is usually impossible to display the complete graph of a function, and so we often display only the more important features of the graph. In **FIGURE 3.1.3(a)**, notice that the graph goes down on its left and right sides. Unless indicated to the contrary, we may assume that there are no major surprises beyond what we have shown and the graph simply continues in the manner indicated. The graph in Figure 3.1.3(a) indicates the so-called **end behavior** or **global behavior** of the function: For a point (x, y) on the graph, the values of the y-coordinate become unbounded in magnitude in the downward or negative direction as the x-coordinate becomes unbounded in magnitude in both the negative and positive directions on the number line. It is convenient to describe this end behavior using the symbols

$$y \to -\infty \text{ as } x \to -\infty \quad \text{and} \quad y \to -\infty \text{ as } x \to \infty. \tag{2}$$

The arrow symbol $\to$ in (2) is read "approaches." Thus, for example, $y \to -\infty$ as $x \to \infty$ is read "y approaches negative infinity as x approaches infinity."

(a) Function (b) Not a function (c) Not a function

FIGURE 3.1.3 Vertical line test

More will be said about this concept of global behavior in Chapter 4. If a graph terminates at either its right or left end, we will indicate this by a dot when clarity demands it. See **FIGURE 3.1.4**. We will use a solid dot to represent the fact that the endpoint is included on the graph and an open dot to signify that the endpoint is not included on the graph.

☐ **Vertical Line Test** From the definition of a function we know that for each x in the domain of f there corresponds only one value $f(x)$ in the range. This means a vertical line that intersects the graph of a function $y = f(x)$ (this is equivalent to choosing an x) can do so in at most one point. Conversely, if *every* vertical line that intersects a graph of an equation does so in at most one point, then the graph is the graph of a function. The last statement is called the **vertical line test** for a function. See Figure 3.1.3(a). On the other hand, if *some* vertical line intersects a graph of an equation more than once, then the graph is not that of a function. See Figures 3.1.3(b) and 3.1.3(c). When a vertical line intersects a graph in several points, the same number x corresponds to different values of y in contradiction to the definition of a function.

FIGURE 3.1.4 Domain and range interpreted graphically

If you have an accurate graph of a function $y = f(x)$, it is often possible to *see* the domain and range of f. In Figure 3.1.4 assume that the colored curve is the entire, or complete, graph of some function f. The domain of f then is the interval $[a, b]$ on the x-axis and the range is the interval $[c, d]$ on the y-axis.

EXAMPLE 6 Example 3 Revisited

From the graph of $f(x) = 4 + \sqrt{x - 3}$ given in **FIGURE 3.1.5**, we can see that the domain and range of f are, respectively, $[3, \infty)$ and $[4, \infty)$. This agrees with the results in Example 3. ≡

As shown in Figure 3.1.3(b), a circle is not the graph of a function. Actually, an equation such as $x^2 + y^2 = 9$ defines (at least) two functions of x. If we solve this equation for y in terms of x we get $y = \pm\sqrt{9 - x^2}$. Because of the single-valued convention for the $\sqrt{}$ symbol, both equations $y = \sqrt{9 - x^2}$ and $y = -\sqrt{9 - x^2}$ define functions. As we saw in Section 2.2, the first equation defines an *upper semicircle* and the second defines a *lower semicircle*. From the graphs shown in **FIGURE 3.1.6**,

FIGURE 3.1.5 Graph of function f in Example 6

(a) Upper semicircle (b) Lower semicircle

FIGURE 3.1.6 These semicircles are graphs of functions

3.1 Functions and Graphs

the domain of $y = \sqrt{9 - x^2}$ is $[-3, 3]$ and the range is $[0, 3]$; the domain and range of $y = -\sqrt{9 - x^2}$ are $[-3, 3]$ and $[-3, 0]$, respectively.

☐ **Intercepts** To graph a function defined by an equation $y = f(x)$, it is usually a good idea to first determine whether the graph of f has any intercepts. Recall that all points on the y-axis are of the form $(0, y)$. Thus, if 0 is in the domain of a function f, the **y-intercept** is the point on the y-axis whose y-coordinate is $f(0)$, in other words, $(0, f(0))$. See FIGURE 3.1.7(a). Similarly, all points on the x-axis have the form $(x, 0)$. This means that to find the x-intercepts of the graph of $y = f(x)$, we determine the values of x that make $y = 0$. That is, we must solve the equation $f(x) = 0$ for x. A number c for which

$$f(c) = 0$$

is referred to as either a **zero** of the function f or a **root** (or **solution**) of the equation $f(x) = 0$. The *real* zeros of a function f are the x-coordinates of the **x-intercepts** of the graph of f. In Figure 3.1.7(b), we have illustrated a function that has three zeros x_1, x_2, and x_3 because $f(x_1) = 0$, $f(x_2) = 0$, and $f(x_3) = 0$. The corresponding three x-intercepts are the points $(x_1, 0)$, $(x_2, 0)$, and $(x_3, 0)$. Of course, the graph of the function may have no intercepts. This is illustrated in Figure 3.1.5.

(a) One y-intercept (b) Three x-intercepts (c) One y-intercept, two x-intercepts

FIGURE 3.1.7 Intercepts of the graph of a function f

More will be said about this in Chapter 4. ▶

A graph does not necessarily have to *cross* a coordinate axis at an intercept, a graph could simply be **tangent** to, or *touch*, an axis. In Figure 3.1.7(c) the graph of $y = f(x)$ is tangent to the x-axis at $(x_1, 0)$. Also, the graph of a function f can have at most one y-intercept since, if 0 is the domain of f, there can correspond only one y-value, namely, $y = f(0)$.

EXAMPLE 7 Intercepts

Find, if possible, the x- and y-intercepts of the given function.

(a) $f(x) = x^2 + 2x - 2$ **(b)** $f(x) = \dfrac{x^2 - 2x - 3}{x}$

Solution (a) Since 0 is in the domain of f, $f(0) = -2$ is the y-coordinate of the y-intercept of the graph of f. The y-intercept is the point $(0, -2)$. To obtain the x-intercepts we must determine whether f has any real zeros, that is, real solutions of the equation $f(x) = 0$. Since the left-hand side of the equation $x^2 + 2x - 2 = 0$ has no obvious factors, we use the quadratic formula to obtain $x = \frac{1}{2}(2 \pm \sqrt{12})$. Since $\sqrt{12} = \sqrt{4 \cdot 3} = 2\sqrt{3}$ the zeros of f are the numbers $1 - \sqrt{3}$ and $1 + \sqrt{3}$. The x-intercepts are the points $(1 - \sqrt{3}, 0)$ and $(1 + \sqrt{3}, 0)$.

(b) Because 0 is not in the domain of f ($f(0) = -3/0$ is not defined), the graph of f possesses no y-intercept. Now since f is a fractional expression, the only way we can have $f(x) = 0$ is to have the numerator equal zero. Factoring the left-hand side of $x^2 - 2x - 3 = 0$ gives $(x + 1)(x - 3) = 0$. Therefore the numbers -1 and 3 are the zeros of f. The x-intercepts are the points $(-1, 0)$ and $(3, 0)$. ≡

☐ **Approximating Zeros** Even when it is obvious that the graph of a function $y = f(x)$ possesses x-intercepts it is not always a straightforward matter to solve the equation $f(x) = 0$. In fact, it is *impossible* to solve some equations exactly; sometimes the best we can do is to **approximate** the zeros of the function. One way of doing this is to obtain a very accurate graph of f.

◀ We will study another way of approximating zeros of a function in Section 4.5.

EXAMPLE 8   Intercepts

With the aid of a graphing utility the graph of the function $f(x) = x^3 - x + 4$ is given in FIGURE 3.1.8. From $f(0) = 4$ we see that the y-intercept is $(0, 4)$. As we see in the figure, there appears to be only one x-intercept with its x-coordinate close to -1.7 or -1.8. But there is no convenient way of finding the exact values of the roots of the equation $x^3 - x + 4 = 0$. We can, however, approximate the real root of this equation with the aid of the *find root* feature of either a graphing calculator or computer algebra system. We find that $x \approx -1.796$ and so the approximate x-intercept is $(-1.796, 0)$. As a check, note that the function value

$$f(-1.796) = (-1.796)^3 - (-1.796) + 4 \approx 0.0028$$

is nearly 0.

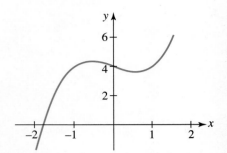

FIGURE 3.1.8 Approximate x-intercept in Example 8

NOTES FROM THE CLASSROOM

When sketching the graph of a function, you should never resort to plotting a lot of points by hand. That is something a graphing calculator or a computer algebra system (CAS) does so well. On the other hand, you should not become dependent on a calculator to obtain a graph. Believe it or not, there are instructors who do not allow the use of graphing calculators on quizzes or tests. Usually there is no objection to your using calculators or computers as an aid in checking homework problems, but in the classroom instructors want to see the product of your own mind, namely, the ability to analyze. So you are strongly encouraged to develop your graphing skills to the point where you are able to quickly sketch by hand the graph of a function from a basic familiarity of types of functions and by plotting a minimum of well-chosen points, and by using the transformations introduced in the next section.

3.1 Exercises
Answers to selected odd-numbered problems begin on page ANS-7.

In Problems 1–6, find the indicated function values.

1. If $f(x) = x^2 - 1$; $f(-5), f(-\sqrt{3}), f(3),$ and $f(6)$
2. If $f(x) = -2x^2 + x$; $f(-5), f(-\frac{1}{2}), f(2),$ and $f(7)$
3. If $f(x) = \sqrt{x + 1}$; $f(-1), f(0), f(3),$ and $f(5)$

4. If $f(x) = \sqrt{2x+4}$; $f(-\tfrac{1}{2}), f(\tfrac{1}{2}), f(\tfrac{5}{2})$, and $f(4)$
5. If $f(x) = \dfrac{3x}{x^2+1}$; $f(-1), f(0), f(1)$, and $f(\sqrt{2})$
6. If $f(x) = \dfrac{x^2}{x^3-2}$; $f(-\sqrt{2}), f(-1), f(0)$, and $f(\tfrac{1}{2})$

In Problems 7 and 8, find

$$f(x), f(2a), f(a^2), f(-5x), f(2a+1), \text{ and } f(x+h)$$

for the given function f and simplify as much as possible.

7. $f(\ ) = -2(\ )^2 + 3(\ )$
8. $f(\ ) = (\ )^3 - 2(\ )^2 + 20$

9. For what values of x is $f(x) = 6x^2 - 1$ equal to 23?
10. For what values of x is $f(x) = \sqrt{x-4}$ equal to 4?

In Problems 11–20, find the domain of the given function f.

11. $f(x) = \sqrt{4x-2}$
12. $f(x) = \sqrt{15-5x}$
13. $f(x) = \dfrac{10}{\sqrt{1-x}}$
14. $f(x) = \dfrac{2x}{\sqrt{3x-1}}$
15. $f(x) = \dfrac{2x-5}{x(x-3)}$
16. $f(x) = \dfrac{x}{x^2-1}$
17. $f(x) = \dfrac{1}{x^2-10x+25}$
18. $f(x) = \dfrac{x+1}{x^2-4x-12}$
19. $f(x) = \dfrac{x}{x^2-x+1}$
20. $f(x) = \dfrac{x^2-9}{x^2-2x-1}$

In Problems 21–26, use the sign-chart method to find the domain of the given function f.

21. $f(x) = \sqrt{25-x^2}$
22. $f(x) = \sqrt{x(4-x)}$
23. $f(x) = \sqrt{x^2-5x}$
24. $f(x) = \sqrt{x^2-3x-10}$
25. $f(x) = \sqrt{\dfrac{3-x}{x+2}}$
26. $f(x) = \sqrt{\dfrac{5-x}{x}}$

In Problems 27–30, determine whether the graph in the figure is the graph of a function.

27.

FIGURE 3.1.9 Graph for Problem 27

28.

FIGURE 3.1.10 Graph for Problem 28

CHAPTER 3 FUNCTIONS AND GRAPHS

29.

FIGURE 3.1.11 Graph for Problem 29

30.

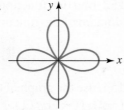

FIGURE 3.1.12 Graph for Problem 30

In Problems 31–34, use the graph of the function f given in the figure to find its domain and range.

31.

FIGURE 3.1.13 Graph for Problem 31

32.

FIGURE 3.1.14 Graph for Problem 32

33.

FIGURE 3.1.15 Graph for Problem 33

34.

FIGURE 3.1.16 Graph for Problem 34

In Problems 35–42, find the zeros of the given function f.

35. $f(x) = 5x + 6$
36. $f(x) = -2x + 9$
37. $f(x) = x^2 - 5x + 6$
38. $f(x) = x^2 - 2x - 1$
39. $f(x) = x(3x - 1)(x + 9)$
40. $f(x) = x^3 - x^2 - 2x$
41. $f(x) = x^4 - 1$
42. $f(x) = 2 - \sqrt{4 - x^2}$

In Problems 43–50, find the x- and y-intercepts, if any, of the graph of the given function f. Do not graph.

43. $f(x) = \frac{1}{2}x - 4$
44. $f(x) = x^2 - 6x + 5$
45. $f(x) = 4(x - 2)^2 - 1$
46. $f(x) = (2x - 3)(x^2 + 8x + 16)$
47. $f(x) = \dfrac{x^2 + 4}{x^2 - 16}$
48. $f(x) = \dfrac{x(x + 1)(x - 6)}{x + 8}$
49. $f(x) = \frac{3}{2}\sqrt{4 - x^2}$
50. $f(x) = \frac{1}{2}\sqrt{x^2 - 2x - 3}$

In Problems 51 and 52, find two functions $y = f_1(x)$ and $y = f_2(x)$ defined by the given equation. Find the domain of the functions f_1 and f_2.

51. $x = y^2 - 5$

52. $x^2 - 4y^2 = 16$

In Problems 53 and 54, use the graph of the function f given in the figure to estimate the values of $f(-3), f(-2), f(-1), f(1), f(2),$ and $f(3)$. Estimate the y-intercept.

53.
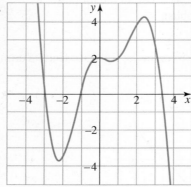
FIGURE 3.1.17 Graph for Problem 53

54.
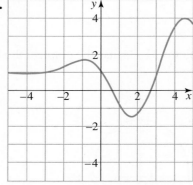
FIGURE 3.1.18 Graph for Problem 54

In Problems 55 and 56, use the graph of the function f given in the figure to estimate the values of $f(-2), f(-1.5), f(0.5), f(1), f(2),$ and $f(3.2)$. Estimate the x-intercepts.

55.
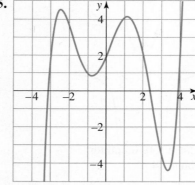
FIGURE 3.1.19 Graph for Problem 55

56.

FIGURE 3.1.20 Graph for Problem 56

57. Factorial Function In your study of mathematics some of the functions that you will encounter have as their domain the set of positive integers n. The factorial function $f(n) = n!$ is defined as the product of the first n positive integers, that is,
$$f(n) = n! = 1 \cdot 2 \cdot 3 \cdots (n-1) \cdot n.$$
(a) Evaluate $f(2), f(3), f(5),$ and $f(7)$.
(b) Show that $f(n+1) = f(n) \cdot (n+1)$.
(c) Simplify $f(n+2)/f(n)$.

58. A Sum Function Another function of a positive integer n gives the sum of the first n squared positive integers:
$$S(n) = \frac{1}{6}n(n+1)(2n+1) = 1^2 + 2^2 + \cdots + n^2.$$
(a) Find the value of the sum $1^2 + 2^2 + \cdots + 99^2 + 100^2$.
(b) Find n such that $300 < S(n) < 400$. [*Hint*: Use a calculator.]

CHAPTER 3 FUNCTIONS AND GRAPHS

For Discussion

59. Determine an equation of a function $y = f(x)$ whose domain is (a) $[3, \infty)$, (b) $(3, \infty)$.
60. Determine an equation of a function $y = f(x)$ whose range is (a) $[3, \infty)$, (b) $(3, \infty)$.
61. What is the only point that can be both an x- and a y-intercept for the graph of a function $y = f(x)$?
62. Consider the function $f(x) = \dfrac{x-1}{x^2-1}$. After factoring the denominator and canceling a common factor we can write $f(x) = \dfrac{1}{x+1}$. Discuss: Is $x = 1$ in the domain of $f(x) = \dfrac{1}{x+1}$?

3.2 Symmetry and Transformations

☰ **Introduction** In this section we discuss two aids in sketching graphs of functions quickly and accurately. If you determine in advance that the graph of a function possesses *symmetry*, then you can cut your work in half. In addition, sketching a graph of a complicated-looking function is expedited if you recognize that the required graph is actually a *transformation* of the graph of a simpler function. This latter graphing aid is based on your prior knowledge of the graphs of some basic functions.

☐ **Power Functions** A function of the form

$$f(x) = x^n$$

where n represents a real number is called a **power function**. The domain of a power function depends on the power n. For example, we have already seen in Section 3.1 for $n = 2, n = \frac{1}{2}$, and $n = -1$, respectively, that

- the domain of $f(x) = x^2$ is the set R of real numbers or $(-\infty, \infty)$,
- the domain of $f(x) = x^{1/2} = \sqrt{x}$ is $[0, \infty)$, and
- the domain of $f(x) = x^{-1} = \dfrac{1}{x}$ is the set R of real numbers except $x = 0$.

Simple power functions, or modified versions of these functions, occur so often in problems that you do not want to spend valuable time plotting their graphs. We suggest that you know (memorize) the short catalogue of graphs of power functions given in FIGURE 3.2.1. You already know that the graph in part (a) of that figure is a **line** and may know that the graph in part (b) is called a **parabola**.

☐ **Symmetry** In Section 2.2 we discussed symmetry of a graph with respect to the y-axis, the x-axis, and the origin. Of those three types of symmetries, the graph of a function can be symmetric with respect to the y-axis or with respect to the origin, but the graph of a nonzero function *cannot* be symmetric with respect to the x-axis. See Problem 43 in Exercises 3.2. If the graph of a function is symmetric with respect to the

◀ Can you explain *why* the graph of a function cannot have symmetry with respect to the x-axis?

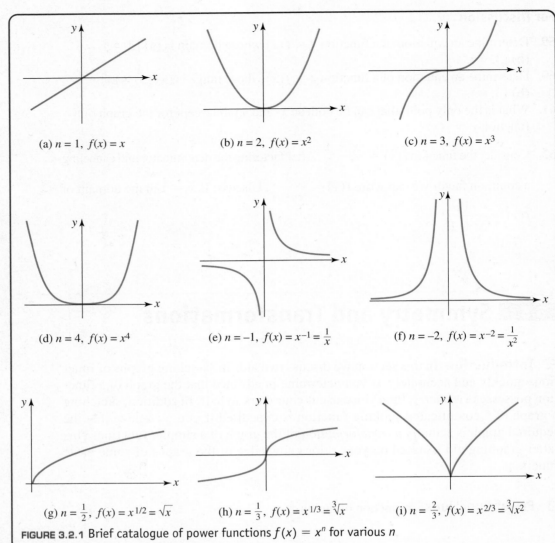

FIGURE 3.2.1 Brief catalogue of power functions $f(x) = x^n$ for various n

y-axis, then as we know the points (x, y) and $(-x, y)$ are on the graph of f. Similarly, if the graph of a function is symmetric with respect to the origin, the points (x, y) and $(-x, -y)$ are on its graph. For functions, the following two tests for symmetry are equivalent to tests (*i*) and (*ii*), respectively, on page 136.

DEFINITION 3.2.1 Even and Odd Functions

Suppose that for every x in the domain of a function f, that $-x$ is also in its domain.

(*i*) A function f is said to be **even** if $f(-x) = f(x)$.
(*ii*) A function f is said to be **odd** if $f(-x) = -f(x)$.

FIGURE 3.2.2 Even function

In **FIGURE 3.2.2**, observe that if f is an even function and

$$f(x) \qquad\qquad f(-x)$$
$$\downarrow \qquad\qquad \downarrow$$

(x, y) is a point on its graph, then necessarily $(-x, y)$ \hfill (1) is also on its graph. Similarly we see in FIGURE 3.2.3 that if f is an odd function and

$$\underset{\downarrow}{f(x)} \qquad \underset{\downarrow}{f(-x) = -f(x)}$$

(x, y) is a point on its graph, then necessarily $(-x, -y)$ \qquad (2)

is on its graph. We have proved the following result.

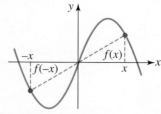

FIGURE 3.2.3 Odd function

THEOREM 3.2.1 Symmetry

(*i*) A function f is even if and only if its graph is symmetric with respect to the y-axis.
(*ii*) A function f is odd if and only if its graph is symmetric with respect to the origin.

Inspection of Figures 3.2.2 and 3.2.3 shows that the graphs, in turn, are symmetric with respect to the y-axis and origin. The function whose graph is given in FIGURE 3.2.4 is neither even nor odd, and so its graph possesses no y-axis or origin symmetry.

In view of Definition 3.2.1 and Theorem 3.2.1 we can determine symmetry of a graph of a function in an algebraic manner.

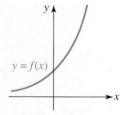

FIGURE 3.2.4 Function is neither odd nor even

EXAMPLE 1 Odd and Even Functions

(a) $f(x) = x^3$ is an odd function since by Definition 3.2.1(*ii*),

$$f(-x) = (-x)^3 = (-1)^3 x^3 = -x^3 = -f(x).$$

This proves what we see in Figure 3.2.1(c), the graph of $f(x) = x^3$ is symmetric with respect to the origin. For example, since $f(1) = 1$, $(1, 1)$ is a point on the graph of $y = x^3$. Because f is an odd function, $f(-1) = -f(1)$ implies $(-1, -1)$ is on the same graph.

(b) $f(x) = x^{2/3}$ is an even function since by Definition 3.2.1(*i*) and the laws of exponents

$$f(-x) = (-x)^{2/3} = (-1)^{2/3} x^{2/3} = \overset{\text{cube root of } -1 \text{ is } -1}{(\sqrt[3]{-1})^2 x^{2/3}} = (-1)^2 x^{2/3} = x^{2/3} = f(x).$$

In Figure 3.2.1(i), we see that the graph of f is symmetric with respect to the y-axis. For example, since $f(8) = 8^{2/3} = 4$, $(8, 4)$ is a point on the graph of $y = x^{2/3}$. Because f is an even function, $f(-8) = f(8)$ implies $(-8, 4)$ is also on the same graph.

(c) $f(x) = x^3 + 1$ is neither even nor odd. From

$$f(-x) = (-x)^3 + 1 = -x^3 + 1$$

we see that $f(-x) \neq f(x)$, and $f(-x) \neq -f(x)$. Hence the graph of f is neither symmetric with respect to the y-axis nor symmetric with respect to the origin. ≡

The graphs in Figure 3.2.1, with part (g) the only exception, possess either y-axis or origin symmetry. The functions in Figures 3.2.1(b), (d), (f), and (i) are even, whereas the functions in Figures 3.2.1(a), (c), (e), and (h) are odd.

Often we can sketch the graph of a function by applying a certain transformation to the graph of a simpler function (such as those given in Figure 3.2.1). We will consider two kinds of graphical transformations, rigid and nonrigid.

3.2 Symmetry and Transformations

☐ **Rigid Transformations** A **rigid transformation** of a graph is one that changes only the *position* of the graph in the *xy*-plane but not its shape. For example, the circle $(x - 2)^2 + (y - 3)^2 = 1$ with center (2, 3) and radius $r = 1$, has *exactly* the same shape as the circle $x^2 + y^2 = 1$ with center at the origin. Thus we can think of the graph of $(x - 2)^2 + (y - 3)^2 = 1$ as the graph of $x^2 + y^2 = 1$ shifted horizontally 2 units to the right followed by an upward vertical shift of 3 units. For the graph of a function $y = f(x)$ we examine four kinds of shifts or translations.

THEOREM 3.2.2 Vertical and Horizontal Shifts

Suppose $y = f(x)$ is a function and c is a positive constant. Then the graph of

(*i*) $y = f(x) + c$ is the graph of f shifted vertically **up** c units,
(*ii*) $y = f(x) - c$ is the graph of f shifted vertically **down** c units,
(*iii*) $y = f(x + c)$ is the graph of f shifted horizontally to the **left** c units,
(*iv*) $y = f(x - c)$ is the graph of f shifted horizontally to the **right** c units.

FIGURE 3.2.5 Graph of $y = f(x)$

Consider the graph of a function $y = f(x)$ given in FIGURE 3.2.5. The shifts of this graph described in (*i*)–(*iv*) of Theorem 3.2.2 are the graphs in red in parts (a)–(d) of FIGURE 3.2.6. If (x, y) is a point on the graph of $y = f(x)$ and the graph of f is shifted, say, upward by $c > 0$ units, then $(x, y + c)$ is a point on the new graph. In general, the *x*-coordinates do not change as a result of a vertical shift. See Figures 3.2.6(a) and 3.2.6(b). Similarly, in a horizontal shift the *y*-coordinates of points on the shifted graph are the same as on the original graph. See Figures 3.2.6(c) and 3.2.6(d).

(a) Vertical shift up

(b) Vertical shift down

(c) Horizontal shift left

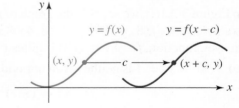
(d) Horizontal shift right

FIGURE 3.2.6 Vertical and horizontal shifts of the graph of $y = f(x)$ by an amount $c > 0$.

EXAMPLE 2   Vertical and Horizontal Shifts

The graphs of $y = x^2 + 1$, $y = x^2 - 1$, $y = (x + 1)^2$, and $y = (x - 1)^2$ are obtained from the blue graph of $f(x) = x^2$ in FIGURE 3.2.7(a) by shifting this graph, in turn, 1 unit up (Figure 3.2.7(b)), 1 unit down (Figure 3.2.7(c)), 1 unit to the left (Figure 3.2.7(d)), and 1 unit to the right (Figure 3.2.7(e)).

(a) Starting point

(b) Shift up

(c) Shift down

(d) Shift left

(e) Shift right

FIGURE 3.2.7 Shifted graphs in red in Example 2

☐ **Combining Shifts** In general, the graph of a function

$$y = f(x \pm c_1) \pm c_2, \quad (3)$$

where c_1 and c_2 are positive constants, combines a horizontal shift (left or right) with a vertical shift (up or down). For example, the graph of $y = f(x - c_1) + c_2$ is the graph of $y = f(x)$ shifted c_1 units to the right and then c_2 units up.

◀ The order in which the shifts are done is irrelevant. We could do the upward shift first followed by the shift to the right.

EXAMPLE 3 Graph Shifted Horizontally and Vertically

Graph $y = (x + 1)^2 - 1$.

Solution From the preceding paragraph we identify in (3) the form $y = f(x + c_1) - c_2$ with $c_1 = 1$ and $c_2 = 1$. Thus, the graph of $y = (x + 1)^2 - 1$ is the graph of $f(x) = x^2$ shifted 1 unit to the left followed by a downward shift of 1 unit. The graph is given in **FIGURE 3.2.8**.

FIGURE 3.2.8 Shifted graph in Example 3

From the graph in Figure 3.2.8 we see immediately that the range of the function $y = (x + 1)^2 - 1 = x^2 + 2x$ is the interval $[-1, \infty)$ on the y-axis. Note also that the graph has x-intercepts $(0, 0)$ and $(-2, 0)$; you should verify this by solving $x^2 + 2x = 0$. Also, if you reexamine Figure 3.1.5 in Section 3.1 you will see that the graph of $y = 4 + \sqrt{x - 3}$ is the graph of the square root function $f(x) = \sqrt{x}$ (Figure 3.2.1(g)) shifted 3 units to the right and then 4 units up.

Another way of rigidly transforming a graph of a function is by a **reflection** in a coordinate axis.

THEOREM 3.2.3 Reflections

Suppose $y = f(x)$ is a function. Then the graph of

(i) $y = -f(x)$ is the graph of f reflected in the **x-axis**,
(ii) $y = f(-x)$ is the graph of f reflected in the **y-axis**.

In part (a) of **FIGURE 3.2.9** we have reproduced the graph of a function $y = f(x)$ given in Figure 3.2.5. The reflections of this graph described in parts (i) and (ii) of Theorem 3.2.3 are illustrated in Figures 3.2.9(b) and 3.2.9(c). If (x, y) denotes a point on the graph of $y = f(x)$, then the point $(x, -y)$ is on the graph of $y = -f(x)$ and $(-x, y)$ is on the graph of $y = f(-x)$. Each of these reflections is a mirror image of the graph of $y = f(x)$ in the respective coordinate axis.

Reflection or mirror image

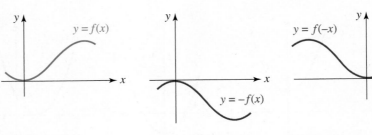

(a) Starting point (b) Reflection in x-axis (c) Reflection in y-axis

FIGURE 3.2.9 Reflections in the coordinate axes

EXAMPLE 4 Reflections

Graph (a) $y = -\sqrt{x}$ (b) $y = \sqrt{-x}$.

Solution The starting point is the graph of $f(x) = \sqrt{x}$ given in **FIGURE 3.2.10(a)**.

(a) The graph of $y = -\sqrt{x}$ is the reflection of the graph of $f(x) = \sqrt{x}$ in the x-axis. Observe in Figure 3.2.10(b) that since $(1, 1)$ is on the graph of f, the point $(1, -1)$ is on the graph of $y = -\sqrt{x}$.

(b) The graph of $y = \sqrt{-x}$ is the reflection of the graph of $f(x) = \sqrt{x}$ in the y-axis. Observe in Figure 3.2.10(c) that since $(1, 1)$ is on the graph of f, the point $(-1, 1)$ is on the graph of $y = \sqrt{-x}$. The function $y = \sqrt{-x}$ looks a little strange, but bear in mind that its domain is determined by the requirement that $-x \geq 0$, or equivalently $x \leq 0$, and so the reflected graph is defined on the interval $(-\infty, 0]$.

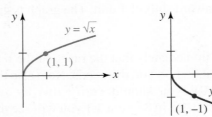

(a) Starting point (b) Reflection in x-axis (c) Reflection in y-axis

FIGURE 3.2.10 Reflected graphs in red in Example 4

If a function f is even, then $f(-x) = f(x)$ shows that a reflection in the y-axis would give precisely the same graph. If a function is odd, then from $f(-x) = -f(x)$ we see that a reflection of the graph of f in the y-axis is identical to the graph of f reflected in the x-axis. In **FIGURE 3.2.11** the blue curve is the graph of the odd function $f(x) = x^3$; the red curve is the graph of $y = f(-x) = (-x)^3 = -x^3$. Notice that if the blue curve is reflected in either the y-axis or the x-axis, we get the red curve.

□ **Nonrigid Transformations** If a function f is multiplied by a constant $c > 0$, the shape of the graph is changed but retains, *roughly*, its original shape. The graph of $y = cf(x)$ is the graph of $y = f(x)$ distorted vertically; the graph of f is either stretched (or elongated) vertically or is compressed (or flattened) vertically depending on the value of c. Stretching or compressing a graph are examples of **nonrigid transformations**.

FIGURE 3.2.11 Reflection of an odd function in y-axis

THEOREM 3.2.4 Vertical Stretches and Compressions

Suppose $y = f(x)$ is a function and c a positive constant. Then the graph of $y = cf(x)$ is the graph of f

(i) vertically stretched by a factor of c units if $c > 1$,
(ii) vertically compressed by a factor of c units if $0 < c < 1$.

FIGURE 3.2.12 Vertical stretch of the graph of $f(x) = x$

If (x, y) represents a point on the graph of f, then the point (x, cy) is on the graph of cf. The graphs of $y = x$ and $y = 3x$ are compared in FIGURE 3.2.12; the y-coordinate of a point on the graph of $y = 3x$ is 3 times as large as the y-coordinate of the point with the same x-coordinate on the graph of $y = x$. The comparison of the graphs of $y = 10x^2$ (blue graph) and $y = \frac{1}{10}x^2$ (red graph) in FIGURE 3.2.13 is a little more dramatic; the graph of $y = \frac{1}{10}x^2$ exhibits considerable vertical flattening, especially in a neighborhood of the origin. Note that c is positive in this discussion. To sketch the graph of $y = -10x^2$ we think of it as $y = -(10x^2)$, which means we first stretch the graph of $y = x^2$ vertically by a factor of 10 units, and then reflect that graph in the x-axis.

The next example illustrates shifting, reflecting, and stretching of a graph.

FIGURE 3.2.13 Vertical stretch (blue) and vertical compression (red) of the graph of $f(x) = x^2$

EXAMPLE 5 Combining Transformations

Graph $y = 2 - 2\sqrt{x - 3}$.

Solution You should recognize that the given function consists of four transformations of the basic function $f(x) = \sqrt{x}$:

We start with the graph of $f(x) = \sqrt{x}$ in FIGURE 3.2.14(a). Then stretch this graph vertically by a factor of 2 to obtain $y = 2\sqrt{x}$ in Figure 3.2.14(b). Reflect this second graph in the x-axis to obtain $y = -2\sqrt{x}$ in Figure 3.2.14(c). Shift this third graph 3 units to the right to obtain $y = -2\sqrt{x - 3}$ in Figure 3.2.14(d). Finally, shift the fourth graph upward 2 units to obtain $y = 2 - 2\sqrt{x - 3}$ in Figure 3.2.14(e). Note that the point $(0, 0)$ on the graph of $f(x) = \sqrt{x}$ remains fixed in the vertical stretch and the reflection in the x-axis, but under the first (horizontal) shift $(0, 0)$ moves to $(3, 0)$ and under the second (vertical) shift $(3, 0)$ moves to $(3, 2)$.

    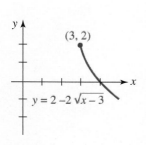

(a) Starting point   (b) Vertical stretch   (c) Reflection in x-axis   (d) Shift right   (e) Shift up

FIGURE 3.2.14 Graph of $y = 2 - 2\sqrt{x - 3}$ in Example 5 is given in part (e)

3.2 Symmetry and Transformations

3.2 Exercises
Answers to selected odd-numbered problems begin on page ANS-7.

In Problems 1–10, use (1) and (2) to determine whether the given function $y = f(x)$ is even, odd, or neither even nor odd. Do not graph.

1. $f(x) = 4 - x^2$
2. $f(x) = x^2 + 2x$
3. $f(x) = x^3 - x + 4$
4. $f(x) = x^5 + x^3 + x$
5. $f(x) = 3x - \dfrac{1}{x}$
6. $f(x) = \dfrac{x}{x^2 + 1}$
7. $f(x) = 1 - \sqrt{1 - x^2}$
8. $f(x) = \sqrt[3]{x^3 + x}$
9. $f(x) = |x^3|$
10. $f(x) = x|x|$

In Problems 11–14, classify the function $y = f(x)$ whose graph is given as even, odd, or neither even nor odd.

11.
FIGURE 3.2.15 Graph for Problem 11

12.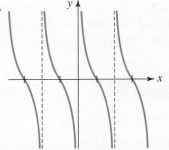
FIGURE 3.2.16 Graph for Problem 12

13.
FIGURE 3.2.17 Graph for Problem 13

14.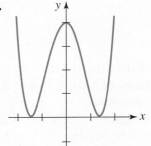
FIGURE 3.2.18 Graph for Problem 14

In Problems 15–18, complete the graph of the given function $y = f(x)$ if **(a)** f is an even function and **(b)** f is an odd function.

15.
FIGURE 3.2.19 Graph for Problem 15

16.
FIGURE 3.2.20 Graph for Problem 16

17.

FIGURE 3.2.21 Graph for Problem 17

18.

FIGURE 3.2.22 Graph for Problem 18

In Problems 19 and 20, suppose that $f(-2) = 4$ and $f(3) = 7$. Determine $f(2)$ and $f(-3)$.

19. If f is an even function

20. If f is an odd function

In Problems 21 and 22, suppose that $g(-1) = -5$ and $g(4) = 8$. Determine $g(1)$ and $g(-4)$.

21. If g is an odd function

22. If g is an even function

In Problems 23–32, the points $(-2, 1)$ and $(3, -4)$ are on the graph of the function $y = f(x)$. Find the corresponding points on the graph obtained by the given transformations.

23. The graph of f shifted up 2 units
24. The graph of f shifted down 5 units
25. The graph of f shifted to the left 6 units
26. The graph of f shifted to the right 1 unit
27. The graph of f shifted up 1 unit and to the left 4 units
28. The graph of f shifted down 3 units and to the right 5 units
29. The graph of f reflected in the y-axis
30. The graph of f reflected in the x-axis
31. The graph of f stretched vertically by a factor of 15 units
32. The graph of f compressed vertically by a factor of $\frac{1}{4}$ unit, then reflected in the x-axis

In Problems 33–36, use the graph of the function $y = f(x)$ given in the figure to graph the following functions

(a) $y = f(x) + 2$ (b) $y = f(x) - 2$
(c) $y = f(x + 2)$ (d) $y = f(x - 5)$
(e) $y = -f(x)$ (f) $y = f(-x)$

33.

FIGURE 3.2.23 Graph for Problem 33

34.

FIGURE 3.2.24 Graph for Problem 34

35.

FIGURE 3.2.25 Graph for Problem 35

36.

FIGURE 3.2.26 Graph for Problem 36

In Problems 37 and 38, use the graph of the function $y = f(x)$ given in the figure to graph the following functions

(a) $y = f(x) + 1$
(b) $y = f(x) - 1$
(c) $y = f(x + \pi)$
(d) $y = f(x - \pi/2)$
(e) $y = -f(x)$
(f) $y = f(-x)$
(g) $y = 3f(x)$
(h) $y = -\frac{1}{2}f(x)$

37.

FIGURE 3.2.27 Graph for Problem 37

38.

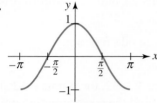

FIGURE 3.2.28 Graph for Problem 38

In Problems 39–42, find the equation of the final graph after the given transformations are applied to the graph of $y = f(x)$.

39. The graph of $f(x) = x^3$ shifted up 5 units and right 1 unit
40. The graph of $f(x) = x^{2/3}$ stretched vertically by a factor of 3 units, then shifted right 2 units
41. The graph of $f(x) = x^4$ reflected in the x-axis, then shifted left 7 units
42. The graph of $f(x) = 1/x$ reflected in the y-axis, then shifted left 5 units and down 10 units

For Discussion

43. Explain why the graph of a function $y = f(x)$ cannot be symmetric with respect to the x-axis.
44. What points, if any, on the graph of $y = f(x)$ remain fixed, that is, the same on the resulting graph after a vertical stretch or compression? After a reflection in the x-axis? After a reflection in the y-axis?
45. Discuss the relationship between the graphs of $y = f(x)$ and $y = f(|x|)$.
46. Discuss the relationship between the graphs of $y = f(x)$ and $y = f(cx)$, where $c > 0$ is a constant. Consider two cases: $0 < c < 1$ and $c > 1$.
47. Review the graphs of $y = x$ and $y = 1/x$ in Figure 3.2.1. Then discuss how to obtain the graph of the reciprocal function $y = 1/f(x)$ from the graph of $y = f(x)$. Sketch the graph of $y = 1/f(x)$ for the function f whose graph is given in Figure 3.2.26.
48. In terms of transformations of graphs, describe the relationship between the graph of the function $y = f(cx)$, c a constant, and the graph of $y = f(x)$. Consider two cases $c > 1$ and $0 < c < 1$. Illustrate your answers with several examples.

3.3 Linear and Quadratic Functions

☰ Introduction When n is a nonnegative integer, the power function $f(x) = x^n$ is just a special case of a class of functions called **polynomial functions**. A polynomial function is a function of the form

$$f(x) = a_n x^n + a_{n-1} x^{n-1} + \cdots + a_2 x^2 + a_1 x + a_0, \qquad (1)$$

◀ Polynomial functions are considered in depth in Chapter 4.

where n is a nonnegative integer. The three functions considered in this section, $f(x) = a_0$, $f(x) = a_1 x + a_0$, and $f(x) = a_2 x^2 + a_1 x + a_0$, are polynomial functions. In the definitions that follow we change the coefficients of these functions to more convenient symbols.

DEFINITION 3.3.1 Constant Function

A **constant function** $y = f(x)$ is a function of the form

$$f(x) = a, \qquad (2)$$

where a is any constant.

DEFINITION 3.3.2 Linear Function

A **linear function** $y = f(x)$ is a function of the form

$$f(x) = ax + b, \qquad (3)$$

where $a \neq 0$ and b are constants.

In the form $y = a$ we know from Section 2.3 that the graph of a constant function is simply a horizontal line. Similarly, when written as $y = ax + b$ we recognize a linear function as the slope-intercept form of a line with the symbol a playing the part of the slope m. Hence the graph of every linear function is a nonhorizontal line with slope. The **domain** of a constant function as well as a linear function is the set of real numbers $(-\infty, \infty)$.

The squaring function $y = x^2$ that played an important role in Section 3.2 is a member of a family of functions called **quadratic functions**.

DEFINITION 3.3.3 Quadratic Function

A **quadratic function** $y = f(x)$ is a function of the form

$$f(x) = ax^2 + bx + c, \qquad (4)$$

where $a \neq 0$, b, and c are real constants.

☐ **Graphs** The graph of any quadratic function is called a **parabola**. The graph of a quadratic function has the same basic shape of the squaring function $y = x^2$ shown in **FIGURE 3.3.1**. In the examples that follow we will see that the graphs of quadratic functions $f(x) = ax^2 + bx + c$ are simply transformations of the graph of $y = x^2$:

- The graph of $f(x) = ax^2$, $a > 0$, is the graph of $y = x^2$ **stretched** vertically when $a > 1$, and **compressed** vertically when $0 < a < 1$.

FIGURE 3.3.1 Graph of simplest parabola

- The graph of $f(x) = ax^2$, $a < 0$, is the graph of $y = ax^2$, $a > 0$, **reflected** in the x-axis.
- The graph of $f(x) = ax^2 + bx + c$, $b \neq 0$, is the graph of $y = ax^2$ **shifted** horizontally or vertically.

From the first two items in the bulleted list, we conclude that the graph of a quadratic function opens upward (as in Figure 3.3.1) if $a > 0$ and opens downward if $a < 0$.

EXAMPLE 1 Stretch, Compression, and Reflection

(a) The graphs of $y = 4x^2$ and $y = \frac{1}{10}x^2$ are, respectively, a vertical stretch and a vertical compression of the graph of $y = x^2$. The graphs of these functions are shown in FIGURE 3.3.2(a); the graph of $y = 4x^2$ is shown in red, the graph of $y = \frac{1}{10}x^2$ is green, and the graph of $y = x^2$ is blue.

(b) The graphs of $y = -4x^2$, $y = -\frac{1}{10}x^2$, and $y = -x^2$ are obtained from the graphs of the functions in part (a) by reflecting their graphs in the x-axis. See Figure 3.3.2(b).

≡

(a) Red graph is a vertical stretch of blue graph; green graph is a vertical compression of blue graph

(b) Reflections in x-axis

FIGURE 3.3.2 Graphs of quadratic functions in Example 1

□ **Vertex and Axis** If the graph of a quadratic function opens upward $a > 0$ (or downward $a < 0$), the lowest (highest) point (h, k) on the parabola is called its **vertex**. All parabolas are symmetric with respect to a vertical line through the vertex (h, k). The line $x = h$ is called the **axis of symmetry** or simply the **axis** of the parabola. See FIGURE 3.3.3.

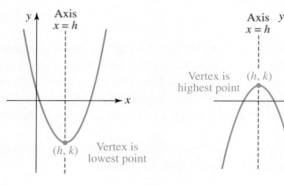

(a) $y = ax^2 + bx + c$, $a > 0$ (b) $y = ax^2 + bx + c$, $a < 0$

FIGURE 3.3.3 Vertex and axis of a parabola

□ **Standard Form** The vertex of a parabola can be determined by recasting the equation $f(x) = ax^2 + bx + c$ into the **standard form**

$$f(x) = a(x - h)^2 + k. \tag{5}$$

See Sections 1.3 and 2.2. ▶

The form (5) is obtained from the equation (4) by *completing the square* in x. Recall, completing the square in (4) starts with factoring the number a from all terms involving the variable x:

$$f(x) = ax^2 + bx + c$$
$$= a\left(x^2 + \frac{b}{a}x\right) + c.$$

CHAPTER 3 FUNCTIONS AND GRAPHS

Within the parentheses we add and subtract the square of one-half the coefficient of x:

$$\text{square } \tfrac{b}{2a} \downarrow$$

$$f(x) = a\left(x^2 + \frac{b}{a}x + \frac{b^2}{4a^2} - \frac{b^2}{4a^2}\right) + c \quad \leftarrow \text{terms in color add to 0}$$

$$= a\left(x^2 + \frac{b}{a}x + \frac{b^2}{4a^2}\right) - \frac{b^2}{4a} + c \quad \leftarrow \text{note that } a \cdot \left(-\frac{b^2}{4a^2}\right) = -\frac{b^2}{4a} \quad (6)$$

$$= a\left(x + \frac{b}{2a}\right)^2 + \frac{4ac - b^2}{4a}.$$

The last expression is equation (5) with the identifications $h = -b/2a$ and $k = (4ac - b^2)/4a$. If $a > 0$, then necessarily $a(x - h)^2 \geq 0$. Hence $f(x)$ in (5) is a minimum when $(x - h)^2 = 0$, that is, for $x = h$. A similar argument shows that if $a < 0$ in (5), $f(x)$ is a maximum value for $x = h$. Thus (h, k) is the vertex of the parabola. The equation of the axis of the parabola is $x = h$ or $x = -b/2a$.

We strongly suggest that you *do not memorize* the result in the last line of (6), but practice completing the square each time. However, if memorization is permitted by your instructor to save time, then the vertex can also be found by computing the coordinates of the point

$$\left(-\frac{b}{2a}, f\left(-\frac{b}{2a}\right)\right). \quad (7)$$

☐ **Intercepts** The graph of (4) always has a **y-intercept** since 0 is in the domain of f. From $f(0) = c$ we see that the y-intercept of a quadratic function is $(0, c)$. To determine whether the graph has x-intercepts we must solve the equation $f(x) = 0$. The last equation can be solved either by factoring or by using the quadratic formula. Recall, a quadratic equation $ax^2 + bx + c = 0$, $a \neq 0$, has the solutions

$$x_1 = \frac{-b - \sqrt{b^2 - 4ac}}{2a}, \quad x_2 = \frac{-b + \sqrt{b^2 - 4ac}}{2a}.$$

We distinguish three cases according to the algebraic sign of the discriminant $b^2 - 4ac$.

- If $b^2 - 4ac > 0$, then there are two distinct real solutions x_1 and x_2. The parabola crosses the x-axis at $(x_1, 0)$ and $(x_2, 0)$.
- If $b^2 - 4ac = 0$, then there is a single real solution x_1. The vertex of the parabola is located on the x-axis at $(x_1, 0)$. The parabola is tangent to, or touches, the x-axis at this point.
- If $b^2 - 4ac < 0$, then there are no real solutions. The parabola does not cross the x-axis.

As the next example shows, a reasonable sketch of a parabola can be obtained by plotting the intercepts and the vertex.

EXAMPLE 2 **Graph Using Intercepts and Vertex**

Graph $f(x) = x^2 - 2x - 3$.

Solution Since $a = 1 > 0$ we know that the parabola will open upward. From $f(0) = -3$ we get the y-intercept $(0, -3)$. To see whether there are any x-intercepts we solve $x^2 - 2x - 3 = 0$. By factoring,

$$(x + 1)(x - 3) = 0$$

FIGURE 3.3.4 Parabola in Example 2

we find the real solutions $x = -1, x = 3$ and so the x-intercepts are $(-1, 0)$ and $(3, 0)$. To locate the vertex we complete the square:

$$f(x) = (x^2 - 2x + 1) - 1 - 3 = (x^2 - 2x + 1) - 4.$$

Thus the standard form is $f(x) = (x - 1)^2 - 4$. With the identifications $h = 1$ and $k = -4$, we conclude that the vertex is $(1, -4)$. Using this information we draw a parabola through these four points as shown in **FIGURE 3.3.4**.

One last observation. By finding the vertex we automatically determine the range of a quadratic function. In our current example, $y = -4$ is the smallest number in the range of f and so the range of f is the interval $[-4, \infty)$ on the y-axis. ≡

EXAMPLE 3   Vertex Is the x-intercept

Graph $f(x) = -4x^2 + 12x - 9$.

Solution The graph of this quadratic function is a parabola that opens downward because $a = -4 < 0$. To complete the square we start by factoring -4 from the two x-terms:

$$\begin{aligned}
f(x) &= -4x^2 + 12x - 9 \\
&= -4(x^2 - 3x) - 9 \\
&= -4\left(x^2 - 3x + \frac{9}{4} - \frac{9}{4}\right) - 9 \\
&= -4\left(x^2 - 3x + \frac{9}{4}\right) - 9 + 9 \quad \leftarrow 9 = (-4) \cdot \left(-\frac{9}{4}\right) \text{ from the preceding line} \\
&= -4\left(x^2 - 3x + \frac{9}{4}\right).
\end{aligned}$$

FIGURE 3.3.5 Parabola in Example 3

Thus the standard form is $f(x) = -4\left(x - \frac{3}{2}\right)^2$. With $h = \frac{3}{2}$ and $k = 0$ we see that the vertex is $\left(\frac{3}{2}, 0\right)$. The y-intercept is $(0, f(0)) = (0, -9)$. Solving $-4x^2 + 12x - 9 = -4\left(x - \frac{3}{2}\right)^2 = 0$, we see that there is only one x-intercept, namely, $\left(\frac{3}{2}, 0\right)$. Of course, this was to be expected because the vertex $\left(\frac{3}{2}, 0\right)$ is on the x-axis. As shown in **FIGURE 3.3.5** a rough sketch can be obtained from these two points alone. The parabola is tangent to the x-axis at $\left(\frac{3}{2}, 0\right)$. ≡

EXAMPLE 4   Using (7) to Find the Vertex

Graph $f(x) = x^2 + 2x + 4$.

Solution The graph is a parabola that opens upward because $a = 1 > 0$. For the sake of illustration we will use (7) this time to find the vertex. With $b = 2, -b/2a = -2/2 = -1$ and

$$f(-1) = (-1)^2 + 2(-1) + 4 = 3,$$

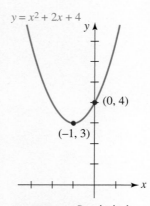

FIGURE 3.3.6 Parabola in Example 4

the vertex is $(-1, f(-1)) = (-1, 3)$. Now the y-intercept is $(0, f(0)) = (0, 4)$ but the quadratic formula shows that the equation $f(x) = 0$ or $x^2 + 2x + 4 = 0$ has no real solutions. Therefore the graph has no x-intercepts. Since the vertex is above the x-axis and the parabola opens upward, the graph must lie entirely above the x-axis. See **FIGURE 3.3.6**. ≡

☐ **Graphs by Transformations** The standard form (5) clearly describes how the graph of any quadratic function is constructed from the graph of $y = x^2$ starting with a non-rigid transformation followed by two rigid transformations:

- $y = ax^2$ is the graph of $y = x^2$ stretched or compressed vertically.
- $y = a(x - h)^2$ is the graph of $y = ax^2$ shifted $|h|$ units horizontally.
- $y = a(x - h)^2 + k$ is the graph of $y = a(x - h)^2$ shifted $|k|$ units vertically.

FIGURE 3.3.7 illustrates the horizontal and vertical shifting in the case where $a > 0$, $h > 0$, and $k > 0$.

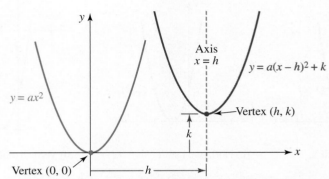

FIGURE 3.3.7 The red graph is obtained by shifting the blue graph h units to the right and k units upward.

EXAMPLE 5   Horizontally Shifted Graphs

Compare the graphs of **(a)** $y = (x - 2)^2$ and **(b)** $y = (x + 3)^2$.

Solution The blue dashed graph in **FIGURE 3.3.8** is the graph of $y = x^2$. Matching the given functions with (6) shows in each case that $a = 1$ and $k = 0$. This means that neither graph undergoes a vertical stretch or a compression, and neither graph is shifted vertically.

(a) With the identification $h = 2$, the graph of $y = (x - 2)^2$ is the graph of $y = x^2$ shifted horizontally 2 units to the right. The vertex $(0, 0)$ for $y = x^2$ becomes the vertex $(2, 0)$ for $y = (x - 2)^2$. See the red graph in Figure 3.3.8.

(b) With the identification $h = -3$, the graph of $y = (x + 3)^2$ is the graph of $y = x^2$ shifted horizontally $|-3| = 3$ units to the left. The vertex $(0, 0)$ for $y = x^2$ becomes the vertex $(-3, 0)$ for $y = (x + 3)^2$. See the green graph in Figure 3.3.8.

FIGURE 3.3.8 Shifted graphs in Example 5

3.3 Linear and Quadratic Functions

EXAMPLE 6 Shifted Graph

Graph $y = 2(x - 1)^2 - 6$.

Solution The graph is the graph of $y = x^2$ stretched vertically upward, followed by a horizontal shift to the right of 1 unit, followed by a vertical shift downward of 6 units. In **FIGURE 3.3.9** you should note how the vertex $(0, 0)$ on the graph of $y = x^2$ is moved to $(1, -6)$ on the graph of $y = 2(x - 1)^2 - 6$ as a result of these transformations. You should also follow by transformations how the point $(1, 1)$ shown in Figure 3.3.9(a) ends up as $(2, -4)$ in Figure 3.3.9(d).

(a) Basic parabola

(b) Vertical stretch

(c) Horizontal shift

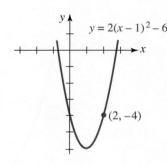
(d) Vertical shift

FIGURE 3.3.9 Graphs in Example 6

❑ **Graphical Solution of Inequalities** Graphs can be of help in solving certain inequalities when a sign chart is not useful because the quadratic does factor conveniently. For example, the quadratic function in Example 6 is equivalent to $y = 2x^2 - 4x - 4$. Were we required to solve the inequality $2x^2 - 4x - 4 \geq 0$ we see in Figure 3.3.9(d) that $y \geq 0$ to the left of the x-intercept on the negative x-axis and to the right of the x-intercept on the positive x-axis. The x-coordinates of these intercepts, obtained by solving $2x^2 - 4x - 4 = 0$ by the quadratic formula are $1 - \sqrt{3}$ and $1 + \sqrt{3}$. Thus the solution of $2x^2 - 4x - 4 \geq 0$ is the union of intervals $(-\infty, 1 - \sqrt{3}] \cup [1 + \sqrt{3}, \infty)$.

❑ **Increasing–Decreasing Functions** We have seen in Figures 2.3.2(a) and 2.3.2(b) that if $a > 0$ (which, as we have just seen plays the part of m), the values of a linear function $f(x) = ax + b$ increase as x-increases, whereas for $a < 0$, the values $f(x)$ decrease as x increases. The notions of increasing and decreasing can be extended to *any* function. The ability to determine intervals over which a function f is either increasing or decreasing plays an important role in applications of calculus.

(a) $f(x_1) < f(x_2)$

(b) $f(x_1) > f(x_2)$

FIGURE 3.3.10 Function f is increasing on $[a, b]$ in (a); is decreasing on $[a, b]$ in (b)

DEFINITION 3.3.4 Increasing/Decreasing

Suppose $y = f(x)$ is a function defined on an interval, and x_1 and x_2 are any two numbers in the interval such that $x_1 < x_2$. Then the function f is

(*i*) **increasing** on the interval if $f(x_1) < f(x_2)$, (8)
(*ii*) **decreasing** on the interval if $f(x_1) > f(x_2)$. (9)

In **FIGURE 3.3.10(a)** the function f is increasing on the interval $[a, b]$, whereas f is decreasing on $[a, b]$ in Figure 3.3.10(b). A linear function $f(x) = ax + b$, increases on the interval $(-\infty, \infty)$ for $a > 0$, and decreases on the interval $(-\infty, \infty)$ for $a < 0$. Similarly, if $a > 0$, then the quadratic function f in (5) is decreasing on the interval

$(-\infty, h]$ and increasing on the interval $[h, \infty)$. If $a < 0$, we have just the opposite, that is, f is increasing on $(-\infty, h]$ followed by decreasing on $[h, \infty)$. Reinspection of Figure 3.3.6 shows that $f(x) = x^2 + 2x + 4$ is decreasing on the interval $(-\infty, -1]$ and increasing on the interval $[-1, \infty)$. In general, if h is the x-coordinate of the vertex of a quadratic function f, then f changes either from increasing to decreasing or from decreasing to increasing at $x = h$. For this reason, the vertex (h, k) of the graph of a quadratic function is also called a **turning point** for the graph of f.

☐ **Freely Falling Object** In rough terms, an equation or a function that is constructed using certain assumptions about some real-world situation or phenomenon with the intent to describe that phenomenon is said to be a **mathematical model**. Suppose an object, such as a ball, is either thrown straight upward (downward) or simply dropped from an initial height s_0. Then if the positive direction is taken to be upward, a mathematical model for the height $s(t)$ of the object aboveground is given by the quadratic function

$$s(t) = \frac{1}{2}gt^2 + v_0 t + s_0, \qquad (10)$$

where g is the acceleration due to gravity (-32 ft/s^2 or -9.8 m/s^2), v_0 is the initial velocity imparted to the object, and t is time measured in seconds. See **FIGURE 3.3.11**. If the object is dropped, then $v_0 = 0$. An assumption in the derivation of (10) is that the motion takes place close to the surface of the Earth and so the retarding effect of air resistance is ignored. Also, the velocity of the object while it is in the air is given by the linear function

$$v(t) = gt + v_0. \qquad (11)$$

See Problems 59–62 in Exercises 3.3.

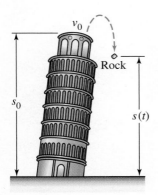

FIGURE 3.3.11 Rock thrown upward from an initial height s_0

3.3 Exercises
Answers to selected odd-numbered problems begin on page ANS-8.

In Problems 1 and 2, find a linear function (3) that satisfies both of the given conditions.

1. $f(-1) = 5, f(1) = 6$ **2.** $f(-1) = 1 + f(2), f(3) = 4f(1)$

In Problems 3–6, find the point of intersection of the graphs of the given linear functions. Sketch both lines.

3. $f(x) = -2x + 1, g(x) = 4x + 6$ **4.** $f(x) = 2x + 5, g(x) = \frac{3}{2}x + 5$
5. $f(x) = 4x + 7, g(x) = \frac{1}{3}x + \frac{10}{3}$ **6.** $f(x) = 2x - 10, g(x) = -3x$

In Problems 7–12, for the given function compute the quotient $\dfrac{f(x + h) - f(x)}{h}$, where h is a constant.

7. $f(x) = -9x + 12$ **8.** $f(x) = \frac{4}{3}x - 5$
9. $f(x) = -x^2 + x$ **10.** $f(x) = 5x^2 - 7x$
11. $f(x) = x^2 - 4x + 2$ **12.** $f(x) = -2x^2 + 5x - 3$

In Problems 13–18, sketch the graph of the given quadratic function f.

13. $f(x) = 2x^2$ **14.** $f(x) = -2x^2$
15. $f(x) = 2x^2 - 2$ **16.** $f(x) = 2x^2 + 5$
17. $f(x) = -2x^2 + 1$ **18.** $f(x) = -2x^2 - 3$

3.3 Linear and Quadratic Functions

In Problems 19–30, consider the given quadratic function f.

(a) Find all intercepts of the graph of f.
(b) Express the function f in standard form.
(c) Find the vertex and axis of symmetry.
(d) Sketch the graph of f.

19. $f(x) = x(x + 5)$
20. $f(x) = -x^2 + 4x$
21. $f(x) = (3 - x)(x + 1)$
22. $f(x) = (x - 2)(x - 6)$
23. $f(x) = x^2 - 3x + 2$
24. $f(x) = -x^2 + 6x - 5$
25. $f(x) = 4x^2 - 4x + 3$
26. $f(x) = -x^2 + 6x - 10$
27. $f(x) = -\frac{1}{2}x^2 + x + 1$
28. $f(x) = x^2 - 2x - 7$
29. $f(x) = x^2 - 10x + 25$
30. $f(x) = -x^2 + 6x - 9$

In Problems 31 and 32, find the maximum or the minimum value of the function f. Give the range of the function f.

31. $f(x) = 3x^2 - 8x + 1$
32. $f(x) = -2x^2 - 6x + 3$

In Problems 33–36, find the largest interval on which the function f is increasing and the largest interval on which f is decreasing.

33. $f(x) = \frac{1}{3}x^2 - 25$
34. $f(x) = -(x + 10)^2$
35. $f(x) = -2x^2 - 12x$
36. $f(x) = x^2 + 8x - 1$

In Problems 37–42, describe in words how the graph of the given function f can be obtained from the graph of $y = x^2$ by rigid or nonrigid transformations.

37. $f(x) = (x - 10)^2$
38. $f(x) = (x + 6)^2$
39. $f(x) = -\frac{1}{3}(x + 4)^2 + 9$
40. $f(x) = 10(x - 2)^2 - 1$
41. $f(x) = (-x - 6)^2 - 4$
42. $f(x) = -(1 - x)^2 + 1$

In Problems 43–48, the given graph is the graph of $y = x^2$ shifted/reflected in the xy-plane. Write an equation of the graph.

43.

FIGURE 3.3.12 Graph for Problem 43

44.

FIGURE 3.3.13 Graph for Problem 44

45.

FIGURE 3.3.14 Graph for Problem 45

46.

FIGURE 3.3.15 Graph for Problem 46

47.

FIGURE 3.3.16 Graph for Problem 47

48.

FIGURE 3.3.17 Graph for Problem 48

In Problems 49 and 50, find a quadratic function $f(x) = ax^2 + bx + c$ that satisfies the given conditions.

49. f has the values $f(0) = 5$, $f(1) = 10$, and $f(-1) = 4$
50. Graph passes through $(2, -1)$, zeros of f are 1 and 3

In Problems 51 and 52, find a quadratic function in standard form $f(x) = a(x - h)^2 + k$ that satisfies the given conditions.

51. The vertex of the graph of f is $(1, 2)$, graph passes through $(2, 6)$
52. The maximum value of f is 10, axis of symmetry is $x = -1$, and y-intercept is $(0, 8)$

In Problems 53–56, sketch the region in the xy-plane that is bounded between the graphs of the given functions. Find the points of intersection of the graphs.

53. $y = -x + 4$, $y = x^2 + 2x$
54. $y = 2x - 2$, $y = 1 - x^2$
55. $y = x^2 + 2x + 2$, $y = -x^2 - 2x + 2$
56. $y = x^2 - 6x + 1$, $y = -x^2 + 2x + 1$

57. (a) Express the square of the distance d from the point (x, y) on the graph of $y = 2x$ to the point $(5, 0)$ shown in FIGURE 3.3.18 as a function of x.
    (b) Use the function in part (a) to find the point (x, y) that is closest to $(5, 0)$.

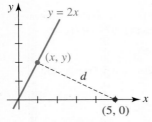

FIGURE 3.3.18 Distance in Problem 57

Miscellaneous Applications

58. **Shooting an Arrow** As shown in FIGURE 3.3.19, an arrow that is shot at a 45° angle with the horizontal travels along a parabolic arc defined by the equation $y = ax^2 + x + c$. Use the fact that the arrow is launched at a vertical height of 6 ft and travels a horizontal distance of 200 ft to find the coefficients a and c. What is the maximum height attained by the arrow?

FIGURE 3.3.19 Arrow in Problem 58

59. **Shooting Another Arrow** An arrow is shot vertically upward with an initial velocity of 64 ft/s from a point 6 ft above the ground. See FIGURE 3.3.20.
    (a) Find the height $s(t)$ and the velocity $v(t)$ of the arrow at time $t \geq 0$.
    (b) What is the maximum height attained by the arrow? What is the velocity of the arrow at the time the arrow attains its maximum height?
    (c) At what time does the arrow fall back to the 6-ft level? What is its velocity at this time?

FIGURE 3.3.20 Arrow in Problem 59

3.3 Linear and Quadratic Functions

60. **How High** The height above ground of a toy rocket launched upward from the top of a building is given by $s(t) = -16t^2 + 96t + 256$.
    (a) What is the height of the building?
    (b) What is the maximum height attained by the rocket?
    (c) Find the time when the rocket strikes the ground.

61. **Impact Velocity** A ball is dropped from the roof of a building that is 122.5 m above ground level.
    (a) What is the height and velocity of the ball at $t = 1$ s?
    (b) At what time does the ball hit the ground?
    (c) What is the impact velocity of the ball when it hits the ground?

62. **A True Story, but...** A few years ago a newspaper in the Midwest reported that an escape artist was planning to jump off a bridge into the Mississippi River wearing 70 lb of chains and manacles. The newspaper article stated that the height of the bridge was 48 ft and predicted that the escape artist's impact velocity on hitting the water would be 85 mi/h. Assuming that he simply dropped from the bridge, then his height (in feet) and velocity (in feet/second) t seconds after jumping off the bridge are given by the functions $s(t) = -16t^2 + 48$ and $v(t) = -32t$, respectively, determine whether the newspaper's estimate of his impact velocity was accurate.

63. **Thermometers** The functional relationship between degrees Celsius T_C and degrees Fahrenheit T_F is linear.
    (a) Express T_F as a function of T_C if (0°C, 32°F) and (60°C, 140°F) are on the graph of T_F.
    (b) Show that 100°C is equivalent to the Fahrenheit boiling point 212°F. See **FIGURE 3.3.21**.

64. **Thermometers—Continued** The functional relationship between degrees Celsius T_C and temperatures measured in kelvin units T_K is linear.
    (a) Express T_K as a function of T_C if (0°C, 273 K) and (27°C, 300 K) are on the graph of T_K.
    (b) Express the boiling point 100°C in kelvin units. See Figure 3.3.21.
    (c) Absolute zero is defined to be 0 K. What is 0 K in degrees Celsius?
    (d) Express T_K as a linear function of T_F.
    (e) What is 0 K in degrees Fahrenheit?

FIGURE 3.3.21 Thermometers in Problems 63 and 64

65. **Simple Interest** In simple interest, the amount A accrued over time is the linear function $A = P + Prt$, where P is the principal, t is measured in years, and r is the annual interest rate (expressed as a decimal). Compute A after 20 years if the principal is $P = \$1000$, and the annual interest rate is 3.4%. At what time is $A = \$2200$?

66. **Linear Depreciation** Straight line, or linear, depreciation consists of an item losing all its initial worth of A dollars over a period of n years by an amount A/n each year. If an item costing $20,000 when new is depreciated linearly over 25 years, determine a linear function giving its value V after x years, where $0 \le x \le 25$. What is the value of the item after 10 years?

67. **Spread of a Disease** One mathematical model for the spread of a flu virus assumes that within a population of P persons the rate at which a disease spreads is jointly proportional to the number D of persons already carrying the disease and the number $P - D$ of persons not yet infected. Mathematically, the model is given by the quadratic function

$$R(D) = kD(P - D),$$

where $R(D)$ is the rate of spread of the flu virus (in cases per day) and $k > 0$ is a constant of proportionality.

Spreading a virus

(a) Show that if the population P is a constant, then the disease spreads most rapidly when exactly one-half the population is carrying the flu.
(b) Suppose that in a town of 10,000 persons, 125 are sick on Sunday, and 37 new cases occur on Monday. Estimate the constant k.
(c) Use the result of part (b) to estimate the number of new cases on Tuesday. [*Hint*: The number of persons carrying the flu on Monday is $162 = 125 + 37$.]
(d) Estimate the number of new cases on Wednesday, Thursday, Friday, and Saturday.

For Discussion

68. Consider the linear function $f(x) = \frac{5}{2}x - 4$. If x is changed by 1 unit, how many units will y change? If x is changed by 2 units? If x is changed by n (n a positive integer) units?
69. Consider the interval $[x_1, x_2]$ and the linear function $f(x) = ax + b$, $a \neq 0$. Show that
$$f\left(\frac{x_1 + x_2}{2}\right) = \frac{f(x_1) + f(x_2)}{2},$$
and interpret this result geometrically for $a > 0$.
70. In Problems 60 and 62, what is the domain of the function $s(t)$? [*Hint*: It is *not* $(-\infty, \infty)$.]
71. On the Moon the acceleration due to gravity is one-sixth the acceleration due to gravity on Earth. If a ball is tossed vertically upward from the surface of the Moon, would it attain a maximum height six times that on Earth when the same initial velocity is used? Defend your answer.
72. Suppose the quadratic function $f(x) = ax^2 + bx + c$ has two distinct real zeros. How would you prove that the x-coordinate of the vertex is the midpoint of the line segment between the x-coordinates of the intercepts? Carry out your ideas.

3.4 Piecewise-Defined Functions

≡ **Introduction** A function f may involve two or more expressions or formulas, with each formula defined on different parts of the domain of f. A function defined in this manner is called a **piecewise-defined function**. For example,
$$f(x) = \begin{cases} x^2, & x < 0 \\ x + 1, & x \geq 0, \end{cases}$$
is not two functions, but a single function in which the rule of correspondence is given in two pieces. In this case, one piece is used for the negative real numbers ($x < 0$) and the other part for the nonnegative real numbers ($x \geq 0$); the domain of f is the union of the intervals $(-\infty, 0) \cup [0, \infty) = (-\infty, \infty)$. For example, since $-4 < 0$, the rule indicates that we square the number:
$$f(-4) = (-4)^2 = 16;$$
on the other hand, since $6 \geq 0$ we add 1 to the number:
$$f(6) = 6 + 1 = 7.$$

☐ **Postage Stamp Function** The USPS first-class mailing rates for a letter, a card, or a package provide a real-world illustration of a piecewise-defined function. As of this writing, the postage for sending a letter in a standard-size envelope by first-class mail depends on its weight in ounces:

$$\text{Postage} = \begin{cases} \$0.44, & 0 < \text{weight} \leq 1 \text{ ounce} \\ \$0.61, & 1 < \text{weight} \leq 2 \text{ ounces} \\ \$0.78, & 2 < \text{weight} \leq 3 \text{ ounces} \\ \vdots \\ \$2.92, & 12 < \text{weight} \leq 13 \text{ ounces.} \end{cases} \quad (1)$$

The rule in (1) is a function P consisting of 14 pieces (letters over 13 ounces are sent priority mail). A value $P(w)$ is one of 14 constants; the constant changes depending on the weight w (in ounces) of the letter.* For example,

$$P(0.5) = \$0.44, P(1.7) = \$0.61, P(2.2) = \$0.78, P(2.9) = \$0.78,$$
$$\text{and } P(12.1) = \$2.92.$$

The domain of the function P is the union of the intervals:

$$(0, 1] \cup (1, 2] \cup (2, 3] \cup \cdots \cup (12, 13] = (0, 13].$$

EXAMPLE 1   Graph of a Piecewise-Defined Function

Graph the piecewise-defined function

$$f(x) = \begin{cases} -1, & x < 0, \\ 0, & x = 0, \\ x + 1, & x > 0. \end{cases} \quad (2)$$

FIGURE 3.4.1 Graph of piecewise-defined function in Example 1

Solution Although the domain of f consists of all real numbers $(-\infty, \infty)$, each piece of the function is defined on a different part of this domain. We draw

- the horizontal line $y = -1$ for $x < 0$,
- the point $(0, 0)$ for $x = 0$, and
- the line $y = x + 1$ for $x > 0$.

The graph is given in FIGURE 3.4.1. ≡

The solid dot at the origin in Figure 3.4.1 indicates that the function in (2) is defined at $x = 0$ only by $f(0) = 0$; the open dots indicate that the formulas corresponding to $x < 0$ and to $x > 0$ do not define f at $x = 0$. Since we are making up a function, consider the definition

$$g(x) = \begin{cases} -1, & x \leq 0, \\ x + 1, & x > 0. \end{cases} \quad (3)$$

FIGURE 3.4.2 Graph of function g defined in (3)

The graph of g shown in FIGURE 3.4.2 is very similar to the graph of (2), but (2) and (3) are not the same function because $f(0) = 0$ but $g(0) = -1$.

---

*Not shown in (1) is the fact that the postage of a letter whose weight falls in the interval $(3, 4]$ is determined by whether its weight falls in $(3, 3.5]$ or $(3.5, 4]$. This is the only interval that is so divided.

☐ **Greatest Integer Function** We consider next a piecewise-defined function that is similar to the "postage stamp" function (1) in that both are examples of *step functions*; each function is constant on an interval and then jumps to another constant value on the next abutting interval. This new function, which has many notations, will be denoted here by $f(x) = [\![x]\!]$, and is defined by the rule

$$[\![x]\!] = n, \text{ where } n \text{ is an integer satisfying } n \leq x < n + 1. \quad (4)$$

The function f is called the **greatest integer function** because (4), translated into words, means that

$f(x)$ is the greatest integer n that is less than or equal to x.

For example,

$f(6) = 6$ since $6 \leq x = 6$, $f(-1.5) = -2$ since $-2 \leq x = -1.5$,
$f(0.4) = 0$ since $0 \leq x = 0.4$, $f(7.6) = 7$ since $7 \leq x = 7.6$,
$f(\pi) = 3$ since $3 \leq x = \pi$, $f(-\sqrt{2}) = -2$ since $-2 \leq x = -\sqrt{2}$,

and so on. The domain of f is the set of real numbers and consists of the union of an infinite number of disjoint intervals, in other words, $f(x) = [\![x]\!]$ is a piecewise-defined function given by

$$f(x) = [\![x]\!] = \begin{cases} \vdots \\ -2, & -2 \leq x < -1 \\ -1, & -1 \leq x < 0 \\ 0, & 0 \leq x < 1 \\ 1, & 1 \leq x < 2 \\ 2, & 2 \leq x < 3 \\ \vdots \end{cases} \quad (5)$$

The range of f is the set of integers. A portion of the graph of f is given on the closed interval $[-2, 5]$ in **FIGURE 3.4.3**.

In computer science the greatest integer function $f(x) = [\![x]\!]$ is known as the **floor function** and is denoted by $f(x) = \lfloor x \rfloor$. See Problems 47, 48, and 53 in Exercises 3.4.

FIGURE 3.4.3 Greatest integer function

EXAMPLE 2 Shifted Graph

Graph $y = [\![x - 2]\!]$.

Solution The function is $y = f(x - 2)$, where $f(x) = [\![x]\!]$. Thus the graph in Figure 3.4.3 is shifted horizontally 2 units to the right. Note in Figure 3.4.3 that if n is an integer, then $f(n) = [\![n]\!] = n$. But in **FIGURE 3.4.4**, for $x = n$, $y = n - 2$. ≡

☐ **Continuous Functions** The graph of a **continuous function** has no holes, finite gaps, or infinite breaks. While the formal definition of continuity of a function is an important topic of discussion in calculus, in this course it suffices to think in informal terms. A continuous function is often characterized by saying that its graph can be drawn "without lifting pencil from paper." Parts (a)–(c) of **FIGURE 3.4.5** illustrate functions that are *not* continuous, or **discontinuous**, at $x = 2$. The function

$$f(x) = \frac{x^2 - 4}{x - 2} = x + 2, \quad x \neq 2,$$

in Figure 3.4.5(a) has a hole in its graph (there is no point $(2, f(2))$); the function $f(x) = \dfrac{|x - 2|}{x - 2}$ in Figure 3.4.5(b) has a finite gap or jump in its graph at $x = 2$; the

FIGURE 3.4.4 Shifted graph in Example 2

3.4 Piecewise-Defined Functions

(a) Hole in graph   (b) Finite gap in graph   (c) Infinite break in graph   (d) No holes, gaps, or breaks

FIGURE 3.4.5 Discontinuous functions (a)–(c); continuous function (d)

function $f(x) = \dfrac{1}{x-2}$ in Figure 3.4.5(c) has an infinite break in its graph at $x = 2$. The function $f(x) = x^3 - 3x + 2$ is continuous; its graph given in Figure 3.4.5(d) has no holes, gaps, or infinite breaks.

You should be aware that constant functions, linear functions, and quadratic functions are continuous. Piecewise-defined functions can be continuous or discontinuous. The functions given in (2), (3), and (4) are discontinuous.

☐ **Absolute-Value Function** The function $y = |x|$ is called the **absolute-value function**. To obtain the graph, we graph its two pieces consisting of perpendicular half lines:

$$y = |x| = \begin{cases} -x, & \text{if } x < 0 \\ x, & \text{if } x \geq 0. \end{cases} \qquad (6)$$

See **FIGURE 3.4.6(a)**. Since $y \geq 0$ for all x, another way of graphing (6) is simply to sketch the line $y = x$ and then reflect in the x-axis that portion of the line that is below the x-axis. See Figure 3.4.6(b). The domain of (6) is the set of real numbers $(-\infty, \infty)$, and as is seen in Figure 3.4.6(a), the absolute-value function is an even function, decreasing on the interval $(-\infty, 0)$, increasing on the interval $(0, \infty)$, and is continuous.

In some applications we are interested in the graph of the absolute value of an arbitrary function $y = f(x)$, in other words, $y = |f(x)|$. Since $|f(x)|$ is nonnegative for all numbers x in the domain of f, the graph of $y = |f(x)|$ does not extend below the x-axis. Moreover, the definition of the absolute value of $f(x)$,

$$|f(x)| = \begin{cases} -f(x), & \text{if } f(x) < 0 \\ f(x), & \text{if } f(x) \geq 0, \end{cases} \qquad (7)$$

(a) $y = |x|$

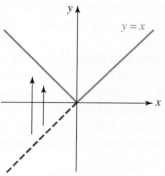

Reflect this portion of $y = x$ in the x-axis

(b)

FIGURE 3.4.6 Absolute-value function (6)

shows that we must negate $f(x)$ whenever $f(x)$ is negative. There is no need to worry about solving the inequalities in (7); to obtain the graph of $y = |f(x)|$, we can proceed just as we did in Figure 3.4.6(b): Carefully draw the graph of $y = f(x)$ and then reflect in the x-axis all portions of the graph that are below the x-axis.

EXAMPLE 3   Absolute Value of a Function

Graph $y = |-3x + 2|$.

Solution We first draw the graph of the linear function $f(x) = -3x + 2$. Note that since the slope is negative, f is decreasing and its graph crosses the x-axis at $(\frac{2}{3}, 0)$. We dash the graph for $x > \frac{2}{3}$ since that portion is below the x-axis. Finally, we reflect that portion upward in the x-axis to obtain the solid blue v-shaped graph in **FIGURE 3.4.7**. Since $f(x) = x$ is a simple linear function, it is not surprising that the graph of the absolute value of any linear function $f(x) = ax + b, a \neq 0$, will result in a graph similar to that of the absolute-value function shown in Figure 3.4.6(a).

FIGURE 3.4.7 Graph of function in Example 3

EXAMPLE 4   Absolute Value of a Function

Graph $y = |-x^2 + 2x + 3|$.

Solution As in Example 3 we begin by drawing the graph of the function $f(x) = -x^2 + 2x + 3$ by finding its intercepts $(-1, 0)$, $(3, 0)$, $(0, 3)$ and, since f is a quadratic function, its vertex $(1, 4)$. Observe in **FIGURE 3.4.8(a)** that $y < 0$ for $x < -1$ and for $x > 3$. These two portions of the graph of f are reflected in the x-axis to obtain the graph of $y = |-x^2 + 2x + 3|$ given in Figure 3.4.8(b).

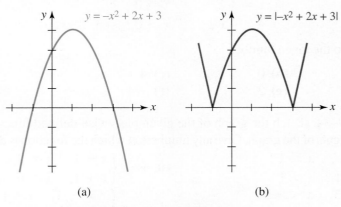

FIGURE 3.4.8 Graph of function in Example 4

3.4   Exercises   Answers to selected odd-numbered problems begin on page ANS-9.

In Problems 1–4, find the indicated values of the given piecewise-defined function f.

1. $f(x) = \begin{cases} \dfrac{x^2 - 4}{x - 2}, & x \neq 2 \\ 4, & x = 2 \end{cases}$; $f(0), f(2), f(-7)$

2. $f(x) = \begin{cases} \dfrac{x^4 - 1}{x^2 - 1}, & x \neq \pm 1 \\ 3, & x = -1 \\ 5, & x = 1 \end{cases}$; $f(-1), f(1), f(3)$

3. $f(x) = \begin{cases} x^2 + 2x, & x \geq 1 \\ -x^3, & x < 1 \end{cases}$; $f(1), f(0), f(-2), f(\sqrt{2})$

4. $f(x) = \begin{cases} 0, & x < 0 \\ x, & 0 < x < 1 \\ x + 1, & x \geq 1 \end{cases}$; $f(-\tfrac{1}{2}), f(\tfrac{1}{3}), f(4), f(6.2)$

5. If the piecewise-defined function f is defined by

$$f(x) = \begin{cases} 1, & x \text{ a rational number} \\ 0, & x \text{ an irrational number,} \end{cases}$$

find each of the following values.
  (a) $f(\tfrac{1}{3})$ (b) $f(-1)$ (c) $f(\sqrt{2})$
  (d) $f(1.\overline{12})$ (e) $f(5.72)$ (f) $f(\pi)$

6. What is the y-intercept of the graph of the function f in Problem 5?

7. Determine the values of x for which the piecewise-defined function

$$f(x) = \begin{cases} x^3 + 1, & x < 0 \\ x^2 - 2, & x \geq 0, \end{cases}$$

is equal to the given number.
  (a) 7   (b) 0   (c) -1
  (d) -2 (e) 1   (f) -7

8. Determine the values of x for which the piecewise-defined function

$$f(x) = \begin{cases} x + 1, & x < 0 \\ 2, & x = 0 \\ x^2, & x > 0, \end{cases}$$

is equal to the given number.
  (a) 1   (b) 0   (c) 4
  (d) $\tfrac{1}{2}$ (e) 2   (f) -4

In Problems 9–34, sketch the graph of the given piecewise-defined function. Find any x- and y-intercepts of the graph. Give any numbers at which the function is discontinuous.

9. $y = \begin{cases} -x, & x \leq 1 \\ -1, & x > 1 \end{cases}$ 10. $y = \begin{cases} x - 1, & x < 0 \\ x + 1, & x \geq 0 \end{cases}$

11. $y = \begin{cases} -3, & x < -3 \\ x, & -3 \leq x \leq 3 \\ 3, & x > 3 \end{cases}$ 12. $y = \begin{cases} -x^2 - 1, & x < 0 \\ 0, & x = 0 \\ x^2 + 1, & x > 0 \end{cases}$

13. $y = [\![x + 2]\!]$ 14. $y = 2 + [\![x]\!]$
15. $y = -[\![x]\!]$ 16. $y = [\![-x]\!]$
17. $y = |x + 3|$ 18. $y = -|x - 4|$
19. $y = 2 - |x|$ 20. $y = -1 - |x|$
21. $y = -2 + |x + 1|$ 22. $y = 1 - \tfrac{1}{2}|x - 2|$
23. $y = -|5 - 3x|$ 24. $y = |2x - 5|$
25. $y = |x^2 - 1|$ 26. $y = |4 - x^2|$
27. $y = |x^2 - 2x|$ 28. $y = |-x^2 - 4x + 5|$
29. $y = ||x| - 2|$ 30. $y = |\sqrt{x} - 2|$
31. $y = |x^3 - 1|$ 32. $y = |[\![x]\!]|$

33. $y = \begin{cases} 1, & x < 0 \\ |x - 1|, & 0 \leq x \leq 2 \\ 1, & x > 2 \end{cases}$ 34. $y = \begin{cases} -x, & x < 0 \\ 1 - |x - 1|, & 0 \leq x \leq 2 \\ x - 2, & x > 2 \end{cases}$

CHAPTER 3 FUNCTIONS AND GRAPHS

35. Without graphing, give the range of the function $f(x) = (-1)^{[\![x]\!]}$.
36. Compare the graphs of $y = 2[\![x]\!]$ and $y = [\![2x]\!]$.

In Problems 37–40, find a piecewise-defined formula for the function f whose graph is given. Assume that the domain of f is $(-\infty, \infty)$.

37.
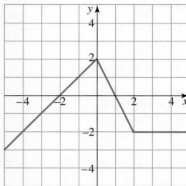
FIGURE 3.4.9 Graph for Problem 37

38.
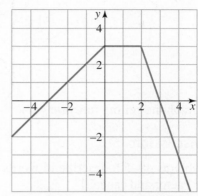
FIGURE 3.4.10 Graph for Problem 38

39.
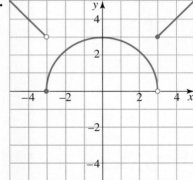
FIGURE 3.4.11 Graph for Problem 39

40.
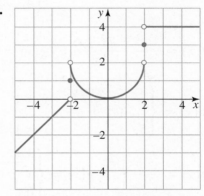
FIGURE 3.4.12 Graph for Problem 40

In Problems 41 and 42, sketch the graph of $y = |f(x)|$.

41. f is the function whose graph is given in FIGURE 3.4.9.
42. f is the function whose graph is given in FIGURE 3.4.10.

In Problems 43 and 44, use the definition of absolute value and express the given function f as a piecewise-defined function.

43. $f(x) = \dfrac{|x|}{x}$

44. $f(x) = \dfrac{x-3}{|x-3|}$

In Problems 45 and 46, find the value of the constant k such that the given piecewise-defined function f is continuous at $x = 2$. That is, the graph of f has no holes, gaps, or breaks in its graph at $x = 2$.

45. $f(x) = \begin{cases} \frac{1}{2}x + 1, & x \leq 2 \\ kx, & x > 2 \end{cases}$

46. $f(x) = \begin{cases} kx + 2, & x < 2 \\ x^2 + 1, & x \geq 2 \end{cases}$

47. The **ceiling function** $g(x) = \lceil x \rceil$ is defined to be the least integer n that is greater than or equal to x. Fill in the blanks

$$g(x) = \lceil x \rceil = \begin{cases} \vdots \\ \underline{}, & -3 < x \le -2 \\ \underline{}, & -2 < x \le -1 \\ \underline{}, & -1 < x \le 0 \\ \underline{}, & 0 < x \le 1 \\ \underline{}, & 1 < x \le 2 \\ \underline{}, & 2 < x \le 3 \\ \vdots \end{cases}$$

48. Graph the ceiling function $g(x) = \lceil x \rceil$ defined in Problem 47.

For Discussion

In Problems 49–52, describe in words how the graphs of the given functions differ. [*Hint*: Factor and cancel.]

49. $f(x) = \dfrac{x^2 - 9}{x - 3}, \quad g(x) = \begin{cases} \dfrac{x^2 - 9}{x - 3}, & x \ne 3 \\ 4, & x = 3 \end{cases}, \quad h(x) = \begin{cases} \dfrac{x^2 - 9}{x - 3}, & x \ne 3 \\ 6, & x = 3 \end{cases}$

50. $f(x) = -\dfrac{x^2 - 7x + 6}{x - 1}, \quad g(x) = \begin{cases} -\dfrac{x^2 - 7x + 6}{x - 1}, & x \ne 1 \\ 8, & x = 1 \end{cases},$

$h(x) = \begin{cases} \dfrac{x^2 - 7x + 6}{x - 1}, & x \ne 1 \\ 5, & x = 1 \end{cases}$

51. $f(x) = \dfrac{x^4 - 1}{x^2 - 1}, \quad g(x) = \begin{cases} \dfrac{x^4 - 1}{x^2 - 1}, & x \ne 1 \\ 0, & x = 1 \end{cases}, \quad h(x) = \begin{cases} \dfrac{x^4 - 1}{x^2 - 1}, & x \ne 1 \\ 2, & x = 1 \end{cases}$

52. $f(x) = \dfrac{x^3 - 8}{x - 2}, \quad g(x) = \begin{cases} \dfrac{x^3 - 8}{x - 2}, & x \ne 2 \\ 5, & x = 2 \end{cases}, \quad h(x) = \begin{cases} \dfrac{x^3 - 8}{x - 2}, & x \ne 2 \\ 12, & x = 2 \end{cases}$

53. Using the notion of a reflection of a graph in an axis, express the ceiling function $g(x) = \lceil x \rceil$, defined in Problem 47, in terms of the floor function $f(x) = \lfloor x \rfloor$ (see page 187).

54. Discuss how to graph the function $y = |x| + |x - 3|$. Carry out your ideas.

3.5 Combining Functions

≡ **Introduction** Two functions f and g can be combined in several ways to create new functions. In this section we examine two such ways in which functions can be combined: through arithmetic operations, and through the operation of function composition.

☐ **Arithmetic Combinations** Two functions can be combined through the familiar four arithmetic operations of addition, subtraction, multiplication, and division.

> **DEFINITION 3.5.1** Arithmetic Combinations
>
> If f and g are two functions, then the **sum** $f + g$, the **difference** $f - g$, the **product** fg, and the **quotient** f/g are defined as follows:
>
> $$(f + g)(x) = f(x) + g(x), \tag{1}$$
> $$(f - g)(x) = f(x) - g(x), \tag{2}$$
> $$(fg)(x) = f(x)g(x), \tag{3}$$
> $$\left(\frac{f}{g}\right)(x) = \frac{f(x)}{g(x)}, \text{ provided } g(x) \neq 0. \tag{4}$$

EXAMPLE 1   Sum, Difference, Product, and Quotient

Consider the functions $f(x) = x^2 + 4x$ and $g(x) = x^2 - 9$. From (1)–(4) of Definition 3.5.1 we can produce four new functions:

$$(f + g)(x) = f(x) + g(x) = (x^2 + 4x) + (x^2 - 9) = 2x^2 + 4x - 9,$$
$$(f - g)(x) = f(x) - g(x) = (x^2 + 4x) - (x^2 - 9) = 4x + 9,$$
$$(fg)(x) = f(x)g(x) = (x^2 + 4x)(x^2 - 9) = x^4 + 4x^3 - 9x^2 - 36x,$$

and

$$\left(\frac{f}{g}\right)(x) = \frac{f(x)}{g(x)} = \frac{x^2 + 4x}{x^2 - 9}.$$

≡

☐ **Domain of an Arithmetic Combination** When combining two functions arithmetically it is necessary that both f and g be defined at a same number x. Hence the **domain** of the functions $f + g$, $f - g$, and fg is the set of real numbers that are *common* to both domains; that is, the domain is the *intersection* of the domain of f with the domain of g. In the case of the quotient f/g, the domain is also the intersection of the two domains, *but* we must also exclude any values of x for which the denominator $g(x)$ is zero. In Example 1 the domain of f and the domain of g is the set of real numbers $(-\infty, \infty)$, and so the domain of $f + g$, $f - g$, and fg is also $(-\infty, \infty)$. However, since $g(-3) = 0$ and $g(3) = 0$, the domain of the quotient $(f/g)(x)$ is $(-\infty, \infty)$ with $x = 3$ and $x = -3$ excluded, in other words, $(-\infty, -3) \cup (-3, 3) \cup (3, \infty)$. In summary, if the domain of f is the set X_1 and the domain of g is the set X_2, then

- the domain of $f + g$, $f - g$, and fg is the intersection $X_1 \cap X_2$, and
- the domain of f/g is the set $\{x \mid x \in X_1 \cap X_2, g(x) \neq 0\}$.

EXAMPLE 2   Domain of $f + g$

By solving the inequality $1 - x \geq 0$, it is seen that the domain of $f(x) = \sqrt{1 - x}$ is the interval $(-\infty, 1]$. Similarly, the domain of the function $g(x) = \sqrt{x + 2}$ is the interval $[-2, \infty)$. Hence, the domain of the sum

$$(f + g)(x) = f(x) + g(x) = \sqrt{1 - x} + \sqrt{x + 2}$$

is the intersection $(-\infty, 1] \cap [-2, \infty)$. You should verify this result by sketching these intervals on the number line and show that this intersection, or the set of numbers common to both domains, is the closed interval $[-2, 1]$.

≡

☐ **Composition of Functions** Another method of combining functions f and g is called **function composition**. To illustrate the idea, let's suppose that for a given x in the domain of g the function value $g(x)$ is a number in the domain of the function f. This means we are able to evaluate f at $g(x)$, in other words, $f(g(x))$. For example, suppose $f(x) = x^2$ and $g(x) = x + 2$. Then for $x = 1$, $g(1) = 3$, and since 3 is the domain of f, we can write $f(g(1)) = f(3) = 3^2 = 9$. Indeed, for these two particular functions it turns out that we can evaluate f at any function value $g(x)$, that is,

$$f(g(x)) = f(x + 2) = (x + 2)^2.$$

The resulting function, called the composition of f and g, is defined next.

DEFINITION 3.5.2 Function Composition

If f and g are two functions, then the **composition** of f and g, denoted by $f \circ g$, is the function defined by

$$(f \circ g)(x) = f(g(x)). \tag{5}$$

The **composition** of g and f, denoted by $g \circ f$, is the function defined by

$$(g \circ f)(x) = g(f(x)). \tag{6}$$

When computing a composition such as $(f \circ g)(x) = f(g(x))$ be sure to substitute $g(x)$ for every x that appears in $f(x)$. See part (a) of the next example.

EXAMPLE 3 Two Compositions

If $f(x) = x^2 + 3x - 1$ and $g(x) = 2x^2 + 1$, find **(a)** $(f \circ g)(x)$ and **(b)** $(g \circ f)(x)$.

Solution (a) For emphasis we replace x by the set of parentheses ( ) and write f in the form

$$f(\ ) = (\ )^2 + 3(\ ) - 1.$$

Thus to evaluate $(f \circ g)(x)$ we fill each set of parentheses with $g(x)$. We find

$$\begin{aligned}(f \circ g)(x) &= f(g(x)) = f(2x^2 + 1) \\ &= (2x^2 + 1)^2 + 3(2x^2 + 1) - 1 \\ &= 4x^4 + 4x^2 + 1 + 3 \cdot 2x^2 + 3 \cdot 1 - 1 \\ &= 4x^4 + 10x^2 + 3.\end{aligned}$$

← use $(a + b)^2 = a^2 + 2ab + b^2$ and the distributive law

(b) In this case write g in the form

$$g(\ ) = 2(\ )^2 + 1.$$

Then

$$\begin{aligned}(g \circ f)(x) &= g(f(x)) = g(x^2 + 3x - 1) \\ &= 2(x^2 + 3x - 1)^2 + 1 \\ &= 2(x^4 + 6x^3 + 7x^2 - 6x + 1) + 1 \\ &= 2 \cdot x^4 + 2 \cdot 6x^3 + 2 \cdot 7x^2 - 2 \cdot 6x + 2 \cdot 1 + 1 \\ &= 2x^4 + 12x^3 + 14x^2 - 12x + 3.\end{aligned}$$

← use $(a + b + c)^2 = ((a + b) + c)^2$
$= (a + b)^2 + 2(a + b)c + c^2$ etc.

≡

Parts (a) and (b) of Example 3 illustrate that function composition is not commutative. That is, in general

$$f \circ g \neq g \circ f.$$

The next example shows that a function can be composed with itself.

EXAMPLE 4 f Composed with f

If $f(x) = 5x - 1$, then the composition $f \circ f$ is given by

$$(f \circ f)(x) = f(f(x)) = f(5x - 1) = 5(5x - 1) - 1 = 25x - 6.$$

EXAMPLE 5 Writing a Function as a Composition

Express $F(x) = \sqrt{6x^3 + 8}$ as the composition of two functions f and g.

Solution If we define f and g as $f(x) = \sqrt{x}$ and $g(x) = 6x^3 + 8$, then

$$F(x) = (f \circ g)(x) = f(g(x)) = f(6x^3 + 8) = \sqrt{6x^3 + 8}.$$

There are other solutions to Example 5. For instance, if the functions f and g are defined by $f(x) = \sqrt{6x + 8}$ and $g(x) = x^3$, then observe

$$(f \circ g)(x) = f(x^3) = \sqrt{6x^3 + 8}.$$

☐ **Domain of a Composition** As stated in the introductory example to this discussion, to evaluate the composition $(f \circ g)(x) = f(g(x))$ the number $g(x)$ must be in the domain of f. For example, the domain $f(x) = \sqrt{x}$ is $x \geq 0$ and the domain of $g(x) = x - 2$ is the set of real numbers $(-\infty, \infty)$. Observe we cannot evaluate $f(g(1))$ because $g(1) = -1$ and -1 is not in the domain of f. In order to substitute $g(x)$ into $f(x)$, $g(x)$ must satisfy the inequality that defines the domain of f, namely, $g(x) \geq 0$. This last inequality is the same as $x - 2 \geq 0$ or $x \geq 2$. The domain of the composition $f(g(x)) = \sqrt{g(x)} = \sqrt{x - 2}$ is $[2, \infty)$, which is only a portion of the original domain $(-\infty, \infty)$ of g. In general:

- The domain of the composition $f \circ g$ consists of the numbers x in the domain of g such that the function values $g(x)$ are in the domain of f.

◄ Read this sentence several times.

EXAMPLE 6 Domain of a Composition

Consider the function $f(x) = \sqrt{x - 3}$. From the requirement that $x - 3 \geq 0$ we see that whatever number x is substituted into f must satisfy $x \geq 3$. Now suppose $g(x) = x^2 + 2$ and we want to evaluate $f(g(x))$. Although the domain of g is the set of all real numbers, in order to substitute $g(x)$ into $f(x)$ we require that x be a number in that domain so that $g(x) \geq 3$. From **FIGURE 3.5.1** we see that the last inequality is satisfied whenever $x \leq -1$ or $x \geq 1$.

In other words, the domain of the composition

$$f(g(x)) = f(x^2 + 2) = \sqrt{(x^2 + 2) - 3} = \sqrt{x^2 - 1}$$

is $(-\infty, -1] \cup [1, \infty)$.

FIGURE 3.5.1 Domain of $(f \circ g)(x)$ in Example 6

In certain applications a quantity y is given as a function of a variable x, which in turn is a function of another variable t. By means of function composition we can express y as a function of t. The next example illustrates the idea; the symbol V plays the part of y and r plays the part of x.

EXAMPLE 7 Inflating a Balloon

A weather balloon is being inflated with a gas. If the radius of the balloon is increasing at a rate of 5 cm/s, express the volume of the balloon as a function of time t in seconds.

Solution Let's assume that as the balloon is inflated, its shape is that of a sphere. If r denotes the radius of the balloon, then $r(t) = 5t$. Since the volume of a sphere is $V = \frac{4}{3}\pi r^3$, the composition is $(V \circ r)(t) = V(r(t)) = V(5t)$ or

$$V = \frac{4}{3}\pi(5t)^3 = \frac{500}{3}\pi t^3.$$

Weather balloon

☐ **Transformations** The rigid and nonrigid transformations that were considered in Section 3.2 are examples of the operations on functions discussed in this section. For $c > 0$ a constant, the rigid transformations defined by $y = f(x) + c$ and $y = f(x) - c$ are the *sum* and *difference*, respectively, of the function $f(x)$ and the constant function $g(x) = c$. The nonrigid transformation $y = cf(x)$ is the *product* of $f(x)$ and the constant function $g(x) = c$. The rigid transformations defined by $y = f(x + c)$ and $y = f(x - c)$ are *compositions* of $f(x)$ with the linear functions $g(x) = x + c$ and $g(x) = x - c$, respectively.

☐ **Difference Quotient** Suppose points P and Q are two distinct points on the graph of $y = f(x)$ with coordinates $(x, f(x))$ and $(x + h, f(x + h))$, respectively. Then as shown in FIGURE 3.5.2, the slope of the secant line through P and Q is

$$m_{sec} = \frac{\text{rise}}{\text{run}} = \frac{f(x + h) - f(x)}{(x + h) - x}$$

or

$$m_{sec} = \frac{f(x + h) - f(x)}{h}. \tag{7}$$

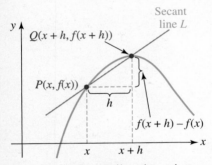

FIGURE 3.5.2 Secant line through two points on a graph

The expression in (7) is called a **difference quotient** and is very important in the study of calculus. The symbol h is just a real number and as seen in Figure 3.5.2 represents an increment or a change in x. The computation of (7) is essentially a *three-step process* and these steps involve only precalculus mathematics: algebra and trigonometry. Getting over the hurdles of algebraic or trigonometric manipulations in these steps is your primary goal. If you are preparing for calculus, we recommend that you be able to carry out the calculation of (7) for functions involving

▶ Review Section R.6 for $(a + b)^n$ for $n = 2$ and 3

▶ Review rational expressions in Section R.8

▶ Review rationalization of numerators and denominators in Section R.4

- positive integer powers of x such as x^n for $n = 1, 2,$ and 3,
- division of functions such as $\frac{1}{x}$ and $\frac{x}{x + 1}$, and
- radicals such as $\sqrt{x}$.

See Problems 47–60 in Exercises 3.5.

EXAMPLE 8 Difference Quotient

(a) Compute the difference quotient (7) for the function $y = x^2 + 2$.
(b) Find the slope of the secant line through the points $(2, f(2))$ and $(2.5, f(2.5))$.

Solution (a) The initial step is the computation of $f(x + h)$. For the given function we write $f(\ ) = (\ )^2 + 2$. The idea is to substitute $x + h$ into the parentheses and carry out the required algebra:

$$\begin{aligned} f(x + h) &= (x + h)^2 + 2 \\ &= (x^2 + 2xh + h^2) + 2 \\ &= x^2 + 2xh + h^2 + 2. \end{aligned}$$

The second step, the computation of the difference $f(x + h) - f(x)$, is the most important step and should be simplified as much as possible. In many of the problems that you will be required to do in calculus you will be able to factor h from the difference $f(x + h) - f(x)$:

$$\begin{aligned} f(x + h) - f(x) &= (x^2 + 2xh + h^2 + 2) - (x^2 + 2) \\ &= x^2 + 2xh + h^2 + 2 - x^2 - 2 \\ &= 2xh + h^2 \\ &= h(2x + h) \quad \leftarrow \text{notice the factor of } h \end{aligned}$$

The computation of the difference quotient $\dfrac{f(x + h) - f(x)}{h}$ is now straightforward. We use the results from the preceding step:

$$\frac{f(x + h) - f(x)}{h} = \frac{\cancel{h}(2x + h)}{\cancel{h}} = 2x + h. \quad \leftarrow \text{cancel the } h\text{'s}$$

Thus the slope m_{sec} of the secant line is

$$m_{\text{sec}} = 2x + h.$$

(b) For the given points we identify $x = 2$ and the change in x as $h = 2.5 - 2 = 0.5$. Therefore the slope of the secant line that passes through $(2, f(2))$ and $(2.5, f(2.5))$ is given by $m_{\text{sec}} = 2(2) + 0.5$ or $m_{\text{sec}} = 4.5$. ≡

3.5 Exercises
Answers to selected odd-numbered problems begin on page ANS-10.

In Problems 1–8, find the functions $f + g$, $f - g$, fg and f/g and give their domains.

1. $f(x) = x^2 + 1$, $g(x) = 2x^2 - x$
2. $f(x) = x^2 - 4$, $g(x) = x + 3$
3. $f(x) = x$, $g(x) = \sqrt{x - 1}$
4. $f(x) = x - 2$, $g(x) = \dfrac{1}{x + 8}$
5. $f(x) = 3x^3 - 4x^2 + 5x$, $g(x) = (1 - x)^2$
6. $f(x) = \dfrac{4}{x - 6}$, $g(x) = \dfrac{x}{x - 3}$
7. $f(x) = \sqrt{x + 2}$, $g(x) = \sqrt{5 - 5x}$
8. $f(x) = \dfrac{1}{x^2 - 9}$, $g(x) = \dfrac{\sqrt{x + 4}}{x}$

9. Fill in the table.

x	0	1	2	3	4
$f(x)$	-1	2	10	8	0
$g(x)$	2	3	0	1	4
$(f \circ g)(x)$					

10. Fill in the table where g is an odd function.

x	0	1	2	3	4
$f(x)$	-2	-3	0	-1	-4
$g(x)$	9	7	-6	-5	13
$(g \circ f)(x)$					

3.5 Combining Functions

In Problems 11–14, find the functions $f \circ g$ and $g \circ f$ and give their domains.

11. $f(x) = x^2 + 1$, $g(x) = \sqrt{x-1}$
12. $f(x) = x^2 - x + 5$, $g(x) = -x + 4$
13. $f(x) = \dfrac{1}{2x-1}$, $g(x) = x^2 + 1$
14. $f(x) = \dfrac{x+1}{x}$, $g(x) = \dfrac{1}{x}$

In Problems 15–20, find the functions $f \circ g$ and $g \circ f$.

15. $f(x) = 2x - 3$, $g(x) = \tfrac{1}{2}(x + 3)$
16. $f(x) = x - 1$, $g(x) = x^3$
17. $f(x) = x + \dfrac{1}{x^2}$, $g(x) = \dfrac{1}{x}$
18. $f(x) = \sqrt{x-4}$, $g(x) = x^2$
19. $f(x) = x + 1$, $g(x) = x + \sqrt{x-1}$
20. $f(x) = x^3 - 4$, $g(x) = \sqrt[3]{x+3}$

In Problems 21–24, find $f \circ f$ and $f \circ (1/f)$.

21. $f(x) = 2x + 6$
22. $f(x) = x^2 + 1$
23. $f(x) = \dfrac{1}{x^2}$
24. $f(x) = \dfrac{x+4}{x}$

In Problems 25 and 26, find $(f \circ g \circ h)(x) = f(g(h(x)))$.

25. $f(x) = \sqrt{x}$, $g(x) = x^2$, $h(x) = x - 1$
26. $f(x) = x^2$, $g(x) = x^2 + 3x$, $h(x) = 2x$

27. For the functions $f(x) = 2x + 7$, $g(x) = 3x^2$, find $(f \circ g \circ g)(x)$.
28. For the functions $f(x) = -x + 5$, $g(x) = -4x^2 + x$, find $(f \circ g \circ f)(x)$.

In Problems 29 and 30, find $(f \circ f \circ f)(x) = f(f(f(x)))$.

29. $f(x) = 2x - 5$
30. $f(x) = x^2 - 1$

In Problems 31–34, find functions f and g such that $F(x) = f \circ g$.

31. $F(x) = (x^2 - 4x)^5$
32. $F(x) = \sqrt{9x^2 + 16}$
33. $F(x) = (x - 3)^2 + 4\sqrt{x-3}$
34. $F(x) = 1 + |2x + 9|$

In Problems 35 and 36, sketch the graphs of the compositions $f \circ g$ and $g \circ f$.

35. $f(x) = |x| - 2$, $g(x) = |x - 2|$
36. $f(x) = [\![x - 1]\!]$, $g(x) = |x|$

37. Consider the function $y = f(x) + g(x)$, where $f(x) = x$ and $g(x) = -[\![x]\!]$. Fill in the blanks and then sketch the graph of the sum $f + g$ over the indicated intervals.

$$y = \begin{cases} \vdots \\ \underline{\qquad}, & -3 \leq x < -2 \\ \underline{\qquad}, & -2 \leq x < -1 \\ \underline{\qquad}, & -1 \leq x < 0 \\ \underline{\qquad}, & 0 \leq x < 1 \\ \underline{\qquad}, & 1 \leq x < 2 \\ \underline{\qquad}, & 2 \leq x < 3 \\ \vdots \end{cases}$$

38. Consider the function $y = f(x) + g(x)$, where $f(x) = |x|$ and $g(x) = [\![x]\!]$. Proceed as in Problem 37 and then sketch the graph of the sum $f + g$.

In Problems 39 and 40, sketch the graph of the sum $f + g$.

39. $f(x) = |x - 1|$, $g(x) = |x|$
40. $f(x) = x$, $g(x) = |x|$

In Problems 41 and 42, sketch the graph of the product fg.

41. $f(x) = x, \quad g(x) = |x|$

42. $f(x) = x, \quad g(x) = [\![x]\!]$

In Problems 43 and 44, sketch the graph of the reciprocal $1/f$.

43. $f(x) = |x|$

44. $f(x) = x - 3$

In Problems 45 and 46, **(a)** find the points of intersection of the graphs of the given functions.
(b) Find the vertical distance d between the graphs on the interval I determined by the x-coordinates of their points of intersection.
(c) Use the concept of a vertex of a parabola to find the maximum value of d on the interval I.

45.

46.

FIGURE 3.5.3 Graph for Problem 45

FIGURE 3.5.4 Graph for Problem 46

In Problems 47–58, **(a)** compute the difference quotient $\dfrac{f(x+h) - f(x)}{h}$ for the given function.
(b) Find the slope of the secant line through the two points $(3, f(3))$, $(3.1, f(3.1))$.

47. $f(x) = -4x^2$

48. $f(x) = x^2 - x$

49. $f(x) = 3x^2 - x + 7$

50. $f(x) = 2x^2 + x - 1$

51. $f(x) = x^3 + 5x - 4$

52. $f(x) = 2x^3 + x^2$

53. $f(x) = \dfrac{1}{4 - x}$

54. $f(x) = \dfrac{3}{2x - 4}$

55. $f(x) = \dfrac{x}{x - 1}$

56. $f(x) = \dfrac{2x + 3}{x + 5}$

57. $f(x) = x + \dfrac{1}{x}$

58. $f(x) = \dfrac{1}{x^2}$

In Problems 59 and 60, compute the difference quotient $\dfrac{f(x+h) - f(x)}{h}$ for the given function. Use appropriate algebra in order to cancel the h in the denominator.

59. $f(x) = 2\sqrt{x}$

60. $f(x) = \sqrt{2x + 1}$

Miscellaneous Applications

61. For the Birds A bird-watcher sights a bird 100 ft due east of her position. If the bird is flying due south at a rate of 500 ft/min, express the distance d from the bird-watcher to the bird as a function of time t. Find the distance 5 minutes after the sighting. See FIGURE 3.5.5.

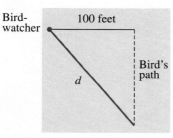

FIGURE 3.5.5 Bird-watcher in Problem 61

62. **Bacteria** A certain bacteria when cultured grows in a circular shape. The radius of the circle, measured in centimeters, is given by the mathematical model

$$r(t) = 4 - \frac{4}{t^2 + 1},$$

where time t is measured in hours.
 (a) Express the area covered by the bacteria as a function of time t.
 (b) Express the circumference of the area covered as a function of time t.

For Discussion

63. Suppose $f(x) = x^2 + 1$ and $g(x) = \sqrt{x}$. Discuss: Why is the domain of

$$(f \circ g)(x) = f(g(x)) = (\sqrt{x})^2 + 1 = x + 1$$

not $(-\infty, \infty)$?

64. Suppose $f(x) = \dfrac{2}{x - 1}$ and $g(x) = \dfrac{5}{x + 3}$. Discuss: Why is the domain of

$$(f \circ g)(x) = f(g(x)) = \frac{2}{g(x) - 1} = \frac{2}{\dfrac{5}{x + 3} - 1} = \frac{2x + 6}{2 - x}$$

not $\{x \mid x \neq 2\}$?

65. Find the error in the following reasoning: If $f(x) = 1/(x - 2)$ and $g(x) = 1/\sqrt{x + 1}$, then

$$\left(\frac{f}{g}\right)(x) = \frac{1/(x - 2)}{1/\sqrt{x + 1}} = \frac{\sqrt{x + 1}}{x - 2} \quad \text{and so} \quad \left(\frac{f}{g}\right)(-1) = \frac{\sqrt{0}}{-3} = 0.$$

66. Suppose $f_1(x) = \sqrt{x + 2}$, $f_2(x) = \dfrac{x}{\sqrt{x(x - 10)}}$, and $f_3(x) = \dfrac{x + 1}{x}$. What is the domain of the function $y = f_1(x) + f_2(x) + f_3(x)$?

67. Suppose $f(x) = x^3 + 4x$, $g(x) = x - 2$, and $h(x) = -x$. Discuss: Without actually graphing, how are the graphs of $f \circ g$, $g \circ f$, $f \circ h$, and $h \circ f$ related to the graph of f?

68. The domain of each piecewise-defined function,

$$f(x) = \begin{cases} x, & x < 0 \\ x + 1, & x \geq 0, \end{cases}$$

$$g(x) = \begin{cases} x^2, & x \leq -1 \\ x - 2, & x > -1, \end{cases}$$

is $(-\infty, \infty)$. Discuss how to find $f + g$, $f - g$, and fg. Carry out your ideas.

69. Discuss how the graph of $y = \frac{1}{2}\{f(x) + |f(x)|\}$ is related to the graph of $y = f(x)$. Illustrate your ideas using $f(x) = x^2 - 6x + 5$.

70. Discuss:
 (a) Is the sum of two even functions f and g even?
 (b) Is the sum of two odd functions f and g odd?
 (c) Is the product of an even function f with an odd function g even, odd, or neither?
 (d) Is the product of an odd function f with an odd function g even, odd, or neither?

71. The product fg of two linear functions with real coefficients, $f(x) = ax + b$ and $g(x) = cx + d$, is a quadratic function. Discuss: Why must the graph of this quadratic function have at least one x-intercept?
72. Make up two different functions f and g so that the domain of $F(x) = f \circ g$ is $[-2, 0) \cup (0, 2]$.

3.6 Inverse Functions

≡ **Introduction** Recall that a function f is a rule of correspondence that assigns to each value x in its domain X, a single or unique value y in its range. This rule does not preclude having the same number y associated with several *different* values of x. For example, for $f(x) = x^2 + 1$, the value $y = 5$ occurs at either $x = -2$ or $x = 2$. On the other hand, for the function $g(x) = x^3$, the value $y = 64$ occurs only at $x = 4$. Indeed, for every value y in the range of the function $g(x) = x^3$, there corresponds only one value of x in its domain. Functions of this last kind are given a special name.

DEFINITION 3.6.1 One-to-One Function

A function f is said to be **one-to-one** if each number in the range of f is associated with exactly one number in its domain X.

☐ **Horizontal Line Test** Interpreted geometrically, this means that a horizontal line ($y =$ constant) can intersect the graph of a one-to-one function in at most one point. Furthermore, if *every* horizontal line that intersects the graph of a function does so in at most one point, then the function is necessarily one-to-one. A function is *not* one-to-one if *some* horizontal line intersects its graph more than once.

EXAMPLE 1   Horizontal Line Test

The graphs of the functions $f(x) = x^2 + 1$ and $g(x) = x^3$, and a horizontal line $y = c$ intersecting the graphs of f and g, are shown in **FIGURE 3.6.1**. Figure 3.6.1(a) indicates that there are two numbers x_1 and x_2 in the domain of f for which $f(x_1) = f(x_2) = c$. Inspection of Figure 3.6.1(b) shows that for every horizontal line $y = c$ intersecting the graph, there is only one number x_1 in the domain of g such that $g(x_1) = c$. Hence the function f is not one-to-one, whereas the function g is one-to-one. ≡

(a) Not one-to-one

(b) One-to-one

FIGURE 3.6.1 Two types of functions

A one-to-one function can be defined in several different ways. Based on the preceding discussion, the following statement should make sense.

- *A function f is **one-to-one** if and only if $f(x_1) = f(x_2)$ implies $x_1 = x_2$ for all x_1 and x_2 in the domain of f.*   (1)

Stated in a negative way, (1) indicates that a function f is *not* one-to-one if different numbers x_1 and x_2 (that is, $x_1 \neq x_2$) can be found in the domain of f such that $f(x_1) = f(x_2)$. You will see this formulation of the one-to-one concept when we solve certain kinds of equations in Chapter 5.

You should consider (1) as a way of determining whether a function f is one-to-one without the benefit of a graph.

EXAMPLE 2   Checking for One-to-One

(a) Consider the function $f(x) = x^4 - 8x + 6$. Observe that $f(0) = f(2) = 6$ but since $0 \neq 2$ we can conclude that f is not one-to-one.

(b) Consider the function $f(x) = \dfrac{1}{2x - 3}$, and let x_1 and x_2 be numbers in the domain of f. If we assume $f(x_1) = f(x_2)$, that is, $\dfrac{1}{2x_1 - 3} = \dfrac{1}{2x_2 - 3}$, then by taking the reciprocal of both sides we see

$$2x_1 - 3 = 2x_2 - 3 \quad \text{implies} \quad 2x_1 = 2x_2 \quad \text{or} \quad x_1 = x_2.$$

From (1) we conclude that f is one-to-one. ≡

☐ **Inverse of a One-to-One Function** Suppose f is a one-to-one function with domain X and range Y. Since every number y in Y corresponds to precisely one number x in X, the function f must actually determine a "reverse" function f^{-1} whose domain is Y and range is X. As shown in **FIGURE 3.6.2**, f and f^{-1} must satisfy

$$f(x) = y \quad \text{and} \quad f^{-1}(y) = x. \tag{2}$$

The equations in (2) are actually the compositions of the functions f and f^{-1}:

$$f(f^{-1}(y)) = y \quad \text{and} \quad f^{-1}(f(x)) = x. \tag{3}$$

The function f^{-1} is called the **inverse** of f or the **inverse function** for f. Following convention that each domain element be denoted by the symbol x, the first equation in (3) is rewritten as $f(f^{-1}(x)) = x$. We summarize the two results given in (3).

Note of Caution: The symbol f^{-1} does *not* mean the reciprocal $1/f$. The number -1 is *not* an exponent.

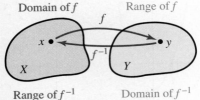

FIGURE 3.6.2 Functions f and f^{-1}

DEFINITION 3.6.2   Inverse Function

Let f be a one-to-one function with domain X and range Y. The **inverse** of f is the function f^{-1} with domain Y and range X for which

$$f(f^{-1}(x)) = x \text{ for every } x \text{ in } Y, \tag{4}$$

and

$$f^{-1}(f(x)) = x \text{ for every } x \text{ in } X. \tag{5}$$

Of course, if a function f is not one-to-one, then it has no inverse function.

☐ **Properties** Before we actually examine methods for finding the inverse of a one-to-one function f, let's list some important properties about f and its inverse f^{-1}.

THEOREM 3.6.1   Properties of Inverse Functions

(*i*) The domain of f^{-1} = range of f.
(*ii*) The range of f^{-1} = domain of f.
(*iii*) $y = f(x)$ is equivalent to $x = f^{-1}(y)$.
(*iv*) An inverse function f^{-1} is one-to-one.
(*v*) The inverse of f^{-1} is f, that is, $(f^{-1})^{-1} = f$.
(*vi*) The inverse of f is unique.

☐ **First Method for Finding f^{-1}** We will consider two ways of finding the inverse of a one-to-one function f. Both methods require that you solve an equation; the first method begins with (4).

EXAMPLE 3   Inverse of a Function

(a) Find the inverse of $f(x) = \dfrac{1}{2x - 3}$.

(b) Find the domain and range of f^{-1}. Find the range of f.

Solution (a) We proved in part (b) of Example 2 that f is one-to-one. To find the inverse of f using (4), we must substitute the symbol $f^{-1}(x)$ wherever x appears in f and then set the expression $f(f^{-1}(x))$ equal to x:

$$f(f^{-1}(x)) = \boxed{\dfrac{1}{2f^{-1}(x) - 3} = x.}$$

solve this equation for $f^{-1}(x)$

By taking the reciprocal of both sides of the equation in the blue outline box we get

$$2f^{-1}(x) - 3 = \dfrac{1}{x}$$

$$2f^{-1}(x) = 3 + \dfrac{1}{x} = \dfrac{3x + 1}{x}. \quad \leftarrow \text{common denominator}$$

Dividing both sides of the last equation by 2 yields the inverse of f:

$$f^{-1}(x) = \dfrac{3x + 1}{2x}.$$

(b) Inspection of f reveals that its domain is the set of real numbers except $\tfrac{3}{2}$, that is, $\{x \mid x \neq \tfrac{3}{2}\}$. Moreover, from the inverse just found we see that the domain of f^{-1} is $\{x \mid x \neq 0\}$. Because range of f^{-1} = domain of f we then know that the range of f^{-1} is $\{y \mid y \neq \tfrac{3}{2}\}$. From domain of f^{-1} = range of f we have also discovered that the range of f is $\{y \mid y \neq 0\}$. ≡

☐ **Second Method for Finding f^{-1}** The inverse of a function f can be found in a different manner. If f^{-1} is the inverse of f, then $x = f^{-1}(y)$. Thus we need only do the following two things:

- Solve $y = f(x)$ for the symbol x in terms of y (if possible). This gives $x = f^{-1}(y)$.
- Relabel the variable x as y and the variable y as x. This gives $y = f^{-1}(x)$.

EXAMPLE 4   Inverse of a Function

Find the inverse of $f(x) = x^3$.

Solution In Example 1 we saw that this function was one-to-one. To begin, we rewrite the function as $y = x^3$. Solving for x then gives $x = y^{1/3}$. Next we relabel variables to obtain $y = x^{1/3}$. Thus $f^{-1}(x) = x^{1/3}$ or equivalently $f^{-1}(x) = \sqrt[3]{x}$. ≡

Finding the inverse of a one-to-one function $y = f(x)$ is sometimes difficult and at times impossible. For example, it can be shown that the function $f(x) = x^3 + x + 3$ is one-to-one and so has an inverse f^{-1}, but solving the equation $y = x^3 + x + 3$ for x is difficult for everyone (including your instructor). Nevertheless, since f only involves positive integer powers of x, its domain is $(-\infty, \infty)$. If you investigate f graphically you are led to the fact that the range of f is also $(-\infty, \infty)$. Consequently the domain and range of f^{-1} are $(-\infty, \infty)$. Even though we don't know f^{-1} explicitly it makes complete sense to talk about the values such as $f^{-1}(3)$ and $f^{-1}(5)$. In the case of $f^{-1}(3)$ note that $f(0) = 3$. This means that $f^{-1}(3) = 0$. Can you figure out the value of $f^{-1}(5)$?

☐ **Graphs of f and f^{-1}** Suppose that (a, b) represents any point on the graph of a one-to-one function f. Then $f(a) = b$ and

$$f^{-1}(b) = f^{-1}(f(a)) = a$$

implies that (b, a) is a point on the graph of f^{-1}. As shown in **FIGURE 3.6.3(a)**, the points (a, b) and (b, a) are reflections of each other in line $y = x$. This means that the line $y = x$ is the perpendicular bisector of the line segment from (a, b) to (b, a). Because each point on one graph is the reflection of a corresponding point on the other graph, we see in Figure 3.6.3(b) that the graphs of f^{-1} and f are **reflections** of each other in the line $y = x$. We also say that the graphs of f^{-1} and f are **symmetric** with respect to the line $y = x$.

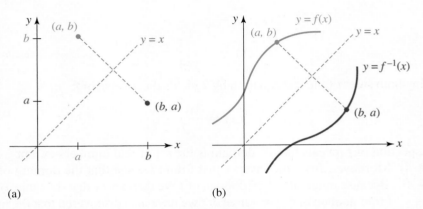

FIGURE 3.6.3 Graphs of f and f^{-1} are reflections in the line $y = x$

EXAMPLE 5 Graphs of f and f^{-1}

In Example 4 we saw that the inverse of $y = x^3$ is $y = x^{1/3}$. In **FIGURE 3.6.4(a)** and Figure 3.6.4(b) we show the graphs of these functions; in Figure 3.6.4(c) the graphs are superimposed on the same coordinate system to illustrate that the graphs are reflections of each other in the line $y = x$.

FIGURE 3.6.4 Graphs of f and f^{-1} in Example 5

Every linear function $f(x) = ax + b$, $a \neq 0$, is one-to-one.

EXAMPLE 6 Inverse of a Function

Find the inverse of the linear function $f(x) = 5x - 7$.

Solution Since the graph of $y = 5x - 7$ is a nonhorizontal line, it follows from the horizontal line test that f is a one-to-one function. To find f^{-1} solve $y = 5x - 7$ for x:

$$5x = y + 7 \quad \text{implies} \quad x = \frac{1}{5}y + \frac{7}{5}.$$

Relabeling the two variables in the last equation gives $y = \frac{1}{5}x + \frac{7}{5}$. Therefore $f^{-1}(x) = \frac{1}{5}x + \frac{7}{5}$. The graphs of f and f^{-1} are compared in FIGURE 3.6.5.

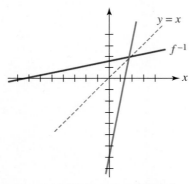

FIGURE 3.6.5 Graphs of f and f^{-1} in Example 6

Every quadratic function $f(x) = ax^2 + bx + c$, $a \neq 0$, is not one-to-one.

☐ **Restricted Domains** For a function f that is not one-to-one, it may be possible to restrict its domain in such a manner so that the new function consisting of f defined on this restricted domain is one-to-one and so has an inverse. In most cases we want to restrict the domain so that the new function retains its original range. The next example illustrates this concept.

EXAMPLE 7 Restricted Domain

In Example 1 we showed graphically that the quadratic function $f(x) = x^2 + 1$ is not one-to-one. The domain of f is $(-\infty, \infty)$, and as seen in FIGURE 3.6.6(a), the range of f is $[1, \infty)$. Now by defining $f(x) = x^2 + 1$ only on the interval $[0, \infty)$, we see two things in Figure 3.6.6(b): The range of f is preserved and $f(x) = x^2 + 1$ confined to the domain $[0, \infty)$ passes the horizontal line test, in other words, is one-to-one. The inverse of this new one-to-one function is obtained in the usual manner. Solving $y = x^2 + 1$ implies

$$x^2 = y - 1 \quad \text{and} \quad x = \pm\sqrt{y - 1} \quad \text{and so} \quad y = \pm\sqrt{x - 1}.$$

The appropriate algebraic sign in the last equation is determined from the fact that the domain and range of f^{-1} are $[1, \infty)$ and $[0, \infty)$, respectively. This forces us to choose $f^{-1}(x) = \sqrt{x - 1}$ as the inverse of f. See Figure 3.6.6(c).

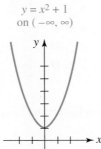

$y = x^2 + 1$
on $(-\infty, \infty)$

(a) Not a one-to-one function

$y = x^2 + 1$
on $[0, \infty)$

(b) One-to-one function

$y = \sqrt{x - 1}$
on $[1, \infty)$

(c) Inverse of function in part (b)

FIGURE 3.6.6 Inverse function in Example 7

3.6 Exercises
Answers to selected odd-numbered problems begin on page ANS-10.

In Problems 1–6, the graph of a function f is given. Use the horizontal line test to determine whether f is one-to-one.

1.

FIGURE 3.6.7 Graph for Problem 1

2.
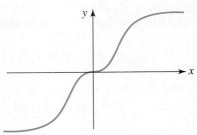
FIGURE 3.6.8 Graph for Problem 2

3.

FIGURE 3.6.9 Graph for Problem 3

4.

FIGURE 3.6.10 Graph for Problem 4

5.

FIGURE 3.6.11 Graph for Problem 5

6.

FIGURE 3.6.12 Graph for Problem 6

In Problems 7–10, sketch the graph of the given piecewise-defined function f to determine whether it is one-to-one.

7. $f(x) = \begin{cases} x - 2, & x < 0 \\ \sqrt{x}, & x \geq 0 \end{cases}$

8. $f(x) = \begin{cases} -\sqrt{-x}, & x < 0 \\ \sqrt{x}, & x \geq 0 \end{cases}$

9. $f(x) = \begin{cases} -x - 1, & x < 0 \\ x^2, & x \geq 0 \end{cases}$

10. $f(x) = \begin{cases} x^2 + x, & x < 0 \\ x^2 - x, & x \geq 0 \end{cases}$

In Problems 11–14, proceed as in Example 2(a) to show that the given function f is not one-to-one.

11. $f(x) = x^2 - 6x$

12. $f(x) = (x - 2)(x + 1)$

13. $f(x) = \dfrac{x^2}{4x^2 + 1}$

14. $f(x) = |x + 10|$

In Problems 15–18, proceed as in Example 2(b) to show that the given function f is one-to-one.

15. $f(x) = \dfrac{2}{5x + 8}$

16. $f(x) = \dfrac{2x - 5}{x - 1}$

17. $f(x) = \sqrt{4 - x}$

18. $f(x) = \dfrac{1}{x^3 + 1}$

In Problems 19 and 20, the given function f is one-to-one. Without finding f^{-1} find its domain and range.

19. $f(x) = 4 + \sqrt{x}$

20. $f(x) = 5 - \sqrt{x + 8}$

In Problems 21 and 22, the given function f is one-to-one. The domain and range of f is indicated. Find f^{-1} and give its domain and range.

21. $f(x) = \dfrac{2}{\sqrt{x}}$, $x > 0, y > 0$

22. $f(x) = 2 + \dfrac{3}{\sqrt{x}}$, $x > 0, y > 2$

In Problems 23–28, the given function f is one-to-one. Find f^{-1}. Sketch the graph of f and f^{-1} on the same coordinate axes.

23. $f(x) = -2x + 6$
24. $f(x) = -2x + 1$
25. $f(x) = x^3 + 2$
26. $f(x) = 1 - x^3$
27. $f(x) = 2 - \sqrt{x}$
28. $f(x) = \sqrt{x - 7}$

In Problems 29–32, the given function f is one-to-one. Find f^{-1}. Proceed as in Example 3(b) and find the domain and range of f^{-1}. Then find the range of f.

29. $f(x) = \dfrac{1}{2x - 1}$

30. $f(x) = \dfrac{2}{5x + 8}$

31. $f(x) = \dfrac{7x}{2x - 3}$

32. $f(x) = \dfrac{1 - x}{x - 2}$

In Problems 33–36, the given function f is one-to-one. Without finding f^{-1}, find the point on the graph of f^{-1} corresponding to the indicated value of x in the domain of f.

33. $f(x) = 2x^3 + 2x$; $x = 2$

34. $f(x) = 8x - 3$; $x = 5$

35. $f(x) = x + \sqrt{x}$; $x = 9$

36. $f(x) = \dfrac{4x}{x + 1}$; $x = \dfrac{1}{2}$

In Problems 37 and 38, sketch the graph of f^{-1} from the graph of f.

37.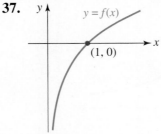

FIGURE 3.6.13 Graph for Problem 37

38.

FIGURE 3.6.14 Graph for Problem 38

In Problems 39 and 40, sketch the graph of f from the graph of f^{-1}.

39.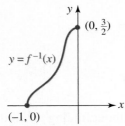

FIGURE 3.6.15 Graph for Problem 39

40.

FIGURE 3.6.16 Graph for Problem 40

In Problems 41–44, the function f is not one-to-one on the given domain but is one-to-one on the restricted domain (the second interval). Find the inverse of the one-to-one function and give its domain. Sketch the graph of f on the restricted domain and the graph of f^{-1} on the same coordinate axes.

41. $f(x) = 4x^2 + 2, (-\infty, \infty); \ [0, \infty)$ **42.** $f(x) = (3 - 2x)^2, (-\infty, \infty); \ [\frac{3}{2}, \infty)$
43. $f(x) = \frac{1}{2}\sqrt{4 - x^2}, [-2, 2]; \ [0, 2]$ **44.** $f(x) = \sqrt{1 - x^2}, [-1, 1]; \ [0, 1]$

In Problems 45 and 46, verify that $f(f^{-1}(x)) = x$ and $f^{-1}(f(x)) = x$.

45. $f(x) = 5x - 10, f^{-1}(x) = \frac{1}{5}x + 2$ **46.** $f(x) = \dfrac{1}{x + 1}, f^{-1}(x) = \dfrac{1 - x}{x}$

For Discussion

47. Suppose f is a continuous function that is increasing (or decreasing) for all x in its domain. Explain why f is necessarily one-to-one.

48. Explain why the graph of a one-to-one function f can have at most one x-intercept.

49. The function $f(x) = |2x - 4|$ is not one-to-one. How should the domain of f be restricted so that the new function has an inverse? Find f^{-1} and give its domain and range. Sketch the graph of f on the restricted domain and the graph of f^{-1} on the same coordinate axes.

50. The equation $y = \sqrt[3]{x} - \sqrt[3]{y}$ defines a one-to-one function $y = f(x)$. Find $f^{-1}(x)$.

51. What property do the one-to-one functions $y = f(x)$ shown in FIGURE 3.6.17 have in common? Find two more explicit functions with this same property. Be very clear about what this property has to do with f^{-1}.

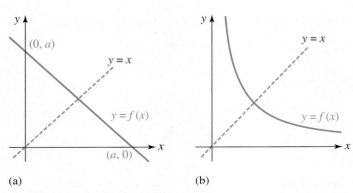

FIGURE 3.6.17 Graphs for Problem 51

3.7 Building a Function from Words

Introduction In subsequent courses in mathematics there are instances when you will be expected to translate the words that describe a problem into mathematical symbols and then set up or construct either an *equation* or a *function*.

In this section we focus on problems that involve functions. We begin with a verbal description about the product of two numbers.

EXAMPLE 1 Product of Two Numbers

The sum of two nonnegative numbers is 15. Express the product of one and the square of the other as a function of one of the numbers.

Solution We first represent the two numbers by the symbols x and y and recall that "nonnegative" means that $x \geq 0$ and $y \geq 0$. The first sentence then says that $x + y = 15$; this is *not* the function we are seeking. The second sentence describes the function we want; it is called "the product." Let's denote "the product" by the symbol P. Now P is the product of one of the numbers, say, x and the square of the other, that is, y^2:

$$P = xy^2. \qquad (1)$$

No, we are not finished because P is supposed to be a "function of *one* of the numbers." We now use the fact that the numbers x and y are related by $x + y = 15$. From this last equation we substitute $y = 15 - x$ into (1) to obtain the desired result:

$$P(x) = x(15 - x)^2. \qquad (2) \equiv$$

Here is a symbolic summary of the analysis of the problem given in Example 1:

$$
\underbrace{\text{The sum of two nonnegative numbers is 15.}}_{\text{let the numbers be } x \geq 0 \text{ and } y \geq 0,\ x+y=15} \text{ Express the } \underbrace{\text{product}}_{P} \text{ of } \underbrace{\text{one}}_{x} \text{ and the } \underbrace{\text{square of the other}}_{y^2} \text{ as a } \underbrace{\text{function of one of the numbers.}}_{\text{use } x} \qquad (3)
$$

Notice that the second sentence is vague about which number is squared. This means that it really doesn't matter; (1) could also be written as $P = yx^2$. Also, we could have used $x = 15 - y$ in (1) to arrive at $P(y) = (15 - y)y^2$. In a course such as calculus it would not have mattered whether we worked with $P(x)$ or with $P(y)$ because by finding *one* of the numbers we automatically find the other from the equation $x + y = 15$. This last equation is commonly called a **constraint**. A constraint not only defines the relationship between the variables x and y but often puts a limitation on how x and y can vary. As we see in the next example, the constraint helps in determining the domain of the function that you have just constructed.

EXAMPLE 2 Example 1 Continued

What is the domain of the function $P(x)$ in (2)?

Solution Taken out of the context of the statement of the problem in Example 1, one would have to conclude from the discussion on page 157 of Section 3.1 that the domain of the cubic polynomial function

$$P(x) = x(15 - x)^2 = 225x - 30x^2 + x^3$$

is the set of real numbers $(-\infty, \infty)$. *But* in the context of the original problem, the numbers were to be nonnegative. From the requirement that $x \geq 0$ *and* $y = 15 - x \geq 0$ we get $x \geq 0$ and $x \leq 15$, which means that x must satisfy the simultaneous inequality $0 \leq x \leq 15$. Using interval notation, the domain of the product function P in (2) is the closed interval $[0, 15]$. ≡

Another way of looking at the conclusion of Example 2 is this: The constraint $x + y = 15$ dictates that $y = 15 - x$. Thus *if* x were allowed to be larger than 15 (say, $x = 17.5$), then $y = 15 - x$ would be a negative number that contradicts the initial assumption that $y \geq 0$.

Invariably whenever word problems are discussed in a mathematics class, students often react with groans, ambivalence, and dismay. While not guaranteeing anything, the following suggestions might help you to get through the problems in Exercises 3.7. These problems are especially important if your future plans include taking a course in calculus.

GUIDELINES FOR BUILDING A FUNCTION

(*i*) At least try to develop a positive attitude.
(*ii*) Try to be neat and organized.
(*iii*) Read the problem slowly. Then read the problem several more times.
(*iv*) Whenever possible, sketch a curve or a picture and identify given quantities in your sketch. Keep your sketch simple.
(*v*) If the description of the function indicates two variables, say x and y, then look for a constraint or relationship between the variables (such as $x + y = 15$ in Example 1). Use the constraint to eliminate one of the variables to express the required function in terms of one variable.

EXAMPLE 3   Area of a Rectangle

A rectangle has two vertices on the x-axis and two vertices on the semicircle whose equation is $y = \sqrt{25 - x^2}$. See **FIGURE 3.7.1(a)**. Express the area of the rectangle as a function of x.

Solution If (x, y), $x > 0$, $y > 0$, denotes the vertex of the rectangle on the circle in the first quadrant, then as shown in Figure 3.7.1(b) the area A is length × width, or

$$A = (2x) \times y = 2xy. \tag{4}$$

The constraint in this problem is the equation $y = \sqrt{25 - x^2}$ of the semicircle. We use the constraint equation to eliminate y in (4) and obtain the area of the rectangle

$$A(x) = 2x\sqrt{25 - x^2}. \tag{5}$$ ≡

Were we again to consider the function $A(x)$ out of the context of the problem in Example 3, its domain would be $[-5, 5]$. Because we assumed that $x > 0$ we can take the domain of $A(x)$ in (4) to be the open interval $(0, 5)$.

FIGURE 3.7.1 Rectangle in Example 3

EXAMPLE 4   Amount of Fencing

A rancher intends to mark off a rectangular plot of land that will have an area of 1000 m². The plot will be fenced and divided into two equal portions by an additional fence parallel to two sides. Express the amount of fencing used to enclose the plot of land as a function of the length of one side of the plot.

Solution A sketch of the land enclosed by the fence is a rectangle with a line drawn down its middle, similar to that given in **FIGURE 3.7.2**. As shown in the figure, let $x > 0$ be the length of the rectangular plot of land and let $y > 0$ denote its width. If the symbol F represents this amount, then the sum of the lengths of the *five* portions—two horizontal and three vertical—of the fence is

$$F = 2x + 3y. \tag{6}$$

Since we want F to be a function of the length of one side of the plot of land we must eliminate either x or y from (6). Because the fenced-in land is to have an area of 1000 m², x and y are related by the constraint $xy = 1000$. From the last equation we get $y = 1000/x$, which can be used to eliminate y in (6). Thus, the amount of fence F as a function of x is $F(x) = 2x + 3(1000/x)$ or

$$F(x) = 2x + \frac{3000}{x}. \tag{7}$$

Since x represents a physical dimension that satisfies $xy = 1000$, we conclude that it is positive. But other than that, there is no restriction on x. Notice that if the positive number x is close to 0 then $y = 1000/x$ is very large, whereas if x is taken to be a very large number, then y is close to 0. Thus the domain of $F(x)$ is $(0, \infty)$. ≡

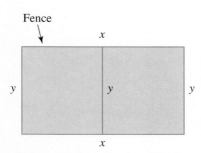

FIGURE 3.7.2 Rectangular plot of land in Example 4

If a problem involves triangles, you should study the problem carefully and determine whether the Pythagorean theorem, similar triangles, or trigonometry is applicable (see Section 8.2).

EXAMPLE 5   Length of a Ladder

A 10-ft wall stands 5 ft from a building. A ladder, supported by the wall, touches a building as shown in **FIGURE 3.7.3**. Express the length of the ladder as a function of x shown in the figure.

Solution Let L denote the length of the ladder. With x and y defined in Figure 3.7.3, we see that there are two right triangles; the larger triangle has three sides with lengths L, y, and $x + 5$ and the smaller triangle has two sides of lengths x and 10. Now the ladder is the hypotenuse of the larger right triangle, so by the Pythagorean theorem,

$$L^2 = (x + 5)^2 + y^2. \tag{8}$$

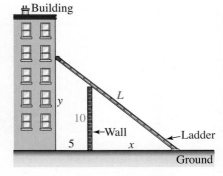

FIGURE 3.7.3 Ladder in Example 5

The right triangles in Figure 3.7.3 are similar because they both contain a right angle and share the common acute angle the ladder makes with the ground. We then use the fact that the ratios of corresponding sides of similar triangles are equal. This enables us to write the constraint

$$\frac{y}{x + 5} = \frac{10}{x}.$$

Solving the last equation for y in terms of x gives $y = 10(x + 5)/x$, and so (8) becomes

$$L^2 = (x + 5)^2 + \left(\frac{10(x + 5)}{x}\right)^2$$

$$= (x + 5)^2\left(1 + \frac{100}{x^2}\right) \quad \leftarrow \text{factoring } (x + 5)^2$$

$$= (x + 5)^2\left(\frac{x^2 + 100}{x^2}\right). \quad \leftarrow \text{common denominator}$$

3.7 Building a Function from Words

Taking the square root gives us L as a function of x,

$$L(x) = \frac{x+5}{x}\sqrt{x^2+100}. \quad \leftarrow \text{square root of a product is the product of the square roots}$$

The domain of the function $L(x)$ is $(0, \infty)$.

EXAMPLE 6   Distance to a Point

Express the distance from a point (x, y) in the first quadrant on the circle $x^2 + y^2 = 1$ to the point $(2, 4)$ as a function of x.

Solution Let d represent the distance from (x, y) to $(2, 4)$. See **FIGURE 3.7.4**. Then from the distance formula, (2) of Section 2.1,

$$d = \sqrt{(x-2)^2 + (y-4)^2} = \sqrt{x^2 + y^2 - 4x - 8y + 20}. \quad (9)$$

The constraint in this problem is the equation of the circle $x^2 + y^2 = 1$. From this we can immediately replace $x^2 + y^2$ in (9) by the number 1. Moreover, using the constraint to write $y = \sqrt{1-x^2}$ allows us to eliminate y in (9). Thus the distance d as a function of x is

$$d(x) = \sqrt{21 - 4x - 8\sqrt{1-x^2}}. \quad (10)$$

Because (x, y) is a point on the circle in the first quadrant the variable x can range from 0 to 1; that is, the domain of the function in (10) is the closed interval $[0, 1]$.

FIGURE 3.7.4 Distance d in Example 6

NOTES FROM THE CLASSROOM

You should not get the impression that every problem that requires you to set up a function from a verbal description must have a constraint. In Problems 11–16 of Exercises 3.7 the required function can be set up using just one variable.

3.7 Exercises
Answers to selected odd-numbered problems begin on page ANS-11.

In Problems 1–40, translate the words into an appropriate function. Give the domain of the function.

1. The product of two positive numbers is 50. Express their sum as a function of one of the numbers.
2. Express the sum of a nonzero number and its reciprocal as a function of the number.
3. The sum of two nonnegative numbers is 1. Express the sum of the square of one and twice the square of the other as a function of one of the numbers.
4. Let m and n be positive integers. The sum of two nonnegative numbers is S. Express the product of the mth power of one and the nth power of the other as a function of one of the numbers.

5. A rectangle has a perimeter of 200 in. Express the area of the rectangle as a function of the length of one of its sides.
6. A rectangle has an area of 400 in². Express the perimeter of the rectangle as a function of the length of one of its sides.
7. Express the area of the rectangle shaded in FIGURE 3.7.5 as a function of x.
8. Express the length of the line segment containing the point $(2, 4)$ shown in FIGURE 3.7.6 as a function of x.
9. Express the distance from a point (x, y) on the graph of $x + y = 1$ to the point $(2, 3)$ as a function of x.
10. Express the distance from a point (x, y) on the graph of $y = 4 - x^2$ to the point $(0, 1)$ as a function of x.
11. Express the perimeter of a square as a function of its area A.
12. Express the area of a circle as a function of its diameter d.
13. Express the diameter of a circle as a function of its circumference C.
14. Express the volume of a cube as a function of the area A of its base.
15. Express the area of an equilateral triangle as a function of its height h.
16. Express the area of an equilateral triangle as a function of the length s of one of its sides.
17. A wire of length x is bent into the shape of a circle. Express the area of the circle as a function of x.
18. A wire of length L is cut x units from one end. One piece of the wire is bent into a square and the other piece is bent into a circle. Express the sum of the areas as a function of x.
19. A tree is planted 30 ft from the base of a street lamp that is 25 ft tall. Express the length of the tree's shadow as a function of its height. See FIGURE 3.7.7.
20. The frame of a kite consists of six pieces of lightweight plastic. The outer frame of the kite consists of four precut pieces; two pieces of length 2 ft and two pieces of length 3 ft. Express the area of the kite as a function of x, where $2x$ is the length of the horizontal crossbar piece shown in FIGURE 3.7.8.

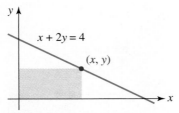

FIGURE 3.7.5 Rectangle in Problem 7

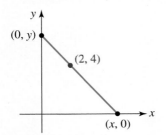

FIGURE 3.7.6 Line segment in Problem 8

FIGURE 3.7.7 Tree in Problem 19

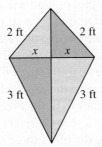

FIGURE 3.7.8 Kite in Problem 20

21. A company wants to construct an open rectangular box with a volume of 450 in³ so that the length of its base is 3 times its width. Express the surface area of the box as a function of the width.
22. A conical tank, with vertex down, has a radius of 5 ft and a height of 15 ft. Water is pumped into the tank. Express the volume of the water as a function of its depth. [*Hint*: The volume of a cone is $V = \frac{1}{3}\pi r^2 h$. Although the tank is a three-dimensional object, examine it in cross section as a two-dimensional triangle. See FIGURE 3.7.9.]

FIGURE 3.7.9 Conical tank in Problem 22

FIGURE 3.7.10 Cars in Problem 23

23. Car A passes point O heading east at a constant rate of 40 mi/h; car B passes the same point 1 hour later heading north at a constant rate of 60 mi/h. Express the distance between the cars as a function of time t, where t is measured starting when car B passes point O. See FIGURE 3.7.10.

24. At time $t = 0$ (measured in hours), two airliners with a vertical separation of 1 mile, pass each other going in opposite directions. If the planes are flying horizontally at rates of 500 mi/h and 550 mi/h, express the horizontal distance between them as a function of t. [*Hint*: Distance = rate × time.]

25. The swimming pool shown in FIGURE 3.7.11 is 3 ft deep at the shallow end, 8 ft deep at the deepest end, 40 ft long, 30 ft wide, and the bottom is an inclined plane. Water is pumped into the pool. Express the volume of the water in the pool as a function of height h of the water above the deep end. [*Hint*: The volume will be a piecewise-defined function with domain defined by $0 \leq h \leq 8$.]

26. U.S. Postal Service regulations for parcel post stipulate that the length plus girth (the perimeter of one end) of a package must not exceed 108 inches. Express the volume of the package as a function of the width x shown in FIGURE 3.7.12. [*Hint*: Assume that the length plus girth equals 108.]

FIGURE 3.7.11 Swimming pool in Problem 25

FIGURE 3.7.12 Package in Problem 26

27. Consider all rectangles that have the same perimeter p. (Here p represents a constant.) Express the area of such a rectangle as a function of the length of one side.

28. The length of a rectangle is x, its height is y, and its perimeter is 20 inches. Express the length of the diagonal of the rectangle as a function of the length x.

29. A rectangular plot of land will be fenced into three equal portions by two dividing fences parallel to two sides. If the area to be enclosed is 4000 m², express the amount of fence needed as a function of the length of the plot of land.

30. A rectangular plot of land will be fenced into three equal portions by two dividing fences parallel to two sides. If the total fence to be used is 8000 m, express the area of the fenced land as a function.

31. A rancher wishes to build a rectangular corral with an area of 128,000 ft² with one side along a straight river. The fence along the river costs $1.50 per foot, whereas along the other three sides the fence costs $2.50 per foot. Express the cost of construction as a function of the length of fence along the river. [*Hint*: Along the river the cost of x ft of fence is $1.50x$.]

32. Express the area of the colored triangular region shown in FIGURE 3.7.13 as a function of x.

33. An open rectangular box is to be constructed with a square base and a volume of 32,000 cm³. Express the amount of material used in its construction as a function of x. See FIGURE 3.7.14.

34. A closed rectangular box is to be constructed with a square base. The material for the top costs $2 per square foot, whereas the material for the remaining sides costs $1 per square foot. The total cost to construct each box is $36. Express the volume of the box as a function of the length of one side of the base.

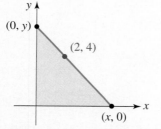

FIGURE 3.7.13 Triangular region in Problem 32

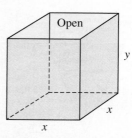

FIGURE 3.7.14 Box in Problem 33

35. A rain gutter with a rectangular cross section is made from a 1-ft × 20-ft piece of metal by bending up equal amounts from the 1-ft side. See FIGURE 3.7.15. Express the capacity of the gutter as a function of x. [*Hint*: Capacity = volume.]
36. A juice can is to be made in the form of a right circular cylinder and have a volume of 32 in^3. See FIGURE 3.7.16(a). Express the amount of material used in its construction as a function of the radius r of the circular cylinder. [*Hint*: Material = total surface area of can = area of top + area of bottom + area of lateral side. If the circular top and bottom covers are removed and the cylinder is cut straight up its side and flattened out, the result is the rectangle shown in Figure 3.7.16(c).]

FIGURE 3.7.15 Rain gutter in Problem 35

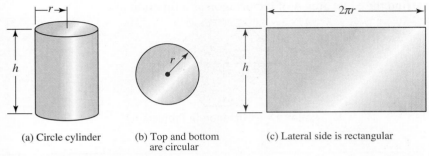

(a) Circle cylinder (b) Top and bottom are circular (c) Lateral side is rectangular

FIGURE 3.7.16 Juice can in Problem 36

37. A printed page will have 2-inch margins of white space on the sides and 1-inch margins of white space on the top and bottom. See FIGURE 3.7.17. The area of the printed portion is 32 in^2. Express the area of the page as a function of the height of the printed portion.

FIGURE 3.7.17 Printed page in Problem 37

38. Many medications are packaged in capsules as shown in the accompanying photo. Assume that a capsule is formed by adjoining two hemispheres to the ends of a right circular cylinder as shown in FIGURE 3.7.18. The total volume of the capsule is to be 0.007 in^3. Express the amount of material used in its construction as a function of the radius r of each hemisphere. [*Hint*: The volume of a sphere is $\frac{4}{3}\pi r^3$ and its surface area is $4\pi r^2$.]

Capsules

39. A 20-ft long water trough has ends in the form of isosceles triangles with sides that are 4 ft long. Express the volume of the trough as a function of the length x shown in FIGURE 3.7.19. [*Hint*: A *right cylinder* is not necessarily a *circular cylinder* where the top and bottom are circles. The top and bottom of a right cylinder are the same but could be a triangle, a pentagon, a trapezoid, and so on. The volume of a right cylinder is the area of the base × the height.]

FIGURE 3.7.18 Model of a capsule in Problem 38

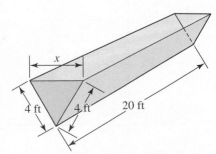

FIGURE 3.7.19 Water trough in Problem 39

3.7 Building a Function from Words

FIGURE 3.7.20 The bird in Problem 40

40. Some birds fly more slowly over water than over land. A bird flies at constant rates 6 km/h over water and 10 km/h over land. Use the information in FIGURE 3.7.20 to express the total flying time between the shore of one island and its nest on the shore of another island in terms of x. [*Hint*: Distance = rate × time.]

For Discussion

41. In Problem 19, what happens to the length of the tree's shadow as its height approaches 25 ft?
42. In an engineering text, the area of the octagon shown in FIGURE 3.7.21 is given as $A = 3.31r^2$. Show that this formula is actually an approximation to the area; that is, find the exact area A of the octagon as a function of r.

FIGURE 3.7.21 Octagon in Problem 42

3.8 Least Squares Line

Introduction
When performing experiments, we often tabulate data in the form of ordered pairs $(x_1, y_1), (x_2, y_2), \ldots, (x_n, y_n)$, with each x_i distinct. Given the data, it is then often desirable to be able to extrapolate or predict y from x by finding a mathematical model—that is, a function that approximates or "fits" the data. In other words, we want a function f such that

$$f(x_1) \approx y_1, f(x_2) \approx y_2, \ldots, f(x_n) \approx y_n.$$

Naturally, we do not want just any function but a function that fits the data as closely as possible. In the discussion that follows we shall confine our attention to the problem of finding a linear polynomial $y = mx + b$ or a straight line that "best fits" the data $(x_1, y_1), (x_2, y_2), \ldots, (x_n, y_n)$. The procedure for finding this linear function is known as **the method of least squares**.

EXAMPLE 1 Fitting a Line to Data

Consider the data (1, 1), (2, 2), (3, 4), (4, 6), (5, 5) shown in FIGURE 3.8.1(a). Looking at Figure 3.8.1(b) and seeing that the line $y = x + 1$ passes through two of the data points, we might take this line as the one that best fits the data.

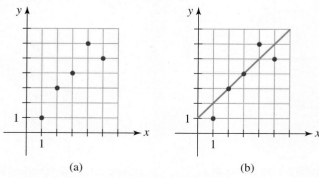

FIGURE 3.8.1 Data in (a); a line fitting data in (b)

Obviously we need something better than a visual guess to determine the linear function $y = f(x)$ as in Example 1. We need a criterion that defines the concept of "best fit" or, as it is sometimes called, the "goodness of fit."

If we try to match the data points with the line $y = mx + b$, then we wish to find m and b that satisfy the system of equations

$$\begin{aligned} y_1 &= mx_1 + b \\ y_2 &= mx_2 + b \\ &\vdots \\ y_n &= mx_n + b. \end{aligned} \tag{1}$$

The system of equations (1) is an example of an **overdetermined system**; that is, a system in which the number of equations is greater than the number of unknowns. We do not expect such a system to have a solution unless, of course, the data points all lie on the same line.

☐ **Least Squares Line** If the data points are $(x_1, y_1), (x_2, y_2), \ldots, (x_n, y_n)$, then one manner of determining how well the linear function $f(x) = mx + b$ fits the data is to measure the vertical distances between the points and the graph of f:

$$e_i = |y_i - f(x_i)|, \quad i = 1, 2, \ldots, n.$$

We can think of each e_i as the **error** in approximating the data value y_i by the function value $f(x_i)$. See **FIGURE 3.8.2**. Intuitively, the function f will fit the data well if the sum of all the e_i is a minimum. Actually, a more convenient approach to the problem is to find a linear function f so that the *sum of the squares* of all the e_i is a minimum. We shall define the solution of the system (1) to be those coefficients m and b that minimize the expression

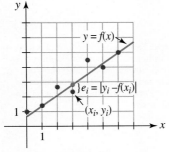

FIGURE 3.8.2 Error in approximating y_i by $f(x_i)$

$$\begin{aligned} E &= e_1^2 + e_2^2 + \cdots + e_n^2 \\ &= [y_1 - f(x_1)]^2 + [y_2 - f(x_2)]^2 + \cdots + [y_n - f(x_n)]^2 \\ &= [y_1 - (mx_1 + b)]^2 + [y_2 - (mx_2 + b)]^2 + \cdots + [y_n - (mx_n + b)]^2. \end{aligned} \tag{2}$$

The expression E is called the **sum of the square errors**. The line $y = mx + b$ that minimizes the sum of the square errors (2) is called the **least squares line** or the **line of best fit** for the data $(x_1, y_1), (x_2, y_2), \ldots, (x_n, y_n)$.

☐ **Summation Notation** Before proceeding to find the least squares line it is convenient to introduce a shorthand notation for sums of numbers. Writing out sums such as (1) can become very tedious. Suppose a_k denotes a real number that depends on an integer k. The sum of n such real numbers a_k, $a_1 + a_2 + a_3 + \cdots + a_n$ is denoted by the symbol $\sum_{k=1}^{n} a_k$, that is,

$$\sum_{k=1}^{n} a_k = a_1 + a_2 + a_3 + \cdots + a_n. \tag{3}$$

◂ Summation notation is covered in greater detail in Section 13.2.

Since Σ is the capital Greek letter sigma, (3) is called **summation notation** or **sigma notation**. The integer k is called the **index of summation** and takes on consecutive integer values starting with $k = 1$ and ending with $k = n$. For example, the sum of the first 100 squared positive integers,

$$1^2 + 2^2 + 3^2 + 4^2 + \cdots + 98^2 + 99^2 + 100^2$$

can be written compactly as

$$\sum_{k=1}^{100} k^2.$$

sum ends with this number (100)
sum starts with this number ($k=1$)

Written using sigma notation the sum of the square errors (2) can be written compactly as

$$E = \sum_{i=1}^{n} [y_i - f(x_i)]^2 \qquad (4)$$
$$= \sum_{i=1}^{n} [y_i - mx_i - b]^2.$$

The problem remains now, can one find a slope m and a y-coordinate b so that (4) is minimum? The answer is: Yes. Although we shall forego the details (which require calculus), the values of m and b that yield the minimum value of E are given by

$$m = \frac{n\sum_{i=1}^{n} x_i y_i - \sum_{i=1}^{n} x_i \sum_{i=1}^{n} y_i}{n\sum_{i=1}^{n} x_i^2 - \left(\sum_{i=1}^{n} x_i\right)^2}, \quad b = \frac{\sum_{i=1}^{n} x_i^2 \sum_{i=1}^{n} y_i - \sum_{i=1}^{n} x_i y_i \sum_{i=1}^{n} x_i}{n\sum_{i=1}^{n} x_i^2 - \left(\sum_{i=1}^{n} x_i\right)^2}. \qquad (5)$$

Don't panic. Evaluating these formulas just requires arithmetic. Also, most graphing calculators can either compute these sums or give you m and b after entering the data. See your owners' manual or the *SRM* that accompanies this text.

EXAMPLE 2 Least Squares Line

Find the least squares line for the data in Example 1. Calculate the sum of the square errors E for this line and the line $y = x + 1$.

Solution From the data (1, 1), (2, 2), (3, 4), (4, 6), (5, 5) we identify $x_1 = 1$, $x_2 = 2$, $x_3 = 3$, $x_4 = 4$, $x_5 = 5$, $y_1 = 1$, $y_2 = 3$, $y_3 = 4$, $y_4 = 6$, and $y_5 = 5$. With these values and $n = 5$, we have

$$\sum_{i=1}^{5} x_i y_i = 68, \quad \sum_{i=1}^{5} x_i = 15, \quad \sum_{i=1}^{5} y_i = 19, \quad \sum_{i=1}^{5} x_i^2 = 55.$$

Substituting these values into the formulas in (5) yields $m = 1.1$ and $b = 0.5$. Thus, the least squares line is $y = 1.1x + 0.5$. By way of comparison, **FIGURE 3.8.3** shows the data, the line $y = x + 1$ in blue and the least squares line $y = 1.1x + 0.5$ in red. ≡

FIGURE 3.8.3 Least squares line (red) in Example 2

For the least squares line $f(x) = 1.1x + 0.5$ found in Example 2 the sum of the square errors (4) is

$$E = \sum_{i=1}^{n} [y_i - mx_i - b]^2$$
$$= [1 - f(1)]^2 + [3 - f(2)]^2 + [4 - f(3)]^2 + [6 - f(4)]^2 + [5 - f(5)]^2$$
$$= [1 - 1.6]^2 + [3 - 2.7]^2 + [4 - 3.8]^2 + [6 - 4.9]^2 + [5 - 6]^2 = 2.7.$$

For the line $y = x + 1$ that we guessed in Example 1 that also passed through two of the data points, we find the sum of the square errors is $E = 3.0$.

Note ▶

In Section 12.6 we will examine another method for obtaining a least squares lines.

It is possible to generalize the least squares technique. For example, we might want to fit the given data to a quadratic polynomial $f(x) = ax^2 + bx + c$ instead of a linear polynomial.

3.8 Exercises

Answers to selected odd-numbered problems begin on page ANS-11.

In Problems 1–6, find the least squares line for the given data.

1. (2, 1), (3, 2), (4, 3), (5, 2)
2. (0, −1), (1, 3), (2, 5), (3, 7)
3. (1, 1), (2, 1.5), (3, 3), (4, 4.5), (5, 5)
4. (0, 0), (2, 1.5), (3, 3), (4, 4.5), (5, 5)
5. (0, 2), (1, 3), (2, 5), (3, 5), (4, 9), (5, 8), (6, 10)
6. (1, 2), (2, 2.5), (3, 1), (4, 1.5), (5, 2), (6, 3.2), (7, 5)

Miscellaneous Applications

7. In an experiment the correspondence given in the table was found between temperature T (in °C) and kinematic viscosity v (in Centistokes) of an oil with a certain additive. Find the least squares line $v = mT + b$. Use this line to estimate the viscosity of the oil at $T = 140$ and $T = 160$.

T	20	40	60	80	100	120
v	220	200	180	170	150	135

8. In an experiment the correspondence given in the table was found between temperature T (in °C) and electrical resistance R (in $M\Omega$). Find the least squares line $R = mT + b$. Use this line to estimate the resistance at $T = 700$.

T	400	450	500	550	600	650
R	0.47	0.90	2.0	3.7	7.5	15

Calculator/CAS Problems

9. (a) A set of data points can be approximated by a least squares *polynomial* of degree n. Learn the syntax for the CAS you have at hand, to obtain a least squares line (linear polynomial), a least squares quadratic, and a least squares cubic to fit the data

$$(-5.5, 0.8), (-3.3, 2.5), (-1.2, 3.8), (0.7, 5.2), (2.5, 5.6), (3.8, 6.5).$$

   (b) Use a CAS to superimpose the plots of the data and the least squares line obtained in part (a) on the same coordinate axes. Repeat for the plots of the data and the least squares quadratic and then the data and the least squares cubic.

10. Use the U.S. census data (in millions) from the year 1900 through 2000

1900	1920	1940	1960	1980	2000
75.994575	105.710620	131.669275	179.321750	226.545805	281.421906

and a least squares line to predict the U.S. population in the year 2020.

CONCEPTS REVIEW

You should be able to give the meaning of each of the following concepts.

Function:
   domain of
   range of
   zero of
Independent variable
Dependent variable
Graphs:
   x-intercept
   y-intercept
Vertical line test
Power function
Monomial

Polynomial function
Linear function
Quadratic function
Odd function
Even function
Rigid transformation:
   vertical shift
   horizontal shift
Reflections
Nonrigid transformations:
   stretches
   compressions

Piecewise-defined function:
   greatest integer function
Continuous function
Arithmetic combinations of functions
Function composition
Difference quotient
One-to-one function
Horizontal line test
Inverse function
Restricted domain
Least squares line:
   sum of the square errors

CHAPTER 3 Review Exercises

Answers to selected odd-numbered problems begin on page ANS-11.

A. True/False

In Problems 1–22, answer true or false:

1. If $(4, 0)$ is on the graph of f, then $(1, 0)$ must be on the graph of $y = \frac{1}{4}f(x)$. ___
2. The graph of a function can have only one y-intercept. ___
3. If f is a function such that $f(a) = f(b)$, then $a = b$. ___
4. No nonzero function f can be symmetric with respect to the x-axis. ___
5. The domain of $f(x) = (x - 1)^{1/3}$ is $(-\infty, \infty)$. ___
6. If $f(x) = x$ and $g(x) = \sqrt{x + 2}$, then the domain of g/f is $[-2, \infty)$. ___
7. A function f is one-to-one if it never takes on the same value twice. ___
8. The domain of the function $y = \sqrt{-x}$ is $(-\infty, 0]$. ___
9. The graph of $y = \sqrt{-x}$ is a reflection of the graph of $f(x) = \sqrt{x}$ in the y-axis. ___
10. A point of intersection of the graphs of f and f^{-1} must lie on the line $y = x$. ___
11. The one-to-one function $f(x) = 1/x$ has the property that $f = f^{-1}$. ___
12. The function $f(x) = 2x^2 + 16x - 2$ decreases on the interval $[-7, -5]$. ___
13. No even function defined on $(-a, a)$, $a > 0$, can be one-to-one. ___
14. All odd functions are one-to-one. ___
15. If a function f is one-to-one, then $f^{-1}(x) = \dfrac{1}{f(x)}$. ___
16. If f is an increasing function on an interval containing $x_1 < x_2$, then $f(x_1) < f(x_2)$. ___
17. The function $f(x) = |x| - 1$ is decreasing on the interval $[0, \infty)$. ___
18. For function composition, $f \circ (g + h) = f \circ g + f \circ h$. ___
19. If the y-intercept for the graph of a function f is $(0, 1)$, then the y-intercept for the graph of $y = 5 - 3f(x)$ is $(0, 2)$. ___
20. If f is a linear function, then $f(x_1 + x_2) = f(x_1) + f(x_2)$. ___
21. The function $f(x) = x^2 + 2x + 1$ is one-to-one on the restricted domain $[-1, \infty)$. The range of f^{-1} is also $[-1, \infty)$. ___
22. The function $f(x) = x^3 + 2x + 5$ is one-to-one. The point $(8, 1)$ is on the graph of f^{-1}. ___

B. Fill in the Blanks

In Problems 1–20, fill in the blanks.

1. If $f(x) = \dfrac{2x^3 - 1}{x^2 + 2}$, then $(\tfrac{1}{2}, \underline{})$ is a point on the graph of f.
2. If $f(x) = \dfrac{Ax}{10x - 2}$ and $f(2) = 3$, then $A = $ \underline{}.
3. The domain of the function $f(x) = \dfrac{1}{\sqrt{5 - x}}$ is \underline{}.
4. The range of the function $f(x) = |x| - 10$ is \underline{}.
5. The zeros of the function $f(x) = \sqrt{x^2 - 2x}$ are \underline{}.
6. If the graph of f is symmetric with respect to the y-axis, $f(-x) = $ \underline{}.
7. If f is an odd function such that $f(-2) = 2$, then $f(2) = $ \underline{}.
8. The graph of a linear function for which $f(-1) = 1$ and $f(1) = 5$ has slope $m = $ \underline{}.
9. A linear function whose intercepts are $(-1, 0)$ and $(0, 4)$ is $f(x) = $ \underline{}.
10. If the graph of $y = |x - 2|$ is shifted 4 units to the left, then its equation is \underline{}.
11. The x- and y-intercepts of the parabola $f(x) = x^2 - 2x - 1$ are \underline{}.
12. The range of the function $f(x) = -x^2 + 6x - 21$ is \underline{}.
13. The quadratic function $f(x) = ax^2 + bx + c$ for which $f(0) = 7$ and whose only x-intercept is $(-2, 0)$ is $f(x) = $ \underline{}.
14. If $f(x) = x + 2$ and $g(x) = x^2 - 2x$, then $(f \circ g)(-1) = $ \underline{}.
15. The vertex of the graph of $f(x) = x^2$ is $(0, 0)$. Therefore, the vertex of the graph of $y = -5(x - 10)^2 + 2$ is \underline{}.
16. Given that $f^{-1}(x) = \sqrt{x - 4}$ is the inverse of a one-to-one function f. Without finding f, the domain of f is \underline{} and the range of f is \underline{}.
17. The x-intercept of a one-to-one function f is $(5, 0)$ and so the y-intercept of f^{-1} is \underline{}.
18. The inverse of $f(x) = \dfrac{x - 5}{2x + 1}$ is $f^{-1} = $ \underline{}.
19. For $f(x) = [\![x + 2]\!] - 4$, $f(-5.3) = $ \underline{}.
20. If f is a one-to-one function such that $f(1) = 4$, then $f(f^{-1}(4)) = $ \underline{}.

C. Review Exercises

In Problems 1 and 2, identify two functions f and g so that $h = f \circ g$.

1. $h(x) = \dfrac{(3x - 5)^2}{x^2}$
2. $h(x) = 4(x + 1) - \sqrt{x + 1}$
3. Write the equation of each new function if the graph of $f(x) = x^3 - 2$ is
    (a) shifted to the left 3 units
    (b) shifted down 5 units
    (c) shifted to the right 1 unit and up 2 units
    (d) reflected in the x-axis
    (e) reflected in the y-axis
    (f) vertically stretched by a factor of 3
4. The graph of a function f with domain $(-\infty, \infty)$ is shown in FIGURE 3.R.1. Sketch the graph of the following functions.
    (a) $y = f(x) - \pi$
    (b) $y = f(x - 2)$
    (c) $y = f(x + 3) + \pi/2$
    (d) $y = -f(x)$
    (e) $y = f(-x)$
    (f) $y = 2f(x)$

FIGURE 3.R.1 Graph for Problem 4

In Problems 5 and 6, use the graph of the one-to-one function f in Figure 3.R.1.

5. Give the domain and range of f^{-1}.
6. Sketch the graph of f^{-1}.
7. Express $y = x - |x| + |x - 1|$ as a piecewise-defined function. Sketch the graph of the function.
8. Sketch the graph of the function $y = [\![x]\!] + [\![-x]\!]$. Give the numbers at which the function is discontinuous.

In Problems 9 and 10, by examining the graph of the function f give the domain of the function g.

9. $f(x) = x^2 - 6x + 10$, $g(x) = \sqrt{x^2 - 6x + 10}$
10. $f(x) = -x^2 + 7x - 6$, $g(x) = \dfrac{1}{\sqrt{-x^2 + 7x - 6}}$

In Problems 11 and 12, the given function f is one-to-one. Find f^{-1}.

11. $f(x) = (x + 1)^3$
12. $f(x) = x + \sqrt{x}$

In Problems 13–16, compute $\dfrac{f(x + h) - f(x)}{h}$ for the given function.

13. $f(x) = -3x^2 + 16x + 12$
14. $f(x) = x^3 - x^2$
15. $f(x) = \dfrac{-1}{2x^2}$
16. $f(x) = x + 4\sqrt{x}$

17. **Area** Express the area of the shaded region in FIGURE 3.R.2 as a function of h.
18. **Parabolic Arch** Determine a quadratic function that describes the parabolic arch shown in FIGURE 3.R.3.
19. **Diameter of a Cube** The diameter d of a cube is the distance between opposite vertices as shown in FIGURE 3.R.4. Express the diameter d as a function of the length s of a side of the cube. [*Hint*: First express the length y of the diagonal in terms of s.]
20. **Inscribed Cylinder** A circular cylinder of height h is inscribed in a sphere of radius 1 as shown in FIGURE 3.R.5. Express the volume of the cylinder as a function of h.

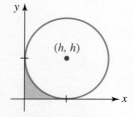

FIGURE 3.R.2 Circle in Problem 17

FIGURE 3.R.3 Arch in Problem 18

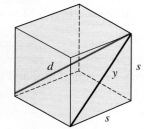

FIGURE 3.R.4 Cube in Problem 19

FIGURE 3.R.5 Inscribed cylinder in Problem 20

21. **Distance from Home** A baseball diamond is a square 90 ft on a side. See FIGURE 3.R.6. After a player hits a home run, he jogs around the bases at a rate of 6 ft/s.
    (a) As the player jogs between home base and first base, express his distance from home base as a function of time t, where $t = 0$ corresponds to the time he left home base—that is, $0 \leq t \leq 15$.
    (b) As the player jogs between home base and first base, express his distance from second base as a function of time t, where $0 \leq t \leq 15$.

FIGURE 3.R.6 Baseball player in Problem 21

22. **Area Again** Consider the four circles shown in FIGURE 3.R.7. Express the area of the shaded region between them as a function of h.
23. **Yet More Area** The running track shown as the black curve in FIGURE 3.R.8 is to consist of two parallel straight parts and two congruent semicircular parts. The length of the track is to be 2 km. Express the area of the rectangular plot of land (the dark green rectangle) enclosed by the running track as a function of the radius of a semicircular end.
24. **Cost of Construction** A pipeline is to be constructed from a refinery across a swamp to storage tanks. See FIGURE 3.R.9. The cost of construction is $25,000 per mile over the swamp and $20,000 per mile over land. Express the cost of construction of the pipeline as a function of x shown in the figure.

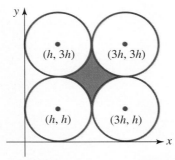

FIGURE 3.R.7 Circles in Problem 22

FIGURE 3.R.8 Running track in Problem 23

FIGURE 3.R.9 Pipeline in Problem 24

4 Polynomial and Rational Functions

In This Chapter

- **4.1** Polynomial Functions
- **4.2** Division of Polynomial Functions
- **4.3** Zeros and Factors of Polynomial Functions
- **4.4** Real Zeros of Polynomial Functions
- **4.5** Approximating Real Zeros
- **4.6** Rational Functions
- Chapter 4 Review Exercises

The curve formed by the cables holding up the roadway of a suspension bridge is described by a polynomial function.

A Bit of History The following problem perplexed mathematicians for centuries: For a general polynomial function $f(x) = a_n x^n + a_{n-1} x^{n-1} + \cdots + a_1 x + a_0$ of degree n, find a formula, or a procedure, that expresses the zeros of f in terms of its coefficients. We saw in Chapter 3 that in the case of a second-degree ($n = 2$), or *quadratic*, polynomial, the zeros of f are expressed in terms of the coefficients by means of the quadratic formula. The problems for third-degree ($n = 3$) polynomials were solved in the sixteenth century through the pioneering work of the Italian mathematician **Niccolò Fontana** (1499–1557).

Around 1540 another Italian mathematician, **Lodovico Ferrari** (1522–1565), first discovered an algebraic formula for determining the zeros of fourth-degree ($n = 4$) polynomial functions. But for the next 284 years no one discovered any formulas for the zeros of general polynomials of degree $n \geq 5$. For good reason! In 1824 the Norwegian mathematician **Niels Henrik Abel** (1802–1829) proved it was impossible to find such formulas for the zeros of all general polynomials of degrees $n \geq 5$ in terms of their coefficients. Over the years it was *assumed*, but never proved, that a polynomial function f of degree n had at most n zeros. It was a truly outstanding achievement when the German mathematician **Carl Friedrich Gauss** (1777–1855) proved in 1799 that every polynomial function f of degree n has *exactly* n zeros. As we will see in this chapter, these zeros may be real or complex numbers.

4.1 Polynomial Functions

Introduction In Chapter 3 we graphed functions such as $y = 3$, $y = 2x - 1$, $y = 5x^2 - 2x + 4$, and $y = x^3$. As we saw in the introduction to Section 3.3, these functions, in which the variable x is raised to *nonnegative integer powers*, are examples of a more general type of function called a **polynomial function**. Our goal in this section is to present some general guidelines for graphing such functions. But first we repeat the definition of a polynomial given in Section 3.3.

DEFINITION 4.1.1 Polynomial Function

A **polynomial function** $y = f(x)$ is a function of the form
$$f(x) = a_n x^n + a_{n-1} x^{n-1} + \cdots + a_2 x^2 + a_1 x + a_0, \tag{1}$$
where the coefficients $a_n, a_{n-1}, \ldots, a_2, a_1, a_0$ are constants and n is a nonnegative integer.

The **domain** of any polynomial function f is the set of all real numbers $(-\infty, \infty)$.

The following functions are *not* polynomials:

$$y = 5x^2 - 3x^{-1} \quad \text{and} \quad y = 2x^{1/2} - 4.$$

(not a nonnegative integer)

The function

$$y = 8x^5 - \tfrac{1}{2}x^4 - 10x^3 + 7x^2 + 6x + 4$$

(nonnegative integer powers)

is a polynomial, where we interpret the number 4 as the coefficient of x^0. Since 0 is a nonnegative integer, a constant function such as $y = 3$ is a polynomial function since it is the same as $y = 3x^0$.

☐ **Terminology** Polynomial functions are classified by their degree. The highest power of x in a polynomial is said to be its **degree**. So if $a_n \neq 0$, then we say that $f(x)$ in (1) has **degree n**. The number a_n in (1) is called the **leading coefficient** and a_0 is called the **constant term** of the polynomial.

For example,

$$f(x) = 3x^5 - 4x^3 - 3x + 8$$

(degree 5; leading coefficient; constant term)

is a polynomial function of degree 5. We have already studied special polynomial functions in Section 3.3. Polynomial functions of degrees $n = 0$, $n = 1$, $n = 2$, and $n = 3$ are, respectively,

$$\begin{aligned} f(x) &= a_0, & &\text{constant function} \\ f(x) &= a_1 x + a_0, & &\text{linear function} \\ f(x) &= a_2 x^2 + a_1 x + a_0, & &\text{quadratic function} \\ f(x) &= a_3 x^3 + a_2 x^2 + a_1 x + a_0, & &\text{cubic function.} \end{aligned}$$

▶ We considered linear and quadratic functions in Section 3.3.

Polynomials of degrees $n = 4$ and $n = 5$ are, in turn, commonly referred to as **quartic** and **quintic functions**. The constant function $f(x) = 0$ is called the **zero polynomial**.

☐ **Graphs** Recall, the graph of a constant function $f(x) = a_0$ is a **horizontal line**, the graph of a linear function $f(x) = a_1x + a_0$, $a_1 \neq 0$, is a **line with slope $m = a_1$**, and the graph of a quadratic function $f(x) = a_2x^2 + a_1x + a_0$, $a_2 \neq 0$, is a **parabola**. See Section 3.3. Such descriptive statements cannot be made about the graph of a higher-degree polynomial function. What is the shape of the graph of a fifth-degree polynomial function? It turns out that the graph of a polynomial function of degree $n \geq 3$ can have several possible shapes. In general, graphing a polynomial function f of degree $n \geq 3$ often demands the use of either calculus or a graphing utility. However, we will see in the discussion that follows that by determining

- shifting,
- end behavior,
- symmetry,
- intercepts, and
- local behavior

of the function we can, in some instances, quickly sketch a reasonable graph of a higher-degree polynomial function while keeping point plotting to a minimum. Before elaborating on each of these concepts we return to the notion of a power function first introduced in Section 3.2.

☐ **Power Function** A special case of the power function is the **single-term polynomial function** or **monomial**,

◀ See Section R.6 for the definition of a monomial

$$f(x) = x^n, \quad n \text{ a positive integer.} \tag{2}$$

The graphs of (2) for degrees $n = 1, 2, 3, 4, 5$, and 6 are given in **FIGURE 4.1.1**. The interesting fact about (2) is that all the graphs for n odd are basically the same; the notable characteristics are that the graphs are symmetric about the origin and become increasingly flatter near the origin as the degree n increases. See Figures 4.1.1(a)–(c). A similar observation is true for the graphs of (2) for n even, except of course, the graphs are symmetric with respect to the y-axis. See Figures 4.1.1(d)–(f).

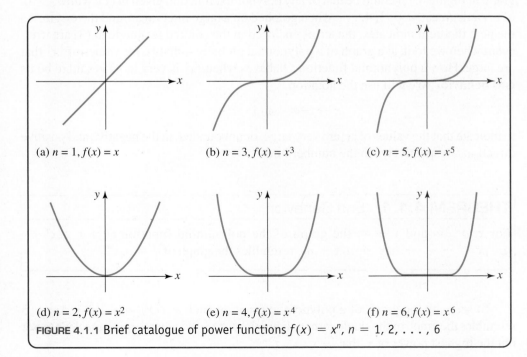

(a) $n = 1, f(x) = x$ (b) $n = 3, f(x) = x^3$ (c) $n = 5, f(x) = x^5$

(d) $n = 2, f(x) = x^2$ (e) $n = 4, f(x) = x^4$ (f) $n = 6, f(x) = x^6$

FIGURE 4.1.1 Brief catalogue of power functions $f(x) = x^n$, $n = 1, 2, \ldots, 6$

☐ **Shifted Graphs** Recall from Section 3.2 that for $c > 0$, the graphs of polynomial functions of the form

$$y = ax^n + c, \quad y = ax^n - c$$

and

$$y = a(x + c)^n, \quad y = a(x - c)^n$$

can be obtained by vertical and horizontal shifts of the graph of $y = ax^n$. Also, if the leading coefficient a is positive, the graph of $y = ax^n$ is either a vertical stretch or a vertical compression of the graph of the basic single-term polynomial function $f(x) = x^n$. When a is negative we also carry out a reflection in the x-axis.

$y = -(x+2)^3 - 1$

FIGURE 4.1.2 Reflected and shifted graph in Example 1

EXAMPLE 1 Graphing a Shifted Polynomial Function

The graph of $y = -(x + 2)^3 - 1$ is the graph of $f(x) = x^3$ reflected in the x-axis, shifted 2 units to the left, and then shifted vertically downward 1 unit. First review Figure 4.1.1(b) and then see **FIGURE 4.1.2**. ≡

☐ **End Behavior** The knowledge of the shape of a single-term polynomial function $f(x) = x^n$ is important for another reason. First, examine the computer-generated graphs given in **FIGURES 4.1.3** and **4.1.4**. Although the graph in Figure 4.1.3 certainly resembles the graphs in Figures 4.1.1(b) and 4.1.1(c), and the graph in Figure 4.1.4 resembles the graphs in Figures 4.1.1(d)–(f), the functions graphed in these two figures are *not* power functions $f(x) = x^n$, n odd, or $f(x) = x^n$, n even. We will not tell you at this point what the specific functions are except to say that they were both graphed on the interval $[-1000, 1000]$. The point is this: The function whose graph is given in Figure 4.1.3 could be almost *any* polynomial function

FIGURE 4.1.3 Mystery graph #1

$$f(x) = a_n x^n \boxed{+ a_{n-1}x^{n-1} + \cdots + a_1 x + a_0} \tag{3}$$

$a_n > 0$, of *odd* degree n, $n = 3, 5, \ldots$ when graphed on $[-1000, 1000]$. Similarly, the graph in Figure 4.1.4 could be that of any polynomial function given in (1), with $a_n > 0$, of *even* degree n, $n = 2, 4, \ldots$ when graphed on a large interval around the origin. As the next theorem indicates, the terms enclosed in the colored rectangle in (3) are irrelevant when we look at a graph of a polynomial globally—that is, for values of $|x|$ that are large. How a polynomial function f behaves when $|x|$ is very large is said to be its **end behavior**. We will use the notation

$$x \to -\infty \quad \text{and} \quad x \to \infty,$$

to indicate that the values of $|x|$ are very large, or unbounded, in the negative and positive directions, respectively, on the number line.

FIGURE 4.1.4 Mystery graph #2

THEOREM 4.1.1 End Behavior

For $x \to -\infty$ and $x \to \infty$, the graph of the polynomial function $f(x) = a_n x^n + a_{n-1}x^{n-1} + \cdots + a_2 x^2 + a_1 x + a_0$, resembles the graph of $y = a_n x^n$.

To see why the graph of a polynomial function such as $f(x) = -2x^3 + 4x^2 + 5$ resembles the graph of the single-term polynomial $y = -2x^3$ when $|x|$ is large, let's factor out the highest power of x, that is, x^3:

$$f(x) = x^3\left(-2 + \frac{4}{x} + \frac{5}{x^3}\right). \tag{4}$$

both these terms become negligible when $|x|$ is large

By letting $|x|$ increase without bound, that is, $x \to -\infty$ or $x \to \infty$, both $4/x$ and $5/x^3$ can be made as close to 0 as we want. Thus when the values of $|x|$ are large, the values of the function f in (4) are closely approximated by the values of $y = -2x^3$.

There can be only four types of end behavior for a polynomial function f. Although two of the end behaviors are already illustrated in Figures 4.1.3 and 4.1.4, we include them again in the pictorial summary given in **FIGURE 4.1.5**. To interpret the red arrows in Figure 4.1.5 let's examine Figure 4.1.5(a). The position and direction of the left arrow (left arrow points down) indicates that as $x \to -\infty$, the values $f(x)$ are negative and large in magnitude. Stated another way, the graph is heading downward as $x \to -\infty$. Similarly, the position and direction of the right arrow (right arrow points up) indicates that the graph is heading upward as $x \to \infty$.

Pictorial Summary of End Behavior of Polynomial Functions

FIGURE 4.1.5 End behavior of a polynomial function f depends on its degree n and on the sign of its leading coefficient a_n

☐ **Local Behavior** The gaps between the arrows in Figure 4.1.5 correspond to some interval around the origin. In these gaps the graph of f exhibits **local behavior**; in other words, the graph of f shows the characteristics of a polynomial function of a particular degree. This local behavior includes the x- and y-intercepts of the graph, the behavior of the graph at an x-intercept, the turning points of the graph, and observable symmetry of the graph (if any). A **turning point** is a point $(c, f(c))$ at which the graph of a polynomial function f changes direction; that is, the function f changes from increasing to decreasing or vice versa. The graph of a polynomial function of degree n can have up to $n - 1$ turning points. In calculus a turning point corresponds to a **relative**, or **local**, **extremum** of f. A relative extremum is classified as either a **maximum** or a **minimum**. If $(c, f(c))$ is a turning point, then in some neighborhood of $x = c$, $f(c)$ is either the *largest* (relative maximum) or *smallest* (relative minimum) function value. If $f(c)$ is a relative maximum, then at $(c, f(c))$ the graph of a polynomial function f must change from increasing immediately to the left of $x = c$ to decreasing immediately to the right of $x = c$, whereas if $f(c)$ is a relative minimum, then at $(c, f(c))$ the function f changes from decreasing to increasing. These concepts are illustrated in Example 2.

☐ **Symmetry** It is easy to tell by inspection those polynomial functions whose graphs possess symmetry with respect to either the *y*-axis or the origin. The words *even* and *odd* functions have special meaning for polynomial functions. Recall, an even function is one for which $f(-x) = f(x)$ and an odd function is one for which $f(-x) = -f(x)$. These two conditions hold for polynomial functions in which the powers of x are all even integers and all odd integers, respectively. For example,

even powers: $f(x) = 5x^4 - 7x^2$, f is even function

odd powers: $f(x) = 10x^5 + 7x^3 + 4x$, f is odd function

mixed powers: $f(x) = -3x^7 + 2x^4 + x^3 + 2$, f is neither even nor odd

A function such as $f(x) = 3x^6 - x^4 + 6$ is an even function because the obvious powers are even integers; the constant term 6 is actually $6x^0$, and 0 is an even nonnegative integer.

☐ **Intercepts** The graph of every polynomial function f passes through the *y*-axis since $x = 0$ is the domain of the function. The *y*-intercept is the point $(0, f(0))$. Recall, a number c is a **zero** of a function f if $f(c) = 0$. In this discussion we assume c is a real zero. If $x - c$ is a factor of a polynomial function f, that is, $f(x) = (x - c)q(x)$, where $q(x)$ is another polynomial, then clearly $f(c) = 0$ and the corresponding point on the graph is $(c, 0)$. Thus the real zeros of a polynomial function are the *x*-coordinates of the *x*-intercepts of its graph. If $(x - c)^m$ is a factor of f, where $m > 1$ is a positive integer, and $(x - c)^{m+1}$ is *not* a factor of f, then c is said to be a **repeated zero**, or more precisely, a **zero of multiplicity** m. For example, $f(x) = x^2 - 10x + 25$ is equivalent to $f(x) = (x - 5)^2$. Hence 5 is a repeated zero or a zero of multiplicity 2. When $m = 1$, c is called a **simple zero**. For example, $-\frac{1}{3}$ and $\frac{1}{2}$ are simple zeros of $f(x) = 6x^2 - x - 1$ since f can be written as $f(x) = 6(x + \frac{1}{3})(x - \frac{1}{2})$. The behavior of the graph of f at an *x*-intercept $(c, 0)$ depends on whether c is a simple zero, or a zero of multiplicity $m > 1$, where m is either an even or an odd integer.

- If c is a simple zero, then the graph of f passes directly through the *x*-axis at $(c, 0)$. See **FIGURE 4.1.6(a)**.
- If c is a zero of odd multiplicity $m = 3, 5, \ldots$, then the graph of f passes through the *x*-axis but is flattened at $(c, 0)$. See Figure 4.1.6(b).
- If c is a zero of even multiplicity $m = 2, 4, \ldots$, then the graph of f does not pass through the *x*-axis but is tangent to, or touches, that axis at $(c, 0)$. See Figure 4.1.6(c).

In the case when c is either a simple zero or a zero of odd multiplicity $m = 3, 5, \ldots$, $f(x)$ changes sign *at* $(c, 0)$, whereas if c is a zero of even multiplicity $m = 2, 4, \ldots$, $f(x)$ does not change sign at $(c, 0)$. We note that in Figure 4.1.6 we have assumed that the polynomial function f has a positive leading coefficient. If the sign of the leading coefficient of f is negative, then the graphs in Figure 4.1.6 would be reflected in the *x*-axis. For example, at a zero of even multiplicity and negative leading coefficient the graph of f would be tangent to the *x*-axis from below that axis.

(a) Simple zero

(b) Zero of odd multiplicity $m = 3, 5, \ldots$

(c) Zero of even multiplicity $m = 2, 4, \ldots$

FIGURE 4.1.6 *x*-intercepts of a polynomial function $f(x)$ with a positive leading coefficient

EXAMPLE 2 Graphing a Polynomial Function

Graph $f(x) = x^3 - 9x$.

Solution Here are some of the things that we look at to sketch the graph of f.

End Behavior: By ignoring all terms but the first, we see that the graph of f resembles the graph of $y = x^3$ for large $|x|$. That is, the graph goes down to the left as $x \to -\infty$ and up to the right as $x \to \infty$, as illustrated in Figure 4.1.5(a).

Symmetry: Since all the powers are odd integers, f is an odd function. The graph of f is symmetric with respect to the origin.

Intercepts: $f(0) = 0$ and so the y-intercept is $(0, 0)$. Setting $f(x) = 0$ we see that we must solve $x^3 - 9x = 0$. Factoring,

$$\overset{\text{difference of two squares}}{x(x^2 - 9) = 0} \quad \text{or} \quad x(x - 3)(x + 3) = 0,$$

shows that the zeros of f are $x = 0$ and $x = \pm 3$. The x-intercepts are $(0, 0)$, $(-3, 0)$, and $(3, 0)$.

The Graph: From left to right, the graph rises (f is increasing) from the third quadrant and passes straight through $(-3, 0)$ since -3 is a simple zero. Although the graph is rising as it passes through this intercept it must turn back downward (f decreasing) at some point in the second quadrant to get through the intercept $(0, 0)$. Since the graph is symmetric with respect to the origin, its behavior is just the opposite in the first and fourth quadrants. See FIGURE 4.1.7. ≡

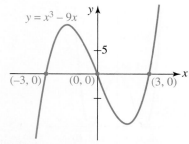

FIGURE 4.1.7 Graph of function in Example 2

In Example 2, the graph of f has two turning points. On the interval $[-3, 0]$ there is a relative maximum and on the interval $[0, 3]$ there is a relative minimum. We made no attempt to locate these points precisely; this is something that would, in general, require techniques from calculus. The best we can do using precalculus mathematics to refine the graph is to resort to plotting additional points on the intervals of interest. By the way, $f(x) = x^3 - 9x$ is the function whose graph on the interval $[-1000, 1000]$ is given in Figure 4.1.3.

EXAMPLE 3 Graphing a Polynomial Function

Graph $f(x) = (1 - x)(x + 1)^2$.

Solution Multiplying out, f is the same as $f(x) = -x^3 - x^2 + x + 1$.

End Behavior: From the preceding line we see that the graph of f resembles the graph of $y = -x^3$ for large $|x|$, just the opposite of the end behavior of the function in Example 1. See Figure 4.1.5(b).

Symmetry: As we see from $f(x) = -x^3 - x^2 + x + 1$, there are both even and odd powers of x present. Hence f is neither even nor odd; its graph possesses no y-axis or origin symmetry.

Intercepts: $f(0) = 1$ so the y-intercept is $(0, 1)$. From the original factored form of the function $f(x)$, we see that $(-1, 0)$ and $(1, 0)$ are the x-intercepts of its graph.

The Graph: From left to right, the graph falls (f decreasing) from the second quadrant and then, because -1 is a zero of multiplicity 2, the graph is tangent to the x-axis at $(-1, 0)$. The graph then rises (f increasing) as it passes through the y-intercept $(0, 1)$. At some point within the interval $[-1, 1]$ the graph turns downward (f decreasing) and, since 1 is a simple zero, passes through the x-axis at $(1, 0)$ heading downward into the fourth quadrant. See FIGURE 4.1.8. ≡

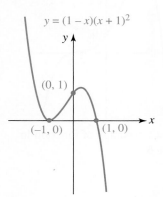

FIGURE 4.1.8 Graph of function in Example 3

In Example 3, there are again two turning points. It should be clear that the point $(-1, 0)$ is a turning point (f changes from decreasing to increasing at this point) and $f(-1) = 0$ is a relative minimum of f. There is a turning point (f changes from increasing to decreasing at this point) on the interval $[-1, 1]$ and corresponds to a relative maximum of f.

EXAMPLE 4 Zeros of Multiplicity Two

Graph $f(x) = x^4 - 4x^2 + 4$.

Solution Before proceeding, note that the right-hand side of f is a perfect square. That is, $f(x) = (x^2 - 2)^2$. Since $x^2 - 2 = (x - \sqrt{2})(x + \sqrt{2})$, by the laws of exponents we can write

$$f(x) = (x - \sqrt{2})^2(x + \sqrt{2})^2. \quad (5)$$

End Behavior: Inspection of $f(x)$ shows that its graph resembles the graph of $y = x^4$ for large $|x|$. That is, the graph goes up to the left as $x \to -\infty$ and up to the right as $x \to \infty$ as shown in Figure 4.1.5(c).

Symmetry: Because $f(x)$ contains only even powers of x, it is an even function and so its graph is symmetric with respect to the y-axis.

Intercepts: $f(0) = 4$ so the y-intercept is $(0, 4)$. Inspection of (5) shows the x-intercepts are $(-\sqrt{2}, 0)$ and $(\sqrt{2}, 0)$.

The Graph: From left to right, the graph falls from the second quadrant and then, because $-\sqrt{2}$ is a zero of multiplicity 2, the graph touches the x-axis at $(-\sqrt{2}, 0)$. The graph then rises from this point to the y-intercept $(0, 4)$. We then use the y-axis symmetry to finish the graph in the first quadrant. See FIGURE 4.1.9.

FIGURE 4.1.9 Graph of function in Example 4

In Example 4, the graph of f has three turning points. From the even multiplicity of the zeros, along with the y-axis symmetry, it can be deduced that the x-intercepts $(-\sqrt{2}, 0)$ and $(\sqrt{2}, 0)$ are turning points and $f(-\sqrt{2}) = 0$ and $f(\sqrt{2}) = 0$ are relative minima, and that the y-intercept $(0, 4)$ is a turning point and $f(0) = 4$ is a relative maximum.

EXAMPLE 5 Zeros of Multiplicity Three

Graph $f(x) = -(x + 4)(x - 2)^3$.

Solution

End Behavior: Inspection of f shows that its graph resembles the graph of $y = -x^4$ for large $|x|$. This end behavior of f is shown in Figure 4.1.5(d).

Symmetry: The function f is neither even nor odd. It is straightforward to show that $f(-x) \neq f(x)$ and $f(-x) \neq -f(x)$.

Intercepts: $f(0) = (-4)(-2)^3 = 32$ so the y-intercept is $(0, 32)$. From the factored form of $f(x)$, we see that $(-4, 0)$ and $(2, 0)$ are the x-intercepts.

The Graph: From left to right, the graph rises from the third quadrant and then, because -4 is a simple zero, the graph of f passes directly through the x-axis at $(-4, 0)$. Somewhere within the interval $[-4, 0]$ the function f must change from increasing to decreasing to enable its graph to pass through the y-intercept $(0, 32)$. After its graph passes through the y-intercept, the function f continues to decrease but, since 2 is a zero of multiplicity 3, its graph flattens as it passes through $(2, 0)$ heading downward into the fourth quadrant. See FIGURE 4.1.10.

FIGURE 4.1.10 Graph of function in Example 5

Note in Example 5 that since f is of degree 4, its graph could have up to three turning points. But as can be seen from Figure 4.1.10, the graph of f possesses only one turning point and corresponds to a relative maximum of f.

EXAMPLE 6 Zeros of Multiplicity Two and Three

Graph $f(x) = (x - 3)(x - 1)^2(x + 2)^3$.

Solution The function f is of degree 6 and so its end behavior resembles the graph of $y = x^6$ for large $|x|$. See Figure 4.1.5(c). Also, the function f is neither even nor odd; its graph possesses no y-axis or origin symmetry. The y-intercept is $(0, f(0)) = (0, -24)$. From the factors of f we see that x-intercepts of the graph are $(-2, 0)$, $(1, 0)$, and $(3, 0)$. Since -2 is a zero of multiplicity 3, the graph of f is flattened as it passes through the point $(-2, 0)$. Since 1 is a zero of multiplicity 2, the graph of f is tangent to the x-axis at $(1, 0)$. Since 3 is a simple zero, the graph of f passes directly through the x-axis at $(3, 0)$. Putting all these facts together we obtain the graph in FIGURE 4.1.11. ≡

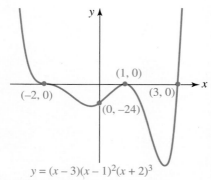

FIGURE 4.1.11 Graph of function in Example 6

In Example 6, since the function f is of degree 6 its graph could have up to five turning points. But as the graph in Figure 4.1.11 shows there are only three turning points. Two of these points correspond to relative minima and $f(1) = 0$ the remaining point, which is $(1, 0)$, is a relative maximum.

4.1 Exercises Answers to selected odd-numbered problems begin on page ANS-12.

In Problems 1–8, proceed as in Example 1 and use transformations to sketch the graph of the given polynomial function.

1. $y = x^3 - 3$
2. $y = -(x + 2)^3$
3. $y = (x - 2)^3 + 2$
4. $y = 3 - (x + 2)^3$
5. $y = (x - 5)^4$
6. $y = x^4 - 1$
7. $y = 1 - (x - 1)^4$
8. $y = 4 + (x + 1)^4$

In Problems 9–12, determine whether the given polynomial function f is even, odd, or neither even nor odd. Do not graph.

9. $f(x) = -2x^3 + 4x$
10. $f(x) = x^6 - 5x^2 + 7$
11. $f(x) = x^5 + 4x^3 + 9x + 1$
12. $f(x) = x^3(x + 2)(x - 2)$

In Problems 13–18, match the given graph with one of the polynomial functions in (a)–(f).

(a) $f(x) = x^2(x - 1)^2$
(b) $f(x) = -x^3(x - 1)$
(c) $f(x) = x^3(x - 1)^3$
(d) $f(x) = -x(x - 1)^3$
(e) $f(x) = -x^2(x - 1)$
(f) $f(x) = x^3(x - 1)^2$

13.

FIGURE 4.1.12 Graph for Problem 13

14.

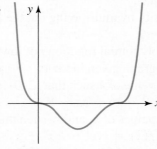

FIGURE 4.1.13 Graph for Problem 14

15.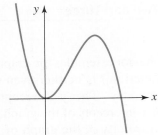

FIGURE 4.1.14 Graph for Problem 15

16.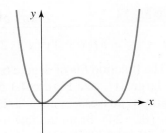

FIGURE 4.1.15 Graph for Problem 16

17.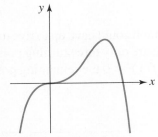

FIGURE 4.1.16 Graph for Problem 17

18.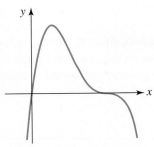

FIGURE 4.1.17 Graph for Problem 18

In Problems 19–40, proceed as in Example 2 and sketch the graph of the given polynomial function f.

19. $f(x) = x^3 - 4x$
20. $f(x) = 9x - x^3$
21. $f(x) = -x^3 + x^2 + 6x$
22. $f(x) = x^3 + 7x^2 + 12x$
23. $f(x) = (x + 1)(x - 2)(x - 4)$
24. $f(x) = (2 - x)(x + 2)(x + 1)$
25. $f(x) = x^4 - 4x^3 + 3x^2$
26. $f(x) = x^2(x - 2)^2$
27. $f(x) = (x^2 - x)(x^2 - 5x + 6)$
28. $f(x) = x^2(x^2 + 3x + 2)$
29. $f(x) = (x^2 - 1)(x^2 + 9)$
30. $f(x) = x^4 + 5x^2 - 6$
31. $f(x) = -x^4 + 2x^2 - 1$
32. $f(x) = x^4 - 6x^2 + 9$
33. $f(x) = x^4 + 3x^3$
34. $f(x) = x(x - 2)^3$
35. $f(x) = x^5 - 4x^3$
36. $f(x) = (x - 2)^5 - (x - 2)^3$
37. $f(x) = 3x(x + 1)^2(x - 1)^2$
38. $f(x) = (x + 1)^2(x - 1)^3$
39. $f(x) = -\frac{1}{2}x^2(x + 2)^3(x - 2)^2$
40. $f(x) = x(x + 1)^2(x - 2)(x - 3)$

FIGURE 4.1.18 Graph for Problem 41

FIGURE 4.1.19 Graph for Problem 42

41. The graph of $f(x) = x^3 - 3x$ is given in **FIGURE 4.1.18**.
    (a) Use the figure to obtain the graph of $g(x) = f(x) + 2$.
    (b) Using only the graph obtained in part (a) write an equation, in *factored* form, for $g(x)$.
    Then verify by multiplying out the factors that your equation for $g(x)$ is the same as $f(x) + 2 = x^3 - 3x + 2$.
42. Find a polynomial function f of lowest possible degree whose graph is consistent with the graph given in **FIGURE 4.1.19**.
43. Find the value of k such that $(2, 0)$ is an x-intercept for the graph of
    $f(x) = kx^5 - x^2 + 5x + 8$.
44. Find the values of k_1 and k_2 such that $(-1, 0)$ and $(1, 0)$ are x-intercepts for the graph of $f(x) = k_1x^4 - k_2x^3 + x - 4$.
45. Find the value of k such that $(0, 10)$ is the y-intercept for the graph of
    $f(x) = x^3 - 2x^2 + 14x - 3k$.

CHAPTER 4 POLYNOMIAL AND RATIONAL FUNCTIONS

46. Consider the polynomial function $f(x) = (x - 2)^{n+1}(x + 5)$, where n is a positive integer. For what values of n does the graph of f touch, but does not cross, the x-axis at $(2, 0)$?
47. Consider the polynomial function $f(x) = (x - 1)^{n+2}(x + 1)$, where n is a positive integer. For what values of n does the graph of f cross the x-axis at $(1, 0)$?
48. Consider the polynomial function $f(x) = (x - 5)^{2m}(x + 1)^{2n-1}$, where m and n are positive integers.
    (a) For what values of m does the graph of f cross the x-axis at $(5, 0)$?
    (b) For what values of n does the graph of f cross the x-axis at $(-1, 0)$?

Miscellaneous Applications

49. **Constructing a Box** An open box can be made from a rectangular piece of cardboard by cutting a square of length x from each corner and bending up the sides. See FIGURE 4.1.20. If the cardboard measures 30-cm by 40-cm, show that the volume of the resulting box is given by

$$V(x) = x(30 - 2x)(40 - 2x).$$

Sketch the graph of $V(x)$ for $x > 0$. What is the domain of the function V?

FIGURE 4.1.20 Box in Problem 49

50. **Another Box** In order to hold its shape, the box in Problem 49 will require tape or some other fastener at the corners. An open box that holds itself together can be made by cutting out a square of length x from each corner of a rectangular piece of cardboard, cutting on the solid line, and folding on the dashed lines, as shown in FIGURE 4.1.21. Find a polynomial function $V(x)$ that gives the volume of the resulting box if the original cardboard measures 30-cm by 40-cm. Sketch the graph of $V(x)$ for $x > 0$.

FIGURE 4.1.21 Box in Problem 50

For Discussion

51. Examine Figure 4.1.5. Then discuss whether there can exist cubic polynomial functions that have no real zeros.
52. Suppose a polynomial function f has three zeros, -3, 2, and 4, and has the end behavior that its graph goes down to the left as $x \to -\infty$ and down to the right as $x \to \infty$. Discuss possible equations for f.

Calculator/Computer Problems

In Problems 53 and 54, use a graphing utility to examine the graph of the given polynomial function on the indicated intervals.

53. $f(x) = -(x - 8)(x + 10)^2$; $[-15, 15]$, $[-100, 100]$, $[-1000, 1000]$
54. $f(x) = (x - 5)^2(x + 5)^2$; $[-10, 10]$, $[-100, 100]$, $[-1000, 1000]$

4.2 Division of Polynomial Functions

≡ **Introduction** In this section we will see that the method for dividing two polynomial functions $f(x)$ and $g(x)$ is quite similar to division of positive integers. Moreover, division of a polynomial function $f(x)$ by a linear polynomial $g(x) = x - c$ is particularly useful because it provides us with a way of evaluating the function f at the number c without having to compute any powers of c.

We begin with a review of the terminology of fractions.

□ **Terminology** If $p > 0$ and $s > 0$ are integers such that $p \geq s$, then p/s is called an **improper fraction**. By dividing p by s, we obtain unique integers q and r that satisfy

$$\frac{p}{s} = q + \frac{r}{s} \quad \text{or} \quad p = sq + r, \tag{1}$$

where $0 \leq r < s$. The number p is called the **dividend**, s is the **divisor**, q the **quotient**, and r is the **remainder**. For example, consider the improper fraction $\frac{1052}{23}$. Performing long division gives

$$\begin{array}{r} 45 \;\; \leftarrow \text{quotient} \\ \text{divisor} \rightarrow 23 \overline{)1052} \;\; \leftarrow \text{dividend} \\ \underline{92} \;\;\; \leftarrow \text{subtract} \\ 132 \;\;\; \\ \underline{115} \;\;\; \\ 17 \;\; \leftarrow \text{remainder} \end{array} \tag{2}$$

The result in (2) can be written as $\frac{1052}{23} = 45 + \frac{17}{23}$, where $\frac{17}{23}$ is a **proper fraction** since the numerator is less than the denominator, in other words, the fraction is less than 1. If we multiply this result by the divisor 23 we obtain the special way of writing the dividend m illustrated in the second equation in (1):

$$1052 = 23 \cdot \underset{\underset{\text{divisor}}{\uparrow}}{45} + \underset{}{17}. \tag{3}$$

(quotient ↓, remainder ↓)

□ **Division of Polynomials** If the degree of a polynomial $f(x)$ is greater than or equal to the degree of the polynomial $g(x)$, then $f(x)/g(x)$ is also called an **improper fraction**. A result analogous to (1) is called the **Division Algorithm** for polynomials.

THEOREM 4.2.1 Division Algorithm

Let $f(x)$ and $g(x) \neq 0$ be polynomials where the degree of $f(x)$ is greater than or equal to the degree of $g(x)$. Then there exist unique polynomials $q(x)$ and $r(x)$ such that

$$\frac{f(x)}{g(x)} = q(x) + \frac{r(x)}{g(x)} \quad \text{or} \quad f(x) = g(x)q(x) + r(x), \tag{4}$$

where $r(x)$ has degree less than the degree of $g(x)$.

The polynomial $f(x)$ is called the **dividend**, $g(x)$ the **divisor**, $q(x)$ the **quotient**, and $r(x)$ the **remainder**. Because $r(x)$ has degree less than the degree of $g(x)$, the rational expression $r(x)/g(x)$ is called a **proper fraction**.

Observe in (4) when $r(x) = 0$, then $f(x) = g(x)q(x)$ and so the divisor $g(x)$ is a factor of $f(x)$. In this case, we say that $f(x)$ is **divisible** by $g(x)$ or, in older terminology, $g(x)$ **divides evenly** into $f(x)$.

EXAMPLE 1 Division of Two Polynomials

Use long division to find the quotient $q(x)$ and remainder $r(x)$ when the polynomial $f(x) = 3x^3 - x^2 - 2x + 6$ is divided by the polynomial $g(x) = x^2 + 1$.

Solution By long division,

$$\begin{array}{r}
3x - 1 \quad \leftarrow \text{quotient}\\
\text{divisor} \to x^2 + 1 \overline{)3x^3 - x^2 - 2x + 6} \quad \leftarrow \text{dividend}\\
\underline{3x^3 + 0x^2 + 3x} \quad \leftarrow \text{subtract}\\
-x^2 - 5x + 6\\
\underline{-x^2 + 0x - 1}\\
-5x + 7 \quad \leftarrow \text{remainder}
\end{array} \qquad (5)$$

The result of the division (5) can be written

$$\frac{3x^3 - x^2 - 2x + 6}{x^2 + 1} = 3x - 1 + \frac{-5x + 7}{x^2 + 1}.$$

Thus the quotient is $3x - 1$ and the remainder is $-5x + 7$. If we multiply both sides of the last equation by the divisor $x^2 + 1$ we get the second form given in (4):

$$3x^3 - x^2 - 2x + 6 = (x^2 + 1)(3x - 1) + (-5x + 7). \qquad (6) \equiv$$

If the divisor $g(x)$ is a linear polynomial $x - c$, it follows from the Division Algorithm that the *degree* of the remainder r is 0, that is to say, r is a constant. Thus (4) becomes

$$f(x) = (x - c)q(x) + r. \qquad (7)$$

When the number $x = c$ is substituted into (7), we discover an alternative way of evaluating a polynomial function:

$$f(c) = (c - c)\overset{\overset{0}{\downarrow}}{q(c)} + r = r.$$

The foregoing result is called the **Remainder Theorem**.

THEOREM 4.2.2 Remainder Theorem

If a polynomial $f(x)$ is divided by a linear polynomial $x - c$, then the remainder r is the value of $f(x)$ at $x = c$, that is, $f(c) = r$.

EXAMPLE 2 Finding the Remainder

Use the Remainder Theorem to find r when $f(x) = 4x^3 - x^2 + 4$ is divided by $x - 2$.

Solution From the Remainder Theorem the remainder r is the value of the function f evaluated at $x = 2$:

$$r = f(2) = 4(2)^3 - (2)^2 + 4 = 32. \qquad (8) \equiv$$

Example 2, where a remainder r is determined by calculating a function value $f(c)$, is more interesting than important. What *is* important is the reverse problem: Determine the function value $f(c)$ by finding the remainder r by division of f by $x - c$. The next two examples illustrate this concept.

EXAMPLE 3 Evaluation by Division

Use the Remainder Theorem to find $f(c)$ for $f(x) = x^5 - 4x^3 + 2x - 10$ when $c = -3$.

Solution The value $f(-3)$ is the remainder when $f(x) = x^5 - 4x^3 + 2x - 10$ is divided by $x - (-3) = x + 3$. For the purposes of long division we must account for the missing x^4 and x^2 terms by rewriting the dividend as

$$f(x) = x^5 + 0x^4 - 4x^3 + 0x^2 + 2x - 10.$$

Then,

$$\begin{array}{r}
x^4 - 3x^3 + 5x^2 - 15x + 47 \\
x + 3 \overline{) x^5 + 0x^4 - 4x^3 + 0x^2 + 2x - 10} \\
\underline{x^5 + 3x^4} \\
-3x^4 - 4x^3 + 0x^2 + 2x - 10 \\
\underline{-3x^4 - 9x^3} \\
5x^3 + 0x^2 + 2x - 10 \\
\underline{5x^3 + 15x^2} \\
-15x^2 + 2x - 10 \\
\underline{-15x^2 - 45x} \\
47x - 10 \\
\underline{47x + 141} \\
-151
\end{array}$$

(9)

The remainder r in the division is the value of the function f at $x = -3$, that is, $f(-3) = -151$. ≡

☐ **Synthetic Division** After working through Example 3 one could justifiably ask the question: Why would anyone want to calculate the value of a polynomial function f by division? The answer is: We would not bother to do this were it not for **synthetic division**. Synthetic division is a shorthand method of dividing a polynomial $f(x)$ by a *linear polynomial* $x - c$; it does not require writing down the various powers of the variable x but only the coefficients of these powers in the dividend $f(x)$ (which must include all 0 coefficients). It is also a very efficient and quick way of evaluating $f(c)$ since the process utilizes only the arithmetic operations of multiplication and addition. No exponentiations such as 2^3 and 2^2 in (8) are involved.

For example, consider the long division:

$$\begin{array}{r}
4x^2 - 3x + 7 \\
x - 2 \overline{) \boxed{4x^3} - 11x^2 + \boxed{13x} - \boxed{5}} \\
(-)\underline{\boxed{4x^3} - 8x^2} \\
\boxed{-3x^2} + \boxed{13x} - \boxed{5} \\
(-)\underline{\boxed{-3x^2} + 6x} \\
\boxed{7x} - \boxed{5} \\
(-)\underline{\boxed{7x} - 14} \\
\boxed{9} \leftarrow \text{remainder}
\end{array}$$

(10)

We can make several observations about (10):

- Below the long-division sign, each column contains terms of the same degree.
- Each rectangle contains identical terms.
- The coefficients of the quotient and the constant remainder are circled.
- Each line marked by $(-)$ is subtracted from the preceding line.

If we delete variables and the observed repetitions in (10), we have

$$x - 2 \rightarrow \quad -2 \overline{\smash{\big)}\, \textcircled{4} \quad -11 \quad 13 \quad -5}$$
$$(-) \quad \underline{\; -8 \;}$$
$$\textcircled{-3}$$
$$(-) \quad \underline{\; 6 \;}$$
$$\textcircled{7}$$
$$(-) \quad \underline{\; -14 \;}$$
$$\textcircled{9} \; \leftarrow \text{remainder} \qquad (11)$$

As indicated by the arrows in (11), this can be written more compactly as

$$-2 \overline{\smash{\big)}\, \textcircled{4} \quad -11 \quad 13 \quad -5}$$
$$(-) \quad \underline{\quad -8 \quad 6 \quad -14}$$
$$\textcircled{-3} \quad \textcircled{7} \quad \textcircled{9} \leftarrow \text{remainder} \qquad (12)$$

If we bring down the leading coefficient 4 of the dividend, as indicated by the red arrow in (12), then the coefficients of the quotient and remainder are on one line.

$$-2\overline{\smash{\big)}\,4 \quad -11 \quad 13 \quad -5} \quad \leftarrow \text{first row}$$
$$(-) \quad \underline{\quad\; -8 \quad 6 \quad -14} \quad \leftarrow \text{second row} \qquad (13)$$
$$4 \quad \textcircled{-3} \quad \textcircled{7} \quad |\textcircled{9} = r \leftarrow \text{third row}$$

Now observe that each number in the second row of (13) can be found by multiplying the number in the third row of the preceding column by -2, and that each number in the third row is obtained by subtracting each number in the second row from the corresponding number in the first row.

Finally, we can avoid the subtraction in (13) by multiplying by $+2$ (the additive inverse of -2) and adding the second row to the first. This is illustrated below:

$$\text{coefficients of the dividend } f(x) = 4x^3 - 11x^2 + 13x - 5$$
$$\text{divisor: } x - 2 \quad 2 \,|\, 4 \quad -11 \quad 13 \quad -5 \quad \leftarrow \text{first row}$$
$$(+) \quad \underline{\quad\;\; 8 \quad -6 \quad 14} \quad \leftarrow \text{second row} \qquad (14)$$
$$4 \quad -3 \quad 7 \quad |9 = r \quad \leftarrow \text{third row}$$
$$\text{coefficients of the quotient } q(x) = 4x^2 - 3x + 7$$

The procedure of synthetic division for dividing $f(x)$, a polynomial of degree $n > 0$, by $x - c$ is summarized as follow:

GUIDELINES FOR SYNTHETIC DIVISION

(i) Write c followed by the coefficients of $f(x)$. Be sure to include any coefficients that are 0, including the constant term.

(ii) Bring down the first coefficient of $f(x)$ to the third row.

(iii) Multiply this number by c and write the product directly under the second coefficient of $f(x)$. Then add the two numbers in this column and write the sum beneath them in the third row.

(iv) Multiply this sum by c and write the product in the second row of the next column. Then add the two numbers in this column and write the sum beneath them in the third row.

(v) Repeat the preceding step as many times as possible.

(vi) The last number in the third row is the constant remainder r; the numbers preceding it in the third row are the coefficients of $q(x)$, the quotient polynomial of degree $n - 1$.

EXAMPLE 4   Example 3 Revisited

Use synthetic division to divide $f(x) = x^5 - 4x^3 + 2x - 10$ when $x + 3$.

Solution The synthetic division of f by $x + 3 = x - (-3)$ is

$$\begin{array}{r|rrrrrr} -3 & 1 & 0 & -4 & 0 & 2 & -10 \\ & & -3 & 9 & -15 & 45 & -141 \\ \hline & 1 & -3 & 5 & -15 & 47 & \boxed{-151} = r \end{array} \quad (15)$$

The quotient is the fourth-degree polynomial $q(x) = x^4 - 3x^3 + 5x^2 - 15x + 47$ and the constant remainder is $r = -151$. Also, the value of the function f at -3 is the remainder $f(-3) = -151$.

EXAMPLE 5   Using Synthetic Division to Evaluate a Function

Use the Remainder Theorem to find $f(2)$ for

$$f(x) = -3x^6 + 4x^5 + x^4 - 8x^3 - 6x^2 + 9.$$

> Using synthetic division we can find $f(2)$ without computing the powers of 2: 2^6, 2^5, 2^4, 2^3, and 2^2.

Solution We will use synthetic division to find the remainder r in the division of f by $x - 2$. We begin by writing down all the coefficients in $f(x)$, including 0 as the coefficient of x. From

$$\begin{array}{r|rrrrrrr} 2 & -3 & 4 & 1 & -8 & -6 & 0 & 9 \\ & & -6 & -4 & -6 & -28 & -68 & -136 \\ \hline & -3 & -2 & -3 & -14 & -34 & -68 & \boxed{-127} = r \end{array}$$

we see that $f(2) = -127$.

EXAMPLE 6   Using Synthetic Division to Evaluate a Function

Use synthetic division to evaluate $f(x) = x^3 - 7x^2 + 13x - 15$ at $x = 5$.

Solution From the synthetic division

$$\begin{array}{r|rrrr} 5 & 1 & -7 & 13 & -15 \\ & & 5 & -10 & 15 \\ \hline & 1 & -2 & 3 & \boxed{0} = r \end{array}$$

we see that $f(5) = 0$.

The result in Example 6 that $f(5) = 0$ shows that 5 is a zero of the given function f. Moreover, we have found additionally that f is evenly divisible by the linear polynomial $x - 5$, or put another way, $x - 5$ is a factor of f. The synthetic division shows that $f(x) = x^3 - 7x^2 + 13x - 15$ is equivalent to

$$f(x) = (x - 5)(x^2 - 2x + 3).$$

In the next section we will further explore the use of the Division Algorithm and Remainder Theorem as a help in finding zeros and factors of a polynomial function.

4.2 Exercises
Answers to selected odd-numbered problems begin on page ANS-12.

In Problems 1–10, use long division to find the quotient $q(x)$ and remainder $r(x)$ when the polynomial $f(x)$ is divided by the given polynomial $g(x)$. In each case write your answer in the form $f(x) = g(x)q(x) + r(x)$.

1. $f(x) = 8x^2 + 4x - 7;\quad g(x) = x^2$
2. $f(x) = x^2 + 2x - 3;\quad g(x) = x^2 + 1$
3. $f(x) = 5x^3 - 7x^2 + 4x + 1;\quad g(x) = x^2 + x - 1$
4. $f(x) = 14x^3 - 12x^2 + 6;\quad g(x) = x^2 - 1$
5. $f(x) = 2x^3 + 4x^2 - 3x + 5;\quad g(x) = (x + 2)^2$
6. $f(x) = x^3 + x^2 + x + 1;\quad g(x) = (2x + 1)^2$
7. $f(x) = 27x^3 + x - 2;\quad g(x) = 3x^2 - x$
8. $f(x) = x^4 + 8;\quad g(x) = x^3 + 2x - 1$
9. $f(x) = 6x^5 + 4x^4 + x^3;\quad g(x) = x^3 - 2$
10. $f(x) = 5x^6 - x^5 + 10x^4 + 3x^2 - 2x + 4;\quad g(x) = x^2 + x - 1$

In Problems 11–16, proceed as in Example 2 and use the Remainder Theorem to find r when $f(x)$ is divided by the given linear polynomial.

11. $f(x) = 2x^2 - 4x + 6;\quad x - 2$
12. $f(x) = 3x^2 + 7x - 1;\quad x + 3$
13. $f(x) = x^3 - 4x^2 + 5x + 2;\quad x - \frac{1}{2}$
14. $f(x) = 5x^3 + x^2 - 4x - 6;\quad x + 1$
15. $f(x) = x^4 - x^3 + 2x^2 + 3x - 5;\quad x - 3$
16. $f(x) = 2x^4 - 7x^2 + x - 1;\quad x + \frac{3}{2}$

In Problems 17–22, proceed as in Example 3 and use the Remainder Theorem to find $f(c)$ for the given value of c.

17. $f(x) = 4x^2 - 10x + 6;\quad c = 2$
18. $f(x) = 6x^2 + 4x - 2;\quad c = \frac{1}{4}$
19. $f(x) = x^3 + 3x^2 + 6x + 6;\quad c = -5$
20. $f(x) = 15x^3 + 17x^2 - 30;\quad c = \frac{1}{5}$
21. $f(x) = 3x^4 - 5x^2 + 20;\quad c = \frac{1}{2}$
22. $f(x) = 14x^4 - 60x^3 + 49x^2 - 21x + 19;\quad c = 1$

In Problems 23–32, use synthetic division to find the quotient $q(x)$ and remainder $r(x)$ when $f(x)$ is divided by the given linear polynomial.

23. $f(x) = 2x^2 - x + 5;\quad x - 2$
24. $f(x) = 4x^2 - 8x + 6;\quad x - \frac{1}{2}$
25. $f(x) = x^3 - x^2 + 2;\quad x + 3$
26. $f(x) = 4x^3 - 3x^2 + 2x + 4;\quad x - 7$
27. $f(x) = x^4 + 16;\quad x - 2$
28. $f(x) = 4x^4 + 3x^3 - x^2 - 5x - 6;\quad x + 3$
29. $f(x) = x^5 + 56x^2 - 4;\quad x + 4$
30. $f(x) = 2x^6 + 3x^3 - 4x^2 - 1;\quad x + 1$
31. $f(x) = x^3 - (2 + \sqrt{3})x^2 + 3\sqrt{3}x - 3;\quad x - \sqrt{3}$
32. $f(x) = x^8 - 3^8;\quad x - 3$

In Problems 33–38, use synthetic division and the Remainder Theorem to find $f(c)$ for the given value of c.

33. $f(x) = 4x^2 - 2x + 9;\quad c = -3$
34. $f(x) = 3x^4 - 5x^2 + 27;\quad c = \frac{1}{2}$
35. $f(x) = 14x^4 - 60x^3 + 49x^2 - 21x + 19;\quad c = 1$

36. $f(x) = 3x^5 + x^2 - 16;\quad c = -2$
37. $f(x) = 2x^6 - 3x^5 + x^4 - 2x + 1;\quad c = 4$
38. $f(x) = x^7 - 3x^5 + 2x^3 - x + 10;\quad c = 5$

In Problems 39 and 40, use long division to find a value of k such that $f(x)$ is divisible by $g(x)$.

39. $f(x) = x^4 + x^3 + 3x^2 + kx - 4;\quad g(x) = x^2 - 1$
40. $f(x) = x^5 - 3x^4 + 7x^3 + kx^2 + 9x - 5;\quad g(x) = x^2 - x + 1$

In Problems 41 and 42, use synthetic division to find a value of k such that $f(x)$ is divisible by $g(x)$.

41. $f(x) = kx^4 + 2x^2 + 9k;\quad g(x) = x - 1$
42. $f(x) = x^3 + kx^2 - 2kx + 4;\quad g(x) = x + 2$

43. Find a value of k such that the remainder in the division of $f(x) = 3x^2 - 4kx + 1$ by $g(x) = x + 3$ is $r = -20$.

44. When $f(x) = x^2 - 3x - 1$ is divided by $x - c$, the remainder is $r = 3$. Determine c.

4.3 Zeros and Factors of Polynomial Functions

≡ Introduction In Section 3.1 we saw that a zero of a function f is a number c for which $f(c) = 0$. A zero c of a function f can be a *real* or a *complex* number. Recall from Section 1.4 that a **complex number** is a number of the form $z = a + bi$, where $i^2 = -1$, and a and b are real numbers. The number a is called the **real part** of z and b is called the **imaginary part** of z. The symbol i is called the **imaginary unit** and it is common practice to write it as $i = \sqrt{-1}$. If $z = a + bi$ is a complex number, then $\bar{z} = a - bi$ is called its **conjugate**. Thus the simple polynomial function $f(x) = x^2 + 1$ has two complex zeros since the solutions of $x^2 + 1 = 0$ are $\pm\sqrt{-1}$, that is, i and $-i$.

In this section we explore the connection between the zeros of a polynomial function f, the operation of division, and the factors of f.

EXAMPLE 1 A Real Zero

Consider the polynomial function $f(x) = 2x^3 - 9x^2 + 6x - 1$. The real number $\frac{1}{2}$ is a zero of the function since

$$f\left(\tfrac{1}{2}\right) = 2\left(\tfrac{1}{2}\right)^3 - 9\left(\tfrac{1}{2}\right)^2 + 6\left(\tfrac{1}{2}\right) - 1$$
$$= \tfrac{2}{8} - \tfrac{9}{4} + 3 - 1$$
$$= \tfrac{1}{4} - \tfrac{9}{4} + \tfrac{8}{4} = 0.$$

≡

EXAMPLE 2 A Complex Zero

See (4) in Section R.6. ▶

Consider the polynomial function $f(x) = x^3 - 5x^2 + 8x - 6$. The complex number $1 + i$ is a zero of the function. To verify this we use the binomial expansion of $(a + b)^3$ and the fact that $i^2 = -1$ and $i^3 = -i$:

$$f(1+i) = (1+i)^3 - 5(1+i)^2 + 8(1+i) - 6$$
$$= (1^3 + 3 \cdot 1^2 \cdot i + 3 \cdot 1 \cdot i^2 + i^3) - 5(1^2 + 2i + i^2) + 8(1+i) - 6$$
$$= (-2 + 2i) - 5(2i) + (2 + 8i)$$
$$= (-2 + 2) + (10 - 10)i = 0 + 0i = 0.$$

☐ **Factor Theorem** We can now relate the notion of a zero of a polynomial function f with division of polynomials. From the Remainder Theorem we know that when $f(x)$ is divided by the linear polynomial $x - c$ the remainder is $r = f(c)$. If c is a zero of f, then $f(c) = 0$ implies $r = 0$. From the form of the Division Algorithm given in (4) of Section 4.2 we can then write f as

$$f(x) = (x - c)q(x). \tag{1}$$

Thus, if c is a zero of a polynomial function f, then $x - c$ is a factor of $f(x)$. Conversely, if $x - c$ is a factor of $f(x)$, then f has the form given in (1). In this case, we see immediately that $f(c) = (c - c)q(c) = 0$. These results are summarized in the **Factor Theorem** given next.

THEOREM 4.3.1 Factor Theorem

A number c is a zero of a polynomial function f if and only if $x - c$ is a factor of $f(x)$.

If a polynomial function f is of degree n and $(x - c)^m$, $m \leq n$, is a factor of $f(x)$, then c is said to be a **zero of multiplicity** m. When $m = 1$, c is a **simple zero**. Equivalently, we say that the number c is a **root of multiplicity** m of the equation $f(x) = 0$. We have already examined the graphical significance of simple and repeated real zeros of a polynomial function f in Section 4.1. See Figure 4.1.6.

■ EXAMPLE 3   **Factors of a Polynomial**

Determine whether

(a) $x + 1$ is a factor of $f(x) = x^4 - 5x^2 + 6x - 1$,
(b) $x - 2$ is a factor of $f(x) = x^3 - 3x^2 + 4$.

Solution We use synthetic division to divide $f(x)$ by the given linear term.
(a) From the division

$$\begin{array}{r|rrrrr} -1 & 1 & 0 & -5 & 6 & -1 \\ & & -1 & 1 & 4 & -10 \\ \hline & 1 & -1 & -4 & 10 & \boxed{-11} = r = f(-1) \end{array}$$

we see that $f(-1) = -11 \neq 0$ and so -1 is not a zero of f. We conclude that $x - (-1) = x + 1$ is not a factor of $f(x)$.
(b) From the division

$$\begin{array}{r|rrrr} 2 & 1 & -3 & 0 & 4 \\ & & 2 & -2 & -4 \\ \hline & 1 & -1 & -2 & \boxed{0} = r = f(2) \end{array}$$

we see that $f(2) = 0$. This means that 2 is a zero and that $x - 2$ is a factor of $f(x)$. From the division we see that the quotient is $q(x) = x^2 - x - 2$ and so $f(x) = (x - 2)(x^2 - x - 2)$.

☐ **Number of Zeros** In Example 6 of Section 4.1 we graphed the polynomial function

$$f(x) = (x - 3)(x - 1)^2(x + 2)^3. \qquad (2)$$

The number 3 is a zero of multiplicity 1, or a simple zero of f, the number 1 is a zero of multiplicity 2, and -2 is a zero of multiplicity 3. Although the function f has three *distinct* zeros (different from one another), nevertheless, it is standard practice to say that f has *six zeros* because we count the multiplicities of each zero. Hence for the function f in (2), the number of zeros is $1 + 2 + 3 = 6$. The question

How many zeros does a polynomial function f have?

is answered next.

THEOREM 4.3.2 Fundamental Theorem of Algebra

A polynomial function f of degree $n > 0$ has at least one zero.

Theorem 4.3.2, first proved by the German mathematician **Carl Friedrich Gauss** (1777–1855) in 1799, is considered one of the major milestones in the history of mathematics. At first reading, this theorem does not appear to say much, but when combined with the Factor Theorem, the Fundamental Theorem of Algebra shows:

Every polynomial function f of degree $n > 0$ has exactly n zeros. $\qquad (3)$

Of course if a zero is repeated, say it has multiplicity k, we count that zero k times. To prove (3), we know from the Fundamental Theorem of Algebra that f has a zero; call it c_1. By the Factor Theorem we can write

$$f(x) = (x - c_1)q_1(x), \qquad (4)$$

where q_1 is a polynomial function of degree $n - 1$. If $n - 1 \neq 0$, then in like manner we know that q_1 must have a zero; call it c_2 and so (4) becomes

$$f(x) = (x - c_1)(x - c_2)q_2(x),$$

where q_2 is a polynomial function of degree $n - 2$. If $n - 2 \neq 0$, we continue and arrive at

$$f(x) = (x - c_1)(x - c_2)(x - c_3)q_3(x), \qquad (5)$$

and so on. Eventually we arrive at a factorization of $f(x)$ with n linear factors and the last factor $q_n(x)$ is of degree 0. In other words, $q_n(x) = a_n$, where a_n is a constant—specifically, a_n is the leading coefficient of f. We have arrived at the *complete* factorization of $f(x)$.

THEOREM 4.3.3 Complete Linear Factorization

Let $c_1, c_2, \ldots, c_n$ be the n (not necessarily distinct) zeros of the polynomial function of degree $n > 0$:

$$f(x) = a_n x^n + a_{n-1} x^{n-1} + \cdots + a_2 x^2 + a_1 x + a_0.$$

Then $f(x)$ can be written as a product of n linear factors

$$f(x) = a_n(x - c_1)(x - c_2) \cdots (x - c_n). \qquad (6)$$

Bear in mind that some or all the zeros $c_1, \ldots, c_n$ in (6) may be complex numbers $a + bi$, where $b \neq 0$.

In the case of a second-degree, or quadratic, polynomial function $f(x) = ax^2 + bx + c$, where the coefficients a, b, and c are real numbers, the zeros c_1 and c_2 of f can be found using the quadratic formula:

$$c_1 = \frac{-b - \sqrt{b^2 - 4ac}}{2a} \quad \text{and} \quad c_2 = \frac{-b + \sqrt{b^2 - 4ac}}{2a}. \qquad (7)$$

The results in (7) tell the whole story about the zeros of the quadratic function:

- the zeros are real and distinct when $b^2 - 4ac > 0$,
- real with multiplicity two when $b^2 - 4ac = 0$, and
- complex and distinct when $b^2 - 4ac < 0$.

It follows from (6) that the complete linear factorization of a quadratic polynomial function $f(x) = ax^2 + bx + c$ is

$$f(x) = a(x - c_1)(x - c_2). \qquad (8)$$

EXAMPLE 4   Example 1 Revisited

In Example 1 we demonstrated that $\frac{1}{2}$ is a zero of $f(x) = 2x^3 - 9x^2 + 6x - 1$. We now know that $x - \frac{1}{2}$ is a factor of $f(x)$ and that $f(x)$ has three zeros. The synthetic division

$$\begin{array}{c|cccc} \frac{1}{2} & 2 & -9 & 6 & -1 \\ & & 1 & -4 & 1 \\ \hline & 2 & -8 & 2 & \boxed{0} = r \end{array}$$

again demonstrates that $\frac{1}{2}$ is a zero of $f(x)$ (the 0 remainder is the value of $f(\frac{1}{2})$) and, additionally, gives us the quotient $q(x)$ obtained in the division of $f(x)$ by $x - \frac{1}{2}$, that is, $f(x) = (x - \frac{1}{2})(2x^2 - 8x + 2)$. As shown in (8) we can now factor the quadratic quotient $q(x) = 2x^2 - 8x + 2$ by finding the roots of $2x^2 - 8x + 2 = 0$ from the quadratic formula:

$$\downarrow \sqrt{48} = \sqrt{16 \cdot 3} = \sqrt{16}\sqrt{3} = 4\sqrt{3}$$

$$x = \frac{-(-8) \pm \sqrt{(-8)^2 - 4(2)(2)}}{4} = \frac{8 \pm \sqrt{48}}{4} = \frac{8 \pm 4\sqrt{3}}{4}$$

$$= \frac{\cancel{4}(2 \pm \sqrt{3})}{\cancel{4}} = 2 \pm \sqrt{3}.$$

Thus the remaining zeros of $f(x)$ are the irrational numbers $2 + \sqrt{3}$ and $2 - \sqrt{3}$. With the identification of the leading coefficient as $a_3 = 2$, it follows from (8) that the complete linear factorization of $f(x)$ is then

$$f(x) = 2(x - \tfrac{1}{2})(x - (2 + \sqrt{3}))(x - (2 - \sqrt{3}))$$
$$= 2(x - \tfrac{1}{2})(x - 2 - \sqrt{3})(x - 2 + \sqrt{3}). \qquad \equiv$$

EXAMPLE 5   Using Synthetic Division

Find the complete linear factorization of

$$f(x) = x^4 - 12x^3 + 47x^2 - 62x + 26$$

given that 1 is a zero of f of multiplicity two.

Solution We know that $x - 1$ is a factor of $f(x)$ so by the division

$$\begin{array}{r|rrrrr} 1 & 1 & -12 & 47 & -62 & 26 \\ & & 1 & -11 & 36 & -26 \\ \hline & 1 & -11 & 36 & -26 & \boxed{0} = r \end{array}$$

we find
$$f(x) = (x - 1)(x^3 - 11x^2 + 36x - 26).$$

Since 1 is a zero of multiplicity 2, $x - 1$ must also be a factor of the quotient $q(x) = x^3 - 11x^2 + 36x - 26$. By the division,

$$\begin{array}{r|rrrr} 1 & 1 & -11 & 36 & -26 \\ & & 1 & -10 & 26 \\ \hline & 1 & -10 & 26 & \boxed{0} = r \end{array}$$

we conclude that $q(x)$ can be written $= (x - 1)(x^2 - 10x + 26)$. Therefore,

$$f(x) = (x - 1)^2(x^2 - 10x + 26).$$

The remaining two zeros, found by solving $x^2 - 10x + 26 = 0$ by the quadratic formula, are the complex numbers $5 + i$ and $5 - i$. Since the leading coefficient is $a_4 = 1$ the complete linear factorization of $f(x)$ is

$$\begin{aligned} f(x) &= (x - 1)^2(x - (5 + i))(x - (5 - i)) \\ &= (x - 1)^2(x - 5 - i)(x - 5 + i). \end{aligned}$$

≡

EXAMPLE 6   Finding a Polynomial Function

Find a polynomial function f of degree 3, with zeros 1, -4, and 5 such that its graph possesses the y-intercept $(0, 5)$.

Solution Because we have three zeros 1, -4, and 5 we know $x - 1$, $x + 4$, and $x - 5$ are factors of f. However, the function we seek is *not*

$$f(x) = (x - 1)(x + 4)(x - 5). \tag{9}$$

The reason for this is that any nonzero constant multiple of f is a different polynomial with the same zeros. Notice too, that the function in (9) gives $f(0) = 20$ but we want $f(0) = 5$. To that end we must assume that f has the form

$$f(x) = a_3(x - 1)(x + 4)(x - 5), \tag{10}$$

where a_3 is some real constant. Using (10), $f(0) = 5$ gives

$$f(0) = a_3(0 - 1)(0 + 4)(0 - 5) = 20a_3 = 5$$

and so $a_3 = \frac{5}{20} = \frac{1}{4}$. The desired function is then

$$f(x) = \tfrac{1}{4}(x - 1)(x + 4)(x - 5).$$

≡

☐ **Conjugate Pairs** In the introduction to this section was saw that i and $-i$ are complex zeros of $f(x) = x^2 + 1$. Likewise, in Example 5 we showed that $5 + i$ and $5 - i$ are complex zeros of $f(x) = x^4 - 12x^3 + 47x^2 - 62x + 26$. In each of these two cases the complex zeros of the polynomial function are conjugate pairs. In other words, one complex zero is the conjugate of the other. This is no coincidence; complex zeros of polynomials with *real* coefficients *always* appear in conjugate pairs. To see this we use the properties that the conjugate of a sum of complex numbers z_1 and z_2 is the sum of the conjugates $\bar{z}_1$ and $\bar{z}_2$ and the conjugate of a nonnegative integer power of a complex number z is the power of the conjugate $\bar{z}$:

$$\overline{z_1 + z_2} = \bar{z}_1 + \bar{z}_2 \quad \text{and} \quad \overline{z^n} = \bar{z}^n. \tag{11}$$

See Problems 80–82 in Exercises 1.4.

Let $f(x) = a_n x^n + a_{n-1} x^{n-1} + \cdots + a_1 x + a_0$, where the coefficients a_i, $i = 0, 1, 2, \ldots, n$, are real numbers. If z denotes a complex zero of f, then we have

$$a_n z^n + a_{n-1} z^{n-1} + \cdots + a_1 z + a_0 = 0.$$

Taking the conjugate of both sides of this equation gives

$$\overline{a_n z^n + a_{n-1} z^{n-1} + \cdots + a_1 z + a_0} = \bar{0}.$$

Now using (11) and the fact that the conjugate of any real number a_i, $i = 0, 1, 2, \ldots, n$, is itself, we obtain

$$a_n \bar{z}^n + a_{n-1} \bar{z}^{n-1} + \cdots + a_1 \bar{z} + a_0 = 0.$$

This means that $f(\bar{z}) = 0$ and, therefore, $\bar{z}$ is a zero of $f(x)$ whenever z is a zero. We state this result as the next theorem.

THEOREM 4.3.4 Conjugate Zeros Theorem

Let $f(x)$ be a polynomial function of degree $n > 1$ with real coefficients. If z is a complex zero of $f(x)$, then its conjugate $\bar{z}$ is also a zero of $f(x)$.

EXAMPLE 7 Example 2 Revisited

Find the complete linear factorization of

$$f(x) = x^3 - 5x^2 + 8x - 6.$$

Solution In Example 2 we demonstrated that $1 + i$ is a complex zero of $f(x) = x^3 - 5x^2 + 8x - 6$. Because the coefficients of f are real numbers we conclude that another zero is the conjugate of $1 + i$, namely, $1 - i$. Thus we know two factors of $f(x)$; $x - (1 + i)$ and $x - (1 - i)$. Carrying out the multiplication, we find

$$(x - 1 - i)(x - 1 + i) = x^2 - 2x + 2.$$

Thus we can write

$$f(x) = (x - 1 - i)(x - 1 + i)q(x) = (x^2 - 2x + 2)q(x).$$

We determine $q(x)$ by performing the *long division* of $f(x)$ by $x^2 - 2x + 2$ (we can't use synthetic division because we are not dividing by a linear factor). From

$$\begin{array}{r}
x - 3 \\
x^2 - 2x + 2 \overline{\smash{\big)}\, x^3 - 5x^2 + 8x - 6} \\
\underline{x^3 - 2x^2 + 2x } \\
-3x^2 + 6x - 6 \\
\underline{-3x^2 + 6x - 6} \\
0
\end{array}$$

we see that the complete linear factorization of $f(x)$ is

$$f(x) = (x - 1 - i)(x - 1 + i)(x - 3).$$

The three zeros of $f(x)$ are $1 + i$, $1 - i$, and 3.

4.3 Exercises
Answers to selected odd-numbered problems begin on page ANS-12.

In Problems 1–6, determine whether the indicated real number is a zero of the given polynomial function f. If yes, find all other zeros and then give the complete factorization of $f(x)$.

1. 1; $f(x) = 4x^3 - 9x^2 + 6x - 1$
2. $\frac{1}{2}$; $f(x) = 2x^3 - x^2 + 32x - 16$
3. 5; $f(x) = x^3 - 6x^2 + 6x + 5$
4. 3; $f(x) = x^3 - 3x^2 + 4x - 12$
5. $-\frac{2}{3}$; $f(x) = 3x^3 - 10x^2 - 2x + 4$
6. -2; $f(x) = x^3 - 4x^2 - 2x + 20$

In Problems 7–10, verify that each of the indicated numbers are zeros of the given polynomial function f. Find all other zeros and then give the complete factorization of $f(x)$.

7. $-3, 5$; $f(x) = 4x^4 - 8x^3 - 61x^2 + 2x + 15$
8. $\frac{1}{4}, \frac{3}{2}$; $f(x) = 8x^4 - 30x^3 + 23x^2 + 8x - 3$
9. $1, -\frac{1}{3}$ (multiplicity 2); $f(x) = 9x^4 + 69x^3 - 29x^2 - 41x - 8$
10. $-\sqrt{5}, \sqrt{5}$; $f(x) = 3x^4 + x^3 - 17x^2 - 5x + 10$

In Problems 11–16, use synthetic division to determine whether the indicated linear polynomial is a factor of the given polynomial function f. If yes, find all other zeros and then give the complete factorization of $f(x)$.

11. $x - 5$; $f(x) = 2x^2 + 6x - 25$
12. $x + \frac{1}{2}$; $f(x) = 10x^2 - 27x + 11$
13. $x - 1$; $f(x) = x^3 + x - 2$
14. $x + \frac{1}{2}$; $f(x) = 2x^3 - x^2 + x + 1$
15. $x - \frac{1}{3}$; $f(x) = 3x^3 - 3x^2 + 8x - 2$
16. $x - 2$; $f(x) = x^3 - 6x^2 - 16x + 48$

In Problems 17–20, use division to show that the indicated polynomial is a factor of the given polynomial function f. Find all other zeros and then give the complete factorization of $f(x)$.

17. $(x - 1)(x - 2)$; $f(x) = x^4 - 3x^3 + 6x^2 - 12x + 8$
18. $x(3x - 1)$; $f(x) = 3x^4 - 7x^3 + 5x^2 - x$
19. $(x - 1)^2$; $f(x) = 2x^4 + x^3 - 5x^2 - x + 3$
20. $(x + 3)^2$; $f(x) = x^4 - 4x^3 - 22x^2 + 84x + 261$

In Problems 21–26, verify that the indicated complex number is a zero of the given polynomial function f. Proceed as in Example 7 to find all other zeros and then give the complete factorization of $f(x)$.

21. $2i$; $f(x) = 3x^3 - 5x^2 + 12x - 20$
22. $\frac{1}{2}i$; $f(x) = 12x^3 + 8x^2 + 3x + 2$
23. $-1 + i$; $f(x) = 5x^3 + 12x^2 + 14x + 4$
24. $-i$; $f(x) = 4x^4 - 8x^3 + 9x^2 - 8x + 5$
25. $1 + 2i$; $f(x) = x^4 - 2x^3 - 4x^2 + 18x - 45$
26. $1 + i$; $f(x) = 6x^4 - 11x^3 + 9x^2 + 4x - 2$

In Problems 27–32, find a polynomial function f with real coefficients of the indicated degree that possesses the given zeros.

27. degree 4; $2, 1, -3$ (multiplicity 2)
28. degree 5; $-4i, -\frac{1}{3}, \frac{1}{2}$ (multiplicity 2)
29. degree 5; $3 + i, 0$ (multiplicity 3)
30. degree 4; $5i, 2 - 3i$
31. degree 2; $1 - 6i$
32. degree 2; $4 + 3i$

In Problems 33–36, find the zeros of the given polynomial function f. State the multiplicity of each zero.

33. $f(x) = x(4x - 5)^2(2x - 1)^3$
34. $f(x) = x^4 + 6x^3 + 9x^2$
35. $f(x) = (9x^2 - 4)^2$
36. $f(x) = (x^2 + 25)(x^2 - 5x + 4)^2$

In Problems 37 and 38, find the value(s) of k such that the indicated number is a zero of $f(x)$. Then give the complete factorization of $f(x)$.

37. 3; $f(x) = 2x^3 - 2x^2 + k$
38. 1; $f(x) = x^3 + 5x^2 - k^2x + k$

In Problems 39 and 40, find a polynomial function f of the indicated degree whose graph is given in the figure.

39. degree 3

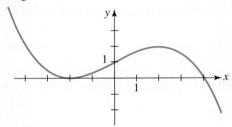

FIGURE 4.3.1 Graph for Problem 39

40. degree 5

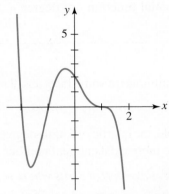

FIGURE 4.3.2 Graph for Problem 40

For Discussion

41. Discuss:
  (a) For what positive integer values of n is $x - 1$ a factor of $f(x) = x^n - 1$?
  (b) For what positive integer values of n is $x + 1$ a factor of $f(x) = x^n + 1$?
42. Suppose f is a polynomial function of degree 3 with real coefficients. Why can't $f(x)$ have three complex zeros? Put another way, why must at least one zero of a cubic polynomial function be a real number? Can you generalize this result?
43. What is the smallest degree that a polynomial function f with real coefficients can have such that $1 + i$ is a complex zero of multiplicity 2? Of multiplicity 3?
44. Let $z = a + bi$. Show that $z + \bar{z}$ and $z\bar{z}$ are real numbers.
45. Let $z = a + bi$. Use the results of Problem 44 to show that

$$f(x) = (x - z)(x - \bar{z})$$

is a polynomial function with real coefficients.
46. Try to prove or disprove the following proposition.

If f is an odd polynomial function, then the graph of f passes through the origin.

4.4 Real Zeros of Polynomial Functions

≡ **Introduction** In the preceding section we saw, as a consequence of the Fundamental Theorem of Algebra, that a polynomial function f of degree n has n zeros when the multiplicities of the zeros are counted. We saw too that a zero of a polynomial function could be either a real or a complex number. In this section we confine our attention to *real zeros* of polynomial functions with real coefficients.

☐ **Real Zeros** If a polynomial function f of degree $n > 0$ has m (not necessarily distinct) real zeros $c_1, c_2, \ldots, c_m$, then by the Factor Theorem each of the linear polynomials $x - c_1, x - c_2, \ldots, x - c_m$ are factors of $f(x)$. That is,

$$f(x) = (x - c_1)(x - c_2) \cdots (x - c_m)q(x),$$

where $q(x)$ is a polynomial. Thus n, the degree of f, must be greater than or possibly equal to m, $(m \leq n)$, the number of real zeros when each is counted according to its multiplicity. Using slightly different words, we restate the last sentence.

THEOREM 4.4.1 Number of Real Zeros

A polynomial function f of degree $n > 0$ has at most n real zeros (not necessarily distinct).

Let's summarize some facts about real zeros of a polynomial function f of degree n:

- *f may not have any real zeros.*

For example, the quartic or fourth-degree polynomial function $f(x) = x^4 + 9$ has no real zeros since there exists no real number x satisfying $x^4 + 9 = 0$ or $x^4 = -9$:

- *f may have m real zeros where $m < n$.*

For example, the third-degree polynomial function $f(x) = (x - 1)(x^2 + 1)$ has one real zero:

- *f may have n real zeros.*

For example, by factoring the third-degree polynomial function $f(x) = x^3 - x$ as $f(x) = x(x^2 - 1) = x(x + 1)(x - 1)$ we see that it has three real zeros:

- *f has at least one real zero when the degree n is odd.*

The last property is a consequence of the fact that complex zeros of a polynomial function f with real coefficients must appear in conjugate pairs and so f cannot have an odd number of complex zeros. Thus if we write down an arbitrary cubic polynomial function such as $f(x) = x^3 + x + 1$, we know that f cannot have just one complex zero nor can it have three complex zeros. Put another way, $f(x) = x^3 + x + 1$ either has exactly one real zero or it has exactly three real zeros. You may have to think about the following property:

- *If the coefficients of $f(x)$ are positive and the constant term $a_0 \neq 0$, then any real zeros of f must be negative.*

☐ **Finding Real Zeros** It is one thing to talk about the existence of real and complex zeros of a polynomial function; it is an entirely different problem to actually find these zeros. As pointed out in the chapter introduction, the problem of finding a *formula* that expresses the zeros of a general nth degree polynomial function f in terms of its coefficients perplexed mathematicians for centuries. We have seen in Sections 1.3 and 3.3 that in the case of a second-degree, or quadratic, polynomial function $f(x) = ax^2 + bx + c$, where the coefficients a, b, and c are real numbers, the zeros c_1 and c_2 of f can be found using the quadratic formula. The problem of finding zeros of third-degree, or cubic, polynomial functions was solved in the sixteenth century by the Italian mathematician **Niccolò Fontana** (1499–1557), also known as Tartaglia—"the stammerer." Around 1540 another Italian mathematician, **Lodovico Ferrari** (1522–1565) discovered an algebraic formula for determining the zeros for fourth-degree, or quartic, polynomial functions. Since these formulas are complicated and difficult to use, they are seldom discussed in elementary courses. Then in 1824, at age 22, the Norwegian mathematician **Niels Henrik Abel** (1802–1829), proved it was impossible to find such formulas for the zeros of all general polynomials of degrees $n \geq 5$ in terms of their coefficients.

Niels Henrik Abel

☐ **Rational Zeros** Real zeros of a polynomial function are either rational or irrational numbers. A rational number is a number of the form p/s, where p and s are integers and $s \neq 0$. An irrational number is one that is not rational. For example, $\frac{1}{4}$ and -9 are rational numbers, but $\sqrt{2}$ and π are irrational; that is, neither $\sqrt{2}$ nor π can be written as a fraction p/s where p and s are integers. So how do we find real zeros for polynomial functions of degree $n > 2$? First, the bad news: For an irrational real zero, we *may* have to be content to use an accurate graph to "eyeball" its location on the x-axis, or use an equation solver of a calculator or one of the many sophisticated analytical methods developed over the years to *approximate* the zero. Now, the good news: We can always find the *rational* real zeros of *any* polynomial function with rational coefficients. We have already seen that synthetic division is a useful method for determining whether a given number c is a zero of a polynomial function $f(x)$. When the remainder in the division of $f(x)$ by $x - c$ is $r = 0$, we have found a zero of the polynomial function f, since $r = f(c) = 0$. For example, $\frac{2}{3}$ is a zero of $f(x) = 18x^3 - 15x^2 + 14x - 8$, since

$$\begin{array}{c|cccc} \frac{2}{3} & 18 & -15 & 14 & -8 \\ & & 12 & -2 & 8 \\ \hline & 18 & -3 & 12 & \boxed{0} = r \end{array}$$

Hence by the Factor Theorem, both $x - \frac{2}{3}$ and the quotient $18x^2 - 3x + 12$ are factors of f and so we can write the polynomial function as the product

$$\begin{aligned} f(x) &= \left(x - \tfrac{2}{3}\right)(18x^2 - 3x + 12) \quad \leftarrow \text{factor 3 from the quadratic polynomial} \\ &= \left(x - \tfrac{2}{3}\right)(3)(6x^2 - x + 4) \\ &= (3x - 2)(6x^2 - x + 4). \end{aligned} \qquad (1)$$

As discussed in the preceding section, if we can factor the polynomial to the point where the remaining factor is a quadratic polynomial, we can then find the remaining two zeros by the quadratic formula. For this example, the factorization in (1) is as far as we can go using real numbers since the zeros of the quadratic factor $6x^2 - x + 4$ are complex (verify). But the indicated multiplication in (1) illustrates something important about rational zeros. The leading coefficient 18 and the constant term -8 of $f(x)$ are obtained from the products

$$\overset{-8}{\overbrace{(3x - 2)(6x^2 - x + 4).}}_{18}$$

Thus we see that the denominator 3 of the rational zero $\frac{2}{3}$ is a *factor* of the leading coefficient 18 of $f(x) = 18x^3 - 15x^2 + 14x - 8$, and the numerator 2 of the rational zero is a factor of the constant term -8.

This example illustrates the following general principle for determining the rational zeros of a polynomial function. Read the following theorem carefully; the coefficients of f are not only real numbers—they must be *integers*.

4.4 Real Zeros of Polynomial Functions

> **THEOREM 4.4.2** Rational Zeros Theorem
>
> Let p/s be a rational number in lowest terms and a zero of the polynomial function
> $$f(x) = a_n x^n + a_{n-1} x^{n-1} + \cdots + a_2 x^2 + a_1 x + a_0,$$
> where the coefficients $a_n, a_{n-1}, \ldots, a_2, a_1, a_0$ are integers with $a_n \neq 0$. Then p is an integer factor of the constant term a_0 and s is an integer factor of the leading coefficient a_n.

The Rational Zeros Theorem deserves to be read several times. Note that the theorem *does not* assert that a polynomial function f with integer coefficients *must* have a rational zero; rather, it states that *if* a polynomial function f with integer coefficients has a rational zero p/s, then necessarily:

$$\frac{p}{s}. \quad \begin{array}{l} \leftarrow \text{is an integer factor of the constant term } a_0 \\ \leftarrow \text{is an integer factor of the lead coefficient } a_n \end{array}$$

By forming all possible quotients of each integer factor of a_0 to each integer factor of a_n, we can construct a list of *potential* rational zeros of f.

EXAMPLE 1   Rational Zeros

Find all rational zeros of $f(x) = 3x^4 - 10x^3 - 3x^2 + 8x - 2$.

Solution We identify the constant term $a_0 = -2$ and leading coefficient $a_4 = 3$ and list all the integer factors of a_0 and a_4, respectively:

$$p: \pm 1, \pm 2,$$
$$s: \pm 1, \pm 3.$$

Now we form a list of all possible rational zeros p/s by dividing all the factors of p by ± 1 and then by ± 3:

$$\frac{p}{s}: \pm 1, \pm 2, \pm \tfrac{1}{3}, \pm \tfrac{2}{3}. \tag{2}$$

We know that the given fourth-degree polynomial function f has four zeros; if any of these zeros is a real number and is rational, then it must appear in the list (2).

To determine which, if any, of the numbers in (2) are zeros, we could use direct substitution into $f(x)$. Synthetic division, however, is usually a more efficient means of evaluating $f(x)$. We begin by testing -1:

$$\begin{array}{r|rrrrr} -1 & 3 & -10 & -3 & 8 & -2 \\ & & -3 & 13 & -10 & 2 \\ \hline & 3 & -13 & 10 & -2 & \boxed{0} = r \end{array} \tag{3}$$

The zero remainder shows $r = f(-1) = 0$ and so -1 is a zero of f. Hence $x - (-1) = x + 1$ is a factor of f. Using the quotient found in (3) we can write

$$f(x) = (x + 1)(3x^3 - 13x^2 + 10x - 2). \tag{4}$$

From (4) we see that any other rational zero of f must be a zero of the quotient $3x^3 - 13x^2 + 10x - 2$. Since the latter polynomial is of lower degree, it will be easier to use synthetic division on it rather than on $f(x)$ to check the next rational zero. At this point in the process you should check to see whether the zero just found is a repeated zero. This is done by determining whether the found zero is also a zero of the quotient. A quick check, using synthetic division, shows that -1 is *not* a repeated zero of f since it is not a zero of $3x^3 - 13x^2 + 10x - 2$. So we move on and determine whether the number 1 is a rational zero of f. Indeed, it is *not* because the division

$$\begin{array}{r|rrrr} 1 & 3 & -13 & 10 & -2 \\ & & 3 & -10 & 0 \\ \hline & 3 & -10 & 0 & \boxed{-2} = r \end{array} \quad \leftarrow \text{coefficients of the quotient} \qquad (5)$$

shows that the remainder is $r = -2 \neq 0$. Checking $\frac{1}{3}$, we have

$$\begin{array}{r|rrrr} \frac{1}{3} & 3 & -13 & 10 & -2 \\ & & 1 & -4 & 2 \\ \hline & 3 & -12 & 6 & \boxed{0} = r \end{array} \qquad (6)$$

Thus $\frac{1}{3}$ is a zero. At this point we can stop using synthetic division since (6) indicates that the remaining factor of f is the quadratic polynomial $3x^2 - 12x + 6$. From the quadratic formula we find that the remaining real zeros are $2 + \sqrt{2}$ and $2 - \sqrt{2}$. Therefore the given polynomial function f has two rational zeros, -1 and $\frac{1}{3}$, and two irrational zeros, $2 + \sqrt{2}$ and $2 - \sqrt{2}$. ≡

If you have access to technology, your selection of rational numbers to test in Example 1 can be motivated by a graph of the function $f(x) = 3x^4 - 10x^3 - 3x^2 + 8x - 2$. With the aid of a graphing utility we obtain the graphs in **FIGURE 4.4.1**. In Figure 4.4.1(a) it would appear that f has at least three real zeros. But by "zooming in" on the graph on the interval $[0, 1]$, Figure 4.4.1(b) reveals that f actually has four real zeros: one negative and three positive. Thus, once you have determined one negative rational zero of f you may disregard all other negative numbers as potential zeros.

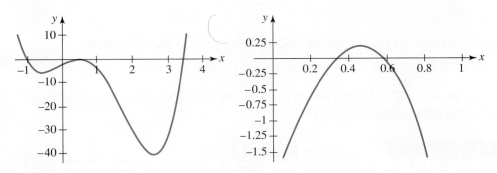

(a) Graph of f on the interval $[-1, 4]$ (b) Zoom-in of graph on the interval $[0, 1]$

FIGURE 4.4.1 Graph of function f in Example 1

EXAMPLE 2   Complete Linear Factorization

Since the function f in Example 1 is of degree 4 and we have found four real zeros we can give its complete factorization. Using the leading coefficient $a_4 = 3$, it follows from (6) of Section 4.3 that

$$\begin{aligned} f(x) &= 3(x + 1)(x - \tfrac{1}{3})(x - (2 - \sqrt{2}))(x - (2 + \sqrt{2})) \\ &= 3(x + 1)(x - \tfrac{1}{3})(x - 2 + \sqrt{2})(x - 2 - \sqrt{2}). \end{aligned}$$

≡

4.4 Real Zeros of Polynomial Functions

EXAMPLE 3 Rational Zeros

Find all rational zeros of $f(x) = x^4 + 4x^3 + 5x^2 + 4x + 4$.

Solution In this case the constant term is $a_0 = 4$ and the leading coefficient is $a_4 = 1$. The integer factors of a_0 and a_4 are, respectively:

$$p: \pm 1, \pm 2, \pm 4,$$
$$s: \pm 1.$$

The list of all possible rational zeros p/s is

$$\frac{p}{s}: \pm 1, \pm 2, \pm 4.$$

Since all the coefficients of f are positive, substituting a positive number from the foregoing list into $f(x)$ can never result in $f(x) = 0$. Thus the only numbers that are potential rational zeros are -1, -2, and -4. From the synthetic division

$$\begin{array}{r|rrrrr} -1 & 1 & 4 & 5 & 4 & 4 \\ & & -1 & -3 & -2 & -2 \\ \hline & 1 & 3 & 2 & 2 & \boxed{2} = r \end{array}$$

we see that -1 is not a zero. However, from

$$\begin{array}{r|rrrrr} -2 & 1 & 4 & 5 & 4 & 4 \\ & & -2 & -4 & -2 & -4 \\ \hline & 1 & 2 & 1 & 2 & \boxed{0} = r \end{array}$$

we see that -2 is a zero. We now test to see whether -2 is a repeated zero. Using the coefficients in the quotient,

$$\begin{array}{r|rrrr} -2 & 1 & 2 & 1 & 2 \\ & & -2 & 0 & -2 \\ \hline & 1 & 0 & 1 & \boxed{0} = r \end{array}$$

it follows that -2 is a zero of multiplicity 2. So far we have shown that

$$f(x) = (x + 2)^2(x^2 + 1).$$

Since the zeros of $x^2 + 1$ are the complex conjugates i and $-i$, we can conclude that -2 is the only real rational zero of $f(x)$.

EXAMPLE 4 No Rational Zeros

Consider the polynomial function $f(x) = x^5 - 4x - 1$. The only possible rational zeros are -1 and 1, and it is easy to see that neither $f(-1)$ nor $f(1)$ are 0. Thus f has no rational zeros. Since f is of odd degree we know that it has at least one real zero and so that zero must be an irrational number. With the aid of a graphing utility we obtain the graph in FIGURE 4.4.2. Note in the figure, the graph to the right of $x = 2$ cannot turn back down *and* the graph to the left of $x = -2$ cannot turn back up so that graph crosses the x-axis five times since that shape of the graph would be inconsistent with the end behavior of f. Thus we can conclude that the function f possesses three irrational real zeros and two complex conjugate zeros. The best we can do here is to approximate these zeros. Using a computer algebra system such as *Mathematica* we can approximate both the real and the complex zeros. We find these approximations to be -1.34, -0.25, 1.47, $0.061 + 1.42i$, and $0.061 - 1.42i$.

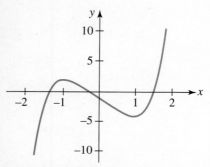

FIGURE 4.4.2 Graph of f in Example 4

Although the Rational Zeros Theorem requires that the coefficients of a polynomial function f be integers, in some circumstances we can apply the theorem to a polynomial function with some rational *noninteger* coefficients. The next example illustrates the concept.

EXAMPLE 5 Noninteger Coefficients

Find the rational zeros of $f(x) = \frac{5}{6}x^4 - \frac{23}{12}x^3 + \frac{10}{3}x^2 - 3x - \frac{3}{4}$.

Solution By multiplying f by the least common denominator 12 of all the rational coefficients we obtain a new function g with integer coefficients:

$$g(x) = 10x^4 - 23x^3 + 40x^2 - 36x - 9.$$

In other words, $g(x) = 12f(x)$. If c is a zero of the function g, then c is also zero of f because $g(c) = 12f(c) = 0$ implies $f(c) = 0$. After working through the numbers in the list of potential rational zeros

$$\frac{p}{s}: \pm 1, \pm 3, \pm 9, \pm \tfrac{1}{2}, \pm \tfrac{3}{2}, \pm \tfrac{9}{2}, \pm \tfrac{1}{5}, \pm \tfrac{3}{5}, \pm \tfrac{9}{5}, \pm \tfrac{1}{10}, \pm \tfrac{3}{10}, \pm \tfrac{9}{10},$$

we find that $-\frac{1}{5}$ and $\frac{3}{2}$ are zeros of g, and hence are zeros of f. ≡

4.4 Exercises
Answers to selected odd-numbered problems begin on page ANS-13.

In Problems 1–20, find all rational zeros of the given polynomial function f.

1. $f(x) = 5x^3 - 3x^2 + 8x + 4$
2. $f(x) = 2x^3 + 3x^2 - x + 2$
3. $f(x) = x^3 - 8x - 3$
4. $f(x) = 2x^3 - 7x^2 - 17x + 10$
5. $f(x) = 4x^4 - 7x^2 + 5x - 1$
6. $f(x) = 8x^4 - 2x^3 + 15x^2 - 4x - 2$
7. $f(x) = x^4 + 2x^3 + 10x^2 + 14x + 21$
8. $f(x) = 3x^4 + 5x^2 + 1$
9. $f(x) = 6x^4 - 5x^3 - 2x^2 - 8x + 3$
10. $f(x) = x^4 + 2x^3 - 2x^2 - 6x - 3$
11. $f(x) = x^4 + 6x^3 - 7x$
12. $f(x) = x^5 - 2x^2 - 12x$
13. $f(x) = x^5 + x^4 - 5x^3 + x^2 - 6x$
14. $f(x) = 128x^6 - 2$
15. $f(x) = \frac{1}{2}x^3 - \frac{9}{4}x^2 + \frac{17}{4}x - 3$
16. $f(x) = 0.2x^3 - x + 0.8$
17. $f(x) = 2.5x^3 + x^2 + 0.6x + 0.1$
18. $f(x) = \frac{3}{4}x^3 + \frac{9}{4}x^2 + \frac{5}{3}x + \frac{1}{3}$
19. $f(x) = 6x^4 + 2x^3 - \frac{11}{6}x^2 - \frac{1}{3}x + \frac{1}{6}$
20. $f(x) = x^4 + \frac{5}{2}x^3 + \frac{3}{2}x^2 - \frac{1}{2}x - \frac{1}{2}$

In Problems 21–30, find all real zeros of the given polynomial function f. Then factor $f(x)$ using only real numbers.

21. $f(x) = 8x^3 + 5x^2 - 11x + 3$
22. $f(x) = 6x^3 + 23x^2 + 3x - 14$
23. $f(x) = 10x^4 - 33x^3 + 66x - 40$
24. $f(x) = x^4 - 2x^3 - 23x^2 + 24x + 144$
25. $f(x) = x^5 + 4x^4 - 6x^3 - 24x^2 + 5x + 20$
26. $f(x) = 18x^5 + 75x^4 + 47x^3 - 52x^2 - 11x + 3$
27. $f(x) = 4x^5 - 8x^4 - 24x^3 + 40x^2 - 12x$
28. $f(x) = 6x^5 + 11x^4 - 3x^3 - 2x^2$
29. $f(x) = 16x^5 - 24x^4 + 25x^3 + 39x^2 - 23x + 3$
30. $f(x) = x^6 - 12x^4 + 48x^2 - 64$

In Problems 31–36, find all real solutions of the given equation.

31. $2x^3 + 3x^2 + 5x + 2 = 0$
32. $x^3 - 3x^2 = -4$
33. $2x^4 + 7x^3 - 8x^2 - 25x - 6 = 0$
34. $9x^4 + 21x^3 + 22x^2 + 2x - 4 = 0$
35. $x^5 - 2x^4 + 2x^3 - 4x^2 + 5x - 2 = 0$
36. $8x^4 - 6x^3 - 7x^2 + 6x - 1 = 0$

In Problems 37 and 38, find a polynomial function f of the indicated degree with integer coefficients that possesses the given rational zeros.

37. degree 4; $-4, \frac{1}{3}, 1, 3$
38. degree 5; $-2, -\frac{2}{3}, \frac{1}{2}, 1$ (multiplicity two)

In Problems 39 and 40, find a cubic polynomial function f that satisfies the given conditions.

39. rational zeros 1 and 2, $f(0) = 1$ and $f(-1) = 4$
40. rational zero $\frac{1}{2}$, irrational zeros $1 + \sqrt{3}$ and $1 - \sqrt{3}$, coefficient of x is 2

Miscellaneous Applications

41. **Construction of a Box** A box with no top is made from a square piece of cardboard by cutting square pieces from each corner and then folding up the sides. See FIGURE 4.4.3. The length of one side of the cardboard is 10 inches. Find the length of one side of the squares that were cut from the corners if the volume of the box is 48 in^3.

42. **Deflection of a Beam** A cantilever beam 20 ft long with a load of 600 lb at its right end is deflected by an amount $d(x) = \frac{1}{16,000}(60x^2 - x^3)$, where d is measured in inches and x in feet. See FIGURE 4.4.4. Find x when the deflection is 0.1215 in. When the deflection is 1 in.

FIGURE 4.4.3 Box in Problem 41

FIGURE 4.4.4 Cantilever beam in Problem 42

For Discussion

43. Discuss: What is the maximum number of times the graphs of the polynomial functions

$$f(x) = a_3x^3 + a_2x^2 + a_1x + a_0 \quad \text{and} \quad g(x) = b_2x^2 + b_1x + b_0$$

can intersect?

44. Consider the polynomial function $f(x) = x^n + a_{n-1}x^{n-1} + \cdots + a_1x + a_0$, where the coefficients $a_{n-1}, \ldots, a_1, a_0$ are nonzero even integers. Discuss why -1 and 1 cannot be zeros of $f(x)$.

45. A polynomial function is a continuous function; that is, the graph of a polynomial function has no breaks in it. If $f(x) = 4x^3 - 11x^2 + 14x - 6$, show that $f(0)f(1) < 0$. Explain why this shows that $f(x)$ has a zero in the interval $[0, 1]$. Find the zero.

46. If the leading coefficient of a polynomial function with integer coefficients is 1, what can be said about the possible rational zeros?

47. If k is a prime number* such that $k > 2$, what are the possible rational zeros of

$$f(x) = 6x^4 - 9x^2 + k?$$

---

*A **prime number** is a positive integer greater than 1 whose only positive integer factors are itself and the number 1. The first five prime numbers are 2, 3, 5, 7, and 11.

48. Show that $\sqrt[3]{7}$ is a zero of the polynomial function $f(x) = x^3 - 7$. Explain why this proves that $\sqrt[3]{7}$ is an irrational number.

4.5 Approximating Real Zeros

Introduction A polynomial function f is a **continuous** function. Recall from Section 3.4, this means that the graph of $y = f(x)$ has no breaks, gaps, or holes in it. The following result is a direct consequence of continuity.

THEOREM 4.5.1 Intermediate Value Theorem

Suppose $y = f(x)$ is a continuous function on the closed interval $[a, b]$. If $f(a) \neq f(b)$ for $a < b$, and if N is any number between $f(a)$ and $f(b)$, then there exists a number c in the open interval (a, b) for which $f(c) = N$.

As we see in **FIGURE 4.5.1**, the Intermediate Value Theorem simply states that a continuous function $f(x)$ takes on all values between the numbers $f(a)$ and $f(b)$. In particular, if the function values $f(a)$ and $f(b)$ have opposite signs, then by identifying $N = 0$, we can say that there is at least one number in the open interval (a, b) for which $f(c) = 0$. In other words,

If either $f(a) > 0$, $f(b) < 0$ or $f(a) < 0$, $f(b) > 0$, then $f(x)$ has at least one zero c in the interval (a, b). (1)

FIGURE 4.5.1 $f(x)$ takes on all values between $f(a)$ and $f(b)$

The plausibility of this conclusion is illustrated in **FIGURE 4.5.2**.

 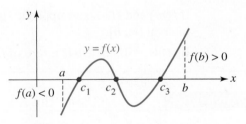

(a) One zero c in (a, b) (b) Three zeros c_1, c_2, c_3 in (a, b)

FIGURE 4.5.2 Locating zeros using the Intermediate Value Theorem

EXAMPLE 1   Using the Intermediate Value Theorem

Consider the polynomial function $f(x) = x^3 - 3x - 1$. From the data in the accompanying table we conclude from (1) that f has a real zero in each of the

x	-2	-1	0	1	2
$f(x)$	-3	1	-1	-3	1

opposite signs

FIGURE 4.5.3 Graph of function in Example 1

FIGURE 4.5.4 If $f(a)$ and $f(b)$ have opposite signs, then a zero c of f must lie either in $[a, m]$ or in $[m, b]$

intervals $[-2, -1]$, $[-1, 0]$, and $[1, 2]$. But using Theorem 4.4.2, we can verify that f has no rational zeros and so the three real zeros of f are irrational numbers. As seen in FIGURE 4.5.3 the graph of f crosses the line $y = 0$ (the x-axis) 3 times.

In the next example we will obtain an approximation to one of the irrational zeros in Example 1 using a technique called the **bisection method**.

☐ **Bisection Method** The basic idea of this method starts with the assumption that a function f is continuous and $f(a)$ and $f(b)$ have opposite signs. From this we know that there exists a number c in (a, b) for which $f(c) = 0$. Then the midpoint $m = (a + b)/2$ of the interval $[a, b]$ is an approximation to c. If $m = (a + b)/2$ is *not* a zero of f, then there is a zero c in an interval (either the open interval (a, m) or the open interval (m, b)) that is one-half the length of the original interval $[a, b]$. If, say, c lies in (m, b) as shown in FIGURE 4.5.4, we then divide this shorter interval in half: Either the new midpoint is a zero or the zero c lies in an interval that is one-fourth the length of the interval $[a, b]$. Continuing in this manner, we can locate the zero c of f in successively shorter intervals. We will then take the midpoints of these intervals as approximations to the zero c. Using this method, we see in Figure 4.5.4 that the error in an approximation to a zero in an interval is less than one-half the length of the interval.

We summarize the discussion as follows:

GUIDELINES FOR APPROXIMATING A ZERO

Let f be a polynomial function such that $f(a)$ and $f(b)$ have opposite signs.

(*i*) Divide the interval $[a, b]$ in half by finding its midpoint $m = (a + b)/2$.
(*ii*) Compute $f(m)$.
(*iii*) If $f(a)$ and $f(m)$ have opposite signs, then f has a zero in the interval $[a, m]$.

If $f(m)$ and $f(b)$ have opposite signs, then f has a zero in the interval $[m, b]$.

If $f(m) = 0$, then m is a zero of f.

EXAMPLE 2 Using the Bisection Method

Find an approximation to the zero of $f(x) = x^3 - 3x - 1$ in the interval $[1, 2]$ that is accurate to three decimal places.

Solution Recall from Example 1 that $f(1) < 0$ and $f(2) > 0$. Now to obtain the desired accuracy, we must have the error less than 0.0005.* The first approximation to the zero in $[1, 2]$ is

$$m_1 = \frac{1 + 2}{2} = 1.5 \quad \text{with error} < \tfrac{1}{2}(2 - 1) = 0.5.$$

Since $f(1.5) = -2.15 < 0$, the zero lies in $[1.5, 2]$.

---

*If we want an approximation that is accurate to two decimal places, we calculate the midpoints m_i, $i = 1, 2, \ldots, n$ until the error becomes less than 0.005.

The second approximation to the zero is

$$m_2 = \frac{1.5 + 2}{2} = 1.75 \quad \text{with error} \ < \tfrac{1}{2}(2 - 1.5) = 0.25.$$

Since $f(1.75) = -0.89065 < 0$, the zero lies in $[1.75, 2]$.

The third approximation to the zero is

$$m_3 = \frac{1.75 + 2}{2} = 1.875 \quad \text{with error} \ < \tfrac{1}{2}(2 - 1.75) = 0.125.$$

Continuing in this manner we eventually find

$$m_{11} = 1.879395 \quad \text{with error} \ < 0.0005.$$

Thus the number 1.879 is an approximation to the zero of f in $[1, 2]$ that is accurate to three decimal places.

In Example 2 we leave the approximation to the zeros of $f(x) = x^3 - 3x - 1$ in the intervals $[-2, -1]$ and $[-1, 0]$ as exercises.

If $f(a)$ and $f(b)$ have the same sign, the polynomial function f could still have one or more zeros in the interval $[a, b]$. See **FIGURE 4.5.5**.

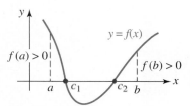

FIGURE 4.5.5 $f(a)$ and $f(b)$ are positive, yet there are two zeros in $[a, b]$

◀ Note of Caution

4.5 Exercises
Answers to selected odd-numbered problems begin on page ANS-13.

In Problems 1 and 2, find an approximation that is accurate to three decimal places to the zero of $f(x) = x^3 - 3x - 1$ in the given interval.

1. $[-2, -1]$ **2.** $[-1, 0]$

In Problems 3–6, use the bisection method to approximate to an accuracy of three decimal places the zero(s) indicated by the graph of the given function.

3. $f(x) = x^3 - x^2 + 4$ **4.** $f(x) = -x^3 - x + 11$

FIGURE 4.5.6 Graph for Problem 3

FIGURE 4.5.7 Graph for Problem 4

5. $f(x) = x^4 - 4x^3 + 10$

FIGURE 4.5.8 Graph for Problem 5

6. $f(x) = 3x^5 - 5x^3 + 1$

FIGURE 4.5.9 Graph for Problem 6

In Problems 7 and 8, use the bisection method to approximate to an accuracy of three decimal places the *x*-coordinates of the point(s) of intersection of the given graphs.

7.

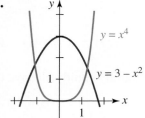

FIGURE 4.5.10 Graph for Problem 7

8.

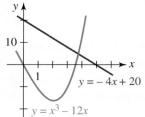

FIGURE 4.5.11 Graph for Problem 8

Miscellaneous Applications

FIGURE 4.5.12 Floating ball in Problem 9

9. Sinking Wooden Ball A spherical wooden ball of radius *r* is placed in water. To determine the depth *h* to which the ball will sink, we equate the weight of the displaced water with the weight of the ball (Archimedes' principle):

$$\frac{\pi}{3}\rho_w h^2 (3r - h) = \frac{4\pi}{3}\rho_b r^3,$$

where ρ_w and ρ_b are the densities of water and wood, respectively. See **FIGURE 4.5.12**. Suppose $\rho_b = 0.4\rho_w$ and $r = 2$ in. Use the bisection method to approximate to an accuracy of two decimal places the depth *h* to which a wooden ball will sink.

10. Sag of a Cable The length *L* of a cable between two vertical supports of a suspension bridge is given by

$$L = r + \frac{8}{3r}s^2 - \frac{32}{5r^3}s^4,$$

FIGURE 4.5.13 Suspension bridge in Problem 10

where *r* is the span of the supports and *s* is the sag of the cable between the supports. See **FIGURE 4.5.13**. If $r = 400$ ft and $L = 404$ ft, use the bisection method to approximate the sag *s* of the cable to an accuracy of two decimal places. [*Hint*: Consider the interval [20, 30].]

4.6 Rational Functions

Introduction Many functions are built up out of polynomial functions by means of arithmetic operations and function composition (see Section 3.5). In this section we construct a class of functions by forming the quotient of two polynomial functions.

As rational numbers are built out of integers, so too a **rational function** is built out of polynomial functions.

DEFINITION 4.6.1 Rational Function

A **rational function** $y = f(x)$ is a function of the form

$$f(x) = \frac{P(x)}{Q(x)}, \tag{1}$$

where P and Q are polynomial functions.

For example, the following three functions are rational functions:

$$y = \frac{x}{x^2 + 5}, \qquad y = \frac{x^3 - x + 7}{x + 3} \;\;\text{(polynomial)}, \qquad y = \frac{1}{x}.$$

The function

$$y = \frac{\sqrt{x}}{x^2 - 1} \quad \leftarrow \text{not a polynomial}$$

is not a rational function. In (1) we cannot allow the denominator to be zero. So the **domain** of a rational function $f(x) = P(x)/Q(x)$ is the set of all real numbers *except* those numbers for which the denominator $Q(x)$ is zero. For example, the domain of the rational function $f(x) = (2x^3 - 1)/(x^2 - 9)$ is $\{x \mid x \neq -3, x \neq 3\}$ or $(-\infty, -3) \cup (-3, 3) \cup (3, \infty)$. It goes without saying that we also disallow the zero polynomial $Q(x) = 0$ as a denominator.

☐ **Graphs** Graphing a rational function f is a little more complicated than graphing a polynomial function because in addition to paying attention to

- intercepts,
- symmetry, and
- shifting/reflecting/stretching of known graphs, you should also keep an eye on the domain of f, and
- the degrees of $P(x)$ and $Q(x)$.

The latter two topics are important in determining whether a graph of a rational function possesses *asymptotes*.

The y-intercept is the point $(0, f(0))$, provided the number 0 is in the domain of f. For example, the graph of the rational function $f(x) = (1 - x)/x$ does not cross the y-axis since $f(0)$ is not defined. If the polynomials $P(x)$ and $Q(x)$ have no common factors, then the x-intercepts of the graph of the rational function $f(x) = P(x)/Q(x)$ are the points whose x-coordinates are the real zeros of the numerator $P(x)$. In other words, the only way we can have $f(x) = P(x)/Q(x) = 0$ is to have $P(x) = 0$. The graph of a rational function f is symmetric with respect to the y-axis if $f(-x) = f(x)$, and symmetric with respect to the origin if $f(-x) = -f(x)$. Since it is easy to spot an even or an odd polynomial function (see page 230), here is an easy way to determine symmetry of the graph of a rational function. We again assume that $P(x)$ and $Q(x)$ have no common factors.

- The quotient of two even functions is even. (2)
- The quotient of two odd functions is even. (3)
- The quotient of an even and an odd function is odd. (4)

See Problem 48 in Exercises 4.6.

We have already seen the graphs of two simple rational functions, $y = 1/x$ and $y = 1/x^2$ in Figures 3.2.1(e) and 3.2.1(f). You are encouraged to review those graphs at this time. Note that $P(x) = 1$ is an even function and $Q(x) = x$ is an odd function, so $y = 1/x$ is an odd function by (4). On the other hand, $P(x) = 1$ is an even function and $Q(x) = x^2$ is an even function, so $y = 1/x^2$ is an even function by (2).

EXAMPLE 1 Shifted Reciprocal Function

Graph the function $f(x) = \dfrac{2}{x-1}$.

Solution The graph possesses no symmetry since $Q(x) = x - 1$ is neither even nor odd. Since $f(0) = -2$, the y-intercept is $(0, -2)$. Because $P(x) = 2$ is never 0, there are no x-intercepts. You might also recognize that the graph of this rational function is the graph of the reciprocal function $y = 1/x$ stretched vertically by a factor of 2 and shifted 1 unit to the right. The point $(1, 1)$ is on the graph of $y = 1/x$; in **FIGURE 4.6.1**, after the vertical stretch and horizontal shift, the corresponding point on the graph of $y = 2/(x - 1)$ is $(2, 2)$.

The vertical line $x = 1$ and the horizontal line $y = 0$ (the equation of the x-axis) are of special importance for this graph.

The vertical dashed line $x = 1$ in Figure 4.6.1 is the y-axis in Figure 3.2.1(e) shifted 1 unit to the right. Although the number 1 is not in the domain of the given function, we can evaluate f at values of x that are *near* 1. For example, you should verify that

x	0.999	1.001
$f(x)$	-2000	2000

(5)

FIGURE 4.6.1 Graph of function in Example 1

The table in (5) shows that for values of x close to 1 the corresponding function values $f(x)$ are large in absolute value. On the other hand, for values of x for which $|x|$ is large, the corresponding function values $f(x)$ are near 0. For example, you should verify that

x	-999	1001
$f(x)$	-0.002	0.002

(6)

Geometrically, as x approaches 1, the graph of the function approaches the vertical line $x = 1$, and as $|x|$ increases without bound the graph of the function approaches the horizontal line $y = 0$.

□ **Notation** To indicate that x is approaching a number a, we use the notation
- $x \to a^-$ to mean that x is approaching a from the *left*, that is, through numbers that are less than a,
- $x \to a^+$ to mean that x is approaching a from the *right*, that is, through numbers that are greater than a, and
- $x \to a$ to mean that x is approaching a from both the *left* and the *right*.

We also use the infinity symbols and the notation
- $x \to -\infty$ to mean that x becomes *unbounded in the negative direction*, and
- $x \to \infty$ to mean that x becomes *unbounded in the positive direction*.

Similar interpretations are given to the symbols $f(x) \to -\infty$ and $f(x) \to \infty$. These notational devices are a convenient way of describing the behavior of a function either near a number $x = a$ or as x increases to the right or decreases to the left. Thus, in Example 1 it is apparent from (5) and Figure 4.6.1 that

◀ This notation was used in the discussion of *end behavior* in Section 4.1.

$$f(x) \to -\infty \text{ as } x \to 1^- \quad \text{and} \quad f(x) \to \infty \text{ as } x \to 1^+.$$

In words, the notation in the preceding line signifies that the function values are decreasing without bound as x approaches 1 from the left, and the function values are increasing without bound as x approaches 1 from the right. From (6) and Figure 4.6.1 it should also be apparent that

$$f(x) \to 0 \quad \text{as} \quad x \to -\infty \quad \text{and} \quad f(x) \to 0 \quad \text{as} \quad x \to \infty.$$

□ **Asymptotes** In Figure 4.6.1, the vertical line whose equation is $x = 1$ is called a **vertical asymptote** for the graph of f, and the horizontal line whose equation is $y = 0$ is called a **horizontal asymptote** for the graph of f.

In this section we will examine three types of asymptotes that correspond to the three types of lines studied in Section 2.3: *vertical lines*, *horizontal lines*, and *slant* (or oblique) *lines*. The characteristic of any asymptote is that the graph of a function f must get close to, or approach, the line.

DEFINITION 4.6.2 Vertical Asymptote

A line $x = a$ is said to be a **vertical asymptote** for the graph of a function f if at least one of the following six statements is true:

$$\begin{aligned} f(x) \to -\infty &\text{ as } x \to a^-, & f(x) \to \infty &\text{ as } x \to a^-, \\ f(x) \to -\infty &\text{ as } x \to a^+, & f(x) \to \infty &\text{ as } x \to a^+, \\ f(x) \to -\infty &\text{ as } x \to a, & f(x) \to \infty &\text{ as } x \to a. \end{aligned} \quad (7)$$

FIGURE 4.6.2 illustrates four of the six possibilities listed in (7) for the unbounded behavior of a function f near a vertical asymptote $x = a$. If the function exhibits the *same kind of unbounded behavior from both sides of* $x = a$, then we write

$$f(x) \to \infty \quad \text{as} \quad x \to a, \qquad (8)$$

or

$$f(x) \to -\infty \quad \text{as} \quad x \to a. \qquad (9)$$

In Figure 4.6.2(d) we see that $f(x) \to \infty$ as $x \to a^-$ and $f(x) \to \infty$ as $x \to a^+$, and so we write $f(x) \to \infty$ as $x \to a$.

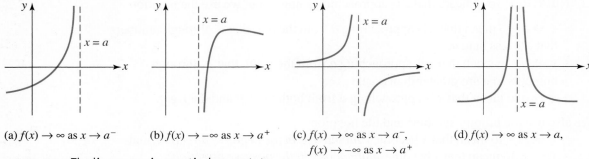

(a) $f(x) \to \infty$ as $x \to a^-$ (b) $f(x) \to -\infty$ as $x \to a^+$ (c) $f(x) \to \infty$ as $x \to a^-$, $f(x) \to -\infty$ as $x \to a^+$ (d) $f(x) \to \infty$ as $x \to a$,

FIGURE 4.6.2 The line $x = a$ is a vertical asymptote

If $x = a$ is a vertical asymptote for the graph of a *rational function* $f(x) = P(x)/Q(x)$, then the function values $f(x)$ become unbounded as x approaches a from *both sides*, that is, from the right $(x \to a^+)$ *and* from the left $(x \to a^-)$. The graphs in Figures 4.6.2(c) and 4.6.2(d) (or the reflection of these graphs in the x-axis) are typical graphs of a rational function with a single vertical asymptote. As can be seen from these figures, a rational function with a vertical asymptote is a **discontinuous function**. There is an infinite break in each graph at $x = a$. As seen in Figures 4.6.2(c) and 4.6.2(d) a single vertical asymptote divides the xy-plane into two regions; within each region there is a single piece or **branch** of the graph of the rational function f.

DEFINITION 4.6.3 Horizontal Asymptote

A line $y = c$ is said to be a **horizontal asymptote** for the graph of a function f if

$$f(x) \to c \text{ as } x \to -\infty \quad \text{or} \quad f(x) \to c \text{ as } x \to \infty. \tag{10}$$

Remember this. ▶

In **FIGURE 4.6.3** we have illustrated some typical horizontal asymptotes. We note, in conjunction with Figure 4.6.3(d) that, in general, the graph of a function can have at most *two* horizontal asymptotes but the graph of a *rational function* $f(x) = P(x)/Q(x)$ can have at most *one*. If the graph of a rational function f possesses a horizontal asymptote $y = c$, then as shown in Figure 4.6.3(c),

$$f(x) \to c \text{ as } x \to -\infty \quad \text{and} \quad f(x) \to c \text{ as } x \to \infty.$$

The last line is a mathematical description of the **end behavior** of the graph of a rational function with a horizontal asymptote. Also, the graph of a function can *never* cross a vertical asymptote but, as suggested in Figure 4.6.3(a), a graph can cross a horizontal asymptote.

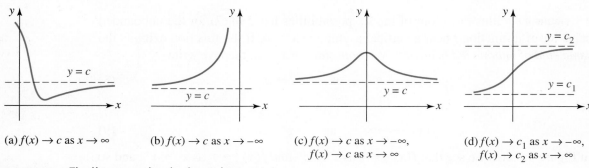

(a) $f(x) \to c$ as $x \to \infty$ (b) $f(x) \to c$ as $x \to -\infty$ (c) $f(x) \to c$ as $x \to -\infty$, $f(x) \to c$ as $x \to \infty$ (d) $f(x) \to c_1$ as $x \to -\infty$, $f(x) \to c_2$ as $x \to \infty$

FIGURE 4.6.3 The line $y = c$ is a horizontal asymptote

DEFINITION 4.6.4 Slant Asymptote

A line $y = mx + b$, $m \neq 0$, is said to be a **slant asymptote** for the graph of a function f if

$$f(x) \to mx + b \text{ as } x \to -\infty \quad \text{or} \quad f(x) \to mx + b \text{ as } x \to \infty. \quad (11)$$

◀ A slant asymptote is also called an oblique asymptote.

The notation in (11) means that the graph of f possesses a slant asymptote whenever the function values $f(x)$ become closer and closer to the values of y on the line $y = mx + b$ as x becomes large in absolute value. Another way of stating (11) is: A line $y = mx + b$ is a slant asymptote for the graph of f if the vertical distance $d(x)$ between points with the same x-coordinate on the two graphs satisfies

$$d(x) = f(x) - (mx + b) \to 0 \text{ as } x \to -\infty \text{ or as } x \to \infty.$$

See **FIGURE 4.6.4**. We note that if a graph of a rational function $f(x) = P(x)/Q(x)$ possesses a slant asymptote it can have vertical asymptotes, but the graph *cannot* have a horizontal asymptote.

On a practical level, vertical and horizontal asymptotes of the graph of a rational function f can be determined by inspection. So for the sake of discussion let us suppose that

$$f(x) = \frac{P(x)}{Q(x)} = \frac{a_n x^n + a_{n-1} x^{n-1} + \cdots + a_1 x + a_0}{b_m x^m + b_{m-1} x^{m-1} + \cdots + b_1 x + b_0}, \quad a_n \neq 0, \quad b_m \neq 0, \quad (12)$$

represents a general rational function.

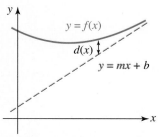

FIGURE 4.6.4 Slant asymptote is $y = mx + b$

☐ **Finding Vertical Asymptotes** Let us assume that the polynomial functions $P(x)$ and $Q(x)$ in (12) have no common factors. In that case:

- If a is a real number such that $Q(a) = 0$, then the line $x = a$ is a vertical asymptote for the graph of f.

Since $Q(x)$ is a polynomial function of degree m, it can have up to m real zeros, and so the graph of a rational function f can have up to m vertical asymptotes. If the graph of a rational function f has, say, k ($k \leq m$) vertical asymptotes, then the k vertical lines divide the xy-plane into $k + 1$ regions. Thus the graph of this rational function would have $k + 1$ branches.

EXAMPLE 2   Vertical Asymptotes

(a) Inspection of the rational function $f(x) = \dfrac{2x + 1}{x^2 - 4}$ shows that the denominator $Q(x) = x^2 - 4 = (x + 2)(x - 2) = 0$ at $x = -2$ and $x = 2$. These are equations of vertical asymptotes for the graph of f. The graph of f has three branches: one to the left of the line $x = -2$, one between the lines $x = -2$ and $x = 2$, and one to the right of the line $x = 2$.

(b) The graph of the rational function $f(x) = \dfrac{1}{x^2 + x + 4}$ has no vertical asymptotes since $Q(x) = x^2 + x + 4 \neq 0$ for all real numbers. ≡

☐ **Finding Horizontal Asymptotes** When we discussed end behavior of a polynomial function $P(x)$ of degree n, we pointed out that $P(x)$ behaves like $y = a_n x^n$, that is, $P(x) \approx a_n x^n$, for values of x large in absolute value. As a consequence, we see from

$$f(x) = \frac{a_n x^n \boxed{+ a_{n-1}x^{n-1} + \cdots + a_1 x + a_0}}{b_m x^m \boxed{+ b_{m-1}x^{m-1} + \cdots + b_1 x + b_0}}$$

lower powers of x are irrelevant as $x \to \pm\infty$

that $f(x)$ behaves like $y = \dfrac{a_n}{b_m} x^{n-m}$ because $f(x) \approx \dfrac{a_n x^n}{b_m x^m} = \dfrac{a_n}{b_m} x^{n-m}$ for $x \to \pm\infty$. Therefore:

If $n = m$, $\quad f(x) \approx \dfrac{a_n}{b_m} x^{\overset{0}{\overset{\downarrow}{n-n}}} \to \dfrac{a_n}{b_m} \quad$ as $x \to \pm\infty$. $\hfill (13)$

If $n < m$, $\quad f(x) \approx \dfrac{a_n}{b_m} x^{\overset{\text{negative}}{\overset{\downarrow}{n-m}}} = \dfrac{a_n}{b_m} \dfrac{1}{x^{m-n}} \to 0 \quad$ as $x \to \pm\infty$. $\hfill (14)$

If $n > m$, $\quad f(x) \approx \dfrac{a_n}{b_m} x^{\overset{\text{positive}}{\overset{\downarrow}{n-m}}} \to \infty \quad$ as $x \to \pm\infty$. $\hfill (15)$

From (13), (14), and (15) we glean the following three facts about horizontal asymptotes for the graph of $f(x) = P(x)/Q(x)$.

- If degree of $P(x) = $ degree of $Q(x)$, then $y = a_n/b_m$ (the quotient of the leading coefficients) is a horizontal asymptote. $\hfill (16)$
- If degree of $P(x) < $ degree of $Q(x)$, then $y = 0$ is a horizontal asymptote. $\hfill (17)$
- If degree of $P(x) > $ degree of $Q(x)$, then the graph of f has *no* horizontal asymptote. $\hfill (18)$

EXAMPLE 3   Horizontal Asymptotes

Determine whether the graph of each of the following rational functions possesses a horizontal asymptote.

(a) $f(x) = \dfrac{3x^2 + 4x - 1}{8x^2 + x}$ **(b)** $f(x) = \dfrac{4x^3 + 7x + 8}{2x^4 + 3x^2 - x + 6}$ **(c)** $f(x) = \dfrac{5x^3 + x^2 + 1}{2x + 3}$

Solution (a) Since the degree of the numerator $3x^2 + 4x - 1$ is the same as the degree of the denominator $8x^2 + x$ (both degrees are 2) we see from (13) that

$$f(x) \approx \frac{3}{8} x^{2-2} = \frac{3}{8} \quad \text{as } x \to \pm\infty.$$

As summarized in (16), $y = \frac{3}{8}$ is a horizontal asymptote for the graph of f.

(b) Since the degree of the numerator $4x^3 + 7x + 8$ is 3 and the degree of the denominator $2x^4 + 3x^2 - x + 6$ is 4 (and $3 < 4$) we see from (14) that

$$f(x) \approx \frac{4}{2} x^{3-4} = \frac{2}{x} \to 0 \quad \text{as } x \to \pm\infty.$$

As summarized in (17), $y = 0$ (the x-axis) is a horizontal asymptote for the graph of f.

(c) Since the degree of the numerator $5x^3 + x^2 - 1$ is 3 and the degree of the denominator $2x + 3$ is 1 (and $3 > 1$) we see from (15) that

$$f(x) \approx \frac{5}{2}x^{3-1} = \frac{5}{2}x^2 \to \infty \quad \text{as } x \to \pm\infty.$$

As summarized in (18), the graph of f has *no* horizontal asymptote. ≡

In the graphing examples that follow we will assume that $P(x)$ and $Q(x)$ in (12) have no common factors.

EXAMPLE 4   Graph of a Rational Function

Graph the function $f(x) = \dfrac{3-x}{x+2}$.

Solution Here are some things we look at to sketch the graph of f.

Symmetry: No symmetry. $P(x) = 3 - x$ and $Q(x) = x + 2$ are neither even nor odd.

Intercepts: $f(0) = \frac{3}{2}$ and so the y-intercept is $\left(0, \frac{3}{2}\right)$. Setting $P(x) = 0$ or $3 - x = 0$ implies 3 is a zero of P. The single x-intercept is $(3, 0)$.

Vertical Asymptotes: Setting $Q(x) = 0$ or $x + 2 = 0$ gives $x = -2$. The line $x = -2$ is a vertical asymptote.

Branches: Because there is only a single vertical asymptote, the graph of f consists of two distinct branches, one to the left of $x = -2$ and one to the right of $x = -2$.

Horizontal Asymptote: The degree of P and the degree of Q are the same (namely, 1) and so the graph of f has a horizontal asymptote. By rewriting f as

$$f(x) = \frac{-x+3}{x+2}$$

we see that the ratio of leading coefficients is $-1/1 = -1$. From (16) we see that the line $y = -1$ is a horizontal asymptote.

The Graph: We draw the vertical and horizontal asymptotes using red dashed lines. The right branch of the graph of f is drawn through the intercepts $\left(0, \frac{3}{2}\right)$ and $(3, 0)$ in such a manner that it approaches both asymptotes. The left branch is drawn *below* the horizontal asymptote $y = -1$. Were we to draw this branch above the horizontal asymptote it would have to be near the horizontal asymptote from above and near the vertical asymptote from the left. In order to do this the branch of the graph would have to cross the x-axis, but since there are no more x-intercepts this is impossible. See **FIGURE 4.6.5**. ≡

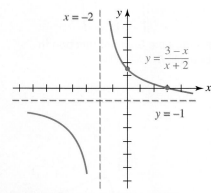

FIGURE 4.6.5 Graph of function in Example 4

EXAMPLE 5   Using Transformations in Example 4

Long division and rigid transformations can sometimes be an aid in graphing rational functions. Note that if we carry out the long division for the function f in Example 4, then we see that

$$f(x) = \frac{3-x}{x+2} \quad \text{is the same as} \quad f(x) = -1 + \frac{5}{x+2}.$$

Thus, starting with the graph of $y = 1/x$, we stretch it vertically by a factor of 5. Then shift the graph of $y = 5/x$ two units to the left. Finally, shift $y = 5/(x+2)$ one unit vertically downward. You should verify that the net result is the graph in Figure 4.6.5. ≡

EXAMPLE 6 Graph of a Rational Function

Graph the function $f(x) = \dfrac{x}{1-x^2}$.

Solution

Symmetry: Since $P(x) = x$ is odd and $Q(x) = 1 - x^2$ is even, the quotient $P(x)/Q(x)$ is odd. The graph of f is symmetric with respect to the origin.

Intercepts: $f(0) = 0$ and so the y-intercept is $(0, 0)$. Setting $P(x) = x = 0$ gives $x = 0$. Thus the only intercept is $(0, 0)$.

Vertical Asymptotes: Setting $Q(x) = 0$ or $1 - x^2 = 0 = 0$ gives $x = -1$ and $x = 1$. The lines $x = -1$ and $x = 1$ are vertical asymptotes.

Branches: Because there are two vertical asymptotes, the graph of f consists of three distinct branches, one to the left of the line $x = -1$, one between the lines $x = -1$ and $x = 1$, and one to the right of the line $x = 1$.

Horizontal Asymptote: Since the degree of the numerator x is 1 and the degree of the denominator $1 - x^2$ is 2 (and $1 < 2$) it follows from (14) and (17) that $y = 0$ is a horizontal asymptote for the graph of f.

The Graph: We can plot the graph for $x \geq 0$ and then use symmetry to obtain the remaining part of the graph of $x < 0$. We begin by drawing the vertical asymptotes using dashed lines. The half-branch of the graph of f on the interval $[0, 1)$ is drawn starting at $(0, 0)$. The function f must then increase because $P(x) = x > 0$ and $Q(x) = 1 - x^2 > 0$ indicates that $f(x) > 0$ for $0 < x < 1$. This implies that near the vertical asymptote $x = 1$, $f(x) \to \infty$ as $x \to 1^-$. The branch of the graph for $x > 1$ is drawn below the horizontal asymptote $y = 0$ since $P(x) = x > 0$ and $Q(x) = 1 - x^2 < 0$ imply $f(x) < 0$. Thus $f(x) \to -\infty$ as $x \to 1^+$ and $f(x) \to 0$ as $x \to \infty$. The remainder of the graph for $x < 0$ is obtained by reflecting the graph for $x < 0$ through the origin. See **FIGURE 4.6.6**.

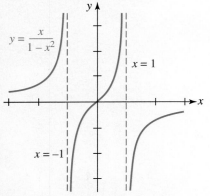

FIGURE 4.6.6 Graph of function in Example 6

EXAMPLE 7 Graph of a Rational Function

Graph the function $f(x) = \dfrac{x}{1+x^2}$.

Solution The given function f is similar to the function in Example 6 in that f is an odd function, $(0, 0)$ is the only intercept of its graph, and its graph has the horizontal asymptote. However, note that since $1 + x^2 > 0$ for all real numbers, there are no vertical asymptotes. Thus there are no branches; the graph is one continuous curve. For $x \geq 0$, the graph passes through $(0, 0)$ and then must increase since $f(x) > 0$ for $x > 0$. Also, f must attain a relative maximum and then decrease in order to satisfy the condition $f(x) \to 0$ as $x \to \infty$. As mentioned in Section 4.1, the exact location of this relative maximum can be obtained through calculus techniques. Finally, we reflect the portion of the graph for $x > 0$ through the origin. The graph must look something like that shown in **FIGURE 4.6.7**.

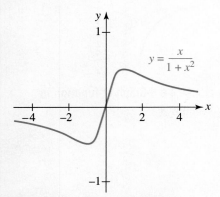

FIGURE 4.6.7 Graph of function in Example 7

□ **Finding Slant Asymptotes** Let us again assume that the polynomials $P(x)$ and $Q(x)$ in (11) have no common factors. In that case we can recognize the existence of a slant asymptote in the following manner:

- If the degree of $P(x)$ is precisely one greater than the degree of $Q(x)$, that is, if the degree of $Q(x)$ is m and the degree of $P(x)$ is $n = m + 1$, then the graph of f possesses a slant asymptote.

We find the slant asymptote by division. Using long division to divide $P(x)$ by $Q(x)$ yields a quotient that is a linear polynomial $mx + b$ and a polynomial remainder $R(x)$:

$$f(x) = \frac{P(x)}{Q(x)} = \overset{\text{quotient}}{mx+b} + \overset{\text{remainder}}{\frac{R(x)}{Q(x)}}. \qquad (19)$$

Because the degree of $R(x)$ must be less than the degree of the divisor $Q(x)$, we have $R(x)/Q(x) \to 0$ as $x \to -\infty$ and as $x \to \infty$, and consequently

$$f(x) \to mx + b \quad \text{as} \quad x \to -\infty \quad \text{and} \quad f(x) \to mx + b \quad \text{as} \quad x \to \infty.$$

In other words, an equation of the slant asymptote is $y = mx + b$, where $mx + b$ is the quotient in (19).

When the denominator $Q(x)$ is a *linear* polynomial, we can use synthetic division to carry out the long division.

EXAMPLE 8 Graph with a Slant Asymptote

Graph the function $f(x) = \dfrac{x^2 - x - 6}{x - 5}$.

Solution

Symmetry: No symmetry. $P(x) = x^2 - x - 6$ and $Q(x) = x - 5$ are neither even nor odd.

Intercepts: $f(0) = \frac{6}{5}$ and so the y-intercept is $(0, \frac{6}{5})$. Setting $P(x) = 0$ or $x^2 - x - 6 = 0$ or $(x + 2)(x - 3) = 0$ shows that -2 and 3 are zeros of $P(x)$. The x-intercepts are $(-2, 0)$ and $(3, 0)$.

Vertical Asymptotes: Setting $Q(x) = 0$ or $x - 5 = 0$ gives $x = 5$. The line $x = 5$ is a vertical asymptote.

Branches: The graph of f consists of two branches, one to the left of $x = 5$ and one to the right of $x = 5$.

Horizontal Asymptote: None.

Slant Asymptote: Since the degree of $P(x) = x^2 - x - 6$ (which is 2) is exactly one greater than the degree of $Q(x) = x - 5$ (which is 1), the graph of $f(x)$ has a slant asymptote. To find it, we divide $P(x)$ by $Q(x)$. Because $Q(x)$ is a linear polynomial we can use synthetic division:

$$\begin{array}{r|rrr} 5 & 1 & -1 & -6 \\ & & 5 & 20 \\ \hline & 1 & 4 & \underline{|14} \end{array}$$

Recall, the latter notation means that

$$\underset{\downarrow}{y = x + 4 \text{ is the slant asymptote}}$$

$$\frac{x^2 - x - 6}{x - 5} = x + 4 + \frac{14}{x - 5}.$$

Note again that $14/(x - 5) \to 0$ as $x \to \pm\infty$. Hence the line $y = x + 4$ is a slant asymptote.

The Graph: Using the foregoing information we obtain the graph in **FIGURE 4.6.8**. The asymptotes are the dashed red lines in the figure. ≡

FIGURE 4.6.8 Graph of function in Example 8

EXAMPLE 9 Graph with a Slant Asymptote

By inspection it should be apparent that the graph of the rational function $f(x) = \dfrac{x^3 - 8x + 12}{x^2 + 1}$ possesses a slant asymptote but no vertical asymptotes. Since the denominator is a quadratic polynomial we resort to long division to obtain

$$\frac{x^3 - 8x + 12}{x^2 + 1} = x + \frac{-9x + 12}{x^2 + 1}.$$

The slant asymptote is the line $y = x$. The graph has no symmetry. The y-intercept is $(0, 12)$. The lack of vertical asymptotes indicates that the function f is continuous; its graph consists of an unbroken curve. Because the numerator is a polynomial of odd degree we know that it has at least one real zero. Since $x^3 - 8x + 12 = 0$ has no rational roots, we can use the approximation method of Section 4.5 or graphical techniques to show that the equation possesses only one real irrational root. Thus the x-intercept is approximately $(-3.4, 0)$. The graph of f is given in FIGURE 4.6.9. Notice in that figure, the graph of f crosses the slant asymptote shown as a red dashed line.

FIGURE 4.6.9 Graph of function in Example 9

☐ **Hole in a Graph** We assumed throughout the foregoing discussion of asymptotes of rational functions that the polynomial functions $P(x)$ and $Q(x)$ in (1) have no common factors. We now know that if a is a real number such that $Q(a) = 0$, and $P(x)$ and $Q(x)$ have no common factors, then the line $x = a$ is a vertical asymptote for the graph of f. Because Q is a polynomial function it follows from the Factor Theorem that $Q(x) = (x - a)q(x)$. The assumption that the numerator P and denominator Q have no common factors tells us that $x - a$ is not a factor of P and so $P(a) \neq 0$. When $P(a) = 0$ *and* $Q(a) = 0$, then $x = a$ *may not* be a vertical asymptote. For example, when a is a *simple zero* of both P and Q, then $x = a$ is *not* a vertical asymptote for the graph of $f(x) = P(x)/Q(x)$. To see this, we know from the Factor Theorem that if $P(a) = 0$ and $Q(a) = 0$ then $x - a$ is a common factor of P and Q and so $P(x) = (x - a)p(x)$ and $Q(x) = (x - a)q(x)$, where p and q are polynomials such that $p(a) \neq 0$ and $q(a) \neq 0$. After cancelling

$$f(x) = \frac{P(x)}{Q(x)} = \frac{(x - a)p(x)}{(x - a)q(x)} = \frac{p(x)}{q(x)}, \quad x \neq a,$$

we see that $f(x)$ is undefined at a, but the function values $f(x)$ do not become unbounded as $x \to a^-$ or as $x \to a^+$ because $q(x)$ is not approaching 0. As an example, we saw in Section 3.4 that the graph of the rational function

$$f(x) = \frac{x^2 - 4}{x - 2} = \frac{(x - 2)(x + 2)}{x - 2} = x + 2, \quad x \neq 2,$$

is basically a straight line. But since $f(2)$ is undefined there is no point $(2, f(2))$ on the line. Instead there is a **hole** in the graph at the point $(2, 4)$. See Figure 3.4.5(a).

EXAMPLE 10 Graph with a Hole

Graph the function $f(x) = \dfrac{x^2 - 2x - 3}{x^2 - 1}$.

Solution Although $x^2 - 1 = 0$ for $x = -1$ and $x = 1$, only $x = 1$ is a vertical asymptote. Note that the numerator $P(x)$ and denominator $Q(x)$ have the common factor $x + 1$ that we cancel provided $x \neq -1$:

equality is true for $x \neq -1$
$$f(x) = \frac{(x+1)(x-3)}{(x+1)(x-1)} = \frac{x-3}{x-1}. \tag{20}$$

Thus we see from (20) that there is no infinite break in the graph at $x = -1$. We graph $y = \frac{x-3}{x-1}$, $x \neq -1$, by observing that the y-intercept is $(0, 3)$, an x-intercept is $(3, 0)$, a vertical asymptote is $x = 1$, and a horizontal asymptote is $y = 1$. The graph of this function has two branches, but the branch to the left of the vertical asymptote $x = 1$ has a hole in it corresponding to the point $(-1, 2)$. See FIGURE 4.6.10.

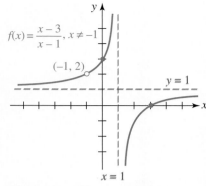

FIGURE 4.6.10 Graph of function in Example 10

NOTES FROM THE CLASSROOM

When asked whether they have ever heard the statement

An asymptote is a line that the graph approaches but does not cross,

a surprising number of students will raise their hands. First, let's make it clear that the statement is false; a graph *can* cross a horizontal asymptote and *can* cross a slant asymptote. A graph can never cross a vertical asymptote $x = a$ since the function is inherently undefined at $x = a$. We can even find the points where a graph crosses a horizontal or slant asymptote. For example, the rational function $f(x) = \frac{x^2 + 2x}{x^2 - 1}$ has the horizontal asymptote $y = 1$. Determining whether the graph of f crosses the horizontal line $y = 1$ is equivalent to asking whether $y = 1$ is in the range of the function f. Setting $f(x)$ equal to 1, that is,

$$\frac{x^2 + 2x}{x^2 - 1} = 1 \quad \text{implies} \quad x^2 + 2x = x^2 - 1 \quad \text{and} \quad x = -\tfrac{1}{2}.$$

Since $x = -\tfrac{1}{2}$ is in the domain of f, the graph of f crosses the horizontal asymptote at $\left(-\tfrac{1}{2}, f\left(-\tfrac{1}{2}\right)\right) = \left(-\tfrac{1}{2}, 1\right)$. Observe in Example 9 that we can find the point where the slant asymptote crosses the graph of $y = x$ by solving $f(x) = x$. You should verify that the point of intersection is $\left(\tfrac{4}{3}, \tfrac{4}{3}\right)$. See Problems 31–36 in Exercises 4.6.

4.6 Exercises
Answers to selected odd-numbered problems begin on page ANS-13.

In Problems 1 and 2, use a calculator to fill in the given table for the rational function $f(x) = \frac{2x}{x-3}$.

1. $x = 3$ is a vertical asymptote for the graph of f

x	3.1	3.01	3.001	3.0001	3.00001
$f(x)$					
x	2.9	2.99	2.999	2.9999	2.99999
$f(x)$					

2. $y = 2$ is a horizontal asymptote for the graph of f

x	10	100	1000	10,000	100,000
$f(x)$					
x	-10	-100	-1000	$-10,000$	$-100,000$
$f(x)$					

In Problems 3–22, find the vertical and horizontal asymptotes for the graph of the given rational function. Find the x- and y-intercepts of the graph. Sketch the graph of f.

3. $f(x) = \dfrac{1}{x-2}$

4. $f(x) = \dfrac{4}{x+3}$

5. $f(x) = \dfrac{x}{x+1}$

6. $f(x) = \dfrac{x}{2x-5}$

7. $f(x) = \dfrac{4x-9}{2x+3}$

8. $f(x) = \dfrac{2x+4}{x-2}$

9. $f(x) = \dfrac{1-x}{x+1}$

10. $f(x) = \dfrac{2x-3}{x}$

11. $f(x) = \dfrac{1}{(x-1)^2}$

12. $f(x) = \dfrac{4}{(x+2)^3}$

13. $f(x) = \dfrac{1}{x^3}$

14. $f(x) = \dfrac{8}{x^4}$

15. $f(x) = \dfrac{x}{x^2-1}$

16. $f(x) = \dfrac{x^2}{x^2-4}$

17. $f(x) = \dfrac{1}{x(x-2)}$

18. $f(x) = \dfrac{1}{x^2-2x-8}$

19. $f(x) = \dfrac{1-x^2}{x^2}$

20. $f(x) = \dfrac{16}{x^2+4}$

21. $f(x) = \dfrac{-2x^2+8}{(x-1)^2}$

22. $f(x) = \dfrac{x(x-5)}{x^2-9}$

In Problems 23–30, find the vertical and slant asymptotes for the graph of the given rational function. Find the x- and y-intercepts of the graph. Sketch the graph of f.

23. $f(x) = \dfrac{x^2-9}{x}$

24. $f(x) = \dfrac{x^2-3x-10}{x}$

25. $f(x) = \dfrac{x^2}{x+2}$

26. $f(x) = \dfrac{x^2-2x}{x+2}$

27. $f(x) = \dfrac{x^2-2x-3}{x-1}$

28. $f(x) = \dfrac{-(x-1)^2}{x+2}$

29. $f(x) = \dfrac{x^3-8}{x^2-x}$

30. $f(x) = \dfrac{5x(x+1)(x-4)}{x^2+1}$

In Problems 31–34, find the point where the graph of f crosses its horizontal asymptote. Sketch the graph of f.

31. $f(x) = \dfrac{x-3}{x^2+3}$

32. $f(x) = \dfrac{(x-3)^2}{x^2-5x}$

33. $f(x) = \dfrac{4x(x-2)}{(x-3)(x+4)}$

34. $f(x) = \dfrac{2x^2}{x^2+x+1}$

In Problems 35 and 36, find the point where the graph of f crosses its slant asymptote. Use a graphing utility to obtain the graph of f and the slant asymptote in the same coordinate plane.

35. $f(x) = \dfrac{x^3 - 3x^2 + 2x}{x^2 + 1}$

36. $f(x) = \dfrac{x^3 + 2x - 4}{x^2}$

In Problems 37–40, find a rational function that satisfies the given conditions. There is no unique answer.

37. vertical asymptote: $x = 2$
horizontal asymptote: $y = 1$
x-intercept: $(5, 0)$

38. vertical asymptote: $x = 1$
horizontal asymptote: $y = -2$
y-intercept: $(0, -1)$

39. vertical asymptotes: $x = -1$, $x = 2$
horizontal asymptote: $y = 3$
x-intercept: $(3, 0)$

40. vertical asymptote: $x = 4$
slant asymptote: $y = x + 2$

In Problems 41–44, find the asymptotes and any holes in the graph of the given rational function. Find the x- and y-intercepts of the graph. Sketch the graph of f.

41. $f(x) = \dfrac{x^2 - 1}{x - 1}$

42. $f(x) = \dfrac{x - 1}{x^2 - 1}$

43. $f(x) = \dfrac{x + 1}{x(x^2 + 4x + 3)}$

44. $f(x) = \dfrac{x^3 + 8}{x + 2}$

Miscellaneous Applications

45. Parallel Resistors A 5-ohm resistor and a variable resistor are placed in parallel as shown in FIGURE 4.6.11. The resulting resistance R (in ohms) is related to the resistance r (in ohms) of the variable resistor by the equation

$$R = \dfrac{5r}{5 + r}.$$

Sketch the graph of R as a function of r for $r > 0$. What is the resulting resistance R as r becomes very large?

FIGURE 4.6.11 Parallel resistors in Problem 45

46. Power The electrical power P produced by a certain source is given by

$$P = \dfrac{E^2 r}{R^2 + 2Rr + r^2},$$

where E is the voltage of the source, R is the resistance of the source, and r is the resistance in the circuit. Sketch the graph of P as a function of r using the values $E = 5$ volts and $R = 1$ ohm.

47. Illumination Intensity The intensity of illumination from a light source at any point is directly proportional to the strength of the source and inversely proportional to the square of the distance from the source. Given two sources of strengths 16 units and 2 units that are 100 cm apart, as shown in FIGURE 4.6.12, the intensity I at any point P between them is given by

$$I(x) = \dfrac{16}{x^2} + \dfrac{2}{(100 - x)^2},$$

where x is the distance from the 16-unit source. Sketch the graph of $I(x)$ for $0 < x < 100$. Describe the behavior of $I(x)$ as $x \to 0^+$. As $x \to 100^-$.

FIGURE 4.6.12 Two light sources in Problem 47

For Discussion

48. Suppose $f(x) = P(x)/Q(x)$. Prove the symmetry rules (2), (3), and (4), for rational functions.

49. Construct a rational function $f(x) = P(x)/Q(x)$ whose graph crosses its slant asymptote twice. Assume $P(x)$ and $Q(x)$ have no common factors.

50. Construct a rational function $f(x) = P(x)/Q(x)$ whose graph has two holes in it.

CONCEPTS REVIEW

You should be able to give the meaning of each of the following concepts.

Polynomial function:
   degree
   leading term
   constant term
Graphs:
   symmetry
   intercepts
   turning point
   end behavior
Zero of a polynomial function:
   simple
   of multiplicity m

Proper fraction
Improper fraction
Division Algorithm
Remainder Theorem
Synthetic division
Factor Theorem
Fundamental Theorem of Algebra
Complete linear factorization
Conjugate Zeros Theorem
Rational Zeros Theorem
Continuous function
Intermediate Value Theorem

Bisection method
Rational function:
   branches of a graph
Asymptotes:
   vertical asymptote
   horizontal asymptote
   slant asymptote
Hole in a graph

CHAPTER 4 Review Exercises

Answers to selected odd-numbered problems begin on page ANS-14.

A. True/False

In Problems 1–18, answer true or false.

1. $f(x) = 2x^3 - 8x^{-2} + 5$ is not a polynomial. ___
2. $f(x) = x + \dfrac{1}{x}$ is a rational function. ___
3. The graph of a polynomial function can have no holes in it. ___
4. A polynomial function of degree 4 has exactly four real zeros. ___
5. When a polynomial of degree greater than 1 is divided by $x + 2$, the remainder is always a constant. ___
6. If the coefficients a, b, c, and d of the polynomial function $f(x) = ax^3 + bx^2 + cx + d$ are positive integers, then f has no positive real zeros. ___
7. The polynomial equation $2x^7 = 1 - x$ has a solution in the interval $[0, 1]$. ___
8. The graph of the rational function $f(x) = (x^2 + 1)/x$ has a slant asymptote. ___
9. The graph of the polynomial function $f(x) = 4x^6 + 3x^2$ is symmetric with respect to the y-axis. ___
10. The graph of a polynomial function that is an odd function must pass through the origin. ___
11. An asymptote is a line that the graph of a function approaches but never crosses. ___
12. The point $\left(\frac{1}{3}, \frac{7}{4}\right)$ is on the graph of $f(x) = \dfrac{2x + 4}{3 - x}$. ___
13. The graph of a rational function $f(x) = P(x)/Q(x)$ has a slant asymptote when the degree of P is greater than the degree of Q. ___
14. If $3 - 4i$ is a zero of a polynomial function $f(x)$ with real coefficients, then $3 + 4i$ is also a zero of $f(x)$. ___
15. A polynomial function must have at least one rational zero. ___
16. If $1 + i$ is a zero of $f(x) = x^3 - 6x^2 + 10x - 8$, then 4 is also a zero of f. ___
17. The continuous function $f(x) = x^6 + 2x^5 - x^2 + 1$ has a real zero within the interval $[-1, 1]$. ___
18. When $f(x) = x^{50} + 1$ is divided by $g(x) = x - 1$, the remainder is $f(1) = 2$. ___

B. Fill in the Blanks

In Problems 1–12, fill in the blanks.

1. The graph of the polynomial function $f(x) = x^3(x - 1)^2(x - 5)$ is tangent to the x-axis at _____ and passes through the x-axis at _____.
2. A third-degree polynomial function with zeros 1 and $3i$ is _____.
3. The end behavior of the graph of $f(x) = x^2(x + 3)(x - 5)$ resembles the graph of the power function $f(x) = $ _____.
4. The polynomial function $f(x) = x^4 - 3x^3 + 17x^2 - 2x + 2$ has _____ (how many) possible rational zeros.
5. For $f(x) = kx^2(x - 2)(x - 3), f(-1) = 8$ if $k = $ _____.
6. The y-intercept of the graph of the rational function $f(x) = \dfrac{2x + 8}{x^2 - 5x + 4}$ is _____.
7. The vertical asymptotes for the graph of the rational function $f(x) = \dfrac{2x + 8}{x^2 - 5x + 4}$ are _____.
8. The x-intercepts of the graph of the rational function $f(x) = \dfrac{x^3 - x}{4 - 2x^3}$ are _____.
9. The horizontal asymptote for the graph of the rational function $f(x) = \dfrac{x^3 - x}{4 - 2x^3}$ is _____.
10. A rational function whose graph has the horizontal asymptote $y = 1$ and an x-intercept $(3, 0)$ is _____.
11. The graph of the rational function $f(x) = \dfrac{x^n}{x^3 + 1}$, where n is a nonnegative integer, has the horizontal asymptote $y = 0$ when $n = $ _____.
12. The graph of the polynomial function $f(x) = 3x^5 - 4x^2 + 5x - 2$ has at most _____ turning points.

C. Review Exercises

In Problems 1 and 2, use long division to divide $f(x)$ by $g(x)$.

1. $f(x) = 6x^5 - 4x^3 + 2x^2 + 4, g(x) = 2x^2 - 1$
2. $f(x) = 15x^4 - 2x^3 + 8x + 6, g(x) = 5x^3 + x + 2$

In Problems 3 and 4, use synthetic division to divide $f(x)$ by $g(x)$.

3. $f(x) = 7x^4 - 6x^2 + 9x + 3, g(x) = x - 2$
4. $f(x) = 4x^3 + 7x^2 - 8x, g(x) = x + 1$
5. Without actually performing the division, determine the remainder when $f(x) = 5x^3 - 4x^2 + 6x - 9$ is divided by $g(x) = x + 3$.
6. Use synthetic division and the Remainder Theorem to find $f(c)$ for
$$f(x) = x^6 - 3x^5 + 2x^4 + 3x^3 - x^2 + 5x - 1$$
when $c = 2$.
7. Determine the values of the positive integer n such that $f(x) = x^n + c^n$ is divisible by $g(x) = x + c$.
8. Suppose that
$$f(x) = 36x^{98} - 40x^{25} + 18x^{14} - 3x^7 + 40x^4 + 5x^2 - x + 2$$
is divided by $g(x) = x - 1$. What is the remainder?

9. List, but do not test, all possible rational zeros of
$$f(x) = 8x^4 + 19x^3 + 31x^2 + 38x - 15.$$

10. Find the complete factorization of $f(x) = 12x^3 + 16x^2 + 7x + 1$.

In Problems 11 and 12, verify that each of the indicated numbers is a zero of the given polynomial function $f(x)$. Find all other zeros and then give the complete factorization of $f(x)$.

11. 2; $f(x) = (x - 3)^3 + 1$

12. -1; $f(x) = (x + 2)^4 - 1$

In Problems 13–16, find the real value of k so that the given condition is satisfied.

13. The remainder in the division of $f(x) = x^4 - 3x^3 - x^2 + kx - 1$ by $g(x) = x - 4$ is $r = 5$.
14. $x + \frac{1}{2}$ is a factor of $f(x) = 8x^2 - 4kx + 9$.
15. $x - k$ is a factor of $f(x) = 2x^3 + x^2 + 2x - 12$.
16. The graph of $f(x) = \dfrac{x - k}{x^2 + 5x + 6}$ has a hole at $x = k$.

In Problems 17 and 18, find a polynomial function f of indicated degree whose graph is given in the figure.

17. fifth degree

FIGURE 4.R.1 Graph for Problem 17

18. sixth degree

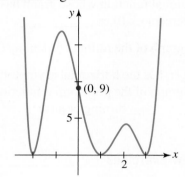

FIGURE 4.R.2 Graph for Problem 18

In Problems 19–28, match the given rational function with one of the graphs (a)–(j).

(a)

FIGURE 4.R.3

(b)

FIGURE 4.R.4

(c)

FIGURE 4.R.5

(d)

FIGURE 4.R.6

(e)

FIGURE 4.R.7

(f)

FIGURE 4.R.8

(g)

FIGURE 4.R.9

(h)

FIGURE 4.R.10

(i)

FIGURE 4.R.11

(j)

FIGURE 4.R.12

19. $f(x) = \dfrac{2x}{x^2 + 1}$

20. $f(x) = \dfrac{x^2 - 1}{x^2 + 1}$

21. $f(x) = \dfrac{2x}{x - 2}$

22. $f(x) = 2 - \dfrac{1}{x^2}$

23. $f(x) = \dfrac{x}{(x - 2)^2}$

24. $f(x) = \dfrac{(x - 1)^2}{x - 2}$

25. $f(x) = \dfrac{x^2 - 10}{2x - 4}$

26. $f(x) = \dfrac{-x^2 + 5x - 5}{x - 2}$

27. $f(x) = \dfrac{2x}{x^3 + 1}$

28. $f(x) = \dfrac{3}{x^2 + 1}$

In Problems 29 and 30, find the asymptotes for the graph of the given rational function. Find the x- and y-intercepts of the graph. Sketch the graph of f.

29. $f(x) = \dfrac{x + 2}{x^2 + 2x - 8}$

30. $f(x) = \dfrac{-x^3 + 2x^2 + 9}{x^2}$

31. Discuss: Without graphing or attempting to find them, why does the polynomial function

$$f(x) = 4x^{10} + 9x^6 + 5x^4 + 13x^2 + 3$$

have no real zeros?

32. Sketch the graph of $f(x) = \dfrac{x - 1}{x(x - 4)}$. From the graph of f, find the solution of the inequality

$$\dfrac{x - 1}{x(x - 4)} \geq 0.$$

33. (a) Show that the number $\sqrt{2} + \sqrt{3}$ is a zero of the polynomial function
$f(x) = x^4 - 10x^2 + 1$.
(b) Discuss: How does part (a) prove that $\sqrt{2} + \sqrt{3}$ is an irrational number?

34. Consider the polynomial function $f(x) = x^3 - i$, where $i = \sqrt{-1}$.
(a) Verify that each of the three numbers $-i, \dfrac{\sqrt{3}}{2} + \dfrac{1}{2}i, -\dfrac{\sqrt{3}}{2} + \dfrac{1}{2}i$ is zero of the function f.
(b) Note in part (b) that each zero is not the conjugate of either of the two remaining zeros. Explain why this does not violate the Conjugate Zero Theorem in Section 4.3.

5 Exponential and Logarithmic Functions

In This Chapter

- **5.1** Exponential Functions
- **5.2** Logarithmic Functions
- **5.3** Exponential and Logarithmic Equations
- **5.4** Exponential and Logarithmic Models
- **5.5** Hyperbolic Functions

Chapter 5 Review Exercises

The curve that describes the shape of a telephone wire hanging between two poles is the graph of a hyperbolic function.

A Bit of History In this chapter we will consider two types of functions that appear often in applications: exponential and logarithmic functions.

The wealthy English political and religious polemicist **John Napier** (1550–1617) is remembered today principally for one of his mathematical diversions, the logarithm. We will see in Section 5.2 that logarithms are essentially exponents and that there are two important types. The invention of the *natural logarithm* is often attributed to Napier. His friend and collaborator, the English mathematician **Henry Briggs** (1561–1631), devised the base ten or *common logarithm*. The word *logarithm* comes from two Greek words: *logos*, which means reasoning or reckoning, and *arithmos*, which means number. Thus a logarithm is a "reckoning number." For several hundred years logarithms were used primarily as an aid in performing complex and tedious arithmetic computations. A calculator prior to 1967 was an analog instrument called the slide rule. The operations performed on a slide rule were based entirely on the properties of logarithms. With the invention of the electronic handheld calculator in 1966 and the evolution of the personal computer, the slide rule and the use of logarithms as a pencil-and-paper computing method have gone the way of the dinosaurs.

5.1 Exponential Functions

☰ **Introduction** In the preceding chapters we considered functions such as $f(x) = x^2$, that is, a function with a variable base x and constant power or exponent 2. We now examine functions having a constant base b and a variable exponent x.

DEFINITION 5.1.1 Exponential Function

If $b > 0$ and $b \neq 1$, then an **exponential function** $y = f(x)$ is a function of the form
$$f(x) = b^x. \tag{1}$$
The number b is called the **base** and x is called the **exponent**.

The **domain** of an exponential function f defined in (1) is the set of all real numbers $(-\infty, \infty)$.

In (1) the base b is restricted to positive numbers in order to guarantee that b^x is always a real number. For example, with this restriction we avoid complex numbers such as $(-4)^{1/2}$. Also, the base $b = 1$ is of little interest to us since it can be shown that f is the constant function $f(x) = 1^x = 1$. Moreover, for $b > 0$, we have $f(0) = b^0 = 1$.

☐ **Exponents** As just mentioned, the domain of an exponential function (1) is the set of all real numbers. This means that the exponent x can be either a rational or an irrational number. For example, if the base $b = 3$ and the exponent x is a *rational number*, say, $x = \frac{1}{5}$ and $x = 1.4$, then

$$3^{1/5} = \sqrt[5]{3} \quad \text{and} \quad 3^{1.4} = 3^{14/10} = 3^{7/5} = \sqrt[5]{3^7}.$$

For an exponent x that is an *irrational number*, b^x is defined, but its precise definition is beyond the scope of this text. We can, however, suggest a procedure for defining a number such as $3^{\sqrt{2}}$. From the decimal representation $\sqrt{2} = 1.414213562\ldots$ we see that the rational numbers

$$1, 1.4, 1.41, 1.414, 1.4142, 1.41421, \ldots$$

are successively better approximations to $\sqrt{2}$. By using these rational numbers as exponents, we would expect that the numbers

$$3^1, 3^{1.4}, 3^{1.41}, 3^{1.414}, 3^{1.4142}, 3^{1.41421}, \ldots$$

are then successively better approximations to $3^{\sqrt{2}}$. In fact, this can be shown to be true with a precise definition of b^x for an irrational value of x. But on a practical level, we can use the $\boxed{y^x}$ key on a calculator to obtain the approximation 4.728804388 to $3^{\sqrt{2}}$.

▶ The precise definition of b^x for x an irrational number requires the concept of a *limit*. The notion of a limit of a function is the backbone of differential and integral calculus.

☐ **Laws of Exponents** As we saw in Chapter 1, the laws of exponents are stated first for integer exponents and then for rational exponents. Since b^x can be defined for all real numbers x when $b > 0$, it can be proved that these same **laws of exponents** hold for all real number exponents. If $a > 0$, $b > 0$ and x, x_1, and x_2 denote real numbers, then

THEOREM 5.1.1 Laws of Exponents

If $a > 0$, $b > 0$ and x, x_1, and x_2 denote real numbers, then

(i) $b^{x_1} \cdot b^{x_2} = b^{x_1 + x_2}$

(ii) $\dfrac{b^{x_1}}{b^{x_2}} = b^{x_1 - x_2}$

(iii) $\dfrac{1}{b^x} = b^{-x}$

(iv) $(b^{x_1})^{x_2} = b^{x_1 x_2}$

(v) $(ab)^x = a^x b^x$

(vi) $\left(\dfrac{a}{b}\right)^x = \dfrac{a^x}{b^x}$.

EXAMPLE 1 Rewriting a Function

At times, we will use the laws of exponents to rewrite a function in a different form. For example, neither $f(x) = 2^{3x}$ nor $g(x) = 4^{-2x}$ has the precise form of the exponential function defined in (1). However, by the laws of exponents given in Theorem 5.1.1, f can be rewritten as $f(x) = 8^x$ ($b = 8$ in (1)), and g can be recast as $g(x) = \left(\frac{1}{16}\right)^x$ ($b = \frac{1}{16}$ in (1)). The details are shown below:

$$f(x) = 2^{3x} \overset{\text{by }(iv)}{=} (2^3)^x \overset{\text{form is now }b^x}{=} 8^x$$

$$g(x) = 4^{-2x} \overset{\text{by }(iv)}{=} (4^{-2})^x \overset{\text{by }(iii)}{=} \left(\dfrac{1}{4^2}\right)^x \overset{\text{form is now }b^x}{=} \left(\dfrac{1}{16}\right)^x.$$

≡

☐ **Graphs** We distinguish two types of graphs for (1) depending on whether the base b satisfies $b > 1$ or $0 < b < 1$. The next two examples illustrate, in turn, the graphs of $f(x) = 3^x$ and $f(x) = \left(\frac{1}{3}\right)^x$. Before graphing, we can make some intuitive observations about both functions. Since the bases $b = 3$ and $b = \frac{1}{3}$ are positive, the values of 3^x and $\left(\frac{1}{3}\right)^x$ are *positive* for every real number x. As a consequence, there are no real numbers x_1 and x_2 for which 3^{x_1} and $\left(\frac{1}{3}\right)^{x_2}$ are zero. Graphically, this means that the graphs of $f(x) = 3^x$ and $f(x) = \left(\frac{1}{3}\right)^x$ have no x-intercepts. Also, $3^0 = 1$ and $\left(\frac{1}{3}\right)^0 = 1$, and so $f(0) = 1$ in each case. This means that the graphs of $f(x) = 3^x$ and $f(x) = \left(\frac{1}{3}\right)^x$ have the same y-intercept $(0, 1)$.

EXAMPLE 2 Graph for $b > 1$

Graph the function $f(x) = 3^x$.

Solution We first construct a table of some function values corresponding to preselected values of x. As shown in **FIGURE 5.1.1**, we plot the corresponding points obtained from the table and connect them with a continuous curve. The graph shows that f is an increasing function on the interval $(-\infty, \infty)$.

x	-3	-2	-1	0	1	2
$f(x)$	$\frac{1}{27}$	$\frac{1}{9}$	$\frac{1}{3}$	1	3	9

≡

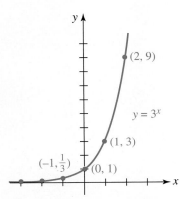

FIGURE 5.1.1 Graph of function in Example 2

FIGURE 5.1.2 Graph of function in Example 3

EXAMPLE 3   Graph for $0 < b < 1$

Graph the function $f(x) = \left(\frac{1}{3}\right)^x$.

Solution Proceeding as in Example 2, we construct a table of some function values corresponding to preselected values of x. Note, for example, by the laws of exponents

$$f(-2) = \left(\tfrac{1}{3}\right)^{-2} = (3^{-1})^{-2} = 3^2 = 9.$$

As shown in **FIGURE 5.1.2**, we plot the corresponding points obtained from the table and connect them with a continuous curve. In this case the graph shows that f is a decreasing function on the interval $(-\infty, \infty)$.

x	-3	-2	-1	0	1	2
$f(x)$	27	9	3	1	$\tfrac{1}{3}$	$\tfrac{1}{9}$

Review Theorem 3.2.3 in Section 3.2 for reflections in the x- and y-axes. ▶

□ **Reflections** Exponential functions with bases satisfying $0 < b < 1$, such as $b = \tfrac{1}{3}$, are frequently written in an alternative manner. We note that $y = \left(\tfrac{1}{3}\right)^x$ is the same as $y = 3^{-x}$. From this last result we see that the graph of $y = 3^{-x}$ is simply the graph of $y = 3^x$ reflected in the y-axis.

□ **Horizontal Asymptote** **FIGURE 5.1.3** illustrates the two general shapes that the graph of an exponential function $f(x) = b^x$ can have; but there is one more important aspect of all such graphs. Observe in Figure 5.1.3 that for $b > 1$,

$$f(x) = b^x \to 0 \quad \text{as} \quad x \to -\infty, \quad \leftarrow \text{blue graph}$$

whereas for $0 < b < 1$,

$$f(x) = b^x \to 0 \quad \text{as} \quad x \to \infty. \quad \leftarrow \text{red graph}$$

In other words, the line $y = 0$ (the x-axis) is a **horizontal asymptote** for both types of exponential graphs.

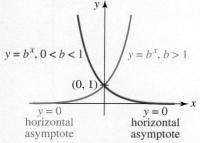

FIGURE 5.1.3 f increasing for $b > 1$; f decreasing for $0 < b < 1$

□ **Properties** The following list summarizes some of the important properties of the exponential function $f(x) = b^x$. Reexamine the graphs in Figures 5.1.1–5.1.3 as you read this list.

PROPERTIES OF THE EXPONENTIAL FUNCTION

(i) The domain of f is the set of real numbers, that is, $(-\infty, \infty)$.
(ii) The range of f is the set of positive real numbers, that is, $(0, \infty)$.
(iii) The y-intercept of f is $(0, 1)$. The graph of f has no x-intercepts.
(iv) The function f is increasing for $b > 1$ and decreasing for $0 < b < 1$.
(v) The x-axis, that is, $y = 0$, is a horizontal asymptote for the graph of f.
(vi) The function f is continuous on $(-\infty, \infty)$.
(vii) The function f is one-to-one.

Although the graphs $y = b^x$ in the case, say, when $b > 1$, all share the same basic shape and all pass through the same point $(0, 1)$, there are subtle differences. The larger the base b the more steeply the graph rises as x increases. In **FIGURE 5.1.4** we compare the graphs of $y = 5^x$, $y = 3^x$, $y = 2^x$, and $y = (1.2)^x$ in green, blue, gold, and red, respectively, on the same coordinate axes. We see from its graph that the values of $y = (1.2)^x$ increase slowly as x increases. For example, for $y = (1.2)^x$, $f(3) = (1.2)^3 = 1.728$, whereas, for $y = 5^x$, $f(3) = 5^3 = 125$.

The fact that (1) is a one-to-one function, follows from the horizontal line test discussed in Section 3.6. Note in Figures 5.1.1–5.1.4 that a horizontal line can cross or intersect an exponential graph in at most one point.

Of course, we can obtain other kinds of graphs by rigid and nonrigid transformations, or when an exponential function is combined with other functions by either an arithmetic operation or by function composition. In the next several examples we examine variations of the exponential graph.

FIGURE 5.1.4 Graphs of $y = b^x$ for $b = 1.2, 2, 3, 5$

EXAMPLE 4   Horizontally Shifted Graph

Graph the function $f(x) = 3^{x+2}$.

Solution From the discussion in Section 3.2 you should recognize that the graph of $f(x) = 3^{x+2}$ is the graph of $y = 3^x$ shifted 2 units to the left. Recall, since the shift is a rigid transformation to the left, the points on the graph of $f(x) = 3^{x+2}$ are the points on the graph of $y = 3^x$ moved horizontally 2 units to the left. This means that the y-coordinates of points (x, y) on the graph of $y = 3^x$ remain unchanged but 2 is subtracted from all the x-coordinates of the points. Thus we see from **FIGURE 5.1.5** that the points $(0, 1)$ and $(2, 9)$ on the graph of $y = 3^x$ are moved, in turn, to the points $(-2, 1)$ and $(0, 9)$ on the graph of $f(x) = 3^{x+2}$. ≡

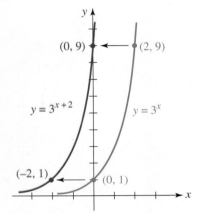

FIGURE 5.1.5 Shifted graph in Example 4

The function $f(x) = 3^{x+2}$ in Example 4 can be rewritten, if desired, as $f(x) = 9 \cdot 3^x$. By (i) of the laws of exponents, $3^{x+2} = 3^2 3^x = 9 \cdot 3^x$. In this manner we can reinterpret the graph of $f(x) = 3^{x+2}$ as a vertical stretch of the graph of $y = 3^x$ by a factor of 9. For example, $(1, 3)$ is on the graph of $y = 3^x$, whereas $(1, 9 \cdot 3) = (1, 27)$ is on the graph of $f(x) = 3^{x+2}$.

☐ **The Number e** Most every student of mathematics has heard of, and has likely worked with, the famous irrational number $\pi = 3.141592654\ldots$. Recall, an irrational number is a nonrepeating and nonterminating decimal. In calculus and applied mathematics the irrational number

$$e = 2.718281828459\ldots$$

arguably plays a role more important than the number π. The usual definition of the number e is the number that the function $f(x) = (1 + 1/x)^x$ approaches as we let x become large without bound in the positive direction, that is,

$$\left(1 + \frac{1}{x}\right)^x \to e \quad \text{as} \quad x \to \infty.$$

See Problems 39 and 40 in Exercises 5.1.

☐ **The Natural Exponential Function** When the base in (1) is chosen to be $b = e$, the function

$$f(x) = e^x \qquad (2)$$

is called the **natural exponential function**. Since $b = e > 1$ and $b = 1/e < 1$, the graphs of $y = e^x$ and $y = e^{-x}$ (or $y = (1/e)^x = 1/e^x$) are given in **FIGURE 5.1.6**.

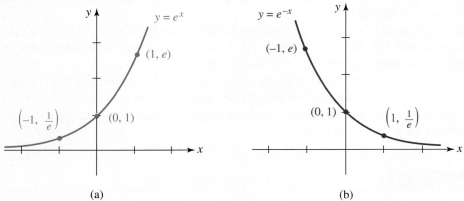

(a)   (b)

FIGURE 5.1.6 Graphs of the natural exponential function (in (a)) and its reciprocal (in (b))

On the face of it, the natural exponential function (2) possesses no noticeable graphical characteristic that distinguishes it from, say, the function $f(x) = 3^x$, and has no special properties other than the ones given in (*i*)–(*vii*) above. Questions as to why (2) is a "natural" and frankly, the most important exponential function, can only be answered fully in courses in calculus and beyond. We will explore some of the importance of the number e in Sections 5.3 and 5.4.

EXAMPLE 5   Reflection and Vertical Shift

Graph the function $f(x) = 2 - e^{-x}$. State the range.

Solution We first draw the graph of $y = e^{-x}$ as shown in **FIGURE 5.1.7(a)**. Then we reflect the first graph in the *x*-axis to obtain the graph of $y = -e^{-x}$ in Figure 5.1.7(b). Finally, the graph in Figure 5.1.7(c) is obtained by shifting the graph in part (b) upward 2 units.

The *y*-intercept $(0, -1)$ of $y = -e^{-x}$ when shifted upward 2 units returns us to the original *y*-intercept in Figure 5.1.7(a). Finally, because of the vertical shift the horizontal asymptote, which was $y = 0$ in parts (a) and (b) of the figure, becomes $y = 2$ in Figure 5.1.7(c). From the last graph we can conclude that the range of the function $f(x) = 2 - e^{-x}$ is the set of real numbers defined by $y < 2$, that is, the interval $(-\infty, 2)$ on the *y*-axis.

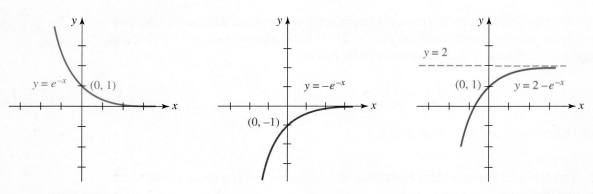

(a) Start with graph of $y = e^{-x}$ (b) Graph in (a) reflected in *x*-axis   (c) Graph in (b) shifted upward 2 units

FIGURE 5.1.7 Graph of function in Example 5

In the next example we graph the function composition of the natural exponential function $y = e^x$ with the simple quadratic polynomial function $y = -x^2$.

EXAMPLE 6 A Function Composition

Graph the function $f(x) = e^{-x^2}$.

Solution Because $f(0) = e^{-0^2} = e^0 = 1$, the y-intercept of the graph is $(0, 1)$. Also, $f(x) \neq 0$ since $e^{-x^2} \neq 0$ for every real number x. This means that the graph of f has no x-intercepts. Then from

$$f(-x) = e^{-(-x)^2} = e^{-x^2} = f(x)$$

we conclude that f is an even function and so its graph is symmetric with respect to the y-axis. Lastly, observe that

$$f(x) = \frac{1}{e^{x^2}} \to 0 \quad \text{as} \quad x \to \infty.$$

By symmetry we can also conclude that $f(x) \to 0$ as $x \to -\infty$. This shows that $y = 0$ is a horizontal asymptote for the graph of f. The graph of f is given in FIGURE 5.1.8.

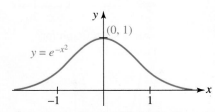

FIGURE 5.1.8 Graph of function in Example 6

Bell-shaped graphs such as that given in Figure 5.1.8 are very important in the study of probability and statistics.

5.1 Exercises Answers to selected odd-numbered problems begin on page ANS-14.

In Problems 1–12, sketch the graph of the given function f. Find the y-intercept and the horizontal asymptote of the graph. State whether the function is increasing or decreasing.

1. $f(x) = \left(\frac{3}{4}\right)^x$
2. $f(x) = \left(\frac{4}{3}\right)^x$
3. $f(x) = -2^x$
4. $f(x) = -2^{-x}$
5. $f(x) = 2^{x+1}$
6. $f(x) = 2^{2-x}$
7. $f(x) = -5 + 3^x$
8. $f(x) = 2 + 3^{-x}$
9. $f(x) = 3 - \left(\frac{1}{5}\right)^x$
10. $f(x) = 9 - e^x$
11. $f(x) = -1 + e^{x-3}$
12. $f(x) = -3 - e^{x+5}$

In Problems 13–16, find an exponential function $f(x) = b^x$ such that the graph of f passes through the given point.

13. $(3, 216)$
14. $(-1, 5)$
15. $(-1, e^2)$
16. $(2, e)$

In Problems 17 and 18, determine the range of the given function.

17. $f(x) = 5 + e^{-x}$
18. $f(x) = 4 - 2^{-x}$

In Problems 19 and 20, find the x- and y-intercepts of the graph of the given function. Do not graph.

19. $f(x) = xe^x + 10e^x$
20. $f(x) = x^2 2^x - 2^x$

In Problems 21–24, use a graph to solve the given inequality.

21. $2^x > 16$
22. $e^x \leq 1$
23. $e^{x-2} < 1$
24. $\left(\frac{1}{2}\right)^x \geq 8$

In Problems 25 and 26, use the graph in Figure 5.1.8 to sketch the graph of the given function f.

25. $f(x) = e^{-(x-3)^2}$
26. $f(x) = 3 - e^{-(x+1)^2}$

In Problems 27 and 28, use $f(-x) = f(x)$ to demonstrate that the given function is even. Sketch the graph of f.

27. $f(x) = e^{x^2}$
28. $f(x) = e^{-|x|}$

In Problems 29–32 use the graphs obtained in Problems 27 and 28 as an aid in sketching the graph of the given function f.

29. $f(x) = 1 - e^{x^2}$
30. $f(x) = 2 + 3e^{|x|}$
31. $f(x) = -e^{|x-3|}$
32. $f(x) = e^{(x+2)^2}$

33. Show that $f(x) = 2^x + 2^{-x}$ is an even function. Sketch the graph of f.
34. Show that $f(x) = 2^x - 2^{-x}$ is an odd function. Sketch the graph of f.

In Problems 35 and 36, sketch the graph of the given piecewise-defined function f.

35. $f(x) = \begin{cases} -e^x, & x < 0 \\ -e^{-x}, & x \geq 0 \end{cases}$
36. $f(x) = \begin{cases} e^{-x}, & x \leq 0 \\ -e^x, & x > 0 \end{cases}$

37. Find an equation of the red line in FIGURE 5.1.9.
38. Find the total area of the shaded region in FIGURE 5.1.10.

FIGURE 5.1.9 Graph for Problem 37

FIGURE 5.1.10 Graph for Problem 38

Calculator Problems

39. Use a calculator to fill in the given table.

x	10	100	1000	10,000	100,000	1,000,000
$(1 + 1/x)^x$						

40. (a) Use a graphing utility to graph the functions $f(x) = (1 + 1/x)^x$ and $g(x) = e$ on the same set of coordinate axes. Use the intervals $(0, 10]$, $(0, 100]$, and $(0, 1000]$. Describe the behavior of f for large values of x. In graphical terms, what is the constant function $g(x) = e$?

(b) Graph the function f in part (a) on the interval $[-10, 0)$. Superimpose that graph with the graph of f on $(0, 10]$ obtained in part (a). Is f a continuous function?

In Problems 41 and 42, use a graphing utility as an aid in determining the x-coordinates of the points of intersection of the graphs of the functions f and g.

41. $f(x) = x^2,\ g(x) = 2^x$
42. $f(x) = x^3,\ g(x) = 3^x$

For Discussion

43. Suppose $2^t = a$ and $6^t = b$. Using the laws of exponents given in this section, answer the following questions in terms of the symbols a and b.
  (a) What does 12^t equal?
  (b) What does 3^t equal?
  (c) What does 6^{-t} equal?
  (d) What does 6^{3t} equal?
  (e) What does $2^{-3t}2^{7t}$ equal?
  (f) What does 18^t equal?

44. Discuss: What does the graph of $y = e^{e^x}$ look like? Do not use a calculator.

5.2 Logarithmic Functions

Introduction Since an exponential function $y = b^x$ is one-to-one, we know that it has an inverse function. To find this inverse, we interchange the variables x and y to obtain $x = b^y$. This last formula defines y as a function of x:

y is that exponent of the base b that produces x.

By replacing the word *exponent* with the word *logarithm*, we can rephrase the preceding line as

y is that logarithm of the base b that produces x.

This last line is abbreviated by the notation $y = \log_b x$ and is called the logarithmic function.

DEFINITION 5.2.1 Logarithmic Function

The **logarithmic function** with base $b > 0$, $b \neq 1$, is defined by

$$y = \log_b x \quad \text{if and only if} \quad x = b^y. \tag{1}$$

For $b > 0$ there is no real number y for which b^y can be either 0 or negative. It then follows from $x = b^y$ that $x > 0$. In other words, the **domain** of a logarithmic function $y = \log_b x$ is the set of positive real numbers $(0, \infty)$.

For emphasis, all that is being said in the preceding sentences is:

The logarithmic expression $y = \log_b x$ and the exponential expression $x = b^y$ are equivalent.

That is, both symbols mean the same thing. As a consequence, within a specific context such as solving a problem, we can use whichever form happens to be more convenient. The following table lists several examples of equivalent logarithmic and exponential statements.

Logarithmic Form	Exponential Form
$\log_3 9 = 2$	$9 = 3^2$
$\log_8 2 = \frac{1}{3}$	$2 = 8^{1/3}$
$\log_{10} 0.001 = -3$	$0.001 = 10^{-3}$
$\log_b 5 = -1$	$5 = b^{-1}$

☐ **Graphs** Recall from Section 3.6 that the graph of an inverse function can be obtained by reflecting the graph of the original function in the line $y = x$. This technique was used to obtain the red graphs from the blue graphs in **FIGURE 5.2.1**. As you inspect the two graphs in Figure 5.2.1(a) and in Figure 5.2.1(b), remember that the domain $(-\infty, \infty)$ and range $(0, \infty)$ of $y = b^x$ become, in turn, the range $(-\infty, \infty)$ and domain $(0, \infty)$ of $y = \log_b x$. Also note that the y-intercept $(0, 1)$ for the exponential function (blue graphs) becomes the x-intercept $(1, 0)$ for the logarithmic function (red graphs).

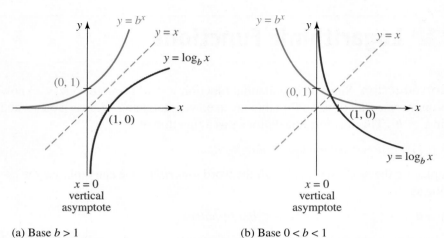

(a) Base $b > 1$ (b) Base $0 < b < 1$

FIGURE 5.2.1 Graphs of logarithmic functions

☐ **Vertical Asymptote** When the exponential function is reflected in the line $y = x$, the horizontal asymptote $y = 0$ for the graph of $y = b^x$ becomes a vertical asymptote for the graph of $y = \log_b x$. In Figure 5.2.1 we see that for $b > 1$,

$$\log_b x \to -\infty \quad \text{as} \quad x \to 0^+, \quad \leftarrow \text{red graph in (a)}$$

whereas for $0 < b < 1$,

$$\log_b x \to \infty \quad \text{as} \quad x \to 0^+. \quad \leftarrow \text{red graph in (b)}$$

From (7) of Section 4.6 we conclude that $x = 0$, which is the equation of the y-axis, is a **vertical asymptote** for the graph of $y = \log_b x$.

☐ **Properties** The following list summarizes some of the important properties of the logarithmic function $f(x) = \log_b x$.

PROPERTIES OF THE LOGARITHMIC FUNCTION

(*i*) The domain of f is the set of positive real numbers, that is, $(0, \infty)$.
(*ii*) The range of f is the set of real numbers, that is, $(-\infty, \infty)$.
(*iii*) The x-intercept of f is $(1, 0)$. The graph of f has no y-intercept.
(*iv*) The function f is increasing for $b > 1$ and decreasing for $0 < b < 1$.
(*v*) The y-axis, that is, $x = 0$, is a vertical asymptote for the graph of f.
(*vi*) The function f is continuous on $(0, \infty)$.
(*vii*) The function f is one-to-one.

We would like to call attention to the third entry in the foregoing list for special emphasis:

$$\log_b 1 = 0 \quad \text{since} \quad b^0 = 1. \tag{2}$$

Also,
$$\log_b b = 1 \quad \text{since} \quad b^1 = b. \tag{3}$$

Thus, in addition to $(1, 0)$ the graph of any logarithmic function (1) with base b also contains the point $(b, 1)$. The equivalence of $y = \log_b x$ and $x = b^y$ also yields two sometimes-useful identities. By substituting $y = \log_b x$ into $x = b^y$, and then $x = b^y$ into $y = \log_b x$ gives

$$x = b^{\log_b x} \quad \text{and} \quad y = \log_b b^y. \tag{4}$$

For example, from (4), $8^{\log_8 10} = 10$ and $\log_{10} 10^5 = 5$.

EXAMPLE 1   Logarithmic Graph for $b > 1$

Graph $f(x) = \log_{10}(x + 10)$.

Solution This is the graph of $y = \log_{10} x$, which has the shape shown in Figure 5.2.1(a), shifted 10 units to the left. To reinforce the fact that the domain of a logarithmic function $y = \log_{10} x$ is the set of positive real numbers, that is, $x > 0$, we can obtain the domain of $f(x) = \log_{10}(x + 10)$ by replacing x by $x + 10$ and requiring that $x + 10 > 0$ or $x > -10$. In interval notation, the domain of f is $(-10, \infty)$. In the short accompanying table, we have chosen convenient values of x in order to plot a few points.

x	-9	0	90
$f(x)$	0	1	2

Notice,

$$f(-9) = \log_{10} 1 = 0 \quad \leftarrow \text{by (2)}$$
$$f(0) = \log_{10} 10 = 1. \quad \leftarrow \text{by (3)}$$

The vertical asymptote $x = 0$ for the graph of $y = \log_{10} x$ becomes $x = -10$ for the shifted graph. This asymptote is the red dashed vertical line in **FIGURE 5.2.2**. ≡

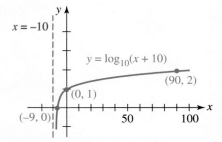

FIGURE 5.2.2 Graph of function in Example 1

☐ **Natural Logarithm** Logarithms with base $b = 10$ are called **common logarithms** and logarithms with base $b = e$ are called **natural logarithms**. Furthermore, it is customary to write the natural logarithm

$$\log_e x \quad \text{as} \quad \ln x.$$

The symbol "$\ln x$" is usually read phonetically as "ell-en of x." Since $b = e > 1$, the graph of $y = \ln x$ has the characteristic logarithmic shape shown in Figure 5.2.1(a). See **FIGURE 5.2.3**. For base $b = e$, (1) becomes

$$y = \ln x \quad \text{if and only if} \quad x = e^y. \tag{5}$$

The analogs of (2) and (3) for the natural logarithm are

$$\ln 1 = 0 \quad \text{since} \quad e^0 = 1, \tag{6}$$
$$\ln e = 1 \quad \text{since} \quad e^1 = e. \tag{7}$$

The identities in (4) become

$$x = e^{\ln x} \quad \text{and} \quad y = \ln e^y. \tag{8}$$

For example, from (8), $e^{\ln 13} = 13$.

Common and natural logarithms can be found on all calculators.

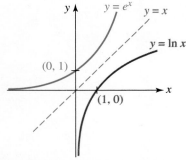

FIGURE 5.2.3 Graph of the natural logarithm is shown in red

☐ **Laws of Logarithms** The laws of exponents given in Theorem 5.1.1 can be restated in an equivalent manner as the laws of logarithms. To see this, suppose we write $M = b^{x_1}$ and $N = b^{x_2}$. Then by (1), $x_1 = \log_b M$ and $x_2 = \log_b N$.

Product: By (*i*) of Theorem 5.1.1, $MN = b^{x_1 + x_2}$. Expressed as a logarithm this is $x_1 + x_2 = \log_b MN$. Substituting for x_1 and x_2 gives

$$\log_b M + \log_b N = \log_b MN.$$

Quotient: By (*ii*) of Theorem 5.1.1, $M/N = b^{x_1 - x_2}$. Expressed as a logarithm this is $x_1 - x_2 = \log_b(M/N)$. Substituting for x_1 and x_2 gives

$$\log_b M - \log_b N = \log_b(M/N).$$

Power: By (*iv*) of Theorem 5.1.1, $M^c = b^{cx_1}$. Expressed as a logarithm this is $cx_1 = \log_b M^c$. Substituting for x_1 gives

$$c \log_b M = \log_b M^c.$$

For convenience and future reference, we summarize these product, quotient, and power laws of logarithms next.

THEOREM 5.2.1 Laws of Logarithms

For any base $b > 0$, $b \neq 1$, and positive numbers M and N:

(*i*) $\log_b MN = \log_b M + \log_b N$

(*ii*) $\log_b\left(\dfrac{M}{N}\right) = \log_b M - \log_b N$

(*iii*) $\log_b M^c = c \log_b M$, for c any real number.

EXAMPLE 2   Using the Laws of Logarithms

Simplify and write as a single logarithm

$$\tfrac{1}{2}\ln 36 + 2\ln 4 - \ln 4.$$

Solution There are several ways to approach this problem. Note, for example, that the second and third terms can be combined arithmetically as

$$2\ln 4 - \ln 4 = \ln 4. \quad \leftarrow \text{analogous to } 2x - x = x$$

Alternatively, we can use law (*iii*) followed by law (*ii*) to combine these terms:

$$\begin{aligned} 2\ln 4 - \ln 4 &= \ln 4^2 - \ln 4 \\ &= \ln 16 - \ln 4 \\ &= \ln \tfrac{16}{4} \\ &= \ln 4. \end{aligned}$$

Hence,
$$\begin{aligned} \tfrac{1}{2}\ln 36 + 2\ln 4 - \ln 4 &= \ln(36)^{1/2} + \ln 4 \quad \leftarrow \text{by (\emph{iii}) of Theorem 5.2.1} \\ &= \ln 6 + \ln 4 \\ &= \ln 24. \quad \leftarrow \text{by (\emph{i}) of Theorem 5.2.1} \end{aligned}$$

EXAMPLE 3   Rewriting Logarithmic Expressions

Use the laws of logarithms to rewrite each expression and evaluate.

(a) $\ln\sqrt{e}$ **(b)** $\ln 5e$ **(c)** $\ln\dfrac{1}{e}$

Solution (a) Since $\sqrt{e} = e^{1/2}$ we have from (*iii*) of the laws of logarithms:

$$\ln\sqrt{e} = \ln e^{1/2} = \tfrac{1}{2}\ln e = \tfrac{1}{2}. \quad \leftarrow \text{from (7), } \ln e = 1$$

(b) From (*i*) of the laws of logarithms and a calculator:

$$\ln 5e = \ln 5 + \ln e = \ln 5 + 1 \approx 2.6094.$$

(c) From (*ii*) of the laws of logarithms:

$$\ln\frac{1}{e} = \ln 1 - \ln e = 0 - 1 = -1. \quad \leftarrow \text{from (6) and (7)}$$

Note that (*iii*) of the laws of logarithms can also be used here:

$$\ln\frac{1}{e} = \ln e^{-1} = (-1)\ln e = -1. \quad \leftarrow \ln e = 1$$

EXAMPLE 4 Value of a Logarithm

If $\log_b 2 = 0.4307$ and $\log_b 3 = 0.6826$, then find $\log_b \sqrt[3]{18}$.

Solution We begin by rewriting $\sqrt[3]{18}$ as $(18)^{1/3}$. Then by the laws of logarithms

$$\begin{aligned}
\log_b(18)^{1/3} &= \tfrac{1}{3}\log_b 18 &&\leftarrow \text{by (\textit{iii}) of Theorem 5.2.1}\\
&= \tfrac{1}{3}\log_b(2 \cdot 3^2)\\
&= \tfrac{1}{3}[\log_b 2 + \log_b 3^2] &&\leftarrow \text{by (\textit{i}) of Theorem 5.2.1}\\
&= \tfrac{1}{3}[\log_b 2 + 2\log_b 3] &&\leftarrow \text{by (\textit{iii}) of Theorem 5.2.1}\\
&= \tfrac{1}{3}[0.4307 + 2(0.6826)]\\
&= 0.5986.
\end{aligned}$$

NOTES FROM THE CLASSROOM

(*i*) Students often struggle with the concept of a *logarithm*. It may help if you repeat to yourself a few dozen times, "A logarithm is an exponent." It may also help if you begin reading a statement such as $3 = \log_{10} 1000$ as "3 is the exponent of 10 that. . . ."

(*ii*) Be *very* careful applying the laws of logarithms. The logarithm does *not* distribute over addition,

$$\log_b(M + N) \neq \log_b M + \log_b N.$$

In other words, the exponent of a sum is not the sum of the exponents.

Also,

$$\frac{\log_b M}{\log_b N} \neq \log_b M - \log_b N.$$

In general, there is no way that we can rewrite either

$$\log_b(M + N) \quad \text{or} \quad \frac{\log_b M}{\log_b N}.$$

(*iii*) In calculus, the first step in a procedure known as *logarithmic differentiation* requires the student to take the natural logarithm of both sides of a complicated function such as $y = \dfrac{x^{10}\sqrt{x^2 + 5}}{\sqrt[3]{8x^3 + 2}}$. The idea is to use the laws of logarithms to transform powers into constant multiples, products into sums, and quotients into differences. See Problems 61–64 in Exercises 5.2.

(*iv*) You may see different notations for the natural exponential function and for the natural logarithm. For example, on some calculators you may see $y = \exp x$ instead of $y = e^x$. In the computer algebra system *Mathematica* the natural exponential function is written Exp[*x*] and the natural logarithm is written Log[*x*].

5.2 Exercises
Answers to selected odd-numbered problems begin on page ANS-15.

In Problems 1–6, rewrite the given exponential expression as an equivalent logarithmic expression.

1. $4^{-1/2} = \frac{1}{2}$
2. $9^0 = 1$
3. $10^4 = 10{,}000$
4. $10^{0.3010} = 2$
5. $t^{-s} = v$
6. $(a + b)^2 = a^2 + 2ab + b^2$

In Problems 7–12, rewrite the given logarithmic expression as an equivalent exponential expression.

7. $\log_2 128 = 7$
8. $\log_5 \frac{1}{25} = -2$
9. $\log_{\sqrt{3}} 81 = 8$
10. $\log_{16} 2 = \frac{1}{4}$
11. $\log_b u = v$
12. $\log_b b^2 = 2$

In Problems 13–18, find the exact value of the given logarithm.

13. $\log_{10}(0.0000001)$
14. $\log_4 64$
15. $\log_2(2^2 + 2^2)$
16. $\log_9 \frac{1}{3}$
17. $\ln e^e$
18. $\ln(e^4 e^9)$

In Problems 19–22, find the exact value of the given expression.

19. $10^{\log_{10} 6^2}$
20. $25^{\log_5 8}$
21. $e^{-\ln 7}$
22. $e^{\frac{1}{2}\ln \pi}$

In Problems 23 and 24, find a logarithmic function $f(x) = \log_b x$ such that the graph of f passes through the given point.

23. $(49, 2)$
24. $\left(4, \frac{1}{3}\right)$

In Problems 25–32, find the domain of the given function f. Find the x-intercept and the vertical asymptote of the graph. Use transformations to graph the given function f.

25. $f(x) = -\log_2 x$
26. $f(x) = -\log_2(x + 1)$
27. $f(x) = \log_2(-x)$
28. $f(x) = \log_2(3 - x)$
29. $f(x) = 3 - \log_2(x + 3)$
30. $f(x) = 1 - 2\log_4(x - 4)$
31. $f(x) = -1 + \ln x$
32. $f(x) = 1 + \ln(x - 2)$

In Problems 33 and 34, use a graph to solve the given inequality.
33. $\ln(x + 1) < 0$
34. $\log_{10}(x + 3) > 1$

35. Show that $f(x) = \ln|x|$ is an even function. Rewrite f as a piecewise-defined function and sketch its graph. Find the x-intercepts and the vertical asymptote of the graph.
36. Use the graph obtained in Problem 35 to sketch the graph of $y = \ln|x - 2|$. Find the x-intercept and the vertical asymptote of the graph.

In Problems 37 and 38, sketch the graph of the given function f.

37. $f(x) = |\ln x|$
38. $f(x) = |\ln(x + 1)|$

In Problems 39–42, find the domain of the given function f.

39. $f(x) = \ln(2x - 3)$
40. $f(x) = \ln(3 - x)$
41. $f(x) = \ln(9 - x^2)$
42. $f(x) = \ln(x^2 - 2x)$

In Problems 43–48, use the laws of logarithms to rewrite the given expression as one logarithm.

43. $\log_{10} 2 + 2\log_{10} 5$
44. $\frac{1}{2}\log_5 49 - \frac{1}{3}\log_5 8 + 13\log_5 1$
45. $\ln(x^4 - 4) - \ln(x^2 + 2)$
46. $\ln\left(\dfrac{x}{y}\right) - 2\ln x^3 - 4\ln y$
47. $\ln 5 + \ln 5^2 + \ln 5^3 - \ln 5^6$
48. $5\ln 2 + 2\ln 3 - 3\ln 4$

In Problems 49–60, use $\log_b 4 = 0.6021$ and $\log_b 5 = 0.6990$ to evaluate the given logarithm. Round your answer to four decimal places.

49. $\log_b 2$
50. $\log_b 20$
51. $\log_b 64$
52. $\log_b 625$
53. $\log_b \sqrt{5}$
54. $\log_b \frac{5}{4}$
55. $\log_b \sqrt[3]{4}$
56. $\log_b 80$
57. $\log_b 0.8$
58. $\log_b 3.2$
59. $\log_4 b$
60. $\log_5 5b$

In Problems 61–64, use the laws of logarithms so that $\ln y$ contains no products, quotients, or powers.

61. $y = \dfrac{x^{10}\sqrt{x^2 + 5}}{\sqrt[3]{8x^3 + 2}}$
62. $y = \sqrt{\dfrac{(2x + 1)(3x + 2)}{4x + 3}}$
63. $y = \dfrac{(x^3 - 3)^5(x^4 + 3x^2 + 1)^8}{\sqrt{x}(7x + 5)^9}$
64. $y = 64x^6\sqrt{x + 1}\,\sqrt[3]{x^2 + 2}$

For Discussion

65. In science it is sometimes useful to display data using logarithmic coordinates. Which of the following equations determines the graph shown in FIGURE 5.2.4?
  (i) $y = 2x + 1$
  (ii) $y = e + x^2$
  (iii) $y = ex^2$
  (iv) $x^2 y = e$

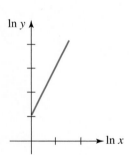

FIGURE 5.2.4 Graph for Problem 65

66. (a) Use a graphing utility to obtain the graph of the function $f(x) = \ln(x + \sqrt{x^2 + 1})$.
  (b) Show that f is an odd function, that is, $f(-x) = -f(x)$.
67. If $a > 0$ and $b > 0$, $a \neq b$, then $\log_a x$ is a constant multiple of $\log_b x$. That is, $\log_a x = k\log_b x$. Find k.
68. Show that $(\log_{10} e)(\log_e 10) = 1$. Can you generalize this result?
69. Discuss: How can the graphs of the given function be obtained from the graph of $f(x) = \ln x$ by means of a rigid transformation (a shift or a reflection)?
  (a) $y = \ln 5x$
  (b) $y = \ln \dfrac{x}{4}$
  (c) $y = \ln x^{-1}$
  (d) $y = \ln(-x)$

70. Find the vertical asymptotes for the graph of $f(x) = \ln\left(\dfrac{x - 3}{x}\right)$. Sketch the graph of f. Do not use a calculator.
71. Using correct mathematical notation, rewrite the statement:

  c is the exponent of 5 that gives the number N,

  in two different, but equivalent, ways.
72. Find the zeros of the function $f(x) = 5 - \log_2|-x + 4|$. Check your answers.

5.3 Exponential and Logarithmic Equations

Introduction Since exponential and logarithmic functions appear in the context of many different applications, we are often called upon to solve equations that involve these functions. While we postpone applications until Section 5.4, we examine in the present section some of the ways that can be used to solve a variety of exponential and logarithmic equations.

Solving Equations Here is a brief list of equation-solving strategies.

SOLVING EXPONENTIAL AND LOGARITHMIC EQUATIONS

(*i*) Rewrite an exponential expression as a logarithmic expression.
(*ii*) Rewrite a logarithmic expression as an exponential expression.
(*iii*) Use the one-to-one properties of b^x and $\log_b x$.
(*iv*) For equations for the form $a^{x_1} = b^{x_2}$, where $a \neq b$, take the natural logarithm of both sides of the equality and simplify using (*iii*) of the laws of logarithms given in Section 5.2.

Of course, this list is not comprehensive and does not reflect the fact that in solving equations involving exponential and logarithmic functions we may also have to employ standard algebraic procedures such as *factoring* and using the *quadratic formula*.

In the first two examples we use the equivalence

$$y = \log_b x \quad \text{if and only if} \quad x = b^y \tag{1}$$

to toggle between logarithmic and exponential expressions.

EXAMPLE 1   Rewriting an Exponential Expression

Solve $e^{10k} = 7$ for k.

Solution We use (1), with $b = e$, to rewrite the given exponential expression as a logarithmic expression:

$$e^{10k} = 7 \quad \text{means} \quad 10k = \ln 7.$$

Therefore, with the aid of a calculator

$$k = \frac{1}{10}\ln 7 \approx 0.1946.$$

EXAMPLE 2   Rewriting a Logarithmic Expression

Solve $\log_2 x = 5$ for x.

Solution We use (1), with $b = 2$, to rewrite the logarithmic expression in its equivalent exponential form:

$$x = 2^5 = 32.$$

□ **One-to-One Properties** Recall from (1) of Section 3.6 that a one-to-one function f possesses the property that if $f(x_1) = f(x_2)$, then necessarily $x_1 = x_2$. We have seen in Sections 5.1 and 5.2 that both the exponential function $y = b^x, b > 0, b \neq 1$, and the logarithmic function $y = \log_b x$ are one-to-one. As a consequence we have:

$$\text{If } b^{x_1} = b^{x_2}, \text{ then } x_1 = x_2. \tag{2}$$
$$\text{If } \log_b x_1 = \log_b x_2, \text{ then } x_1 = x_2. \tag{3}$$

EXAMPLE 3 Using the One-to-One Property (2)

Solve $2^{x-3} = 8^{x+1}$ for x.

Solution Observe on the right-hand side of the given equality that 8 can be written as a power of 2, that is, $8 = 2^3$. Furthermore, by (iv) of the laws of exponents given in Theorem 5.1.1,

$$8^{x+1} = (2^{\overset{\downarrow}{3}})^{\overset{\downarrow}{x+1}} = 2^{3x+3}.$$
$$\quad\quad\quad\text{multiply exponents}$$

Thus, the equation is the same as

$$2^{x-3} = 2^{3x+3}.$$

From the one-to-one property (2) it follows that the exponents are equal, that is, $x - 3 = 3x + 3$. Solving for x then gives $2x = -6$ or $x = -3$. You are encouraged to check this answer by substituting -3 for x in the original equation. ≡

EXAMPLE 4 Using the One-to-One Property (2)

Solve $7^{2(x+1)} = 343$ for x.

Solution By noting that $343 = 7^3$, we have the same base on both sides of the equality:

$$7^{2(x+1)} = 7^3.$$

Thus by (2) we can equate exponents and solve for x:

$$2(x + 1) = 3$$
$$2x + 2 = 3$$
$$2x = 1$$
$$x = \frac{1}{2}.$$
≡

EXAMPLE 5 Using the One-to-One Property (3)

Solve $\ln 2 + \ln(4x - 1) = \ln(2x + 5)$ for x.

Solution By (i) of the laws of logarithms in Theorem 5.2.1, the left-hand side of the equation can be written

$$\ln 2 + \ln(4x - 1) = \ln 2(4x - 1) = \ln(8x - 2).$$

The original equation is then

$$\ln(8x - 2) = \ln(2x + 5).$$

Since two logarithms with the same base are equal, it follows immediately from the one-to-one property (3) that

$$8x - 2 = 2x + 5 \quad \text{or} \quad 6x = 7 \quad \text{or} \quad x = \frac{6}{7}.$$
≡

5.3 Exponential and Logarithmic Equations

For logarithmic equations, especially of the kind in Example 5, you should get accustomed to checking your answer by substituting it back into the original equation. It is possible for a logarithmic equation to have an **extraneous solution**.

EXAMPLE 6   An Extraneous Solution

Solve $\log_2 x + \log_2(x - 2) = 3$.

Solution We start using again that the sum of logarithms on the left-hand side of the equation is the logarithm of a product:

$$\log_2 x(x - 2) = 3.$$

With $b = 2$ we use (1) to rewrite the last equation in the equivalent exponential form

$$x(x - 2) = 2^3.$$

By ordinary algebra we then have

$$x^2 - 2x = 8$$
$$x^2 - 2x - 8 = 0$$
$$(x - 4)(x + 2) = 0.$$

From the last equation we conclude that either $x = 4$ or $x = -2$. However, we must rule out $x = -2$ as a solution. In other words, the number $x = -2$ is an extraneous solution because, when substituted into the original equation, the very first term, $\log_2(-2)$, is not defined. Thus the only solution of the given equation is $x = 4$.

Check:
$$\log_2 4 + \log_2 2 = \log_2 2^2 + \log_2 2$$
$$= \log_2 2^3 = 3\log_2 2 = 3 \cdot 1 = 3.$$ ≡

When we use the phrase "take the logarithm of both sides of an equality" we are actually using the property that if M and N are two positive numbers such that $M = N$, then $\log_b M = \log_b N$.

EXAMPLE 7   Taking the Natural Logarithm of Both Sides

Solve $e^{2x} = 3^{x-4}$.

Solution Since the bases of the exponential expression on each side of the equality are different, one way to proceed is to take the natural logarithm (the common logarithm could also be used) of both sides. From the equality

$$\ln e^{2x} = \ln 3^{x-4}$$

and (*iii*) of the laws of logarithms in Theorem 5.2.1, we get

$$2x \ln e = (x - 4)\ln 3.$$

Now using $\ln e = 1$ and the distributive law, the last equation becomes

$$2x = x\ln 3 - 4\ln 3.$$

Gathering the terms involving the symbol x to one side of the equality then gives

$$\underbrace{2x - x\ln 3}_{\text{factor } x \text{ out of these terms}} = -4\ln 3 \quad \text{or} \quad (2 - \ln 3)x = -4\ln 3 \quad \text{or} \quad x = \frac{-4\ln 3}{2 - \ln 3}.$$

You are encouraged to verify the calculation that $x \approx -4.8752$.   ≡

EXAMPLE 8 Using the Quadratic Formula

Solve $5^x - 5^{-x} = 2$.

Solution Because $5^{-x} = 1/5^x$, the equation is

$$5^x - \frac{1}{5^x} = 2.$$

Multiplying both sides of the foregoing equation by 5^x then gives

$$(5^x)^2 - 1 = 2(5^x) \quad \text{or} \quad (5^x)^2 - 2(5^x) - 1 = 0.$$

If we let $X = 5^x$, then the last equation can be interpreted as a quadratic equation $X^2 - 2X - 1 = 0$. Using the quadratic formula to solve for X yields

$$X = \frac{2 \pm \sqrt{4+4}}{2} = 1 \pm \sqrt{2} \quad \text{or} \quad 5^x = 1 \pm \sqrt{2}.$$

Because $1 - \sqrt{2}$ is a negative number there are no real solutions of $5^x = 1 - \sqrt{2}$ and so

$$5^x = 1 + \sqrt{2}. \qquad (4)$$

Now by taking the natural logarithm of both sides of the equality we obtain

$$\ln 5^x = \ln(1 + \sqrt{2})$$
$$x \ln 5 = \ln(1 + \sqrt{2})$$
$$x = \frac{\ln(1 + \sqrt{2})}{\ln 5}. \qquad (5)$$

Using the $\boxed{\ln}$ key of a calculator, the division yields $x \approx 0.548$. ≡

☐ **Change of Base** In (4) of Example 8 it follows from (1) that a perfectly valid solution of the equation $5^x - 5^{-x} = 2$ is $x = \log_5(1 + \sqrt{2})$. But from a computational viewpoint (that is, expressing x as a number), the last answer is not desirable since no calculator has a logarithmic function with base 5. But by equating $x = \log_5(1 + \sqrt{2})$ with the result in (5) we have discovered that logarithm with base 5 can be expressed in terms of the natural logarithm:

$$\log_5(1 + \sqrt{2}) = \frac{\ln(1 + \sqrt{2})}{\ln 5}. \qquad (6)$$

The result given in (6) is just a special case of a more general result known as the **change-of-base formula**.

THEOREM 5.3.1 Change-of-Base Formula

If $a \neq 1$, $b \neq 1$, and M are positive numbers, then

$$\log_a M = \frac{\log_b M}{\log_b a}. \qquad (7)$$

PROOF: If we let $y = \log_a M$, then from (1) $a^y = M$. Then

$$\log_b a^y = \log_b M$$
$$y \log_b a = \log_b M \quad \leftarrow \text{by (iii) of Theorem 5.2.1}$$
$$y = \frac{\log_b M}{\log_b a} \quad \leftarrow \text{by assumption } y = \log_a M$$
$$\log_a M = \frac{\log_b M}{\log_b a}. \qquad ≡$$

5.3 Exponential and Logarithmic Equations

In order to obtain the numerical value of a logarithm using a calculator, we usually choose $b = 10$ or $b = e$ in (7):

$$\log_a M = \frac{\log_{10} M}{\log_{10} a} \quad \text{or} \quad \log_a M = \frac{\ln M}{\ln a}. \tag{8}$$

EXAMPLE 9 Changing the Base

Find the numerical value of $\log_2 50$.

Solution We can use either formula in (8). If we choose the first formula in (8) with $M = 50$ and $a = 2$, we have

$$\log_2 50 = \frac{\log_{10} 50}{\log_{10} 2}.$$

Using the $\boxed{\log}$ key to calculate the two common logarithms and then dividing yields the approximation

$$\log_2 50 \approx 5.6439.$$

Alternatively, the second formula in (8) gives the same result:

$$\log_2 50 = \frac{\ln 50}{\ln 2} \approx 5.6439. \qquad \equiv$$

We can check the answer in Example 9 on a calculator by using the $\boxed{y^x}$ key. You are urged to verify that $2^{5.6439} \approx 50$.

EXAMPLE 10 Changing the Base

Find the x in the domain of $f(x) = 6^x$ for which $f(x) = 73$.

Solution We must find a solution of the equation $6^x = 73$. One way of proceeding is to rewrite the exponential expression as an equivalent logarithmic expression:

$$x = \log_6 73.$$

Then with the identification $a = 6$ it follows from the second equation in (8) and the aid of a calculator that

$$x = \log_6 73 = \frac{\ln 73}{\ln 6} \approx 2.3946.$$

You should verify that $f(2.3946) = 6^{2.3946} \approx 73$. $\qquad \equiv$

5.3 Exercises
Answers to selected odd-numbered problems begin on page ANS-15.

In Problems 1–20, solve the given exponential equation.

1. $5^{x-2} = 1$
2. $3^x = 27^{x^2}$
3. $10^{-2x} = \dfrac{1}{10{,}000}$
4. $27^x = \dfrac{9^{2x-1}}{3^x}$
5. $e^{5x-2} = 30$
6. $\left(\dfrac{1}{e}\right)^x = e^3$

7. $2^x \cdot 3^x = 36$

8. $\dfrac{4^x}{3^x} = \dfrac{9}{16}$

9. $2^{x^2} = 8^{2x-3}$

10. $\tfrac{1}{4}(10^{-2x}) = 25(10^x)$

11. $5 - 10^{2x} = 0$

12. $7^{-x} = 9$

13. $3^{2(x-1)} = 7^2$

14. $\left(\tfrac{1}{2}\right)^{-x+2} = 8(2^{x-1})^3$

15. $\dfrac{1}{3} = (2^{|x|-2} - 1)^{-1}$

16. $\left(\tfrac{1}{3}\right)^x = 9^{1-2x}$

17. $5^{|x|-1} = 25$

18. $(e^2)^{x^2} - \dfrac{1}{e^{5x+3}} = 0$

19. $4^x = 5^{2x+1}$

20. $3^{x+4} = 2^{x-16}$

In Problems 21–40, solve the given logarithmic equation.

21. $\log_3 5x = \log_3 160$

22. $\ln(10 + x) = \ln(3 + 4x)$

23. $\ln x = \ln 5 + \ln 9$

24. $3 \log_8 x = \log_8 36 + \log_8 12 - \log_8 2$

25. $\log_{10} \dfrac{1}{x^2} = 2$

26. $\log_3 \sqrt{x^2 + 17} = 2$

27. $\log_2(\log_3 x) = 2$

28. $\log_5 |1 - x| = 1$

29. $\log_3 81^x - \log_3 3^{2x} = 3$

30. $\dfrac{\log_2 8^x}{\log_2 \tfrac{1}{4}} = \dfrac{1}{2}$

31. $\log_{10} x = 1 + \log_{10} \sqrt{x}$

32. $\log_2(x - 3) - \log_2(2x + 1) = -\log_2 4$

33. $\log_2 x + \log_2(10 - x) = 4$

34. $\log_8 x + \log_8 x^2 = 1$

35. $\log_6 2x - \log_6(x + 1) = 0$

36. $\log_{10} 54 - \log_{10} 2 = 2 \log_{10} x - \log_{10} \sqrt{x}$

37. $\log_9 \sqrt{10x + 5} - \dfrac{1}{2} = \log_9 \sqrt{x + 1}$

38. $\log_{10} x^2 + \log_{10} x^3 + \log_{10} x^4 - \log_{10} x^5 = \log_{10} 16$

39. $\ln 3 + \ln(2x - 1) = \ln 4 + \ln(x + 1)$

40. $\ln(x + 3) + \ln(x - 4) - \ln x = \ln 3$

In Problems 41–50, either use factoring or the quadratic formula to solve the given equation.

41. $(5^x)^2 - 26(5^x) + 25 = 0$

42. $64^x - 10(8^x) + 16 = 0$

43. $\log_4 x^2 = (\log_4 x)^2$

44. $(\log_{10} x)^2 + \log_{10} x = 2$

45. $(5^x)^2 - 2(5^x) - 1 = 0$

46. $2^{2x} - 12(2^x) + 35 = 0$

47. $(\ln x)^2 + \ln x = 2$

48. $(\log_{10} 2x)^2 = \log_{10}(2x)^2$

49. $2^x + 2^{-x} = 2$

50. $10^{2x} - 103(10^x) + 300 = 0$

In Problems 51–56, find the x-intercepts of the graph of the given function.

51. $f(x) = e^{x+4} - e$

52. $f(x) = 1 - \tfrac{1}{5}(0.1)^x$

53. $f(x) = 4^{x-1} - 3$

54. $f(x) = -3^{2x} + 5$

55. $f(x) = x^3 8^x + 5x^2 8^x + 6x 8^x$

56. $f(x) = \dfrac{2^x - 6 + 2^{3-x}}{x + 2}$

In Problems 57–62, graph the given functions. Determine the approximate x-coordinates of the points of intersection of their graphs.

57. $f(x) = 4e^x,\ \ g(x) = 3^{-x}$

58. $f(x) = 2^x,\ \ g(x) = 3 - 2^x$

59. $f(x) = 3^{x^2},\ \ g(x) = 2(3^x)$

60. $f(x) = \dfrac{1}{3} \cdot 2^{x^2},\ \ g(x) = 2^{x^2} - 1$

61. $f(x) = \log_{10} \dfrac{10}{x},\ \ g(x) = \log_{10} x$

62. $f(x) = \log_{10} \dfrac{x}{2},\ \ g(x) = \log_2 x$

In Problems 63–66, solve the given equation.

63. $x^{\ln x} = e^9$
64. $x^{\log_{10} x} = \dfrac{1000}{x^2}$
65. $\log_x 81 = 2$
66. $\log_5 125^x = -2$

In Problems 67 and 68, use the natural logarithm to find x in the domain of the given function for which f takes on the indicated value.

67. $f(x) = 6^x$; $f(x) = 51$
68. $f(x) = \left(\tfrac{1}{2}\right)^x$; $f(x) = 7$

For Discussion

69. Discuss: How would you find the x-intercepts of the graph of the function $f(x) = e^{2x} - 5e^x + 4$?

70. A Curiosity The logarithm developed by John Napier was actually

$$10^7 \log_{1/e}\left(\frac{x}{10^7}\right).$$

Express this logarithm in terms of the natural logarithm.

5.4 Exponential and Logarithmic Models

≡ **Introduction** In this section we consider some **mathematical models** utilizing exponential or logarithmic functions. Roughly speaking, a mathematical model is a mathematical description of something that we will call a *system*. To construct a mathematical model we start with a set of reasonable assumptions about the system that we are trying to describe. These assumptions include any empirical laws that are applicable to the system. The end result could be a description as simple as a single function.

☐ **Exponential Models** In the physical sciences, the exponential expression Ce^{kt}, where C and k are constants, frequently appears in mathematical models of systems that change with time t. As a consequence, mathematical models are often used to predict a future state of a system. For example, extremely complicated mathematical models are used to predict the weather over various regions of the country for, say, the next week.

☐ **Population Growth** In one model of a growing population, it is assumed that the *rate* of growth of the population is proportional to the *number present* at time t. If $P(t)$ denotes the population or number present at time t, then with the aid of calculus it can be shown that this assumption gives rise to

$$P(t) = P_0 e^{kt}, \quad k > 0, \tag{1}$$

where t is time, and P_0 and k are constants. The function (1) is used to describe the growth of populations of bacteria, small animals, and, in some rare circumstances, humans. Setting $t = 0$ gives $P(0) = P_0$, and so P_0 is called the **initial population**. The constant $k > 0$ is called the **growth constant** or **growth rate**. Since e^{kt}, $k > 0$, is an increasing function on the interval $[0, \infty)$, the model in (1) describes uninhibited growth.

EXAMPLE 1 Bacterial Growth

It is known that the doubling time* of *E. Coli* bacteria, which reside in the large intestine (colon) of healthy people, is just 20 minutes. Use the exponential growth model (1) to find the number of *E. Coli* bacteria in a culture after 6 hours.

Solution Let us use hours as our unit of time, so that 20 min $= \frac{1}{3}$ h. Because the initial number of *E. Coli* in the culture is not specified, we will simply denote the initial size of the culture as P_0. Now using (1), a function interpretation of the first sentence in this example is $P(\frac{1}{3}) = 2P_0$. This means $P_0 e^{k/3} = 2P_0$ or $e^{k/3} = 2$. Solving this last equation for k gives the growth constant

$$\frac{k}{3} = \ln 2 \quad \text{or} \quad k = 3\ln 2 \approx 2.0794.$$

E. Coli bacteria

◂ When working problems such as this, be sure to store the value of k in the memory of your calculator.

A model for the size of the culture after t hours is then $P(t) = P_0 e^{2.0794t}$. Setting $t = 6$ gives $P(6) = P_0 e^{2.0794(6)} \approx 262{,}144 P_0$. Put another way, if the culture consists of only *one* bacterium at $t = 0$, then (with $P_0 = 1$) the model predicts that there will be 262,144 cells 6 hours later. ≡

In the early nineteenth century the English clergyman and economist Thomas R. Malthus used the growth model (1) to predict the world population. For specific values of P_0 and k, the function values $P(t)$ were actually reasonable approximations to the world population for a period of time during the nineteenth century. Since $P(t)$ is an increasing function, Malthus predicted that the future population growth would surpass the world's ability to produce food. As a consequence he also predicted wars and worldwide famine. More a doomsayer than a seer, Malthus failed to foresee that the food supply would keep pace with the increased population through simultaneous advances in science and technology.

In 1840, a more realistic model for predicting human populations in small countries was advanced by the Belgian mathematician/biologist **P. F. Verhulst** (1804–1849). The so-called **logistic function**

$$P(t) = \frac{K}{1 + ce^{rt}}, \quad r < 0, \tag{2}$$

Thomas R. Malthus (1776-1834)

where K, c, and r are constants, has over the years proved to be an accurate growth model for populations of protozoa, bacteria, fruit flies, water fleas, and animals confined to limited spaces. In contrast to uninhibited growth of the Malthusian model (1), (2) exhibits bounded growth. More specifically, the population predicted by (2) will not increase beyond the number K, called the **carrying capacity** of the system. For $r < 0$, $e^{rt} \to 0$ and $P(t) \to K$ as $t \to \infty$. You are asked to graph a special case of (2) in Problem 7 in Exercises 5.4.

☐ **Radioactive Decay** Element 88, better known as **radium**, was discovered by Pierre and Marie Curie in 1898. Radium is a radioactive element, which means that a radium atom spontaneously **decays**, or disintegrates, by emitting radiation in the form of alpha particles, beta particles, and gamma rays. When an atom disintegrates in this manner, its nucleus is transmuted into a nucleus of another element. For example, the nucleus of an atom of the most stable isotope of radium, Ra-226, is transmuted into the nucleus of a radon atom Rn-222. Radon is a heavy, odorless, colorless, and highly dangerous radioactive gas that usually originates in the ground. Because it can penetrate a

Pierre and Marie Curie

---

*In biology the doubling time is sometimes referred to as the **generation time**.

sealed concrete floor, radon frequently accumulates in the basements of some new and highly insulated homes. Some medical organizations have claimed that after cigarette smoking, exposure to radon gas is the second leading cause of lung cancer.

If it is assumed that the rate of decay of a radioactive substance is proportional to the amount remaining or present at time t, then we arrive at basically the same model as in (1). The important difference is that $k < 0$. If $A(t)$ represents the amount of the decaying substance that remains at time t, then

$$A(t) = A_0 e^{kt}, \quad k < 0, \tag{3}$$

where A_0 is the initial amount of the substance present, that is, $A(0) = A_0$. The constant $k < 0$ in (3) is called the **decay constant** or **decay rate**.

EXAMPLE 2   Decay of Radium

Suppose there are 20 grams of radium on hand initially. After t years the amount remaining is modeled by the function $A(t) = 20e^{-0.000418t}$. Find the amount of radium remaining after 100 years. What percent of the original 20 grams has decayed after 100 years?

Solution Using a calculator, we find that after 100 years there remains

$$A(100) = 20e^{-0.000418(100)} \approx 19.18 \text{ g}.$$

Thus, only

$$\frac{20 - 19.18}{20} \times 100\% \approx 4.1\%$$

of the initial 20 grams has decayed.

FIGURE 5.4.1 Time T is the half-life

□ **Half-Life** The half-life of a radioactive substance is the time T it takes for one-half of a given amount of that element to disintegrate and change into a new element. See FIGURE 5.4.1. Half-life is a measure of the stability of an element; that is, the shorter the half-life, the more unstable the element. For example, the half-life of the highly radioactive strontium 90, Sr-90, produced in nuclear explosions, is 29 days, whereas the half-life of the uranium isotope U-238 is 4,560,000 years. The half-life of californium, Cf-244, first discovered in 1950, is only 45 minutes. Polonium, Po-213, has a half-life of 0.000001 second.

EXAMPLE 3   Half-Life of Radium

Use the exponential model in Example 2 to determine the half-life of radium.

Solution If $A(t) = 20e^{-0.000418t}$, then we must find the time T for which

$$A(T) = \tfrac{1}{2}(20) = 10.$$

From $20e^{-0.000418T} = 10$ we get $e^{-0.000418T} = \tfrac{1}{2}$. By rewriting the last expression in the logarithmic form $-0.000418\,T = \ln \tfrac{1}{2}$ we can solve for T:

$$T = \frac{\ln \tfrac{1}{2}}{-0.000418} \approx 1660 \text{ years}.$$

A careful reading of Example 3 reveals that the initial amount present plays no part in the actual calculation of the half-life. Since the solution of $A(T) = A_0 e^{-0.000418T} = \tfrac{1}{2} A_0$

leads to $e^{-0.000418T} = \frac{1}{2}$, we see that T is independent of A_0. In other words, the half-life of 1 gram, 20 grams, or 10,000 grams of radium is the same. It takes about 1660 years for one-half of *any* given quantity of radium to transmute into radon.

Medications also have half-lives. In this case, the half-life of a drug is the time T that it takes for the body to eliminate, by metabolism or excretion, one-half of the amount of the drug taken. For example, the most popular NSAIDs (non-steroidal anti-inflammatory drugs such as aspirin and ibuprofen) taken for the relief of continuing pain, have relatively short half-lives of a few hours and as a consequence must be taken several times a day. The NSAID naproxen has a longer half-life and is usually taken once every 12 hours. See Problem 31 in Exercises 5.4.

Ibuprofen is an NSAID

☐ **Carbon Dating** The approximate age of fossils of once-living matter can sometimes be determined by a method known as **carbon dating**. The radioactive isotope of carbon, carbon-14 or C-14, is formed presumably at a constant rate in the atmosphere by the interaction of cosmic rays on nitrogen-14. The carbon-dating method, invented by the chemist Willard Libby around 1950, is based on the fact that a plant or an animal absorbs C-14 through the process of breathing and eating, and ceases to absorb C-14 when it dies. As the next example shows, the carbon-dating procedure is based on the knowledge that the half-life of C-14 is about 5730 years. Carbon-14 decays back to the original nitrogen-14.

Willard Libby (1908–1980)

Libby won the 1960 Nobel Prize in chemistry for his work. Libby's method has been used to date wooden furniture found in Egyptian tombs, the Dead Sea scrolls written on papyrus and animal skin, the famous linen Shroud of Turin, and a recently discovered copy of the Gnostic Gospel of Judas written on papyrus.

The Psalms scroll

EXAMPLE 4 Carbon Dating a Fossil

A fossilized bone is found to contain $\frac{1}{1000}$ of the initial amount of C-14 that the organism contained while it was alive. Determine the approximate age of the fossil.

Solution If A_0 denotes an initial amount A_0, measured in grams, of C-14 in the organism, then t years after its death there are $A(t) = A_0 e^{kt}$ grams remaining. When $t = 5730$, $A(5730) = \frac{1}{2} A_0$, and so $\frac{1}{2} A_0 = A_0 e^{5730k}$. Solving this last equation for the decay constant k gives

$$e^{5730k} = \frac{1}{2} \quad \text{and so} \quad k = \frac{\ln \frac{1}{2}}{5730} \approx -0.00012097.$$

Hence a model for the amount of C-14 remaining is $A(t) = A_0 e^{-0.00012097t}$. Using this model, we now solve $A(t) = \frac{1}{1000} A_0$ for t:

$$A_0 e^{-0.00012097t} = \frac{1}{1000} A_0 \quad \text{implies} \quad t = \frac{\ln \frac{1}{1000}}{-0.00012097} \approx 57{,}100 \text{ years.} \quad \equiv$$

The age determined in the last example is actually beyond the border of accuracy for the carbon-14-dating method. After 9 half-lives of the isotope, or about 52,000 years, about 99.7% of carbon-14 has decayed making its measurement in a fossil nearly impossible.

☐ **Newton's Law of Cooling/Warming** Suppose an object or body is placed in a medium (air, water, etc.) that is held at constant temperature T_m, called the **ambient temperature**. If the initial temperature T_0 of the body or object at the moment it is

5.4 Exponential and Logarithmic Models

▶ We take this moment to correspond to the time $t = 0$.

placed into the medium is greater than the ambient temperature T_m, then the body will cool. On the other hand, if T_0 is less than T_m, then it will warm up. For example, in an office kept at, say, 70°F, a steaming cup of coffee will cool off, whereas a glass of ice water will warm up. The usual cooling/warming assumption is that the rate at which an object cools/warms is proportional to the difference $T(t) - T_m$, where $T(t)$ represents the temperature of the object at time t. In either case, cooling or warming, this assumption leads to $T(t) - T_m = (T_0 - T_m)e^{kt}$, where k is a negative constant. Observe that since $e^{kt} \to 0$ for $k < 0$, the last expression is consistent with one's intuitive expectation that $T(t) - T_m \to 0$, or equivalently $T(t) \to T_m$, as $t \to \infty$ (the coffee cools to room temperature; the ice water warms to room temperature). Solving for $T(t)$ we obtain a function for the temperature of the object,

$$T(t) = T_m + (T_0 - T_m)e^{kt}, \quad k < 0. \tag{4}$$

The mathematical model in (4), named after its discoverer, is called **Newton's law of cooling/warming**. Note that $T(0) = T_0$.

EXAMPLE 5   Cooling of a Cake

A cake is removed from an oven where the temperature was 350°F into a kitchen where the temperature is 75°F. One minute later the temperature of the cake is measured to be 300°F. Assume that the temperature of the cake in the kitchen is given by (4).

(a) What is the temperature of the cake after 6 minutes?
(b) At what time is the temperature of the cake 80°F?
(c) Graph $T(t)$.

Cake will cool off to room temperature

Solution (a) When the cake is removed from the oven its temperature is also 350°F, that is, $T_0 = 350$. The ambient temperature is the temperature of the kitchen $T_m = 75$. Thus (4) becomes $T(t) = 75 + 275e^{kt}$. The measurement that $T(1) = 300$ is the condition that determines the value of k. From $T(1) = 75 + 275e^k = 300$ we find

$$e^k = \frac{225}{275} = \frac{9}{11} \quad \text{or} \quad k = \ln\frac{9}{11} \approx -0.2007.$$

The mathematical model $T(t) = 75 + 275e^{-0.2007t}$ then predicts that the temperature of the cake in 6 minutes after it is removed from the oven will be

$$T(6) = 75 + 275e^{-0.2007(6)} \approx 157.5°. \tag{5}$$

(b) To determine when the temperature of the cake will be 80°F, we solve the equation $T(t) = 80$ for t. Rewriting $T(t) = 75 + 275e^{-0.2007t} = 80$ as

$$e^{-0.2007t} = \frac{5}{275} = \frac{1}{55} \quad \text{we find} \quad t = \frac{\ln\frac{1}{55}}{-0.2007} \approx 20 \text{ min}.$$

FIGURE 5.4.2 Graph of $T(t)$ in Example 5

(c) With the aid of a graphing utility we obtain the graph of $T(t)$ shown in blue in FIGURE 5.4.2. Since $T(t) = 75 + 275e^{-0.2007t} \to 75$ as $t \to \infty$, $T = 75$, shown in red in Figure 5.4.2, is a horizontal asymptote for the graph of $T(t) = 75 + 275e^{-0.2007t}$. ≡

☐ **Compound Interest** Investments such as savings accounts pay an annual rate of interest that can be compounded annually, quarterly, monthly, weekly, daily, and so on. In general, if a principal of P dollars is invested at an annual rate r of interest that is compounded n times a year, then the amount S accrued at the end of t years is given by

$$S = P\left(1 + \frac{r}{n}\right)^{nt}. \tag{6}$$

S is called the **future value** of the principal P. If the number n is increased without bound, then interest is said to be **compounded continuously**. To find the future value of P in this case, we let $m = n/r$. Then $n = mr$ and

$$\left(1 + \frac{r}{n}\right)^{nt} = \left(1 + \frac{1}{m}\right)^{mrt} = \left[\left(1 + \frac{1}{m}\right)^m\right]^{rt}.$$

Since $n \to \infty$ implies that $m \to \infty$, we see from page 283 of Section 5.1 that $(1 + 1/m)^m \to e$. The right-hand side of (6) becomes

$$P\left[\left(1 + \frac{1}{m}\right)^m\right]^{rt} \to P[e]^{rt} \quad \text{as} \quad m \to \infty.$$

Thus, if an annual rate r of interest is compounded continuously, the future value S of a principal P in t years is

$$S = Pe^{rt}. \tag{7}$$

EXAMPLE 6   Comparison of Future Values

Suppose that $1000 is deposited in a savings account whose annual rate of interest is 3%. Compare the future value of this principal in 10 years **(a)** if interest is compounded monthly and **(b)** if interest is compounded continuously.

Solution (a) Since there are 12 months in a year, we identify $n = 12$. Furthermore, with $P = 1000$, $r = 0.03$, and $t = 10$, (6) becomes

$$S = 1000\left(1 + \frac{0.03}{12}\right)^{12(10)} = 1000(1.0025)^{120} \approx \$1{,}349.35.$$

(b) From (7),

$$S = 1000e^{(0.03)(10)} = 1000e^{0.3} \approx \$1{,}349.86.$$

Thus over 10 years we have gained only $0.51 by compounding continuously rather than monthly.

☐ **Logarithmic Models** Probably the most famous application of the base 10 logarithm, or common logarithm, is the **Richter scale**. In 1935, the American seismologist Charles F. Richter devised a logarithmic scale for comparing the energies of different earthquakes. The magnitude M of an earthquake is defined by

$$M = \log_{10}\frac{A}{A_0}, \tag{8}$$

where A is the amplitude of the largest seismic wave of the earthquake and A_0 is a reference amplitude that corresponds to the magnitude $M = 0$. The number M is calculated to one decimal place. Earthquakes of magnitude 6 or greater are considered potentially destructive.

EXAMPLE 7   Comparing Intensities

The earthquake on December 26, 2004, off the west coast of Northern Sumatra, which spawned a tsunami causing over 200,000 deaths, was initially classified as a 9.3 on the Richter scale. On March 28, 2005, an aftershock in the same area was classified as an 8.7 on the Richter scale. How many times more intense was the 2004 earthquake?

Charles F. Richter (1900–1985)

Solution From (8) we have

$$9.3 = \log_{10}\left(\frac{A}{A_0}\right)_{2004} \quad \text{and} \quad 8.7 = \log_{10}\left(\frac{A}{A_0}\right)_{2005}.$$

This means, in turn, that

$$\left(\frac{A}{A_0}\right)_{2004} = 10^{9.3} \quad \text{and} \quad \left(\frac{A}{A_0}\right)_{2005} = 10^{8.7}.$$

Now, since $9.3 = 0.6 + 8.7$, it follows from the laws of exponents that

$$\left(\frac{A}{A_0}\right)_{2004} = 10^{9.3} = 10^{0.6} 10^{8.7} = 10^{0.6}\left(\frac{A}{A_0}\right)_{2005} \approx 3.98\left(\frac{A}{A_0}\right)_{2005}.$$

Thus the original earthquake in 2004 was approximately 4 times as intense as the aftershock in 2005. ≡

You can see from Example 7 that if, say, one earthquake is a 6.0 and another is a 4.0 on the Richter scale, then the 6.0 earthquake is $10^2 = 100$ times more intense than the 4.0 earthquake.

☐ **pH of a Solution** In chemistry, the hydrogen potential, or **pH**, of a solution is defined as

$$\text{pH} = -\log_{10}[\text{H}^+], \tag{9}$$

where the symbol $[\text{H}^+]$ denotes the concentration of hydrogen ions in a solution measured in moles per liter. The pH scale was invented in 1909 by the Danish biochemist Søren Sørensen. Solutions are classified according to their pH value as *acidic*, *base*, or *neutral*. A solution with a pH in the range $0 < \text{pH} < 7$ is said to be acid; when $\text{pH} > 7$, the solution is base (or alkaline). In the case when $\text{pH} = 7$, the solution is neutral. Water, if uncontaminated by other solutions or by acid rain, is an example of a neutral solution, whereas undiluted lemon juice is highly acid and has a pH in the range $\text{pH} \le 3$. A solution with $\text{pH} = 6$ is ten times more acidic than a neutral solution. See Problems 47–50 in Exercises 5.4.

As the next example illustrates, pH values are usually calculated to one decimal place.

Søren Sørensen (1868–1939)

EXAMPLE 8 pH of Human Blood

The concentration of hydrogen ions in the blood of a healthy person is found to be $[\text{H}^+] = 3.98 \times 10^{-8}$ moles/liter. Find the pH of blood.

Solution From (9) and the laws of logarithms (Theorem 5.2.1),

$$\begin{aligned}\text{pH} &= -\log_{10}[3.98 \times 10^{-8}] \\ &= -[\log_{10} 3.98 + \log_{10} 10^{-8}] \\ &= -[\log_{10} 3.98 - 8 \log_{10} 10] \quad \leftarrow \log_{10} 10 = 1 \\ &= -[\log_{10} 3.98 - 8].\end{aligned}$$

With the help of the base 10 log key on a calculator, we find that

$$\text{pH} \approx -[0.5999 - 8] \approx 7.4.$$

≡

Human blood in usually a base solution. The pH values of blood usually fall within the rather narrow range $7.2 < \text{pH} < 7.6$. A person with a blood pH outside these limits can suffer illness and even death.

5.4 Exercises
Answers to selected odd-numbered problems begin on page ANS-15.

Population Growth

1. After 2 hours the number of bacteria in a culture is observed to have doubled.
   (a) Find an exponential model (1) for the number of bacteria in the culture at time t.
   (b) Find the number of bacteria present in the culture after 5 hours.
   (c) Find the time that it takes the culture to grow to 20 times its initial size.

2. A model for the number of bacteria in a culture after t hours is given by (1).
   (a) Find the growth constant k if it is known that after 1 hour the colony has expanded to 1.5 times its initial population.
   (b) Find the time that it takes for the culture to quadruple in size.

3. A model for the population in a small community is given by $P(t) = 1500e^{kt}$. If the initial population increases by 25% in 10 years, what will the population be in 20 years?

4. A model for the population in a small community after t years is given by (1).
   (a) If the initial population has doubled in 5 years, how long will it take to triple? To quadruple?
   (b) If the population of the community in part (a) is 10,000 after 3 years, what was the initial population?

5. A model for the number of bacteria in a culture after t hours is given by $P(t) = P_0 e^{kt}$. After 3 hours it is observed that 400 bacteria are present. After 10 hours 2000 bacteria are present. What was the initial number of bacteria?

6. In genetic research a small colony of *drosophila* (small two-winged fruit flies) is grown in a laboratory environment. After 2 days it is observed that the population of flies in the colony has increased to 200. After 5 days the colony has 400 flies.
   (a) Find a model $P(t) = P_0 e^{kt}$ for the population of the fruit-fly colony after t days.
   (b) What will be the population of the colony in 10 days?
   (c) When will the population of the colony be 5000 fruit flies?

7. A student sick with a flu virus returns to an isolated college campus of 2000 students. The number of students infected with the flu t days after the student's return is predicted by the logistic function
$$P(t) = \frac{2000}{1 + 1999e^{-0.8905t}}.$$
   (a) According to this model, how many students will be infected with the flu after 5 days?
   (b) How long will it take for one-half of the student population to become infected?
   (c) How many students does the model predict will become infected after a very long period of time?
   (d) Sketch a graph of $P(t)$.

8. In 1920, Pearl and Reed proposed a logistic model for the population of the United States based on the years 1790, 1850, and 1910. The logistic function they proposed was
$$P(t) = \frac{2930.3009}{0.014854 + e^{-0.0313395t}},$$
where P is measured in thousands and t represents the number of years past 1780.

(a) The model agrees quite well with the census figures between 1790 and 1910. Determine the population figures for 1790, 1850, and 1910.
(b) What does this model predict for the population of the United States after a very long time? How does this prediction compare with the 2000 census population of 281 million?

Radioactive Decay and Half-Life

9. Initially 200 milligrams of a radioactive substance was present. After 6 hours the mass had decreased by 3%. Construct an exponential model $A(t) = A_0 e^{kt}$ for the amount remaining of the decaying substance after t hours. Find the amount remaining after 24 hours.
10. Determine the half-life of the substance in Problem 9.
11. Do this problem without using the exponential model (3). Initially there are 400 grams of a radioactive substance on hand. If the half-life of the substance is 8 hours, give an educated guess of how much remains (approximately) after 17 hours. After 23 hours. After 33 hours.
12. Construct an exponential model $A(t) = A_0 e^{kt}$ for the amount remaining of the decaying substance in Problem 11. Compare the predicted values $A(17)$, $A(23)$, and $A(33)$ with your guesses.
13. Iodine 131, used in nuclear medicine procedures, is radioactive and has a half-life of 8 days. Find the decay constant k for iodine 131. If the amount remaining of an initial sample after t days is given by the exponential model $A(t) = A_0 e^{kt}$, how long will it take for 95% of the sample to decay?
14. The amount remaining of a radioactive substance after t hours is given by $A(t) = 100 e^{kt}$. After 12 hours, the initial amount has decreased by 7%. How much remains after 48 hours? What is the half-life of the substance?
15. The half-life of polonium 210, Po-210, is 140 days. If $A(t) = A_0 e^{kt}$ represents the amount of Po-210 remaining after t days, what is the amount remaining after 80 days? After 300 days?
16. Strontium 90 is a dangerous radioactive substance found in acid rain. As such it can make its way into the food chain by polluting the grass in a pasture on which milk cows graze. The half-life of strontium 90 is 29 years.
    (a) Find an exponential model (3) for the amount remaining after t years.
    (b) Suppose a pasture is found to contain Str-90 that is 3 times a safe level A_0. How long will it be before the pasture can be used again for grazing cows?

Charcoal drawing in Problem 17

Carbon Dating

17. Charcoal drawings were discovered on walls and ceilings in a cave in Lascaux, France. Determine the approximate age of the drawings, if it was found that 86% of C-14 in a piece of charcoal found in the cave had decayed through radioactivity.
18. Analysis on an animal bone fossil at an archeological site reveals that the bone has lost between 90% and 95% of C-14. Give an interval for the possible ages of the bone.
19. The shroud of Turin shows the negative image of the body of a man who appears to have been crucified. It is believed by many to be the burial shroud of Jesus of Nazareth. In 1988 the Vatican granted permission to have the shroud carbon dated. Several independent scientific laboratories analyzed the cloth and the consensus opinion was that the shroud is approximately 660 years old, an age consistent with its historical appearance. This age has been disputed by many scholars. Using this age, determine what percentage of the original amount of C-14 remained in the cloth as of 1988.

Shroud image in Problem 19

20. In 1991 hikers found a preserved body of a man partially frozen in a glacier in the Austrian Alps. Through carbon-dating techniques it was found that the body of Ötzi—the iceman, as he came to be called—contained 53% as much C-14 as found in a living person. What is the approximate date of his death?

The iceman in Problem 20

Newton's Law of Cooling/Warming

21. Suppose a pizza is removed from an oven at 400°F into a kitchen whose temperature is a constant 80°F. Three minutes later the temperature of the pizza is found to be 275°F.
    (a) What is the temperature $T(t)$ of the pizza after 5 minutes?
    (b) Determine the time when the temperature of the pizza is 150°F.
    (c) After a very long period of time, what is the approximate temperature of the pizza?

22. A glass of cold water is removed from a refrigerator whose interior temperature is 39°F into a room maintained at 72°F. One minute later the temperature of the water is 43°F. What is the temperature of the water after 10 minutes? After 25 minutes?

23. A thermometer is brought from the outside, where the air temperature is −20°F, into a room where the air temperature is a constant 70°F. After 1 minute inside the room the thermometer reads 0°F. How long will it take for the thermometer to read 60°F?

24. A thermometer is taken from inside a house to the outside, where the air temperature is 5°F. After 1 minute outside the thermometer reads 59°F, and after 5 minutes it reads 32°F. What is the temperature inside the house?

25. A dead body was found within a closed room of a house where the temperature was a constant 70°F. At the time of discovery, the core temperature of the body was determined to be 85°F. One hour later a second measurement showed that the core temperature of the body was 80°F. Assume that the time of death corresponds to $t = 0$ and that the core temperature at that time was 98.6°F. Determine how many hours elapsed before the body was found.

26. Repeat Problem 25 if evidence indicated that the dead person was running a fever of 102°F at the time of death.

Thermometer in Problem 24

Compound Interest

27. Suppose that 1¢ is deposited in a savings account paying 1% annual interest compounded continuously. How much money will have accrued in the account after 2000 years? What is the future value of 1¢ in 2000 years if the account pays 2% annual interest compounded continuously?

28. Suppose that $100,000 is invested at an annual interest rate of 5%. Use (6) and (7) to compare the future values of that amount in 1 year by completing the following table.

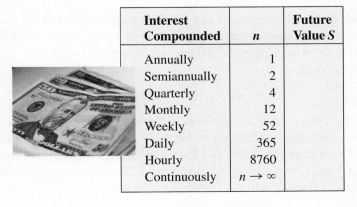

Interest Compounded	n	Future Value S
Annually	1	
Semiannually	2	
Quarterly	4	
Monthly	12	
Weekly	52	
Daily	365	
Hourly	8760	
Continuously	$n \to \infty$	

29. Suppose that $5000 is deposited in a savings account paying 6% annual interest compounded continuously. How much interest will be earned in 8 years?

30. **Present Value** If (7) is solved for P, that is, $P = Se^{-rt}$, we obtain the amount that should be invested now at an annual rate r of interest in order to be worth S dollars after t years. We say that P is the **present value** of the amount S. What is the present value of $100,000 at an annual rate of 3% compounded continuously for 30 years?

Miscellaneous Exponential Models

31. **Effective Half-life** Radioactive substances are removed from living organisms by two processes: natural physical decay and biological metabolism. Each process contributes to an effective half-life E that is defined by

$$1/E = 1/P + 1/B,$$

where P is the physical half-life of the radioactive substance and B is the biological half-life.
  (a) Radioactive iodine, I-131, is used to treat hyperthyroidism (overactive thyroid). It is known that for human thyroids, $P = 8$ days and $B = 24$ days. Find the effective half-life of I-131 in the thyroid.
  (b) Suppose the amount of I-131 in the human thyroid after t days is modeled by $A(t) = A_0 e^{kt}$, $k < 0$. Use the effective half-life found in part (a) to determine the percentage of radioactive iodine remaining in the human thyroid gland 2 weeks after its ingestion.

32. **Newton's Law of Cooling Revisited** The rate at which a body cools also depends on its exposed surface area S. If S is a constant, then a modification of (4) is

$$T(t) = T_m + (T_0 - T_m)e^{kSt}, \quad k < 0.$$

Suppose two cups A and B are filled with coffee at the same time. Initially the temperature of the coffee is 150°F. The exposed surface area of the coffee in cup B is twice the surface area of the coffee in cup A. After 30 min, the temperature of the coffee in cup A is 100°F. If $T_m = 70$°F, what is the temperature of the coffee in cup B after 30 min?

33. **Series Circuit** In a simple series circuit consisting of a constant voltage E, an inductance of L henries, and a resistance of R ohms, it can be shown that the current $I(t)$ is given by

$$I(t) = \frac{E}{R}(1 - e^{-(R/L)t}).$$

Solve for t in terms of the other symbols.

34. **Drug Concentration** Under some conditions the concentration of a drug at time t after injection is given by

$$C(t) = \frac{a}{b} + \left(C_0 - \frac{a}{b}\right)e^{-bt}.$$

Here a and b are positive constants and C_0 is the concentration of the drug at $t = 0$. Determine the steady-state concentration of a drug, that is, the limiting value of $C(t)$ as $t \to \infty$. Determine the time t at which $C(t)$ is one-half the steady-state concentration.

Richter Scale

Marina district in San Francisco, 1989

35. Two of the most devastating earthquakes in the San Francisco Bay area occurred in 1906 along the San Andreas fault and in 1989 in the Santa Cruz Mountains near Loma Prieta peak. The 1906 and 1989 earthquakes measured 8.5 and 7.1 on the Richter scale, respectively. How much greater was the intensity of the 1906 earthquake compared to the 1989 earthquake?

36. How much greater was the intensity of the 2004 Northern Sumatra earthquake (Example 7) compared to the 1964 Alaskan earthquake of magnitude 8.9?
37. If an earthquake has a magnitude 4.2 on the Richter scale, what is the magnitude on the Richter scale of an earthquake that has an intensity 20 times greater? [*Hint*: First solve the equation $10^x = 20$.]
38. Show that the Richter scale defined in (8) of this section can be written

$$M = \frac{\ln A - \ln A_0}{\ln 10}.$$

pH of a Solution

In Problems 39–42, determine the pH of a solution with the given hydrogen-ion concentration $[H^+]$.

39. 10^{-6} 40. 4×10^{-7} 41. 2.8×10^{-8} 42. 5.1×10^{-5}

In Problems 43–46, determine the hydrogen-ion concentration $[H^+]$ of a solution with the given pH.

43. 3.3 44. 7.3 45. 6.6 46. 8.1

In Problems 47–50, determine how many more times acidic the first substance is compared to the second substance.

47. lemon juice: pH = 2.3; vinegar, pH = 3.3
48. battery acid: pH = 1; lye, pH = 13
49. clean rain: pH = 5.6; acidic rain, pH = 3.8
50. NaOH: $[H^+] = 10^{-14}$; HCl, $[H^+] = 1$

Miscellaneous Logarithmic Models

51. **Richter Scale and Energy** Charles Richter working with Beno Gutenberg developed the model

$$M = \tfrac{2}{3}[\log_{10} E - 11.8]$$

that relates the Richter magnitude M of an earthquake and its seismic energy E (measured in ergs). Calculate the seismic energy E of the 2004 Northern Sumatra earthquake where $M = 9.3$.

52. **Intensity Level** The **intensity level** b of a sound measured in decibels (dB) is defined by

$$b = 10 \log_{10} \frac{I}{I_0}, \tag{10}$$

where I is the **intensity of the sound** measured in watts/cm² and $I_0 = 10^{-16}$ watts/cm² is the intensity of the faintest sound that can be heard (0 dB). Use (10) and complete the following table.

Sound	Intensity I (watts/cm²)	Intensity Level b (dB)
Whisper	10^{-14}	
Conversation	10^{-11}	
TV commercials	10^{-10}	
Smoke alarm	10^{-9}	
Jet plane taking off	10^{-7}	
Rock band	10^{-4}	

53. **Threshold of Pain** The threshold of pain is generally taken to be around 140 dB. Find the intensity of sound I corresponding to 140 dB.
54. **Intensity Levels** The intensity of sound I is inversely proportional to the square of the distance d from its source, that is,

$$I = \frac{k}{d^2}, \tag{11}$$

where k is the constant of proportionality. Suppose d_1 and d_2 are distances from a source of sound, and that the corresponding intensity levels of the sounds are b_1 and b_2. Use (11) in (10) to show that b_1 and b_2 are related by

$$b_2 = b_1 + 20 \log_{10} \frac{d_1}{d_2}. \tag{12}$$

55. **Intensity Level** When a plane P_1, flying at an altitude of 1500 ft, passed over a point on the ground its intensity level b_1 was measured as 70 dB. Use (12) to find the intensity level b_2 of a second plane P_2, flying at an altitude of 2600 ft, when it passed over the same point.
56. **Talking Politics** At a distance of 4 ft, the intensity level of an animated political conversation is 60 dB. Use (12) to find the intensity level 14 ft from the conversation.
57. **Pupil of the Eye** An empirical model devised by DeGroot and Gebhard relates the diameter d of the pupil of the eye (measured in millimeters, mm) to the luminance B of light source (measured in millilambert's, mL):

$$\log_{10} d = 0.8558 - 0.000401(8.1 + \log_{10} B)^3.$$

See **FIGURE 5.4.3**.

FIGURE 5.4.3 Pupil diameter in Problem 57

(a) The average luminance of clear sky is approximately $B = 255$ mL. Find the corresponding pupil diameter.
(b) The luminance of the Sun varies from approximately $B = 190{,}000$ mL at sunrise to $B = 51{,}000{,}000$ mL at noon. Find the corresponding pupil diameters.
(c) Find the luminance B corresponding to a pupil diameter of 7 mm.

58. **Body Surface Area** Medical researchers use the empirical mathematical model

$$\log_{10} A = -2.144 + (0.425)\log_{10} m + (0.725)\log_{10} h$$

to estimate body surface area A (measured in square meters), given a person's mass m (in kilograms) and height h (in centimeters).
(a) Estimate the body surface area of a person whose mass is $m = 70$ kg and who is $h = 175$ cm tall.
(b) Determine your mass and height and estimate your own body surface area.

5.5 Hyperbolic Functions

≡ **Introduction** We have already seen in Section 5.4 the usefulness of the exponential function e^x in various mathematical models. As a further application, consider a long rope or a flexible wire, such as a telephone wire hanging only under its own weight between two fixed supports or telephone poles. It can be shown that under certain conditions the hanging wire assumes the shape of the graph of the function

$$f(x) = c\frac{e^{x/c} + e^{-x/c}}{2}. \tag{1}$$

The symbol c stands for a positive constant that depends on the physical characteristics of the wire. Functions such as (1), consisting of certain combinations of e^x and e^{-x}, appear in so many applications that mathematicians have given them names.

☐ **Hyperbolic Functions** In particular, when $c = 1$ in (1), the resulting function $f(x) = \dfrac{e^x + e^{-x}}{2}$ is called the **hyperbolic cosine**.

Telephone wires

DEFINITION 5.5.1 Hyperbolic Functions

For any real number x, the **hyperbolic sine** of x, denoted $\sinh x$ is

$$\sinh x = \frac{e^x - e^{-x}}{2}, \qquad (2)$$

and the **hyperbolic cosine** of x, denoted $\cosh x$, is

$$\cosh x = \frac{e^x + e^{-x}}{2}. \qquad (3)$$

Analogous to the trigonometric functions $\tan x$, $\cot x$, $\sec x$, and $\csc x$ that are defined in terms of $\sin x$ and $\cos x$, there are four additional hyperbolic functions $\tanh x$, $\coth x$, $\text{sech}\, x$, and $\text{csch}\, x$ that are defined in terms of $\sinh x$ and $\cosh x$:

$$\tanh x = \frac{\sinh x}{\cosh x} = \frac{e^x - e^{-x}}{e^x + e^{-x}} \quad \text{and} \quad \coth x = \frac{1}{\tanh x} = \frac{e^x + e^{-x}}{e^x - e^{-x}} \qquad (4)$$

$$\text{sech}\, x = \frac{1}{\cosh x} = \frac{2}{e^x + e^{-x}} \quad \text{and} \quad \text{csch}\, x = \frac{1}{\sinh x} = \frac{2}{e^x - e^{-x}}. \qquad (5)$$

Gateway arch in St. Louis, MO

☐ **Graphs** The graph of the hyperbolic cosine, shown in FIGURE 5.5.1, is called a **catenary**. The word *catenary* derives from the Latin word for a chain, *catena*. The shape of the famous 630-ft tall Gateway arch in St. Louis, Missouri, is an inverted catenary. Compare the shape in Figure 5.5.1 with that in the accompanying photo. The graph of $y = \sinh x$ is given in FIGURE 5.5.2.

FIGURE 5.5.1 Graph of $y = \cosh x$

FIGURE 5.5.2 Graph of $y = \sinh x$

5.5 Hyperbolic Functions

The graphs of the hyperbolic tangent, cotangent, secant, and cosecant are given in **FIGURE 5.5.3**. Observe that $y = 1$ and $y = -1$ are horizontal asymptotes for the graphs of $y = \tanh x$ and $y = \coth x$ and that $x = 0$ is a vertical asymptote for the graphs of $y = \coth x$ and $y = \operatorname{csch} x$.

(a) $y = \tanh x$ (b) $y = \coth x$ (c) $y = \operatorname{sech} x$ (d) $y = \operatorname{csch} x$

FIGURE 5.5.3 Graphs of the hyperbolic tangent (a), cotangent (b), secant (c), and cosecant (d)

☐ **Identities** Although the hyperbolic functions are not periodic, they possess identities that are very similar to trigonometric identities. Analogous to the basic Pythagorean identity of trigonometry $\cos^2 x + \sin^2 x = 1$, for the hyperbolic sine and cosine we have

$$\cosh^2 x - \sinh^2 x = 1. \qquad (6)$$

See Problems 1–6 in Exercises 5.5.

5.5 Exercises
Answers to selected odd-numbered problems begin on page ANS-16.

In Problems 1–6, use the definitions of $\cosh x$ and $\sinh x$ in (2) and (3) to verify the given identity.

1. $\cosh^2 x - \sinh^2 x = 1$
2. $1 - \tanh^2 x = \operatorname{sech}^2 x$
3. $\cosh(-x) = \cosh x$
4. $\sinh(-x) = -\sinh x$
5. $\sinh 2x = 2 \sinh x \cosh x$
6. $\cosh 2x = \cosh^2 x + \sinh^2 x$

7. **(a)** If $\sinh x = -\frac{3}{2}$, use the identity given in Problem 1 to find the value of $\cosh x$.
   (b) Use the result of part (a) to find the numerical values of $\tanh x$, $\coth x$, $\operatorname{sech} x$, and $\operatorname{csch} x$.
8. **(a)** If $\tanh x = \frac{1}{2}$, use the identity given in Problem 2 to find the value of $\operatorname{sech} x$.
   (b) Use the result of part (a) to find the numerical values of $\tanh x$, $\coth x$, $\operatorname{sech} x$, and $\operatorname{csch} x$.
9. As can be seen in Figure 5.5.2, the hyperbolic sine function $y = \sinh x$ is one-to-one. Use (2) in Definition 5.5.1 in the form $e^x - 2y - e^{-x} = 0$ to show that the inverse hyperbolic sine $\sinh^{-1} x$ is given by

$$\sinh^{-1} x = \ln\left(x + \sqrt{x^2 + 1}\right).$$

10. The function $y = \cosh x$ on the restricted domain $[0, \infty)$ is one-to-one. Proceed as in Problem 9 to find the inverse of $y = \cosh x$, $x \geq 0$.

CONCEPTS REVIEW

You should be able to give the meaning of each of the following concepts.

Exponential function:
  base $b > 0, b \neq 1$
  increasing, $b > 1$
  decreasing, $b < 1$
Laws of exponents
Inverse of exponential
  function
The number e
Natural exponential function

Logarithmic function:
  base $b > 0, b \neq 1$
Laws of logarithms
Common logarithm
Natural logarithm
Change-of-base formula
Growth and decay:
  growth constant
  decay constant

Half-life
Carbon dating
Continuous compound interest
pH of a solution
Richter scale
Hyperbolic functions:
  cosine
  sine

CHAPTER 5 Review Exercises

Answers to selected odd-numbered problems begin on page ANS-16.

A. True/False

In Problems 1–14, answer true or false.

1. $y = \ln x$ and $y = e^x$ are inverse functions. _____
2. The point $(b, 1)$ is on the graph of $f(x) = \log_b x$. _____
3. $y = 10^{-x}$ and $y = (0.1)^x$ are the same function. _____
4. If $f(x) = e^{x^2} - 1$, then $f(x) = 1$ when $x = \pm \ln \sqrt{2}$. _____
5. $4^{x/2} = 2^x$ _____
6. $\dfrac{2^{x^2}}{2^x} = 2^x$ _____
7. $2^x + 2^{-x} = (2 + 2^{-1})^x$ _____
8. $2^{3+3x} = 8^{1+x}$ _____
9. $-\ln 2 = \ln\left(\tfrac{1}{2}\right)$ _____
10. $\ln \dfrac{e^a}{e^b} = a - b$ _____
11. $\ln(\ln e) = 1$ _____
12. $\ln \sqrt{43} = \dfrac{\ln 43}{2}$ _____
13. $\ln(e + e) = 1 + \ln 2$ _____
14. $\log_6(36)^{-1} = -2$ _____

B. Fill in the Blanks

In Problems 1–22, fill in the blanks.

1. The graph of $y = 6 - e^{-x}$ has the y-intercept _____ and horizontal asymptote $y =$ _____.
2. The x-intercept of the graph of $y = -10 + 10^{5x}$ is _____.
3. The graph of $y = \ln(x + 4)$ has the x-intercept _____ and vertical asymptote $x =$ _____.
4. The y-intercept of the graph of $y = \log_8(x + 2)$ is _____.
5. $\log_5 2 - \log_5 10 =$ _____
6. $6\ln e + 3\ln \dfrac{1}{e} =$ _____
7. $e^{3\ln 10} =$ _____
8. $10^{\log_{10} 4.89} =$ _____
9. $\log_4(4 \cdot 4^2 \cdot 4^3) =$ _____
10. $\dfrac{\log_5 625}{\log_5 125} =$ _____
11. If $\log_3 N = -2$, then $N =$ _____.
12. If $\log_b 6 = \tfrac{1}{2}$, then $b =$ _____.

13. If $\ln e^3 = y$, then $y = $ _____.
14. If $\ln 3 + \ln(x - 1) = \ln 2 + \ln x$, then $x = $ _____.
15. If $-1 + \ln(x - 3) = 0$, then $x = $ _____.
16. If $\ln(\ln x) = 1$, then $x = $ _____.
17. If $100 - 20e^{-0.15t} = 35$, then to four rounded decimals $t = $ _____.
18. If $3^x = 5$, then $3^{-2x} = $ _____.
19. $f(x) = 4^{3x} = (\underline{})^x$
20. $f(x) = (e^2)^{x/6} = (\underline{})^x$
21. If the graph of $y = e^{x-2} + C$ passes through $(2, 9)$, then $C = $ _____.
22. By rigid transformations, the point $(0, 1)$ on the graph of $y = e^x$ is moved to the point _____ on the graph of $y = 4 + e^{x-3}$.

C. Exercises

In Problems 1 and 2, rewrite the given exponential expression as an equivalent logarithmic expression.

1. $5^{-1} = 0.2$
2. $\sqrt[3]{512} = 8$

In Problems 3 and 4, rewrite the given logarithmic expression as an equivalent exponential expression.

3. $\log_9 27 = 1.5$
4. $\log_6 (36)^{-2} = -4$

In Problems 5–12, solve for x.

5. $2^{1-x} = 8$
6. $3^{2x} = 81$
7. $e^{1-2x} = e^2$
8. $e^{x^2} - e^5 e^{x-1} = 0$
9. $2^{1-x} = 7$
10. $3^x = 7^{x-1}$
11. $e^{x+2} = 6$
12. $3e^x = 4e^{-3x}$

In Problems 13 and 14, solve for the indicated variable.

13. $P = Se^{-rm}$; for m
14. $P = \dfrac{K}{1 + ce^{rt}}$; for t

In Problems 15 and 16, graph the given functions on the same coordinate axes.

15. $y = 4^x$, $y = \log_4 x$
16. $y = \left(\dfrac{1}{2}\right)^x$, $y = \log_{1/2} x$
17. Match the letter of the graph in FIGURE 5.R.1 with the appropriate function.
    (i) $f(x) = b^x$, $b > 2$
    (ii) $f(x) = b^x$, $1 < b < 2$
    (iii) $f(x) = b^x$, $\frac{1}{2} < b < 1$
    (iv) $f(x) = b^x$, $0 < b < \frac{1}{2}$
18. In FIGURE 5.R.2, fill in the blanks for the coordinates of the points on each graph.

In Problems 19 and 20, find the slope of the secant line L given in each figure.

19.

20.

FIGURE 5.R.3 Graph for Problem 19

FIGURE 5.R.4 Graph for Problem 20

FIGURE 5.R.1 Graphs for Problem 17

FIGURE 5.R.2 Graphs for Problem 18

In Problems 21–26, match each of the following functions with one of the given graphs.

(i) $y = \ln(x - 2)$
(ii) $y = 2 - \ln x$
(iii) $y = 2 + \ln(x + 2)$
(iv) $y = -2 - \ln(x + 2)$
(v) $y = -\ln(2x)$
(vi) $y = 2 + \ln(-x + 2)$

21.
FIGURE 5.R.5 Graph for Problem 21

22.
FIGURE 5.R.6 Graph for Problem 22

23.
FIGURE 5.R.7 Graph for Problem 23

24.
FIGURE 5.R.8 Graph for Problem 24

25.
FIGURE 5.R.9 Graph for Problem 25

26.
FIGURE 5.R.10 Graph for Problem 26

In Problems 27 and 28, in words describe the graph of the function f in terms of a transformation of the graph of $y = \ln x$.

27. $f(x) = \ln ex$

28. $f(x) = \ln x^3$

29. Find a function $f(x) = Ae^{kx}$ if $(0, 5)$ and $(6, 1)$ are points on the graph of f.
30. Find a function $f(x) = A10^{kx}$ if $f(3) = 8$ and $f(0) = \frac{1}{2}$.
31. Find a function $f(x) = a + b^x$, $0 < b < 1$, if $f(1) = 5.5$ and the graph of f has a horizontal asymptote $y = 5$.
32. Find a function $f(x) = a + \log_3(x - c)$ if $f(11) = 10$ and the graph of f has a vertical asymptote $x = 2$.
33. **Doubling Time** If the initial number of bacteria present in a culture doubles after 9 hours, how long will it take for the number of bacteria in the culture to double again?
34. **Got Bait?** A commercial fishing lake is stocked with 10,000 fingerlings. Find a model $P(t) = P_0 e^{kt}$ for the fish population of the lake at time t if the owner of the lake estimates that there will be 5000 fish left after 6 months. After how many months does the model predict that there will be 1000 fish left?
35. **Radioactive Decay** Tritium, an isotope of hydrogen, has a half-life of 12.5 years. How much of an initial quantity of this element remains after 50 years?
36. **Old Bones** It is found that 97% of C-14 has been lost in a human skeleton found at an archeological site. What is the approximate age of the skeleton?

37. **Wishful Thinking** A person facing retirement invests $650,000 in a savings account. She wants the account to be worth $1,000,000 in 10 years. What annual rate r of interest compounded continuously will achieve this dream?
38. **Light Intensity** According to the **Beer–Lambert–Bouguer law**, the intensity I (measured in lumens) of a vertical beam of light passing through a transparent substance decreases according to the exponential function $I(x) = I_0 e^{kx}$, $k < 0$, where I_0 is the intensity of the incident beam and x is the depth measured in meters. If the intensity of light 1 meter below the surface of water is 30% of I_0, what is the intensity 3 meters below the surface?
39. The graph of the function $y = ae^{-be^{-cx}}$ is called a **Gompertz curve**. Solve for x in terms of the other symbols.
40. If $a > 0$ and $b > 0$, then show that $\log_{a^2} b^2 = \log_a b$.

6 Right Triangle Trigonometry

In This Chapter

- **6.1** Angles and Their Measure
- **6.2** Right Triangle Trigonometry
- **6.3** Trigonometric Functions of Special Angles
- **6.4** Trigonometric Functions of General Angles

Chapter 6 Review Exercises

A slice of pizza is an example of a circular sector. See Problem 73 in Exercises 6.1.

A Bit of History The early rudiments of trigonometry are found in the milieu of mathematics that sprang from the works of Greek, Egyptian, Indian, Islamic, and Chinese mathematicians. The word *trigonometry* is derived from two Greek words: *trigon* meaning triangle and *metra* meaning measurement. Thus the name *trigonometry* refers to the various relationships between the angles of a triangle and its sides. The Greek astronomer and mathematician **Hipparchus**, who lived in the second century BCE, was one of the principal developers of trigonometry. The tables of "chords" that he constructed were precursors to tables of the values of the trigonometric functions that appeared in all trigonometry texts prior to the appearance of the hand-held calculator. The first European mathematician to define trigonometric functions directly in terms of right triangles instead of circles, with tables for all six trigonometric functions, was the Austrian mathematician and astronomer **Georg Joachim von Lauchen** (1514–1574), aka, **Georg Joachim Rheticus**. Rheticus is also remembered as the sole student of **Nicolaus Copernicus** (1473–1543) and was the first advocate of the Copernican heliocentric theory of the solar system.

We begin this chapter with a discussion of angles and two forms of measuring an angle. In Section 6.2 we get down to the business of defining the trigonometric functions for acute angles in a right triangle. In Section 6.4 we expand these definitions to general angles.

6.1 Angles and Their Measure

(a) Two half-rays

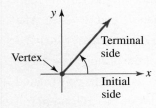

(b) Standard position

FIGURE 6.1.1 Initial and terminal sides of an angle

≡ Introduction We begin our study of trigonometry by discussing angles and two methods of measuring them: degrees and radians. As we will see in Section 7.1, it is the radian measure of an angle that enables us to define trigonometric functions on sets of real numbers.

☐ **Angles** An **angle** is formed by two half-rays, or half-lines, which have a common endpoint, called the **vertex**. We designate one ray the **initial side** of the angle and the other the **terminal side**. It is useful to consider the angle as having been formed by a rotation from the initial side to the terminal side as shown in **FIGURE 6.1.1(a)**. An angle is said to be in **standard position** if its vertex is placed at the origin of a rectangular coordinate system with its initial side coinciding with the positive x-axis, as shown in Figure 6.1.1(b).

☐ **Degree Measure** The **degree** measure of an angle is based on the assignment of 360 degrees (written 360°) to the angle formed by one complete counterclockwise rotation, as shown in **FIGURE 6.1.2**. Other angles are then measured in terms of a 360° angle, with a 1° angle being formed by $\frac{1}{360}$ of a complete rotation. If the rotation is counterclockwise, the measure will be *positive*; if clockwise, the measure is *negative*. For example, the angle in **FIGURE 6.1.3(a)** obtained by one-fourth of a complete counterclockwise rotation will be

$$\tfrac{1}{4}(360°) = 90°.$$

Shown in Figure 6.1.3(b) is the angle formed by three-fourths of a complete clockwise rotation. This angle has measure

$$\tfrac{3}{4}(-360°) = -270°.$$

FIGURE 6.1.2 Angle of 360 degrees

(a) 90° angle (b) −270° angle

FIGURE 6.1.3 Positive measure in (a); negative measure in (b)

FIGURE 6.1.4 Three coterminal angles

☐ **Coterminal Angles** Comparison of Figure 6.1.3(a) with Figure 6.1.3(b) shows that the terminal side of a 90° angle coincides with the terminal side of a −270° angle. When two angles in standard position have the same terminal sides we say they are **coterminal**. For example, the angles θ, $\theta + 360°$, and $\theta - 360°$ shown in **FIGURE 6.1.4** are coterminal. In fact, the addition of any integer multiple of 360° to a given angle results in a coterminal angle. Conversely, any two coterminal angles have degree measures that differ by an integer multiple of 360°.

EXAMPLE 1 Angles and Coterminal Angles

For a 960° angle:
(a) Locate the terminal side and sketch the angle.
(b) Find a coterminal angle between 0° and 360°.
(c) Find a coterminal angle between −360° and 0°.

Solution (a) We first determine how many full rotations are made in forming this angle. Dividing 960 by 360 we obtain a quotient of 2 and a remainder of 240. Equivalently we can write
$$960 = 2(360) + 240.$$
Thus, this angle is formed by making two counterclockwise rotations before completing $\frac{240}{360} = \frac{2}{3}$ of another rotation. As illustrated in **FIGURE 6.1.5(a)**, the terminal side of 960° lies in the third quadrant.

(b) Figure 6.1.5(b) shows that the angle 240° is coterminal with a 960° angle.

(c) Figure 6.1.5(c) shows that the angle $-120°$ is coterminal with a 960° angle.

FIGURE 6.1.5 Angles in (b) and (c) are coterminal with the angle in (a)

☐ **Minutes and Seconds** With calculators it is convenient to represent fractions of degrees by decimals, such as 42.23°. Traditionally, however, fractions of degrees were expressed in **minutes** and **seconds**, where

$$1° = 60 \text{ minutes (written } 60'\text{)}^* \tag{1}$$
and
$$1' = 60 \text{ seconds (written } 60''\text{)}. \tag{2}$$

For example, an angle of 7 degrees, 30 minutes, and 5 seconds is expressed as 7°30′5″. Some calculators have a special DMS key for converting an angle given in decimal degrees to Degrees, Minutes, and Seconds (DMS notation), and vice versa. The following examples show how to perform these conversions by hand.

EXAMPLE 2   **Using (1) and (2)**

Convert:

(a) 86.23° to degrees, minutes, and seconds,
(b) 17°47′13″ to decimal notation.

Solution In each case we will use (1) and (2).

(a) Since 0.23° represents $\frac{23}{100}$ of 1° and 1° = 60′, we have
$$86.23° = 86° + 0.23°$$
$$= 86° + (0.23)(60')$$
$$= 86° + 13.8'.$$

Now $13.8' = 13' + 0.8'$, so we must convert 0.8′ to seconds. Since 0.8′ represents $\frac{8}{10}$ of 1′ and 1′ = 60″, we have
$$86° + 13' + 0.8' = 86° + 13' + (0.8)(60'')$$
$$= 86° + 13' + 48''.$$

Hence, $86.23° = 86°13'48''$.

---

*The use of the number 60 as a base dates back to the Babylonians. Another example of the use of this base in our culture in the measurement of time (1 hour = 60 minutes and 1 minute = 60 seconds).

(b) Since $1° = 60'$, it follows that $1' = \left(\frac{1}{60}\right)°$. Similarly, $1'' = \left(\frac{1}{60}\right)' = \left(\frac{1}{3600}\right)°$. Thus we have

$$\begin{aligned} 17°47'13'' &= 17° + 47' + 13'' \\ &= 17° + 47\left(\tfrac{1}{60}\right)° + 13\left(\tfrac{1}{3600}\right)° \\ &\approx 17° + 0.7833° + 0.0036° \\ &= 17.7869°. \end{aligned}$$

≡

☐ **Radian Measure** Another measure for angles is **radian measure**, which is generally used in almost all applications of trigonometry that involve calculus. The radian measure of an angle θ is based on the length of an arc on a circle. If we place the vertex of the angle θ at the center of a circle of radius r, then θ is called a **central angle**. As we know, an angle θ in standard position can be viewed as having been formed by the initial side rotating from the positive x-axis to the terminal side. The region formed by the initial and terminal sides with a central angle θ is called a **sector** of a circle. As shown in **FIGURE 6.1.6(a)**, if the initial side of θ traverses a distance s along the circumference of the circle, then the **radian measure of θ** is defined by

$$\theta = \frac{s}{r}. \tag{3}$$

In the case when the terminal side of θ traverses an arc of length s along the circumference of the circle equal to the radius r of the circle, then we see from (3) that the measure of the angle θ is **1 radian**. See Figure 6.1.6(b).

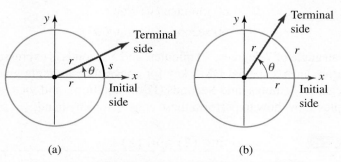

(a)   (b)

FIGURE 6.1.6 Central angle in (a); angle of 1 radian in (b)

FIGURE 6.1.7 Concentric circles

The definition given in (3) does not depend on the size of the circle. To see this, all we need do is to draw another circle centered at the vertex of θ of radius r' and subtended arc length s'. See **FIGURE 6.1.7**. Because the two circular sectors are similar the ratios s/r and s'/r' are equal. Therefore, regardless of which circle we use, we obtain the same radian measure for θ.

In equation (3) any convenient unit of length may be used for s and r, but the same unit must be used for *both* s and r. Thus,

$$\theta(\text{in radians}) = \frac{s(\text{units of length})}{r(\text{units of length})}$$

appears to be a "dimensionless" quantity. For example, if $s = 6$ in. and $r = 2$ in., then the radian measure of the angle is

$$\theta = \frac{4 \text{ in.}}{2 \text{ in.}} = 2,$$

where 2 is simply a real number. This is the reason why sometimes the word *radians* is omitted when an angle is measured in radians. We will come back to this idea in Section 7.1.

One complete rotation of the initial side of θ will traverse an arc equal in length to the circumference of the circle $2\pi r$. It follows from (3) that

$$\text{one rotation} = \frac{s}{r} = \frac{2\pi r}{r} = 2\pi \text{ radians.}$$

We have the same convention as before: An angle formed by a counterclockwise rotation is considered positive, whereas an angle formed by a clockwise rotation is negative. In **FIGURE 6.1.8** we illustrate angles in standard position of $\pi/2$, $-\pi/2$, π, and 3π radians, respectively. Note that the angle of $\pi/2$ radians shown in (a) is obtained by one-fourth of a complete counterclockwise rotation; that is

$$\frac{1}{4}(2\pi \text{ radians}) = \frac{\pi}{2} \text{ radians.}$$

The angle shown in Figure 6.1.8(b), obtained by one-fourth of a complete clockwise rotation, is $-\pi/2$ radians. The angle shown in Figure 6.1.8(c) is coterminal with the angle shown in Figure 6.1.8(d). In general, the addition of any integer multiple of 2π radians to an angle measured in radians results in a coterminal angle. Conversely, any two coterminal angles measured in radians will differ by an integer multiple of 2π.

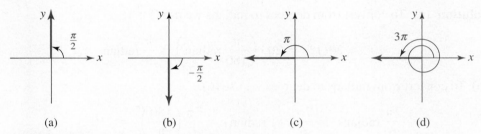

FIGURE 6.1.8 Angles measured in radians

EXAMPLE 3 A Coterminal Angle

Find an angle between 0 and 2π radians that is coterminal with $\theta = 11\pi/4$ radians. Sketch the angle.

Solution Since $2\pi < 11\pi/4 < 3\pi$, we subtract the equivalent of one rotation, or 2π radians, to obtain

$$\frac{11\pi}{4} - 2\pi = \frac{11\pi}{4} - \frac{8\pi}{4} = \frac{3\pi}{4}.$$

Alternatively, we can proceed as in part (a) of Example 1 and divide: $11\pi/4 = 2\pi + 3\pi/4$. Thus, an angle of $3\pi/4$ radians is coterminal with θ, as illustrated in **FIGURE 6.1.9**.

FIGURE 6.1.9 Coterminal angles in Example 3

☐ **Conversion Formulas** Although many scientific calculators have keys that convert between degree and radian measure, there is an easy way to remember the relationship between the two measures. Since the circumference of a unit circle is 2π, one complete rotation has measure 2π radians as well as $360°$. It follows that $360° = 2\pi$ radians or

$$180° = \pi \text{ radians.} \qquad (4)$$

If we interpret (4) as $180(1°) = \pi(1 \text{ radian})$, then we obtain the following two formulas for converting between degree and radian measure.

CONVERSION BETWEEN DEGREES AND RADIANS

$$1° = \frac{\pi}{180} \text{ radian} \tag{5}$$

$$1 \text{ radian} = \left(\frac{180}{\pi}\right)° \tag{6}$$

Using a calculator to carry out the divisions in (5) and (6), we find that

$$1° \approx 0.0174533 \text{ radian} \quad \text{and} \quad 1 \text{ radian} \approx 57.29578°.$$

EXAMPLE 4 Conversion Between Degree and Radians

Convert:

(a) 20° to radians, **(b)** $7\pi/6$ radians to degrees, **(c)** 2 radians to degrees.

Solution (a) To convert from degrees to radians we use (5):

$$20° = 20(1°) = 20 \cdot \left(\frac{\pi}{180} \text{ radian}\right) = \frac{\pi}{9} \text{ radian}.$$

(b) To convert from radians to degrees we use (6):

$$\frac{7\pi}{6} \text{ radians} = \frac{7\pi}{6} \cdot (1 \text{ radian}) = \frac{7\pi}{6}\left(\frac{180}{\pi}\right)° = 210°.$$

(c) We again use (6):

$$2 \text{ radians} = 2 \cdot (1 \text{ radian}) = 2 \cdot \left(\frac{180}{\pi}\right)° = \left(\frac{360}{\pi}\right)° \approx \overbrace{114.59°}^{\text{approximate answer rounded to two decimal places}}.$$

Table 6.1.1 provides the radian and degree measure of the most commonly used angles.

TABLE 6.1.1

Degrees	0	30	45	60	90	180
Radians	0	$\frac{\pi}{6}$	$\frac{\pi}{4}$	$\frac{\pi}{3}$	$\frac{\pi}{2}$	π

☐ **Terminology** You may recall from geometry that a 90° angle is called a **right angle** and a 180° angle is called a **straight angle**. In radian measure, $\pi/2$ is a right angle and π is a straight angle. An **acute angle** has measure between 0° and 90° (or between 0 and $\pi/2$ radians); and an **obtuse angle** has measure between 90° and 180° (or between $\pi/2$ and π radians). Two acute angles are said to be **complementary** if their sum is 90° (or $\pi/2$ radians). Two positive angles are **supplementary** if their sum is 180° (or π radians). The angle 180° (or π radians) is a **straight angle**. An angle whose terminal side coincides with a coordinate axis is called a **quadrantal angle**. For example, 90° (or $\pi/2$ radians) is a quadrantal angle. A triangle that contains a right angle is called a **right triangle**. The lengths a, b, and c of the sides of a right triangle satisfy the Pythagorean relationship $a^2 + b^2 = c^2$, where c is the length of the side opposite the right angle (the hypotenuse).

EXAMPLE 5 Complementary and Supplementary Angles

(a) Find the angle that is complementary to $\theta = 74.23°$.
(b) Find the angle that is supplementary to $\phi = \pi/3$ radians.

Solution (a) Since two angles are complementary if their sum is 90°, we find the angle that is complementary to $\theta = 74.23°$ is

$$90° - \theta = 90° - 74.23° = 15.77°.$$

(b) Since two angles are supplementary if their sum is π radians, we find the angle that is supplementary to $\phi = \pi/3$ radians is

$$\pi - \phi = \pi - \frac{\pi}{3} = \frac{3\pi}{3} - \frac{\pi}{3} = \frac{2\pi}{3} \quad \text{radians.} \qquad \equiv$$

☐ **Arc Length** In many applications it is necessary to find the length s of the arc subtended by a central angle θ in a circle of radius r. See **FIGURE 6.1.10**. From the definition of radian measure given in (3),

$$\theta \text{ (in radians)} = \frac{s}{r}.$$

By multiplying both sides of the last equation by r we obtain the **arc length formula** $s = r\theta$. We summarize the result.

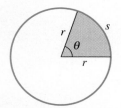

FIGURE 6.1.10 Length of arc s determined by a central angle θ

THEOREM 6.1.1 Arc Length Formula

For a circle of radius r, a central angle of θ radians subtends an arc of length

$$s = r\theta. \qquad (7)$$

EXAMPLE 6 Finding Arc Length

Find the arc length subtended by a central angle of **(a)** 2 radians in a circle of radius 6 inches, **(b)** 30° in a circle of radius 12 feet.

Solution (a) From the arc length formula (7) with $\theta = 2$ radians and $r = 6$ inches, we have $s = r\theta = 2 \cdot 6 = 12$. So the arc length is 12 inches.
(b) We must first express 30° in radians. Recall that $30° = \pi/6$ radians. Then from the arc length formula (7) we have $s = r\theta = (12)(\pi/6) = 2\pi$. So the arc length is $2\pi \approx 6.28$ feet. $\equiv$

◀ Students often apply the arc length formula incorrectly by using degree measure. Remember $s = r\theta$ is valid only if θ is measured in radians.

6.1 Exercises Answers to selected odd-numbered problems begin on page ANS-16.

In Problems 1–16, draw the given angle in standard position. Bear in mind that the lack of a degree symbol (°) in an angular measurement indicates that the angle is measured in radians.

1. 60°
2. −120°
3. 135°
4. 150°
5. 1140°
6. −315°
7. −240°
8. −210°

9. $\dfrac{\pi}{3}$ **10.** $\dfrac{5\pi}{4}$ **11.** $\dfrac{7\pi}{6}$ **12.** $-\dfrac{2\pi}{3}$

13. $-\dfrac{\pi}{6}$ **14.** -3π **15.** 3 **16.** 4

In Problems 17–20, express the given angle in decimal notation.

17. 10°39′17″ **18.** 143°7′2″ **19.** 5°10′ **20.** 10°25′

In Problems 21–24, express the given angle in terms of degrees, minutes, and seconds.

21. 210.78° **22.** 15.45° **23.** 30.81° **24.** 110.5°

In Problems 25–32, convert from degrees to radians.

25. 10° **26.** 15° **27.** 45° **28.** 215°
29. 270° **30.** −120° **31.** −230° **32.** 540°

In Problems 33–40, convert from radians to degrees.

33. $\dfrac{2\pi}{9}$ **34.** $\dfrac{11\pi}{6}$ **35.** $\dfrac{2\pi}{3}$ **36.** $\dfrac{5\pi}{12}$

37. $\dfrac{5\pi}{4}$ **38.** 7π **39.** 3.1 **40.** 12

In Problems 41–44, find the angle between 0° and 360° that is coterminal with the given angle.

41. 875° **42.** 400° **43.** −610° **44.** −150°

45. Find the angle between −360° and 0° that is coterminal with the angle in Problem 41.

46. Find the angle between −360° and 0° that is coterminal with the angle in Problem 43.

In Problems 47–52, find the angle between 0 and 2π that is coterminal with the given angle.

47. $-\dfrac{9\pi}{4}$ **48.** $\dfrac{17\pi}{2}$ **49.** 5.3π

50. $-\dfrac{9\pi}{5}$ **51.** −4 **52.** 7.5

53. Find the angle between -2π and 0 radians that is coterminal with the angle in Problem 47.

54. Find the angle between -2π and 0 radians that is coterminal with the angle in Problem 49.

In Problems 55–62, find an angle that is **(a)** complementary and **(b)** supplementary to the given angle, or state why no such angle can be found.

55. 48.25° **56.** 93° **57.** 98.4° **58.** 63.08°

59. $\dfrac{\pi}{4}$ **60.** $\dfrac{\pi}{6}$ **61.** $\dfrac{2\pi}{3}$ **62.** $\dfrac{5\pi}{6}$

CHAPTER 6 RIGHT TRIANGLE TRIGONOMETRY

63. Find both the degree and the radian measures of the angle formed by (a) three-fifths of a counterclockwise rotation and (b) five and one-eighth clockwise rotations.
64. Find both the degree and the radian measures of the obtuse angle formed by the hands of a clock (a) at 8:00, (b) at 1:00, and (c) at 7:30.
65. Find both the degree and the radian measures of the angle through which the hour hand on a clock rotates in 2 hours.
66. Answer the question in Problem 65 for the minute hand.
67. The Earth rotates on its axis once every 24 hours. How long does it take the Earth to rotate through an angle of (a) 240° and (b) $\pi/6$ radians?
68. The planet Mercury completes one rotation on its axis every 59 days. Through what angle (measured in degrees) does it rotate in (a) 1 day, (b) 1 hour, and (c) 1 minute?

Planet Mercury in Problem 68

69. Find the arc length subtended by a central angle of 3 radians in a circle of (a) radius 3 and (b) radius 5.
70. Find the arc length subtended by a central angle of 30° in a circle of (a) radius 2 and (b) radius 4.
71. Find the measure of a central angle θ in a circle of radius 5 if θ subtends an arc of length 7.5. Give θ in (a) radians and (b) degrees.
72. Find the measure of a central angle θ in a circle of radius 1 if θ subtends an arc of length $\pi/3$. Give θ in (a) radians and (b) degrees.
73. Show that the area A of a sector formed by a central angle of θ radians in a circle of radius r is given by $A = \frac{1}{2}r^2\theta$. See Figure 6.1.10. [*Hint*: Use the proportionality property from geometry that the ratio of the area A of a circular sector to the total area πr^2 of the circle equals the ratio of the central angle θ to one complete revolution 2π.]
74. What is the area of the red shaded circular band shown in FIGURE 6.1.11 if θ is measured (a) in radians and (b) in degrees? [*Hint*: Use the result of Problem 73.]

FIGURE 6.1.11 Circular band in Problem 74

Miscellaneous Applications

75. **Angular and Linear Speed** If we divide (7) by time t we get the relationship $v = r\omega$, where $v = s/t$ is called the **linear speed** of a point on the circumference of a circle and $\omega = \theta/t$ is called the **angular speed** of the point. A communications satellite is placed in a circular geosynchronous orbit 35,786 km above the surface of the Earth. The time it takes the satellite to make one full revolution around the Earth is 23 hours, 56 minutes, 4 seconds and the radius of the Earth is 6378 km. See FIGURE 6.1.12.
    (a) What is the angular speed of the satellite in rad/s?
    (b) What is the linear speed of the satellite in km/s?

FIGURE 6.1.12 Satellite in Problem 75

76. **Pendulum Clock** A clock pendulum is 1.3 m long and swings back and forth along a 15-cm arc. Find (a) the central angle and (b) the area of the sector through which the pendulum sweeps in one swing. [*Hint*: To answer part (b), use the result of Problem 73.]
77. **Sailing at Sea** A nautical mile is defined as the arc length subtended on the surface of the Earth by an angle of measure 1 minute. If the diameter of the Earth is 7927 miles, find how many statute (land) miles there are in a nautical mile.
78. **Circumference of the Earth** Around 230 BCE **Eratosthenes** calculated the circumference of the Earth from the following observations. At noon on the longest day of the year, the Sun was directly overhead in Syene, while it was inclined 7.2° from the vertical in Alexandria. He believed the two cities to be on the same longitudinal line and assumed that the rays of the Sun are parallel. Thus he con-

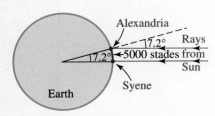

FIGURE 6.1.13 Earth in Problem 78

Yo-yo in Problems 79 and 80

cluded that the arc from Syene to Alexandria was subtended by a central angle of 7.2° at the center of the Earth. See FIGURE 6.1.13. At that time the distance from Syene to Alexandria was measured as 5000 stades. If one stade = 559 feet, find the circumference of the Earth in (a) stades and (b) miles. Show that Eratosthenes' data gives a result that is within 7% of the correct value if the polar diameter of the Earth is 7900 miles (to the nearest mile).

79. **Circular Motion of a Yo-Yo** A yo-yo is whirled around in a circle at the end of its 100-cm string.
    (a) If it makes six revolutions in 4 seconds, find its rate of turning, or angular speed, in radians per second.
    (b) Find the speed at which the yo-yo travels in centimeters per second; that is its linear speed.

80. **More Yo-Yos** If there is a knot in the yo-yo string described in Problem 79 at a point 40 cm from the yo-yo, find (a) the angular speed of the knot and (b) the linear speed.

81. **Circular Motion of a Tire** An automobile with 26-in. diameter tires is traveling at a rate of 55 mi/h.
    (a) Find the number of revolutions per minute that its tires are making.
    (b) Find the angular speed of its tires in radians per minute.

82. **Diameter of the Moon** The average distance from the Earth to the Moon as given by NASA is 238,855 miles. If the angle subtended by the Moon at the eye of an observer on Earth is 0.52°, then what is the approximate diameter of the Moon? FIGURE 6.1.14 is not to scale.

FIGURE 6.1.14 The curved red arc represents the approximate diameter of the Moon

6.2 Right Triangle Trigonometry

≡ **Introduction** As we said in the introduction to this chapter, the word *trigonometry* refers to the measurement of triangles. In this section we define the six trigonometric functions—**sine**, **cosine**, **tangent**, **cotangent**, **secant**, and **cosecant**—as ratios of the lengths of the sides of a right triangle. The names of these functions are abbreviated as **sin**, **cos**, **tan**, **cot**, **sec**, and **csc**, respectively.

We begin with some terminology and a definition.

☐ **Terminology** In FIGURE 6.2.1 we have drawn a right triangle with sides labeled a, b, and c (indicating their respective lengths) and one of the two acute angles denoted by θ. From the Pythagorean theorem we know that $a^2 + b^2 = c^2$. The side opposite the right angle is called the **hypotenuse**; the remaining sides are referred to as the **legs** of the triangle. The legs labeled a and b are, in turn, said to be the **side adjacent** to the angle θ

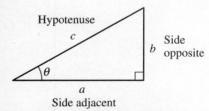

FIGURE 6.2.1 Defining the trigonometric function of the angle θ

and the **side opposite** the angle θ. We will also use the abbreviations **hyp**, **adj**, and **opp** to denote the lengths of these sides.

We next define the six **trigonometric functions of an acute angle**.

DEFINITION 6.2.1 Trigonometric Functions

The trigonometric functions of an acute angle θ in a right triangle are

$$\sin\theta = \frac{\text{opp}}{\text{hyp}} = \frac{b}{c} \qquad \cos\theta = \frac{\text{adj}}{\text{hyp}} = \frac{a}{c}$$

$$\tan\theta = \frac{\text{opp}}{\text{adj}} = \frac{b}{a} \qquad \cot\theta = \frac{\text{adj}}{\text{opp}} = \frac{a}{b} \qquad (1)$$

$$\sec\theta = \frac{\text{hyp}}{\text{adj}} = \frac{c}{a} \qquad \csc\theta = \frac{\text{hyp}}{\text{opp}} = \frac{c}{b}$$

☐ **Domains** The domain of each of these trigonometric functions is the set of all acute angles. In Section 6.4 we will extend these domains to include angles other than acute angles. Then in Chapter 7 we will see how the trigonometric functions can be defined with domains consisting of real numbers rather than angles.

The values of the six trigonometric functions depend only on the size of the angle θ and not on the size of the right triangle. To see this, consider the two right triangles shown in **FIGURE 6.2.2**. Because the right triangles have the same acute angle θ they are similar and thus the ratios of the corresponding sides are equal. For example, from the red triangle in Figure 6.2.2(a) we have

$$\sin\theta = \frac{\text{opp}}{\text{hyp}} = \frac{b}{c},$$

whereas in the smaller blue triangle in Figure 6.2.2(b) we find that

$$\sin\theta = \frac{\text{opp}}{\text{hyp}} = \frac{b'}{c'}.$$

But since the red triangle is similar to the blue triangle we must have

$$\frac{b'}{c'} = \frac{b}{c}.$$

In other words, we get the same value for $\sin\theta$ regardless of which right triangle we use to compute it. A similar argument can be made for the remaining five trigonometric functions.

(a)

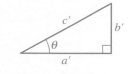
(b)

FIGURE 6.2.2 Similar triangles

EXAMPLE 1   Values of the Trigonometric Functions

Find the exact values of the six trigonometric functions of the acute angle θ in the right triangle shown in **FIGURE 6.2.3**.

Solution Matching Figure 6.2.3 with Figure 6.2.1 we see that the side opposite the angle θ has length 8 and the side adjacent to θ has length 15; that is, $b = 8$ and $a = 15$. From the Pythagorean theorem the hypotenuse $c = \text{hyp}$ is

$$c^2 = 8^2 + 15^2 = 289 \quad \text{and so} \quad c = \sqrt{289} = 17.$$

FIGURE 6.2.3 Right triangle in Example 1

Thus from (1) the values of the six trigonometric functions are

$$\sin\theta = \frac{\text{opp}}{\text{hyp}} = \frac{b}{c} = \frac{8}{17}, \qquad \cos\theta = \frac{\text{adj}}{\text{hyp}} = \frac{a}{c} = \frac{15}{17},$$

$$\tan\theta = \frac{\text{opp}}{\text{adj}} = \frac{a}{b} = \frac{8}{15}, \qquad \cot\theta = \frac{\text{adj}}{\text{opp}} = \frac{b}{a} = \frac{15}{8},$$

$$\sec\theta = \frac{\text{hyp}}{\text{adj}} = \frac{c}{a} = \frac{17}{15}, \qquad \csc\theta = \frac{\text{hyp}}{\text{opp}} = \frac{c}{b} = \frac{17}{8}.$$

≡

☐ **Quotient and Reciprocal Identities** There are many important relationships among the trigonometric functions. The basic ones listed next are referred to as the **fundamental identities** and should be memorized.

Quotient Identities:

$$\tan\theta = \frac{\sin\theta}{\cos\theta}, \qquad \cot\theta = \frac{\cos\theta}{\sin\theta} \qquad (2)$$

Reciprocal Identities:

$$\sec\theta = \frac{1}{\cos\theta}, \qquad \csc\theta = \frac{1}{\sin\theta}, \qquad \cot\theta = \frac{1}{\tan\theta}. \qquad (3)$$

Identities (2) and (3) can be obtained from Definition 6.2.1. For example, the first of the quotient identities is verified as follows:

$$\frac{\sin\theta}{\cos\theta} = \frac{\text{opp/hyp}}{\text{adj/hyp}} = \frac{\text{opp}}{\text{adj}} = \tan\theta.$$

The others can be verified in a similar manner. Using these identities, we can find the values of all six trigonometric functions once we know the values of $\sin\theta$ and $\cos\theta$.

EXAMPLE 2 Using (2) and (3)

Given $\sin\theta = \frac{4}{5}$ and $\cos\theta = \frac{3}{5}$, find the values of the remaining four trigonometric functions.

Solution From the fundamental identities, we have

$$\tan\theta = \frac{\sin\theta}{\cos\theta} = \frac{\frac{4}{5}}{\frac{3}{5}} = \frac{4}{3} \Bigg\} \leftarrow \text{from (2)}$$

$$\sec\theta = \frac{1}{\cos\theta} = \frac{1}{\frac{3}{5}} = \frac{5}{3}$$

$$\csc\theta = \frac{1}{\sin\theta} = \frac{1}{\frac{4}{5}} = \frac{5}{4} \Bigg\} \leftarrow \text{from (3)}$$

$$\cot\theta = \frac{1}{\tan\theta} = \frac{1}{\frac{4}{3}} = \frac{3}{4}.$$

Although we computed $\cot\theta$ using the reciprocal identity in (3), we could have also computed $\cot\theta$ using the quotient identity in (2).

≡

EXAMPLE 3 Using a Right Triangle

Given $\cos\theta = \frac{1}{3}$ and $\tan\theta = 2\sqrt{2}$. Find $\sin\theta$.

Solution We can obtain $\sin\theta$ by multiplying the first identity in (2) by $\cos\theta$:

$$\sin\theta = \cos\theta \tan\theta = \frac{1}{3} \cdot 2\sqrt{2} = \frac{2\sqrt{2}}{3}.$$

≡

The next example illustrates that if just one trigonometric function value of an acute angle is known, it is possible to find the other five function values by drawing an appropriate triangle.

EXAMPLE 4 Using a Right Triangle

If θ is an acute angle and $\sin\theta = \frac{2}{7}$, find the values of the other trigonometric functions of θ.

FIGURE 6.2.4 Right triangle in Example 3

Solution We sketch a right triangle with an acute angle θ satisfying $\sin\theta = \frac{2}{7}$, by making opp $= 2$ and hyp $= 7$ as shown in FIGURE 6.2.4. From the Pythagorean theorem we have

$$2^2 + (\text{adj})^2 = 7^2 \quad \text{so that} \quad (\text{adj})^2 = 7^2 - 2^2 = 45.$$

Thus, adj $= \sqrt{45} = 3\sqrt{5}$.

The values of the remaining five trigonometric functions are obtained from the definitions in (1):

$$\cos\theta = \frac{\text{adj}}{\text{hyp}} = \frac{3\sqrt{5}}{7}, \qquad \sec\theta = \frac{\text{hyp}}{\text{adj}} = \frac{7}{3\sqrt{5}} = \frac{7\sqrt{5}}{15},$$

$$\tan\theta = \frac{\text{opp}}{\text{adj}} = \frac{2}{3\sqrt{5}} = \frac{2\sqrt{5}}{15}, \qquad \cot\theta = \frac{\text{adj}}{\text{opp}} = \frac{3\sqrt{5}}{2},$$

$$\csc\theta = \frac{\text{hyp}}{\text{opp}} = \frac{7}{2}.$$

≡

☐ **Cofunctions** The use of the terminology sine and cosine, tangent and cotangent, secant and cosecant is a result of the following observation. As shown in FIGURE 6.2.5, if the two acute angles of a right triangle ABC are labeled α and β and a is the length of the side opposite α, b is the length of the side opposite β, and c is the length of the side opposite the right angle, then by Definition 6.2.1,

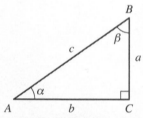

FIGURE 6.2.5 Acute angles α and β in a right triangle

$$\sin\alpha = \frac{a}{c} = \cos\beta, \qquad \cos\alpha = \frac{b}{c} = \sin\beta,$$

$$\tan\alpha = \frac{a}{b} = \cot\beta, \qquad \cot\alpha = \frac{b}{a} = \tan\beta,$$

$$\sec\alpha = \frac{c}{b} = \csc\beta, \qquad \csc\alpha = \frac{c}{a} = \sec\beta.$$

Because the sum of the angles in any triangle is $180°$ (or π radians), the acute angles α and β in a right triangle are complementary. Thus the cosine of an acute angle equals the sine of the complementary angle, the cotangent of an acute angle equals the tangent of the complementary angle, the cosecant of an acute angle equals the secant of the complementary angle, and conversely. For this reason we say that sine and cosine, tangent and cotangent, and secant and cosecant are **cofunctions** of one another. We can summarize the discussion in one simple sentence:

- *Cofunctions of complementary angles are equal.* (4)

☐ **Cofunction Identities** If α and β are the acute angles in the triangle in Figure 6.2.5, then

$$\alpha + \beta = \frac{\pi}{2} \quad \text{or} \quad \alpha = \frac{\pi}{2} - \beta.$$

Because $\cos\beta = \sin\alpha$ we obtain

$$\cos\beta = \sin\left(\frac{\pi}{2} - \beta\right).$$

The last expression is one of six cofunction identities.

Cofunction Identities:

$$\cos\theta = \sin\left(\frac{\pi}{2} - \theta\right) \qquad \cot\theta = \tan\left(\frac{\pi}{2} - \theta\right) \qquad \csc\theta = \sec\left(\frac{\pi}{2} - \theta\right)$$
$$\sin\theta = \cos\left(\frac{\pi}{2} - \theta\right) \qquad \tan\theta = \cot\left(\frac{\pi}{2} - \theta\right) \qquad \sec\theta = \csc\left(\frac{\pi}{2} - \theta\right) \qquad (5)$$

or equivalently

$$\cos\theta = \sin(90° - \theta) \qquad \cot\theta = \tan(90° - \theta) \qquad \csc\theta = \sec(90° - \theta)$$
$$\sin\theta = \cos(90° - \theta) \qquad \tan\theta = \cot(90° - \theta) \qquad \sec\theta = \csc(90° - \theta). \qquad (6)$$

In (5) and (6) it is understood that θ is measured in radians and degrees, respectively.

EXAMPLE 5 Using (5) and (6)

From (5):

complementary angles

(a) $\cot\dfrac{\pi}{6} = \tan\left(\dfrac{\pi}{2} - \dfrac{\pi}{6}\right) = \tan\dfrac{\pi}{3}$

(b) $\cos\dfrac{\pi}{4} = \sin\left(\dfrac{\pi}{2} - \dfrac{\pi}{4}\right) = \sin\dfrac{\pi}{4}.$

From (6):

(c) $\csc 27° = \sec(90° - 27°) = \sec 63°$

(d) $\cot 15° = \tan(90° - 15°) = \tan 75°.$

☐ **Pythagorean Identities** If just *one* trigonometric function value of an acute angle is known, it is possible to find the values of the other five functions without using the relationships in (1). Because the triangle in Figure 6.2.1 is a right triangle, the Pythagorean theorem relates the lengths of the sides of the triangle by

$$a^2 + b^2 = c^2.$$

If we divide this last result by c^2, we obtain $a^2/c^2 + b^2/c^2 = 1$ or

$$\left(\frac{a}{c}\right)^2 + \left(\frac{b}{c}\right)^2 = 1. \qquad (7)$$

Similarly, if we divide $a^2 + b^2 = c^2$ by a^2 and b^2 we obtain, in turn,

$$1 + \left(\frac{b}{a}\right)^2 = \left(\frac{c}{a}\right)^2 \qquad (8)$$

and

$$\left(\frac{a}{b}\right)^2 + 1 = \left(\frac{c}{b}\right)^2. \qquad (9)$$

Using the appropriate definition in (1) in the results (7), (8), and (9) yield another set of important identities.

Pythagorean Identities:

$$\sin^2\theta + \cos^2\theta = 1 \qquad (10)$$
$$1 + \tan^2\theta = \sec^2\theta \qquad (11)$$
$$\cot^2\theta + 1 = \csc^2\theta. \qquad (12)$$

In (10), (11), and (12), the square of the trigonometric functions are written $(\sin\theta)^2 = \sin^2\theta$, $(\cos\theta)^2 = \cos^2\theta$, $(\tan\theta)^2 = \tan^2\theta$, and so on.

EXAMPLE 6 Using (11)

If θ is an acute angle and $\tan\theta = \sqrt{5}$, find the value $\cos\theta$.

Solution There are several ways of solving this problem. One way is to use the Pythagorean identity (11):

$$\sec^2\theta = \tan^2\theta + 1 = (\sqrt{5})^2 + 1 = 5 + 1 = 6$$

and so $\sec\theta = \sqrt{6}$. Because $\sec\theta = 1/\cos\theta$ we have $\cos\theta = 1/\sec\theta$. Therefore $\cos\theta = 1/\sqrt{6} = \sqrt{6}/6$. ≡

NOTES FROM THE CLASSROOM

As we will see in Section 7.4, *all* the identities given in this section hold for any angle θ (not just acute angles).

6.2 Exercises
Answers to selected odd-numbered problems begin on page ANS-17.

In Problems 1–10, find the values of the six trigonometric functions of the angle θ in the given triangle.

1.

FIGURE 6.2.6 Triangle for Problem 1

2.

FIGURE 6.2.7 Triangle for Problem 2

3.

FIGURE 6.2.8 Triangle for Problem 3

4.

FIGURE 6.2.9 Triangle for Problem 4

5.

FIGURE 6.2.10 Triangle for Problem 5

6.

FIGURE 6.2.11 Triangle for Problem 6

7.

FIGURE 6.2.12 Triangle for Problem 7

8.

FIGURE 6.2.13 Triangle for Problem 8

9.

FIGURE 6.2.14 Triangle for Problem 9

10.

FIGURE 6.2.15 Triangle for Problem 10

In Problems 11–20, use the identities given in this section to find the values of the four remaining trigonometric functions at the acute angle θ.

11. $\sin\theta = \dfrac{2}{\sqrt{13}}$, $\cos\theta = \dfrac{3}{\sqrt{13}}$

12. $\sin\theta = \dfrac{1}{\sqrt{10}}$, $\cos\theta = \dfrac{3}{\sqrt{10}}$

13. $\sin\theta = \dfrac{2}{7}$, $\cos\theta = \dfrac{3\sqrt{5}}{7}$

14. $\sin\theta = \dfrac{5}{\sqrt{26}}$, $\cos\theta = \dfrac{1}{\sqrt{26}}$

15. $\sin\theta = \dfrac{1}{\sqrt{65}}$, $\tan\theta = \dfrac{1}{8}$

16. $\cos\theta = \dfrac{5}{\sqrt{29}}$, $\cot\theta = \dfrac{5}{2}$

17. $\csc\theta = \dfrac{5}{3}$, $\sec\theta = \dfrac{5}{4}$

18. $\sin\theta = \dfrac{1}{\sqrt{50}}$, $\cot\theta = 7$

19. $\cos\theta = \dfrac{1}{3}$, $\csc\theta = \dfrac{3}{2\sqrt{2}}$

20. $\sin\theta = \dfrac{1}{\sqrt{50}}$, $\cot\theta = 7$

In Problems 21–28, find the value of the remaining trigonometric functions by drawing an appropriate triangle.

21. $\sin\theta = \dfrac{12}{13}$

22. $\cos\theta = \dfrac{2}{\sqrt{5}}$

23. $\sec\theta = \dfrac{2}{\sqrt{3}}$

24. $\csc\theta = \sqrt{10}$

25. $\tan\theta = \dfrac{2}{5}$

26. $\cot\theta = \dfrac{1}{7}$

27. $\sec\theta = \dfrac{7}{3}$

28. $\tan\theta = 3$

29. If $\cos 75° = \frac{1}{4}(\sqrt{6} - \sqrt{2})$, find the exact value of $\sin 15°$.
30. If $\cos 75° = \frac{1}{4}(\sqrt{6} - \sqrt{2})$, find the exact value of $\sec 75°$.
31. If $\tan(\pi/8) = \sqrt{2} - 1$, find the exact value of $\cot(3\pi/8)$.
32. If $\tan(\pi/8) = \sqrt{2} - 1$, find the exact value of $\tan(3\pi/8)$.

In Problems 33–46, use the identities of this section to find the exact value of the given trigonometric expression. Do not use a calculator.

33. $3\sin^2\dfrac{\pi}{12} + 3\cos^2\dfrac{\pi}{12}$
34. $\sin^2 35° + \sin^2 55°$
35. $1 + \cos^2 18° + \sin^2 18°$
36. $1 + \tan^2 33° - \sec^2 33°$
37. $\tan^2\dfrac{\pi}{8} - \sec^2\dfrac{\pi}{8}$
38. $-4\csc^2 13° + 4\cot^2 13°$
39. $\dfrac{\sin 10°}{\sin 80°} - \dfrac{\sin 10°}{\cos 10°}$
40. $\sec 20° - \csc 70°$
41. $5\cot 41° \cot 49°$
42. $\tfrac{1}{2}\cos 11° \sec 11°$
43. $\sin 28° \cot 28° \csc 62°$
44. $10\sin\dfrac{\pi}{3}\cot\dfrac{\pi}{3}\sec\dfrac{\pi}{3}$
45. $\sin 10° \cos 80° + \cos 10° \sin 80°$
46. $\tan 30° \cot 60° - \sec 30° \csc 60°$

In Problems 47–54, given that $\cos 30° = \sqrt{3}/2$. Use the identities of this section to find the exact value of the given trigonometric function. Do not use a calculator.

47. $\sin 30°$
48. $\cos 60°$
49. $\tan 60°$
50. $\cot 30°$
51. $\sec 30°$
52. $\csc 30°$
53. $\cos 30° \tan 30°$
54. $\tan 30° + \cot 60°$

6.3 Trigonometric Functions of Special Angles

≡ **Introduction** The angles $30°(\pi/6$ radian$)$, $45°(\pi/4$ radian$)$, and $60°(\pi/3$ radians$)$ are considered to be special angles because they occur so often in the study of trigonometry and its use in calculus. Thus, it is to your advantage to know the *exact values* of the sine and cosine of each of these angles. In the discussion that follows, we derive these values using some results from plane geometry.

□ **Values of sin 45° and cos 45°** To find the values of the sine and cosine functions of a 45° angle, we consider the isosceles right triangle with two equal sides of length 1 shown in **FIGURE 6.3.1**. From plane geometry we know that the acute angles in this triangle are equal; therefore, each acute angle measures 45°. We can find the length of the hypotenuse by using the Pythagorean theorem:

$$(\text{hyp})^2 = (1)^2 + (1)^2 = 2 \quad \text{gives} \quad \text{hyp} = \sqrt{2}.$$

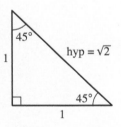

FIGURE 6.3.1 Isosceles right triangle

Thus from (1) of Section 6.2 we obtain

$$\sin 45° = \dfrac{\text{opp}}{\text{hyp}} = \dfrac{1}{\sqrt{2}} = \dfrac{\sqrt{2}}{2} \tag{1}$$

and

$$\cos 45° = \dfrac{\text{adj}}{\text{hyp}} = \dfrac{1}{\sqrt{2}} = \dfrac{\sqrt{2}}{2}. \tag{2}$$

☐ **Values of sin 30° and cos 30°** To find the values of the trigonometric functions of 30° and 60° angles, we consider the equilateral triangle AOB with sides of length 2 shown in FIGURE 6.3.2(a). From plane geometry we know that the three angles of an equilateral triangle each measure 60°. As shown in Figure 6.3.2(b), if we bisect the angle at O, then CO is the perpendicular bisector of AB. It follows that

$$\angle AOC = \tfrac{1}{2}\angle AOB = \tfrac{1}{2}(60°) = 30°,$$
$$\overline{AC} = \tfrac{1}{2}\overline{AB} = \tfrac{1}{2}(2) = 1 \quad \text{and} \quad \angle ACO = 90°.$$

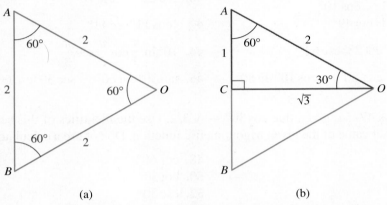

FIGURE 6.3.2 Equilateral triangle in (a); two congruent right triangles in (b)

If we apply the Pythagorean theorem to the red right triangle ACO in Figure 6.3.2(b), we get $(\overline{CO})^2 + 1^2 = 2^2$. Solving for $\overline{CO}$ we get $\overline{CO} = \sqrt{3}$. Therefore, from the right triangle ACO and (1) of Section 6.2, we obtain the following values:

$$\sin 30° = \frac{\text{opp}}{\text{hyp}} = \frac{1}{2} \tag{3}$$

$$\cos 30° = \frac{\text{adj}}{\text{hyp}} = \frac{\sqrt{3}}{2}. \tag{4}$$

☐ **Values of sin 60° and cos 60°** Now using the 60° angle in the red right triangle ACO in Figure 6.3.2(b) we identify opp $= \sqrt{3}$, adj $= 1$, and hyp $= 2$.

Therefore

$$\sin 60° = \frac{\text{opp}}{\text{hyp}} = \frac{\sqrt{3}}{2} \tag{5}$$

$$\cos 60° = \frac{\text{adj}}{\text{hyp}} = \frac{1}{2}. \tag{6}$$

☐ **Cofunctions** We did not have to use a right triangle to derive the values in (5) and (6). Recall, from Section 6.2 we showed that cofunctions of complementary angles are equal. Thus, (5) and (6) follow immediately from the results in (3) and (4):

$$\sin 60° = \cos 30° = \frac{\sqrt{3}}{2}$$

$$\cos 60° = \sin 30° = \frac{1}{2}.$$

EXAMPLE 1   Values of the Other Trigonometric Functions

Find the values of $\tan(\pi/6)$, $\cot(\pi/6)$, $\sec(\pi/6)$, and $\csc(\pi/6)$.

Solution The angle $30°$ is equivalent to $\pi/6$ radian. Using the quotient and reciprocal identities in Section 6.2 along with the results in (3) and (4) we get

$$\tan\frac{\pi}{6} = \frac{\sin(\pi/6)}{\cos(\pi/6)} = \frac{1/2}{\sqrt{3}/2} = \frac{1}{\sqrt{3}} = \frac{\sqrt{3}}{3}$$

$$\cot\frac{\pi}{6} = \frac{1}{\tan(\pi/6)} = \frac{1}{1/\sqrt{3}} = \sqrt{3}$$

$$\sec\frac{\pi}{6} = \frac{1}{\cos(\pi/6)} = \frac{1}{\sqrt{3}/2} = \frac{2}{\sqrt{3}} = \frac{2\sqrt{3}}{3}$$

$$\csc\frac{\pi}{6} = \frac{1}{\sin(\pi/6)} = \frac{1}{1/2} = 2.$$ ≡

We leave finding the values $\tan\theta$, $\cot\theta$, $\sec\theta$, and $\csc\theta$ for $\theta = \pi/4$ and $\theta = \pi/3$ as exercises. See Problems 1 and 2 in Exercises 6.3.

Table 6.3.1 summarizes the values of the sine, cosine, and tangent functions that we have just determined for the special angles $30°$, $45°$, and $60°$. As mentioned in the introduction to this section, these function values are used so frequently that we feel that they should be committed to memory. Knowing these values and the fundamental identities we discussed earlier will enable you to determine any of the trigonometric functions for these special angles.

TABLE 6.3.1

θ (degrees)	θ (radians)	$\sin\theta$	$\cos\theta$	$\tan\theta$
$30°$	$\dfrac{\pi}{6}$	$\dfrac{1}{2}$	$\dfrac{\sqrt{3}}{2}$	$\dfrac{\sqrt{3}}{3}$
$45°$	$\dfrac{\pi}{4}$	$\dfrac{\sqrt{2}}{2}$	$\dfrac{\sqrt{2}}{2}$	1
$60°$	$\dfrac{\pi}{3}$	$\dfrac{\sqrt{3}}{2}$	$\dfrac{1}{2}$	$\sqrt{3}$

EXAMPLE 2   Find the Exact Values

Find the exact value of the given trigonometric expression.

(a) $\sin^2\dfrac{\pi}{4} - \cos\dfrac{\pi}{3}$ (b) $\cos 30° \tan 60°$ (c) $2 + 4\sin\dfrac{\pi}{3} - 6\cos\dfrac{\pi}{6}$

Solution In each case we will use the information in Table 6.3.1.

(a) $\sin^2\dfrac{\pi}{4} - \cos\dfrac{\pi}{3} = \left(\dfrac{\sqrt{2}}{2}\right)^2 - \dfrac{1}{2} = \dfrac{2}{4} - \dfrac{1}{2} = \dfrac{1}{2} - \dfrac{1}{2} = 0$

(b) $\cos 30° \tan 60° = \dfrac{\sqrt{3}}{2} \cdot \sqrt{3} = \dfrac{3}{2}$

(c) $2 + 4\sin\dfrac{\pi}{3} - 6\cos\dfrac{\pi}{6} = 2 + 4 \cdot \dfrac{\sqrt{3}}{2} - 6 \cdot \dfrac{\sqrt{3}}{2} = 2 + 2\sqrt{3} - 3\sqrt{3} = 2 - \sqrt{3}$ ≡

☐ **Use of a Calculator** Approximations to the values of the trigonometric functions can be obtained using a scientific calculator. But before using a calculator to find trigonometric function values of angles measured in *radians*, you must set the calculator in the radian mode. If the angles are measured in degrees, then you must select the degree mode before making your calculations. Also, if the angles are given in degrees, minutes, and seconds, they must be converted to decimal form first. Scientific calculators have keys labeled [sin], [cos], and [tan] for computing the values of these functions. To obtain the values of csc, sec, or cot, we can use the [sin], [cos], or [tan] keys with the reciprocal key [1/x]. The following example illustrates the process.

EXAMPLE 3 Using a Calculator

Use a calculator to approximate each of the following.

(a) $\sin 45°$ **(b)** $\cos 8°15'$ **(c)** $\sec 0.23$ **(d)** $\cot \dfrac{\pi}{7}$

Solution **(a)** First we make sure that the calculator is set in degree mode. Then we enter 45 and use the [sin] key to obtain

$$\sin 45° \approx 0.7071068,$$

which is a seven-decimal-place approximation to the exact value $\sqrt{2}/2$ given in (1).

(b) Since the angle is given in degrees and minutes, we must first convert it to decimal form: $8°15' = 8° + \left(\frac{15}{60}\right)° = 8.25°$. Now with the calculator set in *degree mode*, we enter 8.25 and use the [cos] key to obtain

$$\cos 8°15' = \cos 8.25° \approx 0.9896514.$$

(c) Since degrees are not indicated, we recognize that this angle is measured in radians. To evaluate $\sec 0.23$, we will use the fundamental identity $\sec \theta = 1/\cos \theta$. With the calculator set in radian mode, we enter 0.23, use the [cos] key, and then take the reciprocal of the result by pressing the [1/x] key. Thus we have

$$\sec 0.23 = \frac{1}{\cos 0.23} \approx 1.0270458.$$

(d) We observe that this angle is measured in radians and set the calculator accordingly. We first enter π, divide by 7, use the [tan] key, and then the [1/x] key to obtain

$$\cot \frac{\pi}{7} = \frac{1}{\tan \dfrac{\pi}{7}} \approx 2.0765214. \qquad \equiv$$

6.3 Exercises Answers to selected odd-numbered problems begin on page ANS-17.

In Problems 1 and 2, use the results of this section to find the values of $\tan \theta$, $\cot \theta$, $\sec \theta$, and $\csc \theta$ for the given angle.

1. $45°$ **2.** $\pi/3$

In Problems 3–22, find the exact value of the given trigonometric expression. Do not use a calculator.

3. $\cos^2 \dfrac{\pi}{3}$

4. $\tan^2 \dfrac{\pi}{6}$

5. $\sec 45° \csc 45°$

6. $\sin 60° \cos 30°$

7. $\sin \dfrac{\pi}{4} \cot \dfrac{\pi}{4}$

8. $6 \sec \dfrac{\pi}{3} \csc \dfrac{\pi}{6}$

9. $9 \sec 45° \csc 45°$

10. $\tan 60° \cot 30°$

11. $\sin \dfrac{\pi}{3} \cos \dfrac{\pi}{4} + \cos \dfrac{\pi}{3} \sin \dfrac{\pi}{4}$

12. $\cos \dfrac{\pi}{3} \cos \dfrac{\pi}{6} - \sin \dfrac{\pi}{3} \sin \dfrac{\pi}{6}$

13. $6 \tan 30° + 7 \tan 60°$

14. $3 \sin \dfrac{\pi}{4} - 5 \cos \dfrac{\pi}{4}$

15. $\tan 45° - \cot 45°$

16. $\sec^2 \dfrac{\pi}{4} + 4 \csc^2 \dfrac{\pi}{3}$

17. $\dfrac{8 \sin (\pi/4)}{\sec (\pi/3)}$

18. $\dfrac{2 - \sqrt{2} \sin(\pi/4)}{\cos(\pi/4)}$

19. $\sin^2 30° + \cos^2 45°$

20. $2 + \cot^2 30° - 10 \csc^2 30°$

21. $\dfrac{\tan(\pi/4) - \tan(\pi/6)}{1 + \tan(\pi/4)\tan(\pi/6)}$

22. $\dfrac{\tan(\pi/3) + \tan(\pi/4)}{1 - \tan(\pi/3)\tan(\pi/4)}$

In Problems 23–32, use a calculator to find the approximate values of the six trigonometric functions of the given angle. Round your answer to four decimal places.

23. $17°$

24. $82°$

25. $14.3°$

26. $34.75°$

27. $71°30'15''$

28. $46°15'8''$

29. $\dfrac{\pi}{5}$

30. $\dfrac{\pi}{10}$

31. 0.6725

32. 1.24

For Discussion

33. Without a calculator, find the exact value of the product

$$\tan \dfrac{\pi}{180} \cdot \tan \dfrac{2\pi}{180} \cdot \tan \dfrac{3\pi}{180} \cdots \tan \dfrac{89\pi}{180}.$$

6.4 Trigonometric Functions of General Angles

≡ **Introduction** Until now we have defined the trigonometric functions only for acute angles. However, many applications of trigonometry involve angles that are not acute. Consequently, it is necessary to extend the definition of the six trigonometric functions in (1) of Section 6.2 to general angles. Naturally, we want the extended definition to agree with the earlier definition whenever the angle is acute. To accomplish this we proceed in the following manner.

FIGURE 6.4.1 An acute angle

Let θ be an acute angle in standard position and choose a point $P(x, y)$ on the terminal side of θ. If we let $r = d(O, P) = \sqrt{x^2 + y^2}$, we see in FIGURE 6.4.1 that x, y, and r are the lengths of the sides of a right triangle. With $y = $ opp, $x = $ adj, and $r = $ hyp, and we have from Definition 6.2.1,

$$\sin \theta = \frac{y}{r}, \quad \cos \theta = \frac{x}{r}, \quad \text{and} \quad \tan \theta = \frac{y}{x}. \qquad (1)$$

The expressions in (1) provide us with a model on which to base our extended definition for *any* angle θ in standard position, such as those illustrated in FIGURE 6.4.2.

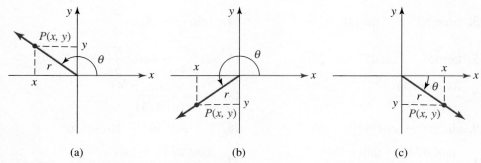

(a) (b) (c)

FIGURE 6.4.2 Angles that are not acute

We now have the following definition of the **trigonometric functions of a general angle**.

DEFINITION 6.4.1 Trigonometric Functions

Let θ be any angle in standard position, and let $P(x, y)$ be any point other than $(0, 0)$ on the terminal side of θ. If $r = \sqrt{x^2 + y^2}$ is the distance between $(0, 0)$ and $P(x, y)$, then the trigonometric functions are defined to be

$$\sin \theta = \frac{y}{r} \qquad \cos \theta = \frac{x}{r}$$
$$\tan \theta = \frac{y}{x} \qquad \cot \theta = \frac{x}{y} \qquad (2)$$
$$\sec \theta = \frac{r}{x} \qquad \csc \theta = \frac{r}{y}$$

provided no denominator is 0.

It can be shown by using similar triangles that the values of the six trigonometric functions depend only on the angle θ and not on which point $P(x, y)$ is chosen on the terminal side of θ. The justification of this statement is like the one made for acute angles on page 329.

☐ **Domains** A trigonometric function defined in (2) will be undefined if its denominator is zero. Since $P(x, y) \neq (0, 0)$, then $r = \sqrt{x^2 + y^2}$ is never zero. Thus the domains of the sine and the cosine functions consist of all angles θ. However, the tangent and the secant functions are undefined if the terminal side of θ lies on the y-axis because then $x = 0$. Therefore, the domains of $\tan \theta$ and $\sec \theta$ consist of all angles θ *except* those having radian measure $\pm \pi/2, \pm 3\pi/2, \pm 5\pi/2$, and so on. Using set notation,

◀ The angles are odd multiples of $\pi/2$.

and the fact that an odd integer can be written as $2n + 1$, n an integer, the domains of the tangent and the secant functions are

$$\{\theta \,|\, \theta \neq (2n + 1)\pi/2, n = 0, \pm 1, \pm 2, \ldots \}$$

or

$$\{\theta \,|\, \theta \neq (2n + 1)90°, n = 0, \pm 1, \pm 2, \ldots \}.$$

The cotangent and the cosecant functions are not defined for angles with their terminal sides on the x-axis because then $y = 0$. Thus the domains of $\cot \theta$ and $\csc \theta$ consist of all angles θ *except* those having radian measure 0, $\pm \pi$, $\pm 2\pi$, $\pm 3\pi$, and so on; that is, $\{\theta \,|\, \theta \neq n\pi, n = 0, \pm 1, \pm 2, \ldots \}$ or $\{\theta \,|\, \theta \neq 180°n, n = 0, \pm 1, \pm 2, \ldots \}$.

◀ The angles are integer multiples of π.

Since $r = \sqrt{x^2 + y^2}$, it follows that $|x| \leq r$ and $|y| \leq r$, or equivalently, $|x/r| \leq 1$ and $|y/r| \leq 1$. Therefore, as before

$$|\sin \theta| \leq 1 \quad \text{and} \quad |\cos \theta| \leq 1. \tag{3}$$

Similarly, because $|r/x| \geq 1$ and $|r/y| \geq 1$ we have

$$|\csc \theta| \geq 1 \quad \text{and} \quad |\sec \theta| \geq 1. \tag{4}$$

The inequalities in (3) and (4) hold for every θ in the domain of each of these functions.

EXAMPLE 1 Values of the Trigonometric Functions

Find the exact values of the six trigonometric functions of the angle θ if θ is in standard position and the terminal side of θ contains the point $P(-3, 1)$.

Solution The terminal side of the obtuse angle θ is sketched in **FIGURE 6.4.3**. With the identifications $x = -3$, $y = 1$, and

$$r = \sqrt{x^2 + y^2} = \sqrt{(-3)^2 + (1)^2} = \sqrt{10},$$

we have from (2),

$$\sin \theta = \frac{1}{\sqrt{10}} = \frac{\sqrt{10}}{10}, \qquad \cos \theta = \frac{-3}{\sqrt{10}} = -\frac{3\sqrt{10}}{10},$$

$$\tan \theta = \frac{1}{-3} = -\frac{1}{3}, \qquad \cot \theta = \frac{-3}{1} = -3,$$

$$\sec \theta = \frac{\sqrt{10}}{-3} = -\frac{\sqrt{10}}{3}, \qquad \csc \theta = \frac{\sqrt{10}}{1} = \sqrt{10}.$$

FIGURE 6.4.3 Angle θ in Example 1

EXAMPLE 2 Values of the Trigonometric Functions

Find the values of the six trigonometric functions of θ if $\theta = -\pi/2$.

Solution First we place θ in standard position as shown in **FIGURE 6.4.4**. According to Definition 6.4.1, we can choose *any* point $P(x, y)$ on the terminal side of θ. For convenience, we select $P(0, -1)$ so that $x = 0$, $y = -1$, and $r = \sqrt{x^2 + y^2} = 1$. Thus,

$$\sin\left(-\frac{\pi}{2}\right) = \frac{-1}{1} = -1, \qquad \cos\left(-\frac{\pi}{2}\right) = \frac{0}{1} = 0,$$

$$\cot\left(-\frac{\pi}{2}\right) = \frac{0}{-1} = 0, \qquad \csc\left(-\frac{\pi}{2}\right) = \frac{1}{-1} = -1.$$

However, the expressions $\tan \theta = y/x$ and $\sec \theta = r/x$ are undefined for $\theta = -\pi/2$ since $x = 0$.

FIGURE 6.4.4 Angle θ in Example 2

☐ **Algebraic Signs** Depending on the quadrant in which the terminal side of θ lies, one or both coordinates of $P(x, y)$ may be negative. Since $r = \sqrt{x^2 + y^2}$ is *always* *positive*, each of the six trigonometric functions of θ has negative as well as positive values. For example, $\sin\theta = y/r$ is positive if the terminal side of θ lies in quadrants I or II (where y is positive), and $\sin\theta = y/r$ is negative if the terminal side of θ lies in quadrants III or IV (where y is negative). **FIGURE 6.4.5** summarizes the algebraic signs of the six trigonometric functions defined in (2). For convenience, if the terminal side of θ lies in quadrant II, we will refer to θ as a quadrant II angle or say that θ is in quadrant II. We will use similar terminology when we refer to angles with terminal sides in quadrants I, III, or IV.

II	I
$\cos\theta < 0 \quad \sin\theta > 0$	$\cos\theta > 0 \quad \sin\theta > 0$
$\tan\theta < 0 \quad \cot\theta < 0$	$\tan\theta > 0 \quad \cot\theta > 0$
$\sec\theta < 0 \quad \csc\theta > 0$	$\sec\theta > 0 \quad \csc\theta > 0$
$\cos\theta < 0 \quad \sin\theta < 0$	$\cos\theta > 0 \quad \sin\theta < 0$
$\tan\theta > 0 \quad \cot\theta > 0$	$\tan\theta < 0 \quad \cot\theta < 0$
$\sec\theta < 0 \quad \csc\theta < 0$	$\sec\theta > 0 \quad \csc\theta < 0$
III	IV

FIGURE 6.4.5 Algebraic signs of the six trigonometric functions

EXAMPLE 3 Using Figure 6.4.5

In which quadrant does the terminal side of θ lie if $\sin\theta > 0$ and $\tan\theta < 0$?

Solution From Figure 6.4.5 we see that the sine function is positive for angles in quadrants I and II and the tangent function is negative in quadrants II and IV, the terminal side of θ must lie in quadrant II. ≡

FIGURE 6.4.6 An arbitrary angle θ

☐ **Pythagorean Identities—Revisited** The reciprocal, quotient, and Pythagorean identities for acute angles given in Section 6.2 also hold for general angles. For example, to derive the **Pythagorean identities** let θ be any angle in standard position. As shown in **FIGURE 6.4.6**, we let $P(x, y)$ be any point other than the origin on the terminal side of θ. If we again let $r = d(O, P) = \sqrt{x^2 + y^2}$, then we have $x^2 + y^2 = r^2$. Dividing both sides of the last equation by r^2 gives

$$\frac{x^2}{r^2} + \frac{y^2}{r^2} = 1 \quad \text{or} \quad \left(\frac{x}{r}\right)^2 + \left(\frac{y}{r}\right)^2 = 1.$$

Recognizing that $x/r = \cos\theta$ and $y/r = \sin\theta$, we obtain the basic Pythagorean identity

$$\sin^2\theta + \cos\theta^2 = 1. \tag{5}$$

In (5) we have followed the convention that $\sin^2\theta$ be written first. If both sides of (5) are divided, in turn, by $\cos^2\theta$ and $\sin^2\theta$ we obtain

$$1 + \tan^2\theta = \sec^2\theta \tag{6}$$

and
$$\cot^2\theta + 1 = \csc^2\theta. \tag{7}$$

Formulas (5), (6), and (7) are identical to (10), (11), and (12) in Section 6.2. But unlike the latter formulas, the trigonometric functions in (5), (6), and (7) are

- valid for all angles for which the functions are defined, and
- the values of the functions can have negative values.

We will encounter the Pythagorean identities one more time (in Chapter 7) when we show that each of the trigonometric functions can be defined for real numbers instead of angles.

EXAMPLE 4 Using (5)

Given that $\cos\theta = \frac{1}{3}$ and that θ is a quadrant IV angle, find the exact values of the remaining five trigonometric functions of θ.

Solution Substituting $\cos\theta = \frac{1}{3}$ into (5) gives

$$\sin^2\theta + \left(\frac{1}{3}\right)^2 = 1$$

$$\sin^2\theta = 1 - \frac{1}{9} = \frac{8}{9}.$$

Since the terminal side of θ is in quadrant IV, $\sin\theta$ is *negative*. Therefore, we must select the negative square root of $\frac{8}{9}$:   ◀ See Figure 6.4.5.

$$\sin\theta = -\sqrt{\frac{8}{9}} = -\frac{2\sqrt{2}}{3}.$$

Now using

$$\tan\theta = \frac{\sin\theta}{\cos\theta}, \quad \cot\theta = \frac{1}{\tan\theta},$$
$$\sec\theta = \frac{1}{\cos\theta}, \quad \csc\theta = \frac{1}{\sin\theta},$$

we find that the values of the remaining four functions are

$$\tan\theta = \frac{-2\sqrt{2}/3}{1/3} = -2\sqrt{2}, \quad \cot\theta = \frac{1}{-2\sqrt{2}} = -\frac{\sqrt{2}}{4},$$

$$\sec\theta = \frac{1}{1/3} = 3, \quad \csc\theta = \frac{1}{-2\sqrt{2}/3} = -\frac{3\sqrt{2}}{4}. \quad \equiv$$

EXAMPLE 5 Using (6)

Given $\tan\theta = -2$ and $\sin\theta > 0$, find the exact values of the remaining five trigonometric functions of θ.

Solution Letting $\tan\theta = -2$ in the identity $1 + \tan^2\theta = \sec^2\theta$, we find

$$\sec^2\theta = 1 + (-2)^2 = 5.$$

Since $\tan\theta$ is negative in quadrants II and IV and $\sin\theta$ is positive in quadrants I and II, the terminal side of θ must lie in quadrant II. Thus we must take

$$\sec\theta = -\sqrt{5}.$$

From $\sec\theta = 1/\cos\theta$, it follows that

$$\cos\theta = \frac{1}{\sec\theta} = \frac{1}{-\sqrt{5}} = -\frac{\sqrt{5}}{5}.$$

Using $\tan\theta = \sin\theta/\cos\theta$, we obtain

$$\sin\theta = \cos\theta\,\tan\theta = \left(-\frac{\sqrt{5}}{5}\right)(-2) = \frac{2\sqrt{5}}{5}.$$

Then

$$\csc\theta = \frac{1}{\sin\theta} = \frac{1}{2\sqrt{5}/5} = \frac{\sqrt{5}}{2}$$

and

$$\cot\theta = \frac{1}{\tan\theta} = \frac{1}{-2} = -\frac{1}{2}. \quad \equiv$$

In Section 6.3 we found exact values for the six trigonometric functions of the special angles of 30°, 45°, and 60° (or $\pi/6$, $\pi/4$, and $\pi/3$, respectively, in radian measure). These values can be used to determine the exact trigonometric function values of certain nonacute angles by means of a **reference angle**.

> **DEFINITION 6.4.2** Reference Angle
>
> Let θ be an angle in standard position such that its terminal side does not lie on a coordinate axis. The **reference angle** θ' for θ is defined to be the acute angle formed by the terminal side of θ and the x-axis.

FIGURE 6.4.7 illustrates this definition for angles with the terminal sides in each of the four quadrants.

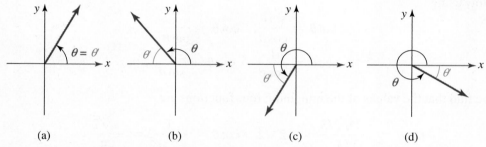

FIGURE 6.4.7 An angle θ (red) and its reference angle θ' (blue)

EXAMPLE 6   Reference Angles

Find the reference angle for each angle θ.

(a) $\theta = 40°$ **(b)** $\theta = \dfrac{2\pi}{3}$ **(c)** $\theta = 210°$ **(d)** $\theta = -\dfrac{9\pi}{4}$

Solution (a) From **FIGURE 6.4.8(a)** we see that $\theta' = 40°$.
(b) From Figure 6.4.8(b), $\theta' = \pi - \theta = \pi - 2\pi/3 = \pi/3$.
(c) From Figure 6.4.8(c), $\theta' = \theta - 180° = 210° - 180° = 30°$.
(d) Since $\theta = -9\pi/4$ is coterminal with

$$-\frac{9\pi}{4} + 2\pi = -\frac{\pi}{4},$$

we find that $\theta' = \pi/4$. See Figure 6.4.8(d).

FIGURE 6.4.8 Reference angles in Example 6

☐ Property of Reference Angles
The usefulness of reference angles in evaluating trigonometric functions is a result of the following property:

- *The absolute value of any trigonometric function of an angle θ equals the value of that function for the reference angle θ'.*

For instance, $|\sin\theta| = \sin\theta'$, $|\cos\theta| = \cos\theta'$, and so on.

We will verify the foregoing property for the sine function. If the terminal side of θ lies in quadrant I, then $\theta = \theta'$ and $\sin\theta$ is positive, so

$$\sin\theta' = \sin\theta = |\sin\theta|.$$

From **FIGURE 6.4.9**, we see that if θ is a quadrant II, III, or IV angle, then we have

$$\sin\theta' = \frac{|y|}{r} = \left|\frac{y}{r}\right| = |\sin\theta|,$$

where $P(x, y)$ is any point on the terminal side of θ and $r = \sqrt{x^2 + y^2}$.

FIGURE 6.4.9 Reference angles

We can now describe a step-by-step procedure for determining a trigonometric function value of any angle θ.

FINDING THE VALUE OF A TRIGONOMETRIC FUNCTION

Suppose θ represents any angle.

(*i*) Find the reference angle θ'.
(*ii*) Determine the value of the trigonometric function for θ'.
(*iii*) Select the correct algebraic sign of the value in (*ii*) by considering in which quadrant the terminal side of the angle θ lies.

EXAMPLE 7 Finding Values Using Reference Angles

Find the exact values of $\sin\theta$, $\cos\theta$, and $\tan\theta$ for each of the following angles.

(a) $\theta = \dfrac{2\pi}{3}$ **(b)** $\theta = 210°$ **(c)** $\theta = -\dfrac{9\pi}{4}$

Solution We use the procedure just discussed along with Table 6.3.1 of Section 6.3.

(a) In part (b) of Example 6 we found the reference angle for $\theta = 2\pi/3$ to be $\theta' = \pi/3$. Now we know that $\sin(\pi/3) = \sqrt{3}/2$, $\cos(\pi/3) = 1/2$, and $\tan(\pi/3) = \sqrt{3}$. Because $\theta = 2\pi/3$ is a quadrant II angle, where the sine is positive but the cosine and the tangent are negative, we conclude

$$\sin\frac{2\pi}{3} = \frac{\sqrt{3}}{2}, \quad \cos\frac{2\pi}{3} = -\frac{1}{2}, \quad \text{and} \quad \tan\frac{2\pi}{3} = -\sqrt{3}.$$

(b) Referring to part (c) of Example 6, we see that the reference angle is $\theta' = 30°$. Using the property of reference angles and the fact that the terminal side of $\theta' = 210°$ lies in quadrant III, we get

$$\left.\begin{array}{l} \sin 210° = -\sin 30° = -\dfrac{1}{2}, \\[4pt] \cos 210° = -\cos 30° = -\dfrac{\sqrt{3}}{2}, \\[4pt] \tan 210° = \tan 30° = \dfrac{\sqrt{3}}{3}. \end{array}\right\} \leftarrow \text{see Figure 6.4.5 for the correct algebraic signs}$$

(c) From part (d) of Example 6 we know that the reference angle $\theta' = \pi/4$. Since $\theta = -9\pi/4$ is a quadrant IV angle, it follows that

$$\sin\left(-\frac{9\pi}{4}\right) = -\sin\frac{\pi}{4} = -\frac{\sqrt{2}}{2},$$

$$\cos\left(-\frac{9\pi}{4}\right) = \cos\frac{\pi}{4} = \frac{\sqrt{2}}{2},$$

$$\tan\left(-\frac{9\pi}{4}\right) = -\tan\frac{\pi}{4} = -1.$$

EXAMPLE 8 Finding Angles

Find all angles θ satisfying $0° \leq \theta < 360°$ such that $\sin\theta = \frac{1}{2}$.

Solution From what we know about the special angles $30°, 60°,$ and $90°$, we know that $\theta = 30°$ is one solution. Using $30°$ as a reference angle in the second quadrant, as shown in **FIGURE 6.4.10**, we find $\theta = 150°$ as a second solution. Since the sine function is negative for angles in quadrants III and IV, there are no additional solutions satisfying $0° \leq \theta < 360°$.

FIGURE 6.4.10 Solutions in Example 8

EXAMPLE 9 Finding Angles

Find all angles θ satisfying $0 \leq \theta < 2\pi$ such that $\cos\theta = -\sqrt{2}/2$.

Solution Since the given value of the cosine function is negative, we first determine the reference angle θ' such that $\cos\theta' = \sqrt{2}/2$. From Section 6.3 we know that $\theta' = \pi/4$. Since the cosine function is negative for angles in quadrants II and III, we position the reference angle $\theta' = \pi/4$ as shown in **FIGURE 6.4.11**. We then obtain $\theta = 3\pi/4$ and $\theta = 5\pi/4$ as solutions.

FIGURE 6.4.11 Solutions in Example 9

NOTES FROM THE CLASSROOM

In this section we have purposely avoided using calculators. For a complete understanding of trigonometry, it is essential that you master the concepts and be able to perform *without the aid of a calculator* the types of calculations and simplifications we have discussed. The exercise set that follows should be worked without the use of a calculator.

6.4 Exercises Answers to selected odd-numbered problems begin on page ANS-17.

We recommend that you do not use a calculator in solving any of the problems that follow.

In Problems 1–10, evaluate the six trigonometric functions of the angle θ if θ is in standard position and the terminal side of θ contains the given point.

1. $(6, 8)$
2. $(-1, 2)$
3. $(5, -12)$
4. $(-8, -15)$
5. $(0, 2)$
6. $(-3, 0)$
7. $(-2, 3)$
8. $(5, -1)$
9. $(-\sqrt{2}, -1)$
10. $(\sqrt{3}, \sqrt{2})$

In Problems 11–18, find the quadrant in which the terminal side of an angle θ lies if θ satisfies the given conditions.

11. $\sin\theta < 0$ and $\tan\theta > 0$
12. $\cos\theta > 0$ and $\sin\theta < 0$
13. $\tan\theta < 0$ and $\sec\theta < 0$
14. $\sec\theta < 0$ and $\csc\theta < 0$
15. $\cot\theta > 0$ and $\sin\theta > 0$
16. $\csc\theta > 0$ and $\cot\theta < 0$
17. $\sin\theta > 0$ and $\cos\theta < 0$
18. $\tan\theta < 0$ and $\csc\theta > 0$

In Problems 19–28, the value of one of the trigonometric functions of an angle θ is given. From the given value and the additional information, determine the values of the five remaining trigonometric functions of θ.

19. $\sin\theta = \frac{1}{4}$, θ is in quadrant II
20. $\cos\theta = -\frac{2}{5}$, θ is in quadrant II
21. $\tan\theta = 3$, θ is in quadrant III
22. $\cot\theta = 2$, θ is in quadrant III
23. $\csc\theta = -10$, θ is in quadrant IV
24. $\sec\theta = 3$, θ is in quadrant IV
25. $\sin\theta = -\frac{1}{5}$, $\cos\theta > 0$
26. $\cos\theta = -\frac{2}{3}$, $\sin\theta < 0$
27. $\tan\theta = 8$, $\sec\theta > 0$
28. $\sec\theta = -4$, $\csc\theta < 0$

29. If $\cos\theta = \frac{3}{10}$, find all possible values of $\sin\theta$.
30. If $\sin\theta = -\frac{2}{7}$, find all possible values of $\cos\theta$.
31. If $2\sin\theta - \cos\theta = 0$, find all possible values of $\sin\theta$ and $\cos\theta$.
32. If $\cot\theta = \frac{3}{4}$, find all possible values of $\csc\theta$.
33. If $\sec\theta = -5$, find all possible values of $\sin\theta$ and $\cos\theta$.
34. If $3\cos\theta = \sin\theta$, find all possible values of $\tan\theta$, $\cot\theta$, $\sec\theta$, and $\csc\theta$.

35. Complete the following table.

θ (degrees)	θ (radians)	sin θ	cos θ	tan θ
0°	0	0	1	0
30°	π/6	1/2	√3/2	√3/3
45°	π/4	√2/2	√2/2	1
60°	π/3	√3/2	1/2	√3
90°	π/2	1	0	—
120°	2π/3	√3/2	−1/2	−√3
135°	3π/4			
150°	5π/6			
180°	π			
210°	7π/6	−1/2	−√3/2	√3/3
225°	5π/4			
240°	4π/3			
270°	3π/2			
300°	5π/3			
315°	7π/4			
330°	11π/6			
360°	2π			

36. Complete the following table.

θ (degrees)	θ (radians)	csc θ	sec θ	cot θ
0°	0	—	1	—
30°	π/6	2	2√3/3	√3
45°	π/4	√2	√2	1
60°	π/3	2√3/3	2	√3/3
90°	π/2	1	—	0
120°	2π/3			
135°	3π/4			
150°	5π/6			
180°	π			
210°	7π/6			
225°	5π/4			
240°	4π/3			
270°	3π/2			
300°	5π/3			
315°	7π/4			
330°	11π/6			
360°	2π			

In Problems 37–52, find the exact value of the given expression.

37. $\cos 5\pi$
38. $\sin\left(-\dfrac{7\pi}{6}\right)$
39. $\cot \dfrac{13\pi}{6}$
40. $\tan \dfrac{9\pi}{2}$
41. $\sin\left(-\dfrac{4\pi}{3}\right)$
42. $\cos \dfrac{23\pi}{4}$
43. $\csc\left(-\dfrac{\pi}{6}\right)$
44. $\tan \dfrac{23\pi}{4}$
45. $\sec(-120°)$
46. $\csc 495°$
47. $\sin 150°$
48. $\cos(-45°)$
49. $\tan 405°$
50. $\sin 315°$
51. $\cot(-720°)$
52. $\sec(-300°)$

In Problems 53–58, find all angles θ, where $0 \le \theta < 360°$, satisfying the given condition.

53. $\tan \theta = \sqrt{3}$
54. $\sin \theta = -\dfrac{1}{2}$
55. $\cos \theta = -\dfrac{\sqrt{2}}{2}$
56. $\sec \theta = \dfrac{2\sqrt{3}}{3}$
57. $\csc \theta = -1$
58. $\cot \theta = -\dfrac{1}{\sqrt{3}}$

In Problems 59–64, find all angles θ, where $0 \le \theta < 2\pi$, satisfying the given condition.

59. $\sin \theta = 0$
60. $\cos \theta = -1$
61. $\sec \theta = -\sqrt{2}$
62. $\csc \theta = 2$
63. $\cot \theta = -\sqrt{3}$
64. $\tan \theta = 1$

Miscellaneous Applications

65. **Free Throw** Under certain conditions the maximum height y attained by a basketball released from a height h at an angle α measured from the horizontal with an initial velocity v_0 is given by

$$y = h + (v_0^2 \sin^2 \alpha)/2g,$$

where g is the acceleration due to gravity. Compute the maximum height reached by a free throw if $h = 2.15$ m, $v_0 = 8$ m/s, $\alpha = 64.47°$, and $g = 9.81$ m/s^2.

66. **Putting the Shot** The range of a shot put released from a height h above the ground with an initial velocity v_0 at an angle θ to the horizontal can be approximated by

$$R = \dfrac{v_0 \cos \phi}{g}\left(v_0 \sin \phi + \sqrt{v_0^2 \sin^2 \phi + 2gh}\,\right),$$

where g is the acceleration due to gravity.
(a) If $v_0 = 13.7$ m/s, $\phi = 40°$, and $g = 9.81$ m/s^2, compare the ranges achieved for the release heights $h = 2.0$ m and $h = 2.4$ m.
(b) Explain why an increase in h yields an increase in R if the other parameters are held fixed.
(c) What does this imply about the advantage that height gives a shot-putter?

Free throw

67. Acceleration Due to Gravity Because of its rotation, the Earth bulges at the equator and is flattened at the poles. As a result, the acceleration due to gravity actually varies with latitude θ. Satellite studies have shown that the acceleration due to gravity g_{sat} is approximated by the function

$$g_{\text{sat}} = 978.0309 + 5.18552 \sin^2\theta - 0.00570 \sin^2 2\theta.$$

(a) Find g_{sat} at the equator ($\theta = 0°$), (b) at the north pole, and (c) at 45° north latitude.

For Discussion

68. Is there an angle θ such that $\cos\theta = \frac{4}{3}$? Explain.
69. Is there an angle θ such that $2\csc\theta = 1$? Explain.
70. Discuss how it is possible to determine without the aid of a calculator that both $\sin 4$ and $\cos 4$ are negative.
71. Let L be a nonvertical line that passes through the origin and makes an angle θ measured counterclockwise from the positive x-axis. Prove that the slope m of the line L is $\tan\theta$. See **FIGURE 6.4.12**.

FIGURE 6.4.12 Line through origin in Problem 71

CONCEPTS REVIEW   *You should be able to give the meaning of each of the following concepts.*

Initial side of an angle
Terminal side of an angle
Standard position of an angle
Coterminal angles
Minutes
Seconds
Degree measure of an angle
Central angle
Radian measure of an angle
Acute angle
Complementary angles
Obtuse angle

Straight angle
Quadrantal angle
Supplementary angles
Right angle
Arc length
Conversion:
   degrees to radians
   radians to degrees
Reference angle
Right triangles:
   side adjacent
   side opposite

hypotenuse
Trigonometric functions:
   of acute angles
   of general angles
Quotient identities
Reciprocal identities
Pythagorean identities
Cofunctions
Cofunction identities

CHAPTER 6   Review Exercises   *Answers to selected odd-numbered problems begin on page ANS-18.*

A. True/False

In Problems 1–10, answer true or false.

1. $\sin(\pi/6) = \cos(\pi/3)$ _____
2. $\sin(\pi/2) = \sin(5\pi/2)$ _____
3. $\sin\frac{1}{2} = 30°$ _____
4. $\tan\pi = 0$ _____
5. $|\csc\theta| \leq 1$ _____
6. $\sin^2\theta + \sin^2(90° - \theta) = 1$ _____
7. The angles 120° and $-240°$ are coterminal. _____
8. If $\tan\theta = \frac{2}{5}$, then $\sin\theta = 2$ and $\cos\theta = 5$. _____
9. If $\sec\theta = \sqrt{7}$, then $\cos\theta = \sqrt{7}/7$. _____
10. 30′ is equivalent to 0.5°. _____

B. Fill in the Blanks

In Problems 1–10, fill in the blanks.

1. The complement of the acute angle 23° is _____.
2. An angle of 1° in standard position is formed by _____ of a complete counterclockwise rotation of its terminal side.
3. An angle in standard position that has negative measure was formed by a _____ rotation of its terminal side.
4. A central angle θ of a circle of radius 8 inches subtends an arc of 12 inches; the measure of the angle θ in radians is _____.
5. If θ is an acute angle measured in degrees such that $\sin\theta = \frac{2}{3}$, then the exact value of $3\cos(90° - \theta) = $ _____.
6. $\sec 51°/\csc 49° = $ _____
7. If θ is an acute angle measured in degrees such that $\cot\theta = 2$, then the exact value of $\cot\theta + \cot(90° - \theta) = $ _____.
8. If θ is an acute angle such that $\tan\theta = \sqrt{3}$, then $\theta = $ _____.
9. The reference angle for $4\pi/3$ is _____.
10. π radians = _____ degrees.

C. Review Exercises

In Problems 1–4, draw the given angle in standard position.

1. $-5\pi/6$ 2. $7\pi/3$ 3. $225°$ 4. $-450°$

In Problems 5–8, convert the given angle to radian measure.

5. $-120°$ 6. $1°$ 7. $48.3°$ 8. $14°14'$

In Problems 9–12, convert the given angle to decimal degrees.

9. $\pi/9$ 10. $78°15'$ 11. 2.3 12. $7\pi/3$

In Problems 13–16, convert the given angle to degrees, minutes, and seconds.

13. $70.5°$ 14. $170.15°$ 15. 3.1 16. $\pi/10$

In Problems 17 and 18, find two positive and two negative angles that are coterminal with the given angle.

17. $85°$ 18. $7\pi/6$

In Problems 19–22, evaluate the six trigonometric functions of the angle θ if θ is in standard position and the terminal side of θ contains the given point.

19. $(-1, 2)$ 20. $(4, 7)$ 21. $(-0.5, -0.3)$ 22. $(\sqrt{2}, \sqrt{5})$

In Problems 23–28, the value of one of the trigonometric functions of an angle θ is given. From the given value and the additional information, determine the values of the five remaining trigonometric functions of θ.

23. $\cos\theta = -\frac{1}{7}$, θ is in quadrant III
24. $\sin\theta = \frac{2}{3}$, θ is in quadrant II
25. $\cot\theta = -5$, θ is in quadrant IV
26. $\sec\theta = 15$, $\sin\theta < 0$
27. $\csc\theta = -7$, $\tan\theta > 0$
28. $\tan\theta = \frac{1}{9}$, $\sec\theta < 0$

29. If $\cot\theta = -4$, find all possible values of $\sin\theta$, $\cos\theta$, $\tan\theta$, $\sec\theta$, and $\csc\theta$.
30. If $4\sin\theta = 3\cos\theta$, find all possible values of $\sin\theta$, $\cos\theta$, $\tan\theta$, $\sec\theta$, and $\csc\theta$.

In Problems 31–34, find the exact value of the given expression. Do not use a calculator.

31. $\sin\left(-\dfrac{7\pi}{4}\right)$ **32.** $\csc\dfrac{13\pi}{6}$

33. $\tan 495°$ **34.** $\sin 330°$

In Problems 35–38, find all angles θ, such that $0° \leq \theta < 360°$, satisfying the given condition. Do not use a calculator.

35. $\sin\theta = \dfrac{\sqrt{2}}{2}$ **36.** $\tan\theta = -\dfrac{\sqrt{3}}{3}$

37. $\sec\theta = -2$ **38.** $\csc\theta = -\sqrt{2}$

In Problems 39–42, find all angles θ, such that $0 \leq \theta < 2\pi$, satisfying the given condition. Do not use a calculator.

39. $\sin\theta = \dfrac{\sqrt{3}}{2}$ **40.** $\csc\theta = -1$

41. $\cot\theta = -1$ **42.** $\cos\theta = \dfrac{1}{2}$

7 Unit Circle Trigonometry

In This Chapter

- **7.1** The Circular Functions
- **7.2** Graphs of Sine and Cosine Functions
- **7.3** Graphs of Other Trigonometric Functions
- **7.4** Special Identities
- **7.5** Inverse Trigonometric Functions
- **7.6** Trigonometric Equations

Chapter 7 Review Exercises

The shape of a plucked guitar string, fixed at both ends, can be described by trigonometric functions of a real variable.

A Bit of History The discussion in Section 6.4 leads directly to a more analytical approach to trigonometry where the cosine and sine are defined as the x- and y-coordinates, respectively, of a point (x, y) on a unit circle. It is this interpretation of the sine and cosine that enables us to define the trigonometric functions of a real number instead of an angle. It is this last approach to trigonometry that is used in calculus and in advanced applications of trigonometry. Moreover, a trigonometric function of a real number can then be graphed as we would an ordinary function $y = f(x)$, where the variable x represents a real number in the domain of f.

From a historical viewpoint, it is not known who made this important leap from sines and cosines of angles to sines and cosines of real numbers.

7.1 The Circular Functions

≡ **Introduction** In Chapter 6 we considered trigonometric functions of *angles* measured either in degrees or in radians. For calculus and the sciences it is necessary to consider trigonometric functions whose domains consist of *real numbers* rather than angles. The transition from angles to real numbers is made by recognizing that to each real number t, there corresponds an angle of measure t radians. As we see next, this correspondence can be visualized using a circle with radius 1 centered at the origin in a rectangular coordinate system. This circle is called the **unit circle**. From Section 2.2 it follows that the equation of the unit circle is $x^2 + y^2 = 1$. In this section the focus will be on the sine and cosine functions. The other four trigonometric functions will be considered in detail in Section 7.3.

We now consider a **central angle** t in standard position; that is, an angle with its vertex at the center of a circle and initial side coinciding with the positive x-axis. From the definition of radian measure, (3) of Section 6.1, the angle t is defined to be $t = s/r$, the ratio of the subtended arc of length s to the radius r of the circle. For the unit circle shown in FIGURE 7.1.1, $r = 1$, and so $t = s/1$ or $t = s$. In other words:

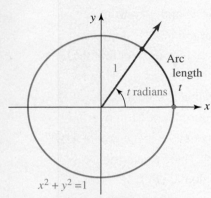

FIGURE 7.1.1 Unit circle

- *On a unit circle, the radian measure of an angle of t radians is equal to the measure t of the subtended arc.*

It follows that for each real number t, the terminal side of an angle of t radians in standard position has traversed a distance of $|t|$ units along the circumference of the unit circle—counterclockwise if $t > 0$, clockwise if $t < 0$. This association of each real number t with an angle of t radians is illustrated in FIGURE 7.1.2.

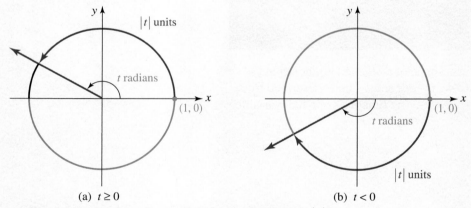

(a) $t \geq 0$ (b) $t < 0$

FIGURE 7.1.2 Angle of t radians subtends an arc of length $|t|$ units

☐ **Trigonometric Functions of Real Numbers** We are now in a position to define **trigonometric functions** of a real number. Before proceeding we need the following important definition.

DEFINITION 7.1.1 Values of the Trigonometric Functions

The value of a trigonometric function at a real number t is defined to be its value at an angle of t radians, provided that value exists.

CHAPTER 7 UNIT CIRCLE TRIGONOMETRY

For example, the sine of the real number $\pi/6 = 0.62359\ldots$ is simply the sine of the angle $\pi/6$ radian that, as we know, is $\frac{1}{2}$. Thus there is really nothing new in evaluating the trigonometric functions of a real number.

The unit circle is very helpful in describing the trigonometric functions of real numbers. If $P(t)$ denotes the point of intersection of the terminal side of the angle t with the unit circle $x^2 + y^2 = 1$ and $P(x, y)$ are the rectangular coordinates of this point, then from (1) of Section 6.4 we have

$$\sin t = \frac{y}{r} = \frac{y}{1} = y \quad \text{and} \quad \cos t = \frac{x}{r} = \frac{x}{1} = x.$$

These definitions, along with the definitions of the remaining four trigonometric functions, are summarized next.

DEFINITION 7.1.2 Trigonometric Functions

Let t be any real number and $P(t) = P(x, y)$ be the point of intersection on the unit circle with the terminal side of the angle of t radians in standard position. Then the six trigonometric functions of the real number t are

$$\sin t = y \qquad \cos t = x$$
$$\tan t = \frac{y}{x} \qquad \cot t = \frac{x}{y} \qquad (1)$$
$$\sec t = \frac{1}{x} \qquad \csc t = \frac{1}{y}.$$

From the first line in (1) of Definition 7.1.2 we see immediately that

- For any real number t, the **cosine** and **sine** of t are the x- and y-coordinates, respectively, of the point P of intersection of the terminal side of the angle of t radians (in standard position) with the unit circle.

See **FIGURE 7.1.3**.

As we will soon see, a number of important properties of the sine and cosine functions can be obtained from this result. Because of the role played by the unit circle in this discussion, the trigonometric functions (1) are often referred to as the **circular functions**.

A number of properties of the sine and cosine functions follow from the fact that $P(t) = (\cos t, \sin t)$ lies on the unit circle. For instance, the coordinates of $P(t)$ must satisfy the equation of the circle:

$$x^2 + y^2 = 1.$$

Substituting $x = \cos t$ and $y = \sin t$ into the foregoing equation gives the familiar result $\cos^2 t + \sin^2 t = 1$. This relationship between the sine and cosine functions is the most fundamental of trigonometric identities, the **Pythagorean identity**. Bear in mind this identity is not just valid for angles as discussed in Sections 6.2 and 6.4; we see now that it is valid for all real numbers t.

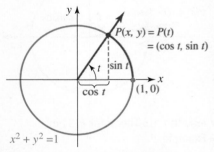

FIGURE 7.1.3 Coordinates of $P(t)$ are $(\cos t, \sin t)$

THEOREM 7.1.1 Pythagorean Identity

For all real numbers t,

$$\sin^2 t + \cos^2 t = 1. \qquad (2)$$

☐ **Bounds on the Values of Sine and Cosine** A number of properties of the sine and cosine functions follow from the fact that $P(t) = P(x, y)$ lies on the unit circle. For instance, it follows that

$$-1 \leq x \leq 1 \quad \text{and} \quad -1 \leq y \leq 1.$$

Since $x = \cos t$ and $y = \sin t$, the foregoing inequalities are equivalent to

$$-1 \leq \cos t \leq 1 \quad \text{and} \quad -1 \leq \sin t \leq 1. \tag{3}$$

The inequalities in (3) can also be expressed using absolute values as $|\cos t| \leq 1$ and $|\sin t| \leq 1$. Thus, for example, there is no real number t for which $\sin t = \frac{3}{2}$.

☐ **Domain and Range** The observations in (3) indicate that both $\cos t$ and $\sin t$ can be any number in the interval $[-1, 1]$. Thus we have the sine and cosine functions,

$$f(t) = \sin t \quad \text{and} \quad g(t) = \cos t,$$

respectively, each with domain the set R of all real numbers and range the interval $[-1, 1]$. The domains and ranges of the other four trigonometric functions will be discussed in Section 7.3.

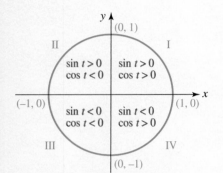

FIGURE 7.1.4 Algebraic signs of $\sin t$ and $\cos t$ in the four quadrants

☐ **Signs of the Circular Functions** The signs of the function values $\sin t$ and $\cos t$ are determined by the quadrant in which the point $P(t)$ lies, and conversely. For example, if $\sin t$ and $\cos t$ are both negative, then the point $P(t)$ and terminal side of the corresponding angle of t radians must lie in quadrant III. FIGURE 7.1.4 displays the signs of the cosine and sine functions in each of the four quadrants.

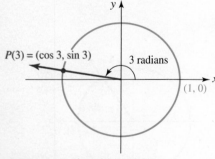

FIGURE 7.1.5 The point $P(3)$ in Example 1

EXAMPLE 1   Sine and Cosine of a Real Number

Use a calculator to approximate $\sin 3$ and $\cos 3$ and give a geometric interpretation of these values.

Solution From a calculator set in *radian mode*, we obtain $\cos 3 \approx -0.9899925$ and $\sin 3 \approx 0.1411200$. These values represent the x and y coordinates, respectively, of the point of intersection $P(3)$ of the terminal side of the angle of 3 radians in standard position, with the unit circle. As shown in FIGURE 7.1.5, this point lies in the second quadrant because $\pi/2 < 3 < \pi$. This would also be expected in view of Figure 7.1.4 since $\cos 3$, the x-coordinate, is *negative* and $\sin 3$, the y-coordinate, is *positive*. ≡

☐ **Periodicity** In Section 6.1 we saw that the angles of t radians and $t \pm 2\pi$ radians are coterminal. Thus they determine the same point $P(x, y)$ on the unit circle. Therefore

$$\sin t = \sin(t \pm 2\pi) \quad \text{and} \quad \cos t = \cos(t \pm 2\pi). \tag{4}$$

In other words, the sine and cosine functions repeat their values every 2π units. It also follows that for any integer n:

$$\begin{aligned} \sin(t + 2n\pi) &= \sin t \\ \cos(t + 2n\pi) &= \cos t. \end{aligned} \tag{5}$$

DEFINITION 7.1.3   Periodic Functions

A nonconstant function f is said to be **periodic** if there is a positive number p such that

$$f(t) = f(t + p) \tag{6}$$

for every t in the domain of f. If p is the smallest positive number for which (6) is true, then p is called the **period** of the function f.

The equations in (4) imply that the sine and cosine functions are periodic with period $p \leq 2\pi$. To see that the period of $\sin t$ is actually 2π, we observe that there is only one point on the unit circle with y-coordinate 1, namely, $P(\pi/2) = (\cos(\pi/2), \sin(\pi/2)) = (0, 1)$. Therefore,

$$\sin t = 1 \quad \text{only for} \quad t = \frac{\pi}{2}, \frac{\pi}{2} \pm 2\pi, \frac{\pi}{2} \pm 4\pi,$$

and so on. Thus the smallest possible positive value of p is 2π. In summary, the sine function $f(t) = \sin t$ and cosine function $g(t) = \cos t$ are periodic with **period** 2π; that is, $f(t) = f(t + 2\pi)$ and $g(t) = g(t + 2\pi)$, respectively. For future reference, we have

$$\sin(t + 2\pi) = \sin t \quad \text{and} \quad \cos(t + 2\pi) = \cos t \qquad (7)$$

for every real number t.

EXAMPLE 2 Using Periodicity

Evaluate **(a)** $\sin(7\pi/3)$ **(b)** $\cos(13\pi/3)$.

Solution **(a)** Because $7\pi/3$ is greater than 2π and can be written

$$\frac{7\pi}{3} = 2\pi + \frac{\pi}{3},$$

it follows from $\sin(t + 2\pi) = \sin t$ with $t = \pi/3$, that

$$\sin\frac{7\pi}{3} = \sin\left(\frac{\pi}{3} + 2\pi\right) = \sin\frac{\pi}{3} = \frac{\sqrt{3}}{2}. \quad \leftarrow \text{See Table 6.3.1} \qquad \blacktriangleleft \text{See the first equation in (7).}$$

(b) Because

$$\frac{19\pi}{3} = 6\pi + \frac{\pi}{3},$$

it follows from $\cos(t + 2n\pi) = \cos t$ with $n = 3$ and $t = \pi/3$, that

$$\cos\frac{19\pi}{3} = \cos\left(\frac{\pi}{3} + 6\pi\right) = \cos\frac{\pi}{3} = \frac{1}{2}. \qquad \blacktriangleleft \text{See the second equation in (7).} \quad \equiv$$

☐ **Odd–Even Properties** The symmetry of the unit circle endows the circular functions with several additional properties. For any real number t, the points $P(t)$ and $P(-t)$ on the unit circle are located on the terminal side of an angle of t and $-t$ radians, respectively. These two points will always be symmetric with respect to the x-axis. **FIGURE 7.1.6** illustrates the situation for a point $P(t)$ lying in the first quadrant: the x-coordinates of the two points are identical, but the y-coordinates have equal magnitudes but opposite signs. The same symmetries will hold regardless of which quadrant contains $P(t)$. Thus, for $f(t) = \sin t$ and $g(t) = \cos t$ and any real number t, $f(-t) = -f(t)$ and $g(-t) = g(t)$, respectively. Applying the definitions of **odd** and **even functions** from Section 3.2 we have the following result.

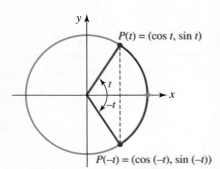

FIGURE 7.1.6 Coordinates of $P(t)$ and $P(-t)$

THEOREM 7.1.2 Odd and Even Functions

The sine function $f(t) = \sin t$ is **odd** and the cosine function $g(t) = \cos t$ is **even**; that is, for every real number t,

$$\sin(-t) = -\sin t \quad \text{and} \quad \cos(-t) = \cos t. \qquad (8)$$

EXAMPLE 3 Using the Odd–Even Properties

Find exact values of $\sin t$ and $\cos t$ for the real number $t = -\pi/6$.

Solution From (8) we have

$$\overbrace{\sin\left(-\frac{\pi}{6}\right)}^{\text{sine is an odd function}} = -\sin\frac{\pi}{6} = -\frac{1}{2}, \quad \leftarrow \text{See Table 6.3.1}$$

and

$$\overbrace{\cos\left(-\frac{\pi}{6}\right)}^{\text{cosine is an even function}} = \cos\frac{\pi}{6} = \frac{\sqrt{3}}{2}.$$

Note that the signs of the answers are consistent with the fact that the terminal side of the angle $-\pi/6$ radian lies in quadrant IV. ≡

The following additional properties of the sine and cosine functions can be verified by considering the symmetries of appropriately chosen points on the unit circle. We first saw the results in (*i*) and (*ii*) in the next theorem stated for acute angles in (5) of Section 6.2.

THEOREM 7.1.3 Additional Properties

For all real numbers t,

(*i*) $\cos\left(\dfrac{\pi}{2} - t\right) = \sin t$ (*ii*) $\sin\left(\dfrac{\pi}{2} - t\right) = \cos t$

(*iii*) $\cos(t + \pi) = -\cos t$ (*iv*) $\sin(t + \pi) = -\sin t$

(*v*) $\cos(\pi - t) = -\cos t$ (*vi*) $\sin(\pi - t) = \sin t$

FIGURE 7.1.7 Geometric justification of (*i*) and (*ii*) of Theorem 7.1.3

For example, to justify properties (*i*) and (*ii*) of Theorem 7.1.3 for $0 < t < \pi/2$, consider **FIGURE 7.1.7**. Since the points $P(t)$ and $P(\pi/2 - t)$ are symmetric with respect to the line $y = x$, we can obtain the coordinates of $P(\pi/2 - t)$ by interchanging the coordinates of $P(t)$. Thus,

$$\cos t = x = \sin\left(\frac{\pi}{2} - t\right) \quad \text{and} \quad \sin t = y = \cos\left(\frac{\pi}{2} - t\right).$$

In Section 7.4 we will use properties (*i*) and (*ii*) to justify two important formulas for the sine function.

EXAMPLE 4 Using Theorem 7.1.3

In Table 6.3.1 in Section 6.3 we saw that $\cos(\pi/3) = \sin(\pi/6)$. This result is a special case of property (*i*) of Theorem 7.1.3; with $t = \pi/3$ we see that

$$\sin\frac{\pi}{6} = \overbrace{\sin\left(\frac{\pi}{2} - \frac{\pi}{3}\right)}^{\text{using property (}i\text{) of Theorem 7.1.3}} = \cos\frac{\pi}{3}. \qquad \equiv$$

☐ **Reference Angle—Revisited** As we noted at the beginning of this section, for each real number t there is a unique angle of t radians in standard position that determines the point $P(t)$, with coordinates $(\cos t, \sin t)$, on the unit circle. As shown in **FIGURE 7.1.8**,

CHAPTER 7 UNIT CIRCLE TRIGONOMETRY

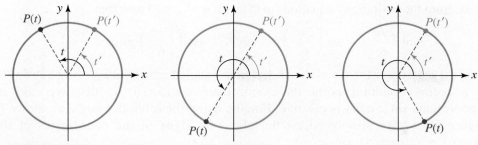

FIGURE 7.1.8 Reference angle t' is an acute angle

the terminal side of any angle of t radians (with $P(t)$ not on an axis) will form an acute angle with the x-axis. We can then locate an angle of t' radians in the first quadrant that is congruent to this acute angle. The angle of t' radians is called the **reference angle** for the real number t. Because of the symmetry of the unit circle, the coordinates of $P(t')$ will be equal *in absolute value* to the respective coordinates of $P(t)$. Hence

$$\sin t = \pm \sin t' \quad \text{and} \quad \cos t = \pm \cos t'.$$

As the following examples will show, reference angles can be used to find the trigonometric function values of any integer multiples of $\pi/6$, $\pi/4$, and $\pi/3$.

EXAMPLE 5   Using a Reference Angle

Find the exact values of $\sin t$ and $\cos t$ for the given real number:
(a) $t = 5\pi/3$ **(b)** $t = -3\pi/4$.

Solution In each part we begin by finding the reference angle corresponding to the given real number t.

(a) From **FIGURE 7.1.9** we find that an angle of $t = 5\pi/3$ radians determines a point $P(5\pi/3)$ in the fourth quadrant and has the reference angle $t' = \pi/3$ radians. After adjusting the signs of the coordinates of $P(\pi/3) = (1/2, \sqrt{3}/2)$ to obtain the fourth-quadrant point $P(5\pi/3) = (1/2, -\sqrt{3}/2)$, we find that

$$\sin \frac{5\pi}{3} = -\sin \overbrace{\frac{\pi}{3}}^{\text{reference angle}} = -\frac{\sqrt{3}}{2} \quad \text{and} \quad \cos \frac{5\pi}{3} = \cos \frac{\pi}{3} = \frac{1}{2}.$$

FIGURE 7.1.9 Reference angle in part (a) of Example 5

(b) The point $P(-3\pi/4)$ lies in the third quadrant and has reference angle $\pi/4$ as shown in **FIGURE 7.1.10**. Therefore

$$\sin\left(-\frac{3\pi}{4}\right) = -\sin\frac{\pi}{4} = -\frac{\sqrt{2}}{2} \quad \text{and} \quad \cos\left(-\frac{3\pi}{4}\right) = -\cos\frac{\pi}{4} = -\frac{\sqrt{2}}{2}. \equiv$$

Sometimes, in order to find the trigonometric values of multiples of our basic fractions of π we must use periodicity or the even–odd function properties in addition to reference angles.

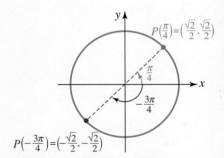

FIGURE 7.1.10 Reference angle in part (b) of Example 5

EXAMPLE 6   Using Periodicity and a Reference Angle

Find the exact values of the coordinates of $P(29\pi/6)$ on the unit circle.

Solution The point $P(29\pi/6)$ has coordinates $(\cos(29\pi/6), \sin(29\pi/6))$. We begin by observing that $29\pi/6$ is greater than 2π, and so we must rewrite $29\pi/6$ as an integer multiple of 2π plus a number less than 2π. By division we have

$$\frac{29\pi}{6} = 4\pi + \frac{5\pi}{6} = 2(2\pi) + \frac{5\pi}{6}.$$

7.1 The Circular Functions

FIGURE 7.1.11 Reference angle in Example 6

Next, from the periodicity equations in (5) with $n = 2$ we know that

$$\sin\left(\frac{29\pi}{6}\right) = \sin\left(\frac{5\pi}{6}\right) \quad \text{and} \quad \cos\left(\frac{29\pi}{6}\right) = \cos\left(\frac{5\pi}{6}\right).$$

Next we see from **FIGURE 7.1.11** that the reference angle for $5\pi/6$ is $\pi/6$. Since $P(5\pi/6)$ is a second-quadrant point its x-coordinate $\cos(5\pi/6)$ is negative and its y-coordinate $\sin(5\pi/6)$ is positive. Finally, using the reference angle as shown in Figure 7.1.11 we simply adjust the algebraic signs of the coordinates of the $P(\pi/6) = (\cos(\pi/6), \sin(\pi/6))$:

$$\cos\frac{29\pi}{6} = \cos\frac{5\pi}{6} = -\cos\frac{\pi}{6} = -\frac{\sqrt{3}}{2}$$

and

$$\sin\frac{29\pi}{6} = \sin\frac{5\pi}{6} = \sin\frac{\pi}{6} = \frac{1}{2}.$$

Thus, $P(29\pi/6) = (-\sqrt{3}/2, 1/2)$. ≡

7.1 Exercises
Answers to selected odd-numbered problems begin on page ANS-18.

In Problems 1–8, for the given real number t, **(a)** locate the point $P(t) = (\cos t, \sin t)$ on the unit circle and **(b)** find the exact values of the coordinates of $P(t)$. Do not use a calculator.

1. $\dfrac{7\pi}{6}$
2. $-\dfrac{4\pi}{3}$
3. $-\dfrac{\pi}{2}$
4. 2π
5. $\dfrac{5\pi}{3}$
6. $-\dfrac{3\pi}{2}$
7. $-\dfrac{11\pi}{6}$
8. $\dfrac{5\pi}{4}$

In Problems 9–16, for the given real number t, **(a)** locate the point $P(t) = (\cos t, \sin t)$ on the unit circle and **(b)** use a calculator to approximate the coordinates of $P(t)$.

9. 1.3
10. −4.4
11. −7.2
12. 0.5
13. 6.1
14. 3.2
15. −2.6
16. 15.3

In Problems 17–24, use periodicity of $\sin t$ and $\cos t$ to find the exact value of the given trigonometric function. Do not use a calculator.

17. $\sin\dfrac{13\pi}{6}$
18. $\cos\dfrac{61\pi}{3}$
19. $\cos\dfrac{9\pi}{4}$
20. $\sin\left(-\dfrac{5\pi}{3}\right)$
21. $\cos 9\pi$
22. $\sin 20\pi$
23. $\sin\dfrac{7\pi}{2}$
24. $\cos\dfrac{27\pi}{4}$

In Problems 25–30, justify the given statement by one of the properties of $\sin t$ and $\cos t$ given in this section.

25. $\sin \pi = \sin 3\pi$
26. $\cos(\pi/4) = \sin(\pi/4)$
27. $\sin(-3 - \pi) = -\sin(3 + \pi)$
28. $\cos 16.8\pi = \cos 14.8\pi$
29. $\cos 0.43 = \cos(-0.43)$
30. $\cos(2.5 + \pi) = -\cos 2.5$

CHAPTER 7 UNIT CIRCLE TRIGONOMETRY

31. Given that $\cos t = -\frac{2}{5}$ and that $P(t)$ is a point on the unit circle in the second quadrant, find $\sin t$.
32. Given that $\sin t = \frac{1}{4}$ and that $P(t)$ is a point on the unit circle in the second quadrant, find $\cos t$.
33. Given that $\sin t = -\frac{2}{3}$ and that $P(t)$ is a point on the unit circle in the third quadrant, find $\cos t$.
34. Given that $\cos t = \frac{3}{4}$ and that $P(t)$ is a point on the unit circle in the fourth quadrant, find $\sin t$.

In Problems 35–38, the y-coordinate of the point $P(5\pi/8)$ on the unit circle is $\frac{1}{2}\sqrt{2 + \sqrt{2}}$. Find the exact value of the given trigonometric function. Do not use a calculator.

35. $\cos\dfrac{5\pi}{8}$
36. $\sin\left(\dfrac{5\pi}{8} - 2\pi\right)$
37. $\sin\left(-\dfrac{5\pi}{8}\right)$
38. $\cos\left(-\dfrac{5\pi}{8}\right)$

In Problems 39–42, use the unit circle to determine all real numbers t for which the given equality is true.

39. $\sin t = \sqrt{2}/2$
40. $\cos t = -\frac{1}{2}$
41. $\cos t = -1$
42. $\sin t = -1$

For Discussion

43. Suppose f is a periodic function with period p. Show that $F(x) = f(ax)$, $a > 0$, is periodic with period p/a.

7.2 Graphs of Sine and Cosine Functions

☰ **Introduction** One way to further your understanding of the trigonometric functions is to examine their graphs. In this section we consider the graphs of the sine and cosine functions.

☐ **Graphs of Sine and Cosine** In Section 7.1 we saw that the domain of the sine function $f(t) = \sin t$ is the set of real numbers $(-\infty, \infty)$ and the interval $[-1, 1]$ is its range. Since the sine function has period 2π, we begin by sketching its graph on the interval $[0, 2\pi]$. We obtain a rough sketch of the graph given in **FIGURE 7.2.1(b)** by considering various positions of the point $P(t)$ on the unit circle, as shown in Figure 7.2.1(a). As t varies from 0 to $\pi/2$, the value $\sin t$ increases from 0 to its maximum value 1. But as t varies from $\pi/2$ to $3\pi/2$, the value $\sin t$ decreases from 1 to its minimum value -1. We note that $\sin t$ changes from positive to negative at $t = \pi$. For t between $3\pi/2$ and 2π, we see that the corresponding values of $\sin t$ increase from -1 to 0. The graph of *any* periodic function over an interval of length equal to its period is said to be one **cycle** of its graph. In the case of the sine function, the graph over the interval $[0, 2\pi]$ in Figure 7.2.1(b) is one cycle of the graph of $f(t) = \sin t$.

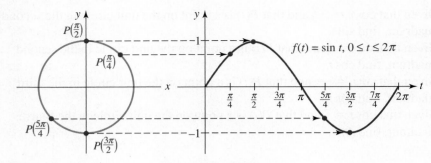

(a) Unit circle   (b) One cycle of sine graph

FIGURE 7.2.1 Points $P(t)$ on the unit circle corresponding to points on the graph

Note: Change of symbols ▶

From this point on we will switch to the traditional symbols x and y when graphing trigonometric functions. Thus, $f(t) = \sin t$ will either be written $f(x) = \sin x$ or simply $y = \sin x$.

The graph of a periodic function is easily obtained by repeatedly drawing one cycle of its graph. In other words, the graph of $y = \sin x$ on, say, the intervals $[-2\pi, 0]$ and $[2\pi, 4\pi]$ is the same as that given in Figure 7.2.1(b). Recall from Section 7.1 that the sine function is an odd function since $f(-x) = \sin(-x) = -\sin x = -f(x)$. Thus, as can be seen in **FIGURE 7.2.2**, the graph of $y = \sin x$ is symmetric with respect to the origin.

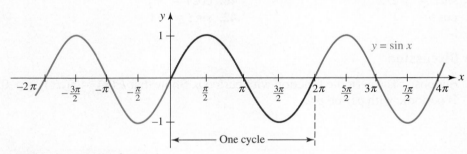

FIGURE 7.2.2 Graph of $y = \sin x$

By working again with the unit circle we can obtain one cycle of the graph of the cosine function $g(x) = \cos x$ on the interval $[0, 2\pi]$. In contrast to the graph of $f(x) = \sin x$ where $f(0) = f(2\pi) = 0$, for the cosine function we have $g(0) = g(2\pi) = 1$. **FIGURE 7.2.3** shows one cycle (in red) of $y = \cos x$ on $[0, 2\pi]$ along with the extension of that cycle (in blue) to the adjacent intervals $[-2\pi, 0]$ and $[2\pi, 4\pi]$. We see from this figure that the graph of the cosine function is symmetric with respect to the y-axis. This is a consequence of the fact that g is an even function: $g(-x) = \cos(-x) = \cos x = g(x)$.

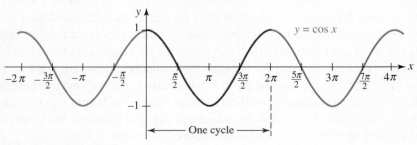

FIGURE 7.2.3 Graph of $y = \cos x$

☐ **Properties of the Sine and Cosine Functions** In this and subsequent courses in mathematics it is important that you know the x-coordinates of the x-intercepts of the sine and cosine graphs, in other words, the zeros of $f(x) = \sin x$ and $g(x) = \cos x$. From the sine graph in Figure 7.2.2 we see that the zeros of the sine function, or the numbers for which $\sin x = 0$, are $x = 0, \pm\pi, \pm 2\pi, \pm 3\pi, \ldots$. These numbers are integer multiples of π. From the cosine graph in Figure 7.2.3 we see that $\cos x = 0$ when $x = \pm\pi/2, \pm 3\pi/2, \pm 5\pi/2, \ldots$. These numbers are odd-integer multiples of $\pi/2$, that is, $x = (2n + 1)\pi/2$, where n is an integer. Using the distributive law, the zeros of $g(x) = \cos x$ are often written as $x = \pi/2 + n\pi$. The following list summarizes some of the important properties of the sine and cosine functions that are apparent from their graphs.

◀ An odd integer can be written as $2n + 1$, where n is an integer.

PROPERTIES OF THE SINE AND COSINE FUNCTIONS

- The domain of $f(x) = \sin x$ and the domain of $g(x) = \cos x$ is the set of real numbers, that is, $(-\infty, \infty)$.
- The range of $f(x) = \sin x$ and the range of $g(x) = \cos x$ is the interval $[-1, 1]$ on the y-axis.
- The zeros of $f(x) = \sin x$ are $x = n\pi$, n an integer. The zeros of $g(x) = \cos x$ are $x = (2n + 1)\pi/2$, n an integer.
- The graph of $f(x) = \sin x$ is symmetric with respect to the origin. The graph of $g(x) = \cos x$ is symmetric with respect to the y-axis.
- The functions $f(x) = \sin x$ and $g(x) = \cos x$ are continuous on the interval $(-\infty, \infty)$.

As we did in Chapter 3 we can obtain variations of the basic sine and cosine graphs through rigid and nonrigid transformations. For the remainder of the discussion we will consider graphs of functions of the form

$$y = A\sin(Bx + C) + D \quad \text{or} \quad y = A\cos(Bx + C) + D, \quad (1)$$

where A, B, C, and D are real constants.

☐ **Graphs of $y = A \sin x + D$ and $y = A \cos x + D$** We begin by considering the special cases of (1):

$$y = A \sin x \quad \text{and} \quad y = A \cos x.$$

For $A > 0$ graphs of these functions are either a vertical stretch or a vertical compression of the graphs of $y = \sin x$ or $y = \cos x$. For $A < 0$ the graphs are also reflected in the x-axis. For example, as **FIGURE 7.2.4** shows we obtain the graph of $y = 2 \sin x$ by stretching the graph of $y = \sin x$ vertically by a factor of 2. Note that the maximum and minimum values of $y = 2 \sin x$ occur at the same x-values as the maximum and minimum values of $y = \sin x$. In general, the maximum distance from any point on the graph of $y = A \sin x$ or $y = A \cos x$ to the x-axis is $|A|$. The number $|A|$ is called the **amplitude** of the functions or of their graphs. The amplitude of the basic functions $y = \sin x$ and $y = \cos x$ is $|A| = 1$. In general, if a periodic function f is continuous, then over a closed interval of length equal to its period, f has both a maximum value M and a minimum value m. The amplitude is defined by

$$\text{amplitude} = \frac{1}{2}[M - m]. \quad (2)$$

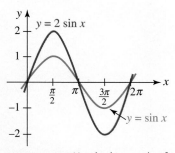

FIGURE 7.2.4 Vertical stretch of $y = \sin x$

7.2 Graphs of Sine and Cosine Functions

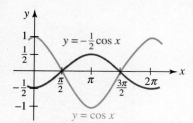

FIGURE 7.2.5 Graph of function in Example 1

FIGURE 7.2.6 Graph of $y = 2\sin x$ shifted up 1 unit

EXAMPLE 1 Vertically Compressed Cosine Graph

Graph $y = -\frac{1}{2}\cos x$.

Solution The graph of $y = -\frac{1}{2}\cos x$ is the graph of $y = \cos x$ compressed vertically by a factor of $\frac{1}{2}$ and then reflected in the x-axis. With the identification $A = -\frac{1}{2}$ we see that the amplitude of the function is $|A| = \left|-\frac{1}{2}\right| = \frac{1}{2}$. The graph of $y = -\frac{1}{2}\cos x$ on the interval $[0, 2\pi]$ is shown in red in FIGURE 7.2.5. ≡

The graphs of

$$y = A\sin x + D \quad \text{and} \quad y = A\cos x + D$$

are the respective graphs of $y = A\sin x$ and $y = A\cos x$ shifted vertically, up for $D > 0$ and down for $D < 0$. For example, the graph of $y = 1 + 2\sin x$ is the graph of $y = 2\sin x$ (Figure 7.2.4) shifted up 1 unit. The amplitude of the graph of either $y = A\sin x + D$ or $y = A\cos x + D$ is still $|A|$. Observe in FIGURE 7.2.6, the maximum of $y = 1 + 2\sin x$ is $y = 3$ at $x = \pi/2$ and the minimum is $y = -1$ at $x = 3\pi/2$. From (2), the amplitude of $y = 1 + 2\sin x$ is then $\frac{1}{2}[3 - (-1)] = 2$.

By interpreting x as a placeholder we can find the x-coordinates of the x-intercepts of the graphs of sine and cosine functions of the form $y = A\sin Bx$ and $y = A\cos Bx$ (considered next). For example, to solve $\sin 2x = 0$, we use the fact that the zeros of $f(x) = \sin x$ are $x = n\pi$, where n is an integer. We simply replace x by $2x$ to obtain

$$2x = n\pi \quad \text{so that} \quad x = \tfrac{1}{2}n\pi, n = 0, \pm 1, \pm 2, \ldots;$$

that is, $\sin 2x = 0$ for $x = 0, \pm\tfrac{1}{2}\pi, \pm\tfrac{2}{2}\pi = \pi, \pm\tfrac{3}{2}\pi, \pm\tfrac{4}{2}\pi = 2\pi$, and so on. See FIGURE 7.2.7.

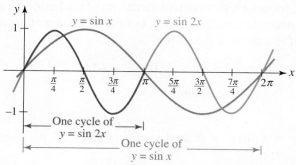

FIGURE 7.2.7 Comparison of the graphs of $y = \sin x$ and $y = \sin 2x$

☐ **Graphs of $y = A\sin Bx$ and $y = A\cos Bx$** We now consider the graph of $y = \sin Bx$, for $B > 0$. The function has amplitude 1 since $A = 1$. Since the period of $y = \sin x$ is 2π, a cycle of the graph of $y = \sin Bx$ begins at $x = 0$ and will start to repeat its values when $Bx = 2\pi$. In other words, a cycle of the function $y = \sin Bx$ is completed on the interval defined by $0 \le Bx \le 2\pi$. Dividing the last inequality by B shows that the **period** of the function $y = \sin Bx$ is $2\pi/B$ and that the graph over the interval $[0, 2\pi/B]$ is one **cycle** of its graph. For example, the period of $y = \sin 2x$ is $2\pi/2 = \pi$, and therefore one cycle of the graph is completed on the interval $[0, \pi]$. Figure 7.2.7 shows that two cycles of the graph of $y = \sin 2x$ (in red and blue) are completed on the interval $[0, 2\pi]$, whereas the graph of $y = \sin x$ (in green) has completed only one cycle. In terms of transformations, we can characterize the cycle of $y = \sin 2x$ on $[0, \pi]$ as a **horizontal compression** of the cycle of $y = \sin x$ on $[0, 2\pi]$.

◀ Careful here; $\sin 2x \ne 2\sin x$

In summary, the graphs of

$$y = A \sin Bx \quad \text{and} \quad y = A \cos Bx$$

for $B > 0$, each have amplitude $|A|$ and period $2\pi/B$.

EXAMPLE 2 Horizontally Compressed Cosine Graph

Find the period of $y = \cos 4x$ and graph the function.

Solution Since $B = 4$, we see that the period of $y = \cos 4x$ is $2\pi/4 = \pi/2$. We conclude that the graph of $y = \cos 4x$ is the graph of $y = \cos x$ compressed horizontally. To graph the function, we draw one cycle of the cosine graph with amplitude 1 on the interval $[0, \pi/2]$ and then use periodicity to extend the graph. FIGURE 7.2.8 shows four complete cycles of $y = \cos 4x$ (the basic cycle in red and the extended graph in blue) and one cycle of $y = \cos x$ (in green) on $[0, 2\pi]$. Notice that $y = \cos 4x$ attains its minimum at $x = \pi/4$ since $\cos 4(\pi/4) = \cos \pi = -1$ and its maximum at $x = \pi/2$ since $\cos 4(\pi/2) = \cos 2\pi = 1$. ≡

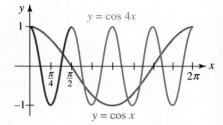

FIGURE 7.2.8 Graph of function in Example 2

If $B < 0$ in either $y = A \sin Bx$ or $y = A \cos Bx$, we can use the odd/even properties, (8) of Section 7.1, to rewrite the function with positive B. This is illustrated in the next example.

EXAMPLE 3 Horizontally Stretched Sine Graph

Find the amplitude and period of $y = \sin(-\tfrac{1}{2}x)$. Graph the function.

Solution Since we require $B > 0$, we use $\sin(-x) = -\sin x$ to rewrite the function as

$$y = \sin(-\tfrac{1}{2}x) = -\sin \tfrac{1}{2} x.$$

With the identification $A = -1$, the amplitude is seen to be $|A| = |-1| = 1$. Now with $B = \tfrac{1}{2}$ we find that the period is $2\pi/\tfrac{1}{2} = 4\pi$. Hence we can interpret the cycle of $y = -\sin \tfrac{1}{2} x$ on $[0, 4\pi]$ as a horizontal stretch and a reflection (in the x-axis because $A < 0$) of the cycle of $y = \sin x$ on $[0, 2\pi]$. FIGURE 7.2.9 shows that on the interval $[0, 4\pi]$ the graph of $y = -\sin \tfrac{1}{2} x$ (in blue) completes one cycle, whereas the graph of $y = \sin x$ (in green) completes two cycles. ≡

FIGURE 7.2.9 Graph of function in Example 3

☐ **Graphs of $y = A \sin(Bx + C)$ and $y = A \cos(Bx + C)$** We have seen that the basic graphs of $y = \sin x$ and $y = \cos x$ can, in turn, be stretched or compressed vertically,

$$y = A \sin x \quad \text{and} \quad y = A \cos x,$$

shifted vertically,

$$y = A \sin x + D \quad \text{and} \quad y = A \cos x + D,$$

and stretched or compressed horizontally,

$$y = A \sin Bx + D \quad \text{and} \quad y = A \cos Bx + D.$$

The graphs of

$$y = A \sin(Bx + C) + D \quad \text{and} \quad y = A \cos(Bx + C) + D,$$

are the graphs of $y = A \sin Bx + D$ and $y = A \cos Bx + D$ shifted horizontally.

In the remaining discussion we are going to focus on the graphs of $y = A \sin(Bx + C)$ and $y = A \cos(Bx + C)$. For example, we know from Section 3.2 that the graph of $y = \cos(x - \pi/2)$ is the basic cosine graph shifted to the right. In

FIGURE 7.2.10 the graph of $y = \cos(x - \pi/2)$ (in red) on the interval $[0, 2\pi]$ is one cycle of $y = \cos x$ on the interval $[-\pi/2, 3\pi/2]$ (in blue) shifted horizontally $\pi/2$ units to the right. Similarly, the graphs of $y = \sin(x + \pi/2)$ and $y = \sin(x - \pi/2)$ are the basic sine graph shifted $\pi/2$ units to the left and to the right, respectively. See FIGURES 7.2.11 and 7.2.12.

FIGURE 7.2.10 Horizontally shifted cosine graph

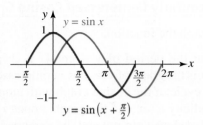

FIGURE 7.2.11 Horizontally shifted sine graph

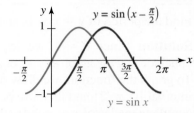

FIGURE 7.2.12 Horizontally shifted sine graph

By comparing the red graphs in Figures 7.2.10–7.2.12 with the graphs in Figures 7.2.2 and 7.2.3 we see that

- the cosine graph shifted $\pi/2$ units to the right is the sine graph,
- the sine graph shifted $\pi/2$ units to the left is the cosine graph, and
- the sine graph shifted $\pi/2$ units to the right is the cosine graph reflected in the x-axis.

In other words, we have graphically verified the identities

$$\cos\left(x - \frac{\pi}{2}\right) = \sin x, \quad \sin\left(x + \frac{\pi}{2}\right) = \cos x, \quad \text{and} \quad \sin\left(x - \frac{\pi}{2}\right) = -\cos x. \quad (3)$$

We now consider the graph of $y = A\sin(Bx + C)$, for $B > 0$. Since the values of $\sin(Bx + C)$ range from -1 to 1, it follows that $A\sin(Bx + C)$ varies between $-A$ and A. That is, the **amplitude** of $y = A\sin(Bx + C)$ is $|A|$. Also, as $Bx + C$ varies from 0 to 2π, the graph will complete one cycle. By solving $Bx + C = 0$ and $Bx + C = 2\pi$, we find that one cycle is completed as x varies from $-C/B$ to $(2\pi - C)/B$. Therefore, the function $y = A\sin(Bx + C)$ has the **period**

$$\frac{2\pi - C}{B} - \left(-\frac{C}{B}\right) = \frac{2\pi}{B}.$$

Moreover, if $f(x) = A\sin Bx$, then

$$f\left(x + \frac{C}{B}\right) = A\sin B\left(x + \frac{C}{B}\right) = A\sin(Bx + C). \quad (4)$$

The result in (4) shows that the graph of $y = A\sin(Bx + C)$ can be obtained by shifting the graph of $f(x) = A\sin Bx$ horizontally a distance $|C|/B$. If $C < 0$ the shift is to the right, whereas if $C > 0$ the shift is to the left. The number $|C|/B$ is called the **phase shift** of the graph of $y = A\sin(Bx + C)$.

EXAMPLE 4 Equation of a Shifted Cosine Graph

The graph of $y = 10\cos 4x$ is shifted $\pi/12$ units to the right. Find its equation.

Solution By writing $f(x) = 10\cos 4x$ and using (4), we find

$$f\left(x - \frac{\pi}{12}\right) = 10\cos 4\left(x - \frac{\pi}{12}\right) \quad \text{or} \quad y = 10\cos\left(4x - \frac{\pi}{3}\right).$$

In the last equation we would identify $C = -\pi/3$. The phase shift is $\pi/12$.

As a practical matter the phase shift of $y = A\sin(Bx + C)$ can be obtained by factoring the number B from $Bx + C$:

$$y = A\sin(Bx + C) = A\sin B\left(x + \frac{C}{B}\right).$$

For convenience the foregoing information is summarized next.

◀ Note

> **SHIFTED SINE AND COSINE GRAPHS**
>
> The graphs of
>
> $$y = A\sin(Bx + C) \quad \text{and} \quad y = A\cos(Bx + C), \quad B > 0,$$
>
> are, respectively, the graphs of $y = A\sin Bx$ and $y = A\cos Bx$ shifted horizontally by $|C|/B$. The shift is to the right if $C < 0$ and to the left if $C > 0$. The number $|C|/B$ is called the **phase shift**. The **amplitude** of each graph is $|A|$ and the **period** of each graph is $2\pi/B$.

EXAMPLE 5   Horizontally Shifted Sine Graph

Graph $y = 3\sin(2x - \pi/3)$.

Solution For purposes of comparison we will first graph $y = 3\sin 2x$. The amplitude of $y = 3\sin 2x$ is $|A| = 3$ and its period is $2\pi/2 = \pi$. Thus one cycle of $y = 3\sin 2x$ is completed on the interval $[0, \pi]$. Then we extend this graph to the adjacent interval $[\pi, 2\pi]$ as shown in blue in FIGURE 7.2.13. Next, we rewrite $y = 3\sin(2x - \pi/3)$ by factoring 2 from $2x - \pi/3$:

$$y = 3\sin\left(2x - \frac{\pi}{3}\right) = 3\sin 2\left(x - \frac{\pi}{6}\right).$$

From the last form we see that the phase shift is $\pi/6$. The graph of the given function, shown in red in Figure 7.2.13, is obtained by shifting the graph of $y = 3\sin 2x$ to the right $\pi/6$ units. Remember, this means that if (x, y) is a point on the blue graph, then $(x + \pi/6, y)$ is the corresponding point on the red graph. For example, $x = 0$ and $x = \pi$ are the x-coordinates of two x-intercepts of the blue graph. Thus $x = 0 + \pi/6 = \pi/6$ and $x = \pi + \pi/6 = 7\pi/6$ are x-coordinates of the x-intercepts of the red or shifted graph. These numbers are indicated by the black arrows in Figure 7.2.13. ≡

FIGURE 7.2.13 Graph of function in Example 5

EXAMPLE 6   Horizontally Shifted Graphs

Determine the amplitude, the period, the phase shift, and the direction of horizontal shift for each of the following functions.

(a) $y = 15\cos\left(5x - \dfrac{3\pi}{2}\right)$ **(b)** $y = -8\sin\left(2x + \dfrac{\pi}{4}\right)$

Solution (a) We first make the identifications $A = 15$, $B = 5$, and $C = -3\pi/2$. Thus the amplitude is $|A| = 15$ and the period is $2\pi/B = 2\pi/5$. The phase shift can be computed either by $(|-3\pi|/2)/5 = 3\pi/10$ or by rewriting the function as

$$y = 15\cos 5\left(x - \frac{3\pi}{10}\right).$$

The last form indicates that the graph of $y = 15\cos(5x - 3\pi/2)$ is the graph of $y = 15\cos 5x$ shifted $3\pi/10$ units to the right.

7.2 Graphs of Sine and Cosine Functions

(b) Since $A = -8$ the amplitude is $|A| = |-8| = 8$. With $B = 2$ the period is $2\pi/2 = \pi$. By factoring 2 from $2x + \pi/4$, we see from

$$y = -8\sin\left(2x + \frac{\pi}{4}\right) = -8\sin 2\left(x + \frac{\pi}{8}\right)$$

that the phase shift is $\pi/8$. The graph of $y = -8\sin(2x + \pi/4)$ is the graph of $y = -8\sin 2x$ shifted $\pi/8$ units to the left. ≡

EXAMPLE 7   Horizontally Shifted Cosine Graph

Graph $y = 2\cos(\pi x + \pi)$.

Solution The amplitude of $y = 2\cos \pi x$ is $|A| = 2$ and the period is $2\pi/\pi = 2$. Thus one cycle of $y = 2\cos \pi x$ is completed on the interval $[0, 2]$. In **FIGURE 7.2.14** two cycles of the graph of $y = 2\cos \pi x$ (in blue) are shown. The x-coordinates of the x-intercepts of this graph are the values of x for which $\cos \pi x = 0$. The last equation implies $\pi x = (2n + 1)\pi/2$ or $x = (2n + 1)/2$, n an integer. In other words, for $n = 0, -1, 1, -2, 2, -3, \ldots$ we get $x = \pm\frac{1}{2}, \pm\frac{3}{2}, \pm\frac{5}{2}$, and so on. Now by rewriting the given function $y = 2\cos(\pi x + \pi)$ as

$$y = 2\cos \pi(x + 1)$$

we see the phase shift is 1. The graph of $y = 2\cos(\pi x + \pi)$ (in red) in Figure 7.2.14, is obtained by shifting the graph of $y = 2\cos \pi x$ to the left 1 unit. This means that the x-intercepts are the same for both graphs. ≡

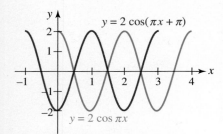

FIGURE 7.2.14 Graph of function in Example 7

EXAMPLE 8   Alternating Current

The current I (in amperes) in a wire of an alternating-current circuit is given by $I(t) = 30\sin 120\pi t$, where t is time measured in seconds. Sketch one cycle of the graph. What is the maximum value of the current?

Solution The graph has amplitude 30 and period $2\pi/120\pi = \frac{1}{60}$. Therefore, we sketch one cycle of the basic sine curve on the interval $[0, \frac{1}{60}]$, as shown in **FIGURE 7.2.15**. From the figure it is evident that the maximum value of the current is $I = 30$ amperes and occurs at $t = \frac{1}{240}$ second since

$$I\left(\tfrac{1}{240}\right) = 30\sin\left(120\pi \cdot \tfrac{1}{240}\right) = 30\sin\frac{\pi}{2} = 30.$$ ≡

FIGURE 7.2.15 Graph of current in Example 8

7.2 Exercises   Answers to selected odd-numbered problems begin on page ANS-18.

In Problems 1–6, use the techniques of shifting, stretching, compressing, and reflecting to sketch at least one cycle of the graph of the given function.

1. $y = \frac{1}{2} + \cos x$
2. $y = -1 + \cos x$
3. $y = 2 - \sin x$
4. $y = 3 + 3\sin x$
5. $y = -2 + 4\cos x$
6. $y = 1 - 2\sin x$

In Problems 7–10, the given figure shows one cycle of a sine or cosine graph. From the figure determine A and D and write an equation of the form $y = A\sin x + D$ or $y = A\cos x + D$ for the graph.

7.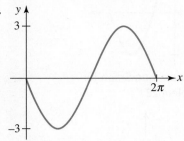

FIGURE 7.2.16 Graph for Problem 7

8.

FIGURE 7.2.17 Graph for Problem 8

9.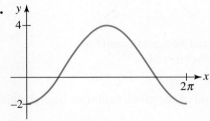

FIGURE 7.2.18 Graph for Problem 9

10.

FIGURE 7.2.19 Graph for Problem 10

In Problems 11–16, find the x-intercepts for the graph of the given function. Do not graph.

11. $y = \sin \pi x$

12. $y = -\cos 2x$

13. $y = 10\cos\dfrac{x}{2}$

14. $y = 3\sin(-5x)$

15. $y = \sin\left(x - \dfrac{\pi}{4}\right)$

16. $y = \cos(2x - \pi)$

In Problems 17 and 18, find the x-intercepts of the graph of the given function on the interval $[0, 2\pi]$. Then find all intercepts using periodicity.

17. $y = -1 + \sin x$

18. $y = 1 - 2\cos x$

In Problems 19–24, the given figure shows one cycle of a sine or cosine graph. From the figure determine A and B and write an equation of the form $y = A\sin Bx$ or $y = A\cos Bx$ for the graph.

19.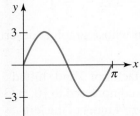

FIGURE 7.2.20 Graph for Problem 19

20.

FIGURE 7.2.21 Graph for Problem 20

7.2 Graphs of Sine and Cosine Functions

21.

FIGURE 7.2.22 Graph for Problem 21

22.

FIGURE 7.2.23 Graph for Problem 22

23.

FIGURE 7.2.24 Graph for Problem 23

24.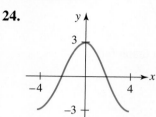

FIGURE 7.2.25 Graph for Problem 24

In Problems 25–32, find the amplitude and period of the given function. Sketch at least one cycle of the graph.

25. $y = 4\sin \pi x$

26. $y = -5\sin \dfrac{x}{2}$

27. $y = -3\cos 2\pi x$

28. $y = \dfrac{5}{2}\cos 4x$

29. $y = 2 - 4\sin x$

30. $y = 2 - 2\sin \pi x$

31. $y = 1 + \cos \dfrac{2x}{3}$

32. $y = -1 + \sin \dfrac{\pi x}{2}$

In Problems 33–42, find the amplitude, period, and phase shift of the given function. Sketch at least one cycle of the graph.

33. $y = \sin\left(x - \dfrac{\pi}{6}\right)$

34. $y = \sin\left(3x - \dfrac{\pi}{4}\right)$

35. $y = \cos\left(x + \dfrac{\pi}{4}\right)$

36. $y = -2\cos\left(2x - \dfrac{\pi}{6}\right)$

37. $y = 4\cos\left(2x - \dfrac{3\pi}{2}\right)$

38. $y = 3\sin\left(2x + \dfrac{\pi}{4}\right)$

39. $y = 3\sin\left(\dfrac{x}{2} - \dfrac{\pi}{3}\right)$

40. $y = -\cos\left(\dfrac{x}{2} - \pi\right)$

41. $y = -4\sin\left(\dfrac{\pi}{3}x - \dfrac{\pi}{3}\right)$

42. $y = 2\cos\left(-2\pi x - \dfrac{4\pi}{3}\right)$

In Problems 43 and 44, write an equation of the function whose graph is described in words.

43. The graph of $y = \cos x$ is vertically stretched up by a factor of 3 and shifted down by 5 units. One cycle of $y = \cos x$ on $[0, 2\pi]$ is compressed to $[0, \pi/3]$ and then the compressed cycle is shifted horizontally $\pi/4$ units to the left.

44. One cycle of $y = \sin x$ on $[0, 2\pi]$ is stretched to $[0, 8\pi]$ and then the stretched cycle is shifted horizontally $\pi/12$ units to the right. The graph is also compressed vertically by a factor of $\dfrac{3}{4}$ and then reflected in the x-axis.

CHAPTER 7 UNIT CIRCLE TRIGONOMETRY

In Problems 45–48, find horizontally shifted sine and cosine functions so that each function satisfies the given conditions. Graph the functions.

45. Amplitude 3, period $2\pi/3$, shifted by $\pi/3$ units to the right
46. Amplitude $\frac{2}{3}$, period π, shifted by $\pi/4$ units to the left
47. Amplitude 0.7, period 0.5, shifted by 4 units to the right
48. Amplitude $\frac{5}{4}$, period 4, shifted by $1/2\pi$ units to the left

In Problems 49 and 50, graphically verify the given identity.

49. $\cos(x + \pi) = -\cos x$ **50.** $\sin(x + \pi) = -\sin x$

Miscellaneous Applications

51. Pendulum The angular displacement θ of a pendulum from the vertical at time t seconds is given by $\theta(t) = \theta_0 \cos \omega t$, where θ_0 is the initial displacement at $t = 0$ seconds. See **FIGURE 7.2.26**. For $\omega = 2$ rad/s and $\theta_0 = \pi/10$, sketch two cycles of the resulting function.

52. Current In a certain kind of electrical circuit, the current I measured in amperes at time t seconds is given by

$$I(t) = 10\cos\left(120\pi t + \frac{\pi}{3}\right).$$

Sketch two cycles of the graph of I as a function of time t.

53. Depth of Water The depth d of water at the entrance to a small harbor at time t is modeled by a function of the form

$$d(t) = A\sin B\left(t - \frac{\pi}{2}\right) + C,$$

where A is one-half the difference between the high- and low-tide depths, $2\pi/B$, $B > 0$, is the tidal period, and C is the average depth. Assume that the tidal period is 12 hours, the depth at high tide is 18 feet, and the depth at low tide is 6 feet. Sketch two cycles of the graph of d.

54. Fahrenheit Temperature Suppose that

$$T(t) = 50 + 10\sin\frac{\pi}{12}(t - 8),$$

$0 \le t \le 24$, is a mathematical model of the Fahrenheit temperature at t hours after midnight on a certain day of the week.
(a) What is the temperature at 8 AM?
(b) At what time(s) does $T(t) = 60$?
(c) Sketch the graph of T.
(d) Find the maximum and minimum temperatures and the times at which they occur.

FIGURE 7.2.26 Pendulum in Problem 51

Calculator Problems

In Problems 55–58, use a calculator to investigate whether the given function is periodic.

55. $f(x) = \sin\left(\frac{1}{x}\right)$ **56.** $f(x) = \frac{1}{\sin 2x}$
57. $f(x) = 1 + (\cos x)^2$ **58.** $f(x) = x\sin x$

For Discussion

59. The function $f(x) = \sin\frac{1}{2}x + \sin 2x$ is periodic. What is the period of f?
60. Discuss and then sketch the graphs of $y = |\sin x|$ and $y = |\cos x|$.

7.3 Graphs of Other Trigonometric Functions

≡ **Introduction** Four additional trigonometric functions are defined in terms of quotients and reciprocals of the sine and cosine functions. In this section we will consider the properties and graphs of these new functions.

We begin with a definition that follows directly from (1) of Section 7.1.

DEFINITION 7.3.1 Four More Trigonometric Functions

The **tangent**, **cotangent**, **secant**, and **cosecant** functions are denoted by $\tan x$, $\cot x$, $\sec x$, and $\csc x$, respectively, and are defined as follows:

$$\tan x = \frac{\sin x}{\cos x}, \qquad \cot x = \frac{\cos x}{\sin x}, \tag{1}$$

$$\sec x = \frac{1}{\cos x}, \qquad \csc x = \frac{1}{\sin x}. \tag{2}$$

Note that the tangent and cotangent functions are related by

$$\cot x = \frac{\cos x}{\sin x} = \frac{1}{\frac{\sin x}{\cos x}} = \frac{1}{\tan x}.$$

In view of the definitions in (2) and the foregoing result, $\cot x$, $\sec x$, and $\csc x$ are referred to as the **reciprocal functions**.

☐ **Domain and Range** Because the functions in (1) and (2) are quotients, the **domain** of each function consists of the set of real numbers except those numbers for which the denominator is zero. We have seen in Section 7.2 that $\cos x = 0$ for $x = (2n + 1)\pi/2$, $n = 0, \pm 1, \pm 2, \ldots$, and so

- the domain of $\tan x$ and of $\sec x$ is $\{x \mid x \neq (2n + 1)\pi/2, n = 0, \pm 1, \pm 2, \ldots\}$.

Similarly, since $\sin x = 0$ for $x = n\pi$, $n = 0, \pm 1, \pm 2, \ldots$, it follows that

- the domain of $\cot x$ and of $\csc x$ is $\{x \mid x \neq n\pi, n = 0, \pm 1, \pm 2, \ldots\}$.

We know that the values of the sine and cosine are bounded, that is, $|\sin x| \leq 1$ and $|\cos x| \leq 1$. From these last inequalities we have

$$|\sec x| = \left|\frac{1}{\cos x}\right| = \frac{1}{|\cos x|} \geq 1 \tag{3}$$

and

$$|\csc x| = \left|\frac{1}{\sin x}\right| = \frac{1}{|\sin x|} \geq 1. \tag{4}$$

Recall, an inequality such as (3) means that $\sec x \geq 1$ or $\sec x \leq -1$. Hence the range of the secant function is $(-\infty, -1] \cup [1, \infty)$. The inequality in (4) implies that the cosecant function has the same range $(-\infty, -1] \cup [1, \infty)$. When we consider the graphs of the tangent and cotangent functions we will see that they have the same range: $(-\infty, \infty)$.

If we interpret x as an angle, then FIGURE 7.3.1 illustrates the algebraic signs of the tangent, cotangent, secant, and cosecant functions in each of the four quadrants. This is easily verified using the signs of the sine and cosine functions displayed in Figure 7.1.4.

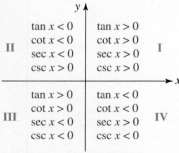

FIGURE 7.3.1 Signs of tan x, cot x, sec x, and csc x in the four quadrants

EXAMPLE 1 Example 3 of Section 7.1 Revisited

Find $\tan x$, $\cot x$, $\sec x$, and $\csc x$ for $x = -\pi/6$.

Solution In Example 3 of Section 7.1 we saw that

$$\sin\left(-\frac{\pi}{6}\right) = -\sin\frac{\pi}{6} = -\frac{1}{2} \quad \text{and} \quad \cos\left(-\frac{\pi}{6}\right) = \cos\frac{\pi}{6} = \frac{\sqrt{3}}{2}.$$

Therefore, by the definitions in (1) and (2):

$$\tan\left(-\frac{\pi}{6}\right) = \frac{-1/2}{\sqrt{3}/2} = -\frac{1}{\sqrt{3}}, \quad \cot\left(-\frac{\pi}{6}\right) = \frac{\sqrt{3}/2}{-1/2} = -\sqrt{3} \quad \leftarrow \text{we could also use } \cot x = 1/\tan x$$

$$\sec\left(-\frac{\pi}{6}\right) = \frac{1}{\sqrt{3}/2} = \frac{2}{\sqrt{3}}, \quad \csc\left(-\frac{\pi}{6}\right) = \frac{1}{-1/2} = -2.$$

Table 7.3.1 summarizes some important values of the tangent, cotangent, secant, and cosecant and was constructed using values of the sine and cosine from Section 6.3. A dash in the table indicates the trigonometric function is not defined at that particular value of x.

TABLE 7.3.1

x	0	$\frac{\pi}{6}$	$\frac{\pi}{4}$	$\frac{\pi}{3}$	$\frac{\pi}{2}$
$\tan x$	0	$\frac{1}{\sqrt{3}}$	1	$\sqrt{3}$	–
$\cot x$	–	$\sqrt{3}$	1	$\frac{1}{\sqrt{3}}$	0
$\sec x$	1	$\frac{2}{\sqrt{3}}$	$\sqrt{2}$	2	–
$\csc x$	–	2	$\sqrt{2}$	$\frac{2}{\sqrt{3}}$	1

☐ **Periodicity** Because the cosine and sine functions are 2π periodic necessarily the secant and cosecant function have the same period. But from Theorem 7.1.3 of Section 7.1 we have

$$\tan(x + \pi) = \underbrace{\frac{\sin(x + \pi)}{\cos(x + \pi)}}_{(iii) \text{ of Theorem 7.1.3}} = \overbrace{\frac{-\sin x}{-\cos x}}^{(iv) \text{ of Theorem 7.1.3}} = \tan x. \qquad (5)$$

◀ Also, see Problems 49 and 50 in Exercises 7.2.

Thus (5) implies that $\tan x$ and $\cot x$ are periodic with a period $p \leq \pi$. In the case of the tangent function, $\tan x = 0$ only if $\sin x = 0$, that is, only if $x = 0, \pm\pi, \pm 2\pi$, and so on. Therefore, the smallest positive number p for which $\tan(x + p) = \tan x$ is $p = \pi$. The cotangent function, since it is the reciprocal of the tangent function, has the same period.

In summary, the secant and cosecant functions are periodic with **period 2π**:

$$\sec(x + 2\pi) = \sec x \quad \text{and} \quad \csc(x + 2\pi) = \csc x. \qquad (6)$$

The tangent and cotangent function are periodic with **period π**:

$$\tan(x + \pi) = \tan x \quad \text{and} \quad \cot(x + \pi) = \cot x. \qquad (7)$$

Of course it is understood that (6) and (7) hold for every real number x for which the functions are defined.

☐ **Graphs of $y = \tan x$ and $y = \cot x$** The numbers that make the denominators of $\tan x$, $\cot x$, $\sec x$, and $\csc x$ equal to zero correspond to **vertical asymptotes** of their graphs. For example, we encourage you to verify using a calculator that

$$\tan x \to -\infty \text{ as } x \to -\frac{\pi}{2}^+ \quad \text{and} \quad \tan x \to \infty \text{ as } x \to \frac{\pi}{2}^-.$$

In other words, $x = -\pi/2$ and $x = \pi/2$ are vertical asymptotes. The graph of $y = \tan x$ on the interval $(-\pi/2, \pi/2)$ given in **FIGURE 7.3.2** is one **cycle** of the graph of $y = \tan x$. Using periodicity we extend the cycle in Figure 7.3.2 to adjacent intervals of length π as shown in **FIGURE 7.3.3**. The x-intercepts of the graph of the tangent function are $(0, 0), (\pm\pi, 0), (\pm 2\pi, 0), \ldots$ and the vertical asymptotes of the graph are $x = \pm\pi/2, \pm 3\pi/2, \pm 5\pi/2, \ldots$.

FIGURE 7.3.2 One cycle of the graph of $y = \tan x$

◀ This is a good time to review (7) of Section 4.6.

7.3 Graphs of Other Trigonometric Functions

FIGURE 7.3.3 Graph of of $y = \tan x$

FIGURE 7.3.4 Graph of $y = \cot x$

The graph of $y = \cot x$ is similar to the graph of the tangent function and is given in FIGURE 7.3.4. In this case, the graph of $y = \cot x$ on the interval $(0, \pi)$ is one **cycle** of the graph of $y = \cot x$. The x-intercepts of the graph of the cotangent function are $(\pm \pi/2, 0), (\pm 3\pi/2, 0), (\pm 5\pi/2, 0), \ldots$ and the vertical asymptotes of the graph are the vertical lines $x = 0, \pm \pi, \pm 2\pi, \pm 3\pi, \ldots$.

Note that the graphs of $y = \tan x$ and $y = \cot x$ are symmetric with respect to the origin.

THEOREM 7.3.1 ODD FUNCTIONS

The tangent function $f(x) = \tan x$ and the cotangent function $g(x) = \cot x$ are **odd** functions, that

$$\tan(-x) = -\tan x \quad \text{and} \quad \cot(-x) = -\cot x \qquad (8)$$

for every real number x for which the functions are defined.

☐ **Graphs of sec x and csc x** For both $y = \sec x$ and $y = \csc x$ we know that $|y| \geq 1$ and so no portion of their graphs can appear in the horizontal strip $-1 < y < 1$ of the Cartesian plane. Hence the graphs of $y = \sec x$ and $y = \csc x$ have no x-intercepts. As we have already seen, $y = \sec x$ and $y = \csc x$ have period 2π. The vertical asymptotes for the graph of $y = \sec x$ are the same as $y = \tan x$, namely, $x = \pm \pi/2, \pm 3\pi/2, \pm 5\pi/2, \ldots$. Because $y = \cos x$ is an even function so is $y = \sec x = 1/\cos x$. The graph of $y = \sec x$ is symmetric with respect to the y-axis. On the other hand, the vertical asymptotes for the graph of $y = \csc x$ are the same as $y = \cot x$, namely, $x = 0, \pm \pi, \pm 2\pi, \pm 3\pi, \ldots$. Because $y = \sin x$ is an odd function so is $y = \csc x = 1/\sin x$. The graph of $y = \csc x$ is symmetric with respect to the origin. One cycle of the graph of $y = \sec x$ on $[0, 2\pi]$ is extended to the interval $[-2\pi, 0]$ by periodicity (or y-axis symmetry) in FIGURE 7.3.5. Similarly, in FIGURE 7.3.6 we extend one cycle of $y = \csc x$ on $(0, 2\pi)$ to the interval $(-2\pi, 0)$ by periodicity (or origin symmetry).

FIGURE 7.3.5 Graph of $y = \sec x$

FIGURE 7.3.6 Graph of $y = \csc x$

☐ **Transformations and Graphs** Like the sine and cosine graphs, rigid and nonrigid transformations can be applied to the graphs of $y = \tan x$, $y = \cot x$, $y = \sec x$, and $y = \csc x$. For example, a function such as $y = A \tan(Bx + C) + D$ can be analyzed in the following manner:

$$\underset{\substack{\uparrow \\ \text{horizontal stretch/compression} \\ \text{by changing period}}}{\overset{\substack{\text{vertical stretch/compression/reflection} \downarrow \qquad\qquad\qquad \downarrow \text{vertical shift} \\ }}{y = A \tan(Bx + C) + D.}} \qquad (9)$$

CHAPTER 7 UNIT CIRCLE TRIGONOMETRY

If $B > 0$, then the period of

$$y = A\tan(Bx + C) \quad \text{and} \quad y = A\cot(Bx + C) \text{ is } \pi/B, \qquad (10)$$

whereas the period of

$$y = A\sec(Bx + C) \quad \text{and} \quad y = A\csc(Bx + C) \text{ is } 2\pi/B. \qquad (11)$$

As we see in (9) the number A in each case can be interpreted as either a vertical stretch or compression of a graph. However, you should be aware of the fact that the functions in (10) and (11) have no amplitude, because none of the functions have a maximum *and* a minimum value.

◀ Of the six trigonometric functions, only the sine and cosine functions have an amplitude.

EXAMPLE 2 Comparison of Graphs

Find the period, x-intercepts, and vertical asymptotes for the graph of $y = \tan 2x$. Graph the function on $[0, \pi]$.

Solution With the identification $B = 2$, we see from (10) that the period is $\pi/2$. Since $\tan 2x = \sin 2x/\cos 2x$, the x-intercepts of the graph occur at the zeros of $\sin 2x$. From the properties of the sine function given in Section 7.2, we know that $\sin 2x = 0$ for

$$2x = n\pi \quad \text{so that} \quad x = \tfrac{1}{2}n\pi, n = 0, \pm 1, \pm 2, \ldots.$$

That is, $x = 0, \pm \pi/2, \pm 2\pi/2 = \pi, \pm 3\pi/2, \pm 4\pi/2 = 2\pi$, and so on. The x-intercepts are $(0, 0), (\pm \pi/2, 0), (\pm \pi, 0), (\pm 3\pi/2, 0), \ldots$. The vertical asymptotes of the graph occur at zeros of $\cos 2x$. Moreover, the numbers for which $\cos 2x = 0$ are found in the following manner:

$$2x = (2n + 1)\frac{\pi}{2} \quad \text{so that} \quad x = (2n + 1)\frac{\pi}{4}, n = 0, \pm 1, \pm 2, \ldots.$$

That is, the vertical asymptotes are $x = \pm \pi/4, \pm 3\pi/4, \pm 5\pi/4, \ldots$. On the interval $[0, \pi]$ the graph of $y = \tan 2x$ has three intercepts, $(0, 0), (\pi/2, 0), (\pi, 0)$, and two vertical asymptotes, $x = \pi/4$ and $x = 3\pi/4$. In **FIGURE 7.3.7** we have compared the graphs of $y = \tan x$ and $y = \tan 2x$ on the interval. The graph of $y = \tan 2x$ is a horizontal compression of the graph of $y = \tan x$.

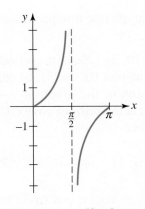

(a) $y = \tan x$ on $[0, \pi]$

(b) $y = \tan 2x$ on $[0, \pi]$

FIGURE 7.3.7 Graphs of functions in Example 2

EXAMPLE 3 Comparisons of Graphs

Compare one cycle of the graphs of $y = \tan x$ and $y = \tan(x - \pi/4)$.

Solution The graph of $y = \tan(x - \pi/4)$ is the graph of $y = \tan x$ shifted horizontally $\pi/4$ units to the right. The intercept $(0, 0)$ for the graph of $y = \tan x$ is shifted to $(\pi/4, 0)$ on the graph of $y = \tan(x - \pi/4)$. The vertical asymptotes $x = -\pi/2$ and $x = \pi/2$ for the graph of $y = \tan x$ are shifted to $x = -\pi/4$ and $x = 3\pi/4$ for the graph of $y = \tan(x - \pi/4)$. In **FIGURES 7.3.8(a)** and 7.3.8(b) we see, respectively, a cycle of the graph of $y = \tan x$ on the interval $(-\pi/2, \pi/2)$ is shifted to the right to yield a cycle of the graph of $y = \tan(x - \pi/4)$ on the interval $(-\pi/4, 3\pi/4)$.

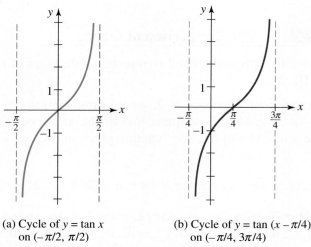

(a) Cycle of $y = \tan x$ on $(-\pi/2, \pi/2)$

(b) Cycle of $y = \tan(x - \pi/4)$ on $(-\pi/4, 3\pi/4)$

FIGURE 7.3.8 Graph of functions in Example 3

As we did in the analysis of the graphs of $y = A\sin(Bx + C)$ and $y = A\cos(Bx + C)$ we can determine the amount of horizontal shift for graphs of functions such as $y = A\tan(Bx + C)$ and $y = A\sec(Bx + C)$ by factoring the number $B > 0$ from $Bx + C$.

EXAMPLE 4 Two Shifts and Two Compressions

Graph $y = 2 - \frac{1}{2}\sec(3x - \pi/2)$.

Solution Let's break down the analysis of the graph into four parts, namely, by transformations.

(*i*) One cycle of the graph of $y = \sec x$ occurs on $[0, 2\pi]$. Since the period of $y = \sec 3x$ is $2\pi/3$, one cycle of its graph occurs on the interval $[0, 2\pi/3]$. In other words, the graph of $y = \sec 3x$ is a horizontal compression of the graph of $y = \sec x$. Since $\sec 3x = 1/\cos 3x$ the vertical asymptotes occur at the zeros of $\cos 3x$. Using the zeros of the cosine function given in Section 7.2 we find

$$3x = (2n+1)\frac{\pi}{2} \quad \text{or} \quad x = (2n+1)\frac{\pi}{6}, n = 0, \pm 1, \pm 2, \ldots.$$

FIGURE 7.3.9(a) shows two cycles of the graph $y = \sec 3x$; one cycle on $[-2\pi/3, 0]$ and another on $[0, 2\pi/3]$. Within those intervals the vertical asymptotes are $x = -\pi/2, x = -\pi/6, x = \pi/6,$ and $x = \pi/2$.

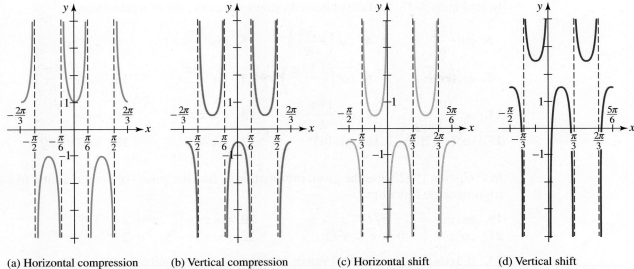

(a) Horizontal compression (b) Vertical compression and reflection in x-axis (c) Horizontal shift (d) Vertical shift

FIGURE 7.3.9 Graph of function in Example 4

(*ii*) The graph of $y = -\frac{1}{2}\sec 3x$ is the graph of $y = \sec 3x$ compressed vertically by a factor of $\frac{1}{2}$ and then reflected in the x-axis. See Figure 7.3.9(b).

(*iii*) By factoring 3 from $3x - \pi/2$, we see from

$$y = -\frac{1}{2}\sec\left(3x - \frac{\pi}{2}\right) = -\frac{1}{2}\sec 3\left(x - \frac{\pi}{6}\right)$$

that the graph of $y = -\frac{1}{2}\sec(3x - \pi/2)$ is the graph of $y = -\frac{1}{2}\sec 3x$ shifted $\pi/6$ units to the right. By shifting the two intervals $[-2\pi/3, 0]$ and $[0, 2\pi/3]$ in Figure 7.3.9(b) to the right $\pi/6$ units, we see in Figure 7.3.9(c) two cycles of $y = -\frac{1}{2}\sec(3x - \pi/2)$ on the intervals $[-\pi/2, \pi/6]$ and $[\pi/6, 5\pi/6]$. The vertical asymptotes $x = -\pi/2$, $x = -\pi/6$, $x = \pi/6$, and $x = \pi/2$ shown in Figure 7.3.9(b) are shifted to $x = -\pi/3$, $x = 0$, $x = \pi/3$, and $x = 2\pi/3$. Observe that the y-intercept $(0, -\frac{1}{2})$ in Figure 7.3.9(b) is now moved to $(\pi/6, -\frac{1}{2})$ in Figure 7.3.9(c).

(*iv*) Finally, we obtain the graph $y = 2 - \frac{1}{2}\sec(3x - \pi/2)$ in Figure 7.3.9(d) by shifting the graph of $y = -\frac{1}{2}\sec(3x - \pi/2)$ in Figure 7.3.9(c) 2 units upward. ≡

7.3 Exercises
Answers to selected odd-numbered problems begin on page ANS-19.

In Problems 1 and 2, complete the given table.

1.

x	$\frac{2\pi}{3}$	$\frac{3\pi}{4}$	$\frac{5\pi}{6}$	π	$\frac{7\pi}{6}$	$\frac{5\pi}{4}$	$\frac{4\pi}{3}$	$\frac{3\pi}{2}$	$\frac{5\pi}{3}$	$\frac{7\pi}{4}$	$\frac{11\pi}{6}$	2π
$\tan x$												
$\cot x$												

2.

x	$\frac{2\pi}{3}$	$\frac{3\pi}{4}$	$\frac{5\pi}{6}$	π	$\frac{7\pi}{6}$	$\frac{5\pi}{4}$	$\frac{4\pi}{3}$	$\frac{3\pi}{2}$	$\frac{5\pi}{3}$	$\frac{7\pi}{4}$	$\frac{11\pi}{6}$	2π
$\sec x$												
$\csc x$												

In Problems 3–18, find the indicated value without the use of a calculator.

3. $\cot \dfrac{13\pi}{6}$
4. $\csc\left(-\dfrac{3\pi}{2}\right)$
5. $\tan \dfrac{9\pi}{2}$
6. $\sec 7\pi$
7. $\csc\left(-\dfrac{\pi}{3}\right)$
8. $\cot\left(-\dfrac{13\pi}{3}\right)$
9. $\tan \dfrac{23\pi}{4}$
10. $\tan\left(-\dfrac{5\pi}{6}\right)$
11. $\sec \dfrac{10\pi}{3}$
12. $\cot \dfrac{17\pi}{6}$
13. $\csc 5\pi$
14. $\sec \dfrac{29\pi}{4}$
15. $\sec(-120°)$
16. $\tan 405°$
17. $\csc 495°$
18. $\cot(-720°)$

In Problems 19–22, use the given information to find the values of the remaining five trigonometric functions.

19. $\tan x = -2, \; \pi/2 < x < \pi$
20. $\cot x = \tfrac{1}{2}, \; \pi < x < 3\pi/2$
21. $\csc x = \tfrac{4}{3}, \; 0 < x < \pi/2$
22. $\sec x = -5, \; \pi/2 < x < \pi$

23. If $3\cos x = \sin x$, find all values of $\tan x$, $\cot x$, $\sec x$, and $\csc x$.
24. If $\csc x = \sec x$, find all values of $\tan x$, $\cot x$, $\sin x$, and $\cos x$.

In Problems 25–32, find the period, x-intercepts, and the vertical asymptotes of the given function. Sketch at least one cycle of the graph.

25. $y = \tan \pi x$
26. $y = \tan \dfrac{x}{2}$
27. $y = \cot 2x$
28. $y = -\cot \dfrac{\pi x}{3}$
29. $y = \tan\left(\dfrac{x}{2} - \dfrac{\pi}{4}\right)$
30. $y = \dfrac{1}{4}\cot\left(x - \dfrac{\pi}{2}\right)$
31. $y = -1 + \cot \pi x$
32. $y = \tan\left(x + \dfrac{5\pi}{6}\right)$

In Problems 33–40, find the period and the vertical asymptotes of the given function. Sketch at least one cycle of the graph.

33. $y = -\sec x$
34. $y = 2\sec \dfrac{\pi x}{2}$
35. $y = 3\csc \pi x$
36. $y = -2\csc \dfrac{x}{3}$
37. $y = \sec\left(3x - \dfrac{\pi}{2}\right)$
38. $y = \csc x(4x + \pi)$
39. $y = 3 + \csc\left(2x + \dfrac{\pi}{2}\right)$
40. $y = -1 + \sec(x - 2\pi)$

In Problems 41 and 42, use the graphs of $y = \tan x$ and $y = \sec x$ to find numbers A and C for which the given equality is true.

41. $\cot x = A\tan(x + C)$
42. $\csc x = A\sec(x + C)$

For Discussion

43. Using a calculator in radian mode, compare the values of $\tan 1.57$ and $\tan 1.58$. Explain the difference in these values.
44. Using a calculator in radian mode, compare the values of $\cot 3.14$ and $\cot 3.15$.
45. Can $9\csc x = 1$ for any real number x?
46. Can $7 + 10\sec x = 0$ for any real number x?
47. For which real numbers x is (a) $\sin x \leq \csc x$? (b) $\sin x < \csc x$?

48. For which real numbers x is (a) $\sec x \leq \cos x$? (b) $\sec x < \cos x$?
49. Discuss and then sketch the graphs of $y = |\sec x|$ and $y = |\csc x|$.
50. Use Definition 3.2.1 to prove Theorem 7.3.1, that is, $f(x) = \tan x$ and $g(x) = \cot x$ are odd functions.

7.4 Special Identities

≡ Introduction In this section we will examine identities for trigonometric functions. We have already seen some of these identities, such as the Pythagorean identities, in earlier sections. A **trigonometric identity** is an equation or formula involving trigonometric functions that is valid for all angles or real numbers for which both sides of the equality are defined. There are *numerous* trigonometric identities, but we are going to develop only those of special importance in courses in mathematics and science.

☐ **Pythagorean Identities Revisited** In Sections 6.2 and 6.4 we saw that the sine and cosine of an angle θ are related by the fundamental identity $\sin^2\theta + \cos^2\theta = 1$. We saw that by dividing this identity, in turn, by $\cos^2\theta$ and then by $\sin^2\theta$ we obtain two more identities, one relating $\tan^2\theta$ to $\sec^2\theta$ and the other relating $\cot^2\theta$ to $\csc^2\theta$. These so-called **Pythagorean identities** are also valid for a real number x as well as to an angle θ measured in degrees or in radians. Also, see (2) in Section 7.1.

The Pythagorean identities are so basic to trigonometry that we give them again for future reference.

THEOREM 7.4.1 Pythagorean Identities

For x a real number for which the functions are defined,

$$\sin^2 x + \cos^2 x = 1 \tag{1}$$
$$1 + \tan^2 x = \sec^2 x \tag{2}$$
$$1 + \cot^2 x = \csc^2 x. \tag{3}$$

☐ **Trigonometric Substitutions** In calculus it is often useful to make use of trigonometric substitution to change the form of certain algebraic expressions involving radicals. Generally, this is done using the Pythagorean identities. The following example illustrates the technique.

EXAMPLE 1   Rewriting a Radical

Rewrite $\sqrt{a^2 - x^2}$ as a trigonometric expression without radicals by means of the substitution $x = a\sin\theta$, $a > 0$ and $-\pi/2 \leq \theta \leq \pi/2$.

Solution If $x = a\sin\theta$, then

$$\begin{aligned}\sqrt{a^2 - x^2} &= \sqrt{a^2 - (a\sin\theta)^2} \\ &= \sqrt{a^2 - a^2\sin^2\theta} \\ &= \sqrt{a^2(1 - \sin^2\theta)} \quad \leftarrow \text{now use (1) of Theorem 7.4.1} \\ &= \sqrt{a^2\cos^2\theta}.\end{aligned}$$

Since $a > 0$ and $\cos\theta \geq 0$ for $-\pi/2 \leq \theta \leq \pi/2$, the original radical is the same as

$$\sqrt{a^2 - x^2} = \sqrt{a^2\cos^2\theta} = a\cos\theta. \qquad \equiv$$

☐ **Sum and Difference Formulas** The **sum** and **difference formulas** for the cosine and sine functions are identities that reduce $\cos(x_1 + x_2)$, $\cos(x_1 - x_2)$, $\sin(x_1 + x_2)$, and $\sin(x_1 - x_2)$ to expressions that involve $\cos x_1$, $\cos x_2$, $\sin x_1$, and $\sin x_2$. We will derive the formula for $\cos(x_1 - x_2)$ first, and then we will use that result to obtain the others.

For convenience, let us suppose that x_1 and x_2 represent angles measured in radians. As shown in **FIGURE 7.4.1(a)**, let d denote the distance between $P(x_1)$ and $P(x_2)$. If we place the angle $x_1 - x_2$ in standard position as shown in Figure 7.4.1(b), then d is also the distance between $P(x_1 - x_2)$ and $P(0)$. Equating the squares of these distances gives

$$(\cos x_1 - \cos x_2)^2 + (\sin x_1 - \sin x_2)^2 = (\cos(x_1 - x_2) - 1)^2 + \sin^2(x_1 - x_2)$$

or
$$\cos^2 x_1 - 2\cos x_1 \cos x_2 + \cos^2 x_2 + \sin^2 x_1 - 2\sin x_1 \sin x_2 + \sin^2 x_2$$
$$= \cos^2(x_1 - x_2) - 2\cos(x_1 - x_2) + 1 + \sin^2(x_1 - x_2).$$

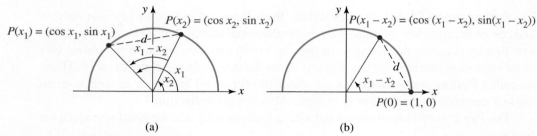

(a) (b)

FIGURE 7.4.1 The difference of two angles

In view of (1),
$$\cos^2 x_1 + \sin^2 x_1 = 1, \quad \cos^2 x_2 + \sin^2 x_2 = 1, \quad \cos^2(x_1 - x_2) + \sin^2(x_1 - x_2) = 1$$
and so the preceding equation simplifies to
$$\cos(x_1 - x_2) = \cos x_1 \cos x_2 + \sin x_1 \sin x_2.$$

This last result can be put to work immediately to find the cosine of the sum of two angles. Since $x_1 + x_2$ can be rewritten as the difference $x_1 - (-x_2)$,
$$\cos(x_1 + x_2) = \cos(x_1 - (-x_2))$$
$$= \cos x_1 \cos(-x_2) + \sin x_1 \sin(-x_2).$$

By the even–odd identities, $\cos(-x_2) = \cos x_2$ and $\sin(-x_2) = -\sin x_2$, it follows that the last line is the same as
$$\cos(x_1 + x_2) = \cos x_1 \cos x_2 - \sin x_1 \sin x_2.$$

The two results just obtained are summarized next.

THEOREM 7.4.2 Sum and Difference Formulas for the Cosine

For all real numbers x_1 and x_2,
$$\cos(x_1 + x_2) = \cos x_1 \cos x_2 - \sin x_1 \sin x_2 \qquad (4)$$
$$\cos(x_1 - x_2) = \cos x_1 \cos x_2 + \sin x_1 \sin x_2. \qquad (5)$$

EXAMPLE 2 Cosine of a Sum

Evaluate $\cos(7\pi/12)$.

Solution We have no way of evaluating $\cos(7\pi/12)$ directly. However, observe that

$$\frac{7\pi}{12} \text{ radians } = 105° = 60° + 45° = \frac{\pi}{3} + \frac{\pi}{4}.$$

Because $7\pi/12$ radians is a second-quadrant angle we know that the value of $\cos(7\pi/12)$ is negative. Proceeding, the sum formula (4) gives

$$\cos\frac{7\pi}{12} = \overbrace{\cos\left(\frac{\pi}{3} + \frac{\pi}{4}\right) = \cos\frac{\pi}{3}\cos\frac{\pi}{4} - \sin\frac{\pi}{3}\sin\frac{\pi}{4}}^{\text{this is (4) of Theorem 7.4.2}}$$

$$= \frac{1}{2}\frac{\sqrt{2}}{2} - \frac{\sqrt{3}}{2}\frac{\sqrt{2}}{2} = \frac{\sqrt{2}}{4}(1 - \sqrt{3}).$$

Using $\sqrt{2}\sqrt{3} = \sqrt{6}$ this result can also be written as $\cos(7\pi/12) = (\sqrt{2} - \sqrt{6})/4$. Since $\sqrt{6} > \sqrt{2}$, we see that $\cos(7\pi/12) < 0$ as expected. ≡

To obtain the corresponding sum/difference identities for the sine function we will make use of two identities:

$$\cos\left(x - \frac{\pi}{2}\right) = \sin x \quad \text{and} \quad \sin\left(x - \frac{\pi}{2}\right) = -\cos x. \tag{6}$$

These identities were first presented in Section 6.2 as cofunction identities and then rediscovered in Section 7.2 by shifting the graphs of the cosine and sine. However, both results in (6) can now be proved using (5):

$$\cos\left(x - \frac{\pi}{2}\right) = \cos x \cos\frac{\pi}{2} + \sin x \sin\frac{\pi}{2} = \cos x \cdot 0 + \sin x \cdot 1 = \sin x \qquad \blacktriangleleft \text{This proves the first equation in (6).}$$

$$\cos x = \cos\left(\overset{\text{zero}}{\overset{\downarrow}{\frac{\pi}{2} - \frac{\pi}{2}}} + x\right) = \cos\left(\frac{\pi}{2} - \left(\frac{\pi}{2} - x\right)\right) \qquad \blacktriangleleft \text{This proves the second equation in (6).}$$

$$= \cos\frac{\pi}{2}\cos\left(\frac{\pi}{2} - x\right) + \sin\frac{\pi}{2}\sin\left(\frac{\pi}{2} - x\right)$$

$$= 0 \cdot \cos\left(\frac{\pi}{2} - x\right) + 1 \cdot \sin\left(\frac{\pi}{2} - x\right)$$

$$= -\sin\left(x - \frac{\pi}{2}\right).$$

We put the first equation in (6) to work immediately by writing the sine of the sum $x_1 + x_2$ as

$$\sin(x_1 + x_2) = \cos\left((x_1 + x_2) - \frac{\pi}{2}\right)$$

$$= \cos\left(x_1 + \left(x_2 - \frac{\pi}{2}\right)\right) = \cos x_1 \cos\left(x_2 - \frac{\pi}{2}\right) - \sin x_1 \sin\left(x_2 - \frac{\pi}{2}\right) \quad \leftarrow \text{by (4)}$$

$$= \cos x_1 \sin x_2 - \sin x_1 (-\cos x_2). \quad \leftarrow \text{by (6)}$$

The preceding line is traditionally written as

$$\sin(x_1 + x_2) = \sin x_1 \cos x_2 + \cos x_1 \sin x_2.$$

To obtain the sine of the difference $x_1 - x_2$, we use again $\cos(-x_2) = \cos x_2$ and $\sin(-x_2) = -\sin x_2$:

$$\sin(x_1 - x_2) = \sin(x_1 + (-x_2)) = \sin x_1 \cos(-x_2) + \cos x_1 \sin(-x_2)$$
$$= \sin x_1 \cos x_2 - \cos x_1 \sin x_2.$$

THEOREM 7.4.3 Sum and Difference Formulas for the Sine

For all real numbers x_1 and x_2,

$$\sin(x_1 + x_2) = \sin x_1 \cos x_2 + \cos x_1 \sin x_2, \tag{7}$$
$$\sin(x_1 - x_2) = \sin x_1 \cos x_2 - \cos x_1 \sin x_2. \tag{8}$$

EXAMPLE 3 Sine of a Sum

Evaluate $\sin(7\pi/12)$.

Solution We proceed as in Example 2, except we use the sum formula (7):

$$\sin\frac{7\pi}{12} = \overbrace{\sin\left(\frac{\pi}{3} + \frac{\pi}{4}\right) = \sin\frac{\pi}{3}\cos\frac{\pi}{4} + \cos\frac{\pi}{3}\sin\frac{\pi}{4}}^{\text{this is (7) of Theorem 7.4.3}}$$
$$= \frac{\sqrt{3}}{2}\frac{\sqrt{2}}{2} + \frac{1}{2}\frac{\sqrt{2}}{2} = \frac{\sqrt{2}}{4}(1 + \sqrt{3}).$$

As in Example 2, the result can be rewritten as $\sin(7\pi/12) = (\sqrt{2} + \sqrt{6})/4$.

Since we know the value of $\cos(7\pi/12)$ from Example 2 we can also compute the value of $\sin(7\pi/12)$ using the Pythagorean identity (1):

$$\sin^2\frac{7\pi}{12} + \cos^2\frac{7\pi}{12} = 1.$$

We solve for $\sin(7\pi/12)$ and take the positive square root:

$$\sin\frac{7\pi}{12} = \sqrt{1 - \cos^2\frac{7\pi}{12}} = \sqrt{1 - \left[\frac{\sqrt{2}}{4}(1 - \sqrt{3})\right]^2}$$
$$= \sqrt{\frac{4 + 2\sqrt{3}}{8}} = \frac{\sqrt{2 + \sqrt{3}}}{2}. \tag{9}$$

Although the number in (9) does not look like the result obtained in Example 3, the values are the same. See Problem 62 in Exercises 7.4.

There are sum and difference formulas for the tangent function as well. We can derive the sum formula using the sum formulas for the sine and cosine as follows:

$$\tan(x_1 + x_2) = \frac{\sin(x_1 + x_2)}{\cos(x_1 + x_2)} = \frac{\sin x_1 \cos x_2 + \cos x_1 \sin x_2}{\cos x_1 \cos x_2 - \sin x_1 \sin x_2}. \tag{10}$$

We now divide the numerator and denominator of (10) by $\cos x_1 \cos x_2$ (assuming that x_1 and x_2 are such that $\cos x_1 \cos x_2 \neq 0$),

$$\tan(x_1 + x_2) = \frac{\dfrac{\sin x_1 \cos x_2}{\cos x_1 \cos x_2} + \dfrac{\cos x_1 \sin x_2}{\cos x_1 \cos x_2}}{\dfrac{\cos x_1 \cos x_2}{\cos x_1 \cos x_2} - \dfrac{\sin x_1 \sin x_2}{\cos x_1 \cos x_2}} = \frac{\tan x_1 + \tan x_2}{1 - \tan x_1 \tan x_2}. \tag{11}$$

The derivation of the difference formula for $\tan(x_1 - x_2)$ is obtained in a similar manner. We summarize the two results.

THEOREM 7.4.4 Sum and Difference Formulas for the Tangent

For all real numbers x_1 and x_2 for which the functions are defined,

$$\tan(x_1 + x_2) = \frac{\tan x_1 + \tan x_2}{1 - \tan x_1 \tan x_2} \qquad (12)$$

$$\tan(x_1 - x_2) = \frac{\tan x_1 - \tan x_2}{1 + \tan x_1 \tan x_2}. \qquad (13)$$

EXAMPLE 4 Tangent of a Sum

Evaluate $\tan(\pi/12)$.

Solution If we think of $\pi/12$ as an angle in radians, then

$$\frac{\pi}{12} \text{ radians} = 15° = 45° - 30° = \frac{\pi}{4} - \frac{\pi}{6} \text{ radians}.$$

It follows from formula (13):

$$\tan\frac{\pi}{12} = \tan\left(\frac{\pi}{4} - \frac{\pi}{6}\right) = \overbrace{\frac{\tan\frac{\pi}{4} - \tan\frac{\pi}{6}}{1 + \tan\frac{\pi}{4}\tan\frac{\pi}{6}}}^{\text{this is (13) of Theorem 7.4.4}}$$

$$= \frac{1 - \frac{1}{\sqrt{3}}}{1 + 1 \cdot \frac{1}{\sqrt{3}}} = \frac{\sqrt{3} - 1}{\sqrt{3} + 1} \quad \leftarrow \text{this is the answer but we can simplify the expression}$$

$$= \frac{\sqrt{3} - 1}{\sqrt{3} + 1} \cdot \frac{\sqrt{3} - 1}{\sqrt{3} - 1} \quad \leftarrow \text{rationalizing the denominator}$$

$$= \frac{(\sqrt{3} - 1)^2}{2} = \frac{4 - 2\sqrt{3}}{2} = \frac{2(2 - \sqrt{3})}{2} = 2 - \sqrt{3}. \qquad \equiv$$

◀ You should rework this example using $\pi/12 = \pi/3 - \pi/4$ to see that the result is the same.

Strictly speaking, we really do not need the identities for $\tan(x_1 \pm x_2)$ since we can always compute $\sin(x_1 \pm x_2)$ and $\cos(x_1 \pm x_2)$ using (4)–(8) and then proceed as in (10), that is, from the quotient $\sin(x_1 \pm x_2)/\cos(x_1 \pm x_2)$.

☐ **Double-Angle Formulas** Many useful trigonometric formulas can be derived from the sum and difference formulas. The **double-angle formulas** express the cosine and sine of $2x$ in terms of the cosine and sine of x.

If we set $x_1 = x_2 = x$ in (4) and use $\cos(x + x) = \cos 2x$, then

$$\cos 2x = \cos x \cos x - \sin x \sin x = \cos^2 x - \sin^2 x.$$

Similarly, by setting $x_1 = x_2 = x$ in (7) and using $\sin(x + x) = \sin 2x$, then

$$\sin 2x = \overset{\downarrow}{\sin x \cos x} + \overset{\downarrow}{\cos x \sin x} = 2 \sin x \cos x.$$

these two terms are equal

We summarize the last two results.

THEOREM 7.4.5 Double-Angle Formulas for the Cosine and Sine

For any real number x,

$$\cos 2x = \cos^2 x - \sin^2 x \qquad (14)$$

$$\sin 2x = 2\sin x \cos x. \qquad (15)$$

EXAMPLE 5 Using the Double-Angle Formulas

If $\sin x = -\frac{1}{4}$ and $\pi < x < 3\pi/2$, find the exact values of $\cos 2x$ and $\sin 2x$.

Solution First, we compute $\cos x$ using $\sin^2 x + \cos^2 x = 1$. Since $\pi < x < 3\pi/2$, $\cos x < 0$ and so we choose the negative square root:

$$\cos x = -\sqrt{1 - \sin^2 x} = -\sqrt{1 - \left(-\frac{1}{4}\right)^2} = -\frac{\sqrt{15}}{4}.$$

From the double-angle formula (14),

$$\cos 2x = \cos^2 x - \sin^2 x$$
$$= \left(-\frac{\sqrt{15}}{4}\right)^2 - \left(-\frac{1}{4}\right)^2$$
$$= \frac{15}{16} - \frac{1}{16} = \frac{14}{16} = \frac{7}{8}.$$

Finally, from the double-angle formula (15),

$$\sin 2x = 2\sin x \cos x = 2\left(-\frac{1}{4}\right)\left(-\frac{\sqrt{15}}{4}\right) = \frac{\sqrt{15}}{8}. \qquad \equiv$$

The formula in (14) has two useful alternative forms. By (1), we know that $\sin^2 x = 1 - \cos^2 x$. Substituting the last expression into (14) yields $\cos 2x = \cos^2 x - (1 - \cos^2 x)$ or

$$\cos 2x = 2\cos^2 x - 1. \qquad (16)$$

On the other hand, if we substitute $\cos^2 x = 1 - \sin^2 x$ into (14) we get

$$\cos 2x = 1 - 2\sin^2 x. \qquad (17)$$

☐ **Half-Angle Formulas** The alternative forms of (16) and (17) of the double-angle formula (14) are the source of two **half-angle formulas**. Solving (16) and (17) for $\cos^2 x$ and $\sin^2 x$ gives, respectively,

$$\cos^2 x = \frac{1}{2}(1 + \cos 2x) \quad \text{and} \quad \sin^2 x = \frac{1}{2}(1 - \cos 2x). \qquad (18)$$

By replacing the symbol x in (18) by $x/2$ and using $2(x/2) = x$, we obtain the following formulas.

THEOREM 7.4.6 Half-Angle Formulas for the Cosine and Sine

For any real number x,

$$\cos^2 \frac{x}{2} = \frac{1}{2}(1 + \cos x) \qquad (19)$$

$$\sin^2 \frac{x}{2} = \frac{1}{2}(1 - \cos x). \qquad (20)$$

EXAMPLE 6 Using the Half-Angle Formulas

Find the exact values of $\cos(5\pi/8)$ and $\sin(5\pi/8)$.

Solution If we let $x = 5\pi/4$, then $x/2 = 5\pi/8$ and formulas (19) and (20) yield, respectively

$$\cos^2\left(\frac{5\pi}{8}\right) = \frac{1}{2}\left(1 + \cos\frac{5\pi}{4}\right) = \frac{1}{2}\left[1 + \left(-\frac{\sqrt{2}}{2}\right)\right] = \frac{2 - \sqrt{2}}{4},$$

and $\quad \sin^2\left(\dfrac{5\pi}{8}\right) = \dfrac{1}{2}\left(1 - \cos\dfrac{5\pi}{4}\right) = \dfrac{1}{2}\left[1 - \left(-\dfrac{\sqrt{2}}{2}\right)\right] = \dfrac{2 + \sqrt{2}}{4}.$

Because $5\pi/8$ radians is a second-quadrant angle, $\cos(5\pi/8) < 0$ and $\sin(5\pi/8) > 0$. Therefore, we take the negative square root for the value of the cosine,

$$\cos\left(\frac{5\pi}{8}\right) = -\sqrt{\frac{2 - \sqrt{2}}{4}} = -\frac{\sqrt{2 - \sqrt{2}}}{2},$$

and the positive square root for the value of the sine,

$$\sin\left(\frac{5\pi}{8}\right) = \sqrt{\frac{2 + \sqrt{2}}{4}} = \frac{\sqrt{2 + \sqrt{2}}}{2}. \qquad \equiv$$

NOTES FROM THE CLASSROOM

(i) Should you memorize all the identities presented in this section? You should consult your instructor about this, but in the opinion of the authors, you should at the very least memorize formulas (1)–(8), (14), (15), and the two formulas in (18).

(ii) If you eventually enroll in a calculus course, check the title of your text. If it has the words *Early Transcendentals* in its title, then your knowledge of the graphs and properties of the trigonometric functions will come into play almost immediately. The sum identities (4) and (7) are used in differential calculus to compute functions known as derivatives of $\sin x$ and $\cos x$. Identities are especially useful in integral calculus. Replacing a radical by a trigonometric function as illustrated in Example 1 in this section is a standard technique for evaluating some types of integrals. Also, to evaluate integrals of $\cos^2 x$ and $\sin^2 x$ you would use the half-angle formulas in the form given in (18):

$$\cos^2 x = \tfrac{1}{2}(1 + \cos 2x) \quad \text{and} \quad \sin^2 x = \tfrac{1}{2}(1 - \cos 2x).$$

At some point in your study of integral calculus you may be required to evaluate integrals of products such as

$$\sin 2x \sin 5x \quad \text{and} \quad \sin 10x \cos 4x.$$

One way of doing this is to use the sum/difference formulas to devise an identity that converts these products into either a sum of sines or a sum of cosines. See Problems 66–70 in Exercises 7.4.

7.4 Exercises
Answers to selected odd-numbered problems begin on page ANS-20.

In Problems 1–8, proceed as in Example 1 and rewrite the given expression as a trigonometric expression without radicals by making the indicated substitution. Assume that $a > 0$.

1. $\sqrt{a^2 - x^2}$, $\quad x = a\cos\theta$, $\quad 0 \leq \theta \leq \pi$
2. $\sqrt{a^2 + x^2}$, $\quad x = a\tan\theta$, $\quad -\pi/2 < \theta < \pi/2$
3. $\sqrt{x^2 - a^2}$, $\quad x = a\sec\theta$, $\quad 0 \leq \theta < \pi/2$
4. $\sqrt{16 - 25x^2}$, $\quad x = \frac{4}{5}\sin\theta$, $\quad -\pi/2 \leq \theta \leq \pi/2$
5. $\dfrac{x}{\sqrt{9 - x^2}}$, $\quad x = 3\sin\theta$, $\quad -\pi/2 < \theta < \pi/2$
6. $\dfrac{\sqrt{x^2 - 3}}{x^2}$, $\quad x = \sqrt{3}\sec\theta$, $\quad 0 < \theta < \pi/2$
7. $\dfrac{1}{\sqrt{7 + x^2}}$, $\quad x = \sqrt{7}\tan\theta$, $\quad -\pi/2 < \theta < \pi/2$
8. $\dfrac{\sqrt{5 - x^2}}{x}$, $\quad x = \sqrt{5}\cos\theta$, $\quad 0 \leq \theta \leq \pi$

In Problems 9–30, use a sum or difference formula to find the exact value of the given expression.

9. $\cos\dfrac{\pi}{12}$
10. $\sin\dfrac{\pi}{12}$
11. $\sin 75°$
12. $\cos 75°$
13. $\sin\dfrac{7\pi}{12}$
14. $\cos\dfrac{11\pi}{12}$
15. $\tan\dfrac{5\pi}{12}$
16. $\cos\left(-\dfrac{5\pi}{12}\right)$
17. $\sin\left(-\dfrac{\pi}{12}\right)$
18. $\tan\dfrac{11\pi}{12}$
19. $\sin\dfrac{11\pi}{12}$
20. $\tan\dfrac{7\pi}{12}$
21. $\cos 165°$
22. $\sin 165°$
23. $\tan 165°$
24. $\cos 195°$
25. $\sin 195°$
26. $\tan 195°$
27. $\cos 345°$
28. $\sin 345°$
29. $\cos\dfrac{13\pi}{12}$
30. $\tan\dfrac{17\pi}{12}$

In Problems 31–34, use a double-angle formula to write the given expression as a single trigonometric function of twice the angle.

31. $2\cos\beta\sin\beta$
32. $\cos^2 2t - \sin^2 2t$
33. $1 - 2\sin^2\dfrac{\pi}{5}$
34. $2\cos^2\left(\dfrac{19}{2}x\right) - 1$

CHAPTER 7 UNIT CIRCLE TRIGONOMETRY

In Problems 35–40, use the given information to find the exact values of (a) $\cos 2x$, (b) $\sin 2x$, and (c) $\tan 2x$.

35. $\sin x = \sqrt{2}/3, \quad \pi/2 < x < \pi$
36. $\cos x = \sqrt{3}/5, \quad 3\pi/2 < x < 2\pi$
37. $\tan x = \frac{1}{2}, \quad \pi < x < 3\pi/2$
38. $\csc x = -3, \quad \pi < x < 3\pi/2$
39. $\sec x = -\frac{13}{5}, \quad \pi/2 < x < \pi$
40. $\cot x = \frac{4}{3}, \quad 0 < x < \pi/2$

In Problems 41–48, use a half-angle formula to find the exact value of the given expression.

41. $\cos(\pi/12)$
42. $\sin(\pi/8)$
43. $\sin(3\pi/8)$
44. $\tan(\pi/12)$
45. $\cos 67.5°$
46. $\sin 15°$
47. $\csc(13\pi/12)$
48. $\sec(-3\pi/8)$

In Problems 49–54, use the given information to find the exact values of (a) $\cos(x/2)$, (b) $\sin(x/2)$, and (c) $\tan(x/2)$.

49. $\sin x = \frac{12}{13}, \quad \pi/2 < x < \pi$
50. $\cos x = \frac{4}{5}, \quad 3\pi/2 < x < 2\pi$
51. $\tan x = 2, \quad \pi < x < 3\pi/2$
52. $\csc x = 9, \quad 0 < x < \pi/2$
53. $\sec x = \frac{3}{2}, \quad 0 < x < 90°$
54. $\cot x = -\frac{1}{4}, \quad 90° < x < 180°$

55. If $P(x_1)$ and $P(x_2)$ are points in quadrant II on the terminal side of the angles x_1 and x_2, respectively, with $\cos x_1 = -\frac{1}{3}$ and $\sin x_2 = \frac{2}{3}$, find
   (a) $\sin(x_1 + x_2)$
   (b) $\cos(x_1 + x_2)$
   (c) $\sin(x_1 - x_2)$
   (d) $\cos(x_1 - x_2)$.
56. If x_1 is a quadrant II angle, x_2 is a quadrant III angle, $\sin x_1 = \frac{8}{17}$, and $\tan x_2 = \frac{3}{4}$, find
   (a) $\sin(x_1 + x_2)$, (b) $\sin(x_1 - x_2)$, (c) $\cos(x_1 + x_2)$, and (d) $\cos(x_1 - x_2)$.

Miscellaneous Applications

57. **Mach Number** The ratio of the speed of an airplane to the speed of sound is called the Mach number M of the plane. If $M > 1$, the plane makes sound waves that form a (moving) cone, as shown in **FIGURE 7.4.2**. A sonic boom is heard at the intersection of the cone with the ground. If the vertex angle of the cone is θ, then

$$\sin\frac{\theta}{2} = \frac{1}{M}.$$

If $\theta = \pi/6$, find the exact value of the Mach number.

FIGURE 7.4.2 Airplane in Problem 57

58. **Cardiovascular Branching** A mathematical model for blood flow in a large blood vessel predicts that the optimal values of the angles θ_1 and θ_2, which represent the (positive) angles of the smaller daughter branches (vessels) with respect to the axis of the parent branch, are given by

$$\cos\theta_1 = \frac{A_0^2 + A_1^2 - A_2^2}{2A_0A_1} \quad \text{and} \quad \cos\theta_2 = \frac{A_0^2 - A_1^2 + A_2^2}{2A_0A_2},$$

where A_0 is the cross-sectional area of the parent branch and A_1 and A_2 are the cross-sectional areas of the daughter branches. See **FIGURE 7.4.3**. Let $\psi = \theta_1 + \theta_2$ be the junction angle, as shown in the figure.
   (a) Show that

$$\cos\psi = \frac{A_0^2 - A_1^2 - A_2^2}{2A_1A_2}.$$

FIGURE 7.4.3 Branching of a large blood vessel in Problem 58

FIGURE 7.4.4 Isosceles triangle in Problem 59

(b) Show that for the optimal values of θ_1 and θ_2, the cross-sectional area of the daughter branches, $A_1 + A_2$, is greater than or equal to that of the parent branch. Therefore, the blood must slow down in the daughter branches.

59. **Area of a Triangle** Show that the area of an isosceles triangle with equal sides of length x is $A = \frac{1}{2}x^2 \sin\theta$, where θ is the angle formed by the two equal sides. See **FIGURE 7.4.4**. [*Hint*: Consider $\theta/2$ as shown in the figure.]

60. **Putting the Shot** We saw in Problem 66 in Exercises 6.4 that when a projectile, such as a shot put, is released from a height h, upward at an angle ϕ with velocity v_0, the range R at which it strikes the ground is given by

$$R = \frac{v_0 \cos\phi}{g}\left(v_0 \sin\phi + \sqrt{v_0^2 \sin^2\phi + 2gh}\right),$$

where g is the acceleration due to gravity. See **FIGURE 7.4.5**. It can be shown that the maximum range R_{max} is achieved if the angle ϕ satisfies the equation

$$\cos 2\phi = \frac{gh}{v_0^2 + gh}.$$

Show that
$$R_{max} = \frac{v_0\sqrt{v_0^2 + 2gh}}{g},$$

by using the expressions for R and $\cos 2\phi$ and the half-angle formulas for the sine and the cosine with $t = 2\phi$.

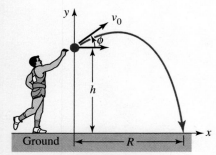

FIGURE 7.4.5 Projectile in Problem 60

For Discussion

61. Discuss: Why would you expect to get an error message from your calculator when you try to evaluate

$$\frac{\tan 35° + \tan 55°}{1 - \tan 35° \tan 55°}?$$

62. In Example 3 we showed that $\sin\dfrac{7\pi}{12} = \dfrac{\sqrt{2} + \sqrt{6}}{4}$. Following the example, we then showed that $\sin\dfrac{7\pi}{12} = \dfrac{\sqrt{2 + \sqrt{3}}}{2}$. Demonstrate that these answers are equivalent.

63. Discuss: How would you express $\sin 3\theta$ in terms of $\sin\theta$? Carry out your ideas.
64. In Problem 55, in what quadrants do the points $P(x_1 + x_2)$ and $P(x_1 - x_2)$ lie?
65. In Problem 56, in which quadrant does the terminal side of $x_1 + x_2$ lie? The terminal side of $x_1 - x_2$?
66. Use the sum/difference formulas (4), (5), (7), and (8) to derive the **product-to-sum formulas**:

$$\sin x_1 \sin x_2 = \tfrac{1}{2}[\cos(x_1 - x_2) - \cos(x_1 + x_2)]$$
$$\cos x_1 \cos x_2 = \tfrac{1}{2}[\cos(x_1 - x_2) + \cos(x_1 + x_2)]$$
$$\sin x_1 \cos x_2 = \tfrac{1}{2}[\sin(x_1 + x_2) + \sin(x_1 - x_2)].$$

In Problems 67–70, use a product-to-sum formula in Problem 66 to write the given product as a sum of sines or a sum of cosines.

67. $\cos 4\theta \cos 3\theta$
68. $\sin\dfrac{3t}{2}\cos\dfrac{t}{2}$
69. $\sin 2x \sin 5x$
70. $\sin 10x \cos 4x$

In Problems 71 and 72, use one of the formulas in Problem 66 to find the exact value of the given expression.

71. $\sin 15° \sin 45°$

72. $\sin 75° \cos 15°$

73. Prove the double-angle formula for the tangent function:
$$\tan 2x = \frac{2\tan x}{1 - \tan^2 x}.$$

74. Prove the half-angle formula for the tangent function:
$$\tan^2 \frac{x}{2} = \frac{1 - \cos x}{1 + \cos x}.$$

In Problems 75 and 76, prove the alternative half-angle formulas for the tangent function. [*Hint*: In Problem 75, multiply the numerator and denominator of $\dfrac{\sin(x/2)}{\cos(x/2)}$ by $2\sin(x/2)$ and then look at (15) and (20).]

75. $\tan \dfrac{x}{2} = \dfrac{1 - \cos x}{\sin x}$

76. $\tan \dfrac{x}{2} = \dfrac{\sin x}{1 + \cos x}$

77. Discuss: Why are the formulas in Problems 75 and 76 more useful than the formula in Problem 74?

7.5 Inverse Trigonometric Functions

≡ **Introduction** Although we can find the values of the trigonometric functions of real numbers or angles, in many applications we must do the reverse: Given the value of a trigonometric function, find a corresponding angle or number. This suggests we consider inverse trigonometric functions. Before we define the inverse trigonometric functions, let's recall from Section 3.6 some of the properties of a one-to-one function f and its inverse f^{-1}.

◀ Recall, a function f is one-to-one if every y in its range corresponds to exactly one x in its domain.

☐ **Properties of Inverse Functions** If $y = f(x)$ is a one-to-one function, then there is a unique inverse function f^{-1} with the following properties:

> **PROPERTIES OF INVERSE FUNCTIONS**
> - The domain of f^{-1} = range of f.
> - The range of f^{-1} = domain of f.
> - $y = f(x)$ is equivalent to $x = f^{-1}(y)$.
> - The graphs of f and f^{-1} are reflections in the line $y = x$.
> - $f(f^{-1}(x)) = x$ for x in the domain of f^{-1}.
> - $f^{-1}(f(x)) = x$ for x in the domain of f.

Inspection of the graphs of the various trigonometric functions clearly shows that *none* of these functions are one-to-one. In Section 3.6 we discussed the fact that if a function f is not one-to-one, it may be possible to restrict the function to a portion of its domain where it is one-to-one. Then we can define an inverse for f on that restricted domain. Normally, when we restrict the domain, we make sure to preserve the entire range of the original function.

See Example 7 in Section 3.6.

☐ **Arcsine Function** From FIGURE 7.5.1(a) we see that the function $y = \sin x$ on the closed interval $[-\pi/2, \pi/2]$ takes on all values in its range $[-1, 1]$. Notice that any horizontal line drawn to intersect the red portion of the graph can do so at most once. Thus the sine function on this restricted domain is one-to-one and has an inverse. There are two commonly used notations to denote the inverse of the function shown in Figure 7.5.1(b):

$$\arcsin x \quad \text{or} \quad \sin^{-1} x,$$

and are read **arcsine of x** and **inverse sine of x**, respectively.

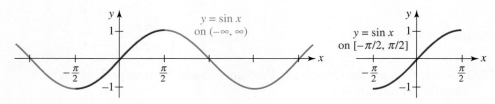

(a) Not a one-to-one function (b) A one-to-one function

FIGURE 7.5.1 Restricting the domain of $y = \sin x$ to produce a one-to-one function

In **FIGURE 7.5.2(a)** we have reflected the portion of the graph of $y = \sin x$ on the interval $[-\pi/2, \pi/2]$ (the red graph in Figure 7.5.1(b)) about the line $y = x$ to obtain the graph of $y = \arcsin x$ (in blue). For clarity, we have reproduced this blue graph in Figure 7.5.2(b). As this curve shows, the domain of the arcsine function is $[-1, 1]$ and the range is $[-\pi/2, \pi/2]$.

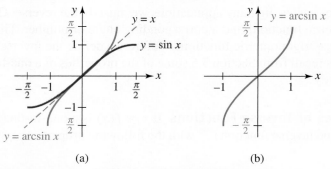

FIGURE 7.5.2 Graph of $y = \arcsin x$ is the blue curve

DEFINITION 7.5.1 Arcsine Function

The **arcsine function**, or **inverse sine function**, is defined by

$$y = \arcsin x \quad \text{if and only if} \quad x = \sin y, \tag{1}$$

where $-1 \leq x \leq 1$ and $-\pi/2 \leq y \leq \pi/2$.

In other words:

> The arcsine of the number x is that number y (or radian-measured angle) between $-\pi/2$ and $\pi/2$ whose sine is x.

When using the notation $\sin^{-1}x$ it is important to realize that "–1" is not an exponent; rather, it denotes an inverse function. The notation $\arcsin x$ has an advantage over the notation $\sin^{-1}x$ in that there is no "–1" and hence no potential for misinterpretation; moreover, the prefix "arc" refers to an angle—*the* angle whose sine is x. But since $y = \arcsin x$ and $y = \sin^{-1}x$ are used interchangeably in calculus and in applications, we will continue to alternate their use so that you become comfortable with both notations.

◀ **Note of Caution:**
$(\sin x)^{-1} = \dfrac{1}{\sin x} \neq \sin^{-1}x$

EXAMPLE 1   Evaluating the Inverse Sine Function

Find **(a)** $\arcsin\tfrac{1}{2}$, **(b)** $\sin^{-1}(-\tfrac{1}{2})$, and **(c)** $\sin^{-1}(-1)$.

Solution (a) If we let $y = \arcsin\tfrac{1}{2}$, then by (1) we must find the number y (or radian-measured angle) that satisfies $\sin y = \tfrac{1}{2}$ and $-\pi/2 \leq y \leq \pi/2$. Since $\sin(\pi/6) = \tfrac{1}{2}$ and $\pi/6$ satisfies the inequality $-\pi/2 \leq y \leq \pi/2$ it follows that $y = \pi/6$.

(b) If we let $y = \sin^{-1}(-\tfrac{1}{2})$, then $\sin y = -\tfrac{1}{2}$. Since we must choose y such that $-\pi/2 \leq y \leq \pi/2$, we find that $y = -\pi/6$.

(c) Letting $y = \sin^{-1}(-1)$, we have that $\sin y = -1$ and $-\pi/2 \leq y \leq \pi/2$. Hence $y = -\pi/2$.

≡

In parts (b) and (c) of Example 1 we were careful to choose y so that $-\pi/2 \leq y \leq \pi/2$. For example, it is a common error to think that because $\sin(3\pi/2) = -1$, then necessarily $\sin^{-1}(-1)$ can be taken to be $3\pi/2$. Remember: If $y = \sin^{-1}x$, then y is subject to the restriction $-\pi/2 \leq y \leq \pi/2$ and $3\pi/2$ does not satisfy this inequality.

◀ Read this paragraph several times.

EXAMPLE 2   Evaluating a Composition

Without using a calculator, find $\tan(\sin^{-1}\tfrac{1}{4})$.

Solution We must find the tangent of the angle of t radians with sine equal to $\tfrac{1}{4}$, that is, $\tan t$, where $t = \sin^{-1}\tfrac{1}{4}$. The angle t is shown in **FIGURE 7.5.3**. Since

$$\tan t = \frac{\sin t}{\cos t} = \frac{\tfrac{1}{4}}{\cos t},$$

we want to determine the value of $\cos t$. From Figure 7.5.3 and the Pythagorean identity $\sin^2 t + \cos^2 t = 1$, we see that

$$\left(\frac{1}{4}\right)^2 + \cos^2 t = 1 \quad \text{or} \quad \cos t = \frac{\sqrt{15}}{4}.$$

Hence we have

$$\tan t = \frac{1/4}{\sqrt{15}/4} = \frac{1}{\sqrt{15}} = \frac{\sqrt{15}}{15},$$

and so

$$\tan\left(\sin^{-1}\frac{1}{4}\right) = \tan t = \frac{\sqrt{15}}{15}.$$

≡

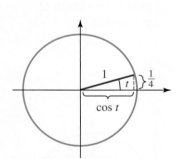

FIGURE 7.5.3 The angle $x = \sin^{-1}\tfrac{1}{4}$ in Example 2

☐ **Arccosine Function** If we restrict the domain of the cosine function to the closed interval $[0, \pi]$, the resulting function is one-to-one and thus has an inverse. We denote this inverse by

$$\arccos x \quad \text{or} \quad \cos^{-1} x,$$

which gives us the following definition.

DEFINITION 7.5.2 Arccosine Function

The **arccosine function**, or **inverse cosine function**, is defined by

$$y = \arccos x \quad \text{if and only if} \quad x = \cos y, \qquad (2)$$

where $-1 \leq x \leq 1$ and $0 \leq y \leq \pi$.

The graphs shown in FIGURE 7.5.4 illustrate how the function $y = \cos x$ restricted to the interval $[0, \pi]$ becomes a one-to-one function. The inverse of the function shown in Figure 7.5.4(b) is $y = \arccos x$.

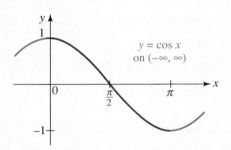
(a) Not a one-to-one function

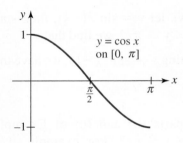
(b) A one-to-one function

FIGURE 7.5.4 Restricting the domain of $y = \cos x$ to produce a one-to-one function

By reflecting the graph of the one-to-one function in Figure 7.5.4(b) in the line $y = x$ we obtain the graph of $y = \arccos x$ shown in FIGURE 7.5.5.

Note that the figure clearly shows that the domain and range of $y = \arccos x$ are $[-1, 1]$ and $[0, \pi]$, respectively.

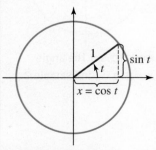
FIGURE 7.5.5 Graph of $y = \arccos x$

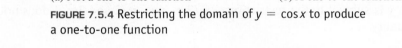

EXAMPLE 3   **Evaluating the Inverse Cosine Function**

Find **(a)** $\arccos\left(\sqrt{2}/2\right)$ and **(b)** $\cos^{-1}\left(-\sqrt{3}/2\right)$.

Solution (a) If we let $y = \arccos(\sqrt{2}/2)$, then $\cos y = \sqrt{2}/2$ and $0 \leq y \leq \pi$. Thus $y = \pi/4$.
(b) Letting $y = \cos^{-1}(-\sqrt{3}/2)$, we have that $\cos y = -\sqrt{3}/2$, and we must find y such that $0 \leq y \leq \pi$. Therefore, $y = 5\pi/6$ since $\cos(5\pi/6) = -\sqrt{3}/2$. ≡

EXAMPLE 4   **Evaluating the Compositions of Functions**

Write $\sin(\cos^{-1} x)$ as an algebraic expression in x.

Solution In FIGURE 7.5.6 we have constructed an angle of t radians with cosine equal to x. Then $t = \cos^{-1} x$, or $x = \cos t$, where $0 \leq t \leq \pi$. Now to find $\sin(\cos^{-1} x) = \sin t$, we use the identity $\sin^2 t + \cos^2 t = 1$. Thus

FIGURE 7.5.6 The angle $t = \cos^{-1} x$ in Example 4

$$\sin^2 t + x^2 = 1$$
$$\sin^2 t = 1 - x^2$$
$$\sin t = \sqrt{1 - x^2}$$
$$\sin(\cos^{-1} x) = \sqrt{1 - x^2}.$$

We use the positive square root of $1 - x^2$, since the range of $\cos^{-1} x$ is $[0, \pi]$, and the sine of an angle t in the first or second quadrant is positive. ≡

☐ **Arctangent Function** If we restrict the domain of $\tan x$ to the open interval $(-\pi/2, \pi/2)$, then the resulting function is one-to-one and thus has an inverse. This inverse is denoted by

$$\arctan x \quad \text{or} \quad \tan^{-1} x.$$

DEFINITION 7.5.3 Arctangent Function

The **arctangent**, or **inverse tangent**, function is defined by

$$y = \arctan x \quad \text{if and only if} \quad x = \tan y, \qquad (3)$$

where $-\infty < x < \infty$ and $-\pi/2 < y < \pi/2$.

The graphs shown in **FIGURE 7.5.7** illustrate how the function $y = \tan x$ restricted to the open interval $(-\pi/2, \pi/2)$ becomes a one-to-one function.

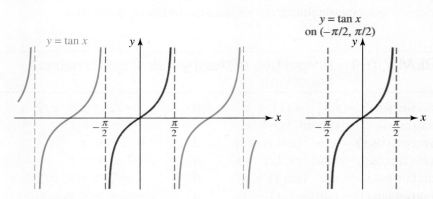

(a) Not a one-to-one function (b) A one-to-one function

FIGURE 7.5.7 Restricting the domain of $y = \tan x$ to produce a one-to-one function

By reflecting the graph of the one-to-one function in Figure 7.5.7(b) in the line $y = x$ we obtain the graph of $y = \arctan x$ shown in **FIGURE 7.5.8**. We see in the figure that the domain and range of $y = \arctan x$ are, in turn, the intervals $(-\infty, \infty)$ and $(-\pi/2, \pi/2)$.

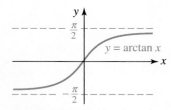

FIGURE 7.5.8 Graph of $y = \arctan x$

EXAMPLE 5 **Evaluating the Inverse Tangent Function**

Find $\tan^{-1}(-1)$.

Solution If $\tan^{-1}(-1) = y$, then $\tan y = -1$, where $-\pi/2 < y < \pi/2$. It follows that $\tan^{-1}(-1) = y = -\pi/4$. ≡

EXAMPLE 6 Evaluating the Compositions of Functions

Without using a calculator, find $\sin\left(\arctan\left(-\frac{5}{3}\right)\right)$.

Solution If we let $t = \arctan\left(-\frac{5}{3}\right)$, then $\tan t = -\frac{5}{3}$. The Pythagorean identity $1 + \tan^2 t = \sec^2 t$ can be used to find $\sec t$:

$$1 + \left(-\frac{5}{3}\right)^2 = \sec^2 t$$

$$\sec t = \sqrt{\frac{25}{9} + 1} = \sqrt{\frac{34}{9}} = \frac{\sqrt{34}}{3}.$$

In the preceding line we take the positive square root because $t = \arctan\left(-\frac{5}{3}\right)$ is in the interval $(-\pi/2, \pi/2)$ (the range of the arctangent function) and the secant of an angle t in the first or fourth quadrant is positive. Also, from $\sec t = \sqrt{34}/3$ we find the value of $\cos t$ from the reciprocal identity:

$$\cos t = \frac{1}{\sec t} = \frac{1}{\sqrt{34}/3} = \frac{3}{\sqrt{34}}.$$

Finally, we can use the identity $\tan t = \sin t/\cos t$ in the form $\sin t = \tan t \cos t$ to compute $\sin\left(\arctan\left(-\frac{5}{3}\right)\right)$. It follows that

$$\sin t = \tan t \cos t = \left(-\frac{5}{3}\right)\left(\frac{3}{\sqrt{34}}\right) = -\frac{5}{\sqrt{34}}.$$

☐ **Properties of the Inverses** Recall from Section 3.6 that $f^{-1}(f(x)) = x$ and $f(f^{-1}(x)) = x$ hold for any function f and its inverse under suitable restrictions on x. Thus for the inverse trigonometric functions, we have the following properties.

THEOREM 7.5.1 Properties of the Inverse Trigonometric Functions

(i) $\arcsin(\sin x) = \sin^{-1}(\sin x) = x$ if $-\pi/2 \leq x \leq \pi/2$
(ii) $\sin(\arcsin x) = \sin(\sin^{-1} x) = x$ if $-1 \leq x \leq 1$
(iii) $\arccos(\cos x) = \cos^{-1}(\cos x) = x$ if $0 \leq x \leq \pi$
(iv) $\cos(\arccos x) = \cos(\cos^{-1} x) = x$ if $-1 \leq x \leq 1$
(v) $\arctan(\tan x) = \tan^{-1}(\tan x) = x$ if $-\pi/2 < x < \pi/2$
(vi) $\tan(\arctan x) = \tan(\tan^{-1} x) = x$ if $-\infty < x < \infty$

EXAMPLE 7 Using the Inverse Properties

Without using a calculator, evaluate:

(a) $\sin^{-1}\left(\sin\frac{\pi}{12}\right)$ **(b)** $\cos\left(\cos^{-1}\frac{1}{3}\right)$ **(c)** $\tan^{-1}\left(\tan\frac{3\pi}{4}\right)$.

Solution In each case we use the properties of the inverse trigonometric functions given in Theorem 7.5.1.

(a) Because $\pi/12$ satisfies $-\pi/2 \leq x \leq \pi/2$ it follows from property (i) that

$$\sin^{-1}\left(\sin\frac{\pi}{12}\right) = \frac{\pi}{12}.$$

(b) By property *(iv)*, $\cos(\cos^{-1}\frac{1}{3}) = \frac{1}{3}$.

(c) In this case we *cannot* apply property *(v)*, since $3\pi/4$ is not in the interval $(-\pi/2, \pi/2)$. If we first evaluate $\tan(3\pi/4) = -1$, then we have

$$\tan^{-1}\left(\tan\frac{3\pi}{4}\right) \overset{\text{see Example 5}}{=} \tan^{-1}(-1) = -\frac{\pi}{4}.$$

In the next section we illustrate how inverse trigonometric functions can be used to solve trigonometric equations.

☐ **Postscript—The Other Inverse Trig Functions** The functions $\cot x$, $\sec x$, and $\csc x$ also have inverses when their domains are suitably restricted. See Problems 49–51 in Exercises 7.5. Because these functions are not used as often as arctan, arccos, and arcsin, most scientific calculators do not have keys for them. However, any calculator that computes arcsin, arccos, and arctan can be used to obtain values for **arccsc**, **arcsec**, and **arccot**. Unlike the fact that $\sec x = 1/\cos x$, we note that $\sec^{-1}x \neq 1/\cos^{-1}x$; rather, $\sec^{-1}x = \cos^{-1}(1/x)$ for $|x| \geq 1$. Similar relationships hold for $\csc^{-1}x$ and $\cot^{-1}x$. See Problems 56–58 in Exercises 7.5.

7.5 Exercises
Answers to selected odd-numbered problems begin on page ANS-20.

In Problems 1–14, find the exact value of the given trigonometric expression. Do not use a calculator.

1. $\sin^{-1}0$
2. $\tan^{-1}\sqrt{3}$
3. $\arccos(-1)$
4. $\arcsin\dfrac{\sqrt{3}}{2}$
5. $\arccos\frac{1}{2}$
6. $\arctan(-\sqrt{3})$
7. $\sin^{-1}\left(-\dfrac{\sqrt{3}}{2}\right)$
8. $\cos^{-1}\dfrac{\sqrt{3}}{2}$
9. $\tan^{-1}1$
10. $\sin^{-1}\dfrac{\sqrt{2}}{2}$
11. $\arctan\left(-\dfrac{\sqrt{3}}{3}\right)$
12. $\arccos\left(-\frac{1}{2}\right)$
13. $\sin^{-1}\left(-\dfrac{\sqrt{2}}{2}\right)$
14. $\arctan 0$

In Problems 15–32, find the exact value of the given trigonometric expression. Do not use a calculator.

15. $\sin(\cos^{-1}\frac{3}{5})$
16. $\cos(\sin^{-1}\frac{1}{3})$
17. $\tan(\arccos(-\frac{2}{3}))$
18. $\sin(\arctan\frac{1}{4})$
19. $\cos(\arctan(-2))$
20. $\tan(\sin^{-1}(-\frac{1}{6}))$
21. $\csc(\sin^{-1}\frac{3}{5})$
22. $\sec(\tan^{-1}4)$
23. $\sin(\sin^{-1}\frac{1}{5})$
24. $\cos(\cos^{-1}(-\frac{4}{5}))$
25. $\tan(\tan^{-1}1.2)$
26. $\sin(\arcsin 0.75)$

27. $\arcsin\left(\sin\dfrac{\pi}{16}\right)$ **28.** $\arccos\left(\cos\dfrac{2\pi}{3}\right)$

29. $\tan^{-1}(\tan\pi)$ **30.** $\sin^{-1}\left(\sin\dfrac{5\pi}{6}\right)$

31. $\cos^{-1}\left(\cos\left(-\dfrac{\pi}{4}\right)\right)$ **32.** $\arctan\left(\tan\dfrac{\pi}{7}\right)$

In Problems 33–40, write the given expression as an algebraic expression in x.

33. $\sin(\tan^{-1}x)$ **34.** $\cos(\tan^{-1}x)$
35. $\tan(\arcsin x)$ **36.** $\sec(\arccos x)$
37. $\cot(\sin^{-1}x)$ **38.** $\cos(\sin^{-1}x)$
39. $\csc(\arctan x)$ **40.** $\tan(\arccos x)$

In Problems 41–48, sketch the graph of the given function.

41. $y = \arctan|x|$ **42.** $y = \dfrac{\pi}{2} - \arctan x$

43. $y = |\arcsin x|$ **44.** $y = \sin^{-1}(x+1)$
45. $y = 2\cos^{-1}x$ **46.** $y = \cos^{-1}2x$
47. $y = \arccos(x-1)$ **48.** $y = \cos(\arcsin x)$

49. The **arccotangent** function can be defined by $y = \text{arccot } x$ (or $y = \cot^{-1}x$) if and only if $x = \cot y$, where $0 < y < \pi$. Graph $y = \text{arccot } x$, and give the domain and the range of this function.

50. The **arccosecant** function can be defined by $y = \text{arccsc } x$ (or $y = \csc^{-1}x$) if and only if $x = \csc y$, where $-\pi/2 \leq y \leq \pi/2$ and $y \neq 0$. Graph $y = \text{arccsc } x$, and give the domain and the range of this function.

51. One definition of the **arcsecant** function is $y = \text{arcsec } x$ (or $y = \sec^{-1}x$) if and only if $x = \sec y$, where $0 \leq y \leq \pi$ and $y \neq \pi/2$. (See Problem 52 for an alternative definition.) Graph $y = \text{arcsec } x$, and give the domain and the range of this function.

52. An alternative definition of the arcsecant function can be made by restricting the domain of the secant function to $[0, \pi/2) \cup [\pi, 3\pi/2)$. Under this restriction, define the arcsecant function. Graph $y = \text{arcsec } x$, and give the domain and the range of this function.

53. Using the definition of the arccotangent function from Problem 49, for what values of x is it true that **(a)** $\cot(\text{arccot } x) = x$ and **(b)** $\text{arccot}(\cot x) = x$?

54. Using the definition of the arccosecant function from Problem 50, for what values of x is it true that **(a)** $\csc(\text{arccsc } x) = x$ and **(b)** $\text{arccsc}(\csc x) = x$?

55. Using the definition of the arcsecant function from Problem 51, for what values of x is it true that **(a)** $\sec(\text{arcsec } x) = x$ and **(b)** $\text{arcsec}(\sec x) = x$?

56. Verify that $\text{arccot } x = \dfrac{\pi}{2} - \arctan x$, for all real numbers x.

57. Verify that $\text{arccsc } x = \arcsin(1/x)$ for $|x| \geq 1$.
58. Verify that $\text{arcsec } x = \arccos(1/x)$ for $|x| \geq 1$.

In Problems 59–64, use the results of Problems 56–58 and a calculator to find the indicated value.

59. $\cot^{-1}0.75$ **60.** $\csc^{-1}(-1.3)$
61. $\text{arccsc}(-1.5)$ **62.** $\text{arccot}(-0.3)$
63. $\text{arcsec}(-1.2)$ **64.** $\sec^{-1}2.5$

Miscellaneous Applications

65. Projectile Motion The departure angle θ for a bullet to hit a target at a distance R (assuming that the target and the gun are at the same height) satisfies

$$R = \frac{v_0^2 \sin 2\theta}{g},$$

where v_0 is the muzzle velocity and g is the acceleration due to gravity. If the target is 800 ft from the gun and the muzzle velocity is 200 ft/s, find the departure angle. Use $g = 32$ ft/s^2. [*Hint*: There are two solutions.]

66. Olympic Sports For the Olympic event, the hammer throw, it can be shown that the maximum distance is achieved for the release angle θ (measured from the horizontal) that satisfies

$$\cos 2\theta = \frac{gh}{v_0^2 + gh},$$

where h is the height of the hammer above the ground at release, v_0 is the initial velocity, and g is the acceleration due to gravity. For $v_0 = 13.7$ m/s and $h = 2.25$ m, find the optimal release angle. Use $g = 9.81$ m/s^2.

67. Highway Design In the design of highways and railroads, curves are banked to provide centripetal force for safety. The optimal banking angle θ is given by $\tan \theta = v^2/Rg$, where v is the speed of the vehicle, R is the radius of the curve, and g is the acceleration due to gravity. See **FIGURE 7.5.9**. As the formula indicates, for a given radius there is no one correct angle for all speeds. Consequently, curves are banked for the average speed of the traffic over them. Find the correct banking angle for a curve of radius 600 ft on a country road where speeds average 30 mi/h. Use $g = 32$ ft/s^2. [*Hint*: Use consistent units.]

68. Highway Design—Continued If μ is the coefficient of friction between the car and the road, then the maximum velocity v_m that a car can travel around a curve without slipping is given by $v_m^2 = gR\tan(\theta + \tan^{-1}\mu)$, where θ is the banking angle of the curve. Find v_m for the country road in Problem 67 if $\mu = 0.26$.

69. Geology Viewed from the side, a volcanic cinder cone usually looks like an isosceles trapezoid. See **FIGURE 7.5.10**. Studies of cinder cones less than 50,000 years old indicate that cone height H_{co} and crater width W_{cr} are related to the cone width W_{co} by the equations $H_{co} = 0.18W_{co}$ and $W_{cr} = 0.40W_{co}$. If $W_{co} = 1.00$, use these equations to determine the base angle ϕ of the trapezoid in Figure 7.5.10.

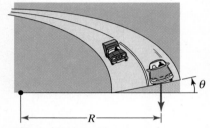

FIGURE 7.5.9 Banked curve in Problem 67

Volcanic cone

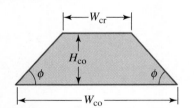

FIGURE 7.5.10 Volcanic cinder cone in Problem 69

For Discussion

70. Using a calculator set in radian mode, evaluate arctan (tan 1.8), arccos (cos 1.8), and arcsin (sin 1.8). Explain the results.

71. Using a calculator set in radian mode, evaluate $\tan^{-1}(\tan(-1))$, $\cos^{-1}(\cos(-1))$, and $\sin^{-1}(\sin(-1))$. Explain the results.

72. In Section 7.2 we saw that the graphs of $y = \sin x$ and $y = \cos x$ are related by shifting and reflecting. Justify the identity

$$\arcsin x + \arccos x = \frac{\pi}{2},$$

for all x in $[-1, 1]$, by finding a similar relationship between the graphs of $y = \arcsin x$ and $y = \arccos x$.

73. With a calculator set in radian mode determine which of the following inverse trigonometric evaluations result in an error message: **(a)** $\sin^{-1}(-2)$, **(b)** $\cos^{-1}(-2)$, **(c)** $\tan^{-1}(-2)$. Explain.

74. Discuss: Can any periodic function be one-to-one?

75. Show that $\arcsin \frac{3}{5} + \arcsin \frac{5}{13} = \arcsin \frac{56}{65}$. [*Hint*: See (7) of Section 7.4.]

7.6 Trigonometric Equations

Introduction In Section 7.4 we considered trigonometric identities, which are equations involving trigonometric functions that are satisfied by all values of the variable for which both sides of the equality are defined. In this section we examine **conditional trigonometric equations**, that is, equations that are true for only certain values of the variable. We discuss techniques for finding those values of the variable (if any) that satisfy the equation.

We begin by considering the problem of finding all real numbers x that satisfy $\sin x = \sqrt{2}/2$. Interpreted as the x-coordinates of the points of intersection of the graphs of $y = \sin x$ and $y = \sqrt{2}/2$, FIGURE 7.6.1 shows that there exists infinitely many solutions of the equation $\sin x = \sqrt{2}/2$:

$$\ldots, -\frac{7\pi}{4}, \frac{\pi}{4}, \frac{9\pi}{4}, \frac{17\pi}{4}, \ldots \tag{1}$$

$$\ldots, -\frac{5\pi}{4}, \frac{3\pi}{4}, \frac{11\pi}{4}, \frac{19\pi}{4}, \ldots \tag{2}$$

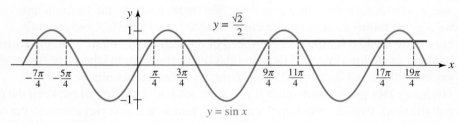

FIGURE 7.6.1 Graphs of $y = \sin x$ and $y = \frac{\sqrt{2}}{2}$

Note that in each of the lists (1) and (2), two successive solutions differ by $2\pi = 8\pi/4$. This is a consequence of the periodicity of the sine function. It is common for trigonometric equations to have an infinite number of solutions because of the periodicity of the trigonometric functions. In general, to obtain solutions of an equation such as $\sin x = \sqrt{2}/2$, it is more convenient to use a unit circle and reference angles rather than a graph of the trigonometric function. We illustrate this approach in the following example.

EXAMPLE 1   Using the Unit Circle

Find all real numbers x satisfying $\sin x = \sqrt{2}/2$.

Solution If $\sin x = \sqrt{2}/2$, the reference angle for x is $\pi/4$ radian. Since the value of $\sin x$ is positive, the terminal side of the angle x lies in either the first or second quadrant. Thus, as shown in FIGURE 7.6.2, the only solutions between 0 and 2π are

$$x = \frac{\pi}{4} \quad \text{and} \quad x = \frac{3\pi}{4}.$$

Since the sine function is periodic with period 2π, all of the remaining solutions can be obtained by adding integer multiples of 2π to these solutions. The two solutions are

$$x = \frac{\pi}{4} + 2n\pi \quad \text{and} \quad x = \frac{3\pi}{4} + 2n\pi, \tag{3}$$

FIGURE 7.6.2 Unit circle in Example 1

where n is an integer The numbers that you see in (1) and (2) correspond, respectively, to letting $n = -1$, $n = 0$, $n = 1$, and $n = 2$ in the first and second formulas in (3). ≡

When we are faced with a more complicated equation, such as

$$4\sin^2 x - 8\sin x + 3 = 0,$$

the basic approach is to solve for a single trigonometric function (in this case, it would be $\sin x$) by using methods similar to those for solving algebraic equations.

EXAMPLE 2 Solving a Trigonometric Equation by Factoring

Find all solutions of $4\sin^2 x - 8\sin x + 3 = 0$.

Solution We first observe that this is a quadratic equation in $\sin x$, and that it factors as

$$(2\sin x - 3)(2\sin x - 1) = 0.$$

This implies that either

$$\sin x = \frac{3}{2} \quad \text{or} \quad \sin x = \frac{1}{2}.$$

The first equation has no solution since $|\sin x| \leq 1$. As we see in FIGURE 7.6.3 the two angles between 0 and 2π for which $\sin x$ equals $\frac{1}{2}$ are

$$x = \frac{\pi}{6} \quad \text{and} \quad x = \frac{5\pi}{6}.$$

Therefore, by the periodicity of the sine function, the solutions are

$$x = \frac{\pi}{6} + 2n\pi \quad \text{and} \quad x = \frac{5\pi}{6} + 2n\pi,$$

where n is an integer. ≡

FIGURE 7.6.3 Unit circle in Example 2

EXAMPLE 3 Checking for Lost Solutions

Find all solutions of

$$\sin x = \cos x. \tag{4}$$

Solution In order to work with a single trigonometric function, we divide both sides of the equation by $\cos x$ to obtain

$$\tan x = 1. \tag{5}$$

Equation (5) is equivalent to (4) *provided* that $\cos x \neq 0$. We observe that if $\cos x = 0$, then as we have seen in Section 7.2, $x = (2n + 1)\pi/2 = \pi/2 + n\pi$, for n an integer. By the sum formula for the sine,

$$\overset{\text{see (7) of Section 7.4}}{\underset{\downarrow}{\sin\left(\frac{\pi}{2} + n\pi\right)}} = \sin\frac{\pi}{2}\overset{(-1)^n}{\underset{\downarrow}{\cos n\pi}} + \cos\frac{\pi}{2}\overset{0}{\underset{\downarrow}{\sin n\pi}} = (-1)^n \neq 0,$$

we see that these values of x do not satisfy the original equation. Thus we will find *all* the solutions to (4) by solving equation (5).

Now $\tan x = 1$ implies that the reference angle for x is $\pi/4$ radian. Since $\tan x = 1 > 0$, the terminal side of the angle of x radians can lie either in the first or in the third quadrant, as shown in FIGURE 7.6.4. Thus the solutions are

$$x = \frac{\pi}{4} + 2n\pi \quad \text{and} \quad x = \frac{5\pi}{4} + 2n\pi,$$

◄ $\cos 0 = 1$, $\cos \pi = -1$, $\cos 2\pi = 1$, $\cos 3\pi = -1$, and so on. In general, $\cos n\pi = (-1)^n$, where n is an integer.

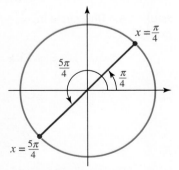

FIGURE 7.6.4 Unit circle in Example 3

7.6 Trigonometric Equations

where n is an integer. We can see from Figure 7.6.4 that these two sets of numbers can be written more compactly as

$$x = \frac{\pi}{4} + n\pi,$$

where n is an integer.

▶ This also follows from the fact that $\tan x$ is π-periodic.

☐ **Losing Solutions** When solving an equation, if you divide by an expression containing a variable, you may lose some solutions of the original equation. For example, in algebra a common mistake in solving equations such as $x^2 = x$ is to divide by x to obtain $x = 1$. But by writing $x^2 = x$ as $x^2 - x = 0$ or $x(x - 1) = 0$ we see that in fact $x = 0$ or $x = 1$. To prevent the loss of a solution you must determine the values that make the expression zero and check to see whether they are solutions of the original equation. Note that in Example 3, when we divided by $\cos x$, we took care to check that no solutions were lost.

Whenever possible, it is preferable to avoid dividing by a variable expression. As illustrated with the algebraic equation $x^2 = x$, this can frequently be accomplished by collecting all nonzero terms on one side of the equation and then factoring (something we could not do in Example 3). Example 4 illustrates this technique.

EXAMPLE 4 Solving a Trigonometric Equation by Factoring

Solve
$$2\sin x \cos^2 x = -\frac{\sqrt{3}}{2}\cos x. \qquad (6)$$

Solution To avoid dividing by $\cos x$, we write the equation as

$$2\sin x \cos^2 x + \frac{\sqrt{3}}{2}\cos x = 0$$

and factor:
$$\cos x \left(2\sin x \cos x + \frac{\sqrt{3}}{2}\right) = 0.$$

Thus either
$$\cos x = 0 \qquad \text{or} \qquad 2\sin x \cos x + \frac{\sqrt{3}}{2} = 0.$$

Since the cosine is zero for all odd multiples of $\pi/2$, the solutions of $\cos x = 0$ are

$$x = (2n + 1)\frac{\pi}{2} = \frac{\pi}{2} + n\pi,$$

where n is an integer.

In the second equation we replace $2\sin x \cos x$ by $\sin 2x$ from the double-angle formula for the sine function to obtain an equation with a single trigonometric function:

$$\sin 2x + \frac{\sqrt{3}}{2} = 0 \qquad \text{or} \qquad \sin 2x = -\frac{\sqrt{3}}{2}.$$

Thus the reference angle for $2x$ is $\pi/3$. Since the sine is negative, the angle $2x$ must be in either the third quadrant or the fourth quadrant. As **FIGURE 7.6.5** illustrates, either

$$2x = \frac{4\pi}{3} + 2n\pi \qquad \text{or} \qquad 2x = \frac{5\pi}{3} + 2n\pi.$$

FIGURE 7.6.5 Unit circle in Example 4

▶ See (15) in Section 7.4.

Dividing by 2 gives

$$x = \frac{2\pi}{3} + n\pi \quad \text{or} \quad x = \frac{5\pi}{6} + n\pi.$$

Therefore, all solutions of (6) are

$$x = \frac{\pi}{2} + n\pi, \quad x = \frac{2\pi}{3} + n\pi, \quad \text{and} \quad x = \frac{5\pi}{6} + n\pi,$$

where n is an integer.

In Example 4 had we simplified the equation by dividing by $\cos x$ and not checked to see whether the values of x for which $\cos x = 0$ satisfied equation (6), we would have lost the solutions $x = \pi/2 + n\pi$, where n is an integer.

EXAMPLE 5 Using a Trigonometric Identity

Solve $3\cos^2 x - \cos 2x = 1$.

Solution We observe that the given equation involves both the cosine of x and the cosine of $2x$. Consequently, we use the double-angle formula for the cosine in the form

$$\cos 2x = 2\cos^2 x - 1 \quad \leftarrow \text{See (16) of Section 7.4}$$

to replace the equation by an equivalent equation that involves $\cos x$ only. We find that

$$3\cos^2 x - (2\cos^2 x - 1) = 1 \quad \text{becomes} \quad \cos^2 x = 0.$$

Therefore, $\cos x = 0$, and the solutions are

$$x = (2n + 1)\frac{\pi}{2} = \frac{\pi}{2} + n\pi,$$

where n is an integer.

So far in this section we have viewed the variable in the trigonometric equation as representing either a real number or an angle measured in radians. If the variable represents an angle measured in degrees, the technique for solving is the same.

EXAMPLE 6 Equation When the Angle Is in Degrees

Solve $\cos 2\theta = -\frac{1}{2}$, where θ is an angle measured in degrees.

Solution Since $\cos 2\theta = -\frac{1}{2}$, the reference angle for 2θ is $60°$ and the angle 2θ must be in either the second or the third quadrant. FIGURE 7.6.6 illustrates that either $2\theta = 120°$ or $2\theta = 240°$. Any angle that is coterminal with one of these angles will also satisfy $\cos 2\theta = -\frac{1}{2}$. These angles are obtained by adding any integer multiple of $360°$ to $120°$ or to $240°$:

$$2\theta = 120° + 360°n \quad \text{or} \quad 2\theta = 240° + 360°n,$$

where n is an integer. Dividing by 2 the last line yields the two solutions

$$\theta = 60° + 180°n \quad \text{and} \quad \theta = 120° + 180°n.$$

FIGURE 7.6.6 Unit circle in Example 6

☐ **Extraneous Solutions** The next example shows that by squaring a trigonometric equation we may introduce extraneous solutions. In other words, the resulting equation after squaring may *not* be equivalent to the original.

EXAMPLE 7 Extraneous Roots

Find all solutions of $1 + \tan\alpha = \sec\alpha$, where α is an angle measured in degrees.

Solution The equation does not factor, but we see that if we square both sides, we can use a fundamental identity to obtain an equation involving a single trigonometric function:

$$(1 + \tan\alpha)^2 = (\sec\alpha)^2$$
$$1 + 2\tan\alpha + \tan^2\alpha = \sec^2\alpha \qquad \leftarrow \text{See (2) of Section 7.4.}$$
$$1 + 2\tan\alpha + \tan^2\alpha = 1 + \tan^2\alpha$$
$$2\tan\alpha = 0$$
$$\tan\alpha = 0.$$

The values of α in $[0°, 360°)$ for which $\tan\alpha = 0$ are

$$\alpha = 0° \quad \text{and} \quad \alpha = 180°.$$

Since we squared each side of the original equation, we may have introduced extraneous solutions. Therefore, it is important that we check all solutions in the original equation. Substituting $\alpha = 0°$ into $1 + \tan\alpha = \sec\alpha$, we obtain the *true* statement $1 + 0 = 1$. But after substituting $\alpha = 180°$, we obtain the *false* statement $1 + 0 = -1$. Therefore, $180°$ is an extraneous solution and $\alpha = 0°$ is the only solution in the interval $[0°, 360°)$. Thus, all the solutions of the equation are given by

$$\alpha = 0° + 360°n = 360°n,$$

where n is an integer. For $n \neq 0$, these are the angles that are coterminal with $0°$. ≡

Recall from Section 3.1 that to find the x-intercepts of the graph of a function $y = f(x)$ we find the zeros of f, that is, we must solve the equation $f(x) = 0$. The following example makes use of this fact.

EXAMPLE 8 Intercepts of a Graph

Find the first three x-intercepts of the graph of $f(x) = \sin 2x \cos x$ on the positive x-axis.

Solution We must solve $f(x) = 0$, that is, $\sin 2x \cos x = 0$. It follows that either $\sin 2x = 0$ or $\cos x = 0$.

From $\sin 2x = 0$, we obtain $2x = n\pi$, where n is an integer, or $x = n\pi/2$, where n is an integer. From $\cos x = 0$, we find $x = \pi/2 + n\pi$, where n is an integer. Then for $n = 2$, $x = n\pi/2$ gives $x = \pi$, whereas for $n = 0$ and $n = 1$, $x = \pi/2 + n\pi$ gives $x = \pi/2$ and $x = 3\pi/2$, respectively. Thus the first three x-intercepts on the positive x-axis are $(\pi/2, 0)$, $(\pi, 0)$, and $(3\pi/2, 0)$. ≡

☐ **Using Inverse Functions** So far all of the trigonometric equations have had solutions that were related by reference angles to the special angles 0, $\pi/6$, $\pi/4$, $\pi/3$, or $\pi/2$. If this is not the case, we will see in the next example how to use inverse trigonometric functions and a calculator to find solutions.

EXAMPLE 9 Solving Equations Using Inverse Functions

Find the solutions of $4\cos^2 x - 3\cos x - 2 = 0$ in the interval $[0, \pi]$.

Solution We recognize that this is a quadratic equation in $\cos x$. Since the left-hand side of the equation does not readily factor, we apply the quadratic formula to obtain

$$\cos x = \frac{3 \pm \sqrt{41}}{8}.$$

At this point we can discard the value $(3 + \sqrt{41})/8 \approx 1.18$, because $\cos x$ cannot be greater than 1. We then use the inverse cosine function (and the aid of a calculator) to solve the remaining equation:

$$\cos x = \frac{3 - \sqrt{41}}{8} \quad \text{which implies} \quad x = \cos^{-1}\left(\frac{3 - \sqrt{41}}{8}\right) \approx 2.01. \quad \equiv$$

Of course in Example 9, had we attempted to compute $\cos^{-1}[(3 + \sqrt{41})/8]$ with a calculator, we would have received an error message.

7.6 Exercises   Answers to selected odd-numbered problems begin on page ANS-20.

In Problems 1–6, find all solutions of the given trigonometric equation if x represents an angle measured in radians.

1. $\sin x = \sqrt{3}/2$
2. $\cos x = -\sqrt{2}/2$
3. $\sec x = \sqrt{2}$
4. $\tan x = -1$
5. $\cot x = -\sqrt{3}$
6. $\csc x = 2$

In Problems 7–12, find all solutions of the given trigonometric equation if x represents a real number.

7. $\cos x = -1$
8. $2\sin x = -1$
9. $\tan x = 0$
10. $\sqrt{3}\sec x = 2$
11. $-\csc x = 1$
12. $\sqrt{3}\cot x = 1$

In Problems 13–18, find all solutions of the given trigonometric equation if θ represents an angle measured in degrees.

13. $\csc \theta = 2\sqrt{3}/3$
14. $2\sin \theta = \sqrt{2}$
15. $1 + \cot \theta = 0$
16. $\sqrt{3}\sin \theta = \cos \theta$
17. $\sec \theta = -2$
18. $2\cos \theta + \sqrt{2} = 0$

In Problems 19–46, find all solutions of the given trigonometric equation if x is a real number and θ is an angle measured in degrees.

19. $\cos^2 x - 1 = 0$
20. $2\sin^2 x - 3\sin x + 1 = 0$
21. $3\sec^2 x = \sec x$
22. $\tan^2 x + (\sqrt{3} - 1)\tan x - \sqrt{3} = 0$
23. $2\cos^2 \theta - 3\cos \theta - 2 = 0$
24. $2\sin^2 \theta - \sin \theta - 1 = 0$
25. $\cot^2 \theta + \cot \theta = 0$
26. $2\sin^2 \theta + (2 - \sqrt{3})\sin \theta - \sqrt{3} = 0$
27. $\cos 2x = -1$
28. $\sec 2x = 2$
29. $2\sin 3\theta = 1$
30. $\tan 4\theta = -1$
31. $\cot(x/2) = 1$
32. $\csc(\theta/3) = -1$
33. $\sin 2x + \sin x = 0$
34. $\cos 2x + \sin^2 x = 1$
35. $\cos 2\theta = \sin \theta$
36. $\sin 2\theta + 2\sin \theta - 2\cos \theta = 2$
37. $\sin^4 x - 2\sin^2 x + 1 = 0$
38. $\tan^4 \theta - 2\sec^2 \theta + 3 = 0$
39. $\sec x \sin^2 x = \tan x$
40. $\dfrac{1 + \cos \theta}{\cos \theta} = 2$
41. $\sin \theta + \cos \theta = 1$
42. $\sin x + \cos x = 0$
43. $\sqrt{\dfrac{1 + 2\sin x}{2}} = 1$
44. $\sin x + \sqrt{\sin x} = 0$
45. $\cos \theta - \sqrt{\cos \theta} = 0$
46. $\cos \theta \sqrt{1 + \tan^2 \theta} = 1$

In Problems 47–54, find the first three x-intercepts of the graph of the given function on the positive x-axis.

47. $f(x) = -5\sin(3x + \pi)$
48. $f(x) = 2\cos\left(x + \dfrac{\pi}{4}\right)$
49. $f(x) = 2 - \sec\dfrac{\pi}{2}x$
50. $f(x) = 1 + \cos\pi x$
51. $f(x) = \sin x + \tan x$
52. $f(x) = 1 - 2\cos\left(x + \dfrac{\pi}{3}\right)$
53. $f(x) = \sin x - \sin 2x$
54. $f(x) = \cos x + \cos 3x$
[*Hint*: Write $3x = x + 2x$.]

In Problems 55–58, by graphing determine whether the given equation has any solutions.

55. $\tan x = x$ [*Hint*: Graph $y = \tan x$ and $y = x$ on the same set of axes.]
56. $\sin x = x$
57. $\cot x - x = 0$
58. $\cos x + x + 1 = 0$

In Problems 59–64, using a inverse trigonometric function find the solutions of the given equation in the indicated interval. Round your answers to two decimal places.

59. $20\cos^2 x + \cos x - 1 = 0$, $[0, \pi]$
60. $3\sin^2 x - 8\sin x + 4 = 0$, $[-\pi/2, \pi/2]$
61. $\tan^2 x + \tan x - 1 = 0$, $(-\pi/2, \pi/2)$
62. $3\sin 2x + \cos x = 0$, $[-\pi/2, \pi/2]$
63. $5\cos^3 x - 3\cos^2 x - \cos x = 0$, $[0, \pi]$
64. $\tan^4 x - 3\tan^2 x + 1 = 0$, $(-\pi/2, \pi/2)$

Miscellaneous Applications

65. Isosceles Triangle From Problem 59 in Exercises 7.4, the area of the isosceles triangle with vertex angle θ as shown in Figure 7.4.4 is given by $A = \tfrac{1}{2}x^2 \sin\theta$. If the length x is 4, what value of θ will give a triangle with area 4?

66. Circular Motion An object travels in a circular path centered at the origin with constant angular speed. The y-coordinate of the object at any time t seconds is given by $y = 8\cos(\pi t - \pi/12)$. At what time(s) does the object cross the x-axis?

67. Mach Number Use Problem 57 in Exercises 7.4 to find the vertex angle of the cone of sound waves made by an airplane flying at Mach 2.

68. Alternating Current An electric generator produces a 60-cycle alternating current given by $I(t) = 30\sin 120\pi\left(t - \tfrac{7}{36}\right)$, where $I(t)$ is the current in amperes at t seconds. Find the smallest positive value of t for which the current is 15 amperes.

69. Electrical Circuits If the voltage given by $V = V_0 \sin(\omega t + \alpha)$ is impressed on a series circuit, an alternating current is produced. If $V_0 = 110$ volts, $\omega = 120\pi$ radians per second, and $\alpha = -\pi/6$, when is the voltage equal to zero?

70. Refraction of Light Consider a ray of light passing from one medium (such as air) into another medium (such as a crystal). Let ϕ be the angle of incidence and θ the angle of refraction. As shown in **FIGURE 7.6.7**, these angles are measured from a vertical line. According to **Snell's law**, there is a constant c that depends on the two mediums, such that $\dfrac{\sin\phi}{\sin\theta} = c$. Assume that for light passing from air into a crystal, $c = 1.437$. Find ϕ and θ such that the angle of incidence is twice the angle of refraction.

FIGURE 7.6.7 Light rays in Problem 70

71. Snow Cover On the basis of data collected from 1966 to 1980, the extent of snow cover S in the northern hemisphere, measured in millions of square kilometers, can be modeled by the function

$$S(w) = 25 + 21\cos\frac{\pi}{26}(w - 5),$$

where w is the number of weeks past January 1.
  (a) How much snow cover does this formula predict for April Fool's Day? (Round w to the nearest integer.)
  (b) In which week does the formula predict the least amount of snow cover?
  (c) What month does this fall in?

CONCEPTS REVIEW

You should be able to give the meaning of each of the following concepts.

Circular functions:
  unit circle
  central angle
  reference angle
Periodic functions:
  period of sine
  period of cosine
  period of tangent
  period of cotangent
  period of secant
  period of cosecant

Graphs of trignometric functions:
  cycle
  amplitude
  phase shift
Identities:
  Pythagorean
  odd-even
Special formulas:
  addition
  subtraction
  double-angle
  half-angle

Inverse trigonometric functions:
  arcsine
  arccosine
  arctangent
Graphs of inverse trigonometric functions:
  arcsine
  arccosine
  arctangent
Trigonometric equations

CHAPTER 7 Review Exercises

Answers to selected odd-numbered problems begin on page ANS-21.

A. True/False

In Problems 1–20, answer true or false.

1. If $\tan t = \frac{3}{4}$, then $\sin t = 3$ and $\cos t = 4$. ____
2. In a right triangle, if $\sin\theta = \frac{11}{61}$, then $\cot\theta = \frac{60}{11}$. ____
3. $\sec(-\pi) = \csc\left(\dfrac{3\pi}{2}\right)$ ____
4. There is no angle t such that $\sec t = \frac{1}{2}$. ____
5. $\sin(2\pi - t) = -\sin t$ ____
6. $1 + \sec^2\theta = \tan^2\theta$ ____
7. $(-2, 0)$ is an x-intercept of the graph of $y = 3\sin(\pi x/2)$. ____
8. $(2\pi/3, -1/\sqrt{3})$ is a point on the graph of $y = \cot x$. ____
9. The range of the function $y = \csc x$ is $(-\infty, -1] \cup [1, \infty)$. ____
10. The graph of $y = \csc x$ does not intersect the y-axis. ____
11. The line $x = \pi/2$ is a vertical asymptote for the graph of $y = \tan x$. ____
12. If $\tan(x + 2\pi) = 0.3$, then $\tan x = 0.3$. ____
13. For the function $f(x) = -2\sin x$, the range is defined by $-2 \le y \le 2$. ____
14. $\sin 20x = 2\sin 10x \cos 10x$ ____

15. The graph of $y = \sin(2x - \pi/3)$ is the graph of $y = \sin 2x$ shifted $\pi/3$ units to the right. ___
16. The graphs $y = 3\sin(-2x)$ and $y = -3\cos(2x - \pi/2)$ are the same. ___
17. Since $\tan(5\pi/4) = 1$, then $\arctan(1) = 5\pi/4$. ___
18. $\tan 8\pi = \tan 9\pi$ ___
19. The function $f(x) = \arcsin x$ is not periodic. ___
20. $\arcsin(\frac{1}{2}) = 30°$ ___

B. Fill in the Blanks

In Problems 1–14, fill in the blanks.

1. If $\sin u = \frac{3}{5}$, $0 < u < \pi/2$, and $\cos v = 1/\sqrt{5}$, $3\pi/2 < v < 2\pi$, then $\cos(u + v) = $ _____.
2. The y-intercept for the graph of the function $y = 2\sec(x + \pi)$ is _____.
3. The period of the function $y = 2\sin\frac{\pi}{3}x$ is _____.
4. The first vertical asymptote for the graph of $y = \tan\left(x - \frac{\pi}{4}\right)$ to the right of the y-axis is _____.
5. The phase shift for the graph of $y = 5\cos(3x - 4\pi)$ is _____.
6. If $\sin t = \frac{1}{6}$, then $\cos\left(t - \frac{\pi}{2}\right) = $ _____.
7. The amplitude of $y = -10\cos\left(\frac{\pi}{3}x\right)$ is _____.
8. $\cos\left(\frac{\pi}{6} - \frac{5\pi}{4}\right) = $ _____.
9. The exact value of $\arccos\left(\cos\frac{9\pi}{5}\right) = $ _____.
10. The period of the function $y = \tan 4x$ is _____.
11. The fifth x-intercept on the positive x-axis for the graph of the function $y = \sin \pi x$ is _____.
12. If $P(t) = \left(-\frac{1}{3}, \frac{2\sqrt{2}}{3}\right)$ is a point on the unit circle, then $\sin 2t = $ _____.
13. If $\cos x = \frac{\sqrt{2}}{3}$, where $3\pi/2 < x < 2\pi$, then the exact values of $\sin\frac{x}{2} = $ _____, $\cos\frac{x}{2} = $ _____, $\sin 2x = $ _____, and $\cos 2x = $ _____.
14. From the results in Problem 13, we find $\tan\frac{x}{2} = $ _____ and $\tan 2x = $ _____.

C. Review Exercises

In Problems 1–4, graph the given functions. Give the amplitude, the period, and the phase shift where appropriate.

1. $y = 5(1 + \sin x)$
2. $y = -\frac{4}{3}\cos x$
3. $y = 10\cos\left(-3x + \frac{\pi}{2}\right)$
4. $y = -4\cos\left(\frac{1}{4}x - \pi\right)$

In Problems 5–10, find all solutions of the given equation in the interval $[0, 2\pi]$.

5. $\cos x \sin x - \cos x + \sin x - 1 = 0$
6. $\cos x - \sin x = 0$
7. $4\sin^2 x - 1 = 0$
8. $\sin x = 2\tan x$
9. $\cos 4x = -1$
10. $\tan x - 3\cot x = 2$

In Problems 11 and 12, find the solutions of the given equation in the interval $(-\pi/2, \pi/2)$. Round your solutions to two decimal places.

11. $3\cos 2x + \sin x = 0$
12. $\tan^4 x + \tan^2 x - 1 = 0$

In Problems 13–20, find the indicated value without using a calculator.

13. $\cos^{-1}(-\tfrac{1}{2})$
14. $\arcsin(-1)$
15. $\cot(\cos^{-1}\tfrac{13}{4})$
16. $\cos(\arcsin\tfrac{2}{5})$
17. $\sin^{-1}(\sin \pi)$
18. $\cos(\arccos 0.42)$
19. $\sin(\arccos(\tfrac{5}{13}))$
20. $\arctan(\cos \pi)$

In Problems 21 and 22, write the given expression as an algebraic expression in x.

21. $\sin(\arccos x)$
22. $\sec(\tan^{-1} x)$

In Problems 23–26, give two examples of the indicated trigonometric function such that each has the given properties.

23. Sine function with period 4 and amplitude 6
24. Cosine function with period π, amplitude 4, and phase shift $\tfrac{1}{2}$
25. Sine function with period $\pi/2$, amplitude 3, and phase shift $\pi/4$
26. Tangent function whose graph completes one cycle on the interval $(-\pi/8, \pi/8)$

In Problems 27–30, the given graph can be interpreted as a rigid/nonrigid transformation of the graph of $y = \sin x$ and of the graph of $y = \cos x$. Find an equation of the graph using the sine function. Then find an equation of the same graph using the cosine function.

27.

FIGURE 7.R.1 Graph for Problem 27

28.

FIGURE 7.R.2 Graph for Problem 28

29.

FIGURE 7.R.3 Graph for Problem 29

30.

FIGURE 7.R.4 Graph for Problem 30

8 Applications of Trigonometry

In This Chapter

- **8.1** Solving Right Triangles
- **8.2** Applications of Right Triangles
- **8.3** Law of Sines
- **8.4** Law of Cosines
- **8.5** Simple Harmonic Motion
- **8.6** Trigonometric Form of Complex Numbers
- **8.7** Powers and Roots of Complex Numbers
- Chapter 8 Review Exercises

The wine rack shown is constructed from right triangles.

A Bit of History Trigonometry was developed from early efforts to advance the study of astronomy by predicting the paths and positions of the heavenly bodies and to improve accuracy in navigation and the reckoning of time and calendars.

Much of the mathematics developed in the eighteenth century arose in an attempt to describe certain physical phenomena. For example, what is the shape of a sail under the pressure of the wind? What is the shape of a vibrating elastic string (for example, a violin string or guitar string) fixed at both ends? Answers to such questions frequently involve the trigonometric functions.

Ernst Florens Chladni (1756–1827), a German physicist, applied mathematician, and musician developed an interesting technique for studying the vibration of elastic plates. After sprinkling fine sand on drumlike surfaces, he drew a violin bow along the edges to start them vibrating. The French Academy of Sciences offered a prize for the best mathematical description of this phenomenon. In 1816 **Marie-Sophie Germain** (1776–1831), a largely self-taught French mathematician, won the Paris Academy's prize with her paper on elasticity.

In this chapter we put to use the trigonometric functions developed in the last two chapters to solve various kinds of mathematical and applied problems.

8.1 Solving Right Triangles

Introduction The expression "to solve a triangle" means that we wish to find the length of each side and the measure of each angle in the triangle. We can solve any right triangle if we know either two sides or one acute angle and one side. As the examples in this section will show, sketching and labeling the triangle is an essential part of the solution process. It will be our general practice to label a right triangle as shown in FIGURE 8.1.1. The three vertices will be denoted by A, B, and C with C at the vertex of the right angle. We denote the angles at A and B by the Greek letters α and β (*alpha* and *beta*) and the lengths of the sides opposite these angles by a and b, respectively. The length of the side opposite the right angle at the vertex C is denoted by c.

Applications of right triangle trigonometry in fields such as surveying and navigation involve solving right triangles. We will consider these applications in the next section.

FIGURE 8.1.1 Standard labeling for a right triangle

EXAMPLE 1 Solving a Right Triangle

Solve the right triangle having a hypotenuse of length $4\sqrt{3}$ and one 60° angle.

Solution First we make a sketch of the triangle and label it as shown in FIGURE 8.1.2. We wish to find a, b, and β. Since α and β are complementary angles, $\alpha + \beta = 90°$ yields

$$\beta = 90° - \alpha = 90° - 60° = 30°.$$

We are given the length of the hypotenuse, namely hyp $= 4\sqrt{3}$. To find a, the length of the side opposite the angle $\alpha = 60°$, we select the sine function. From $\sin\alpha = \text{opp/hyp}$, we obtain

$$\sin 60° = \frac{a}{4\sqrt{3}} \quad \text{or} \quad a = 4\sqrt{3}\sin 60°.$$

Since $\sin 60° = \sqrt{3}/2$, we have

$$a = 4\sqrt{3}\sin 60° = 4\sqrt{3}\left(\frac{\sqrt{3}}{2}\right) = 6.$$

FIGURE 8.1.2 Right triangle in Example 1

Finally, for the length b of the side adjacent to the 60° angle, we select the cosine function. From $\cos\alpha = \text{adj/hyp}$, we obtain

$$\cos 60° = \frac{b}{4\sqrt{3}} \quad \text{or} \quad b = 4\sqrt{3}\cos 60°.$$

Because $\cos 60° = \frac{1}{2}$, we find

$$b = 4\sqrt{3}\cos 60° = 4\sqrt{3}(\tfrac{1}{2}) = 2\sqrt{3}.$$

In Example 1 once we determined a, we could have found b by using either the Pythagorean theorem or the tangent function. In general, there are usually several ways to solve a triangle.

☐ **Use of a Calculator** If angles other than 30°, 45°, or 60° are involved in a problem, we obtain approximations of the desired trigonometric function values from a calculator. For the remainder of this chapter, whenever an approximation is used, we will round the final results to the nearest hundredth unless the problem specifies otherwise. To take full advantage of the calculator's accuracy, *store* the computed values of the

Some calculators have a key labeled ▶ **stor** or **sto**.

trigonometric functions in the calculator for subsequent calculations. If, instead, a rounded version of a displayed value is written down and then later keyed back into the calculator, the accuracy of the final result may be diminished.

In many applications we are able to determine the value of one of the trigonometric functions, say $\sin\theta$, and we wish to find θ. We can do so by using the inverse trigonometric functions and a calculator. Recall, the inverse trigonometric functions are denoted by $\sin^{-1}$ or arcsin, $\cos^{-1}$ or arccos, $\tan^{-1}$ or arctan, and so on. Scientific calculators have keys labeled $\boxed{\sin^{-1}}$, $\boxed{\cos^{-1}}$, and $\boxed{\tan^{-1}}$ or $\boxed{\text{arcsin}}$, $\boxed{\text{arccos}}$, and $\boxed{\text{arctan}}$. (Read the manual for your calculator if you need more explanation.)

◀ On HP scientific calculators the notation asin, acos, and atan is used.

EXAMPLE 2 Find θ

Use a calculator to find an acute angle θ measured in **(a)** degrees and **(b)** radians for which $\cos\theta = 0.5$.

Solution (a) We must first set the calculator in degree mode. We enter 0.5 and then press the appropriate key. The resulting display is 60. Thus, $\theta = 60°$.
(b) With the calculator set in radian mode, we enter 0.5 and then press the same key; the display reads 1.0471976. Thus for $\cos\theta = 0.5$, $\theta \approx 1.0471976$. Note that 1.0471976 is a decimal approximation of $\pi/3$.

EXAMPLE 3 Solving a Right Triangle

Solve the right triangle with legs of length 4 and 5.

Solution After sketching and labeling the triangle as shown in **FIGURE 8.1.3**, we see that we need to find c, α, and β. From the Pythagorean theorem, the hypotenuse c is given by

$$c = \sqrt{5^2 + 4^2} = \sqrt{41} \approx 6.40.$$

To find β, we will use $\tan\beta = \text{opp/adj}$. (By choosing to work with the given quantities, we avoid error due to previous approximations.) Thus we have

$$\tan\beta = \tfrac{4}{5} = 0.8.$$

The inverse tangent key of a calculator set in degree mode yields $\beta \approx 38.66°$. Since α and β are complementary angles, we use $\alpha = 90° - \beta$ to obtain $\alpha \approx 51.34°$.

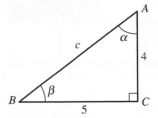

FIGURE 8.1.3 Right triangle in Example 3

8.1 Exercises Answers to selected odd-numbered problems begin on page ANS-21.

In Problems 1–12, find the indicated unknowns. Each problem refers to the triangle shown in **FIGURE 8.1.4**.

1. $a = 4$, $\beta = 27°$; b, c
2. $c = 10$, $\beta = 49°$; a, b
3. $b = 8$, $\beta = 34.33°$; a, c
4. $c = 25$, $\alpha = 50°$; a, b
5. $b = 1.5$, $c = 3$; α, β, a
6. $a = 5$, $b = 2$; α, β, c
7. $a = 4$, $b = 10$; α, β, c
8. $b = 4$, $\alpha = 58°$; a, c
9. $a = 9$, $c = 12$; α, β, b
10. $b = 3$, $c = 6$; α, β, a
11. $b = 20$, $\alpha = 23°$; a, c
12. $a = 11$, $\alpha = 33.5°$; b, c

FIGURE 8.1.4 Triangle for Problems 1–12

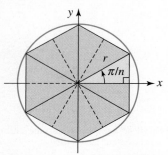

FIGURE 8.1.5 Inscribed *n*-gon for Problem 13

For Discussion

13. **(a)** A regular *n*-gon is an *n*-sided polygon inscribed in a circle; the polygon is formed by *n* equally spaced points on the circle. Suppose the polygon shown in FIGURE 8.1.5 represents a regular *n*-gon inscribed in a circle of radius *r*. Use right triangle trigonometry to show that the area $A(n)$ of the *n*-gon is given by

$$A(n) = \frac{n}{2}r^2\sin\left(\frac{2\pi}{n}\right).$$

(b) Compute A_{100}, A_{500}, and A_{1000}.
(c) Discuss: Is $A(n)$ approaching a number as $n \to \infty$?

8.2 Applications of Right Triangles

≡ **Introduction** Right triangle trigonometry can be used to solve many practical problems, particularly those involving lengths, heights, and distances.

EXAMPLE 1 Finding the Height of a Tree

A kite is caught in the top branches of a tree. If the 90-ft kite string makes an angle of 22° with the ground, estimate the height of the tree by finding the distance from the kite to the ground.

Solution Let *h* denote the height of the kite. From FIGURE 8.2.1 we see that

$$\frac{h}{90} = \sin 22° \quad \text{or} \quad h = 90\sin 22°.$$

A calculator set in degree mode gives $h \approx 33.71$ ft.

FIGURE 8.2.1 Tree in Example 1

EXAMPLE 2 Length of a Saw Cut

A carpenter cuts the end of a 4-in.-wide board on a 25° bevel from the vertical, starting at a point $1\frac{1}{2}$ in. from the end of the board. Find the lengths of the diagonal cut and the remaining side. See FIGURE 8.2.2.

Solution Let *x*, *y*, and *z* be the (unknown) dimensions, as labeled in Figure 8.2.2. It follows from the definition of the tangent function that

$$\frac{x}{4} = \tan 25° \quad \text{so therefore} \quad x = 4\tan 25° \approx 1.87 \text{ in.}$$

To find *y* we observe that

$$\frac{4}{y} = \cos 25° \quad \text{so} \quad y = \frac{4}{\cos 25°} \approx 4.41 \text{ in.}$$

Since $z = \frac{3}{2} + x$ and $x \approx 1.87$ in., we see that $z \approx 1.5 + 1.87 \approx 3.37$ in.

FIGURE 8.2.2 Saw cut in Example 2

☐ **Angles of Elevation and Depression** The angle between an observer's line of sight to an object and the horizontal is given a special name. As FIGURE 8.2.3 illustrates, if the line of sight is to an object above the horizontal, the angle is called an **angle of**

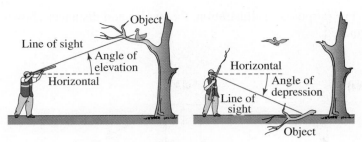

FIGURE 8.2.3 Angles of elevation and depression

elevation, whereas if the line of sight is to an object below the horizontal, the angle is called an **angle of depression**.

EXAMPLE 3   Using Angles of Elevation

A surveyor uses an instrument called a theodolite to measure the angle of elevation between ground level and the top of a mountain. At one point the angle of elevation is measured to be 41°. A half kilometer farther from the base of the mountain, the angle of elevation is measured to be 37°. How high is the mountain?

Solution Let h represent the height of the mountain. FIGURE 8.2.4 shows that there are two right triangles sharing the common side h, so we obtain two equations in two unknowns z and h:

$$\frac{h}{z + 0.5} = \tan 37° \quad \text{and} \quad \frac{h}{z} = \tan 41°.$$

We can solve each of these for h, obtaining, respectively,

$$h = (z + 0.5)\tan 37° \quad \text{and} \quad h = z\tan 41°.$$

Equating the last two results gives an equation from which we can determine the distance z:

$$(z + 0.5)\tan 37° = z\tan 41°.$$

Solving for z gives us

$$z = \frac{-0.5 \tan 37°}{\tan 37° - \tan 41°}.$$

Using $h = z\tan 41°$ we find the height h of the mountain to be

$$h = \frac{-0.5 \tan 37° \tan 41°}{\tan 37° - \tan 41°} \approx 2.83 \text{ km}.$$

FIGURE 8.2.4 Mountain in Example 3

EXAMPLE 4   Glide Path

Most airplanes approach San Francisco International Airport (SFO) on a straight 3° glide path starting at a point 5.5 miles from the field. A few years ago, the FAA experimented with a computerized two-segment approach where a plane approaches the field on a 6° glide path starting at a point 5.5 mi out and then switches to a 3° glide path 1.5 mi from the point of touchdown. The point of this experimental approach was to reduce the noise of the planes over the outlying residential areas. Compare the height of a plane P' using the standard 3° approach with the height of a plane P using the experimental approach when both planes are 5.5 mi from the airport.

Solution For purposes of illustration, the angles and distances shown in FIGURE 8.2.5 are exaggerated.

FIGURE 8.2.5 Glide paths in Example 4

First, suppose y is the height of plane P' on the standard approach when it is 5.5 mi out from the airport. As we see in Figure 8.2.5(a),

$$\frac{y}{5.5} = \tan 3° \quad \text{or} \quad y = 5.5 \tan 3°.$$

Because distances from the airport are measured in miles, we convert y to feet

$$y = 5.5(5280) \tan 3° \text{ ft} \approx 1522 \text{ ft}.$$

Now, suppose z is the height of plane P on the experimental approach when it is 5.5 mi out from the airport. As shown in Figure 8.2.5(b), $z = x + w$, so we use two right triangles to obtain

$$\frac{x}{1.5} = \tan 3° \quad \text{or} \quad x = 1.5 \tan 3°.$$

and

$$\frac{w}{4} = \tan 6° \quad \text{or} \quad w = 4 \tan 6°.$$

Hence the approximate height of plane P at a point 5.5 mi out from the airport is

$$z = x + w$$
$$= 1.5 \tan 3° + 4 \tan 6°$$
$$= 1.5(5280) \tan 3° + 4(5280) \tan 6° \approx 2635 \text{ ft}.$$

In other words, plane P is approximately 1113 ft higher than plane P'. ≡

☐ **Words Into Functions** Section 3.7 was devoted to setting up or constructing functions that were described or expressed in words. As emphasized in that section, this is a task that you will surely face in a course in calculus. Our final example illustrates a recommended procedure of sketching a figure and labeling quantities of interest with appropriate variables.

EXAMPLE 5 Functions That Involve Trigonometry

A plane flying horizontally at an altitude of 2 miles approaches a radar station as shown in FIGURE 8.2.6.

(a) Express the distance d between the plane and the radar station as a function of the angle of elevation θ.

FIGURE 8.2.6 Plane in Example 5

(b) Express the angle of elevation θ of the plane as a function of the horizontal separation x between the plane and the radar station.

Solution As shown in Figure 8.2.6, θ is an acute angle in a right triangle.

(a) We can relate the distance d and the angle θ by $\sin \theta = 2/d$. Solving for d gives

$$d(\theta) = \frac{2}{\sin \theta} \quad \text{or} \quad d(\theta) = 2\csc \theta,$$

where $0 < \theta \leq 90°$.

(b) The horizontal separation x and θ are related by $\tan \theta = 2/x$. We make use of the inverse tangent function to solve for θ:

$$\theta(x) = \tan^{-1} \frac{2}{x},$$

where $0 < x < \infty$.

≡

8.2 Exercises
Answers to selected odd-numbered problems begin on page ANS-21.

1. A building casts a shadow 20 m long. If the angle from the tip of the shadow to a point on top of the building is 69°, how high is the building?
2. Two trees are on opposite sides of a river, as shown in **FIGURE 8.2.7**. A baseline of 100 ft is measured from tree T_1, and from that position the angle β to T_2 is measured to be 29.7°. If the baseline is perpendicular to the line segment between T_1 and T_2, find the distance between the two trees.
3. A 50-ft tower is located on the edge of a river. The angle of elevation between the opposite bank and the top of the tower is 37°. How wide is the river?
4. A surveyor uses a geodometer to measure the straight-line distance from a point on the ground to a point on top of a mountain. Use the information given in **FIGURE 8.2.8** to find the height of the mountain.
5. An observer on the roof of building A measures a 27° angle of depression between the horizontal and the base of building B. The angle of elevation from the same point to the roof of the second building is 41.42°. What is the height of building B if the height of building A is 150 ft? Assume buildings A and B are on the same horizontal plane.
6. Find the height h of a mountain using the information given in **FIGURE 8.2.9**.
7. The top of a 20-ft ladder is leaning against the edge of the roof of a house. If the angle of inclination of the ladder from the horizontal is 51°, what is the approximate height of the house and how far is the bottom of the ladder from the base of the house?
8. An airplane flying horizontally at an altitude of 25,000 ft approaches a radar station located on a 2000-ft-high hill. At one instant in time, the angle between the radar dish pointed at the plane and the horizontal is 57°. What is the straight-line distance in miles between the airplane and the radar station at that particular instant?
9. A 5-mi straight segment of a road climbs a 4000-ft hill. Determine the angle that the road makes with the horizontal.
10. A box has dimensions as shown in **FIGURE 8.2.10**. Find the length of the diagonal between the corners P and Q. What is the angle θ formed between the diagonal and the bottom edge of the box?

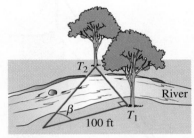

FIGURE 8.2.7 Trees and river in Problem 2

FIGURE 8.2.8 Mountain in Problem 4

FIGURE 8.2.9 Mountain in Problem 6

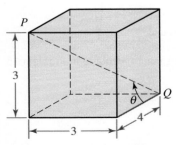

FIGURE 8.2.10 Box in Problem 10

FIGURE 8.2.11 Mountain in Problem 11

11. Observers in two towns A and B on either side of a 12,000-ft mountain measure the angles of elevation between the ground and the top of the mountain. See FIGURE 8.2.11. Assuming that the towns and the mountaintop lie in the same vertical plane, find the horizontal distance between them.

12. A drawbridge* measures 7.5 m from shore to shore, and when completely open it makes an angle of 43° with the horizontal. See FIGURE 8.2.12(a). When the bridge is closed, the angle of depression from the shore to a point on the surface of the water below the opposite end is 27°. See Figure 8.2.12(b). When the bridge is fully open, what is the distance d between the highest point of the bridge and the water below?

(a) Open bridge

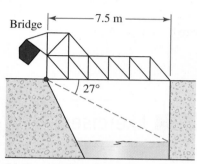
(b) Closed bridge

FIGURE 8.2.12 Drawbridge in Problem 12

FIGURE 8.2.13 Flagpole in Problem 13

13. A flagpole is located at the edge of a sheer 50-ft cliff at the bank of a river of width 40 ft. See FIGURE 8.2.13. An observer on the opposite side of the river measures an angle of 9° between her line of sight to the top of the flagpole and her line of sight to the top of the cliff. Find the height of the flagpole.

14. From an observation site 1000 ft from the base of Mt. Rushmore the angle of elevation to the top of the sculpted head of George Washington is measured to be 80.05°, whereas the angle of elevation to the bottom of his head is 79.946°. Determine the height of George Washington's head.

15. The length of a Boeing 747 airplane is 231 ft. What is the plane's altitude if it subtends an angle of 2° when it is directly above an observer on the ground? See FIGURE 8.2.14.

Bust of George Washington on Mt. Rushmore

16. The height of a gnomon (pin) of a sundial is 4 in. If it casts a 6-in. shadow, what is the angle of elevation of the Sun?

17. Weather radar is capable of measuring both the angle of elevation to the top of a thunderstorm and its range (the horizontal distance to the storm). If the range of a storm is 90 km and the angle of elevation is 4°, can a passenger plane that is able to climb to 10 km fly over the storm?

FIGURE 8.2.14 Airplane in Problem 15

Sundial

---

*The drawbridge shown in Figure 8.2.12, where the span is continuously balanced by a counterweight, is called a *bascule* bridge.

18. Cloud ceiling is the lowest altitude at which solid cloud is present. The cloud ceiling at airports must be sufficiently high for safe takeoffs and landings. At night the cloud ceiling can be determined by illuminating the base of the clouds with a searchlight pointed vertically upward. If an observer is 1 km from the searchlight and the angle of elevation to the base of the illuminated cloud is 8°, find the cloud ceiling. See **FIGURE 8.2.15**. (During the day cloud ceilings are generally estimated by sight. However, if an accurate reading is required, a balloon is inflated so that it will rise at a known constant rate. Then it is released and timed until it disappears into the cloud. The cloud ceiling is determined by multiplying the rate by the time of the ascent; trigonometry is not required for this calculation.)

FIGURE 8.2.15 Searchlight in Problem 18

19. Assuming that the Earth is a sphere, show that $C_\theta = C_e \cos\theta$, where C_θ is the circumference of the parallel of latitude at the latitude angle θ and C_e is the Earth's circumference at the equator. See **FIGURE 8.2.16**.
[*Hint*: $R\cos\theta = r$.]

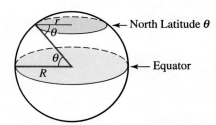
FIGURE 8.2.16 Earth in Problem 19

20. Use Problem 19 and the fact that the radius R of the Earth is 6400 km to find:
    (a) the circumference of the Arctic Circle, which lies at 66°33′ N (66.55° N) latitude, and
    (b) the distance "around the world" at the 58°40′ N (58.67° N) latitude.

21. The distance between the Earth and the Moon varies as the Moon revolves around the Earth. At a particular time the **geocentric parallax** angle shown in **FIGURE 8.2.17** is measured to be 1°. Calculate to the nearest hundred miles the distance between the center of the Earth and the center of the Moon at this instant. Assume that the radius of the Earth is 3963 miles.

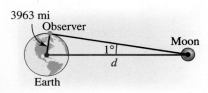
FIGURE 8.2.17 Angle in Problem 21

22. The final length of a volcanic lava flow seems to decrease as the elevation of the lava vent from which it originates increases. An empirical study of Mt. Etna gives the final lava flow length L in terms of elevation h by the formula

$$L = 23 - 0.0053h,$$

Mt. Etna

where L is measured in kilometers and h is measured in meters. Suppose that a Sicilian village at elevation 750 m is on a 10° slope directly below a lava vent at 2500 m. See **FIGURE 8.2.18**. According to the formula, how close will the lava flow get to the village?

FIGURE 8.2.18 Lava flow in Problem 22

23. As shown in **FIGURE 8.2.19**, two tracking stations S_1 and S_2 sight a weather balloon between them at elevation angles α and β, respectively. Express the height h of the balloon in terms of α and β, and the distance c between the tracking stations.

24. An entry in a soapbox derby rolls down a hill. Using the information given in **FIGURE 8.2.20**, find the total distance $d_1 + d_2$ that the soapbox travels.

FIGURE 8.2.19 Weather balloon in Problem 23

25. Find the height and area of the isosceles trapezoid shown in **FIGURE 8.2.21**.

FIGURE 8.2.20 Soapbox in Problem 24

FIGURE 8.2.21 Trapezoid in Problem 25

FIGURE 8.2.22 Rectangles in Problem 26

FIGURE 8.2.23 Searchlight in Problem 28

26. Consider the blue rectangle circumscribed around the red rectangle in FIGURE 8.2.22. With the aid of calculus it can be shown that the area of the blue rectangle is greatest when $\theta = \pi/4$. Find this area in terms of a and b.

In Problems 27–30, proceed as in Example 5 and translate the words into an appropriate function.

27. A tracking telescope, located 1.25 km from the point of a rocket launch, follows a vertically ascending rocket. Express the height h of the rocket as a function of the angle of elevation θ.
28. A searchlight one-half mile offshore illuminates a point P on the shore. Express the distance d from the searchlight to the point of illumination P as a function of the angle θ shown in FIGURE 8.2.23.
29. A statue is placed on a pedestal as shown in FIGURE 8.2.24. Express the viewing angle θ as a function of the distance x from the pedestal.
30. A woman on an island wishes to reach a point R on a straight shore on the mainland from a point P on the island. The point P is 9 mi from the shore and 15 mi from point R. See FIGURE 8.2.25. If the woman rows a boat at a rate of 3 mi/h to a point Q on the mainland, then walks the rest of the way at a rate of 5 mi/h, express the total time it takes the woman to reach point R as a function of the indicated angle θ. [*Hint*: Distance = rate × time.]

FIGURE 8.2.24 Viewing angle in Problem 29

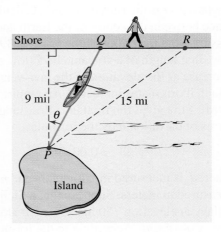

FIGURE 8.2.25 Woman rowing to shore in Problem 30

8.3 Law of Sines

≡ **Introduction** In Section 8.1 we saw how to solve right triangles. In this and the next section we consider two techniques for solving general triangles.

☐ **Law of Sines** Consider the triangle ABC, shown in FIGURE 8.3.1, with angles α, β, and γ, and corresponding opposite sides BC, AC, and AB. If we know the length of one side and two other parts of the triangle, we can then find the remaining three parts. One way of doing this is by the **Law of Sines**.

FIGURE 8.3.1 General triangle

CHAPTER 8 APPLICATIONS OF TRIGONOMETRY

THEOREM 8.3.1 The Law of Sines

Suppose angles α, β, and γ, and opposite sides of length a, b, and c are as shown in Figure 8.3.1. Then

$$\frac{\sin\alpha}{a} = \frac{\sin\beta}{b} = \frac{\sin\gamma}{c}. \qquad (1)$$

PROOF: Although the Law of Sines is valid for any triangle, we will prove it only for acute triangles—that is, a triangle in which all three angles α, β, and γ, are less than 90°. As shown in **FIGURE 8.3.2**, let h be the length of the altitude from vertex A to side BC. Since the altitude is perpendicular to the base BC it determines two right triangles. Consequently, we can write

$$\frac{h}{c} = \sin\beta \quad \text{and} \quad \frac{h}{b} = \sin\gamma. \qquad (2)$$

Thus (2) gives

$$h = c\sin\beta \quad \text{and} \quad h = b\sin\gamma. \qquad (3)$$

Equating the two expressions in (3) gives $c\sin\beta = b\sin\gamma$ so that

$$\frac{\sin\beta}{b} = \frac{\sin\gamma}{c}. \qquad (4)$$

If we use the altitude from the vertex C to the side AB, it follows in the same manner that

$$\frac{\sin\alpha}{a} = \frac{\sin\beta}{b}. \qquad (5)$$

Combining (4) and (5) yields the result in (1). ≡

FIGURE 8.3.2 Acute triangle

EXAMPLE 1   Determining the Parts of a Triangle

Find the remaining parts of the triangle shown in **FIGURE 8.3.3**.

Solution Let $\beta = 20°$, $\alpha = 130°$, and $b = 6$. Because the sum of the angles in a triangle is 180° we have $\gamma + 20° + 130 = 180°$ and so $\gamma = 180° - 20° - 130° = 30°$. From (1) we then see that

$$\frac{\sin 130°}{a} = \frac{\sin 20°}{6} = \frac{\sin 30°}{c}. \qquad (6)$$

We use the first equality in (6) to solve for a:

$$a = 6\frac{\sin 130°}{\sin 20°} \approx 13.44.$$

The second equality in (6) gives c:

$$c = 6\frac{\sin 30°}{\sin 20°} \approx 8.77.$$

≡

FIGURE 8.3.3 Triangle in Example 1

EXAMPLE 2   Height of a Building

A building is situated on the side of a hill that slopes downward at an angle of 15°. The Sun is uphill from the building at an angle of elevation of 42°. Find the building's height if it casts a shadow 36 ft long.

8.3 Law of Sines

FIGURE 8.3.4 Triangle QPS in Example 2

Solution Denote the height of the building on the downward slope by h and construct a right triangle QPS as shown in FIGURE 8.3.4. Now $\alpha + 15° = 42°$ so that $\alpha = 27°$. Since $\triangle QPS$ is a right triangle, $\gamma + 42° = 90°$ gives $\gamma = 90° - 42° = 48°$. From the Law of Sines (1),

$$\frac{\sin 27°}{h} = \frac{\sin 48°}{36} \qquad \text{so} \qquad h = 36 \frac{\sin 27°}{\sin 48°} \approx 21.99 \text{ ft.}$$

In Examples 1 and 2, where we were given *two angles and a side opposite one of these angles*, each triangle had a unique solution. However, this may not always be true for triangles where we know *two sides and an angle opposite one of these sides*. The next example illustrates the latter situation.

EXAMPLE 3 Two Triangles

Find the remaining parts of the triangle with $\beta = 50°$, $b = 5$, and $c = 6$.

Solution From the Law of Sines, we have

$$\frac{\sin 50°}{5} = \frac{\sin \gamma}{6} \qquad \text{or} \qquad \sin \gamma = \frac{6}{5}\sin 50° \approx 0.9193.$$

From a calculator set in degree mode, we obtain $\gamma \approx 66.82°$. At this point it is essential to recall that the sine function is also positive for second quadrant angles. In other words, there is another angle satisfying $0° \leq \gamma \leq 180°$ for which $\sin \gamma \approx 0.9193$. Using $66.82°$ as a reference angle we find the second quadrant angle to be $180° - 66.82° = 113.18°$. Therefore, the two possibilities for γ are $\gamma_1 \approx 66.82°$ and $\gamma_2 \approx 113.18°$. Thus, as shown in FIGURE 8.3.5, there are two possible triangles ABC_1 and ABC_2 satisfying the given three conditions.

To complete the solution of triangle ABC_1 (Figure 8.3.5(a)), we first find $\alpha_1 = 180° - \gamma_1 - \beta$ or $\alpha_1 \approx 63.18°$. To find the side opposite this angle we use

$$\frac{\sin 63.18°}{a_1} = \frac{\sin 50°}{5} \qquad \text{which gives} \qquad a_1 = 5\left(\frac{\sin 63.18°}{\sin 50°}\right)$$

or $a_1 \approx 5.82$.

To complete the solution of triangle ABC_2 (Figure 8.3.5(b)), we find $\alpha_2 = 180° - \gamma_2 - \beta$ or $\alpha_2 \approx 16.82°$. Then from

$$\frac{\sin 16.82°}{a_2} = \frac{\sin 50°}{5} \qquad \text{we find} \qquad a_2 = 5\left(\frac{\sin 16.82°}{\sin 50°}\right)$$

or $a_2 \approx 1.89$.

FIGURE 8.3.5 Triangles in Example 3

□ **Ambiguous Case** When solving triangles, the situation where two sides and an angle opposite one of these sides are given is called the **ambiguous case**. We have just seen in Example 3 that the given information may determine two different triangles. In the ambiguous case other complications can arise. For instance, suppose that the length of sides AB and AC (that is, c and b, respectively) and the angle β in triangle ABC are specified. As shown in FIGURE 8.3.6, we draw the angle β and mark off side AB with length c to locate the vertices A and B. The third vertex C is located on the base by drawing an arc of a circle of radius b (the length of AC) with center A. As shown in FIGURE 8.3.7, there are four possible outcomes of this construction:

- The arc does not intersect the base and no triangle is formed.
- The arc intersects the base in two distinct points C_1 and C_2 and two triangles are formed (as in Example 3).

FIGURE 8.3.6 Horizontal base, the angle β, and side AB

CHAPTER 8 APPLICATIONS OF TRIGONOMETRY

(a) No triangle (b) Two triangles (c) Single triangle (d) Right triangle

FIGURE 8.3.7 Solution possibilities for the ambiguous case in the Law of Sines

- The arc intersects the base in one point and one triangle is formed.
- The arc is tangent to the base and a single right triangle is formed.

EXAMPLE 4 Determining the Parts of a Triangle

Find the remaining parts of the triangle with $\beta = 40°$, $b = 5$, and $c = 9$.

Solution From the Law of Sines (1), we have

$$\frac{\sin 40°}{5} = \frac{\sin \gamma}{9} \quad \text{and so} \quad \sin \gamma = \frac{9}{5}\sin 40° \approx 1.1570.$$

Since the sine of any angle must be between -1 and 1, $\sin \gamma \approx 1.1570$ is impossible. This means the triangle has no solution; the side with length b is not long enough to reach the base. This is the case illustrated in Figure 8.3.7(a). ≡

8.3 Exercises
Answers to selected odd-numbered problems begin on page ANS-21.

In Problems 1–16, use the Law of Sines to solve the triangle. ◀ In Problems 1–16, refer to Figure 8.3.1.

1. $\alpha = 80°, \beta = 20°, b = 7$
2. $\alpha = 60°, \beta = 15°, c = 30$
3. $\beta = 37°, \gamma = 51°, a = 5$
4. $\alpha = 30°, \gamma = 75°, a = 6$
5. $\beta = 72°, b = 12, c = 6$
6. $\alpha = 120°, a = 9, c = 4$
7. $\gamma = 62°, b = 7, c = 4$
8. $\beta = 110°, \gamma = 25°, a = 14$
9. $\gamma = 15°, a = 8, c = 5$
10. $\alpha = 55°, a = 20, c = 18$
11. $\gamma = 150°, b = 7, c = 5$
12. $\alpha = 35°, a = 9, b = 12$
13. $\beta = 30°, a = 10, b = 7$
14. $\alpha = 140°, \gamma = 20°, c = 12$
15. $\alpha = 20°, a = 8, c = 27$
16. $\alpha = 75°, \gamma = 45°, b = 8$

Miscellaneous Applications

17. **Length of a Pool** A 10-ft rope that is available to measure the length between two points A and B at opposite ends of a kidney-shaped swimming pool is not long enough. A third point C is found such that the distance from A to C is 10 ft. It is determined that angle ACB is $115°$ and angle ABC is $35°$. Find the distance from A to B. See **FIGURE 8.3.8**.

18. **Width of a River** Two points A and B lie on opposite sides of a river. Another point C is located on the same side of the river as B at a distance of 230 ft from B. If angle ABC is $105°$ and angle ACB is $20°$, find the distance across the river from A to B.

FIGURE 8.3.8 Pool in Problem 17

FIGURE 8.3.9 Telephone pole in Problem 19

19. **Length of a Telephone Pole** A telephone pole makes an angle of 82° with the level ground. As shown in **FIGURE 8.3.9**, the angle of elevation of the Sun is 76°. Find the length of the telephone pole if its shadow is 3.5 m. (Assume that the tilt of the pole is away from the Sun and in the same plane as the pole and the Sun.)

20. **Not on the Level** A man 5 ft 9 in. tall stands on a sidewalk that slopes down at a constant angle. A vertical street lamp directly behind him causes his shadow to be 25 ft long. The angle of depression from the top of the man to the tip of his shadow is 31°. Find the angle α, as shown in **FIGURE 8.3.10**, that the sidewalk makes with the horizontal.

21. **How High?** If the man in Problem 20 is 20 ft down the sidewalk from the lamppost, find the height of the light above the sidewalk.

22. **Plane with an Altitude** Angles of elevation to an airplane are measured from the top and the base of a building that is 20 m tall. The angle from the top of the building is 38°, and the angle from the base of the building is 40°. Find the altitude of the airplane.

23. **Angle of Drive** The distance from the tee to the green on a particular golf hole is 370 yd. A golfer hits his drive and paces its distance off at 210 yd. From the point where the ball lies, he measures an angle of 160° between the tee and the green. Find the angle of his drive off the tee measured from the dashed line from the tee to the green shown in **FIGURE 8.3.11**.

FIGURE 8.3.10 Sloping sidewalk in Problem 20

FIGURE 8.3.11 Angle of drive in Problem 23

24. In Problem 23, what is the distance from the ball to the green?

8.4 Law of Cosines

≡ **Introduction** Triangles for which we know either *three sides* or *two sides and the included angle* (that is, the angle formed by the given sides) cannot be solved directly using the Law of Sines. The method we consider next can be used to solve triangles in these two cases.

☐ **Pythagorean Theorem** In a right triangle, such as the one shown in **FIGURE 8.4.1**, the length c of the hypotenuse is related to the lengths a and b of the other two sides by the Pythagorean theorem

$$c^2 = a^2 + b^2. \tag{1}$$

FIGURE 8.4.1 Right triangle

This last equation is a special case of a general formula that relates the lengths of the sides of *any* triangle.

☐ **Law of Cosines** A generalization of (1) is called the **Law of Cosines**. Like the Law of Sines, (1) of Section 8.3, the Law of Cosines is valid for any triangle. But for convenience, we will prove the last two equations in (2) of Theorem 8.4.1 using again the acute triangle given in Figure 8.3.2.

THEOREM 8.4.1   The Law of Cosines

Suppose angles α, β, and γ, and opposite sides a, b, and c are as shown in Figure 8.3.1. Then

$$a^2 = b^2 + c^2 - 2bc \cos\alpha$$
$$b^2 = a^2 + c^2 - 2ac \cos\beta \qquad (2)$$
$$c^2 = a^2 + b^2 - 2ab \cos\gamma.$$

PROOF: Let P denote the point where the altitude from the vertex A intersects side BC. Then, since both $\triangle BPA$ and $\triangle CPA$ in **FIGURE 8.4.2** are right triangles we have from (1),

$$c^2 = h^2 + (c\cos\beta)^2 \qquad (3)$$

and
$$b^2 = h^2 + (b\cos\gamma)^2. \qquad (4)$$

Now the length of BC is $a = c\cos\beta + b\cos\gamma$ so that

$$c\cos\beta = a - b\cos\gamma. \qquad (5)$$

Moreover, from (4),

$$h^2 = b^2 - (b\cos\gamma)^2. \qquad (6)$$

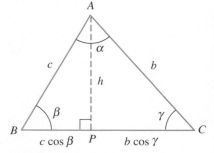

FIGURE 8.4.2 Acute triangle

Substituting (5) and (6) into (3) and simplifying yields the third equation in (2):

$$c^2 = b^2 - (b\cos\gamma)^2 + (a - b\cos\gamma)^2$$
$$= b^2 - b^2\cos^2\gamma + a^2 - 2ab\cos\gamma + b^2\cos^2\gamma$$

or
$$c^2 = a^2 + b^2 - 2ab\cos\gamma. \qquad (7)$$

Note that equation (7) reduces to the Pythagorean theorem (1) when $\gamma = 90°$.

Similarly, if we use $b\cos\gamma = a - c\cos\beta$ and $h^2 = c^2 - (c\cos\beta)^2$ to eliminate $b\cos\gamma$ and h^2 in (4), we obtain the second equation in (2). ≡

EXAMPLE 1   Determining the Parts of a Triangle

Find the remaining parts of the triangle shown in **FIGURE 8.4.3**.

Solution First, if we call the unknown side b and identify $a = 12$, $c = 10$, and $\beta = 26°$, then from the second equation in (2) we can write

$$b^2 = (12)^2 + (10)^2 - 2(12)(10)\cos 26°.$$

Therefore, $b^2 \approx 28.2894$ and so $b \approx 5.32$.

FIGURE 8.4.3 Triangle in Example 1

Next, we use the Law of Cosines to determine the remaining angles in the triangle in Figure 8.4.3. If γ is the angle at the vertex C, then the third equation in (2) gives

$$10^2 = 12^2 + (5.32)^2 - 2(12)(5.32)\cos\gamma \qquad \text{or} \qquad \cos\gamma \approx 0.5663.$$

With the aid of a calculator and the inverse cosine we find $\gamma \approx 55.51°$. Note that since the cosine of an angle between $90°$ and $180°$ is negative, there is no need to consider two possibilities as we did in Example 3 in Section 8.3. Finally, the angle at the vertex A is $\alpha = 180° - \beta - \gamma$ or $\alpha \approx 98.49°$. ≡

In Example 1, observe that after b is found, we know two sides and an angle opposite one of these sides. Hence we could have used the Law of Sines to find the angle γ.

In the next example we consider the case in which the lengths of the three sides of a triangle are given.

8.4 Law of Cosines

FIGURE 8.4.4 Triangle in Example 2

EXAMPLE 2 Determining the Angles in a Triangle

Find the angles α, β, and γ in the triangle shown in FIGURE 8.4.4.

Solution We use Law of Cosines to find the angle opposite the longest side:

$$9^2 = 6^2 + 7^2 - 2(6)(7)\cos\gamma \quad \text{or} \quad \cos\gamma = \tfrac{1}{21}.$$

A calculator then gives $\gamma \approx 87.27°$. Although we could use the Law of Cosines, we choose to find β by the Law of Sines:

$$\frac{\sin\beta}{6} = \frac{\sin 87.27°}{9} \quad \text{or} \quad \sin\beta = \frac{6}{9}\sin 87.27° \approx 0.6659.$$

Since γ is the angle opposite the longest side it is the largest angle in the triangle, so β must be an acute angle. Thus, $\sin\beta \approx 0.6659$ yields $\beta \approx 41.75°$. Finally, from $\alpha = 180° - \beta - \gamma$ we find $\alpha \approx 50.98°$. ≡

☐ **Bearing** In navigation directions are given using bearings. A **bearing** designates the acute angle that a line makes with the north–south line. For example, FIGURE 8.4.5(a) illustrates a bearing of S40°W, meaning 40 degrees west of south. The bearings in Figures 8.4.5(b) and 8.4.5(c) are N65°E and S80°E, respectively.

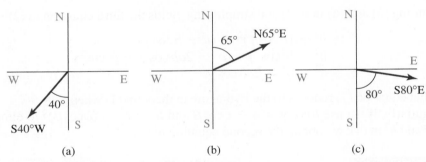

FIGURE 8.4.5 Three examples of bearings

FIGURE 8.4.6 Ships in Example 3

EXAMPLE 3 Bearings of Two Ships

Two ships leave a port at 7:00 AM, one traveling at 12 knots (nautical miles per hour) and the other at 10 knots. If the faster ship maintains a bearing of N47°W and the other ship maintains a bearing of S20°W, what is their separation (to the nearest nautical mile) at 11:00 AM that day?

Solution Since the elapsed time is 4 hours, the faster ship has traveled $4 \cdot 12 = 48$ nautical miles from port and the slower ship $4 \cdot 10 = 40$ nautical miles. Using these distances and the given bearings, we can sketch the triangle (valid at 11:00 AM) shown in FIGURE 8.4.6. In the triangle, c denotes the distance separating the ships and γ is the angle opposite that side. Since $47° + \gamma + 20° = 180°$ we find $\gamma = 113°$. Finally, the Law of Cosines

$$c^2 = 48^2 + 40^2 - 2(48)(40)\cos 113°,$$

gives $c^2 \approx 5404.41$ or $c \approx 73.51$. Thus the distance between the ships (to the nearest nautical mile) is 74 nautical miles. ≡

CHAPTER 8 APPLICATIONS OF TRIGONOMETRY

NOTES FROM THE CLASSROOM

(*i*) An important first step in solving a triangle is determining which of the three approaches we have discussed to use: right triangle trigonometry, the Law of Sines, or the Law of Cosines. The following table describes the various types of problems and gives the most appropriate approach for each. The term *oblique* refers to any triangle that is not a right triangle.

Type of Triangle	Information Given	Technique
Right	Two sides or an angle and a side	Basic definitions of sine, cosine, and tangent; the Pythagorean Theorem
Oblique	Three sides	Law of Cosines
Oblique	Two sides and the included angle	Law of Cosines
Oblique	Two angles and a side	Law of Sines
Oblique	Two sides and an angle opposite one of the sides	Law of Sines (if the given angle is acute; it is an ambiguous case)

(*ii*) Here are some additional bits of advice for solving triangles.
- Students will frequently use the Law of Sines when a right triangle trigonometric function could have been used. A right triangle approach is the simplest and most efficient.
- When three sides are given, check first to see whether the length of the longest side is greater than or equal to the sum of the lengths of the other two sides. If it is, there can be no solution (even though the given information indicates a Law of Cosines approach). This is because the shortest distance between two points is the length of the line segment joining them.
- In applying the Law of Sines, if you obtain a value greater than 1 for the sine of an angle, there is no solution.
- In the ambiguous case of the Law of Sines, when solving for the first unknown angle, you must consider *both the acute angle found from your calculator and its supplement as possible solutions.* The supplement will be a solution if the sum of the supplement and the angle given in the triangle is less than 180°.

8.4 Exercises
Answers to selected odd-numbered problems begin on page ANS-21.

In Problems 1–16, use the Law of Cosines to solve the triangle.

◀ In Problems 1–16, refer to Figure 8.3.1.

1. $\gamma = 65°, a = 5, b = 8$
2. $\beta = 48°, a = 7, c = 6$
3. $a = 8, b = 10, c = 7$
4. $\gamma = 31.5°, a = 4, b = 8$
5. $\gamma = 97.33°, a = 3, b = 6$
6. $a = 7, b = 9, c = 4$
7. $a = 11, b = 9.5, c = 8.2$
8. $\alpha = 162°, b = 11, c = 8$

9. $a = 5, b = 7, c = 10$
10. $a = 6, b = 5, c = 7$
11. $a = 3, b = 4, c = 5$
12. $a = 5, b = 12, c = 13$
13. $a = 6, b = 8, c = 12$
14. $\beta = 130°, a = 4, c = 7$
15. $\alpha = 22°, b = 3, c = 9$
16. $\beta = 100°, a = 22.3, b = 16.1$

Miscellaneous Applications

17. **How Far?** A ship sails due west from a harbor for 22 nautical miles. It then sails S62°W for another 15 nautical miles. How far is the ship from the harbor?
18. **How Far Apart?** Two hikers leave their camp simultaneously, taking bearings of N42°W and S20°E, respectively. If they each average a rate of 5 km/h, how far apart are they after 1 h?
19. **Bearings** On a hiker's map point A is 2.5 in. due west of point B and point C is 3.5 in. from B and 4.2 in. from A, respectively. See FIGURE 8.4.7. Find **(a)** the bearing of A from C, and **(b)** the bearing of B from C.

FIGURE 8.4.7 Triangle in Problem 19

20. **How Long Will It Take?** Two ships leave port simultaneously, one traveling at 15 knots and the other at 12 knots. They maintain bearings of S42°W and S10°E, respectively. After 3 h the first ship runs aground and the second ship immediately goes to its aid.
    (a) How long will it take the second ship to reach the first ship if it travels at 14 knots?
    (b) What bearing should it take?
21. **A Robotic Arm** A two-dimensional robot arm "knows" where it is by keeping track of a "shoulder" angle α and an "elbow" angle β. As shown in FIGURE 8.4.8, this arm has a fixed point of rotation at the origin. The shoulder angle is measured counterclockwise from the x-axis, and the elbow angle is measured counterclockwise from the upper to the lower arm. Suppose that the upper and lower arms are both of length 2 and that the elbow angle β is prevented from "hyperextending" beyond 180°. Find the angles α and β that will position the robot's hand at the point (1, 2).

FIGURE 8.4.8 Robotic arm in Problem 21

22. **Which Way?** Two lookout towers are situated on mountain tops A and B, 4 mi from each other. A helicopter firefighting team is located in a valley at point C, 3 mi from A and 2 mi from B. Using the line between A and B as a reference, a lookout spots a fire at an angle of 40° from tower A and 82° from tower B. See FIGURE 8.4.9. At what angle, measured from CB, should the helicopter fly in order to head directly for the fire?

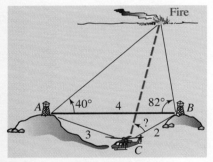

FIGURE 8.4.9 Fire in Problem 22

23. **Making a Kite** For the kite shown in FIGURE 8.4.10, use the Law of Cosines to find the lengths of the two dowels required for the diagonal supports.
24. **Distance Across a Canyon** From the floor of a canyon it takes 62 ft of rope to reach the top of one canyon wall and 86 ft to reach the top of the opposite wall. See FIGURE 8.4.11. If the two ropes make an angle of 123°, what is the distance d from the top of one canyon wall to the other?

FIGURE 8.4.10 Kite in Problem 23

FIGURE 8.4.11 Canyon in Problem 24

For Discussion

25. Use the Law of Cosines to derive **Heron's formula**,*
$$A = \sqrt{s(s-a)(s-b)(s-c)},$$
for the area of a triangle with sides a, b, c where $s = \frac{1}{2}(a+b+c)$.

26. **Garden Plot** Use Heron's formula in Problem 25 to find the area of a triangular garden plot if the lengths of the three sides are 25, 32, and 41 m, respectively.

27. **Corner Lot** Find the area of the irregular corner lot shown in FIGURE 8.4.12. [*Hint*: Divide the lot into two triangular lots as shown and then find the area of each triangle. Use Heron's formula in Problem 25 for the area of the acute triangle.]

28. Use Heron's formula in Problem 25 to find the area of a triangle with vertices located at $(3, 2)$, $(-3, -6)$, and $(0, 6)$ in a rectangular coordinate system.

29. **Blue Man** The effort in climbing a flight of stairs depends largely on the flexing angle of the leading knee. A simplified blue stick-figure model of a person walking up a staircase indicates that the maximum flexing of the knee occurs when the back leg is straight and the hips are directly over the heel of the front foot. See FIGURE 8.4.13. Show that
$$\cos\theta = \left(\frac{R}{a}\right)\sqrt{4 - \left(\frac{T}{a}\right)^2} + \frac{(T/a)^2 - (R/a)^2}{2} - 1,$$
where θ is the knee joint angle, $2a$ is the length of the leg, R is the rise of a single stair step, and T is the width of a step. [*Hint*: Let h be the vertical distance from hip to heel of the leading leg, as shown in the figure. Set up two equations involving h: one by applying the Pythagorean theorem to the right triangle outlined in color and the other by using the Law of Cosines on the angle θ. Then eliminate h and solve for $\cos\theta$.]

FIGURE 8.4.12 Corner lot in Problem 27

FIGURE 8.4.13 Blue man in Problem 29

8.5 Simple Harmonic Motion

≡ **Introduction** Many physical objects vibrate or oscillate in a regular manner, repeatedly moving back and forth over a definite time interval. Some examples are clock pendulums, a mass on a spring, sound waves, strings on a guitar when plucked, the human heart, tides, and alternating current. In this section we will focus on mathematical models of the undamped oscillatory motion of a mass on a spring.

Before proceeding with the main discussion we need to discuss the graph of the sum of constant multiples of $\cos Bx$ and $\sin Bx$, that is, $y = c_1 \cos Bx + c_2 \sin Bx$, where c_1 and c_2 are constants.

☐ **Addition of Two Sinusoidal Functions** In Section 7.2 we examined the graphs of horizontally shifted sine and cosine graphs. It turns out that any linear combination of a sine function and a cosine function of the form
$$y = c_1 \cos Bx + c_2 \sin Bx, \qquad (1)$$

---

*This formula is named after the Greek mathematician Heron but should actually be credited to Archimedes.

where c_1 and c_2 are constants, can be expressed either as a shifted sine function $y = A\sin(Bx + \phi)$, $B > 0$, or as a shifted cosine function $y = A\cos(Bx + \phi)$. Note that in (1) the sine and cosine functions have the same period $2\pi/B$.

EXAMPLE 1 Addition of a Sine and a Cosine

Graph the function $y = \cos 2x - \sqrt{3}\sin 2x$.

Solution Using a graphing utility we have shown in FIGURE 8.5.1 four cycles of the graphs of $y = \cos 2x$ (in red) and $y = -\sqrt{3}\sin 2x$ (in green). It is apparent in FIGURE 8.5.2 that the period of the sum of these two functions is π, the common period of $\cos 2x$ and $\sin 2x$. Also apparent is that the blue graph is a horizontally shifted sine (or cosine) function. Although Figure 8.5.2 suggests that the amplitude of the function $y = \cos 2x - \sqrt{3}\sin 2x$ is 2, the exact phase shift of the graph is certainly is *not* apparent.

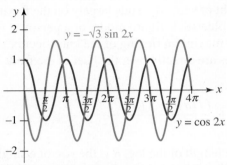

FIGURE 8.5.1 Superimposed graphs of $y = \cos 2x$ and $y = -\sqrt{3}\sin 2x$ in Example 1

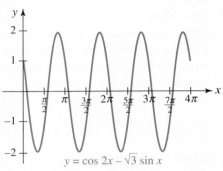

FIGURE 8.5.2 Graph of the sum $y = \cos 2x - \sqrt{3}\sin 2x$ in Example 1

▶ The sine form $y = A\sin(Bx + \phi)$ is slightly easier to use than the cosine form $y = A\cos(Bx + \phi)$.

☐ **Reduction to a Sine Function** We examine only the reduction of (1) to the form $y = A\sin(Bx + \phi)$, $B > 0$.

THEOREM 8.5.1 Reduction of (1) to (2)

For real numbers c_1, c_2, B, and x,
$$c_1 \cos Bx + c_2 \sin Bx = A\sin(Bx + \phi), \tag{2}$$
where A and ϕ are defined by
$$A = \sqrt{c_1^2 + c_2^2}, \tag{3}$$
and
$$\left.\begin{aligned}\sin\phi &= \frac{c_1}{A}\\ \cos\phi &= \frac{c_2}{A}\end{aligned}\right\} \quad \tan\phi = \frac{c_1}{c_2}. \tag{4}$$

PROOF: To prove (2), we use the sum formula (7) of Section 7.4:
$$\begin{aligned}A\sin(Bx + \phi) &= A\sin Bx \cos\phi + A\cos Bx \sin\phi\\ &= (A\sin\phi)\cos Bx + (A\cos\phi)\sin Bx\\ &= c_1 \cos Bx + c_2 \sin Bx\end{aligned}$$

and identify $A\sin\phi = c_1$, $A\cos\phi = c_2$. Thus, $\sin\phi = c_1/A = c_1/\sqrt{c_1^2 + c_2^2}$ and $\cos\phi = c_2/A = c_2/\sqrt{c_1^2 + c_2^2}$.

EXAMPLE 2 Example 1 Revisited

Express $y = \cos 2x - \sqrt{3}\sin 2x$ as a single sine function.

Solution With the identifications $c_1 = 1$, $c_2 = -\sqrt{3}$, and $B = 2$, we have from (3) and (4),

$$A = \sqrt{c_1^2 + c_2^2} = \sqrt{1^2 + (-\sqrt{3})^2} = \sqrt{4} = 2,$$

$$\left. \begin{array}{l} \sin\phi = \dfrac{1}{2} \\ \cos\phi = -\dfrac{\sqrt{3}}{2} \end{array} \right\} \quad \tan\phi = -\dfrac{1}{\sqrt{3}}.$$

Although $\tan\phi = -1/\sqrt{3}$ we cannot blindly assume that $\phi = \tan^{-1}(-1/\sqrt{3})$. The angle we take for ϕ must be consistent with the algebraic signs of $\sin\phi$ and $\cos\phi$. Because $\sin\phi > 0$ and $\cos\phi < 0$ the terminal side of the angle ϕ lies in the second quadrant. But since the range of the inverse tangent function is the interval $(-\pi/2, \pi/2)$, $\tan^{-1}(-1/\sqrt{3}) = -\pi/6$ is a fourth-quadrant angle. The correct angle is found by using the reference angle $\pi/6$ for $\tan^{-1}(-1/\sqrt{3})$ to find the second-quadrant angle

$$\phi = \pi - \frac{\pi}{6} = \frac{5\pi}{6} \text{ radians.}$$

Therefore $y = \cos 2x - \sqrt{3}\sin 2x$ can be rewritten as

$$y = 2\sin\left(2x + \frac{5\pi}{6}\right) \quad \text{or} \quad y = 2\sin 2\left(x + \frac{5\pi}{12}\right).$$

Hence the graph of $y = \cos 2x - \sqrt{3}\sin 2x$ is the graph of $y = 2\sin 2x$, which has amplitude 2, period $2\pi/2 = \pi$, and is shifted $5\pi/12$ units to the left. $\equiv$

☐ **Simple Harmonic Motion** Consider the motion of a mass on a spring as shown in **FIGURE 8.5.3**. In the absence of frictional or damping forces, a mathematical model for the displacement (or directed distance) of the mass measured from a position called the **equilibrium position** is given by the function

$$y(t) = y_0 \cos\omega t + \frac{v_0}{\omega}\sin\omega t. \qquad (5)$$

Oscillatory motion modeled by the function (5) is said to be **simple harmonic motion**. More precisely, we have the following definition.

FIGURE 8.5.3 An undamped spring/mass system exhibits simple harmonic motion

---

DEFINITION 8.5.1 Simple Harmonic Motion

A point moving on a coordinate line whose position at time t is given by

$$y(t) = A\sin(\omega t + \phi) \quad \text{or} \quad y(t) = A\cos(\omega t + \phi), \qquad (6)$$

where A, $\omega > 0$, and ϕ are constants, is said to exhibit **simple harmonic motion**.

Special cases of the trigonometric functions in (6) are $y(t) = A\sin\omega t$, $y(t) = A\cos\omega t$, and $y(t) = c_1\cos\omega t + c_2\sin\omega t$.

☐ **Terminology** The function (5) is said to be the **equation of motion** of the mass. Also, in (5), $\omega = \sqrt{k/m}$, where k is the **spring constant** (an indicator of the stiffness of the spring), m is the **mass** attached to the spring (measured in slugs or kilograms),

y_0 is the **initial displacement** of the mass (measured above or below the equilibrium position), v_0 is the **initial velocity** of the mass, t is **time** measured in seconds, and the **period** p of motion is $p = 2\pi/\omega$ seconds. The number $f = 1/p = 1/(2\pi/\omega) = \omega/2\pi$ is called the **frequency** of motion. The frequency indicates the number of cycles completed by the graph per unit time. For example, if the period of (5) is, say, $p = 2$ seconds, then we know that one cycle of the function is complete in 2 seconds. The frequency $f = 1/p = \frac{1}{2}$ means one-half of a cycle is complete in 1 second.

In the study of simple harmonic motion it is convenient to recast the equation of motion (5) as a single expression involving only the sine function:

$$y(t) = A\sin(\omega t + \phi). \tag{7}$$

The reduction of (5) to the sine function (7) can be done in exactly the same manner as illustrated in Example 2. In this situation we make the following identifications between (2) and (5):

$$c_1 = y_0, \quad c_2 = v_0/\omega, \quad A = \sqrt{c_1^2 + c_2^2}, \quad \text{and} \quad B = \omega.$$

EXAMPLE 3 Equation of Motion

(a) Find the equation of simple harmonic motion (5) for a spring mass system if $m = \frac{1}{16}$ slug, $y_0 = -\frac{2}{3}$ ft, $k = 4$ lb/ft, and $v_0 = \frac{4}{3}$ ft/s.
(b) Find the period and frequency of motion.

Solution (a) We begin with the simple harmonic motion equation (5). Since $k/m = 4/(\frac{1}{16}) = 64$, $\omega = \sqrt{k/m} = 8$, and $v_0/\omega = (\frac{4}{3})/8 = \frac{1}{6}$, therefore (5) becomes

$$y(t) = -\tfrac{2}{3}\cos 8t + \tfrac{1}{6}\sin 8t. \tag{8}$$

(b) The period of motion is $2\pi/8 = \pi/4$ second; the frequency is $4/\pi \approx 1.27$ cycles per second. ≡

EXAMPLE 4 Example 3 Continued

Express the equation of motion (8) as a single sine function (7).

Solution With $c_1 = -\frac{2}{3}$, $c_2 = \frac{1}{6}$, we find the amplitude of motion is

$$A = \sqrt{(-\tfrac{2}{3})^2 + (\tfrac{1}{6})^2} = \tfrac{1}{6}\sqrt{17} \text{ ft}.$$

Then from

$$\left. \begin{array}{l} \sin\phi = -\tfrac{2}{3}/\tfrac{\sqrt{17}}{6} < 0 \\ \cos\phi = \tfrac{1}{6}/\tfrac{\sqrt{17}}{6} > 0 \end{array} \right\} \tan\phi = -4$$

we can see from algebraic signs $\sin\phi < 0$ and $\cos\phi > 0$ that the terminal side of the angle ϕ lies in the fourth quadrant. Hence the correct value of ϕ is $\tan^{-1}(-4) = -1.3258$. The equation of motion is then $y(t) = \frac{1}{6}\sqrt{17}\sin(8t - 1.3258)$. As shown in **FIGURE 8.5.4**, the amplitude of motion is $A = \sqrt{17}/6 \approx 0.6872$. Since we are assuming that that is no resistance to the motion, once the spring/mass system is set in motion the model indicates it stays in motion bouncing back and forth between its maximum displacement $\sqrt{17}/6$ feet above the equilibrium position and a minimum of $-\sqrt{17}/6$ feet below the equilibrium position. ≡

FIGURE 8.5.4 Graph of the equation of motion in Example 4

Only in the two cases, $c_1 > 0$, $c_2 > 0$ or $c_1 < 0$, $c_2 > 0$, can we use $\tan\phi$ in (4) to write $\phi = \tan^{-1}(c_1/c_2)$. (Why?) Correspondingly, ϕ is a first or a fourth-quadrant angle.

8.5 Exercises
Answers to selected odd-numbered problems begin on page ANS-22.

In Problems 1–6, proceed as in Example 2 and reduce the given trigonometric expression to the form $y = A\sin(Bx + \phi)$. Sketch the graph and give the amplitude, the period, and the phase shift.

1. $y = \cos \pi x - \sin \pi x$
2. $y = \sin\dfrac{\pi}{2}x - \sqrt{3}\cos\dfrac{\pi}{2}x$
3. $y = \sqrt{3}\sin 2x + \cos 2x$
4. $y = \sqrt{3}\cos 4x - \sin 4x$
5. $y = \dfrac{\sqrt{2}}{2}(-\sin x - \cos x)$
6. $y = \sin x + \cos x$

In Problems 7–10, proceed as in Examples 3 and 4 and use the given information to express the equation of simple harmonic motion (5) for a spring/mass system in the trigonometric form (7). Give the amplitude, period, and frequency of motion.

7. $m = \tfrac{1}{4}$ slug, $y_0 = \tfrac{1}{2}$ ft, $k = 1$ lb/ft, and $v_0 = \tfrac{3}{2}$ ft/s
8. $m = 1.6$ slug, $y_0 = -\tfrac{1}{3}$ ft, $k = 40$ lb/ft, and $v_0 = -\tfrac{5}{4}$ ft/s
9. $m = 1$ slug, $y_0 = -1$ ft, $k = 16$ lb/ft, and $v_0 = -2$ ft/s
10. $m = 2$ slug, $y_0 = -\tfrac{2}{3}$ ft, $k = 200$ lb/ft, and $v_0 = 5$ ft/s

11. The equation of simple harmonic motion of a spring/mass system is $y(t) = \tfrac{5}{2}\sin(2t - \pi/3)$. Determine the initial displacement y_0 and initial velocity v_0 of the mass. [*Hint*: Use (5).]
12. Use the equation of simple harmonic motion of the spring/mass system given in Problem 11 to find the times for which the mass passes through the equilibrium position $y = 0$.

Miscellaneous Applications

13. **Electrical Circuits** Under certain conditions the current $I(t)$ in an electrical circuit at time t is given by $I(t) = I_0[\sin(\omega t + \theta)\cos\phi + \cos(\omega t + \theta)\sin\phi]$. Express $I(t)$ as a single sine function of the form given in (7). [*Hint*: Review the sum formula in (7) of Theorem 7.4.3.]

8.6 Trigonometric Form of Complex Numbers

◀ A review of Section 1.4 is recommended.

☰ Introduction A complex number $z = a + bi$ is uniquely determined by an *ordered pair* of real numbers (a, b). The first and second entries of the ordered pairs correspond, in turn, with the real and imaginary parts of the complex number. For example, the ordered pair $(2, -3)$ corresponds to the complex number $z = 2 - 3i$. Conversely, the complex number $z = 2 - 3i$ determines the ordered pair $(2, -3)$. The numbers 10, i, and $-5i$ are equivalent to $(10, 0)$, $(0, 1)$, and $(0, -5)$, respectively. In this manner we are able to associate a complex number $z = a + bi$ with a *point* (a, b) in a rectangular coordinate system.

☐ **Complex Plane** Because of the correspondence between a complex number $z = a + bi$ and one and only one point $P(a, b)$ in a coordinate plane we shall use the

FIGURE 8.6.1 Complex plane

FIGURE 8.6.2 The complex numbers in Example 1 interpreted as points

FIGURE 8.6.3 Modulus and argument of a complex number z

terms *complex number* and *point* interchangeably. The coordinate plane illustrated in FIGURE 8.6.1 is called the **complex plane** or simply the **z-plane**. The horizontal or x-axis is called the **real axis** because each point on that axis represents a real number. The vertical or y-axis is called the **imaginary axis** because a point on that axis represents a pure imaginary number.

EXAMPLE 1   Graphing Complex Numbers

Graph the complex numbers

$$z_1 = 5 + 4i, \quad z_2 = -2i, \quad z_3 = -2 - 3i, \quad \text{and} \quad z_4 = -4 + 2i.$$

Solution We identify the complex numbers z_1, z_2, z_3, z_4 with the points $(5, 4)$, $(0, -2)$, $(-2, -3)$, $(-4, 2)$, respectively. These points are, in turn, the red, blue, green, and orange dots in FIGURE 8.6.2.   ≡

□ **Trigonometric Form** If $z = a + bi$ is a nonzero complex number and $P(a, b)$ is its geometric representation, as shown in FIGURE 8.6.3, then the distance from P to the origin is given by $\sqrt{a^2 + b^2}$. This distance is called the **modulus**, **magnitude**, or **absolute value** of z and is denoted by $|z|$,

$$|z| = \sqrt{a^2 + b^2}. \tag{1}$$

For example, if $z = i$, then $|i| = \sqrt{0^2 + 1^2} = \sqrt{1^2} = 1$; if $z = 3 - 4i$, then $|3 - 4i| = \sqrt{3^2 + (-4)^2} = \sqrt{25} = 5$. From (4) of Section 1.4, we know that if $\bar{z} = a - bi$ is the conjugate of $z = a + bi$, then $z\bar{z} = a^2 + b^2$. Hence (1) can also be written as

$$|z| = \sqrt{z\bar{z}}.$$

If we let θ be the angle in standard position whose terminal side passes through $P(a, b)$ and $r = |z|$, then $\cos\theta = a/r$ and $\sin\theta = b/r$, from which we obtain $a = r\cos\theta$ and $b = r\sin\theta$. Substituting these expressions for a and b in $z = a + bi$, we obtain $z = a + bi = (r\cos\theta) + (r\sin\theta)i$ or

$$z = r(\cos\theta + i\sin\theta). \tag{2}$$

We say that (2) is the **trigonometric form**, or **polar form**, of the complex number z. The angle θ is called the **argument** of z and satisfies $\tan\theta = b/a$. However, θ is not necessarily $\arctan(b/a)$ since θ is not restricted to the interval $(-\pi/2, \pi/2)$. See Examples 2 and 3 that follow. Also, the argument θ is *not uniquely determined*, since $\cos\theta = \cos(\theta + 2k\pi)$ and $\sin\theta = \sin(\theta + 2k\pi)$ for any integer k. If $z = a + bi = 0$, then $a = b = 0$. In this case, $r = 0$ and we can take any angle θ as the argument.

EXAMPLE 2   Trigonometric Form

Write the complex numbers in trigonometric form: **(a)** $1 + i$ **(b)** $1 - \sqrt{3}i$.

Solution **(a)** If we identify $a = 1$ and $b = 1$, then the modulus of $1 + i$ is

$$r = |1 + i| = \sqrt{(1)^2 + (1)^2} = \sqrt{2}.$$

Because $\tan\theta = b/a = 1$ and the point $(1, 1)$ lies in the first quadrant, we can take the argument of the complex number to be $\theta = \pi/4$, as shown in FIGURE 8.6.4. Thus,

$$z = \sqrt{2}\left(\cos\frac{\pi}{4} + i\sin\frac{\pi}{4}\right).$$

FIGURE 8.6.4 Complex number in part (a) of Example 2

CHAPTER 8 APPLICATIONS OF TRIGONOMETRY

(b) In this case,
$$r = |1 - \sqrt{3}i| = \sqrt{1^2 + (-\sqrt{3})^2} = \sqrt{4} = 2.$$
From $\tan\theta = -\sqrt{3}/1 = -\sqrt{3}$ and the fact that $(1, -\sqrt{3})$ lies in the fourth quadrant, we take $\theta = \tan^{-1}(-\sqrt{3}) = -\pi/3$, as shown in FIGURE 8.6.5. Thus,
$$z = 2\left[\cos\left(-\frac{\pi}{3}\right) + i\sin\left(-\frac{\pi}{3}\right)\right].$$

FIGURE 8.6.5 Complex number in part (b) of Example 2

◀ Note

Following convention, in the remainder of the discussion as well as in Exercises 8.6 we will take the argument θ of a complex number z either as an angle in measured radians in the interval $[0, 2\pi)$ or an angle measured in degrees that satisfies $0 \le \theta < 360°$. For example, the answer in part (b) of Example 2 can be written in the alternative form
$$z = 2\left(\cos\frac{5\pi}{3} + i\sin\frac{5\pi}{3}\right).$$
The argument of $1 - \sqrt{3}i$ that lies in the interval $[0, 2\pi)$ is $\theta = 5\pi/3$, as is shown in Figure 8.6.5.

EXAMPLE 3 Trigonometric Form

Express the complex number
$$z = 2\sqrt{2}\left(\cos\frac{7\pi}{4} + i\sin\frac{7\pi}{4}\right)$$
in the standard form $z = a + bi$.

Solution By using the reference angle concept discussed in Section 7.1, we find $\cos(7\pi/4) = \sqrt{2}/2$ and $\sin(7\pi/4) = -\sqrt{2}/2$. Therefore,
$$z = 2\sqrt{2}\left(\cos\frac{7\pi}{4} + i\sin\frac{7\pi}{4}\right) = 2\sqrt{2}\left(\frac{\sqrt{2}}{2} - i\frac{\sqrt{2}}{2}\right)$$
or $z = 2 - 2i$.

EXAMPLE 4 Trigonometric Form

Find the trigonometric form of $z = -4 + 5i$.

Solution The modulus of $z = -4 + 5i$ is
$$r = |-4 + 5i| = \sqrt{16 + 25} = \sqrt{41}.$$

Because the point $(-4, 5)$ lies in the second quadrant, we must take care to adjust the value of the angle obtained from $\tan\theta = -\frac{5}{4}$ and a calculator so that our final answer is a quadrant II angle. See FIGURE 8.6.6. One approach is to use a calculator set in radian mode to obtain the reference angle $\theta' = \tan^{-1}\frac{5}{4} \approx 0.8961$ radian. The desired second-quadrant angle is then $\theta = \pi - \theta' \approx 2.2455$. Thus,
$$z \approx \sqrt{41}(\cos 2.2455 + i\sin 2.2455).$$

Alternatively the foregoing trigonometric form can be written using a degree-measured angle. With the calculator set in degree mode, we would obtain $\theta' \approx 51.34°$ and $\theta = 180° - \theta' \approx 128.66°$, from which it follows that
$$z \approx \sqrt{41}(\cos 128.66° + i\sin 128.66°).$$

FIGURE 8.6.6 Complex number in Example 4

8.6 Trigonometric Form of Complex Numbers

FIGURE 8.6.7 The product in Example 5

EXAMPLE 5   Modulus and Argument of a Product

Find the modulus and the argument of $z_1 z_2$, where $z_1 = 2i$ and $z_2 = 1 + i$.

Solution The product is

$$z_1 z_2 = 2i(1 + i) = -2 + 2i,$$

and hence the modulus is

$$r = |z_1 z_2| = |-2 + 2i| = \sqrt{8} = 2\sqrt{2}.$$

By identifying $a = -2$ and $b = 2$, we have $\tan\theta = -1$. Since θ is a second quadrant angle, we conclude that the argument of $z_1 z_2$ is $\theta = 3\pi/4$. See **FIGURE 8.6.7**.

☐ **Multiplication and Division** In Example 5 notice that the modulus $r = 2\sqrt{2}$ of the product $z_1 z_2$ is the *product* of the modulus $r_1 = 2$ of z_1 and the modulus $r_2 = \sqrt{2}$ of z_2. Also, the argument $\theta = 3\pi/4$ of $z_1 z_2$ is the *sum* of the arguments $\theta_1 = \pi/2$ and $\theta_2 = \pi/4$ of z_1 and z_2, respectively. We have illustrated a particular case of the following theorem, which describes how to multiply and divide complex numbers when they are written in trigonometric form.

THEOREM 8.6.1   Product and Quotient

If $z_1 = r_1(\cos\theta_1 + i\sin\theta_1)$ and $z_2 = r_2(\cos\theta_2 + i\sin\theta_2)$, then

$$z_1 z_2 = r_1 r_2 [\cos(\theta_1 + \theta_2) + i\sin(\theta_1 + \theta_2)] \tag{3}$$

$$\frac{z_1}{z_2} = \frac{r_1}{r_2}[\cos(\theta_1 - \theta_2) + i\sin(\theta_1 - \theta_2)], \quad r_2 \neq 0. \tag{4}$$

PROOF: We will prove only (4) of Theorem 8.6.1; the proof of (3) is very similar. If we multiply the numerator and the denominator of

$$\frac{z_1}{z_2} = \frac{r_1(\cos\theta_1 + i\sin\theta_1)}{r_2(\cos\theta_2 + i\sin\theta_2)}$$

by $\cos\theta_2 - i\sin\theta_2$, we obtain

$$\frac{z_1}{z_2} = \frac{r_1}{r_2} \frac{(\cos\theta_1 + i\sin\theta_1)(\cos\theta_2 - i\sin\theta_2)}{\cos^2\theta_2 + \sin^2\theta_2} \quad \leftarrow \text{denominator equals 1}$$

$$= \frac{r_1}{r_2}(\cos\theta_1 + i\sin\theta_1)(\cos\theta_2 - i\sin\theta_2).$$

Performing the multiplication and then using the difference formulas from Section 7.4, we have

$$\frac{z_1}{z_2} = \frac{r_1}{r_2}[\overbrace{(\cos\theta_1\cos\theta_2 + \sin\theta_1\sin\theta_2)}^{\text{see (5) of Theorem 7.4.2}} + i\overbrace{(\sin\theta_1\cos\theta_2 - \cos\theta_1\sin\theta_2)}^{\text{see (8) of Theorem 7.4.3}}]$$

$$= \frac{r_1}{r_2}[\cos(\theta_1 - \theta_2) + i\sin(\theta_1 - \theta_2)].$$

EXAMPLE 6 Product and Quotient

If $z_1 = 4(\cos 75° + i \sin 75°)$ and $z_2 = \frac{1}{2}(\cos 45° + i \sin 45°)$, find **(a)** $z_1 z_2$ **(b)** z_1/z_2. Express each answer in the standard form $a + bi$.

Solution (a) From (3) of Theorem 8.6.1 we can write the product as

$$z_1 z_2 = 4 \cdot \overbrace{\frac{1}{2}}^{\text{multiply moduli}} [\cos(\overbrace{75° + 45°}^{\text{add arguments}}) + i\sin(75° + 45°)]$$

$$= 2[\cos 120° + i\sin 120°] = 2\left[-\frac{1}{2} + \frac{\sqrt{3}}{2}i\right]$$

and so $z_1 z_2 = -1 + \sqrt{3}i$.

(b) Now from (4) of Theorem 8.6.1 the quotient is

$$\frac{z_1}{z_2} = \overbrace{\frac{4}{\frac{1}{2}}}^{\text{divide moduli}} [\cos(\overbrace{75° - 45°}^{\text{subtract arguments}}) + i\sin(75° - 45°)]$$

$$= 8[\cos 30° + i\sin 30°] = 8\left[\frac{\sqrt{3}}{2} + \frac{1}{2}i\right]$$

or $z_1/z_2 = 4\sqrt{3} + 4i$. ≡

8.6 Exercises
Answers to selected odd-numbered problems begin on page ANS-22.

In Problems 1–10, graph the given complex number(s) and evaluate and graph the indicated complex number.

1. $z_1 = 2 + 5i$; $\bar{z}_1$
2. $z_1 = -8 - 4i$; $\frac{1}{4}\bar{z}_1$
3. $z_1 = 1 + i, z_2 = 2 - 2i$; $z_1 + z_2$
4. $z_1 = 4i, z_2 = -4 + i$; $z_1 - z_2$
5. $z_1 = 6 - 3i, z_2 = -i$; $\bar{z}_1 + z_2$
6. $z_1 = 5 + 2i, z_2 = -1 + 2i$; $z_1 + \bar{z}_2$
7. $z_1 = -2i, z_2 = 1 - i$; $z_1 z_2$
8. $z_1 = 1 + i, z_2 = 2 - i$; $z_1 z_2$
9. $z_1 = 2\sqrt{3} + 2i, z_2 = 1 - \sqrt{3}i$; $\frac{z_1}{z_2}$
10. $z_1 = i, z_2 = 1 - i$; $\frac{z_1}{z_2}$

In Problems 11–22, find the modulus and an argument of the given complex number.

11. $z = \frac{1}{2} - \frac{\sqrt{3}}{2}i$
12. $z = 4 + 3i$
13. $z = \sqrt{2} - 4i$
14. $z = -5 + 2i$
15. $z = \frac{3}{4} - \frac{1}{4}i$
16. $z = -8 - 2i$
17. $z = 3 + 3i$
18. $z = -1 - i$
19. $z = \sqrt{3} + i$
20. $z = 2 - 2\sqrt{3}i$
21. $z = 2 - i$
22. $z = 4 + 8i$

In Problems 23–32, write the given complex number in trigonometric form.

23. $z = -4i$
24. $z = 15i$
25. $z = 5\sqrt{3} + 5i$
26. $z = 3 + i$
27. $z = -2 + 5i$
28. $z = 2 + 2\sqrt{3}i$

29. $z = 3 - 5i$
30. $z = -10 + 6i$
31. $z = -2 - 2i$
32. $z = 1 - i$

In Problems 33–42, write the given complex number in the standard form $z = a + bi$. Do not use a calculator.

33. $z = \sqrt{2}\left(\cos\dfrac{\pi}{4} + i\sin\dfrac{\pi}{4}\right)$
34. $z = 6\left(\cos\dfrac{7\pi}{4} + i\sin\dfrac{7\pi}{4}\right)$
35. $z = 10(\cos 210° + i\sin 210°)$
36. $z = \sqrt{5}(\cos 420° + i\sin 420°)$
37. $z = 2(\cos 30° + i\sin 30°)$
38. $z = 7\left(\cos\dfrac{7\pi}{12} + i\sin\dfrac{7\pi}{12}\right)$
39. $z = \cos\dfrac{5\pi}{3} + i\sin\dfrac{5\pi}{3}$
40. $z = \dfrac{3}{2}\left(\cos\dfrac{5\pi}{4} + i\sin\dfrac{5\pi}{4}\right)$
41. $z = 4[\cos(\tan^{-1}2) + i\sin(\tan^{-1}2)]$
42. $z = 20\left[\cos\left(\tan^{-1}\dfrac{3}{5}\right) + i\sin\left(\tan^{-1}\dfrac{3}{5}\right)\right]$

In Problems 43–48, find $z_1 z_2$ and z_1/z_2 in trigonometric form by first writing z_1 and z_2 in trigonometric form.

43. $z_1 = 3i,\ z_2 = 6 + 6i$
44. $z_1 = 1 + i,\ z_2 = -1 + i$
45. $z_1 = 1 + \sqrt{3}i,\ z_2 = 2\sqrt{3} + 2i$
46. $z_1 = 5i,\ z_2 = -10i$
47. $z_1 = \sqrt{3} + i,\ z_2 = 5 - 5i$
48. $z_1 = -\sqrt{2} + \sqrt{2}i,\ z_2 = \dfrac{5\sqrt{2}}{2} + \dfrac{5\sqrt{2}}{2}i$

In Problems 49–52, find $z_1 z_2$ and z_1/z_2. Write the answer in the standard form $z = a + bi$.

49. $z_1 = \sqrt{6}\left(\cos\dfrac{\pi}{3} + i\sin\dfrac{\pi}{3}\right),\ z_2 = \sqrt{2}\left(\cos\dfrac{7\pi}{4} + i\sin\dfrac{7\pi}{4}\right)$
50. $z_1 = 10\left(\cos\dfrac{7\pi}{6} + i\sin\dfrac{7\pi}{6}\right),\ z_2 = \dfrac{1}{2}\left(\cos\dfrac{\pi}{6} + i\sin\dfrac{\pi}{6}\right)$
51. $z_1 = 3\left(\cos\dfrac{\pi}{4} + i\sin\dfrac{\pi}{4}\right),\ z_2 = 4\left(\cos\dfrac{15\pi}{8} + i\sin\dfrac{15\pi}{8}\right)$
52. $z_1 = \cos 57° + i\sin 57°,\ z_2 = 7(\cos 73° + i\sin 73°)$

8.7 Powers and Roots of Complex Numbers

Introduction The trigonometric form of a product $z_1 z_2$ given in (3) of Theorem 8.6.1 of the last section also gives a means of computing *powers* of a complex number, that is, z^n, where n is a positive integer. In this section we also show how to find the n distinct nth *roots* of a complex number z.

We begin the discussion with an example.

☐ **Powers of a Complex Number** Suppose $z = 1 + i$ and we wish to compute z^3. Of course, there are several ways of proceeding. We can carry out the multiplications zz and $(zz)z$ using the standard forms of the numbers or we can treat the number $z = 1 + i$ as a binomial and use the binomial expansion

See (*iii*) of Definition 1.4.2 on page 95. ▶

$$(a + b)^3 = a^3 + 3a^2 b + 3ab^2 + b^3$$

CHAPTER 8 APPLICATIONS OF TRIGONOMETRY

with $a = 1$ and $b = i$. Alternatively, we can use the trigonometric form

$$z = 1 + i = \sqrt{2}\left(\cos\frac{\pi}{4} + i\sin\frac{\pi}{4}\right).$$

With $z = z_1 = z_2$ in (3) of Theorem 8.6.1, we obtain the square of z:

$$z^2 = z \cdot z = (\sqrt{2})(\sqrt{2})\left[\cos\left(\frac{\pi}{4} + \frac{\pi}{4}\right) + i\sin\left(\frac{\pi}{4} + \frac{\pi}{4}\right)\right]$$

$$= (\sqrt{2})^2\left[\cos 2\left(\frac{\pi}{4}\right) + i\sin 2\left(\frac{\pi}{4}\right)\right].$$

Then (3) of Theorem 8.6.1 also gives

$$z^3 = z^2 \cdot z = (\sqrt{2})^2(\sqrt{2})\left[\cos\left(\frac{2\pi}{4} + \frac{\pi}{4}\right) + i\sin\left(\frac{2\pi}{4} + \frac{\pi}{4}\right)\right]$$

$$= (\sqrt{2})^3\left[\cos 3\left(\frac{\pi}{4}\right) + i\sin 3\left(\frac{\pi}{4}\right)\right] \qquad (1)$$

$$= 2\sqrt{2}\left(\cos\frac{3\pi}{4} + i\sin\frac{3\pi}{4}\right).$$

After simplifying the last expression we get $z^3 = -2 + 2i$.

The result in (1),

$$z^3 = (\sqrt{2})^3\left[\cos 3\left(\frac{\pi}{4}\right) + i\sin 3\left(\frac{\pi}{4}\right)\right], \qquad (2)$$

illustrates a particular case of the following theorem named after the French mathematician **Abraham DeMoivre** (1667–1754). The formal proof of this theorem cannot be given at this point because it requires mathematical induction, which is discussed in Section 13.4.

THEOREM 8.7.1 DeMoivre's Theorem

If $z = r(\cos\theta + i\sin\theta)$ and n is a positive integer, then

$$z^n = r^n(\cos n\theta + i\sin n\theta). \qquad (3)$$

Inspection of (2) shows that the result is DeMoivre's theorem with $z = 1 + i$, $r = \sqrt{2}$ and $\theta = \pi/4$ in blue, and $n = 3$ in red.

EXAMPLE 1   Power of a Complex Number

Evaluate $(\sqrt{3} + i)^8$.

Solution First, the modulus of $\sqrt{3} + i$ is $r = \sqrt{(\sqrt{3})^2 + 1^2} = 2$. Then from $\tan\theta = 1/\sqrt{3}$, an argument of the number is $\theta = \pi/6$ since $(\sqrt{3}, 1)$ lies in quadrant I. Hence from DeMoivre's theorem,

$$(\sqrt{3} + i)^8 = 2^8\left[\cos 8\left(\frac{\pi}{6}\right) + i\sin 8\left(\frac{\pi}{6}\right)\right] = 256\left(\cos\frac{4\pi}{3} + i\sin\frac{4\pi}{3}\right)$$

$$= 256\left(-\frac{1}{2} - \frac{\sqrt{3}}{2}i\right) = -128 - 128\sqrt{3}i. \qquad \equiv$$

☐ **Roots of a Complex Number** Recall from Section R.4 that -2 and 2 are said to be square roots of the number 4 because $(-2)^2 = 4$ and $2^2 = 4$. In other words, the two square roots of 4 are distinct solutions of the equation $w^2 = 4$. In like manner we say $w = 3$ is a cube root of 27 since $w^3 = 3^3 = 27$. In general, we say that a number $w = a + bi$ is a complex **nth root** of a nonzero complex number z if $w^n = (a + bi)^n = z$, where n is a positive integer. For example, you are urged to verify that $w_1 = \frac{1}{2}\sqrt{2} + \frac{1}{2}\sqrt{2}i$ and $w_2 = -\frac{1}{2}\sqrt{2} - \frac{1}{2}\sqrt{2}i$ are the two square roots of the complex number $z = i$ because $w_1^2 = i$ and $w_2^2 = i$. See also Problem 77 in Exercises 1.4.

We will now demonstrate that there are exactly n solutions of the equation $w^n = z$.

Let the modulus and the argument of w be ρ and ϕ, respectively, so that $w = \rho(\cos\phi + i\sin\phi)$. If w is an nth root of the complex number $z = r(\cos\theta + i\sin\theta)$, then $w^n = z$. DeMoivre's theorem enables us to write the last equation as

$$\rho^n(\cos n\phi + i\sin n\phi) = r(\cos\theta + i\sin\theta).$$

When two complex numbers are equal, their moduli are necessarily equal. Thus we have

$$\rho^n = r \quad \text{or} \quad \rho = r^{1/n}$$

and

$$\cos n\phi + i\sin n\phi = \cos\theta + i\sin\theta.$$

Equating the real and imaginary parts in this equation gives

$$\cos n\phi = \cos\theta, \quad \sin n\phi = \sin\theta,$$

from which it follows that $n\phi = \theta + 2k\pi$, or

$$\phi = \frac{\theta + 2k\pi}{n},$$

where k is any integer. As k takes on the successive integer values $0, 1, 2, \ldots, n-1$, we obtain n distinct roots of z. For $k \geq n$, the values of $\sin\phi$ and $\cos\phi$ repeat the values obtained by letting $k = 0, 1, 2, \ldots, n-1$. To see this, suppose that $k = n + m$, where $m = 0, 1, 2, \ldots$. Then

$$\phi = \frac{\theta + 2(n+m)\pi}{n} = \frac{\theta + 2m\pi}{n} + 2\pi.$$

Since the sine and cosine each have period 2π, we have

$$\sin\phi = \sin\left(\frac{\theta + 2m\pi}{n}\right) \quad \text{and} \quad \cos\phi = \cos\left(\frac{\theta + 2m\pi}{n}\right),$$

and so no new roots are obtained for $k \geq n$. Summarizing these results gives the following theorem.

THEOREM 8.7.2 Complex Roots

If $z = r(\cos\theta + i\sin\theta)$ and n is a positive integer, then n distinct complex nth roots of z are given by

$$w_k = r^{1/n}\left[\cos\left(\frac{\theta + 2k\pi}{n}\right) + i\sin\left(\frac{\theta + 2k\pi}{n}\right)\right], \tag{4}$$

where $k = 0, 1, 2, \ldots, n-1$.

We will denote the n roots by $w_0, w_1, \ldots, w_{n-1}$ corresponding to $k = 0, 1, \ldots, n-1$, respectively, in (4).

EXAMPLE 2 Three Cube Roots

Find the three cube roots of i.

Solution In the trigonometric form for i, $r = 1$ and $\theta = \pi/2$, so that

$$i = \cos\frac{\pi}{2} + i\sin\frac{\pi}{2}.$$

With $n = 3$ we find from (4) of Theorem 8.7.2 that

$$w_k = 1^{1/3}\left[\cos\left(\frac{\pi/2 + 2k\pi}{3}\right) + i\sin\left(\frac{\pi/2 + 2k\pi}{3}\right)\right], \quad k = 0, 1, 2.$$

Now for

$$k = 0, \quad w_0 = \cos\frac{\pi}{6} + i\sin\frac{\pi}{6}$$

$$k = 1, \quad w_1 = \cos\left(\frac{\pi}{6} + \frac{2\pi}{3}\right) + i\sin\left(\frac{\pi}{6} + \frac{2\pi}{3}\right)$$

$$= \cos\frac{5\pi}{6} + i\sin\frac{5\pi}{6}$$

$$k = 2, \quad w_2 = \cos\left(\frac{\pi}{6} + \frac{4\pi}{3}\right) + i\sin\left(\frac{\pi}{6} + \frac{2\pi}{3}\right)$$

$$= \cos\frac{3\pi}{2} + i\sin\frac{3\pi}{2}.$$

Therefore, in standard form the three cube roots of i are $w_0 = \frac{1}{2}\sqrt{3} + \frac{1}{2}i$, $w_1 = -\frac{1}{2}\sqrt{3} + \frac{1}{2}i$, and $w_2 = -i$. $\equiv$

The three cube roots of i found in Example 2 are plotted in **FIGURE 8.7.1**. We note that they are equally spaced around a circle of radius 1 centered at the origin. In general, the n distinct nth roots of a nonzero complex number z are equally spaced on the circumference of the circle of radius $|z|^{1/n}$ with center at the origin.

As the next example shows, the roots of a complex number do not have to be "nice" numbers as in Example 2.

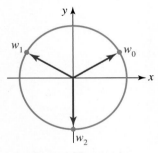

FIGURE 8.7.1 Three cube roots of i in Example 2

EXAMPLE 3 Solving an Equation

Solve the equation $z^4 = 1 + i$.

Solution Solving this equation is equivalent to finding the four complex fourth roots of the number $1 + i$. In this case, the modulus and an argument of $1 + i$ are $r = \sqrt{2}$ and $\theta = \pi/4$, respectively. From (4) with $n = 4$ and the symbol z_k playing the part of w_k we obtain

$$z_k = (\sqrt{2})^{1/4}\left[\cos\left(\frac{\pi/4 + 2k\pi}{4}\right) + i\sin\left(\frac{\pi/4 + 2k\pi}{4}\right)\right]$$

$$= \sqrt[8]{2}\left[\cos\left(\frac{\pi/4 + 2k\pi}{4}\right) + i\sin\left(\frac{\pi/4 + 2k\pi}{4}\right)\right], \quad k = 0, 1, 2, 3,$$

$$k = 0, \quad z_0 = \sqrt[8]{2}\left(\cos\frac{\pi}{16} + i\sin\frac{\pi}{16}\right)$$

$$k = 1, \quad z_1 = \sqrt[8]{2}\left(\cos\frac{9\pi}{16} + i\sin\frac{9\pi}{16}\right)$$

$$k = 2, \quad z_2 = \sqrt[8]{2}\left(\cos\frac{17\pi}{16} + i\sin\frac{17\pi}{16}\right)$$
$$k = 3, \quad z_3 = \sqrt[8]{2}\left(\cos\frac{25\pi}{16} + i\sin\frac{25\pi}{16}\right).$$

With the aid of a calculator we find the approximate standard forms,

$$z_0 \approx 1.0696 + 0.2127i$$
$$z_1 \approx -0.2127 + 1.0696i$$
$$z_2 \approx -1.0696 - 0.2127i$$
$$z_3 \approx 0.2127 - 1.0696i.$$

As shown in **FIGURE 8.7.2** the four roots lie on a circle centered at the origin of radius $r = \sqrt[8]{2} \approx 1.09$ and are spaced at equal angular intervals of $2\pi/4 = \pi/2$ radians beginning with the root whose argument is $\pi/16$.

FIGURE 8.7.2 Four fourth roots of $1 + i$ in Example 3

8.7 Exercises
Answers to selected odd-numbered problems begin on page ANS-22.

In Problems 1–10, use DeMoivre's theorem to calculate the given power. Write your answer in the standard form $z = a + bi$. If necessary, use a calculator.

1. $\left(\cos\dfrac{5\pi}{8} + i\sin\dfrac{5\pi}{8}\right)^{24}$
2. $\left(\cos\dfrac{\pi}{10} + i\sin\dfrac{\pi}{10}\right)^{5}$
3. $\left[\sqrt{2}\left(\cos\dfrac{\pi}{16} + i\sin\dfrac{\pi}{16}\right)\right]^{4}$
4. $\left[\sqrt{3}\left(\cos\dfrac{7\pi}{16} + i\sin\dfrac{7\pi}{16}\right)\right]^{4}$
5. $[\sqrt{3}(\cos 21° + i\sin 21°)]^{10}$
6. $\left[\sqrt{3}\left(\cos\dfrac{\pi}{24} + i\sin\dfrac{\pi}{24}\right)\right]^{8}$
7. $[\sqrt{5}(\cos 13.5° + i\sin 13.5°)]^{6}$
8. $[2(\cos 67° + i\sin 67°)]^{3}$
9. $[3.2(\cos 12° + i\sin 12°)]^{3}$
10. $[\tfrac{1}{2}(\cos 24° + i\sin 24°)]^{5}$

In Problems 11 and 12, use (3) of this section and (4) of Section 8.6 to simplify the given complex number. Write your answer in the standard form $z = a + bi$.

11. $\dfrac{\left[2\left(\cos\dfrac{\pi}{16} + i\sin\dfrac{\pi}{16}\right)\right]^{10}}{\left[4\left(\cos\dfrac{3\pi}{8} + i\sin\dfrac{3\pi}{8}\right)\right]^{3}}$

12. $\dfrac{\left(\cos\dfrac{\pi}{9} + i\sin\dfrac{\pi}{9}\right)^{12}}{\left[\dfrac{1}{2}\left(\cos\dfrac{\pi}{6} + i\sin\dfrac{\pi}{6}\right)\right]^{5}}$

In Problems 13–24, use the trigonometric form of a complex number along with DeMoivre's theorem to calculate the given power. Write your answer in the standard form $z = a + bi$.

13. i^{30}
14. $-i^{15}$
15. $(1 + i)^{6}$
16. $(1 - i)^{9}$
17. $(-2 + 2i)^{4}$
18. $(-4 - 4i)^{3}$
19. $(\sqrt{3} + i)^{5}$
20. $(-\sqrt{3} + i)^{10}$
21. $\left(\dfrac{\sqrt{2}}{2} - \dfrac{\sqrt{6}}{2}i\right)^{9}$
22. $\left(\dfrac{\sqrt{3}}{6} + \dfrac{1}{2}i\right)^{8}$
23. $(1 + 2i)^{4}$
24. $(\tfrac{1}{2} + \tfrac{1}{2}i)^{20}$

In Problems 25–34, find the indicated roots. Write your answer in the standard form $z = a + bi$.

25. The three cube roots of -8
26. The three cube roots of 1
27. The four fourth roots of i
28. The two square roots of i
29. The four fourth roots of $-1 - \sqrt{3}i$
30. The two square roots of $-1 + \sqrt{3}i$
31. The two square roots of $1 + i$
32. The three cube roots of $-2\sqrt{3} + 2i$
33. The six sixth roots of $64(\cos 54° + i \sin 54°)$
34. The two square roots of $81\left(\cos\dfrac{5\pi}{3} + i\sin\dfrac{5\pi}{3}\right)$

In Problems 35 and 36, find the indicated roots. Proceed as in Example 2 and plot these roots on an appropriate circle.

35. The six sixth roots of 1
36. The eight eighth roots of 1
37. For what positive integers n will $(\sqrt{2}/2 + \sqrt{2}i/2)^n$ be equal to 1? Equal to i? Equal to $-\sqrt{2}/2 - \sqrt{2}i/2$? Equal to $\sqrt{2}/2 + \sqrt{2}i/2$?
38. (a) Verify that $(4 + 3i)^2 = 7 + 24i$.
    (b) Use part (a) to find the two values of $(7 + 24i)^{1/2}$.

In Problems 39–42, solve the given equation. Write your answer in the standard form $z = a + bi$.

39. $z^4 + 1 = 0$
40. $x^3 - 125i = 0$
41. $x^2 + 8 + 8\sqrt{3}i = 0$
42. $z^2 - 8z + 18 = 8i$

For Discussion

43. DeMoivre's theorem implies
$$(\cos\theta + i\sin\theta)^2 = \cos 2\theta + i\sin 2\theta.$$
Use this information to derive trigonometric identities for $\cos 2\theta$ and $\sin 2\theta$ by multiplying out the left-hand side of the equation and then equating real and imaginary parts.
44. Use a procedure analogous to that outlined in Problem 43 to find trigonometric identities of $\cos 3\theta$ and $\sin 3\theta$.

CONCEPTS REVIEW
You should be able to give the meaning of each of the following concepts.

Solving a right triangle	Equation of motion:	Trigonometric form of a complex number:
Law of Sines:	amplitude	modulus
ambiguous case	period	argument
Law of Cosines	frequency	product
Bearing	Complex plane	quotient
Simple harmonic motion	Imaginary axis	DeMoivre's theorem
	Real axis	Roots of a complex number

CHAPTER 8 Review Exercises
Answers to selected odd-numbered problems begin on page ANS-23.

A. True/False

In Problems 1–6, answer true or false.

1. If $w^5 = z$, then w is a fifth root of z. _____
2. The argument of $-\frac{\sqrt{2}}{2} - \frac{\sqrt{2}}{2}i$ is $5\pi/4$. _____
3. The complex number $-\frac{\sqrt{2}}{2} - \frac{\sqrt{2}}{2}i$ is a square root of i. _____
4. If the modulus of a complex number is $\sqrt{3}$, then the modulus of z^4 is 9. _____
5. $\sin t + \cos t = \sqrt{2}\sin\left(t + \frac{\pi}{4}\right)$ _____
6. The Pythagorean theorem is a special case of the Law of Cosines. _____

B. Fill in the Blanks

In Problems 1–10, fill in the blanks.

1. To solve a triangle in which you know two angles and a side opposite one of these angles, you would first use the Law of _____.
2. The ambiguous case refers to solving a triangle when _____ are given.
3. To solve a triangle in which you know two sides and the included angle, you would first use the Law of _____.
4. In standard form, $z = 0.6(\cos 200° + i\sin 200°) = $ _____.
5. In trigonometric form, $z = -5 = $ _____.
6. If $z_1 = \frac{1}{2}(\cos 20° + i\sin 20°)$ and $z_2 = 2(\cos 15° + i\sin 15°)$, then the argument of $z_1 z_2$ is _____.
7. For the complex numbers in Problem 6, the trigonometric form of z_1/z_2 is _____.
8. For the complex numbers in Problem 6, the trigonometric form of z_1^4 is _____.
9. The modulus of $\frac{1}{2} + \frac{\sqrt{3}}{2}i$ is _____.
10. $\left(\frac{1}{2} + \frac{\sqrt{3}}{2}i\right)^{483} = $ _____.

C. Exercises

In Problems 1–4, solve the triangle satisfying the given conditions.

1. $\alpha = 30°, \beta = 70°, b = 10$
2. $\gamma = 145°, a = 25, c = 20$
3. $\alpha = 51°, b = 20, c = 10$
4. $a = 4, b = 6, c = 3$

FIGURE 8.R.1 Cliff in Problem 5

5. A surveyor 100 m from the base of an overhanging cliff measures a 28° angle of elevation from that point to the top of the cliff. See FIGURE 8.R.1. If the cliff makes an angle of 65° with the horizontal ground, determine its height h.

6. A rocket is launched from ground level at an angle of elevation of 43°. If the rocket hits a drone target plane flying at 20,000 ft, find the horizontal distance between the rocket launch site and the point directly beneath the plane. What is the straight-line distance between the rocket launch site and the target plane?

7. A competition water skier leaves a ramp at point R and lands at point S. See FIGURE 8.R.2. A judge at point J measures an $\angle RJS$ as 47°. If the distance from the ramp to the judge is 110 ft, find the length of the jump. Assume that $\angle SRJ$ is 90°.

FIGURE 8.R.2 Water skier in Problem 7

8. The angle between two sides of a parallelogram is 40°. If the lengths of the sides are 5 and 10 cm, find the lengths of the two diagonals.
9. A weather satellite orbiting the equator of the Earth at a height of $H = 36{,}000$ km spots a thunderstorm to the north at P at an angle of $\theta = 6.5°$ from its vertical. See FIGURE 8.R.3.
   (a) Given that the Earth's radius is approximately $R = 6370$ km, find the latitude ϕ of the thunderstorm.
   (b) Show that angles θ and ϕ are related by
   $$\tan\theta = \frac{R\sin\phi}{H + R(1-\cos\phi)}.$$

FIGURE 8.R.3 Satellite in Problem 9

10. It can be shown that a basketball of diameter d approaching the basket from an angle θ to the horizontal will pass through a hoop of diameter D if $D\sin\theta > d$, where $0° \le \theta \le 90°$. See FIGURE 8.R.4. If the basketball has diameter 24.6 cm and the hoop has diameter 45 cm, what range of approach angles θ will result in a basket?

FIGURE 8.R.4 Basketball in Problem 10

11. Each of the 24 NAVSTAR Global Positioning System (GPS) satellites orbits the Earth at an altitude of $h = 20{,}200$ km. Using this network of satellites, an inexpensive handheld GPS receiver can determine its position on the surface of the Earth to within 10 m. Find the greatest distance s (in km) on the surface of the Earth that can be observed from a single GPS satellite. See FIGURE 8.R.5. Take the radius of the Earth to be 6370 km. [*Hint*: Find the central angle θ subtended by s.]

12. An airplane flying horizontally at a speed of 400 miles per hour is climbing at an angle of 6° from the horizontal. When the airplane passes directly over a car traveling 60 miles per hour, it is 2 miles above the car. Assuming that the airplane and the car remain in the same vertical plane, find the angle of elevation from the car to the airplane after 30 minutes.

FIGURE 8.R.5 GPS satellite in Problem 11

13. A house measures 45 ft from front to back. The roof measures 32 ft from the front of the house to the peak and 18 ft from the peak to the back of the house. See FIGURE 8.R.6. Find the angles of elevation of the front and back parts of the roof.

14. The angle between two sides of a parallelogram is 40°. If the lengths of the sides are 5 and 10 cm, find the lengths of the two diagonals.

In Problems 15–22, translate the words into an appropriate function.

15. A 20-ft-long water trough has ends in the form of isosceles triangles with sides that are 4 ft long. See Figure 3.7.19 in Exercises 3.7. As shown in FIGURE 8.R.7, let θ denote the angle between the vertical and one of the sides of a triangular end. Express the volume of the trough as a function of 2θ.

16. A person driving a car approaches a freeway sign as shown in FIGURE 8.R.8. Let θ be her viewing angle of the sign and let x represent her horizontal distance (measured in feet) to that sign. Express θ as a function of x.

FIGURE 8.R.6 House in Problem 13

FIGURE 8.R.7 End of water trough in Problem 15

FIGURE 8.R.8 Freeway sign in Problem 16

FIGURE 8.R.9 Plank in Problem 17

FIGURE 8.R.10 Pasture in Problem 18

17. As shown in FIGURE 8.R.9, a plank is supported by a sawhorse so that one end rests on the ground and the other end rests against a building. Express the length of the plank as a function of the indicated angle θ.

18. A farmer wishes to enclose a pasture in the form of a right triangle using 2000 ft of fencing on hand. See FIGURE 8.R.10. Show that the area of the pasture as a function of the indicated angle θ is

$$A(\theta) = \frac{1}{2}\cot\theta \cdot \left(\frac{2000}{1 + \cot\theta + \csc\theta}\right)^2.$$

19. Express the volume of the box shown in FIGURE 8.R.11 as a function of the indicated angle θ.

20. A corner of an 8.5-in. $\times$ 11-in. piece of paper is folded over to the other edge of the paper as shown in FIGURE 8.R.12. Express the length L of the crease as a function of the angle θ shown in the figure.

FIGURE 8.R.11 Box in Problem 19

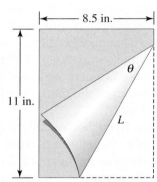

FIGURE 8.R.12 Folded paper in Problem 20

FIGURE 8.R.13 Gutter in Problem 21

FIGURE 8.R.14 Pipe in Problem 22

21. A gutter is to be made from a sheet of metal 30 cm wide by turning up the edges of width 10 cm along each side so that the sides make equal angles ϕ with the vertical. See FIGURE 8.R.13. Express the cross-sectional area of the gutter as a function of the angle ϕ.

22. A metal pipe is to be carried horizontally around a right-angled corner from a hallway 8 feet wide into a hallway that is 6 feet wide. See FIGURE 8.R.14. Express the length L of the pipe as a function of the angle θ shown in the figure.

In Problems 23–26, find the modulus and the argument of the given complex number and write the number in trigonometric form.

23. $3 - 3i$
24. $\sqrt{3} - i$
25. $-2 + 3i$
26. $1 + 4i$

In Problems 27 and 28, write the given complex number in the standard form $z = a + bi$.

27. $z = 4\left(\cos\dfrac{4\pi}{3} + i\sin\dfrac{4\pi}{3}\right)$
28. $z = 2(\cos 330° + i\sin 330°)$

In Problems 29 and 30, find $z_1 z_2$ and z_1/z_2 in trigonometric form by first writing z_1 and z_2 in trigonometric form.

29. $z_1 = -1 - i,\ z_2 = 1 + \sqrt{3}i$
30. $z_1 = \sqrt{3} - i,\ z_2 = 2 - 2i$

In Problems 31 and 32, use DeMoivre's theorem to calculate the given power.

31. $(-1 - i)^7$

32. $\left[4\left(\cos\dfrac{5\pi}{6} + i\sin\dfrac{5\pi}{6} \right) \right]^{10}$

In Problems 33 and 34, find the indicated roots.

33. The five fifth roots of 32
34. The three cube roots of $-i$

35. Factor the polynomial $x^3 - 27i$.
36. **(a)** Verify that $(2 + 2i)^2 = 8i$.
  (b) Use the result in part (a) to find all solutions of the equation
    $z^2 - 8z + 16 = 8i$.

9 Topics in Analytic Geometry

In This Chapter

9.1 The Parabola
9.2 The Ellipse
9.3 The Hyperbola
9.4 Rotation of Axes
9.5 Parametric Equations
Chapter 9 Review Exercises

Planets, asteroids, and some comets revolve around the Sun in elliptical orbits

A Bit of History **Hypatia** is the first woman in the history of mathematics about whom we have considerable knowledge. Born in Alexandria (circa 370 CE), she was renowned as a mathematician, philosopher, and prophetess. Her life and untimely death at the hands of a fanatical mob are romanticized in an 1853 novel by Charles Kingsley (*Hypatia, or New Foes with Old Faces*, Chicago: W. B. Conkey, 1853). Among her writings is *On the Conics of Apollonius*, which popularized Apollonius' work on curves that can be obtained by intersecting a cone with a plane: the circle, *parabola*, *ellipse*, and *hyperbola*. With the close of the Greek period, interest in the conic sections waned and, after Hypatia, study of these curves was neglected for over 1000 years. In the seventeenth century, **Galileo Galilei** (1564–1642) showed that in the absence of air resistance the path of a projectile follows a parabolic arc. About the same time, the astronomer, astrologer, and mathematician **Johannes Kepler** (1571–1630) hypothesized that the orbits of planets about the Sun are ellipses with the Sun at one focus. This was later verified by Newton, using the methods of the newly discovered calculus. Kepler also experimented with the reflecting properties of parabolic mirrors; these investigations sped the development of reflecting telescopes. The Greeks had known little of these practical applications. They had studied the conics for their beauty and fascinating properties.

In this chapter, we examine both the ancient properties and the modern applications of these curves.

9.1 The Parabola

Hypatia

≡ Introduction As mentioned in the introduction to this chapter, the Greek mathematician **Hypatia** popularized **Apollonius'** (200 BCE) work on curves that can be obtained by intersecting a double-napped cone with a plane: the circle, ellipse, parabola, and hyperbola. See FIGURE 9.1.1.

FIGURE 9.1.1 Conic sections

Solar system

But the Greeks knew little of the practical applications of these **conic sections**. They had studied the conics for their beauty and fascinating properties. In the first three sections of this chapter, we will examine both the ancient properties and the modern applications of these curves. Rather than using a cone, we will see how the parabola, ellipse, and hyperbola are defined by means of distance. Using a rectangular coordinate system and the distance formula, we obtain equations for the conics. Each of these equations will be in the form of a quadratic equation in variables x and y:

$$Ax^2 + Bxy + Cy^2 + Dx + Ey + F = 0,$$

where A, B, C, D, E, and F are constants. We have already studied the special case $y = ax^2 + bx + c$ of the foregoing equation in Section 3.3.

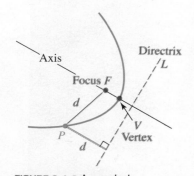
FIGURE 9.1.2 A parabola

DEFINITION 9.1.1 Parabola

A parabola is the set of points $P(x, y)$ in the plane that are equidistant from a fixed line L, called the **directrix**, and a fixed point F, called the **focus**.

A parabola is shown in FIGURE 9.1.2. The line through the focus perpendicular to the directrix is called the **axis** of the parabola. The point of intersection of the parabola and the axis is called the **vertex**, denoted by the symbol V in Figure 9.1.2.

FIGURE 9.1.3 Parabola with vertex (0, 0) and focus on the y-axis

□ **Parabola with Vertex (0, 0)** To describe a parabola analytically, we use a rectangular coordinate system where the directrix is a horizontal line $y = -c$, where $c > 0$, and the focus is the point $F(0, c)$. Then we see that the axis of the parabola is along the y-axis, as FIGURE 9.1.3 shows. The origin is necessarily the vertex, since it lies on the axis c units from both the focus and the directrix. The distance from a point $P(x, y)$ to the directrix is

$$y - (-c) = y + c.$$

Using the distance formula, the distance from P to the focus F is

$$d(P, F) = \sqrt{(x - 0)^2 + (y - c)^2}.$$

CHAPTER 9 TOPICS IN ANALYTIC GEOMETRY

From the definition of the parabola it follows that $d(P, F) = y + c$, or

$$\sqrt{(x-0)^2 + (y-c)^2} = y + c.$$

By squaring both sides and simplifying, we obtain

$$x^2 + (y-c)^2 = (y+c)^2$$
$$x^2 + y^2 - 2cy + c^2 = y^2 + 2cy + c^2$$

or
$$x^2 = 4cy. \tag{1}$$

Equation (1) is referred to as the **standard form** of the equation of a parabola with focus $(0, c)$, directrix $y = -c$, $c > 0$, and vertex $(0, 0)$. The graph of any parabola with standard form (1) is symmetric with respect to the y-axis.

Equation (1) does not depend on the assumption that $c > 0$. However, the direction in which the parabola opens does depend on the sign of c. Specifically, if $c > 0$, the parabola opens *upward* as in Figure 9.1.3; if $c < 0$, the parabola opens *downward*.

If the focus of a parabola is assumed to lie on the x-axis at $F(c, 0)$ and the directrix is $x = -c$, then the x-axis is the axis of the parabola and the vertex is $(0, 0)$. If $c > 0$, the parabola opens to the right; if $c < 0$, it opens to the left. In either case, the **standard form** of the equation is

$$y^2 = 4cx. \tag{2}$$

The graph of any parabola with standard form (2) is symmetric with respect to the x-axis.

A summary of all this information for equations (1) and (2) is given in **FIGURE 9.1.4** and **FIGURE 9.1.5**, respectively. You may be surprised to see in Figure 9.1.4(b) that the directrix above the x-axis is labeled $y = -c$ and the focus on the negative y-axis has coordinates $F(0, c)$. Bear in mind that in this case the assumption is that $c < 0$ and so $-c > 0$. A similar remark holds for Figure 9.1.5(b).

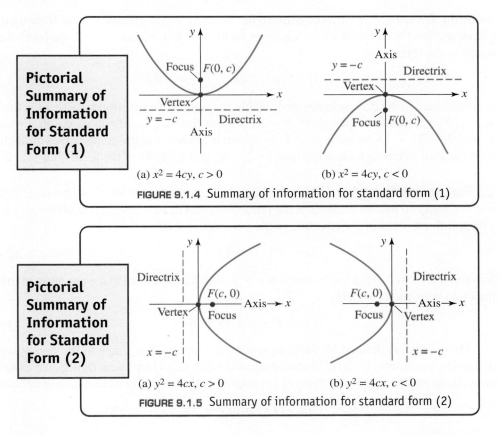

FIGURE 9.1.4 Summary of information for standard form (1)

FIGURE 9.1.5 Summary of information for standard form (2)

FIGURE 9.1.6 Graph of equation in Example 1

Graphing tip for equations (1) and (2).

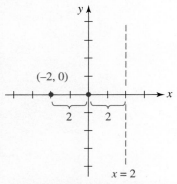

FIGURE 9.1.7 Directrix and focus in Example 2

FIGURE 9.1.8 Graph of parabola in Example 2

EXAMPLE 1 The Simplest Parabola

We first encountered the graph of $y = x^2$ in Section 2.2. By comparing this equation with (1) we see

$$x^2 = \underset{\underset{4c}{\uparrow}}{1} \cdot y$$

and so $4c = 1$ or $c = \frac{1}{4}$. Therefore the graph of $y = x^2$ is a parabola with vertex at the origin, focus at $\left(0, \frac{1}{4}\right)$, and directrix $y = -\frac{1}{4}$. These details are indicated in the graph in FIGURE 9.1.6. ≡

Knowing the basic parabolic shape, all we need to know to sketch a *rough* graph of either equation (1) or (2) is the fact that the graph passes through its vertex $(0, 0)$ and the direction in which the parabola opens. To add more accuracy to the graph it is convenient to use the number c determined by the standard form equation to plot two additional points. Note that if we choose $y = c$ in (1), then $x^2 = 4c^2$ implies $x = \pm 2c$. Thus $(2c, c)$ and $(-2c, c)$ lie on the graph of $x^2 = 4cy$. Similarly, the choice $x = c$ in (2) implies $y = \pm 2c$, and so $(c, 2c)$ and $(c, -2c)$ are points on the graph of $y^2 = 4cx$. The *line segment* through the focus with endpoints $(2c, c)$, $(-2c, c)$ for equations with standard form (1), and $(c, 2c)$, $(c, -2c)$ for equations with standard form (2) is called the **focal chord**. For example, in Figure 9.1.6, if we choose $y = \frac{1}{4}$, then $x^2 = \frac{1}{4}$ implies $x = \pm\frac{1}{2}$. Endpoints of the horizontal focal chord for $y = x^2$ are $\left(-\frac{1}{2}, \frac{1}{4}\right)$ and $\left(\frac{1}{2}, \frac{1}{4}\right)$.

EXAMPLE 2 Finding an Equation of a Parabola

Find the equation in standard form of the parabola with directrix $x = 2$ and focus $(-2, 0)$. Graph.

Solution In FIGURE 9.1.7 we have graphed the directrix and the focus. We see from their placement that the equation we seek is of the form $y^2 = 4cx$. Since $c = -2$, the parabola opens to the left and so

$$y^2 = 4(-2)x \quad \text{or} \quad y^2 = -8x.$$

As mentioned in the discussion preceding this example, if we substitute $x = c$, or in this case $x = -2$, into the equation $y^2 = -8x$ we can find two points on its graph. From $y^2 = -8(-2) = 16$ we get $y = \pm 4$. As shown in FIGURE 9.1.8, the graph passes through $(0, 0)$ as well as through the endpoints $(-2, -4)$ and $(-2, 4)$ of the focal chord. ≡

☐ **Parabola with Vertex (h, k)** Suppose that a parabola is shifted both horizontally and vertically so that its vertex is at the point (h, k) and its axis is the vertical line $x = h$. The **standard form** of the equation of the parabola is then

$$(x - h)^2 = 4c(y - k). \tag{3}$$

Similarly, if its axis is the horizontal line $y = k$, the standard form of the equation of the parabola with vertex (h, k) is

$$(y - k)^2 = 4c(x - h). \tag{4}$$

The parabolas defined by these equations are identical in shape to the parabolas defined by equations (1) and (2) because equations (3) and (4) represent rigid transformations (shifts up, down, left, and right) of the graphs of (1) and (2). For example, the parabola

$$(x + 1)^2 = 8(y - 5)$$

has vertex $(-1, 5)$. Its graph is the graph of $x^2 = 8y$ shifted horizontally 1 unit to the left followed by an upward vertical shift of 5 units.

For each of the equations, (1) and (2) or (3) and (4), the *distance* from the vertex to the focus, as well as the distance from the vertex to the directrix, is $|c|$.

EXAMPLE 3   Find an Equation of a Parabola

Find the equation in standard form of the parabola with vertex $(-3, -1)$ and directrix $y = 3$.

Solution We begin by graphing the vertex at $(-3, -1)$ and the directrix $y = 3$. From **FIGURE 9.1.9** we can see that the parabola must open downward, and so its standard form is (3). This fact, plus the observation that the vertex lies 4 units below the directrix, indicates that the appropriate solution of $|c| = 4$ is $c = -4$. Substituting $h = -3$, $k = -1$, and $c = -4$ into (3) gives

$$[x - (-3)]^2 = 4(-4)[y - (-1)] \quad \text{or} \quad (x + 3)^2 = -16(y + 1).$$

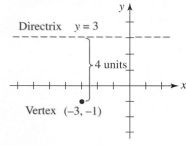

FIGURE 9.1.9 Vertex and directrix in Example 3

EXAMPLE 4   Find Everything

Find the vertex, focus, directrix, intercepts, and graph of the parabola

$$y^2 - 4y - 8x - 28 = 0. \tag{5}$$

Solution In order to write the equation in one of the standard forms we complete the square in y:

$$y^2 - 4y + 4 = 8x + 28 + 4 \quad \leftarrow \text{add 4 to both sides}$$
$$(y - 2)^2 = 8x + 32.$$

Thus the standard form of equation (5) is $(y - 2)^2 = 8(x + 4)$. Comparing this equation with (4) we conclude that the vertex is $(-4, 2)$ and that $4c = 8$ or $c = 2$. Thus the parabola opens to the right. From $c = 2 > 0$, the focus is 2 units to the right of the vertex at $(-4 + 2, 2)$ or $(-2, 2)$. The directrix is the vertical line 2 units to the left of the vertex, $x = -4 - 2$ or $x = -6$. Knowing the parabola opens to the right from the point $(-4, 2)$ also tells us that the graph has intercepts. To find the x-intercept we set $y = 0$ in (5) and find immediately that $x = -\frac{28}{8} = -\frac{7}{2}$. The x-intercept is $\left(-\frac{7}{2}, 0\right)$. To find the y-intercepts we set $x = 0$ in (5) and find from the quadratic formula that $y = 2 \pm 4\sqrt{2}$ or $y \approx 7.66$ and $y \approx -3.66$. The y-intercepts are $(0, 2 - 4\sqrt{2})$ and $(0, 2 + 4\sqrt{2})$. Putting all this information together we get the graph in **FIGURE 9.1.10**.

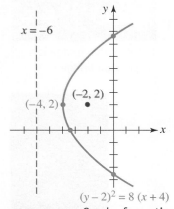

FIGURE 9.1.10 Graph of equation in Example 4

☐ **Applications of the Parabola** The parabola has many interesting properties that make it suitable for certain applications. Reflecting surfaces are often designed to take advantage of a reflection property of parabolas. Such surfaces, called **paraboloids**, are three-dimensional and are formed by rotating a parabola about its axis. As illustrated in **FIGURE 9.1.11(a)**, rays of light (or electronic signals) from a point source located at the focus of a parabolic reflecting surface will be reflected along lines parallel to the axis. This is the idea behind the design of searchlights, some flashlights, and on-location satellite dishes. Conversely, if the incoming rays of light are parallel to the axis of a parabola, they will be reflected off the surface along lines passing through the focus. See Figure 9.1.11(b). Beams of light from a distant object such as a galaxy are essentially parallel, and so when these beams enter a reflecting telescope they are reflected by the parabolic mirror to the focus, where a camera is usually placed to capture the image over time. A parabolic home satellite dish operates on the same principle as the

Searchlights

200 inch reflecting telescope at Mt. Palomar

TV satellite dish

reflecting telescope; the digital signal from a TV satellite is captured at the focus of the dish by a receiver.

(a) Rays emitted at focus are reflected as parallel rays

(b) Incoming rays reflected to focus

FIGURE 9.1.11 Parabolic reflecting surface

Parabolas are also important in the design of suspension bridges. It can be shown that if the weight of the bridge is distributed uniformly along its length, then a support cable in the shape of a parabola will bear the load evenly.

The trajectory of an obliquely launched projectile—say, a basketball thrown from the free-throw line—will travel in a parabolic arc.

Tuna, which prey on smaller fish, have been observed swimming in schools of 10–20 fish arrayed approximately in a parabolic shape. One possible explanation for this is that the smaller fish caught in the school of tuna will try to escape by "reflecting" off the parabola. See **FIGURE 9.1.12**. As a result, they are concentrated at the focus and become easy prey for the tuna.

The Brooklyn bridge is a suspension bridge

The ball travels in a parabolic arc

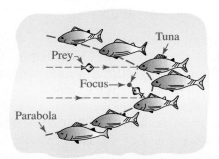

FIGURE 9.1.12 Tuna hunting in a parabolic arc

9.1 Exercises
Answers to selected odd-numbered problems begin on page ANS-23.

In Problems 1–24, find the vertex, focus, directrix, and axis of the given parabola. Graph the parabola.

1. $y^2 = 4x$
2. $y^2 = \frac{7}{2}x$
3. $y^2 = -\frac{4}{3}x$
4. $y^2 = -10x$
5. $x^2 = -16y$
6. $x^2 = \frac{1}{10}y$
7. $x^2 = 28y$
8. $x^2 = -64y$
9. $(y - 1)^2 = 16x$
10. $(y + 3)^2 = -8(x + 2)$
11. $(x + 5)^2 = -4(y + 1)$
12. $(x - 2)^2 + y = 0$
13. $y^2 + 12y - 4x + 16 = 0$
14. $x^2 + 6x + y + 11 = 0$

15. $x^2 + 5x - \frac{1}{4}y + 6 = 0$
16. $x^2 - 2x - 4y + 17 = 0$
17. $y^2 - 8y + 2x + 10 = 0$
18. $y^2 - 4y - 4x + 3 = 0$
19. $4x^2 = 2y$
20. $3(y - 1)^2 = 9x$
21. $-2x^2 + 12x - 8y - 18 = 0$
22. $4y^2 + 16y - 6x - 2 = 0$
23. $6y^2 - 12y - 24x - 42 = 0$
24. $3x^2 + 30x - 8y + 75 = 0$

In Problems 25–44, find an equation of the parabola that satisfies the given conditions.

25. Focus $(0, 7)$, directrix $y = -7$
26. Focus $(0, -5)$, directrix $y = 5$
27. Focus $(-4, 0)$, directrix $x = 4$
28. Focus $\left(\frac{3}{2}, 0\right)$, directrix $x = -\frac{3}{2}$
29. Focus $\left(\frac{5}{2}, 0\right)$, vertex $(0, 0)$
30. Focus $(0, -10)$, vertex $(0, 0)$
31. Focus $(2, 3)$, directrix $y = -3$
32. Focus $(1, -7)$, directrix $x = -5$
33. Focus $(-1, 4)$, directrix $x = 5$
34. Focus $(-2, 0)$, directrix $y = \frac{3}{2}$
35. Focus $(1, 5)$, vertex $(1, -3)$
36. Focus $(-2, 3)$, vertex $(-2, 5)$
37. Focus $(8, -3)$, vertex $(0, -3)$
38. Focus $(1, 2)$, vertex $(7, 2)$
39. Vertex $(0, 0)$, directrix $y = -\frac{7}{4}$
40. Vertex $(0, 0)$, directrix $x = 6$
41. Vertex $(5, 1)$, directrix $y = 7$
42. Vertex $(-1, 4)$, directrix $x = 0$
43. Vertex $(0, 0)$, through $(-2, 8)$, axis along the y-axis
44. Vertex $(0, 0)$, through $\left(1, \frac{1}{4}\right)$, axis along the x-axis

In Problems 45–48, find the x- and y-intercepts of the given parabola.

45. $(y + 4)^2 = 4(x + 1)$
46. $(x - 1)^2 = -2(y - 1)$
47. $x^2 + 2y - 18 = 0$
48. $y^2 - 8y - x + 15 = 0$

Miscellaneous Applications

49. **Spotlight** A large spotlight is designed so that a cross section through its axis is a parabola and the light source is at the focus. Find the position of the light source if the spotlight is 4 ft across at the opening and 2 ft deep.
50. **Reflecting Telescope** A reflecting telescope has a parabolic mirror that is 20 ft across at the top and 4 ft deep at the center. Where should the eyepiece be located?
51. **Light Ray** Suppose that a light ray emanating from the focus of the parabola $y^2 = 4x$ strikes the parabola at $(1, -2)$. What is the equation of the reflected ray?
52. **Suspension Bridge** Suppose that two towers of a suspension bridge are 350 ft apart and the vertex of the parabolic cable is tangent to the road midway between the towers. If the cable is 1 ft above the road at a point 20 ft from the vertex, find the height of the towers above the road.
53. **Another Suspension Bridge** Two 75-ft towers of a suspension bridge with a parabolic cable are 250 ft apart. The vertex of the parabola is tangent to the road midway between the towers. Find the height of the cable above the roadway at a point 50 ft from one of the towers.
54. **Drain Pipe** Assume that the water gushing from the end of a horizontal pipe follows a parabolic arc with the vertex at the end of the pipe. The pipe is 20 m above the ground. At a point 2 m below the end of the pipe, the horizontal distance from the water to a vertical line through the end of the pipe is 4 m. See FIGURE 9.1.13. Where does the water strike the ground?
55. **A Bull's-Eye** A dart thrower releases a dart 5 ft above the ground. The dart is thrown horizontally and follows a parabolic path. It hits the ground $10\sqrt{10}$ ft from the dart thrower. At a distance of 10 ft from the dart thrower, how high should a bull's-eye be placed in order for the dart to hit it?

FIGURE 9.1.13 Pipe in Problem 54

FIGURE 9.1.14 Focal width in Problem 57

56. **Path of a Projectile** The vertical position of a projectile is given by the equation $y = -16t^2$ and the horizontal position by $x = 40t$ for $t \geq 0$. By eliminating t between the two equations, show that the path of the projectile is a parabolic arc. Graph the path of the projectile.

57. **Focal Width** The focal width of a parabola is the length of the focal chord, that is, the line segment through the focus perpendicular to the axis, with endpoints on the parabola. See **FIGURE 9.1.14**.
   (a) Find the focal width of the parabola $x^2 = 8y$.
   (b) Show that the focal width of the parabola $x^2 = 4cy$ and $y^2 = 4cx$ is $4|c|$.

58. **Parabolic Orbit** The orbit of a comet is a parabola with the Sun at the focus. When the comet is 50,000,000 km from the Sun, the line from the comet to the Sun is perpendicular to the axis of the parabola. Use the result of Problem 57(b) to write an equation of the comet's path. (A comet with a parabolic path will not return to the solar system.)

For Discussion

59. **Reflecting Surfaces** Suppose that two parabolic reflecting surfaces face one another (with foci on a common axis). Any sound emitted at one focus will be reflected off the parabolas and concentrated at the other focus. **FIGURE 9.1.15** shows the paths of two typical sound waves. Using the definition of a parabola on page 448, show that all waves will travel the same distance. (*Note*: This result is important for the following reason. If the sound waves traveled paths of different lengths, then the waves would arrive at the second focus at different times. The result would be interference rather than clear sound.)

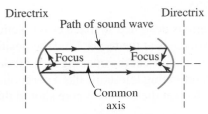

FIGURE 9.1.15 Parabolic reflecting surfaces in Problem 59

60. The point closest to the focus is the vertex. How would you go about proving this? Carry out your ideas.

61. For the comet in Problem 58, use the result of Problem 60 to determine the shortest distance between the Sun and the comet.

9.2 The Ellipse

≡ **Introduction** The ellipse occurs frequently in astronomy. For example, the paths of the planets around the Sun are elliptical with the Sun located at one focus. Similarly, communication satellites, the Hubble space telescope, and the international space station revolve around the Earth in elliptical orbits with the Earth at one focus. In this section we define the ellipse and study some of its properties and applications.

DEFINITION 9.2.1 Ellipse

An **ellipse** is the set of points $P(x, y)$ in the plane such that the sum of the distances between P and two fixed points F_1 and F_2 is constant. The fixed points F_1 and F_2 are called **foci** (plural for **focus**). The midpoint of the line segment joining points F_1 and F_2 is called the **center** of the ellipse.

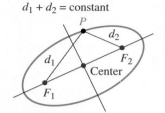

FIGURE 9.2.1 An ellipse

As shown in FIGURE 9.2.1, if P is a point on the ellipse and if $d_1 = d(F_1, P)$ and $d_2 = d(F_2, P)$ are the distances from the foci to P, then the preceding definition asserts that

$$d_1 + d_2 = k, \tag{1}$$

where $k > 0$ is some constant.

On a practical level, equation (1) suggests a way of generating an ellipse. FIGURE 9.2.2 shows that if a string of length k is attached to a piece of paper by two tacks, then an ellipse can be traced out by inserting a pencil against the string and moving it in such a manner that the string remains taut.

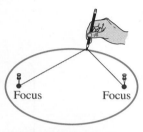

FIGURE 9.2.2 A way to draw an ellipse

□ **Ellipse with Center (0, 0)** We now derive an equation of the ellipse. For algebraic convenience, let us choose $k = 2a > 0$ and put the foci on the x-axis with coordinates $F_1(-c, 0)$ and $F_2(c, 0)$ as shown in FIGURE 9.2.3. It follows from (1) that

$$\sqrt{(x + c)^2 + y^2} + \sqrt{(x - c)^2 + y^2} = 2a \tag{2}$$

or

$$\sqrt{(x + c)^2 + y^2} = 2a - \sqrt{(x - c)^2 + y^2}.$$

We square both sides of the second equation in (2) and simplify,

$$(x + c)^2 + y^2 = 4a^2 - 4a\sqrt{(x - c)^2 + y^2} + (x - c)^2 + y^2$$
$$a\sqrt{(x - c)^2 + y^2} = a^2 - cx.$$

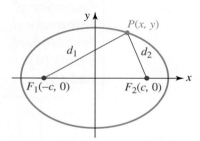

FIGURE 9.2.3 Ellipse with center (0, 0) and foci on the x-axis

Squaring a second time gives

$$a^2[(x - c)^2 + y^2] = a^4 - 2a^2cx + c^2x^2$$

or

$$(a^2 - c^2)x^2 + a^2y^2 = a^2(a^2 - c^2). \tag{3}$$

Referring to Figure 9.2.3, we see that the points F_1, F_2, and P form a triangle. Because the sum of the lengths of any two sides of a triangle is greater than the remaining side, we must have $2a > 2c$ or $a > c$. Hence, $a^2 - c^2 > 0$. When we let $b^2 = a^2 - c^2$, then (3) becomes $b^2x^2 + a^2y^2 = a^2b^2$. Dividing this last equation by a^2b^2 gives

$$\frac{x^2}{a^2} + \frac{y^2}{b^2} = 1. \tag{4}$$

Equation (4) is called the **standard form** of the equation of an ellipse centered at $(0, 0)$ with foci $(-c, 0)$ and $(c, 0)$, where c is defined by $b^2 = a^2 - c^2$ and $a > b > 0$.

If the foci are placed on the y-axis, then a repetition of the above analysis leads to

$$\frac{x^2}{b^2} + \frac{y^2}{a^2} = 1. \tag{5}$$

Equation (5) is called the **standard form** of the equation of an ellipse centered at $(0, 0)$ with foci $(0, -c)$ and $(0, c)$, where c is defined by $b^2 = a^2 - c^2$ and $a > b > 0$.

☐ **Major and Minor Axes** The **major axis** of an ellipse is the line segment through its center, containing the foci, and with endpoints on the ellipse. For an ellipse with standard equation (4) the major axis is horizontal, whereas for (5) the major axis is vertical. The line segment through the center, perpendicular to the major axis, and with endpoints on the ellipse is called the **minor axis**. The two endpoints of the major axis are called the **vertices** of the ellipse. For (4) the vertices are the x-intercepts. Setting $y = 0$ in (4) gives $x = \pm a$. The vertices are then $(-a, 0)$ and $(a, 0)$. For (5) the vertices are the y-intercepts $(0, -a)$ and $(0, a)$. For equation (4), the endpoints of the minor axis are $(0, -b)$ and $(0, b)$; for (5) the endpoints are $(-b, 0)$ and $(b, 0)$. For either (4) or (5), the **length of the major axis** is $a - (-a) = 2a$; the length of the minor axis is $2b$. Since $a > b$, the major axis of an ellipse is always longer than its minor axis.

A summary of all this information for equations (4) and (5) is given in **FIGURE 9.2.4**.

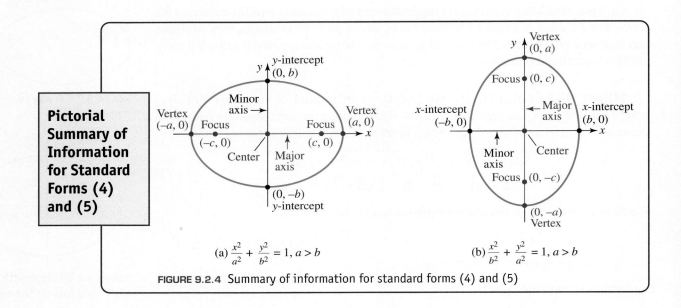

FIGURE 9.2.4 Summary of information for standard forms (4) and (5)

EXAMPLE 1 Vertices and Foci

Find the vertices and foci of the ellipse whose equation is $3x^2 + y^2 = 9$. Graph.

Solution By dividing both sides of the equality by 9 the standard form of the equation is

$$\frac{x^2}{3} + \frac{y^2}{9} = 1.$$

We see that $9 > 3$ and so we identify the equation with (5). From $a^2 = 9$ and $b^2 = 3$, we see that $a = 3$ and $b = \sqrt{3}$. The major axis is vertical with endpoints $(0, -3)$ and $(0, 3)$. The minor axis is horizontal with endpoints $(-\sqrt{3}, 0)$ and $(\sqrt{3}, 0)$. Of course, the vertices are also the y-intercepts and the endpoints of the minor axis are the x-intercepts. Now, to find the foci we use $b^2 = a^2 - c^2$ or $c^2 = a^2 - b^2$ to write $c = \sqrt{a^2 - b^2}$. With $a = 3$, $b = \sqrt{3}$, we get $c = \sqrt{9 - 3} = \sqrt{6}$. Hence, the foci are on the y-axis at $(0, -\sqrt{6})$ and $(0, \sqrt{6})$. The graph is given in **FIGURE 9.2.5**. ≡

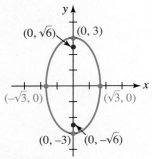

FIGURE 9.2.5 Ellipse in Example 1

EXAMPLE 2 Finding an Equation of an Ellipse

Find an equation of the ellipse with a focus $(2, 0)$ and an x-intercept $(5, 0)$.

Solution Since the given focus is on the x-axis, we can find an equation in standard form (4). Consequently, $c = 2$, $a = 6$, $a^2 = 25$, and $b^2 = a^2 - c^2$ or $b^2 = 5^2 - 2^2 = 21$. The desired equation is

$$\frac{x^2}{25} + \frac{y^2}{21} = 1.$$

≡

☐ **Ellipse with Center (h, k)** When the center is at (h, k), the **standard form** for the equation of an ellipse is either

$$\frac{(x - h)^2}{a^2} + \frac{(y - k)^2}{b^2} = 1 \tag{6}$$

or

$$\frac{(x - h)^2}{b^2} + \frac{(y - k)^2}{a^2} = 1. \tag{7}$$

The ellipses defined by these equations are identical in shape to the ellipses defined by equations (4) and (5) since equations (6) and (7) represent rigid transformations of the graphs of (4) and (5). For example, the ellipse

$$\frac{(x - 1)^2}{9} + \frac{(y + 3)^2}{16} = 1$$

has center $(1, -3)$. Its graph is the graph of $x^2/9 + y^2/16 = 1$ shifted horizontally 1 unit to the right followed by a downward vertical shift of 3 units.

It is not a good idea to memorize formulas for the vertices and foci of an ellipse with center (h, k). Everything is the same as before, a, b, and c are positive and $a > b$, $a > c$. You can locate vertices, foci, and endpoints of the minor axis using the fact that a is the distance from the center to a vertex, b is the distance from the center to an endpoint on the minor axis, and c is the distance from the center to a focus. Also, we still have $c^2 = a^2 - b^2$.

EXAMPLE 3 Ellipse Centered at (h, k)

Find the vertices and foci of the ellipse $4x^2 + 16y^2 - 8x - 96y + 84 = 0$. Graph.

Solution To write the given equation in one of the standard forms (6) or (7) we must complete the square in x and in y. Recall, in order to complete the square we want the coefficients of the quadratic terms x^2 and y^2 to be 1. To do this we factor 4 from both x^2 and x and factor 16 from both y^2 and y:

$$4(x^2 - 2x \quad) + 16(y^2 - 6y \quad) = -84.$$

Then from

$$\underbrace{4(x^2 - 2x + 1) + 16(y^2 - 6y + 9)}_{4 \cdot 1 \text{ and } 16 \cdot 9 \text{ are added to both sides}} = -84 + 4 \cdot 1 + 16 \cdot 9$$

we obtain

$$4(x - 1)^2 + 16(y - 3)^2 = 64$$

or

$$\frac{(x - 1)^2}{16} + \frac{(y - 3)^2}{4} = 1. \tag{8}$$

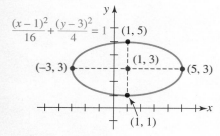

FIGURE 9.2.6 Ellipse in Example 3

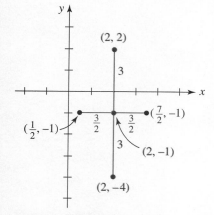

FIGURE 9.2.7 Graphical interpretation of data in Example 4

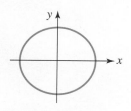

(a) e close to zero

(b) e close to 1

FIGURE 9.2.8 Effect of eccentricity on the shape of an ellipse

From (8) we see that the center of the ellipse is $(1, 3)$. Since the last equation has the standard form (6) we identify $a^2 = 16$ or $a = 4$ and $b^2 = 4$ or $b = 2$. The major axis is horizontal and lies on the horizontal line $y = 3$ passing through $(1, 3)$. This is the red horizontal dashed line segment in FIGURE 9.2.6. By measuring $a = 4$ units to the left and then to the right of the center along the line $y = 3$ we arrive at the vertices $(-3, 3)$ and $(5, 3)$. By measuring $b = 2$ units both down and up the vertical line $x = 1$ through the center we arrive at the endpoints of the minor axis $(1, 1)$ and $(1, 5)$. The minor axis is the black dashed vertical line segment in Figure 9.2.6. Because $c^2 = a^2 - b^2 = 16 - 4 = 12$, $c = 2\sqrt{3}$. Finally, by measuring $c = 2\sqrt{3}$ units to the left and right of the center along $y = 3$ we obtain the foci $(1 - 2\sqrt{3}, 3)$ and $(1 + 2\sqrt{3}, 3)$.

EXAMPLE 4 Finding an Equation of an Ellipse

Find an equation of the ellipse with center $(2, -1)$, vertical major axis of length 6, and minor axis of length 3.

Solution The length of the major axis is $2a = 6$; hence $a = 3$. Similarly, the length of the minor axis is $2b = 3$, so $b = \frac{3}{2}$. By sketching the center and the axes, we see from FIGURE 9.2.7 that the vertices are $(2, 2)$ and $(2, -4)$ and the endpoints of the minor axis are $(\frac{1}{2}, -1)$ and $(\frac{7}{2}, -1)$. Because the major axis is vertical, the standard equation of this ellipse is

$$\frac{(x-2)^2}{\left(\frac{3}{2}\right)^2} + \frac{(y-(-1))^2}{3^2} = 1 \quad \text{or} \quad \frac{(x-2)^2}{\frac{9}{4}} + \frac{(y+1)^2}{9} = 1.$$

☐ **Eccentricity** Associated with each conic section is a number e called its **eccentricity**. The eccentricity of an ellipse is defined to be

$$e = \frac{c}{a},$$

where $c = \sqrt{a^2 - b^2}$. Since $0 < \sqrt{a^2 - b^2} < a$, the eccentricity of an ellipse satisfies $0 < e < 1$.

EXAMPLE 5 Example 3 Revisited

Determine the eccentricity of the ellipse in Example 3.

Solution In the solution of Example 3 we found that $a = 4$ and $c = 2\sqrt{3}$. Hence, the eccentricity of the ellipse is $e = (2\sqrt{3})/4 = \sqrt{3}/2 \approx 0.87$.

Eccentricity is an indicator of the shape of an ellipse. When $e \approx 0$, that is, e is close to zero, the ellipse is nearly circular, and when $e \approx 1$ the ellipse is flattened or elongated. To see this, observe that if e is close to 0, it follows from $e = \sqrt{a^2 - b^2}/a$ that $c = \sqrt{a^2 - b^2} \approx 0$ and consequently $a \approx b$. As you can see from the standard equations in (4) and (5), this means that the shape of the ellipse is close to circular. Also, because c is the distance from the center of the ellipse to a focus, the two foci are close together near the center. See FIGURE 9.2.8(a). On the other hand, if $e \approx 1$ or $\sqrt{a^2 - b^2}/a \approx 1$, then $c = \sqrt{a^2 - b^2} \approx a$ and so $b \approx 0$. Also, $c \approx a$ means that the foci are far apart; each focus is close to a vertex. Thus, the ellipse is elongated as shown in Figure 9.2.8(b).

☐ **Applications of the Ellipse** Ellipses have a reflection property analogous to the one discussed in Section 9.1 for the parabola. It can be shown that if a light or sound source is placed at one focus of an ellipse, then all rays or waves will be reflected off the ellipse to the other focus. See FIGURE 9.2.9. For example, if a pool table is constructed in the form of an ellipse with a pocket at one focus, then any shot originating at the other focus will never miss the pocket. Similarly, if a ceiling is elliptical with two foci on (or near) the floor, but considerably distant from each other, then anyone whispering at one focus will be heard at the other. Some famous "whispering galleries" are the Statuary Hall at the Capitol in Washington, DC, the Mormon Tabernacle in Salt Lake City, and St. Paul's Cathedral in London.

Using his law of universal gravitation, Isaac Newton was the first to prove Kepler's first law of planetary motion: The orbit of each planet about the Sun is an ellipse with the Sun at one focus.

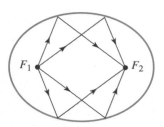

FIGURE 9.2.9 Reflection property of an ellipse

EXAMPLE 6   Eccentricity of Earth's Orbit

The perihelion distance of the Earth (the least distance between the Earth and the Sun) is approximately 9.16×10^7 miles, and its aphelion distance (the greatest distance between the Earth and the Sun) is approximately 9.46×10^7 miles. What is the eccentricity of Earth's orbit?

Statuary Hall in Washington, DC

Solution Let us assume that the orbit of the Earth is as shown in FIGURE 9.2.10. From the figure we see that

$$a - c = 9.16 \times 10^7$$
$$a + c = 9.46 \times 10^7.$$

Solving this system of equations gives $a = 9.31 \times 10^7$ and $c = 0.15 \times 10^7$. Thus the eccentricity $e = c/a$ is

$$e = \frac{0.15 \times 10^7}{9.31 \times 10^7} \approx 0.016.$$ ≡

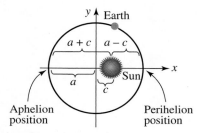

FIGURE 9.2.10 Graphical interpretation of data in Example 6

The orbits of seven of the eight planets have eccentricities less than 0.1 and, hence, the orbits are not far from circular. Mercury is the exception. The orbit of the well-known dwarf planet Pluto has the eccentricity 0.25. Many of the asteroids and comets have highly eccentric orbits. The orbit of the asteroid Hildago is one of the most eccentric, with $e = 0.66$. Another notable case is the orbit of Comet Halley. See Problem 43 in Exercises 9.2.

9.2 Exercises   Answers to selected odd-numbered problems begin on page ANS-24.

In Problems 1–20, find the center, foci, vertices, endpoints of the minor axis, and eccentricity of the given ellipse. Graph the ellipse.

1. $\dfrac{x^2}{25} + \dfrac{y^2}{9} = 1$

2. $\dfrac{x^2}{16} + \dfrac{y^2}{4} = 1$

3. $x^2 + \dfrac{y^2}{16} = 1$

4. $\dfrac{x^2}{4} + \dfrac{y^2}{10} = 1$

5. $9x^2 + 16y^2 = 144$

6. $2x^2 + y^2 = 4$

7. $9x^2 + 4y^2 = 36$

8. $x^2 + 4y^2 = 4$

9. $\dfrac{(x-1)^2}{49} + \dfrac{(y-3)^2}{36} = 1$

10. $\dfrac{(x+1)^2}{25} + \dfrac{(y-2)^2}{36} = 1$

11. $(x+5)^2 + \dfrac{(y+2)^2}{16} = 1$ 12. $\dfrac{(x-3)^2}{64} + \dfrac{(y+4)^2}{81} = 1$
13. $4x^2 + (y+\tfrac{1}{2})^2 = 4$ 14. $36(x+2)^2 + (y-4)^2 = 72$
15. $5(x-1)^2 + 3(y+2)^2 = 45$ 16. $6(x-2)^2 + 8y^2 = 48$
17. $25x^2 + 9y^2 - 100x + 18y - 116 = 0$ 18. $9x^2 + 5y^2 + 18x - 10y - 31 = 0$
19. $x^2 + 3y^2 + 18y + 18 = 0$ 20. $12x^2 + 4y^2 - 24x - 4y + 1 = 0$

In Problems 21–40, find an equation of the ellipse that satisfies the given conditions.

21. Vertices $(\pm 5, 0)$, foci $(\pm 3, 0)$
22. Vertices $(\pm 9, 0)$, foci $(\pm 2, 0)$
23. Vertices $(0, \pm 3)$, foci $(0, \pm 1)$
24. Vertices $(0, \pm 7)$, foci $(0, \pm 3)$
25. Vertices $(0, \pm 3)$, endpoints of minor axis $(\pm 1, 0)$
26. Vertices $(\pm 4, 0)$, endpoints of minor axis $(0, \pm 2)$
27. Vertices $(-3, -3)$, $(5, -3)$, endpoints of minor axis $(1, -1)$, $(1, -5)$
28. Vertices $(1, -6)$, $(1, 2)$, endpoints of minor axis $(-2, -2)$, $(4, -2)$
29. One focus $(0, -2)$, center at origin, $b = 3$
30. One focus $(1, 0)$, center at origin, $a = 3$
31. Foci $(\pm\sqrt{2}, 0)$, length of minor axis 6
32. Foci $(0, \pm\sqrt{5})$, length of major axis 16
33. Foci $(0, \pm 3)$, passing through $(-1, 2\sqrt{2})$
34. Vertices $(\pm 5, 0)$, passing through $(\sqrt{5}, 4)$
35. Vertices $(\pm 4, 1)$, passing through $(2\sqrt{3}, 2)$
36. Center $(1, -1)$, one focus $(1, 1)$, $a = 5$
37. Center $(1, 3)$, one focus $(1, 0)$, one vertex $(1, -1)$
38. Center $(5, -7)$, length of vertical major axis 8, length of minor axis 6
39. Endpoints of minor axis $(0, 5)$, $(0, -1)$, one focus $(6, 2)$
40. Endpoints of major axis $(2, 4)$, $(13, 4)$, one focus $(4, 4)$

41. The orbit of the planet Mercury is an ellipse with the Sun at one focus. The length of the major axis of this orbit is 72 million miles and the length of the minor axis is 70.4 million miles. What is the least distance (perihelion) between Mercury and the Sun? What is the greatest distance (aphelion)?
42. What is the eccentricity of the orbit of Mercury in Problem 41?
43. The orbit of Comet Halley is an ellipse whose major axis is 3.34×10^9 miles long, and whose minor axis is 8.5×10^8 miles long. What is the eccentricity of the comet's orbit?
44. A satellite orbits the Earth in an elliptical path with the center of the Earth at one focus. It has a minimum altitude of 200 mi and a maximum altitude of 1000 mi above the surface of the Earth. If the radius of the Earth is 4000 mi, what is an equation of the satellite's orbit?

Miscellaneous Applications

45. **Archway** A semielliptical archway has a vertical major axis. The base of the arch is 10 ft across and the highest part of the arch is 15 ft. Find the height of the arch above the point on the base of the arch 3 ft from the center.
46. **Gear Design** An elliptical gear rotates about its center and is always kept in mesh with a circular gear that is free to move horizontally. See **FIGURE 9.2.11**. If the origin of the xy-coordinate system is placed at the center of the ellipse, then

FIGURE 9.2.11 Elliptical and circular gears in Problem 46

the equation of the ellipse in its present position is $3x^2 + 9y^2 = 24$. The diameter of the circular gear equals the length of the minor axis of the elliptical gear. Given that the units are centimeters, how far does the center of the circular gear move horizontally during the rotation from one vertex of the elliptical gear to the next?

47. **Carpentry** A carpenter wishes to cut an elliptical top for a coffee table from a rectangular piece of wood that is 4-ft by 3-ft utilizing the entire length and width available. If the ellipse is to be drawn using the string-and-tack method illustrated in Figure 9.2.2, how long should the piece of string be and where should the tacks be placed?

48. **Park Design** The Ellipse is a park in Washington, DC. It is bounded by an elliptical path with a major axis of length 458 m and a minor axis of length 390 m. Find the distance between the foci of this ellipse.

49. **Whispering Gallery** Suppose that a room is constructed on a flat elliptical base by rotating a semiellipse 180° about its major axis. Then, by the reflection property of the ellipse, anything whispered at one focus will be distinctly heard at the other focus. If the height of the room is 16 ft and the length is 40 ft, find the location of the whispering and listening posts.

50. **Focal Width** The focal width of the ellipse is the length of a focal chord, that is, a line segment perpendicular to the major axis, through a focus with endpoints on the ellipse. See **FIGURE 9.2.12**.
    (a) Find the focal width of the ellipse $x^2/9 + y^2/4 = 1$.
    (b) Show that, in general, the focal width of the ellipse $x^2/a^2 + y^2/b^2 = 1$ is $2b^2/a$.

51. Find an equation of the ellipse with foci $(0, 2)$ and $(8, 6)$ and fixed distance sum $2a = 12$. [*Hint*: Here the major axis is neither horizontal nor vertical; thus none of the standard forms from this section apply. Use the definition of the ellipse.]

52. Proceed as in Problem 51, and find an equation of the ellipse with foci $(-1, -3)$ and $(-5, 7)$ and fixed distance sum $2a = 20$.

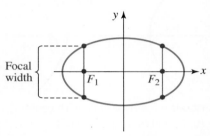

FIGURE 9.2.12 Focal width in Problem 50

For Discussion

53. The graph of the ellipse $x^2/4 + (y - 1)^2/9 = 1$ is shifted 4 units to the right. What are the center, foci, vertices, and endpoints of the minor axis for the shifted graph?

54. The graph of the ellipse $(x - 1)^2/9 + (y - 4)^2 = 1$ is shifted 5 units to the left and 3 units up. What are the center, foci, vertices, and endpoints of the minor axis for the shifted graph?

55. In engineering the eccentricity of an ellipse is often expressed only in terms of a and b. Show that $e = \sqrt{1 - b^2/a^2}$.

9.3 The Hyperbola

≡ **Introduction** The definition of a hyperbola is basically the same as the definition of the ellipse with only one exception: the word *sum* is replaced by the word *difference*.

DEFINITION 9.3.1 Hyperbola

A **hyperbola** is the set of points $P(x, y)$ in the plane such that the difference of the distances between P and two fixed points F_1 and F_2 is constant. The fixed points F_1 and F_2 are called **foci** (plural for **focus**). The midpoint of the line segment joining points F_1 and F_2 is called the **center** of the hyperbola.

FIGURE 9.3.1 A hyperbola

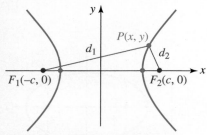

FIGURE 9.3.2 Hyperbola with center (0, 0) and foci on the *x*-axis

As shown in FIGURE 9.3.1, a hyperbola consists of two **branches**. If P is a point on the hyperbola, then

$$|d_1 - d_2| = k, \tag{1}$$

where $d_1 = d(F_1, P)$ and $d_2 = d(F_2, P)$.

☐ **Hyperbola with Center (0, 0)** Proceeding as for the ellipse, we place the foci on the *x*-axis at $F_1(-c, 0)$ and $F_2(c, 0)$ as shown in FIGURE 9.3.2 and choose the constant k to be $2a$ for algebraic convenience. It follows from (1) that

$$d_1 - d_2 = \pm 2a. \tag{2}$$

As drawn in Figure 9.3.2, P is on the right branch of the hyperbola and so $d_1 - d_2 = 2a > 0$. If P is on the left branch, then the difference is $-2a$. Writing (2) as

$$\sqrt{(x + c)^2 + y^2} - \sqrt{(x - c)^2 + y^2} = \pm 2a$$

or

$$\sqrt{(x + c)^2 + y^2} = \pm 2a + \sqrt{(x - c)^2 + y^2}$$

we square, simplify, and square again:

$$(x + c)^2 + y^2 = 4a^2 \pm 4a\sqrt{(x - c)^2 + y^2} + (x - c)^2 + y^2$$
$$\pm a\sqrt{(x - c)^2 + y^2} = cx - a^2$$
$$a^2[(x - c)^2 + y^2] = c^2x^2 - 2a^2cx + a^4$$
$$(c^2 - a^2)x^2 - a^2y^2 = a^2(c^2 - a^2). \tag{3}$$

From Figure 9.3.2, we see that the triangle inequality gives

$$d_1 < d_2 + 2c \quad \text{and} \quad d_2 < d_1 + 2c,$$

or

$$d_1 - d_2 < 2c \quad \text{and} \quad d_2 - d_1 < 2c.$$

Using $d_1 - d_2 = \pm 2a$ the last two inequalities imply that $2a < 2c$ or $a < c$. Since $c < a < 0$, $c^2 - a^2$ is a positive constant. If we let $b^2 = c^2 - a^2$, (3) becomes $b^2x^2 - a^2y^2 = a^2b^2$ or, after dividing by a^2b^2,

$$\frac{x^2}{a^2} - \frac{y^2}{b^2} = 1. \tag{4}$$

Equation (4) is called the **standard form** of the equation of a hyperbola centered at $(0, 0)$ with foci $(-c, 0)$ and $(c, 0)$, where c is defined by $b^2 = c^2 - a^2$.

When the foci lie on the *y*-axis, a repetition of the foregoing algebra leads to

$$\frac{y^2}{a^2} - \frac{x^2}{b^2} = 1. \tag{5}$$

Equation (5) is the **standard form** of the equation of a hyperbola centered at $(0, 0)$ with foci $(0, -c)$ and $(0, c)$. Here again, $c > a$ and $b^2 = c^2 - a^2$.

Note of Caution ▶ For the hyperbola (unlike the ellipse) bear in mind that in (4) and (5) there is no relationship between the relative sizes of a and b; rather, a^2 is always the denominator of the *positive term* and the intercepts *always* have $\pm a$ as a coordinate.

☐ **Transverse and Conjugate Axes** The line segment with endpoints on the hyperbola and lying on the line through the foci is called the **transverse axis**; its endpoints are called the **vertices** of the hyperbola. For the hyperbola described by equation (4), the transverse axis lies on the *x*-axis. Therefore, the coordinates of the vertices are the *x*-intercepts. Setting $y = 0$ gives $x^2/a^2 = 1$, or $x = \pm a$. Thus, as shown in FIGURE 9.3.3(a) the vertices are $(-a, 0)$ and $(a, 0)$; the **length of the transverse axis** is $2a$. Notice that by setting $y = 0$ in (4), we get $-y^2/b^2 = 1$ or $y^2 = -b^2$, which has no real solutions. Hence the graph of any equation in that form has no *y*-intercepts. Nonetheless, the numbers $\pm b$ are important. The line segment through the center of the hyperbola perpendicular to the transverse axis and with endpoints $(0, -b)$ and $(0, b)$ is called the **conjugate axis**. Similarly, the graph of an equation in standard form (5) has no x-intercepts. The conjugate axis for (5) is the line segment with endpoints $(-b, 0)$ and $(b, 0)$.

This information for equations (4) and (5) is summarized in Figure 9.3.3.

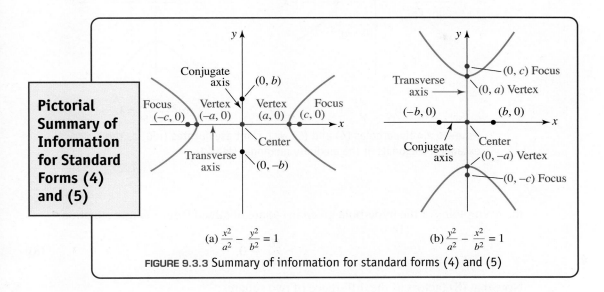

Pictorial Summary of Information for Standard Forms (4) and (5)

(a) $\dfrac{x^2}{a^2} - \dfrac{y^2}{b^2} = 1$ (b) $\dfrac{y^2}{a^2} - \dfrac{x^2}{b^2} = 1$

FIGURE 9.3.3 Summary of information for standard forms (4) and (5)

☐ **Asymptotes** Every hyperbola possesses a pair of **slant asymptotes** that pass through its center. These asymptotes are indicative of end behavior, and as such are an invaluable aid in sketching the graph of a hyperbola. Solving (4) for y in terms of x gives

$$y = \pm \frac{b}{a} x \sqrt{1 - \frac{a^2}{x^2}}.$$

As $x \to -\infty$ or as $x \to \infty$, $a^2/x^2 \to 0$, and thus $\sqrt{1 - a^2/x^2} \to 1$. Therefore, for large values of $|x|$, points on the graph of the hyperbola are close to the points on the lines

$$y = \frac{b}{a} x \quad \text{and} \quad y = -\frac{b}{a} x. \tag{6}$$

By a similar analysis we find that the slant asymptotes for (5) are

$$y = \frac{a}{b} x \quad \text{and} \quad y = -\frac{a}{b} x. \tag{7}$$

Each pair of asymptotes intersect at the origin, which is the center of the hyperbola. Note, too, in **FIGURE 9.3.4(a)** that the asymptotes are simply the *extended diagonals* of a rectangle of width $2a$ (the length of the transverse axis) and height $2b$ (the length of the conjugate axis); in Figure 9.3.4(b) the asymptotes are the extended diagonals of a rectangle of width $2b$ and height $2a$. This rectangle is referred to as the **auxiliary rectangle**.

We recommend that you *do not* memorize the equations in (6) and (7). There is an easy method for obtaining the asymptotes of a hyperbola. For example, since $y = \pm \dfrac{b}{a} x$ is equivalent to

$$\frac{x^2}{a^2} = \frac{y^2}{b^2}$$

9.3 The Hyperbola

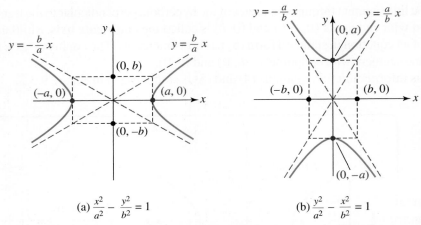

(a) $\dfrac{x^2}{a^2} - \dfrac{y^2}{b^2} = 1$ (b) $\dfrac{y^2}{a^2} - \dfrac{x^2}{b^2} = 1$

FIGURE 9.3.4 Hyperbolas (4) and (5) with slant asymptotes (red) as the extended diagonals of the auxiliary rectangle (black)

the asymptotes of the hyperbola given in (4) are obtained from a single equation

$$\dfrac{x^2}{a^2} - \dfrac{y^2}{b^2} = 0. \tag{8}$$

Note that (8) factors as the difference of two squares:

$$\left(\dfrac{x}{a} - \dfrac{y}{b}\right)\left(\dfrac{x}{a} + \dfrac{y}{b}\right) = 0.$$

This is a mnemonic, or memory device. It ▶ has no geometric significance.

Setting each factor equal to zero and solving for y gives an equation of an asymptote. You do not even have to memorize (8) because it is simply the left-hand side of the standard form of the equation of a hyperbola given in (4). In like manner, to obtain the asymptotes for (5) just replace 1 by 0 in the standard form, factor $y^2/a^2 - x^2/b^2 = 0$, and solve for y.

EXAMPLE 1 Hyperbola Centered at (0, 0)

Find the vertices, foci, and asymptotes of the hyperbola $9x^2 - 25y^2 = 225$. Graph.

Solution We first put the equation into standard form by dividing the left-hand side by 225:

$$\dfrac{x^2}{25} - \dfrac{y^2}{9} = 1. \tag{9}$$

From this equation we see that $a^2 = 25$ and $b^2 = 9$, and so $a = 5$ and $b = 3$. Therefore the vertices are $(-5, 0)$ and $(5, 0)$. Since $b^2 = c^2 - a^2$ implies $c^2 = a^2 + b^2$, we have $c^2 = 34$, and so the foci are $(-\sqrt{34}, 0)$ and $(\sqrt{34}, 0)$. To find the slant asymptotes we use the standard form (9) with 1 replaced by 0:

$$\dfrac{x^2}{25} - \dfrac{y^2}{9} = 0 \quad \text{factors as} \quad \left(\dfrac{x}{5} - \dfrac{y}{3}\right)\left(\dfrac{x}{5} + \dfrac{y}{3}\right) = 0.$$

Setting each factor equal to zero and solving for y gives the asymptotes $y = \pm 3x/5$. We plot the vertices and graph the two lines through the origin. Both branches of the hyperbola must become arbitrarily close to the asymptotes as $x \to \pm\infty$. See FIGURE 9.3.5.

FIGURE 9.3.5 Hyperbola in Example 1

EXAMPLE 2   Finding an Equation of a Hyperbola

Find an equation of the hyperbola with vertices $(0, -4)$, $(0, 4)$ and asymptotes $y = -\frac{1}{2}x$, $y = \frac{1}{2}x$.

Solution The center of the hyperbola is $(0, 0)$. This is revealed by the fact that the asymptotes intersect at the origin. Moreover, the vertices are on the y-axis and are 4 units on either side of the origin. Thus the equation we seek is of form (5). From (7) or Figure 9.3.4(b), the asymptotes must be of the form $y = \pm\frac{a}{b}x$ so that $a/b = 1/2$. From the given vertices we identify $a = 4$, and so

$$\frac{4}{b} = \frac{1}{2} \quad \text{implies} \quad b = 8.$$

The equation of the hyperbola is then

$$\frac{y^2}{4^2} - \frac{x^2}{8^2} = 1 \quad \text{or} \quad \frac{y^2}{16} - \frac{x^2}{64} = 1. \qquad \equiv$$

☐ **Hyperbola with Center (h, k)** When the center of the hyperbola is (h, k) the **standard form** analogs of equations (4) and (5) are, in turn,

$$\frac{(x - h)^2}{a^2} - \frac{(y - k)^2}{b^2} = 1 \qquad (10)$$

and

$$\frac{(y - k)^2}{a^2} - \frac{(x - h)^2}{b^2} = 1. \qquad (11)$$

As in (4) and (5) the numbers a^2, b^2, and c^2 are related by $b^2 = c^2 - a^2$.

You can locate vertices and foci using the fact that a is the distance from the center to a vertex, and c is the distance from the center to a focus. The slant asymptotes for (10) can be obtained by factoring

$$\frac{(x - h)^2}{a^2} - \frac{(y - k)^2}{b^2} = 0 \quad \text{as} \quad \left(\frac{x - h}{a} - \frac{y - k}{b}\right)\left(\frac{x - h}{a} + \frac{y - k}{b}\right) = 0.$$

Similarly, the asymptotes for (11) can be obtained from factoring

$$\frac{(y - k)^2}{a^2} - \frac{(x - h)^2}{b^2} = 0,$$

setting each factor equal to zero and solving for y in terms of x. As a check on your work, remember that (h, k) must be a point that lies on each asymptote.

EXAMPLE 3   Hyperbola Centered at (h, k)

Find the center, vertices, foci, and asymptotes of the hyperbola $4x^2 - y^2 - 8x - 4y - 4 = 0$. Graph.

Solution Before completing the square in x and y, we factor 4 from the two x-terms and factor -1 from the two y-terms so that the leading coefficient in each expression is 1. Then we have

$$\begin{aligned} 4(x^2 - 2x) + (-1)(y^2 + 4y) &= 4 \\ 4(x^2 - 2x + 1) - (y^2 + 4y + 4) &= 4 + 4 \cdot 1 + (-1) \cdot 4 \\ 4(x - 1)^2 - (y + 2)^2 &= 4 \\ \frac{(x - 1)^2}{1} - \frac{(y + 2)^2}{4} &= 1. \end{aligned}$$

We see now that the center is $(1, -2)$. Since the term in the standard form involving x has the positive coefficient, the transverse axis is horizontal along the line $y = -2$, and we identify $a = 1$ and $b = 2$. The vertices are 1 unit to the left and to the right of the center at $(0, -2)$ and $(2, -2)$, respectively. From $b^2 = c^2 - a^2$, we have

$$c^2 = a^2 + b^2 = 1 + 4 = 5,$$

and so $c = \sqrt{5}$. Hence the foci are $\sqrt{5}$ units to the left and the right of the center $(1, -2)$ at $(1 - \sqrt{5}, -2)$ and $(1 + \sqrt{5}, -2)$.

To find the asymptotes, we solve

$$\frac{(x-1)^2}{1} - \frac{(y+2)^2}{4} = 0 \quad \text{or} \quad \left(x - 1 - \frac{y+2}{2}\right)\left(x - 1 + \frac{y+2}{2}\right) = 0$$

for y. From $y + 2 = \pm 2(x - 1)$ we find that the asymptotes are $y = -2x$ and $y = 2x - 4$. Observe that by substituting $x = 1$, both equations give $y = -2$, which means that both lines pass through the center. We then locate the center, plot the vertices, and graph the asymptotes. As shown in FIGURE 9.3.6, the graph of the hyperbola passes through the vertices and becomes closer and closer to the asymptotes as $x \to \pm\infty$.

FIGURE 9.3.6 Hyperbola in Example 3

EXAMPLE 4 Finding an Equation of a Hyperbola

Find an equation of the hyperbola with center $(2, -3)$, passing through the point $(4, 1)$, and having one vertex $(2, 0)$.

Solution Since the distance from the center to one vertex is a, we have $a = 3$. From the location of the center and the vertex, it follows that the transverse axis is vertical and lies along the line $x = 2$. Therefore, the equation of the hyperbola must be of form (11):

$$\frac{(y+3)^2}{3^2} - \frac{(x-2)^2}{b^2} = 1, \tag{12}$$

where b^2 is yet to be determined. Since the point $(4, 1)$ is on the graph on the hyperbola, its coordinates must satisfy equation (12). From

$$\frac{(1+3)^2}{3^2} - \frac{(4-2)^2}{b^2} = 1$$

$$\frac{16}{9} - \frac{4}{b^2} = 1$$

$$\frac{7}{9} = \frac{4}{b^2}$$

we find $b^2 = \frac{36}{7}$. We conclude that the desired equation is

$$\frac{(y+3)^2}{3^2} - \frac{(x-2)^2}{\frac{36}{7}} = 1.$$

(a) e close to 1

(b) e much greater than 1

FIGURE 9.3.7 Effect of eccentricity on the shape of a hyperbola

☐ **Eccentricity** Like the ellipse, the equation that defines the **eccentricity** of a hyperbola is $e = c/a$. Except in this case the number c is given by $c = \sqrt{a^2 + b^2}$. Since $0 < a < \sqrt{a^2 + b^2}$, the eccentricity of an ellipse satisfies $e > 1$. As with the ellipse, the magnitude of the eccentricity of a hyperbola is an indicator of its shape. FIGURE 9.3.7 shows examples of two extreme cases: $e \approx 1$ and e much bigger than 1.

EXAMPLE 5 Eccentricity of a Hyperbola

Find the eccentricity of the hyperbola $\dfrac{y^2}{2} - \dfrac{(x-1)^2}{36} = 1$.

Solution Identifying $a^2 = 2$ and $b^2 = 36$, we get $c^2 = 2 + 36 = 38$. Thus the eccentricity of the given hyperbola is

$$e = \frac{c}{a} = \frac{\sqrt{38}}{\sqrt{2}} = \sqrt{19} \approx 4.4.$$

We conclude that the hyperbola is one whose branches open widely as in Figure 9.3.7(b). ≡

☐ **Applications of the Hyperbola** The hyperbola has several important applications involving sounding techniques. In particular, several navigational systems utilize hyperbolas as follows. Two fixed radio transmitters at a known distance from each other transmit synchronized signals. The difference in reception times by a navigator determines the difference $2a$ of the distances from the navigator to the two transmitters. This information locates the navigator somewhere on the hyperbola with foci at the transmitters and fixed difference in distances from the foci equal to $2a$. By using two sets of signals obtained from a single master station paired with each of two second stations, the long-range navigation system LORAN locates a ship or plane at the intersection of two hyperbolas. See FIGURE 9.3.8.

FIGURE 9.3.8 The idea behind LORAN

The next example illustrates the use of a hyperbola in another situation involving sounding techniques.

EXAMPLE 6 Locating a Big Blast

The sound of a dynamite blast is heard at different times by two observers at points A and B. Knowing that the speed of sound is approximately 1100 ft/s or 335 m/s, it is determined that the blast occurred 1000 meters closer to point A than to point B. If A and B are 2600 meters apart, show that the location of the blast lies on a branch of a hyperbola. Find an equation of the hyperbola.

Solution In FIGURE 9.3.9, we have placed the points A and B on the x-axis at $(1300, 0)$ and $(-1300, 0)$, respectively. If $P(x, y)$ denotes the location of the blast, then

$$d(P, B) - d(P, A) = 1000.$$

From the definition of the hyperbola on page 461 and the derivation following it, we see that this is the equation for the right branch of a hyperbola with fixed distance difference $2a = 100$ and $c = 1300$. Thus the equation has the form

$$\frac{x^2}{a^2} - \frac{y^2}{b^2} = 1, \quad \text{where } x \geq 0,$$

or after solving for x,

$$x = a\sqrt{1 + \frac{y^2}{b^2}}.$$

With $a = 500$ and $c = 1300$, $b^2 = (1300)^2 - (500)^2 = (1200)^2$. Substituting in the foregoing equation gives

$$x = 500\sqrt{1 + \frac{y^2}{(1200)^2}} \quad \text{or} \quad x = \frac{5}{12}\sqrt{(1200)^2 + y^2}. \quad \equiv$$

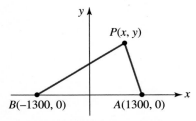

FIGURE 9.3.9 Graph in Example 6

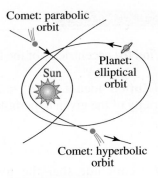

(a) Sonic footprint (b) Cassegrain telescope (c) Orbits around the Sun

FIGURE 9.3.10 Applications of hyperbolas

To find the exact location of the blast in Example 6 we would need another observer hearing the blast at a third point C. Knowing the time between when this observer hears the blast and when the observer at A hears the blast, we find a second hyperbola. The actual point of detonation is a point of intersection of the two hyperbolas.

There are many other applications of the hyperbola. As shown in **FIGURE 9.3.10(a)**, a plane flying at a supersonic speed parallel to level ground leaves a hyperbolic sonic "footprint" on the ground. Like the parabola and ellipse, a hyperbola also possesses a reflecting property. The Cassegrain reflecting telescope shown in Figure 9.3.10(b) utilizes a convex hyperbolic secondary mirror to reflect a ray of light back through a hole to an eyepiece (or camera) behind the parabolic primary mirror. This telescope construction makes use of the fact that a beam of light directed along a line through one focus of a hyperbolic mirror will be reflected on a line through the other focus.

Orbits of objects in the universe can be parabolic, elliptic, or hyperbolic. When an object passes close to the Sun (or a planet), it is not necessarily captured by the gravitational field of the larger body. Under certain conditions, the object picks up a fractional amount of orbital energy of this much larger body and the resulting "slingshot-effect" orbit of the object as it passes the Sun is hyperbolic. See Figure 9.3.10(c).

9.3 Exercises
Answers to selected odd-numbered problems begin on page ANS-25.

In Problems 1–20, find the center, foci, vertices, asymptotes, and eccentricity of the given hyperbola. Graph the hyperbola.

1. $\dfrac{x^2}{16} - \dfrac{y^2}{25} = 1$
2. $\dfrac{x^2}{4} - \dfrac{y^2}{4} = 1$
3. $\dfrac{y^2}{64} - \dfrac{x^2}{9} = 1$
4. $\dfrac{y^2}{6} - 4x^2 = 1$
5. $4x^2 - 16y^2 = 64$
6. $5x^2 - 5y^2 = 25$
7. $y^2 - 5x^2 = 20$
8. $9x^2 - 16y^2 + 144 = 0$
9. $\dfrac{(x-5)^2}{4} - \dfrac{(y+1)^2}{49} = 1$
10. $\dfrac{(x+2)^2}{10} - \dfrac{(y+4)^2}{25} = 1$
11. $\dfrac{(y-4)^2}{36} - x^2 = 1$
12. $\dfrac{(y-\frac{1}{4})^2}{4} - \dfrac{(x+3)^2}{9} = 1$
13. $25(x-3)^2 - 5(y-1)^2 = 125$
14. $10(x+1)^2 - 2(y-\frac{1}{2})^2 = 100$
15. $8(x+4)^2 - 5(y-7)^2 + 40 = 0$
16. $9(x-1)^2 - 81(y-2)^2 = 9$
17. $5x^2 - 6y^2 - 20x + 12y - 16 = 0$
18. $16x^2 - 25y^2 - 256x - 150y + 399 = 0$
19. $4x^2 - y^2 - 8x + 6y - 4 = 0$
20. $2y^2 - 9x^2 - 18x + 20y + 5 = 0$

In Problems 21–44, find an equation of the hyperbola that satisfies the given conditions.

21. Foci $(\pm 5, 0)$, $a = 3$
22. Foci $(\pm 10, 2)$, $b = 2$
23. Foci $(0, \pm 4)$, one vertex $(0, -2)$
24. Foci $(0, \pm 3)$, one vertex $\left(0, -\frac{3}{2}\right)$
25. Foci $(\pm 4, 0)$, length of transverse axis 6
26. Foci $(0, \pm 7)$, length of transverse axis 10
27. Center $(0, 0)$, one vertex $\left(0, \frac{5}{2}\right)$, one focus $(0, -3)$
28. Center $(0, 0)$, one vertex $(7, 0)$, one focus $(9, 0)$
29. Center $(0, 0)$, one vertex $(-2, 0)$, one focus $(-3, 0)$
30. Center $(0, 0)$, one vertex $(1, 0)$, one focus $(5, 0)$
31. Vertices $(0, \pm 8)$, asymptotes $y = \pm 2x$
32. Foci $(0, \pm 3)$, asymptotes $y = \pm\frac{3}{2}x$
33. Vertices $(\pm 2, 0)$, asymptotes $y = \pm\frac{4}{3}x$
34. Foci $(\pm 5, 0)$, asymptotes $y = \pm\frac{3}{5}x$
35. Center $(1, -3)$, one focus $(1, -6)$, one vertex $(1, -5)$
36. Center $(2, 3)$, one focus $(0, 3)$, one vertex $(3, 3)$
37. Foci $(-4, 2)$, $(2, 2)$, one vertex $(-3, 2)$
38. Vertices $(2, 5)$, $(2, -1)$, one focus $(2, 7)$
39. Vertices $(\pm 2, 0)$, passing through $(2\sqrt{3}, 4)$
40. Vertices $(0, \pm 3)$, passing through $\left(\frac{16}{5}, 5\right)$
41. Center $(-1, 3)$, one vertex $(-1, 4)$, passing through $(-5, 3 + \sqrt{5})$
42. Center $(3, -5)$, one vertex $(3, -2)$, passing through $(1, -1)$
43. Center $(2, 4)$, one vertex $(2, 5)$, one asymptote $2y - x - 6 = 0$
44. Eccentricity $\sqrt{10}$, endpoints of conjugate axis $(-5, 4)$, $(-5, 10)$

45. Three points are located at $A(-10, 16)$, $B(-2, 0)$, and $C(2, 0)$, where the units are kilometers. An artillery gun is known to lie on the line segment between A and C, and using sounding techniques it is determined that the gun is 2 km closer to B than to C. Find the point where the gun is located.

46. It can be shown that a ray of light emanating from one focus of a hyperbola will be reflected back along the line from the opposite focus. See FIGURE 9.3.11. A light ray from the left focus of the hyperbola $x^2/16 - y^2/20 = 1$ strikes the hyperbola at $(-6, -5)$. Find an equation of the reflected ray.

47. Find an equation of the hyperbola with foci $(0, -2)$ and $(8, 4)$ and fixed distance difference $2a = 8$. [*Hint*: See Problem 51 in Exercises 9.2.]

48. **Focal Width** The **focal width** of a hyperbola is the length of a focal chord, that is, a line segment, perpendicular to the line containing the transverse axis and through a focus, with endpoints on the hyperbola. See FIGURE 9.3.12.
    (a) Find the focal width of the hyperbola $x^2/4 - y^2/9 = 1$.
    (b) Show that, in general, the focal width of the hyperbola $x^2/a^2 - y^2/b^2 = 1$ is $2b^2/a$.

For Discussion

49. **Sub Hunting** Two sonar detectors are located at a distance d from one another. Suppose that a sound (such as a sneeze aboard a submarine) is heard at the two detectors with a time delay h between them. See FIGURE 9.3.13. Assume that sound travels in straight lines to the two detectors with speed v.
    (a) Explain why h cannot be larger than d/v.
    (b) Explain why, for given values of d, v, and h, the source of the sound can be determined to lie on one branch of a hyperbola. [*Hint*: Where do you suppose that the foci might be?]

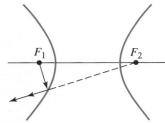

FIGURE 9.3.11 Reflecting property in Problem 46

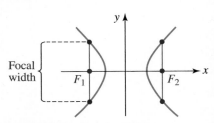

FIGURE 9.3.12 Focal width in Problem 48

FIGURE 9.3.13 Sonic detectors in Problem 49

(c) Find an equation for the hyperbola in part (b), assuming that the detectors are at the points $(0, d/2)$ and $(0, -d/2)$. Express the answer in the standard form $y^2/a^2 - x^2/b^2 = 1$.

50. The hyperbolas

$$\frac{x^2}{a^2} - \frac{y^2}{b^2} = 1 \quad \text{and} \quad \frac{y^2}{b^2} - \frac{x^2}{a^2} = 1$$

are said to be **conjugates** of each other.

(a) Find the equation of the hyperbola that is conjugate to

$$\frac{x^2}{25} - \frac{y^2}{144} = 1.$$

(b) Discuss how the graphs of conjugate hyperbolas are related.

51. A **rectangular hyperbola** is one for which the asymptotes are perpendicular.
    (a) Show that $y^2 - x^2 + 5y + 3x = 1$ is a rectangular hyperbola.
    (b) Which of the hyperbolas given in Problems 1–20 are rectangular?

9.4 Rotation of Axes

☰ **Introduction** In the introduction to Section 9.1 we pointed out that equations of conic sections are special cases of the general second-degree equation

$$Ax^2 + Bxy + Cy^2 + Dx + Ey + F = 0. \tag{1}$$

When $B = 0$, we obtain the standard forms of equations of circles, parabolas, ellipses, and hyperbolas studied in preceding sections from equations of the form

$$Ax^2 + Cy^2 + Dx + Ey + F = 0 \tag{2}$$

by completion of the square. Since each standard form is a second-degree equation we must have $A \neq 0$ or $C \neq 0$ in (2).

In addition to the familiar conics, equation (1) could also represent two intersecting lines, one line, a single point, two parallel lines, or no graph at all. These are referred to as the **degenerate cases** of equation (1). See Problems 29–31 in Exercises 9.4.

When $B \neq 0$, we will see in the discussion that follows that it is possible to remove the xy-term in equation (1) by a **rotation of axes**. In other words, it is always possible to select the angle of rotation θ so that any equation of the form (1) can be transformed into an equation in x' and y' with no $x'y'$-term:

$$A'(x')^2 + C'(y')^2 + D'x' + E'y' + F' = 0. \tag{3}$$

Proceeding as we would for equation (2), we can recast (3) into a standard form and thereby enabling us to identify the conic and graph it in the $x'y'$-coordinate plane.

☐ **Rotation of Axes** We begin with an xy-coordinate system with origin O and rotate the x- and y-axes about O through an angle θ, as shown in FIGURE 9.4.1. In their rotated position we will denote the x- and y-axes by the symbols x' and y', respectively. In this manner, any point P in the plane has two sets of coordinates: (x, y) in terms of the original xy-coordinate system and (x', y') in terms of the $x'y'$-coordinate system. It is a

FIGURE 9.4.1 Rotated axes in red

straightforward exercise in trigonometry to show that the xy-coordinates of P can be converted to the new $x'y'$-coordinates by

$$x' = x\cos\theta + y\sin\theta$$
$$y' = -x\sin\theta + y\cos\theta. \quad (4)$$

Conversely, by solving (4) for x and y we obtain a set of equations that allow us to convert the $x'y'$-coordinates of P to xy-coordinates:

$$x = x'\cos\theta - y'\sin\theta$$
$$y = x'\sin\theta + y'\cos\theta. \quad (5)$$

See Problem 32 in Exercises 9.4. Of the two sets of **rotation equations**, (4) and (5), the set given in (5) is the more important for our purposes.

EXAMPLE 1   Coordinates

Suppose that the x-axis is rotated by an angle of $60°$. Find
(a) the $x'y'$-coordinates of the point whose xy-coordinates are $(4, 4)$,
(b) the xy-coordinates of the point whose $x'y'$-coordinates are $(3, -5)$.

Solutions (a) The point $(4, 4)$ is indicated by the black dot in FIGURE 9.4.2. With $\theta = 60°$, $x = 4$, and $y = 4$ the equations in (4) give

$$x' = 4\cos 60° + 4\sin 60° = 4\left(\frac{1}{2}\right) + 4\left(\frac{\sqrt{3}}{2}\right) = 2 + 2\sqrt{3}$$

$$y' = -4\sin 60° + 4\cos 60° = -4\left(\frac{\sqrt{3}}{2}\right) + 4\left(\frac{1}{2}\right) = 2 - 2\sqrt{3}.$$

The $x'y'$-coordinates of $(4, 4)$ are $(2 + 2\sqrt{3}, 2 - 2\sqrt{3})$ or approximately $(5.46, -1.46)$.
(b) The point $(3, -5)$ is indicated by the red dot in Figure 9.4.2. With $\theta = 60°, x' = 3$, and $y' = -5$ the equations in (5) give

$$x = 3\cos 60° - (-5)\sin 60° = 3\left(\frac{1}{2}\right) + 5\left(\frac{\sqrt{3}}{2}\right) = \frac{3 + 5\sqrt{3}}{2}$$

$$y = 3\sin 60° + (-5)\cos 60° = 3\left(\frac{\sqrt{3}}{2}\right) - 5\left(\frac{1}{2}\right) = \frac{3\sqrt{3} - 5}{2}.$$

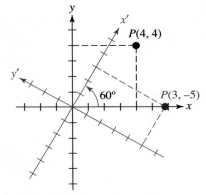

FIGURE 9.4.2 Rotated axes in Example 1

Thus the xy-coordinates of $(3, -5)$ are $\left(\frac{1}{2}(3 + 5\sqrt{3}), \frac{1}{2}(3\sqrt{3} - 5)\right)$ or approximately $(5.83, 0.10)$.

Using the rotation equations in (5) it is possible to determine an angle of rotation θ so that any equation of form (1) where $B \neq 0$, can be transformed into an equation in x' and y' with no $x'y'$-term. Substituting the equations in (5) for x and y in (1),

$$A(x'\cos\theta - y'\sin\theta)^2 + B(x'\cos\theta - y'\sin\theta)(x'\sin\theta + y'\cos\theta) + C(x'\sin\theta + y'\cos\theta)^2$$
$$+ D(x'\cos\theta - y'\sin\theta) + E(x'\sin\theta + y'\cos\theta) + F = 0,$$

and simplifying, we discover that the resulting equation can be written

$$A'(x')^2 + B'x'y' + C'(y')^2 + D'x' + E'y' + F' = 0. \quad (6)$$

The coefficients A', B', C', D', E', F' depend on A, B, C, D, E, F and on $\sin\theta, \cos\theta$. In particular, the coefficient of the $x'y'$-term in (6) is

$$B' = 2(C - A)\sin\theta\cos\theta + B(\cos^2\theta - \sin^2\theta).$$

Thus, in order to eliminate the $x'y'$-term in (6), we can select any angle θ so that $B' = 0$, that is,

$$2(C - A)\sin\theta\cos\theta + B(\cos^2\theta - \sin^2\theta) = 0.$$

See (14) and (15) in Section 7.4. ▶ By the double-angle formulas for sine and cosine, the last equation is equivalent to

$$(C - A)\sin 2\theta + B\cos 2\theta = 0 \quad \text{or} \quad \cot 2\theta = \frac{A - C}{B}.$$

We have proved the following result.

THEOREM 9.4.1 Elimination of the xy-Term

The xy-term can be eliminated from the general second-degree equation

$$Ax^2 + Bxy + Cy^2 + Dx + Ey + F = 0,$$

where $B \neq 0$, by a rotation of axes through an angle θ that satisfies

$$\cot 2\theta = \frac{A - C}{B}. \tag{7}$$

Although equation (7) possesses an infinite number of solutions, it suffices to take a solution such that $0° < \theta < 90°$. This inequality comes from the three cases:

- If $\cot 2\theta = 0$, then $2\theta = 90°$ and therefore $\theta = 45°$.
- If $\cot 2\theta > 0$, then we can take $0° < 2\theta < 90°$ or $0° < \theta < 45°$.
- If $\cot 2\theta < 0$, then $90° < 2\theta < 180°$ or $45° < \theta < 90°$.

EXAMPLE 2 An $x'y'$-Equation

The simple equation $xy = 1$ can be written in terms of x' and y' without the product xy. By comparing $xy = 1$ with equation (1) we see that $A = 0$, $C = 0$, and $B = 1$. Thus (7) shows $\cot 2\theta = 0$. Using $\theta = 45°$, $\cos 45° = \sin 45° = \sqrt{2}/2$, the rotation equations in (5) become

$$x = x'\cos 45° - y'\sin 45° = \frac{\sqrt{2}}{2}(x' - y')$$

$$y = x'\sin 45° + y'\cos 45° = \frac{\sqrt{2}}{2}(x' + y').$$

Substituting these expressions for x and y in $xy = 1$, we obtain

$$\frac{\sqrt{2}}{2}(x' - y') \cdot \frac{\sqrt{2}}{2}(x' + y') = 1$$

or

$$\frac{x'^2}{2} - \frac{y'^2}{2} = 1.$$

We recognize this as the standard equation of a hyperbola with vertices on the x'-axis at the $x'y'$-points $(\pm\sqrt{2}, 0)$. The asymptotes of the hyperbola are $y' = -x'$ and $y' = x'$ (which are simply the original x- and y-axes). See **FIGURE 9.4.3**. ≡

FIGURE 9.4.3 Rotated axes in Example 2

We do not actually have to determine the value of θ if we simply want to obtain an $x'y'$-equation to identify the conic. When $\cot 2\theta \neq 0$, it is clear that in order to use (5) we only need to know both $\sin\theta$ and $\cos\theta$. To do this, we use the value of $\cot 2\theta$ to find the value of $\cos 2\theta$ and then use the half-angle formulas

$$\sin\theta = \sqrt{\frac{1 - \cos 2\theta}{2}} \quad \text{and} \quad \cos\theta = \sqrt{\frac{1 + \cos 2\theta}{2}}. \qquad (8)$$

However, if we wish to sketch the conic, then we need to find θ to determine the position of the x' and y' axes. The next example illustrates these ideas.

EXAMPLE 3 Eliminating the *xy*-Term

After a suitable rotation of axes, identify and sketch the graph of

$$5x^2 + 3xy + y^2 = 44.$$

Solutions With $A = 5$, $B = 3$, and $C = 1$, (7) shows that the desired rotation angle satisfies

$$\cot 2\theta = \frac{5 - 1}{3} = \frac{4}{3}. \qquad (9)$$

From the discussion following (7), since $\cot 2\theta$ is positive, we can choose 2θ such that $0 < \theta < 45°$. From the identity $1 + \cot^2 2\theta = \csc^2 2\theta$ we find $\csc 2\theta = \frac{5}{3}$ and so $\sin 2\theta = \frac{3}{5}$. Then $\cot 2\theta = \cos 2\theta/\sin 2\theta = \frac{4}{3}$ yields $\cos 2\theta = \frac{4}{5}$. Now from the half-angle formulas in (8), we find

$$\sin\theta = \sqrt{\frac{1 - \cos 2\theta}{2}} = \sqrt{\frac{1 - \frac{4}{5}}{2}} = \frac{1}{\sqrt{10}}$$

$$\cos\theta = \sqrt{\frac{1 + \cos 2\theta}{2}} = \sqrt{\frac{1 + \frac{4}{5}}{2}} = \frac{3}{\sqrt{10}}. \qquad (10)$$

Thus, the equations in (5) become

$$x = \frac{3}{\sqrt{10}}x' - \frac{1}{\sqrt{10}}y' = \frac{1}{\sqrt{10}}(3x' - y')$$

$$y = \frac{1}{\sqrt{10}}x' + \frac{3}{\sqrt{10}}y' = \frac{1}{\sqrt{10}}(x' + 3y').$$

Substituting these into the given equation, we have

$$5\left(\frac{1}{\sqrt{10}}\right)^2(3x' - y')^2 + 3\frac{1}{\sqrt{10}}(3x' - y') \cdot \frac{1}{\sqrt{10}}(x' + 3y') + \left(\frac{1}{\sqrt{10}}\right)^2(x' + 3y')^2 = 44$$

$$\frac{5}{10}(9x'^2 - 6x'y' + y'^2) + \frac{3}{10}(3x'^2 + 8x'y' - 3y'^2) + \frac{1}{10}(x'^2 + 6x'y' + 9y'^2) = 44$$

$$45x'^2 - 30x'y' + 5y'^2 + 9x'^2 + 24x'y' - 9y'^2 + x'^2 + 6x'y' + 9y'^2 = 440.$$

The last equation simplifies to

$$\frac{x'^2}{8} + \frac{y'^2}{88} = 1. \qquad (11)$$

We recognize this as the standard equation of an ellipse. Now from (10) we have $\sin\theta = 1/\sqrt{10}$ and so with the aid of a calculator we find $\theta \approx 18.4°$. This rotation angle is shown in **FIGURE 9.4.4** and we use the new axes to sketch the ellipse.

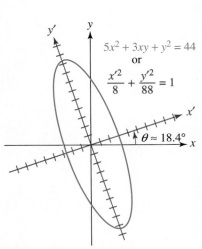

FIGURE 9.4.4 Rotated axes in Example 3

Don't be misled by the last two examples. After using the rotation equations in (5) the conic section may not be immediately identifiable without some extra work. For example, after an appropriate rotation of axes, the equation

$$11x^2 + 16\sqrt{2}xy + 19y^2 - 24\sqrt{3}x - 24\sqrt{6}y + 45 = 0 \quad (12)$$

is transformed into

$$9x'^2 - 24x' + y'^2 + 15 = 0.$$

After completing the square in x', $(3x' - 4)^2 + y'^2 = 1$, we recognize that the equation (12) defines an ellipse.

☐ **Identifying Conics Without Rotation** If for the sake of discussion, we simply wish to identify a conic defined by an equation of the form given in (1), we can do so by examining its coefficients. All we need do is calculate the **discriminant** $B^2 - 4AC$ of the equation.

THEOREM 9.4.2 Identifying Conics

Excluding the degenerate cases, the graph of the second-degree equation (1) is

(*i*) a **parabola** when $B^2 - 4AC = 0$,
(*ii*) an **ellipse** when $B^2 - 4AC < 0$, or
(*iii*) a **hyperbola** when $B^2 - 4AC > 0$.

EXAMPLE 4   Identification

Identify the conic defined by the given equation
(a) $9x^2 + 12xy + 4y^2 + 2x - 3y = 0$ **(b)** $3x^2 - 5y^2 + 8x - y + 2 = 0$

Solutions **(a)** With $A = 9$, $B = 12$, $C = 4$ the discriminant

$$B^2 - 4AC = (12)^2 - 4(9)(4) = 144 - 144 = 0$$

indicates that the equation defines a parabola.
(b) With $A = 3$, $B = 0$, $C = -5$ the discriminant is

$$B^2 - 4AC = (0)^2 - 4(3)(-5) = 60 > 0.$$

The equation defines a hyperbola. ≡

9.4   Exercises   Answers to selected odd-numbered problems begin on page ANS-25.

In Problems 1–4, use (4) to find the $x'y'$-coordinates of the given xy-point. Use the specified angle of rotation θ.

1. $(6, 2), \theta = 45°$ **2.** $(-2, 8), \theta = 30°$
3. $(-1, -1), \theta = 60°$ **4.** $(5, 3), \theta = 15°$

In Problems 5–10, use (5) to find the xy-coordinates of the given $x'y'$-point. Use the specified angle of rotation θ.

5. $(2, -8), \theta = 30°$ **6.** $(-5, 7), \theta = 45°$
7. $(0, 4), \theta = \dfrac{\pi}{2}$ **8.** $(3, 0), \theta = \dfrac{\pi}{3}$
9. $(4, 6), \theta = 15°$ **10.** $(1, 1), \theta = 75°$

In Problems 11–16, use rotation of axes to eliminate the xy-term in the given equation. Identify the conic and graph.

11. $x^2 + xy + y^2 = 4$
12. $2x^2 - 3xy - 2y^2 = 5$
13. $x^2 - 2xy + y^2 = 8x + 8y$
14. $3x^2 + 4xy = 16$
15. $x^2 + 4xy - 2y^2 - 6 = 0$
16. $x^2 + 4xy + 4y^2 = 16\sqrt{5}x - 8\sqrt{5}y$

In Problems 17–20, use rotation of axes to eliminate the xy-term in the given equation. Identify the conic.

17. $4x^2 - 4xy + 7y^2 + 12x + 6y - 9 = 0$
18. $-x^2 + 6\sqrt{3}xy + 5y^2 - 8\sqrt{3}x + 8y = 12$
19. $8x^2 - 8xy + 2y^2 + 10\sqrt{5}x = 5$
20. $x^2 - xy + y^2 - 4x - 4y = 20$

21. Given $3x^2 + 2\sqrt{3}xy + y^2 + 2x - 2\sqrt{3}y = 0$.
    (a) By rotation of axes show that the graph of the equation is a parabola.
    (b) Find the $x'y'$-coordinates of the focus. Use this information to find the xy-coordinates of the focus.
    (c) Find an equation of the directrix in terms of the $x'y'$-coordinates. Use this information to find an equation of the directrix in terms of the xy-coordinates.

22. Given $13x^2 - 8xy + 7y^2 = 30$.
    (a) By rotation of axes show that the graph of the equation is an ellipse.
    (b) Find the $x'y'$-coordinates of the foci. Use this information to find the xy-coordinates of the foci.
    (c) Find the xy-coordinates of the vertices.

In Problems 23–28, use the discriminant to determine the type of graph without actually graphing.

23. $x^2 - 3xy + y^2 = 5$
24. $2x^2 - 2xy + 2y^2 = 1$
25. $4x^2 - 4xy + y^2 - 6 = 0$
26. $x^2 + \sqrt{3}xy - \frac{1}{2}y^2 = 0$
27. $x^2 + xy + y^2 - x + 2y + 1 = 0$
28. $3x^2 + 2\sqrt{3}xy + y^2 - 2x + 2\sqrt{3}y - 4 = 0$

For Discussion

29. In (2), show that if A and C have the same signs, then the graph of the equation is either an ellipse, a circle, or a point, or does not exist. Give an example of each type of equation.
30. In (2), show that if A and C have opposite signs, then the graph of the equation is either a hyperbola or a pair of intersecting lines. Give an example of each type of equation.
31. In (2), show that if either $A = 0$ or $C = 0$, then the graph of the equation is either a parabola, two parallel lines, or one line, or does not exist. Give an example of each type of equation.
32. (a) Use FIGURE 9.4.5 to show that
$$x = r\cos\phi, \quad y = r\sin\phi$$
and
$$x' = r\cos(\phi - \theta), \quad y' = r\sin(\phi - \theta).$$
(b) Use the results from part (a) to derive the rotation equations in (4).
(c) Use (4) to find the rotation equations in (5).

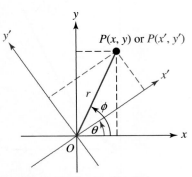

FIGURE 9.4.5 Rotated axes in Problem 32

9.5 Parametric Equations

Introduction Rectangular equations and polar equations are not the only, and often not the most convenient, ways of describing curves in the coordinate plane. In this section we will consider a different way of representing a curve that is important in many applications. Let's consider one example. The motion of a particle along a curve, in contrast to a straight line, is called *curvilinear motion*. If it is assumed that a golf ball is hit off the ground, perfectly straight (no hook or slice), and that its path stays in a coordinate plane as shown in FIGURE 9.5.1, then it can be shown that its x- and y-coordinates at time t are given by

$$x = (v_0 \cos \theta_0)t, \quad y = -\tfrac{1}{2}gt^2 + (v_0 \sin \theta_0)t, \tag{1}$$

where θ_0 is the launch angle, v_0 is its initial velocity, and $g = 32$ ft/s² is the acceleration due to gravity. These equations, which give the golf ball's position in the coordinate plane at time t, are said to be **parametric equations**. The third variable t in (1) is called a **parameter** and is restricted to some interval defined by $0 \le t \le T$, where $t = 0$ gives the origin $(0, 0)$, and $t = T$ is the time the ball hits the ground.

In general, a curve in a coordinate plane can be *defined* in terms of parametric equations.

FIGURE 9.5.1 Fore!

DEFINITION 9.5.1 Plane Curve

A **plane curve** is a set C of ordered pairs $(f(t), g(t))$, where f and g are functions defined on a common interval I. The equations

$$x = f(t), \quad y = g(t), \quad \text{for } t \text{ in } I,$$

are called **parametric equations** for C. The variable t is called a **parameter**.

It is also common practice to refer to $x = f(t), y = g(t)$, for t in I, as a **parameterization** for C.

The **graph** of a plane curve C is the set of all points (x, y) in the coordinate plane corresponding to the ordered pairs $(f(t), g(t))$. Hereafter, we will refer to a plane curve as a **curve** or as a **parameterized curve**.

EXAMPLE 1 Graph of a Parametric Curve

Graph the curve C that has the parametric equations

$$x = t^2, \quad y = t^3, \quad -1 \le t \le 2.$$

Solution As shown in the accompanying table, for any choice of t in the interval $[-1, 2]$, we obtain a single ordered pair (x, y). By connecting the points with a curve, we obtain the graph in FIGURE 9.5.2.

FIGURE 9.5.2 Curve in Example 1

t	-1	$-\tfrac{1}{2}$	0	$\tfrac{1}{2}$	1	$\tfrac{3}{2}$	2
x	1	$\tfrac{1}{4}$	0	$\tfrac{1}{4}$	1	$\tfrac{9}{4}$	4
y	-1	$-\tfrac{1}{8}$	0	$\tfrac{1}{8}$	1	$\tfrac{27}{8}$	8

In Example 1, if we think in terms of motion and t as time, then as t increases from -1 to 2, a point P defined as (t^2, t^3) starts from $(1, -1)$, advances up the lower branch to the origin $(0, 0)$, passes to the upper branch, and finally stops at $(4, 8)$. In general, as we plot points corresponding to *increasing values* of the parameter, the curve C is traced out by $(f(t), g(t))$ in a certain *direction* indicated by the arrowheads on the curve in Figure 9.5.2. This direction is called the **orientation** of the curve C.

A parameter need have no relation to time. When the interval I over which f and g in (1) are defined is a closed interval $[a, b]$, we say that $(f(a), g(a))$ is the **initial point** of the curve C and that $(f(b), g(b))$ is its **terminal point**. In Example 1 the initial point is $(1, -1)$ and the terminal point is $(4, 8)$. If the terminal point is the same as the initial point, that is,

$$(f(a), g(a)) = (f(b), g(b)),$$

then C is a **closed curve**. If C is closed but does not cross itself, then it is called a **simple closed curve**. In FIGURE 9.5.3, A and B represent the initial and terminal points, respectively.

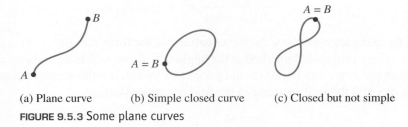

(a) Plane curve (b) Simple closed curve (c) Closed but not simple

FIGURE 9.5.3 Some plane curves

The next example illustrates a simple closed curve.

EXAMPLE 2   A Parameterization of a Circle

Find a parameterization for the circle $x^2 + y^2 = a^2$.

Solution The circle has center at the origin and radius a. If t represents the central angle, that is, an angle with vertex at the origin and initial side coinciding with the positive x-axis, then as shown in FIGURE 9.5.4 the equations

$$x = a\cos t, \quad y = a\sin t, \quad 0 \le t \le 2\pi \tag{2}$$

give every point P on the circle. For example, at $t = 0$ we get $x = a$ and $y = 0$; in other words, the initial point is $(a, 0)$. The terminal point corresponds to $t = 2\pi$ and is also $(a, 0)$. Since the initial and terminal points are the same, this proves the obvious: the curve C defined by the parametric equations in (2) is a closed curve. Note the orientation of C in Figure 9.5.4; as t increases from 0 to 2π, the point P traces out C in a counterclockwise direction.

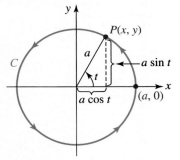

FIGURE 9.5.4 Circle in Example 2

In Example 2, the upper *semicircle* $x^2 + y^2 = a^2$, $0 \le y \le a$, is defined parametrically by restricting the parameter interval to $[0, \pi]$,

$$x = a\cos t, \quad y = a\sin t, \quad 0 \le t \le \pi.$$

Observe that when $t = \pi$, the terminal point is now $(-a, 0)$. On the other hand, if we wish to describe *two* complete counterclockwise revolutions around the circle, we again modify the parameter interval by writing

$$x = a\cos t, \quad y = a\sin t, \quad 0 \le t \le 4\pi.$$

☐ **Eliminating the Parameter** Given a set of parametric equations, we sometimes desire to eliminate or clear the parameter to obtain a rectangular equation for the curve. There is no well-defined method of eliminating the parameter; the method is dictated by the parametric equations. For example, to eliminate the parameter in (2), we simply square x and y and add the two equations:

$$x^2 + y^2 = a^2\cos^2 t + a^2\sin^2 t \quad \text{implies} \quad x^2 + y^2 = a^2$$

since $\sin^2 t + \cos^2 t = 1$. In the next example, we illustrate a substitution method.

EXAMPLE 3　　Eliminating the Parameter

(a) In the first equation in (1) we can solve for t in terms of x and then substitute $t = x/(v_0 \cos \theta_0)$ into the second equation. This gives an equation involving only the variables x and y:

$$y = -\frac{g}{2(v_0 \cos \theta_0)^2} x^2 + (\tan \theta_0) x.$$

Since v_0, θ_0, and g are constants, the last equation has the form $y = ax^2 + bx$ and so the trajectory of any projectile launched at the angle $0 < \theta_0 < \pi/2$ is a parabolic arc.

(b) In Example 1, we can eliminate the parameter by solving the second equation for t in terms of y and then substituting in the first equation. We find

$$t = y^{1/3} \quad \text{and so} \quad x = (y^{1/3})^2 = y^{2/3}.$$

The curve shown in Figure 9.5.2 is only a portion of the graph of $x = y^{2/3}$. For $-1 \leq t \leq 2$, we have correspondingly $-1 \leq y \leq 8$. Thus, a rectangular equation for the curve in Example 1 is given by $x = y^{2/3}$, $-1 \leq y \leq 8$.　　≡

▶ A curve C can have many different parameterizations.

A curve C can have more than one parameterization. For example, an alternative parameterization for the circle in Example 2 is

$$x = a\cos 2t, \quad y = a\sin 2t, \quad 0 \leq t \leq \pi.$$

Note that the parameter interval is now $[0, \pi]$. We see that as t increases from 0 to π, the new angle $2t$ increases from 0 to 2π.

EXAMPLE 4　　Alternative Parameterizations

(a) Consider the curve C that has the parametric equations $x = t$, $y = 2t^2$, $-\infty < t < \infty$. We can eliminate the parameter by using $t = x$ and substituting in $y = 2t^2$. This gives the rectangular equation $y = 2x^2$ that we recognize as a parabola. Moreover, since $-\infty < t < \infty$ is equivalent to $-\infty < x < \infty$, the point $(t, 2t^2)$ traces out the complete parabola $y = 2x^2$, $-\infty < x < \infty$.

(b) An alternative parameterization of the curve C in part (a) is given by $x = t^3/4$, $y = t^6/8$, $-\infty < t < \infty$. Using $t^3 = 4x$ and substituting in $y = t^6/8$ or $y = (t^3 \cdot t^3)/8$ gives $y = (4x)^2/8 = 2x^2$. Moreover, $-\infty < t < \infty$ implies $-\infty < t^3 < \infty$ and so $-\infty < x < \infty$.　　≡

We note in Example 4 that a point on the curve C need not correspond to the same value of the parameter in each set of parametric equations for C. For example, $(1, 2)$ is obtained for $t = 1$ in $x = t$, $y = 2t^2$, but $t = \sqrt[3]{4}$ yields $(1, 2)$ in $x = t^3/4$, $y = t^6/8$.

EXAMPLE 5 Example 4 Revisited

One has to be careful when working with parametric equations. Eliminating the parameter in $x = t^2$, $y = 2t^4$, $-\infty < t < \infty$, would seem to yield the same parabola $y = 2x^2$ as in Example 4. However, this is *not* the case because for any value of t satisfying $-\infty < t < \infty$, we have $t^2 \geq 0$ and so $x \geq 0$. In other words, the set of equations $x = t^2$, $y = 2t^4$, $-\infty < t < \infty$ is a parametric representation of only the right-hand branch of the parabola, that is, $y = 2x^2$, $0 \leq x < \infty$.

◀ Proceed with caution when eliminating the paramater.

EXAMPLE 6 Eliminating the Parameter

Consider the curve C defined parametrically by

$$x = \sin t, \quad y = \cos 2t, \quad 0 \leq t \leq \pi/2.$$

Eliminate the parameter and obtain a rectangular equation for C.

Solution Using the double-angle formula $\cos 2t = \cos^2 t - \sin^2 t$, we can write

$$\begin{aligned} y &= \cos^2 t - \sin^2 t \\ &= (1 - \sin^2 t) - \sin^2 t \\ &= 1 - 2\sin^2 t \quad \leftarrow \text{substitute } \sin t = x \\ &= 1 - 2x^2. \end{aligned}$$

Now the curve C described by the parametric equations does not consist of the complete parabola, that is, $y = 1 - 2x^2$, $-\infty < x < \infty$. See **FIGURE 9.5.5(a)**. For $0 \leq t \leq \pi/2$ we have $0 \leq \sin t \leq 1$ and $-1 \leq \cos 2t \leq 1$. This means that C is only that portion of the parabola for which the coordinates of a point $P(x, y)$ satisfy $0 \leq x \leq 1$ and $-1 \leq y \leq 1$. The curve C, along with its orientation, is shown in Figure 9.5.5(b). A rectangular equation for C is $y = 1 - 2x^2$ with the restricted domain $0 \leq x \leq 1$.

(a) $y = 1 - 2x^2$

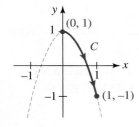

(b) $x = \sin t, y = \cos 2t$, $0 \leq t \leq \pi/2$

FIGURE 9.5.5 Curve C in Example 6

☐ **Intercepts** We can get intercepts of a curve C without finding its rectangular equation. For instance, in Example 6 we can find the x-intercept by finding the value of t in the parameter interval for which $y = 0$. The equation $\cos 2t = 0$ yields $2t = \pi/2$ so that $t = \pi/4$. The corresponding point at which C crosses the x-axis is $(\sqrt{2}/2, 0)$. Similarly, the y-intercept of C is found by solving $x = 0$. From $\sin t = 0$ we immediately conclude $t = 0$ and so the y-intercept is $(0, 1)$.

☐ **Applications of Parametric Equations** Cycloidal curves were a popular topic of study by mathematicians in the seventeenth century. Suppose a point $P(x, y)$, marked on a circle of radius a, is at the origin when its diameter lies along the y-axis. As the circle rolls along the x-axis, the point P traces out a curve C that is called a **cycloid**. See **FIGURE 9.5.6**.*

(a) Circle rolling on x-axis

(b) Point P on the circle traces out this curve

FIGURE 9.5.6 Curve in red is a cycloid

---

*For an animation of the rolling circle go to http://mathworld.wolfram.com/Cycloid.html.

FIGURE 9.5.7 Sliding bead

Two problems were extensively studied in the seventeenth century. Consider a flexible (frictionless) wire fixed at points A and B and a bead free to slide down the wire starting at P. See **FIGURE 9.5.7**. Is there a particular shape of the wire so that, regardless of where the bead starts, the time to slide down the wire to B will be the same? Also, what would the shape of the wire be so that the bead slides from P to B in the shortest time? The so-called **tautochrone** (same time) and **brachistochrone** (least time) were shown to be an inverted half-arch of a cycloid.

EXAMPLE 7 Parameterization of a Cycloid

Find a parameterization for the cycloid shown in Figure 9.5.6(b).

Solution A circle of radius a whose diameter initially lies along the y-axis rolls along the x-axis without slipping. We take as a parameter the angle θ (in radians) through which the circle has rotated. The point $P(x, y)$ starts at the origin, which corresponds to $\theta = 0$. As the circle rolls through an angle θ, its distance from the origin is the arc $PE = \overline{OE} = a\theta$. From **FIGURE 9.5.8** we then see that the x-coordinate of P is

$$x = \overline{OE} - \overline{QE} = a\theta - a\sin\theta.$$

Now the y-coordinate of P is seen to be

$$y = \overline{CE} - \overline{CD} = a - a\cos\theta.$$

Hence, parametric equations for the cycloid are

$$x = a\theta - a\sin\theta, \quad y = a - a\cos\theta.$$

As shown in Figure 9.5.6(a), one arch of a cycloid is generated by one rotation of the circle and corresponds to the parameter interval $0 \leq \theta \leq 2\pi$.

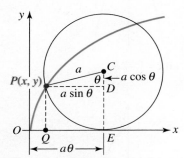

FIGURE 9.5.8 The angle θ is the parameter for the cycloid

☐ **Parameterizations of Rectangular Curves** A curve C described by a continuous function $y = f(x)$ can always be parameterized by simply letting $x = t$. Parametric equations for C are then

$$x = t, \quad y = f(t). \tag{3}$$

EXAMPLE 8 Parameterization of a Rectangular Curve

Find a parameterization of $f(x) = \sqrt{x - 1}$.

Solution If we let $x = t$, then a parameterization of the curve defined by f is

$$x = t, \quad y = \sqrt{t - 1}, \quad t \geq 1.$$

Alternative Solution A curve can have many different parameterizations. If we let $x - 1 = t$, then another parameterization of the curve defined by f is

$$x = t + 1, \quad y = \sqrt{t}, \quad t \geq 0.$$

NOTES FROM THE CLASSROOM

In this section we have focused on **plane curves**, curves C defined parametrically in two dimensions. In the study of multivariable calculus you will see curves and surfaces in three dimensions that are defined by means of parametric equations. For example, a **space curve** C consists of a set of ordered triples $(f(t), g(t), h(t))$, where f, g, and h are defined on a common interval. Parametric equations for

C are $x = f(t), y = g(t), z = h(t)$. For example, the **circular helix** such as shown in FIGURE 9.5.9 is a space curve whose parametric equations are

$$x = a\cos t, \quad y = a\cos t, z = bt, \quad t \geq 0. \tag{4}$$

Surfaces in three dimensions can be represented by a set of parametric equations involving *two* parameters, $x = f(u, v), y = g(u, v), z = h(u, v)$.

For example, the **circular helicoid** shown in FIGURE 9.5.10 arises from the study of minimal surfaces and is defined by the set of parametric equations similar to those in (4):

$$x = u\cos v, \quad y = u\sin v, \quad z = bv,$$

where b is a constant. The circular helicoid has a circular helix as its boundary. You might recognize the helicoid as the model for the rotating curved blade in machinery such as post-hole diggers, ice augers, and snow blowers.

FIGURE 9.5.9 Circular helix

DNA is a double helix

Helical antenna

FIGURE 9.5.10 Circular helicoid

9.5 Exercises
Answers to selected odd-numbered problems begin on page ANS-26.

In Problems 1 and 2, fill in the table for the given set of parametric equations. Find the x- and y-intercepts. Sketch the curve and indicate its orientation.

1. $x = t + 2, y = 3 + \frac{1}{2}t, -\infty < t < \infty$

t	-3	-2	-1	0	1	2	3
x							
y							

2. $x = 2t + 1, y = t^2 + t, -\infty < t < \infty$

t	-3	-2	-1	0	1	2	3
x							
y							

In Problems 3–10, sketch the curve that has the given set of parametric equations.

3. $x = t - 1, y = 2t - 1, \quad -1 \le t \le 5$
4. $x = t^2 - 1, y = 3t, \quad -2 \le t \le 3$
5. $x = \sqrt{t}, y = 5 - t, \quad t \ge 0$
6. $x = t^3 + 1, y = t^2 - 1, \quad -2 \le t \le 2$
7. $x = 3\cos t, y = 5\sin t, \quad 0 \le t \le 2\pi$
8. $x = 3 + 2\sin t, y = 4 + \sin t, \quad -\pi/2 \le t \le \pi/2$
9. $x = e^t, y = e^{3t}, \quad 0 \le t \le \ln 2$
10. $x = -e^t, y = e^{-t}, \quad t \ge 0$

In Problems 11–18, eliminate the parameter from the given set of parametric equations and obtain a rectangular equation that has the same graph.

11. $x = t^2, y = t^4 + 3t^2 - 1$
12. $x = t^3 + t + 4, y = -2(t^3 + t)$
13. $x = \cos 2t, y = \sin t, \quad -\pi/2 \le t \le \pi/2$
14. $x = e^t, y = \ln t, \quad t > 0$
15. $x = t^3, y = 3\ln t, \quad t > 0$
16. $x = \tan t, y = \sec t, \quad -\pi/2 < t < \pi/2$
17. $x = 4\cos t, y = 2\sin t, \quad 0 \le t \le 2\pi$
18. $x = -1 + \cos t, y = 2 + \sin t, \quad 0 \le t \le 2\pi$

In Problems 19–24, graphically show the difference between the given curves.

19. $y = x$ and $x = \sin t, y = \sin t$
20. $y = x^2$ and $x = -\sqrt{t}, y = t$
21. $y = \frac{1}{4}x^2 - 1$ and $x = 2t, y = t^2 - 1, \quad -1 \le t \le 2$
22. $y = -x^2$ and $x = e^t, y = -e^{2t}, \quad t \ge 0$
23. $x^2 - y^2 = 1$ and $x = \cosh t, y = \sinh t$ ← See (2) and (3) in Section 5.5.
24. $y = 2x - 2$ and $x = t^2 - 1, y = 2t^2 - 4$

In Problems 25–28, graphically show the difference between the given curves. Assume that $a > 0$ and $b > 0$.

25. $x = a\cos t, y = a\sin t, \quad 0 \le t \le \pi$
    $x = a\sin t, y = a\cos t, \quad 0 \le t \le \pi$
26. $x = a\cos t, y = b\sin t, \quad a > b, \pi \le t \le 2\pi$
    $x = a\sin t, y = b\cos t, \quad a > b, \pi \le t \le 2\pi$
27. $x = a\cos t, y = a\sin t, \quad -\pi/2 \le t \le \pi/2$
    $x = a\cos 2t, y = a\sin 2t, \quad -\pi/2 \le t \le \pi/2$
28. $x = a\cos\frac{t}{2}, y = a\sin\frac{t}{2}, \quad 0 \le t \le \pi$
    $x = a\cos\left(-\frac{t}{2}\right), y = a\sin\left(-\frac{t}{2}\right), \quad -\pi \le t \le 0$

In Problems 29 and 30, find the x- and y-intercepts of the given curves.

29. $x = t^2 - 2t, y = t + 1, \quad -2 \le t < 4$
30. $x = t^2 + t, y = t^2 + t - 6, \quad -5 \le t < 5$

31. Show that parametric equations for a line through (x_1, y_1) and (x_2, y_2) are
$$x = x_1 + (x_2 - x_1)t, \quad y = y_1 + (y_2 - y_1)t, \quad -\infty < t < \infty.$$
What do these equations represent when $0 \le t \le 1$?

32. (a) Use the result of Problem 31 to find parametric equations of the line through $(-2, 5)$ to $(4, 8)$.
(b) Eliminate the parameter in part (a) to obtain a rectangular equation for the line.
(c) Find parametric equations for the line segment with $(-2, 5)$ as the initial point and $(4, 8)$ as the terminal point.

33. A famous golfer can generate a club head speed of approximately 130 mi/h or $v_0 = 190$ ft/s. If the golf ball leaves the ground at an angle $\theta_0 = 45°$, use (1) to find parametric equations for the path of the ball. What are the coordinates of the ball at $t = 2$ s?

34. Use the parametric equations obtained in Problem 33 to determine:
(a) How long the golf ball is in the air,
(b) Its maximum height, and
(c) The horizontal distance that the golf ball travels.

35. As shown in FIGURE 9.5.11, a piston is attached by means of a rod of length L to a circular crank mechanism of radius r. Parameterize the coordinates of the point P in terms of the angle ϕ shown in the figure.

36. Consider a circle of radius a, which is tangent to the x-axis at the origin. Let B be a point on the horizontal line $y = 2a$ and let the line segment OB cut the circle at point A. As shown in FIGURE 9.5.12, the projection of AB on the vertical gives the line segment BP. Using the angle θ in the figure as a parameter, find parametric equations of the curve traced by the point P as A varies around the circle. The curve, more historically famous than useful, is called the **witch of Agnesi**.*

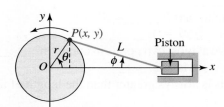

FIGURE 9.5.11 Crank mechanism in Problem 35

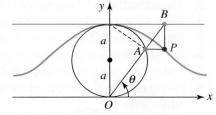

FIGURE 9.5.12 Witch of Agnesi in Problem 36

Calculator/Computer Problems

In Problems 37–42, use a graphing utility to obtain the graph of the given set of parametric equations.

37. $x = 4\sin 2t, \ y = 2\sin t, \quad 0 \le t \le 2\pi$
38. $x = 6\cos 3t, \ y = 4\sin 2t, \quad 0 \le t \le 2\pi$
39. $x = 6\sin 4t, \ y = 4\sin t, \quad 0 \le t \le 2\pi$
40. $x = \cos t + t\sin t, \ y = \sin t - t\cos t, \quad 0 \le t \le 3\pi$
41. $x = 4\cos t - \cos 4t, \ y = 4\sin t - \sin 4t, \quad 0 \le t \le 2\pi$
42. $x = \cos^3 t, \ y = \sin^3 t, \quad 0 \le t \le 2\pi$

---

*No, the curve has nothing to do with witches and goblins. This curve, called *versoria*, which is Latin for a kind of rope, was included in a text on analytic geometry written in 1748 by the Italian mathematician Maria Agnesi. A translator of the text confused *versoria* with the Italian word *versiera*, which means *female goblin*. In English, *female goblin* became a *witch*.

CONCEPTS REVIEW

You should be able to give the meaning of each of the following concepts.

Conic section
Parabola:
   focus
   directrix
   axis
   vertex
Ellipse:
   foci
   center
   major axis
   minor axis
   vertices

Hyperbola:
   foci
   center
   branches
   transverse axis
   vertices
   conjugate axis
   asymptotes
Eccentricity of a conic
Standard form of equations

Rotation of axes:
   rotation equations
Discriminant
Plane curve
Parametric equations
Parameter
Orientation of a curve
Closed curve
Simple closed curve
Elimination of the parameter

CHAPTER 9 Review Exercises

Answers to selected odd-numbered problems begin on page ANS-26.

A. True/False

In Problems 1–20, answer true or false.

1. The axis of the parabola $x^2 = -4y$ is vertical. _____
2. The foci of an ellipse lie on its graph. _____
3. The eccentricity of a parabola is $e = 0$. _____
4. The minor axis of an ellipse bisects the major axis. _____
5. The point $(-2, 5)$ is on the ellipse $x^2/8 + y^2/50 = 1$. _____
6. The graphs of $y = x^2$ and $y^2 - x^2 = 1$ have at most two points in common. _____
7. The eccentricity of the hyperbola $x^2 - y^2 = 1$ is $\sqrt{2}$. _____
8. For an ellipse, the length of the major axis is always greater than the length of the minor axis. _____
9. The vertex and focus are both on the axis of symmetry of a parabola. _____
10. The asymptotes for $(x - h)^2/a^2 - (y - k)^2/b^2 = 1$ must pass through (h, k). _____
11. An ellipse with eccentricity $e = 0.01$ is nearly circular. _____
12. The transverse axis of the hyperbola $x^2/9 - y^2/49 = 1$ is vertical. _____
13. The two hyperbolas $x^2 - y^2/25 = 1$ and $y^2/25 - x^2 = 1$ have the same pair of slant asymptotes. _____
14. The graph of the curve $x = t^2$, $y = t^4 + 1$, $-\infty < t < \infty$ is the same as the graph of $y = x^2 + 1$. _____
15. If P is a point on a parabola, then the perpendicular distance between P and the directrix equals the distance between P and the vertex. _____
16. The curve with parametric equations $x = 1 + \cos t$, $y = 1 + \sin t$, $0 \leq t \leq 2\pi$, is a circle of radius 1 centered at $(1, 1)$. _____
17. The curve with parametric equations $x = e^t$, $y = e^t - 5$, $-\infty < t < \infty$ is the same as the graph of the rectangular equation $y = x - 5$. _____
18. The curve with parametric equations $x = t^3 - 1$, $y = t^2 + 1$, $-\infty < t < \infty$ has no x-intercept. _____
19. The hyperbola $(x - 1)^2 - (y + 1)^2 = 1$ has no y-intercept(s). _____
20. The two sets of parametric equations $x = \cos t$, $y = \sin t$, $0 \leq t \leq 2\pi$ and $x = \sin t$, $y = \cos t$, $0 \leq t \leq 2\pi$ define the same circle. _____

B. Fill in the Blanks

In Problems 1–20, fill in the blanks.

1. An equation in the standard form $y^2 = 4cx$ of a parabola with focus $(5, 0)$ is _____.
2. An equation in the standard form $x^2 = 4cy$ of a parabola through $(2, 6)$ is _____.
3. A rectangular equation of a parabola with focus $(1, -3)$ and directrix $y = -7$ is _____.
4. The directrix and vertex of a parabola are $x = -3$ and $(-1, -2)$, respectively. The focus of the parabola is _____.
5. The focus and directrix of a parabola are $\left(0, \tfrac{1}{4}\right)$ and $y = -\tfrac{1}{4}$, respectively. The vertex of the parabola is _____.
6. The vertex and focus of the parabola $8(x + 4)^2 = y - 2$ are _____.
7. The eccentricity of a parabola is $e =$ _____.
8. The center and vertices of the ellipse $\dfrac{(x-2)^2}{16} + \dfrac{(y+5)^2}{4} = 1$ are _____.
9. The center and vertices of the hyperbola $y^2 - \dfrac{(x+3)^2}{4} = 1$ are _____.
10. The asymptotes of the hyperbola $y^2 - (x-1)^2 = 1$ are _____.
11. The y-intercepts of the hyperbola $y^2 - (x-1)^2 = 1$ are _____.
12. The eccentricity of the hyperbola $x^2 - \tfrac{1}{2}y^2 = 1$ is _____.
13. If the graph of an ellipse is very elongated, then its eccentricity e is close to _____. (Fill in with 0 or 1.)
14. The line segment with endpoints on a hyperbola and lying on the line through its foci is called _____.
15. The equations $x = t + 2$, $y = 3 + \tfrac{1}{2}t$, $-\infty < t < \infty$, are a parametric representation of a _____.
16. The point on the curve defined by the parametric equations $x = -4 + 2\cos t$, $y = 2 + \sin t$, $-\infty < t < \infty$, corresponding to $t = 5\pi/2$ is _____.
17. The x-intercept of the graph of the rectangular equation for the curve defined by the parametric equations $x = e^t$, $y = e^t - 5$, $-\infty < t < \infty$, is _____.
18. Because $B^2 - 4AC$ _____ (fill in with < 0, $= 0$, or > 0), the conic $3x^2 - xy - y^2 + 1 = 0$ is a _____.
19. The y-intercepts of the curve defined by the parametric equations $x = t^2 - 4$, $y = t^3 - 3t$, $-\infty < t < \infty$, are _____.
20. The center of a hyperbola with asymptotes $y = -\tfrac{5}{4}x + \tfrac{3}{2}$ and $y = \tfrac{5}{4}x + \tfrac{13}{2}$ is _____.

C. Review Exercises

In Problems 1–4, find the vertex, focus, directrix, and axis of the given parabola. Graph the parabola.

1. $(y - 3)^2 = -8x$
2. $8(x + 4)^2 = y - 2$
3. $x^2 - 2x + 4y + 1 = 0$
4. $y^2 + 10y + 8x + 41 = 0$

In Problems 5–8, find an equation of the parabola satisfying the given conditions.

5. Focus $(1, -3)$, directrix $y = -7$
6. Focus $(3, -1)$, vertex $(0, -1)$
7. Vertex $(1, 2)$, vertical axis, passing through $(4, 5)$
8. Vertex $(-1, -4)$, directrix $x = 2$

In Problems 9–12, find the center, vertices, and foci of the given ellipse. Graph the ellipse.

9. $\dfrac{x^2}{3} + \dfrac{(y+5)^2}{25} = 1$
10. $\dfrac{(x-2)^2}{16} + \dfrac{(y+5)^2}{4} = 1$
11. $4x^2 + y^2 + 8x - 6y + 9 = 0$
12. $5x^2 + 9y^2 - 20x + 54y + 56 = 0$

In Problems 13–16, find an equation of the ellipse that satisfies the given conditions.

13. Endpoints of minor axis $(0, \pm 4)$, foci $(\pm 5, 0)$
14. Foci $(2, -1 \pm \sqrt{2})$, one vertex $(2, -1 + \sqrt{6})$
15. Vertices $(\pm 2, -2)$, passing through $(1, -2 + \tfrac{1}{2}\sqrt{3})$
16. Center $(2, 4)$, one focus $(2, 1)$, one vertex $(2, 0)$

In Problems 17–20, find the center, vertices, foci, and asymptotes of the given hyperbola. Graph the hyperbola.

17. $(x - 1)(x + 1) = y^2$
18. $y^2 - \dfrac{(x+3)^2}{4} = 1$
19. $9x^2 - y^2 - 54x - 2y + 71 = 0$
20. $16y^2 - 9x^2 - 64y - 80 = 0$

In Problems 21–24, find an equation of the hyperbola that satisfies the given conditions.

21. Center $(0, 0)$, one vertex $(6, 0)$, one focus $(8, 0)$
22. Foci $(2, \pm 3)$, one vertex $(2, -\tfrac{3}{2})$
23. Foci $(\pm 2\sqrt{5}, 0)$, asymptotes $y = \pm 2x$
24. Vertices $(-3, 2)$ and $(-3, 4)$, one focus $(-3, 3 + \sqrt{2})$

In Problems 25 and 26, perform a suitable rotation of axes so that the resulting $x'y'$-equation has no $x'y'$-term. Sketch the graph.

25. $xy = -8$
26. $8x^2 - 4xy + 5y^2 = 36$
27. Find an equation of the ellipse $4x^2 + y^2 = 4$ when its center is translated to the point $(-5, 2)$.
28. Carefully describe the graphs of the given functions.
    (a) $f(x) = \sqrt{36 - 9x^2}$
    (b) $f(x) = -\sqrt{36 + 9x^2}$
29. **Distance from a Satellite** A satellite orbits the planet Neptune in an elliptical orbit with the center of the planet at one focus. If the length of the major axis of the orbit is 2×10^9 m and the length of the minor axis is 6×10^8 m, find the maximum distance between the satellite and the center of the planet.
30. **Mirror, Mirror . . .** A parabolic mirror has a depth of 7 cm at its center and the distance across the top of the mirror is 20 cm. Find the distance from the vertex to the focus.

Parabolic mirror

10 Polar Coordinates

In This Chapter

10.1 Polar Coordinates

10.2 Graphs of Polar Equations

10.3 Conic Sections in Polar Coordinates

10.4 Vectors in the Plane

10.5 Dot Product

Chapter 10 Review Exercises

The sliced view of a nautilus shell shows an approximation to a curve called a logarithmic spiral. See Problem 31 in Exercises 10.2.

A Bit of History The rectangular or Cartesian coordinate system is one of many different coordinate systems that have been developed over the years to serve different purposes. In mathematical applications a problem that is intractable in rectangular coordinates may become tractable in a different coordinate system. In this chapter our primary study is of the polar coordinate system—a two-dimensional coordinate system in which points are specified by a directed distance from the origin and an angle measured from the positive x-axis.

Although bits and pieces of the concept of polar coordinates can be traced back to the Greek mathematicians **Hipparchus** (c. 190–120 BCE) and **Archimedes** (c. 287–212 BCE), the independent introduction of polar coordinates into the mathematical milieu of the seventeenth century is attributed to the Belgian Jesuit priest **Grégoire de Saint-Vincent** (1584–1667) and the Italian theologian and mathematician **Bonaventura Cavalieri** (1598–1647). However, it was the Italian mathematician **Gregorio Fontana** (1735–1803) who actually coined the term *polar coordinates*. This term appeared in English for the first time in the 1816 translation of **Sylvestre François Lacroix's** (French, 1765–1843) *Differential and Integral Calculus* by the English mathematician **George Peacock** (1791–1858).

10.1 Polar Coordinates

≡ **Introduction** So far we have used the rectangular coordinate system to specify a point P in the plane. We can regard this system as a grid of horizontal and vertical lines. The coordinates (a, b) of a point P are determined by the intersection of two lines: one line $x = a$ is perpendicular to the horizontal reference line called the x-axis, and the other $y = b$ is perpendicular to the vertical reference line called the y-axis. See FIGURE 10.1.1(a). Another system for locating points in the plane is the **polar coordinate system**.

☐ **Terminology** To set up a **polar coordinate system**, we use a system of circles centered at a point O, called the **pole**, and straight lines or rays emanating from O. We take as a reference axis a horizontal half-line directed to the right of the pole and call it the **polar axis**. By specifying a directed (signed) distance r from O and an angle θ whose initial side is the polar axis and whose terminal side is the ray OP, we label the point P by (r, θ). We say that the ordered pair (r, θ) are the **polar coordinates** of P. See Figures 10.1.1(b) and 10.1.1(c).

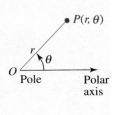

(a) Rectangular coordinate system (b) Polar coordinate system (c) Polar coordinates of P

FIGURE 10.1.1 Comparison of rectangular and polar coordinates of a point P

Although the measure of the angle θ can be either in degrees or radians, in calculus radian measure is used almost exclusively. Consequently, we shall use only radian measure in this discussion.

In the polar coordinate system we adopt the following conventions.

DEFINITION 10.1.1 Conventions in Polar Coordinates

(i) Angles $\theta > 0$ are measured counterclockwise from the polar axis, whereas angles $\theta < 0$ are measured clockwise.
(ii) To graph a point $(-r, \theta)$, where $-r < 0$, measure $|r|$ units along the ray $\theta + \pi$.
(iii) The coordinates of the pole O are $(0, \theta)$, where θ is any angle.

EXAMPLE 1   Plotting Polar Points

Plot the points whose polar coordinates are given.
(a) $(4, \pi/6)$ (b) $(2, -\pi/4)$ (c) $(-3, 3\pi/4)$

Solution (a) Measure 4 units along the ray $\pi/6$ as shown in FIGURE 10.1.2(a).

(b) Measure 2 units along the ray $-\pi/4$. See Figure 10.1.2(b).

(c) Measure 3 units along the ray $3\pi/4 + \pi = 7\pi/4$. Equivalently, we can measure 3 units along the ray $3\pi/4$ extended *backward* through the pole. Note carefully in Figure 10.1.2(c) that the point $(-3, 3\pi/4)$ is not in the same quadrant as the terminal side of the given angle.

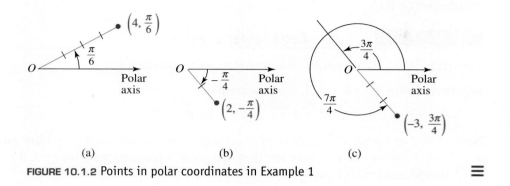

(a) (b) (c)

FIGURE 10.1.2 Points in polar coordinates in Example 1

In contrast to the rectangular coordinate system, the description of a point in polar coordinates is not unique. This is an immediate consequence of the fact that

$$(r, \theta) \quad \text{and} \quad (r, \theta + 2n\pi), \quad n \text{ an integer,}$$

are equivalent. To compound the problem, negative values of r can be used.

EXAMPLE 2 Equivalent Polar Points

The following coordinates are some alternative representations of the point $(2, \pi/6)$:

$$(2, 13\pi/6), \quad (2, -11\pi/6), \quad (-2, 7\pi/6), \quad (-2, -5\pi/6).$$

☐ **Conversion of Polar Coordinates to Rectangular** By superimposing a rectangular coordinate system on a polar coordinate system, as shown in **FIGURE 10.1.3**, we can convert a polar description of a point to rectangular coordinates by using

$$x = r\cos\theta, \quad y = r\sin\theta. \tag{1}$$

These conversion formulas hold true for any values of r and θ in an equivalent polar representation of (r, θ).

FIGURE 10.1.3 Relating polar and rectangular coordinates

EXAMPLE 3 Polar to Rectangular

Convert $(2, \pi/6)$ in polar coordinates to rectangular coordinates.

Solution With $r = 2$, $\theta = \pi/6$, we have from (1),

$$x = 2\cos\frac{\pi}{6} = 2\left(\frac{\sqrt{3}}{2}\right) = \sqrt{3}$$

$$y = 2\sin\frac{\pi}{6} = 2\left(\frac{1}{2}\right) = 1.$$

Thus, $(2, \pi/6)$ is equivalent to $(\sqrt{3}, 1)$ in rectangular coordinates.

☐ **Conversion of Rectangular Coordinates to Polar** It should be evident from Figure 10.1.3 that x, y, r, and θ are also related by

$$r^2 = x^2 + y^2, \quad \tan\theta = \frac{y}{x}. \tag{2}$$

The equations in (2) are used to convert the rectangular coordinates (x, y) to the polar coordinates (r, θ).

EXAMPLE 4 Rectangular to Polar

Convert $(-1, 1)$ in rectangular coordinates to polar coordinates.

Solution With $x = -1$, $y = 1$, we have from (2)

$$r^2 = 2 \quad \text{and} \quad \tan\theta = -1.$$

Now, $r^2 = 2$ or $r = \pm\sqrt{2}$, and two of many angles that satisfy $\tan\theta = -1$ are $3\pi/4$ and $7\pi/4$. From **FIGURE 10.1.4** we see that two polar representations for $(-1, 1)$ are $(\sqrt{2}, 3\pi/4)$ and $(-\sqrt{2}, 7\pi/4)$. ≡

FIGURE 10.1.4 Point in Example 4

In Example 4, observe that we cannot pair just *any* angle θ and *any* value r that satisfy (2); these solutions must also be consistent with (1). Because the points $(-\sqrt{2}, 3\pi/4)$ and $(\sqrt{2}, 7\pi/4)$ lie in the fourth quadrant, they are not polar representations of the second-quadrant point $(-1, 1)$.

There are instances in calculus when a rectangular equation must be expressed as a polar equation $r = f(\theta)$. The next example shows how to do this using the conversion formulas in (1).

EXAMPLE 5 Rectangular Equation to Polar Equation

Find a polar equation that has the same graph as the circle $x^2 + y^2 = 8x$.

Solution Substituting $x = r\cos\theta$, $y = r\sin\theta$, into the given equation we find

$$r^2\cos^2\theta + r^2\sin^2\theta = 8r\cos\theta$$
$$r^2(\cos^2\theta + \sin^2\theta) = 8r\cos\theta \quad \leftarrow \cos^2\theta + \sin^2\theta = 1$$
$$r(r - 8\cos\theta) = 0.$$

The last equation implies that

$$r = 0 \quad \text{or} \quad r = 8\cos\theta.$$

Since $r = 0$ determines only the pole O, we conclude that a polar equation of the circle is $r = 8\cos\theta$. Note that the circle $x^2 + y^2 = 8x$ passes through the origin since $x = 0$ and $y = 0$ satisfy the equation. Relative to the polar equation $r = 8\cos\theta$ of the circle, the origin or pole corresponds to the polar coordinates $(0, \pi/2)$. ≡

EXAMPLE 6 Rectangular Equation to Polar Equation

Find a polar equation that has the same graph as the parabola $x^2 = 8(2 - y)$.

Solution We replace x and y in the given equation by $x = r\cos\theta$, $y = r\sin\theta$ and solve for r in terms of θ:

$$r^2\cos^2\theta = 8(2 - r\sin\theta)$$
$$r^2(1 - \sin^2\theta) = 16 - 8r\sin\theta$$
$$r^2 = r^2\sin^2\theta - 8r\sin\theta + 16 \quad \leftarrow \text{right side is a perfect square}$$
$$r^2 = (r\sin\theta - 4)^2$$
$$r = \pm(r\sin\theta - 4).$$

Solving for r gives two equations,

$$r = \frac{4}{1 + \sin\theta} \quad \text{or} \quad r = \frac{-4}{1 - \sin\theta}.$$

Now recall that, by convention (ii) in Definition 10.1.1, (r, θ) and $(-r, \theta + \pi)$ represent the same point. You should verify that if (r, θ) is replaced by $(-r, \theta + \pi)$ in the second of these two equations, we obtain the first equation. In other words, the equations are equivalent and so we may simply take the polar equation of the parabola to be $r = 4/(1 + \sin\theta)$. ≡

EXAMPLE 7   Polar Equation to Rectangular Equation

Find a rectangular equation that has the same graph as the polar equation $r^2 = 9\cos 2\theta$.

Solution First, we use the trigonometric identity for the cosine of a double angle:

$$r^2 = 9(\cos^2\theta - \sin^2\theta). \quad \leftarrow \cos 2\theta = \cos^2\theta - \sin^2\theta$$

Then, from $r^2 = x^2 + y^2$, $\cos\theta = x/r$, $\sin\theta = y/r$, we have

$$x^2 + y^2 = 9\left(\frac{x^2}{x^2 + y^2} - \frac{y^2}{x^2 + y^2}\right) \quad \text{or} \quad (x^2 + y^2)^2 = 9(x^2 - y^2). \quad \equiv$$

The next section will be devoted to graphing polar equations.

10.1 Exercises   Answers to selected odd-numbered problems begin on page ANS-27.

In Problems 1–6, plot the point with the given polar coordinates.

1. $(3, \pi)$ **2.** $(2, -\pi/2)$ **3.** $\left(-\frac{1}{2}, \pi/2\right)$
4. $(-1, \pi/6)$ **5.** $(-4, -\pi/6)$ **6.** $\left(\frac{2}{3}, 7\pi/4\right)$

In Problems 7–12, find alternative polar coordinates that satisfy

(a) $r > 0, \theta < 0$ (b) $r > 0, \theta > 2\pi$
(c) $r < 0, \theta > 0$ (d) $r < 0, \theta < 0$

for each point with the given polar coordinates.

7. $(2, 3\pi/4)$ **8.** $(5, \pi/2)$ **9.** $(4, \pi/3)$
10. $(3, \pi/4)$ **11.** $(1, \pi/6)$ **12.** $(3, 7\pi/6)$

In Problems 13–18, find the rectangular coordinates for each point with the given polar coordinates.

13. $\left(\frac{1}{2}, 2\pi/3\right)$ **14.** $(-1, 7\pi/4)$ **15.** $(-6, -\pi/3)$
16. $(\sqrt{2}, 11\pi/6)$ **17.** $(4, 5\pi/4)$ **18.** $(-5, \pi/2)$

In Problems 19–24, find polar coordinates that satisfy

(a) $r > 0, -\pi < \theta \leq \pi$ (b) $r < 0, -\pi < \theta \leq \pi$

for each point with the given rectangular coordinates.

19. $(-2, -2)$ **20.** $(0, -4)$ **21.** $(1, -\sqrt{3})$
22. $(\sqrt{6}, \sqrt{2})$ **23.** $(7, 0)$ **24.** $(1, 2)$

In Problems 25–30, sketch the region on the plane that consists of points (r, θ) whose polar coordinates satisfy the given conditions.

25. $2 \leq r < 4, 0 \leq \theta \leq \pi$
26. $2 < r \leq 4$
27. $0 \leq r \leq 2, -\pi/2 \leq \theta \leq \pi/2$
28. $r \geq 0, \pi/4 < \theta < 3\pi/4$
29. $-1 \leq r \leq 1, 0 \leq \theta \leq \pi/2$
30. $-2 \leq r < 4, \pi/3 \leq \theta \leq \pi$

In Problems 31–40, find a polar equation that has the same graph as the given rectangular equation.

31. $y = 5$
32. $x + 1 = 0$
33. $y = 7x$
34. $3x + 8y + 6 = 0$
35. $y^2 = -4x + 4$
36. $x^2 - 12y - 36 = 0$
37. $x^2 + y^2 = 36$
38. $x^2 - y^2 = 1$
39. $x^2 + y^2 + x = \sqrt{x^2 + y^2}$
40. $x^3 + y^3 - xy = 0$

In Problems 41–52, find a rectangular equation that has the same graph as the given polar equation.

41. $r = 2 \sec \theta$
42. $r \cos \theta = -4$
43. $r = 6 \sin 2\theta$
44. $2r = \tan \theta$
45. $r^2 = 4 \sin 2\theta$
46. $r^2 \cos 2\theta = 16$
47. $r + 5 \sin \theta = 0$
48. $r = 2 + \cos \theta$
49. $r = \dfrac{2}{1 + 3\cos\theta}$
50. $r(4 - \sin \theta) = 10$
51. $r = \dfrac{5}{3\cos\theta + 8\sin\theta}$
52. $r = 3 + 3 \sec \theta$

For Discussion

53. How would you express the distance d between two points (r_1, θ_1) and (r_2, θ_2) in terms of their polar coordinates?

54. You know how to find a rectangular equation of a line through two points with rectangular coordinates. How would you find a polar equation of a line through two points with polar coordinates (r_1, θ_1) and (r_2, θ_2)? Carry out your ideas by finding a polar equation of the line through $(3, 3\pi/4)$ and $(1, \pi/4)$. Find the polar coordinates of the x- and y-intercepts of the line.

55. In rectangular coordinates the x-intercepts of the graph of a function $y = f(x)$ are determined from the solutions of the equation $f(x) = 0$. In the next section we will graph polar equations $r = f(\theta)$. What is the significance of the solutions of the equation $f(\theta) = 0$?

10.2 Graphs of Polar Equations

≡ **Introduction** The graph of a polar equation $r = f(\theta)$ is the set of points P with *at least* one set of polar coordinates that satisfies the equation. Since it is most likely that your classroom does not have a polar coordinate grid, to facilitate graphing and discussion of graphs of a polar equation $r = f(\theta)$, we will, as in the preceding section, superimpose a rectangular coordinate over the polar coordinate system.

We begin with some simple polar graphs.

EXAMPLE 1 A Circle Centered at the Origin

Graph $r = 3$.

Solution Since θ is not specified, the point $(3, \theta)$ lies on the graph of $r = 3$ for any value of θ and is 3 units from the origin. We see in FIGURE 10.2.1 that the graph is the circle of radius 3 centered at the origin.

Alternatively, we know from (2) of Section 10.1 that $r = \pm\sqrt{x^2 + y^2}$ so that $r = 3$ yields the familiar rectangular equation $x^2 + y^2 = 3^2$ of a circle of radius 3 centered at the origin. ≡

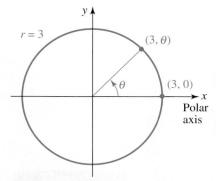

FIGURE 10.2.1 Circle in Example 1

□ **Circles Centered at the Origin** In general, if a is any nonzero constant, the polar graph of

$$r = a \qquad (1)$$

is a circle of radius $|a|$ with center at the origin.

EXAMPLE 2 A Ray Through the Origin

Graph $\theta = \pi/4$.

Solution Since r is not specified, the point $(r, \pi/4)$ lies on the graph for any value of r. If $r > 0$, then this point lies on the half-line in the first quadrant; if $r < 0$, then the point lies on the half-line in the third quadrant. For $r = 0$, the point $(0, \pi/4)$ is the pole or origin. Therefore, the polar graph of $\theta = \pi/4$ is the line through the origin that makes an angle of $\pi/4$ with the polar axis or positive x-axis. See FIGURE 10.2.2. ≡

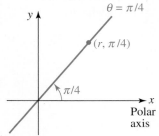

FIGURE 10.2.2 Line in Example 2

□ **Lines Through the Origin** In general, if α is any nonzero real constant, the polar graph of

$$\theta = \alpha \qquad (2)$$

is a line through the origin that makes an angle of α radians with the polar axis.

EXAMPLE 3 A Spiral

Graph $r = \theta$.

Solution As $\theta \geq 0$ increases, r increases and the points (r, θ) wind around the pole in a counterclockwise manner. This is illustrated by the blue portion of the graph in FIGURE 10.2.3. The red portion of the graph is obtained by plotting points for $\theta < 0$.

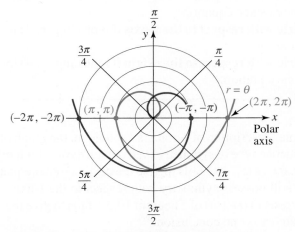

FIGURE 10.2.3 Graph of equation in Example 3

☐ **Spirals** Many graphs in polar coordinates are given special names. The graph in Example 3 is a special case of

$$r = a\theta, \tag{3}$$

where a is a constant. A graph of this equation is called a **spiral of Archimedes**. You are asked to graph other types of spiral curves in Problems 31 and 32 in Exercises 10.2.

In addition to basic point plotting, symmetry can often be utilized to graph a polar equation.

Symmetries of a snowflake

☐ **Symmetry** As shown in FIGURE 10.2.4, a polar graph can have three types of symmetry. A polar graph is **symmetric with respect to the y-axis** if whenever (r, θ) is a point on the graph, $(r, \pi - \theta)$ is also a point on the graph. A polar graph is **symmetric with respect to the x-axis** if whenever (r, θ) is a point on the graph, $(r, -\theta)$ is also a point on the graph. Finally, a polar graph is **symmetric with respect to the origin** if whenever (r, θ) is on the graph, $(-r, \theta)$ is also a point on the graph.

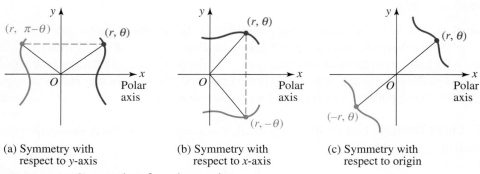

(a) Symmetry with respect to y-axis

(b) Symmetry with respect to x-axis

(c) Symmetry with respect to origin

FIGURE 10.2.4 Symmetries of a polar graph

We have the following tests for symmetries in polar coordinates.

THEOREM 10.2.1 Tests for Symmetry

The graph of a polar equation is:

(*i*) **symmetric with respect to the y-axis** if replacing (r, θ) by $(r, \pi - \theta)$ results in the same equation;

(*ii*) **symmetric with respect to the x-axis** if replacing (r, θ) by $(r, -\theta)$ results in the same equation;

(*iii*) **symmetric with respect to the origin** if replacing (r, θ) by $(-r, \theta)$ results in the same equation.

▶ In rectangular coordinates the description of a point is unique. Hence, in rectangular coordinates if a test for a particular type of symmetry fails, then we can definitely say that the graph does not possess that symmetry.

Because the polar description of a point is not unique, the graph of a polar equation may still have a particular type of symmetry even though the test for it may fail. For example, if replacing (r, θ) by $(r, -\theta)$ fails to give the original polar equation, the graph of that equation may still possess symmetry with respect to the x-axis. Therefore, if one of the replacement tests in (*i*)–(*iii*) of Theorem 10.2.1 fails to give the same polar equation, the best we can say is "no conclusion."

EXAMPLE 4 Graphing a Polar Equation

Graph $r = 1 - \cos\theta$.

Solution One way of graphing this equation is to plot a few well-chosen points corresponding to $0 \leq \theta \leq 2\pi$. As the following table shows

θ	0	$\pi/4$	$\pi/2$	$3\pi/4$	π	$5\pi/4$	$3\pi/2$	$7\pi/4$	2π
r	0	0.29	1	1.71	2	1.71	1	0.29	0

as θ advances from $\theta = 0$ to $\theta = \pi/2$, r increases from $r = 0$ (the origin) to $r = 1$. See **FIGURE 10.2.5(a)**. As θ advances from $\theta = \pi/2$ to $\theta = \pi$, r continues to increase from $r = 1$ to its maximum value of $r = 2$. See Figure 10.2.5(b). Then, for $\theta = \pi$ to $\theta = 3\pi/2$, r begins to decrease from $r = 2$ to $r = 1$. For $\theta = 3\pi/2$ to $\theta = 2\pi$, r continues to decrease and we end up again at the origin $r = 0$. See Figures 10.2.5(c) and 10.2.5(d).

By taking advantage of symmetry we could have simply plotted points for $0 \leq \theta \leq \pi$. From the trigonometric identity for the cosine function $\cos(-\theta) = \cos\theta$ it follows from (*ii*) of Theorem 10.2.1 that the graph of $r = 1 - \cos\theta$ is symmetric with respect to the x-axis. We can obtain the complete graph of $r = 1 - \cos\theta$ by reflecting in the x-axis that portion of the graph given in Figure 10.2.5(b).

FIGURE 10.2.5 Graph of equation in Example 4

☐ **Cardioids** The polar equation in Example 4 is a member of a family of equations that all have a "heart-shaped" graph that passes through the origin. A graph of any polar equation of the form

$$r = a \pm a\sin\theta \quad \text{or} \quad r = a \pm a\cos\theta \quad (4)$$

is called a **cardioid**. The only difference in the graph of these four equations is their symmetry with respect to the y-axis ($r = a \pm a\sin\theta$) or symmetry with respect to the x-axis ($r = a \pm a\cos\theta$). See **FIGURE 10.2.6**.

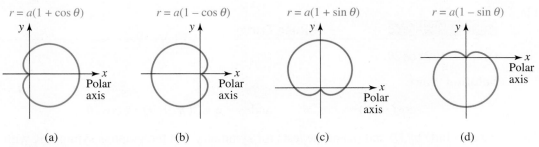

FIGURE 10.2.6 Cardioids

10.2 Graphs of Polar Equations

By knowing the basic shape and orientation of a cardioid, you can obtain a quick and accurate graph by plotting the four points corresponding to $\theta = 0$, $\theta = \pi/2$, $\theta = \pi$, and $\theta = 3\pi/2$.

☐ **Limaçons** Cardioids are special cases of polar curves known as **limaçons**:

$$r = a \pm b\sin\theta \quad \text{or} \quad r = a \pm b\cos\theta. \tag{5}$$

The shape of a limaçon depends on the relative magnitudes of a and b. Let us assume that $a > 0$ and $b > 0$. For $a/b < 1$, we get a **limaçon with an interior loop** as shown in FIGURE 10.2.7(a). When $a = b$ or equivalently $a/b = 1$ we get a **cardioid**. For $1 < a/b < 2$, we get a **dimpled limaçon** as shown in Figure 10.2.7(b). For $a/b \geq 2$, the curve is called a **convex limaçon**. See Figure 10.2.7(c).

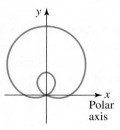
(a) Limaçon with interior loop

(b) Dimpled limaçon

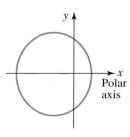
(c) Convex limaçon

FIGURE 10.2.7 Three kinds of limaçons

EXAMPLE 5   A Limaçon

The graph of $r = 3 - \sin\theta$ is a convex limaçon, since $a = 3$, $b = 1$, and $a/b = 3 > 2$. ≡

EXAMPLE 6   A Limaçon

The graph of $r = 1 + 2\cos\theta$ is a limaçon with an interior loop, since $a = 1$, $b = 2$, and $a/b = \frac{1}{2} < 1$. For $\theta \geq 0$, notice in FIGURE 10.2.8 the limaçon starts at $\theta = 0$ or $(3, 0)$. The graph passes through the y-axis at $(1, \pi/2)$ and then enters the origin $(r = 0)$ for the first angle for which $r = 0$ or $1 + 2\cos\theta = 0$ or $\cos\theta = -\frac{1}{2}$. This implies that $\theta = 2\pi/3$. At $\theta = \pi$, the curve passes through $(-1, \pi)$. The remainder of the graph can then be completed using the fact that it is symmetric with respect to the x-axis. ≡

FIGURE 10.2.8 Graph of equation in Example 6

EXAMPLE 7   A Rose Curve

Graph $r = 2\cos 2\theta$.

Solution Since

$$\cos(-2\theta) = \cos 2\theta \quad \text{and} \quad \cos 2(\pi - \theta) = \cos 2\theta$$

we conclude by (*i*) and (*ii*) of the tests for symmetry that the graph is symmetric with respect to both the x- and the y-axes. A moment of reflection should convince you that we need only consider $0 \leq \theta \leq \pi/2$. Using the data in the following table, we see that the dashed portion of the graph given in FIGURE 10.2.9 is that completed by symmetry. The graph is called a **rose curve with four petals**.

θ	0	$\pi/12$	$\pi/6$	$\pi/4$	$\pi/3$	$5\pi/12$	$\pi/2$
r	2	1.7	1	0	-1	-1.7	12

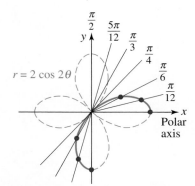

FIGURE 10.2.9 Graph of equation in Example 7

☐ **Rose Curves** In general, if n is a positive integer, the graphs of

$$r = a\sin n\theta \quad \text{or} \quad r = a\cos n\theta, \quad n \geq 2 \tag{6}$$

are called **rose curves**, although as you can see in FIGURE 10.2.10 the curve looks more like a daisy. When n is odd, the number of **loops** or **petals** of the curve is n; if n is even, the curve has $2n$ petals. To graph a rose curve we can start by graphing one petal. To begin, we find an angle θ for which r is a maximum. This gives the center line of the petal. We then find corresponding values of θ for which the rose curve enters the origin ($r = 0$). To complete the graph we use the fact that the center lines of the petals are spaced $2\pi/n$ radians ($360/n$ degrees) apart if n is odd, and $2\pi/2n = \pi/n$ radians ($180/n$ degrees) apart if n is even. In Figure 10.2.10 we have drawn the graph of $r = a\sin 5\theta$, $a > 0$. The spacing between the center lines of the five petals is $2\pi/5$ radians ($72°$).

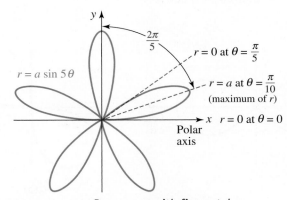

FIGURE 10.2.10 Rose curve with five petals

In Example 5 in Section 10.1 we saw that the polar equation $r = 8\cos\theta$ is equivalent to the rectangular equation $x^2 + y^2 = 8x$. By completing the square in x in the rectangular equation, we recognize

$$(x - 4)^2 + y^2 = 16$$

as a circle of radius 4 centered at $(4, 0)$ on the x-axis. Polar equations such as $r = 8\cos\theta$ or $r = 8\sin\theta$ are circles and are also special cases of rose curves. See FIGURE 10.2.11.

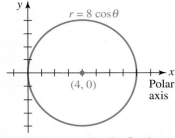

FIGURE 10.2.11 Graph of polar equation $r = 8\cos\theta$

☐ **Circles with Centers on an Axis** When $n = 1$ in (6) we get

$$r = a\sin\theta \quad \text{or} \quad r = a\cos\theta, \tag{7}$$

which are polar equations of circles passing through the origin with diameter $|a|$ with centers $(a/2, 0)$ on the x-axis ($r = a\cos\theta$), or with centers $(0, a/2)$ on the y-axis ($r = a\sin\theta$). FIGURE 10.2.12 illustrates the graphs of the equations in (7) in the cases when $a > 0$ and $a < 0$.

10.2 Graphs of Polar Equations

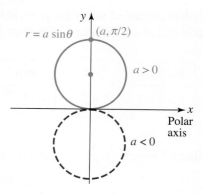

(a) Centers on the x-axis (b) Centers on the y-axis

FIGURE 10.2.12 Circles through the origin with centers on an axis

☐ **Lemniscates** If n is a positive integer, the graphs of

$$r^2 = a\cos 2\theta \quad \text{or} \quad r^2 = a\sin 2\theta \tag{8}$$

where $a > 0$, are called **lemniscates**. By (*iii*) of the tests for symmetry you can see the graphs of both of the equations in (8) are symmetric with respect to the origin. Moreover, by (*ii*) of the tests for symmetry the graph of $r^2 = a\cos 2\theta$ is symmetric with respect to the x-axis. **FIGURES 10.2.13(a)** and 10.2.13(b) show typical graphs of the equations $r^2 = a\cos 2\theta$ and $r^2 = a\sin 2\theta$, respectively.

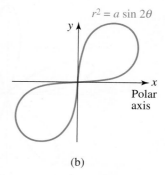

(a)   (b)

FIGURE 10.2.13 Lemniscates

☐ **Points of Intersection** In rectangular coordinates we can find the points (x, y) where the graphs of two functions $y = f(x)$ and $y = g(x)$ intersect by equating the y-values. The real solutions of the equation $f(x) = g(x)$ correspond to *all* the x-coordinates of the points where the graphs intersect. In contrast, problems may arise in polar coordinates when we try the same method to determine where the graphs of two polar equations $r = f(\theta)$ and $r = g(\theta)$ intersect.

EXAMPLE 8   Intersecting Circles

FIGURE 10.2.14 shows that the circles $r = \sin\theta$ and $r = \cos\theta$ have two points of intersection. By equating the r values, the equation $\sin\theta = \cos\theta$ leads to $\theta = \pi/4$. Substituting this value into either equation yields $r = \sqrt{2}/2$. Thus we have found only a single polar point $(\sqrt{2}/2, \pi/4)$ where the graphs intersect. From the figure, it is apparent that the graphs also intersect at the origin. But the problem here is that the origin or pole is $(0, \pi/2)$ on the graph of $r = \cos\theta$ but is $(0, 0)$ on the graph of $r = \sin\theta$. This situation is analogous to the curves reaching the same point at different times. ≡

FIGURE 10.2.14 Intersecting circles in Example 8

☐ **Rotation of Polar Graphs** In Section 3.2 we saw that if $y = f(x)$ is the rectangular equation of a function, then the graphs of $y = f(x - c)$ and $y = f(x + c)$, $c > 0$, are obtained by *shifting* the graph of f horizontally c units to the right and to the left, respectively. In contrast, if $r = f(\theta)$ is a polar equation, then the graphs of $f(\theta - \gamma)$ and $f(\theta + \gamma)$, where $\gamma > 0$, can be obtained by *rotating* the graph of f by an amount γ. Specifically:

- The graph of $r = f(\theta - \gamma)$ is the graph of $r = f(\theta)$ rotated *counterclockwise* about the origin by an amount γ.
- The graph of $r = f(\theta + \gamma)$ is the graph of $r = f(\theta)$ rotated *clockwise* about the origin by an amount γ.

For example, the graph of the cardioid $r = a(1 + \cos\theta)$ is shown in Figure 10.2.6(a). The graph of $r = a(1 + \cos(\theta - \pi/2))$ is the graph of $r = a(1 + \cos\theta)$ rotated counterclockwise about the origin by an amount $\pi/2$. Its graph then must be that given in Figure 10.2.6(c). This makes sense, because the difference formula of the cosine gives

$$r = a[1 + \cos(\theta - \pi/2)] = a[1 + \cos\theta\cos(\pi/2) + \sin\theta\sin(\pi/2)] = a(1 + \sin\theta).$$

◀ See the identity in (5) of Section 7.4.

Similarly, rotating $r = a(1 + \cos\theta)$ counterclockwise about the origin by an amount π gives the equation

$$r = a[1 + \cos(\theta - \pi)] = a[1 + \cos\theta\cos\pi + \sin\theta\sin\pi] = a(1 - \cos\theta)$$

whose graph is given in Figure 10.2.6(b). As another example, take a look again at Figure 10.2.13. From

$$r^2 = a\cos 2\left(\theta - \frac{\pi}{4}\right) = a\cos\left(2\theta - \frac{\pi}{2}\right) = a\sin 2\theta$$

we see that the graph of the lemniscate in Figure 10.2.13(b) is the graph in Figure 10.2.13(a) rotated counterclockwise about the origin by an amount $\pi/4$.

EXAMPLE 9   **Rotated Polar Graphs**

Graph $r = 1 + 2\sin(\theta + \pi/4)$.

Solution The graph of the given equation is the graph of the limaçon $r = 1 + 2\sin\theta$ rotated clockwise about the origin by an amount $\pi/4$. In **FIGURE 10.2.15** the blue graph is that of $r = 1 + 2\sin\theta$ and the red graph is the rotated graph. ≡

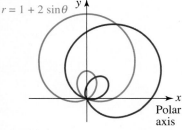

FIGURE 10.2.15 Graphs of polar equations in Example 9

NOTES FROM THE CLASSROOM

(*i*) Example 8 illustrates one of several frustrating difficulties of working in polar coordinates:

A point can be on the graph of a polar equation even though its coordinates do not satisfy the equation.

You should verify that $(2, \pi/2)$ is an alternative polar description of the point $(-2, 3\pi/2)$. Moreover, verify that $(-2, 3\pi/2)$ is a point on the graph of $r = 1 + 3\sin\theta$ by showing that the coordinates satisfy the equation. However, note that the alternative coordinates $(2, \pi/2)$ do not satisfy the equation.

(*ii*) It is sometimes convenient to use parametric equations to plot the graphs of polar equations. This can be done using the conversion formulas

$x = r\cos\theta$, $y = r\sin\theta$. If $r = f(\theta)$, $\alpha \leq \theta \leq \beta$ describes a polar graph C, then a parameterization for C is given by

$$x = f(\theta)\cos\theta, \quad y = f(\theta)\sin\theta, \quad \alpha \leq \theta \leq \beta. \tag{9}$$

See Problems 47–50 in Exercises 10.2.

(*iii*) The four-petal rose curve obtained in Example 7 is obtained by plotting r for θ-values satisfying $0 \leq \theta < 2\pi$. See **FIGURE 10.2.16**. Do not assume this is true for every rose curve. Indeed, the five-petal rose curve discussed in Figure 10.2.10 is obtained using θ-values satisfying $0 \leq \theta < \pi$. In general, a rose curve $r = a\sin n\theta$ or $r = a\cos n\theta$ is traced out exactly once for $0 \leq \theta < 2\pi$ if n is even and once for $0 \leq \theta < \pi$ if n is odd.

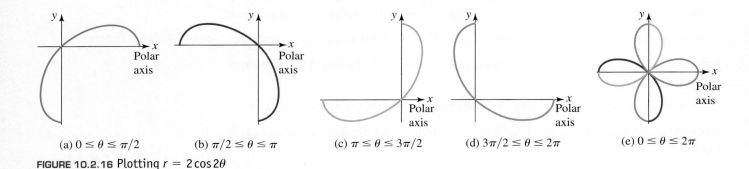

(a) $0 \leq \theta \leq \pi/2$ (b) $\pi/2 \leq \theta \leq \pi$ (c) $\pi \leq \theta \leq 3\pi/2$ (d) $3\pi/2 \leq \theta \leq 2\pi$ (e) $0 \leq \theta \leq 2\pi$

FIGURE 10.2.16 Plotting $r = 2\cos 2\theta$

10.2 Exercises
Answers to selected odd-numbered problems begin on page ANS-27.

In Problems 1–30, identify by name the graph of the given polar equation. Then sketch the graph of the equation.

1. $r = 6$
2. $r = -1$
3. $\theta = \pi/3$
4. $\theta = 5\pi/6$
5. $r = 2\theta, \theta \leq 0$
6. $r = 3\theta, \theta \geq 0$
7. $r = 1 + \cos\theta$
8. $r = 5 - 5\sin\theta$
9. $r = 2(1 + \sin\theta)$
10. $2r = 1 - \cos\theta$
11. $r = 1 - 2\cos\theta$
12. $r = 2 + 4\sin\theta$
13. $r = 4 - 3\sin\theta$
14. $r = 3 + 2\cos\theta$
15. $r = 4 + \cos\theta$
16. $r = 4 - 2\sin\theta$
17. $r = \sin 2\theta$
18. $r = 3\sin 4\theta$
19. $r = 3\cos 3\theta$
20. $r = 2\sin 3\theta$
21. $r = \cos 5\theta$
22. $r = 2\sin 9\theta$
23. $r = 6\cos\theta$
24. $r = -2\cos\theta$
25. $r = -3\sin\theta$
26. $r = 5\sin\theta$
27. $r^2 = 4\sin 2\theta$
28. $r^2 = 4\cos 2\theta$
29. $r^2 = -25\cos 2\theta$
30. $r^2 = -9\sin 2\theta$

In Problems 31 and 32, the graph of the given equation is a spiral. Sketch its graph.

31. $r = 2^\theta, \theta \geq 0$ (logarithmic)
32. $r\theta = \pi, \theta > 0$ (hyperbolic)

In Problems 33–38, find an equation of the given polar graph.

33.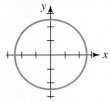
FIGURE 10.2.17 Graph for Problem 33

34.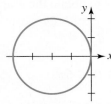
FIGURE 10.2.18 Graph for Problem 34

35.
FIGURE 10.2.19 Graph for Problem 35

36.
FIGURE 10.2.20 Graph for Problem 36

37.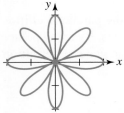
FIGURE 10.2.21 Graph for Problem 37

38.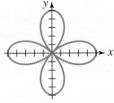
FIGURE 10.2.22 Graph for Problem 38

In Problems 39–42, find the points of intersection of the graphs of the given pair of polar equations.

39. $r = 2$
$r = 4\sin\theta$

40. $r = \sin\theta$
$r = \sin 2\theta$

41. $r = 1 - \cos\theta$
$r = 1 + \cos\theta$

42. $r = 3 - 3\cos\theta$
$r = 3\cos\theta$

Calculator/CAS Problems

43. Use a graphing utility to obtain the graph of the **bifolium** $r = 4\sin\theta\cos^2\theta$ and the circle $r = \sin\theta$ on the same axes. Find all points of intersection of the graphs.

44. Use a graphing utility to verify that the cardioid $r = 1 + \cos\theta$ and the lemniscate $r^2 = 4\cos\theta$ intersect at four points. Find these points of intersection of the graphs.

In Problems 45 and 46, the graphs of the equations (a)–(d) represent a rotation of the graph of the given equation. Try sketching these graphs by hand. If you have difficulties, then use a calculator or CAS.

45. $r = 1 + \sin\theta$
   (a) $r = 1 + \sin(\theta - \pi/2)$
   (b) $r = 1 + \sin(\theta + \pi/2)$
   (c) $r = 1 + \sin(\theta - \pi/6)$
   (d) $r = 1 + \sin(\theta + \pi/4)$

46. $r = 2 + 4\cos\theta$
   (a) $r = 2 + 4\cos(\theta + \pi/6)$
   (b) $r = 2 + 4\cos(\theta - 3\pi/2)$
   (c) $r = 2 + 4\cos(\theta + \pi)$
   (d) $r = 2 + 4\cos(\theta - \pi/8)$

In Problems 47–50, use (9) to parameterize the curve whose polar equation is given. Use a graphing utility to obtain the graph of the resulting set of parametric equations.

47. $r = 2\sin\dfrac{\theta}{2}, 0 \leq \theta \leq 4\pi$
48. $r = 2\sin\dfrac{\theta}{4}, 0 \leq \theta \leq 8\pi$
49. $r = 2\cos\dfrac{\theta}{5}, 0 \leq \theta \leq 6\pi$
50. $r = 2\cos\dfrac{3\theta}{2}, 0 \leq \theta \leq 6\pi$

For Discussion

In Problems 51 and 52, suppose $r = f(\theta)$ is a polar equation. Graphically interpret the given property.

51. $f(-\theta) = f(\theta)$ (even function)
52. $f(-\theta) = -f(\theta)$ (odd function)

10.3 Conic Sections in Polar Coordinates

≡ **Introduction** In Chapter 9 we derived equations for the parabola, ellipse, and hyperbola using the distance formula in rectangular coordinates. By using polar coordinates and the concept of eccentricity, we can now give one general definition of a conic section that encompasses all three curves.

DEFINITION 10.3.1 Conic Section

Let L be a fixed line in the plane, and let F be a point not on the line. A **conic section** is the set of points P in the plane for which the distance from P to F divided by the distance from P to L is a constant.

The fixed line L is called a **directrix**, and the point F is a **focus**. The fixed constant is the **eccentricity** e of the conic. As **FIGURE 10.3.1** shows, the point P lies on the conic if and only if

$$\frac{d(P, F)}{d(P, Q)} = e, \qquad (1)$$

where Q denotes the foot of the perpendicular from P to L. In (1), if

- $e = 1$, the conic is a **parabola**,
- $0 < e < 1$, the conic is an **ellipse**, and
- $e > 1$, the conic is a **hyperbola**.

FIGURE 10.3.1 Geometric interpretation of (1)

□ **Polar Equations of Conics** Equation (1) is readily interpreted using polar coordinates. Suppose F is placed at the pole and L is p units ($p > 0$) to the left of F perpendicular to the extended polar axis. We see from **FIGURE 10.3.2** that (1) written as $d(P, F) = e d(P, Q)$ is the same as

$$r = e(p + r\cos\theta) \quad \text{or} \quad r - er\cos\theta = ep. \qquad (2)$$

Solving for r yields

$$r = \frac{ep}{1 - e\cos\theta}. \qquad (3)$$

FIGURE 10.3.2 Polar coordinate interpretation of (1)

To see that (3) yields the familiar equations of the conics, let us superimpose a rectangular coordinate system on the polar coordinate system with origin at the pole and the positive *x*-axis coinciding with the polar axis. We then express the first equation in (2) in rectangular coordinates and simplify:

$$\pm\sqrt{x^2 + y^2} = ex + ep$$
$$x^2 + y^2 = e^2x^2 + 2e^2px + e^2p^2$$
$$(1 - e^2)x^2 - 2e^2px + y^2 = e^2p^2. \tag{4}$$

Choosing $e = 1$, (4) becomes

$$-2px + y^2 = p^2 \quad \text{or} \quad y^2 = 2p\left(x + \tfrac{1}{2}p\right).$$

The last equation is the standard form of a parabola whose axis is the *x*-axis, vertex is at $\left(-\tfrac{1}{2}p, 0\right)$ and, consistent with the placement of *F*, whose focus is at the origin.

It is a good exercise in algebra to show that (3) yields standard form equations of an ellipse in the case $0 < e < 1$ and a hyperbola in the case $e > 1$. See Problem 43 in Exercises 10.3. Thus, depending on the value of *e*, the polar equation (3) can have three possible graphs as shown in FIGURE 10.3.3.

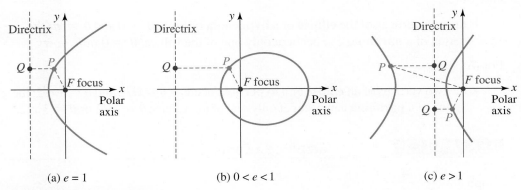

(a) $e = 1$ (b) $0 < e < 1$ (c) $e > 1$

FIGURE 10.3.3 Graphs of equation (3)

If we had placed the focus *F* to the *left* of the directrix in our derivation of the polar equation (3), then the equation $r = ep/(1 + e\cos\theta)$ would be obtained. When the directrix *L* is chosen parallel to the polar axis (that is, horizontal), then the equation of the conic is found to be either $r = ep/(1 - e\sin\theta)$ or $r = ep/(1 + e\sin\theta)$. A summary of the preceding discussion is given next.

THEOREM 10.3.1 Polar Equations of Conics

Any polar equation of the form

$$r = \frac{ep}{1 \pm e\cos\theta} \tag{5}$$

or

$$r = \frac{ep}{1 \pm e\sin\theta} \tag{6}$$

is a conic section with focus at the origin and axis along a coordinate axis. The axis of the conic section is along the *x*-axis for equations of the form (5) and along the *y*-axis for equations of the form (6). The conic is a **parabola** if $e = 1$, an **ellipse** if $0 < e < 1$, and a **hyperbola** if $e > 1$.

EXAMPLE 1 Identifying Conics

Identify each of the following conics **(a)** $r = \dfrac{2}{1 - 2\sin\theta}$ **(b)** $r = \dfrac{3}{4 + \cos\theta}$.

Solution **(a)** A term-by-term comparison of the given equation with the polar form $r = ep/(1 - e\sin\theta)$ enables us to make the identification $e = 2$. Hence the conic is a hyperbola.

(b) In order to identify the conic section, we divide the numerator and the denominator of the given equation by 4. This puts the equation into the form

$$r = \dfrac{\frac{3}{4}}{1 + \frac{1}{4}\cos\theta}.$$

Then by comparison with $r = ep/(1 + e\cos\theta)$ we see that $e = \frac{1}{4}$. Hence the conic is an ellipse. ≡

☐ **Graphs** A rough graph of a conic defined by (5) or (6) can be obtained by knowing the orientation of its axis, finding the *x*- and *y*-intercepts, and finding the vertices. In the case of (5),

- the two vertices of the **ellipse** or a **hyperbola** occur at $\theta = 0$ and $\theta = \pi$; the vertex of a **parabola** can occur at only one of the values: $\theta = 0$ or $\theta = \pi$.

For (6),

- the two vertices of an **ellipse** or a **hyperbola** occur at $\theta = \pi/2$ and $\theta = 3\pi/2$; the vertex of a **parabola** can occur at only one of the values: $\theta = \pi/2$ or $\theta = 3\pi/2$.

EXAMPLE 2 Graphing a Conic

Graph $r = \dfrac{4}{3 - 2\sin\theta}$.

Solution By writing the equation as $r = \dfrac{\frac{4}{3}}{1 - \frac{2}{3}\sin\theta}$ we see that the eccentricity is $e = \frac{2}{3}$ and so the conic is an ellipse. Moreover, because the equation is of the form given in (6), we know that the axis of the ellipse is vertical along the *y*-axis. Now in view of the discussion preceding this example, we obtain:

$$\text{vertices: } (4, \pi/2), \left(\tfrac{4}{5}, 3\pi/2\right)$$
$$\text{x-intercepts: } \left(\tfrac{4}{3}, 0\right), \left(\tfrac{4}{3}, \pi\right).$$

The graph of the equation is given in **FIGURE 10.3.4**. ≡

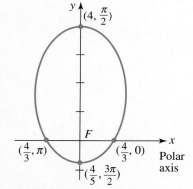

FIGURE 10.3.4 Graph of polar equation in Example 2

EXAMPLE 3 Graphing a Conic

Graph $r = \dfrac{1}{1 - \cos\theta}$.

Solution Inspection of the equation shows that it is of the form given in (5) with $e = 1$. Hence the conic section is a parabola whose axis is horizontal along the *x*-axis. Since *r* is undefined at $\theta = 0$, the vertex of the parabola occurs at $\theta = \pi$:

$$\text{vertex: } \left(\tfrac{1}{2}, \pi\right)$$
$$\text{y-intercepts: } (1, \pi/2), (1, 3\pi/2).$$

The graph of the equation is given in **FIGURE 10.3.5**. ≡

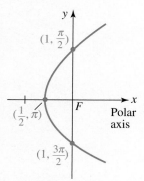

FIGURE 10.3.5 Graph of polar equation in Example 3

EXAMPLE 4 Graphing a Conic

Graph $r = \dfrac{2}{1 + 2\cos\theta}$.

Solution From (5) we see that $e = 2$ and so the conic section is a hyperbola whose axis is horizontal along the x-axis. The vertices, the endpoints of the transverse axis of the hyperbola, occur at $\theta = 0$ and at $\theta = \pi$:

$$\text{vertices: } \left(\tfrac{2}{3}, 0\right), (-2, \pi)$$
$$y\text{-intercepts: } (2, \pi/2), (2, 3\pi/2).$$

The graph of the equation is given in FIGURE 10.3.6.

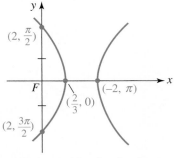

FIGURE 10.3.6 Graph of polar equation in Example 4

☐ **Rotated Conics** We saw in Section 10.2 that graphs of $r = f(\theta - \gamma)$ and $r = f(\theta + \gamma), \gamma > 0$, are rotations of the graph of the polar equation $r = f(\theta)$ about the origin by an amount γ. Thus

$$\left.\begin{array}{l} r = \dfrac{ep}{1 \pm e\cos(\theta - \gamma)} \\[6pt] r = \dfrac{ep}{1 \pm e\sin(\theta - \gamma)} \end{array}\right\} \begin{array}{l}\text{conics rotated}\\ \text{counterclockwise}\\ \text{about the origin}\end{array} \qquad \left.\begin{array}{l} r = \dfrac{ep}{1 \pm e\cos(\theta + \gamma)} \\[6pt] r = \dfrac{ep}{1 \pm e\sin(\theta + \gamma)} \end{array}\right\} \begin{array}{l}\text{conics rotated}\\ \text{clockwise}\\ \text{about the origin}\end{array}$$

EXAMPLE 5 Rotated Conic

In Example 2 we saw that the graph of $r = \dfrac{4}{3 - 2\sin\theta}$ is an ellipse with major axis along the y-axis. This is the blue graph in FIGURE 10.3.7. The graph of $r = \dfrac{4}{3 - 2\sin(\theta - 2\pi/3)}$ is the red graph in Figure 10.3.7 and is a counterclockwise rotation of the blue graph by the amount $2\pi/3$ radians (or 120°) about the origin. The major axis of the red graph lies along the line $\theta = 7\pi/6$.

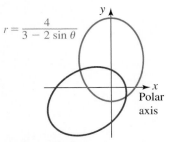

FIGURE 10.3.7 Graphs of polar equations in Example 5

☐ **Applications** Equations of the type in (5) and (6) are well suited to describe a closed orbit of satellite around the Sun (Earth or Moon) since such an orbit is an ellipse with the Sun (Earth or Moon) at one focus. Suppose that an equation of the orbit is given by $r = ep/(1 - e\cos\theta), 0 < e < 1$, and r_p is the value of r at perihelion (perigee or perilune) and r_a is the value of r at aphelion (apogee or apolune). These are the points in the orbit, occurring on the x-axis, at which the satellite is closest and farthest, respectively, from the Sun (Earth or Moon). See FIGURE 10.3.8. It is left as an exercise to show that the eccentricity e of the orbit is related to r_p and r_a by

$$e = \frac{r_a - r_p}{r_a + r_p}. \tag{7}$$

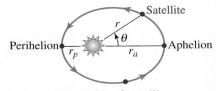

FIGURE 10.3.8 Orbit of satellite

EXAMPLE 6 Polar Equation of an Orbit

Find a polar equation of the orbit of the planet Mercury around the Sun if $r_p = 2.85 \times 10^7$ miles and $r_a = 4.36 \times 10^7$ miles.

Solution From (7), the eccentricity of Mercury's orbit is

$$e = \frac{4.36 \times 10^7 - 2.85 \times 10^7}{4.36 \times 10^7 + 2.85 \times 10^7} = 0.21.$$

Mercury is the closest planet to the Sun

Hence
$$r = \frac{0.21p}{1 - 0.21\cos\theta}. \tag{8}$$

All we need to do now is to solve for the quantity $0.21p$. To do this we use the fact that aphelion occurs at $\theta = 0$:

$$4.36 \times 10^7 = \frac{0.21p}{1 - 0.21}.$$

The last equation yields $0.21p = 3.44 \times 10^7$. Substituting this value of $0.21p$ in (8) yields the polar equation

$$r = \frac{3.44 \times 10^7}{1 - 0.21\cos\theta}$$

of Mercury's orbit.   ≡

10.3 Exercises
Answers to selected odd-numbered problems begin on page ANS-28.

In Problems 1–10, determine the eccentricity, identify the conic, and sketch its graph.

1. $r = \dfrac{2}{1 - \sin\theta}$
2. $r = \dfrac{2}{2 - \cos\theta}$
3. $r = \dfrac{16}{4 + \cos\theta}$
4. $r = \dfrac{5}{2 + 2\sin\theta}$
5. $r = \dfrac{4}{1 + 2\sin\theta}$
6. $r = \dfrac{-4}{\cos\theta - 1}$
7. $r = \dfrac{18}{3 - 6\cos\theta}$
8. $r = \dfrac{4\csc\theta}{3\csc\theta + 2}$
9. $r = \dfrac{6}{1 - \cos\theta}$
10. $r = \dfrac{2}{2 + 5\cos\theta}$

In Problems 11–14, determine the eccentricity e of the given conic. Then convert the polar equation to a rectangular equation and verify that $e = c/a$.

11. $r = \dfrac{6}{1 + 2\sin\theta}$
12. $r = \dfrac{10}{2 - 3\cos\theta}$
13. $r = \dfrac{12}{3 - 2\cos\theta}$
14. $r = \dfrac{2\sqrt{3}}{\sqrt{3} + \sin\theta}$

In Problems 15–20, find a polar equation of the conic with focus at the origin that satisfies the given conditions.

15. $e = 1$, directrix $x = 3$
16. $e = \frac{3}{2}$, directrix $y = 2$
17. $e = \frac{2}{3}$, directrix $y = -2$
18. $e = \frac{1}{2}$, directrix $x = 4$
19. $e = 2$, directrix $x = 6$
20. $e = 1$, directrix $y = -2$

21. Find a polar equation of the conic in Problem 15 if the graph is rotated clockwise about the origin by an amount $2\pi/3$.
22. Find a polar equation of the conic in Problem 16 if the graph is rotated counterclockwise about the origin by an amount $\pi/6$.

In Problems 23–28, find a polar equation of the parabola with focus at the origin and the given vertex.

23. $(\frac{3}{2}, 3\pi/2)$
24. $(2, \pi)$
25. $(\frac{1}{2}, \pi)$
26. $(2, 0)$
27. $(\frac{1}{4}, 3\pi/2)$
28. $(\frac{3}{2}, \pi/2)$

In Problems 29–32, find the polar coordinates of the vertex or vertices of the given rotated conic.

29. $r = \dfrac{4}{1 + \cos(\theta - \pi/4)}$
30. $r = \dfrac{5}{3 + 2\cos(\theta - \pi/3)}$
31. $r = \dfrac{10}{2 - \sin(\theta + \pi/6)}$
32. $r = \dfrac{6}{1 + 2\sin(\theta + \pi/3)}$

Miscellaneous Applications

33. **Perigee Distance** A communications satellite is 12,000 km above the Earth at its apogee. The eccentricity of its elliptical orbit is 0.2. Use (7) to find its perigee distance.
34. **Orbit** Find a polar equation $r = ep/(1 - e\cos\theta)$ of the orbit of the satellite in Problem 33.
35. **Earth's Orbit** Find a polar equation of the orbit of the Earth around the Sun if $r_p = 1.47 \times 10^8$ km and $r_a = 1.52 \times 10^8$ km.
36. **Comet Halley** (a) The eccentricity of the elliptical orbit of Comet Halley is 0.97 and the length of the major axis of its orbit is 3.34×10^9 mi. Find a polar equation of its orbit of the form $r = ep/(1 - e\cos\theta)$.
    (b) Use the equation in part (a) to obtain r_p and r_a for the orbit of Comet Halley.

Next visit of Comet Halley to the Solar System will be in 2061

Calculator/CAS Problems

The orbital characteristics (eccentricity, perigee, and major axis) of a satellite near the Earth gradually degrade over time due to many small forces acting on the satellite other than the gravitational force of the Earth. These forces include atmospheric drag, the gravitational attractions of the Sun and the Moon, and magnetic forces. Approximately once a month tiny rockets are activated for a few seconds in order to "boost" the orbital characteristics back into the desired range. Rockets are turned on longer to a major change in the orbit of a satellite. The most fuel efficient way to move from an inner orbit to an outer orbit, called a **Hohmann transfer**, is to add velocity in the direction of flight at the time the satellite reaches perigee on the inner orbit, follow the Hohmann transfer ellipse halfway around to its apogee, and add velocity again to achieve the outer orbit. A similar process (subtracting velocity at apogee on the outer orbit and subtracting velocity at perigee on the Hohmann transfer orbit) moves a satellite from an outer orbit to an inner orbit.

In Problems 37–40, use a graphing calculator or CAS to superimpose the graphs of the given three polar equations on the same coordinate axes. Print out your result and use a colored pencil to trace out the Hohmann transfer.

37. Inner orbit $r = \dfrac{24}{1 + 0.2\cos\theta}$, Hohmann transfer $r = \dfrac{32}{1 + 0.6\cos\theta}$,

    outer orbit $r = \dfrac{56}{1 + 0.3\cos\theta}$

38. Inner orbit $r = \dfrac{5.5}{1 + 0.1\cos\theta}$, Hohmann transfer $r = \dfrac{7.5}{1 + 0.5\cos\theta}$, outer orbit $r = \dfrac{13.5}{1 + 0.1\cos\theta}$

39. Inner orbit $r = 9$, Hohmann transfer $r = \dfrac{15.3}{1 + 0.7\cos\theta}$, outer orbit $r = 51$

40. Inner orbit $r = \dfrac{73.5}{1 + 0.05\cos\theta}$, Hohmann transfer $r = \dfrac{77}{1 + 0.1\cos\theta}$, outer orbit $r = \dfrac{84.7}{1 + 0.01\cos\theta}$

In Problems 41 and 42, use a calculator or CAS to superimpose the graphs of the given two polar equations on the same coordinate axes.

41. $r = \dfrac{4}{4 + 3\cos\theta}$; $r = \dfrac{4}{4 + 3\cos(\theta - \pi/3)}$

42. $r = \dfrac{2}{1 - \sin\theta}$; $r = \dfrac{2}{1 - \sin(\theta + 3\pi/4)}$

For Discussion

43. Show that (3) yields standard form equations of an ellipse in the case $0 < e < 1$ and a hyperbola in the case $e < 1$.
44. Use the equation $r = ed/(1 - e\cos\theta)$ to derive the result in (7).

10.4 Vectors in the Plane

Introduction In order to describe certain physical quantities accurately, we must have two pieces of information: a magnitude and a direction. For example, when we discuss the flight of an airplane both its speed and its heading are important. Quantities that involve both magnitude and direction are represented by **vectors**. In this section we will survey some basic definitions and operations on vectors that lie in the coordinate plane. We also draw on trigonometry from Chapters 6–8.

Terminology In science, mathematics, and engineering we distinguish two important quantities: *scalars* and *vectors*. A **scalar** is simply a real number and is generally represented by a lowercase italicized letter, such as a, k, or x. Scalars may be used to represent magnitudes and may have specific units attached to them; for example, 80 feet, 10 lb, or 20°Celsius.

On the other hand, a **vector**, or **displacement vector**, may be thought of as an arrow or directed line segment (a line with a direction specified by an arrowhead) connecting points A and B in 2-space. The *tail* of the arrow is called the **initial point** and the *tip* of the arrow is called the **terminal point**. As shown in FIGURE 10.4.1, a vector is usually denoted by a bold-faced letter such as $\mathbf{v}$ or, if we wish to emphasize the initial and terminal points A and B, by the symbol $\overrightarrow{AB}$.

FIGURE 10.4.1 Directed line segment in 2-space

☐ **Magnitude and Direction** The length of the directed line segment is called the **magnitude** of the vector $\overrightarrow{AB}$ and is represented by $|\overrightarrow{AB}|$. Two vectors $\overrightarrow{AB}$ and $\overrightarrow{CD}$ are said to be **equal**, written $\overrightarrow{AB} = \overrightarrow{CD}$, if they have both the same magnitude and the same direction, as shown in FIGURE 10.4.2. Thus vectors can be translated from one position to another so long as neither the magnitude nor the direction is changed.

Because we can move a vector provided its magnitude and its direction are unchanged, we can place the initial point at the origin. Then, as shown in FIGURE 10.4.3, the terminal point P will have rectangular coordinates (x, y). Conversely, every ordered pair of real numbers (x, y) determines a vector $\overrightarrow{OP}$, where P has rectangular coordinates (x, y). Thus we have a one-to-one correspondence between vectors and ordered pairs of real numbers. We say that $\mathbf{v} = \overrightarrow{OP}$ is the **position vector** of the point $P(x, y)$ and is written

$$\overrightarrow{OP} = \langle x, y \rangle.$$

In general, any vector in the plane can be identified with a unique position vector $\mathbf{v} = \langle a, b \rangle$. The numbers a and b are said to be the **components** of the position vector $\mathbf{v}$ and the notation $\langle a, b \rangle$ is called the **component form of a vector**.

Since the magnitude of $\langle a, b \rangle$ is the distance from (a, b) to the origin, we define the **magnitude** $|\mathbf{v}|$ of the vector $\mathbf{v} = \langle a, b \rangle$ to be

$$|\mathbf{v}| = \sqrt{a^2 + b^2}. \tag{1}$$

The **zero vector**, denoted by $\mathbf{0}$, is defined by $\mathbf{0} = \langle 0, 0 \rangle$. The magnitude of the zero vector is zero. The zero vector is not assigned any direction. Let $\mathbf{v} = \langle x, y \rangle$ be a nonzero vector. If θ is an angle in standard position formed by $\mathbf{v}$ and the positive x-axis, as shown in FIGURE 10.4.4, then θ is called a **direction angle** for $\mathbf{v}$. We note that any angle coterminal with θ is also a direction angle for $\mathbf{v}$. Thus a vector $\mathbf{v}$ can be specified by giving either its components $\mathbf{v} = \langle x, y \rangle$ or its magnitude $|\mathbf{v}|$ and a direction angle. From trigonometry, we have the following relationships between the components, magnitude, and the direction angle of a vector $\mathbf{v}$.

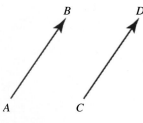
$\overrightarrow{AB} = \overrightarrow{CD}$
FIGURE 10.4.2 Equal vectors

FIGURE 10.4.3 Position vector

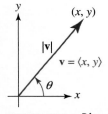
FIGURE 10.4.4 Direction angle of a vector

DEFINITION 10.4.1 Direction Angle

For any nonzero vector $\mathbf{v} = \langle x, y \rangle$ with direction angle θ:

$$\cos\theta = \frac{x}{|\mathbf{v}|}, \quad \sin\theta = \frac{y}{|\mathbf{v}|}, \quad \tan\theta = \frac{y}{x}, \quad x \neq 0, \tag{2}$$

where $|\mathbf{v}| = \sqrt{x^2 + y^2}$.

EXAMPLE 1   **Direction Angle**

Sketch each of the following vectors. Find the magnitude and the smallest positive direction angle θ of each vector.
(a) $\mathbf{v} = \langle -2, 2 \rangle$ (b) $\mathbf{u} = \langle 0, 3 \rangle$ (c) $\mathbf{w} = \langle 1, -\sqrt{3} \rangle$

Solution The three vectors are sketched using different colors in FIGURE 10.4.5.
(a) From (1),

$$|\mathbf{v}| = \sqrt{(-2)^2 + 2^2} = \sqrt{8} = 2\sqrt{2},$$

and from (2)

$$\tan\theta = \frac{2}{-2} = -1.$$

FIGURE 10.4.5 Vectors in Example 1

As we see in Figure 10.4.5, θ is a second-quadrant angle and so we conclude that $\theta = 3\pi/4$.

(b) The magnitude of **u** is $|\mathbf{u}| = \sqrt{0^2 + 3^2} = 3$ and from Figure 10.4.5 we see immediately that $\theta = \pi/2$.

(c) The magnitude of **w** is $|\mathbf{w}| = \sqrt{1^2 + (-\sqrt{3})^2} = \sqrt{4} = 2$. Since $\tan\theta = -\sqrt{3}$ and θ is a fourth-quadrant angle, we choose $\theta = 5\pi/3$. ≡

☐ **Vector Arithmetic** Vectors can be combined with other vectors by the arithmetic operation of addition. In addition, vectors can be combined with scalars through multiplication. Using the component form of a vector we give next the algebraic definitions of the **sum** of two vectors, the **scalar multiple** of a vector, and **equality** of two vectors.

DEFINITION 10.4.2 Operations on Vectors

Let $\mathbf{u} = \langle a, b \rangle$ and $\mathbf{v} = \langle c, d \rangle$ be vectors, and let k be a real number. Then we define the

Sum: $\mathbf{u} + \mathbf{v} = \langle a + c, b + d \rangle$ (3)
Scalar Multiple: $k\mathbf{u} = \langle ka, kb \rangle$ (4)
Equality: $\mathbf{u} = \mathbf{v}$ if and only if $a = c, b = d$ (5)

☐ **Subtraction** Using (4), we define the **negative** of a vector **u** by

$$-\mathbf{u} = (-1)\mathbf{u} = \langle -a, -b \rangle.$$

We can then define **subtraction**, or the **difference**, of two vectors as

$$\mathbf{u} - \mathbf{v} = \mathbf{u} + (-\mathbf{v}) = \langle a - c, b - d \rangle. \tag{6}$$

EXAMPLE 2   Addition, Subtraction, and Scalar Multiplication

If $\mathbf{u} = \langle 2, 1 \rangle$ and $\mathbf{v} = \langle -1, 5 \rangle$ find $4\mathbf{u}$, $\mathbf{u} + \mathbf{v}$, and $3\mathbf{u} - 2\mathbf{v}$.

Solution From the definitions of addition, subtraction, and scalar multiples of vectors, we find

$$4\mathbf{u} = 4\langle 2, 1 \rangle = \langle 8, 4 \rangle \quad \leftarrow \text{from (4)}$$
$$\mathbf{u} + \mathbf{v} = \langle 2, 1 \rangle + \langle -1, 5 \rangle = \langle 1, 6 \rangle \quad \leftarrow \text{from (3)}$$
$$3\mathbf{u} - 2\mathbf{v} = 3\langle 2, 1 \rangle - 2\langle -1, 5 \rangle = \langle 6, 3 \rangle - \langle -2, 10 \rangle = \langle 8, -7 \rangle. \leftarrow \text{from (4) and (6)} \equiv$$

Operations (3), (4), and (6) possess the following properties.

THEOREM 10.4.1 Properties of Vector Operations

(i) $\mathbf{u} + \mathbf{v} = \mathbf{v} + \mathbf{u}$ (ii) $\mathbf{u} + (\mathbf{v} + \mathbf{w}) = (\mathbf{u} + \mathbf{v}) + \mathbf{w}$
(iii) $k(\mathbf{u} + \mathbf{v}) = k\mathbf{u} + k\mathbf{v}$ (iv) $(k_1 + k_2)\mathbf{u} = k_1\mathbf{u} + k_2\mathbf{u}$
(v) $k_1(k_2\mathbf{u}) = (k_1 k_2)\mathbf{u}$ (vi) $\mathbf{u} + \mathbf{0} = \mathbf{u}$
(vii) $\mathbf{u} + (-\mathbf{u}) = \mathbf{0}$ (viii) $0\mathbf{u} = \mathbf{0}$
(ix) $1\mathbf{u} = \mathbf{u}$ (x) $|k\mathbf{u}| = |k||\mathbf{u}|$

You should recognize properties (*i*) and (*ii*) as the commutative and associative laws of addition, respectively.

□ **Geometric Interpretations** The sum $\mathbf{u} + \mathbf{v}$ of two vectors can readily be interpreted geometrically in the plane using the concept of a position vector. If $\mathbf{u} = \langle a, b \rangle$ and $\mathbf{v} = \langle c, d \rangle$, then the three vectors $\mathbf{u}, \mathbf{v}$, and $\mathbf{u} + \mathbf{v}$ can be represented by directed line segments from the origin to the points $A(a, b)$, $B(c, d)$, and $C(a + c, b + d)$, respectively. As shown in FIGURE 10.4.6, if the vector $\mathbf{v}$ is translated so that its initial point is A, then its terminal point will be C. Thus a geometric representation of the sum $\mathbf{u} + \mathbf{v}$ can be obtained by placing the initial point of $\mathbf{v}$ on the terminal point of $\mathbf{u}$ and drawing the vector from the initial point of $\mathbf{u}$ to the terminal point of $\mathbf{v}$. By examining the coordinates of the quadrilateral $OACB$ in Figure 10.4.6, we see that it is a parallelogram formed by the vectors $\mathbf{u}$ and $\mathbf{v}$, with $\mathbf{u} + \mathbf{v}$ as one of its diagonals.

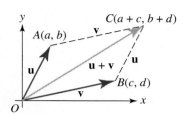

FIGURE 10.4.6 Sum of two vectors $\mathbf{u}$ and $\mathbf{v}$

We now consider a scalar multiple of the vector $\mathbf{v} = \langle x, y \rangle$. Let k be any real number, then

$$|k\mathbf{v}| = \sqrt{(kx)^2 + (ky)^2} = \sqrt{k^2(x^2 + y^2)}$$
$$= \sqrt{k^2}\sqrt{x^2 + y^2} = |k|\sqrt{x^2 + y^2} = |k||\mathbf{v}|.$$

We have derived the property of scalar multiplication given in property (x) of Theorem 10.4.1, that is,

$$|k\mathbf{v}| = |k||\mathbf{v}|. \tag{7}$$

This property states that in the scalar multiplication of a vector $\mathbf{v}$ by a real number k, the magnitude of $\mathbf{v}$ is multiplied by $|k|$. As shown in FIGURE 10.4.7, if $k > 0$, the direction of $\mathbf{v}$ does not change; but if $k < 0$, the direction of $\mathbf{v}$ is reversed. In particular, $\mathbf{v}$ and its negative $-\mathbf{v}$ have the same length but opposite direction.

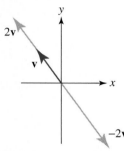

FIGURE 10.4.7 Scalar multiples of vector $\mathbf{v}$

The geometric interpretation of the difference $\mathbf{u} - \mathbf{v}$ of two vectors is obtained by observing that $\mathbf{u} = \mathbf{v} + (\mathbf{u} - \mathbf{v})$. Thus, $\mathbf{u} - \mathbf{v}$ is the vector that when added to $\mathbf{v}$ yields $\mathbf{u}$. As we see in FIGURE 10.4.8, the initial point of $\mathbf{u} - \mathbf{v}$ will be at the terminal point of $\mathbf{v}$, and the terminal point of $\mathbf{u} - \mathbf{v}$ coincides with the terminal point of $\mathbf{u}$. Hence the vector $\mathbf{u} - \mathbf{v}$ is one diagonal of the parallelogram determined by $\mathbf{u}$ and $\mathbf{v}$, with $\mathbf{u} + \mathbf{v}$ being the other diagonal. See FIGURE 10.4.9.

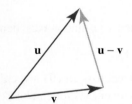

FIGURE 10.4.8 Difference of two vectors $\mathbf{u}$ and $\mathbf{v}$

FIGURE 10.4.9 Sum and difference of vectors $\mathbf{u}$ and $\mathbf{v}$ as diagonals of a parallelogram

EXAMPLE 3 Sum and Difference

Let $\mathbf{u} = \langle -1, 1 \rangle$ and $\mathbf{v} = \langle 3, 2 \rangle$. Sketch the geometric interpretations of $\mathbf{u} + \mathbf{v}$ and $\mathbf{u} - \mathbf{v}$.

Solution To interpret these vectors geometrically, we form the parallelogram determined by the vectors $\mathbf{u}$ and $\mathbf{v}$ and identify the diagonals $\mathbf{u} + \mathbf{v}$ and $\mathbf{u} - \mathbf{v}$, as shown in FIGURE 10.4.10. ≡

FIGURE 10.4.10 Sum and difference of vectors in Example 3

EXAMPLE 4 Scalar Multiples

Let $\mathbf{v} = \langle 1, 2 \rangle$. The geometric interpretations of scalar multiples $2\mathbf{v}$ and $-\mathbf{v}$ are shown in FIGURE 10.4.11. ≡

FIGURE 10.4.11 Scalar multiples of a vector in Example 4

10.4 Vectors in the Plane

☐ **Unit Vectors** Any vector with magnitude 1 is called a **unit vector**. We can obtain a unit vector **u** in the same direction as a nonzero vector **v** by multiplying **v** by the positive scalar $k = 1/|\mathbf{v}|$ (reciprocal of its magnitude). In this case we say that

$$\mathbf{u} = \left(\frac{1}{|\mathbf{v}|}\right)\mathbf{v} = \frac{\mathbf{v}}{|\mathbf{v}|} \tag{8}$$

is the **normalization** of the vector **v**. It follows from (7) that the normalization of a vector **v** is a unit vector because

$$|\mathbf{u}| = \left|\frac{1}{|\mathbf{v}|}\mathbf{v}\right| = \frac{1}{|\mathbf{v}|}|\mathbf{v}| = 1.$$

EXAMPLE 5 Unit Vector

Given $\mathbf{v} = \langle 2, -1 \rangle$, find a unit vector **(a)** in the same direction as **v**, and **(b)** in the opposite direction of **v**.

Solution First, we find the magnitude of the vector **v**:

$$|\mathbf{v}| = \sqrt{4 + (-1)^2} = \sqrt{5}.$$

(a) From (8), a unit vector in the same direction as **v** is then

$$\mathbf{u} = \frac{1}{\sqrt{5}}\mathbf{v} = \frac{1}{\sqrt{5}}\langle 2, -1 \rangle = \left\langle \frac{2}{\sqrt{5}}, \frac{-1}{\sqrt{5}} \right\rangle.$$

(b) A unit vector in the opposite direction of **v** is the negative of **u**:

$$-\mathbf{u} = \left\langle -\frac{2}{\sqrt{5}}, \frac{1}{\sqrt{5}} \right\rangle.$$

≡

FIGURE 10.4.12 The **i** and **j** vectors

☐ **i, j Vectors** The unit vectors in the direction of the positive x- and y-axes, denoted by

$$\mathbf{i} = \langle 1, 0 \rangle \quad \text{and} \quad \mathbf{j} = \langle 0, 1 \rangle, \tag{9}$$

are of special importance. See **FIGURE 10.4.12**. The unit vectors in (9) are called the **standard basis vectors** for the vectors in the plane because every vector can be expressed in terms of **i** and **j**. To see why this is so we use the definition of scalar multiplication to rewrite $\mathbf{u} = \langle a, b \rangle$ as

or
$$\mathbf{u} = \langle a, 0 \rangle + \langle 0, b \rangle = a\langle 1, 0 \rangle + b\langle 0, 1 \rangle$$
$$\mathbf{u} = \langle a, b \rangle = a\mathbf{i} + b\mathbf{j}.$$

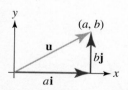

FIGURE 10.4.13 A vector **u** is a linear combination of **i** and **j**

As shown in **FIGURE 10.4.13**, since **i** and **j** are unit vectors, the vectors $a\mathbf{i}$ and $b\mathbf{j}$ are horizontal and vertical vectors of length and $|a|$ and $|b|$, respectively. For this reason, a is called the **horizontal component** of **u**, and b is called the **vertical component**. The vector $a\mathbf{i} + b\mathbf{j}$ is often referred to as a **linear combination** of **i** and **j**. Using this notation for the vectors $\mathbf{u} = a\mathbf{i} + b\mathbf{j}$ and $\mathbf{v} = c\mathbf{i} + d\mathbf{j}$, we can write the definition of the sum, difference, and scalar multiples of **u** and **v** in the following manner:

Sum: $(a\mathbf{i} + b\mathbf{j}) + (c\mathbf{i} + d\mathbf{j}) = (a + c)\mathbf{i} + (b + d)\mathbf{j}$ (10)
Difference: $(a\mathbf{i} + b\mathbf{j}) - (c\mathbf{i} + d\mathbf{j}) = (a - c)\mathbf{i} + (b - d)\mathbf{j}$ (11)
Scalar multiple: $k(a\mathbf{i} + b\mathbf{j}) = (ka)\mathbf{i} + (kb)\mathbf{j}$ (12)

EXAMPLE 6 Difference of Vectors

If $\mathbf{u} = 3\mathbf{i} + \mathbf{j}$ and $\mathbf{v} = 5\mathbf{i} - 2\mathbf{j}$, find $4\mathbf{u} - 2\mathbf{v}$.

Solution We use (12) followed by (11) to obtain

$$\begin{aligned}4\mathbf{u} - 2\mathbf{v} &= 4(3\mathbf{i} + \mathbf{j}) - 2(5\mathbf{i} - 2\mathbf{j}) \\ &= (12\mathbf{i} + 4\mathbf{j}) - (10\mathbf{i} - 4\mathbf{j}) \\ &= (12 - 10)\mathbf{i} + (4 - (-4))\mathbf{j} \\ &= 2\mathbf{i} + 8\mathbf{j}.\end{aligned}$$

≡

☐ **Trigonometric Form of a Vector** There is yet another way of representing vectors. For a nonzero vector $\mathbf{v} = \langle x, y \rangle$ with direction angle θ, we see from (2) that $x = |\mathbf{v}|\cos\theta$ and $y = |\mathbf{v}|\sin\theta$. Thus,

$$\mathbf{v} = x\mathbf{i} + y\mathbf{j} = |\mathbf{v}|\cos\theta\,\mathbf{i} + |\mathbf{v}|\sin\theta\,\mathbf{j},$$

or

$$\mathbf{v} = |\mathbf{v}|(\cos\theta\,\mathbf{i} + \sin\theta\,\mathbf{j}). \tag{13}$$

This latter representation is called the **trigonometric form** of the vector $\mathbf{v}$.

EXAMPLE 7 Trigonometric Form

Express $\mathbf{v} = \sqrt{3}\mathbf{i} - 3\mathbf{j}$ in trigonometric form.

Solution To write $\mathbf{v}$ in trigonometric form, we must find the magnitude $|\mathbf{v}|$ and its direction angle θ. From (1) we find

$$|\mathbf{v}| = \sqrt{(\sqrt{3})^2 + (-3)^2} = \sqrt{12} = 2\sqrt{3}$$

and from (2)

$$\tan\theta = -\frac{3}{\sqrt{3}} = -\sqrt{3}.$$

To determine θ, we sketch $\mathbf{v}$ and observe that the terminal side of the angle θ lies in the fourth quadrant. See FIGURE 10.4.14. Thus, with $|\mathbf{v}| = 2\sqrt{3}$ and $\theta = 5\pi/3$, (13) gives the trigonometric form of $\mathbf{v}$:

$$\mathbf{v} = 2\sqrt{3}\left(\cos\frac{5\pi}{3}\mathbf{i} + \sin\frac{5\pi}{3}\mathbf{j}\right).$$

≡

FIGURE 10.4.14 Vector and direction angle in Example 7

EXAMPLE 8 Velocity as a Vector

Given that an airplane is flying at 200 mi/h on a bearing of N20°E, express its velocity as a vector.

Solution The desired velocity vector $\mathbf{v}$ is shown in FIGURE 10.4.15. Since $\theta = 90° - 20° = 70°$ and $|\mathbf{v}| = 200$, we have

$$\mathbf{v} = 200(\cos 70°\mathbf{i} + \sin 70°\mathbf{j}) \approx 68.4\mathbf{i} + 187.9\mathbf{j}.$$

≡

FIGURE 10.4.15 Velocity vector in Example 8

In Example 8 we see that velocity is a vector quantity. The magnitude $|\mathbf{v}|$ of the velocity $\mathbf{v}$ is a scalar quantity called **speed**.

In physics it is shown that when two forces act simultaneously at the same point P on an object, the object reacts as though a single force equal to the vector sum of the two forces is acting on the object at P. This single force is called the **resultant force**.

10.4 Vectors in the Plane

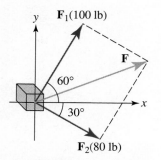

FIGURE 10.4.16 Resultant force (gold) in Example 9

EXAMPLE 9 Resultant Force

Two people push on a crate with forces $\mathbf{F}_1$ and $\mathbf{F}_2$, whose magnitudes and directions are shown in **FIGURE 10.4.16**. Find the magnitude and the direction of the resultant force.

Solution From the figure, we see that the direction angles for the two forces $\mathbf{F}_1$ and $\mathbf{F}_2$ are $\theta_1 = 60°$ and $\theta_2 = 330°$, respectively. Thus,

$$\mathbf{F}_1 = 100(\cos 60°\mathbf{i} + \sin 60°\mathbf{j}) = 50\mathbf{i} + 50\sqrt{3}\mathbf{j}$$

and

$$\mathbf{F}_2 = 80(\cos 330°\mathbf{i} + \sin 330°\mathbf{j}) = 40\sqrt{3}\mathbf{i} - 40\mathbf{j}.$$

The resultant force $\mathbf{F}$ can then be found by vector addition:

$$\mathbf{F} = \mathbf{F}_1 + \mathbf{F}_2 = (50\mathbf{i} + 50\sqrt{3}\mathbf{j}) + (40\sqrt{3}\mathbf{i} - 40\mathbf{j})$$
$$= (50 + 40\sqrt{3})\mathbf{i} + (50\sqrt{3} - 40)\mathbf{j}.$$

Thus the magnitude $|\mathbf{F}|$ of the resultant force is

$$|\mathbf{F}| = \sqrt{(50 + 40\sqrt{3})^2 + (50\sqrt{3} - 40)^2} \approx 128.06.$$

If θ is a direction angle for $\mathbf{F}$, then we know from (2) that

$$\tan \theta = \frac{50\sqrt{3} - 40}{50 + 40\sqrt{3}} \approx 0.3907.$$

Since θ is a first-quadrant angle, we find with the help of a calculator that $\theta \approx 21.34°$. ≡

NOTES FROM THE CLASSROOM

You should not draw the conclusion from the preceding discussion that all vector quantities can be pictured as arrows. Many applications of vectors in advanced mathematics do not lend themselves to this interpretation. However, for purposes in this text, we find this interpretation both convenient and useful.

10.4 Exercises   Answers to selected odd-numbered problems begin on page ANS-29.

In Problems 1–8, sketch the given vector. Find the magnitude and the smallest positive direction angle of each vector.

1. $\langle \sqrt{3}, -1 \rangle$
2. $\langle 4, -4 \rangle$
3. $\langle 5, 0 \rangle$
4. $\langle -2, 2\sqrt{3} \rangle$
5. $-4\mathbf{i} + 4\sqrt{3}\mathbf{j}$
6. $\mathbf{i} - \mathbf{j}$
7. $-10\mathbf{i} + 10\mathbf{j}$
8. $-3\mathbf{j}$

In Problems 9–14, find $\mathbf{u} + \mathbf{v}$, $\mathbf{u} - \mathbf{v}$, $-3\mathbf{u}$, and $3\mathbf{u} - 4\mathbf{v}$.

9. $\mathbf{u} = \langle 2, 3 \rangle$, $\mathbf{v} = \langle 1, -1 \rangle$
10. $\mathbf{u} = \langle 4, -2 \rangle$, $\mathbf{v} = \langle 10, 2 \rangle$
11. $\mathbf{u} = \langle -4, 2 \rangle$, $\mathbf{v} = \langle 4, 1 \rangle$
12. $\mathbf{u} = \langle -1, -5 \rangle$, $\mathbf{v} = \langle 8, 7 \rangle$
13. $\mathbf{u} = \langle -5, -7 \rangle$, $\mathbf{v} = \langle \frac{1}{2}, -\frac{1}{4} \rangle$
14. $\mathbf{u} = \langle 0.1, 0.2 \rangle$, $\mathbf{v} = \langle -0.3, 0.4 \rangle$

In Problems 15–20, find $\mathbf{u} - 4\mathbf{v}$ and $2\mathbf{u} + 5\mathbf{v}$.

15. $\mathbf{u} = \mathbf{i} - 2\mathbf{j}, \mathbf{v} = 8\mathbf{i} + 3\mathbf{j}$
16. $\mathbf{u} = \mathbf{j}, \mathbf{v} = 4\mathbf{i} - \mathbf{j}$
17. $\mathbf{u} = \frac{1}{2}\mathbf{i} - \frac{3}{2}\mathbf{j}, \mathbf{v} = 2\mathbf{i}$
18. $\mathbf{u} = 2\mathbf{i} - 3\mathbf{j}, \mathbf{v} = 3\mathbf{i} - 2\mathbf{j}$
19. $\mathbf{u} = 0.2\mathbf{i} + 0.1\mathbf{j}, \mathbf{v} = -1.4\mathbf{i} - 2.1\mathbf{j}$
20. $\mathbf{u} = 5\mathbf{i} - 10\mathbf{j}, \mathbf{v} = -10\mathbf{i}$

In Problems 21–24, sketch the vectors $\mathbf{u} + \mathbf{v}$ and $\mathbf{u} - \mathbf{v}$.

21. $\mathbf{u} = 2\mathbf{i} + 3\mathbf{j}, \mathbf{v} = -\mathbf{i} + 2\mathbf{j}$
22. $\mathbf{u} = -4\mathbf{i} + \mathbf{j}, \mathbf{v} = 2\mathbf{i} + 2\mathbf{j}$
23. $\mathbf{u} = 5\mathbf{i} - \mathbf{j}, \mathbf{v} = 4\mathbf{i} - 3\mathbf{j}$
24. $\mathbf{u} = 2\mathbf{i} - 7\mathbf{j}, \mathbf{v} = -7\mathbf{i} - 3\mathbf{j}$

In Problems 25–28, sketch the vectors $2\mathbf{v}$ and $-2\mathbf{v}$.

25. $\mathbf{v} = \langle -2, 1 \rangle$
26. $\mathbf{v} = \langle 4, 7 \rangle$
27. $\mathbf{v} = 3\mathbf{i} - 5\mathbf{j}$
28. $\mathbf{v} = -\frac{1}{2}\mathbf{i} + \frac{3}{2}\mathbf{j}$

In Problems 29–32, if $\mathbf{u} = 3\mathbf{i} - \mathbf{j}$ and $\mathbf{v} = 2\mathbf{i} + 4\mathbf{j}$, find the horizontal and the vertical components of the indicated vector.

29. $2\mathbf{u} - \mathbf{v}$
30. $3(\mathbf{u} + \mathbf{v})$
31. $\mathbf{v} - 4\mathbf{u}$
32. $4(\mathbf{u} + 3\mathbf{v})$

In Problems 33–36, express the given vector **(a)** in trigonometric form and **(b)** as a linear combination of the unit vectors $\mathbf{i}$ and $\mathbf{j}$.

33. $\langle -\sqrt{2}, \sqrt{2} \rangle$
34. $\langle 7, 7\sqrt{3} \rangle$
35. $\langle -3\sqrt{3}, 3 \rangle$
36. $\langle -4, -4 \rangle$

In Problems 37–40, find a unit vector **(a)** in the same direction as $\mathbf{v}$, and **(b)** in the opposite direction of $\mathbf{v}$.

37. $\mathbf{v} = \langle 2, 2 \rangle$
38. $\mathbf{v} = \langle -3, 4 \rangle$
39. $\mathbf{v} = \langle 0, -5 \rangle$
40. $\mathbf{v} = \langle 1, -\sqrt{3} \rangle$

In Problems 41 and 42, normalize the given vector when $\mathbf{v} = \langle 2, 8 \rangle$ and $\mathbf{w} = \langle 3, 4 \rangle$.

41. $\mathbf{v} + \mathbf{w}$
42. $2\mathbf{v} - 3\mathbf{w}$

43. Two forces $\mathbf{F}_1$ and $\mathbf{F}_2$ of magnitudes 4 N and 7 N, respectively, act on a point. If the angle between the forces is 47°, find the magnitude of the resultant force $\mathbf{F}$ and the angle between $\mathbf{F}_1$ and $\mathbf{F}$.

44. The resultant $\mathbf{F}$ of two forces $\mathbf{F}_1$ and $\mathbf{F}_2$ has a magnitude of 100 lb and direction as shown in FIGURE 10.4.17. If $\mathbf{F}_1 = -200\mathbf{i}$, find the horizontal and the vertical components of $\mathbf{F}_2$.

FIGURE 10.4.17 Resultant in Problem 44

Miscellaneous Applications

45. Resultant Force A mass weighing 10 lb is hanging from a rope. A 2-lb force is applied horizontally to the weight, moving the rope from its horizontal position. See FIGURE 10.4.18. Find the resultant of this force and the force due to gravity.

46. Resultant Force A small boat is pulled along a canal by two tow ropes on opposite sides of the canal. The angle between the ropes is 50°. If one rope is pulled with a force of 250 lb and the other with a force of 400 lb, find the magnitude of the resultant force and the angle it makes with the 250-lb force.

47. Actual Direction The current in a river that is 0.5 mi across is 6 mi/h. A swimmer heads out from shore perpendicular to the current at 2 mi/h. In what direction is the swimmer actually going?

FIGURE 10.4.18 Hanging mass in Problem 45

48. **What Direction?** As a freight train, traveling at 10 mi/h, passes a landing, a mail sack is tossed out perpendicular to the train with a velocity of 15 feet per second. In what direction does it slide on the landing?
49. **What Is the Speed?** In order for an airplane to fly due north at 300 mi/h, it must set a course 10° west of north (N10°W) because of a strong wind blowing due east. What is the speed of the wind?
50. **Getting One's Bearings** A hiker walks 1.0 mi to the northeast, then 1.5 mi to the east, and then 2.0 mi to the southeast. What are the hiker's distance and bearing from the starting point? [*Hint*: Each part of the journey can be represented by a vector. Find the vector sum.]

10.5 Dot Product

☰ **Introduction** In the last section we considered two kinds of vector operations on vectors, addition and scalar multiplication, that produced another vector. We now consider a special kind of product between vectors that originated in the study of mechanics. This product, known as the **dot product**, or **inner product**, of vectors **u** and **v** is denoted by **u** · **v** and is a real number, or scalar, defined in terms of the components of the vectors.

DEFINITION 10.5.1 Dot Product

In 2-space the **dot product** of two vectors $\mathbf{u} = \langle a, b \rangle$ and $\mathbf{v} = \langle c, d \rangle$ is the number

$$\mathbf{u} \cdot \mathbf{v} = ac + bd. \tag{1}$$

EXAMPLE 1 Dot Product Using (1)

Suppose $\mathbf{u} = \langle -2, 5 \rangle$, $\mathbf{v} = \langle \frac{1}{2}, 4 \rangle$, and $\mathbf{w} = \langle 8, -1 \rangle$. Find
(a) **u** · **v**   (b) **w** · **u**   (c) **v** · **w**.

Solution It follows from (1) that
(a) $\mathbf{u} \cdot \mathbf{v} = \langle -2, 5 \rangle \cdot \langle \frac{1}{2}, 4 \rangle = (-2)(\frac{1}{2}) + (5)(4) = -1 + 20 = 19$
(b) $\mathbf{w} \cdot \mathbf{u} = \langle 8, -1 \rangle \cdot \langle -2, 5 \rangle = (8)(-2) + (-1)(5) = -16 - 5 = -21$
(c) $\mathbf{v} \cdot \mathbf{w} = \langle \frac{1}{2}, 4 \rangle \cdot \langle 8, -1 \rangle = (\frac{1}{2})(8) + (4)(-1) = 4 - 4 = 0.$ ☰

☐ **Properties** The dot product possesses the following properties.

THEOREM 10.5.1 Properties of the Dot Product

(*i*) $\mathbf{u} \cdot \mathbf{v} = 0$ if $\mathbf{u} = \mathbf{0}$ or $\mathbf{v} = \mathbf{0}$
(*ii*) $\mathbf{u} \cdot \mathbf{v} = \mathbf{v} \cdot \mathbf{u}$ ← commutative law
(*iii*) $\mathbf{u} \cdot (\mathbf{v} + \mathbf{w}) = \mathbf{u} \cdot \mathbf{v} + \mathbf{u} \cdot \mathbf{w}$ ← distributive law
(*iv*) $\mathbf{u} \cdot (k\mathbf{v}) = (k\mathbf{u}) \cdot \mathbf{v} = k(\mathbf{u} \cdot \mathbf{v})$, k a scalar
(*v*) $\mathbf{v} \cdot \mathbf{v} \geq 0$
(*vi*) $\mathbf{v} \cdot \mathbf{v} = |\mathbf{v}|^2$

PROOF: We prove parts (*ii*) and (*vi*). The remaining proofs are straightforward and left for the reader. To prove part (*ii*) we let $\mathbf{u} = \langle a, b \rangle$ and $\mathbf{v} = \langle c, d \rangle$. Then

$$\mathbf{u} \cdot \mathbf{v} = \langle a, b \rangle \cdot \langle c, d \rangle$$
$$= ac + bd$$
$$= ca + db \quad \longleftarrow \begin{cases} \text{since multiplication} \\ \text{of real numbers is} \\ \text{commutative} \end{cases}$$
$$= \langle c, d \rangle \cdot \langle a, b \rangle = \mathbf{v} \cdot \mathbf{u}.$$

To prove part (*vi*) we note that

$$\mathbf{v} \cdot \mathbf{v} = \langle c, d \rangle \cdot \langle c, d \rangle = c^2 + d^2 = |\mathbf{v}|^2.$$

≡

EXAMPLE 2 Dot Products

Let $\mathbf{u} = \langle 3, 2 \rangle$ and $\mathbf{v} = \langle -4, -5 \rangle$. Find
(a) $(\mathbf{u} \cdot \mathbf{v})\mathbf{u}$ (b) $\mathbf{u} \cdot (\frac{1}{2}\mathbf{v})$ (c) $|\mathbf{v}|$.

Solution (a) From (1),

$$\mathbf{u} \cdot \mathbf{v} = \langle 3, 2 \rangle \cdot \langle -4, -5 \rangle$$
$$= 3(-4) + 2(-5)$$
$$= -22.$$

Because $\mathbf{u} \cdot \mathbf{v}$ is a scalar we have from (4) of Section 10.4,

$$(\mathbf{u} \cdot \mathbf{v})\mathbf{u} = (-22)\langle 3, 2 \rangle = \langle -66, -44 \rangle.$$

(b) From (*iv*) of Theorem 10.5.1 and part (a),

$$\mathbf{u} \cdot (\tfrac{1}{2}\mathbf{v}) = \tfrac{1}{2}(\mathbf{u} \cdot \mathbf{v}) = \tfrac{1}{2}(-22) = -11.$$

(c) Part (*vi*) of Theorem 10.5.1 relates the magnitude of a vector with the dot product. From (1) we have

$$\mathbf{v} \cdot \mathbf{v} = \langle -4, -5 \rangle \cdot \langle -4, -5 \rangle$$
$$= (-4)(-4) + (-5)(-5)$$
$$= 41.$$

Therefore,

$$\mathbf{v} \cdot \mathbf{v} = |\mathbf{v}|^2 \quad \text{implies} \quad |\mathbf{v}| = \sqrt{\mathbf{v} \cdot \mathbf{v}} = \sqrt{41}.$$

≡

☐ **Alternative Form** The dot product of two vectors can also be expressed in terms of the lengths of the vectors and the angle between them.

THEOREM 10.5.2 Alternative Form of the Dot Product

The dot product of two vectors $\mathbf{u}$ and $\mathbf{v}$ is

$$\mathbf{u} \cdot \mathbf{v} = |\mathbf{u}||\mathbf{v}| \cos\theta, \qquad (2)$$

where θ is the angle between the vectors such that $0 \leq \theta \leq \pi$.

◀ This more geometric form is what is generally used as the definition of the dot product in a physics course.

PROOF: Suppose θ is the angle between the vectors $\mathbf{u} = a\mathbf{i} + b\mathbf{j}$ and $\mathbf{v} = c\mathbf{i} + d\mathbf{j}$. Then the vector

$$\mathbf{w} = \mathbf{v} - \mathbf{u} = (c - a)\mathbf{i} + (d - b)\mathbf{j}$$

FIGURE 10.5.1 The vector **w** in the proof of Theorem 10.5.2

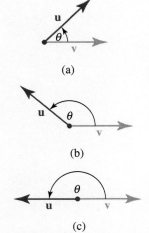

FIGURE 10.5.2 The angle θ in the dot product

FIGURE 10.5.3 Angle between the vector in Example 3

is the third side of the triangle indicated in **FIGURE 10.5.1**. By the Law of Cosines, (2) of Section 8.4, we can write

$$|\mathbf{w}|^2 = |\mathbf{v}|^2 + |\mathbf{u}|^2 - 2|\mathbf{v}||\mathbf{u}|\cos\theta \quad \text{or} \quad |\mathbf{v}||\mathbf{u}|\cos\theta = \tfrac{1}{2}(|\mathbf{v}|^2 + |\mathbf{u}|^2 - |\mathbf{w}|^2). \quad (3)$$

Using $|\mathbf{u}|^2 = a^2 + b^2$, $|\mathbf{v}|^2 = c^2 + d^2$,
and $|\mathbf{w}|^2 = |\mathbf{v} - \mathbf{u}|^2 = (c-a)^2 + (d-b)^2$,

the right-hand side of the second equation in (3) simplifies to $ac + bd$. Since this is the definition of the dot product given in (1), we see that $|\mathbf{u}||\mathbf{v}|\cos\theta = \mathbf{u}\cdot\mathbf{v}$. ≡

☐ **Angle Between Vectors** **FIGURE 10.5.2** illustrates three cases of the angle θ in (2). If the vectors **u** and **v** are not parallel, then θ is the *smaller* of the two possible angles between them. Solving for $\cos\theta$ in (2) and then using the definition of the dot product in (1) we have a formula for the cosine of the angle between two vectors:

$$\cos\theta = \frac{\mathbf{u}\cdot\mathbf{v}}{|\mathbf{u}||\mathbf{v}|} = \frac{ac + bd}{|\mathbf{u}||\mathbf{v}|}. \quad (4)$$

EXAMPLE 3   **Angle Between Two Vectors**

Find the angle between $\mathbf{u} = 2\mathbf{i} + 5\mathbf{j}$ and $\mathbf{v} = 5\mathbf{i} - 4\mathbf{j}$.

Solution We have $|\mathbf{u}| = \sqrt{\mathbf{u}\cdot\mathbf{u}} = \sqrt{29}$, $|\mathbf{v}| = \sqrt{\mathbf{v}\cdot\mathbf{v}} = \sqrt{41}$, and $\mathbf{u}\cdot\mathbf{v} = -10$. Hence, (4) gives

$$\cos\theta = \frac{-10}{\sqrt{29}\sqrt{41}} \approx -0.2900,$$

and so $\theta = \cos^{-1}\left(\dfrac{-10}{\sqrt{29}\sqrt{41}}\right) \approx 1.8650$ radians or $\theta = 108.86°$. See **FIGURE 10.5.3**. ≡

☐ **Orthogonal Vectors** If **u** and **v** are nonzero vectors, then Theorem 10.5.2 implies that

(*i*) $\mathbf{u}\cdot\mathbf{v} > 0$ if and only if θ is acute,
(*ii*) $\mathbf{u}\cdot\mathbf{v} < 0$ if and only if θ is obtuse, and
(*iii*) $\mathbf{u}\cdot\mathbf{v} = 0$ if and only if $\cos\theta = 0$.

But in the last case, the only number in the interval $[0, 2\pi]$ for which $\cos\theta = 0$ is $\theta = \pi/2$. When $\theta = \pi/2$, we say that the vectors are **orthogonal** or **perpendicular**. Thus, we are led to the following result.

THEOREM 10.5.3   Criterion for Orthogonal Vectors

Two nonzero vectors **u** and **v** are orthogonal if and only if $\mathbf{u}\cdot\mathbf{v} = 0$.

As seen in Figure 10.4.12 the standard basis vectors **i** and **j** are orthogonal. Moreover, because $\mathbf{i} = \langle 1, 0\rangle$, $\mathbf{j} = \langle 0, 1\rangle$ we have

$$\mathbf{i}\cdot\mathbf{j} = \langle 1, 0\rangle\cdot\langle 0, 1\rangle = (1)(0) + (0)(1) = 0,$$

and so from Theorem 10.5.3 the vectors **i** and **j** are orthogonal. Inspection of the result in part (c) of Example 1 shows that the two vectors $\mathbf{v} = \langle \tfrac{1}{2}, 4\rangle$, and $\mathbf{w} = \langle 8, -1\rangle$ are orthogonal.

EXAMPLE 4 Orthogonal Vectors

If $\mathbf{u} = \langle 4, 6 \rangle$ and $\mathbf{v} = \langle -3, 2 \rangle$, then

$$\mathbf{u} \cdot \mathbf{v} = (4)(-3) + (6)(2) = -12 + 12 = 0.$$

From Theorem 10.5.3, we conclude that $\mathbf{u}$ and $\mathbf{v}$ are orthogonal. See **FIGURE 10.5.4**. ≡

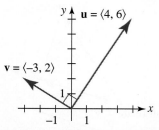

FIGURE 10.5.4 Orthogonal vectors in Example 4

☐ **Component of u on v** Parts (*ii*), (*iii*), and (*vi*) of Theorem 10.5.1 enable us to express the components of a vector $\mathbf{u} = a\mathbf{i} + b\mathbf{j}$ in terms of a dot product:

$$\mathbf{i} \cdot \mathbf{u} = \mathbf{u} \cdot \mathbf{i} = (a\mathbf{i} + b\mathbf{j}) \cdot \mathbf{i} = a\overbrace{(\mathbf{i} \cdot \mathbf{i})}^{1} + b\overbrace{(\mathbf{j} \cdot \mathbf{i})}^{0} = a.$$

That is, $\mathbf{u} \cdot \mathbf{i} = a$. Similarly, $\mathbf{u} \cdot \mathbf{j} = b$. Symbolically, we write these components of $\mathbf{u}$ as

$$\text{comp}_\mathbf{i}\mathbf{u} = \mathbf{u} \cdot \mathbf{i} \quad \text{and} \quad \text{comp}_\mathbf{j}\mathbf{u} = \mathbf{u} \cdot \mathbf{j}. \quad (5)$$

We shall now see that the procedure indicated in (5) carries over to finding the **component of u on a vector v**. Note that in either of the two cases shown in **FIGURE 10.5.5**,

$$\text{comp}_\mathbf{v}\mathbf{u} = |\mathbf{u}|\cos\theta. \quad (6)$$

In Figure 10.5.5(a), $\text{comp}_\mathbf{v}\mathbf{u} > 0$ since $0 < \theta \leq \pi/2$, whereas in Figure 10.5.5(b), $\text{comp}_\mathbf{v}\mathbf{u} < 0$ since $\pi/2 < \theta \leq \pi$. Now, by writing (6) as

$$\text{comp}_\mathbf{v}\mathbf{u} = \frac{|\mathbf{u}||\mathbf{v}|\cos\theta}{|\mathbf{v}|} = \frac{\mathbf{u} \cdot \mathbf{v}}{|\mathbf{v}|},$$

we see that

$$\text{comp}_\mathbf{v}\mathbf{u} = \mathbf{u} \cdot \left(\frac{\mathbf{v}}{|\mathbf{v}|}\right). \quad (7)$$

In other words:

To find the component of vector $\mathbf{u}$ on vector $\mathbf{v}$, we dot $\mathbf{u}$ with a unit vector in the direction of $\mathbf{v}$.

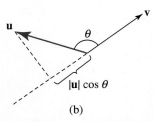

FIGURE 10.5.5 Component of vector $\mathbf{u}$ on vector $\mathbf{v}$

EXAMPLE 5 Component of a Vector on Another

Let $\mathbf{u} = 2\mathbf{i} + 3\mathbf{j}$ and $\mathbf{v} = \mathbf{i} + \mathbf{j}$. Find **(a)** $\text{comp}_\mathbf{v}\mathbf{u}$, and **(b)** $\text{comp}_\mathbf{u}\mathbf{v}$.

Solution (a) We first form a unit vector in the direction of $\mathbf{v}$:

$$|\mathbf{v}| = \sqrt{2} \quad \text{so} \quad \frac{\mathbf{v}}{|\mathbf{v}|} = \frac{1}{\sqrt{2}}(\mathbf{i} + \mathbf{j}).$$

Then from (7) we have

$$\text{comp}_\mathbf{v}\mathbf{u} = (2\mathbf{i} + 3\mathbf{j}) \cdot \frac{1}{\sqrt{2}}(\mathbf{i} + \mathbf{j}) = \frac{5}{\sqrt{2}}.$$

(b) By modifying (7) accordingly, we have

$$\text{comp}_\mathbf{u}\mathbf{v} = \mathbf{v} \cdot \overbrace{\left(\frac{\mathbf{u}}{|\mathbf{u}|}\right)}^{\text{unit vector in the direction of }\mathbf{u}}.$$

Then $|\mathbf{u}| = \sqrt{13}$ so $\dfrac{\mathbf{u}}{|\mathbf{u}|} = \dfrac{1}{\sqrt{13}}(2\mathbf{i} + 3\mathbf{j}),$

and $\text{comp}_\mathbf{u}\mathbf{v} = (\mathbf{i} + \mathbf{j}) \cdot \dfrac{1}{\sqrt{13}}(2\mathbf{i} + 3\mathbf{j}) = \dfrac{5}{\sqrt{13}}.$ ≡

□ **Projection of u onto v** The *projection* of a vector **u** in any of the directions determined by **i** and **j**, is the *vector* formed by multiplying the component of $\mathbf{u} = a\mathbf{i} + b\mathbf{j}$ in the specified direction with a unit vector in that direction; for example,

$$\text{proj}_\mathbf{i}\mathbf{u} = (\text{comp}_\mathbf{i}\mathbf{u})\mathbf{i} = (\mathbf{u} \cdot \mathbf{i})\mathbf{i} = a\,\mathbf{i},$$

and so on. **FIGURE 10.5.6** shows the general case of the **projection of u onto v**:

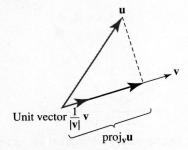

FIGURE 10.5.6 Projection of vector **u** onto vector **v**

$$\text{proj}_\mathbf{v}\mathbf{u} = (\text{comp}_\mathbf{v}\mathbf{u})\overset{\text{unit vector}}{\left(\frac{\mathbf{v}}{|\mathbf{v}|}\right)}. \tag{8}$$

That is,

To find the projection of vector **u** *onto a vector* **v**, *we multiply a unit vector in the direction of* **v** *by the component of* **u** *on* **v**.

If desired, the result in (8) can be expressed in terms of two dot products. Using (7)

$$\text{proj}_\mathbf{v}\mathbf{u} = \left(\mathbf{u} \cdot \overset{\text{scalar}}{\left(\frac{\mathbf{v}}{|\mathbf{v}|}\right)}\right)\overset{\text{unit vector}}{\left(\frac{\mathbf{v}}{|\mathbf{v}|}\right)} = \left(\frac{\mathbf{u} \cdot \mathbf{v}}{|\mathbf{v}|^2}\right)\mathbf{v}$$

or

$$\text{proj}_\mathbf{v}\mathbf{u} = \left(\frac{\mathbf{u} \cdot \mathbf{v}}{\mathbf{v} \cdot \mathbf{v}}\right)\mathbf{v}. \quad \leftarrow |\mathbf{v}|^2 = \mathbf{v} \cdot \mathbf{v} \text{ by } (vi) \text{ of Theorem 10.5.1}$$

EXAMPLE 6   Projection of u onto v

Find the projection of $\mathbf{u} = 4\mathbf{i} + \mathbf{j}$ onto the vector $\mathbf{v} = 2\mathbf{i} + 3\mathbf{j}$. Graph.

Solution First, we find the component of **u** on **v**. A unit vector in the direction of **v** is

$$\frac{1}{\sqrt{13}}(2\mathbf{i} + 3\mathbf{j})$$

and so the component of **u** on **v** is the number

$$\text{comp}_\mathbf{v}\mathbf{u} = (4\mathbf{i} + \mathbf{j}) \cdot \frac{1}{\sqrt{13}}(2\mathbf{i} + 3\mathbf{j}) = \frac{11}{\sqrt{13}}.$$

Thus, from (8)

FIGURE 10.5.7 Projection of **u** onto **v** in Example 6

The graph of this vector is shown in gold color in **FIGURE 10.5.7**.   ≡

□ **Physical Interpretation of the Dot Product** When a constant force of magnitude F moves an object a distance d in the same direction of the force, the work done is defined to be

$$W = Fd. \tag{9}$$

However, if a constant force **F** applied to a body acts at an angle θ to the direction of motion, then the work done by **F** is defined to be the product of the component of **F** in the direction of the displacement and the distance $|\mathbf{d}|$ that the body moves:

$$W = (|\mathbf{F}|\cos\theta)|\mathbf{d}| = |\mathbf{F}||\mathbf{d}|\cos\theta.$$

See **FIGURE 10.5.8**. It follows from Theorem 10.5.2 that if $\mathbf{F}$ causes a displacement $\mathbf{d}$ of a body, then the work done is

$$W = \mathbf{F} \cdot \mathbf{d}. \qquad (10)$$

Note that (10) reduces to (9) when $\theta = 0$.

EXAMPLE 7 Work Done by a Force at an Angle

Find the work done by a constant force $\mathbf{F} = 2\mathbf{i} + 4\mathbf{j}$ on a block that moves from $P_1(1, 1)$ to $P_2(4, 6)$. Assume that $|\mathbf{F}|$ is measured in pounds and $|\mathbf{d}|$ is measured in feet.

Solution The displacement of the block is given by

$$\mathbf{d} = \overrightarrow{P_1P_2} = \overrightarrow{OP_2} - \overrightarrow{OP_1} = 3\mathbf{i} + 5\mathbf{j}.$$

It follows from (10) that the work done is

$$W = \mathbf{F} \cdot \mathbf{d} = (2\mathbf{i} + 4\mathbf{j}) \cdot (3\mathbf{i} + 5\mathbf{j}) = 26 \text{ ft-lb}.$$

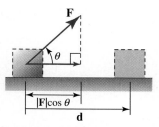

FIGURE 10.5.8 Work done by a force acting at an angle θ to the direction of motion

10.5 Exercises Answers to selected odd-numbered problems begin on page ANS-29.

In Problems 1–4, find the dot product $\mathbf{u} \cdot \mathbf{v}$.

1. $\mathbf{u} = \langle 4, 2 \rangle$, $\mathbf{v} = \langle 3, -1 \rangle$
2. $\mathbf{u} = \langle 1, -2 \rangle$, $\mathbf{v} = \langle 4, 0 \rangle$
3. $\mathbf{u} = 3\mathbf{i} - 2\mathbf{j}$, $\mathbf{v} = \mathbf{i} + \mathbf{j}$
4. $\mathbf{u} = 4\mathbf{i}$, $\mathbf{v} = -3\mathbf{j}$

In Problems 5–18, $\mathbf{u} = \langle 2, -3 \rangle$, $\mathbf{v} = \langle -1, 5 \rangle$, and $\mathbf{w} = \langle 3, -2 \rangle$. Find the indicated scalar or vector.

5. $\mathbf{u} \cdot \mathbf{v}$
6. $\mathbf{v} \cdot \mathbf{w}$
7. $\mathbf{u} \cdot \mathbf{w}$
8. $\mathbf{v} \cdot \mathbf{v}$
9. $\mathbf{w} \cdot \mathbf{w}$
10. $\mathbf{u} \cdot (\mathbf{v} + \mathbf{w})$
11. $\mathbf{u} \cdot (4\mathbf{v})$
12. $\mathbf{v} \cdot (\mathbf{u} - \mathbf{w})$
13. $(-\mathbf{v}) \cdot (\frac{1}{2}\mathbf{w})$
14. $(2\mathbf{v}) \cdot (3\mathbf{w})$
15. $\mathbf{u} \cdot (\mathbf{u} + \mathbf{v} + \mathbf{w})$
16. $(2\mathbf{u}) \cdot (\mathbf{u} - 2\mathbf{v})$
17. $\left(\dfrac{\mathbf{u} \cdot \mathbf{v}}{\mathbf{v} \cdot \mathbf{v}}\right)\mathbf{v}$
18. $(\mathbf{w} \cdot \mathbf{v})\mathbf{u}$

In Problems 19 and 20, find the dot product $\mathbf{u} \cdot \mathbf{v}$ if the smaller angle between $\mathbf{u}$ and $\mathbf{v}$ is as given.

19. $|\mathbf{u}| = 10$, $|\mathbf{v}| = 5$, $\theta = \pi/4$
20. $|\mathbf{u}| = 6$, $|\mathbf{v}| = 12$, $\theta = \pi/6$

In Problems 21–24, find the angle between the given pair of vectors. Round your answer to two decimal places.

21. $\langle 1, 4 \rangle$, $\langle 2, -1 \rangle$
22. $\langle 3, 5 \rangle$, $\langle -4, -2 \rangle$
23. $\mathbf{i} - \mathbf{j}$, $3\mathbf{i} + \mathbf{j}$
24. $2\mathbf{i} - \mathbf{j}$, $4\mathbf{i} + \mathbf{j}$

In Problems 25–28, determine whether the given vectors are orthogonal.

25. $\mathbf{u} = \langle -5, -4 \rangle$, $\mathbf{v} = \langle -6, 8 \rangle$
26. $\mathbf{u} = \langle 3, -2 \rangle$, $\mathbf{v} = \langle -6, -9 \rangle$
27. $4\mathbf{i} - 5\mathbf{j}$, $\mathbf{i} + \frac{4}{5}\mathbf{j}$
28. $\frac{1}{2}\mathbf{i} + \frac{3}{4}\mathbf{j}$, $-\frac{2}{5}\mathbf{i} + \frac{4}{5}\mathbf{j}$

In Problems 29 and 30, find a scalar c so that the given vectors are orthogonal.

29. $\mathbf{u} = 2\mathbf{i} - c\mathbf{j}$, $\mathbf{v} = 3\mathbf{i} + 2\mathbf{j}$
30. $\mathbf{u} = 4c\mathbf{i} - 8\mathbf{j}$, $\mathbf{v} = c\mathbf{i} + 2\mathbf{j}$

31. Verify that the vector
$$\mathbf{w} = \mathbf{v} - \frac{\mathbf{v} \cdot \mathbf{u}}{|\mathbf{u}|^2}\mathbf{u}$$
is orthogonal to the vector $\mathbf{u}$.

32. Find a scalar c so that the angle between the vectors $\mathbf{u} = \mathbf{i} + c\mathbf{j}$ and $\mathbf{u} = \mathbf{i} + \mathbf{j}$ is $45°$.

In Problems 33–36, $\mathbf{u} = \langle 1, -1 \rangle$ and $\mathbf{v} = \langle 2, 6 \rangle$. Find the indicated number.

33. $\text{comp}_\mathbf{v} \mathbf{u}$
34. $\text{comp}_\mathbf{u} \mathbf{v}$
35. $\text{comp}_\mathbf{u}(\mathbf{v} - \mathbf{u})$
36. $\text{comp}_{2\mathbf{v}}(\mathbf{u} + \mathbf{v})$

In Problems 37 and 38, find (a) $\text{proj}_\mathbf{v}\mathbf{u}$, and (b) $\text{proj}_\mathbf{u}\mathbf{v}$.

37. $\mathbf{u} = -5\mathbf{i} + 5\mathbf{j}$, $\mathbf{v} = -3\mathbf{i} + 4\mathbf{j}$
38. $\mathbf{u} = 4\mathbf{i} + 2\mathbf{j}$, $\mathbf{v} = -3\mathbf{i} + \mathbf{j}$

In Problems 39 and 40, $\mathbf{u} = 4\mathbf{i} + 3\mathbf{j}$ and $\mathbf{v} = -\mathbf{i} + \mathbf{j}$. Find the indicated vector.

39. $\text{proj}_{\mathbf{u}+\mathbf{v}}\mathbf{u}$
40. $\text{proj}_{\mathbf{u}-\mathbf{v}}\mathbf{v}$

41. A sled is pulled horizontally over ice by a rope attached to its front. A 20-lb force acting at an angle of $60°$ with the horizontal moves the sled 100 ft. Find the work done.

42. A block with weight $\mathbf{w}$ is pulled along a frictionless horizontal surface by a constant force $\mathbf{F}$ of magnitude 30 lb in the direction by the vector $\mathbf{d}$. See FIGURE 10.5.9.
(a) What is the work done by the weight $\mathbf{w}$?
(b) What is the work done by the force $\mathbf{F}$ if $\mathbf{d} = 4\mathbf{i} + 3\mathbf{j}$?

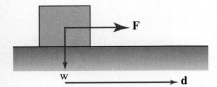

FIGURE 10.5.9 Block in Problem 42

43. A constant force $\mathbf{F}$ of magnitude 3 lb is applied to the block shown in FIGURE 10.5.10. The force $\mathbf{F}$ has the same direction as the vector $\mathbf{u} = 3\mathbf{i} + 4\mathbf{j}$. Find the work done in the direction of motion if the block moves from $P_1(3, 1)$ to $P_2(9, 3)$. Assume distance is measured in feet.

FIGURE 10.5.10 Block in Problem 43

For Discussion

44. Suppose $\text{proj}_\mathbf{v}\mathbf{u} = \mathbf{0}$, where $\mathbf{0}$ is the zero vector. Then what can be said about the vectors $\mathbf{u}$ and $\mathbf{v}$?

CONCEPTS REVIEW

You should be able to give the meaning of each of the following concepts.

Polar coordinate system:
  pole
  polar axis
Conversion formulas:
  polar to rectangular
  rectangular to polar
Graphs of polar equations:
  circles
  lines through origin
  spirals
  cardioids
  limaçons
  lemniscates
  rose curves

Tests for symmetry
Conic section:
  directrix
  focus
  eccentricity
Polar equations of conics
Rotated conics
Vector:
  geometric
  component form
Scalar
Negative of a vector
Position vector
Magnitude of a vector

Scalar multiple
Direction angle of a vector
Unit vector
Normalization
Basis vectors
Trigonometric form
Dot product
Angle between two vectors
Orthogonal vectors
Component of a vector onto another
Projection of a vector onto another

CHAPTER 10 Review Exercises
Answers to selected odd-numbered problems begin on page ANS-29.

A. True/False

In Problems 1–14, answer true or false.

1. Rectangular coordinates of a point in the plane are unique. ____
2. The graph of the polar equation $r = 5\sec\theta$ is a line. ____
3. $(3, \pi/6)$ and $(-3, -5\pi/6)$ are polar coordinates of the same point. ____
4. The graph of the ellipse $r = \dfrac{90}{15 - \sin\theta}$ is nearly circular. ____
5. The graph of the rose curve $r = 5\sin 6\theta$ has six petals. ____
6. The graph of $r = 2 + 4\sin\theta$ is a limaçon with an interior loop. ____
7. The graph of the polar $r^2 = 4\sin 2\theta$ is symmetric with respect to the origin. ____
8. The graphs of the cardioids $r = 3 + 3\cos\theta$ and $r = -3 + 3\cos\theta$ are the same. ____
9. The point $(4, 3\pi/2)$ is not on the graph of $r = 4\cos 2\theta$, because its coordinates do not satisfy the equation. ____
10. The vector $\mathbf{v} = \langle \sqrt{3}, \sqrt{5} \rangle$ is twice as long as the vector $\mathbf{u} = \langle -1, 1 \rangle$. ____
11. If $\mathbf{u}$ is a unit vector, then $\mathbf{u} \cdot \mathbf{u} = 1$. ____
12. If $\mathbf{u}$ and $\mathbf{v}$ are unit vectors, then $\mathbf{u} + \mathbf{v}$ and $\mathbf{u} - \mathbf{v}$ are orthogonal. ____
13. The terminal side of the angle θ is always in the same quadrant as the point (r, θ). ____
14. The eccentricity e of a parabola satisfies $0 < e < 1$. ____

B. Fill in the Blanks

In Problems 1–14, fill in the blanks.

1. The rectangular coordinates of the point with polar coordinates $(-\sqrt{2}, 5\pi/4)$ are _____.
2. Approximate polar coordinates of the point with rectangular coordinates $(-1, 3)$ are _____.
3. Polar coordinates of the point with rectangular coordinates $(0, -10)$ are _____.
4. On the graph of the polar equation $r = 4\cos\theta$, two pairs of coordinates of the pole or origin are _____.
5. The radius of the circle $r = \cos\theta$ is _____.
6. If $a > 0$, the center of the circle $r = -2a\sin\theta$ is _____.
7. The conic section $r = \dfrac{1}{2 + 5\cos\theta}$ is a _____.
8. In polar coordinates, the graph of $\theta = \pi/3$ is a _____.
9. The name of the polar graph of $r = 2 + \sin\theta$ is _____.
10. $r = \dfrac{12}{2 + \cos\theta}$, center _____, foci _____, vertices _____.
11. The angle between the vectors $\mathbf{u} = 5\mathbf{i}$ and $\mathbf{v} = -2\mathbf{j}$ is _____.
12. If $|\mathbf{u}| = 4$, $|\mathbf{v}| = 3$ and the angle between $\mathbf{u}$ and $\mathbf{v}$ is $\theta = 2\pi/3$, then $\mathbf{u} \cdot \mathbf{v} = $ _____.
13. A unit vector in the same direction as $\mathbf{v} = 12\mathbf{i} + 5\mathbf{j}$ is _____.
14. If $\mathbf{u} = \langle -3, 2 \rangle$ and $\mathbf{v} = \langle -1, 1 \rangle$, then $\text{proj}_\mathbf{v}\mathbf{u} = $ _____.

C. Review Exercises

In Problems 1 and 2, find a rectangular equation that has the same graph as the given polar equation.

1. $r = \cos\theta + \sin\theta$
2. $r(\cos\theta + \sin\theta) = 1$

In Problems 3 and 4, find a polar equation that has the same graph as the given rectangular equation.

3. $x^2 + y^2 - 4y = 0$
4. $(x^2 + y^2 - 2x)^2 = 9(x^2 + y^2)$

5. Determine the rectangular coordinates of the vertices of the ellipse whose polar equation is $r = 2/(2 - \sin\theta)$.
6. Find a polar equation of the hyperbola with focus at the origin, vertices (in rectangular coordinates) $(0, -\frac{4}{3})$ and $(0, -4)$, and eccentricity 2.

In Problems 7 and 8, find polar coordinates satisfying (a) $r > 0$, $-\pi < \theta \le \pi$, and (b) $r < 0$, $-\pi < \theta \le \pi$, for each point given in rectangular coordinates.

7. $(\sqrt{3}, -\sqrt{3})$
8. $\left(-\frac{1}{4}, \frac{1}{4}\right)$

In Problems 9–20, identify and sketch the graph of the given polar equation.

9. $r = 5$
10. $\theta = -\pi/3$
11. $r = 5\sin\theta$
12. $r = -2\cos\theta$
13. $r = 4 - 4\cos\theta$
14. $r = 1 + \sin\theta$
15. $r = 2 + \sin\theta$
16. $r = 1 - 2\cos\theta$
17. $r = \sin 3\theta$
18. $r = 3\sin 4\theta$
19. $r = \dfrac{8}{3 - 2\cos\theta}$
20. $r = \dfrac{1}{1 + \cos\theta}$

In Problems 21 and 22, give an equation of the given polar graph.

21.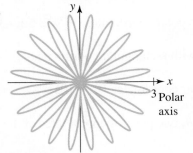

FIGURE 10.R.1 Graph for Problem 21

22.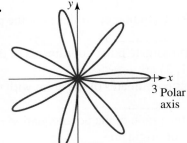

FIGURE 10.R.2 Graph for Problem 22

In Problems 23 and 24, the graph of the given polar equation is rotated about the origin by the indicated amount. (a) Find a polar equation of the new graph. (b) Find a rectangular equation for the new graph.

23. $r = 2\cos\theta$; counterclockwise, $\pi/4$
24. $r = 1/(1 + \cos\theta)$; clockwise, $\pi/6$
25. (a) Show that the graph of the polar equation

$$r = a\sin\theta + b\cos\theta$$

for $a \ne 0$ and $b \ne 0$, is a circle.
(b) Determine the center and radius of the circle in part (a).

26. (a) Find a rectangular equation that has the same graph as the given polar equation: $r\cos\theta = 1$, $r\cos(\theta - \pi/3) = 1$, $r = 1$. Sketch the graph of each equation.
   (b) How are the graphs of $r\cos\theta = 1$ and $r\cos(\theta - \pi/3) = 1$ related?
   (c) Show that rectangular point $\left(\frac{1}{2}, \frac{\sqrt{3}}{2}\right)$ is on the graphs of $r\cos(\theta - \pi/3) = 1$ and $r = 1$.
   (d) Use the information in parts (a) and (c) to explain how the graphs of $r\cos(\theta - \pi/3) = 1$ and $r = 1$ are related.

In Problems 27–42, $\mathbf{u} = -2\mathbf{i} + 3\mathbf{j}$, $\mathbf{v} = \mathbf{i} + \mathbf{j}$, and $\mathbf{w} = \mathbf{i} - 4\mathbf{j}$. Find the indicated vector or scalar.

27. $-5\mathbf{u} + 3\mathbf{v}$
28. $\mathbf{u} - 10\mathbf{v}$
29. $\mathbf{u} + (2\mathbf{v} + 3\mathbf{w})$
30. $4\mathbf{u} - (3\mathbf{v} + \mathbf{w})$
31. $(\mathbf{u} \cdot \mathbf{v})\mathbf{w} + (\mathbf{w} \cdot \mathbf{v})\mathbf{u}$
32. $(\mathbf{u} - \mathbf{v}) \cdot (\mathbf{v} + \mathbf{w})$
33. $\text{comp}_\mathbf{w} \mathbf{v}$
34. $\text{comp}_\mathbf{u}(-\mathbf{v})$
35. $\text{proj}_\mathbf{v}(2\mathbf{u})$
36. $\text{proj}_\mathbf{w}(\mathbf{u} + \mathbf{v})$
37. $|\mathbf{u}| + |2\mathbf{v}|$
38. $|\mathbf{u} + \mathbf{v}|$
39. Trigonometric form of $2\mathbf{v}$
40. Horizontal component of $-2(\mathbf{u} + \mathbf{w})$
41. A unit vector in the opposite direction of $\mathbf{w}$
42. The angle between $\mathbf{v}$ and $\mathbf{w}$

11 Systems of Equations and Inequalities

In This Chapter

- **11.1** Systems of Linear Equations
- **11.2** Systems of Nonlinear Equations
- **11.3** Partial Fractions
- **11.4** Systems of Inequalities
- **11.5** Linear Programming

Chapter 11 Review Exercises

Student successfully solving a system of linear equations

A Bit of History Many of the mathematical concepts considered in this text are several hundred years old. In this chapter we have a rare opportunity to examine, albeit briefly, a topic that has its origins in the twentieth century. *Linear programming,* like many other branches of mathematics, originated in an attempt to solve practical problems. But unlike the mathematics of earlier centuries, which was often rooted in the sciences of physics and astronomy, linear programming developed from an effort to solve problems in business, manufacturing, shipping, economics, and military planning. In these problems it was usually necessary to find the optimal values (that is, maximum and minimum values) of a linear function when certain restrictions were placed on the variables. There was no general mathematical procedure for solving this kind of problem until **George B. Dantzig** (1914–2005) published his *simplex method* in 1947. Dantzig and his colleagues in the U.S. Air Force developed this method for finding optimal values of a linear function by investigating certain problems in transportation and military logistical planning. The word "programming," it should be pointed out, does not refer to computer programming but rather to a program of action.

Although we will not study the simplex method itself, we will see in Section 11.5 that linear-programming problems involving two variables can be solved in a geometric manner.

11.1 Systems of Linear Equations

Introduction Recall from Section 2.3 that a **linear equation in two variables** x and y is any equation that can be put in the form $ax + by = c$, where a and b are real numbers not both zero. In general, a **linear equation in n variables** $x_1, x_2, \ldots, x_n$ is an equation of the form

$$a_1 x_1 + a_2 x_2 + \cdots + a_n x_n = b, \tag{1}$$

where the real numbers $a_1, a_2, \ldots, a_n$ are not all zero. The number b is called the **constant term** of the equation. The equation in (1) is also called a **first-degree equation** in that the exponent of each of the n variables is 1. In this and the next section we examine solution methods for systems of equations.

☐ **Terminology** A **system of equations** consists of two or more equations with each equation containing at least one variable. If each equation in a system is linear, we say that it is a **system of linear equations** or simply a **linear system**. Whenever possible, we will use the familiar symbols x, y, and z to represent variables in a system. For example,

$$\begin{cases} 2x + y - z = 0 \\ x + 3y + z = 2 \\ -x - y + 5z = 14 \end{cases} \tag{2}$$

is a linear system of three equations in three variables. The brace in (2) is just a way of reminding us that we are trying to solve a system of equations and that the equations must be dealt with simultaneously. A **solution** of a system of n equations in n variables consists of values of the variables that satisfy each equation in the system. A solution of such a system is also written as an **ordered n-tuple**. For example, as we see $x = 2$, $y = -1$, and $z = 3$ satisfy each equation in the linear system (2):

$$\begin{cases} 2x + y - z = 0 \\ x + 3y + z = 2 \\ -x - y + 5z = 14 \end{cases} \xrightarrow[\text{and } z = 3]{\substack{\text{substituting} \\ x = 2, y = -1,}} \begin{cases} 2 \cdot 2 + (-1) - 3 = 4 - 4 = 0 \\ 2 + 3(-1) + 3 = 5 - 3 = 2 \\ -2 - (-1) + 5 \cdot 3 = 16 - 2 = 14 \end{cases}$$

and so these values constitute a solution. Alternatively, this solution can be written as the **ordered triple** $(2, -1, 3)$. To **solve** a system of equations we find all solutions of the system. Often to solve a system of equations we perform operations on the system to transform it into an equivalent set of equations. Two systems of equations are said to be **equivalent** if they have precisely the same **solution sets**.

☐ **Linear Systems in Two Variables** The simplest linear system consists of two equations in two variables:

$$\begin{cases} a_1 x + b_1 y = c_1 \\ a_2 x + b_2 y = c_2. \end{cases} \tag{3}$$

Because the graph of a linear equation $ax + by = c$ is a straight line, the system determines two straight lines in the xy-plane.

☐ **Consistent and Inconsistent Systems** As shown in **FIGURE 11.1.1** there are three possible cases for the graphs of the equations in system (3):

- The lines intersect in a single point.   ← Figure 11.1.1(a)
- The equations describe coincident lines.   ← Figure 11.1.1(b)
- The two lines are parallel.   ← Figure 11.1.1(c)

FIGURE 11.1.1 Two lines in the plane

In these three cases we say, respectively:

- The system is **consistent** and the equations are **independent**. The system has exactly one solution, that is, the ordered pair of real numbers corresponding to the point of intersection of the lines.
- The system is **consistent**, but the equations are **dependent**. The system has infinitely many solutions, that is, all the ordered pairs of real numbers corresponding to the points on the one line.
- The system is **inconsistent**. The lines are parallel and so there are no solutions.

For example, the equations in the linear system

$$\begin{cases} x - y = 0 \\ x - y = 3 \end{cases}$$

are parallel lines as in Figure 11.1.1(c). Hence the system is inconsistent.

To solve a system of linear equations, we can use either the method of substitution or the method of elimination.

☐ **Method of Substitution** The first solution technique considered is called the **method of substitution**.

> **METHOD OF SUBSTITUTION**
>
> (*i*) Use one of the equations in the system to solve for one variable in terms of the other variables.
> (*ii*) Substitute this expression into the other equations.
> (*iii*) If one of the equations obtained in step (*ii*) contains one variable, then solve it. Otherwise repeat (*i*) until one equation in one variable is obtained.
> (*iv*) Finally, use back-substitution to find the values of the remaining variables.

EXAMPLE 1 Method of Substitution

Solve the system

$$\begin{cases} 3x + 4y = -5 \\ 2x - y = 4. \end{cases}$$

Solution Solving the second equation for y yields

$$y = 2x - 4.$$

We substitute this expression into the first equation and solve for x:

$$3x + 4(2x - 4) = -5 \quad \text{or} \quad 11x = 11 \quad \text{or} \quad x = 1.$$

Back substitution ▶ We then substitute this value back into the first equation:

$$3(1) + 4y = -5 \quad \text{or} \quad 4y = -8 \quad \text{or} \quad y = -2.$$

Solution written as an ordered pair. ▶ Thus the only solution of the system is $(1, -2)$. The system is consistent and the equations are independent. ≡

☐ **Linear Systems in Three Variables** In calculus it is shown that the graph of a **linear equation in three variables**,

$$ax + by + cz = d,$$

where a, b, and c are not all zero, determines a *plane* in three-dimensional space. As we have seen in (2), a solution of a system of three equations in three unknowns

$$\begin{cases} a_1x + b_1y + c_1z = d_1 \\ a_2x + b_2y + c_2z = d_2 \\ a_3x + b_3y + c_3z = d_3 \end{cases} \quad (4)$$

is an ordered triple of the form (x, y, z); an ordered triple of numbers represents a point in three-dimensional space. The intersection of the three planes described by the system (4) may be

- a single point,
- infinitely many points, or
- no points.

As before, to each of these cases we apply the terms *consistent and independent*, *consistent and dependent*, and *inconsistent*, respectively. Each is illustrated in FIGURE 11.1.2.

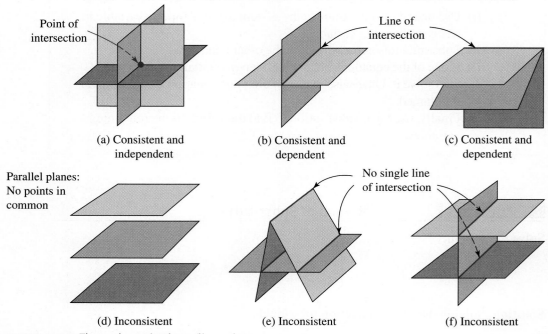

(a) Consistent and independent
(b) Consistent and dependent
(c) Consistent and dependent
(d) Inconsistent
(e) Inconsistent
(f) Inconsistent

FIGURE 11.1.2 Three planes in three dimensions

☐ **Method of Elimination** The next method that we illustrate uses **elimination operations**. When applied to a system of equations, these operations yield an equivalent system of equations.

> **METHOD OF ELIMINATION**
>
> (*i*) Interchange any two equations in a system.
> (*ii*) Multiply an equation by a nonzero constant.
> (*iii*) Add a nonzero constant multiple of an equation in a system to another equation in the same system.

We often add a nonzero constant multiple of one equation to other equations in a system with the intention of eliminating a variable from those equations.

For convenience, we represent these operations by the following symbols, where the letter E stands for the word *equation*:

$E_i \leftrightarrow E_j$: Interchange the ith equation with the jth equation.
kE_i: Multiply the ith equation by a constant k.
$kE_i + E_j$: Multiply the ith equation by k and add to the jth equation.

Reading a linear system from the top, E_1 represents the first equation, E_2 represents the second equation, and so on.

Using the method of elimination it is possible to reduce the system (4) of three linear equations in three variables to an equivalent system in triangular form,

$$\begin{cases} a_1'x + b_1'y + c_1'z = d_1' \\ b_2'y + c_2'z = d_2' \\ c_3'z = d_3'. \end{cases}$$

A solution of the system (if one exists) can be readily obtained by **back-substitution**. The next example illustrates the procedure.

EXAMPLE 2 Elimination and Back Substitution

Solve the system

$$\begin{cases} x + 2y + z = -6 \\ 4x - 2y - z = -4 \\ 2x - y + 3z = 19. \end{cases}$$

Solution We begin by eliminating x from the second and third equations:

$$\begin{matrix} x + 2y + z = -6 \\ 4x - 2y - z = -4 \\ 2x - y + 3z = 19 \end{matrix} \quad \xrightarrow[-2E_1 + E_3]{-4E_1 + E_2} \quad \begin{cases} x + 2y + z = -6 \\ -10y - 5z = 20 \\ -5y + z = 31. \end{cases} \quad (5)$$

We then eliminate y from the third equation and obtain an equivalent system in triangular form:

$$\begin{matrix} x + 2y + z = -6 \\ -10y - 5z = 20 \\ -5y + z = 31 \end{matrix} \quad \xrightarrow{-\frac{1}{2}E_2 + E_3} \quad \begin{cases} x + 2y + z = -6 \\ -10y - 5z = 20 \\ \frac{7}{2}z = 21. \end{cases} \quad (6)$$

We arrive at another triangular form that is equivalent to the original system by multiplying the third equation by $\frac{2}{7}$:

$$\left.\begin{array}{r} x + 2y + z = -6 \\ -10y - 5z = 20 \\ \frac{7}{2}z = 21 \end{array}\right\} \xrightarrow{\frac{2}{7}E_3} \begin{cases} x + 2y + z = -6 \\ y + \frac{1}{2}z = -2 \\ z = 6. \end{cases}$$

From this last system it is evident that $z = 6$. Using this value and substituting back into the second equation gives

$$y = -\tfrac{1}{2}z - 2 = -\tfrac{1}{2}(6) - 2 = -5.$$

Finally, by substituting $y = -5$ and $z = 6$ back into the first equation, we obtain

$$x = -2y - z - 6 = -2(-5) - 6 - 6 = -2.$$

Therefore the solution of the system is $(-2, -5, 6)$.

▶ The answer indicates that the three planes intersect at a point as in Figure 11.1.2 (a).

EXAMPLE 3   Elimination and Back Substitution

Solve the system

$$\begin{cases} x + y + z = 2 \\ 5x - 2y + 2z = 0 \\ 8x + y + 5z = 6. \end{cases} \qquad (7)$$

Solution Using the first equation to eliminate the variable x from the second and third equations, we get the equivalent system

$$\left.\begin{array}{r} x + y + z = 2 \\ 5x - 2y + 2z = 0 \\ 8x + y + 5z = 6 \end{array}\right\} \xrightarrow[-8E_1 + E_3]{-5E_1 + E_2} \begin{cases} x + y + z = 2 \\ -7y - 3z = -10 \\ -7y - 3z = -10. \end{cases}$$

This system, in turn, is equivalent to the system in triangular form:

$$\left.\begin{array}{r} x + y + z = 2 \\ -7y - 3z = -10 \\ -7y - 3z = -10 \end{array}\right\} \xrightarrow[-E_2 + E_3]{-E_2} \begin{cases} x + y + z = 2 \\ 7y + 3z = 10 \\ 0z = 0. \end{cases} \qquad (8)$$

In this system we cannot determine unique values for x, y, and z. At best we can solve for two variables in terms of the remaining variable. For example, from the second equation in (8), we obtain y in terms of z:

$$y = -\tfrac{3}{7}z + \tfrac{10}{7}.$$

Substituting this equation for y in the first equation for x gives

$$x + \left(-\tfrac{3}{7}z + \tfrac{10}{7}\right) + z = 2 \quad \text{or} \quad x = -\tfrac{4}{7}z + \tfrac{4}{7}.$$

Thus in the solutions for y and x, we can choose z *arbitrarily*. If we denote z by the symbol α, where α represents a real number, then the solutions of the system are all ordered triples of the form $\left(-\tfrac{4}{7}\alpha + \tfrac{4}{7}, -\tfrac{3}{7}\alpha + \tfrac{10}{7}, \alpha\right)$. We emphasize that for any real number α, we obtain a solution of (7). For example, by choosing α to be, say, 0, 1, and 2, we obtain the solutions $\left(\tfrac{4}{7}, \tfrac{10}{7}, 0\right)$, $(0, 1, 1)$, and $\left(-\tfrac{4}{7}, \tfrac{4}{7}, 2\right)$, respectively. In other words, the system is consistent and has infinitely many solutions.

▶ The answer indicates that the two planes intersect in a line as in Figure 11.1.2 (b).

In Example 3 there is nothing special about solving (8) for x and y in terms of z. For instance, by solving (8) for x and z in terms of y, we obtain the solution $\left(\frac{4}{3}\beta - \frac{4}{3}, \beta, -\frac{7}{3}\beta + \frac{10}{3}\right)$, where β is any real number. Note that by setting β equal to $\frac{10}{7}$, 1, and $\frac{4}{7}$, we get the same solutions in Example 3 corresponding, in turn, to $\alpha = 0$, $\alpha = 1$, and $\alpha = 2$.

EXAMPLE 4   No Solution

Solve the system
$$\begin{cases} 2x - y - z = 0 \\ 2x + 3y = 1 \\ 8x - 3z = 4. \end{cases}$$

Solution The elimination method,

$$\left.\begin{aligned} 2x - y - z &= 0 \\ 2x + 3y &= 1 \\ 8x - 3z &= 4 \end{aligned}\right\} \xrightarrow[-4E_1 + E_3]{-E_1 + E_2} \begin{cases} 2x - y - z = 0 \\ 4y + z = 1 \\ 4y + z = 4 \end{cases}$$

$$\left.\begin{aligned} 2x - y - z &= 0 \\ 4y + z &= 1 \\ 4y - z &= 4 \end{aligned}\right\} \xrightarrow{-E_2 + E_3} \begin{cases} 2x - y - z = 0 \\ 4y + z = 1 \\ 0z = 3, \end{cases}$$

shows that the last equation $0z = 3$ is *never* satisfied for any number z since $0 \neq 3$. Thus, the system is inconsistent and so has no solutions. ≡

EXAMPLE 5   Elimination and Back Substitution

A force of smallest magnitude P is applied to a 300-lb block on an inclined plane in order to keep it from sliding down the plane. See **FIGURE 11.1.3**. If the coefficient of friction between the block and the surface is 0.4, then the magnitude of the frictional force is $0.4N$, where N is the magnitude of the normal force exerted on the block by the plane. Since the system is in equilibrium, the horizontal and the vertical components of the forces must be zero:

$$\begin{cases} P\cos 30° + 0.4N\cos 30° - N\cos 60° = 0 \\ P\sin 30° + 0.4N\sin 30° + N\cos 60° - 300 = 0. \end{cases}$$

Solve this system for P and N.

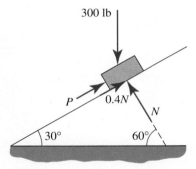

FIGURE 11.1.3 Inclined plane in Example 5

Solution Using $\sin 30° = \cos 60° = \frac{1}{2}$ and $\sin 60° = \cos 30° = \sqrt{3}/2$, we simplify the system above to

$$\begin{cases} \sqrt{3}P + (0.4\sqrt{3} - 1)N = 0 \\ P + (0.4 + \sqrt{3})N = 600. \end{cases}$$

By elimination,

$$\left.\begin{aligned} \sqrt{3}P + (0.4\sqrt{3} - 1)N &= 0 \\ P + (0.4 + \sqrt{3})N &= 600 \end{aligned}\right\} \xrightarrow{E_1 - \sqrt{3}E_2} \begin{cases} \sqrt{3}P + (0.4\sqrt{3} - 1)N = 0 \\ -4N = -600\sqrt{3} \end{cases}$$

$$\left.\begin{aligned} \sqrt{3}P + (0.4\sqrt{3} - 1)N &= 0 \\ -4N &= -600\sqrt{3} \end{aligned}\right\} \xrightarrow{-\frac{1}{4}E_2} \begin{cases} \sqrt{3}P + (0.4\sqrt{3} - 1)N = 0 \\ N = 150\sqrt{3}. \end{cases}$$

The second equation of the last system gives $N = 150\sqrt{3} \approx 259.81$ lb. By substituting this value into the first equation we obtain $P = 150(1 - 0.4\sqrt{3}) \approx 46.08$ lb.

☐ **Homogeneous Systems** A linear system in which all the constant terms are zero, such as

$$\begin{cases} a_1 x + b_1 y = 0 \\ a_2 x + b_2 y = 0 \end{cases} \quad (9)$$

or

$$\begin{cases} a_1 x + b_1 y + c_1 z = 0 \\ a_2 x + b_2 y + c_2 z = 0 \\ a_3 x + b_3 y + c_3 z = 0, \end{cases} \quad (10)$$

is said to be **homogeneous**. Note that systems (9) and (10) have the solutions $(0, 0)$ and $(0, 0, 0)$, respectively. A solution of a system of equations in which each of its variables is zero is called the **zero solution** or the **trivial solution**. Because a homogeneous linear system always possesses at least the zero solution, such a system is *always consistent*. In addition to the zero solution, however, there *may* exist infinitely many nonzero solutions. These solutions can be found by proceeding exactly as in Example 3.

▶ A homogeneous system is consistent even in the case where the linear system consists of m equations in n variables, where $m \neq n$.

EXAMPLE 6   A Homogeneous System

The same steps used to solve the system in Example 3 can be used to solve the related homogeneous system

$$\begin{cases} x + 2y + z = 0 \\ 5x - 2y + 2z = 0 \\ 8x + 2y + 5z = 0. \end{cases}$$

In this case the elimination steps yield

$$\begin{cases} x + y + z = 0 \\ 7y + 3z = 0 \\ 0z = 0. \end{cases}$$

Choosing $z = \alpha$, where α is a real number, we find from the second equation of the last system that $y = -\frac{3}{7}\alpha$. Then using the first equation, we obtain $x = -\frac{4}{7}\alpha$. Thus, the solutions of the system consist of all ordered triples of the form $\left(-\frac{4}{7}\alpha, -\frac{3}{7}\alpha, \alpha\right)$. Note that for $\alpha = 0$, we obtain the trivial solution $(0, 0, 0)$ but for, say, $\alpha = -7$, we obtain the nontrivial solution $(4, 3, -7)$.

The discussion in this section is also applicable to systems of n linear equations in n variables for $n > 3$. See Problems 25 and 26 in Exercises 11.1.

11.1 Exercises   Answers to selected odd-numbered problems begin on page ANS-30.

In Problems 1–26, solve the given linear system. State whether the system is consistent, with independent or dependent equations, or whether it is inconsistent.

1. $\begin{cases} 2x + y = 2 \\ 3x - 2y = -4 \end{cases}$ **2.** $\begin{cases} 2x - 2y = 1 \\ 3x + 5y = 11 \end{cases}$

3. $\begin{cases} 4x - y + 1 = 0 \\ x + 3y + 9 = 0 \end{cases}$
4. $\begin{cases} x - 4y + 1 = 0 \\ 3x + 2y - 1 = 0 \end{cases}$

5. $\begin{cases} x - 2y = 6 \\ -0.5x + y = 1 \end{cases}$
6. $\begin{cases} 6x - 4y = 9 \\ -3x + 2y = -4.5 \end{cases}$

7. $\begin{cases} x - y = 2 \\ x + y = 1 \end{cases}$
8. $\begin{cases} 2x + y = 4 \\ 2x + y = 0 \end{cases}$

9. $\begin{cases} -x - 2y + 4 = 0 \\ 5x + 10y - 20 = 0 \end{cases}$
10. $\begin{cases} 7x - 3y - 14 = 0 \\ x + y - 1 = 0 \end{cases}$

11. $\begin{cases} x + y - z = 0 \\ x - y + z = 2 \\ 2x + y - 4z = -8 \end{cases}$
12. $\begin{cases} x + y + z = 8 \\ x - 2y + z = 4 \\ x + y - z = -4 \end{cases}$

13. $\begin{cases} 2x + 6y + z = -2 \\ 3x + 4y - z = 2 \\ 5x - 2y - 2z = 0 \end{cases}$
14. $\begin{cases} x + 7y - 4z = 1 \\ 2x + 3y + z = -3 \\ -x - 18y + 13z = 2 \end{cases}$

15. $\begin{cases} 2x + y + z = 1 \\ x - y + 2z = 5 \\ 3x + 4y - z = -2 \end{cases}$
16. $\begin{cases} x + y - 5z = -1 \\ 4x - y + 3z = 1 \\ 5x - 5y + 21z = 5 \end{cases}$

17. $\begin{cases} x - 5y + z = 0 \\ 10x + y + 3z = 0 \\ 4x + 2y - 5z = 0 \end{cases}$
18. $\begin{cases} -5x + y + z = 0 \\ 4x - y = 0 \\ 2x - y + 2z = 0 \end{cases}$

19. $\begin{cases} x - 3y = 22 \\ y + 6z = -3 \\ \frac{1}{3}x + 2z = 3 \end{cases}$
20. $\begin{cases} 2x - z = 12 \\ x + y = 7 \\ 5x + 4z = -9 \end{cases}$

21. $\begin{cases} -x + 3y + 2z = 2 \\ \frac{1}{2}x - \frac{3}{2}y - z = -1 \\ -\frac{1}{3}x + y + \frac{2}{3}z = \frac{2}{3} \end{cases}$
22. $\begin{cases} x + 6y + z = 9 \\ 3x + y - 2z = 7 \\ -6x + 3y + 7z = -2 \end{cases}$

23. $\begin{cases} x + y - z = 0 \\ 2x + 2y - 2z = 1 \\ 5x + 5y - 5z = 2 \end{cases}$
24. $\begin{cases} x + y + z = 4 \\ 2x - y + 2z = 11 \\ 4x + 3y - 6z = -18 \end{cases}$

25. $\begin{cases} 2x - y + 3z - w = 8 \\ x + y - z + w = 3 \\ x - y + 5z - 3w = -1 \\ 6x + 2y + z - w = -2 \end{cases}$
26. $\begin{cases} x - 2y + z - 3w = 0 \\ 8x - 8y - z - 5w = 16 \\ -x - y + 3w = -6 \\ 4x - 7y + 3z - 10w = 2 \end{cases}$

In Problems 27–30, solve the given system.

27. $\begin{cases} \dfrac{1}{x} - \dfrac{1}{y} = \dfrac{1}{6} \\ \dfrac{4}{x} + \dfrac{3}{y} = 3 \end{cases}$
28. $\begin{cases} \dfrac{1}{x} - \dfrac{1}{y} + \dfrac{2}{z} = 3 \\ \dfrac{2}{x} + \dfrac{1}{y} - \dfrac{4}{z} = -1 \\ \dfrac{3}{x} + \dfrac{1}{y} + \dfrac{1}{z} = \dfrac{5}{2} \end{cases}$

29. $\begin{cases} 3\log_{10} x + \log_{10} y = 2 \\ 5\log_{10} x + 2\log_{10} y = 1 \end{cases}$
30. $\begin{cases} \cos x - \sin y = 1 \\ 2\cos x + \sin y = -1 \end{cases}$

11.1 Systems of Linear Equations

FIGURE 11.1.4 Cables in Problem 31

31. The magnitudes T_1 and T_2 of the tensions in the two cables shown in FIGURE 11.1.4 satisfy the system of equations

$$\begin{cases} T_1\cos 25° - T_2\cos 15° = 0 \\ T_1\sin 25° + T_2\sin 15° - 200 = 0. \end{cases}$$

Find T_1 and T_2.

32. If we change the direction of the frictional force in Figure 11.1.3 of Example 5, then the system of equations becomes

$$\begin{cases} P\cos 30° - 0.4N\cos 30° - N\cos 60° = 0 \\ P\sin 30° - 0.4N\sin 30° + N\sin 60° - 300 = 0. \end{cases}$$

In this case, P represents the magnitude of the force that is just enough to start the block up the plane. Find P and N.

Miscellaneous Applications

33. **Speed** An airplane flies 3300 mi from Hawaii to California in 5.5 h with a tailwind. From California to Hawaii, flying against a wind of the same velocity, the trip takes 6 h. Determine the speed of the plane and the speed of the wind.
34. **How Many Coins?** A person has 20 coins, consisting of dimes and quarters, which total $4.25. Determine how many of each coin the person has.
35. **Number of Gallons** A 100-gal tank is full of water in which 50 lb of salt is dissolved. A second tank contains 200 gal of water with 75 lb of salt. How much should be removed from both tanks and mixed together in order to make a solution of 90 gal with $\frac{4}{9}$ lb of salt per gallon?
36. **Playing with Numbers** The sum of three numbers is 20. The difference of the first two numbers is 5, and the third number is 4 times the sum of the first two. Find the numbers.
37. **How Long?** Three pumps P_1, P_2, and P_3 working together can fill a tank in 2 hours. Pumps P_1 and P_2 can fill the same tank in 3 hours, whereas pumps P_2 and P_3 can fill it in 4 hours. Determine how long it would take each pump working alone to fill the tank.
38. **Parabola Through Three Points** The parabola $y = ax^2 + bx + c$ passes through the points $(1, 10)$, $(-1, 12)$, and $(2, 18)$. Find a, b, and c.
39. **Area** Find the area of the right triangle shown in FIGURE 11.1.5.
40. **Current** According to Kirchhoff's law of voltages, the currents i_1, i_2, and i_3 in the parallel circuit shown in FIGURE 11.1.6 satisfy the equations

$$\begin{cases} i_1 + 2(i_1 - i_2) + 0i_3 = 6 \\ 3i_2 + 4(i_2 - i_3) + 2(i_2 - i_1) = 0 \\ 2i_3 + 4(i_3 - i_2) + 0i_1 = 12. \end{cases}$$

Solve for i_1, i_2, and i_3.

FIGURE 11.1.5 Triangle in Problem 39

FIGURE 11.1.6 Circuit in Problem 40

41. **The A, B, C's** When Beth graduated from college, she had completed 40 courses, in which she received grades of A, B, and C. Her final GPA (grade point average) was 3.125. Her GPA in only those courses in which she received grades of A and B was 3.8. Assume that A, B, and C grades are worth four points, three points, and two points, respectively. Determine the number of A's, B's, and C's that Beth received.

42. **Conductivity** Cosmic rays are deflected toward the poles by the Earth's magnetic field, so that only the most energetic rays can penetrate the equatorial regions. See **FIGURE 11.1.7**. As a result, the ionization rate, and hence the conductivity σ of the stratosphere, is greater near the poles than it is near the equator. Conductivity can be approximated by the formula

$$\sigma = (A + B \sin^4 \phi)^{1/2},$$

where ϕ is latitude and A and B are constants that must be chosen to fit the physical data. Balloon measurements made in the southern hemisphere indicated a conductivity of approximately 3.8×10^{-12} siemens/meter at $35.5°$ south latitude and 5.6×10^{-12} siemens/meter at $51°$ south latitude. (A *siemen* is the reciprocal of an *ohm*, which is a unit of electrical resistance.) Determine the constants A and B. What is the conductivity at $42°$ south latitude?

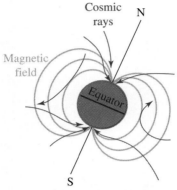

FIGURE 11.1.7 Earth's magnetic field in Problem 42

For Discussion

43. Determine conditions on a_1, a_2, b_1, and b_2 so that the linear system (9) has only the trivial solution.
44. Determine a value of k such that the linear system

$$\begin{cases} 2x - 3y = 10 \\ 6x - 9y = k \end{cases}$$

is **(a)** inconsistent, and **(b)** dependent.

45. Devise a system of two linear equations whose solution is $(2, -5)$.

11.2 Systems of Nonlinear Equations

☰ Introduction As **FIGURE 11.2.1** illustrates, the graphs of the parabolas $y = x^2 - 4x$ and $y = -x^2 + 8$ intersect at two points. Thus the coordinates of the points of intersection must satisfy *both* equations,

$$\begin{cases} y = x^2 - 4x \\ y = -x^2 + 8. \end{cases} \quad (1)$$

Recall from Sections 2.3 and 11.1 that any equation that can be put in the form $ax + by + c = 0$ is called a **linear equation** in two variables. A **nonlinear equation** is simply one that is not linear. For example, in system (1) both equations $y = x^2 - 4x$ and $y = -x^2 + 8$ are nonlinear. A system of equations in which at least one of the equations is nonlinear will be referred to as a **system of nonlinear equations** or simply a **nonlinear system**.

In the examples that follow, will use the *methods of substitution* and *elimination* introduced in Section 11.1 to solve nonlinear systems.

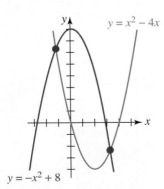

FIGURE 11.2.1 Intersection of two parabolas

EXAMPLE 1 Solution of (1)

Find the solutions of the system (1).

Solution Since the first equation already expresses y in terms of x, we substitute this expression for y into the second equation to get a single equation in one variable:

$$x^2 - 4x = -x^2 + 8.$$

Simplifying the last equation we get a quadratic equation $x^2 - 2x - 4 = 0$ that we solve using the quadratic formula: $x = 1 - \sqrt{5}$ and $x = 1 + \sqrt{5}$. We then substitute each of these numbers *back* into the first equation in (1) to solve for the corresponding values of y. This gives

$$y = (1 - \sqrt{5})^2 - 4(1 - \sqrt{5}) = 2 + 2\sqrt{5}$$
and
$$y = (1 + \sqrt{5})^2 - 4(1 + \sqrt{5}) = 2 - 2\sqrt{5}.$$

Thus, $(1 - \sqrt{5}, 2 + 2\sqrt{5})$ and $(1 + \sqrt{5}, 2 - 2\sqrt{5})$ are solutions of the system. ≡

EXAMPLE 2 Solving a Nonlinear System

Find the solutions of the system

$$\begin{cases} x^4 - 2(10^{2y}) - 3 = 0 \\ x - 10^y = 0. \end{cases}$$

Solution From the second equation, we have $x = 10^y$, and therefore $x^2 = 10^{2y}$. Substituting this last result into the first equation gives

$$x^4 - 2x^2 - 3 = 0,$$
or
$$(x^2 - 3)(x^2 + 1) = 0.$$

Since $x^2 + 1 > 0$ for all real numbers x, it follows that $x^2 = 3$ or $x = \pm\sqrt{3}$. But $x = 10^y > 0$ for all y; therefore, we must take $x = \sqrt{3}$. Solving $\sqrt{3} = 10^y$ for y gives

$$y = \log_{10}\sqrt{3} \quad\text{or}\quad y = \tfrac{1}{2}\log_{10}3.$$

Hence, $x = \sqrt{3}, y = \tfrac{1}{2}\log_{10}3$ is the only solution of the system. ≡

▶ Solution written by specifying values of the variables.

EXAMPLE 3 Dimensions of a Rectangle

Consider a rectangle in the first quadrant bounded by the x- and y-axes and the graph of $y = 20 - x^2$. See **FIGURE 11.2.2**. Find the dimensions of such a rectangle if its area is 16 square units.

Solution Let (x, y) be the coordinates of the point P on the graph of $y = 20 - x^2$ shown in the figure. Then the

$$\text{area of the rectangle} = xy \quad\text{or}\quad 16 = xy.$$

Thus we obtain the system of equations

$$\begin{cases} xy = 16 \\ y = 20 - x^2. \end{cases}$$

FIGURE 11.2.2 Rectangle in Example 3

The first equation of the system yields $y = 16/x$. After substituting this expression for y in the second equation, we get

$$\frac{16}{x} = 20 - x^2 \quad \leftarrow \text{multiply this equation by } x$$

or $\quad 16 = 20x - x^3 \quad$ or $\quad x^3 - 20x + 16 = 0.$

Now from the Rational Zeros Theorem in Section 4.4 the only possible rational roots of the last equation are $\pm 1, \pm 2, \pm 4, \pm 8,$ and ± 16. Testing these numbers by synthetic division eventually shows that

$$\begin{array}{r|rrrr} 4 & 1 & 0 & -20 & 16 \\ & & 4 & 16 & -16 \\ \hline & 1 & 4 & -4 & \boxed{0 = r} \end{array}$$

and so 4 is a solution. But the division above gives the factorization

$$x^3 - 20x + 16 = (x - 4)(x^2 + 4x - 4).$$

Applying the quadratic formula to $x^2 + 4x - 4 = 0$ reveals two more real roots:

$$x = \frac{-4 \pm \sqrt{32}}{2} = -2 \pm 2\sqrt{2}.$$

The positive number $-2 + 2\sqrt{2}$ is another solution. Since dimensions are positive, we reject the negative number $-2 - 2\sqrt{2}$. In other words, there are two rectangles with area 16 square units.

To find y, we use $y = 16/x$. If $x = 4$, then $y = 4$, and if $x = -2 + 2\sqrt{2} \approx 0.83$, then $y = 16/(-2 + 2\sqrt{2}) \approx 19.31$. Thus the dimensions of the two rectangles are

$$4 \times 4 \quad \text{and} \quad 0.83 \times 19.31 \text{ (approximately)}. \quad \equiv$$

Note: In Example 3 observe that the equation $16 = 20x - x^3$ was obtained by multiplying the equation preceding it by x. Remember, when equations are multiplied by a variable, there is the possibility of introducing an extraneous solution. To make sure that this is not the case, you should check each solution.

EXAMPLE 4 Solving a Nonlinear System

Find the solutions of the system

$$\begin{cases} x^2 + y^2 = 4 \\ -2x^2 + 7y^2 = 7. \end{cases}$$

Solution In preparation for eliminating an x^2-term, we begin by multiplying the first equation by 2. The system

$$\begin{cases} 2x^2 + 2y^2 = 8 \\ -2x^2 + 7y^2 = 7 \end{cases} \tag{2}$$

is equivalent to the given system. Now, by adding the first equation of this last system to the second, we obtain yet another system equivalent to the original system. In this case, we have eliminated x^2 from the second equation:

$$\begin{cases} 2x^2 + 2y^2 = 8 \\ \qquad\quad 9y^2 = 15. \end{cases}$$

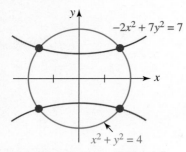

FIGURE 11.2.3 Intersection of a circle and a hyperbola in Example 4

From the last equation, we see that $y = \pm\frac{1}{3}\sqrt{15}$. Substituting these two values of y into $x^2 + y^2 = 4$ then gives

$$x^2 + \tfrac{15}{9} = 4 \quad \text{or} \quad x^2 = \tfrac{21}{9}$$

so that $x = \pm\frac{1}{3}\sqrt{21}$. Thus, $\left(\frac{1}{3}\sqrt{21}, \frac{1}{3}\sqrt{15}\right)$, $\left(-\frac{1}{3}\sqrt{21}, \frac{1}{3}\sqrt{15}\right)$, $\left(\frac{1}{3}\sqrt{21}, -\frac{1}{3}\sqrt{15}\right)$, and $\left(-\frac{1}{3}\sqrt{21}, -\frac{1}{3}\sqrt{15}\right)$ are all solutions. The graphs of the given equations and the four points corresponding to the ordered pairs are indicated by the red dots in FIGURE 11.2.3. ≡

In Example 4 we note that the system can also be solved by the substitution method by substituting, say, $y^2 = 4 - x^2$ into the second equation.

In the next example, we use the third elimination operation to simplify the system *before* applying the substitution method.

EXAMPLE 5   Solving a Nonlinear System

Find the solutions of the system

$$\begin{cases} x^2 - 2x + y^2 = 0 \\ x^2 - 2y + y^2 = 0. \end{cases}$$

Solution By multiplying the first equation by -1 and adding the result to the second, we eliminate x^2 and y^2 from that equation:

$$\begin{cases} x^2 - 2x + y^2 = 0 \\ \phantom{x^2 - {}} 2x - 2y = 0. \end{cases}$$

The second equation of the latter system implies that $y = x$. Substituting this expression into the first equation then yields

$$x^2 - 2x + x^2 = 0 \quad \text{or} \quad 2x(x - 1) = 0.$$

It follows that $x = 0$, $x = 1$ and, correspondingly, $y = 0$, $y = 1$. Thus the solutions of the system are $(0, 0)$ and $(1, 1)$. ≡

By completing the square in x and y, we can write the system in Example 5 as

$$\begin{cases} (x - 1)^2 + y^2 = 1 \\ x^2 + (y - 1)^2 = 1. \end{cases}$$

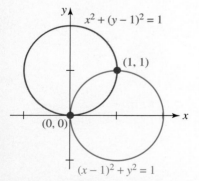

FIGURE 11.2.4 Intersecting circles in Example 5

From this system we see that both equations describe circles of radius $r = 1$. The circles and their points of intersection are illustrated in FIGURE 11.2.4.

11.2 Exercises Answers to selected odd-numbered problems begin on page ANS-30.

In Problems 1–6, determine graphically whether the given nonlinear system has any solutions.

1. $\begin{cases} x = 5 \\ x = y^2 \end{cases}$

2. $\begin{cases} y = 3 \\ (x + 1)^2 + y^2 = 10 \end{cases}$

3. $\begin{cases} -x^2 + y = -1 \\ x^2 + y = 4 \end{cases}$

4. $\begin{cases} x + y = 5 \\ x^2 + y^2 = 1 \end{cases}$

5. $\begin{cases} x^2 + y^2 = 1 \\ x^2 - 4x + y^2 = -3 \end{cases}$

6. $\begin{cases} y = 2^x - 1 \\ y = \log_2(x + 2) \end{cases}$

In Problems 7–42, solve the given nonlinear system.

7. $\begin{cases} y = x \\ y^2 = x + 2 \end{cases}$
8. $\begin{cases} y = 3x \\ x^2 + y^2 = 4 \end{cases}$
9. $\begin{cases} y = 2x - 1 \\ y = x^2 \end{cases}$
10. $\begin{cases} x + y = 1 \\ x^2 - 2y = 0 \end{cases}$
11. $\begin{cases} 64x + y = 1 \\ x^3 - y = -1 \end{cases}$
12. $\begin{cases} y - x = 3 \\ x^2 + y^2 = 9 \end{cases}$
13. $\begin{cases} x = \sqrt{y} \\ x^2 = \dfrac{6}{y} + 1 \end{cases}$
14. $\begin{cases} y = 2\sqrt{2}x^2 \\ y = \sqrt{x} \end{cases}$
15. $\begin{cases} xy = 1 \\ x + y = 1 \end{cases}$
16. $\begin{cases} xy = 3 \\ x + y = 4 \end{cases}$
17. $\begin{cases} xy = 5 \\ x^2 + y^2 = 10 \end{cases}$
18. $\begin{cases} xy = 1 \\ x^2 = y^2 + 2 \end{cases}$
19. $\begin{cases} 16x^2 - y^4 = 16y \\ y^2 + y = x^2 \end{cases}$
20. $\begin{cases} x^3 + 3y = 26 \\ y = x(x + 1) \end{cases}$
21. $\begin{cases} x^2 - y^2 = 4 \\ 2x^2 + y^2 = 1 \end{cases}$
22. $\begin{cases} 3x^2 + 2y^2 = 4 \\ x^2 + 4y^2 = 1 \end{cases}$
23. $\begin{cases} x^2 + y^2 = 4 \\ x^2 - 4x + y^2 - 2y = 4 \end{cases}$
24. $\begin{cases} x^2 + y^2 - 6y = -9 \\ x^2 + 4x + y^2 = -1 \end{cases}$
25. $\begin{cases} x^2 + y^2 = 5 \\ y = x^2 - 5 \end{cases}$
26. $\begin{cases} y = x(x^2 - 6x + 8) \\ y + 4 = (x - 2)^2 \end{cases}$
27. $\begin{cases} (x - y)^2 = 4 \\ (x + y)^2 = 12 \end{cases}$
28. $\begin{cases} (x - y)^2 = 0 \\ (x + y)^2 = 1 \end{cases}$
29. $\begin{cases} y = \sin x \\ y = \cos x \end{cases}$
30. $\begin{cases} y = \cos x \\ 2y \tan x = \sqrt{3} \end{cases}$
31. $\begin{cases} 2y \sin x = 1 \\ y = 2 \sin x \end{cases}$
32. $\begin{cases} y = \sin 2x \\ y = \sin x \end{cases}$
33. $\begin{cases} y = \log_{10} x \\ y^2 = 5 + 4\log_{10} x \end{cases}$
34. $\begin{cases} x + \log_{10} y = 2 \\ y + 15 = 10^x \end{cases}$
35. $\begin{cases} \log_{10}(x^2 + y)^2 = 8 \\ y = 2x + 1 \end{cases}$
36. $\begin{cases} \log_{10} x = y - 5 \\ 7 = y - \log_{10}(x + 6) \end{cases}$
37. $\begin{cases} x = 3^y \\ x = 9^y - 20 \end{cases}$
38. $\begin{cases} y = 2^{x^2} \\ \sqrt{5}x = \log_2 y \end{cases}$
39. $\begin{cases} 2x + \lambda = 0 \\ 2y + \lambda = 0 \\ xy - 3 = 0 \end{cases}$
40. $\begin{cases} -2x + \lambda = 0 \\ y - y\lambda = 0 \\ y^2 - x = 0 \end{cases}$
41. $\begin{cases} y^2 = 2x\lambda \\ 2xy = 2y\lambda \\ x^2 + y^2 - 1 = 0 \end{cases}$
42. $\begin{cases} 8x + 5y = 2xy\lambda \\ 5x = x^2\lambda \\ x^2y - 1000 = 0 \end{cases}$

Miscellaneous Applications

43. **Dimensions of a Corral** The perimeter of a rectangular corral is 260 ft and its area is 4000 ft². What are its dimensions?
44. **Inscribed Rectangle** Find the dimensions of the rectangle(s) with area 10 cm² inscribed in the triangle consisting of the blue line and the two coordinate axes shown in FIGURE 11.2.5.

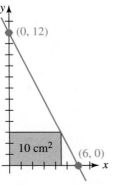

FIGURE 11.2.5 Rectangle in Problem 44

11.2 Systems of Nonlinear Equations

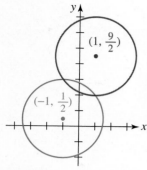

FIGURE 11.2.6 Circles in Problem 46

FIGURE 11.2.7 Rectangle in Problem 47

FIGURE 11.2.8 Open box in Problem 49

45. Sum of Areas The sum of the radii of two circles is 8 cm. Find the radii if the sum of the areas of the circles is 32π cm^2.

46. Intersecting Circles Find the two points of intersection of the circles shown in FIGURE 11.2.6 if the radius of each circle is $\frac{5}{2}$.

47. Golden Ratio The **golden ratio** for the rectangle shown in FIGURE 11.2.7 is defined by

$$\frac{x}{y} = \frac{y}{x+y}.$$

This ratio is often used in architecture and in paintings. Find the dimensions of a rectangular sheet of paper containing 100 in^2 that satisfy the golden ratio.

48. Length The hypotenuse of a right triangle is 20 cm. Find the lengths of the remaining two sides if the shorter side is one-half the length of the longer side.

49. Topless Box A box is to be made with a square base and no top. See FIGURE 11.2.8. The volume of the box is to be 32 ft^3, and the combined areas of the sides and bottom are to be 68 ft^2. Find the dimensions of the box.

50. Dimensions of a Cylinder The volume of a right circular cylinder is 63π in^3, and its height h is 1 in. greater than twice its radius r. Find the dimensions of the cylinder.

For Discussion

51. A **tangent to an ellipse** is defined exactly as it was for the circle, namely, a straight line that touches the ellipse at only one point (x_1, y_1). See Problem 42 in Exercises 2.3. It can be shown (see Problem 52) that an equation of the tangent line at a given point (x_1, y_1) on an ellipse $x^2/a^2 + y^2/b^2 = 1$ is

$$\frac{xx_1}{a^2} + \frac{yy_1}{b^2} = 1. \tag{3}$$

(a) Find the equation of the tangent line to the ellipse $x^2/50 + y^2/8 = 1$ at the point $(5, -2)$.
(b) Write your answer in the form of $y = mx + b$.
(c) Sketch the ellipse and the tangent line.

52. In this problem, you are guided through the steps to derive equation (3).
(a) An alternative form of the equation $x^2/a^2 + y^2/b^2 = 1$ is

$$b^2x^2 + a^2y^2 = a^2b^2.$$

Since the point (x_1, y_1) is on the ellipse, its coordinates must satisfy the foregoing equation:

$$b^2x_1^2 + a^2y_1^2 = a^2b^2.$$

Show that

$$b^2(x^2 - x_1^2) + a^2(y^2 - y_1^2) = 0.$$

(b) Using the point-slope form of a line, the tangent line at (x_1, y_1) is $y - y_1 = m(x - x_1)$. Use substitution in the system

$$\begin{cases} b^2(x^2 - x_1^2) + a^2(y^2 - y_1^2) = 0 \\ y - y_1 = m(x - x_1) \end{cases}$$

to show that

$$b^2(x^2 - x_1^2) + a^2m^2(x - x_1)^2 + 2a^2my_1(x - x_1) = 0. \tag{4}$$

The last equation is a quadratic equation in x. Explain why x_1 is a repeated root or a root of multiplicity 2.

(c) By factoring, (4) becomes

$$(x - x_1)[(b^2(x + x_1) + a^2m^2(x - x_1) + 2a^2my_1] = 0.$$

and so we must have

$$b^2(x + x_1) + a^2m^2(x - x_1) + 2a^2my_1 = 0.$$

Use the last equation to find the slope m of the tangent line at (x_1, y_1). Finish the problem by finding the equation of the tangent as given in (3).

11.3 Partial Fractions

≡ Introduction When two rational functions, say, $f(x) = \dfrac{2}{x + 5}$ and $g(x) = \dfrac{1}{x + 1}$ are added, the terms are combined by means of a common denominator:

$$\frac{2}{x + 5} + \frac{1}{x + 1} = \frac{2}{x + 5}\left(\frac{x + 1}{x + 1}\right) + \frac{1}{x + 1}\left(\frac{x + 5}{x + 5}\right). \tag{1}$$

Adding numerators on the right-hand side of (1) yields the single rational expression

$$\frac{2x + 7}{(x + 5)(x + 1)}. \tag{2}$$

An important procedure in the study of integral calculus requires that we be able to reverse the process; in other words, starting with a rational expression such as (2) break it down, or *decompose* it, into simpler component fractions $2/(x + 5)$ and $1/(x + 1)$ called **partial fractions**.

☐ **Terminology** The algebraic process for breaking down a rational expression such as (2) into partial fractions is known as **partial fraction decomposition**. For convenience we will assume that the rational function $P(x)/Q(x)$, $Q(x) \neq 0$, is a **proper fraction** or **proper rational expression**; that is, the degree of $P(x)$ is less than the degree of $Q(x)$. We will also assume once again that the polynomials $P(x)$ and $Q(x)$ have no common factors.

In the discussion that follows we consider four cases of partial fraction decomposition of $P(x)/Q(x)$. The cases depend on the factors in the denominator $Q(x)$. When the polynomial $Q(x)$ is factored as a product of $(ax + b)^n$ and $(ax^2 + bx + c)^m$, $n = 1, 2, \ldots, m = 1, 2, \ldots$, where the coefficients a, b, c are real numbers and the quadratic polynomial $ax^2 + bx + c$ is **irreducible** over the real numbers (that is, does not factor using real numbers), the rational expression $P(x)/Q(x)$ can be decomposed into a sum of partial fractions of the form

$$\frac{C_k}{(ax + b)^k} \quad \text{and} \quad \frac{A_k x + B_k}{(ax^2 + bx + c)^k}.$$

CASE 1: $Q(x)$ Contains Only Nonrepeated Linear Factors

We state the following fact from algebra without proof. If the denominator can be factored completely into linear factors,

$$Q(x) = (a_1 x + b_1)(a_2 x + b_2) \cdots (a_n x + b_n),$$

where all the $a_i x + b_i$, $i = 1, 2, \ldots, n$ are distinct (that is, no two factors are the same), then unique real constants $C_1, C_2, \ldots, C_n$ can be found such that

$$\frac{P(x)}{Q(x)} = \frac{C_1}{a_1 x + b_1} + \frac{C_2}{a_2 x + b_2} + \cdots + \frac{C_n}{a_n x + b_n}. \tag{3}$$

In practice we will use the letters $A, B, C, \ldots$ in place of the subscripted coefficients $C_1, C_2, C_3, \ldots$. The next example illustrates this first case.

EXAMPLE 1   Distinct Linear Factors

To decompose $\dfrac{2x + 1}{(x - 1)(x + 3)}$ into individual partial fractions we make the assumption, based on the form given in (3), that the rational function can be written as

$$\frac{2x + 1}{(x - 1)(x + 3)} = \frac{A}{x - 1} + \frac{B}{x + 3}. \tag{4}$$

We now clear (4) of fractions; this can be done by either combining the terms on the right-hand side of the equality over a least common denominator and equating numerators or by simply multiplying both sides of the equality by the denominator $(x - 1)(x + 3)$ on the left-hand side. Either way, we arrive at

$$2x + 1 = A(x + 3) + B(x - 1). \tag{5}$$

Multiplying out the right-hand side of (5) and grouping by powers of x gives

$$2x + 1 = A(x + 3) + B(x - 1) = (A + B)x + (3A - B). \tag{6}$$

Each of the equations (5) and (6) is an identity, which means that the equality is true for *all* real values of x. As a consequence, the coefficients of x on the left-hand side of (6) must be the same as the coefficients of the corresponding powers of x on the right-hand side, that is,

$$2x + 1x^0 = (A + B)x + (3A - B)x^0.$$

with the indicated coefficients equal.

The result is a system of two linear equations in two variables A and B:

$$\begin{cases} 2 = A + B \\ 1 = 3A - B. \end{cases} \tag{7}$$

By adding the two equations we get $3 = 4A$ and so we find that $A = \tfrac{3}{4}$. Substituting this value into either equation in (7) then yields $B = \tfrac{5}{4}$. Hence the desired decomposition is

$$\frac{2x + 1}{(x - 1)(x + 3)} = \frac{\tfrac{3}{4}}{x - 1} + \frac{\tfrac{5}{4}}{x + 3}.$$

You are encouraged to verify the foregoing result by combining the terms on the right-hand side of the last equation by means of a common denominator. ≡

☐ **A Shortcut Worth Knowing** If the denominator contains, say, three linear factors such as in $\dfrac{4x^2 - x + 1}{(x - 1)(x + 3)(x - 6)}$, then the partial fraction decomposition looks like this

$$\frac{4x^2 - x + 1}{(x - 1)(x + 3)(x - 6)} = \frac{A}{x - 1} + \frac{B}{x + 3} + \frac{C}{x - 6}.$$

CHAPTER 11 SYSTEMS OF EQUATIONS AND INEQUALITIES

By following the same steps as in Example 1, we would find that the analog of (7) is now three equations in the three unknowns A, B, and C. The point is this: The more linear factors in the denominator the larger the system of equations we must solve. There is a procedure worth learning that can cut down on some of the algebra. To illustrate, let's return to the identity (5). Since the equality is true for every value of x, it holds for $x = 1$ and $x = -3$, *the zeros of the denominator.* Setting $x = 1$ in (5) gives $3 = 4A$, from which it follows immediately that $A = \frac{3}{4}$. Similarly, by setting $x = -3$ in (5), we obtain $-5 = (-4)B$ or $B = \frac{5}{4}$.

CASE 2: $Q(x)$ Contains Repeated Linear Factors

If the denominator $Q(x)$ contains a repeated linear factor $(ax + b)^n$, $n > 1$, then unique real constants $C_1, C_2, \ldots, C_n$ can be found such that the partial fraction decomposition of $P(x)/Q(x)$ contains the terms

$$\frac{C_1}{ax + b} + \frac{C_2}{(ax + b)^2} + \cdots + \frac{C_n}{(ax + b)^n}. \tag{8}$$

EXAMPLE 2   Repeated Linear Factors

To decompose $\dfrac{6x - 1}{x^3(2x - 1)}$ into partial fractions we first observe that the denominator consists of the repeated linear factor x and the nonrepeated linear factor $2x - 1$. Based on the forms in (3) and (8) we assume that

$$\frac{6x - 1}{x^3(2x - 1)} = \overbrace{\frac{A}{x} + \frac{B}{x^2} + \frac{C}{x^3}}^{\text{according to Case 2}} + \overbrace{\frac{D}{2x - 1}}^{\text{according to Case 1}}. \tag{9}$$

Multiplying (9) by $x^3(2x - 1)$ clears it of fractions and yields

$$6x - 1 = Ax^2(2x - 1) + Bx(2x - 1) + C(2x - 1) + Dx^3 \tag{10}$$

or $\quad 6x - 1 = (2A + D)x^3 + (-A + 2B)x^2 + (-B + 2C)x - C. \tag{11}$

Now the zeros of the denominator in the original expression are $x = 0$ and $x = \frac{1}{2}$. If we then set $x = 0$ and $x = \frac{1}{2}$ in (10), we find, in turn, that $C = 1$ and $D = 16$. Because the denominator of the original expression has only two distinct zeros, we can find A and B by equating the corresponding coefficients of x^3 and x^2 in (11):

$$0 = 2A + D, \qquad 0 = -A + 2B.$$

◀ The coefficients of x^3 and x^2 on the left-hand side of (11) are both 0.

Using the known value of D, the first equation yields $A = -D/2 = -8$. The second then gives $B = A/2 = -4$. The partial fraction decomposition is

$$\frac{6x - 1}{x^3(2x - 1)} = -\frac{8}{x} - \frac{4}{x^2} + \frac{1}{x^3} + \frac{16}{2x - 1}. \qquad \equiv$$

CASE 3: $Q(x)$ Contains Nonrepeated Irreducible Quadratic Factors

If the denominator $Q(x)$ contains nonrepeated irreducible quadratic factors $a_i x^2 + b_i x + c_i$, then unique real constants $A_1, A_2, \ldots, A_n, B_1, B_2, \ldots, B_n$ can be found such that the partial fraction decomposition of $P(x)/Q(x)$ contains the terms

$$\frac{A_1 x + B_1}{a_1 x^2 + b_1 x + c_1} + \frac{A_2 x + B_2}{a_2 x^2 + b_2 x + c_2} + \cdots + \frac{A_n x + B_n}{a_n x^2 + b_n x + c_n}. \tag{12}$$

11.3 Partial Fractions

EXAMPLE 3   Irreducible Quadratic Factors

To decompose $\dfrac{4x}{(x^2 + 1)(x^2 + 2x + 3)}$ into partial fractions we first observe that the quadratic polynomials $x^2 + 1$ and $x^2 + 2x + 3$ are irreducible over the real numbers. Hence by (12) we assume that

> Use the quadratic formula. For either factor you will find that $b^2 - 4ac < 0$.

$$\frac{4x}{(x^2 + 1)(x^2 + 2x + 3)} = \frac{Ax + B}{x^2 + 1} + \frac{Cx + D}{x^2 + 2x + 3}.$$

After clearing fractions in the preceding line, we find

$$\begin{aligned} 4x &= (Ax + B)(x^2 + 2x + 3) + (Cx + D)(x^2 + 1) \\ &= (A + B)x^3 + (2A + B + D)x^2 + (3A + 2B + C)x + (3B + D). \end{aligned}$$

Because the denominator of the original fraction has no real zeros, we have no recourse except to form a system of equations by comparing coefficients of all powers of x:

$$\begin{cases} 0 = A + C \\ 0 = 2A + B + D \\ 4 = 3A + 2B + C \\ 0 = 3B + D. \end{cases}$$

Using $C = -A$ and $D = -3B$ from the first and fourth equations we can eliminate C and D in the second and third equations:

$$\begin{cases} 0 = A - B \\ 2 = A + B. \end{cases}$$

Solving this simpler system of equations yields $A = 1$ and $B = 1$. Hence, $C = -1$ and $D = -3$. The partial fraction decomposition is

$$\frac{4x}{(x^2 + 1)(x^2 + 2x + 3)} = \frac{x + 1}{x^2 + 1} - \frac{x + 3}{x^2 + 2x + 3}. \qquad \equiv$$

CASE 4: $Q(x)$ **Contains Repeated Irreducible Quadratic Factors**

If the denominator $Q(x)$ contains a repeated irreducible quadratic factor $(ax^2 + bx + c)^n$, $n > 1$, then unique real constants $A_1, A_2, \ldots, A_n, B_1, B_2, \ldots, B_n$ can be found such that the partial fraction decomposition of $P(x)/Q(x)$ contains the terms

$$\frac{A_1 x + B_1}{ax^2 + bx + c} + \frac{A_2 x + B_2}{(ax^2 + bx + c)^2} + \cdots + \frac{A_n x + B_n}{(ax^2 + bx + c)^n}. \qquad (13)$$

EXAMPLE 4   Repeated Quadratic Factor

Decompose $\dfrac{x^2}{(x^2 + 4)^2}$ into partial fractions.

Solution The denominator contains only the repeated irreducible quadratic factor $x^2 + 4$. As indicated in (13) we assume a decomposition of the form

$$\frac{x^2}{(x^2 + 4)^2} = \frac{Ax + B}{x^2 + 4} + \frac{Cx + D}{(x^2 + 4)^2}.$$

are inequalities in two variables. A **solution** of an inequality in two variables is any ordered pair of real numbers (x_0, y_0) that satisfies the inequality—that is, results in a true statement—when x_0 and y_0 are substituted for x and y, respectively. A **graph** of the solution set of an inequality in two variables is made up of all points in the plane whose coordinates satisfy the inequality.

Many results obtained in higher-level mathematics courses are valid only in a specialized region either in the xy-plane or in three-dimensional space, and these regions are often defined by means of **systems of inequalities** in two or three variables. In this section we consider only systems of inequalities involving two variables x and y.

We begin with linear inequalities in two variables.

☐ **Half-Planes** A **linear inequality in two variables** x and y is any inequality that has one of the forms

$$ax + by + c < 0, \quad ax + by + c > 0, \tag{1}$$
$$ax + by + c \leq 0, \quad ax + by + c \geq 0. \tag{2}$$

Since the inequalities in (1) and (2) have infinitely many solutions, the notation

$$\{(x, y) \mid ax + by + c < 0\}, \quad \{(x, y) \mid ax + by + c \geq 0\},$$

and so on, is used to denote a set of solutions. Geometrically, each of these sets describes a **half-plane**. As shown in FIGURE 11.4.1, the graph of the linear equation $ax + by + c = 0$ divides the xy-plane into two regions, or half-planes. One of these half-planes is the graph of the set of solutions of the linear inequality. If the inequality is strict, as in (1), then we draw the graph of $ax + by + c = 0$ as a dashed line, because the points on the line are not in the set of solutions of the inequality. See Figure 11.4.1(a). On the other hand, if the inequality is nonstrict, as in (2), the set of solutions includes the points satisfying $ax + by + c = 0$, and so we draw the graph of the equation as a solid line. See Figure 11.4.1(b).

(a)

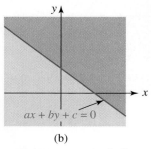

(b)

FIGURE 11.4.1 A single line determines two half-planes

EXAMPLE 1 Graph of a Linear Inequality

Graph the linear inequality $2x - 3y \geq 12$.

Solution First, we graph the line $2x - 3y = 12$, as shown in FIGURE 11.4.2(a). Solving the given inequality for y gives

$$y \leq \tfrac{2}{3}x - 4. \tag{3}$$

Since the y-coordinate of any point (x, y) on the graph of $2x - 3y \geq 12$ must satisfy (3), we conclude that the point (x, y) must lie on or below the graph of the line. This solution set is the region that is shaded blue in Figure 11.4.2(b).

Alternatively, we know that the set

$$\{(x, y) \mid 2x - 3y - 12 \geq 0\}$$

describes a half-plane. Thus we can determine whether the graph of the inequality includes the region above or below the line $2x - 3y = 12$ by determining whether a test point not on the line, such as $(0, 0)$, satisfies the original inequality. Substituting $x = 0$, $y = 0$ into $2x - 3y \geq 12$ gives $0 \geq 12$. This false statement implies that the graph of the inequality is the region on the other side of the line $2x - 3y = 12$, that is, the side that does *not* contain the origin. Note that the blue half-plane in Figure 11.4.2(b) does not contain the point $(0, 0)$. ≡

(a)

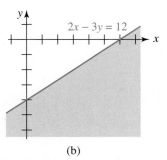

(b)

FIGURE 11.4.2 Half-plane in Example 1

In general, given a linear inequality of the forms in (1) or (2), we can graph the solutions by proceeding in the following manner.

- Graph the line $ax + by + c = 0$.
- Select a **test point** not on this line.
- Shade the half-plane containing the test point if its coordinates satisfy the original inequality. If they do not satisfy the inequality, shade the other half-plane.

EXAMPLE 2 Graph of a Linear Inequality

Graph the linear inequality $3x + y - 2 < 0$.

Solution In FIGURE 11.4.3 we draw the graph of $3x + y = 2$ as a dashed line, since it will not be part of the solution set of the inequality. Then we select $(0, 0)$ as a test point that is not on the line. Because substituting $x = 0$, $y = 0$ into $3x + y - 2 < 0$ gives the true statement $-2 < 0$ we shade that region of the plane containing the origin. ≡

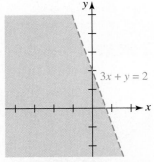

FIGURE 11.4.3 Half-plane in Example 2

☐ **Systems of Inequalities** We say (x_0, y_0) is a **solution of a system of inequalities** when it is a member of the set of solutions *common* to all inequalities. In other words, the **solution set** of a system of inequalities is the intersection of the solution sets of the individual inequalities in the system.

In the next two examples we graph the solution set of a system of linear inequalities.

EXAMPLE 3 System of Linear Inequalities

Graph the system of linear inequalities

$$\begin{cases} x \geq 1 \\ y \leq 2. \end{cases}$$

Solution The sets

$$\{(x, y) \mid x \geq 1\} \quad \text{and} \quad \{(x, y) \mid y \leq 2\}$$

denote the sets of solutions for each inequality. These sets are illustrated in FIGURE 11.4.4 by the blue and the red shading, respectively. The solutions of the given system are the ordered pairs in the intersection

$$\{(x, y) \mid x \geq 1\} \cap \{(x, y) \mid y \leq 2\} = \{(x, y) \mid x \geq 1 \text{ and } y \leq 2\}.$$

This last set is the region of darker color (overlapping red and blue colors) shown in the figure. ≡

FIGURE 11.4.4 Solution set in Example 3

EXAMPLE 4 System of Linear Inequalities

Graph the system of linear inequalities

$$\begin{cases} x + y \leq 1 \\ -x + 2y \geq 4. \end{cases} \quad (4)$$

Solution Substitution of $(0, 0)$ into the first inequality in (4) gives the true statement $0 \leq 1$, which implies that the graph of the solutions of $x + y \leq 1$ is the half-plane *below* (and including) the line $x + y = 1$. This is the shaded blue region in FIGURE 11.4.5(a). Similarly, substituting $(0, 0)$ into the second inequality gives the false statement $0 \geq 4$, and so the graph of the solutions of $-x + 2y \geq 4$ is the half-plane *above* (and including) the line $-x + 2y = 4$. This is the shaded red region in Figure 11.4.5(b). The graph of the solutions of the system of inequalities is then the intersection of the graphs of these two solution sets. This intersection is the darker region of overlapping colors shown in Figure 11.4.5(c).

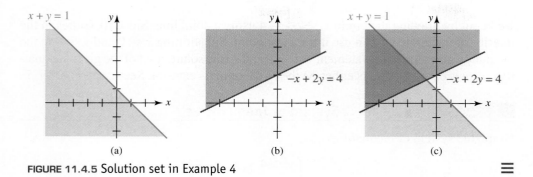

FIGURE 11.4.5 Solution set in Example 4

Often we are interested in the solutions of a system of linear inequalities subject to the restrictions that $x \geq 0$ and $y \geq 0$. This means that the graph of the solutions is a subset of the set consisting of the points in the first quadrant and on the nonnegative coordinate axes. For example, inspection of Figure 11.4.5(c) reveals that the system of inequalities (4) subject to the added requirements that $x \geq 0$, $y \geq 0$, has no solutions.

EXAMPLE 5   System of Linear Inequalities

The graph of the solutions of the system of linear inequalities

$$\begin{cases} -2x + y \leq 2 \\ x + 2y \leq 8 \end{cases}$$

is the region shown in FIGURE 11.4.6(a). The graph of the solutions of

$$\begin{cases} -2x + y \leq 2 \\ x + 2y \leq 8 \\ x \geq 0, \ y \geq 0 \end{cases}$$

is the region in the first quadrant along with portions of the two lines and portions of the coordinate axes illustrated in Figure 11.4.6(b).

FIGURE 11.4.6 Solution set in Example 5

☐ **Nonlinear Inequalities** Graphing **nonlinear inequalities** in two variables x and y is basically the same as graphing linear inequalities. In the next example we again utilize the notion of a test point.

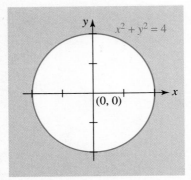

FIGURE 11.4.7 Solution set in Example 6

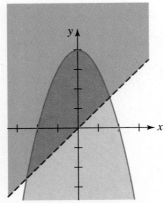

FIGURE 11.4.8 Solution set in Example 7

EXAMPLE 6 Graph of a Nonlinear Inequality

To graph the nonlinear inequality

$$x^2 + y^2 - 4 \geq 0$$

we begin by drawing the circle $x^2 + y^2 = 4$ using a solid line. Since $(0, 0)$ lies in the interior of the circle we can use it for a test point. Substituting $x = 0$ and $y = 0$ in the inequality gives the false statement $-4 \geq 0$ and so the solution set of the given inequality consists of all the points either on the circle or in its exterior. See **FIGURE 11.4.7**. ≡

EXAMPLE 7 System of Inequalities

Graph the system of inequalities

$$\begin{cases} y \leq 4 - x^2 \\ y > x. \end{cases}$$

Solution Substitution of the coordinates of $(0, 0)$ into the first inequality gives the true statement $0 \leq 4$ and so the graph of $y \leq 4 - x^2$ is the shaded blue region in **FIGURE 11.4.8** below the parabola $y = 4 - x^2$. Note that we cannot use $(0, 0)$ as a test point for the second inequality since $(0, 0)$ is a point on the line $y = x$. However, if we use $(1, 2)$ as a test point, the second inequality gives the true statement $2 > 1$. Thus the graph of the solutions of $y > x$ is the shaded red half-plane above the line $y = x$ in Figure 11.4.8. The line itself is dashed because of the strict inequality. The intersection of these two colored regions is the darker region in the figure. ≡

11.4 Exercises Answers to selected odd-numbered problems begin on page ANS-30.

In Problems 1–12, graph the given inequality.

1. $x + 3y \geq 6$
2. $x - y \leq 4$
3. $x + 2y < -x + 3y$
4. $2x + 5y > x - y + 6$
5. $-y \geq 2(x + 3) - 5$
6. $x \geq 3(x + 1) + y$
7. $y \geq (x - 1)^2$
8. $x^2 + \frac{1}{4}y^2 < 1$
9. $y - 1 \leq \sqrt{x}$
10. $y \geq \sqrt{x + 1}$
11. $y \geq |x + 2|$
12. $xy \geq 3$

In Problems 13–36, graph the given system of inequalities.

13. $\begin{cases} y \leq x \\ x \geq 2 \end{cases}$
14. $\begin{cases} y \geq x \\ y \geq 0 \end{cases}$
15. $\begin{cases} x - y > 0 \\ x + y > 1 \end{cases}$
16. $\begin{cases} x + y < 1 \\ -x + y < 1 \end{cases}$
17. $\begin{cases} x + 2y \leq 4 \\ -x + 2y \geq 6 \\ x \geq 0 \end{cases}$
18. $\begin{cases} 4x + y \geq 12 \\ -2x + y \leq 0 \\ y \geq 0 \end{cases}$
19. $\begin{cases} x - 3y > -9 \\ x \geq 0, y \geq 0 \end{cases}$
20. $\begin{cases} x + y > 4 \\ x \geq 0, y \geq 0 \end{cases}$
21. $\begin{cases} y < x + 2 \\ 1 \leq x \leq 3 \\ y \geq 1 \end{cases}$
22. $\begin{cases} 4y > x \\ x \geq 2 \\ y \leq 5 \end{cases}$

23. $\begin{cases} x + y \leq 4 \\ y \geq -x \\ y \leq 2x \end{cases}$

24. $\begin{cases} 2x + 3y \geq 6 \\ x - y \geq -6 \\ 2x + y \leq 6 \end{cases}$

25. $\begin{cases} -2x + y \leq 2 \\ x + 3y \leq 10 \\ x - y \leq 5 \\ x \geq 0, y \geq 0 \end{cases}$

26. $\begin{cases} -x + y \leq 0 \\ -x + 3y \geq 0 \\ x + y - 8 \geq 0 \\ y - 2 \leq 0 \end{cases}$

27. $\begin{cases} x^2 + y^2 \geq 1 \\ \frac{1}{9}x^2 + \frac{1}{4}y^2 \leq 1 \end{cases}$

28. $\begin{cases} x^2 + y^2 \leq 25 \\ x + y \geq 5 \end{cases}$

29. $\begin{cases} y \leq x^2 + 1 \\ y \geq -x^2 \end{cases}$

30. $\begin{cases} x^2 + y^2 \leq 4 \\ y \leq x^2 - 1 \end{cases}$

31. $\begin{cases} y \geq |x| \\ x^2 + y^2 \leq 2 \end{cases}$

32. $\begin{cases} y \leq e^x \\ y \geq x - 1 \\ x \geq 0 \end{cases}$

33. $\begin{cases} \frac{1}{9}x^2 - \frac{1}{4}y^2 \geq 1 \\ y \geq 0 \end{cases}$

34. $\begin{cases} y < \ln x \\ y > 0 \end{cases}$

35. $\begin{cases} y \leq x^3 + 1 \\ x \geq 0 \\ x \leq 1 \\ y \geq 0 \end{cases}$

36. $\begin{cases} y \geq x^4 \\ y \leq 2 \\ x \geq -1 \\ x \leq 1 \end{cases}$

In Problems 37–40, find a system of linear inequalities whose graph is the region shown in the figure.

37.

FIGURE 11.4.9 Region for Problem 37

38.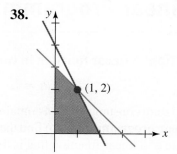

FIGURE 11.4.10 Region for Problem 38

39.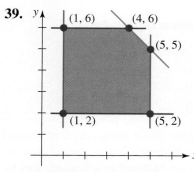

FIGURE 11.4.11 Region for Problem 39

40.

FIGURE 11.4.12 Region for Problem 40

For Discussion

In Problems 41 and 42, graph the given inequality.

41. $-1 \leq x + y \leq 1$
42. $-x \leq y \leq x$

Project

FIGURE 11.4.13 Envelope in Problem 43

43. Ancient History and USPS Some years ago the restrictions on first-class envelope size were a bit more confusing than they are today. Consider the rectangular envelope of length x and height y shown in FIGURE 11.4.13 and the following postal regulation of November 1978:

All first-class items weighing one ounce or less and all single-piece third-class items weighing two ounces or less are subject to an extra mailing fee when the height is greater than $6\frac{1}{8}$ in., or the length is greater than $11\frac{1}{2}$ in., or the length is less than 1.3 times the height, or the length is greater than 2.5 times the height.

In parts (a)–(c) assume that the weight specification is satisfied.
(a) Using x and y, interpret the above regulation as a system of linear inequalities.
(b) Graph the region that describes envelope sizes that are *not* subject to an extra mailing fee.
(c) Under this regulation does an envelope of length 8 in. and height 4 in. require an extra fee?
(d) Do some research and compare the 2010 first class regulation against the one just given.

11.5 Linear Programming

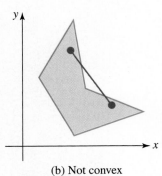

FIGURE 11.5.1 Polygons in the plane

≡ **Introduction** A **linear function in two variables** is a function of the form

$$F(x, y) = ax + by + c, \quad (1)$$

where a, b, and c are constants, with a domain a subset of the Cartesian plane. The basic problem in **linear programming** is to find the maximum (largest) value or the minimum (smallest) value of a linear function that is defined on a set determined by a system of linear inequalities. In this section we will discuss a way of finding a maximum or minimum value of F.

☐ **Terminology** A typical linear-programming problem is given by

Maximize: $F(x, y) = 5x + 10y$

subject to: $\begin{cases} x + 2y \leq 6 \\ 3x + y \leq 9 \\ x \geq 0, y \geq 0. \end{cases}$ (2)

In this context, F is called the **objective function** and the linear inequalities are called **constraints**. Any ordered pair of real numbers (x_0, y_0) that satisfies all the constraints is said to be a **feasible solution** of the problem. The set of feasible solutions will be denoted by S. It can be shown that for any two points in the graph of S, the line segment joining them lies entirely in the graph. Any set in the plane having this property is called **convex**. FIGURE 11.5.1(a) illustrates a convex polygon; Figure 11.5.1(b) illustrates a polygon that is not convex. The corner points of the convex set S determined by the constraints are called **vertices**.

Throughout this discussion we will be concerned with the graph of the set S of feasible solutions of a system of linear inequalities in which $x \geq 0$ and $y \geq 0$. We state the following theorem without proof.

THEOREM 11.5.1 Maximum and Minimum Values

Let the function $F(x, y) = ax + by + c$ be defined on a set S in the plane. If the graph of S is a convex polygon, then F has both a maximum and a minimum value on S, and each of these values is attained at a vertex of S.

EXAMPLE 1 Finding Max and Min Values

Find the maximum and minimum values of the objective function in (2).

Solution We first graph the set S of feasible solutions and find all vertices by solving the appropriate simultaneous equations. For example, the vertex $\left(\frac{12}{5}, \frac{9}{5}\right)$ is obtained by solving the system of equations:
$$\begin{cases} x + 2y = 6 \\ 3x + y = 9. \end{cases}$$

See FIGURE 11.5.2. It follows from Theorem 11.5.1 that the maximum and minimum values of F occur at vertices. From the accompanying table, we see that the maximum value of the objective function is
$$F\left(\tfrac{12}{5}, \tfrac{9}{5}\right) = 10\left(\tfrac{12}{5}\right) + 15\left(\tfrac{9}{5}\right) = 51.$$
The minimum value is $F(0, 0) = 0$.

Vertex	Value of F
(0, 3)	45
(0, 0)	0
(3, 0)	30
$\left(\tfrac{12}{5}, \tfrac{9}{5}\right)$	51

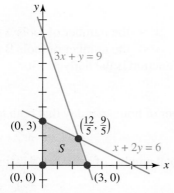

FIGURE 11.5.2 Convex set for Example 1

EXAMPLE 2 Finding Max and Min Values

Find the maximum and minimum values of
$$F(x, y) = 6x + y + 1$$
subject to
$$\begin{cases} x + y \leq 10 \\ 1 \leq x \leq 5 \\ 2 \leq y \leq 6. \end{cases}$$

Solution In FIGURE 11.5.3 the graph of S determined by the constraints is given and the vertices are labeled. As we see from the accompanying table, the maximum value of F is

$$F(5, 5) = 6(5) + 5 + 1 = 36.$$

The minimum value is

$$F(1, 2) = 6(1) + 2 + 1 = 9.$$

Vertex	Value of F
(1, 6)	13
(1, 2)	9
(5, 2)	33
(5, 5)	36
(4, 6)	31

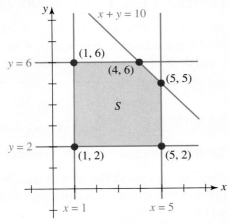

FIGURE 11.5.3 Convex set for Example 2

EXAMPLE 3   Maximum Profit

A small tool manufacturing company has two forges F_1 and F_2 each of which, because of maintenance requirements, can operate at most 20 h per day. The company makes two types of tools, A and B. Tool A must spend 1 h in forge F_1 and 3 h in forge F_2. Tool B must spend 2 h in forge F_1 and 1 h in forge F_2. The company makes a $20 profit on tool A and a $10 profit on tool B. Determine the number of each type of tool the company must make in order to maximize its daily profit.

Solution Let

$x =$ the number of tools A produced each day, and
$y =$ the number of tools B produced each day.

The objective function is the daily profit

$$P(x, y) = 20x + 10y.$$

The total number of hours per day that both tools spend in forge F_1 must satisfy

$$1 \cdot x + 2 \cdot y \leq 20.$$

Similarly, the total number of hours per day that both tools spend in forge F_2 must satisfy

$$3 \cdot x + 1 \cdot y \leq 20.$$

Thus we wish to

Maximize: $P(x, y) = 20x + 10y$

subject to: $\begin{cases} x + 2y \leq 20 \\ 3x + y \leq 20 \\ x \geq 0, y \geq 0. \end{cases}$

The graph of S determined by the constraints is shown in FIGURE 11.5.4. From the table accompanying Figure 11.5.4, we see that the maximum daily profit is

$$P(4, 8) = 20(4) + 10(8) = 160.$$

That is, when the company makes four of tool A and eight of tool B each day, its maximum daily profit is $160.

Vertex	Value of P
(0, 10)	100
(0, 0)	0
$(\frac{20}{3}, 0)$	133.33
(4, 8)	160

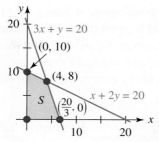

FIGURE 11.5.4 Convex set for Example 3

EXAMPLE 4   Minimum Cost

In his spare time John does piecework at home making pairs of mittens, scarfs, and stocking caps. During the winter he produces a total of 300 of these items per month. He has a standing monthly order at a large outdoor mail-order company for 50 to 100 pairs of mittens, for at least 100 scarfs, and for at least 70 stocking caps. The costs of the material used are $0.20 for each pair of mittens, $0.40 for each scarf, and $0.50 for each stocking cap. Determine the number of each item that should be made each month in order to minimize his monthly total cost.

Solution Let x and y denote the number of pairs of mittens and scarfs, respectively, supplied by John to the mail-order company each month. The number of stocking caps supplied each month is then $300 - x - y$. The objective function is the total monthly costs

$$C(x, y) = 0.2x + 0.4y + 0.5(300 - x - y),$$
or
$$C(x, y) = -0.3x - 0.1y + 150.$$

The constraints are

$$\begin{cases} x \geq 0, y \geq 0 \\ 300 - x - y \geq 0 \\ 50 \leq x \leq 100 \\ y \geq 100 \\ 300 - x - y \geq 70. \end{cases}$$

The last inequality in this system of inequalities is equivalent to $x + y \leq 230$. The graph of the S of feasible solutions determined by the constraints and the vertices of the set is shown in FIGURE 11.5.5. From the accompanying table, we see that $C(100, 300) = 107$ is the minimum. Thus John should make 100 pairs of mittens, 130 scarfs, and 70 stocking caps each month in order to minimize his total costs.

Vertex	Value of C
(50, 100)	125
(50, 180)	117
(100, 130)	107
(100, 100)	110

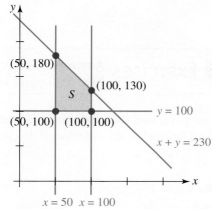

FIGURE 11.5.5 Convex set for Example 4

You should not get the impression from the preceding discussion that an objective function must have both a maximum and a minimum. If the graph of S is not a convex polygon, then the conclusion of Theorem 11.5.1 may not be true. Depending on the constraints, it could happen that a linear function F has a minimum but no maximum (or vice versa). However, it can be shown that if F has a minimum (or a maximum) on S, then it is attained at a vertex of the region. Also, the maximum (or minimum) of an objective function may occur at more than one vertex.

EXAMPLE 5 Minimum but No Maximum

Consider the linear function

$$F(x, y) = 5x + 20y$$

subject to
$$\begin{cases} 3x + 4y \geq 12 \\ 2x + y \geq 4 \\ x \geq 0, y \geq 0. \end{cases}$$

Inspection of **FIGURE 11.5.6** shows that the graph of the set S of the solutions of the constraints is not a convex polygon. It can be proved that

$$F(4, 0) = 20$$

is a minimum. However, in this case, the objective function has no maximum since $F(x, y)$ can be increased without bound simply by increasing x ($x \geq 4$) or by increasing y ($y \geq 4$).

Vertex	Value of F
$(0, 4)$	80
$(\frac{4}{5}, \frac{12}{5})$	52
$(4, 0)$	20

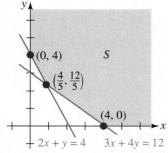

FIGURE 11.5.6 Convex set for Example 5

11.5 Exercises   Answers to selected odd-numbered problems begin on page ANS-31.

In Problems 1–12, find the maximum and minimum values of the given linear function F on the set S defined by the constraints. In each problem assume that $x \geq 0$ and $y \geq 0$.

1. $F(x, y) = 4x + 7$

$$\begin{cases} x \leq 3, y \geq 1 \\ y \leq x \end{cases}$$

2. $F(x, y) = 20x - 3y$

$$\begin{cases} y \leq 4 \\ x + y \geq 3 \\ x - y \leq 0 \end{cases}$$

3. $F(x, y) = 5x + 8y$

$$\begin{cases} x \leq 2 \\ x + y \leq 3 \\ x - y \geq 0 \end{cases}$$

4. $F(x, y) = 3x + 6y$

$$\begin{cases} x \geq 2, y \leq 4 \\ x + 3y \geq 6 \\ 2x - y \leq -2 \\ 3x + 2y \leq 18 \end{cases}$$

5. $F(x, y) = 3x + 6y$

$$\begin{cases} x + 3y \geq 6 \\ 2x - y \leq -2 \\ 3x + 2y \leq 18 \end{cases}$$

6. $F(x, y) = 8x + 12y$

$$\begin{cases} x - 4y \leq -6 \\ -3x + y \leq -4 \\ 2x + 3y \leq 21 \end{cases}$$

7. $F(x, y) = x + 4y$

$$\begin{cases} x + y \geq 1 \\ 2x - y \geq -1 \\ 2x + y \leq 5 \end{cases}$$

8. $F(x, y) = x + y$

$$\begin{cases} y \leq 5 \\ x - y \leq 3 \end{cases}$$

9. $F(x, y) = 3x + 6y$

$$\begin{cases} x \leq 4, y \leq 5 \\ 2x + y \leq 10 \\ x + 2y \leq 10 \end{cases}$$

10. $F(x, y) = x - 4y$

$$\begin{cases} 3x + y \geq 12 \\ 3x - 2y \leq 6 \\ 3x + 4y \leq 30 \\ 3x - 2y \geq 3 \end{cases}$$

11. $F(x, y) = 4x + 2y + 25$

$$\begin{cases} -x + 2y \geq 0 \\ x + y \leq 10 \\ -3x + y \leq 5 \end{cases}$$

12. $F(x, y) = 2x + 3y + 6$

$$\begin{cases} x \leq 8, y \leq 5 \\ 3x + 2y \geq 8 \\ -x + 5y \leq 20 \end{cases}$$

In Problems 13–16, the given objective function F subject to the constraints has a minimum value. Find this value. Explain why F has no maximum value. Assume that $x \geq 0$ and $y \geq 0$.

13. $F(x, y) = 6x + 4y$

$$\begin{cases} 2x - y \geq 6 \\ 2x + 5y \geq 10 \end{cases}$$

14. $F(x, y) = 10x + 20y$

$$\begin{cases} x \geq 3, y \geq 1 \\ y \leq x \end{cases}$$

15. $F(x, y) = 10x + 10y$

$$\begin{cases} x + 2y \geq 5 \\ 2x + y \geq 6 \end{cases}$$

16. $F(x, y) = 10x + 5y$

$$\begin{cases} x + y \geq 4 \\ x + 2y \geq 6 \\ x + 4y \geq 8 \end{cases}$$

Miscellaneous Applications

17. **How Many?** A company manufactures satellite radios and portable DVD players. It makes a $10 profit on each radio and a $40 profit on a DVD player. Due to limited production facilities, the total number of radios and DVD players that the company can manufacture in one month is at most 350. Because of the availability of parts, the company can manufacture at most 300 radios and 100 DVD players each month. Determine how many satellite radios and DVD players the company should produce each month in order to maximize its profit.

18. **Making Money** A woman has up to $10,000 that she wishes to invest in two kinds of certificates of deposit, A and B, that return, respectively, 6.5% and 7.5% annually. She wants to invest at most three times as much in B as in A. Find the

maximum annual return if no more than $6000 can be invested in B and no more than $5000 in A.

How much of each?

19. **Out-of-Pocket Cost** A patient is informed that her daily intake of vitamins must be at least 6 units of A, 4 units of B, and 18 units of C, but no more than 12 units of A, 8 units of B, and 56 units of C. She finds that a drugstore sells two brands, X and Y, of multiple vitamins containing the necessary vitamins. One capsule of brand X contains 1 unit of A, 1 unit of B, and 7 units of C, and costs 5 cents. One capsule of brand Y contains 3 units of A, 1 unit of B, and 2 units of C, and costs 6 cents. How may capsules of each brand should the patient take each day in order to minimize her cost?

20. **Cost of Doing Business** An insurance company uses two computers, an IBC 490 and a CDM 500. Each hour the IBC can process 8 units (1 unit = 1000) of medical claims, 1 unit of life insurance claims, and 2 units of car insurance claims. Each hour the CDM can process 2 units of medical claims, 1 unit of life insurance claims, and 7 units of car insurance claims. The company finds it necessary to process at least 16 units of medical claims, at least 5 units of life insurance claims, and at least 20 units of car insurance claims per day. If it costs the company $100 an hour to run the IBC and $200 an hour to run the CDM, at most how many hours should each computer be run each day in order to keep the company's cost per day at a minimum? What is the minimum cost? Is there a maximum cost per day?

21. **Making a Profit** Kerry's Warehouse has a supply of 1300 pairs of designer jeans and 1700 pairs of generic-brand jeans, which are to be shipped out to two stores: an upscale location and a discount outlet. The warehouse receives a profit of $14.25 per pair on the designer jeans and $12.50 per pair on the generic-brand jeans at the upscale location. The corresponding profits at the discount outlet are $4.80 and $3.40 per pair. The upscale location, however, can carry at most 1800 pairs of jeans, whereas the discount outlet has room for at most 2500 pairs. Find the number of pairs of both designer jeans and generic-brand jeans that the warehouse should ship to each store in order to maximize its total profit. What is the maximum profit?

22. **Minimize the Cost** Joan's manufacturing company makes three kinds of custom sports cars; A, B, and C. Her company turns out a total of 100 cars each year. The number of B cars made must not exceed three times the number of A cars made, but their combined yearly production must be at least 20. The number of C cars made each year must be at least 32. Each A car costs $9000 to build. The B and C cars cost $6000 and $8000, respectively. How many of each kind of car should be built in order to minimize the company's yearly production cost? What is the minimum cost?

23. **A Healthy Diet** Moose living in a national park in Michigan eat both aquatic plants, which are high in sodium but low in energy content (they contain a lot of water), and terrestrial plants, which are high in energy content but contain essentially no sodium. Experiments have shown that a moose can get about 0.8 mj (megajoules) of energy from 1 kg of aquatic plants and about 3.2 mj of energy from 1 kg of terrestrial plants. It has been estimated that an adult moose needs to eat at least 17 kg of aquatic plants daily in order to satisfy its sodium needs. It has also been estimated that the moose rumen (the first stomach) is incapable of digesting more than 33 kg of food daily. Find the daily intakes of both aquatic and terrestrial plants that will maximize a moose's energy intake, subject to its sodium requirement and rumen capacity.

Moose eating aquatic plants

CONCEPTS REVIEW

You should be able to give the meaning of each of the following concepts.

Linear equation
Systems of linear equations
Solution of a linear system
Equivalent linear systems
Consistent system
Inconsistent system
Method of substitution
Method of elimination
Back substitution
Homogeneous systems

Trivial solution
Proper rational expression
Improper rational expression
Partial fraction decomposition:
    irreducible quadratic factor
Linear inequality
Solution of an inequality:
    half-plane
System of linear inequalities:
    test point

Nonlinear inequality
System of nonlinear inequalities
Graph of a solution set
Linear programming:
    objective function
    feasible solution
    constraints
    convex set
    vertices of a convex set

CHAPTER 11 Review Exercises Answers to selected odd-numbered problems begin on page ANS-31.

A. True/False

In Problems 1–10, answer true or false.

1. The graphs of $2x + 7y = 6$ and $x^4 + 8xy - 3y^6 = 0$ intersect at $(-4, 2)$. _____
2. The homogeneous system
$$\begin{cases} x + 2y - 3z = 0 \\ x + y + z = 0 \\ -2x - 4y + 6z = 0 \end{cases}$$
possesses only the zero solution $(0, 0, 0)$. _____
3. The system
$$\begin{cases} y = mx \\ x^2 + y^2 = k \end{cases}$$
always has two solutions when $m \neq 0$ and $k > 0$. _____
4. The system
$$\begin{cases} y = \ln x \\ y = 1 - x \end{cases}$$
has no solution. _____
5. The nonlinear systems
$$\begin{cases} y = \sqrt{x} \\ y = \sqrt{4-x} \end{cases} \quad \text{and} \quad \begin{cases} y^2 = x \\ y^2 = 4 - x \end{cases}$$
are equivalent. _____
6. $(1, -2)$ is a solution of the inequality $4x - 3y + 5 \leq 0$. _____
7. The origin is in the half-plane determined by $4x - 3y < 6$. _____
8. The system of inequalities
$$\begin{cases} x + y > 4 \\ x + y < -1 \end{cases}$$
has no solutions. _____

9. The system of nonlinear equations

$$\begin{cases} x^2 + y^2 = 25 \\ x^2 - y = 5 \end{cases}$$

has exactly three solutions. ____

10. The form of the partial fraction decomposition of $\dfrac{1}{x^2(x+1)^2}$ is

$$\dfrac{A}{x^2} + \dfrac{B}{(x+1)^2}. \text{\_\_\_\_}$$

B. Fill in the Blanks

In Problems 1–10, fill in the blanks.

1. The system

$$\begin{cases} x - y = 5 \\ 2x - 2y = 1 \end{cases}$$

is _____ (consistent or inconsistent).

2. The system

$$\begin{cases} x - 2y = 3 \\ -\tfrac{1}{2}x + y = b \end{cases}$$

is consistent for $b =$ _____.

3. The graph of a single linear inequality in two variables represents a _____ in the plane.

4. In words, describe the graph of the inequality $1 \leq x - y \leq 4$. _____

5. To decompose

$$\dfrac{x^3}{(x+1)(x+2)}$$

into partial fractions we begin with _____.

6. The solution of the system

$$\begin{cases} 3x + y + z = 2 \\ y + 2z = 1 \\ 4z = -8 \end{cases}$$

is _____.

7. If the system of two linear equations in two variables has an infinite number of solutions, then the equations are said to be _____.

8. If the graph of $y = ax^2 + bx$ passes through $(1, 1)$ and $(2, 1)$, then $a =$ _____ and $b =$ _____.

9. The graph of the system

$$\begin{cases} x^2 + y^2 \leq 25 \\ y - 1 > 0 \\ x + 1 < 0 \end{cases}$$

lies in the _____ quadrant.

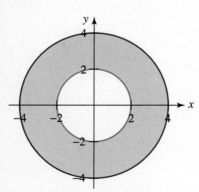

FIGURE 11.R.1 Graph for Problem 10

10. A system of inequalities whose graph is given in FIGURE 11.R.1 is _____.

C. Review Exercises

In Problems 1–14, solve the given system of equations.

1. $\begin{cases} x^2 - 4x + y = 5 \\ x + y = -1 \end{cases}$
2. $\begin{cases} 101y = 10^x + 10^{-x} \\ y - 10^x = 0 \end{cases}$
3. $\begin{cases} 4x^2 + y^2 = 16 \\ x^2 + 4y^2 = 16 \end{cases}$
4. $\begin{cases} xy = 12 \\ -\dfrac{1}{x} + \dfrac{1}{y} = \dfrac{1}{3} \end{cases}$
5. $\begin{cases} y - \log_{10} x = 0 \\ y^2 - 4\log_{10} x + 4 = 0 \end{cases}$
6. $\begin{cases} x^2 y = 63 \\ y = 16 - x^2 \end{cases}$
7. $\begin{cases} 2\ln x + \ln y = 3 \\ 5\ln x + 2\ln y = 8 \end{cases}$
8. $\begin{cases} x^2 + y^2 = 4 \\ xy = 1 \end{cases}$
9. $\begin{cases} x + y + z = 0 \\ x + 2y + 3z = 0 \\ x - y - z = 0 \end{cases}$
10. $\begin{cases} x + 5y - 6z = 1 \\ 4x - y + 2z = 4 \\ 2x - 11y + 14z = 2 \end{cases}$
11. $\begin{cases} 2x + y - z = 7 \\ x + y + z = -2 \\ 4x + 2y + 2z = -6 \end{cases}$
12. $\begin{cases} 5x - 2y = 10 \\ x + 4y = 4 \end{cases}$
13. $\begin{cases} x = 4 + \log_2 y \\ y = 8^x \end{cases}$
14. $\begin{cases} y = |x| \\ -x + y = 1 \end{cases}$

15. **Playing with Numbers** In a two-digit number, the units digit is 1 greater than 3 times the tens digit. When the digits are reversed, the new number is 45 more than the old number. Find the old number.
16. **Lengths** A right triangle has an area of 24 cm². If its hypotenuse has a length of 10 cm, find the lengths of the remaining two sides of the triangle.
17. **Got a Wire Cutter?** A wire 1 m long is cut into two pieces. One piece is bent into a circle and the other piece is bent into a square. The sum of the areas of the circle and the square is $\tfrac{1}{16}$ m². What are the length of the side of the square and the radius of the circle?
18. **Coordinates** Find the coordinates of the point P of intersection of the line and the parabola shown in FIGURE 11.R.2.

FIGURE 11.R.2 Graphs for Problem 18

In Problems 19–22, find the partial fraction decomposition of the given rational expression.

19. $\dfrac{2x - 1}{x(x^2 + 2x - 3)}$
20. $\dfrac{1}{x^4(x^2 + 5)}$
21. $\dfrac{x^2}{(x^2 + 4)^2}$
22. $\dfrac{x^5 - x^4 + 2x^3 + 5x - 1}{(x - 1)^2}$

In Problems 23–28, graph the given system of inequalities.

23. $\begin{cases} y - x \le 0 \\ y + x \le 0 \\ y \ge -1 \end{cases}$
24. $\begin{cases} x + y \le 4 \\ 2x - 3y \ge -6 \\ 3x - 2y \le 12 \end{cases}$
25. $\begin{cases} x + y \le 5 \\ x + y \ge 1 \\ -x + y \le 7 \end{cases}$
26. $\begin{cases} 1 \le x \le 4 \\ 2 \le y \le 6 \\ x + y \ge 5 \\ -x + y \le 9 \end{cases}$

27. $\begin{cases} x^2 + y^2 \leq 4 \\ x^2 + y^2 - 4y \leq 0 \end{cases}$

28. $\begin{cases} y \leq -x^2 - x + 6 \\ y \geq x^2 - 2x \end{cases}$

In Problems 29–32, use the functions $y = x^2$ and $y = 2 - x$ to form a system of inequalities whose graph is given in the figure.

29.

FIGURE 11.R.3 Graph for Problem 29

30.

FIGURE 11.R.4 Graph for Problem 30

31.

FIGURE 11.R.5 Graph for Problem 31

32.

FIGURE 11.R.6 Graph for Problem 32

33. Find the maximum and minimum values of

$$F(x, y) = 100x - 40y$$

subject to

$$\begin{cases} -x + 3y \leq 5 \\ x + y \geq 3 \\ 2x - y \leq 7 \\ x \geq 0, 1 \leq y \leq 3. \end{cases}$$

34. The function $F(x, y) = 20x + 5y$ subject to the constraints

$$\begin{cases} 4x + 5y \geq 20 \\ 3x + y \geq 10 \\ x \geq 0, y \geq 0 \end{cases}$$

has a minimum value. What is it? Explain why the function has no maximum value.

35. **Total Return** On a certain small farm, farming an acre of corn requires 6 h of labor and $36 of capital, whereas farming an acre of oats requires 2 h of labor and $18 of capital. Suppose the farmer has 12 acres of land, 48 h of labor, and $360 of capital available. If the return on the corn is $40 per acre and on oats is $20 per acre, how many acres of each should the farmer plant in order to maximize the total return (including unused capital)?

12 Matrices and Determinants

In This Chapter

12.1 Introduction to Matrices

12.2 Algebra of Matrices

12.3 Determinants

12.4 Inverse of a Matrix

12.5 Linear Systems: Augmented Matrices

12.6 Linear Systems: Matrix Inverses

12.7 Linear Systems: Determinants

12.8 Cryptography

Chapter 12 Review Exercises

A rectangular array of numbers or symbols is called a matrix.

A Bit of History The focus of this chapter will be on three topics: **matrices**, **determinants**, and **systems of equations**. We will see how the first two concepts can be used to solve systems of n linear equations in n unknowns.

Matrices were the invention of the eminent English mathematicians **Arthur Cayley** (1821–1895) and **James Joseph Sylvester** (1814–1897). As with many mathematical inventions, the theory and algebra of matrices grew as an offshoot of Cayley's primary mathematical interest and investigations. Cayley was a child prodigy in mathematics and excelled in that subject as a student at Trinity College at Cambridge. However, for want of a job in mathematics, he became a lawyer at the age of 28. After enduring 14 years in this profession, he was offered a professorship of mathematics at Cambridge in 1863, where he was influential in getting the first women admitted to Cambridge. Arthur Cayley also invented the concept of n-dimensional geometry and made many significant contributions to the theory of determinants. During 1881–1882, Cayley taught in the United States at Johns Hopkins University. Sylvester also taught at Johns Hopkins University from 1877 to 1883.

12.1 Introduction to Matrices

≡ Introduction The method of solution, and the solution itself, of a system of linear equations does not depend in any way on what symbols are used as variables. In Example 2 of Section 11.1, we saw that the solution of the system

$$\begin{cases} x + 2y + z = -6 \\ 4x - 2y - z = -4 \\ 2x - y + 3z = 19 \end{cases} \quad (1)$$

is $x = -2$, $y = -5$, $z = 6$ or as an ordered triple $(-2, -5, 6)$. This same ordered triple is also a solution of

$$\begin{cases} u + 2v + w = -6 \\ 4u - 2v - w = -4 \\ 2u - v + 3w = 19 \end{cases} \quad \text{and} \quad \begin{cases} r + 2s + t = -6 \\ 4r - 2s - t = -4 \\ 2r - s + 3t = 19. \end{cases}$$

The point is this: The solution of a system of linear equations depends only on the coefficients and constants that appear in the system and not on what symbols are used to represent the variables. We will see that we could solve (1) by suitable manipulations on the *array* of numbers

$$\begin{bmatrix} 1 & 2 & 1 & -6 \\ 4 & -2 & -1 & -4 \\ 2 & -1 & 3 & 19 \end{bmatrix}. \quad (2)$$

In (2) the first, second, and third columns represent the coefficients of x, y, and z, respectively in (1), and the last column consists of the constants to the right of the equality sign in (1).

Before examining this idea we need to develop a mathematical system whose elements are arrays of numbers. A rectangular array such as (2) is called a **matrix**.

DEFINITION 12.1.1 Matrix

A **matrix** is a rectangular array of numbers:

$$A = \begin{bmatrix} a_{11} & a_{12} & \cdots & a_{1n} \\ a_{21} & a_{22} & \cdots & a_{2n} \\ \vdots & \vdots & \cdots & \vdots \\ a_{m1} & a_{m2} & \cdots & a_{mn} \end{bmatrix}. \quad (3)$$

☐ **Terminology** If there are m rows and n columns, we say that the **dimension**, or **size**, of the matrix is $m \times n$, and we refer to it as an "m by n matrix," or simply, as a **rectangular matrix**. The matrix in (3) is an $m \times n$ matrix. An $n \times n$ matrix is called a **square matrix** and is said to be of **order** n. The **entry**, or **element**, in the ith row and jth column of an $m \times n$ matrix A is denoted by the symbol a_{ij}. Thus the entry in, say, the third row and fourth column of a matrix A is a_{34}.

EXAMPLE 1 Dimensions

(a) The dimensions of the matrices

$$A = \begin{bmatrix} 1 & 2 & 3 \\ 4 & 5 & 6 \end{bmatrix} \quad \text{and} \quad B = \begin{bmatrix} -1 \\ 0 \\ \frac{1}{2} \end{bmatrix}$$

are, respectively, 2×3 and 3×1.

(b) The 2×2 matrix

$$C = \begin{bmatrix} \sqrt{2} & 0 \\ -3 & \pi \end{bmatrix}$$

is also referred to as a square matrix of order 2. ≡

A $1 \times n$ matrix consists of one row and n columns and is called a **row matrix** or **row vector**. On the other hand, an $m \times 1$ matrix has m rows and one column and is naturally called a **column matrix** or **column vector**. The 3×1 matrix B in part (a) of Example 1 is a column matrix.

☐ **Matrix Notation** To save time and space in writing, it is convenient to use a special notation for a general matrix. An $m \times n$ matrix A is often abbreviated as $A = (a_{ij})_{m \times n}$.

EXAMPLE 2 Find the Matrix

Determine the matrix $A = (a_{ij})_{3 \times 2}$ if $a_{ij} = i + j$ for each i and j.

Solution To obtain the entry in the first row and first column, we let $i = 1$ and $j = 1$; that is, $a_{11} = 1 + 1 = 2$. The remaining entries are obtained in a similar fashion:

$$A = \begin{bmatrix} 1+1 & 1+2 \\ 2+1 & 2+2 \\ 3+1 & 3+2 \end{bmatrix} = \begin{bmatrix} 2 & 3 \\ 3 & 4 \\ 4 & 5 \end{bmatrix}.$$ ≡

The entries $a_{11}, a_{22}, a_{33}, \ldots$ in a square matrix are said to be on the **main diagonal** of the matrix. For example, the main diagonal entries of the square matrices

$$A = \begin{bmatrix} 2 & 4 & 6 \\ 5 & 6 & 0 \\ -2 & 7 & 9 \end{bmatrix} \quad \text{and} \quad B = \begin{bmatrix} 16 & -5 \\ 8 & 3 \end{bmatrix}$$

are highlighted in red.

☐ **Equality** Two matrices are **equal** if they have the same dimension and if their corresponding entries are equal.

DEFINITION 12.1.2 Equality

If $A = (a_{ij})_{m \times n}$ and $B = (b_{ij})_{m \times n}$, then $A = B$ if and only if $a_{ij} = b_{ij}$ for all i and j.

EXAMPLE 3 Equality of Two Matrices

From Definition 12.1.2, we have the equality

$$\begin{bmatrix} 1 & \frac{1}{2} & 2 & -3 \\ 0 & -\pi & \sqrt{2} & 4+1 \end{bmatrix} = \begin{bmatrix} \frac{100}{100} & 0.5 & \sqrt{4} & -3 \\ 0 & (-1)\pi & \sqrt{2} & 5 \end{bmatrix}$$

but

$$\begin{bmatrix} 1 & 2 & 3 \\ 4 & 0 & 1 \end{bmatrix} \neq \begin{bmatrix} 1 & 2 & 3 \\ 1 & 0 & 4 \end{bmatrix},$$

since the corresponding entries in the second row are not all equal. Also,

$$\begin{bmatrix} 1 & 1 \\ 1 & 1 \end{bmatrix} \neq \begin{bmatrix} 1 & 1 & 1 \\ 1 & 1 & 1 \end{bmatrix},$$

since the matrices do not have the same dimension. ≡

EXAMPLE 4 Matrix Equation

Solve for x and y if

$$\begin{bmatrix} -1 & 2 \\ x^3 & 0 \end{bmatrix} = \begin{bmatrix} 2y+1 & 2 \\ 8 & 0 \end{bmatrix}.$$

Solution From Definition 12.1.2, we equate corresponding entries. It follows that

$$-1 = 2y + 1 \quad \text{and} \quad x^3 = 8.$$

Solving these equations gives $y = -1$ and $x = 2$. ≡

DEFINITION 12.1.3 Transpose of a Matrix

The transpose of the $m \times n$ matrix A in (3) is the $n \times m$ matrix A^T given by

$$A^T = \begin{bmatrix} a_{11} & a_{21} & \cdots & a_{m1} \\ a_{12} & a_{22} & \cdots & a_{m2} \\ \vdots & \vdots & \cdots & \vdots \\ a_{1n} & a_{2n} & \cdots & a_{mn} \end{bmatrix}.$$

In other words, the rows of a matrix A become the columns of its transpose A^T.

EXAMPLE 5 Transpose

Find the transpose of **(a)** $A = \begin{bmatrix} 3 & 2 & -1 & 10 \\ 6 & 5 & 2 & 9 \\ 2 & 1 & 4 & 8 \end{bmatrix}$, and **(b)** $B = \begin{bmatrix} 5 & 3 \end{bmatrix}$.

Solution (a) Since A is a 3×4 matrix its transpose A^T will be a 4×3 matrix. In forming the transpose, we write the first row as the first column, the second row as the second column, and so on. Hence,

$$A^T = \begin{bmatrix} 3 & 6 & 2 \\ 2 & 5 & 1 \\ -1 & 2 & 4 \\ 10 & 9 & 8 \end{bmatrix}.$$

(b) The transpose of a row matrix becomes a column matrix:

$$B^T = \begin{bmatrix} 5 \\ 3 \end{bmatrix}.$$

≡

As we will see in Section 12.4, the transpose of a square matrix is particularly useful.

12.1 Exercises
Answers to selected odd-numbered problems begin on page ANS-32.

In Problems 1–10, state the dimension of the given matrix.

1. $\begin{bmatrix} 1 & 2 & 5 \\ 0 & -1 & 0 \\ 3 & 4 & 2 \end{bmatrix}$

2. $\begin{bmatrix} 4 & 7 \\ 0 & -6 \end{bmatrix}$

3. $\begin{bmatrix} 4 & 0 \\ -1 & 2 \\ 3 & 6 \end{bmatrix}$

4. $\begin{bmatrix} 2 & 8 & 6 & 0 \\ 3 & 1 & 7 & 5 \\ -1 & 4 & 0 & 1 \end{bmatrix}$

5. $[8]$

6. $[0 \quad 5 \quad -7]$

7. $\begin{bmatrix} 1 \\ 4 \\ -6 \end{bmatrix}$

8. $\begin{bmatrix} 1 & 5 & 6 & 8 \\ 0 & 4 & 3 & -1 \end{bmatrix}$

9. $(a_{ij})_{5 \times 7}$

10. $(a_{ij})_{6 \times 6}$

In Problems 11–16, suppose that a matrix $A = (a_{ij})_{3 \times 4}$ is defined by

$$A = \begin{bmatrix} -1 & 2 & 7 & 10 \\ 0 & -3 & \frac{1}{2} & -9 \\ \frac{1}{4} & 5 & 11 & 27 \end{bmatrix}.$$

Find the indicated number.

11. a_{13}
12. a_{32}
13. a_{24}
14. a_{33}
15. $2a_{11} + 5a_{31}$
16. $a_{23} - 4a_{33}$

In Problems 17–22, determine the matrix $(a_{ij})_{2 \times 3}$ satisfying the given condition.

17. $a_{ij} = i - j$
18. $a_{ij} = ij$
19. $a_{ij} = ij^2$
20. $a_{ij} = 2i + 3j$
21. $a_{ij} = \dfrac{4i}{j}$
22. $a_{ij} = i^j$

In Problems 23–26, determine whether the given matrices are equal.

23. $\begin{bmatrix} |-4| & 3^2 \\ 2 & \sqrt{2} \end{bmatrix}, \begin{bmatrix} 4 & 9 \\ 2 & 1.4 \end{bmatrix}$

24. $[0 \quad 0], [0 \quad 0 \quad 0]$

25. $\begin{bmatrix} 0 & |4-5| & 0 \\ \frac{7}{2} & -4 & 6 \end{bmatrix}, \begin{bmatrix} 1-1 & 1 & 0 \\ 3.5 & -4 & \frac{12}{2} \end{bmatrix}$

26. $\begin{bmatrix} 1 & 0 \\ 2 & 3 \end{bmatrix}, \begin{bmatrix} 1 & 2 \\ 0 & 3 \end{bmatrix}$

In Problems 27–32, solve for the variables.

27. $\begin{bmatrix} 1 & 2 & 0 \\ x & -y & 3 \end{bmatrix} = \begin{bmatrix} -z & 2 & 0 \\ 3 & 4 & 3 \end{bmatrix}$

28. $\begin{bmatrix} x+y & 2 \\ x-y & 3 \end{bmatrix} = \begin{bmatrix} 6 & 2 \\ 4 & 3 \end{bmatrix}$

29. $\begin{bmatrix} w+1 & 10+x \\ 3y-2 & x-4 \end{bmatrix} = \begin{bmatrix} 2 & 2x+1 \\ y-5 & 4z \end{bmatrix}$

30. $\begin{bmatrix} 1 & x \\ y & -x \end{bmatrix} = \begin{bmatrix} y & y \\ y & -x \end{bmatrix}$

31. $\begin{bmatrix} 0 & 4 \\ x^2 & y^2 \end{bmatrix} = \begin{bmatrix} 0 & 4 \\ 4 & 9 \end{bmatrix}$

32. $\begin{bmatrix} x^2-9x & x \\ 1 & 0 \end{bmatrix} = \begin{bmatrix} 10 & x \\ 1 & y^2-1 \end{bmatrix}$

In Problems 33–38, find the transpose of the given matrix.

33. $\begin{bmatrix} 2 & 1 \\ 5 & \frac{1}{2} \end{bmatrix}$

34. $\begin{bmatrix} -1 & 6 & 7 \end{bmatrix}$

35. $\begin{bmatrix} 1 & 2 & 10 \\ 5 & 6 & 9 \end{bmatrix}$

36. $\begin{bmatrix} 1 & 3 & 4 \\ 0 & 2 & 0 \\ 1 & 0 & 3 \end{bmatrix}$

37. $\begin{bmatrix} 0 & 4 & -2 \\ 0 & 1 & 3 \\ 0 & 4 & -5 \end{bmatrix}$

38. $\begin{bmatrix} 1 & 4 & 8 & 2 \\ 0 & 7 & 3 & 6 \\ -2 & -1 & 2 & 1 \end{bmatrix}$

In Problems 39 and 40, verify that the given matrix is symmetric. An $n \times n$ matrix A is said to be **symmetric** if $A^T = A$.

39. $A = \begin{bmatrix} 2 & 5 \\ 5 & -3 \end{bmatrix}$

40. $A = \begin{bmatrix} 2 & 5 & 2 \\ 5 & 1 & 0 \\ 2 & 0 & -1 \end{bmatrix}$

For Discussion

41. Give an example of a 4×4 matrix that is symmetric. See Problem 39.
42. Suppose A^T is the transpose of the matrix A. Discuss: What is $(A^T)^T$?

12.2 Algebra of Matrices

≡ Introduction In ordinary algebra, we take for granted the fact that two real numbers can be added, subtracted, and multiplied. In matrix algebra, however, two matrices can be added, subtracted, and multiplied only under certain conditions.

☐ **Addition of Matrices** Only matrices that have the same dimensions can be added. If A and B are both $m \times n$ matrices, then their sum $A + B$ is the $m \times n$ matrix formed by adding the corresponding entries in each matrix. In other words, the entry in the first row and first column of A is added to the entry in the first row and first column of B, and so on. Using symbols we have the following definition.

DEFINITION 12.2.1 Sum of Two Matrices

If $A = (a_{ij})_{m \times n}$ and $B = (b_{ij})_{m \times n}$, then their **sum** $A + B$ is the matrix

$$A + B = (a_{ij} + b_{ij})_{m \times n}. \tag{1}$$

If matrices A and B have different dimensions, then they cannot be added.

EXAMPLE 1   **Sum of Two Matrices**

(a) Because the two matrices

$$A = \begin{bmatrix} 1 & 2 & 0 \\ 7 & 3 & -4 \end{bmatrix} \quad \text{and} \quad B = \begin{bmatrix} 3 & 1 & 3 \\ -5 & 0 & 6 \end{bmatrix}$$

have the same dimension (namely, 2×3) we can add them to obtain a third matrix with the same dimension:

$$A + B = \begin{bmatrix} 1+3 & 2+1 & 0+3 \\ 7+(-5) & 3+0 & -4+6 \end{bmatrix} = \begin{bmatrix} 4 & 3 & 3 \\ 2 & 3 & 2 \end{bmatrix}.$$

(b) Because the two matrices

$$A = \begin{bmatrix} 1 & 2 & 0 \\ 1 & 2 & 1 \end{bmatrix} \quad \text{and} \quad B = \begin{bmatrix} -3 & 7 \\ 9 & 1 \end{bmatrix}$$

have different dimensions (respectively, 2×3 and 2×2) we cannot add them. ≡

It follows directly from the properties of real numbers and Definition 12.2.1 that the operation of addition on the set of $m \times n$ matrices possesses the following two familiar properties:

        Commutative Law: $A + B = B + A$,
        Associative Law: $A + (B + C) = (A + B) + C$.

☐ **Additive Identity** A matrix whose entries are all zeros is said to be a **zero matrix** and is denoted by O. If A and O are $m \times n$ matrices, then we have $A + O = O + A = A$ for every $m \times n$ matrix A. We say that the $m \times n$ zero matrix O is the **additive identity** for the set of $m \times n$ matrices. For example, for the set of 3×2 matrices the zero matrix is

$$O = \begin{bmatrix} 0 & 0 \\ 0 & 0 \\ 0 & 0 \end{bmatrix}$$

and

$$\begin{bmatrix} a_{11} & a_{12} \\ a_{21} & a_{22} \\ a_{31} & a_{32} \end{bmatrix} + \begin{bmatrix} 0 & 0 \\ 0 & 0 \\ 0 & 0 \end{bmatrix} = \begin{bmatrix} 0 & 0 \\ 0 & 0 \\ 0 & 0 \end{bmatrix} + \begin{bmatrix} a_{11} & a_{12} \\ a_{21} & a_{22} \\ a_{31} & a_{32} \end{bmatrix} = \begin{bmatrix} a_{11} & a_{12} \\ a_{21} & a_{22} \\ a_{31} & a_{32} \end{bmatrix}.$$

☐ **Scalar Multiple** In the discussion of matrices, real numbers are referred to as **scalars**. If k is a real number, then the **scalar multiple** of a matrix A and a real number k is the matrix kA with each entry equal to the product of the real number k and the corresponding entry in the given matrix.

DEFINITION 12.2.2   Scalar Multiple

If $A = (a_{ij})_{m \times n}$ and k is any real number, then the **scalar multiple** of A by k is

$$kA = (ka_{ij})_{m \times n}. \tag{2}$$

EXAMPLE 2 Scalar Multiple

For $k = 3$ and $A = \begin{bmatrix} 1 & 2 \\ 3 & 4 \end{bmatrix}$, it follows from Definition 12.2.2 that

$$3A = 3\begin{bmatrix} 1 & 2 \\ 3 & 4 \end{bmatrix} = \begin{bmatrix} 3(1) & 3(2) \\ 3(3) & 3(4) \end{bmatrix} = \begin{bmatrix} 3 & 6 \\ 9 & 12 \end{bmatrix}.$$

≡

EXAMPLE 3 Sum of Two Scalar Multiples

Consider the 2×3 matrices $A = \begin{bmatrix} 1 & 0 & 2 \\ -3 & 4 & 5 \end{bmatrix}$ and $B = \begin{bmatrix} 7 & 2 & -1 \\ 4 & 0 & 3 \end{bmatrix}$. Find
(a) $(-1)A + 2B$, and **(b)** $(-1)A + A$.

Solution **(a)** Applying Definition 12.2.2, we have the scalar multiples

$$(-1)A = \begin{bmatrix} -1 & 0 & -2 \\ 3 & -4 & -5 \end{bmatrix} \quad \text{and} \quad 2B = \begin{bmatrix} 14 & 4 & -2 \\ 8 & 0 & 6 \end{bmatrix}.$$

Using the foregoing results, we have from Definition 12.2.1,

$$(-1)A + 2B = \begin{bmatrix} -1+14 & 0+4 & -2+(-2) \\ 3+8 & -4+0 & -5+6 \end{bmatrix} = \begin{bmatrix} 13 & 4 & -4 \\ 11 & -4 & 1 \end{bmatrix}.$$

(b) $(-1)A + A = \begin{bmatrix} -1 & 0 & -2 \\ 3 & -4 & -5 \end{bmatrix} + \begin{bmatrix} 1 & 0 & 2 \\ -3 & 4 & 5 \end{bmatrix}$

$$= \begin{bmatrix} -1+1 & 0+0 & -2+2 \\ 3+(-3) & -4+4 & -5+5 \end{bmatrix} = \begin{bmatrix} 0 & 0 & 0 \\ 0 & 0 & 0 \end{bmatrix}.$$

≡

The following properties of the scalar product are readily established using Definitions 12.2.1 and 12.2.2. If k_1 and k_2 are scalars, then

$$k_1(A + B) = k_1A + k_1B,$$
$$(k_1 + k_2)A = k_1A + k_2A,$$
$$k_1(k_2A) = (k_1k_2)A.$$

☐ **Additive Inverse** As Example 3 shows, the **additive inverse** $-A$ of the matrix A is defined to be the scalar multiple $(-1)A$. Thus, if O is the $m \times n$ zero matrix,

$$A + (-A) = O = (-A) + A$$

for any $m \times n$ matrix A. We use the additive inverse to define **subtraction**, or the **difference** $A - B$, of two $m \times n$ matrices A and B as follows:

$$A - B = A + (-B).$$

From $\quad A + (-B) = (a_{ij} + (-b_{ij}))_{m \times n} = (a_{ij} - b_{ij})_{m \times n},$

we see that the difference is obtained by subtracting the entries in B from the corresponding entries in A.

EXAMPLE 4   Difference

The two row matrices
$$A = [1 \quad 2 \quad 3] \quad \text{and} \quad B = [-2 \quad 7 \quad 4]$$
have the same dimension (namely, 1×3) and so their difference is
$$\begin{aligned} A - B &= [1 \quad 2 \quad 3] - [-2 \quad 7 \quad 4] \\ &= [1 - (-2) \quad 2 - 7 \quad 3 - 4] = [3 \quad -5 \quad -1]. \end{aligned}$$
$\equiv$

EXAMPLE 5   Difference

If $\quad A = \begin{bmatrix} 4 & 5 & -2 \\ 10 & 6 & 8 \\ 9 & -7 & -1 \end{bmatrix} \quad \text{and} \quad B = \begin{bmatrix} 5 & 1 & 3 \\ -1 & 2 & 6 \\ 4 & 9 & -8 \end{bmatrix},$

find $A - B$ and $B - A$.

Solution

$$A - B = \begin{bmatrix} 4 - 5 & 5 - 1 & -2 - 3 \\ 10 - (-1) & 6 - 2 & 8 - 6 \\ 9 - 4 & -7 - 9 & -1 - (-8) \end{bmatrix} = \begin{bmatrix} -1 & 4 & -5 \\ 11 & 4 & 2 \\ 5 & -16 & 7 \end{bmatrix}$$

$$B - A = \begin{bmatrix} 5 - 4 & 1 - 5 & 3 - (-2) \\ -1 - 10 & 2 - 6 & 6 - 8 \\ 4 - 9 & 9 - (-7) & -8 - (-1) \end{bmatrix} = \begin{bmatrix} 1 & -4 & 5 \\ -11 & -4 & -2 \\ -5 & 16 & -7 \end{bmatrix}.$$ $\equiv$

Example 5 illustrates that $B - A = -(A - B)$.

☐ **Multiplication of Two Matrices** In order to find the **product** AB of two matrices A and B, we require that the number of columns in A be equal to the number of rows in B. Suppose $A = (a_{ij})_{m \times n}$ is an $m \times n$ matrix and $B = (b_{ij})_{n \times p}$ is an $n \times p$ matrix. As illustrated below, to find the entry c_{ij} in the product $C = AB$, we pair the numbers in the ith row of A with those in the jth column of B. We then multiply the pairs and add the products, as follows:

$$c_{ij} = a_{i1}b_{1j} + a_{i2}b_{2j} + \cdots + a_{in}b_{nj}, \tag{3}$$

that is,

$$\begin{bmatrix} a_{11} & a_{12} & \cdots & a_{1n} \\ a_{21} & a_{22} & \cdots & a_{2n} \\ \vdots & \vdots & & \vdots \\ a_{i1} & a_{i2} & \cdots & a_{in} \\ \vdots & \vdots & & \vdots \\ a_{m1} & a_{m2} & \cdots & a_{mn} \end{bmatrix} \begin{bmatrix} b_{11} & b_{12} & \cdots & b_{1j} & \cdots & b_{1p} \\ b_{21} & b_{22} & \cdots & b_{2j} & \cdots & b_{2p} \\ \vdots & \vdots & & \vdots & & \vdots \\ b_{n1} & b_{n2} & \cdots & b_{nj} & \cdots & b_{np} \end{bmatrix} = \begin{bmatrix} \\ c_{ij} \\ \\ \end{bmatrix}$$

We say that (3) is the product of the ith row of A and the jth row of B. It follows from (3) that the product AB has m rows and p columns. In other words, the dimension of the product $C = AB$ is determined from the number of rows in A and the number of columns in B:

$$\underset{\text{dimension of product}}{A_{m \times n} B_{n \times p}} = C_{m \times p}.$$

with the inner dimensions "must be equal" and the outer dimensions giving the "dimension of product".

For example, the product of a 2×3 matrix and a 3×3 matrix is a 2×3 matrix:

$$\begin{bmatrix} 0 & 1 & 2 \\ 3 & 4 & 5 \end{bmatrix} \begin{bmatrix} 6 & 7 & 8 \\ 9 & 10 & 11 \\ 12 & 13 & 14 \end{bmatrix} = \begin{bmatrix} c_{11} & c_{12} & c_{13} \\ c_{21} & c_{22} & c_{23} \end{bmatrix}.$$

The entry, say, c_{12} is the product of the first row in A and second column of B:

$$\begin{aligned} c_{12} &= a_{11}b_{12} + a_{12}b_{22} + a_{13}b_{32} \\ &= 0 \cdot 7 + 1 \cdot 10 + 2 \cdot 13 = 36. \end{aligned}$$

We summarize the foregoing discussion with a formal definition.

DEFINITION 12.2.3 Product of Two Matrices

If $A = (a_{ij})_{m \times n}$ and $B = (b_{ij})_{n \times p}$, then the **product** AB is the $m \times p$ matrix $C = (c_{ij})_{m \times p}$, where c_{ij} is the product of the ith row of A and the jth row of B defined by

$$c_{ij} = a_{i1}b_{1j} + a_{i2}b_{2j} + \cdots + a_{in}b_{nj}. \qquad (4)$$

Although at first inspection the definition of the product of two matrices may seem unnatural, it does have many applications. For example, in Section 12.6, it will provide us with a new technique for solving some system of equations.

EXAMPLE 6   Product

If $A = \begin{bmatrix} 6 & -5 \end{bmatrix}$ and $B = \begin{bmatrix} 1 & -2 \\ 3 & 6 \end{bmatrix}$, then by inspection we see that A and B are *conformable* to multiplication in the order AB because the dimension of A is 1×2 and the dimension of B is 2×2. It follows from Definition 12.2.3 that

$$\begin{aligned} AB &= \begin{bmatrix} 6 & -5 \end{bmatrix} \begin{bmatrix} 1 & -2 \\ 3 & 6 \end{bmatrix} \\ &= \begin{bmatrix} 6 \cdot 1 + (-5) \cdot 3 & 6 \cdot (-2) + (-5) \cdot 6 \end{bmatrix} \\ &= \begin{bmatrix} -9 & -42 \end{bmatrix}. \end{aligned}$$

≡

▶ In other words, the number of columns in A equals the number of rows in B.

It is possible for the product AB to exist even though the product BA may not be defined. In Example 6, because the number of columns in B (that is, 2) does not equal the number of rows in A (that is, 1), the product BA is not defined.

EXAMPLE 7 Product

If $A = \begin{bmatrix} 2 & 1 & 5 \\ 3 & 0 & 4 \end{bmatrix}$ and $B = \begin{bmatrix} -1 & 7 & 8 \\ 4 & 6 & 0 \\ 5 & 7 & 3 \end{bmatrix}$, find the product AB.

Solution Using Definition 12.2.3, we have

$$\begin{bmatrix} 2 & 1 & 5 \\ 3 & 0 & 4 \end{bmatrix} \begin{bmatrix} -1 & 7 & 8 \\ 4 & 6 & 0 \\ 5 & 7 & 3 \end{bmatrix}$$

$$= \begin{bmatrix} 2 \cdot (-1) + 1 \cdot 4 + 5 \cdot 5 & 2 \cdot 7 + 1 \cdot 6 + 5 \cdot 7 & 2 \cdot 8 + 1 \cdot 0 + 5 \cdot 3 \\ 3 \cdot (-1) + 0 \cdot 4 + 4 \cdot 5 & 3 \cdot 7 + 0 \cdot 6 + 4 \cdot 7 & 3 \cdot 8 + 0 \cdot 0 + 4 \cdot 3 \end{bmatrix}$$

$$= \begin{bmatrix} 27 & 55 & 31 \\ 17 & 49 & 36 \end{bmatrix}. \quad \equiv$$

EXAMPLE 8 Comparing AB and BA

If $A = \begin{bmatrix} 1 & 2 \\ 3 & 4 \end{bmatrix}$ and $B = \begin{bmatrix} -2 & 0 \\ -1 & 3 \end{bmatrix}$, find the products AB and BA.

Solution Because A and B are both of dimension 2×2 we can form the products AB and BA:

$$AB = \begin{bmatrix} 1 & 2 \\ 3 & 4 \end{bmatrix} \begin{bmatrix} -2 & 0 \\ -1 & 3 \end{bmatrix} = \begin{bmatrix} 1 \cdot (-2) + 2 \cdot (-1) & 1 \cdot 0 + 2 \cdot 3 \\ 3 \cdot (-2) + 4 \cdot (-1) & 3 \cdot 0 + 4 \cdot 3 \end{bmatrix} = \begin{bmatrix} -4 & 6 \\ -10 & 12 \end{bmatrix},$$

$$BA = \begin{bmatrix} -2 & 0 \\ -1 & 3 \end{bmatrix} \begin{bmatrix} 1 & 2 \\ 3 & 4 \end{bmatrix} = \begin{bmatrix} -2 \cdot 1 + 0 \cdot 3 & -2 \cdot 2 + 0 \cdot 4 \\ -1 \cdot 1 + 3 \cdot 3 & -1 \cdot 2 + 3 \cdot 4 \end{bmatrix} = \begin{bmatrix} -2 & -4 \\ 8 & 10 \end{bmatrix}. \quad \equiv$$

Note in Example 8 that even though both products AB and BA are defined, we have $AB \neq BA$. In other words, matrix multiplication is not, in general, commutative. However, it can be demonstrated that matrix multiplication does possess the following properties:

$$\text{Associative Law: } A(BC) = (AB)C,$$
$$\text{Distributive Laws: } \begin{cases} A(B + C) = AB + AC, \\ (A + B)C = AC + BC, \end{cases}$$

provided these sums and products are defined. See Problems 23–26 in Exercises 12.2.

☐ **Identity Matrix** The set of all square matrices of a given order n has a multiplicative identity, that is, there is a unique $n \times n$ matrix I_n such that

$$AI_n = I_nA = A,$$

for any $n \times n$ matrix A. We say that I_n is the **identity matrix of order n**, or simply, the **identity matrix**. It can be shown that each entry on the main diagonal of I_n is 1 and all other entries are 0:

$$I_n = \begin{bmatrix} 1 & 0 & 0 & \cdots & 0 \\ 0 & 1 & 0 & \cdots & 0 \\ 0 & 0 & 1 & \cdots & 0 \\ \vdots & \vdots & \vdots & \cdots & \vdots \\ 0 & 0 & 0 & \cdots & 1 \end{bmatrix}.$$

For example,

$$I_2 = \begin{bmatrix} 1 & 0 \\ 0 & 1 \end{bmatrix} \quad \text{and} \quad I_3 = \begin{bmatrix} 1 & 0 & 0 \\ 0 & 1 & 0 \\ 0 & 0 & 1 \end{bmatrix}$$

are identity matrices of orders 2 and 3, respectively.

EXAMPLE 9   Identity for the Set of 2 × 2 Matrices

Verify that $I_2 = \begin{bmatrix} 1 & 0 \\ 0 & 1 \end{bmatrix}$ is the multiplicative identity for the set of 2 × 2 matrices.

Solution Let $A = \begin{bmatrix} a_{11} & a_{12} \\ a_{21} & a_{22} \end{bmatrix}$ be a 2 × 2 matrix. Then

$$AI_2 = \begin{bmatrix} a_{11} & a_{12} \\ a_{21} & a_{22} \end{bmatrix}\begin{bmatrix} 1 & 0 \\ 0 & 1 \end{bmatrix} = \begin{bmatrix} a_{11}\cdot 1 + a_{12}\cdot 0 & a_{11}\cdot 0 + a_{12}\cdot 1 \\ a_{21}\cdot 1 + a_{22}\cdot 0 & a_{21}\cdot 0 + a_{22}\cdot 1 \end{bmatrix} = \begin{bmatrix} a_{11} & a_{12} \\ a_{21} & a_{22} \end{bmatrix} = A$$

$$I_2 A = \begin{bmatrix} 1 & 0 \\ 0 & 1 \end{bmatrix}\begin{bmatrix} a_{11} & a_{12} \\ a_{21} & a_{22} \end{bmatrix} = \begin{bmatrix} 1\cdot a_{11} + 0\cdot a_{21} & 1\cdot a_{12} + 0\cdot a_{22} \\ 0\cdot a_{11} + 1\cdot a_{21} & 0\cdot a_{12} + 1\cdot a_{22} \end{bmatrix} = \begin{bmatrix} a_{11} & a_{12} \\ a_{21} & a_{22} \end{bmatrix} = A. \equiv$$

NOTES FROM THE CLASSROOM

A $1 \times n$ matrix, or row matrix, and an $m \times 1$ matrix, or column matrix,

$$[b_{11} \quad b_{12} \quad \cdots \quad b_{1n}] \quad \text{and} \quad \begin{bmatrix} a_{11} \\ a_{21} \\ \vdots \\ a_{m1} \end{bmatrix},$$

are also called a **row vector** and a **column vector**, respectively. For example,

$$[1 \quad 2], \quad [9 \quad 1 \quad 5], \quad [-2 \quad \tfrac{1}{2} \quad 4 \quad -3]$$

are row vectors, and

$$\begin{bmatrix} 1 \\ 3 \end{bmatrix}, \quad \begin{bmatrix} 2 \\ -7 \\ 10 \end{bmatrix}, \quad \begin{bmatrix} 0 \\ 0 \\ 1 \\ 0 \end{bmatrix}$$

are column vectors.

Suppose A is a $1 \times n$ row vector and B is an $n \times 1$ column vector, that is, A has the same number of columns as the row matrix B has rows, then the product AB is a 1×1 matrix or a scalar. We also say that AB is the **inner product** of the two matrices. For example, if $A = [4 \quad 8]$ and $B = \begin{bmatrix} 2 \\ -5 \end{bmatrix}$, then their inner product is

$$AB = [4 \quad 8]\begin{bmatrix} 2 \\ -5 \end{bmatrix} = 4 \cdot 2 + 8 \cdot (-5) = -32.$$

That is what is going on in (4): If A is an $m \times n$ matrix and B is an $n \times p$ matrix, then the $m \times p$ matrix AB is formed by taking the inner product of each row vector in A with all the column vectors in B.

See Section 10.5 ▶

12.2 Exercises
Answers to selected odd-numbered problems begin on page ANS-32.

In Problems 1–10, find $A + B$, $A - B$, $4A$, and $3A - 2B$.

1. $A = \begin{bmatrix} 1 & -2 & 3 \\ 0 & 4 & 7 \end{bmatrix}, B = \begin{bmatrix} -7 & 6 & 1 \\ 0 & 2 & 3 \end{bmatrix}$
2. $A = \begin{bmatrix} 1 & 0 \\ 4 & 2 \end{bmatrix}, B = \begin{bmatrix} -3 & 6 \\ 8 & 7 \end{bmatrix}$

3. $A = \begin{bmatrix} 4 & -1 \\ 2 & 8 \end{bmatrix}, B = \begin{bmatrix} -1 & 2 \\ 0 & -2 \end{bmatrix}$
4. $A = \begin{bmatrix} 1 & 0 \\ 3 & 4 \\ -2 & 1 \end{bmatrix}, B = \begin{bmatrix} 2 & 1 \\ 3 & 0 \\ 7 & 5 \end{bmatrix}$

5. $A = \begin{bmatrix} 1 \\ 0 \end{bmatrix}, B = \begin{bmatrix} 6 \\ 2 \end{bmatrix}$

6. $A = \begin{bmatrix} 1 & -6 & 3 & 7 \\ 2 & 0 & 4 & 1 \end{bmatrix}, B = \begin{bmatrix} 4 & 6 & 9 & 1 \\ 5 & 0 & 7 & 2 \end{bmatrix}$

7. $A = [10 \quad 3], B = [4 \quad -5]$
8. $A = [5], B = [2]$

9. $A = \begin{bmatrix} 0 & 0 \\ 0 & 0 \end{bmatrix}, B = \begin{bmatrix} 1 & 2 \\ 3 & 4 \end{bmatrix}$
10. $A = \begin{bmatrix} 5 & 6 \\ -7 & -8 \end{bmatrix}, B = \begin{bmatrix} -5 & -6 \\ 7 & 8 \end{bmatrix}$

In Problems 11–20, find AB and BA, if possible.

11. $A = \begin{bmatrix} 1 & 0 \\ 2 & 4 \end{bmatrix}, B = \begin{bmatrix} 3 & 1 \\ -2 & 5 \end{bmatrix}$
12. $A = \begin{bmatrix} -3 & 6 \\ 2 & 1 \end{bmatrix}, B = \begin{bmatrix} 5 & 0 \\ 6 & 2 \end{bmatrix}$

13. $A = \begin{bmatrix} 4 & 0 & 1 \\ 2 & -3 & 5 \end{bmatrix}, B = \begin{bmatrix} 3 & 2 \\ 1 & 4 \\ 0 & 6 \end{bmatrix}$
14. $A = \begin{bmatrix} 5 & 1 \\ 6 & 2 \\ -3 & 1 \end{bmatrix}, B = \begin{bmatrix} 4 & 0 \\ 1 & 3 \\ 2 & -2 \end{bmatrix}$

15. $A = \begin{bmatrix} 1 & 0 & 2 \\ 0 & 3 & 1 \\ 5 & 0 & 0 \end{bmatrix}, B = \begin{bmatrix} 0 & 0 & 8 \\ 1 & 0 & 3 \\ 2 & 1 & 0 \end{bmatrix}$

16. $A = \begin{bmatrix} 5 & 1 & 4 \\ 3 & 0 & 6 \\ -1 & 2 & 1 \end{bmatrix}, B = \begin{bmatrix} 8 & 2 & -1 \\ 1 & 6 & 0 \\ 4 & 1 & 3 \end{bmatrix}$

17. $A = [1 \quad 2 \quad -3], B = \begin{bmatrix} 1 \\ 1 \\ 0 \end{bmatrix}$
18. $A = \begin{bmatrix} 2 & 2 \\ 0 & 10 \end{bmatrix}, B = \begin{bmatrix} 4 & 1 & 3 \\ \frac{1}{2} & -1 & \frac{5}{2} \end{bmatrix}$

19. $A = \begin{bmatrix} 1 \\ -2 \end{bmatrix}, B = [-3 \quad 4]$
20. $A = [4 \quad 0 \quad 2], B = \begin{bmatrix} 1 \\ 6 \\ 5 \end{bmatrix}$

In Problems 21 and 22, find $A^2 = AA$ for the given matrix A.

21. $A = \begin{bmatrix} 2 & -3 \\ -5 & 4 \end{bmatrix}$
22. $A = \begin{bmatrix} 1 & -1 & 2 \\ -1 & 3 & 0 \\ 3 & 2 & 2 \end{bmatrix}$

In Problems 23–26, find the given matrix if

$$A = \begin{bmatrix} 1 & -2 \\ -2 & 4 \end{bmatrix}, B = \begin{bmatrix} 6 & 3 \\ 2 & 1 \end{bmatrix}, \text{ and } C = \begin{bmatrix} 0 & 2 \\ 3 & 4 \end{bmatrix}.$$

23. $A(BC)$
24. $C(BA)$
25. $A(B + C)$
26. $B(C - A)$

In Problems 27 and 28, write the given sum as a single column matrix.

27. $\begin{bmatrix} 2 & -3 \\ 1 & 4 \end{bmatrix} \begin{bmatrix} -2 \\ 5 \end{bmatrix} - \begin{bmatrix} -1 & 6 \\ -2 & 3 \end{bmatrix} \begin{bmatrix} -7 \\ 2 \end{bmatrix}$

28. $\begin{bmatrix} 1 & -3 & 4 \\ 2 & 5 & -1 \\ 0 & -4 & -2 \end{bmatrix} \begin{bmatrix} 3 \\ 2 \\ -1 \end{bmatrix} + \begin{bmatrix} -1 \\ 1 \\ 4 \end{bmatrix} - \begin{bmatrix} 2 \\ 8 \\ -6 \end{bmatrix}$

In Problems 29–32, give the dimension of the matrix A so that the following products are defined.

29. $\begin{bmatrix} 1 & 2 \\ 3 & 4 \\ 5 & 6 \end{bmatrix} A$

30. $A \begin{bmatrix} 1 & 2 \\ 3 & 4 \\ 5 & 6 \end{bmatrix}$

31. $\begin{bmatrix} 1 & 2 \\ 3 & 4 \end{bmatrix} A \begin{bmatrix} 1 & 2 \\ 3 & 4 \\ 5 & 6 \end{bmatrix}$

32. $\begin{bmatrix} 1 \\ 2 \\ 3 \end{bmatrix} A \begin{bmatrix} 4 \\ 5 \\ 6 \end{bmatrix}$

In Problems 33 and 34, find c_{23} and c_{12} for the matrix $C = 2A - 3B$.

33. $A = \begin{bmatrix} 2 & 3 & -1 \\ -1 & 6 & 0 \end{bmatrix}$, $B = \begin{bmatrix} 4 & -2 & 6 \\ 1 & 3 & -3 \end{bmatrix}$

34. $A = \begin{bmatrix} 1 & -1 & 1 \\ 2 & 2 & 1 \\ 0 & -4 & 1 \end{bmatrix}$, $B = \begin{bmatrix} 2 & 0 & 5 \\ 0 & 4 & 0 \\ 3 & 0 & 7 \end{bmatrix}$

35. Verify that the polynomial in two variables $ax^2 + bxy + cy^2$ is the same as the matrix product

$$[x \quad y] \begin{bmatrix} a & \tfrac{1}{2}b \\ \tfrac{1}{2}b & c \end{bmatrix} \begin{bmatrix} x \\ y \end{bmatrix}.$$

36. Write $\begin{bmatrix} a_{11} & a_{12} \\ a_{21} & a_{22} \end{bmatrix} \begin{bmatrix} x_1 \\ x_2 \end{bmatrix} = \begin{bmatrix} b_1 \\ b_2 \end{bmatrix}$ without matrices.

In Problems 37 and 38, find a 2×2 matrix X that satisfies the given equation.

37. $X + 3 \begin{bmatrix} 5 & -5 \\ 2 & 4 \end{bmatrix} = 10 I_{2 \times 2}$

38. $X + 2 I_{2 \times 2} = \begin{bmatrix} 3 & -1 \\ 1 & 2 \end{bmatrix}$

In Problems 39–42, suppose $A = \begin{bmatrix} 2 & 4 \\ -3 & 2 \end{bmatrix}$ and $B = \begin{bmatrix} 4 & 10 \\ 2 & 5 \end{bmatrix}$. Verify the given property of the transpose by computing the left and right members of the given equality.

39. $(A + B)^T = A^T + B^T$
40. $(A - B)^T = A^T - B^T$
41. $(AB)^T = B^T A^T$
42. $(6A)^T = 6A^T$

Miscellaneous Applications

43. **Spread of a Disease** Two persons X and Y have infectious hepatitis. They have the possibility of contact, and thus of passing on the disease, with four persons $P_1, P_2, P_3,$ and P_4. Let a 2×4 matrix be given as follows:

$$A = \begin{array}{c} \\ X \\ Y \end{array} \begin{array}{c} P_1 \; P_2 \; P_3 \; P_4 \\ \begin{bmatrix} 1 & 0 & 1 & 1 \\ 0 & 1 & 0 & 1 \end{bmatrix} \end{array}.$$

If person X (or Y) has contact with any of the four persons, a 1 is entered in the row labeled X (or Y) in the appropriate column. If X (or Y) has no contact with a particular person, a 0 is entered. Define the contacts between P_1, P_2, P_3, and P_4 with four additional persons P_5, P_6, P_7, and P_8 by

$$B = \begin{array}{c} \\ P_1 \\ P_2 \\ P_3 \\ P_4 \end{array} \begin{array}{cccc} P_1 & P_2 & P_3 & P_4 \\ \left[\begin{array}{cccc} 0 & 1 & 0 & 0 \\ 1 & 0 & 0 & 1 \\ 1 & 1 & 0 & 1 \\ 0 & 1 & 1 & 1 \end{array}\right] \end{array}.$$

Compute the product AB and interpret the entries.

44. Revenue The revenue, in thousands of dollars, for three successive weeks for five stores in a grocery chain are represented by the entries in the following matrices R_1, R_2, and R_3:

$$R_1 = [100 \quad 150 \quad 210 \quad 125 \quad 190], R_2 = 2R_1, \text{ and } R_3 = R_1.$$

Over the same period the costs can be represented by

$$C_1 = [40 \quad 60 \quad 80 \quad 50 \quad 70], \quad C_2 = 1.5C_1, \text{ and } C_3 = C_1.$$

Compute the matrix $4R_1 - 3.5C_1$ and interpret the entries.

45. Retailing A retail store purchased two brands of stereophonic equipment consisting of amplifiers, tuners, and speakers from wholesale outlets. Because of limited quantities, the store must purchase these supplies from three wholesale dealers. Matrix A gives the wholesale price of each piece of equipment in dollars. Matrix B represents the number of units of each piece of equipment purchased. (For example, $b_{11} = 1$ means that one amplifier, one tuner, and one set of speakers of brand 1 is purchased from wholesale dealer 1.)

$$A = \begin{array}{c} \\ \text{Amplifiers} \\ \text{Tuners} \\ \text{Speakers} \end{array} \begin{array}{cc} \text{Price of} & \text{Price of} \\ \text{brand 1} & \text{brand 2} \\ \left[\begin{array}{cc} 200 & 100 \\ 200 & 150 \\ 400 & 300 \end{array}\right] \end{array}$$

$$B = \begin{array}{c} \\ \text{Units of brand 1} \\ \text{Units of brand 2} \end{array} \begin{array}{ccc} \text{Wholesale} & \text{Wholesale} & \text{Wholesale} \\ \text{dealer 1} & \text{dealer 2} & \text{dealer 3} \\ \left[\begin{array}{ccc} 1 & 2 & 3 \\ 2 & 4 & 2 \end{array}\right] \end{array}$$

If

$$C = \begin{array}{cc} \text{State} & \text{City} \\ \text{sales tax} & \text{sales tax} \\ \left[\begin{array}{cc} 0.06 & 0.01 \\ 0.06 & 0.01 \\ 0.06 & 0.01 \end{array}\right], \end{array}$$

find the matrix $P = (AB)C$, and interpret the significance of its entries.

46. Investigative Reporting A television station does a weekly comparison of three supermarkets for the costs of five basic food items. In a particular week, the following matrix gives the per-pound price for each item:

$$\begin{array}{c} \\ \text{Vegetables} \\ \text{Meat} \\ \text{Bread} \\ \text{Cheese} \\ \text{Fruit} \end{array} \begin{array}{ccc} \text{Store 1} & \text{Store 2} & \text{Store 3} \\ \left[\begin{array}{ccc} 0.39 & 0.41 & 0.38 \\ 1.50 & 1.29 & 1.35 \\ 0.72 & 0.68 & 0.70 \\ 1.00 & 0.92 & 0.98 \\ 0.50 & 0.58 & 0.52 \end{array}\right]. \end{array}$$

The number of pounds purchased of each items is given by the matrix

$$[2 \quad 3 \quad 1 \quad 2 \quad 4].$$

By an appropriate matrix multiplication, compare the total costs at the three stores.

47. Inventory A company owns five tire stores. The inventory of tires in store S_1 is given by

$$\begin{array}{c} \text{Brand } X \quad \text{Brand } Y \quad \text{Brand } Z \end{array}$$
$$\begin{array}{c} \text{Belted tires} \\ \text{Radials} \\ \text{Steel-belted radials} \\ \text{Regular tires} \end{array} \begin{bmatrix} 100 & 50 & 40 \\ 80 & 20 & 50 \\ 200 & 60 & 20 \\ 100 & 100 & 100 \end{bmatrix}.$$

Stores S_2 and S_3 each have 3 times as many tires as S_1; store S_4 has $\frac{1}{2}$ as many tires as stores S_1; and store S_5 has 2 times as many tires as S_1. Find the matrix that gives the total inventory of tires that the company owns.

48. Distance Traveled The velocities of cars X, Y, and Z in kilometers per hour are given in the 3×1 matrix

$$A = \begin{bmatrix} 50 \\ 80 \\ 120 \end{bmatrix}.$$

The number of hours each car travels is given in the 1×3 matrix $B = [3 \quad 4 \quad 6]$. Compute the products AB and BA and interpret the entries in each.

49. College Dropouts A college has 3000 students enrolled at the beginning of a particular academic year. A breakdown by classes is given by the matrix

$$\begin{array}{c} \text{Year} \\ \text{Number of students} \end{array} \begin{array}{c} \text{Freshmen} \quad \text{Sophomores} \quad \text{Juniors} \quad \text{Seniors} \\ [1100 \quad 800 \quad 600 \quad 500]. \end{array}$$

It is projected that the percentage of dropouts per class, in any given year, is given by

$$\begin{bmatrix} 0.20 \\ 0.15 \\ 0.05 \\ 0.03 \end{bmatrix}.$$

That is to say, 20% of the freshmen class is expected to leave school before the completion of the year, and so on. Using matrix multiplication, determine the projected total number of dropouts for that particular year.

For Discussion

50. Show by example that, in general, for $n \times n$ matrices A and B,

$$(A - B)(A + B) \neq A^2 - B^2.$$

[*Hint*: Use 2×2 matrices and $A^2 = AA$ and $B^2 = BB$.]

51. Let A and B be 2×2 matrices. Is it true, in general, that

$$(A + B)^2 = A^2 + 2AB + B^2?$$

Explain.

52. Find two different 2×2 matrices A, where $A \neq O$, for which $A^2 = O$.

53. Suppose $A = \begin{bmatrix} 2 & 1 \\ 5 & 3 \end{bmatrix}$. Find a 2×2 matrix B, such that $AB = BA = I_2$.

54. If a, b, and c are real numbers and $c \neq 0$, then $ac = bc$ implies $a = b$. For matrices, $AC = BC$ does not imply $A = B$. Verify this for the matrices

$$A = \begin{bmatrix} 2 & 1 & 4 \\ 3 & 2 & 1 \\ 1 & 3 & 2 \end{bmatrix}, B = \begin{bmatrix} 5 & 1 & 6 \\ 9 & 2 & -3 \\ -1 & 3 & 7 \end{bmatrix}, \text{ and } C = \begin{bmatrix} 0 & 0 & 0 \\ 2 & 3 & 4 \\ 0 & 0 & 0 \end{bmatrix}.$$

12.3 Determinants

≡ **Introduction** To each square matrix A, we can associate a *number* called the **determinant** of A. For example, the determinants of 2×2 and 3×3 matrices

$$\begin{bmatrix} a_{11} & a_{12} \\ b_{21} & b_{22} \end{bmatrix} \quad \text{and} \quad \begin{bmatrix} a_{11} & a_{12} & a_{13} \\ a_{21} & a_{22} & a_{23} \\ a_{31} & a_{32} & a_{33} \end{bmatrix} \quad (1)$$

are written

$$\begin{vmatrix} a_{11} & a_{12} \\ b_{21} & b_{22} \end{vmatrix} \quad \text{and} \quad \begin{vmatrix} a_{11} & a_{12} & a_{13} \\ a_{21} & a_{22} & a_{23} \\ a_{31} & a_{32} & a_{33} \end{vmatrix}; \quad (2)$$

in other words, the brackets are replaced by vertical bars. A determinant of an $n \times n$ matrix is said to be a **determinant of order n** or an **nth-order determinant**. The determinants in (2) are, in turn, determinants of orders 2 and 3. In a discussion, the determinant of a square matrix A is denoted by symbols such as $\det A$ or $|A|$. We will use the former symbol exclusively. Thus, if

$$A = \begin{bmatrix} 4 & 9 \\ 5 & -6 \end{bmatrix}, \text{ then } \det A = \begin{vmatrix} 4 & 9 \\ 5 & -6 \end{vmatrix}.$$

◀ Even though a determinant is a number it is often convenient to think of it as a square array. Thus, for example, second-order and third-order determinants are, in turn, referred to as 2×2 and 3×3 determinants.

We will examine two applications of determinants in Sections 12.4 and 12.7.

☐ **Determinant of a 2 x 2 Matrix** As we just indicated, a determinant is a number. We begin with the definition of the determinant of order 2, that is, the determinant of a 2×2 matrix.

DEFINITION 12.3.1 Determinant of a 2×2 Matrix

If

$$A = \begin{bmatrix} a_{11} & a_{12} \\ a_{21} & a_{22} \end{bmatrix},$$

then $\det A$ is the number

$$\det A = \begin{vmatrix} a_{11} & a_{12} \\ a_{21} & a_{22} \end{vmatrix} = a_{11}a_{22} - a_{12}a_{21}. \quad (3)$$

EXAMPLE 1 Determinant of Order 2

Evaluate the determinant of the matrix

$$A = \begin{bmatrix} 2 & 3 \\ 4 & -5 \end{bmatrix}.$$

Solution From (3),

$$\det A = \begin{vmatrix} 2 & 3 \\ 4 & -5 \end{vmatrix} = 2(-5) - 3(4) = -22. \qquad \equiv$$

As a mnemonic for the formula in (3), remember that the determinant is the difference of the products of the diagonal elements:

$$\begin{vmatrix} a_{11} & a_{12} \\ a_{21} & a_{22} \end{vmatrix} = a_{11}a_{22} - a_{12}a_{21}.$$

Determinants of 2×2 matrices play a fundamental role in the evaluating of determinants of $n \times n$ matrices for $n > 2$. In general, a determinant of an $n \times n$ matrix can be expressed in terms of determinants of $(n-1) \times (n-1)$ matrices, that is, determinants of order $n - 1$. Thus, for example, the determinant of a 3×3 matrix can be expressed in terms of determinants of order 2. In preparation for a method of evaluating a determinant of an $n \times n$ matrix, $n > 2$, we need to introduce the notion of a cofactor determinant.

☐ **Minor and Cofactor** If a_{ij} denotes the entry in the ith row and jth column of a square matrix A, then the **minor** M_{ij} of a_{ij} is defined to be the determinant of the matrix obtained by deleting the ith row and the jth column of A. Thus for the matrix

$$A = \begin{bmatrix} 1 & 5 & 3 \\ 2 & 4 & 5 \\ 1 & 2 & 3 \end{bmatrix}, \qquad (4)$$

the minors of $a_{11} = 1$, $a_{12} = 5$, $a_{22} = 4$, and $a_{32} = 2$ are, in turn, the determinants

$$M_{11} = \begin{vmatrix} 1 & 5 & 3 \\ 2 & 4 & 5 \\ 1 & 2 & 3 \end{vmatrix} = \begin{vmatrix} 4 & 5 \\ 2 & 3 \end{vmatrix} = 4(3) - 5(2) = 2,$$

$$M_{12} = \begin{vmatrix} 1 & 5 & 3 \\ 2 & 4 & 5 \\ 1 & 2 & 3 \end{vmatrix} = \begin{vmatrix} 2 & 5 \\ 1 & 3 \end{vmatrix} = 2(3) - 5(1) = 1,$$

$$M_{22} = \begin{vmatrix} 1 & 5 & 3 \\ 2 & 4 & 5 \\ 1 & 2 & 3 \end{vmatrix} = \begin{vmatrix} 1 & 3 \\ 1 & 3 \end{vmatrix} = 1(3) - 3(1) = 0,$$

$$M_{32} = \begin{vmatrix} 1 & 5 & 3 \\ 2 & 4 & 5 \\ 1 & 2 & 3 \end{vmatrix} = \begin{vmatrix} 1 & 3 \\ 2 & 5 \end{vmatrix} = 1(5) - 3(2) = -1.$$

The **cofactor** A_{ij} of the entry a_{ij} is defined to be the minor M_{ij} multiplied by $(-1)^{i+j}$, that is,

$$A_{ij} = (-1)^{i+j} M_{ij}. \quad (5)$$

◀ $(-1)^{i+j}$ is 1 if $i+j$ is even, $(-1)^{i+j}$ is -1 if $i+j$ is odd.

Thus for the matrix A in (4) the cofactors associated with the foregoing minor determinants are

$$A_{11} = (-1)^{1+1} M_{11} = 2,$$
$$A_{12} = (-1)^{1+2} M_{12} = -1,$$
$$A_{22} = (-1)^{2+2} M_{22} = 0,$$
$$A_{32} = (-1)^{3+2} M_{32} = -(-1) = 1,$$

and so on. For a 3×3 matrix, the coefficient $(-1)^{i+j}$ of the minor M_{ij} follows the pattern

$$\begin{bmatrix} + & - & + \\ - & + & - \\ + & - & + \end{bmatrix}.$$

This "checkerboard" pattern of signs extends to matrices of larger size as well.

EXAMPLE 2 Cofactors

For the matrix

$$A = \begin{bmatrix} -2 & 1 & 0 \\ 5 & -3 & 7 \\ -1 & 6 & -5 \end{bmatrix}$$

find the cofactor of the given entry: **(a)** 0 **(b)** 7 **(c)** -1.

Solution (a) The number 0 is the entry of the first row ($i = 1$) and third column ($j = 3$). From (5) the cofactor of 0 is the determinant

$$A_{13} = (-1)^{1+3} M_{13} = (1) \begin{vmatrix} 5 & -3 \\ -1 & 6 \end{vmatrix} = 30 - 3 = 27.$$

(b) The number 7 is the entry in the second row ($i = 2$) and third column ($j = 3$). Thus the cofactor is

$$A_{23} = (-1)^{2+3} M_{23} = (-1) \begin{vmatrix} -2 & 1 \\ -1 & 6 \end{vmatrix} = (-1) \cdot [-12 - (-1)] = 11.$$

(c) Finally, because -1 is the entry in the third row ($i = 3$) and first column ($j = 1$), its cofactor is

$$A_{31} = (-1)^{3+1} M_{31} = (1) \begin{vmatrix} 1 & 0 \\ -3 & 7 \end{vmatrix} = 7 - 0 = 7. \quad \equiv$$

We are now in a position to evaluate the determinant of any square matrix.

THEOREM 12.3.1 Expansion Theorem

The **determinant** det A of an $n \times n$ matrix A can be evaluated by multiplying each entry in any row (or column) by its cofactor and adding the results.

12.3 Determinants

When we use Theorem 12.3.1 to find the value of the determinant of a square matrix A, we say that we have **expanded the determinant of A by a given row or by a given column**. For example, the expansion of the determinant of the 3×3 matrix in (1) by the first row is

$$\det A = a_{11}A_{11} + a_{12}A_{12} + a_{13}A_{13}$$
$$= a_{11}(-1)^{1+1}\begin{vmatrix} a_{22} & a_{23} \\ a_{32} & a_{33} \end{vmatrix} + a_{12}(-1)^{1+2}\begin{vmatrix} a_{21} & a_{23} \\ a_{31} & a_{33} \end{vmatrix} + a_{13}(-1)^{1+3}\begin{vmatrix} a_{21} & a_{22} \\ a_{31} & a_{32} \end{vmatrix}$$

or

$$\begin{vmatrix} a_{11} & a_{12} & a_{13} \\ a_{21} & a_{22} & a_{23} \\ a_{31} & a_{32} & a_{33} \end{vmatrix} = a_{11}\begin{vmatrix} a_{22} & a_{23} \\ a_{32} & a_{33} \end{vmatrix} - a_{12}\begin{vmatrix} a_{21} & a_{23} \\ a_{31} & a_{33} \end{vmatrix} + a_{13}\begin{vmatrix} a_{21} & a_{22} \\ a_{31} & a_{32} \end{vmatrix}. \qquad (6)$$

■ EXAMPLE 3   Expansion by the First Row

Evaluate the determinant of the 3×3 matrix

$$A = \begin{bmatrix} 6 & 5 & 3 \\ 2 & 4 & 5 \\ 1 & 2 & -3 \end{bmatrix}.$$

Solution Using the expansion by the first row given in (6) we have

$$\det A = \begin{vmatrix} 6 & 5 & 3 \\ 2 & 4 & 5 \\ 1 & 2 & -3 \end{vmatrix} = 6\begin{vmatrix} 4 & 5 \\ 2 & -3 \end{vmatrix} - 5\begin{vmatrix} 2 & 5 \\ 1 & -3 \end{vmatrix} + 3\begin{vmatrix} 2 & 4 \\ 1 & 2 \end{vmatrix}$$
$$= 6 \cdot (-22) - 5 \cdot (-11) + 3 \cdot 0 = -77. \qquad \equiv$$

Theorem 12.3.1 states that the determinant of a square matrix A can be expanded by *any* row or *any* column. For example, the expansion of the determinant of the 3×3 matrix in (1) by, say, the second row gives

$$\det A = a_{21}A_{21} + a_{22}A_{22} + a_{23}A_{23}$$
$$= (-1)^{2+1}a_{21}\begin{vmatrix} a_{12} & a_{13} \\ a_{32} & a_{33} \end{vmatrix} + (-1)^{2+2}a_{22}\begin{vmatrix} a_{11} & a_{13} \\ a_{31} & a_{33} \end{vmatrix} + (-1)^{2+3}a_{23}\begin{vmatrix} a_{11} & a_{12} \\ a_{31} & a_{32} \end{vmatrix}$$
$$= -a_{21}\begin{vmatrix} a_{12} & a_{13} \\ a_{32} & a_{33} \end{vmatrix} + a_{22}\begin{vmatrix} a_{11} & a_{13} \\ a_{31} & a_{33} \end{vmatrix} - a_{23}\begin{vmatrix} a_{11} & a_{12} \\ a_{31} & a_{32} \end{vmatrix}.$$

■ EXAMPLE 4   Example 3 Revisited

The expansion of the determinant in Example 3 by the third column is

$$\det A = \begin{vmatrix} 6 & 5 & 3 \\ 2 & 4 & 5 \\ 1 & 2 & -3 \end{vmatrix} = 3(-1)^{1+3}\begin{vmatrix} 2 & 4 \\ 1 & 2 \end{vmatrix} + 5(-1)^{2+3}\begin{vmatrix} 6 & 5 \\ 1 & 2 \end{vmatrix} + (-3)(-1)^{3+3}\begin{vmatrix} 6 & 5 \\ 2 & 4 \end{vmatrix}$$
$$= 3\begin{vmatrix} 2 & 4 \\ 1 & 2 \end{vmatrix} + 5(-1)\begin{vmatrix} 6 & 5 \\ 1 & 2 \end{vmatrix} + (-3)\begin{vmatrix} 6 & 5 \\ 2 & 4 \end{vmatrix}$$
$$= 3 \cdot 0 + 5 \cdot (-1) \cdot 7 + (-3) \cdot 14 = -77. \qquad \equiv$$

In the expansion of a determinant, since the entries in a row (or column) multiply the cofactors of that row (or column), it makes sense that if a determinant has a row (or column) with several 0 entries that we expand the determinant by that row (or column).

EXAMPLE 5 Determinant of Order 4

Evaluate the determinant of the 4×4 matrix

$$A = \begin{bmatrix} 1 & 0 & 0 & 3 \\ 0 & -1 & 0 & 4 \\ 2 & 3 & 0 & 0 \\ 1 & 5 & -2 & 6 \end{bmatrix}.$$

Solution Since the third column has only one nonzero entry, we expand the det A by the third column:

$$\det A = a_{13}A_{13} + a_{23}A_{23} + a_{33}A_{33} + a_{43}A_{43}$$
$$= (0)A_{13} + (0)A_{23} + (0)A_{33} + (-2)A_{43}$$
$$= (-2)A_{43},$$

where the cofactor A_{43} is

$$A_{43} = (-1)^{4+3} \begin{vmatrix} 1 & 0 & 3 \\ 0 & -1 & 4 \\ 2 & 3 & 0 \end{vmatrix} = (-1) \begin{vmatrix} 1 & 0 & 3 \\ 0 & -1 & 4 \\ 2 & 3 & 0 \end{vmatrix}.$$

Using (6), we expand the last determinant by the first row:

$$A_{43} = (-1)\left((1) \begin{vmatrix} -1 & 4 \\ 3 & 0 \end{vmatrix} - 0 \begin{vmatrix} 0 & 4 \\ 2 & 0 \end{vmatrix} + 3 \begin{vmatrix} 0 & -1 \\ 2 & 3 \end{vmatrix} \right)$$
$$= (-1)(-12 + 0 + 2 \cdot 3)$$
$$= (-1)(-6) = 6.$$

Thus,
$$\det A = (-2)A_{43} = (-2)(6) = -12. \quad \equiv$$

☐ **Properties** Determinants have many special properties, some of which are given in the following theorem.

THEOREM 12.3.2 Properties of Determinants

Suppose A is a square matrix.

(*i*) If every entry in a row (or column) of A is 0, then $\det A = 0$.
(*ii*) If a matrix B is formed by interchanging two rows (or columns) of A, then $\det B = -\det A$.
(*iii*) If a matrix B is formed by multiplying each entry in a row (or column) of A by a real number k, then $\det B = k \det A$.
(*iv*) If two rows (or columns) of A are equal, then $\det A = 0$.
(*v*) If a matrix B is formed by replacing any row (or column) of A by the sum of that row (or column) and k times any other row (or column) of A, then $\det B = \det A$.

It is easy to prove part (*i*) of Theorem 12.3.2 from Theorem 12.3.1: We expand det *A* by the row (or column) that contains all zero entries. As exercises, you will be asked to verify parts (*ii*)–(*v*) of Theorem 12.3.2 for matrices of order 2. See Problems 31–34 in Exercises 12.3.

EXAMPLE 6 Using Theorem 12.3.2

Without expanding, it follows immediately from (*i*) of Theorem 12.3.2 that

$$\text{row of zeros} \rightarrow \begin{vmatrix} 1 & -2 & 3 \\ 0 & 0 & 0 \\ 4 & -8 & 6 \end{vmatrix} = 0.$$

EXAMPLE 7 Using Theorem 12.3.2

It follows from (*ii*) of Theorem 12.3.2 that

$$\begin{vmatrix} 1 & 0 & 2 \\ 3 & 7 & 8 \\ 4 & -1 & 4 \end{vmatrix} = - \begin{vmatrix} 2 & 0 & 1 \\ 8 & 7 & 3 \\ 4 & -1 & 4 \end{vmatrix}$$

(interchanging these two columns yields a minus sign)

by interchanging the first column and the third column.

EXAMPLE 8 Using Theorem 12.3.2

By factoring 2 from each entry in the first row, it follows from (*iii*) of Theorem 12.3.2 that

$$\begin{vmatrix} 4 & 8 & 2 \\ 0 & 3 & 4 \\ -1 & 7 & 8 \end{vmatrix} = \begin{vmatrix} 2\cdot 2 & 2\cdot 4 & 2\cdot 1 \\ 0 & 3 & 4 \\ -1 & 7 & 8 \end{vmatrix} = 2 \begin{vmatrix} 2 & 4 & 1 \\ 0 & 3 & 4 \\ -1 & 7 & 8 \end{vmatrix}.$$

(2 is a common factor of the first row)

EXAMPLE 9 Using Theorem 12.3.2

Because the first and second columns are the same, it follows from (*iv*) of Theorem 12.3.2 that

$$\begin{vmatrix} 2 & 2 & 3 \\ -4 & -4 & 6 \\ 7 & 7 & -5 \end{vmatrix} = 0.$$

(columns are the same)

As the next example shows, using (*v*) of Theorem 12.3.2 can simplify the evaluation of a determinant.

EXAMPLE 10 Using Theorem 12.3.2

Evaluate the determinant of the 3 × 3 matrix

$$A = \begin{bmatrix} 1 & 2 & 5 \\ -2 & 3 & 0 \\ 3 & -5 & 2 \end{bmatrix}. \tag{7}$$

Solution We use (v) of Theorem 12.3.2 to obtain a matrix, with the same determinant, that has a row (or column) with only one nonzero entry. To avoid fractions, it is best to use a row (or column) containing the element 1 or -1, if possible. Thus we will use the first row to introduce zeros into the first column as follows. We multiply the first row by 2 and add the result to the second and obtain

$$\begin{vmatrix} 1 & 2 & 5 \\ 0 & 7 & 10 \\ 3 & -5 & 2 \end{vmatrix}.$$

Now multiplying the first row by -3 and adding the result to the third row gives

$$\begin{vmatrix} 1 & 2 & 5 \\ 0 & 7 & 10 \\ 0 & -11 & -13 \end{vmatrix}. \tag{8}$$

Expanding (8) by the first column, we find that

$$\begin{vmatrix} 1 & 2 & 5 \\ 0 & 7 & 10 \\ 0 & -11 & -13 \end{vmatrix} = (1) \begin{vmatrix} 7 & 10 \\ -11 & -13 \end{vmatrix} = 19.$$

From (v) of Theorem 12.3.2, it follows that the determinant of the matrix given in (7) has the same value; that is, $\det A = 19$. ≡

12.3 Exercises   Answers to selected odd-numbered problems begin on page ANS-32.

In Problems 1–4, find the minor and the cofactor of each element in the given matrix.

1. $\begin{bmatrix} 4 & 0 \\ 3 & -2 \end{bmatrix}$

2. $\begin{bmatrix} 6 & -2 \\ 5 & 1 \end{bmatrix}$

3. $\begin{bmatrix} 1 & -7 & 8 \\ 2 & 1 & 0 \\ -3 & 0 & 5 \end{bmatrix}$

4. $\begin{bmatrix} 4 & -3 & 0 \\ 2 & -1 & 6 \\ -5 & 4 & 1 \end{bmatrix}$

In Problems 5–18, evaluate the determinant of the given matrix. In Problem 10, assume that the numbers a and b are nonzero.

5. $\begin{bmatrix} 3 & -1 \\ 8 & -5 \end{bmatrix}$

6. $\begin{bmatrix} 0 & -1 \\ 8 & 0 \end{bmatrix}$

7. $\begin{bmatrix} 4 & 2 \\ 0 & 3 \end{bmatrix}$

8. $\begin{bmatrix} 3 & -4 \\ 5 & 6 \end{bmatrix}$

9. $\begin{bmatrix} a & -b \\ b & a \end{bmatrix}$

10. $\begin{bmatrix} a & b \\ \dfrac{1}{b} & \dfrac{1}{a} \end{bmatrix}$

11. $\begin{bmatrix} -3 & 4 & 1 \\ 2 & -6 & 1 \\ 6 & 8 & -4 \end{bmatrix}$

12. $\begin{bmatrix} 6 & 2 & 1 \\ 0 & 3 & -4 \\ 1 & 0 & 2 \end{bmatrix}$

13. $\begin{bmatrix} 4 & 6 & 1 \\ 3 & 2 & 3 \\ 0 & -1 & 7 \end{bmatrix}$

14. $\begin{bmatrix} 5 & 4 & 0 \\ 3 & -6 & 1 \\ 2 & 0 & 3 \end{bmatrix}$

15. $\begin{bmatrix} 2 & 0 & 1 & 7 \\ 0 & 3 & 0 & -2 \\ 5 & 1 & 4 & 0 \\ 4 & 0 & -6 & 8 \end{bmatrix}$

16. $\begin{bmatrix} 0 & 1 & 0 & -3 \\ 0 & 0 & 5 & 4 \\ 2 & 2 & 0 & 0 \\ 3 & 1 & -1 & 3 \end{bmatrix}$

17. $\begin{bmatrix} 0 & 0 & 0 & a \\ 0 & c & 0 & 0 \\ 0 & 0 & b & 0 \\ d & 0 & 0 & 0 \end{bmatrix}$

18. $\begin{bmatrix} a & b & c & d \\ 0 & e & f & g \\ 0 & 0 & h & i \\ 0 & 0 & 0 & j \end{bmatrix}$

In Problems 19–26, state why the equality is true without evaluating the given determinant.

19. $\begin{vmatrix} -1 & 2 & -4 \\ 0 & -3 & 1 \\ 8 & 2 & 4 \end{vmatrix} = -\begin{vmatrix} -1 & 2 & -4 \\ 8 & 2 & 4 \\ 0 & -3 & 1 \end{vmatrix}$

20. $\begin{vmatrix} 1 & 3 \\ 4 & 6 \end{vmatrix} = -2\begin{vmatrix} 1 & 3 \\ -2 & -3 \end{vmatrix}$

21. $\begin{vmatrix} 1 & 0 & 0 & 2 \\ 0 & 1 & 0 & 2 \\ 2 & 1 & 0 & 0 \\ 0 & 0 & 0 & 1 \end{vmatrix} = 0$

22. $\begin{vmatrix} 4 & 0 & -2 & 1 \\ 3 & -6 & 5 & 0 \\ 2 & 1 & -1 & 4 \\ 0 & -2 & 1 & 3 \end{vmatrix} = -\begin{vmatrix} 4 & 0 & 2 & 1 \\ 3 & -6 & -5 & 0 \\ 2 & 1 & 1 & 4 \\ 0 & -2 & -1 & 3 \end{vmatrix}$

23. $\begin{vmatrix} 1 & 8 & 2 \\ 3 & 0 & 4 \\ -1 & 2 & 3 \end{vmatrix} = \begin{vmatrix} 5 & 0 & -10 \\ 3 & 0 & 4 \\ -1 & 2 & 3 \end{vmatrix}$

24. $\begin{vmatrix} 1 & 2 & 3 \\ 4 & 5 & 6 \\ 7 & 8 & 9 \end{vmatrix} = \begin{vmatrix} 7 & 8 & 9 \\ -4 & -5 & -6 \\ 1 & 2 & 3 \end{vmatrix}$

25. $\begin{vmatrix} 2 & 6 & 2 \\ 1 & 8 & 1 \\ -5 & 4 & -5 \end{vmatrix} = 0$

26. $\begin{vmatrix} 4 & -5 & 8 \\ 2 & 6 & 4 \\ 1 & 3 & 2 \end{vmatrix} = 0$

In Problems 27–30, use (v) of Theorem 12.3.2 to introduce zeros, as in Example 10, before evaluating the given determinant.

27. $\begin{vmatrix} 1 & 0 & 6 \\ 2 & 4 & 3 \\ -2 & 5 & 2 \end{vmatrix}$

28. $\begin{vmatrix} 3 & -2 & -4 \\ 2 & 0 & 5 \\ -5 & 6 & 3 \end{vmatrix}$

29. $\begin{vmatrix} 6 & -1 & 0 & 4 \\ 3 & 3 & -2 & 0 \\ 0 & 1 & 8 & 6 \\ 2 & 3 & 0 & 4 \end{vmatrix}$

30. $\begin{vmatrix} -5 & 0 & 4 & 2 \\ -9 & 6 & -2 & 18 \\ -2 & 1 & 0 & 3 \\ 0 & 3 & 6 & 8 \end{vmatrix}$

In Problems 31–34, verify the given identity by expanding each determinant.

31. $\begin{vmatrix} c & d \\ a & b \end{vmatrix} = -\begin{vmatrix} a & b \\ c & d \end{vmatrix}$

32. $\begin{vmatrix} ka & kb \\ c & d \end{vmatrix} = k\begin{vmatrix} a & b \\ c & d \end{vmatrix}$

33. $\begin{vmatrix} a & b \\ a & b \end{vmatrix} = 0$

34. $\begin{vmatrix} a & b \\ ka+c & kb+d \end{vmatrix} = \begin{vmatrix} a & b \\ c & d \end{vmatrix}$

35. Prove that
$$\begin{vmatrix} 1 & a & a^2 \\ 1 & b & b^2 \\ 1 & c & c^2 \end{vmatrix} = (b-a)(c-a)(c-b).$$

36. Verify that an equation of the line through the points (x_1, y_1), (x_2, y_2) is given by
$$\begin{vmatrix} x & y & 1 \\ x_1 & y_1 & 1 \\ x_2 & y_2 & 1 \end{vmatrix} = 0.$$

In Problems 37–40, find the values of λ for which $\det(A - \lambda I_n) = 0$. These numbers are called the *eigenvalues* of the matrix A.

37. $A = \begin{bmatrix} 1 & 2 \\ 3 & -4 \end{bmatrix}$

38. $A = \begin{bmatrix} 6 & 3 \\ -11 & -6 \end{bmatrix}$

39. $A = \begin{bmatrix} -1 & 1 & 0 \\ 1 & 2 & 1 \\ 0 & 3 & -1 \end{bmatrix}$

40. $A = \begin{bmatrix} 13 & 0 & 0 \\ 0 & \frac{1}{2} & 0 \\ 0 & 0 & -7 \end{bmatrix}$

41. Without expanding, state why
$$\begin{vmatrix} 1 & 1 & 1 \\ a & b & c \\ b+c & a+c & a+b \end{vmatrix} = 0.$$

42. Verify that $\det(AB) = \det A \cdot \det B$ for the 2×2 matrices
$$A = \begin{bmatrix} 9 & -3 \\ 4 & -2 \end{bmatrix} \quad \text{and} \quad B = \begin{bmatrix} 2 & 5 \\ -1 & 2 \end{bmatrix}.$$

In Problems 43–48, find the value of each determinant given that
$$\begin{vmatrix} a_{11} & a_{12} & a_{13} \\ a_{21} & a_{22} & a_{23} \\ a_{31} & a_{32} & a_{33} \end{vmatrix} = 3.$$

43. $\begin{vmatrix} a_{31} & a_{32} & a_{33} \\ a_{21} & a_{22} & a_{23} \\ a_{11} & a_{12} & a_{13} \end{vmatrix}$

44. $\begin{vmatrix} a_{12} & a_{13} & a_{11} \\ a_{22} & a_{23} & a_{21} \\ a_{32} & a_{33} & a_{31} \end{vmatrix}$

45. $\begin{vmatrix} a_{11} & a_{12} & a_{13} \\ a_{11} & a_{12} & a_{13} \\ a_{21} & a_{22} & a_{23} \end{vmatrix}$

46. $\begin{vmatrix} 2a_{11} & 2a_{12} & 2a_{13} \\ 3a_{21} & 3a_{22} & 3a_{23} \\ 4a_{31} & 4a_{32} & 4a_{33} \end{vmatrix}$

47. $\begin{vmatrix} a_{11} & a_{12} & a_{13} \\ -2a_{31} & -2a_{32} & -2a_{33} \\ -a_{21} & -a_{22} & -a_{23} \end{vmatrix}$

48. $\begin{vmatrix} a_{11}-5a_{21} & a_{12}-5a_{22} & a_{13}-5a_{23} \\ a_{21} & a_{22} & a_{23} \\ a_{31} & a_{32} & a_{33} \end{vmatrix}$

In Problems 49 and 50, solve for x.

49. $\begin{vmatrix} x & 1 & -2 \\ 1 & -1 & 1 \\ -1 & 0 & 2 \end{vmatrix} = 7$

50. $\begin{vmatrix} x & 1 & -1 \\ 1 & x & 1 \\ -1 & x & 2 \end{vmatrix} = 0$

For Discussion

51. Without expanding the determinant of the matrix

$$A = \begin{bmatrix} 1 & 1 & 1 \\ \dfrac{1}{a} & \dfrac{1}{b} & \dfrac{1}{c} \\ bc & ac & ab \end{bmatrix},$$

where a, b, and c are nonzero constants, explain why $\det A = 0$.

52. Let A be a square matrix and let A^T be its transpose. Discuss: Are $\det A$ and $\det A^T$ related in any way?

12.4 Inverse of a Matrix

≡ Introduction
In ordinary algebra, every nonzero real number a has a multiplicative inverse b such that

$$ab = ba = 1,$$

where the number 1 is the multiplicative identity. The number b is the *reciprocal* of the number a, that is, $a^{-1} = 1/a$. Similarly, a matrix A can have a multiplicative inverse, but as we see in the discussion that follows A has to be a certain kind of *square* matrix.

☐ Multiplicative Inverse
If A is an $n \times n$ matrix and if there exists an $n \times n$ matrix B such that

$$AB = BA = I_n, \tag{1}$$

we say that B is the **multiplicative inverse**, or simply the **inverse**, of A. The multiplicative inverse of A is written $B = A^{-1}$. Unlike in the real number system, we note that the symbol A^{-1} does *not* denote the reciprocal of A, that is, A^{-1} is not $1/A$. In matrix theory, $1/A$ is not defined. A square matrix that has a multiplicative inverse is said to be **nonsingular** or **invertible**. When a square matrix A has no inverse, it is said to be **singular** or **noninvertible**.

EXAMPLE 1   Inverse of a Matrix

If $A = \begin{bmatrix} 2 & 5 \\ 1 & 3 \end{bmatrix}$ and $B = \begin{bmatrix} 3 & -5 \\ -1 & 2 \end{bmatrix}$,

then

$$AB = \begin{bmatrix} 2 & 5 \\ 1 & 3 \end{bmatrix}\begin{bmatrix} 3 & -5 \\ -1 & 2 \end{bmatrix} = \begin{bmatrix} 6-5 & -10+10 \\ 3-3 & -5+6 \end{bmatrix} = \begin{bmatrix} 1 & 0 \\ 0 & 1 \end{bmatrix}$$

and

$$BA = \begin{bmatrix} 3 & -5 \\ -1 & 2 \end{bmatrix}\begin{bmatrix} 2 & 5 \\ 1 & 3 \end{bmatrix} = \begin{bmatrix} 6-5 & 15-15 \\ -2+2 & -5+6 \end{bmatrix} = \begin{bmatrix} 1 & 0 \\ 0 & 1 \end{bmatrix}.$$

Because $AB = BA = I_2$ we conclude from (1) that the matrix A is nonsingular and that the inverse A^{-1} of the matrix A is the given matrix B. ≡

☐ **Finding A^{-1}: Method 1**   We can find the inverse of a nonsingular matrix by two different methods. The first method considered uses determinants. We begin with the special case where A is a 2×2 matrix:

$$A = \begin{bmatrix} a_{11} & a_{12} \\ a_{21} & a_{22} \end{bmatrix}.$$

In order for a 2×2 matrix

$$B = \begin{bmatrix} b_{11} & b_{12} \\ b_{21} & b_{22} \end{bmatrix}$$

to be the inverse of A, we must have

$$\begin{bmatrix} a_{11} & a_{12} \\ a_{21} & a_{22} \end{bmatrix}\begin{bmatrix} b_{11} & b_{12} \\ b_{21} & b_{22} \end{bmatrix} = \begin{bmatrix} 1 & 0 \\ 0 & 1 \end{bmatrix}.$$

By multiplication and equality of matrices, we find that b_{11} and b_{21} must satisfy the linear system

$$\begin{cases} a_{11}b_{11} + a_{12}b_{21} = 1 \\ a_{21}b_{11} + a_{22}b_{21} = 0, \end{cases} \qquad (2)$$

while b_{12} and b_{22} must satisfy

$$\begin{cases} a_{11}b_{12} + a_{12}b_{22} = 0 \\ a_{21}b_{12} + a_{22}b_{22} = 1. \end{cases} \qquad (3)$$

Solving these two systems of equations gives

$$b_{11} = \frac{a_{22}}{a_{11}a_{22} - a_{12}a_{21}}, \quad b_{12} = \frac{-a_{12}}{a_{11}a_{22} - a_{12}a_{21}},$$
$$b_{21} = \frac{-a_{21}}{a_{11}a_{22} - a_{12}a_{21}}, \quad b_{12} = \frac{a_{11}}{a_{11}a_{22} - a_{12}a_{21}}.$$
(4)

An inspection of the expressions in (4) reveals that the denominator in each fraction is the value of the determinant of the matrix A, that is

$$\det A = a_{11}a_{22} - a_{12}a_{21}.$$

Thus
$$B = \begin{bmatrix} b_{11} & b_{12} \\ b_{21} & b_{22} \end{bmatrix} = \begin{bmatrix} \dfrac{a_{22}}{\det A} & \dfrac{-a_{12}}{\det A} \\ \dfrac{-a_{21}}{\det A} & \dfrac{a_{11}}{\det A} \end{bmatrix}.$$

This result leads to the following theorem.

THEOREM 12.4.1 Inverse of a 2 x 2 Matrix

Let
$$A = \begin{bmatrix} a_{11} & a_{12} \\ a_{21} & a_{22} \end{bmatrix}.$$

If $\det A \neq 0$, then the multiplicative inverse of A is the matrix

$$A^{-1} = \frac{1}{\det A} \begin{bmatrix} a_{22} & -a_{12} \\ -a_{21} & a_{11} \end{bmatrix}. \tag{5}$$

From the derivation preceding Theorem 12.4.1, we have shown that $AA^{-1} = I_2$. We leave it as an exercise to verify that $A^{-1}A = I_2$, where A^{-1} is given by (5). See Problem 34 in Exercises 12.4.

EXAMPLE 2 Using (5)

Find A^{-1} for
$$A = \begin{bmatrix} 3 & 2 \\ -7 & -4 \end{bmatrix}.$$

Solution First, we calculate the determinant of the matrix:

$$\det A = 3 \cdot (-4) - 2 \cdot (-7) = 2.$$

With the identifications $a_{11} = 3$, $a_{12} = 2$, $a_{21} = -7$, and $a_{22} = -4$ we see from (5) of Theorem 12.4.1 that the inverse of the matrix A is

$$A^{-1} = \frac{1}{2} \begin{bmatrix} -4 & -2 \\ 7 & 3 \end{bmatrix} = \begin{bmatrix} -2 & -1 \\ \tfrac{7}{2} & \tfrac{3}{2} \end{bmatrix}. \qquad \equiv$$

Theorem 12.4.1 is a special case of the next theorem, which we state without proof. Before reading Theorem 12.4.2 you are encouraged to review Definition 12.1.3 on the *transpose* of a matrix.

THEOREM 12.4.2 Inverse of an $n \times n$ Matrix

Let

$$A = \begin{bmatrix} a_{11} & a_{12} & \cdots & a_{1n} \\ a_{21} & a_{22} & \cdots & a_{2n} \\ \vdots & \vdots & \cdots & \vdots \\ a_{n1} & a_{n2} & \cdots & a_{nn} \end{bmatrix} = (a_{ij})_{n \times n}.$$

If $\det A \neq 0$, then the multiplicative inverse of A is the matrix

$$A^{-1} = \frac{1}{\det A} \begin{bmatrix} A_{11} & A_{12} & \cdots & A_{1n} \\ A_{21} & A_{22} & \cdots & A_{2n} \\ \vdots & \vdots & \cdots & \vdots \\ A_{n1} & A_{n2} & \cdots & A_{nn} \end{bmatrix}^T, \qquad (6)$$

where A_{ij} is the cofactor of the entry a_{ij} of A.

The transpose of the matrix of cofactors,

$$\begin{bmatrix} A_{11} & A_{12} & \cdots & A_{1n} \\ A_{21} & A_{22} & \cdots & A_{2n} \\ \vdots & \vdots & \cdots & \vdots \\ A_{n1} & A_{n2} & \cdots & A_{nn} \end{bmatrix}^T = \begin{bmatrix} A_{11} & A_{21} & \cdots & A_{n1} \\ A_{12} & A_{22} & \cdots & A_{n2} \\ \vdots & \vdots & \cdots & \vdots \\ A_{1n} & A_{2n} & \cdots & A_{nn} \end{bmatrix},$$

given in (6) is called the **adjoint** of the matrix A and is denoted by adj A. In the adjoint matrix given in (6), it is important to note that the entries a_{ij} of the matrix A are replaced by their corresponding cofactors A_{ij} and *then* we take the transpose of that matrix. The inverse in (6) can be written

$$A^{-1} = \frac{1}{\det A} \text{ adj } A.$$

EXAMPLE 3   Using (6)

Find A^{-1} for

$$A = \begin{bmatrix} 1 & -2 & 4 \\ -1 & 3 & 2 \\ 5 & 0 & -6 \end{bmatrix}.$$

Solution The determinant of A is $\det A = -86$. Now, for each entry in A the corresponding cofactor is

$$A_{11} = \begin{vmatrix} 3 & 2 \\ 0 & -6 \end{vmatrix} = -18, \quad A_{12} = -\begin{vmatrix} -1 & 2 \\ 5 & -6 \end{vmatrix} = 4, \quad A_{13} = \begin{vmatrix} -1 & 3 \\ 5 & 0 \end{vmatrix} = -15,$$

$$A_{21} = -\begin{vmatrix} -2 & 4 \\ 0 & -6 \end{vmatrix} = -12, \quad A_{22} = \begin{vmatrix} 1 & 4 \\ 5 & -6 \end{vmatrix} = -26, \quad A_{23} = -\begin{vmatrix} 1 & -2 \\ 5 & 0 \end{vmatrix} = -10,$$

$$A_{31} = \begin{vmatrix} -2 & 4 \\ 3 & 2 \end{vmatrix} = -16, \quad A_{32} = -\begin{vmatrix} 1 & 4 \\ -1 & 2 \end{vmatrix} = -6, \quad A_{33} = \begin{vmatrix} 1 & -2 \\ -1 & 3 \end{vmatrix} = -1.$$

Thus by (6) of Theorem 12.4.2, the inverse of A is $-\frac{1}{86}$ times the adjoint of the matrix A:

$$A^{-1} = -\frac{1}{86}\begin{bmatrix} A_{11} & A_{12} & A_{13} \\ A_{21} & A_{22} & A_{23} \\ A_{31} & A_{32} & A_{33} \end{bmatrix}^T = -\frac{1}{86}\begin{bmatrix} -18 & 4 & -15 \\ -12 & -26 & -10 \\ -16 & -6 & -1 \end{bmatrix}^T$$

$$= -\frac{1}{86}\underbrace{\begin{bmatrix} -18 & -12 & -16 \\ 4 & -26 & -6 \\ -15 & -10 & 1 \end{bmatrix}}_{\text{this adj }A} = \begin{bmatrix} \frac{18}{86} & \frac{12}{86} & \frac{16}{86} \\ -\frac{4}{86} & \frac{26}{86} & \frac{6}{86} \\ \frac{15}{86} & \frac{10}{86} & -\frac{1}{86} \end{bmatrix} = \begin{bmatrix} \frac{9}{43} & \frac{6}{43} & \frac{8}{43} \\ -\frac{2}{43} & \frac{13}{43} & \frac{3}{43} \\ \frac{15}{86} & \frac{5}{43} & -\frac{1}{86} \end{bmatrix}. \equiv$$

In Theorems 12.4.1 and 12.4.2, we saw that we could calculate A^{-1} provided $\det A \neq 0$. Conversely, if A^{-1} exists, then it can be shown that $\det A \neq 0$. We conclude:

- An $n \times n$ matrix A is nonsingular if and only if $\det A \neq 0$. (7)

EXAMPLE 4   Using (7)

The matrix

$$A = \begin{bmatrix} 4 & -8 \\ -1 & 2 \end{bmatrix}$$

has no inverse because $\det A = 8 - 8 = 0$. Thus by (7), A is a singular matrix. $\equiv$

It should be apparent that the use of (6) becomes formidable for matrices of order $n > 3$. For example, for a 4×4 matrix we must first compute *sixteen* determinants of order 3. A more efficient method for finding the multiplicative inverse of a matrix makes use of elementary row operations on the matrix.

☐ **Finding A^{-1}: Method 2** For *any* matrix A, the **elementary row operations** on A are defined to be the following three transformations of A.

(*i*) Interchange any two rows.
(*ii*) Multiply any row by a nonzero constant k.
(*iii*) Add a nonzero constant multiple of one row to another row.

Analogous to the notation that we used in Section 11.1 to denote operations on *equations* in a system of linear equations, we will use the following abbreviations for elementary *row* operations. The symbol R, of course, stands for the word *row*:

$R_i \leftrightarrow R_j$: Interchange the ith row with the jth row.
kR_i: Multiply the ith row by a constant k.
$kR_i + R_j$: Multiply the ith row by k and add to the jth row.

We state without proof:

- *The sequence of elementary row operations that transforms an $n \times n$ matrix A into the multiplicative identity I_n is the same sequence of row operations that transforms I_n into A^{-1}.*

By forming the $n \times 2n$ matrix consisting of the entries of A to the left of a vertical bar and the entries of I_n to the right of the vertical bar,

$$\begin{bmatrix} a_{11} & a_{12} & \cdots & a_{1n} & | & 1 & 0 & \cdots & 0 \\ a_{21} & a_{22} & \cdots & a_{2n} & | & 0 & 1 & \cdots & 0 \\ \vdots & \vdots & \cdots & \vdots & | & \vdots & \vdots & \cdots & \vdots \\ a_{n1} & a_{n2} & \cdots & a_{nn} & | & 0 & 0 & \cdots & 1 \end{bmatrix} \quad (8)$$

we then apply a succession or row operations on (8) until we have transformed it into a new matrix,

$$\begin{bmatrix} 1 & 0 & \cdots & 0 & b_{11} & b_{12} & \cdots & b_{1n} \\ 0 & 1 & \cdots & 0 & b_{21} & b_{22} & \cdots & b_{2n} \\ \vdots & \vdots & \cdots & \vdots & \vdots & \vdots & \cdots & \vdots \\ 0 & 0 & \cdots & 1 & b_{n1} & b_{n2} & \cdots & b_{nn} \end{bmatrix},$$

where the matrix to the left of the vertical bar is now I_n. The inverse of A is

$$A^{-1} = \begin{bmatrix} b_{11} & b_{12} & \cdots & b_{1n} \\ b_{21} & b_{22} & \cdots & b_{2n} \\ \vdots & \vdots & \cdots & \vdots \\ b_{n1} & b_{n2} & \cdots & b_{nn} \end{bmatrix}.$$

This procedure is illustrated in the next two examples.

EXAMPLE 5 Using Elementary Row Operations

Use elementary row operations to find A^{-1} for

$$A = \begin{bmatrix} 2 & 3 \\ 1 & 6 \end{bmatrix}.$$

Solution We begin by forming the matrix

$$\begin{bmatrix} 2 & 3 & | & 1 & 0 \\ 1 & 6 & | & 0 & 1 \end{bmatrix}.$$

The idea is to transform the matrix to the left of the vertical line into the matrix I_2. Now,

$$\begin{bmatrix} 2 & 3 & | & 1 & 0 \\ 1 & 6 & | & 0 & 1 \end{bmatrix} \xrightarrow{R_{12}} \begin{bmatrix} 1 & 6 & | & 0 & 1 \\ 2 & 3 & | & 1 & 0 \end{bmatrix}$$

$$\xrightarrow{-2R_1 + R_2} \begin{bmatrix} 1 & 6 & | & 0 & 1 \\ 0 & -9 & | & 1 & -2 \end{bmatrix}$$

$$\xrightarrow{-\frac{1}{9}R_2} \begin{bmatrix} 1 & 6 & | & 0 & 1 \\ 0 & 1 & | & -\frac{1}{9} & \frac{2}{9} \end{bmatrix}$$

$$\xrightarrow{-6R_2 + R_1} \begin{bmatrix} 1 & 0 & | & \frac{2}{3} & -\frac{1}{3} \\ 0 & 1 & | & -\frac{1}{9} & \frac{2}{9} \end{bmatrix}.$$

Because I_2 now appears to the left of the vertical line, we conclude that the matrix to the right of this line is

$$A^{-1} = \begin{bmatrix} \frac{2}{3} & -\frac{1}{3} \\ -\frac{1}{9} & \frac{2}{9} \end{bmatrix}.$$

This result can now be checked by the previous method or by multiplication. Choosing the latter procedure, we see that AA^{-1} and $A^{-1}A$ are, in turn,

$$\begin{bmatrix} 2 & 3 \\ 1 & 6 \end{bmatrix} \begin{bmatrix} \frac{2}{3} & -\frac{1}{3} \\ -\frac{1}{9} & \frac{2}{9} \end{bmatrix} = \begin{bmatrix} \frac{4}{3} - \frac{3}{9} & -\frac{2}{3} + \frac{6}{9} \\ \frac{2}{3} - \frac{6}{9} & -\frac{1}{3} + \frac{12}{9} \end{bmatrix} = \begin{bmatrix} 1 & 0 \\ 0 & 1 \end{bmatrix},$$

$$\begin{bmatrix} \frac{2}{3} & -\frac{1}{3} \\ -\frac{1}{9} & \frac{2}{9} \end{bmatrix} \begin{bmatrix} 2 & 3 \\ 1 & 6 \end{bmatrix} = \begin{bmatrix} \frac{4}{3} - \frac{1}{3} & \frac{6}{3} - \frac{6}{3} \\ -\frac{2}{9} + \frac{2}{9} & -\frac{3}{9} + \frac{12}{9} \end{bmatrix} = \begin{bmatrix} 1 & 0 \\ 0 & 1 \end{bmatrix}. \qquad \equiv$$

EXAMPLE 6 Example 3 Revisited

Use elementary row operations to find A^{-1} for the matrix in Example 3.

Solution We have

$$\left[\begin{array}{rrr|rrr} 1 & -2 & 4 & 1 & 0 & 0 \\ -1 & 3 & 2 & 0 & 1 & 0 \\ 5 & 0 & -6 & 0 & 0 & 1 \end{array}\right] \xrightarrow{R_1+R_2} \left[\begin{array}{rrr|rrr} 1 & -2 & 4 & 1 & 0 & 0 \\ 0 & 1 & 6 & 1 & 1 & 0 \\ 5 & 0 & -6 & 0 & 0 & 1 \end{array}\right]$$

$$\xrightarrow{-5R_1+R_3} \left[\begin{array}{rrr|rrr} 1 & -2 & 4 & 1 & 0 & 0 \\ 0 & 1 & 6 & 1 & 1 & 0 \\ 0 & 10 & -26 & -5 & 0 & 1 \end{array}\right]$$

$$\xrightarrow{2R_2+R_1} \left[\begin{array}{rrr|rrr} 1 & 0 & 16 & 3 & 2 & 0 \\ 0 & 1 & 6 & 1 & 1 & 0 \\ 0 & 10 & -26 & -5 & 0 & 1 \end{array}\right]$$

$$\xrightarrow{-10R_2+R_3} \left[\begin{array}{rrr|rrr} 1 & 0 & 16 & 3 & 2 & 0 \\ 0 & 1 & 6 & 1 & 1 & 0 \\ 0 & 0 & -86 & -15 & -10 & 1 \end{array}\right]$$

$$\xrightarrow{-\frac{1}{86}R_3} \left[\begin{array}{rrr|rrr} 1 & 0 & 16 & 3 & 2 & 0 \\ 0 & 1 & 6 & 1 & 1 & 0 \\ 0 & 0 & 1 & \frac{15}{86} & \frac{10}{86} & -\frac{1}{86} \end{array}\right]$$

$$\xrightarrow{-6R_3+R_2} \left[\begin{array}{rrr|rrr} 1 & 0 & 16 & 3 & 2 & 0 \\ 0 & 1 & 0 & -\frac{4}{86} & \frac{26}{86} & \frac{6}{86} \\ 0 & 0 & 1 & \frac{15}{86} & \frac{10}{86} & -\frac{1}{86} \end{array}\right]$$

$$\xrightarrow{-16R_3+R_1} \left[\begin{array}{rrr|rrr} 1 & 0 & 0 & \frac{18}{86} & \frac{12}{86} & \frac{16}{86} \\ 0 & 1 & 0 & -\frac{4}{86} & \frac{26}{86} & \frac{6}{86} \\ 0 & 0 & 1 & \frac{15}{86} & \frac{10}{86} & -\frac{1}{86} \end{array}\right]$$

As before, we see that

$$A^{-1} = \begin{bmatrix} \frac{18}{86} & \frac{12}{86} & \frac{16}{86} \\ -\frac{4}{86} & \frac{26}{86} & \frac{6}{86} \\ \frac{15}{86} & \frac{10}{86} & -\frac{1}{86} \end{bmatrix} = \frac{1}{86}\begin{bmatrix} 18 & 12 & 16 \\ -4 & 26 & 6 \\ 15 & 10 & -1 \end{bmatrix}.$$

≡

In conclusion, we note that if an $n \times n$ matrix A *cannot* be transformed into the multiplicative identity I_n by elementary row operations, then A is necessarily singular. If, at some point in the applications of the elementary row operations, we find a row of zeros in the matrix on the left side of the vertical line, then the matrix A is singular. For example, from

$$\left[\begin{array}{rr|rr} -2 & 4 & 1 & 0 \\ 1 & -2 & 0 & 1 \end{array}\right] \xrightarrow{\frac{1}{2}R_1+R_2} \left[\begin{array}{rr|rr} -2 & 4 & 1 & 0 \\ 0 & 0 & \frac{1}{2} & 1 \end{array}\right]$$

we see that it is now impossible, using only row operations, to obtain I_2 to the left of the vertical line. Thus the 2×2 matrix

$$A = \begin{bmatrix} -2 & 4 \\ 1 & -2 \end{bmatrix}$$

is singular.

12.4 Exercises
Answers to selected odd-numbered problems begin on page ANS-33.

In Problems 1 and 2, verify that the matrix B is the inverse of the matrix A.

1. $A = \begin{bmatrix} 1 & \frac{1}{2} \\ 2 & \frac{3}{2} \end{bmatrix}$, $B = \begin{bmatrix} 3 & -1 \\ -4 & 2 \end{bmatrix}$

2. $A = \begin{bmatrix} 1 & -1 & 0 \\ 3 & 0 & 2 \\ 1 & 1 & 1 \end{bmatrix}$, $B = \begin{bmatrix} 2 & -1 & 2 \\ 1 & -1 & 2 \\ -3 & 2 & -3 \end{bmatrix}$

In Problems 3–14, use Method 1 of this section to find the multiplicative inverse, if it exists, of the given matrix. Assume all variables are nonzero.

3. $\begin{bmatrix} 3 & 1 \\ 2 & -1 \end{bmatrix}$

4. $\begin{bmatrix} 7 & -4 \\ -5 & 2 \end{bmatrix}$

5. $\begin{bmatrix} 3 & -9 \\ -1 & 3 \end{bmatrix}$

6. $\begin{bmatrix} 0 & 1 \\ -1 & 0 \end{bmatrix}$

7. $\begin{bmatrix} a & a \\ -a & a \end{bmatrix}$

8. $\begin{bmatrix} a & b \\ -b & a \end{bmatrix}$

9. $\begin{bmatrix} 1 & 0 & 1 \\ 0 & 1 & 0 \\ 1 & 1 & 0 \end{bmatrix}$

10. $\begin{bmatrix} 2 & -1 & 1 \\ 1 & 1 & 0 \\ -1 & 2 & -1 \end{bmatrix}$

11. $\begin{bmatrix} 3 & 4 & -7 \\ 2 & 1 & 8 \\ 5 & 5 & 1 \end{bmatrix}$

12. $\begin{bmatrix} -3 & 5 & -1 \\ 2 & 0 & 6 \\ 4 & 1 & -5 \end{bmatrix}$

13. $\begin{bmatrix} 4 & 1 & 0 \\ 0 & 2 & 0 \\ 0 & 0 & -5 \end{bmatrix}$

14. $\begin{bmatrix} 1 & -1 & -2 \\ -4 & 4 & 3 \\ 3 & 2 & 0 \end{bmatrix}$

In Problems 15–26, use Method 2 of this section to find the multiplicative inverse, if it exists, of the given matrix.

15. $\begin{bmatrix} -1 & 4 \\ 2 & 7 \end{bmatrix}$

16. $\begin{bmatrix} 5 & -6 \\ 2 & -4 \end{bmatrix}$

17. $\begin{bmatrix} 6 & 12 \\ -3 & -6 \end{bmatrix}$

18. $\begin{bmatrix} \frac{1}{2} & 1 \\ -1 & 8 \end{bmatrix}$

19. $\begin{bmatrix} 4 & 0 \\ 0 & 5 \end{bmatrix}$

20. $\begin{bmatrix} 3 & 15 \\ 1 & \frac{1}{5} \end{bmatrix}$

21. $\begin{bmatrix} 0 & -1 & -2 \\ 1 & 3 & 0 \\ 4 & 0 & -6 \end{bmatrix}$

22. $\begin{bmatrix} 2 & -3 & 1 \\ 0 & 1 & 2 \\ 0 & -2 & 4 \end{bmatrix}$

23. $\begin{bmatrix} 2 & 0 & -4 \\ 1 & 5 & -1 \\ 3 & 2 & 1 \end{bmatrix}$

24. $\begin{bmatrix} 8 & -3 & 6 \\ 2 & 1 & -1 \\ -2 & 1 & 1 \end{bmatrix}$

12.4 Inverse of a Matrix

25. $\begin{bmatrix} -2 & 0 & 0 & -1 \\ 0 & 3 & 0 & 4 \\ 0 & 1 & -1 & 2 \\ 0 & 0 & 1 & 0 \end{bmatrix}$
26. $\begin{bmatrix} 1 & 0 & 1 & 1 \\ 2 & 3 & 0 & -1 \\ 0 & 4 & 0 & 2 \\ 1 & -1 & 1 & 0 \end{bmatrix}$

27. If $A^{-1} = \begin{bmatrix} 4 & 3 \\ 3 & 2 \end{bmatrix}$, what is A?

28. Find the inverse of $A = \begin{bmatrix} \sin\theta & \cos\theta \\ -\cos\theta & \sin\theta \end{bmatrix}$.

In Problems 29 and 30, suppose $A = \begin{bmatrix} 4 & 2 \\ 3 & 2 \end{bmatrix}$ and $B = \begin{bmatrix} 4 & 8 \\ 2 & 5 \end{bmatrix}$. Verify the given property of the inverse by computing the left and right members of the given equality.

29. $(A^{-1})^{-1} = A$
30. $(AB)^{-1} = B^{-1}A^{-1}$

For Discussion

31. Let

$$A = \begin{bmatrix} a_{11} & 0 & \cdots & 0 \\ 0 & a_{22} & \cdots & 0 \\ \vdots & \vdots & \cdots & \vdots \\ 0 & 0 & \cdots & a_{nn} \end{bmatrix},$$

where $a_{ii} \neq 0$, $i = 1, 2, \ldots, n$. What is A^{-1}?

32. Use the result in Problem 31 to find A^{-1} for the matrix

$$A = \begin{bmatrix} 2 & 0 & 0 \\ 0 & 4 & 0 \\ 0 & 0 & -6 \end{bmatrix}.$$

33. If $\begin{bmatrix} 1 & -2 \\ 3 & -8 \end{bmatrix}\begin{bmatrix} x \\ y \end{bmatrix} = \begin{bmatrix} 2 \\ 3 \end{bmatrix}$, then find x and y.

34. If A^{-1} is given by (5), verify that $A^{-1}A = I_2$.

35. Let A, B, and C be $n \times n$ matrices, where A is nonsingular. Discuss: How would you prove that if $AB = AC$, then $B = C$?

36. In Problem 29, we saw that $(A^{-1})^{-1} = A$ for a nonsingular 2×2 matrix. This result is true for any nonsingular $n \times n$ matrix. Discuss: How would you prove this general result?

37. In Problem 30, we saw that $(AB)^{-1} = B^{-1}A^{-1}$ for two nonsingular 2×2 matrices. This result is true for any two nonsingular $n \times n$ matrices. Discuss: How would you prove this general result?

38. Let A be a 2×2 matrix for which $\det A \neq 0$. Show that $\det A^{-1} = 1/\det A$. This result is true for any nonsingular $n \times n$ matrix.

12.5 Linear Systems: Augmented Matrices

Introduction In Example 2 of Section 11.1, we solved the system of linear equations

$$\begin{cases} x + 2y + z = -6 \\ 4x - 2y - z = -4 \\ 2x - y + 3z = 19 \end{cases} \quad (1)$$

by finding an equivalent system in triangular form:

$$\begin{cases} x + 2y + z = -6 \\ y + \tfrac{1}{2}z = -1 \\ z = 6. \end{cases} \quad (2)$$

The system (2) was obtained from system (1) by a series of operations on the equations that changed the coefficients of the variables and the constants on the right-hand side of the equality. Throughout this procedure the variables acted as "placeholders." Therefore, these same calculations can be simplified by performing operations on the rows of the matrix

$$\begin{bmatrix} 1 & 2 & 1 & | & -6 \\ 4 & -2 & -1 & | & -4 \\ 2 & -1 & 3 & | & 19 \end{bmatrix}. \quad (3)$$

☐ **Augmented Matrices** The matrix in (3) is called the **augmented matrix** of the system (1) and consists of the **coefficient matrix**

$$\begin{bmatrix} 1 & 2 & 1 \\ 4 & -2 & -1 \\ 2 & -1 & 3 \end{bmatrix},$$

augmented by adding a column whose entries consist of the constant terms of the system. The vertical line in an augmented matrix enables us to distinguish the coefficients of the variables in the system from the constant terms of the system.

When the elimination operations introduced in Section 11.1 are applied to a system of equations, we obtain an equivalent system of equations. These elimination operations are analogous to the elementary row operations that we discussed in the preceding section. When elementary **row operations** are applied to an augmented matrix, the result is the augmented matrix of an equivalent system. For this reason, the original matrix and the resulting matrix are said to be **row equivalent**. The procedure of carrying out elementary row operations on a matrix to obtain a row equivalent matrix is called **row reduction**.

☐ **Gaussian Elimination** To solve a system such as (1) using an augmented matrix, we will use either **Gaussian elimination** or **Gauss-Jordan elimination**. In the Gaussian elimination, we row reduce the augmented matrix of the system until we arrive at a row equivalent augmented matrix in **row echelon form**.

> **DEFINITION 12.5.1** Row Echelon Form
>
> A matrix is in **row echelon form** when:
> - (*i*) The first nonzero entry in each nonzero row is the number 1.
> - (*ii*) In consecutive nonzero rows, the first entry 1 in the lower row appears to the right of the 1 in the higher row.
> - (*iii*) Rows in which the entries are all zero appear at the bottom of the matrix.

EXAMPLE 1   Row Echelon Form

The two augmented matrices

$$\begin{bmatrix} 1 & 4 & 6 & | & 1 \\ 0 & 1 & 5 & | & 8 \\ 0 & 0 & 1 & | & 2 \end{bmatrix}, \quad \begin{bmatrix} 1 & 0 & | & 5 \\ 0 & 1 & | & -1 \\ 0 & 0 & | & 0 \end{bmatrix},$$

are in row echelon form, whereas the matrix

$$\begin{bmatrix} 0 & 2 & 4 & | & 5 \\ 0 & 0 & 0 & | & 0 \\ 1 & 4 & 3 & | & 1 \end{bmatrix} \begin{array}{l} \leftarrow \text{violates } (i) \\ \leftarrow \text{violates } (iii) \\ \leftarrow \text{violates } (ii) \end{array}$$

is not in row echelon form.

As we see in Example 1, the row echelon form of an augmented matrix is roughly a *triangular form* with zero entries below a diagonal consisting of 1's.

To reduce an augmented matrix to row echelon form we use the same elementary row operations introduced in Section 12.4:

- $R_i \leftrightarrow R_j$:   Interchange the *i*th row with the *j*th row.
- kR_i: Multiply the *i*th row by a constant *k*.
- $kR_i + R_j$:   Multiply the *i*th row by *k* and add to the *j*th equation.

EXAMPLE 2   System (1) Revisited

Solve the system (1) using Gaussian elimination.

Solution   We begin by using the first row to introduce 0's below the 1 in the first column:

$$\begin{bmatrix} 1 & 2 & 1 & | & -6 \\ 4 & -2 & -1 & | & -4 \\ 2 & -1 & 3 & | & 19 \end{bmatrix} \xrightarrow{-4R_1 + R_2} \begin{bmatrix} 1 & 2 & 1 & | & -6 \\ 0 & -10 & -5 & | & 20 \\ 2 & -1 & 3 & | & 19 \end{bmatrix}$$

$$\xrightarrow{-2R_1 + R_3} \begin{bmatrix} 1 & 2 & 1 & | & -6 \\ 0 & -10 & -5 & | & 20 \\ 0 & -5 & 1 & | & 31 \end{bmatrix}$$

$$\xrightarrow{-\frac{1}{2}R_2 + R_3} \begin{bmatrix} 1 & 2 & 1 & | & -6 \\ 0 & -10 & -5 & | & -20 \\ 0 & 0 & \frac{7}{2} & | & 21 \end{bmatrix}$$

$$\xrightarrow{-\frac{1}{10}R_2} \begin{bmatrix} 1 & 2 & 1 & | & -6 \\ 0 & 1 & \frac{1}{2} & | & -2 \\ 0 & 0 & \frac{7}{2} & | & 21 \end{bmatrix}$$

$$\xrightarrow{\frac{2}{7}R_3} \begin{bmatrix} 1 & 2 & 1 & | & -6 \\ 0 & 1 & \frac{1}{2} & | & -2 \\ 0 & 0 & 1 & | & 6 \end{bmatrix}.$$

Since this final augmented matrix is in row echelon form (and corresponds to the system (2)) we have, in fact, solved the original system. The last row of the matrix implies that $z = 6$. The remaining variables are determined by **back substitution**. Substituting $z = 6$ back into the equation corresponding to the second row of the matrix yields $y = -5$. Finally, substituting $y = -5$ and $z = 6$ into the equation corresponding to the first row yields $x = -2$. Thus the solution is $x = -2, y = -5, z = 6$. ≡

In Example 2, note that we repeated, in order, the elementary row operations corresponding to the operations that were performed on the equations when we solved this system by elimination (see Example 2 in Section 11.1). Thus we are not doing anything new here. We have simply removed the variables and the equality signs from the equations and are relying on the format of the matrix to keep things in order.

EXAMPLE 3 Using Gaussian Elimination

Use Gaussian elimination to solve the system

$$\begin{cases} x + y - 2z = 2 \\ x + 5y + 6z = 7 \\ x + 3y + 3z = 4. \end{cases}$$

Solution We form the augmented matrix of the system and perform elementary row operations until we obtain a row echelon form:

$$\begin{bmatrix} 1 & 1 & -2 & | & 2 \\ 1 & 5 & 6 & | & 7 \\ 1 & 3 & 3 & | & 4 \end{bmatrix} \xrightarrow{-R_1+R_2} \begin{bmatrix} 1 & 1 & -2 & | & 2 \\ 0 & 4 & 8 & | & 5 \\ 1 & 3 & 3 & | & 4 \end{bmatrix}$$

$$\xrightarrow{-R_1+R_3} \begin{bmatrix} 1 & 1 & -2 & | & 2 \\ 0 & 4 & 8 & | & 5 \\ 0 & 2 & 5 & | & 2 \end{bmatrix}$$

$$\xrightarrow{-\frac{1}{2}R_2+R_3} \begin{bmatrix} 1 & 1 & -2 & | & 2 \\ 0 & 4 & 8 & | & 5 \\ 0 & 0 & 1 & | & -\frac{1}{2} \end{bmatrix}$$

$$\xrightarrow{\frac{1}{4}R_2} \begin{bmatrix} 1 & 1 & -2 & | & 2 \\ 0 & 1 & 2 & | & \frac{5}{4} \\ 0 & 0 & 1 & | & -\frac{1}{2} \end{bmatrix}.$$

The last augmented matrix is in row echelon form and corresponds to the system

$$\begin{cases} x + y - 2z = 2 \\ y + 2z = \frac{5}{4} \\ z = -\frac{1}{2}. \end{cases}$$

From the last equation we see immediately that $z = -\frac{1}{2}$. From the second equation we then obtain $y + 2(-\frac{1}{2}) = \frac{5}{4}$, or $y = \frac{9}{4}$. And finally, the first equation gives us $x + \frac{9}{4} - 2(-\frac{1}{2}) = 2$, or $x = -\frac{5}{4}$. Therefore, $x = -\frac{5}{4}, y = \frac{9}{4}, z = -\frac{1}{2}$ is the solution of the system.

EXAMPLE 4   Using Gaussian Elimination

Use Gaussian elimination to solve the system

$$\begin{cases} 2x - 3y + z = 6 \\ x + y - z = -2 \\ 4x - y - z = 2. \end{cases}$$

Solution   We use elementary row operations:

$$\begin{bmatrix} 2 & -3 & 1 & | & 6 \\ 1 & 1 & -1 & | & -2 \\ 4 & -1 & -1 & | & 2 \end{bmatrix} \xrightarrow{R_{12}} \begin{bmatrix} 1 & 1 & 1 & | & -2 \\ 2 & -3 & 1 & | & 6 \\ 4 & -1 & -1 & | & 2 \end{bmatrix}$$

$$\xrightarrow{-2R_1 + R_2} \begin{bmatrix} 1 & 1 & -1 & | & -2 \\ 0 & -5 & 3 & | & 10 \\ 4 & -1 & -1 & | & 2 \end{bmatrix}$$

$$\xrightarrow{-4R_1 + R_3} \begin{bmatrix} 1 & 1 & -1 & | & -2 \\ 0 & -5 & 3 & | & 10 \\ 0 & -5 & 3 & | & 10 \end{bmatrix}$$

$$\xrightarrow{-R_2 + R_3} \begin{bmatrix} 1 & 1 & -1 & | & -2 \\ 0 & -5 & 3 & | & 10 \\ 0 & 0 & 0 & | & 0 \end{bmatrix}$$

$$\xrightarrow{-\frac{1}{5}R_2} \begin{bmatrix} 1 & 1 & -1 & | & -2 \\ 0 & 1 & -\frac{3}{5} & | & -2 \\ 0 & 0 & 0 & | & 0 \end{bmatrix}.$$

The last matrix is the augmented matrix of the system

$$\begin{cases} x + y - z = 12 \\ y - \frac{3}{5}z = -2. \end{cases}$$

Solving for x and y in terms of z, we find

$$\begin{cases} x = \frac{2}{5}z \\ y = \frac{3}{5}z - 2. \end{cases}$$

Thus the given system is consistent, but the equations are dependent. There are infinitely many solutions of the systems obtained by assigning arbitrary real values to z. If we denote z by α, the solutions of the system consist of all x, y, and z defined by $x = \frac{2}{5}\alpha, y = \frac{3}{5}\alpha - 2, z = \alpha$, respectively, where α represents any real number.

The technique discussed in this section is also applicable to systems of m linear equations in n unknowns. In the following example, we consider a system of two equations in three unknowns.

EXAMPLE 5 Using Gaussian Elimination

Use Gaussian elimination to solve the system

$$\begin{cases} x + 2y - 4z = 6 \\ 5x - y + 2z = -3. \end{cases}$$

Solution We have

$$\begin{bmatrix} 1 & 2 & -4 & | & 6 \\ 5 & -1 & 2 & | & -3 \end{bmatrix} \xrightarrow{-5R_1 + R_2} \begin{bmatrix} 1 & 2 & -4 & | & 6 \\ 0 & -11 & 22 & | & -33 \end{bmatrix}$$

$$\xrightarrow{-\frac{1}{11}R_3} \begin{bmatrix} 1 & 2 & -4 & | & 6 \\ 0 & 1 & -2 & | & 3 \end{bmatrix}.$$

From the last matrix in row echelon form, we find

$$\begin{cases} x + 2y - 4z = 6 \\ y - 2z = 3. \end{cases}$$

Using the second equation to eliminate y from the first, we then obtain

$$\begin{cases} x = 0 \\ y - 2z = 3 \end{cases} \quad \text{or} \quad \begin{cases} x = 0 \\ y = 2z + 3. \end{cases}$$

As in Example 4, we may assign any value to z. Hence the solutions of the system are defined by $x = 0$, $y = 2\alpha + 3$, $z = \alpha$, where α is any real number. ≡

☐ **Gauss-Jordan Elimination** In the Gauss-Jordan elimination method, the elementary row operations are continued until we obtain an augmented matrix that is in **reduced row echelon form**. A reduced row echelon matrix has the three properties (*i*)–(*iii*) in Definition 12.5.1, but in addition:

(*iv*) A column containing first entry 1 has zeros everywhere else.

EXAMPLE 6 Reduced Row Echelon Form

(a) The augmented matrices

$$\begin{bmatrix} 1 & 0 & 0 & | & 7 \\ 0 & 1 & 0 & | & -1 \\ 0 & 0 & 0 & | & 0 \end{bmatrix} \quad \text{and} \quad \begin{bmatrix} 0 & 0 & 1 & -6 & 2 & | & 5 \\ 0 & 0 & 0 & 0 & 1 & | & 4 \end{bmatrix}$$

are in reduced row echelon form. You should verify that the four criteria for this form are satisfied.

(b) We have seen in Example 1 that the augmented matrix

$$\begin{bmatrix} 1 & 4 & 6 & | & 1 \\ 0 & 1 & 5 & | & 8 \\ 0 & 0 & 1 & | & 2 \end{bmatrix}$$

is in row echelon form. However, the augmented matrix is not in reduced row echelon form because the remaining entries (indicated in red) in the columns that contain a leading entry 1 are not all zero. ≡

In should be noted that in Gaussian elimination, we stop when we have obtained *an* augmented matrix in row echelon form. In other words, by using different sequences of row operations, we may arrive at different row echelon forms. This method then requires the use of back substitution. In Gauss-Jordan elimination, we stop when we have obtained the augmented matrix in reduced row echelon form. Any sequence of row operations will lead to the same augmented matrix in reduced row echelon form. This method does not require back substitution; the solution of the system will be apparent by inspection of the final matrix. In terms of the equations of the original system our goal is to simply make the coefficient of the first variable in the first equation* equal to 1 and then use multiples of that equation to eliminate the variable from other equations. The process is repeated on the other variables.

EXAMPLE 7   Example 3 Revisited

In solving the system in Example 3,

$$\begin{cases} x + y - 2z = 2 \\ x + 5y + 6z = 7 \\ x + 3y + 3z = 4, \end{cases}$$

we stopped when we obtained a row echelon form. We now start with the last matrix in Example 3. Since the first entries in the second and third rows are 1's, we must, in turn, make the remaining entries in the second and third columns 0's:

$$\begin{bmatrix} 1 & 1 & -2 & | & 2 \\ 1 & 5 & 6 & | & 7 \\ 1 & 3 & 3 & | & 4 \end{bmatrix} \xrightarrow{\text{row operations}} \begin{bmatrix} 1 & 1 & -2 & | & 2 \\ 0 & 1 & 2 & | & \frac{5}{4} \\ 0 & 0 & 1 & | & -\frac{1}{2} \end{bmatrix} \leftarrow \text{last matrix in Example 3}$$

$$\xrightarrow{-R_2 + R_1} \begin{bmatrix} 1 & 0 & -4 & | & \frac{3}{4} \\ 0 & 1 & 2 & | & \frac{5}{4} \\ 0 & 0 & 1 & | & -\frac{1}{2} \end{bmatrix}$$

$$\xrightarrow[4R_3 + R_1]{-2R_3 + R_2} \begin{bmatrix} 1 & 0 & 0 & | & -\frac{5}{4} \\ 0 & 1 & 0 & | & \frac{9}{4} \\ 0 & 0 & 1 & | & -\frac{1}{2} \end{bmatrix}.$$

The last matrix is in reduced row echelon form. Bearing in mind what the matrix means in terms of equations, we see immediately that the solution is $x = -\frac{5}{4}$, $y = \frac{9}{4}$, $z = -\frac{1}{2}$. ≡

EXAMPLE 8   Inconsistent System

Use Gauss-Jordan elimination to solve the system

$$\begin{cases} x + y = 1 \\ 4x - y = -6 \\ 2x - 3y = 8. \end{cases}$$

Solution In the process of applying Gauss-Jordan elimination to the matrix of the system, we stop at

$$\begin{bmatrix} 1 & 1 & | & 1 \\ 4 & -1 & | & -6 \\ 2 & -3 & | & 8 \end{bmatrix} \xrightarrow{\text{row operations}} \begin{bmatrix} 1 & 0 & | & -1 \\ 0 & 1 & | & 2 \\ 0 & 0 & | & 16 \end{bmatrix}.$$

The third row of the last matrix means $0x + 0y = 16$ (or $0 = 16$). Since no numbers x and y can satisfy this equation, we conclude that the system has no solution, that is, it is inconsistent. ≡

---

*We can always interchange equations so that the first equation contains the variable x_1.

EXAMPLE 9 Balancing a Chemical Equation

Balance the chemical equation $C_2H_6 + O_2 \to CO_2 + H_2O$.

Solution We seek positive integers $x, y, z,$ and w so that the balanced equation is

$$xC_2H_6 + yO_2 \to zCO_2 + wH_2O.$$

Because the number of atoms of each element must be the same on each side of the last equation, we obtain a homogeneous system of three equations in four variables:

carbon (C): $\quad 2x = z \qquad\qquad 2x + 0y - z + 0w = 0$
hydrogen (H): $\quad 6x = 2w \quad$ or $\quad 6x + 0y + 0z - 2w = 0$
oxygen (O): $\quad 2y = 2z + w \qquad 0x + 2y - 2z - w = 0.$

Since the last system is homogeneous it must be consistent.
Using elementary row operations, we find

$$\begin{bmatrix} 2 & 0 & -1 & 0 & | & 0 \\ 6 & 0 & 0 & -2 & | & 0 \\ 0 & 2 & -2 & -1 & | & 0 \end{bmatrix} \xrightarrow{\text{row operations}} \begin{bmatrix} 1 & 0 & 0 & -\frac{1}{3} & | & 0 \\ 0 & 1 & 0 & -\frac{7}{6} & | & 0 \\ 0 & 0 & 1 & -\frac{2}{3} & | & 0 \end{bmatrix}$$

◀ See Section 11.1

and so a solution of the system is $x = \frac{1}{3}\alpha, y = \frac{7}{6}\alpha, z = \frac{2}{3}\alpha, w = \alpha$. In this case α must be a positive integer chosen in such a manner so that $x, y, z,$ and w are positive integers. To accomplish this we pick $\alpha = 6$. This gives $x = 2, y = 7, z = 4,$ and $w = 6$. The balanced equation is then

$$2C_2H_6 + 7O_2 \to 4CO_2 + 6H_2O.$$

≡

12.5 Exercises

Answers to selected odd-numbered problems begin on page ANS-33.

In Problems 1–4, write the coefficient matrix and the augmented matrix of the given system of equations.

1. $\begin{cases} 4x - 6y = 1 \\ x + 3y = 7 \end{cases}$

2. $\begin{cases} x - y + z = 1 \\ 2y + 8z = 6 \\ 7x - 3y - 5z = 0 \end{cases}$

3. $\begin{cases} x - y = 1 \\ y + z = 2 \\ 2x - z = 3 \end{cases}$

4. $\begin{cases} x + y + z - w = 2 \\ y - 5z + w = 1 \\ x - 9w = 4 \\ y + 6z = -1 \end{cases}$

In Problems 5–32, solve the given system, or show that no solution exists. Use Gaussian elimination or Gauss-Jordan elimination as directed by your instructor.

5. $\begin{cases} 2x + 3y = 2 \\ -4x + 6y = 0 \end{cases}$

6. $\begin{cases} 2x_1 - x_2 = -3 \\ 3x_1 + 2x_2 = 13 \end{cases}$

7. $\begin{cases} -x + 3y = 4 \\ 3x - 9y = -12 \end{cases}$

8. $\begin{cases} 2x - y = 1 \\ -6x + 3y = -5 \end{cases}$

9. $\begin{cases} x - y + z = -2 \\ 2x + y - z = 8 \\ 2x + 2y + 3z = -3 \end{cases}$

10. $\begin{cases} 2x + y - 2z = 2 \\ x + 3y + z = 0 \\ 4x + 2y + 2z = 1 \end{cases}$

12.5 Linear Systems: Augmented Matrices

11. $\begin{cases} x - y + z = 3 \\ x + y - z = -5 \\ -2x - y + z = 6 \end{cases}$
12. $\begin{cases} x - y + z = 3 \\ -x + y + z = -1 \\ -3x + 3y + z = -5 \end{cases}$

13. $\begin{cases} x - 2y + z = 2 \\ 2x - 4y + 2z = 4 \\ 5x - y + 2z = 13 \end{cases}$
14. $\begin{cases} 12x + 2y - 4z = 26 \\ 2x - y + z = -3 \\ 3x + 6y - 9z = 38 \end{cases}$

15. $\begin{cases} u + v - 3w = 6 \\ 2u - v + 6w = 7 \\ 3u - 9w = 9 \end{cases}$
16. $\begin{cases} 3x - y + z = 1 \\ 3x + 3y + 5z = 3 \\ 3x + y + 3z = 2 \end{cases}$

17. $\begin{cases} x_1 + 2x_2 + x_3 = 1 \\ 2x_1 - x_2 + 4x_3 = 7 \\ 3x_1 + 2x_2 - x_3 = 7 \end{cases}$
18. $\begin{cases} 2x - y + 2z = -1 \\ -x + y + 2z = -2 \\ -4x + 3y + 2z = 0 \end{cases}$

19. $\begin{cases} x - 2y + 3z = 2 \\ 3x - y + 2z = -5 \\ x + 3y + 4z = 1 \end{cases}$
20. $\begin{cases} 2x + y - z = -1 \\ 2x + 5y - 5z = 3 \\ x + y - z = 0 \end{cases}$

21. $\begin{cases} x - 2y - z = 0 \\ x - 2y + 4z = 0 \\ 2x + y + 3z = 0 \end{cases}$
22. $\begin{cases} 4x + y + z = 0 \\ x - 5y + 5z = 0 \\ x + y + z = 0 \end{cases}$

23. $\begin{cases} 5x - 7y + 6z = 0 \\ x - 2y + 6z = 0 \\ x - y - 2z = 0 \end{cases}$
24. $\begin{cases} x + y + z = 0 \\ x + 2y + 3z = 0 \\ -x + 6z = 0 \end{cases}$

25. $\begin{cases} 2x - y + z = 1 \\ y + z - w = 3 \\ 3x - y + w = 1 \\ 2x - z - w = 1 \end{cases}$
26. $\begin{cases} x + y - z + w = 4 \\ -x + y - 2z + w = 3 \\ 2x - y + 3z - 2w = -4 \\ 3x - 2y + z - w = 3 \end{cases}$

27. $\begin{cases} x + 2y + w = 7 \\ 4x + 9y + z + 12w = 21 \\ 3x + 9y + 6z + 21w = 9 \\ 3x + 9y + 6z + 21w = 9 \end{cases}$
28. $\begin{cases} x + y + z + w = 1 \\ 2x - y + 3z - w = 0 \\ 3x + 4z = 1 \\ -x + 2y - 2z + 2w = 3 \end{cases}$

29. $\begin{cases} x - y + 4z = 1 \\ 6x + y - z = 2 \end{cases}$
30. $\begin{cases} 4x + 2y + z = 9 \\ y - z = 2 \end{cases}$

31. $\begin{cases} 2x_1 - 3x_2 = 2 \\ x_1 + 2x_2 = 1 \\ 3x_1 + 2x_2 = -1 \end{cases}$
32. $\begin{cases} x - y + z = 0 \\ x + y - z = 0 \end{cases}$

In Problems 33–38, use the procedure illustrated in Example 9 to balance the given chemical equation.

33. $Na + H_2O \rightarrow NaOH + H_2$
34. $KClO_3 \rightarrow KCl + O_2$
35. $Fe_3O_4 + C \rightarrow Fe + CO$
36. $C_5H_8 + O_2 \rightarrow CO_2 + H_2O$
37. $Cu + HNO_3 \rightarrow Cu(NO_3)_2 + H_2O + NO$
38. $Ca_3(PO_4)_2 + H_3PO_4 \rightarrow Ca(H_2PO_4)_2$

39. Find a quadratic function $f(x) = ax^2 + bx + c$ whose graph passes through the three points $(1, 8)$, $(-1, -4)$, and $(3, 4)$.
40. Find coefficients a, b, and c such that $(1, 1, -1)$, $(-2, -3, 3)$, and $(1, 2, -\frac{3}{2})$ are solutions of the equation $ax + by + cz = 1$.

41. Find x, y, z, and w such that
$$\begin{bmatrix} x+y & 2z+w \\ 5x-3w & y-w \end{bmatrix} = \begin{bmatrix} 2 & 7 \\ 5 & 9 \end{bmatrix}.$$

42. Write a system of equations corresponding to the augmented matrix
$$\begin{bmatrix} 3 & 7 & 9 & 4 & -1 \\ 1 & -6 & 0 & 2 & 10 \\ 7 & 0 & -4 & -5 & 0 \\ 5 & 3 & -1 & 0 & 4 \end{bmatrix}.$$

Miscellaneous Applications

43. Current in a Circuit The currents i_1, i_2, and i_3 in the electrical network shown in FIGURE 12.5.1 can be shown to satisfy the system of linear equations

$$\begin{aligned} i_1 - i_2 - i_3 &= 0 \\ i_1 R + i_2 R_2 &= E \\ i_2 R_2 - i_3 R_3 &= 0, \end{aligned}$$

where R_1, R_2, R_3, and E are positive constants. Use Gauss-Jordan elimination to solve the system when $R_1 = 10$, $R_2 = 20$, $R_3 = 10$, and $E = 12$.

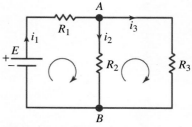

FIGURE 12.5.1 Network in Problem 43

44. Counting the A, B, C's A company has 100 employees divided into three categories A, B, and C. As the following table shows, each employee makes a different contribution to a retirement fund. Following negotiation of a new contract, the monthly contribution of the employees increases by the indicated percentage. The total monthly contribution of $4450 by all employees then increases to $5270 because of the new contract. Use the concept of an augmented matrix to determine the number of employees in each category.

	A	B	C	Total monthly contribution
Monthly contribution to pension fund per employee	$20	$30	$50	$4450
Percent increase per month per employee	10%	10%	20%	$5270

Calculator/CAS Problems

In Problems 45–48, use a calculator or CAS to solve the given system.

45. $\begin{cases} x + y + z = 4.280 \\ 0.2x - 0.1y - 0.5z = -1.978 \\ 4.1x + 0.3y + 0.12z = 1.686 \end{cases}$

46. $\begin{cases} 2.5x + 1.4y + 4.5z = 2.6170 \\ 1.35x + 0.95y + 1.2z = 0.7545 \\ 2.7x + 3.05y - 1.44z = -1.4292 \end{cases}$

47. $\begin{cases} 1.2x + 3.5y - 4.4z + 3.1w = 1.8 \\ 0.2x - 6.1y - 2.3z + 5.4w = -0.6 \\ 3.3x - 3.5y - 2.4z - 0.1w = 2.5 \\ 5.2x + 8.5y - 4.4z - 2.9w = 0 \end{cases}$

48. $\begin{cases} x_1 - x_2 - x_3 + 2x_4 - x_5 = 5 \\ 6x_1 + 9x_2 - 6x_3 + 17x_4 - x_5 = 40 \\ 2x_1 + x_2 - 2x_3 + 5x_4 - x_5 = 12 \\ x_1 + 2x_2 - x_3 + 3x_4 = 7 \\ x_1 + 2x_2 + x_3 + 3x_4 = 1 \end{cases}$

12.6 Linear Systems: Matrix Inverses

≡ **Introduction** In this and the next section we will confine our attention to solving only linear systems with n equations and n variables $x_1, x_2, \ldots, x_n$:

$$\begin{cases} a_{11}x_1 + a_{12}x_2 + \cdots + a_{1n}x_n = b_1 \\ a_{21}x_1 + a_{22}x_2 + \cdots + a_{2n}x_n = b_2 \\ \phantom{a_{11}x_1} \vdots \phantom{+ a_{12}x_2 + \cdots +} \vdots \\ a_{n1}x_1 + a_{n2}x_2 + \cdots + a_{nn}x_n = b_n. \end{cases} \quad (1)$$

If the **determinant of the coefficients** of the variables in the system (1) is denoted by

$$\det A = \begin{vmatrix} a_{11} & a_{12} & \cdots & a_{1n} \\ a_{21} & a_{22} & \cdots & a_{2n} \\ \vdots & \vdots & \cdots & \vdots \\ a_{n1} & a_{n2} & \cdots & a_{nn} \end{vmatrix},$$

then in this and Section 12.7 we solve linear systems such as (1) under the further assumption that $\det A \neq 0$.

☐ **Matrix Form of (1)** Using matrix multiplication and equality, we can write the linear system (1) as the matrix equation

$$\begin{bmatrix} a_{11} & a_{12} & \cdots & a_{1n} \\ a_{21} & a_{22} & \cdots & a_{2n} \\ \vdots & \vdots & \cdots & \vdots \\ a_{n1} & a_{n2} & \cdots & a_{nn} \end{bmatrix} \begin{bmatrix} x_1 \\ x_2 \\ \vdots \\ x_{nn} \end{bmatrix} = \begin{bmatrix} b_1 \\ b_2 \\ \vdots \\ b_n \end{bmatrix}. \quad (2)$$

In other words, if A is the **coefficient matrix** for the system (1), then we can write (1) as

$$AX = B, \quad (3)$$

where

$$A = \begin{bmatrix} a_{11} & a_{12} & \cdots & a_{1n} \\ a_{21} & a_{22} & \cdots & a_{2n} \\ \vdots & \vdots & \cdots & \vdots \\ a_{n1} & a_{n2} & \cdots & a_{nn} \end{bmatrix}, \quad X = \begin{bmatrix} x_1 \\ x_2 \\ \vdots \\ x_n \end{bmatrix}, \quad \text{and} \quad B = \begin{bmatrix} b_1 \\ b_2 \\ \vdots \\ b_n \end{bmatrix}.$$

EXAMPLE 1   Matrix Form of a Linear System

The system of equations

$$\begin{cases} 6x - y = 6 \\ 9x + 2y = -5 \end{cases} \quad (4)$$

can be written as

$$\begin{bmatrix} 6 & -1 \\ 9 & 2 \end{bmatrix} \begin{bmatrix} x \\ y \end{bmatrix} = \begin{bmatrix} 6 \\ -5 \end{bmatrix}. \quad (5) \equiv$$

☐ **Matrix Solution** If the inverse of the coefficient matrix A exists, we can solve system (3) by multiplying both sides of the equation by A^{-1}:

$$A^{-1}(AX) = A^{-1}B$$
$$(A^{-1}A)X = A^{-1}B$$
$$I_n X = A^{-1}B,$$

or
$$X = A^{-1}B. \qquad (6)$$

EXAMPLE 2 Example 1 Revisited

Use the inverse of a matrix to solve the system in (4).

Solution Because the determinant of the coefficient matrix A is not zero,

$$\det A = \begin{vmatrix} 6 & -1 \\ 9 & 2 \end{vmatrix} = 21 \neq 0$$

and the matrix A has a multiplicative inverse, and so

$$\begin{bmatrix} 6 & -1 \\ 9 & 2 \end{bmatrix}\begin{bmatrix} x \\ y \end{bmatrix} = \begin{bmatrix} 6 \\ -5 \end{bmatrix} \quad \text{implies} \quad \begin{bmatrix} x \\ y \end{bmatrix} = \begin{bmatrix} 6 & -1 \\ 9 & 2 \end{bmatrix}^{-1}\begin{bmatrix} 6 \\ -5 \end{bmatrix}.$$

By either of the two methods of Section 12.4, we find that

$$\begin{bmatrix} 6 & -1 \\ 9 & 2 \end{bmatrix}^{-1} = \frac{1}{21}\begin{bmatrix} 2 & 1 \\ -9 & 6 \end{bmatrix}.$$

Hence, it follows from (6) that the solution of (4) is

$$\begin{bmatrix} x \\ y \end{bmatrix} = \begin{bmatrix} 6 & -1 \\ 9 & 2 \end{bmatrix}^{-1}\begin{bmatrix} 6 \\ -5 \end{bmatrix}$$
$$= \frac{1}{21}\begin{bmatrix} 2 & 1 \\ -9 & 6 \end{bmatrix}\begin{bmatrix} 6 \\ -5 \end{bmatrix}$$
$$= \begin{bmatrix} \frac{1}{3} \\ -4 \end{bmatrix}.$$

Therefore, $x = \frac{1}{3}, y = -4$ is the solution of the given system. ≡

EXAMPLE 3 Using a Matrix Inverse

Use the inverse of a matrix to solve the system

$$\begin{cases} 2x + y + z = -1 \\ -x + 3y - z = 4 \\ x - y + z = 0. \end{cases}$$

Solution The given system can be written as

$$\begin{bmatrix} 2 & 1 & 1 \\ -1 & 3 & -1 \\ 1 & -1 & 1 \end{bmatrix}\begin{bmatrix} x \\ y \\ z \end{bmatrix} = \begin{bmatrix} -1 \\ 4 \\ 0 \end{bmatrix}.$$

12.6 Linear Systems: Matrix Inverses

Because the determinant of the coefficient matrix is $\det A = 2 \neq 0$ its inverse is found to be

$$\begin{bmatrix} 2 & 1 & 1 \\ -1 & 3 & -1 \\ 1 & -1 & 1 \end{bmatrix}^{-1} = \tfrac{1}{2}\begin{bmatrix} 2 & -2 & -4 \\ 0 & 1 & 1 \\ -2 & 3 & 7 \end{bmatrix}.$$

Thus by (6) we have

$$\begin{bmatrix} x \\ y \\ z \end{bmatrix} = \begin{bmatrix} 2 & 1 & 1 \\ -1 & 3 & -1 \\ 1 & -1 & 1 \end{bmatrix}^{-1}\begin{bmatrix} -1 \\ 4 \\ 0 \end{bmatrix}$$

$$= \tfrac{1}{2}\begin{bmatrix} 2 & -2 & -4 \\ 0 & 1 & 1 \\ -2 & 3 & 7 \end{bmatrix}\begin{bmatrix} -1 \\ 4 \\ 0 \end{bmatrix}$$

$$= \begin{bmatrix} -5 \\ 2 \\ 7 \end{bmatrix}.$$

The solution of the system is given by $x = -5$, $y = 2$, $z = 7$. ≡

☐ **Least Squares Line—Revisited** In Section 3.8 we saw that if we try to fit a line $y = mx + b$ to a set of data points $(x_1, y_1), (x_2, y_2), \ldots, (x_n, y_n)$, then m and b must satisfy a system of linear equations

$$\begin{aligned} y_1 &= mx_1 + b \\ y_2 &= mx_2 + b \\ &\vdots \\ y_n &= mx_n + b. \end{aligned} \quad (7)$$

In terms of matrices system (7) is $Y = AX$, where

$$Y = \begin{bmatrix} y_1 \\ y_2 \\ \vdots \\ y_n \end{bmatrix}, \quad A = \begin{bmatrix} x_1 & 1 \\ x_2 & 1 \\ \vdots & \vdots \\ x_n & 1 \end{bmatrix}, \quad X = \begin{bmatrix} m \\ b \end{bmatrix}. \quad (8)$$

We define the solution of (7) to be the values of m and b that minimize the sum of square errors

$$E = [y_1 - (mx_1 + b)]^2 + [y_2 - (mx_2 + b)]^2 + \cdots + [y_n - (mx_n + b)]^2$$

and correspondingly $y = mx + b$ is called the **least squares line** or the **line of best fit** for the data. The slope m and constant b are defined by cumbersome quotients of sums given in (5) of Section 3.8. These formulas, derived using calculus, are actually solutions of a linear system in the two variables m and b:

$$\begin{aligned} \left(\sum_{i=1}^{n} x_i^2\right) m + \left(\sum_{i=1}^{n} x_i\right) b &= \sum_{i=1}^{n} x_i y_i \\ \left(\sum_{i=1}^{n} x_i\right) m + \quad\quad nb &= \sum_{i=1}^{n} y_i. \end{aligned} \quad (9)$$

The foregoing system can be written in a special manner using matrices:

$$A^T A X = A^T Y, \qquad (10)$$

where A, X, and Y are defined in (8). Since A is an $n \times 2$ matrix and its transpose A^T is a $2 \times n$ matrix, the dimension of the product $A^T A$ is 2×2. Moreover, unless the data points all lie on the same vertical line, the matrix $A^T A$ is nonsingular. Thus, (10) has the unique solution

$$X = (A^T A)^{-1} A^T Y. \qquad (11)$$

We say that X is the **least squares solution** of the overdetermined system (7).

The next example is Example 2 of Section 3.8, but worked this time using matrices.

EXAMPLE 4   Example 2, Section 3.8

Find the least squares line for the data $(1, 1)$, $(2, 2)$, $(3, 4)$, $(4, 6)$, $(5, 5)$.

Solution The linear equation $y = mx + b$ and the data $(1, 1)$, $(2, 2)$, $(3, 4)$, $(4, 6)$, $(5, 5)$ lead to the overdetermined system

$$\begin{aligned} 1 &= m + b \\ 3 &= 2m + b \\ 4 &= 3m + b \\ 6 &= 4m + b \\ 5 &= 5m + b. \end{aligned} \qquad (12)$$

Written as a matrix equation $Y = AX$, (12) enables us to make the identifications

$$Y = \begin{bmatrix} 1 \\ 3 \\ 4 \\ 6 \\ 5 \end{bmatrix}, \quad A = \begin{bmatrix} 1 & 1 \\ 2 & 1 \\ 3 & 1 \\ 4 & 1 \\ 5 & 1 \end{bmatrix}, \quad X = \begin{bmatrix} m \\ b \end{bmatrix}.$$

Now

$$A^T A = \begin{bmatrix} 55 & 15 \\ 15 & 5 \end{bmatrix} \quad \text{and} \quad (A^T A)^{-1} = \frac{1}{50} \begin{bmatrix} 5 & -15 \\ -15 & 55 \end{bmatrix}$$

and so (11) gives

$$X = \begin{bmatrix} 55 & 15 \\ 15 & 5 \end{bmatrix}^{-1} \begin{bmatrix} 1 & 1 \\ 2 & 1 \\ 3 & 1 \\ 4 & 1 \\ 5 & 1 \end{bmatrix}^T \begin{bmatrix} 1 \\ 3 \\ 4 \\ 6 \\ 5 \end{bmatrix}$$

$$= \frac{1}{50} \begin{bmatrix} 5 & -15 \\ -15 & 55 \end{bmatrix} \begin{bmatrix} 1 & 2 & 3 & 4 & 5 \\ 1 & 1 & 1 & 1 & 1 \end{bmatrix} \begin{bmatrix} 1 \\ 3 \\ 4 \\ 6 \\ 5 \end{bmatrix}$$

$$= \frac{1}{50} \begin{bmatrix} 5 & -15 \\ -15 & 55 \end{bmatrix} \begin{bmatrix} 68 \\ 19 \end{bmatrix}$$

$$= \begin{bmatrix} 1.1 \\ 0.5 \end{bmatrix}.$$

From the entries in the last matrix, the least squares solution of (12) is $m = 1.1$ and $b = 0.5$ and the least squares line is $y = 1.1x + 0.5$.

NOTES FROM THE CLASSROOM

In conclusion, we note that not all *consistent* $n \times n$ systems of linear equations can be solved by the method outlined in Examples 2 and 3. This procedure obviously fails for those linear systems in which the coefficient matrix A has no inverse, that is, for those systems for which $\det A = 0$. However, when $\det A \neq 0$, it can be proved that an $n \times n$ system of linear equations has a *unique* solution.

12.6 Exercises   Answers to selected odd-numbered problems begin on page ANS-33.

In Problems 1–18, write each linear system in the form $AX = B$. Then use an inverse matrix to solve the system.

1. $\begin{cases} x - 2y = 4 \\ 2x + y = 3 \end{cases}$

2. $\begin{cases} 3x - 3y = 12 \\ 2x + 4y = 8 \end{cases}$

3. $\begin{cases} 2x + 3y = 0 \\ -2x + 6y = -3 \end{cases}$

4. $\begin{cases} 2x + y = 0 \\ -x + 4y = 0 \end{cases}$

5. $\begin{cases} 4x + 5y = 1 \\ -2x - 3y = 2 \end{cases}$

6. $\begin{cases} x - y = 1 \\ x + y = 7 \end{cases}$

7. $\begin{cases} x + y = 0 \\ -2x + 3y = -15 \end{cases}$

8. $\begin{cases} 3x + y = 6 \\ 2x - y = -1 \end{cases}$

9. $\begin{cases} x + y + z = 3 \\ 2x - y + z = 5 \\ 3y - 2z = -2 \end{cases}$

10. $\begin{cases} 2x - y + z = 4 \\ x + 3y + z = -1 \\ x - 2y - z = 2 \end{cases}$

11. $\begin{cases} x + y - z = 0 \\ 2x - 3y + z = 7 \\ x + 2y - z = 1 \end{cases}$

12. $\begin{cases} x + y - z = 0 \\ 2x + y + z = 3 \\ 2y + 3z = 12 \end{cases}$

13. $\begin{cases} -x + y + 2z = 3 \\ 2x + 3y + z = -1 \\ 3x + y - z = -5 \end{cases}$

14. $\begin{cases} x + y - z = -2 \\ 2x - 2y + z = 5 \\ 4x + 2y + z = 0 \end{cases}$

15. $\begin{cases} 4x + 8y - z = -1 \\ x - 2y + z = 5 \\ 3x + 2y + 2z = 12 \end{cases}$

16. $\begin{cases} x + z = 0 \\ x + 2y - z = 0 \\ -4x - 3y + 4z = 0 \end{cases}$

17. $\begin{cases} x + y + z + w = 2 \\ x - y + w = -1 \\ 2y + z + w = 3 \\ x + z - w = 1 \end{cases}$

18. $\begin{cases} 2x + y - z + w = 2 \\ x + y - 2z = -1 \\ 2x + 2y + 3w = 11 \\ x + y = 1 \end{cases}$

In Problems 19–22, use an inverse matrix to solve the system

$$\begin{cases} 5x + 9y = b_1 \\ 4x + 7y = b_2 \end{cases}$$

for the given matrix $B = \begin{bmatrix} b_1 \\ b_2 \end{bmatrix}$.

19. $\begin{bmatrix} 2 \\ 4 \end{bmatrix}$ **20.** $\begin{bmatrix} -10 \\ 3 \end{bmatrix}$

21. $\begin{bmatrix} -1 \\ -8 \end{bmatrix}$ **22.** $\begin{bmatrix} 0 \\ 6 \end{bmatrix}$

In Problems 23–28, proceed as in Example 4 and find the least squares line for the given data.

23. (2, 1), (3, 2), (4, 3), (5, 2)
24. (0, −1), (1, 3), (2, 5), (3, 7)
25. (1, 1), (2, 1.5), (3, 3), (4, 4.5), (5, 5)
26. (0, 0), (2, 1.5), (3, 3), (4, 4.5), (5, 5)
27. (0, 2), (1, 3), (2, 5), (3, 5), (4, 9), (5, 8), (6, 10)
28. (1, 2), (2, 2.5), (3, 1), (4, 1.5), (5, 2), (6, 3.2), (7, 5)

Miscellaneous Applications

In Problems 29–31, use an inverse matrix to solve the problem.

29. Playing with Numbers The sum of three numbers is 12. The second number is 1 more than 3 times the first, and the third number is 1 less than 2 times the second. Find the numbers.

30. Counting the A, B, C's A company manufactures three products A, B, and C from materials m_1, m_2, and m_3. The two matrices

$$\begin{array}{c} \\ m_1 \\ m_2 \\ m_3 \end{array} \begin{array}{ccc} A & B & C \end{array} \\ \begin{bmatrix} 1 & 1 & 2 \\ 2 & 3 & 1 \\ 4 & 2 & 3 \end{bmatrix} \quad \text{and} \quad \begin{array}{c} m_1 \\ m_2 \\ m_3 \end{array} \begin{bmatrix} 26 \\ 47 \\ 59 \end{bmatrix}$$

represent, in turn, the number of units of material used in the construction of each product and the number of units of each type of material used in a specific week. Find

$$\begin{bmatrix} x \\ y \\ z \end{bmatrix},$$

where x, y, and z are, respectively, the number of products that are manufactured in that particular week.

31. Probing the Depths This problem shows how the depth of an ocean and the speed of sound in water can be measured by a procedure known as echo sounding. Suppose that an oceanographic vessel emits sonar signals and that the arrival times of the signals reflected from the flat ocean floor are recorded at two trailing sonobuoys. See **FIGURE 12.6.1**. Using the relation $distance = rate \times time$, we see from the figure that $2l_1 = vt_1$ and $2l_2 = vt_2$, where v is the speed of sound in water, t_1 and t_2 are the arrival times of the signals at the two sonobuoys, and l_1 and l_2 are the indicated distances.

FIGURE 12.6.1 Echo sounding procedure in Problem 31

(a) Show that the speed of sound in water v and ocean depth D satisfy the matrix equation

$$\begin{bmatrix} t_1^2 & -4 \\ t_2^2 & -4 \end{bmatrix} \begin{bmatrix} v^2 \\ D^2 \end{bmatrix} = \begin{bmatrix} d_1^2 \\ d_2^2 \end{bmatrix}.$$

[*Hint*: Use the Pythagorean theorem to relate l_1, d_1, and D, and l_2, d_2, and D.]

(b) Solve the matrix equation in part (a) to obtain formulas for v and D in terms of the measurable quantities d_1, d_2, t_1, and t_2.

(c) The sonobuoys, trailing at 1000 and 2000 m, record the arrival times of the reflected signals at 1.4 and 1.8 seconds, respectively. Find the depth of the ocean and the speed of sound in water.

12.7 Linear Systems: Determinants

≡ **Introduction** In some circumstances, we can use determinants to solve systems of n linear equations in n variables.

Suppose the linear equations in the system

$$\begin{cases} a_1 x + b_1 y = c_1 \\ a_2 x + b_2 y = c_2 \end{cases} \tag{1}$$

are independent. If we multiply the first equation by b_2 and the second by $-b_1$, we obtain

$$\begin{cases} a_1 b_2 x + b_1 b_2 y = c_1 b_2 \\ -a_2 b_1 x - b_2 b_1 y = -c_2 b_1. \end{cases}$$

In the last form of the system we can eliminate the y-variable by adding the two equations. Solving for x gives

$$x = \frac{c_1 b_2 - b_1 c_2}{a_1 b_2 - b_1 a_2}. \tag{2}$$

Similarly, by eliminating the x-variable, we find

$$y = \frac{a_1 c_2 - c_1 a_2}{a_1 b_2 - b_1 a_2}. \tag{3}$$

If we denote three 2 × 2 matrices by

$$A = \begin{bmatrix} a_1 & b_1 \\ a_2 & b_2 \end{bmatrix}, \quad A_x = \begin{bmatrix} c_1 & b_1 \\ c_2 & b_2 \end{bmatrix}, \quad A_y = \begin{bmatrix} a_1 & c_1 \\ a_2 & c_2 \end{bmatrix}, \tag{4}$$

where A is the matrix of the coefficients in (1), then the numerators and the common denominator in (2) and (3) can be written as second-order determinants:

$$\det A = \begin{vmatrix} a_1 & b_1 \\ a_2 & b_2 \end{vmatrix}, \quad \det A_x = \begin{vmatrix} c_1 & b_1 \\ c_2 & b_2 \end{vmatrix}, \quad \det A_y = \begin{vmatrix} a_1 & c_1 \\ a_2 & c_2 \end{vmatrix}. \tag{5}$$

The determinants $\det A_x$ and $\det A_y$ in (5) are, in turn, obtained from $\det A$ by replacing the x-coefficients and then the y-coefficients by the constant terms c_1 and c_2 of system (1).

With this notation we can summarize the discussion in a compact fashion.

THEOREM 12.7.1 Two Equations in Two Variables

If $\det A \neq 0$, then the system in (1) has the unique solution

$$x = \frac{\det A_x}{\det A}, \quad y = \frac{\det A_y}{\det A}. \tag{6}$$

EXAMPLE 1 Using (6)

Solve the system

$$\begin{cases} 3x - y = -3 \\ -2x + 4y = 6. \end{cases}$$

Solution Since

$$\det A = \begin{vmatrix} 3 & -1 \\ -2 & 4 \end{vmatrix} = 10 \neq 0,$$

it follows that the system has a unique solution. Continuing, we find

$$\det A_x = \begin{vmatrix} -3 & -1 \\ 6 & 4 \end{vmatrix} = -6, \quad \det A_y = \begin{vmatrix} 3 & -3 \\ -2 & 6 \end{vmatrix} = 12.$$

From (6) the solution is given by

$$x = \frac{-6}{10} = -\frac{3}{5} \quad \text{and} \quad y = \frac{12}{10} = \frac{6}{5}. \quad \equiv$$

In like manner, the solution (6) can be extended to larger systems of linear equations. In particular, for a system of three equations in three variables,

$$\begin{cases} a_1 x + b_1 y + c_1 z = d_1 \\ a_2 x + b_2 y + c_2 z = d_2 \\ a_3 x + b_3 y + c_3 z = d_3, \end{cases} \tag{7}$$

and the determinants analogous to those in (5) are

$$\det A = \begin{vmatrix} a_1 & b_1 & c_1 \\ a_2 & b_2 & c_2 \\ a_3 & b_3 & c_3 \end{vmatrix}, \quad \det A_x = \begin{vmatrix} d_1 & b_1 & c_1 \\ d_2 & b_2 & c_2 \\ d_3 & b_3 & c_3 \end{vmatrix},$$

$$\det A_y = \begin{vmatrix} a_1 & d_1 & c_1 \\ a_2 & d_2 & c_2 \\ a_3 & d_3 & c_3 \end{vmatrix}, \quad \det A_z = \begin{vmatrix} a_1 & b_1 & d_1 \\ a_2 & b_2 & d_2 \\ a_3 & b_3 & d_3 \end{vmatrix}. \tag{8}$$

As in (5), the three determinants $\det A_x$, $\det A_y$, and $\det A_z$ are obtained from the determinant $\det A$ of the coefficients of the system, by replacing the x-, y-, and z-coefficients, respectively, by the constant terms d_1, d_2, and d_3 in each of the three linear equations in the system (7). The solution of the system (7) that is analogous to (6) is given next.

◀ Indicated by the red d_1, d_2, and d_3 in (8).

THEOREM 12.7.2 Three Equations in Three Variables

If $\det A \neq 0$, then the system in (7) has the unique solution

$$x = \frac{\det A_x}{\det A}, \quad y = \frac{\det A_y}{\det A}, \quad z = \frac{\det A_z}{\det A}. \tag{9}$$

The solutions in (6) and (9) are special cases of a more general method known as **Cramer's rule**, named after the Swiss mathematician **Gabriel Cramer** (1704–1752), who was the first to publish these results.

EXAMPLE 2 — Using Cramer's Rule

Solve the system

$$\begin{cases} -x + 2y + 4z = 9 \\ x - y + 6z = -2 \\ 4x + 6y - 2z = -1. \end{cases}$$

Solution We evaluate the four 3×3 determinants in (9) using cofactor expansion. We begin by finding the value of the determinant of the coefficients in the system:

$$\det A = \begin{vmatrix} -1 & 2 & 4 \\ 1 & -1 & 6 \\ 4 & 6 & -2 \end{vmatrix} = 126 \neq 0.$$

The fact that this determinant is nonzero is sufficient to indicate that the system is consistent and has a unique solution. Continuing, we find

$$\det A_x = \begin{vmatrix} 9 & 2 & 4 \\ -2 & -1 & 6 \\ -1 & 6 & -2 \end{vmatrix} = -378, \quad \det A_y = \begin{vmatrix} -1 & 9 & 4 \\ 1 & -2 & 6 \\ 4 & -1 & -2 \end{vmatrix} = 252,$$

$$\det A_z = \begin{vmatrix} -1 & 2 & 9 \\ 1 & -1 & -2 \\ 4 & 6 & -1 \end{vmatrix} = 63.$$

From (9), the solution of the system is then

$$x = \frac{-378}{126} = -3, \quad y = \frac{252}{126} = 2, \quad z = \frac{63}{126} = \frac{1}{2}.$$

When the determinant of the coefficients of the variables in a linear system is 0, Cramer's rule cannot be used. As we see in the next example, this does *not* mean the system has no solution.

EXAMPLE 3 — Consistent System

For the system

$$\begin{cases} 4x - 16y = 3 \\ -x + 4y = -0.75 \end{cases}$$

we see that

$$\begin{vmatrix} 4 & -16 \\ -1 & 4 \end{vmatrix} = 16 - 16 = 0.$$

Although we cannot apply (6), the method of elimination would show us that the system is consistent but that the equations in the system are dependent.

The equations in (9) can be extended to systems of n linear equations in n variables $x_1, x_2, \ldots, x_n$. For the system (1) in Section 12.6, Cramer's rule is

$$x_1 = \frac{\det A_{x_1}}{\det A}, \quad x_2 = \frac{\det A_{x_2}}{\det A}, \ldots, \quad x_n = \frac{\det A_{x_n}}{\det A},$$

provided det $A \neq 0$. As a practical matter, Cramer's rule is seldom used on systems with a large number of equations simply because evaluating the determinants becomes a Herculean task. For large systems, the methods discussed in Section 12.5 are the most efficient means of solution.

12.7 Exercises
Answers to selected odd-numbered problems begin on page ANS-33.

In Problems 1–14, use Cramer's rule to solve the given system.

1. $\begin{cases} x - y = 7 \\ 3x + 2y = 6 \end{cases}$
2. $\begin{cases} -x + 2y = 0 \\ 4x - 2y = 3 \end{cases}$
3. $\begin{cases} 2x - y = -3 \\ -x + 3y = 19 \end{cases}$
4. $\begin{cases} 4x + y = 1 \\ 8x - 2y = 2 \end{cases}$
5. $\begin{cases} 2x - 5y = 5 \\ -x + 10y = -15 \end{cases}$
6. $\begin{cases} -x - 3y = -7 \\ -2x + 6y = -9 \end{cases}$
7. $\begin{cases} x + y - z = 5 \\ 2x - y + 3z = -3 \\ 2x + 3y = -4 \end{cases}$
8. $\begin{cases} 2x + y - z = -1 \\ 3x + 3y + z = 9 \\ x - 2y + 4z = 8 \end{cases}$
9. $\begin{cases} 2x - y + 3z = 13 \\ 3y + z = 5 \\ x - 7y + z = -1 \end{cases}$
10. $\begin{cases} 2x - y - 2z = 4 \\ 4x - y + 2z = -1 \\ 2x + 3y + 8z = 3 \end{cases}$
11. $\begin{cases} x + y + z = 2 \\ 4x - 8y + 3z = -2 \\ 2x - 2y + 2z = 1 \end{cases}$
12. $\begin{cases} 2x - y + z = 0 \\ x + 2y + z = 10 \\ 3x + y = 0 \end{cases}$
13. $\begin{cases} x + y - z + w = 4 \\ -x - y = -1 \\ 2x + y - 3w = -4 \\ 2y + z - 2w = -5 \end{cases}$
14. $\begin{cases} x + y - z = 1 \\ -x + y + z + w = 3 \\ 2x - 2y + w = 4 \\ 2y - z + 3w = 4 \end{cases}$

Miscellaneous Applications

15. **Take Your Vitamins** The U.S. recommended daily allowance (U.S. RDA), in percent of vitamin content per ounce of food groups X, Y, and Z, is given in the following table.

	X	Y	Z
Vitamin A	9	5	4
Vitamin C	3	5	0
Vitamin B_1	24	10	5

Use Cramer's rule to determine how many ounces of each food group one must consume each day in order to get 100% of the daily recommended allowance of vitamin A, 30% of the daily recommended allowance of vitamin B_1, and 200% of the daily recommended allowance of vitamin C. Let x, y, and z be the number of ounces of food groups X, Y, and Z, respectively.

12.8 Cryptography

Introduction The word *cryptography* is a combination of two Greek words: *crypto*, which means "hidden" or "secret," and *grapho*, which means "writing." Cryptography then is the study of making "secret writings" or **codes**. In this section we will consider a system of encoding and decoding messages that requires both the sender of the message and the receiver of the message to know

- a specified rule of correspondence between a set of symbols (such as letters of the alphabet and punctuation marks from which messages are composed) and a set of integers, and
- a specified *nonsingular* matrix A.

☐ **Encoding/Decoding** A natural correspondence between the first 27 nonnegative integers and the letters of the alphabet and a blank space (to separate words) is given by

0	1	2	3	4	5	6	7	8	9	10	11	12	13	14	15	16	17	18	19	20	21	22	23	24	25	26
space	a	b	c	d	e	f	g	h	i	j	k	l	m	n	o	p	q	r	s	t	u	v	w	x	y	z

(1)

Using the correspondence (1) the numerical equivalent of the message

SEND THE DOCUMENT TODAY

is

$$19\ 5\ 14\ 4\ 0\ 20\ 8\ 5\ 0\ 4\ 15\ 3\ 21\ 13\ 5\ 14\ 20\ 0\ 20\ 15\ 4\ 1\ 25. \quad (2)$$

The sender of the message will encode it by means of a nonsingular matrix A and the receiver of the encoded message will decode it by means of the (unique) matrix A^{-1}. The matrix A is called the **encoding matrix** and A^{-1} is called the **decoding matrix**. The numerical message (2) is now written as a matrix M. Since there are 23 symbols in the message we need a matrix that will hold a minimum of 24 entries (an $m \times n$ matrix has mn entries). We choose to write (2) as the 3×8 matrix

$$M = \begin{bmatrix} 19 & 5 & 14 & 4 & 0 & 20 & 8 & 5 \\ 0 & 4 & 15 & 3 & 21 & 13 & 5 & 14 \\ 20 & 0 & 20 & 15 & 4 & 1 & 25 & 0 \end{bmatrix}. \quad (3)$$

Note that the last entry (a_{38}) in the matrix M has been simply padded with a space represented by the number 0. Of course we could have written the message (2) as a 6×4 matrix or a 4×6 matrix but that would require a larger encoding matrix. The choice of a 3×8 matrix allows us to encode the message by means of a 3×3 matrix. The size of the matrices used is a concern only when the encoding and decoding are done by hand rather than by a computer.

The encoding matrix A is chosen, or rather, constructed, so that

- A is nonsingular,
- A has only integer entries, and
- A^{-1} has only integer entries.

The last criterion is not particularly difficult to accomplish. We need only select the integer entries of A in such a manner so that $\det A = \pm 1$. For a 2×2 or a 3×3 matrix we can then find A^{-1} by the formulas in (5) and (6) of Section 12.4. If A has integer entries, then all the cofactors A_{11}, A_{12}, and so on are also integers. For the discussion on hand we choose

$$A = \begin{bmatrix} -1 & 0 & -1 \\ 2 & 3 & 4 \\ 2 & 4 & 5 \end{bmatrix}. \tag{4}$$

You should verify that $\det A = -1$.

The original message is **encoded** by premultiplying the message matrix M by the encoding matrix A; that is, the message is sent as the matrix $B = AM$. Using (3) and (4) the encoded message is

$$B = AM = \begin{bmatrix} -1 & 0 & -1 \\ 2 & 3 & 4 \\ 2 & 4 & 5 \end{bmatrix} \begin{bmatrix} 19 & 5 & 14 & 4 & 0 & 20 & 8 & 5 \\ 0 & 4 & 15 & 3 & 21 & 13 & 5 & 14 \\ 20 & 0 & 20 & 15 & 4 & 1 & 25 & 0 \end{bmatrix}$$

$$= \begin{bmatrix} -39 & -5 & -34 & -19 & -4 & -21 & -33 & -5 \\ 118 & 22 & 153 & 77 & 79 & 83 & 131 & 52 \\ 138 & 26 & 188 & 95 & 104 & 97 & 161 & 66 \end{bmatrix} \tag{5}$$

You should try to imagine the difficulty of decoding the last matrix in (5) without knowledge of A. But the receiver of the message B knows A and its inverse and so the **decoding** is the straightforward computation of premultiplying B by the decoding matrix A^{-1}:

$$AM = B \quad \text{implies} \quad M = A^{-1}B. \tag{6}$$

For the matrix (4) we find from (6) of Section 12.4 that

$$A^{-1} = \begin{bmatrix} 1 & 4 & -3 \\ 2 & 3 & -2 \\ -2 & -4 & 3 \end{bmatrix}.$$

Thus from (6) the decoded message is

$$M = A^{-1}B = \begin{bmatrix} 1 & 4 & -3 \\ 2 & 3 & -2 \\ -2 & -4 & 3 \end{bmatrix} \begin{bmatrix} -39 & -5 & -34 & -19 & -4 & -21 & -33 & -5 \\ 118 & 22 & 153 & 77 & 79 & 83 & 131 & 52 \\ 138 & 26 & 188 & 95 & 104 & 97 & 161 & 66 \end{bmatrix}$$

$$= \begin{bmatrix} 19 & 5 & 14 & 4 & 0 & 20 & 8 & 5 \\ 0 & 4 & 15 & 3 & 21 & 13 & 5 & 14 \\ 20 & 0 & 20 & 15 & 4 & 1 & 25 & 0 \end{bmatrix}$$

or

19 5 14 4 0 20 8 5 0 4 15 3 21 13 5 14 20 0 20 15 4 1 25 0.

By also knowing the original correspondence (1), the receiver translates the numbers into

SEND_THE_DOCUMENT_TODAY_,

where we have emphasized the blank spaces by dashes.

☐ **Observations** Several observations are in order. The correspondence or mapping (1) is one of many such correspondences that can be set up between the letters of the alphabet (we could even include punctuation symbols such as the comma and period) and integers. Using the 26 letters of the alphabet and the blank space we can set up 27! of these kinds of correspondences. (See Section 13.6.) Furthermore, we could have used a 2×2 matrix to encode (2). The size of the message matrix M would then have to be at least 2×12 in order to contain the 23 entries of the message. For example, if the encoding and message matrices are, respectively,

$$A = \begin{bmatrix} 1 & 2 \\ 0 & 1 \end{bmatrix} \quad \text{and} \quad M = \begin{bmatrix} 19 & 5 & 14 & 4 & 0 & 20 & 8 & 5 & 0 & 4 & 15 & 3 \\ 21 & 13 & 5 & 14 & 20 & 0 & 20 & 15 & 4 & 1 & 25 & 0 \end{bmatrix},$$

then the encoded message is

$$B = AM = \begin{bmatrix} 61 & 31 & 24 & 32 & 40 & 20 & 48 & 35 & 8 & 6 & 65 & 3 \\ 21 & 13 & 5 & 14 & 20 & 0 & 20 & 15 & 4 & 1 & 25 & 0 \end{bmatrix}.$$

Using the decoding matrix $A^{-1} = \begin{bmatrix} 1 & -2 \\ 0 & 1 \end{bmatrix}$ we obtain as before

$$M = A^{-1}B = \begin{bmatrix} 1 & -2 \\ 0 & 1 \end{bmatrix} \begin{bmatrix} 61 & 31 & 24 & 32 & 40 & 20 & 48 & 35 & 8 & 6 & 65 & 3 \\ 21 & 13 & 5 & 14 & 20 & 0 & 20 & 15 & 4 & 1 & 25 & 0 \end{bmatrix}$$

$$= \begin{bmatrix} 19 & 5 & 14 & 4 & 0 & 20 & 8 & 5 & 0 & 4 & 15 & 3 \\ 21 & 13 & 5 & 14 & 20 & 0 & 20 & 15 & 4 & 1 & 25 & 0 \end{bmatrix}.$$

There is no particular reason why the numerical message (2) has to be broken down as rows (1×8 row vectors) as in the matrix (3). Alternatively, (2) could be broken down as columns (3×1 column vectors) as shown in the matrix

$$\begin{bmatrix} 19 & 4 & 8 & 4 & 21 & 14 & 20 & 1 \\ 5 & 0 & 5 & 15 & 13 & 20 & 15 & 25 \\ 14 & 20 & 0 & 3 & 5 & 0 & 4 & 0 \end{bmatrix}.$$

Finally, it may be desirable to send the encoded message as letters of the alphabet rather than as numbers. In Problem 13 in Exercises 12.8 we show how to transmit SEND THE DOCUMENT TODAY encoded as

OVTHWFUVJVRWYBWYCZZNWPZL.

12.8 Exercises
Answers to selected odd-numbered problems begin on page ANS-33.

In Problems 1–6, **(a)** use the matrix A and the correspondence (1) to encode the given message, and **(b)** verify your work by decoding the encoded message.

1. $A = \begin{bmatrix} 1 & 2 \\ 1 & 1 \end{bmatrix}$; SEND HELP

2. $A = \begin{bmatrix} 3 & 5 \\ 1 & 2 \end{bmatrix}$; THE MONEY IS HERE

3. $A = \begin{bmatrix} 3 & 5 \\ 2 & 3 \end{bmatrix}$; PHONE HOME

4. $A = \begin{bmatrix} 1 & 2 & 3 \\ 1 & 1 & 2 \\ 0 & 1 & 2 \end{bmatrix}$; MADAME X HAS THE PLANS

5. $A = \begin{bmatrix} 2 & 1 & 1 \\ 1 & 1 & 1 \\ -1 & 1 & 0 \end{bmatrix}$; GO NORTH ON MAIN ST

6. $A = \begin{bmatrix} 5 & 3 & 0 \\ 4 & 3 & -1 \\ 5 & 2 & 2 \end{bmatrix}$; DR JOHN IS THE SPY

In Problems 7–10, use the matrix A and the correspondence (1) to decode the given message.

7. $A = \begin{bmatrix} 8 & 3 \\ 5 & 2 \end{bmatrix}$; $B = \begin{bmatrix} 152 & 184 & 171 & 86 & 212 \\ 95 & 116 & 107 & 56 & 133 \end{bmatrix}$

8. $A = \begin{bmatrix} 2 & -1 \\ 1 & -1 \end{bmatrix}$; $B = \begin{bmatrix} 46 & -7 & -13 & 22 & -18 & 1 & 10 \\ 23 & -15 & -14 & 2 & -18 & -12 & 5 \end{bmatrix}$

9. $A = \begin{bmatrix} 1 & 0 & 1 \\ 0 & 1 & 0 \\ 1 & 0 & 0 \end{bmatrix}$; $B = \begin{bmatrix} 31 & 21 & 21 & 22 & 20 & 9 \\ 19 & 0 & 9 & 13 & 16 & 15 \\ 13 & 1 & 20 & 8 & 0 & 9 \end{bmatrix}$

10. $A = \begin{bmatrix} 2 & 1 & 1 \\ 0 & 0 & -1 \\ 1 & 1 & 1 \end{bmatrix}$; $B = \begin{bmatrix} 36 & 32 & 28 & 61 & 26 & 56 & 10 & 12 \\ -9 & -2 & -18 & -1 & -18 & -25 & 0 & 0 \\ 23 & 27 & 23 & 41 & 26 & 43 & 5 & 12 \end{bmatrix}$

11. Using the correspondence (1), a message M was encoded using a 2×2 matrix to give

$$B = \begin{bmatrix} 17 & 16 & 18 & 5 & 34 & 0 & 34 & 20 & 9 & 5 & 25 \\ -30 & -31 & -32 & -10 & -59 & 0 & -54 & -35 & -13 & -6 & -50 \end{bmatrix}.$$

Decode the message if its first two letters are DA and its last two letters are AY.

12. (a) Using the correspondence

1	2	3	4	5	6	7	8	9	10	11	12	13	14	15	16	17	18	19	20	21	22	23	24	25	26	27
j	k	l	n	m	s	t	u	w	x	g	h	i	o	p	q	r	v	y	z	a	b	c	d	e	f	space

find the numerical equivalent of the message

BUY ALL AVAILABLE STOCK AT MARKET.

(b) Encode the message by *postmultiplying* the message matrix M by

$$A = \begin{bmatrix} 1 & 1 & 0 \\ 1 & 0 & 1 \\ 1 & 1 & -1 \end{bmatrix}.$$

(c) Verify your work by decoding the encoded message in part (b).

13. Consider the matrices A and B defined in (4) and (5), respectively.

(a) Rewrite B as the matrix B' using integers modulo 27.*

---

*For integers a and b, we write $a = b$ (mod 27) if b is the remainder ($0 \le b < 27$) when a is divided by 27. For example, $33 = 6$ (mod 27), $28 = 1$ (mod 27), and so on. Negative integers are handled in the following manner. Since $27 = 0$ (mod 27) then, for example, $25 + 2 = 0$ (mod 27) so that $-25 = 2$ (mod 27) and $-2 = 25$ (mod 27). Also, $-30 = 24$ (mod 27) since $30 + 24 = 54 = 0$ (mod 27).

(b) Verify that the encoded message to be sent in letters is

OVTHWFUVJVRWYBWYCZZNWPZL.

(c) Decode the encoded message by computing $A^{-1}B'$ and rewriting the result using integers modulo 27.

CONCEPTS REVIEW

You should be able to give the meaning of each of the following concepts.

Matrix:
  dimension
  transpose
  scalar multiple
  column
  row
Additive identity
Additive inverse
Sum of matrices
Product of matrices
Multiplicative identity

Determinant:
  minor
  cofactor
  Expansion by cofactors
  Multiplicative inverse
  Nonsingular matrix
  Singular matrix
  Adjoint matrix
  Augmented matrix:
    coefficient matrix
  Row operations

Row equivalent matrices
Row echelon form
Gaussian elimination:
  back substitution
Gauss-Jordan elimination
Reduced row echelon form
Least squares line
Cramer's rule
Cryptography:
  encoding
  decoding

CHAPTER 12 Review Exercises

Answers to selected odd-numbered problems begin on page ANS-33.

A. True/False

In Problems 1–12, answer true or false.

1. If A is a square matrix such that det $A = 0$, then A^{-1} does not exist. ___

2. If A is a 3×3 matrix for which the cofactor $A_{11} = \begin{vmatrix} 0 & 0 \\ 0 & 0 \end{vmatrix}$, then det $A = 0$. ___

3. If A, B, and C are matrices such that $AB = AC$, then $B = C$. ___
4. Every square matrix A has an inverse matrix A^{-1}. ___
5. If $AB = C$ and if C has two columns, then B necessarily has two columns. ___

6. The matrix $A = \begin{bmatrix} 2 & 2 & 4 \\ 5 & 1 & -1 \\ 0 & 0 & 0 \end{bmatrix}$ is singular. ___

7. If A and B are 2×2 singular matrices, then $A + B$ is also singular. ___
8. If A is a nonsingular matrix such that $A^2 = A$, then det $A = 1$. ___

9. If $A = \begin{bmatrix} a & b \\ c & d \end{bmatrix}$ and $ad = bc$, then A^{-1} does not exist. ___

10. $\begin{bmatrix} 0 & a \\ a & 0 \end{bmatrix}^2 = a^2 I_2$. ___

11. The augmented matrix $\begin{bmatrix} 1 & 1 & 1 & | & 2 \\ 0 & 1 & 0 & | & 3 \\ 0 & 0 & 0 & | & 0 \end{bmatrix}$ is in reduced row echelon form. ___

12. If $a \neq 0$ and $b \neq 0$, then
$$\begin{bmatrix} 0 & a \\ b & 0 \end{bmatrix}^{-1} = \begin{bmatrix} 0 & 1/b \\ 1/a & 0 \end{bmatrix}. \quad \underline{}$$

B. Fill in the Blanks

In Problems 1–12, fill in the blanks.

1. If A has 3 rows and 4 columns and B has 4 rows and 6 columns, then the dimension of the product AB is _____.
2. If A has dimension $m \times n$ and B has dimension $n \times m$, then the dimension of AB is _____ and the dimension of BA is _____.
3. If $A = (a_{ij})_{4 \times 5}$, where $a_{ij} = 2i^2 - j^3$, then the entry in the third row and second column is _____.
4. Suppose A is a square matrix and B is the matrix formed by interchanging the first two columns in A. If $\det A = 12$, then $\det B =$ _____.
5. If $A = \begin{bmatrix} 2 & 3 \\ 4 & x \end{bmatrix}$ and $\det A = 2$, then $x =$ _____.
6. If $A = \begin{bmatrix} -4 & 1 & 1 \\ 1 & 5 & -1 \\ 0 & 1 & -3 \end{bmatrix}$ and $K = \begin{bmatrix} 10 \\ -1 \\ 1 \end{bmatrix}$, the number λ for which $AK = \lambda K$ is _____.
7. If $A = \begin{bmatrix} 4 & -6 \\ -8 & 3 \end{bmatrix}$ and $A + B = O$, then $B =$ _____.
8. If $A = (a_{ij})_{4 \times 4}$, where $a_{ij} = 0$, $i \neq j$, then $\det A =$ _____.
9. An example of a 2×2 matrix $A \neq I_2$ such that $A^2 = I_2$ is _____.
10. The minor determinant M_{22} of the matrix $A = \begin{bmatrix} 1 & 0 & 3 \\ 5 & 1 & 1 \\ 1 & 2 & 3 \end{bmatrix}$ is _____.
11. The cofactor of 5 in the matrix $A = \begin{bmatrix} 1 & 0 & 3 \\ 5 & 1 & 1 \\ 1 & 2 & 3 \end{bmatrix}$ is _____.
12. If A and B are $n \times n$ matrices whose corresponding entries in the third column are the same, then $\det(A - B) = 0$ because _____.

C. Review Exercises

1. Solve for the unknowns: $\begin{bmatrix} x - y & 0 \\ 3 & 2 \end{bmatrix} = \begin{bmatrix} 0 & z \\ 3 & x + y \end{bmatrix}$.
2. For

$$A = \begin{bmatrix} 2 & 1 & 0 \\ 3 & -1 & 5 \end{bmatrix} \quad \text{and} \quad B = \begin{bmatrix} -6 & 2 & -1 \\ 4 & 0 & 3 \end{bmatrix},$$

find $A + B$ and $(-2)A + 3B$.

3. The currents in an electrical network satisfy the system of equations

$$\begin{cases} i_1 - i_2 - i_3 - i_4 = 0 \\ i_2 R_1 = E \\ i_2 R_1 - i_3 R_2 = 0 \\ i_3 R_2 - i_4 R_3 = 0, \end{cases}$$

where $R_1, R_2, R_3,$ and E are positive constants. Use Cramer's rule to show that

$$i_1 = E\left(\frac{1}{R_1} + \frac{1}{R_2} + \frac{1}{R_3}\right).$$

4. Two $n \times n$ matrices A and B are said to **anticommute** if $AB = -BA$. Find two different pairs of 2×2 matrices that anticommute.

5. Find all cofactors A_{ij} of the entries a_{ij} for the matrix

$$A = \begin{bmatrix} 2 & -2 & 1 \\ 5 & -1 & 4 \\ 3 & 2 & 6 \end{bmatrix}.$$

6. Show that if A is a nonsingular matrix for which $\det A = 1$, then $A^{-1} = \operatorname{adj} A$.

In Problems 7–10, suppose

$$A = \begin{bmatrix} a_{11} & a_{12} & a_{13} \\ a_{21} & a_{22} & a_{23} \\ a_{31} & a_{32} & a_{33} \end{bmatrix}$$

and $\det A = 2$. Determine the value of the given determinant.

7. $\det(4A)$

8. $\begin{bmatrix} a_{12} & a_{11} & a_{13} \\ a_{22} & a_{21} & a_{23} \\ a_{32} & a_{31} & a_{33} \end{bmatrix}$

9. $\begin{bmatrix} a_{11} + a_{21} & a_{12} + a_{22} & a_{13} + a_{23} \\ a_{21} & a_{22} & a_{23} \\ a_{31} & a_{32} & a_{33} \end{bmatrix}$

10. $\begin{bmatrix} -a_{11} & -a_{12} & -a_{13} \\ 2a_{21} & 2a_{22} & 2a_{23} \\ -3a_{31} & -3a_{32} & -3a_{33} \end{bmatrix}$

In Problems 11 and 12, evaluate the determinant of the given matrix.

11. $\begin{bmatrix} a & 0 & c \\ 0 & e & 0 \\ f & 0 & g \end{bmatrix}$

12. $\begin{bmatrix} 1 & 0 & 1 & 1 \\ 0 & 2 & 0 & 0 \\ 0 & 1 & -1 & 0 \\ -1 & 0 & -2 & 1 \end{bmatrix}$

In Problems 13 and 14, use Theorems 12.4.1 and 12.4.2 to find A^{-1} for the given matrix.

13. $A = \begin{bmatrix} 1 & 5 \\ -2 & 0 \end{bmatrix}$

14. $A = \begin{bmatrix} 3 & 1 & 4 \\ 4 & -2 & -1 \\ 2 & -1 & 3 \end{bmatrix}$

In Problems 15 and 16, use elementary row operations to find A^{-1} for the given matrix.

15. $A = \begin{bmatrix} \frac{1}{2} & \frac{3}{4} \\ -\frac{5}{2} & 2 \end{bmatrix}$

16. $A = \begin{bmatrix} 2 & 1 & 1 \\ 2 & 2 & 2 \\ 2 & 2 & 1 \end{bmatrix}$

In Problems 17 and 18, write the system as an augmented matrix. Solve the system by Gaussian elimination.

17. $\begin{cases} x - y = 0 \\ 2x + y = 1 \end{cases}$

18. $\begin{cases} x + y - z = 3 \\ 2x - y + z = 0 \\ -x + 3y + 2z = 0 \end{cases}$

In Problems 19 and 20, write the system as an augmented matrix. Solve the system by Gauss-Jordan elimination.

19. $\begin{cases} x - y - 2z = 11 \\ 2x + 4y + 5z = -35 \\ 6x - z = -1 \end{cases}$

20. $\begin{cases} x - y = 11 \\ 4x + 3y = -5 \end{cases}$

In Problems 21 and 22, use an inverse matrix to solve the given system.

21. $\begin{cases} 5x + 6y - z = 4 \\ x + y + z = 10 \\ x - y + 2z = 3 \end{cases}$

22. $\begin{cases} 4x - 5y = 8 \\ 6x + 2y = 5 \end{cases}$

In Problems 23 and 24, use Cramer's rule to solve the given system.

23. $\begin{cases} 0.5x - 0.2y = 4 \\ 0.6x + 0.8y = -1 \end{cases}$

24. $\begin{cases} x + y - z = 0 \\ 2x + 3y + 2z = 1 \\ -4x + y + 2z = 2 \end{cases}$

In Problems 25 and 26, expand the given determinant by cofactors other than those of the first row.

25. $\begin{vmatrix} 2 & -3 & 1 \\ 0 & 2 & -4 \\ 0 & 6 & 5 \end{vmatrix}$

26. $\begin{vmatrix} 4 & 3 & -1 \\ 7 & 0 & -4 \\ -1 & 5 & 8 \end{vmatrix}$

27. Find the values of x for which the matrix

$$A = \begin{bmatrix} x & 0 & 1 \\ 0 & 1 & 0 \\ 1 & 1 & x \end{bmatrix}$$

is nonsingular. Find A^{-1}.

28. Prove that the matrix

$$A = \begin{bmatrix} 1 & 3 & 2 \\ -1 & 0 & 2 \\ 2 & 6 & 4 \end{bmatrix}$$

is singular.

In Problems 29–40, suppose

$$A = \begin{bmatrix} 5 & -6 & 7 \end{bmatrix}, \quad B = \begin{bmatrix} 3 \\ 4 \\ -1 \end{bmatrix}, \quad C = \begin{bmatrix} 1 & 2 & 4 \\ 0 & 1 & -1 \\ 3 & 2 & 1 \end{bmatrix}, \text{ and } D = \begin{bmatrix} 1 & 1 \\ 5 & 0 \\ 2 & -1 \end{bmatrix}.$$

Find the indicated matrix if it is defined.

29. $A - B^T$ **30.** $-3C$
31. $A(2B)$ **32.** $5BA$
33. AD **34.** DA
35. $(BA)C$ **36.** $(AB)C$
37. $C^T B$ **38.** $B^T D$
39. $C^{-1}(BA)$ **40.** C^2

41. Show that there exists no 2×2 matrix with real entries such that

$$A^2 = \begin{bmatrix} 0 & 1 \\ 1 & 0 \end{bmatrix}.$$

42. Show that

$$\begin{bmatrix} \cos\theta & -\sin\theta \\ -\sin\theta & -\cos\theta \end{bmatrix}^{-1} = \begin{bmatrix} \cos\theta & -\sin\theta \\ -\sin\theta & -\cos\theta \end{bmatrix}.$$

43. Let A and B be 2×2 matrices. Is it true, in general, that

$$(AB)^2 = A^2 B^2?$$

44. Show that if $A = \begin{bmatrix} -1 & 1 \\ -1 & 0 \end{bmatrix}$, then $A^2 = A^{-1}$, where $A^2 = AA$.

45. Show that

$$\begin{vmatrix} a_{11} & a_{12} \\ a_{21} & a_{22} \end{vmatrix} = \begin{vmatrix} 1 & a & b & c \\ 0 & 1 & d & e \\ 0 & 0 & a_{11} & a_{12} \\ 0 & 0 & a_{21} & a_{22} \end{vmatrix}.$$

46. Suppose that A is the 2×2 matrix

$$A = \begin{bmatrix} -2 & 4 \\ -1 & 3 \end{bmatrix}.$$

(a) Show that the matrix A satisfies the equation $A^2 - A - 2I_2 = O$, where O is the 2×2 zero matrix.

(b) Use the equation in part (a) to show that $A^3 = 3A + 2I_2$ and $A^4 = 5A + 6I_2$. [*Hint*: Multiply the equation by A.]

(c) Use the right-hand side of the appropriate equation in part (b) to compute A^3 and A^4.

47. Show that if X_1 and X_2 are solutions of the homogeneous system of linear equations $AX = O$, then $X_1 + X_2$ is also a solution.

48. Use the matrix $A = \begin{bmatrix} 10 & 1 \\ 9 & 1 \end{bmatrix}$ to encode the message

SATELLITE LAUNCHED ON FRI.

Use the correspondence (1) of Section 12.8.

13 Sequences, Series, and Probability

In This Chapter

- **13.1** Sequences
- **13.2** Series
- **13.3** Convergence of Sequences and Series
- **13.4** Mathematical Induction
- **13.5** The Binomial Theorem
- **13.6** Principles of Counting
- **13.7** Introduction to Probability
- Chapter 13 Review Exercises

In rolling a pair of dice, what is the probability of obtaining a 7? See Example 4 in Section 13.7.

A Bit of History This chapter could have been titled *Discrete Mathematics*, since each of the topics that we will consider—sequences, series, induction, the Binomial Theorem, counting, and probability—depends in a special way only on a set of integers. For example, we will see in Section 13.1 that a sequence is a function whose domain is the set of positive integers. When we study probability in Section 13.7, we leave the world of certitude. In life an event may or may not happen. When tossing, or flipping, an unbiased coin, a head or a tail is equally likely to appear. If we were asked, "Will a head appear when the coin is tossed?" our best reply would be "There is a 50–50 chance of obtaining a head." In more precise mathematical terminology, the probability of a head appearing is $\frac{1}{2}$ (one head out of two possible outcomes).

The year 1654 marks the beginning of probability theory. It was then that the Chevalier de Méré, a member of the French nobility, sent some questions that had occurred to him during gambling to the mathematician, physicist, and philosopher **Blaise Pascal** (1623–1662). Pascal's interest was aroused and he, in turn, corresponded with **Pierre de Fermat** (1601–1665), the leading French mathematician of the day. The resulting exchange of letters contained the basic results on which the subject was founded.

13.1 Sequences

Introduction Most people have heard the phrases "sequence of cards," "sequence of events," and "sequence of car payments." Intuitively, we can describe a **sequence** as a list of objects, events, or numbers that come one after the other, that is, a list of things given in some definite order. The months of the year listed in the order in which they occur,

$$\text{January, February, March, \ldots, December} \qquad (1)$$

and

$$3, 4, 5, \ldots, 12 \qquad (2)$$

are two examples of sequences. Each object in the list is called a **term** of the sequence. The lists in (1) and (2) are **finite sequences**: The sequence in (1) has 12 terms and the sequence in (2) has 10 terms. A sequence such as

$$1, \tfrac{1}{2}, \tfrac{1}{3}, \tfrac{1}{4}, \ldots, \qquad (3)$$

where no last term is indicated, is understood to be an **infinite sequence**. The three dots in (1), (2), and (3) is called an *ellipsis* and indicates that succeeding terms follow the same pattern as that set by the terms given.

Note ▶ In this chapter, unless otherwise specified, we will use the word *sequence* to mean *infinite sequence*.

The terms of a sequence can be put into a one-to-one correspondence with the set N of positive integers. For example, a natural correspondence for the sequence in (3) is

$$\begin{array}{cccc} 1, & \tfrac{1}{2}, & \tfrac{1}{3}, & \tfrac{1}{4}, \ldots \\ \uparrow & \uparrow & \uparrow & \uparrow \ldots \\ 1 & 2 & 3 & 4 \end{array}$$

Because of this correspondence property, we can give a precise mathematical definition of a sequence.

DEFINITION 13.1.1 Sequence

A **sequence** is a function f with domain the set N of positive integers $\{1, 2, 3, \ldots\}$.

You should be aware that in some instances it is convenient to take the domain of a sequence to be the set of nonnegative integers $\{0, 1, 2, 3, \ldots\}$. A **finite sequence** is also a function and its domain is some subset $\{1, 2, 3, \ldots, n\}$ of N.

☐ **Terminology** The elements in the **range** of a sequence are simply the terms of the sequence. We will assume hereafter that the range of a sequence is some set of real numbers. The number $f(1)$ is taken to be the first term of the sequence, the second term is $f(2)$, and, in general, the **nth term** is $f(n)$. Rather than using function notation, we commonly represent the terms of a sequence using subscripts: $f(1) = a_1, f(2) = a_2, \ldots,$ and so on. The nth term $f(n) = a_n$ is also called the **general term** of the sequence. We denote a sequence

$$a_1, a_2, a_3, \ldots, a_n, \ldots$$

by the notation $\{a_n\}$. If we identify the general term in (3) as $1/n$, the sequence $1, \tfrac{1}{2}, \tfrac{1}{3}, \ldots,$ can then be written compactly as $\{1/n\}$.

EXAMPLE 1   Three Sequences

List the first five terms of the given sequence.

(a) $\{2^n\}$ (b) $\left\{\dfrac{1}{n^2}\right\}$ (c) $\left\{\dfrac{(-1)^n n}{n+1}\right\}$

Solution Letting n take on the values 1, 2, 3, 4, and 5, the first five terms of the (infinite) sequences are

(a) $2^1, 2^2, 2^3, 2^4, 2^5, \ldots$ or $2, 4, 8, 16, 32, \ldots$.

(b) $\dfrac{1}{1^2}, \dfrac{1}{2^2}, \dfrac{1}{3^2}, \dfrac{1}{4^2}, \dfrac{1}{5^2}, \ldots$ or $1, \dfrac{1}{4}, \dfrac{1}{9}, \dfrac{1}{16}, \dfrac{1}{25}, \ldots$.

(c) $\dfrac{(-1)^1 \cdot 1}{1+1}, \dfrac{(-1)^2 \cdot 2}{2+1}, \dfrac{(-1)^3 \cdot 3}{3+1}, \dfrac{(-1)^4 \cdot 4}{4+1}, \dfrac{(-1)^5 \cdot 5}{5+1}, \ldots$

or $-\dfrac{1}{2}, \dfrac{2}{3}, -\dfrac{3}{4}, \dfrac{4}{5}, -\dfrac{5}{6}, \ldots$.   ≡

☐ **Sequences Defined Recursively** Instead of giving the general term of a sequence $a_1, a_2, a_3, \ldots, a_n, a_{n+1}, \ldots$, sequences are often defined using a rule or formula in which a_{n+1} is expressed using the preceding terms. For example, if we set $a_1 = 1$ and define successive terms by $a_{n+1} = a_n + 2$ for $n = 1, 2, \ldots$, then

$$a_2 = \overset{\text{given}}{\underset{\downarrow}{a_1}} + 2 = 1 + 2 = 3,$$
$$a_3 = a_2 + 2 = 3 + 2 = 5,$$
$$a_4 = a_3 + 2 = 5 + 2 = 7,$$

and so on. Sequences such as this are said to be defined **recursively**. In this example, the rule that $a_{n+1} = a_n + 2$ is called a **recursion formula**.

EXAMPLE 2   Sequence Defined Recursively

List the first five terms of the sequence defined by $a_1 = 2$ and $a_{n+1} = (n+2)a_n$.

Solution We are given $a_1 = 2$. From the recursion formula we have, respectively, for $n = 1, 2, 3, 4, \ldots$

$$a_2 = (1+2)\overset{\text{given}}{\underset{\downarrow}{a_1}} = 3(2) = 6,$$
$$a_3 = (2+2)a_2 = 4(6) = 24,$$
$$a_4 = (3+2)a_3 = 5(24) = 120,$$
$$a_5 = (4+2)a_4 = 6(120) = 720,$$

and so on. Including $a_1 = 2$ the first five terms of the sequence are

$$2, 6, 24, 120, 720, \ldots.$$ ≡

Of course, if we choose a different value for a_1 in Example 2 we would obtain an entirely different sequence.

For the remainder of this section we will examine two special types of recursively defined sequences.

☐ **Arithmetic Sequence** In the sequence $1, 3, 5, 7, \ldots$, note that each term after the first is obtained by adding the number 2 to the term preceding it. In other words, successive terms in the sequence differ by 2. A sequence of this type is known as an **arithmetic sequence**.

> **DEFINITION 13.1.2** Arithmetic Sequence
>
> A sequence such that the successive terms a_{n+1} and a_n, for $n = 1, 2, 3, \ldots$, have a fixed difference $a_{n+1} - a_n = d$ is called an **arithmetic sequence**. The number d is called the **common difference** of the sequence.

From $a_{n+1} - a_n = d$, we obtain the recursion formula

$$a_{n+1} = a_n + d \tag{4}$$

for an arithmetic sequence with common difference d.

EXAMPLE 3 An Arithmetic Sequence

The first several terms of the recursive sequence defined by $a_1 = 3$ and $a_{n+1} = a_n + 4$ are

$$a_1 = 3$$
$$a_2 = a_1 + 4 = 3 + 4 = 7,$$
$$a_3 = a_2 + 4 = 7 + 4 = 11,$$
$$a_4 = a_3 + 4 = 11 + 4 = 15,$$
$$a_5 = a_4 + 4 = 15 + 4 = 19,$$
$$\vdots$$

or

$$3, 7, 11, 15, 19, \ldots.$$

This is an arithmetic sequence with common difference 4. ≡

If we let a_1 be the first term of an arithmetic sequence having common difference d, we find from the recursion formula (4) that

$$a_2 = a_1 + d$$
$$a_3 = a_2 + d = a_1 + 2d$$
$$a_4 = a_3 + d = a_1 + 3d$$
$$\vdots$$
$$a_n = a_{n-1} + d = a_1 + (n-1)d$$

and so on. In general, an arithmetic sequence with first term a_1 and common difference d is given by

$$\{a_1 + (n-1)d\}. \tag{5}$$

EXAMPLE 4 Arithmetic Sequence Using (5)

A woman decides to jog a particular distance each week according to the following schedule. The first week she will jog 1000 m per day. Each succeeding week she will jog 250 m per day further than she did the preceding week.
(a) How far will she jog per day in the 26th week?
(b) In which week will she jog 10,000 m per day?

Solution The example describes an arithmetic sequence with $a_1 = 1000$ and $d = 250$.
(a) To find the distance the woman jogs per day in the 26th week, we set $n = 26$ and compute a_{26} using (5):

$$a_{26} = 1000 + (26 - 1)(250) = 1000 + 6250 = 7250.$$

Thus she will jog 7250 m per day in the 26th week.

(b) Here we are given $a_n = 10{,}000$ and we need to find n. From (5) we have $10{,}000 = 1000 + (n-1)(250)$ or $9000 = (n-1)(250)$. Solving for n gives

$$n - 1 = \tfrac{9000}{250} = 36 \quad \text{or} \quad n = 37.$$

Therefore, she will jog 10,000 m per day in the 37th week.

EXAMPLE 5   Find the First Term

The common difference in an arithmetic sequence is -2 and the sixth term is 3. Find the first term of the sequence.

Solution From (5) the sixth term of the sequence is

$$a_6 = a_1 + (6-1)d.$$

Setting $a_6 = 3$ and $d = -2$, we have $3 = a_1 + 5(-2)$, or $a_1 = 3 + 10$. Thus the first term is $a_1 = 13$.

Check: The sequence with $a_1 = 13$ and $d = -2$ is $13, 11, 9, 7, 5, 3, \ldots$. The sixth term of this sequence is 3.

☐ **Geometric Sequence** In the sequence $1, 2, 4, 8, \ldots$, each term after the first is obtained by multiplying the term preceding it by the number 2. In this case, we observe that the ratio of a term to the term preceding it is a constant, namely, 2. A sequence of this type is said to be a **geometric sequence**.

DEFINITION 13.1.3   Geometric Sequence

A sequence such that the successive terms a_{n+1} and a_n, for $n = 1, 2, 3, \ldots$, have a fixed ratio $a_{n+1}/a_n = r$, is called a **geometric sequence**. The number r is called the **common ratio** of the sequence.

From $a_{n+1}/a_n = r$, we see that a geometric sequence with a common ratio r is defined recursively by the formula

$$a_{n+1} = a_n r. \tag{6}$$

EXAMPLE 6   Geometric Sequence Using (6)

The sequence defined recursively by $a_1 = 2$ and $a_{n+1} = -3a_n$ is

$$2, -6, 18, -54, \ldots.$$

This is a geometric sequence with common ratio $r = -3$.

If we let $a_1 = a$ be the first term of a geometric sequence with common ratio r, we find from the recursion formula (6) that

$$\begin{aligned}
a_2 &= a_1 r = ar \\
a_3 &= a_2 r = ar^2 \\
a_4 &= a_3 r = ar^3 \\
&\vdots \\
a_n &= a_{n-1} r = ar^{n-1}
\end{aligned}$$

and so on. In general, a geometric sequence with first term a and common ratio r is

$$\{ar^{n-1}\}. \tag{7}$$

EXAMPLE 7 Find the Third Term

Find the third term of a geometric sequence with common ratio $\frac{2}{3}$ and sixth term $\frac{128}{81}$.

Solution We first find a. Since $a_6 = \frac{128}{81}$ and $r = \frac{2}{3}$, we have from (7) that

$$\frac{128}{81} = a\left(\frac{2}{3}\right)^{6-1}.$$

Solving for a, we find

$$a = \frac{\frac{128}{81}}{\left(\frac{2}{3}\right)^5} = \frac{2^7}{3^4}\left(\frac{3^5}{2^5}\right) = 12.$$

Applying (7) again with $n = 3$, we have

$$a_3 = 12\left(\frac{2}{3}\right)^{3-1} = 12\left(\frac{4}{9}\right) = \frac{16}{3}.$$

The third term of the sequence is $a_3 = \frac{16}{3}$.

□ **Compound Interest** An initial amount of money deposited in a savings account is called the **principal** and is denoted by P. Suppose that the annual **rate of interest** for the account is r. If interest is *compounded annually*, then at the end of the first year the interest on P is Pr and the amount A_1 accumulated in the account at the end of the first year is principal plus interest:

$$A_1 = P + Pr = P(1 + r).$$

The interest earned on this amount at the end of the second year is $P(1 + r)r$. If this amount is deposited, then at the end of the second year the account contains

$$A_2 = P(1 + r) + P(1 + r)r$$
$$= P(1 + 2r + r^2) = P(1 + r)^2.$$

Continuing in this fashion, we can construct the following table.

Year	Amount at the End of the Year
1	$P(1 + r)$
2	$P(1 + r)^2$
3	$P(1 + r)^3$
4	$P(1 + r)^4$
⋮	⋮

The amounts in the second column of the table form a geometric sequence with first term $P(1 + r)$ and common ratio $1 + r$. Thus from (7) we conclude that the amount in the savings account at the end of the nth year is $A_n = [P(1 + r)](1 + r)^{n-1}$ or

$$A_n = P(1 + r)^n. \tag{8}$$

EXAMPLE 8 Compound Interest

On January 1, 2010, a principal of $500 was deposited in an account drawing 4% interest compounded annually. Find the amount in the account on January 1, 2024.

Solution We make the identification $P = 500$ and $r = 0.04$. As of January 1, 2024, the principal will have drawn interest for 14 years. Using (8) and a calculator, we find

$$A_{14} = 500(1 + 0.04)^{14}$$
$$= 500(1.04)^{14}$$
$$\approx 865.84.$$

To the nearest dollar amount, the account will contain $866 at the end of 14 years. ≡

13.1 Exercises
Answers to selected odd-numbered problems begin on page ANS-34.

In Problems 1–10, list the first five terms of the given sequence.

1. $\{(-1)^n\}$
2. $\left\{\dfrac{n}{n+3}\right\}$
3. $\{\tfrac{1}{2}n(n+1)\}$
4. $\left\{\dfrac{(-2)^n}{n^2}\right\}$
5. $\left\{\dfrac{1}{n^2+1}\right\}$
6. $\left\{\dfrac{n+1}{n+2}\right\}$
7. $\{n\cos n\pi\}$
8. $\left\{\dfrac{1}{n^3}\sin\dfrac{n\pi}{2}\right\}$
9. $\left\{\dfrac{n+(-1)^n}{1+4n}\right\}$
10. $\{(-1)^{n-1}(1+n)^2\}$

In Problems 11 and 12, list the first six terms of a sequence whose general term is given.

11. $a_n = \begin{cases} -2^n, & n \text{ odd,} \\ n^2, & n \text{ even} \end{cases}$
12. $a_n = \begin{cases} \sqrt{n}, & n \text{ odd,} \\ 1/n, & n \text{ even} \end{cases}$

In Problems 13 and 14, discern a pattern for the given sequence and determine the next three terms.

13. $1, 2, \tfrac{1}{9}, 4, \tfrac{1}{25}, 6, \ldots$
14. $2, 3, 5, 8, 12, 17, \ldots$

In Problems 15–22, list the first five terms of the sequence defined recursively.

15. $a_1 = 3, a_n = \dfrac{(-1)^n}{a_{n-1}}$
16. $a_1 = \tfrac{1}{2}, a_n = (-1)^n(a_{n-1})^2$
17. $a_1 = 0, a_n = 2 + 3a_{n-1}$
18. $a_1 = 2, a_n = \tfrac{1}{3}na_{n-1}$
19. $a_1 = 1, a_n = \dfrac{1}{n}a_{n-1}$
20. $a_1 = 0, a_2 = 1, a_n = a_{n-1} - a_{n-2}$
21. $a_1 = 7, a_{n+1} = a_n + 2$
22. $a_1 = -6, a_{n+1} = \tfrac{2}{3}a_n$

In Problems 23–32, the given sequence is either an arithmetic or a geometric sequence. Find either the common difference or the common ratio. Write the general term and the recursion formula of the sequence.

23. $4, -1, -6, -11, \ldots$
24. $\tfrac{1}{16}, \tfrac{1}{8}, \tfrac{1}{4}, \tfrac{1}{2}, \ldots$
25. $4, -3, \tfrac{9}{4}, -\tfrac{27}{16}, \ldots$
26. $\tfrac{1}{2}, 1, \tfrac{3}{2}, 2, \ldots$

27. $2, -9, -20, -31, \ldots$
28. $-\frac{1}{3}, 1, -3, 9, \ldots$
29. $0.1, 0.01y, 0.001y^2, 0.0001y^3, \ldots$
30. $4x, 7x, 10x, 13x, \ldots$
31. $\frac{3}{8}, -\frac{1}{4}, \frac{1}{6}, -\frac{1}{9}, \ldots$
32. $\log_3 2, \log_3 4, \log_3 8, \log_3 16, \ldots$

33. Find the twentieth term of the sequence $-1, 5, 11, 17, \ldots$.
34. Find the fifteenth term of the sequence $2, 6, 10, 14, \ldots$.
35. Find the fifth term of a geometric sequence with first term 8 and common ratio $r = -\frac{1}{2}$.
36. Find the eighth term of the sequence $\frac{1}{1024}, \frac{1}{128}, \frac{1}{16}, \frac{1}{2}, \ldots$
37. Find the first term of a geometric sequence with third and fourth terms 2 and 8, respectively.
38. Find the first term of an arithmetic sequence with fourth and fifth terms 5 and -3, respectively.
39. Find the seventh term of an arithmetic sequence with first and third terms 357 and 323, respectively.
40. Find the tenth term of a geometric sequence with fifth and sixth terms 2 and 3, respectively.
41. Find an arithmetic sequence whose first term is 4 such that the sum of the second and third terms is 17.
42. Find a geometric sequence whose second term is 1 such that $a_5/a_3 = 64$.
43. If $1000 is invested at 7% interest compounded annually, find the amount in the account after 20 years.
44. Find the amount that must be deposited in an account drawing 5% interest compounded annually in order to have $10,000 in the account 30 years later.
45. At what rate of interest compounded annually should $450 be deposited in order to have $750 in 8 years?
46. At 6% interest compounded annually, how long will it take an initial investment to double?

Miscellaneous Applications

47. **Cookie-Jar Savings** A couple decides to set aside $5 each month the first year of their marriage, $15 each month the second year, $25 each month the third year, and so on, increasing the monthly amount by $10 each year. Find the amount they will set aside each month of the fifteenth year.
48. **Cookie-Jar Savings—Continued** In Problem 47, find a formula for the amount the couple will set aside each month of the nth year.
49. **Population Growth** The population of a certain community is observed to grow geometrically by a factor of $\frac{3}{2}$ each year. If the population at the beginning of the first year is 1000, find the population at the beginning of the eleventh year.
50. **Profit** A small company expects its profits to increase at a rate of $10,000 per year. If its profit after the first year is $6000, how much profit can the company expect after 15 years of operation?
51. **Family Tree** Everyone has two parents. Determine how many great-great-great-grandparents a person will have.
52. **How Many Rabbits?** Besides its famous leaning bell tower, the city of Pisa, Italy, is also noted as the birthplace of **Leonardo Pisano**, aka **Leonardo Fibonacci** (1170–1250). Fibonacci was the first in Europe to introduce the Hindu–Arabic place-valued decimal system and the use of Arabic numerals. His book *Liber Abacci*, published in 1202, is basically a text on how to do arithmetic

Statue of Fibonacci in Pisa, Italy

in this decimal system. But in Chapter 12 of *Liber Abacci*, Fibonacci poses and solves the following problem on the reproduction of rabbits:

> How many pairs of rabbits will be produced in a year beginning with a single pair, if in every month each pair bears a new pair that become productive from the second month on?

Discern the pattern of the solution of this problem and complete the following table.

	Start	After each month
		1 2 3 4 5 6 7 8 9 10 11 12
Adult pairs	1	1 2 3 5 8 13 21
Baby pairs	0	1 1 2 3 5 8 13
Total pairs	1	2 3 5 8 13 21 34

53. Write out five terms, after the initial two, of the sequence defined recursively by $F_{n+1} = F_n + F_{n-1}$, $F_1 = 1$, $F_2 = 1$. The elements in the range of this sequence are called **Fibonacci numbers**. Reexamine Problem 52.

For Discussion

54. Verify that the general term of the sequence defined in Problem 53 is

$$F_n = \frac{1}{\sqrt{5}}\left(\frac{1+\sqrt{5}}{2}\right)^n - \frac{1}{\sqrt{5}}\left(\frac{1-\sqrt{5}}{2}\right)^n$$

by showing that this result satisfies the recursion formula.

55. Find two different values of x such that $-\frac{3}{2}, x, -\frac{8}{27}, \ldots$ is a geometric sequence.

13.2 Series

≡ **Introduction** In the following discussion, we will be concerned with the sum of the terms of a sequence. Of special interest are the sums of the first n terms of arithmetic and geometric sequences. We begin by considering a special notation that is used as a convenient shorthand for an indicated sum of terms.

☐ **Summation Notation** Suppose we are interested in the sum of the first n terms of a sequence $\{a_n\}$. Rather than writing

$$a_1 + a_2 + \cdots + a_n,$$

mathematicians have invented a notation for representing such sums in a compact manner:

$$\sum_{k=1}^{n} a_k = a_1 + a_2 + \cdots + a_n.$$

Because $\sum$ is the capital Greek letter *sigma*, the notation $\sum_{k=1}^{n} a_k$ is referred to as **summation** or **sigma notation** and is read "the sum from $k = 1$ to $k = n$ of a sub k." The subscript k is called the **index of summation** and takes on the successive values $1, 2, \ldots, n$:

sum ends with this number
↓
$$\sum_{k=1}^{n} a_k.$$
↑
sum starts with this number

EXAMPLE 1 Summation Notation

Write out each sum.

(a) $\sum_{k=1}^{4} k^2$ (b) $\sum_{k=1}^{20} (3k+1)$ (c) $\sum_{k=1}^{n} (-1)^{k+1} a_k$

Solution (a) $\sum_{k=1}^{4} k^2 = 1^2 + 2^2 + 3^2 + 4^2 = 1 + 4 + 9 + 16$,

(b) $\sum_{k=1}^{20} (3k+1) = (3(1)+1) + (3(2)+1) + (3(3)+1) + \cdots + (3(20)+1)$

$= 4 + 7 + 10 + \cdots + 61$,

(c) $\sum_{k=1}^{n} (-1)^{k+1} a_k = (-1)^{1+1} a_1 + (-1)^{2+1} a_2 + (-1)^{3+1} a_3 + \cdots + (-1)^{n+1} a_n$

$= a_1 - a_2 + a_3 - \cdots + (-1)^{n+1} a_n.$ ≡

The choice of the letter used as the index of summation is arbitrary. Although we will consistently use the letter k, we note that

$$\sum_{k=1}^{n} a_k = \sum_{j=1}^{n} a_j = \sum_{m=1}^{n} a_m,$$

and so on. Also, as we see in the next example, we may sometimes allow the index of summation to start at a value other than $k = 1$.

EXAMPLE 2 Using Summation Notation

Write $1 - \frac{1}{2} + \frac{1}{4} - \frac{1}{8} + \cdots + \frac{1}{256}$ using summation notation.

Solution We observe that the kth term of the sequence $1, -\frac{1}{2}, \frac{1}{4}, -\frac{1}{8}, \ldots$ can be written as $(-1)^k \frac{1}{2^k}$, where $k = 0, 1, 2, \ldots$. We note too that $\frac{1}{256} = \frac{1}{2^8}$. Therefore,

$$\sum_{k=0}^{8} (-1)^k \frac{1}{2^k} = 1 - \frac{1}{2} + \frac{1}{4} - \frac{1}{8} + \cdots + \frac{1}{256}.$$ ≡

☐ **Properties** Some properties of summation notation are listed in the theorem that follows next.

THEOREM 13.2.1 Properties of Summation Notation

Suppose c is a constant (that is, does not depend on k), then

(i) $\sum_{k=1}^{n} c a_k = c \sum_{k=1}^{n} a_k,$ (ii) $\sum_{k=1}^{n} c = nc,$

(iii) $\sum_{k=1}^{n} (a_k + b_k) = \sum_{k=1}^{n} a_k + \sum_{k=1}^{n} b_k,$ (iv) $\sum_{k=1}^{n} (a_k - b_k) = \sum_{k=1}^{n} a_k - \sum_{k=1}^{n} b_k.$

Property (*i*) of Theorem 13.2.1 is simply factoring a common term from a sum:

$$\sum_{k=1}^{n} ca = ca_1 + ca_2 + \cdots + ca_n = c(a_1 + a_2 + \cdots + a_n) = c \sum_{k=1}^{n} a_k.$$

To understand property (*ii*), consider the following simple examples:

$$\underbrace{2 + 2 + 2}_{\text{three 2's}} = 3 \cdot 2 = 6 \quad \text{and} \quad \underbrace{7 + 7 + 7 + 7}_{\text{four 7's}} = 4 \cdot 7 = 28.$$

Thus, if $a_k = c$ is a real constant for $k = 1, 2, \ldots, n$, then

$$a_1 = c, \quad a_2 = c, \ldots, \quad a_n = c.$$

Consequently,

$$\sum_{k=1}^{n} c = \underbrace{c + c + c + \cdots + c}_{n \text{ terms}} = nc.$$

For example, $\sum_{k=1}^{10} 6 = 10 \cdot 6 = 60$.

☐ **Arithmetic Series** Recall, we saw in (5) of Section 13.1 that an arithmetic sequence could be written as $\{a_1 + (n - 1)d\}$. The addition of the first n terms of an arithmetic sequence,

$$S_n = \sum_{k=1}^{n} (a_1 + (k - 1)d) = a_1 + (a_1 + d) + (a_1 + 2d) + \cdots + (a_1 + (n - 1)d) \quad (1)$$

is called an **arithmetic series**. If we denote the last term of the series in (1) by a_n, then S_n can be written as

$$S_n = (a_n - (n - 1)d) + \cdots + (a_n - 2d) + (a_n - d) + a_n. \quad (2)$$

Reversing the terms in (1), we have

$$S_n = (a_1 + (n - 1)d) + \cdots + (a_1 + 2d) + (a_1 + d) + a_1. \quad (3)$$

Adding (2) and (3) gives

$$2S_n = (a_1 + a_n) + (a_1 + a_n) + \cdots + (a_1 + a_n) + (a_1 + a_n) = n(a_1 + a_n).$$

Thus,

$$S_n = n\left(\frac{a_1 + a_n}{2}\right). \quad (4)$$

In other words, the sum of the first n terms of an arithmetic sequence is the number of terms n times the average of the first term a_1 and the nth term a_n of the sequence.

EXAMPLE 3 **Arithmetic Series**

Find the sum of the first seven terms of the arithmetic sequence $\{5 - 4(n - 1)\}$.

Solution The first term of the sequence is 5 and the seventh term is -19. By identifying $a_1 = 5$, $a_7 = -19$, and $n = 7$ it follows from (4) that the sum of the seven terms in the arithmetic series

$$5 + 1 + (-3) + (-7) + (-11) + (-15) + (-19)$$

is

$$S_7 = 7\left(\frac{5 + (-19)}{2}\right) = 7(-7) = -49. \qquad \equiv$$

EXAMPLE 4 Sum of the First 100 Positive Integers

Find the sum of the first 100 positive integers.

Solution The sequence of positive integers $\{k\}$,

$$1, 2, 3, \ldots,$$

is an arithmetic sequence with common difference 1. Thus, from (4) the value of $S_{100} = 1 + 2 + 3 + \cdots + 100$ is given by

$$S_{100} = 100\left(\frac{1 + 100}{2}\right) = 50(101) = 5050.$$

An alternative form for the sum of an arithmetic series can be obtained by substituting $a_1 + (n - 1)d$ for a_n in (4). We then have

$$S_n = n\left(\frac{2a_1 + (n - 1)d}{2}\right), \tag{5}$$

which expresses the sum of an arithmetic series in terms of the first term, the number of terms, and the common difference.

EXAMPLE 5 Paying Off a Loan

A woman wishes to pay off an interest-free loan of $1300 by paying $10 the first month and increasing her payments by $15 each succeeding month. How many months will it take to pay off the entire loan? Find the amount of the final payment.

Solution The monthly payments form an arithmetic sequence with first term $a_1 = 10$ and common difference $d = 15$. Since the sum of the arithmetic series formed by the sequence of payments is $1300, we let $S_n = 1300$ in (5) and solve for n:

$$1300 = n\left(\frac{2(10) + (n - 1)15}{2}\right)$$
$$= n\left(\frac{5 + 15n}{2}\right)$$
$$2600 = 5n + 15n^2.$$

By dividing by 5 the last equation simplifies to $3n^2 + n - 520 = 0$ or $(3n + 40)(n - 13) = 0$. Thus, $n = -\frac{40}{3}$ or $n = 13$. Since n must be a positive integer, we conclude that it will take 13 months to pay off the loan. The final payment will be

$$a_{13} = 10 + (13 - 1)15 = 10 + 180 = 190 \text{ dollars}.$$

☐ **Geometric Series** The addition of the first n terms of a geometric sequence $\{ar^{n-1}\}$ is

$$S_n = \sum_{k=1}^{n} ar^{k-1} = a + ar + ar^2 + \cdots + ar^{n-1} \tag{6}$$

and is called a **finite geometric series**. Multiplying (6) by the common ratio r gives

$$rS_n = ar + ar^2 + ar^3 + \cdots + ar^n. \tag{7}$$

Subtracting (7) from (6) and simplifying gives

$$S_n - rS_n = (a + ar + ar^2 + \cdots + ar^{n-1}) - (ar + ar^2 + \cdots + ar^n) = a - ar^n,$$

or
$$(1 - r)S_n = a(1 - r^n).$$

Solving this equation for S_n, we obtain a formula for the sum of a geometric series containing n terms:

$$S_n = \frac{a(1 - r^n)}{1 - r}. \tag{8}$$

EXAMPLE 6 Sum of a Geometric Series

Compute the sum $3 + \frac{3}{2} + \frac{3}{4} + \frac{3}{8} + \frac{3}{16} + \frac{3}{32}$.

Solution This geometric series is the sum of the first six terms of the geometric sequence $\{3(\frac{1}{2})^{n-1}\}$. Identifying the first term $a = 3$, the common ratio $r = \frac{1}{2}$, and $n = 6$ in (8), we have

$$S_6 = \frac{3(1 - (\frac{1}{2})^6)}{1 - \frac{1}{2}} = \frac{3(1 - \frac{1}{64})}{\frac{1}{2}} = 6\left(\frac{63}{64}\right) = \frac{189}{32}.$$

EXAMPLE 7 Sum of a Geometric Series

A developer constructed one house in 2002. With his profits, he was able to build two houses in 2003. With the profits from these, he constructed four houses in 2004. Assuming that he is able to continue doubling the number of houses he builds each year, find the total number of houses he will have constructed by the end of 2012.

Solution The total number of houses he constructs in the 11 years from 2002 through 2012 is the sum of the geometric series with first term $a = 1$ and common ratio $r = 2$. From (8) the total number of houses is

$$S_{11} = \frac{1 \cdot (1 - 2^{11})}{1 - 2} = \frac{1 - 2048}{-1} = 2047.$$

We will return to the subject of geometric series in Section 13.3.

13.2 Exercises Answers to selected odd-numbered problems begin on page ANS-34.

In Problems 1–6, compute the given sum.

1. $\sum_{k=1}^{4} (k - 1)^2$
2. $\sum_{k=1}^{3} (-1)^k 2^k$
3. $\sum_{k=0}^{5} (k - k^2)$
4. $\sum_{k=1}^{15} 3$
5. $\sum_{k=2}^{6} (-1)^k \frac{30}{k}$
6. $\sum_{k=0}^{3} (1 - k)^3$

In Problems 7–10, write out the terms of the given sum.

7. $\sum_{k=1}^{5} \sqrt{k}$
8. $\sum_{k=1}^{5} k a_k$
9. $\sum_{k=0}^{3} (-1)^n$
10. $\sum_{k=0}^{4} k^2 f(k)$

In Problems 11–16, write the given series in summation notation.

11. $3 + 5 + 7 + 9 + 11$
12. $\frac{1}{2} + \frac{4}{3} + \frac{9}{4} + \frac{16}{5} + \frac{25}{6} + \frac{36}{7}$
13. $\frac{1}{3} - \frac{1}{6} + \frac{1}{12} - \frac{1}{24} + \frac{1}{48} - \frac{1}{96}$
14. $\frac{3}{5} + \frac{5}{6} + \frac{7}{7} + \frac{9}{8} + \frac{11}{9}$
15. $\frac{3}{2} + \frac{5}{4} + \frac{9}{8} + \frac{17}{16} + \frac{33}{32}$
16. $a_0 + \frac{1}{3}a_2 + \frac{1}{5}a_4 + \frac{1}{7}a_6 + \cdots + \frac{1}{2n+1}a_{2n}$

In Problems 17–22, find the sum of the given arithmetic series.

17. $1 + 4 + 7 + 10 + 13$
18. $131 + 111 + 91 + 71 + 51 + 31$
19. $\sum_{k=1}^{12} [3 + (k-1)8]$
20. $\sum_{k=1}^{20} [-6 + (k-1)3]$
21. $12 + 5 - 2 - \cdots - 100$
22. $-5 - 3 - 1 + \cdots + 25$

In Problems 23–28, find the sum of the given geometric series.

23. $\frac{1}{3} + \frac{1}{9} + \frac{1}{27} + \frac{1}{81}$
24. $7 + 14 + 28 + 56 + 112 + 224$
25. $60 + 6 + 0.6 + 0.06 + 0.006$
26. $1 - \frac{2}{3} + \frac{4}{9} - \frac{8}{27} + \frac{16}{81} - \frac{32}{243}$
27. $\sum_{k=1}^{8} \left(-\frac{1}{2}\right)^{k-1}$
28. $\sum_{k=1}^{5} 4\left(\frac{1}{5}\right)^{k-1}$

29. If $\{a_n\}$ is an arithmetic sequence with $d = 2$ such that $S_{10} = 135$, find a_1 and a_{10}.
30. If $\{a_n\}$ is an arithmetic sequence with $a_1 = 4$ such that $S_8 = 86$, find a_8 and d.
31. Suppose that $a_1 = 5$ and $a_n = 45$ are the first and nth terms, respectively, of an arithmetic series for which $S_n = 2000$. Find n.
32. If $\{a_n\}$ is a geometric sequence with $r = \frac{1}{2}$ such that $S_6 = \frac{65}{8}$, find the first term a.
33. The sum of the first n terms of the geometric sequence $\{2^n\}$ is $S_n = 8190$. Find n.
34. Find the sum of the first 10 terms of the arithmetic sequence

$$y, \frac{x + 3y}{2}, x + 2y, \ldots.$$

35. Find the sum of the first 15 terms of the geometric sequence

$$\frac{x}{y}, -1, \frac{y}{x}, \ldots.$$

36. (a) Find a formula for the sum of the first n positive integers:

$$1 + 2 + 3 + \cdots + n.$$

(b) Reexamine the photo on the opening page for Chapter R. Assume that the numerator of the fraction is the sum of the first 100 positive integers. Find the value of the fraction.

37. Find a formula for the sum of the first n even integers.
38. Find a formula for the sum of the first n odd integers.
39. Use the result obtained in part (a) of Problem 36 to find the sum of the first 1000 positive integers.
40. Use the result obtained in Problem 38 to find the sum of the first 50 odd integers.

Miscellaneous Applications

41. Cookie-Jar Savings A couple decides to set aside $5 each month the first year of their marriage, $15 each month the second year, $25 each month the third year, and so on, increasing the monthly amount by $10 each year. Find the total amount that they will have set aside by the end of the fifteenth year.

42. **Cookie-Jar Savings—Continued** In Problem 41, find a formula for the total amount that the couple will have set aside by the end of the nth year.

43. **Distance Traveled** An automobile accelerating at a constant rate travels 2 m the first second, 6 m the second second, 10 m the third second, and so on, traveling an additional 4 m each second. Find the total distance that the automobile has traveled after 6 seconds.

44. **Total Distance** Find a formula for the total distance traveled by the automobile in Problem 43 after n seconds.

45. **Annuity** If the same amount of money P is invested each year for n years at a rate of interest r compounded annually, then the amount accumulated after the nth payment is given by

$$S = P(1 + r)^{n-1} + P(1 + r)^{n-2} + \cdots + P(1 + r) + P.$$

Such a savings plan is called an **annuity**. Show that the value of the annuity after the nth payment is

$$S = P\left[\frac{(1 + r)^n - 1}{r}\right].$$

46. **Watch the Bouncing Ball** A ball is dropped from an initial height of 15 ft onto a concrete slab. Each time it bounces, it reaches a height of $\frac{2}{3}$ its preceding height. What height does it reach on its third bounce? On its nth bounce? How many times does the ball have to hit the concrete before its height is less than $\frac{1}{2}$ ft? See **FIGURE 13.2.1**.

47. **Total Distance** In Problem 46, find the total distance the ball has traveled up to the time when it hits the concrete slab for the seventh time.

FIGURE 13.2.1 Bouncing ball in Problem 46

48. **Desalinization** A solution of salt water containing 10 kg of salt is passed through a filter that removes 20% of the salt. The resulting solution is then filtered again, removing 20% of the remaining salt. If 20% of the salt is removed during each filtration, find the amount of salt removed from the solution after 10 filtrations.

49. **Drug Accumulation** A patient takes 50 mg of a drug each day and of the amount accumulated, 90% is excreted each day by bodily functions. Determine how much of the drug has accumulated in the body immediately after the eighth dosage.

50. **Pyramid Display** A grocer wants to display canned soup in a pyramid with 20 cans on the bottom row, 19 cans on the next row, 18 on the next row, and so on, with a single can at the top. How many cans of soup are required for the display?

Display of soup cans

For Discussion

51. **A Chess Master** According to legend, the king of a middle eastern country was so taken with the new game of chess that he queried its peasant inventor on how he might reward him. The inventor's modest request was for the sum of the grains of wheat that would fill the chess board according to the rule: 1 grain on the first square, 2 grains on the second square, 4 on the third square, 8 on the fourth square, and so on, for the entire 64 squares. The king immediately acceded to this request. If an average bushel contains 10^6 grains of wheat, how many bushels did the king owe the inventor? Do you think the peasant lived to see his reward?

Chess board

13.3 Convergence of Sequences and Series

☰ Introduction Sequences and series are important and are studied in depth in a typical course in calculus. In that study, we distinguish between sequences that are convergent or are divergent. In the discussion that follows we examine these concepts from an intuitive point of view.

☐ **Convergence** The sequence $\left\{\dfrac{n}{n+1}\right\}$ is an example of a **convergent** sequence. Although it is apparent that the terms of the sequence,

$$\frac{1}{2}, \frac{2}{3}, \frac{3}{4}, \frac{4}{5}, \ldots, \frac{n}{n+1}, \ldots$$

are increasing as n increases, the values $a_n = \dfrac{n}{n+1}$ do not increase without bound. This is because $n < n + 1$ and so

$$\frac{n}{n+1} < 1$$

for all values of n. For example, for $n = 100$, $a_{100} = \frac{100}{101} < 1$. Moreover, it appears that the terms of the sequence can be made closer and closer to 1 by letting the values of n become progressively larger. Using the $\to$ symbol for the word *approach* as we did in earlier chapters, this is written

$$a_n = \frac{n}{n+1} \to 1 \quad \text{as} \quad n \to \infty.$$

We can see the foregoing a little better by dividing the numerator and the denominator of the general term $n/(n+1)$ by n:

$$\frac{n}{n+1} = \frac{1}{1 + \dfrac{1}{n}}.$$

As $n \to \infty$ the term $1/n$ in the denominator gets closer and closer to 0 and so

$$\frac{1}{1 + \dfrac{1}{n}} \to \frac{1}{1 + 0} = 1.$$

We write $\lim\limits_{n \to \infty} \dfrac{n}{n+1} = 1$ and we say that sequence $\left\{\dfrac{n}{n+1}\right\}$ **converges** to 1.

☐ **Notation** In general, if the nth term a_n of a sequence $\{a_n\}$ can be made arbitrarily close to a number L for n sufficiently large we say that the sequence $\{a_n\}$ **converges** to L. We indicate that a sequence is convergent to a number L by writing either

$$a_n \to L \quad \text{as} \quad n \to \infty \quad \text{or} \quad \lim_{n \to \infty} a_n = L.$$

The notions of "arbitrarily close" and "for n sufficiently large" are made precise in a course in calculus. For our purposes, in determining whether a sequence $\{a_n\}$ converges, we will work directly with $\lim\limits_{n \to \infty} a_n$. The symbol "lim" is an abbreviation for the word *limit*.

We summarize the discussion.

DEFINITION 13.3.1 Convergent Sequence

(*i*) A sequence $\{a_n\}$ is said to be **convergent** if
$$\lim_{n\to\infty} a_n = L. \qquad (1)$$

The number L is said to be the **limit of the sequence**.

(*ii*) If $\lim_{n\to\infty} a_n$ does not exist, then the sequence is said to be **divergent**.

If a sequence $\{a_n\}$ converges, then its limit L is a unique number.

If a_n either increases or decreases without bound as $n \to \infty$, then $\{a_n\}$ is necessarily divergent and we write, respectively,

$$\lim_{n\to\infty} a_n = \infty \quad \text{or} \quad \lim_{n\to\infty} a_n = -\infty.$$

◀ In each case, the limits do not exist.

In the first case we say that $\{a_n\}$ **diverges to infinity** and in the second, $\{a_n\}$ **diverges to negative infinity**. For example, the sequence $1, 2, 3, \ldots, n, \ldots$ diverges to infinity.

To determine whether a sequence converges or diverges we often have to rely on analytic procedures (such as algebra) or on previously proven theorems. So in this brief discussion we will accept without proof the following three results:

$$\lim_{n\to\infty} c = c, \text{ where } c \text{ is any real constant,} \qquad (2)$$

$$\lim_{n\to\infty} \frac{1}{n^r} = 0, \text{ where } r \text{ is a positive rational number,} \qquad (3)$$

$$\lim_{n\to\infty} r^n = 0, \text{ for } |r| < 1, r \text{ a nonzero real number.} \qquad (4)$$

EXAMPLE 1 Three Convergent Sequences

(a) The constant sequence $\{\pi\}$,

$$\pi, \pi, \pi, \pi, \ldots$$

converges to π because of (2), $\lim_{n\to\infty} \pi = \pi$.

(b) The sequence $\left\{\dfrac{1}{\sqrt{n}}\right\}$,

$$\frac{1}{\sqrt{1}}, \frac{1}{\sqrt{2}}, \frac{1}{\sqrt{3}}, \frac{1}{\sqrt{4}}, \ldots \quad \text{or} \quad 1, \frac{1}{\sqrt{2}}, \frac{1}{\sqrt{3}}, \frac{1}{2}, \ldots$$

◀ When $n = 1{,}000{,}000$, the laws of exponents shows that $\frac{1}{\sqrt{1{,}000{,}000}} = 0.001$.

converges to 0. With the identification $r = \frac{1}{2}$ in (3), we have

$$\lim_{n\to\infty} \frac{1}{\sqrt{n}} = \lim_{n\to\infty} \frac{1}{n^{1/2}} = 0.$$

(c) The sequence $\{(\tfrac{1}{2})^n\}$,

$$\frac{1}{2}, \frac{1}{2^2}, \frac{1}{2^3}, \frac{1}{2^4}, \ldots \quad \text{or} \quad \frac{1}{2}, \frac{1}{4}, \frac{1}{8}, \frac{1}{18}, \ldots$$

◀ The 20th term of the sequence is approximately $a_{20} \approx 0.00000095$.

converges to 0. With the identifications $r = \frac{1}{2}$ and $|r| = \frac{1}{2} < 1$ in (4), we see that $\lim_{n\to\infty}(\tfrac{1}{2})^n = 0$. ≡

EXAMPLE 2 Divergent Sequences

(a) The sequence $\{(-1)^{n-1}\}$,

$$1, -1, 1, -1, \ldots$$

is divergent. As $n \to \infty$, the terms of the sequence oscillate between 1 and -1. Thus $\lim_{n\to\infty}(-1)^n$ does not exist because $a_n = (-1)^n$ does not approach a *single* constant L for large values of n.

(b) The first four terms of the sequence $\{\left(\frac{5}{2}\right)^n\}$ are

$$\frac{5}{2}, \frac{25}{4}, \frac{125}{8}, \frac{625}{16}, \ldots \quad \text{or} \quad 2.5, 6.25, 15.625, 39.0625, \ldots.$$

▶ The 20th term of the sequence is approximately $a_{20} \approx 90{,}949{,}470.2$.

Because the general term $a_n = \left(\frac{5}{2}\right)^n$ increases without bound as $n \to \infty$, we conclude that $\lim_{n\to\infty} a_n = \infty$; in other words, the sequence diverges to infinity.

Expanding on (4) and part (b) of Example 2, it can be proved that

- *The sequence $\{r^n\}$ converges to 0 for $|r| < 1$, and diverges for $|r| > 1$.* (5)

It is often necessary to manipulate the general term of a sequence to demonstrate convergence of the sequence.

EXAMPLE 3 Convergent Sequence

Determine whether the sequence $\left\{\sqrt{\dfrac{n}{9n+1}}\right\}$ converges.

Solution By dividing the numerator and denominator by n it follows that

$$\frac{1}{9 + \dfrac{1}{n}} \to \frac{1}{9}$$

as $n \to \infty$. Thus, we can write

$$\lim_{n\to\infty} \sqrt{\frac{n}{9n+1}} = \lim_{n\to\infty} \sqrt{\frac{1}{9 + \dfrac{1}{n}}} = \sqrt{\frac{1}{9}} = \frac{1}{3}.$$

The sequence converges to $\frac{1}{3}$.

EXAMPLE 4 Convergent Sequence

Determine whether the sequence $\left\{\dfrac{12e^n - 5}{3e^n + 2}\right\}$ converges.

Solution Since $e > 1$, a fast inspection of the general term may lead you to the false conclusion that the sequence is divergent because $12e^n - 5 \to \infty$ and $3e^n + 2 \to \infty$ as $n \to \infty$. But if we divide the numerator and denominator by e^n and then use $12 - 5e^{-n} \to 12$ and $3 + 2e^{-n} \to 3$ as $n \to \infty$, we can write

$$\lim_{n\to\infty} \frac{12e^n - 5}{3e^n + 2} = \lim_{n\to\infty} \frac{12 - 5e^{-n}}{3 + 2e^{-n}} = \frac{12 - 0}{3 + 0} = 4.$$

▶ Note that $e^{-n} = \left(\frac{1}{e}\right)^n$. Since $1/e < 1$, it follows from (4) that $\left(\frac{1}{e}\right)^n \to 0$ as $n \to \infty$.

The sequence converges to 4.

☐ **Infinite Series** Under certain conditions it is possible to assign a numerical value to an **infinite series**. In Section 13.2 we saw that we could add terms of a sequence using sigma notation. Associated with every sequence $\{a_n\}$ is another sequence called the **sequence of partial sums** $\{S_n\}$, where S_1 is the first term, S_2 is the sum of the first two terms, S_3 is the sum of the first three terms, and so on. In symbols:

sequence: $a_1, a_2, a_3, \ldots, a_n, \ldots$
sequence of partial sums: $a_1, a_1 + a_2, a_1 + a_2 + a_3, \ldots, a_1 + a_2 + a_3 + \cdots + a_n, \ldots$

In other words, the sequence of partial sums for $\{a_n\}$ is the sequence $\{S_n\}$, where the general term can be written $S_n = \sum_{k=1}^{n} a_k$. Just as we can ask whether a sequence $\{a_n\}$ converges, we now ask whether a sequence of partial sums can converge.

This question is answered in the next definition.

DEFINITION 13.3.2 Convergent Infinite Series

(*i*) If $a_1, a_2, a_3, \ldots, a_n, \ldots$ is an **infinite sequence**, we say that

$$\sum_{k=1}^{\infty} a_k = a_1 + a_2 + a_3 + \cdots + a_n + \cdots$$

is an **infinite series**.

(*ii*) An infinite series $\sum_{k=1}^{\infty} a_k$ is said to be **convergent** if the sequence of partial sums $\{S_n\}$ converges, that is,

$$\lim_{n \to \infty} S_n = \lim_{n \to \infty} \sum_{k=1}^{n} a_k = S.$$

The number S is called the **sum** of the infinite series.

(*iii*) If $\lim_{n \to \infty} S_n$ does not exist, the infinite series is said to be **divergent**.

Although the proper place for digging deeper into the above concepts is a course in calculus, we can readily illustrate the notion of convergence of an infinite series using geometric series.

Every student of mathematics knows that

$$0.333 \ldots \tag{6}$$

◀ Suspend, for the sake of illustration, that you know this rational number.

is the decimal representation of a well-known rational number. The decimal in (6) is the same as the infinite series

$$0.3 + 0.03 + 0.003 + \cdots = \tfrac{3}{10} + \tfrac{3}{100} + \tfrac{3}{1000} + \cdots$$
$$= \tfrac{3}{10} + \tfrac{3}{10^2} + \tfrac{3}{10^3} + \cdots = \sum_{k=1}^{\infty} \tfrac{3}{10^k}. \tag{7}$$

If we consider the geometric sequence

$$\frac{3}{10}, \frac{3}{10^2}, \frac{3}{10^3}, \ldots,$$

it is possible to find a formula for the general term of the associated sequence of partial sums:

$$
\begin{aligned}
S_1 &= \tfrac{3}{10} = 0.3 \\
S_2 &= \tfrac{3}{10} + \tfrac{3}{10^2} = 0.33 \\
S_3 &= \tfrac{3}{10} + \tfrac{3}{10^2} + \tfrac{3}{10^3} = 0.333 \\
&\vdots \\
S_n &= \tfrac{3}{10} + \tfrac{3}{10^2} + \tfrac{3}{10^3} + \cdots + \tfrac{3}{10^n} = \overbrace{0.333\ldots 3}^{n\ 3\text{'s}}. \\
&\vdots
\end{aligned}
\tag{8}
$$

In view (8) of Section 13.2 with the identifications $a = \tfrac{3}{10}$ and $r = \tfrac{1}{10}$ we can write the general term S_n of the sequence (8):

$$
S_n = \frac{3}{10} \frac{1 - \left(\tfrac{1}{10}\right)^n}{1 - \tfrac{1}{10}}.
\tag{9}
$$

We now let n increase without bound, that is, $n \to \infty$. From (4) and (5) we know that $\left(\tfrac{1}{10}\right)^n \to 0$ as $n \to \infty$ and so the limit of (9) is

$$
\lim_{n\to\infty} S_n = \lim_{n\to\infty} \frac{3}{10} \frac{1 - \left(\tfrac{1}{10}\right)^n}{1 - \tfrac{1}{10}} = \frac{\tfrac{3}{10}}{\tfrac{9}{10}} = \frac{3}{9} = \frac{1}{3}.
$$

Thus, $\tfrac{1}{3}$ is the sum of the infinite series in (7):

$$
\frac{1}{3} = \sum_{k=1}^{\infty} \frac{3}{10^k} \quad \text{or} \quad \frac{1}{3} = 0.333\ldots.
$$

☐ **Geometric Series** In general, the sum of an **infinite geometric series**

$$
\sum_{k=1}^{\infty} ar^{k-1} = a + ar + ar^2 + \cdots + ar^{n-1} + \cdots
\tag{10}
$$

is defined whenever $|r| < 1$. To see why this is so, recall

$$
S_n = \sum_{k=1}^{n} ar^{k-1} = a + ar + ar^2 + \cdots + ar^{n-1} = \frac{a(1 - r^n)}{1 - r}.
\tag{11}
$$

By letting $n \to \infty$ and using $r^n \to 0$ whenever $|r| < 1$, we see that

$$
\lim_{n\to\infty} S_n = \lim_{n\to\infty} \frac{a(1 - r^n)}{1 - r} = \frac{a}{1 - r}.
$$

Therefore for $|r| < 1$ we define the sum of the infinite geometric series in (10) to be $a/(1 - r)$.

THEOREM 13.3.1 Sum of a Geometric Series

(i) An infinite geometric series $\sum_{k=1}^{\infty} ar^{k-1}$ **converges** for $|r| < 1$. The sum of the series is then

$$
\sum_{k=1}^{\infty} ar^{k-1} = a + ar + ar^2 + \cdots + ar^{n-1} + \cdots = \frac{a}{1 - r}.
\tag{12}
$$

(ii) An infinite geometric series $\sum_{k=1}^{\infty} ar^{k-1}$ **diverges** for $|r| \geq 1$.

A divergent geometric series $\sum_{k=1}^{\infty} ar^{k-1}$ has no sum.

Formula (12) gives a method for converting a repeating decimal to a quotient of integers. We use the fact that

Every repeating decimal is the sum of an infinite geometric series.

Before giving another example of this, let's be clear that a **repeating decimal** is a decimal number that after a finite number of decimal places has a sequence of one or more digits that repeats endlessly.

◀ Recall from Section R.1, a **rational number** is one that is either a terminating decimal or a repeating decimal. An **irrational number** is one that is neither a terminating nor a repeating decimal.

EXAMPLE 5   Repeating Decimal

Write $0.232323\ldots$ as a quotient of integers.

Solution Written as an infinite geometric series, the repeating decimal is the same as

$$\frac{23}{100} + \frac{23}{100^2} + \frac{23}{100^3} + \cdots = \sum_{k=1}^{\infty} \frac{23}{100^k}.$$

With the identifications $a = \frac{23}{100}$ and $|r| = \left|\frac{1}{100}\right| < 1$ it follows from (12) that

$$\sum_{k=1}^{\infty} \frac{23}{100^k} = \frac{\frac{23}{100}}{1 - \frac{1}{100}} = \frac{\frac{23}{100}}{\frac{99}{100}} = \frac{23}{99}.$$

≡

EXAMPLE 6   Repeating Decimal

Write $0.72555\ldots$ as a quotient of integers.

Solution The repeating digit 5 does not appear until the third decimal place so we write the number as the sum of a terminating decimal and a repeating decimal:

$$0.72555\ldots = 0.72 + \overbrace{0.00555}^{\text{geometric series}} \ldots$$

$$= \tfrac{72}{100} + \left(\tfrac{5}{1000} + \tfrac{5}{10{,}000} + \tfrac{5}{100{,}000} + \cdots\right)$$

$$= \tfrac{72}{100} + \left(\tfrac{5}{10^3} + \tfrac{5}{10^4} + \tfrac{5}{10^5} + \cdots\right) \quad \leftarrow a = \tfrac{5}{10^3},\, r = \tfrac{1}{10}$$

$$= \tfrac{72}{100} + \frac{\tfrac{5}{10^3}}{1 - \tfrac{1}{10}} \quad \leftarrow \text{from (12)}$$

$$= \tfrac{72}{100} + \tfrac{5}{900}.$$

Combining the last two rational numbers by a common denominator we find

$$0.72555\ldots = \frac{653}{900}.$$

≡

Every repeating decimal number (rational number) is a geometric series, but do not get the impression that the sum of every convergent geometric series need be a quotient of integers.

EXAMPLE 7   Sum of a Geometric Series

The infinite series $1 - \dfrac{1}{e} + \dfrac{1}{e^2} - \dfrac{1}{e^3} + \cdots$ is a convergent geometric series because $|r| = |-1/e| = 1/e < 1$. By (12) the sum of the series is the number

$$\frac{1}{1-(-1/e)} = \frac{e}{e+1}.$$

≡

13.3 Convergence of Sequences and Series

EXAMPLE 8 A Divergent Geometric Series

The infinite series

$$2 - 3 + \frac{3^2}{2} - \frac{3^3}{2^2} + \cdots$$

is a divergent geometric series because $|r| = \left|-\frac{3}{2}\right| = \frac{3}{2} > 1.$

NOTES FROM THE CLASSROOM

(*i*) When written in terms of summation notation, a geometric series may not be immediately recognizable, or if it is, the values of a and r may not be apparent. For example, to see whether $\sum_{n=3}^{\infty} 4\left(\frac{1}{2}\right)^{n+2}$ is a geometric series it is a good idea to write out two or three terms:

$$\sum_{n=3}^{\infty} 4\left(\frac{1}{2}\right)^{n+2} = \overbrace{4\left(\frac{1}{2}\right)^5}^{a} + \overbrace{4\left(\frac{1}{2}\right)^6}^{ar} + \overbrace{4\left(\frac{1}{2}\right)^7}^{ar^2} + \cdots.$$

From the right side of the last equality, we can make the identifications $a = 4\left(\frac{1}{2}\right)^5$ and $|r| = \frac{1}{2} < 1$. Consequently, the sum of the series is $\dfrac{4\left(\frac{1}{2}\right)^5}{1 - \frac{1}{2}} = \frac{1}{4}$.

If desired, although there is no real need to do this, we can express $\sum_{n=3}^{\infty} 4\left(\frac{1}{2}\right)^{n+2}$ in the more familiar form $\sum_{k=1}^{\infty} ar^{k-1}$ by letting $k = n - 2$. The result is

$$\sum_{n=3}^{\infty} 4\left(\frac{1}{2}\right)^{n+2} = \sum_{k=1}^{\infty} 4\left(\frac{1}{2}\right)^{k+4} = \sum_{k=1}^{\infty} \overbrace{4\left(\frac{1}{2}\right)^5}^{a} \overbrace{\left(\frac{1}{2}\right)^{k-1}}^{r^{k-1}}.$$

(*ii*) In general, it is very difficult to find the sum of a convergent infinite series using the sequence of partial sums. In most cases it is impossible to find a formula for the general term $S_n = \sum_{k=1}^{n} a_k$ of this sequence. The geometric series is, of course, an important exception. But there is another type of infinite series whose sum can be found by finding the limit of the sequence $\{S_n\}$. If interested, see Problems 37 and 38 in Exercises 13.3.

13.3 Exercises
Answers to selected odd-numbered problems begin on page ANS-34.

In Problems 1–20, determine whether the given sequence converges.

1. $\left\{\dfrac{10}{n}\right\}$

2. $\left\{1 + \dfrac{1}{n^2}\right\}$

3. $\left\{\dfrac{1}{5n + 6}\right\}$

4. $\left\{\dfrac{4}{2n + 7}\right\}$

5. $\left\{\dfrac{3n-2}{6n+1}\right\}$
6. $\left\{\dfrac{n}{1-2n}\right\}$
7. $\left\{\dfrac{3n(-1)^{n-1}}{n+1}\right\}$
8. $\left\{\left(-\tfrac{1}{3}\right)^n\right\}$
9. $\left\{\dfrac{n^2-1}{2n}\right\}$
10. $\left\{\dfrac{7n}{n^2+1}\right\}$
11. $\left\{\sqrt{\dfrac{2n+1}{n}}\right\}$
12. $\left\{\dfrac{n}{\sqrt{n+1}}\right\}$
13. $\{\cos n\pi\}$
14. $\{\sin n\pi\}$
15. $\left\{\dfrac{5-2^{-n}}{6+4^{-n}}\right\}$
16. $\left\{\dfrac{2^n}{3^n+1}\right\}$
17. $\left\{\dfrac{10e^n - 3e^{-n}}{2e^n + e^{-n}}\right\}$
18. $\left\{4 + \dfrac{3^n}{2^n}\right\}$
19. $2, \tfrac{2}{3}, \tfrac{2}{9}, \tfrac{2}{27}, \ldots$
20. $1 + \tfrac{1}{2}, \tfrac{1}{2} + \tfrac{1}{3}, \tfrac{1}{3} + \tfrac{1}{4}, \tfrac{1}{4} + \tfrac{1}{5}, \ldots$

In Problems 21–26, write the given repeating decimal as a quotient of integers.

21. $0.222\ldots$
22. $0.555\ldots$
23. $0.616161\ldots$
24. $0.393939\ldots$
25. $1.314314\ldots$
26. $0.5262626\ldots$

In Problems 27–36, determine whether the given infinite geometric series converges. If convergent, find its sum.

27. $2 + 1 + \dfrac{1}{2} + \cdots$
28. $1 + \dfrac{1}{3} + \dfrac{1}{9} + \cdots$
29. $\dfrac{2}{3} - \dfrac{4}{9} + \dfrac{8}{27} - \cdots$
30. $1 + 0.1 + 0.01 + \cdots$
31. $9 + 2 + \dfrac{4}{9} + \cdots$
32. $\dfrac{1}{81} - \dfrac{1}{54} + \dfrac{1}{36} - \cdots$
33. $\sum_{k=1}^{\infty}\dfrac{1}{(\sqrt{3}-\sqrt{2})^{k-1}}$
34. $\sum_{k=1}^{\infty}\left(\dfrac{\sqrt{5}}{1+\sqrt{5}}\right)^{k-1}$
35. $\sum_{k=1}^{\infty}(-3)^k 7^{-k}$
36. $\sum_{k=1}^{\infty}\pi^k\left(\tfrac{1}{3}\right)^{k-1}$

37. The infinite series $\sum_{k=1}^{\infty}\dfrac{1}{k(k+1)}$ is an example of a **telescoping series**. For such series it is possible to find a formula for the general term S_n of the sequence of partial sums.

   (a) Use the partial fraction decomposition
   $$\dfrac{1}{k(k+1)} = \dfrac{1}{k} - \dfrac{1}{k+1}$$
   as an aid in finding a formula for S_n. This will also explain the meaning of the word *telescoping*.

   (b) Use part (a) to find the sum of the infinite series.

38. Use the procedure in part (a) of Problem 37 to find the sum of the infinite series $\sum_{k=1}^{\infty}\dfrac{1}{(k+1)(k+2)}$.

Miscellaneous Applications

39. Distance Traveled A ball is dropped from an initial height of 15 ft onto a concrete slab. Each time the ball bounces, it reaches a height of $\frac{2}{3}$ its preceding height. Use an infinite geometric series to determine the distance the ball travels before it comes to rest.

40. Drug Accumulation A patient takes 15 mg of a drug at the same time each day. If 80% of the drug accumulated is excreted each day by bodily functions, how much of the drug will accumulate in the patient's body after a long period of time, that is, as $n \to \infty$? (Assume that the measurement of the accumulation is made immediately after each dose.)

Calculator/Computer Problems

41. It can be proved that the terms of the sequence $\{a_n\}$ defined recursively by the formula

$$a_{n+1} = \frac{1}{2}\left(a_n + \frac{r}{a_n}\right), \quad r > 0,$$

converges when $a_1 = 1$ and $r = 3$. Use a calculator to find the first 10 terms of the sequence. Conjecture the limit of the sequence.

42. The sequence

$$\{1 + \tfrac{1}{2} + \tfrac{1}{3} + \cdots + \tfrac{1}{n} - \ln n\}$$

is known to converge to a number γ called **Euler's constant**. Calculate at least the first 10 terms of the sequence. Conjecture the limit of the sequence.

For Discussion

43. Use algebra to show that the sequence $\{\sqrt{n}(\sqrt{n+1} - \sqrt{n})\}$ converges.

44. Use the graph of the inverse tangent function to show the sequence $\left\{\dfrac{\pi}{4} - \arctan(n)\right\}$ converges.

45. The infinite series $\sum_{k=1}^{\infty} \dfrac{2^k - 1}{4^k}$ is known to be convergent. Discuss how the sum of the series can be found. State any assumptions that you make.

46. Find the values of x for which the infinite series $\sum_{k=1}^{\infty} \left(\dfrac{x}{2}\right)^{k-1}$ converges.

47. The infinite series

$$1 + 1 + 1 + \cdots$$

is a divergent geometric series with $r = 1$. Note that formula (5) does not yield the general term for the sequence of partial sums. Find a formula for S_n and use that formula to argue that the infinite series is divergent.

48. Consider the rational function $f(x) = 1/(1 - x)$. Show that

$$\frac{1}{1 - x} = 1 + x + x^2 + \cdots.$$

For what values of x is the equality true?

49. Discuss whether the equality $1 = 0.999\ldots$ is true or false.

50. The Trains and the Fly At a specified time two trains T_1 and T_2, 20 miles apart on the same track, start on a collision course at a rate of 10 mi/h. Suppose that at the precise instant the trains start a fly leaves the front of train T_1, flies at a rate of 20 mi/h in a straight line to the front of the engine of train T_2, then flies back to

T_1 at 20 mi/h, then back to T_2, and so on. Use geometric series to find the total distance traversed by the fly when the trains collide (and the fly is squashed). Then use common sense to find the total distance the fly flies. See **FIGURE 13.3.1**.

FIGURE 13.3.1 Trains and fly in Problem 50

FIGURE 13.3.2 Embedded squares in Problem 51

51. **Embedded Squares** In **FIGURE 13.3.2** the square shown in red is 1 unit on a side. A second blue square is constructed inside the first square by connecting the midpoints of the first one. A third green square is constructed by connecting the midpoints of the sides of the second square, and so on.
    (a) Find a formula for the area A_n of the nth inscribed square.
    (b) Make a conjecture about the convergence of the sequence $\{A_n\}$.
    (c) Consider the sequence $\{S_n\}$, where $S_n = A_1 + A_2 + \cdots + A_n$. Calculate the numerical values of the first 10 terms of this sequence.
    (d) Make a conjecture about the convergence of the sequence $\{S_n\}$.

52. **Length of a Polygonal Path** In **FIGURE 13.3.3**, there are twelve blue rays emanating from the origin and the angle between each pair of consecutive rays is 30°. The line segment AP_1 is perpendicular to ray L_1, the line segment P_1P_2 is perpendicular to ray L_2, and so on.
    (a) Show that the length of the red polygonal path $AP_1P_2P_3 \ldots$ is the infinite series

$$AP_1 + P_1P_2 + P_2P_3 + P_3P_4 + \cdots$$
$$= \sin 30° + (\cos 30°)\sin 30° + (\cos 30°)^2 \sin 30° + (\cos 30°)^3 \sin 30° + \cdots.$$

    (b) Find the sum of the infinite series in part (a).

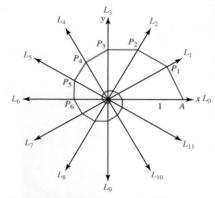

FIGURE 13.3.3 Polygonal path in Problem 52

13.4 Mathematical Induction

≡ Introduction Frequently, a mathematical statement or proposition that depends on the natural numbers or positive integers $N = \{1, 2, 3, \ldots\}$ can be proved using a technique known as **mathematical induction**. Suppose we can show two things:

- *a statement is true for the number* 1; *and*
- *whenever the statement is true for the positive integer k, then it is true for the next positive integer $k + 1$.*

In other words, suppose we can demonstrate that the

$$\boxed{\text{statement is true for } 1} \tag{1}$$

and that the

$$\boxed{\text{statement is true for } k} \text{ implies the } \boxed{\text{statement is true for } k + 1.} \tag{2}$$

What can we conclude from this? From (1) we have that

the statement is true for the number 1,

and by (2) *the statement is true for the number* $1 + 1 = 2$.

In addition, it now follows from (2) that

the statement is true for the number $2 + 1 = 3$,
the statement is true for the number $3 + 1 = 4$,
the statement is true for the number $4 + 1 = 5$,

and so on. Symbolically, we can represent this sequence of implications by

$\boxed{\text{statement is true for 1}} \Rightarrow \boxed{\text{statement is true for 2}} \Rightarrow \boxed{\text{statement is true for 3}} \Rightarrow \cdots$

It seems clear that the statement must be true for *all* positive integers n. This is precisely the assertion of the following principle.

THEOREM 13.4.1 Principle of Mathematical Induction

Let $S(n)$ be a statement involving a positive integer n such that

(*i*) $S(1)$ is true, and
(*ii*) whenever $S(k)$ is true for a positive integer k, then $S(k + 1)$ is also true.

Then $S(n)$ is true for every positive integer.

Although we have stated the Principle of Mathematical Induction as a theorem, it is actually considered to be an *axiom* of the natural numbers.

By way of a physical analogy to the foregoing principle, imagine that we have an endless row of correctly spaced dominoes each standing on its end. Suppose we can demonstrate that whenever a domino (give it a name, say, the kth domino) falls over that its neighboring domino (the $(k + 1)$st domino) also falls over. Then we conclude that all the dominoes must fall over provided we can show one more thing, namely, that the first domino falls over.

We now illustrate the use of induction with several examples. We begin with an example from arithmetic.

Falling dominoes

EXAMPLE 1   Using Mathematical Induction

Prove that the sum of the first n positive integers is given by

$$1 + 2 + 3 + \cdots + n = \frac{n(n + 1)}{2}. \tag{3}$$

Solution Here the statement $S(n)$ is the formula in (3). The first step is to show that $S(1)$ is true, where $S(1)$ is the statement

$$1 = \frac{1 \cdot 2}{2}.$$

Since this is clearly true, condition (*i*) is satisfied.

The next step is to verify condition (*ii*). This requires that from the hypothesis "$S(k)$ is true," we prove that "$S(k + 1)$ is true." Thus we assume that the statement $S(k)$,

$$1 + 2 + 3 + \cdots + k = \frac{k(k + 1)}{2}, \tag{4}$$

is true. From this assumption we wish to demonstrate that $S(k + 1)$,

$$1 + 2 + 3 + \cdots + (k + 1) = \frac{(k + 1)[(k + 1) + 1]}{2}, \tag{5}$$

is also true. Now we can obtain a formula for the sum of the first $k + 1$ positive integers by using the equality (4) and some algebra:

$$1 + 2 + 3 + \cdots + k + (k + 1) = \overbrace{}^{\text{by (4) this equals } \frac{k(k+1)}{2}} + (k+1) = \frac{k(k + 1)}{2} + (k + 1)$$

$$= \frac{k(k + 1) + 2(k + 1)}{2}$$

$$= \frac{(k + 1)(k + 2)}{2}$$

$$= \frac{(k + 1)[(k + 1) + 1]}{2}. \quad \leftarrow \text{this is (5)}$$

Thus we have shown that the statement $S(k + 1)$ is true. It follows from the Principle of Mathematical Induction that $S(n)$ is true for every positive integer n. ≡

In basic algebra we learned how to factor. In particular, from the factorizations

$$x - y = x - y,$$
$$x^2 - y^2 = (x - y)(x + y), \quad \leftarrow \text{See (4) of Section R.7}$$
$$x^3 - y^3 = (x - y)(x^2 + xy + y^2), \quad \leftarrow \text{See (5) of Section R.7}$$
$$x^4 - y^4 = (x^2 - y^2)(x^2 + y^2) = (x - y)(x + y)(x^2 + y^2),$$

a reasonable conjecture is that $x - y$ is always a factor of $x^n - y^n$ for any positive integer n. We now prove that this is so.

EXAMPLE 2   Using Mathematical Induction

Prove that $x - y$ is a factor of $x^n - y^n$ for any positive integer n.

Solution For the statement $S(n)$,

$$x - y \text{ is a factor of } x^n - y^n,$$

we must show that the two conditions (*i*) and (*ii*) are satisfied. For $n = 1$ we have the true statement $S(1)$,

$$x - y \text{ is a factor of } x^1 - y^1.$$

Now assume that $S(k)$,

$$x - y \text{ is a factor of } x^k - y^k,$$

is true. Using this assumption, we must show that $S(k + 1)$ is true; that is, $x - y$ is a factor of $x^{k+1} - y^{k+1}$. To this end we perform a bit of cleverness, namely, let's *subtract* and *add* xy^k to $x^{k+1} - y^{k+1}$:

$$x^{k+1} - y^{k+1} = x^{k+1} - \overbrace{xy^k + xy^k}^{0} - y^{k+1} = x\overbrace{(x^k - y^k)}^{\substack{x - y \text{ is assumed} \\ \text{to be a factor of} \\ \text{this term}}} + y^k\overbrace{(x - y)}^{\substack{\text{here is a} \\ \text{factor of} \\ x - y}}. \quad (6)$$

But by hypothesis, $x - y$ is a factor of $x^k - y^k$. Therefore, $x - y$ is a factor of *both* terms on the right-hand side of (6). It follows that $x - y$ is a factor of the right-hand side, and thus we have shown that the statement $S(k + 1)$,

$$x - y \text{ is a factor of } x^{k+1} - y^{k+1},$$

is true. It follows by the Principle of Mathematical Induction that $x - y$ is a factor of $x^n - y^n$ for any positive integer n. ≡

EXAMPLE 3 Using Mathematical Induction

Prove that $8^n - 1$ is divisible by 7 for all positive integers n.

Solution We let $S(n)$ be the statement "$8^n - 1$ is divisible by 7 for all positive integers n." With $n = 1$ we see that $8^1 - 1 = 7$ is obviously divisible by 7.

Therefore $S(1)$ is true. Now let us assume that $S(k)$ is true; that is, $8^k - 1$ is divisible by 7 for some positive integer k. Using that assumption we must show that $8^{k+1} - 1$ is divisible by 7. Consider

$$8^{k+1} - 1 = 8^k 8 - 1$$
$$= 8^k(1 + 7) - 1 \quad \leftarrow \text{rearrange terms}$$
$$= \underbrace{(8^k - 1)}_{\text{assumed to be divisible by 7}} + \underbrace{7 \cdot 8^k}_{\text{divisible by 7}}.$$

The last equality proves $S(k + 1)$ is true because both $8^k - 1$ and $7 \cdot 8^k$ are divisible by 7. It follows from the Principle of Mathematical Induction that $S(n)$ is true for every positive integer n. ≡

13.4 Exercises   Answers to selected odd-numbered problems begin on page ANS-34.

In Problems 1–20, use the Principle of Mathematical Induction to prove that the given statement is true for every positive integer n.

1. $2 + 4 + 6 + \cdots + 2n = n^2 + n$
2. $1 + 3 + 5 + \cdots + (2n - 1) = n^2$
3. $1^2 + 2^2 + 3^2 + \cdots + n^2 = \frac{1}{6}n(n + 1)(2n + 1)$
4. $1^3 + 2^3 + 3^3 + \cdots + n^3 = \frac{1}{4}n^2(n + 1)^2$
5. $\sum_{k=1}^{n} \frac{1}{2^k} + \frac{1}{2^n} = 1$
6. $\sum_{k=1}^{n} (4k - 5) = n(2n - 3)$
7. $\frac{1}{1 \cdot 2} + \frac{1}{2 \cdot 3} + \frac{1}{3 \cdot 4} + \cdots + \frac{1}{n(n + 1)} = \frac{n}{n + 1}$
8. $\frac{1}{2 \cdot 3} + \frac{1}{3 \cdot 4} + \frac{1}{4 \cdot 5} + \cdots + \frac{1}{(n + 1)(n + 2)} = \frac{n}{2n + 4}$
9. $1 + 4 + 4^2 + \cdots + 4^{n-1} = \frac{1}{3}(4^n - 1)$
10. $10 + 10^2 + 10^3 + \cdots + 10^n = \frac{1}{9}(10^{n+1} - 10)$
11. $n^3 + 2n$ is divisible by 3
12. $n^2 + n$ is divisible by 2
13. 4 is a factor of $5^n - 1$
14. 6 is a factor of $n^3 - n$
15. 7 is a factor of $3^{2n} - 2^n$
16. $x + y$ is a factor of $x^{2n-1} + y^{2n-1}$
17. If $a \geq -1$, then $(1 + a)^n \geq 1 + na$.
18. $2n \leq 2^n$
19. If $r > 1$, then $r^n > 1$.
20. If $0 < r < 1$, then $0 < r^n < 1$.

For Discussion

21. If we assume that

$$2 + 4 + 6 + \cdots + 2n = n^2 + n + 1$$

is true for $n = k$, show that the formula is true for $n = k + 1$. Show, however, that the formula itself is false. Explain why this does not violate the Principle of Mathematical Induction.

13.5 The Binomial Theorem

Introduction When $(a + b)^n$ is expanded for an arbitrary positive integer n, the exponents of a and b follow a definite pattern. For example, from the expansions

$$(a + b)^2 = a^2 + 2ab + b^2$$
$$(a + b)^3 = a^3 + 3a^2b + 3ab^2 + b^3$$
$$(a + b)^4 = a^4 + 4a^3b + 6a^2b^2 + 4ab^3 + b^4,$$

we see that the exponents of a *decrease* by 1, starting with the first term, whereas the exponents of b *increase* by 1, starting with the second term. In the case of $(a + b)^4$, we have

powers decreasing by 1 ↓

$$(a + b)^4 = a^4 + 4a^3b^1 + 6a^2b^2 + 4a^1b^3 + b^4.$$

↑ powers increasing by 1

To extend this pattern, we consider the first and last terms to be multiplied by b^0 and a^0, respectively; that is,

$$(a + b)^4 = a^4b^0 + 4a^3b^1 + 6a^2b^2 + 4a^1b^3 + a^0b^4. \qquad (1)$$

We also note that the sum of the exponents in each of the five terms of the expansion $(a + b)^4$ is 4. For example, in the second term we have $4\overbrace{a^3b^1}^{4=3+1}$.

EXAMPLE 1 Using (1)

Expand $(y^2 - 1)^4$.

Solution With the identifications $a = y^2$ and $b = -1$, it follows from (1) and the laws of exponents that

$$(y^2 - 1)^4 = (y^2 + (-1))^4$$
$$= (y^2)^4 + 4(y^2)^3(-1) + 6(y^2)^2(-1)^2 + 4(y^2)(-1)^3 + (-1)^4$$
$$= y^8 - 4y^6 + 6y^4 - 4y^2 + 1. \qquad \equiv$$

□ **The Coefficients** The coefficients in the expansion of $(a + b)^n$ also follow a pattern. To illustrate, we display the coefficients in the expansions of $(a + b)^0$, $(a + b)^1$, $(a + b)^2$, $(a + b)^3$, and $(a + b)^4$ in a triangular array

$$\begin{array}{ccccccccc} & & & & 1 & & & & \\ & & & 1 & & 1 & & & \\ & & 1 & & 2 & & 1 & & \\ & 1 & & 3 & & 3 & & 1 & \\ 1 & & 4 & & 6 & & 4 & & 1 \end{array} \qquad (2)$$

Observe that each number in the interior of this array is the *sum* of the two numbers directly above it. Thus the next line in the array can be obtained as follows:

$$\begin{array}{ccccccccccc} 1 & & 4 & & 6 & & 4 & & 1 & & \\ & \searrow & \swarrow & & \searrow \swarrow & & \searrow & \swarrow & & \searrow & \\ 1 & & 5 & & 10 & & 10 & & 5 & & 1. \end{array}$$

As you might expect, these numbers are the coefficients of the powers of a and b in the expansion of $(a + b)^5$; that is,

$$(a + b)^5 = 1a^5 + 5a^4b + 10a^3b^2 + 10a^2b^3 + 5ab^4 + 1b^5. \qquad (3)$$

The array obtained by continuing in this manner is called **Pascal's triangle** after the French philosopher and mathematician **Blaise Pascal** (1623–1662).

EXAMPLE 2   Using (3)

Expand $(3 - x)^5$.

Solution From (3), with $a = 3$ and $b = -x$, we can write

$$(3 - x)^5 = (3 + (-x))^5$$
$$= 1(3)^5 + 5(3)^4(-x) + 10(3)^3(-x)^2 + 10(3)^2(-x)^3 + 5(3)(-x)^4 + 1(-x)^5$$
$$= 243 - 405x + 270x^2 - 90x^3 + 15x^4 - x^5. \qquad \equiv$$

□ **Factorial Notation** Before we give a general formula for the expansion of $(a + b)^n$, it will be helpful to introduce **factorial notation**. The symbol $r!$ is defined for any positive integer r as the product

See Problem 57 in Exercises 3.1. ▶

$$r! = r \cdot (r - 1) \cdot (r - 2) \cdots 3 \cdot 2 \cdot 1, \qquad (4)$$

and is read "r factorial." For example, $1! = 1$ and $4! = 4 \cdot 3 \cdot 2 \cdot 1 = 24$. Also, it is convenient to define

$$0! = 1.$$

EXAMPLE 3   A Simplification

Simplify $\dfrac{r!(r + 1)}{(r - 1)!}$, where r is a positive integer.

Solution Using the definition of $r!$ in (4) we can write the numerator as

$$r!(r + 1) = (r + 1)r! = (r + 1)r(r - 1) \cdots 2 \cdot 1 = (r + 1)r(r - 1)!.$$

Thus, $\qquad \dfrac{r!(r + 1)}{(r - 1)!} = \dfrac{(r + 1)r(r - 1)!}{(r - 1)!} = (r + 1)r. \qquad \equiv$

□ **The Binomial Theorem** The general formula for the expansion of $(a + b)^n$ is given in the following result, known as the **Binomial Theorem**.

THEOREM 13.5.1   Binomial Theorem

For any positive integer n,

$$(a + b)^n = a^n + \frac{n}{1!}a^{n-1}b + \frac{n(n - 1)}{2!}a^{n-2}b^2$$
$$+ \cdots + \frac{n(n - 1) \cdots (n - r + 1)}{r!}a^{n-r}b^r + \cdots + b^n. \qquad (5)$$

By paying attention to the increasing powers on b in (5) we see that the expression

$$\frac{n(n-1)\cdots(n-r+1)}{r!}a^{n-r}b^r \quad (6)$$

is the $(r+1)$st term in the expansion of $(a+b)^n$. For $r = 0, 1, \ldots, n$, the numbers

$$\frac{n(n-1)\cdots(n-r+1)}{r!} \quad (7)$$

are called **binomial coefficients** and are, of course, the same as those obtained from Pascal's triangle. Before proving the Binomial Theorem by mathematical induction, we consider some examples.

EXAMPLE 4 Using (5)

Expand $(a+b)^4$.

Solution We use the Binomial Theorem (5) with coefficients given by (7). With $n = 4$ we obtain:

$$(a+b)^4 = a^4 + \frac{4}{1!}a^{4-1}b + \frac{4\cdot 3}{2!}a^{4-2}b^2 + \frac{4\cdot 3\cdot 2}{3!}a^{4-3}b^3 + \frac{4\cdot 3\cdot 2\cdot 1}{4!}b^4$$

$$= a^4 + 4a^3b + \frac{12}{2}a^2b^2 + \frac{24}{6}ab^3 + \frac{24}{24}b^4$$

$$= a^4 + 4a^3b + 6a^2b^2 + 4ab^3 + b^4. \quad \equiv$$

EXAMPLE 5 Finding the Sixth Term

Find the sixth term in the expansion of $(x^2 - 2y)^7$.

Solution Since (6) gives the $(r+1)$st term in the expansion of $(a+b)^n$, the sixth term in the expansion of $(x^2 - 2y)^7$ corresponds to $r = 5$ (that is, $r + 1 = 5 + 1 = 6$). With the identifications $n = 7, r = 5, a = x^2$, and $b = -2y$, it follows from (6) that the sixth term is

$$\frac{7\cdot 6\cdot 5\cdot 4\cdot 3}{5!}(x^2)^{7-5}(-2y)^5 = 21x^4(-32y^5)$$

$$= -672x^4y^5. \quad \equiv$$

☐ **An Alternative Form** The binomial coefficients can be written in a more compact manner using factorial notation. If r is any integer such that $0 \le r \le n$, then

$$n(n-1)\cdots(n-r+1) = \frac{n(n-1)\cdots(n-r+1)}{1}\cdot\overbrace{\frac{(n-r)(n-r-1)\cdots 3\cdot 2\cdot 1}{(n-r)(n-r-1)\cdots 3\cdot 2\cdot 1}}^{\text{this fraction is 1}}$$

$$= \frac{n(n-1)\cdots(n-r+1)(n-r)(n-r-1)\cdots 3\cdot 2\cdot 1}{(n-r)(n-r-1)\cdots 3\cdot 2\cdot 1}$$

$$= \frac{n!}{(n-r)!}.$$

Thus the binomial coefficients of $a^{n-r}b^r$ for $r = 0, 1, \ldots, n$ given in (7) are the same as $n!/r!(n-r)!$. This latter quotient is usually denoted by the symbol $\binom{n}{r}$. That is, the **binomial coefficients** are

$$\binom{n}{r} = \frac{n!}{r!(n-r)!}. \quad (8)$$

Hence the Binomial Theorem (5) can be written in the alternative form

$$(a + b)^n = \binom{n}{0}a^n + \binom{n}{1}a^{n-1}b + \cdots + \binom{n}{r}a^{n-r}b^r + \cdots + \binom{n}{n}b^n. \qquad (9)$$

It is this form that we will use to prove (5).

☐ **Summation Notation** The Binomial Theorem can be expressed in a compact manner by using summation notation. Using (6) and (8), the sums in (5) and (9) can be written as

$$(a + b)^n = \sum_{k=0}^{n} \frac{n(n-1)\cdots(n-k+1)}{k!} a^{n-k}b^k$$

or

$$(a + b)^n = \sum_{k=0}^{n} \binom{n}{k} a^{n-k}b^k,$$

respectively. From these forms it is apparent that since the index of summation starts at 0 and ends at n, a binomial expansion contains $n + 1$ terms.

The following property of the binomial coefficient $\binom{n}{r}$ will play a pivotal role in the proof of the Binomial Theorem. For any integer r, $0 < r \leq n$, we have

$$\binom{n}{r-1} + \binom{n}{r} = \binom{n+1}{r}. \qquad (10)$$

We leave the verification of (10) as an exercise (see Problem 63 in Exercises 13.5).

☐ **Proof of Theorem 13.5.1** We now prove the Binomial Theorem by mathematical induction. Substituting $n = 1$ into (9) gives a true statement,

$$(a + b)^1 = \binom{1}{0}a^1 + \binom{1}{1}b^1 = a + b,$$

since

$$\binom{1}{0} = \frac{1!}{0!1!} = 1 \quad \text{and} \quad \binom{1}{1} = \frac{1!}{1!0!} = 1.$$

This completes the verification of the first condition of the Principle of Mathematical Induction.

For the second condition, we assume that (9) is true for some positive integer $n = k$:

$$(a + b)^k = \binom{k}{0}a^k + \binom{k}{1}a^{k-1}b + \cdots + \binom{k}{r}a^{k-r}b^r + \cdots + \binom{k}{k}b^k. \qquad (11)$$

Using this assumption we then must show that (9) is also true for $n = k + 1$. To do this we multiply both sides of (11) by $(a + b)$:

$$(a+b)(a+b)^k = (a+b)\left[\binom{k}{0}a^k + \binom{k}{1}a^{k-1}b + \cdots + \binom{k}{r}a^{k-r}b^r + \cdots \binom{k}{k}b^k\right]$$

$$= \binom{k}{0}(a^{k+1} + a^k b) + \binom{k}{1}(a^k b + a^{k-1}b^2) + \cdots + \binom{k}{r}(a^{k-r+1}b^r + a^{k-r}b^{r+1}) + \cdots + \binom{k}{k}(ab^k + b^{k+1}) \qquad (12)$$

$$= \binom{k}{0}a^{k+1} + \left[\binom{k}{0} + \binom{k}{1}\right]a^k b + \left[\binom{k}{1} + \binom{k}{2}\right]a^{k-1}b^2 + \cdots + \left[\binom{k}{r-1} + \binom{k}{r}\right]a^{k-r+1}b^r + \cdots + \binom{k}{k}b^{k+1}.$$

Using (10) to rewrite the coefficient of the $(r + 1)$st term in (12) as

$$\binom{k}{r-1} + \binom{k}{r} = \binom{k+1}{r}$$

and the facts that $(a + b)(a + b)^k = (a + b)^{k+1}$,

$$\binom{k}{0} = 1 = \binom{k+1}{0}, \quad \text{and} \quad \binom{k}{k} = 1 = \binom{k+1}{k+1},$$

the last line in (12) becomes

$$(a + b)^{k+1} = \binom{k+1}{0}a^{k+1} + \binom{k+1}{1}a^k b + \cdots + \binom{k+1}{r}a^{k+1-r}b^r + \cdots + \binom{k+1}{k+1}b^{k+1}.$$

Because this is (9) with n replaced by $k + 1$, the proof is complete by the Principle of Mathematical Induction. ≡

13.5 Exercises Answers to selected odd-numbered problems begin on page ANS-35.

In Problems 1–12, evaluate the given expression.

1. $3!$
2. $5!$
3. $\dfrac{2!}{5!}$
4. $\dfrac{6!}{3!}$
5. $3!4!$
6. $0!5!$
7. $\binom{5}{3}$
8. $\binom{6}{3}$
9. $\binom{7}{6}$
10. $\binom{9}{9}$
11. $\binom{4}{1}$
12. $\binom{4}{0}$

In Problems 13–16, simplify the given expression.

13. $\dfrac{n!}{(n-1)!}$
14. $\dfrac{(n-1)!}{(n-3)!}$
15. $\dfrac{n!(n+1)!}{(n+2)!(n+3)!}$
16. $\dfrac{(2n+1)!}{(2n)!}$

In Problems 17–26, use factorial notation to rewrite the given product.

17. $5 \cdot 4 \cdot 3 \cdot 2 \cdot 1$
18. $7 \cdot 6 \cdot 5 \cdot 4 \cdot 3 \cdot 2 \cdot 1$
19. $100 \cdot 99 \cdot 98 \cdots 3 \cdot 2 \cdot 1$
20. $t(t - 1)(t - 2) \cdots 3 \cdot 2 \cdot 1$
21. $(4 \cdot 3 \cdot 2 \cdot 1)(5 \cdot 4 \cdot 3 \cdot 2 \cdot 1)$
22. $(6 \cdot 5 \cdot 4 \cdot 3 \cdot 2 \cdot 1)/(3 \cdot 2 \cdot 1)$
23. $4 \cdot 3$
24. $10 \cdot 9 \cdot 8$
25. $n(n - 1), n \geq 2$
26. $n(n-1)(n-2) \cdots (n-r+1), n \geq r$

In Problems 27–32, answer true or false.

27. $5! = 5 \cdot 4!$
28. $3! + 3! = 6!$
29. $\dfrac{8!}{4!} = 2!$
30. $\dfrac{8!}{4} = 2$
31. $n!(n + 1) = (n + 1)!$
32. $\dfrac{n!}{n} = (n - 1)!$

In Problems 33–42, use the Binomial Theorem to expand the given expression.

33. $(x^2 - 5y^4)^2$
34. $(x^{-1} + y^{-1})^3$
35. $(x^2 - y^2)^3$
36. $(x^{-2} + 1)^4$
37. $(x^{1/2} + y^{1/2})^4$
38. $(3 - y^2)^4$
39. $(x^2 + y^2)^5$
40. $\left(2x + \dfrac{1}{x}\right)^5$
41. $(a - b - c)^3$
42. $(x + y + z)^4$

43. By referring to Pascal's triangle, determine the coefficients in the expansion of $(a + b)^n$ for $n = 6$ and $n = 7$.
44. If $f(x) = x^n$, where n is a positive integer, use the Binomial Theorem to simplify the difference quotient:
$$\frac{f(x + h) - f(x)}{h}.$$

In Problems 45–54, find the indicated term in the expansion of the given expression.

45. Sixth term of $(a + b)^6$
46. Second term of $(x - y)^5$
47. Fourth term of $(x^2 - y^2)^6$
48. Third term of $(x - 5)^5$
49. Fifth term of $(4 + x)^7$
50. Seventh term of $(a - b)^7$
51. Tenth term of $(x + y)^{14}$
52. Fifth term of $(t + 1)^4$
53. Eighth term of $(2 - y)^9$
54. Ninth term of $(3 - z)^{10}$

55. Find the coefficient of the constant term in $(x + 1/x)^{10}$.
56. Find the first five terms in the expansion of $(x^2 - y)^{11}$.
57. Use the first four terms in the expansion of $(1 - 0.01)^5$ to find an approximation to $(0.99)^5$. Compare with the answer obtained from a calculator.
58. Use the first four terms in the expansion of $(1 + 0.01)^{10}$ to find an approximation to $(1.01)^{10}$. Compare with the answer obtained from a calculator.

For Discussion

59. Without adding the terms, determine the value of $\sum_{k=0}^{4}\binom{4}{k}4^k$.
60. If $\sum_{k=0}^{5}\binom{5}{k}x^{5-k} = 0$, what is x?
61. Use the Binomial Theorem to show that
$$\sum_{k=0}^{n}(-1)^k\binom{n}{k} = 0.$$

62. Use the Binomial Theorem to show that
$$\binom{n}{0} + \binom{n}{1} + \cdots + \binom{n}{n} = 2^n.$$

63. Prove that
$$\binom{n}{r-1} + \binom{n}{r} = \binom{n+1}{r}, \qquad 0 < r \leq n.$$

64. Prove that
$$\binom{n}{r+1} = \frac{n-r}{r+1}\binom{n}{r}, \qquad 0 \leq r < n.$$

13.6 Principles of Counting

Introduction A wide variety of practical problems involve counting the number of ways in which something can occur. For example, the telephone prefix at a certain university is 642. If the prefix is followed by four digits, how many telephone numbers are possible before a second prefix is needed? We will be able to solve this problem (see Example 2) and others using the counting techniques discussed in this section.

Tree Diagrams We begin by considering a more abstract problem. How many different arrangements can be made of the three letters a, b, and c using two letters at a time? One way to solve this problem is to list all the possible arrangements. As shown in FIGURE 13.6.1, a **tree diagram** can be used to illustrate all the possibilities. From the point labeled "Start," line segments lead to each of the three possible choices for a first letter. From each of these, a line segment leads to each of the possible choices for a second letter. Each possible arrangement corresponds to a path, or **branch** of the tree, beginning at the "Start" and traveling to the right through the tree. We see that there are 6 different arrangements of the three letters:

$$ab, \quad ac, \quad ba, \quad bc, \quad ca, \quad cb.$$

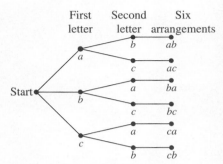

FIGURE 13.6.1 Tree diagram for number of arrangements of a, b, c taken two at a time

Another way to solve the foregoing problem is to recognize that each arrangement consists of a selection of letters to fill the two blank positions indicated by the red lines:

$$\underset{\text{first letter}}{\underline{}} \; \underset{\text{second letter}}{\underline{}}.$$

Any one of the *three* letters a, b, or c can be chosen for the first position. Once this choice is made, any one of the *two* remaining letters can be chosen for the second position. Since each of the three letters for the first position can be associated with either of the remaining two letters, the total number of arrangements is given by the *product*

$$\underset{\text{first letter}}{\underline{3}} \cdot \underset{\text{second letter}}{\underline{2}} = 6.$$

This simple example illustrates the **Fundamental Counting Principle**.

THEOREM 13.6.1 Fundamental Counting Principle

If one event can occur in m different ways and, after it has happened, a second event can occur in n different ways, then the total number of ways in which both events can take place is the product mn.

This Fundamental Counting Principle can be extended to three or more events in an obvious way:

Simply multiply the number of ways each event can occur.

EXAMPLE 1 Number of Outfits

A college student has 5 shirts, 3 pairs of slacks, and 2 pairs of shoes. How many different outfits can he wear consisting of a shirt, a pair of slacks, and a pair of shoes?

Solution Three selections or events are to occur, with 5 choices for the first event (choosing a shirt), 3 choices for the second event (choosing a pair of slacks), and 2 choices for the third event (choosing a pair of shoes). By the Fundamental Counting Principle, the number of different outfits is the product $5 \cdot 3 \cdot 2 = 30$.

We now return to the problem given in the introduction.

EXAMPLE 2   Telephone Numbers

The telephone prefix at a certain university is 642. If the prefix is followed by four digits, how many different telephone numbers are possible before a second prefix is needed?

Solution Four events are to occur: selecting the first digit after the prefix, selecting the second digit after the prefix, and so on. Since repeated digits are allowed in telephone numbers, any one of the 10 digits 0, 1, 2, 3, 4, 5, 6, 7, 8, 9 can be selected for each position. Hence there are $10 \cdot 10 \cdot 10 \cdot 10 = 10{,}000$ possible different phone numbers with the single prefix 642. ≡

EXAMPLE 3   Arrangements of Letters

How many different ways are there to arrange the letters in the word RANDOM?

Solution Since RANDOM has 6 distinct letters, there are 6 events: choosing the first letter, choosing the second letter, and so on. Any one of the 6 letters can be chosen for the first position, then any of the *remaining* 5 letters can be chosen for the second position, then any of the *remaining* 4 letters can be chosen for the third position, and so on. The total number of arrangements is $6 \cdot 5 \cdot 4 \cdot 3 \cdot 2 \cdot 1 = 720$. ≡

☐ **Permutations** A **permutation** is an arrangement that is made by using some or all of the elements of a set *without repetition*. This means that no element of the set appears more than once in the arrangement. For example, 312 is a permutation of the digits in the set $\{1, 2, 3\}$, but 112 is not. In Example 3, each of the rearrangements of the six letters in the word RANDOM (for instance, MODRAN) is a permutation. More generally, we have the following definition.

DEFINITION 13.6.1   Permutation

An ordered arrangement of r elements selected from a set of n distinct elements is called a **permutation** of n elements taken r at a time ($n \geq r$).

☐ **Notation** We will use the symbol $P(n, r)$ to denote the number of permutations of n distinct objects taken r at a time. Using the notation $P(n, r)$, we write the number of permutations of 5 objects taken 3 at a time as $P(5, 3)$.

It is possible to find an explicit formula for $P(n, r)$, that is, the number of permutations of n distinct objects taken r at a time for $0 \leq r \leq n$. For $r \geq 1$, we can think of the process of forming a permutation of n objects taken r at a time as r events: choose the first object, choose the second object, and so on. When we make the first choice, there are n objects available; when we make the second choice, there are $n - 1$ objects; for the third choice, there are $n - 2$ objects; and so on. When we choose the rth object, there are $n - (r - 1)$ objects to choose from. Thus from Theorem 13.6.1,

$$P(n, r) = \overbrace{n(n-1)(n-2)\cdots(n-(r-1))}^{r \text{ factors}}$$

or $$P(n, r) = n(n-1)(n-2)\cdots(n-r+1). \qquad (1)$$

An alternative expression for $P(n, r)$ involving factorial notation can be found by multiplying the right-hand side of (1) by

$$\frac{(n-r)!}{(n-r)!} = 1.$$

The result is

$$P(n, r) = \frac{n(n-1)(n-2)\cdots(n-r+1)\overbrace{(n-r)(n-r-1)\cdots 2\cdot 1}^{(n-r)!}}{(n-r)!},$$

or
$$P(n, r) = \frac{n!}{(n-r)!}. \qquad (2)$$

When $r = n$, formula (2) reduces to

$$P(n, n) = \frac{n!}{0!} = n!,$$

because 0! is defined to be 1. This result is the same as that obtained by using the counting principle in Theorem 13.6.1:

$$P(n, n) = n(n-1)(n-2)\cdots 2\cdot 1 = n!, \qquad (3)$$

since any one of the n objects can be chosen first, any one of the remaining objects can be chosen second, and so on. In Example 3, the number of 6-letter arrangements of the words RANDOM is the number of permutations of the 6 letters taken 6 at a time, that is, $P(6, 6) = 6! = 720$.

If $r = 0$, we define $P(n, 0) = 1$, which is consistent with (2).

EXAMPLE 4   Using (2) and (3)

Evaluate **(a)** $P(5, 3)$, **(b)** $P(5, 1)$, and **(c)** $P(5, 5)$.

Solution In (a) and (b) we use formula (2):

(a) $P(5, 3) = \dfrac{5!}{(5-3)!} = \dfrac{5!}{2!} = \dfrac{5\cdot 4\cdot 3\cdot \overbrace{2\cdot 1}^{2!}}{2!} = 60,$

(b) $P(5, 1) = \dfrac{5!}{(5-1)!} = \dfrac{5!}{4!} = \dfrac{5\cdot \overbrace{4\cdot 3\cdot 2\cdot 1}^{4!}}{4!} = 5.$

(c) From formula (3) we find that

$$P(5, 5) = 5! = 5\cdot 4\cdot 3\cdot 2\cdot 1 = 120.$$

EXAMPLE 5   Awarding Medals

At a track meet 6 athletes are entered in the 100 m dash. In how many ways can gold, silver, and bronze metals be awarded?

Solution We wish to count the number of ways of arranging 3 of the 6 athletes in the winning positions. The solution is given by the number of permutations of 6 things (athletes) taken 3 at a time:

$$P(6, 3) = \frac{6!}{(6-3)!} = \frac{6!}{3!} = 120.$$

This problem can also be solved using the Fundamental Counting Principle. Since there are 3 choices to be made, with 6 athletes available for the gold medal, 5 for the silver, and 4 for the bronze, we find $6\cdot 5\cdot 4 = 120$.

Ten books on a shelf

EXAMPLE 6   Arrangements of Books

How many arrangements are possible for 10 different books on a bookshelf?

Solution We wish to find the number of permutations of 10 objects taken 10 at a time, or $P(10, 10) = 10! = 3,628,800$. ≡

☐ **Combinations** In the preceding discussion we were concerned with the number of ways of arranging or choosing r elements from a set of n elements, where the order in which they were arranged or chosen was considered. However, in certain applications the order of the elements is not important. For example, if a committee of two is to be chosen from the four students Angie, Brandon, Cecilia, and David, the committee formed by choosing Angie and Brandon is the same as the committee formed by choosing Brandon and Angie. A selection of objects in which the order does not make any difference is called a **combination**.

DEFINITION 13.6.2   Combination

A subset of r elements of a set of n distinct elements is called a **combination** of n elements taken r at a time ($n \geq r$).

☐ **Notation** We use the symbol $C(n, r)$ to denote the number of combinations of n distinct objects taken r at a time. By using (2) it is possible to derive a formula for $C(n, r)$. At the beginning of this section we saw that there are 6 arrangements (permutations) of the 3 letters a, b, and c taken 2 at a time:

$$ab \quad ac \quad ba \quad bc \quad ca \quad cb. \tag{4}$$

(with ab/ba, ac/ca, bc/cb marked as the same combination)

In (4) we see that if we disregard the order in which the letters are listed, then there are only 3 combinations of the letters: ab, ac, and bc. Thus, $C(3, 2) = 3$. We see that each of these combinations can be arranged in $2!$ ways to yield the list of permutations in (4). By the Fundamental Counting Principle,

$$P(3, 2) = 6 = 2!C(3, 2).$$

In general, for $0 < r \leq n$, each of the $C(n, r)$ combinations can be rearranged in $r!$ different ways, so that

$$P(n, r) = r!\, C(n, r),$$

or

$$C(n, r) = \frac{P(n, r)}{r!} = \frac{\dfrac{n!}{(n-r)!}}{r!}.$$

Thus,

$$C(n, r) = \frac{n!}{(n-r)!\, r!}. \tag{5}$$

For $r = 0$, we define $C(n, 0) = 1$, which is consistent with formula (5).

Note that $C(n, r)$ is identical to the binomial coefficient $\binom{n}{r}$ in the expansion of $(a + b)^n$, where n is a nonnegative integer. See (7) and (8) of Section 13.5.

EXAMPLE 7 Using Formula (5)

Evaluate **(a)** $C(5, 3)$, **(b)** $C(5, 1)$, and **(c)** $C(5, 5)$.

Solution Using formula (5), we have the following:

(a) $C(5, 3) = \dfrac{5!}{(5-3)!\,3!} = \dfrac{5!}{2!\,3!} = \dfrac{5 \cdot 4 \cdot \overset{3 \cdot 2 \cdot 1}{3!}}{2!\,3!} = 10,$

(b) $C(5, 1) = \dfrac{5!}{(5-1)!\,1!} = \dfrac{5!}{4!\,1!} = \dfrac{5 \cdot \overset{4 \cdot 3 \cdot 2 \cdot 1}{4!}}{4!\,1!} = 5,$

(c) $C(5, 5) = \dfrac{5!}{(5-5)!\,5!} = \dfrac{5!}{0!\,5!} = \dfrac{1}{\underset{1}{0!}} = 1.$

EXAMPLE 8 Number of Card Hands

How many different 5-card hands can be dealt from a deck of 52 cards?

Solution Since a hand is the same regardless of the order of the cards, we are talking about the number of combinations of 52 cards taken 5 at a time. Using (5), the solution is

$$C(52, 5) = \dfrac{52!}{47!\,5!} = \dfrac{52 \cdot 51 \cdot 50 \cdot 49 \cdot 48 \cdot 47!}{47!\,5!}$$

$$= \dfrac{52 \cdot 51 \cdot 50 \cdot 49 \cdot 48}{5!} = 2{,}598{,}960.$$

Note that we cancelled the larger of the two factorials 47! and 5! to simplify the calculations of $C(52, 5)$.

EXAMPLE 9 Organizing a Club

A card club has 8 members.
(a) In how many ways can 3 members be chosen to be president, secretary, and treasurer?
(b) In how many ways can a committee of 3 members be chosen?

Solution In choosing officers, order *does* matter, whereas in choosing a committee, the order of the selection does not affect the resulting committee. Thus in (a) we are counting permutations and in (b) we are counting combinations. We find

(a) $P(8, 3) = \dfrac{8!}{5!} = 336,$

(b) $C(8, 3) = \dfrac{8!}{5!\,3!} = 56.$

In deciding whether to use the formula for $P(n, r)$ or $C(n, r)$, consider the following two informal rules. ◀ **Note of Caution**

- Permutations are involved if you are considering arrangements in which *different orderings* of the same objects *are to be counted*.
- Combinations are involved if you are considering ways of choosing objects in which the *order* of the chosen objects *makes no difference*.

EXAMPLE 10 Choosing Reporters

A college newspaper staff has 6 junior reporters and 8 senior reporters. In how many ways can 2 junior and 3 senior reporters be chosen for a special assignment?

Solution Two events are to occur: the selection of 2 junior reporters and the selection of 3 senior reporters. Because the order in which the 2 junior reporters are chosen makes no difference, we count combinations. Therefore, the number of ways of choosing 2 junior reporters is

$$C(6, 2) = \frac{6!}{4!2!} = 15.$$

Likewise in selecting the 3 senior reporters order does not matter, so we again count combinations:

$$C(8, 3) = \frac{8!}{5!3!} = 56.$$

Thus we choose the junior reporters in 15 ways and, for each of these selections, there are 56 ways of selecting the senior reporters. Applying the Fundamental Counting Principle gives

$$C(6, 2) \cdot C(8, 3) = 15 \cdot 56 = 840$$

ways to make the choices for the special assignment.

EXAMPLE 11 Selecting a Display

A cheese store has 10 varieties of domestic cheese and 8 varieties of imported cheese. In how many ways can a selection of 6 cheeses, consisting of 2 domestic and 4 imported varieties, be placed on a display shelf?

Solution The domestic varieties can be chosen in $C(10, 2)$ ways and the imported varieties in $C(8, 4)$ ways. Thus by the Fundamental Counting Principle the 6 cheeses can be selected in $C(10, 2) \cdot C(8, 4)$ ways. Up to this point in the solution, order has not been important in making the *selection* of the cheeses. Now we observe that *each* selection of 6 cheeses can be placed or *arranged* on the shelf in $P(6, 6)$ ways. Thus the total number of ways the cheese can be displayed is

$$C(10, 2) \cdot C(8, 4) \cdot P(6, 6) = \frac{10!}{8!\,2!} \cdot \frac{8!}{4!\,4!} \cdot \frac{6!}{(6-6)!}$$
$$= 2{,}268{,}000.$$

13.6 Exercises
Answers to selected odd-numbered problems begin on page ANS-35.

In Problems 1–4, use a tree diagram.

1. List all possible arrangements of the letters a, b, and c.
2. If a coin is tossed 4 times, list all possible sequences of heads (H) and tails (T).
3. If a red die and a black die are rolled, list all possible results.
4. If a coin is tossed and then a die is rolled, list all possible results.

In Problems 5–8, use the Fundamental Counting Principle.

5. **Number of Meals** A cafeteria offers 8 salads, 6 entrees, 4 vegetables, and 3 desserts. How many different meals are possible if one item is selected from each category?
6. **Number of Systems** How many different stereo systems consisting of speakers, receiver, and CD player can be purchased if a store carries 6 models of speakers, 4 of receivers, and 2 of compact disc players?
7. **Number of Prefixes** How many different 3-digit telephone prefixes are possible if neither 0 nor 1 can occupy the first position?
8. **Number of License Plates** If a license plate consists of 3 letters followed by 3 digits, how many license plates are possible if the first letter cannot be O or I?

In Problems 9–16, evaluate $P(n, r)$.

9. $P(6, 3)$
10. $P(6, 4)$
11. $P(6, 1)$
12. $P(4, 0)$
13. $P(100, 2)$
14. $P(4, 4)$
15. $P(8, 6)$
16. $P(7, 6)$

In Problems 17–24, evaluate $C(n, r)$.

17. $C(4, 2)$
18. $C(4, 1)$
19. $C(50, 2)$
20. $C(2, 2)$
21. $C(13, 11)$
22. $C(8, 2)$
23. $C(2, 0)$
24. $C(7, 4)$

Miscellaneous Applications

In Problems 25–28, use permutations to solve the given problem.

25. **Family Portrait** In how many ways can a family of four line up in a row to have their family portrait taken?
26. **Volunteer Work** As part of a fund-raising drive, a volunteer is given 5 names to contact. In how many different orders can the volunteer complete the task?
27. **Scrabble** A *Scrabble* game player has the following 7 letters: A, T, E, L, M, Q, F.
    (a) How many different 7-letter "words" can be considered?
    (b) How many different 5-letter "words"?
28. **Politics** From a class of 24, elections are held for president, vice president, secretary, and treasurer. In how many ways can the offices be filled?

A family of four

In Problems 29–32, use combinations to solve the given problem.

29. **Good Luck!** A student must answer any 10 questions on a 12-question exam. In how many different ways can the student select the questions?
30. **Chem Lab** For a chemistry lab class, a student must correctly identify 3 "unknown" samples. In how many ways can the 3 samples be chosen from 10 chemicals?
31. **Volunteers** In how many ways can 5 subjects be chosen from a group of 10 volunteers for a psychology experiment?
32. **Potpourri** In how many ways can 4 herbs be chosen from 8 available herbs to make a potpourri?

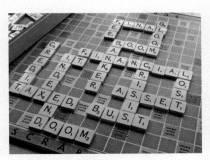

The game *Scrabble*®

In Problems 33–44, use one or more of the techniques discussed in this section to solve the given counting problem.

33. **Spelling Bee** If 10 students enter a spelling bee, in how many different ways can first- and second-place awards be made?
34. **Show Business** A theater company has a repertoire consisting of 8 dramatic skits, 6 comedies, and 4 musical numbers. In how many ways can a program be selected consisting of a dramatic skit followed by either a comedy or a musical number?
35. **Take Your Pick** A pediatrician allows a well-behaved child to select any 2 of 5 small plastic toys to take home. How many different selections of toys are possible?
36. **Tournament Rankings** If 8 teams enter a soccer tournament, in how many different ways can first, second, and third place be decided, assuming ties are not allowed?
37. **Another Jackson Pollock** If 8 colors are available to make an abstract spatter-paint picture, how many different color combinations are possible if only 3 colors are chosen?
38. **Seating Arrangements** Three couples have reserved seats in a row at the theater. In how many different ways can they be seated
    (a) if there are no restrictions?
    (b) if each couple wishes to sit together?
    (c) if the 3 women and 3 men wish to sit together in 2 groups?
39. **Mastermind** In a popular board game that originated in England called *Mastermind*, one player creates a secret "code" by filling 4 slots with any one of 6 colors. How many codes are possible
    (a) if repetitions are not allowed?
    (b) if repetitions are allowed?
    (c) if repetitions and blank slots are allowed?
40. **Super Mastermind** Some advertisements for the game *Super Mastermind* (a more difficult version of the *Mastermind* game described in Problem 39) claim that up to 59,000 codes are possible. If *Super Mastermind* involves filling 5 slots with any one of 8 colors and if blanks and repetitions are allowed, is the claim correct?
41. **Playing with Letters** From 5 different consonants and 3 different vowels, how many 5-letter "words" can be made consisting of 3 different consonants and 2 different vowels?
42. **Defective Lights** A box contains 24 Christmas tree bulbs, 4 of which are defective. In how many ways can 4 bulbs be chosen so that
    (a) all 4 are defective?
    (b) all 4 are good?
    (c) 2 are good and 2 are defective?
    (d) 3 are good and 1 is defective?
43. **More Playing with Letters** How many 3-letter "words" can be made from 4 different consonants and 2 different vowels
    (a) if the middle letter must be a vowel?
    (b) if the first letter cannot be a vowel? Assume that repeated letters are not allowed.
44. **Store Display** A wine store has 12 different California wines and 8 different French wines. In how many ways can a group of 6 wines consisting of 4 California and 2 French wines
    (a) be selected for display?
    (b) be placed in a row on a display shelf?

The game *Mastermind*®

13.7 Introduction to Probability

Introduction As we mentioned in the chapter introduction, the development of the mathematical theory of **probability** was initially motivated by questions arising in the seventeenth century about games of chance. Today, applications of probability are found in medicine, sports, law, business, and many other areas. In this section we present a brief introduction to this fascinating subject.

☐ **Terminology** Consider an experiment that has a finite number of possible results or **outcomes**. The set S of all possible outcomes of a particular experiment is called the **sample space** of the experiment. For our purposes we will assume that each outcome is *equally likely* to occur. Thus, if the experiment consists of tossing, or flipping, a fair coin, there are two possible equally likely outcomes: obtaining a head or obtaining a tail. If the outcome of obtaining a head is denoted by H and the outcome of obtaining a tail is denoted by T, then the sample space of the experiment can be written in set notation as

$$S = \{H, T\}. \tag{1}$$

Any subset E of a sample space S is called an **event**. Generally, an event E is one or more outcomes of an experiment. For example,

$$E = \{H\} \tag{2}$$

is the event of obtaining a head when a coin is tossed.

There are two outcomes in tossing a coin

EXAMPLE 1 Sample Space and Two Events

On a single roll of a fair die, there is an equal chance of obtaining a 1, 2, 3, 4, 5, or 6. Thus the sample space of the experiment of rolling a fair die is the set

$$S = \{1, 2, 3, 4, 5, 6\}. \tag{3}$$

(a) The event E_1 of obtaining a 4 on a roll of the die is the subset $E_1 = \{4\}$ of S.
(b) The event E_2, consisting of obtaining an odd number on a roll of the die, is the subset $E_2 = \{1, 3, 5\}$ of S.

A die showing 4 is one of six possible outcomes

≡

We shall use the notation $n(S)$ to denote the number of outcomes in a sample space S and $n(E)$ to denote the number of outcomes associated with an event E. Thus, in Example 1 we have $n(S) = 6$; in parts (a) and (b) of the example we have $n(E_1) = 1$ and $n(E_2) = 3$, respectively.

The definition of the probability $P(E)$ of an event E is expressed in terms of $n(S)$ and $n(E)$.

DEFINITION 13.7.1 Probability of an Event

Let S be the sample space of an experiment and let E be an event. If each outcome of the experiment is equally likely, then the **probability** of the event E is given by

$$P(E) = \frac{n(E)}{n(S)}, \tag{4}$$

where $n(E)$ and $n(S)$ denote the number of outcomes in the sets E and S, respectively.

EXAMPLE 2 The Probability of Tossing a Head

Find the probability of obtaining a head if a coin is tossed.

Solution From (1) and (2), $E = \{H\}$, $S = \{H, T\}$, and so $n(E) = 1$ and $n(S) = 2$. From (4) of Definition 13.7.1 the probability of obtaining a head is

$$P(E) = \frac{n(E)}{n(S)} = \frac{1}{2}.$$

≡

EXAMPLE 3 Three Probabilities

On a single roll of a fair die, find the probability
(a) of obtaining a 4, (b) of obtaining an odd number, (c) of obtaining a number that is not a 4.

Solution Let the symbols E_1, E_2, and E_3 denote, respectively, the events in parts (a), (b), and (c) of this example. Also, in each part we have $S = \{1, 2, 3, 4, 5, 6\}$.
(a) From part (a) of Example 1, $E_1 = \{4\}$, and so $n(E_1) = 1$ and $n(S) = 6$. From (4), the probability of obtaining a 4 when a die is rolled is then

$$P(E_1) = \frac{n(E_1)}{n(S)} = \frac{1}{6}.$$

(b) From part (b) of Example 1, $E_2 = \{1, 3, 5\}$, so $n(E_2) = 3$ and $n(S) = 6$. Again from (4), the probability of rolling an odd number is

$$P(E_2) = \frac{n(E_2)}{n(S)} = \frac{3}{6} = \frac{1}{2}.$$

(c) The event of obtaining a number that is not a 4 on a roll of a die is the subset $E_3 = \{1, 2, 3, 5, 6\}$ of S. Using $n(E_3) = 5$ and $n(S) = 6$ the probability of obtaining a number that is not a 4 is

$$P(E_3) = \frac{n(E_3)}{n(S)} = \frac{5}{6}.$$

≡

EXAMPLE 4 Probability of 7

Find the probability of obtaining a total of 7 when two dice are rolled.

Solution Since there are 6 numbers on each die, we conclude from the Fundamental Counting Principle of Section 13.6 that there are $6 \cdot 6 = 36$ possible outcomes in the sample space S; that is, $n(S) = 36$. In the accompanying table, we have listed the possible ways of obtaining a total of 7.

One of six possible ways of throwing a 7 on a pair of dice

E = two dice showing a total of 7						
First die	1	2	3	4	5	6
Second die	6	5	4	3	2	1

From the table we see that $n(E) = 6$. Hence from (4) the probability of throwing a 7 with two dice is

$$P(E) = \frac{n(E)}{n(S)} = \frac{6}{36} = \frac{1}{6}.$$

≡

EXAMPLE 5   Using Combinations

A bag contains 5 white marbles and 3 red marbles. A person reaches into the bag and randomly withdraws 3 marbles. What is the probability that all the marbles will be white?

Solution The sample space S of the experiment is the set of all possible combinations of 3 marbles drawn from the 8 marbles in the bag. The number of ways of choosing 3 marbles from a bag of 8 marbles is the number of *combinations* of 8 objects taken 3 at a time; that is, $n(S) = C(8, 3)$. Similarly, the number of ways of choosing 3 white marbles from 5 white marbles is the number of combinations $n(E) = C(5, 3)$. Since the event E is "all marbles are white," we have

$$P(E) = \frac{n(E)}{n(S)} = \frac{C(5,3)}{C(8,3)} = \frac{\frac{5!}{3!2!}}{\frac{8!}{3!5!}} = \frac{5}{28}.$$

☐ **Bounds on the Probability of an Event** Since any event E is a subset of a sample space S, it follows that $0 \leq n(E) \leq n(S)$. By dividing the last inequality by $n(S)$ we see that

$$0 \leq \frac{n(E)}{n(S)} \leq \frac{n(S)}{n(S)}$$

or
$$0 \leq P(E) \leq 1.$$

If $E = S$, then $n(E) = n(S)$ and $P(E) = n(S)/n(S) = 1$; whereas if E has no elements, we take $E = \emptyset$, $n(\emptyset) = 0$, and $P(E) = n(\emptyset)/n(S) = 0/n(S) = 0$. If $P(E) = 1$, then E always happens and E is called a **certain event**. On the other hand, if $P(E) = 0$, then E is an **impossible event**, that is, E never happens.

EXAMPLE 6   Rolling a Die

Suppose a fair die is rolled once.
(a) What is the probability of obtaining a 7?
(b) What is the probability of obtaining a number less than 7?

Solution (a) Because the number 7 is not in the set S of all possible outcomes (3) the event E of "obtaining a 7" is an impossible event; that is, $E = \emptyset$, $n(\emptyset) = 0$. Therefore,

$$P(E) = \frac{n(\emptyset)}{n(S)} = \frac{0}{6} = 0.$$

(b) Because the outcomes of rolling a fair die are all positive integers less than 7 we have $E = \{1, 2, 3, 4, 5, 6\} = S$. Thus E is a certain event and

$$P(E) = \frac{n(E)}{n(S)} = \frac{6}{6} = 1.$$

☐ **Complement of an Event** The set of all outcomes in the sample space S that do not belong to an event E is called the **complement of E** and is denoted by the symbol E'. For example, in rolling a die, if E is the event of "obtaining a 4," then E' is the event of "obtaining any number *except* 4." Because events are sets, we can describe the relationship between an event E and its complement E' using the operations of union and intersection:

$$E \cup E' = S \quad \text{and} \quad E \cap E' = \emptyset.$$

In view of the foregoing properties we can write $n(E) + n(E') = n(S)$. Dividing both sides of the last equality by $n(S)$ we see that the probabilities of E and E' are related by

$$\frac{n(E)}{n(S)} + \frac{n(E')}{n(S)} = \frac{n(S)}{n(S)}$$

or
$$P(E) + P(E') = 1. \tag{5}$$

For instance, the complement of the event $E_1 = \{4\}$ in part (a) of Example 3 is the set $E_1' = E_3 = \{1, 2, 3, 5, 6\}$ in part (c). Observe in accordance with (5), we have $P(E_1) + P(E_3) = P(E_1) + P(E_1') = \frac{1}{6} + \frac{5}{6} = 1$.

The relationship (5) is useful in either of the two forms:

$$P(E) = 1 - P(E') \quad \text{or} \quad P(E') = 1 - P(E). \tag{6}$$

The second of the two formulas in (6) allows us to find the probability of an event if we know the probability of its complement. Sometimes it is easier to calculate $P(E')$ than it is to calculate $P(E)$. Also, it is interesting to note that the equation $P(E) + P(E') = 1$ can be interpreted as saying that *something* must happen.

EXAMPLE 7   Probability of an Ace

If 5 cards are drawn from a well-shuffled 52-card deck without replacement, find the probability of obtaining at least one ace.

Solution We let E be the event of obtaining at least one ace. Since E consists of all 5-card hands that contain 1, 2, 3, or 4 aces, it is actually easier to consider E'; that is, all 5-card hands that contain no aces. The sample space S consists of all possible 5-card hands. From Section 13.6 we have that $n(S) = C(52, 5)$. Since 48 of the 52 cards are *not* aces we find $n(E') = C(48, 5)$. By (4) the probability of drawing 5 cards where none of the cards are aces is given by

$$P(E') = \frac{C(48, 5)}{C(52, 5)} = \frac{1{,}712{,}304}{2{,}598{,}960}.$$

From the first formula in (5) the probability of drawing 5 cards where at least one of them is an ace is

$$P(E) = 1 - P(E') = 1 - \frac{1{,}712{,}304}{2{,}598{,}960} \approx 0.3412. \qquad \equiv$$

Up to this point we have considered the probability of a single event. In the discussion that follows, we examine the probability of two or more events.

☐ **Union of Two Events** Two events E_1 and E_2 are said to be **mutually exclusive** if they have no outcomes, or elements, in common. In other words the events E_1 and E_2 cannot occur at the same time. In terms of sets, E_1 and E_2 are **disjoint** sets; that is, $E_1 \cap E_2 = \emptyset$. Recall, the set $E_1 \cup E_2$ consists of the elements that are in E_1 or in E_2. In this case of mutually exclusive events the number of outcomes in the set $E_1 \cup E_2$ is given by

▶ Review the notions of the union and intersection of two sets in Section R.1

$$n(E_1 \cup E_2) = n(E_1) + n(E_2). \tag{7}$$

By dividing (7) by $n(S)$ we obtain

$$\frac{n(E_1 \cup E_2)}{n(S)} = \frac{n(E_1)}{n(S)} + \frac{n(E_2)}{n(S)}.$$

In view of (4), the foregoing expression is the same as

$$P(E_1 \cup E_2) = P(E_1) + P(E_2). \tag{8}$$

In the next example we return to the results in Example 3.

EXAMPLE 8 Mutually Exclusive Events

On a single roll of a fair die, find the probability of obtaining a 4 or an odd number.

Solution From Example 3 the two events are $E_1 = \{4\}$, $E_2 = \{1, 3, 5\}$, and the sample space is again $S = \{1, 2, 3, 4, 5, 6\}$. The events of rolling a 4 and rolling an odd number are mutually exclusive: $E_1 \cap E_2 = \{4\} \cap \{1, 3, 5\} = \varnothing$. Thus by (8) the probability $P(E_1 \text{ or } E_2)$ of rolling a 4 or an odd number is given by

$$P(E_1 \cup E_2) = P(E_1) + P(E_2) = \frac{1}{6} + \frac{3}{6} = \frac{4}{6} = \frac{2}{3}.$$

Alternative Solution From $E_1 \cup E_2 = \{1, 3, 4, 5\}$, $n(E_1 \cup E_2) = 4$, and so (4) of Definition 13.7.1 yields

$$P(E_1 \cup E_2) = \frac{n(E_1 \cup E_2)}{n(S)} = \frac{4}{6} = \frac{2}{3}.$$

≡

The additive property in (8) extends to the probability of three or more mutually exclusive events. See Problems 31 and 32 in Exercises 13.7.

☐ **Addition Rule** Formula (8) is just a special case of a more general rule. In (8) there were no outcomes in common in the events E_1 and E_2. Of course, this need not be the case. For example, in the experiment of rolling a single fair die, the events $E_1 = \{1\}$ and $E_2 = \{1, 3, 5\}$ are not mutually exclusive because the number 1 is an element in both sets. When two sets E_1 and E_2 have a nonempty intersection, the number of outcomes in $n(E_1 \cup E_2)$ is not given by (7) but rather by the formula

$$n(E_1 \cup E_2) = n(E_1) + n(E_2) - n(E_1 \cap E_2). \tag{9}$$

◀ See Problem 76 in Exercises R.1.

Dividing (9) by $n(S)$ yields

$$\frac{n(E_1 \cup E_2)}{n(S)} = \frac{n(E_1)}{n(S)} + \frac{n(E_2)}{n(S)} - \frac{n(E_1 \cap E_2)}{n(S)}$$

or

$$P(E_1 \cup E_2) = P(E_1) + P(E_2) - P(E_1 \cap E_2). \tag{10}$$

The result in (10) is called the **addition rule** of probability.

EXAMPLE 9 Probability of a Union of Two Events

On a single roll of a fair die, find the probability of obtaining a 1 or an odd number.

Solution The sets are $E_1 = \{1\}$, $E_2 = \{1, 3, 5\}$, and $S = \{1, 2, 3, 4, 5, 6\}$. Now $\{1\} \cap \{1, 3, 5\} = \{1\}$ so that $n(E_1 \cap E_2) = 1$. Thus by (10) the probability of rolling a 1 or an odd number is given by

$$P(E_1 \cup E_2) = P(E_1) + P(E_2) - P(E_1 \cap E_2) = \frac{1}{6} + \frac{3}{6} - \frac{1}{6} = \frac{3}{6} = \frac{1}{2}.$$

Alternative Solution Since E_1 is a subset of E_2, $E_1 \cup E_2 = E_2 = \{1, 3, 5\}$, and $n(E_1 \cup E_2) = 3$. From (4) of Definition 13.7.1,

$$P(E_1 \cup E_2) = \frac{n(E_1 \cup E_2)}{n(S)} = \frac{3}{6} = \frac{1}{2}.$$

≡

Note ▶ It might help if you think of the symbols $P(E_1 \cup E_2)$ and $P(E_1 \cap E_2)$ in (10) as $P(E_1 \text{ or } E_2)$ and $P(E_1 \text{ and } E_2)$, respectively.

EXAMPLE 10   Probability of a Union of Two Events

A single card is drawn from a well-shuffled standard deck. Find the probability of obtaining either an ace or a heart.

Solution As shown in the photo to the left, a standard deck contains 52 cards divided into 4 suits with 13 cards in each suit. Thus the sample space S of this experiment consists of the 52 cards. The event E_1 of drawing an ace consists of the 4 aces and so the probability of drawing an ace is $P(E_1) = \frac{4}{52}$. The event E_2 of drawing a card that is a heart consists of the 13 hearts in that suit and so the probability of drawing a heart is $P(E_2) = \frac{13}{52}$. Since one of the hearts is an ace, $n(E_1 \cap E_2) = 1$, and so $P(E_1 \cap E_2) = \frac{1}{52}$. Therefore, from (10)

$$P(\overbrace{E_1 \cup E_2}^{\text{ace or a heart}}) = P(\overbrace{E_1}^{\text{ace}}) + P(\overbrace{E_2}^{\text{heart}}) - P(\overbrace{E_1 \cap E_2}^{\text{ace and a heart}})$$

$$= \frac{4}{52} + \frac{13}{52} - \frac{1}{52} = \frac{16}{52} = \frac{4}{13}.$$

≡

52-card deck in Example 10

13.7 Exercises   Answers to selected odd-numbered problems begin on page ANS-35.

In Problems 1–4, use set notation to write the sample space S of the given experiment.

1. Two coins are tossed.
2. Three coins are tossed.
3. A die is rolled and then a coin is tossed.
4. Two dice are rolled.

In Problems 5–12, find the probability of the given event.

5. Drawing a face card (jack, queen, or king) from a deck of 52 cards
6. Drawing a heart from a deck of 52 cards
7. Rolling a 2 with a single die
8. Rolling a number less than 3 with a single die
9. Rolling snake eyes (a total of 2) with two dice
10. Rolling a total of 7 or 11 with two dice
11. Obtaining all heads when 3 coins are tossed
12. Obtaining exactly 1 head when 3 coins are tossed

In Problems 13–16, find the probability of obtaining the indicated hand by drawing 5 cards without replacement from a well-shuffled 52-card deck.

13. Four of a kind (such as 4 aces)
14. A straight (5 cards in sequence, such as 4, 5, 6, 7, 8, where an ace can count as a 1 or an ace)
15. A flush (5 cards, all of the same suit)
16. A royal flush (10, jack, queen, king, and ace, all of the same suit)

In Problems 17–20, use the first formula in (5) to find the probability of the given event.

17. Obtaining at least 1 heart if 5 cards are drawn without replacement from a 52-card deck
18. Obtaining at least 1 face card if 5 cards are drawn without replacement from a 52-card deck
19. Obtaining at least 1 head in 10 tosses of a coin
20. Obtaining at least one 6 when 3 dice are rolled

Miscellaneous Applications

21. **Family Planning** Assume that the probability of having a girl equals the probability of having a boy. Find the probability that a family with 4 children has at least 1 girl.
22. **Thank You! OOPS!** After Joshua writes personalized thank-you notes to each of his 3 aunts for their birthday gifts, his sister randomly inserts them into preaddressed envelopes. Find the probability that (a) each aunt receives the correct thank-you note, (b) at least one aunt receives the correct thank-you note.
23. **Now Hiring** Five male and eight female applicants are found to be qualified for 3 identical positions as bank tellers. If 3 of the applicants are selected at random, find the probability that
    (a) only women are hired, (b) at least one woman is hired.
24. **Forming a Committee** A committee of 6 people is to be chosen at random from a group of 4 administrators, 7 faculty members, and 8 staff members. Find the probability that all 4 administrators and no faculty members are on the committee.
25. **Just Guessing** On a 10-question true–false examination, find the probability of scoring 100% if a student guesses the answer for each question.
26. **Got Caramel?** In a box of 20 chocolates of the same shape and appearance, 10 are known to have caramel centers. Four chocolates are selected at random from the box. Find the probability that all four will have caramel centers.

In Problems 27–30, proceed as in Example 8 to find the indicated probability.

27. **A Natural** In the dice game of "craps," a player rolls 2 dice and wins on the first roll if a total of 7 or 11 is obtained. Find the probability of winning on the first roll.
28. **Black or Red** A drawer contains 8 black socks, 4 white socks, and 2 red socks. If 1 sock is drawn at random, find the probability that it is either black or red.
29. **You Want to Bet?** At the beginning of the baseball season, an oddsmaker estimates that the probability of the Dodgers winning the World Series is $\frac{1}{10}$ and the probability of the Mets winning is $\frac{1}{20}$. On the basis of these probabilities determine the probability that either the Dodgers or the Mets will win the World Series.

30. **Trying for a Good Grade** A student estimates that his probability of earning an A in a certain math course is $\frac{3}{10}$, a B is $\frac{2}{5}$, a C is $\frac{1}{5}$, and a D is $\frac{1}{10}$. What is the probability that he earns either an A or a B?
31. **Rolling Dice** Two dice are rolled. Find the probability that the total showing on the dice is at most 4.
32. **Tossing a Coin** A coin is tossed 5 times. Let E_1 be the event of obtaining 3 tails, E_2 be the event of obtaining 4 tails, and E_3 be the event of obtaining 5 tails. Intuitively, which of the following probabilities
    (a) $P(E_1 \text{ or } E_2)$ (b) $P(E_2 \text{ or } E_3)$ (c) $P(E_1 \text{ or } E_2 \text{ or } E_3)$
    is the least number? Now compute each probability in parts (a)–(c).
33. **Raindrops Keep Falling** According to the newspaper there is a 40% probability of rain tomorrow. What is the probability that it will not rain tomorrow?
34. **Will She Lose?** A tennis player believes that she has a 75% chance of winning a tournament. Assuming ties are played off, what does she think the probability of losing is?

In Problems 35 and 36, proceed as in Example 10 to find the indicated probability.

35. **More Rolling Dice** Two dice are rolled. Find the probability that the total showing on the dice is either an even number or a multiple of 3.
36. **Choosing at Random** At ABC Plumbing and Heating Company, 30% of the workers are female, 70% are plumbers, and 40% of the workers are female plumbers. If a worker is chosen at random, find the probability that the worker is either female or a plumber.

For Discussion

37. A 12-sided die can be constructed in the form of a regular dodecahedron; each face of the die is a regular pentagon. See **FIGURE 13.7.1**. When rolled, one of the pentagonal faces will be horizontal to a table top. If each of the numbers from 1 to 6 appears twice on the die, show that the probability of each outcome is the same as that for an ordinary 6-sided die.
38. Suppose a die is a 12-sided regular dodecahedron as in Problem 37, where each face of the die is a regular pentagon. But in this case, suppose that each face bears one of the numbers 1, 2, ..., 12 as shown in **FIGURE 13.7.2**.
    (a) If two such dice are rolled, what is the probability of obtaining a total of 13?
    (b) A total of 8?
    (c) A total of 23?
    (d) What number total is least likely to appear?

FIGURE 13.7.1 12-sided die in Problem 37

FIGURE 13.7.2 12-sided die in Problem 38

39. For the spinner shown in FIGURE 13.7.3, let S be the sample space for a single spin of the spinner. Let B and R be the events that the pointer lands on blue and red, respectively, so that $S = \{B, R\}$. What, if anything, is wrong with the computation $P(B) = n(B)/n(S) = \frac{1}{2}$ for the probability of the pointer landing on blue?

Project

40. **The Birthday Problem** Find the probability that in a group of n people at least 2 people have the same birthday. Assume that a year has 365 days. Consider the three cases
    (a) $n = 10$,   (b) $n = 25$,   (c) $n = 90$.

FIGURE 13.7.3 Spinner in Problem 39

Same birthday

CONCEPTS REVIEW

You should be able to give the meaning of each of the following concepts.

Sequence:
   general term
Recursively defined sequence:
   recursion formula
Arithmetic sequence:
   common difference
Geometric sequence:
   common ratio
Series
Summation notation:
   index of summation
Arithmetic series
Geometric series

Convergence:
   sequence
   infinite series
Divergence:
   sequence
   infinite series
Infinite geometric series
Sum of a geometric series
Repeating decimal as a
   geometric series
Binomial Theorem:
   factorial notation
   binomial coefficient

Tree diagram
Fundamental Counting Principle
Permutation
Combination
Outcomes
Events
Sample space
Probability of an event
Complement of an event
Mutually exclusive events
Addition rule of probability

CHAPTER 13 Review Exercises

Answers to selected odd-numbered problems begin on page ANS-36.

A. True/False

In Problems 1–12, answer true or false.

1. $2(8!) = 16!$ _____
2. $\dfrac{10!}{9!} = 10$ _____
3. $(n - 1)!n = n!$ _____
4. $2^{10} < 10!$ _____
5. There is no constant term in the expansion of $\left(x + \dfrac{1}{x^2}\right)^{20}$. _____
6. There are exactly 100 terms in the expansion of $(a + b)^{100}$. _____
7. A sequence that is defined recursively by $a_{n+1} = (-1)a_n$ is a geometric sequence. _____
8. $\{\ln 5^n\}$ is an arithmetic sequence. _____
9. $\sum_{k=1}^{5} \ln k = \ln 120$ _____
10. $1 = 0.999\ldots$ _____
11. $0! = 1$ _____
12. $P(n, n) = n!$ _____

B. Fill in the Blanks

In Problems 1–12, fill in the blanks.

1. The next three terms in the arithmetic sequence $2x + 1, 2x + 4, \ldots$ are _____.

2. $\dfrac{x}{2} + \dfrac{x^2}{4} + \dfrac{x^3}{6} + \cdots + \dfrac{x^{10}}{20} = \sum_{-}^{-} \underline{\quad}$

3. The fifth term of the sequence $\left\{ \sum_{k=1}^{n} \dfrac{1}{k} \right\}$ is _____.

4. The twentieth term of the arithmetic sequence $-2, 3, 8, \ldots$ is _____.

5. The common ratio r of the geometric sequence $\left\{ \dfrac{2^{n+1}}{5^{n-1}} \right\}$ is _____.

6. $\dbinom{100}{100} = $ _____

7. $\sum_{k=1}^{50}(3 + 2k) = $ _____

8. $\sum_{k=1}^{100}(-1)^k = $ _____

9. $3 - 1 + \tfrac{1}{3} - \tfrac{1}{9} + \cdots = $ _____

10. For $|x| > 1$, $\sum_{k=0}^{\infty} \dfrac{1}{x^k} = $ _____.

11. If E_1 and E_2 are mutually exclusive events such that $P(E_1) = \tfrac{1}{5}$ and $P(E_2) = \tfrac{1}{3}$, then $P(E_1 \cup E_2) = $ _____.

12. If $P(E_1) = 0.3$, $P(E_2) = 0.8$, and $P(E_1 \cap E_2) = 0.7$, then $P(E_1 \cup E_2) = $ _____.

C. Review Exercises

In Problems 1–4, list the first five terms of the given sequence.

1. $\{6 - 3(n - 1)\}$
2. $\{-5 + 4n\}$
3. $\{(-1)^n n\}$
4. $\left\{ \dfrac{(-1)^n 2^n}{n + 3} \right\}$

5. List the first five terms of the sequence defined by $a_1 = 1$, $a_2 = 3$, and $a_n = (n + 1)a_{n-1} + 2$.

6. Find the seventeenth term of the arithmetic sequence with first term 3 and third term 11.

7. Find the first term of the geometric sequence with third term $-\tfrac{1}{2}$ and fourth term 1.

8. Find the sum of the first 30 terms of the sequence defined by $a_1 = 4$ and $a_{n+1} = a_n + 3$.

9. Find the sum of the first 10 terms of the geometric series with first term 2 and common ratio $-\tfrac{1}{2}$.

10. Write $2.515151\ldots$ as an infinite geometric series and express the sum as a quotient of integers.

11. **Best Gift** Determine the best gift from the following choices:
    - **A:** $10 each month for 10 years.
    - **B:** 10¢ the first month, 20¢ the second month, 30¢ the third month, and so on, receiving an increase of 10¢ each month for 10 years.
    - **C:** 1¢ the first month, 2¢ the second month, 4¢ the third month, and so on, doubling the amount received each month for 2 years.

12. **Distance Traveled** The Italian astronomer and physicist **Galileo Galilei** (1564–1642) discovered that the distance a mass moves down an inclined plane in consecutive time intervals is proportional to an odd integer. Therefore, the total distance D that a mass will move down the inclined plane in n seconds is proportional to $1 + 3 + 5 + \cdots + (2n - 1)$. Show that D is proportional to n^2.

13. **Annuity** If an annual rate of interest r is compounded continuously, then the amount S accrued in an annuity immediately after the nth deposit of P dollars is given by

$$S = P + Pe^r + Pe^{2r} + \cdots + Pe^{(n-1)r}.$$

Show that

$$S = P\frac{1 - e^{rn}}{1 - e^r}.$$

14. **Number of Sales** In 2009 a new high-tech firm projects that its sales will double each year for the next 5 years. If its sales in 2009 are $1,000,000, what does it expect its sales to be in 2014?

In Problems 15–20, use the Principle of Mathematical Induction to prove that the given statement is true for every positive integer n.

15. $n^2(n + 1)^2$ is divisible by 4

16. $\sum_{k=1}^{n}(2k + 6) = n(n + 7)$

17. $1(1!) + 2(2!) + \cdots + n(n!) = (n + 1)! - 1$

18. 9 is a factor of $10^{n+1} - 9n - 10$

19. $\left(1 + \frac{1}{1}\right)\left(1 + \frac{1}{2}\right)\left(1 + \frac{1}{3}\right)\cdots\left(1 + \frac{1}{n}\right) = n + 1$

20. $(\cos\theta + i\sin\theta)^n = \cos n\theta + i\sin n\theta$, where $i^2 = -1$

In Problems 21–26, evaluate the given expression.

21. $\dfrac{6!}{4! - 3!}$

22. $\dfrac{6!4!}{10!}$

23. $C(7, 2)$

24. $P(9, 6)$

25. $\dfrac{(n + 3)!}{n!}$

26. $\dfrac{(2n + 1)!}{(2n - 1)!}$

In Problems 27–30, use the Binomial Theorem to expand the given expression.

27. $(a + 4b)^4$

28. $(2y - 1)^6$

29. $(x^2 - y)^5$

30. $(4 - (a + b))^3$

In Problems 31–34, find the indicated term in the expansion of the given expression.

31. Fourth term of $(5a - b^3)^8$

32. Tenth term of $(8y^2 - 2x)^{11}$

33. Fifth term of $(xy^2 + z^3)^{10}$

34. Third term of $\left(\dfrac{10}{a} - 3bc\right)^7$

35. A multiple of x^2 occurs as which term in the expansion of $(x^{1/2} + 1)^{40}$?

36. Solve for x:

$$\sum_{k=0}^{n}\binom{n}{k}x^{2n-2k}(-4)^k = 0.$$

37. If the first term of an infinite geometric series is 10 and the sum of the series is $\frac{25}{2}$, then what is the value of the common ratio r?

38. Consider the sequence $\{a_n\}$ whose first four terms are

$$1,\ 1+\tfrac{1}{2},\ 1+\cfrac{1}{2+\tfrac{1}{2}},\ 1+\cfrac{1}{2+\cfrac{1}{2+\tfrac{1}{2}}},\ \ldots$$

  (a) With $a_1 = 1$, find a recursion formula that defines the sequence.
  (b) What are the fifth and sixth terms of the sequence?

In Problems 39 and 40, conjecture whether the given sequence converges.

39. $\left\{\dfrac{2^n}{n!}\right\}$

40. $\sqrt{3},\ \sqrt{3\sqrt{3}},\ \sqrt{3\sqrt{3\sqrt{3}}},\ \ldots$

41. If a coin is tossed 3 times, use a tree diagram to find all possible sequences of heads (H) and tails (T).

42. List all possible 3-digit numbers using only the digits 2, 4, 6, and 8.

43. **Ice Cream** If 32 different flavors of ice cream are available, in how many ways can a double scoop cone be ordered
   (a) if both scoops must be different flavors?
   (b) if both scoops must be the same flavor?
   [*Hint*: Assume that the order in which the scoops are placed on the cone does not matter.]

44. **More Ice Cream** At a dessert bar there are 3 flavors of ice cream, 6 different toppings, 2 kinds of nuts, and whipped cream. How many different sundaes can be made consisting of 1 flavor of ice cream with 1 topping
   (a) if nuts and whipped cream are required?
   (b) if nuts are optional, but whipped cream is required?
   (c) if both nuts and whipped cream are optional?

45. **Build Your Own** Domingo's Pizza offers 10 extra toppings. How many different pizzas can be made using just 3 of the toppings?

46. **Arrangements** In how many ways can 8 books be arranged on a shelf?

47. **Rearrangements** In making up a scrambled word puzzle, how many rearrangements of the letters in the word *shower* are possible?

48. **Time to Plant** Burtee's seed catalog offers 9 varieties of tomatoes. In how many ways can a gardener choose 3 to order?

49. **Modeling** There are 10 casual and 12 formal outfits to be modeled one at a time in a fashion show. In how many different orders can they be shown
   (a) if all the casual outfits are grouped together and all the formal outfits are grouped together?
   (b) if there are no restrictions on the order?

50. **At the Races** In how many ways can win, place, and show (that is, first-, second-, and third-place finish, respectively) be decided if 10 horses are entered in a race? Assume that there are no ties.

In Problems 51 and 52, use set notation to write the sample space of the given experiment.

51. The spinner in FIGURE 13.R.1 is spun twice.
52. The spinner in Figure 13.R.1 is spun once and then a coin is tossed.
53. **Drawing Cards** If 2 cards are drawn from a well-shuffled 52-card deck, what is the probability that both are black?

FIGURE 13.R.1 Spinner in Problems 51 and 52

54. **Choosing a Pen** Five pens are selected at random from a batch of 100 Pic pens. If 90% of this batch of Pic pens will write the first time, what is the probability that
    (a) all 5 of the pens selected will write the first time?
    (b) none of them will write the first time?
    (c) at least 1 of them will write the first time?
55. **Family Planning** Assume that the probability of giving birth to a female baby equals the probability of giving birth to a male baby. In a family of 4 children, which is more likely: (*i*) all the same sex, (*ii*) 2 of each sex, (*iii*) 3 of one sex and 1 of the other?
56. **Average Young Woman** Statistics indicate that the probability of death in the next year for a 20-year-old female is 0.0006. What is the probability that an "average" 20-year-old female will live through the next year?
57. **Feeling Lucky?** A drawer contains 8 black socks and 4 white socks. If 2 socks are drawn at random, what is the probability that
    (a) a black pair is obtained?
    (b) a white pair is obtained?
    (c) a matching pair is obtained?
58. **Bingo** A Bingo card has 5 rows and 5 columns. See **FIGURE 13.R.2**. Any five of the numbers 1 through 15 appear in the first column (designated B); any five of the numbers 16 through 30 appear in the second column (I); any four of 31 through 45 appear in the third column (N), where the center square marked "FREE" is found; any five of 46 through 60 appear in the fourth column (G); and any five of 61 through 75 appear in the last column (O). How many different Bingo cards are possible? (Consider 2 cards to be different if any 2 corresponding entries are different.)
59. **More Bingo** One version of Bingo requires a player to cover all the numbers on the card as numbers are called out at random. See Problem 58.
    (a) What is the minimum number of calls before there can be a winner in this version?
    (b) Assume that there is a winner at the minimum number of calls obtained in part (a). What is the probability that the card being played is a winning card at that point?

B	I	N	G	O
1	16	33	52	72
12	20	41	47	65
2	22	FREE	55	68
7	30	36	60	74
8	28	45	49	61

FIGURE 13.R.2 Bingo card in Problem 58

60. **Areas** Let $\{A_n\}$ be the sequence of areas of the isosceles triangles shown in **FIGURE 13.R.3**. Find the sum of the infinite series $A_1 + A_2 + A_3 + \cdots$.

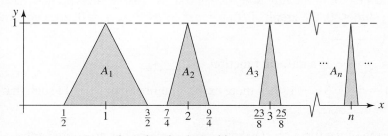

FIGURE 13.R.3 Isosceles triangles in Problem 60

Final Examination

Answers to all questions are found in the *SRM*.

In Problems 1–14, fill in the blanks.

1. Completing the square in x for $2x^2 + 6x + 5$ gives _____.
2. In the binomial expansion of $(1 - 2x)^3$ the coefficient of x^2 is _____.
3. In interval notation, the solution set of $\dfrac{x(x^2 - 9)}{x^2 - 25} \geq 0$ is _____.
4. If $a - 3$ is a negative number, then $|a - 3| = $ _____.
5. If $|5x| = 80$, then $x = $ _____.
6. If (a, b) is a point in the third quadrant, then $(-a, b)$ is a point in the _____ quadrant.
7. Given that the point $(1, 7)$ is on a graph, give the coordinates of another point on the graph if it is:
   (a) symmetric with respect to the x-axis. _____
   (b) symmetric with respect to the y-axis. _____
   (c) symmetric with respect to the origin. _____
8. The lines $6x + 2y = 1$ and $kx - 9y = 5$ are parallel if $k = $ _____. The lines are perpendicular if $k = $ _____.
9. The complete factorization of the function $f(x) = x^3 - 2x^2 - 6x$ is _____.
10. The only potential rational zeros of $f(x) = x^3 + 4x + 2$ are _____.
11. The phase shift of the graph of $y = 5\sin(4x + \pi)$ is _____.
12. If $f(x) = x^4 \arctan(x/2)$, then the exact value of $f(-2)$ is _____.
13. $5\ln 2 - \ln \frac{2}{3} = \ln$ _____.
14. The graph of $y = \ln(2x + 5)$ has the vertical asymptote $x = $ _____.

In Problems 15–32, answer true or false.

15. The absolute value of any real number x is positive. _____
16. The inequality $|x| > -1$ has no solutions. _____
17. For any function f, if $f(a) = f(b)$, then $a = b$. _____
18. The graph of $y = f(x + c)$, $c > 0$, is the graph of $y = f(x)$ shifted c units to the right. _____
19. The points $(1, 3)$, $(3, 11)$, and $(5, 19)$ are collinear. _____
20. The function $f(x) = x^5 - 4x^3 + 2$ is an odd function. _____
21. $x + \frac{1}{4}$ is a factor of the function $f(x) = 64x^4 + 16x^3 + 48x^2 - 36x - 12$. _____
22. If $b^2 - 4ac < 0$, the graph of $f(x) = ax^2 + bx + c$, $a \neq 0$, does not cross the x-axis. _____
23. $f(x) = \dfrac{\sqrt{x}}{2x + 1}$ is a rational function. _____
24. If $f(x) = x^5 + 3x - 1$, then there exists a number c in $[-1, 1]$ such that $f(c) = 0$. _____
25. The graph of the function $f(x) = \dfrac{1}{x - 1} + \dfrac{1}{x - 2}$ has no x-intercepts. _____
26. $x = 0$ is a vertical asymptote for the graph of the rational function $f(x) = \dfrac{x^2 - 2x}{x}$. _____
27. The graph of $y = \cos(x/6)$ is the graph of $y = \cos x$ stretched horizontally. _____

28. $f(x) = \csc x$ is not defined at $x = \pi/2$. _____
29. The function $f(x) = e^{-4x^2}$ is not one-to-one. _____
30. The exponential function $f(x) = (\frac{3}{2})^x$ increases on the interval $(-\infty, \infty)$. _____
31. The domain of the function $f(x) = \ln x + \ln(x - 4)$ is $(4, \infty)$. _____
32. The solutions of the equation $\ln x^2 = \ln 3x$ are $x = 0$ and $x = 3$. _____
33. Match the given interval with the appropriate inequality.

   (i) $[2, 4]$ (ii) $[2, 4)$
   (iii) $(2, 4)$ (iv) $(2, 4]$
   (a) $|x - 3| \leq 1$ (b) $1 < x - 1 \leq 3$
   (c) $-2 < 2 - x \leq 0$ (d) $|x - 3| < 1$

34. Write the solution of the absolute-value inequality $|3x - 1| > 7$ using interval notation.
35. The answer to a problem given in the back of a mathematics text is $1 + \sqrt{3}$, but your answer is $2/(\sqrt{3} - 1)$. Are you correct?
36. In which quadrants in the Cartesian plane is the quotient x/y negative?
37. Which one of the following equations best describes a circle that passes through the origin? The symbols $a, b, c, d,$ and e stand for different nonzero real constants.

   (a) $ax^2 + by^2 + cx + dy + e = 0$ (b) $ax^2 + ay^2 + cx + dy + e = 0$
   (c) $ax^2 + ay^2 + cx + dy = 0$ (d) $ax^2 + by^2 + cx + dy = 0$
   (e) $ax^2 + ay^2 + e = 0$ (f) $ax^2 + ay^2 + cx + e = 0$

38. Match the given rational function f with the most appropriate phrase.

   (i) $f(x) = \dfrac{x^4}{x^2 - 2}$ (ii) $f(x) = \dfrac{x^2}{x^2 + 2}$
   (iii) $f(x) = \dfrac{x^5}{x^2 + 2}$ (iv) $f(x) = \dfrac{x^3}{x^2 + 2}$

   (a) slant asymptote (b) no asymptotes
   (c) horizontal asymptote (d) vertical asymptote

39. What is the range of the rational function $f(x) = \dfrac{10}{x^2 + 1}$?
40. What is the domain of the function $f(x) = \dfrac{\sqrt{x + 2}}{x^2}$?
41. Find an equation of the line that passes through the origin and through the point of intersection of the graphs of $x + y = 1$ and $2x - y = 7$.
42. Find a quadratic function f whose graph has the y-intercept $(0, -6)$ and the vertex of the graph is $(1, 4)$.

For those going on to a course in calculus, you will often be required to rewrite a function either in a simpler form or in a form that is more helpful in solving the problem. In Problems 43–48, rewrite each function by following the given instruction. In calculus you would be expected to recognize what to do from the context of the actual problem.

43. $f(x) = \sqrt{x^6 + 4} - x^3$. Express f as quotient using rationalization and simplification.
44. $f(x) = \dfrac{5x^3 - 4x^2\sqrt{x} + 8}{\sqrt[3]{x}}$. Carry out the indicated division and express each term as a power of x.
45. $f(x) = \dfrac{7x^2 - 7x - 6}{x^3 - x^2}$. Decompose f into partial fractions.

46. $f(x) = \dfrac{1}{1 + \sin x}$. Express f in terms of $\sec x$ and $\tan x$.

47. $f(x) = e^{3\ln x}$. Express f as a power of x.

48. $f(x) = |x^2 - 3x|$. Express f without absolute-value signs.

In Problems 49 and 50, solve the equation $f(x) = 0$ by following the given instruction.

49. $f(x) = x^2 \tfrac{1}{2}(4 - x^2)^{-1/2}(-2x) + 2x\sqrt{4 - x^2}$. Rewrite f as a single expression without negative exponents.

50. $f(x) = 2\sin x \cos x - \sin x$. Find the zeros of f on the interval $[-\pi, \pi]$.

In Problems 51 and 52, compute and simplify the difference quotient

$$\frac{f(x + h) - f(x)}{h}$$

for the given function.

51. $f(x) = \dfrac{3x}{2x + 5}$

52. $f(x) = -x^3 + 10x^2$

53. Consider the trigonometric function $y = -8\sin(\pi x/3)$. What is the amplitude of the function? Give an interval over which one cycle of the graph is completed.

54. If $\tan\theta = \sqrt{5}$ and $\pi < \theta < 3\pi/2$, then what is the value of $\cos\theta$?

55. Suppose $f(x) = \sin x$ and $f(c) = 0.7$. What is the value of

$$2f(-c) + f(c + 2\pi) + f(c - 6\pi)?$$

56. Suppose $f(x) = \sin x$ and $g(x) = \ln x$. Solve $(f \circ g)(x) = 0$.

57. Find the x- and y-intercepts of the parabola whose equation is $(y + 4)^2 = 4(x + 1)$.

58. Find the center, foci, vertices, and endpoints of the minor axis of the ellipse whose equation is

$$x^2 + 2y^2 + 2x - 20y + 49 = 0.$$

59. The slant asymptotes of a hyperbola are $y = -5x + 2$ and $y = 5x - 8$. What is the center of the hyperbola?

60. From a point 220 ft from the base of a cell-phone antenna a person measures a $30°$ angle of inclination from the ground to the top of the antenna. What is the angle of inclination to the top of the antenna if the person moves 100 ft closer to its base?

61. Iodine 131 is radioactive and is used in certain medical procedures. Assume that iodine 131 decays exponentially. If the half-life of I-131 is 8 days, then how much of a 5-gram sample remains at the end of 15 days?

62. The polar coordinate equation $r = 3\cos 4\theta$ is a rose curve with eight petals. Find all radian-measure angles satisfying $0 \leq \theta \leq 2\pi$ for which $|r| = 3$.

63. Solve the linear system

$$\begin{cases} x - 2y + 3z = 1 \\ x + y - z = 5 \\ 4x - 5y + 8z = 8 \end{cases}$$

and interpret the solution geometrically.

FINAL EXAMINATION

64. Solve the equation $\begin{vmatrix} x & 0 & 4 \\ 0 & x & 0 \\ 4 & 0 & x \end{vmatrix} = 0$ for x.

65. Here is a nonlinear system of equations taken from a calculus text:

$$\begin{cases} 2x\lambda = -4 \\ 2y\lambda = 2y \\ x^2 + y^2 = 9. \end{cases}$$

Solve for x, y, and λ.

66. Graph the system of inequalities:

$$\begin{cases} y \leq 2^x \\ 6y - 7x \geq 10 \\ x \geq 0. \end{cases}$$

In Problems 67–70, answer the given question about the sequence

$$128, 64, 32, 16, \ldots.$$

67. What is the eighth term of the sequence?
68. What is the sum S_8 of the first eight terms of the sequence?
69. Is the sequence convergent or divergent?
70. Does the infinite series

$$128 + 64 + 32 + 16 + \cdots$$

have a sum S?

In Problems 71–74, find the nth term a_n of the given sequence.

71. $-2, -1, 0, 1, \ldots$ **72.** $0, 3, 8, 15, \ldots$
73. $1000, -100, 10, -1, \ldots$ **74.** $1, \frac{1}{7}, \frac{1}{13}, \frac{1}{19}, \ldots$

75. If d is a digit (any numeral 0 through 9), find a rational number whose decimal representation is $0.ddd\ldots$.

76. If $A = \begin{bmatrix} 1 & -1 & 1 \\ 1 & 1 & 1 \\ 2 & 2 & 2 \end{bmatrix}$, explain why $\det A = 0$ from the properties of determinants.

77. Find real number coefficients a, b, and c such that the points $(-1, -3)$, $(1, -5)$, and $(2, 0)$ are on the graph of the quadratic function $f(x) = ax^2 + bx + c$.

78. If the points P_1, P_2, and P_3 are collinear as shown in **FIGURE FEX.1**, then find an equation relation relating the distances $d(P_1, P_2)$, $d(P_2, P_3)$, and $d(P_1, P_3)$.

79. Find the values of the six trigonometric functions of the angle θ given in **FIGURE FEX.2**.

FIGURE FEX.1 Graph for Problem 78

FIGURE FEX.2 Triangle for Problem 79

80. If $\sin \theta = \frac{24}{25}$, $90° < \theta < 180°$, then find **(a)** $\sin 2\theta$ **(b)** $\cos 2\theta$ **(c)** $\tan 2\theta$.

Answers to Selected Odd-Numbered Problems

Exercises R.1   Page 9

1. $\{1, 3, 4, 6, 8, 9, 10, 11, 12, 14, 15\}$
3. $\{1, 2, 3, 5, 7, 8, 9, 11, 12, 13, 14\}$
5. $\{1, 8\}$
7. $\{3, 9, 11, 12, 14\}$
9. $\{-1, -2, 1, 2\}$
11. $\{\frac{2}{3}, \frac{4}{3}\}$
13. $\{-2, -1\}$
15. $\{x \mid x = 2n, n \in Z\}$
17. property 3(*i*)
19. property 2(*i*)
21. property 3(*ii*)
23. property 4(*ii*)
25. property 5(*ii*)
27. property 4(*i*)
29. property 2(*ii*)
31. property 5(*i*)
33. property 10(*iv*)
35. property 7(*i*)
37. property 9(*i*)
39. property 8(*ii*)
41. property 16(*i*)
43. property 13
45. $-a$
47. $\dfrac{3+c}{c}$
49. 0
51. (a) greater than

Exercises R.2   Page 15

1. [number line showing points at $-2, -1, -\frac{1}{2}, 0, 1, \frac{4}{3}, 2, 2.5$]
3. $x > 0$
5. $x + y \geq 0$
7. $b \geq 100$
9. $|t - 1| < 50$
11. $-3 < 15$
13. $1.33 < \frac{4}{3}$
15. $3.14 < \pi$
17. $-2 \geq -7$
19. $2.5 \geq \frac{5}{2}$ and $\frac{5}{2} \geq 2.5$ are both correct
21. $\frac{423}{157} \geq 2.6$
23. 7
25. 22
27. $\frac{22}{7}$
29. $\sqrt{5}$
31. $4 - \pi$
33. 4
35. 4
37. $3 - \sqrt{5}$
39. $8 - \sqrt{7}$
41. $2.3 - \sqrt{5}$
43. $2\pi - 6.28$
45. $-h$
47. $2 - x$
49. $x - 2$
51. 0
53. 0
55. -1
57. (a) 4 (b) 5
59. (a) 0.2 (b) 0.7
61. (a) 3 (b) -6.5
63. (a) 3 (b) 0
65. $a = 2, b = 8$
67. $a = -1.5, b = 5.5$
69. $m = 4 + \pi, b = 4 + 2\pi$
71. $a = -3 - \sqrt{2}, b = -3 - \frac{\sqrt{2}}{2}$
73. $(10)(10) = 100$
75. $\pi > 3.14$
77. $\frac{7}{11} = 0.\overline{63}$
79. $\sqrt{2} > 1.4$
81. Natalie lives either $\frac{5}{4}$ miles or $\frac{3}{4}$ miles from Greg.

Exercises R.3   Page 21

1. $\dfrac{1}{8^3}$
3. $(2y)^4$
5. 4^{-5}
7. x^{-3}
9. (a) 81 (b) $\frac{1}{81}$ (c) -81
11. (a) 49 (b) $\frac{1}{49}$ (c) $-\frac{1}{49}$
13. (a) 1 (b) 1 (c) -1
15. $-\frac{3}{2}$
17. $\frac{1}{5}$
19. 0
21. 13
23. 11
25. $-\frac{7}{6}$
27. x^4
29. $-21x^6$
31. 2^5
33. $\dfrac{1}{10^{11}}$
35. $25x^2$
37. 5^6
39. $\dfrac{64x^6}{y^3}$
41. x
43. $49ab$
45. $\dfrac{9y^9}{x}$
47. $a^6 b^{10}$
49. $-x^2 y^4 z^6$
51. negative
53. positive
55. positive
57. $A = s^2$
59. $A = \pi r^2$
61. $V = \pi r^2 h$
63. (a) 1.05×10^6 (b) 1.05×10^{-5}
65. (a) 1.2×10^9 (b) 1.2×10^{-10}
67. (a) $32,500,000$ (b) 0.0000325
69. (a) 0.0000000000000000987 (b) $9,870,000,000,000$
71. 4.9064×10^{-4}
73. 9.533×10^{-6}
75. 8.2500×10^{13}
77. (a) $7,200,000,000,000$ (b) 7.2×10^{12}
79. (a) $32,000,000,000,000,000,000$ (b) 3.2×10^{19}
81. 1.335×10^9
83. (a) $\$14,261,000,000,000$ (b) $\$1.4261 \times 10^{13}$
85. approximately 1.5×10^{19} mi

Exercises R.4   Page 29

1. -5
3. 10
5. 0.1
7. $-\frac{4}{3}$
9. $\dfrac{1}{xy^2}$
11. $\dfrac{xy^2}{z^3}$
13. $0.5x^2z^2$
15. $4ab$
17. $\frac{1}{3}$
19. $\dfrac{1}{7b}$
21. 0.2
23. ab^2
25. xyz^5
27. abc
29. $2rs^3$
31. $\sqrt{2}x^2$
33. $\dfrac{\sqrt{3}}{9}$
35. $\dfrac{\sqrt{x+1}}{x+1}$
37. $\dfrac{-7 + 2\sqrt{10}}{3}$
39. $\dfrac{x + 2\sqrt{xy} + y}{x - y}$
41. $\sqrt[3]{2}$
43. $\dfrac{4\sqrt[3]{(x-1)^2}}{x - 1}$
45. $\dfrac{2}{\sqrt{2(x+h)} + \sqrt{2x}}$
47. $\dfrac{1}{\sqrt{x+h+1} + \sqrt{x+1}}$

49. $5\sqrt{x}$
51. $2\sqrt[3]{2}$
53. $(6x - 3y + 4x^2)\sqrt{2x}$
55. $(1 - a)\sqrt{\dfrac{a}{b}}$
57. $s = \sqrt{A}$
59. $c = \sqrt{a^2 + b^2}$
61. 7.74×10^3 m/s

Exercises R.5 Page 34

1. $(ab)^{1/3}$ **3.** $x^{-4/3}$ **5.** $(x + y)^{1/7}$ **7.** $(x + x^{1/2})^{1/2}$
9. $\sqrt[3]{a^2}$ **11.** $\sqrt[3]{9a^2}$ **13.** $3 + \sqrt[3]{a^2}$ **15.** $\dfrac{3}{\sqrt[3]{a^2}}$
17. (a) 7 (b) $\tfrac{1}{7}$
19. (a) 0.0000128 (b) $\dfrac{1}{0.0000128} = 78{,}125$
21. (a) 2187 (b) $-\dfrac{1}{2187}$
23. $12x^{5/6}$ **25.** $4a^{13/6}$ **27.** $x^{7/8}$ **29.** $a^{1/2}b$
31. $125x^{1/2}y^{3/2}$ **33.** $\dfrac{c^{2/3}}{d^{2/3}}$ **35.** $\dfrac{64x^3 z}{y^4}$ **37.** $\dfrac{9a^2}{b^4}$
39. 1 **41.** $125x^3$ **43.** $\dfrac{b^6 c^{3/2}}{a^4}$ **45.** $p^{1/6}q^{1/2}$
47. $\dfrac{r}{s}$ **49.** $\sqrt[6]{500}$ **51.** $\sqrt[6]{64} = 2$ **53.** $\sqrt[4]{x^3}$
55. $\sqrt[8]{a^3}$ **57.** 8000 **59.** $0.00001x^5 y^{10}$
61. approximately 1.0139 s **63.** 1126.30 ft/s

Exercises R.6 Page 43

1. (a) 30 (b) $\tfrac{15}{4}$ (c) 6
3. (a) -192 (b) $\tfrac{1}{2}$ (c) 0
5. (a) $-\tfrac{5}{2}$ (b) $-\tfrac{3}{4}$ (c) -1
7. (a) 2.6 (b) -0.025 (c) 0.2
9. polynomial; degree 1; leading coefficient 8
11. not a polynomial
13. polynomial; degree 101; leading coefficient 26
15. not a polynomial
17. $3x^5 + x^3 - 8x^2 + 6x - 6$
19. $6y^3 + y^2 - 2y - 7$
21. $-3x^4 + 5x^2 - 1$
23. $3x^7 - 7x^6 + x^5 - x^4 + 2x^2 - 8x - 14$
25. $-2t^3 - 2t^2 - 27t + 46$
27. $2v^2 - 8v^2 - 24v$
29. $y^4 + y^3 - y^2 + 14y - 20$
31. $15a^4 - a^3 b + 8a^2 b^2 - 8ab^3 + 11b^4$
33. $6a^3 - 5a^2 b + 3ab^2 - b^3$ **35.** $5 - 20s$
37. $1 - x^2 + 2x^6 y$ **39.** $x^2 + x - 2$
41. $2r^6 - 13r^3 - 7$ **43.** $10t^2 + 26t - 56$
45. $24x - 2\sqrt{x} - 2$ **47.** $3x^2 + 7.63x + 1.47$
49. $x^2 - \tfrac{13}{24}x - \tfrac{1}{12}$ **51.** $1 + 10b + 25b^2$
53. $50x^2 + 40x + 8$ **55.** $4 - 3x$
57. $y^{-2} - 4x^2$ **59.** $8x^3 - 36x^2 + 54x - 27$
61. $x^6 y^9 + 6x^4 y^6 + 12x^2 y^3 + 8$ **63.** $x^3 + 3x^2 y + 3xy^2 + y^3$
65. $a^3 - 27$ **67.** $729 + y^3$
69. $625x^4 - y^4$ **71.** $x^2 + 2xy + y^2 + 2x + 2y + 1$
73. $x^3 + 3x^2 y + 3xy^2 + y^3 + 3x^2 + 6xy + 3y^2 + 3x + 3y + 1$
75. $x^{4/3} - x^{2/3}$ **77.** $\dfrac{1}{y^6} - \dfrac{1}{x^6}$
79. $x^6 - x^5 - x^3 + x^2$
81. (a) $-4x^3 - 8x^2 + 100x + 200$
 (b) $-8x^2 + 40x + 280$

Exercises R.7 Page 50

1. $2x(6x^2 + x + 3)$ **3.** $(2y - z)(y + 3)$
5. $(3t + s)(5a + b)$ **7.** $xyz(z^2 - y^2 + x^2)$
9. $(p^2 + 1)(2p - 1)$ **11.** $(6x - 5)(6x + 5)$
13. $(2xy - 1)(2xy + 1)$ **15.** $(x - y)(x + y)(x^2 + y^2)$
17. $(x - y)(x + y)(x^2 + y^2)(x^4 + y^4)$
19. $(2xy + 3)(4x^2 y^2 - 6xy + 9)$
21. $(y - 1)(y + 1)(y^2 + y + 1)(y^2 - y + 1)$
23. $(x - 3)(x - 2)$ **25.** $(y + 5)(y + 2)$
27. $(x - 2)(x + 2)(x^2 + 1)$ **29.** $(r + 1)^2$
31. $(x - 2y)(x + y)$ **33.** $(x + 5)^2$
35. $(s - 4t)^2$ **37.** $(2p + 5)(p + 1)$
39. $(6a^2 - 5)(a^2 + 3)$ **41.** $(2x - y)(x - 3y)$
43. $(x^2 + y^2)(x^4 + y^4 - x^2 y^2 + 3x^2 - 3y^2 + 3)$
45. $(x - y)^2$
47. $(y - x)(y + x)(x^4 + y^4 + x^2 y^2 - 3x^2 - 3y^2 + 3)$
49. $(1 - 2v)(1 + 2v)(1 + 4v^2)(1 + 16v^4)$
51. $(x + 2)(x - 1)(x^2 - 2x + 4)(x^2 + x + 1)$
53. $(rs - 2t)(r^2 s^2 + 2rst + 4t^2)$
55. $(a - b)(a + b)^2$ **57.** $(4z - y)(z + 2y)$
59. $(4a - 3b)^2$ **61.** $(x - \sqrt{3})(x + \sqrt{3})$
63. $(\sqrt{5}y - 1)(\sqrt{5}y + 1)$ **65.** $(a + \tfrac{1}{2})^2$
67. $(a - \sqrt{2}b)(a + \sqrt{2}b)$ **69.** $(2\sqrt{6} - x)(2\sqrt{6} + x)$

Exercises R.8 Page 57

1. $\dfrac{x + 1}{x + 4}$ **3.** $\dfrac{z - 3}{z^2 - 3z + 9}$
5. $\dfrac{3x + 5}{2x + 3}$ **7.** $\dfrac{w + 3}{w - 3}$
9. $(x + 2)(x - 1)$ **11.** $b^2(b - 2)(b + 3)(b - 6)$
13. $c(c - 1)(c + 1)^2$ **15.** $x^2(x - 1)(x + 1)^2$
17. 1 **19.** $\dfrac{7z + 1}{7z - 1}$
21. $\dfrac{2x^2 - 2x + 5}{x^2 - 1}$ **23.** $\dfrac{y^2 + x^2}{y^2 - x^2}$
25. $\dfrac{r^2 - 4r + 2}{r^2 - r - 12}$ **27.** $\dfrac{-8x + 2}{2x^2 + 3x - 2}$
29. $\dfrac{t^2 + t - 20}{t^2 + t - 6}$ **31.** $\dfrac{x^2 - 1}{x^2 + x - 1}$
33. $\dfrac{1}{3x - 6}$ **35.** $\dfrac{u + 7}{u + 2}$
37. $\dfrac{x^2}{x^2 + 9x + 20}$ **39.** $\dfrac{q + 1}{q - 4}$
41. $\dfrac{-s + 3}{s + 5}$ **43.** $\dfrac{1 - x^3}{1 + x^3}$
45. $\dfrac{z}{2}$ **47.** $\dfrac{xy}{x - y}$
49. $\dfrac{-2x - h}{x^2(x + h)^2}$ **51.** $\dfrac{ab}{b - a}$
53. $\dfrac{v^2 - u^2}{u^4 v^4}$ **55.** $\dfrac{w\sqrt{u} + u\sqrt{w}}{uw}$
57. $\dfrac{\sqrt{xy} + \sqrt{y}}{\sqrt{xy} + \sqrt{x}}$ **59.** $\dfrac{-3x^2 - 3xh - h^2}{x^3(x + h)^3}$
61. $\dfrac{-10}{(2x - 1)(2x + 2h - 1)}$ **63.** $\dfrac{1 - 3x^2}{2x^{1/2}(x^2 + 1)^2}$
65. $\dfrac{R_1 R_2 R_3}{R_1 R_2 + R_1 R_3 + R_2 R_3}$

Chapter R Review Exercises Page 60

A. 1. false 3. false 5. false 7. false
9. true 11. false 13. false 15. false
17. false 19. true 21. true 23. false
25. true

B. 1. 0, 1, 2 3. irrational
5. left 7. reciprocal or multiplicative inverse
9. $x^{3/4}$ 11. polynomial; 4, 3, 0
13. $|b - a|$ 15. scientific notation
17. disjoint
19. x is the base, 3 is the exponent or power
21. commutative

C. 1. $\{1, 2, 3, 4, 5, 7, 9\}$ 3. $\{2, 4\}$
5. $\{1, 2, 3, 4, 5, 6, 7, 8, 9\}$ 7. $x - y \geq 10$
9. $-1.4 > -\sqrt{2}$ 11. $\frac{2}{3} < 0.67$
13. $3 - \sqrt{8}$ 15. $x^2 + 5$
17. $-(t + 5)$
19. (a) 9.3 (b) 1.15
21. $108u^5v^8$ 23. $\dfrac{x^4}{2y}$
25. $\dfrac{4c^8}{d^6}$ 27. $\dfrac{y^{1/9}}{x}$
29. 25 31. xy^3
33. $\dfrac{-q^3}{p^2}$ 35. $(3 + \sqrt{2})\sqrt{xy}$
37. $x^{1/2}$ 39. 7.023×10^{-7}
41. 5×10^{21}
43. (a) \$52,670,000,000 (b) $\$5.267 \times 10^{10}$
45. $4x^3 - 4x^2 + 9x - 6$ 47. $a^3 - 2a^2 - 5a - 6$
49. $9z^8 - 12z^5 + 4z^2$ 51. $9x^4 - 25y^2$
53. $(3x + 2)(4x - 9)$ 55. $(y - 3)(2x + 3)$
57. $(2x + 5y^2)(4x^2 - 10xy^2 + 25y^4)$
59. $(2t^2 - s)^2$ 61. $\dfrac{3x - 3}{x^2 - 4}$
63. $\dfrac{1}{xy}$ 65. $\dfrac{1}{x}$
67. $\dfrac{r^2 + 2rs}{s^2 + 2rs}$ 69. $\dfrac{-4(3x^2 + 3xh + h^2)}{x^3(x + h)^3}$
71. $\dfrac{20x + 8}{(2x + 1)^{3/4}}$ 73. $\dfrac{2\sqrt{s} - 2\sqrt{t}}{s - t}$
75. $\dfrac{2}{\sqrt{2x + 2h + 3} + \sqrt{2x + 3}}$

Exercises 1.1 Page 70

1. equivalent 3. equivalent
5. not equivalent 7. $\{-7\}$
9. $\{-\frac{1}{5}\}$ 11. $\{1\}$ 13. $\{-\frac{5}{2}\}$ 15. $\{\frac{1}{2}\}$
17. $\{3\}$ 19. $\{0\}$ 21. $\{-3.5\}$ 23. $\{0.5\}$
25. $\{-\frac{1}{2}\}$ 27. $\{1\}$ 29. $\{0\}$ 31. $\{\frac{1}{2}\}$
33. $\{-1\}$ 35. $\{3\}$ 37. $\{1\}$ 39. $\{x \mid x \neq 2\}$
41. $\{9\}$ 43. $\{1\}$ 45. $\varnothing$ 47. $\{4\}$
49. $a = 3$ 51. $c = -\frac{7}{3}$ 53. $a = -1$ 55. $r = C/2\pi$
57. $t = I/Pr$ 59. $P = A/(1 + rt)$
61. $n = (a_n - a + d)/d$ 63. $m_1 = Fr^2/(gm_2)$
65. $R_2 = RR_1/(R_1 - R)$
67. (a) $T_F = \frac{9}{5}T_C + 32$ (b) 23°F, 32°F, 60.8°F, 95°F, 212°F
69. 84 years 71. −68.2875°C

Exercises 1.2 Page 78

1. 12, 38 3. 15, 16, 17 5. 13 years old 7. 9.6%
9. \$4000 11. $183\frac{1}{3}$ mi 13. $\frac{4}{3}$ h
15. Car: 40 km/h; bicycle: 10 km/h
17. 3.75 qt
19. 0.5 ft³ of 10% peat; 1.5 ft³ of 30% peat
21. 20% of \$3.95 per pound; 80% of \$4.20 per pound
23. $\frac{24}{7}$ h ≈ 3.43 h 25. 75 min
27. 10 cm by 15 cm 29. 6 cm, 7 cm, 8 cm
31. 8 cm 33. 47
35. 110 37. 13 nickels; 17 dimes
39. 27 41. 20
43. 10 months
45. (a) It makes no difference.
    (b) Let x be the original cost of an item. Taking a $d\%$ discount first and then adding the $t\%$ sales tax is represented by $\left(x - \dfrac{d}{100}x\right)\left(1 + \dfrac{t}{100}\right)$. Adding the $t\%$ sales tax first and then taking the $d\%$ discount is given by $\left(x + \dfrac{t}{100}x\right)\left(1 - \dfrac{d}{100}\right)$. By factoring, each of these expressions is equal to $\left(1 + \dfrac{t}{100}\right)\left(1 - \dfrac{d}{100}\right)x$.

Exercises 1.3 Page 90

1. $\{-4, 4\}$ 3. $\{-1, \frac{1}{2}\}$
5. $\{-\frac{1}{2}\}$ 7. $\{3, 4\}$
9. $\{-\frac{1}{5}, -\frac{2}{5}\}$ 11. $\{-\frac{1}{4}, \frac{1}{4}\}$
13. $\{-3, 0, 3\}$ 15. $\{-\frac{5}{2}, 0, \frac{5}{2}\}$
17. $\{-\sqrt{17}, \sqrt{17}\}$ 19. $\{-5 - \sqrt{5}, -5 + \sqrt{5}\}$
21. $\{-1 - \sqrt{3}, -1 + \sqrt{3}\}$ 23. $\{-b, b\}$
25. $\{a - b, a + b\}$ 27. $\{-1 - \sqrt{2}, -1 + \sqrt{2}\}$
29. $\{-\frac{3}{2}, -1\}$ 31. $\{1 - \frac{3\sqrt{10}}{10}, 1 + \frac{3\sqrt{10}}{10}\}$
33. $\{2 - \frac{\sqrt{35}}{3}, 2 + \frac{\sqrt{35}}{3}\}$ 35. $\{\frac{1}{3}, 2\}$
37. $\{-\frac{5}{3}\}$ 39. $\{\frac{1}{4} - \frac{\sqrt{105}}{20}, \frac{1}{4} + \frac{\sqrt{105}}{20}\}$
41. no real solutions 43. $\{\frac{1}{2} - \frac{\sqrt{3}}{2}, \frac{1}{2} + \frac{\sqrt{3}}{2}\}$
45. $\{\pm\sqrt{3 - \sqrt{2}}, \pm\sqrt{3 + \sqrt{2}}\}$ 47. $\{\frac{1}{3}, 4\}$
49. $\{2 - 2\sqrt{2}, 2 + 2\sqrt{2}\}$ 51. no real solutions
53. $\{-\frac{\sqrt{2}}{4}, 0, \frac{\sqrt{2}}{4}\}$ 55. $\{-\sqrt{15}, \sqrt{15}\}$
57. $r = \sqrt{\dfrac{V}{\pi h}}$ 59. $r = \dfrac{1}{2}\left(-h + \sqrt{h^2 + \dfrac{2A}{\pi}}\right)$
61. $t = (-v_0 + \sqrt{v_0^2 + 2gs})/g$ 63. 2, 18
65. 4 67. 7, 15
69. Base: 17 cm; altitude: 14 cm 71. 18 m by 20 m
73. 5000 ft² 75. 30 cm, 40 cm, 50 cm
77. 90 mi
79. John drove at 54 mi/h; James drove at 48 mi/h
81. 12 83. 3 m
85. 9 in. by 18 in. 87. 14.7 m

Exercises 1.4 Page 98

1. $-i$ 3. i
5. $-i$ 7. $-i$
9. i 11. $10i$
13. $-3 - \sqrt{3}i$ 15. $-1 + 4i$

17. $2 - 13i$
19. $-9 - 15i$
21. $-11 + 7i$
23. $1 - 5i$
25. $35i$
27. $1 + 4i$
29. $-1 + 12i$
31. -10
33. $-18 - 16i$
35. $15 + 8i$
37. 4
39. $\frac{4}{25} + \frac{3}{25}i$
41. $\frac{20}{41} - \frac{16}{41}i$
43. $\frac{1}{2} + \frac{1}{2}i$
45. $6 - 4i$
47. i
49. $-\frac{6}{53} + \frac{32}{53}i$
51. $\frac{11}{2} + \frac{9}{2}i$
53. $-\frac{6}{13} - \frac{4}{13}i$
55. $9 + i$
57. $x = 2, y = \frac{3}{2}$
59. $x = -9, y = -20$
61. $x = -4, y = -5$
63. $x = \frac{1}{5}, y = -\frac{2}{5}$
65. $\pm 3i$
67. $\pm \frac{\sqrt{10}}{2}i$
69. $\frac{1}{4} - \frac{\sqrt{7}}{4}i, \frac{1}{4} + \frac{\sqrt{7}}{4}i$
71. $-4 - 6i, -4 + 6i$
73. $\frac{1}{8} - \frac{\sqrt{31}}{8}i, \frac{1}{8} + \frac{\sqrt{31}}{8}i$
75. $\pm i, \pm \sqrt{2}i$
77. $\frac{\sqrt{2}}{2} + \frac{\sqrt{2}}{2}i, -\frac{\sqrt{2}}{2} - \frac{\sqrt{2}}{2}i$

Exercises 1.5 Page 104

1. $(-\infty, 0)$;
3. $[5, \infty)$;
5. $(8, 10]$;
7. $[-2, 4]$;
9. $-7 \le x \le 9$
11. $x < 2$
13. $4 < x \le 20$
15. $4x + 4 \ge x$
    $4x + 4 - x \le x - x$ ← by (i)
    $3x + 4 - 4 \le 0 - 4$ ← by (i)
    $\frac{1}{3}(3x) \le \frac{1}{3}(-4)$ ← by (ii)
    $x \le -\frac{4}{3}$
17. $0 < 2(4 - x) < 6$
    $0 < 8 - 2x < 6$
    $-8 + 0 < -8 + 8 - 2x < -8 + 6$ ← by (i)
    $-8 < -2x < -2$
    $-\frac{1}{2}(-8) > -\frac{1}{2}(-2x) > -\frac{1}{2}(-2)$ ← by (iii)
    $4 > x > 1$ or $1 < x < 4$
19. $(-5, \infty)$;
21. $(-\infty, 4]$;
23. $(-\infty, \frac{3}{4})$;
25. $(-\infty, \frac{5}{2}]$;
27. $(-10, 6)$;
29. $(-5, 3)$;
31. $[-10, -8]$;

33. $[0, 6)$;
35. $-2x + 7$
37. 3
39. $2x - 2$
41. 4
43. The number is less than .
45. With the rebate, the larger jar is more economical if it costs less than $5.30.
47. A person can ride 8, 9, 10, 11, 12, 13, 14, 15, or 16 quarter miles.

Exercises 1.6 Page 108

1. $\{-\frac{1}{4}, \frac{3}{4}\}$
3. $\{-\frac{1}{2}, \frac{5}{6}\}$
5. $\{\frac{2}{3}, 2\}$
7. $\{-3, 3\}$
9. $\{1, 1 - \sqrt{2}, 1 + \sqrt{2}\}$
11. $(-\frac{4}{5}, \frac{4}{5})$;
13. $(-\infty, -10) \cup (4, \infty)$;
15. $[3, 4]$;
17. $(-\infty, -1 - \sqrt{2}) \cup [1 - \sqrt{2}, \infty)$;
19. $(-\frac{7}{3}, 3)$;
21. $(4.99, 5.01)$;
23. $|x - 4| < 7$
25. $|x - 5| > 4$
27. $|x + 3| \ge 2$; $(-\infty, -5] \cup [-1, \infty)$
29. $|A_B - A_M| \le 3$
31. $(11.95, 12.05)$
33. $|x - 0.623| \le 0.005$; $[0.618, 0.628]$

Exercises 1.7 Page 114

1. $(-\infty, -5) \cup (3, \infty)$
3. $(2, 6)$
5. $(-\infty, 0] \cup [5, \infty)$
7. $(-3, 3)$
9. $\varnothing$
11. $(-4, 4)$
13. $[-2\sqrt{3}, 2\sqrt{3}]$
15. $\{-3\}$
17. $(-2, 3)$
19. $[-3, 3)$
21. $(-\infty, \frac{1}{3}) \cup (1, \infty)$
23. $(-\infty, \frac{2}{5}) \cup [-\frac{1}{12}, \infty)$
25. $(-\infty, -8)$
27. $\varnothing$
29. $(-5, 0] \cup [1, \infty)$
31. $(-\infty, -1) \cup [1, 2]$
33. $(-3, -2) \cup (-1, \infty)$
35. $[-2, -1] \cup [1, 2]$
37. $(-\infty, -5)$
39. $(-\infty, -4) \cup (-4, -3) \cup (-3, 5) \cup (5, \infty)$
41. $(-\infty, \frac{1}{2} - \frac{\sqrt{5}}{2}) \cup (\frac{1}{2} + \frac{\sqrt{5}}{2}, \infty)$
43. $(-\infty, \frac{5}{2} - \frac{\sqrt{13}}{2}] \cup [\frac{5}{2} + \frac{\sqrt{13}}{2}, \infty)$
45. (a) $(0, 1)$ (b) $(-\infty, 0) \cup (1, \infty)$
47. Yes, because the solution set of $x^2 - 1 \le 0$ is $[-1, 1]$
49. If $x > 0$ is the number, then $x > 3$.
51. $n > 10$
53. If x denotes the width, then $x > 7$.
55. $R > \frac{10}{3}$

Chapter 1 Review Exercises   Page 117

A. 1. false 3. true 5. false 7. true 9. true
B. 1. $x \le 9$ 3. -6
5. $x = 5, x = 15$ 7. $|x - \sqrt{2}| > 3$
9. $-a$
C. 1. $\{-5\}$ 3. $\{2\}$
5. $\{\frac{4}{3}\}$ 7. $\{\frac{1}{2} - \frac{1}{2}i, \frac{1}{2} + \frac{1}{2}i\}$
9. $\{-4, \frac{3}{2}\}$ 11. $\{\frac{3}{4}\}$
13. $\{\frac{3 - \sqrt{15}}{2}, \frac{3 + \sqrt{15}}{2}\}$ 15. $\{\pm 5\sqrt{2}i\}$
17. $\{2, -1 - \sqrt{3}i, -1 + \sqrt{3}i\}$
19. $\{\pm\sqrt{-2 + 2\sqrt{3}}, \pm\sqrt{2 + 2\sqrt{3}i}\}$
21. $\{1\}$ 23. $\{\pm 9\}$
25. $\{7\}$ 27. $(-\infty, -3]$
29. $(4, 12)$ 31. $(-\infty, -10) \cup (10, \infty)$
33. $(-\frac{1}{3}, 3)$ 35. $[-1, \frac{5}{2}]$
37. $(-1, 0) \cup (1, \infty)$ 39. $(0, 1) \cup (1, \infty)$
41. (a) $-6 < x \le 2$ (b) $(-6, 2]$
43. (a) $x \ge -4$ (b) $[-4, \infty)$
45. $b = \dfrac{A - 2ac}{2(a + c)}$ 47. $f = \dfrac{pq}{p + q}$
49. $x = \pm\dfrac{a}{b}\sqrt{b^2 - y^2}$ 51. $10 - 2i$
53. $22 - 7i$ 55. $\frac{1}{5} + \frac{1}{10}i$
57. $-\frac{14}{25} - \frac{23}{25}i$ 59. $x = \frac{16}{3}, y = 4$
61. $x = -2, y = 2$ 63. 15, 18
65. 88   67. 59.1 mi/h
69. 2 m   71. 4 mi/h
73. 5 in.   75. $(2, 2.5)$
77. $10,000 at 6%; $20,000 at 8%   79. 25
81. approximately 0.34 cm
83. (a) $t < 34.5°C$
    (b) No. If $P_v = t = 37°C$, then
    $37 = 30.1 + 0.2(37) - (4.12 - 0.13(37))v$. It follows that
    $v = -\frac{50}{69}$, but it makes no sense to have a negative wind
    speed in this application.
    (c) $t > 4.12/0.13 \approx 31.7°C$
85. (a) $T = \sqrt{D^3/216}$
    (b) approximately 0.19245 h or 11.55 min

Exercises 2.1   Page 128

1.
3.
5. II   7. III   9. II   11. I   13. III   15. IV
17.

19. $(3, 6)$
21.
23.
25.
27. $2\sqrt{5}$
29. 10   31. 5
33. not a right triangle   35. a right triangle
37. an isosceles triangle
39. (a) $2x + y - 5 = 0$
    (b) The points (x, y) lie on the perpendicular bisector of the line
    segment joining A and B.
41. $(6, 8)$ and $(6, -4)$ 43. $(1, \frac{5}{2})$
45. $(-\frac{9}{2}, \frac{5}{2})$ 47. $(3a, -\frac{3}{2}b)$
49. $(5, -1)$ 51. $(-7, -10)$
53. 6   55. $(2, -5)$
57. $(\frac{7}{2}, \frac{13}{2})$, $(4, 7)$, $(\frac{9}{2}, \frac{15}{2})$

Exercises 2.2   Page 136

1. center $(0, 0)$; radius $\sqrt{5}$ 3. center $(0, 3)$; radius 7

5. center $(\frac{1}{2}, \frac{3}{2})$; radius 1   7. center $(0, -4)$; radius 4

9. center $(-1, 2)$; radius 3
11. center $(10, -8)$; radius 6
13. center $(-1, -4)$; radius $\sqrt{\frac{33}{2}}$
15. $x^2 + y^2 = 1$
17. $x^2 + (y - 2)^2 = 2$
19. $(x - 1)^2 + (y - 6)^2 = 8$
21. $x^2 + y^2 = 5$
23. $(x - 5)^2 + (y - 6)^2 = 36$

25.

27.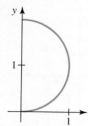

29. $y = 3 + \sqrt{4-x^2}$; $x = \sqrt{4-(y-3)^2}$

31.

33.

35. $(3-\sqrt{13}, 0)$, $(3+\sqrt{13}, 0)$, $(0, -6-2\sqrt{10})$, $(0, -6+2\sqrt{10})$

37. $(0, 0)$; origin

39. $(-1, 0)$, $(0, \frac{1}{2})$; no symmetry

41. $(0, 0)$; x-axis

43. $(-2, 0)$, $(2, 0)$, $(0, 4)$; y-axis

45. $(1-\sqrt{3}, 0)$, $(1+\sqrt{3}, 0)$, $(0, -2)$; no symmetry

47. $(0, 0)$, $(-\sqrt{3}, 0)$, $(\sqrt{3}, 0)$; origin

49. $(0, -4)$, $(0, 4)$; x-axis

51. $(0, -3)$, $(0, 3)$; x-axis, y-axis, and origin

53. $(-\sqrt{7}, 0)$, $(\sqrt{7}, 0)$; origin

55. $(-4, 0)$, $(5, 0)$, $(0, -\frac{10}{3})$; no symmetry

57. $(9, 0)$, $(0, -3)$; no symmetry

59. $(9, 0)$, $(0, 9)$; no symmetry

61. $(-4, 0)$, $(4, 0)$, $(0, -4)$, $(0, 4)$, x-axis, y-axis, and origin

63. x-axis, y-axis, and origin

65. y-axis

67.

69.

71.

Exercises 2.3 Page 144

1. $-\frac{7}{2}$;

3. 5;

5. -1;

7. $-\frac{5}{12}$

9. $\frac{3}{4}$; $(-4, 0)$, $(0, 3)$;

11. $\frac{2}{3}$; $(\frac{9}{2}, 0)$, $(0, -3)$;

13. $-\frac{2}{5}$; $(4, 0)$, $(0, \frac{8}{5})$;

15. $-\frac{2}{3}$; $(\frac{3}{2}, 0)$, $(0, 1)$;

17. $y = \frac{2}{3}x + \frac{4}{3}$

19. $y = 2$

21. $y = -x + 3$

23. $y = -2x + 7$

25. $y = 1$

27. $x = -2$

29. $y = -3x - 2$

31. $x = 5$

33. $y = -4x + 11$

35. $y = -\frac{1}{5}x - 5$

37. (a) and (c) are parallel, (b) and (e) are parallel; (a) and (c) are perpendicular to (b) and (e); (d) is perpendicular to (f)

39. (a) and (d) are perpendicular, (b) and (c) are perpendicular, (e) and (f) are perpendicular

41. $y = x + 3$

Exercises 2.4 Page 149

1. 12

3. 3

5. (a) $F = 40x$ (b) 1.25 ft

7. $s = 16t^2$; 400 ft; 80 ft

9. The length should be $4L$.

11. approximately 2.06 m²

13. $F_L = k\dfrac{I_1 I_2}{r^2}$; F_L is quadrupled

15. $P = k\dfrac{T}{V}$; approximately 3646 ft³

17. July 4th; approximately 2357 cm/s

19. approximately 0.016 cm

Chapter 2 Review Exercises Page 151

A. 1. false **3.** true **5.** true
7. false **9.** true **11.** true
13. true **15.** true **17.** true
19. true **21.** true

B. 1. $-\frac{6}{5}$

3. $\frac{4}{3}$, $(-12, 0)$, $(0, 16)$

5. $\frac{3}{2}$

7. $(4, -3)$ and $(-4, -3)$

9. $(2, -7)$, $2\sqrt{2}$

11. $x_1 = -12$; $y_2 = 9$

13. quadrant II and quadrant IV

15. $x = -2$

17. semicircle

19. $\sqrt{10}$

C. 1. a right triangle

3. $y = \frac{1}{3}x$

5. $y = -\frac{15}{4}x + 8$

7. $(1, 2)$, $(-4, 4)$

9. $x^2 + y^2 = 16$

11. $(x+9)^2 + (y-3)^2 = 100$, $(x-9)^2 + (y+3)^2 = 100$

13. $(0, -4)$
15. $x = 4, x = -2$
17. $x^2 = 4y$
19. $(3, 0), (5, -6)$
21. $x - \sqrt{3}y + 4\sqrt{3} - 7 = 0$
23. (g)
25. (h)
27. (b)
29. (d)

Exercises 3.1 Page 161

1. 24, 2, 8, 35
3. $0, 1, 2, \sqrt{6}$
5. $-\frac{3}{2}, 0, \frac{3}{2}, \sqrt{2}$
7. $-2x^2 + 3x, -8a^2 + 6a, -2a^4 + 3a^2, -50x^2 - 15x,$
$-8a^2 - 2a + 1, -2x^2 - 4xh - 2h^2 + 3x + 3h$
9. $-2, 2$
11. $[\frac{1}{2}, \infty)$
13. $(-\infty, 1)$
15. $\{x \mid x \neq 0, x \neq 3\}$
17. $\{x \mid x \neq 5\}$
19. $(-\infty, \infty)$
21. $[-5, 5]$
23. $(-\infty, 0] \cup [5, \infty)$
25. $(-2, 3]$
27. not a function
29. function
31. $[-4, 4], [0, 5]$
33. $[1, 9], [1, 6]$
35. $-\frac{6}{5}$
37. 2, 3
39. $0, \frac{1}{3}, -9$
41. $-1, 1$
43. $(8, 0), (0, -4)$
45. $(\frac{3}{2}, 0), (\frac{5}{2}, 0), (0, 15)$
47. $(0, -\frac{1}{4})$
49. $(-2, 0), (2, 0), (0, 3)$
51. $f_1(x) = \sqrt{x + 5}, f_2(x) = -\sqrt{x + 5}; [-5, \infty)$
53. $0, -3.4, 0.3, 2, 3.8, 2.9; (0, 2)$
55. $3.6, 2, 3.3, 4.1, 2, -4.1; (-3.2, 0), (2.3, 0), (3.8, 0)$
57. (a) $2; 6; 120; 5040$ (c) $(n + 1)(n + 2)$

Exercises 3.2 Page 172

1. even
3. neither even nor odd
5. odd
7. even
9. even
11. odd
13. neither even nor odd
15. (a) (b)

17. (a) (b)

19. $f(2) = 4, f(-3) = 7$
21. $g(1) = 5, g(-4) = -8$
23. $(-2, 3), (3, -2)$
25. $(-8, 1), (-3, -4)$
27. $(-6, 2), (-1, -3)$
29. $(2, 1), (-3, -4)$
31. $(-2, 15), (3, -60)$
33. (a) (b)

(c) (d)

(e) (f)

35. (a) (b)

(c) (d)

(e) (f)

37. (a) (b)

(c) (d)

(e) (f)

(g)

(h)

23. (a) $(1, 0), (2, 0), (0, 2)$ (b) $y = \left(x - \frac{3}{2}\right)^2 - \frac{1}{4}$
(c) $\left(\frac{3}{2}, -\frac{1}{4}\right), x = \frac{3}{2}$ (d)

39. $y = (x - 1)^3 + 5$ **41.** $y = -(x + 7)^4$

Exercises 3.3   Page 181

1. $f(x) = \frac{1}{2}x + \frac{11}{2}$

3. $\left(-\frac{5}{6}, \frac{8}{3}\right);$

5. $(-1, 3);$

25. (a) $(0, 3)$ (b) $y = 4\left(x - \frac{1}{2}\right)^2 + 2$
(c) $\left(\frac{1}{2}, 2\right), x = \frac{1}{2}$ (d)

7. -9
9. $-2x + 1 - h$
11. $2x - 4 + h$

13.

15.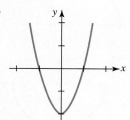

27. (a) $\left(1 - \sqrt{3}, 0\right), \left(1 + \sqrt{3}, 0\right), (0, 1)$
(b) $y = -\frac{1}{2}(x - 1)^2 + \frac{3}{2}$
(c) $\left(1, \frac{3}{2}\right), x = 1$
(d)

17.

29. (a) $(5, 0), (0, 25)$
(b) $y = (x - 5)^2$
(c) $(5, 0), x = 5$
(d)

31. Minimum functional value is $f\left(\frac{4}{3}\right) = -\frac{13}{3}; \left[-\frac{13}{3}, \infty\right)$
33. Increasing on $[0, \infty)$, decreasing on $(-\infty, 0]$
35. Increasing on $(-\infty, -3]$, decreasing on $[-3, \infty)$
37. The graph of $y = x^2$ is shifted horizontally 10 units to the right.
39. The graph of $y = x^2$ is compressed vertically, followed by a reflection in the x-axis, followed by a horizontal shift of 4 units to the left, followed by a vertical shift of 9 units upward.
41. Because $f(x) = (-x - 6)^2 - 4 = (x + 6)^2 - 4$ the graph of f is the graph of $y = x^2$ shifted horizontally 6 units to the left, followed by a vertical shift of 4 units downward.

19. (a) $(0, 0), (-5, 0)$ (b) $y = \left(x + \frac{5}{2}\right)^2 - \frac{25}{4}$
(c) $\left(-\frac{5}{2}, -\frac{25}{4}\right), x = -\frac{5}{2}$ (d)

43. $y = (x + 2)^2$ **45.** $y = -x^2 - 1$
47. $y = -(x - 1)^2 + 5$ **49.** $f(x) = 2x^2 + 3x + 5$
51. $f(x) = 4(x - 1)^2 + 2$

21. (a) $(-1, 0), (3, 0), (0, 3)$ (b) $y = -(x - 1)^2 + 4$
(c) $(1, 4), x = 1$ (d)

53. $(-4, 8), (1, 3),$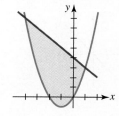

55. $(-2, 2), (0, 2),$

57. (a) $d^2 = 5x^2 - 10x + 25$ (b) $(1, 2)$
59. (a) $s(t) = -16t^2 + 64t + 6$, $v(t) = -32t + 64$
 (b) 70 ft, 0 ft/s (c) 4 s, −64 ft/s
61. (a) 117.6 m, −9.8 m/s (b) in 5 seconds (c) −49 m/s
63. (a) $T_F = \frac{9}{5}T_C + 32$
65. $1680; approximately 35.3 years
67. (a) The graph of $R(D) = -kD^2 + kPD$ is a parabola with vertex at $-b/2a = (-kP)/(-2k) = P/2$. Since k is positive, the graph opens downward, and so $R(D)$ is a maximum at this value. Since $R(D)$ measures the rate at which the disease spreads, we conclude that the disease spreads most rapidly when exactly one-half the population is infected.
 (b) 3×10^{-5}
 (c) approximately 48
 (d) approximately 62, 79, 102, and 130

Exercises 3.4 Page 189

1. 2, 4, −5
3. 3, 0, 8, $2 + 2\sqrt{2}$
5. (a) 1 (b) 1 (c) 0 (d) 1 (e) 1 (f) 0
7. (a) 3 (b) $-1, \sqrt{2}$ (c) $\sqrt[3]{-2}, 1$ (d) $\sqrt[3]{-3}, 0$
 (e) $\sqrt{3}$ (f) -2
9. (0, 0); continuous; **11.** (0, 0); continuous;

13. x-intercepts are the points $(x, 0)$, where $-2 \leq x < -1$; y-intercept is $(0, 2)$; the function is discontinuous at every integer value of x;

15. x-intercepts are the points $(x, 0)$, where $0 \leq x < 1$; y-intercept is $(0, 0)$; the function is discontinuous at every integer value of x;

17. $(-3, 0)$, $(0, 3)$; continuous;

19. $(-2, 0)$, $(2, 0)$, $(0, 2)$; continuous;

21. $(-3, 0)$, $(1, 0)$, $(0, -1)$; continuous;

23. $(\frac{5}{3}, 0)$, $(0, -5)$; continuous;

25. $(-1, 0)$, $(1, 0)$, $(0, 1)$; continuous;

27. $(0, 0)$, $(2, 0)$; continuous;

29. $(-2, 0)$, $(2, 0)$, $(0, 2)$; continuous;

31. $(1, 0)$, $(0, 1)$; continuous;

33. $(1, 0)$, $(0, 1)$; continuous;

35. $\{-1, 1\}$

37. $f(x) = \begin{cases} x + 2, & x < 0 \\ -2x + 2, & 0 \leq x < 2 \\ -2, & x \geq 2 \end{cases}$

39. $f(x) = \begin{cases} -x, & x < -3 \\ \sqrt{9 - x^2}, & -3 \leq x < 3 \\ x, & x \geq 3 \end{cases}$

41.

43. $f(x) = \begin{cases} 1, & x > 0 \\ -1, & x < 0 \end{cases}$

45. $k = 1$

47. $g(x) = \lceil x \rceil = \begin{cases} -2, & -3 < x \leq -2 \\ -1, & -2 < x \leq -1 \\ 0, & -1 < x \leq 0 \\ 1, & 0 < x \leq 1 \\ 2, & 1 < x \leq 2 \\ 3, & 2 < x \leq 3 \end{cases}$

Exercises 3.5 Page 197

1. $(f + g)(x) = 3x^2 - x + 1$, domain: $(-\infty, \infty)$
   $(f - g)(x) = -x^2 + x + 1$, domain: $(-\infty, \infty)$
   $(fg)(x) = 2x^4 - x^3 + 2x^2 - x$, domain: $(-\infty, \infty)$
   $(f/g)(x) = (x^2 + 1)/(2x^2 - x)$, domain: real numbers except $x = 0$ and $x = \frac{1}{2}$

3. $(f + g)(x) = x + \sqrt{x - 1}$, domain: $[1, \infty)$,
   $(f - g)(x) = x - \sqrt{x - 1}$, domain: $[1, \infty)$,
   $(fg)(x) = x\sqrt{x - 1}$, domain: $[1, \infty)$,
   $(f/g)(x) = x/\sqrt{x - 1}$, domain: $[1, \infty)$

5. $(f + g)(x) = 3x^3 - 3x^2 + 3x + 1$, domain: $(-\infty, \infty)$,
   $(f - g)(x) = 3x^3 - 5x^2 + 7x - 1$, domain: $(-\infty, \infty)$,
   $(fg)(x) = 3x^5 - 10x^4 + 16x^3 - 14x^2 + 5x$, domain: $(-\infty, \infty)$,
   $(f/g)(x) = (3x^3 - 4x^2 + 5x)/(1 - x)^2$, domain: real numbers except $x = 1$

7. $(f + g)(x) = \sqrt{x + 2} + \sqrt{5 - 5x}$, domain: $[-2, 1]$,
   $(f - g)(x) = \sqrt{x + 2} - \sqrt{5 - 5x}$, domain: $[-2, 1]$,
   $(fg)(x) = \sqrt{5(x + 2)(1 - x)}$, domain: $[-2, 1]$,
   $\left(\dfrac{f}{g}\right)(x) = \sqrt{\dfrac{x + 2}{5 - 5x}}$, domain: $[-2, 1]$

9. $10, 8, -1, 2, 0$

11. $(f \circ g)(x) = x$, domain: $[1, \infty)$;
    $(g \circ f)(x) = \sqrt{x^2} = |x|$, domain: $(-\infty, \infty)$

13. $(f \circ g)(x) = \dfrac{1}{2x^2 + 1}$, domain: $(-\infty, \infty)$;
    $(g \circ f)(x) = \dfrac{4x^2 - 4x + 2}{4x^2 - 4x + 1}$, domain: real numbers except $x = \frac{1}{2}$

15. $(f \circ g)(x) = x$, $(g \circ f)(x) = x$

17. $(f \circ g)(x) = \dfrac{x^3 + 1}{x}$, $(g \circ f)(x) = \dfrac{x^2}{x^3 + 1}$

19. $(f \circ g)(x) = x + 1 + \sqrt{x - 1}$,
    $(g \circ f)(x) = x + 1 + \sqrt{x}$

21. $(f \circ f)(x) = 4x + 18$, $\left(f \circ \dfrac{1}{f}\right)(x) = \dfrac{6x + 19}{x + 3}$

23. $(f \circ f)(x) = x^4$, $\left(f \circ \dfrac{1}{f}\right)(x) = \dfrac{1}{x^4}$

25. $(f \circ g \circ h)(x) = |x - 1|$
27. $(f \circ g \circ g)(x) = 54x^4 + 7$
29. $(f \circ f \circ f)(x) = 8x - 35$
31. $f(x) = x^5, g(x) = x^2 - 4x$
33. $f(x) = x^2 + 4\sqrt{x}, g(x) = x - 3$

35.

37. $y = \begin{cases} \vdots \\ x + 3, & -3 \leq x < -2 \\ x + 2, & -2 \leq x < -1 \\ x + 1, & -1 \leq x < 0 \\ x, & 0 \leq x < 1 \\ x - 1, & 1 \leq x < 2 \\ x - 2, & 2 \leq x < 3 \\ \vdots \end{cases}$

39.

41.

43.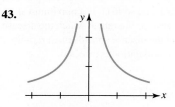

45. (a) $(-2, 3), (1, 0)$ (b) $d = -x^2 - x + 2$
    (c) $\frac{9}{4}$
47. (a) $-8x - 4h$ (b) -24.4
49. (a) $6x + 3h - 1$ (b) 17.3
51. (a) $3x^2 + 3xh + h^2 + 5$ (b) 32.91
53. (a) $\dfrac{1}{(4 - x)(4 - x - h)}$ (b) $\frac{10}{9}$
55. (a) $\dfrac{-1}{(x - 1)(x + h - 1)}$ (b) $-\frac{5}{21}$
57. (a) $1 - \dfrac{1}{x(x + h)}$ (b) $\frac{83}{93}$
59. $\dfrac{2}{\sqrt{x + h} + \sqrt{x}}$
61. $d = \sqrt{10,000 + 250,000t^2}$; approximately 2502 ft

Exercises 3.6 Page 206

1. not one-to-one
3. not one-to-one
5. one-to-one
7. one-to-one;

9. not one-to-one;

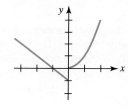

19. domain is $[4, \infty)$; range is $[0, \infty)$
21. $f^{-1}(x) = \dfrac{4}{x^2}, x > 0, y > 0$
23. $f^{-1}(x) = -\frac{1}{2}x + 3$,

25. $f^{-1}(x) = \sqrt[3]{x - 2}$,

27. $f^{-1}(x) = (2-x)^2, x \leq 2,$

29. $f^{-1}(x) = \dfrac{x+1}{2x}$, domain of f^{-1} is the set of real numbers except $x = 0$; range of f^{-1} is the set of real numbers except $y = \tfrac{1}{2}$; range of f is the set of real numbers except $y = 0$

31. $f^{-1}(x) = \dfrac{3x}{2x-7}$, domain of f^{-1} is the set of real numbers except $x = \tfrac{7}{2}$; range of f^{-1} is the set of real numbers except $y = \tfrac{3}{2}$; range of f is the set of real numbers except $y = \tfrac{7}{2}$

33. $(20, 2)$ **35.** $(12, 9)$

37.   **39.**

41. $f^{-1}(x) = \tfrac{1}{2}\sqrt{x-2}, x \geq 2;$ **43.** $f^{-1}(x) = 2\sqrt{1-x^2}, [0,1];$

Exercises 3.7   Page 212

1. $S(x) = x + \dfrac{50}{x}, (0, \infty)$
3. $S(x) = 3x^2 - 4x + 2, [0, 1]$
5. $A(x) = 100x - x^2, [0, 100]$
7. $A(x) = 2x - \tfrac{1}{2}x^2, [0, 4]$
9. $d(x) = \sqrt{2x^2 + 8}, (-\infty, \infty)$
11. $P(A) = 4\sqrt{A}, (0, \infty)$
13. $d(C) = \dfrac{1}{\pi}C, (0, \infty)$
15. $A(h) = \dfrac{1}{\sqrt{3}}h^2, (0, \infty)$
17. $A(x) = \dfrac{1}{4\pi}x^2, (0, \infty)$
19. $s(h) = \dfrac{30h}{25-h}, [0, 25)$
21. $S(w) = 3w^2 + \dfrac{1200}{w}, (0, \infty)$
23. $d(t) = 20\sqrt{13t^2 + 8t + 4}, (0, \infty)$
25. $V(h) = \begin{cases} 120h^2, & 0 \leq h < 5 \\ 1200h - 3000, & 5 \leq h \leq 8 \end{cases}, [0, 8]$

27. $A(x) = \tfrac{1}{2}xp - x^2, [0, \tfrac{1}{2}p]$
29. $F(x) = 2x + \dfrac{16{,}000}{x}, (0, \infty)$
31. $C(x) = 4x + \dfrac{640{,}000}{x}, (0, \infty)$
33. $A(x) = x^2 + \dfrac{128{,}000}{x}, (0, \infty)$
35. $V(x) = 20x - 40x^2, [0, \tfrac{1}{2}]$
37. $A(y) = 40 + 4y + \dfrac{64}{y}, (0, \infty)$
39. $V(x) = 5x\sqrt{64 - x^2}, [0, 8]$

Exercises 3.8   Page 219

1. $y = 0.4x + 0.6$
3. $y = 1.1x - 0.3$
5. $y = 1.3571x + 1.9286$
7. $v = -0.8357T + 234.333; 117.335, 100.621$
9. (a) $y = 0.5996x + 4.3665; y = -0.0232x^2 + 0.5618x + 4.5942;$
   $y = 0.00079x^3 - 0.0212x^2 + 0.5498x + 4.5840$

Chapter 3 Review Exercises   Page 220

A. 1. false **3.** false
5. true **7.** true
9. true **11.** true
13. true **15.** false
17. false **19.** true
21. true

B. 1. $-\tfrac{1}{3}$ **3.** $(-\infty, 5)$
5. $x = 0, x = 2$ **7.** -2
9. $f(x) = 4x + 4$
11. $(1 - \sqrt{2}, 0), (1 + \sqrt{2}, 0), (0, -1)$
13. $f(x) = \tfrac{7}{4}(x+2)^2$ **15.** $(10, 2)$
17. $(0, 5)$ **19.** -8

C. 1. $f(x) = x^2, g(x) = \dfrac{3x - 5}{x}$
3. (a) $y = (x+3)^3 - 2$ **(b)** $y = x^3 - 7$ **(c)** $y = (x-1)^3$
  (d) $y = -x^3 + 2$ **(e)** $y = -x^3 - 2$ **(f)** $y = 3x^3 - 6$
5. domain: $(\pi/2, 3\pi/2)$; range: $(-\infty, \infty)$
7. $f(x) = \begin{cases} x + 1, & x < 0 \\ -x + 1, & 0 \leq x < 1 \\ x - 1, & x \geq 1 \end{cases}$

9. From the graph of f it is seen that $f(x) > 0$ for all x. Thus the domain of g is $(-\infty, \infty)$.
11. $f^{-1}(x) = -1 + \sqrt[3]{x}$
13. $-6x - 3h + 16$
15. $\dfrac{2x + h}{2x^2(x+h)^2}$
17. $A(h) = h^2(1 - \pi/4)$
19. $d(s) = \sqrt{3}s$
21. (a) $d(t) = 6t$ **(b)** $d(t) = \sqrt{90^2 + (90 - 6t)^2}$
23. $A(x) = 2x(1 - \pi x)$, where x is the radius of the semicircle

Exercises 4.1 Page 233

1.

3.

5.

7.

9. odd

11. neither even nor odd

13. (f) **15.** (e) **17.** (b)

19.

21.

23.

25.

27.

29.

31.

33.

35.

37.

39.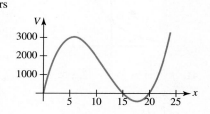

41. (b) $f(x) = (x - 1)^2(x + 2)$

43. $k = -\frac{7}{16}$

45. $k = -\frac{10}{3}$

47. the odd positive integers

49. domain is [0, 15]; graph of V is

Exercises 4.2 Page 241

1. $f(x) = x^2 \cdot 8 + 4x - 7$
3. $f(x) = (x^2 + x - 1) \cdot (5x - 12) + 21x - 11$
5. $f(x) = (x + 2)^2 \cdot (2x - 4) + 5x + 21$
7. $f(x) = (3x^2 - x) \cdot (9x + 3) + 4x - 2$
9. $f(x) = (6x^2 + 4x + 1) \cdot (x^3 - 2) + 12x^2 + 8x + 2$
11. $r = 6$
13. $r = \frac{29}{8}$
15. $r = 76$
17. $f(2) = 2$
19. $f(-5) = -74$
21. $f\left(\frac{1}{2}\right) = \frac{303}{16}$
23. $q(x) = 2x + 3, r = 11$
25. $q(x) = x^2 - 4x + 12, r = -34$
27. $q(x) = x^3 + 2x^2 + 4x + 8, r = 32$
29. $q(x) = x^4 - 4x^3 + 16x^2 - 8x + 32, r = -132$
31. $q(x) = x^2 - 2x + \sqrt{3}, r = 0$
33. $f(-3) = 51$
35. $f(1) = 1$
37. $f(4) = 5369$
39. $k = -1$
41. $k = -\frac{1}{5}$
43. $k = -4$

Exercises 4.3 Page 248

1. $f(x) = 4\left(x - \frac{1}{4}\right)(x - 1)^2$
3. 5 is not a zero
5. $f(x) = 3\left(x + \frac{2}{3}\right)(x - 2 + \sqrt{2})(x - 2 - \sqrt{2})$
7. $f(x) = 4(x + 3)(x - 5)\left(x - \frac{1}{2}\right)\left(x + \frac{1}{2}\right)$
9. $f(x) = 9(x - 1)\left(x + \frac{1}{3}\right)^2(x + 8)$
11. $x - 5$ is not a factor
13. $f(x) = (x - 1)\left(x + \frac{1}{2} + \frac{1}{2}\sqrt{7}i\right)\left(x + \frac{1}{2} - \frac{1}{2}\sqrt{7}i\right)$
15. $x - \frac{1}{3}$ is not a factor
17. $f(x) = (x - 1)(x - 2)(x - 2i)(x + 2i)$
19. $f(x) = 2(x - 1)^2(x + 1)\left(x + \frac{3}{2}\right)$

21. $f(x) = 3(x - \frac{5}{3})(x + 2i)(x - 2i)$
23. $f(x) = 5(x - \frac{2}{5})(x + 1 - i)(x + 1 + i)$
25. $f(x) = (x - 3)(x + 3)(x - 1 + 2i)(x - 1 - 2i)$
27. $f(x) = (x - 2)(x - 1)(x + 3)^2 = x^4 + 3x^3 - 7x^2 - 15x + 18$
29. $f(x) = x^5 - 6x^4 + 10x^3$
31. $f(x) = x^2 - 2x + 37$
33. 0 is a simple zero; $\frac{5}{4}$ is a zero of multiplicity 2; $\frac{1}{2}$ is a zero of multiplicity 3
35. $-\frac{2}{3}$ is a zero of multiplicity 2; $\frac{2}{3}$ is a zero of multiplicity 2
37. $k = -36; f(x) = 2(x - 3)(x + 1 - \sqrt{5}i)(x + 1 + \sqrt{5}i)$
39. $f(x) = -\frac{1}{16}(x - 4)(x + 2)^2$

Exercises 4.4 Page 255

1. $\frac{2}{5}$ 3. 3 5. $\frac{1}{2}$ (multiplicity 2)
7. no rational zeros 9. $\frac{1}{3}, \frac{3}{2}$
11. 0, 1 13. $-3, 0, 2$ 15. $\frac{3}{2}$
17. $-\frac{1}{5}$ 19. $-\frac{1}{2}$ (multiplicity 2); $\frac{1}{3}$ (multiplicity 2)
21. $\frac{3}{8}, -\frac{1}{2} - \frac{1}{2}\sqrt{5}, -\frac{1}{2} + \frac{1}{2}\sqrt{5}$;
$f(x) = (8x - 3)(x + \frac{1}{2} + \frac{1}{2}\sqrt{5})(x + \frac{1}{2} - \frac{1}{2}\sqrt{5})$
23. $\frac{4}{5}, \frac{5}{2}, -\sqrt{2}, \sqrt{2}$;
$f(x) = (5x - 4)(2x - 5)(x + \sqrt{2})(x - \sqrt{2})$
25. $-4, -1, 1, -\sqrt{5}, \sqrt{5}$;
$f(x) = (x + 4)(x + 1)(x - 1)(x + \sqrt{5})(x - \sqrt{5})$
27. $0, 1, 3, -1 - \sqrt{2}, -1 + \sqrt{2}$;
$f(x) = 4x(x - 1)(x - 3)(x + 1 + \sqrt{2})(x + 1 - \sqrt{2})$
29. $-1, \frac{1}{4}$ (multiplicity 2); $f(x) = (x + 1)(4x - 1)^2(x^2 - 2x + 3)$
31. $-\frac{1}{2}$
33. $-\frac{3}{2}, 2, -2 - \sqrt{3}, -2 + \sqrt{3}$
35. 1 (multiplicity 3)
37. $f(x) = 3x^4 - x^3 - 39x^2 + 49x - 12$
39. $f(x) = -\frac{1}{6}(x - 1)(x - 2)(x - 3)$
41. 3 in. or $\frac{1}{2}(7 - \sqrt{33}) \approx 0.63$ in.

Exercises 4.5 Page 259

1. -1.531 3. -1.314
5. $1.611; 3.820$ 7. $-1.141; 1.141$
9. 1.730 in.

Exercises 4.6 Page 271

1.
x	3.1	3.01	3.001	3.0001	3.00001
$f(x)$	62	602	6002	60,002	600,002
x	2.9	2.99	2.999	2.9999	2.99999
$f(x)$	-58	-598	-5998	$-59,998$	$-599,998$

3. Asymptotes: $x = 2, y = 0$
   Intercepts: $(0, -\frac{1}{2})$

5. Asymptotes: $x = -1, y = 1$
   Intercepts: $(0, 0)$

7. Asymptotes: $x = -\frac{3}{2}, y = 2$
   Intercepts: $(\frac{9}{4}, 0), (0, -3)$

9. Asymptotes: $x = -1, y = -1$
   Intercepts: $(1, 0), (0, 1)$

11. Asymptotes: $x = 1, y = 0$
    Intercepts: $(0, 1)$

13. Asymptotes: $x = 0, y = 0$
    Intercepts: none

15. Asymptotes: $x = 1, x = -1, y = 0$
    Intercepts: $(0, 0)$

17. Asymptotes: $x = 0, x = 2, y = 0$
    Intercepts: none

19. Asymptotes: $x = 0, y = -1$
    Intercepts: $(-1, 0), (1, 0)$

21. Asymptotes: $x = 1, y = -2$
    Intercepts: $(-2, 0), (2, 0), (0, 8)$

23. Asymptotes: $x = 0, y = x$
    Intercepts: $(-3, 0), (3, 0)$

25. Asymptotes: $x = -2, y = x - 2$
    Intercepts: $(0, 0)$

27. Asymptotes: $x = 1$, $y = x - 1$
Intercepts: $(3, 0)$, $(-1, 0)$, $(0, 3)$

29. Asymptotes: $x = 1$, $x = 0$, $y = x + 1$
Intercepts: $(2, 0)$

3. $7x^3 + 14x^2 + 22x + 53 + \dfrac{109}{x - 2}$

5. $r = f(-3) = -198$ **7.** n an odd positive integer

9. $\pm 1, \pm 3, \pm 5, \pm 15, \pm \tfrac{1}{2}, \pm \tfrac{3}{2}, \pm \tfrac{5}{2}, \pm \tfrac{15}{2}, \pm \tfrac{1}{4}, \pm \tfrac{3}{4}, \pm \tfrac{5}{4}, \pm \tfrac{15}{4},$
$\pm \tfrac{1}{8}, \pm \tfrac{3}{8}, \pm \tfrac{5}{8}, \pm \tfrac{15}{8}$

11. $f(x) = (x - 2)\left(x - \tfrac{7}{2} + \tfrac{1}{2}\sqrt{3}i\right)\left(x - \tfrac{7}{2} - \tfrac{1}{2}\sqrt{3}i\right)$

13. $k = -\tfrac{21}{2}$ **15.** $k = \tfrac{3}{2}$

17. $f(x) = 3x^2(x + 2)^2(x - 1)$ **19.** (f)

21. (d)   **23.** (h)

25. (c)   **27.** (b)

29. $y = 0$, $x = -4$, $x = 2$, $(-2, 0)$, $\left(0, -\tfrac{1}{4}\right)$,

31. $(3, 0)$;

33. $(4, 4)$;

Exercises 5.1   Page 285

1. $(0, 1)$; $y = 0$; decreasing

3. $(0, -1)$; $y = 0$; decreasing

35. $(-3, -6)$

37. $y = \dfrac{x - 5}{x - 2}$

39. $y = \dfrac{3x(x - 3)}{(x + 1)(x - 2)}$

41. Hole is the graph at $x = 1$
Intercepts: $(-1, 0)$, $(0, 1)$

43. Hole in the graph at $x = -1$
Intercepts: none

5. $(0, 2)$; $y = 0$; increasing

7. $(0, -4)$; $y = -5$; increasing

45. $R \to 5$ as $r \to \infty$

47. $I(x) \to \infty$ as $x \to 0^+$;
$I(x) \to \infty$ as $x \to 100^-$

9. $(0, 2)$; $y = 3$; increasing

11. $(0, -1 + e^{-3})$; $y = -1$; increasing

Chapter 4 Review Exercises   Page 274

A. 1. true   **3.** true
5. true   **7.** true
9. true   **11.** false
13. false   **15.** false
17. true

B. 1. $(1, 0)$; $(0, 0)$, $(5, 0)$ **3.** $f(x) = x^4$
5. $k = \tfrac{2}{3}$ **7.** $x = 1$, $x = 4$
9. $y = -\tfrac{1}{2}$ **11.** $n = 0$, $n = 1$, $n = 2$

C. 1. $3x^3 - \tfrac{1}{2}x + 1 + \dfrac{-\tfrac{1}{2}x + 5}{2x^2 - 1}$

13. $f(x) = 6^x$ **15.** $f(x) = (e^{-2})^x = e^{-2x}$
17. $(5, \infty)$ **19.** $(-10, 0)$, $(0, 10)$
21. $x > 4$ **23.** $x < 2$

25.

27.

ANS-14   ANSWERS TO SELECTED ODD-NUMBERED PROBLEMS

29.

31.

33.

35.

37. $y = \frac{26}{9}x + \frac{29}{9}$

41. the interval $(-3, 3)$
43. $\log_{10} 50$
45. $\ln(x^2 - 2)$
47. $\ln 1 = 0$
49. 0.3011
51. 1.8063
53. 0.3495
55. 0.2007
57. -0.0969
59. 1.6609
61. $\ln y = 10\ln x + \frac{1}{2}\ln(x^2 + 5) - \frac{1}{3}\ln(8x^3 + 2)$
63. $\ln y = 5\ln(x^3 - 3) + 8\ln(x^4 + 3x^2 + 1) - \frac{1}{2}\ln x - 9\ln(7x + 5)$

Exercises 5.2 Page 292

1. $-\frac{1}{2} = \log_4 \frac{1}{2}$
3. $4 = \log_{10} 10{,}000$
5. $-s = \log_t v$
7. $2^7 = 128$
9. $(\sqrt{3})^8 = 81$
11. $b^v = u$
13. -7
15. 3
17. e
19. 36
21. $\frac{1}{7}$
23. $f(x) = \log_7 x$
25. $(0, \infty); (1, 0), x = 0$
27. $(-\infty, 0); (-1, 0), x = 0$

29. $(-3, \infty); (5, 0), x = -3$
31. $(0, \infty); (e, 0), x = 0$

33. $-1 < x < 0$
35. $f(x) = \begin{cases} \ln x, & x > 0 \\ \ln(-x), & x < 0 \end{cases}$, $(-1, 0), (1, 0), x = 0$

37.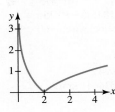

39. the interval $\left(\frac{3}{2}, \infty\right)$

Exercises 5.3 Page 298

1. 2
3. 2
5. 1.0802
7. 2
9. 3
11. 0.3495
13. 2.7712
15. ± 4
17. ± 3
19. -0.8782
21. 32
23. 45
25. $\pm \frac{1}{10}$
27. 81
29. $\frac{3}{2}$
31. 100
33. $2, 8$
35. 1
37. 4
39. $\frac{7}{2}$
41. $0, 2$
43. $1, 16$
45. $\log_5(1 + \sqrt{2}) = \dfrac{\ln(1 + \sqrt{2})}{\ln 5}$
47. e^{-2}, e
49. 0
51. $(-3, 0)$
53. $\left(1 + \frac{\ln 3}{\ln 4}, 0\right) \approx (1.7925, 0)$
55. $(-3, 0), (-2, 0), (0, 0)$
57. approximately -0.6606
59. approximately $-0.4481, 1.5468$

61. $\sqrt{10} \approx 3.1623$
63. e^{-3}, e^3

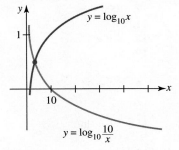

65. 9
67. $x \approx 2.1944$

Exercises 5.4 Page 307

1. (a) $P(t) = P_0 e^{0.3466t}$ (b) $5.66 P_0$ (c) 8.64 h
3. 2344
5. 201

7. (a) 82 (b) 8.53 days (c) 2000
(d)
9. $A(t) = 200e^{-0.005077t}$; 177 mg
11. approximately 100 g, 50 g, 25 g
13. approximately $k = -0.08664$; 34.58 days
15. $0.6730 A_0$; $0.2264 A_0$
17. approximately 16,253 years old
19. approximately 92%
21. (a) 220.2°F (b) 9.2 minutes (c) 80°F
23. 8.74 minutes
25. approximately 1.6 hours
27. $4,851,651.95; 2.35×10^{15}
29. $3080.37 in interest
31. (a) 6 days (b) 19.84%
33. $t = -\dfrac{L}{R}\ln\left(1 - \dfrac{IR}{E}\right)$
35. approximately 25 times stronger
37. 5.5
39. 6
41. 7.6
43. 5×10^{-4}
45. 2.5×10^{-7}
47. 10 times as acidic
49. 158.5 times as acidic
51. 5.62×10^{25} ergs
53. 10^{-2} watts/cm^2
55. 65 dB
57. (a) 2.46 mm (b) 0.79 mm, 0.19 mm (c) 7.7×10^{-6} mL

Exercises 5.5 Page 314

7. (a) $\dfrac{\sqrt{13}}{2}$ (b) $\dfrac{-3}{\sqrt{13}}, -\dfrac{\sqrt{13}}{3}, \dfrac{2}{\sqrt{13}}, -\dfrac{2}{3}$

Chapter 5 Review Exercises Page 315

A. 1. true 3. true 5. true
7. false 9. true 11. false
13. true

B. 1. $(0, 5)$; $y = 6$ 3. $(-3, 0)$; $x = -4$
5. -1 7. 1000
9. 6 11. $\dfrac{1}{9}$
13. 3 15. $3 + e$
17. -7.8577 19. 64
21. 8

C. 1. $\log_5 0.2 = -1$ 3. $9^{1.5} = 27$
5. -2 7. $-\dfrac{1}{2}$
9. $1 - \log_2 7 = 1 - \dfrac{\ln 7}{\ln 2}$ 11. $-2 + \ln 6$
13. $m = -\dfrac{1}{r}\ln(P/S)$ 15.

17. C, D, A, B
19. $\dfrac{3^{1-h} - 3}{h}$
21. (ii)
23. (iv)
25. (iii)
27. upward shift of 1 unit
29. $f(x) = 5e^{(-\frac{1}{6}\ln 5)x} = 5e^{-0.2682x}$
31. $f(x) = 5 + \left(\dfrac{1}{2}\right)^x$
33. After doubling, it will take another 9 hours to double again. In other words, it will take a total of 18 hours for the population to grow to 4 times the initial population.
35. $0.0625 A_0$ or $6\frac{1}{4}$% of the initial quantity A_0
37. 4.3%
39. $x = \dfrac{1}{c}[\ln b - \ln(\ln a - \ln y)]$

Exercises 6.1 Page 325

1.
3.
5.
7.
9.
11.

13.
15.
17. 10.6547°
19. 5.17°
21. 210°46′48″
23. 30°48′36″
25. $\pi/18$
27. $\pi/4$
29. $3\pi/2$
31. $-23\pi/18$
33. 40°
35. 120°
37. 225°
39. 177.62°
41. 155°
43. 110°
45. $-205°$
47. $7\pi/4$
49. 1.3π
51. $2\pi - 4 \approx 2.28$
53. $-\pi/4$
55. (a) 41.75° (b) 131.75°
57. (a) The given angle is greater than 90°. (b) 81.6°
59. (a) $\pi/4$ (b) $3\pi/4$
61. (a) The given angle is greater than $\pi/2$. (b) $\pi/3$
63. (a) 216°; 1.2π (b) $-1845°$; -10.25π
65. because the hour hand moves counterclockwise: $-60°$, $-\pi/3$
67. (a) 16 h (b) 2 h
69. (a) 9 (b) 15
71. (a) 1.5 (b) 85.94°
75. (a) 0.000072921 rad/s (b) 3.074641 km/s
77. 1.15 statute miles

79. (a) 3π rad/s (b) 300π cm/s
81. (a) 711.1 rev/min (b) 4468 rad/min

Exercises 6.2   Page 333

1. $\sin\theta = \frac{4}{5}, \cos\theta = \frac{3}{5}, \tan\theta = \frac{4}{3}, \cot\theta = \frac{3}{4}, \sec\theta = \frac{5}{3}, \csc\theta = \frac{5}{4}$
3. $\sin\theta = \frac{3\sqrt{10}}{10}, \cos\theta = \frac{\sqrt{10}}{10}, \tan\theta = 3, \cot\theta = \frac{1}{3},$
  $\sec\theta = \sqrt{10}, \csc\theta = \frac{\sqrt{10}}{3}$
5. $\sin\theta = \frac{2}{5}, \cos\theta = \frac{\sqrt{21}}{5}, \tan\theta = \frac{2\sqrt{21}}{21}, \cot\theta = \frac{\sqrt{21}}{2},$
  $\sec\theta = \frac{5\sqrt{21}}{21}, \csc\theta = \frac{5}{2}$
7. $\sin\theta = \frac{1}{3}, \cos\theta = \frac{2\sqrt{2}}{3}, \tan\theta = \frac{\sqrt{2}}{4}, \cot\theta = 2\sqrt{2},$
  $\sec\theta = \frac{3\sqrt{2}}{4}, \csc\theta = 3$
9. $\sin\theta = \frac{y}{\sqrt{x^2+y^2}}, \cos\theta = \frac{x}{\sqrt{x^2+y^2}}, \tan\theta = \frac{y}{x}, \cot\theta = \frac{x}{y},$
  $\sec\theta = \frac{\sqrt{x^2+y^2}}{x}, \csc\theta = \frac{\sqrt{x^2+y^2}}{y}$
11. $\tan\theta = \frac{2}{3}, \cot\theta = \frac{3}{2}, \sec\theta = \frac{\sqrt{13}}{3}, \csc\theta = \frac{\sqrt{13}}{2}$
13. $\tan\theta = \frac{2\sqrt{5}}{15}, \cot\theta = \frac{3\sqrt{5}}{2}, \sec\theta = \frac{7\sqrt{5}}{15}, \csc\theta = \frac{7}{2}$
15. $\cos\theta = \frac{8\sqrt{65}}{65}, \cot\theta = 8, \sec\theta = \frac{\sqrt{65}}{8}, \csc\theta = \sqrt{65}$
17. $\sin\theta = \frac{3}{5}, \cos\theta = \frac{4}{5}, \tan\theta = \frac{3}{4}, \cot\theta = \frac{4}{3}$
19. $\sin\theta = \frac{2\sqrt{2}}{3}, \tan\theta = 2\sqrt{2}, \cot\theta = \frac{\sqrt{2}}{4}, \sec\theta = 3$
21. $\cos\theta = \frac{5}{13}, \tan\theta = \frac{12}{5}, \cot\theta = \frac{5}{12}, \sec\theta = \frac{13}{5}, \csc\theta = \frac{13}{12}$
23. $\sin\theta = \frac{1}{2}, \cos\theta = \frac{\sqrt{3}}{2}, \tan\theta = \frac{\sqrt{3}}{3}, \cot\theta = \sqrt{3}, \csc\theta = 2$
25. $\sin\theta = \frac{2\sqrt{29}}{29}, \cos\theta = \frac{5\sqrt{29}}{29}, \cot\theta = \frac{5}{2}, \sec\theta = \frac{\sqrt{29}}{5}, \csc\theta = \frac{\sqrt{29}}{2}$
27. $\sin\theta = \frac{2\sqrt{10}}{7}, \cos\theta = \frac{3}{7}, \tan\theta = \frac{2\sqrt{10}}{3}, \cot\theta = \frac{3\sqrt{10}}{20}, \csc\theta = \frac{7\sqrt{10}}{20}$
29. $\frac{1}{4}(\sqrt{6} - \sqrt{2})$ **31.** $\sqrt{2} - 1$
33. 3   **35.** 2   **37.** -1
39. 0   **41.** 5   **43.** 1
45. 1   **47.** $\frac{1}{2}$ **49.** $\sqrt{3}$
51. $\frac{2\sqrt{3}}{3}$ **53.** $\frac{1}{2}$

Exercises 6.3   Page 338

1. $\tan 45° = 1, \cot 45° = 1, \sec 45° = \sqrt{2}, \csc 45° = \sqrt{2}$
3. $\frac{1}{4}$ **5.** 2   **7.** $\frac{\sqrt{2}}{2}$
9. 18   **11.** $\frac{1}{4}(\sqrt{6} + \sqrt{2})$
13. $9\sqrt{3}$ **15.** 0   **17.** $2\sqrt{2}$
19. $\frac{3}{4}$ **21.** $2 - \sqrt{3}$
23. $\sin 17° = 0.2924, \cos 17° = 0.9563, \tan 17° = 0.3057,$
  $\cot 17° = 3.2709, \sec 17° = 1.0457, \csc 17° = 3.4203$
25. $\sin 14.3° = 0.2470, \cos 14.3° = 0.9690, \tan 14.3° = 0.2549,$
  $\cot 14.3° = 3.9232, \sec 14.3° = 1.0320, \csc 14.3° = 4.0486$
27. $\sin 71.50417° = 0.9483, \cos 71.50417° = 0.3172,$
  $\tan 71.50417° = 2.9894, \cot 71.50417° = 0.3345,$
  $\sec 71.50417° = 3.1522, \csc 71.50417° = 1.0545$
29. $\sin\frac{\pi}{5} = 0.5878, \cos\frac{\pi}{5} = 0.8090, \tan\frac{\pi}{5} = 0.7265,$
  $\cot\frac{\pi}{5} = 1.3764, \sec\frac{\pi}{5} = 1.2361, \csc\frac{\pi}{5} = 1.7013$
31. $\sin 0.6725 = 0.6229, \cos 0.6725 = 0.7823,$
  $\tan 0.6725 = 0.7963, \cot 0.6725 = 1.2558,$
  $\sec 0.6725 = 1.2783, \csc 0.6725 = 1.6053$

Exercises 6.4   Page 347

1. $\sin\theta = \frac{4}{5}, \cos\theta = \frac{3}{5}, \tan\theta = \frac{4}{3}, \csc\theta = \frac{5}{4}, \sec\theta = \frac{5}{3}, \cot\theta = \frac{3}{4}$
3. $\sin\theta = -\frac{12}{13}, \cos\theta = \frac{5}{13}, \tan\theta = -\frac{12}{5}, \csc\theta = -\frac{13}{12},$
  $\sec\theta = \frac{13}{5}, \cot\theta = -\frac{5}{12}$

5. $\sin\theta = 1, \cos\theta = 0, \tan\theta$ is undefined, $\csc\theta = 1, \sec\theta$ is undefined, $\cot\theta = 0$
7. $\sin\theta = \frac{3\sqrt{13}}{13}, \cos\theta = -\frac{2\sqrt{13}}{13}, \tan\theta = -\frac{3}{2}, \csc\theta = \frac{\sqrt{13}}{3},$
  $\sec\theta = -\frac{\sqrt{13}}{2}, \cot\theta = -\frac{2}{3}$
9. $\sin\theta = -\frac{\sqrt{3}}{3}, \cos\theta = -\frac{\sqrt{6}}{3}, \tan\theta = \frac{\sqrt{2}}{2}, \csc\theta = -\sqrt{3},$
  $\sec\theta = -\frac{\sqrt{6}}{2}, \cot\theta = \sqrt{2}$
11. III   **13.** II
15. I   **17.** II
19. $\cos\theta = -\frac{\sqrt{15}}{4}, \tan\theta = -\frac{\sqrt{15}}{15}, \csc\theta = 4, \sec\theta = -\frac{4\sqrt{15}}{15},$
  $\cot\theta = -\sqrt{15}$
21. $\sin\theta = -\frac{3\sqrt{10}}{10}, \cos\theta = -\frac{\sqrt{10}}{10}, \csc\theta = -\frac{\sqrt{10}}{3}, \sec\theta = -\sqrt{10},$
  $\cot\theta = \frac{1}{3}$
23. $\sin\theta = -\frac{1}{10}, \cos\theta = \frac{3\sqrt{11}}{10}, \tan\theta = -\frac{\sqrt{11}}{33}, \sec\theta = \frac{10\sqrt{11}}{33},$
  $\cot\theta = -3\sqrt{11}$
25. $\cos\theta = \frac{2\sqrt{6}}{5}, \tan\theta = -\frac{\sqrt{6}}{12}, \csc\theta = -5, \sec\theta = \frac{5\sqrt{6}}{12},$
  $\cot\theta = -2\sqrt{6}$
27. $\sin\theta = \frac{8\sqrt{65}}{65}, \cos\theta = \frac{\sqrt{65}}{65}, \csc\theta = \frac{\sqrt{65}}{8}, \sec\theta = \sqrt{65}, \cot\theta = \frac{1}{8}$
29. $\pm\frac{\sqrt{91}}{10}$
31. $\sin\theta = \pm\frac{\sqrt{5}}{5}, \cos\theta = \pm\frac{2\sqrt{5}}{5}$
33. $\cos\theta = -\frac{1}{5}, \sin\theta = \pm\frac{2\sqrt{6}}{5}$

35.

θ (radians)	θ (degrees)	$\sin\theta$	$\cos\theta$	$\tan\theta$
0°	0	0	1	0
30°	$\frac{\pi}{6}$	$\frac{1}{2}$	$\frac{\sqrt{3}}{2}$	$\frac{\sqrt{3}}{3}$
45°	$\frac{\pi}{4}$	$\frac{\sqrt{2}}{2}$	$\frac{\sqrt{2}}{2}$	1
60°	$\frac{\pi}{3}$	$\frac{\sqrt{3}}{2}$	$\frac{1}{2}$	$\sqrt{3}$
90°	$\frac{\pi}{2}$	1	0	—
120°	$\frac{2\pi}{3}$	$\frac{\sqrt{3}}{2}$	$-\frac{1}{2}$	$-\sqrt{3}$
135°	$\frac{3\pi}{4}$	$\frac{\sqrt{2}}{2}$	$-\frac{\sqrt{2}}{2}$	-1
150°	$\frac{5\pi}{6}$	$\frac{1}{2}$	$-\frac{\sqrt{3}}{2}$	$-\frac{\sqrt{3}}{3}$
180°	π	0	-1	0
210°	$\frac{7\pi}{6}$	$-\frac{1}{2}$	$-\frac{\sqrt{3}}{2}$	$\frac{\sqrt{3}}{3}$
225°	$\frac{5\pi}{4}$	$-\frac{\sqrt{2}}{2}$	$-\frac{\sqrt{2}}{2}$	1
240°	$\frac{4\pi}{3}$	$-\frac{\sqrt{3}}{2}$	$-\frac{1}{2}$	$\sqrt{3}$
270°	$\frac{3\pi}{2}$	-1	0	—
300°	$\frac{5\pi}{3}$	$-\frac{\sqrt{3}}{2}$	$\frac{1}{2}$	$-\sqrt{3}$
315°	$\frac{7\pi}{4}$	$-\frac{\sqrt{2}}{2}$	$\frac{\sqrt{2}}{2}$	-1
330°	$\frac{11\pi}{6}$	$-\frac{1}{2}$	$\frac{\sqrt{3}}{2}$	$-\frac{\sqrt{3}}{3}$
360°	2π	0	1	0

37. -1 **39.** $\sqrt{3}$
41. $\frac{\sqrt{3}}{2}$ **43.** -2
45. -2 **47.** $\frac{1}{2}$
49. 1   **51.** undefined

53. 60°, 240° **55.** 135°, 225°
57. 270° **59.** 0, π
61. $3\pi/4, 5\pi/4$ **63.** $5\pi/6, 11\pi/6$
65. 4.81 m
67. (a) 978.0309 cm/s² **(b)** 983.21642 cm/s²
  (c) 980.61796 cm/s²

Chapter 6 Review Exercises   Page 350

A. 1. true **3.** false **5.** false
  7. true **9.** true
B. 1. 57° **3.** clockwise **5.** 2
  7. $\frac{5}{2}$ **9.** $\pi/3$
C. 1. **3.**

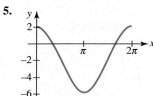

5. $-2\pi/3$ **7.** 0.27π
9. 20° **11.** 131.78°
13. 70°30′ **15.** 177°37′1″
17. 445°, 805°, $-275°$, $-635°$ (Note: Other answers are possible.)
19. $\sin\theta = \frac{2\sqrt{5}}{5}, \cos\theta = -\frac{\sqrt{5}}{5}, \tan\theta = -2, \cot\theta = -\frac{1}{2},$
  $\sec\theta = -\sqrt{5}, \csc\theta = \frac{\sqrt{5}}{2}$
21. $\sin\theta = -\frac{3\sqrt{34}}{34}, \cos\theta = -\frac{5\sqrt{34}}{34}, \tan\theta = \frac{3}{5}, \cot\theta = \frac{5}{3},$
  $\sec\theta = -\frac{\sqrt{34}}{5}, \csc\theta = -\frac{\sqrt{34}}{3}$
23. $\sin\theta = -\frac{4\sqrt{3}}{7}, \tan\theta = 4\sqrt{3}, \cot\theta = \frac{\sqrt{3}}{12}, \sec\theta = -7,$
  $\csc\theta = -\frac{7\sqrt{3}}{12}$
25. $\sin\theta = -\frac{\sqrt{26}}{26}, \cos\theta = \frac{5\sqrt{26}}{26}, \tan\theta = -\frac{1}{5}, \sec\theta = \frac{\sqrt{26}}{5},$
  $\csc\theta = -\sqrt{26}$
27. $\sin\theta = -\frac{1}{7}, \cos\theta = -\frac{4\sqrt{3}}{7}, \tan\theta = \frac{\sqrt{3}}{12}, \cot\theta = 4\sqrt{3},$
  $\sec\theta = -\frac{7\sqrt{3}}{12}$
29. $\sin\theta = \pm\frac{\sqrt{17}}{17}, \cos\theta = \mp\frac{4\sqrt{17}}{17}, \tan\theta = -\frac{1}{4}, \sec\theta = \mp\frac{\sqrt{17}}{4},$
  $\csc\theta = \pm\sqrt{17}$
31. $\frac{\sqrt{2}}{2}$ **33.** -1
35. 45°, 135° **37.** 120°, 240°
39. $\pi/3, 2\pi/3$ **41.** $3\pi/4, 7\pi/4$

Exercises 7.1   Page 360

1. (b) $\left(-\frac{\sqrt{3}}{2}, -\frac{1}{2}\right)$ **3. (b)** $(0, -1)$
5. (b) $\left(\frac{1}{2}, -\frac{\sqrt{3}}{2}\right)$ **7. (b)** $\left(\frac{\sqrt{3}}{2}, \frac{1}{2}\right)$
9. (b) $(0.2675, 0.9636)$ **11. (b)** $(0.6084, -0.7937)$
13. (b) $(0.9833, -0.1822)$ **15. (b)** $(-0.8569, -0.5155)$
17. $\frac{1}{2}$ **19.** $\frac{\sqrt{2}}{2}$
21. -1 **23.** -1
25. $\sin t$ is 2π-periodic **27.** $\sin t$ is an odd function
29. $\cos t$ is an even function **31.** $\frac{\sqrt{21}}{5}$
33. $-\frac{\sqrt{5}}{3}$ **35.** $-\frac{1}{2}\sqrt{2-\sqrt{2}}$
37. $-\frac{1}{2}\sqrt{2+\sqrt{2}}$
39. $t = \pi/4 + 2n\pi, t = 3\pi/4 + 2n\pi$, where n is an integer
41. $t = (2n+1)\pi$, where n is an integer

Exercises 7.2   Page 368

1.

3.

5.

7. $y = -3\sin x$

9. $y = 1 - 3\cos x$ **11.** $(n, 0)$, where n is an integer
13. $((2n+1)\pi, 0)$, where n is an integer
15. $(\pi/4 + n\pi, 0)$, where n is an integer
17. $(\pi/2, 0); (\pi/2 + 2n\pi, 0)$, where n is an integer
19. $y = 3\sin 2x$ **21.** $y = \frac{1}{2}\cos \pi x$
23. $y = -\sin \pi x$
25. amplitude: 4; period: 2 **27.** amplitude: 3; period: 1

29. amplitude: 4; period: 2π **31.** amplitude: 1; period: 3π

33. amplitude: 1; period: 2π; **35.** amplitude: 1; period: 2π;
  phase shift: $\pi/6$ phase shift: $\pi/6$

37. amplitude: 4; period: π; **39.** amplitude: 3; period: 4π;
  phase shift: $3\pi/4$ phase shift: $2\pi/3$

41. amplitude: 4; period: 6; phase shift: 1

43. $y = -5 + 3\cos\left(6x + \dfrac{3\pi}{2}\right)$

25. period: 1;
x-intercepts: $(n, 0)$, where n is an integer;
asymptotes: $x = \dfrac{2n+1}{2}$, n an integer;

45.

$y = 3 \sin(3x - \pi)$

$y = 3 \cos(3x - \pi)$

27. period: $\pi/2$;
x-intercepts: $\left(\dfrac{2n+1}{4}\pi, 0\right)$,
where n is an integer;
asymptotes: $x = n\pi/2$, n an integer;

47.

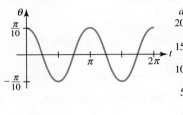
$y = 0.7 \sin[4\pi(x-4)]$ $\quad$ $y = 0.7 \cos[4\pi(x-4)]$

29. period: 2π;
x-intercepts: $\left(\dfrac{\pi}{2} + 2n\pi, 0\right)$,
where n is an integer;
asymptotes: $x = \dfrac{3\pi}{2} + 2n\pi$, n an integer;

51. **53.**

31. period: 1;
x-intercepts: $\left(\dfrac{1}{4} + n, 0\right)$,
where n is an integer;
asymptotes: $x = n$, n an integer;

Exercises 7.3 Page 377

1.

x	$\dfrac{2\pi}{3}$	$\dfrac{3\pi}{4}$	$\dfrac{5\pi}{6}$	π	$\dfrac{7\pi}{6}$	$\dfrac{5\pi}{4}$	$\dfrac{4\pi}{3}$	$\dfrac{3\pi}{2}$	$\dfrac{5\pi}{3}$	$\dfrac{7\pi}{4}$	$\dfrac{11\pi}{6}$	2π
$\tan x$	$-\sqrt{3}$	-1	$-\dfrac{1}{\sqrt{3}}$	0	$\dfrac{1}{\sqrt{3}}$	1	$\sqrt{3}$	—	$-\sqrt{3}$	-1	$-\dfrac{1}{\sqrt{3}}$	0
$\cot x$	$-\dfrac{1}{\sqrt{3}}$	-1	$-\sqrt{3}$	—	$\sqrt{3}$	1	$\dfrac{1}{\sqrt{3}}$	0	$-\dfrac{1}{\sqrt{3}}$	-1	$-\sqrt{3}$	—

3. $\sqrt{3}$
5. undefined
7. $-2/\sqrt{3}$
9. -1
11. -2
13. undefined
15. -2
17. $\sqrt{2}$
19. $\cot x = -\dfrac{1}{2}$, $\sec x = -\sqrt{5}$, $\cos x = -\dfrac{1}{\sqrt{5}}$, $\sin x = \dfrac{2}{\sqrt{5}}$, $\csc x = \dfrac{\sqrt{5}}{2}$
21. $\sin x = \dfrac{3}{4}$, $\cos x = \dfrac{\sqrt{7}}{4}$, $\tan x = \dfrac{3}{\sqrt{7}}$, $\cot x = \dfrac{\sqrt{7}}{3}$, $\sec x = \dfrac{4}{\sqrt{7}}$
23. $\tan x = 3$, $\cot x = \dfrac{1}{3}$, $\sec x = \pm\sqrt{10}$, $\csc x = \pm\dfrac{\sqrt{10}}{3}$

33. period: 2π;
asymptotes: $x = \dfrac{2n+1}{2}\pi$, n an integer;

35. period: 2;
asymptotes: $x = n$, n an integer;

37. period: $2\pi/3$;
asymptotes: $x = n\pi/3$, n an integer;

39. period: π;
asymptotes: $x = \dfrac{2n-1}{4}\pi$, n an integer;

41. $\cot x = -\tan\left(x - \dfrac{\pi}{2}\right)$

Exercises 7.4   Page 386

1. $a\sin\theta$
3. $a\tan\theta$
5. $\tan\theta$
7. $\dfrac{\sqrt{7}}{7}\cos\theta$
9. $\dfrac{\sqrt{2}}{4}(1+\sqrt{3})$
11. $\dfrac{\sqrt{2}}{4}(1+\sqrt{3})$
13. $\dfrac{\sqrt{2}}{4}(1+\sqrt{3})$
15. $2+\sqrt{3}$
17. $\dfrac{\sqrt{2}}{4}(1-\sqrt{3})$
19. $\dfrac{\sqrt{2}}{4}(\sqrt{3}-1)$
21. $-\dfrac{\sqrt{2}}{4}(1+\sqrt{3})$
23. $-2+\sqrt{3}$
25. $\dfrac{\sqrt{2}}{4}(1-\sqrt{3})$
27. $\dfrac{\sqrt{2}}{4}(\sqrt{3}+1)$
29. $-\dfrac{\sqrt{2}}{4}(1+\sqrt{3})$
31. $\sin 2\beta$
33. $\cos\dfrac{2\pi}{5}$
35. (a) $\dfrac{5}{9}$ (b) $-\dfrac{2\sqrt{14}}{9}$ (c) $-\dfrac{2\sqrt{14}}{5}$
37. (a) $\dfrac{3}{5}$ (b) $\dfrac{4}{5}$ (c) $\dfrac{4}{3}$
39. (a) $-\dfrac{119}{169}$ (b) $-\dfrac{120}{169}$ (c) $\dfrac{120}{119}$
41. $\dfrac{\sqrt{2+\sqrt{3}}}{2}$
43. $\dfrac{\sqrt{2+\sqrt{2}}}{2}$
45. $\dfrac{\sqrt{2-\sqrt{2}}}{2}$
47. $\dfrac{-2}{\sqrt{2-\sqrt{3}}}$
49. (a) $\dfrac{2\sqrt{13}}{13}$ (b) $\dfrac{3\sqrt{13}}{13}$ (c) $\dfrac{3}{2}$
51. (a) $-\sqrt{(5-\sqrt{5})/10}$ (b) $\sqrt{(5+\sqrt{5})/10}$
    (c) $-\sqrt{(3+\sqrt{5})/2}$
53. (a) $\dfrac{\sqrt{30}}{6}$ (b) $\dfrac{\sqrt{6}}{6}$ (c) $\dfrac{\sqrt{5}}{5}$
55. (a) $-\dfrac{2}{9}(\sqrt{10}+1)$ (b) $\dfrac{1}{9}(\sqrt{5}-4\sqrt{2})$
    (c) $\dfrac{2}{9}(1-\sqrt{10})$ (d) $\dfrac{1}{9}(\sqrt{5}+4\sqrt{2})$
57. $2\sqrt{2+\sqrt{3}} \approx 3.86$

Exercises 7.5   Page 395

1. 0
3. π
5. $\pi/3$
7. $-\pi/3$
9. $\pi/4$
11. $-\pi/6$
13. $-\pi/4$
15. $\dfrac{4}{5}$
17. $-\sqrt{5}/2$
19. $\sqrt{5}/5$
21. $\dfrac{5}{3}$
23. $\dfrac{1}{5}$
25. 1.2
27. $\pi/16$
29. 0

31. $\pi/4$
33. $\dfrac{x}{\sqrt{1+x^2}}$
35. $\dfrac{x}{\sqrt{1-x^2}}$
37. $\dfrac{\sqrt{1-x^2}}{x}$
39. $\dfrac{\sqrt{1+x^2}}{x}$

41.
43.
45.
47.

49. domain: $(-\infty, \infty)$;
range: $(0, \pi)$

51. domain: $(-\infty, -1] \cup [1, \infty)$;
range: $[0, \pi/2) \cup (\pi/2, \pi]$

53. (a) x in $(-\infty, \infty)$ (b) x in $(0, \pi)$
55. (a) x in $(-\infty, -1] \cup [1, \infty)$ (b) x in $[0, \pi/2) \cup (\pi/2, \pi]$
59. 0.9273
61. -0.7297
63. 2.5559
65. $19.9°, 70.1°$
67. $5.76°$
69. $\phi = 0.5404$ radian $\approx 31°$

Exercises 7.6   Page 403

1. $x = \dfrac{\pi}{3} + 2n\pi$ or $x = \dfrac{2\pi}{3} + 2n\pi$, where n is an integer
3. $x = \dfrac{\pi}{4} + 2n\pi$ or $x = \dfrac{7\pi}{4} + 2n\pi$, where n is an integer
5. $x = \dfrac{5\pi}{6} + n\pi$, where n is an integer
7. $x = \pi + 2n\pi = (2n+1)\pi$, where n is an integer
9. $x = n\pi$, where n is an integer
11. $x = \dfrac{3\pi}{2} + 2n\pi$, where n is an integer
13. $\theta = 60° + 360°n, \theta = 120° + 360°n$, where n is an integer
15. $\theta = 135° + 180°n$, where n is an integer
17. $\theta = 120° + 360°n, \theta = 240° + 360°n$, where n is an integer
19. $x = n\pi$, where n is an integer
21. no solutions
23. $\theta = 120° + 360°n, \theta = 240° + 360°n$, where n is an integer
25. $\theta = 90° + 180°n, \theta = 135° + 180°n$, where n is an integer
27. $x = \dfrac{\pi}{2} + n\pi$, where n is an integer
29. $\theta = 10° + 120°n, \theta = 50° + 120°n$, where n is an integer
31. $x = \dfrac{\pi}{2} + 2n\pi$, where n is an integer
33. $x = n\pi, x = \dfrac{2\pi}{3} + 2n\pi, x = \dfrac{4\pi}{3} + 2n\pi$, where n is an integer

35. $\theta = 30° + 360°n, \theta = 150° + 360°n, \theta = 270° + 360°n$, where n is an integer

37. $x = \dfrac{\pi}{2} + n\pi$, where n is an integer

39. $x = n\pi$, where n is an integer

41. $\theta = 2n\pi, \theta = \dfrac{\pi}{2} + 2n\pi$, where n is an integer

43. $x = \dfrac{\pi}{6} + 2n\pi, x = \dfrac{5\pi}{6} + 2n\pi$, where n is an integer

45. $\theta = 90° + 180°n, \theta = 360°n$, where n is an integer

47. $(\pi/3, 0), (2\pi/3, 0), (\pi, 0)$

49. $(\tfrac{2}{3}, 0), (\tfrac{10}{3}, 0), (\tfrac{14}{3}, 0)$

51. $(\pi, 0), (2\pi, 0), (3\pi, 0)$

53. $(\pi/3, 0), (\pi, 0), (5\pi/3, 0)$

55. The equation has infinitely many solutions.

57. The equation has infinitely many solutions.

59. 1.37, 1.82
61. $-1.02, 0.55$
63. 0.58, 1.57, 1.81
65. 30°, 150°
67. 60°
69. $t = \dfrac{1}{120}(\tfrac{1}{6} + n)$, where n is an integer

71. (a) 36.93 million square kilometers (b) $w = 31$ weeks (c) August

17. 0
19. $\tfrac{12}{13}$
21. $\sqrt{1 - x^2}$
23. $6\sin(\pi x/2); -6\sin(\pi x/2)$
25. $3\sin(4x - \pi); -3\sin(4x + \pi)$
27. $y = -\sin x, y = \cos(x + \pi/2)$
29. $y = 1 + \tfrac{1}{2}\sin(x + \pi/2), y = 1 + \tfrac{1}{2}\cos x$

Exercises 8.1   Page 411

1. $b = 2.04, c = 4.49$
3. $a = 11.71, c = 14.19$
5. $\alpha = 60°, \beta = 30°, a = 2.6$
7. $\alpha = 21.8°, \beta = 68.2°, c = 10.8$
9. $\alpha = 48.6°, \beta = 41.4°, b = 7.9$
11. $a = 8.5, c = 21.7$

Exercises 8.2   Page 415

1. 52.1 m
3. 66.4 ft
5. 409.7 ft
7. height: 15.5 ft; distance: 12.6 ft
9. 8.7°
11. 34,157 ft $\approx$ 6.5 mi
13. 20.2 ft
15. 6617 ft
17. Yes, since the altitude of the storm is approximately 6.3 km
21. 227,100 mi
23. $h = \dfrac{c}{\cot\alpha + \cot\beta}$

25. height is approximately 1.35; area is approximately 4.8
27. $h(\theta) = 1.25\tan\theta$
29. $\theta(x) = \arctan\left(\dfrac{1}{x}\right) - \arctan\left(\dfrac{1}{2x}\right)$, where x is measured in meters

Chapter 7 Review Exercises   Page 405

A. 1. false **3.** true **5.** true **7.** true
9. true **11.** true **13.** true **15.** false
17. false **19.** true

B. 1. $\dfrac{2\sqrt{5}}{5}$ **3.** 6 **5.** $4\pi/3$ **7.** 10
9. $\pi/5$ **11.** $(5, 0)$
13. $\sqrt{(3-\sqrt{2})/6}, -\sqrt{(3+\sqrt{2})/6}, -\dfrac{2\sqrt{14}}{9}, -\dfrac{5}{9}$

C.
1. amplitude: 5; period: 2π
3. amplitude: 10; period: $2\pi/3$; phase shift: $\pi/6$

Exercises 8.3   Page 421

1. $\gamma = 80°, a = 20.16, c = 20.16$
3. $\alpha = 92°, b = 3.01, c = 3.89$
5. $\alpha = 79.61°, \gamma = 28.39°, a = 12.41$
7. no solution
9. $\alpha = 24.46°, \beta = 140.54°, b = 12.28$; $\alpha = 155.54°, \beta = 9.46°, b = 3.18$
11. no solution
13. $\alpha = 45.58°, \gamma = 104.42°, c = 13.56$; $\alpha = 134.42°, \gamma = 15.58°, c = 3.76$
15. no solution **17.** 15.80 ft
19. 9.07 m **21.** 10.35 ft
23. 8.81°

Exercises 8.4   Page 425

1. $\alpha = 37.59°, \beta = 77.41°, c = 7.43$
3. $\alpha = 52.62°, \beta = 83.33°, \gamma = 44.05°$
5. $\alpha = 25°, \beta = 57.67°, c = 7.04$
7. $\alpha = 76.45°, \beta = 57.10°, \gamma = 46.45°$
9. $\alpha = 27.66°, \beta = 40.54°, \gamma = 111.80°$
11. $\alpha = 36.87°, \beta = 53.13°, \gamma = 90°$
13. $\alpha = 26.38°, \beta = 36.34°, \gamma = 117.28°$
15. $\beta = 10.24°, \gamma = 147.76°, a = 6.32$
17. 35.94 nautical miles
19. (a) S33.66°W (b) S2.82°E
21. $\alpha = 119.45°, \beta = 67.98°$
23. 91.77, 176.18

5. $\pi/2, \pi$
7. $\pi/6, 5\pi/6, 7\pi/6, 11\pi/6$
9. $\pi/4$
11. $-0.68, 0.92$
13. $2\pi/3$
15. $\dfrac{3\sqrt{7}}{7}$

Exercises 8.5　Page 431

1. $y = \sqrt{2}\sin(\pi x + 3\pi/4)$; amplitude: $\sqrt{2}$; period: 2; phase shift: $\tfrac{3}{4}$; one cycle of the graph is

3. $y = 2\sin(2x + \pi/6)$; amplitude: 2; period: π; phase shift: $\pi/12$; one cycle of the graph is

5. $y = \sin(x + 5\pi/4)$; amplitude: 1; period: 2π; phase shift: $5\pi/4$; one cycle of the graph is

7. $y = \tfrac{\sqrt{13}}{4}\sin(2t + 0.5880)$; amplitude: $\tfrac{\sqrt{13}}{4}$ feet; period: π seconds; frequency: $1/\pi$ cycle per second
9. $y = \tfrac{\sqrt{5}}{2}\sin(4t + 4.2487)$; amplitude: $\tfrac{\sqrt{5}}{2}$ feet; period: $\pi/2$ seconds; frequency: $2/\pi$ cycle per second
11. $y_0 = -\tfrac{5\sqrt{3}}{4}$, $v_0 = \tfrac{5}{2}$
13. $I(t) = I_0 \sin(\omega t + \theta + \phi)$

Exercises 8.6　Page 435

1. $\bar{z}_1 = 2 - 5i$

3. $z_1 + z_2 = 3 - i$

5. $\bar{z}_1 + z_2 = 6 + 2i$

7. $z_1 z_2 = -2 - 2i$

9. $\dfrac{z_1}{z_2} = 2i$

11. $r = 1, \theta = \dfrac{5\pi}{3}$

13. $r = 3\sqrt{2}, \theta = 289.47°$
15. $r = \dfrac{\sqrt{10}}{4}, \theta = 341.57°$
17. $r = 3\sqrt{2}, \theta = \dfrac{\pi}{4}$
19. $r = 2, \theta = \dfrac{\pi}{6}$
21. $r = \sqrt{5}, \theta = 333.43°$
23. $z = 4\left(\cos\dfrac{3\pi}{2} + i\sin\dfrac{3\pi}{2}\right)$
25. $z = 10\left(\cos\dfrac{\pi}{6} + i\sin\dfrac{\pi}{6}\right)$
27. $z = \sqrt{29}(\cos 111.8° + i\sin 111.8°)$
29. $z = \sqrt{34}(\cos 300.96° + i\sin 300.96°)$
31. $z = 2\sqrt{2}\left(\cos\dfrac{5\pi}{4} + i\sin\dfrac{5\pi}{4}\right)$
33. $1 + i$
35. $-5\sqrt{3} - 5i$
37. $\sqrt{3} + i$
39. $\tfrac{1}{2} - \tfrac{1}{2}\sqrt{3}i$
41. $\tfrac{4}{5}\sqrt{5} + \tfrac{8}{5}\sqrt{5}i$
43. $z_1 z_2 = 18\sqrt{2}\left(\cos\dfrac{3\pi}{4} + i\sin\dfrac{3\pi}{4}\right), \dfrac{z_1}{z_2} = \dfrac{\sqrt{2}}{4}\left(\cos\dfrac{\pi}{4} + i\sin\dfrac{\pi}{4}\right)$
45. $z_1 z_2 = 8\left(\cos\dfrac{\pi}{2} + i\sin\dfrac{\pi}{2}\right), \dfrac{z_1}{z_2} = \dfrac{1}{2}\left(\cos\dfrac{\pi}{6} + i\sin\dfrac{\pi}{6}\right)$
47. $z_1 z_2 = 10\sqrt{2}\left(\cos\dfrac{23\pi}{12} + i\sin\dfrac{23\pi}{12}\right),$
$\dfrac{z_1}{z_2} = \dfrac{\sqrt{2}}{5}\left(\cos\dfrac{5\pi}{12} + i\sin\dfrac{5\pi}{12}\right)$
49. $z_1 z_2 = 2\sqrt{3}\left(\cos\dfrac{\pi}{12} + i\sin\dfrac{\pi}{12}\right) \approx 3.3461 + 0.8966i,$
$\dfrac{z_1}{z_2} = \sqrt{3}\left(\cos\dfrac{7\pi}{12} + i\sin\dfrac{7\pi}{12}\right) \approx -0.4483 + 1.6730i$
51. $z_1 z_2 = 12\left(\cos\dfrac{\pi}{8} + i\sin\dfrac{\pi}{8}\right) \approx 11.0866 + 4.5922i,$
$\dfrac{z_1}{z_2} = \dfrac{3}{4}\left(\cos\dfrac{3\pi}{8} + i\sin\dfrac{3\pi}{8}\right) \approx 0.2870 + 0.6929i$

Exercises 8.7　Page 440

1. -1
3. $2\sqrt{2} + 2\sqrt{2}i$
5. $-\tfrac{243}{2}\sqrt{3} - \tfrac{243}{2}i$
7. approximately $19.5543 + 123.4610i$
9. approximately $26.5099 + 19.2605i$
11. $-16i$
13. -1
15. $-8i$
17. -64
19. $-16\sqrt{3} + 16i$
21. $-16\sqrt{2}$
23. $-7 - 24i$
25. $1 + \sqrt{3}i, -2, 1 - \sqrt{3}i$
27. $0.9239 + 0.3827i, -0.3827 + 0.9239i,$
$-0.9239 - 0.3827i, 0.3827 - 0.9239i$
29. $\sqrt[4]{2}\left(\tfrac{1}{2} + \tfrac{1}{2}\sqrt{3}i\right), \sqrt[4]{2}\left(-\tfrac{1}{2}\sqrt{3} + \tfrac{1}{2}i\right),$
$\sqrt[4]{2}\left(-\tfrac{1}{2} - \tfrac{1}{2}\sqrt{3}i\right), \sqrt[4]{2}\left(\tfrac{1}{2}\sqrt{3} - \tfrac{1}{2}i\right)$
31. $\sqrt[4]{2}(0.9239 + 0.3827i), \sqrt[4]{2}(-0.9239 - 0.3827i)$
33. $1.9754 + 0.3129i, 0.7167 + 1.8672i, -1.2586 + 1.5543i,$
$-1.9754 - 0.3129i, -0.7167 - 1.8672i, 1.2586 - 1.5543i$

35. $1, \frac{1}{2} + \frac{1}{2}\sqrt{3}i, -\frac{1}{2} + \frac{1}{2}\sqrt{3}i, -1, -\frac{1}{2} - \frac{1}{2}\sqrt{3}i, \frac{1}{2} - \frac{1}{2}\sqrt{3}i$

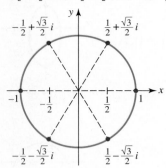

37. $n = 8k, k = 1, 2, 3, \ldots; n = 2 + 8k, k = 0, 1, 2, \ldots;$
$n = 5 + 8k, k = 0, 1, 2, \ldots; n = 1 + 8k, k = 0, 1, 2, \ldots$

39. $\frac{1}{2}\sqrt{2} + \frac{1}{2}\sqrt{2}i, \frac{1}{2}\sqrt{2} - \frac{1}{2}\sqrt{2}i,$
$-\frac{1}{2}\sqrt{2} + \frac{1}{2}\sqrt{2}i, -\frac{1}{2}\sqrt{2} - \frac{1}{2}\sqrt{2}i$

41. $-2 + 2\sqrt{3}i, 2 - 2\sqrt{3}i$

Chapter 8 Review Exercises Page 442

A. 1. true 3. true 5. true
B. 1. Sines 3. Cosines 5. $z = 5(\cos\pi + i\sin\pi)$
  7. $\frac{1}{4}(\cos 5° + i\sin 5°)$ 9. 1
C. 1. $\gamma = 80°, a = 5.32, c = 10.48$
  3. $a = 15.76, \beta = 99.44°, \gamma = 29.56°$
  5. 42.61 m 7. 118 ft
  9. (a) 42.35° 11. 16,927.6 km
  13. front: 18.88°; back: 35.12° 15. $V(\theta) = 160\sin 2\theta$
  17. $L(\theta) = 3\csc\theta + 4\sec\theta$
  19. $V(\theta) = 360 + 75\cot\theta$
  21. $A(\phi) = 100\cos\phi + 50\sin 2\phi$
  23. $r = 3\sqrt{2}, \theta = \frac{7\pi}{4}, 3\sqrt{2}\left(\cos\frac{7\pi}{4} + i\sin\frac{7\pi}{4}\right)$
  25. $r = \sqrt{13}, \theta = 123.69°, \sqrt{13}(\cos 123.69° + i\sin 123.69°)$
  27. $-2 - 2\sqrt{3}i$
  29. $z_1 z_2 = 2\sqrt{2}\left(\cos\frac{19\pi}{12} + i\sin\frac{19\pi}{12}\right),$
   $\frac{z_1}{z_2} = \frac{\sqrt{2}}{2}\left(\cos\frac{11\pi}{12} + i\sin\frac{11\pi}{12}\right)$
  31. $-8 + 8i$
  33. $2, 0.6180 + 1.9021i, -1.6180 + 1.1756i,$
   $-1.6180 - 1.1756i, 0.6180 - 1.9021i$
  35. $\left(x - \frac{3}{2}\sqrt{3} - \frac{3}{2}i\right)\left(x + \frac{3}{2}\sqrt{3} - \frac{3}{2}i\right)(x + 3i)$

Exercises 9.1 Page 452

1. Vertex: $(0, 0)$
   Focus: $(1, 0)$
   Directrix: $x = -1$
   Axis: $y = 0$
   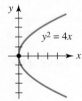

3. Vertex: $(0, 0)$
   Focus: $\left(-\frac{1}{3}, 0\right)$
   Directrix: $x = \frac{1}{3}$
   Axis: $y = 0$

5. Vertex: $(0, 0)$
   Focus: $(0, -4)$
   Directrix: $y = 4$
   Axis: $x = 0$

7. Vertex: $(0, 0)$
   Focus: $(0, 7)$
   Directrix: $y = -7$
   Axis: $x = 0$

9. Vertex: $(0, 1)$
   Focus: $(4, 1)$
   Directrix: $x = -4$
   Axis: $y = 1$

11. Vertex: $(-5, -1)$
    Focus: $(-5, -2)$
    Directrix: $y = 0$
    Axis: $x = -5$

13. Vertex: $(-5, -6)$
    Focus: $(-4, -6)$
    Directrix: $x = -6$
    Axis: $y = -6$

15. Vertex: $\left(-\frac{5}{2}, -1\right)$
    Focus: $\left(-\frac{5}{2}, -\frac{15}{16}\right)$
    Directrix: $y = -\frac{17}{16}$
    Axis: $x = -\frac{5}{2}$
    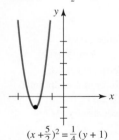

17. Vertex: $(3, 4)$
    Focus: $\left(\frac{5}{2}, 4\right)$
    Directrix: $x = \frac{7}{2}$
    Axis: $y = 4$

19. Vertex: $(0, 0)$
    Focus: $\left(0, \frac{1}{8}\right)$
    Directrix: $y = -\frac{1}{8}$
    Axis: $x = 0$

21. Vertex: $(3, 0)$
    Focus: $(3, -1)$
    Directrix: $y = 1$
    Axis: $x = 3$

23. Vertex: $(-2, 1)$
    Focus: $(-1, 1)$
    Directrix: $x = -3$
    Axis: $y = 1$

25. $x^2 = 28y$
27. $y^2 = -16x$
29. $y^2 = 10x$
31. $(x-2)^2 = 12y$
33. $(y-4)^2 = -12(x-2)$
35. $(x-1)^2 = 32(y+3)$
37. $(y+3)^2 = 32x$
39. $x^2 = 7y$
41. $(x-5)^2 = -24(y-1)$
43. $x^2 = \frac{1}{2}y$
45. $(3, 0), (0, -2), (0, -6)$
47. $(-3\sqrt{2}, 0), (3\sqrt{2}, 0), (0, 9)$
49. at the focus 6 in. from the vertex
51. $y = -2$
53. 27 ft
55. 4.5 ft
57. **(a)** 8

Exercises 9.2   Page 459

1. Center: $(0, 0)$
   Foci: $(\pm 4, 0)$
   Vertices: $(\pm 5, 0)$
   Minor axis endpoints: $(0, \pm 3)$
   Eccentricity: $\frac{4}{5}$

3. Center: $(0, 0)$
   Foci: $(0, \pm\sqrt{15})$
   Vertices: $(0, \pm 4)$
   Minor axis endpoints: $(\pm 1, 0)$
   Eccentricity: $\sqrt{15}/4$

$\frac{x^2}{25} + \frac{y^2}{9} = 1$

$\frac{x^2}{1} + \frac{y^2}{16} = 1$

5. Center: $(0, 0)$
   Foci: $(\pm\sqrt{7}, 0)$
   Vertices: $(\pm 4, 0)$
   Minor axis endpoints: $(0, \pm 3)$
   Eccentricity: $\sqrt{7}/4$

7. Center: $(0, 0)$
   Foci: $(0, \pm\sqrt{5})$
   Vertices: $(0, \pm 3)$
   Minor axis endpoints: $(\pm 2, 0)$
   Eccentricity: $\sqrt{5}/3$

$\frac{x^2}{16} + \frac{y^2}{9} = 1$

$\frac{x^2}{4} + \frac{y^2}{9} = 1$

9. Center: $(1, 3)$
   Foci: $(1 \pm \sqrt{13}, 3)$
   Vertices: $(-6, 3), (8, 3)$
   Minor axis endpoints: $(1, -3), (1, 9)$
   Eccentricity: $\sqrt{13}/7$

$\frac{(x-1)^2}{49} + \frac{(y-3)^2}{36} = 1$

11. Center: $(-5, -2)$
    Foci: $(-5, -2 \pm \sqrt{15})$
    Vertices: $(-5, -6), (-5, 2)$
    Minor axis endpoints: $(-6, -2), (-4, -2)$
    Eccentricity: $\sqrt{15}/4$

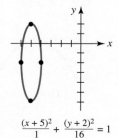

$\frac{(x+5)^2}{1} + \frac{(y+2)^2}{16} = 1$

13. Center: $(0, -\frac{1}{2})$
    Foci: $(0, -\frac{1}{2} \pm \sqrt{3})$
    Vertices: $(0, -\frac{5}{2}), (0, \frac{3}{2})$
    Minor axis endpoints: $(-1, -\frac{1}{2}), (1, -\frac{1}{2})$
    Eccentricity: $\sqrt{3}/2$

$\frac{x^2}{1} + \frac{(y+\frac{1}{2})^2}{4} = 1$

15. Center: $(1, -2)$
    Foci: $(1, -2 \pm \sqrt{6})$
    Vertices: $(1, -2 \pm \sqrt{15})$
    Minor axis endpoints: $(-2, -2), (4, -2)$
    Eccentricity: $\sqrt{\frac{2}{5}}$

$\frac{(x-1)^2}{9} + \frac{(y+2)^2}{15} = 1$

17. Center: $(2, -1)$
    Foci: $(2, -5), (2, 3)$
    Vertices: $(2, -6), (2, 4)$
    Minor axis endpoints: $(-1, -1), (5, -1)$
    Eccentricity: $\frac{4}{5}$

$\frac{(x-2)^2}{9} + \frac{(y+1)^2}{25} = 1$

19. Center: $(0, -3)$
    Foci: $(\pm\sqrt{6}, -3)$
    Vertices: $(-3, -3), (3, -3)$
    Minor axis endpoints: $(0, -3 \pm \sqrt{3})$
    Eccentricity: $\sqrt{6}/3$

$\frac{x^2}{9} + \frac{(y+3)^2}{3} = 1$

21. $\frac{x^2}{25} + \frac{y^2}{16} = 1$
23. $\frac{x^2}{8} + \frac{y^2}{9} = 1$
25. $\frac{x^2}{1} + \frac{y^2}{9} = 1$
27. $\frac{(x-1)^2}{16} + \frac{(y+3)^2}{4} = 1$
29. $\frac{x^2}{9} + \frac{y^2}{13} = 1$
31. $\frac{x^2}{11} + \frac{y^2}{9} = 1$
33. $\frac{x^2}{3} + \frac{y^2}{12} = 1$
35. $\frac{x^2}{16} + \frac{(y-1)^2}{4} = 1$
37. $\frac{(x-1)^2}{7} + \frac{(y-3)^2}{16} = 1$
39. $\frac{x^2}{45} + \frac{(y-2)^2}{9} = 1$

41. least distance is 28.5 million miles; greatest distance is 43.5 million miles
43. approximately 0.97
45. 12 ft
47. The piece of string should be 4 ft long. The tacks should be placed $\sqrt{7}/2$ ft from the center of the rectangle on the major axis of the ellipse.
49. on the major axis, 12 ft to either side from the center of the room
51. $5x^2 - 4xy + 8y^2 - 24x - 48y = 0$

Exercises 9.3 Page 468

1. Center: $(0, 0)$
Foci: $(\pm\sqrt{41}, 0)$
Vertices: $(\pm 4, 0)$
Asymptotes: $y = \pm \frac{5}{4}x$
Eccentricity: $\sqrt{41}/4$

$\frac{x^2}{16} - \frac{y^2}{25} = 1$

3. Center: $(0, 0)$
Foci: $(0, \pm\sqrt{73})$
Vertices: $(0, \pm 8)$
Asymptotes: $y = \pm \frac{8}{3}x$
Eccentricity: $\sqrt{73}/8$

$\frac{y^2}{64} - \frac{x^2}{9} = 1$

5. Center: $(0, 0)$
Foci: $(\pm 2\sqrt{5}, 0)$
Vertices: $(\pm 4, 0)$
Asymptotes: $y = \pm \frac{1}{2}x$
Eccentricity: $\sqrt{5}/2$

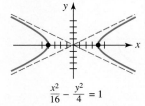

$\frac{x^2}{16} - \frac{y^2}{4} = 1$

7. Center: $(0, 0)$
Foci: $(0, \pm 2\sqrt{6})$
Vertices: $(0, \pm 2\sqrt{5})$
Asymptotes: $y = \pm\sqrt{5}x$
Eccentricity: $\sqrt{\frac{6}{5}}$

$\frac{y^2}{20} - \frac{x^2}{4} = 1$

9. Center: $(5, -1)$
Foci: $(5 \pm \sqrt{53}, -1)$
Vertices: $(3, -1), (7, -1)$
Asymptotes: $y = -1 \pm \frac{7}{2}(x - 5)$
Eccentricity: $\sqrt{53}/2$

$\frac{(x-5)^2}{4} - \frac{(y+1)^2}{49} = 1$

11. Center: $(0, 4)$
Foci: $(0, 4 \pm \sqrt{37})$
Vertices: $(0, -2), (0, 10)$
Asymptotes: $y = 4 \pm 6x$
Eccentricity: $\sqrt{37}/6$

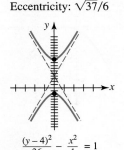

$\frac{(y-4)^2}{36} - \frac{x^2}{1} = 1$

13. Center: $(3, 1)$
Foci: $(3 \pm \sqrt{30}, 1)$
Vertices: $(3 \pm \sqrt{5}, 1)$
Asymptotes: $y = 1 \pm \sqrt{5}(x - 3)$
Eccentricity: $\sqrt{6}$

$\frac{(x-3)^2}{5} - \frac{(y-1)^2}{25} = 1$

15. Center: $(-4, 7)$
Foci: $(-4, 7 \pm \sqrt{13})$
Vertices: $(-4, 7 \pm 2\sqrt{2})$
Asymptotes: $y = 7 \pm \sqrt{\frac{8}{5}}(x + 4)$
Eccentricity: $\sqrt{\frac{13}{8}}$

$\frac{(y-7)^2}{8} - \frac{(x+4)^2}{5} = 1$

17. Center: $(2, 1)$
Foci: $(2 \pm \sqrt{11}, 1)$
Vertices: $(2 \pm \sqrt{6}, 1)$
Asymptotes: $y = 1 \pm \sqrt{\frac{5}{6}}(x - 2)$
Eccentricity: $\sqrt{\frac{11}{6}}$

$\frac{(x-2)^2}{6} - \frac{(y-1)^2}{5} = 1$

19. Center: $(1, 3)$
Foci: $(1, 3 \pm \frac{1}{2}\sqrt{5})$
Vertices: $(1, 2), (1, 4)$
Asymptotes: $y = 3 \pm 2(x - 1)$
Eccentricity: $\sqrt{5}/2$

$\frac{(y-3)^2}{1} - \frac{(x-1)^2}{\frac{1}{4}} = 1$

21. $\frac{x^2}{9} - \frac{y^2}{16} = 1$ **23.** $\frac{y^2}{4} - \frac{x^2}{12} = 1$ **25.** $\frac{x^2}{9} - \frac{y^2}{7} = 1$

27. $\frac{y^2}{\frac{25}{4}} - \frac{x^2}{\frac{11}{4}} = 1$ **29.** $\frac{x^2}{4} - \frac{y^2}{5} = 1$ **31.** $\frac{y^2}{64} - \frac{x^2}{16} = 1$

33. $\frac{x^2}{4} - \frac{y^2}{\frac{64}{9}} = 1$ **35.** $\frac{(y+3)^2}{4} - \frac{(x-1)^2}{5} = 1$

37. $\frac{(x+1)^2}{4} - \frac{(y-2)^2}{5} = 1$ **39.** $\frac{x^2}{4} - \frac{y^2}{8} = 1$

41. $\frac{(y-3)^2}{1} - \frac{(x+1)^2}{4} = 1$ **43.** $\frac{(y-4)^2}{1} - \frac{(x-2)^2}{4} = 1$

45. $(-7, 12)$
47. $7y^2 - 24xy + 24x + 82y + 55 = 0$

Exercises 9.4 Page 474

1. $(4\sqrt{2}, -2\sqrt{2})$ **3.** $\left(-\frac{1}{2} - \frac{\sqrt{3}}{2}, \frac{\sqrt{3}}{2} - \frac{1}{2}\right)$
5. $(4 + \sqrt{3}, 1 - 4\sqrt{3})$ **7.** $(-4, 0)$
9. approximately $(2.31, 6.83)$
11. ellipse rotated $45°$, $3x'^2 + y'^2 = 8$,

13. parabola rotated 45°, $y'^2 = 4\sqrt{2}x'$,

15. hyperbola rotated approximately 27°, $2x'^2 - 3y'^2 = 6$,

17. ellipse, $3(x' + \sqrt{5})^2 + 8y'^2 = 24$
19. parabola, $(y' - 1)^2 = -(x' - \frac{3}{2})$
21. (a) $y' = x'^2$
    (b) $x'y'$-coordinates: $(0, \frac{1}{4})$; xy-coordinates: $(-\frac{1}{8}, \frac{\sqrt{3}}{8})$
    (c) $y' = -\frac{1}{4}$, $2x - 2\sqrt{3}y = 1$
23. hyperbola   25. parabola   27. ellipse

Exercises 9.5   Page 481

1.
t	-3	-2	-1	0	1	2	3
x	-1	0	1	2	3	4	5
y	$\frac{3}{2}$	2	$\frac{5}{2}$	3	$\frac{7}{2}$	4	$\frac{9}{2}$

intercepts: $(-4, 0), (0, 2)$; curve:

3.   5.

7.   9.

11. $y = x^2 + 3x - 1, x \geq 0$ 13. $x = 1 - 2y^2, -1 \leq x \leq 1$

15. $y = \ln x, x > 0$ 17. $\dfrac{x^2}{16} + \dfrac{y^2}{4} = 1$

19.

21.

23.

25.

27.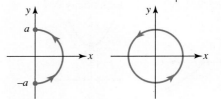

29. $(3, 0); (0, 1), (0, 3)$
31. The line segment between (x_1, y_1) and (x_2, y_2).
33. $x = 95\sqrt{2}\,t, y = -16t^2 + 95\sqrt{2}\,t, t \geq 0$;
    $(190\sqrt{2}, 190\sqrt{2} - 64) \approx (268.70, 204.70)$
35. $x = \pm\sqrt{r^2 - L^2\sin^2\phi}, y = L\sin\phi$

Chapter 9 Review Exercises   Page 484

A. 1. true   3. false   5. true   7. true
   9. true   11. true   13. true   15. false
   17. false   19. false
B. 1. $y^2 = 20x$ 3. $(x - 1)^2 = 8(y + 5)$
   5. $(0, 0)$ 7. 1
   9. $(-3, 0); (-3, -1), (-3, 1)$ 11. $(0, -\sqrt{2}), (0, \sqrt{2})$
   13. 1   15. line
   17. $(5, 0)$ 19. $(0, -2), (0, 2)$
C. 1. Vertex: $(0, 3)$ 3. Vertex: $(1, 0)$
      Focus: $(-2, 3)$ Focus: $(1, -1)$
      Directrix: $x = 2$ Directrix: $y = 1$
      Axis: $y = 3$ Axis: $x = 1$

$(y - 3)^2 = -8x$

$(x - 1)^2 = -4y$

5. $(x-1)^2 = 8(y+5)$ **7.** $(x-1)^2 = 3(y-2)$

9. Center: $(0, -5)$
Vertices: $(0, -10), (0, 0)$
Foci: $(0, -5 - \sqrt{22})$,
$(0, -5 + \sqrt{22})$

11. Center: $(-1, 3)$
Vertices: $(-1, 1), (-1, 5)$
Foci: $(-1, 3 - \sqrt{3})$,
$(-1, 3 + \sqrt{3})$

$\dfrac{x^2}{3} + \dfrac{(y+5)^2}{25} = 1$

$(x+1)^2 + \dfrac{(y-3)^2}{4} = 1$

13. $\dfrac{x^2}{41} + \dfrac{y^2}{16} = 1$ **15.** $\dfrac{x^2}{4} + (y+2)^2 = 1$

17. Center: $(0, 0)$
Vertices: $(-1, 0), (1, 0)$
Foci: $(-\sqrt{2}, 0), (\sqrt{2}, 0)$
Asymptotes: $y = \pm x$

19. Center: $(3, -1)$
Vertices: $(2, -1), (4, -1)$
Foci: $(3 - \sqrt{10}, -1)$,
$(3 + \sqrt{10}, -1)$
Asymptotes:
$y + 1 = \pm 3(x - 3)$

$x^2 - y^2 = 1$

$(x-3)^2 - \dfrac{(y+1)^2}{9} = 1$

21. $\dfrac{x^2}{36} - \dfrac{y^2}{28} = 1$ **23.** $\dfrac{x^2}{4} - \dfrac{y^2}{16} = 1$

25.
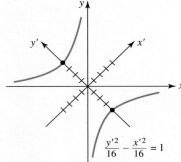
$\dfrac{y'^2}{16} - \dfrac{x'^2}{16} = 1$

27. $(x+5)^2 + \dfrac{(y-2)^2}{4} = 1$ **29.** 1.95×10^9 m

Exercises 10.1 Page 491

1.   **3.**

5. $\left(-4, -\dfrac{\pi}{6}\right)$

7. (a) $(2, -5\pi/4)$ (b) $(2, 11\pi/4)$
(c) $(-2, 7\pi/4)$ (d) $(-2, -\pi/4)$
9. (a) $(4, -5\pi/3)$ (b) $(4, 7\pi/3)$
(c) $(-4, 4\pi/3)$ (d) $(-4, -2\pi/3)$
11. (a) $(1, -11\pi/6)$ (b) $(1, 13\pi/6)$
(c) $(-1, 7\pi/6)$ (d) $(-1, -5\pi/6)$
13. $\left(-\dfrac{1}{4}, \dfrac{\sqrt{3}}{4}\right)$ **15.** $(-3, 3\sqrt{3})$
17. $(-2\sqrt{2}, -2\sqrt{2})$
19. (a) $(2\sqrt{2}, -3\pi/4)$ (b) $(-2\sqrt{2}, \pi/4)$
21. (a) $(2, -\pi/3)$ (b) $(-2, 2\pi/3)$
23. (a) $(7, 0)$ (b) $(-7, \pi)$

25.

27.

29.

31. $r = 5 \csc\theta$ **33.** $\theta = \tan^{-1} 7$
35. $r = 2/(1 + \cos\theta)$ **37.** $r = 6$
39. $r = 1 - \cos\theta$ **41.** $x = 2$
43. $(x^2 + y^2)^3 = 144 x^2 y^2$ **45.** $(x^2 + y^2)^2 = 8xy$
47. $x^2 + y^2 + 5y = 0$ **49.** $8x^2 - 12x - y^2 + 4 = 0$
51. $3x + 8y = 5$

Exercises 10.2 Page 500

1. circle   **3.** line through the pole

5. spiral

7. cardioid

9. cardioid

11. limaçon with interior loop

13. dimpled limaçon

15. convex limaçon

17. rose curve

19. rose curve

21. rose curve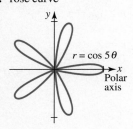

23. circle with center on x-axis

25. circle with center on y-axis

27. lemniscate

29. lemniscate

31.

33. $r = \frac{5}{2}$

35. $r = 4 - 3\cos\theta$

37. $r = 2\cos 4\theta$

39. $(2, \pi/6), (2, 5\pi/6)$

41. $(1, \pi/2), (1, 3\pi/2)$, origin

43. $(0, 0), \left(\frac{\sqrt{3}}{2}, \frac{\pi}{3}\right), \left(\frac{\sqrt{3}}{2}, \frac{2\pi}{3}\right)$

Exercises 10.3 Page 506

1. $e = 1$, parabola
$r = \dfrac{2}{1 - \sin\theta}$

3. $e = \frac{1}{4}$, ellipse
$r = \dfrac{16}{4 + \cos\theta}$

5. $e = 2$, hyperbola
$r = \dfrac{4}{1 + 2\sin\theta}$

7. $e = 2$, hyperbola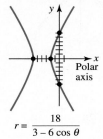
$r = \dfrac{18}{3 - 6\cos\theta}$

9. $e = 1$, parabola
$r = \dfrac{6}{1 - \cos\theta}$

11. $e = 2$, $\dfrac{(y-4)^2}{4} - \dfrac{x^2}{12} = 1$

13. $e = \frac{2}{3}$, $\dfrac{\left(x - \frac{24}{5}\right)^2}{\frac{1296}{25}} + \dfrac{y^2}{\frac{144}{5}} = 1$

15. $r = \dfrac{3}{1 + \cos\theta}$

17. $r = \dfrac{4}{3 - 2\sin\theta}$

19. $r = \dfrac{12}{1 + 2\cos\theta}$

21. $r = \dfrac{3}{1 + \cos(\theta + 2\pi/3)}$

23. $r = \dfrac{3}{1 - \sin\theta}$

25. $r = \dfrac{1}{1 - \cos\theta}$

27. $r = \dfrac{1}{2 - 2\sin\theta}$

29. vertex: $(2, \pi/4)$

31. vertices: $(10, \pi/3)$ and $\left(\frac{10}{3}, 4\pi/3\right)$

33. $r_p = 8000$ km   **35.** $r = \dfrac{1.495 \times 10^8}{1 - 0.0167\cos\theta}$

43. 10.16, 30.26°   **45.** $2\mathbf{i} - 10\mathbf{j}$

47. Actual course makes an angle of 71.6° from the original heading.

49. 52.9 mi/h

Exercises 10.4 Page 514

1. $|\mathbf{v}| = 2, \theta = 11\pi/6$ **3.** $|\mathbf{v}| = 5, \theta = 2\pi$

5. $|\mathbf{v}| = 8, \theta = 2\pi/3$ **7.** $|\mathbf{v}| = 10\sqrt{2}, \theta = 3\pi/4$

9. $\langle 3, 2\rangle, \langle 1, 4\rangle, \langle -6, -9\rangle, \langle 2, 13\rangle$
11. $\langle 0, 3\rangle, \langle -8, 1\rangle, \langle 12, -6\rangle, \langle -28, 2\rangle$
13. $\langle -\tfrac{9}{2}, -\tfrac{29}{4}\rangle, \langle -\tfrac{11}{2}, -\tfrac{27}{4}\rangle, \langle 15, 21\rangle, \langle -17, -20\rangle$
15. $-31\mathbf{i} - 14\mathbf{j}, 42\mathbf{i} + 11\mathbf{j}$
17. $-\tfrac{15}{2}\mathbf{i} - \tfrac{3}{2}\mathbf{j}, 11\mathbf{i} - 3\mathbf{j}$
19. $5.8\mathbf{i} + 8.5\mathbf{j}, -6.6\mathbf{i} - 10.3\mathbf{j}$

21.   **23.**

25.   **27.**

29. horizontal component: 4; vertical component: -6
31. horizontal component: -10; vertical component: 8
33. (a) $2\left(\cos\dfrac{3\pi}{4}\mathbf{i} + \sin\dfrac{3\pi}{4}\mathbf{j}\right)$ (b) $-\sqrt{2}\mathbf{i} + \sqrt{2}\mathbf{j}$
35. (a) $6\left(\cos\dfrac{5\pi}{6}\mathbf{i} + \sin\dfrac{5\pi}{6}\mathbf{j}\right)$ (b) $-3\sqrt{3}\mathbf{i} + 3\mathbf{j}$
37. (a) $\langle 1/\sqrt{2}, 1/\sqrt{2}\rangle$ (b) $\langle -1/\sqrt{2}, -1/\sqrt{2}\rangle$
39. (a) $\langle 0, -1\rangle$ (b) $\langle 0, 1\rangle$ **41.** $\langle \tfrac{5}{13}, \tfrac{12}{13}\rangle$

Exercises 10.5 Page 521

1. 10   **3.** 1   **5.** -17
7. 12   **9.** 13   **11.** -68
13. $\tfrac{13}{2}$ **15.** 8   **17.** $\langle \tfrac{17}{26}, -\tfrac{85}{26}\rangle$
19. $25\sqrt{2}$ **21.** 102.53°   **23.** 63.43°
25. not orthogonal   **27.** orthogonal   **29.** $c = 3$
33. $-\tfrac{\sqrt{10}}{5}$ **35.** $-3\sqrt{2}$
37. (a) $-\tfrac{21}{5}\mathbf{i} + \tfrac{28}{5}\mathbf{j}$ (b) $-\tfrac{7}{2}\mathbf{i} + \tfrac{7}{2}\mathbf{j}$ **39.** $\tfrac{72}{25}\mathbf{i} + \tfrac{96}{25}\mathbf{j}$
41. 1000 ft-lb   **43.** $\tfrac{78}{5}$ ft-lb

Chapter 10 Review Exercises Page 523

A. 1. true   **3.** true
 5. false   **7.** true
 9. false   **11.** true
 13. false
B. 1. $(1, 1)$ **3.** $(10, 3\pi/2)$
 5. $\tfrac{1}{2}$ **7.** hyperbola
 9. convex limaçon   **11.** $\pi/2$ radians or 90°
 13. $\tfrac{12}{13}\mathbf{i} + \tfrac{5}{13}\mathbf{j}$
C. 1. $\left(x - \tfrac{1}{2}\right)^2 + \left(y - \tfrac{1}{2}\right)^2 = \tfrac{1}{2}$ **3.** $r = 4\sin\theta$
 5. $(0, 2), \left(0, -\tfrac{2}{3}\right)$
 7. (a) $(\sqrt{6}, -\pi/4)$ (b) $(-\sqrt{6}, 3\pi/4)$
 9. circle of radius 5;   **11.** circle with center
   centered at the origin   on y-axis

13. cardioid   **15.** convex limaçon

 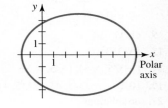

17. rose curve   **19.** ellipse

21. $r = 3\sin 10\theta$
23. (a) $r = 2\cos(\theta - \pi/4)$ (b) $x^2 + y^2 = \sqrt{2}x + \sqrt{2}y$
25. (b) center $(b/2, a/2)$, radius $\frac{1}{2}\sqrt{a^2 + b^2}$
27. $13\mathbf{i} - 12\mathbf{j}$ 29. $3\mathbf{i} - 7\mathbf{j}$
31. $7\mathbf{i} - 13\mathbf{j}$ 33. $\frac{-3}{\sqrt{17}}$
35. $\mathbf{i} + \mathbf{j}$ 37. $\sqrt{13} + 2\sqrt{2}$
39. $2\sqrt{2}\left(\cos\frac{\pi}{4}\mathbf{i} + \sin\frac{\pi}{4}\mathbf{j}\right)$ 41. $-\frac{1}{\sqrt{17}}\mathbf{i} + \frac{4}{\sqrt{17}}\mathbf{j}$

Exercises 11.1 Page 534

1. $(0, 2)$; consistent; independent
3. $(-\frac{12}{13}, -\frac{35}{13})$; consistent; independent
5. no solutions; inconsistent
7. $(\frac{3}{2}, -\frac{1}{2})$; consistent; independent
9. $(-2\alpha + 4, \alpha)$; α a real number; consistent; dependent
11. $(1, 2, 3)$; consistent; independent
13. $(-1, \frac{1}{2}, -3)$; consistent; independent
15. no solutions; inconsistent
17. $(0, 0, 0)$; consistent; independent
19. $(7, -5, \frac{1}{3})$; consistent; independent
21. $(2\alpha + 3\beta - 2, \beta, \alpha)$; α and β real numbers; consistent; dependent
23. no solutions; inconsistent
25. $(1, -2, 4, 8)$; consistent; independent
27. $x = 2, y = 3$
29. $x = 10^3, y = 10^{-7}$
31. $T_1 = \dfrac{200\cos 15°}{\sin 40°} \approx 300.54$, $T_2 = \dfrac{200\cos 25°}{\sin 40°} \approx 281.99$
33. plane: 575 mi/h; wind: 25 mi/h
35. 50 gal from the first tank, 40 gal from the second tank
37. P_1: 4 h; P_2: 12 h; P_3: 6 h
39. 25   41. 20 A's, 5 B's, 15 C's

Exercises 11.2 Page 540

1. two solutions
3. two solutions

5. one solution   7. $(-1, -1), (2, 2)$

9. $(1, 1)$ 11. $(0, 1)$
13. $(\sqrt{3}, 3)$ 15. no solutions
17. $(-\sqrt{5}, \sqrt{5}), (\sqrt{5}, \sqrt{5})$
19. $(0, 0), (-2\sqrt{5}, 4), (2\sqrt{5}, 4), (-2\sqrt{3}, -4), (2\sqrt{3}, -4)$
21. no solutions

23. $(-\frac{2}{5}\sqrt{5}, \frac{4}{5}\sqrt{5}), (\frac{2}{5}\sqrt{5}, -\frac{4}{5}\sqrt{5})$
25. $(-\sqrt{5}, 0), (\sqrt{5}, 0), (-2, -1), (2, -1)$
27. $(-1 - \sqrt{3}, 1 - \sqrt{3}), (-1 + \sqrt{3}, 1 + \sqrt{3}),$
    $(1 - \sqrt{3}, -1 - \sqrt{3}), (1 + \sqrt{3}, -1 + \sqrt{3})$
29. $\left\{\left(\dfrac{\pi}{4} + 2n\pi, \dfrac{\sqrt{2}}{2}\right) \middle| n = 0, \pm 1, \ldots\right\}$
    $\cup \left\{\left(\dfrac{\pi}{4} + (2n+1)\pi, -\dfrac{\sqrt{2}}{2}\right) \middle| n = 0, \pm 1, \ldots\right\}$
31. $\left\{\left(\dfrac{\pi}{6} + 2n\pi, 1\right) \middle| n = 0, \pm 1, \ldots\right\}$
    $\cup \left\{\left(\dfrac{5\pi}{6} + 2n\pi, 1\right) \middle| n = 0, \pm 1, \ldots\right\}$
    $\cup \left\{\left(\dfrac{7\pi}{6} + 2n\pi, -1\right) \middle| n = 0, \pm 1, \ldots\right\}$
    $\cup \left\{\left(\dfrac{11\pi}{6} + 2n\pi, -1\right) \middle| n = 0, \pm 1, \ldots\right\}$
33. $(0.1, -1), (100,000, 5)$ 35. $(-101, -201), (99, 199)$
37. $(5, \log_3 5)$
39. $(\sqrt{3}, \sqrt{3}, -2\sqrt{3}), (-\sqrt{3}, -\sqrt{3}, 2\sqrt{3})$
41. $(1/\sqrt{3}, \sqrt{2/3}, 1/\sqrt{3}), (-1/\sqrt{3}, \sqrt{2/3}, -1/\sqrt{3}),$
    $(1/\sqrt{3}, -\sqrt{2/3}, 1/\sqrt{3}), (-1/\sqrt{3}, -\sqrt{2/3}, -1/\sqrt{3}),$
    $(1, 0, 0), (-1, 0, 0)$
43. 50 ft $\times$ 80 ft   45. each radius is 4 cm
47. approximately 7.9 in. $\times$ 12.7 in.
49. 2 ft $\times$ 2 ft $\times$ 8 ft, or approximately 7.06 ft $\times$ 7.06 ft $\times$ 0.64 ft

Exercises 11.3 Page 548

1. $\dfrac{\frac{1}{2}}{x} - \dfrac{\frac{1}{2}}{x + 2}$ 3. $-\dfrac{6}{x + 1} - \dfrac{3}{x - 5}$
5. $\dfrac{\frac{3}{2}}{x + 1} - \dfrac{10}{x + 2} + \dfrac{\frac{21}{2}}{x + 3}$ 7. $\dfrac{\frac{3}{2}}{x + 4} + \dfrac{\frac{3}{2}}{x - 4}$
9. $\dfrac{5}{x - 3} + \dfrac{9}{(x - 3)^2}$ 11. $-\dfrac{\frac{1}{4}}{x} + \dfrac{\frac{1}{4}}{x^2} + \dfrac{\frac{1}{4}}{x + 2} + \dfrac{\frac{1}{4}}{(x + 2)^2}$
13. $-\dfrac{\frac{11}{27}}{x} - \dfrac{\frac{7}{9}}{x^2} + \dfrac{\frac{1}{3}}{x^3} + \dfrac{\frac{1}{2}}{x - 1} - \dfrac{\frac{5}{54}}{x + 3}$
15. $\dfrac{1}{x - 1} + \dfrac{5x - 2}{x^2 + 9}$ 17. $\dfrac{\frac{36}{7}}{2x - 3} + \dfrac{-\frac{4}{7}x + \frac{26}{7}}{x^2 - x + 1}$
19. $-\dfrac{\frac{7}{4}}{t + 1} + \dfrac{\frac{9}{4}}{t - 1} + \dfrac{-\frac{1}{2}t - 4}{t^2 + 1}$
21. $\dfrac{2x}{x^2 + 2} - \dfrac{x}{x^2 + 1}$ 23. $\dfrac{1}{x^2 + 1} + \dfrac{2x}{(x^2 + 1)^2}$
25. $x^3 + x + \dfrac{\frac{1}{2}}{x - 1} + \dfrac{\frac{1}{2}}{x + 1}$ 27. $\dfrac{1}{2} - \dfrac{\frac{13}{3}}{x + 2} + \dfrac{\frac{13}{6}}{2x + 1}$
29. $x^3 + 2x^2 + 3x + 6 + \dfrac{\frac{64}{5}}{x - 2} + \dfrac{\frac{1}{5}x + \frac{2}{5}}{x^2 + 1}$

Exercises 11.4 Page 552

1.   3.

5.
7.
9.
11.
13.
15.
17. no solutions
19.
21.
23.
25.
27.
29.
31.
33.
35.

37. $\begin{cases} 3x + 2y \geq 12 \\ x + 2y \geq 8 \\ x \geq 0, y \geq 0 \end{cases}$

39. $\begin{cases} x + y \leq 10 \\ 1 \leq x \leq 5 \\ 2 \leq y \leq 6 \end{cases}$

Exercises 11.5 — Page 558

1. minimum: 11; maximum: 33
3. minimum: 0; maximum: $\frac{39}{2}$
5. minimum: 12; maximum: 54
7. minimum: 1; maximum: 13
9. minimum: 0; maximum: 30
11. minimum: 25; maximum: $\frac{175}{3}$
13. minimum: $\frac{68}{3}$. The objective function has no maximum since it can be increased without bound by increasing x.
15. minimum: $\frac{110}{3}$. The objective function has no maximum since it can be increased without bound by increasing x or y.
17. 250 satellite radios; 100 DVD players
19. 3 capsules of X; 1 capsule of Y
21. 1300 pairs of designer jeans and 500 pairs of generic jeans to the upscale store; no designer jeans and 1200 pairs of generic-brand jeans to the discount outlet. The maximum profit is $28,855.
23. 17 kg of aquatic plants and 16 kg of terrestrial plants

Chapter 11 Review Exercises — Page 561

A.
1. true
3. true
5. false
7. true
9. true

B.
1. inconsistent
3. half-plane
5. long division
7. dependent
9. second

C. 1. $(-1, 0), (6, -7)$
3. $\left(-\frac{4\sqrt{5}}{5}, -\frac{4\sqrt{5}}{5}\right), \left(-\frac{4\sqrt{5}}{5}, \frac{4\sqrt{5}}{5}\right), \left(\frac{4\sqrt{5}}{5}, -\frac{4\sqrt{5}}{5}\right), \left(\frac{4\sqrt{5}}{5}, \frac{4\sqrt{5}}{5}\right)$
5. $(100, 2)$ 7. (e^2, e^{-1})
9. $(0, 0, 0)$ 11. $(-1, 4, -5)$
13. $\left(-2, \frac{1}{64}\right)$ 15. 33
17. length of side of square: $\dfrac{4-\pi}{4(4+\pi)} \approx 0.03$; radius of circle: $\dfrac{1}{4+\pi} \approx 0.14$
19. $\dfrac{\frac{1}{3}}{x} + \dfrac{\frac{1}{4}}{x-1} - \dfrac{\frac{7}{12}}{x+3}$ 21. $\dfrac{1}{x^2+4} - \dfrac{4}{(x^2+4)^2}$
23.   25.

27.

29. $\begin{cases} y \geq x^2 \\ y \geq 2-x \end{cases}$

31. $\begin{cases} y \leq x^2 \\ y \leq 2-x \end{cases}$ 33. minimum: 20; maximum: 380

35. 6 acres of corn and 6 acres of oats

Exercises 12.1   Page 569

1. 3×3 3. 3×2
5. 1×1 7. 3×1
9. 5×7 11. 7
13. -9 15. $-\frac{3}{4}$
17. $\begin{bmatrix} 0 & -1 & -2 \\ 1 & 0 & -1 \end{bmatrix}$ 19. $\begin{bmatrix} 1 & 4 & 9 \\ 2 & 8 & 18 \end{bmatrix}$
21. $\begin{bmatrix} 4 & 2 & \frac{4}{3} \\ 8 & 4 & \frac{8}{3} \end{bmatrix}$ 23. not equal
25. equal   27. $x = 3, y = -4, z = -1$
29. $x = 9, y = -\frac{3}{2}, z = \frac{5}{4}, w = 1$ 31. $x = \pm 2, y = \pm 3$
33. $\begin{bmatrix} 2 & 5 \\ 1 & \frac{1}{2} \end{bmatrix}$ 35. $\begin{bmatrix} 1 & 5 \\ 2 & 6 \\ 10 & 9 \end{bmatrix}$
37. $\begin{bmatrix} 0 & 0 & 0 \\ 4 & 1 & 4 \\ -2 & 3 & -5 \end{bmatrix}$

Exercises 12.2   Page 577

1. $\begin{bmatrix} -6 & 4 & 4 \\ 0 & 6 & 10 \end{bmatrix}, \begin{bmatrix} 8 & -8 & 2 \\ 0 & 2 & 4 \end{bmatrix},$ $\begin{bmatrix} 4 & -8 & 12 \\ 0 & 16 & 28 \end{bmatrix}, \begin{bmatrix} 17 & -18 & 7 \\ 0 & 8 & 15 \end{bmatrix}$

3. $\begin{bmatrix} 3 & 1 \\ 2 & 6 \end{bmatrix}, \begin{bmatrix} 5 & -3 \\ 2 & 10 \end{bmatrix}, \begin{bmatrix} 16 & -4 \\ 8 & 32 \end{bmatrix}, \begin{bmatrix} 14 & -7 \\ 6 & 28 \end{bmatrix}$
5. $\begin{bmatrix} 7 \\ 2 \end{bmatrix}, \begin{bmatrix} -5 \\ -2 \end{bmatrix}, \begin{bmatrix} 4 \\ 0 \end{bmatrix}, \begin{bmatrix} -9 \\ -4 \end{bmatrix}$
7. $[14\ \ -2], [6\ \ 8], [40\ \ 12], [22\ \ 19]$
9. $\begin{bmatrix} 1 & 2 \\ 3 & 4 \end{bmatrix}, \begin{bmatrix} -1 & -2 \\ -3 & -4 \end{bmatrix}, \begin{bmatrix} 0 & 0 \\ 0 & 0 \end{bmatrix}, \begin{bmatrix} -2 & -4 \\ -6 & -8 \end{bmatrix}$
11. $\begin{bmatrix} 3 & 1 \\ -2 & 22 \end{bmatrix}, \begin{bmatrix} 5 & 4 \\ 8 & 20 \end{bmatrix}$
13. $\begin{bmatrix} 12 & 14 \\ 3 & 22 \end{bmatrix}, \begin{bmatrix} 16 & -6 & 13 \\ 12 & -12 & 21 \\ 12 & -18 & 30 \end{bmatrix}$
15. $\begin{bmatrix} 4 & 2 & 8 \\ 5 & 1 & 9 \\ 0 & 0 & 40 \end{bmatrix}, \begin{bmatrix} 40 & 0 & 0 \\ 16 & 0 & 2 \\ 2 & 3 & 5 \end{bmatrix}$
17. AB is not defined; $\begin{bmatrix} 1 & 2 & -3 \\ 0 & 0 & 0 \end{bmatrix}$
19. $\begin{bmatrix} -3 & 4 \\ 6 & -8 \end{bmatrix}, [-11]$ 21. $\begin{bmatrix} 19 & -18 \\ -30 & 31 \end{bmatrix}$
23. $\begin{bmatrix} 3 & 8 \\ -6 & -16 \end{bmatrix}$ 25. $\begin{bmatrix} -4 & -5 \\ 8 & 10 \end{bmatrix}$
27. $\begin{bmatrix} -38 \\ -2 \end{bmatrix}$ 29. $2 \times n$
31. 2×3 33. $c_{23} = 9, c_{12} = 12$
37. $\begin{bmatrix} -5 & 15 \\ -6 & -2 \end{bmatrix}$
43. $\begin{bmatrix} 1 & 3 & 1 & 2 \\ 1 & 1 & 1 & 2 \end{bmatrix}$. An entry of 1 means that $P_5, P_6, P_7,$ or P_8 has had only one secondary contact with either X or Y. An entry of 2 means two secondary contacts with either X or Y, and so on.
45. $\begin{bmatrix} 120 & 20 \\ 144 & 24 \\ 288 & 48 \end{bmatrix}$. The three entries in the first column represent (in dollars) the state sales tax paid by the retail store on all amplifiers, tuners, and speakers, respectively. The entries in the second column represent the city sales tax paid by the retail store on all amplifiers, tuners, and speakers, respectively.
47. $\begin{bmatrix} 950 & 475 & 380 \\ 760 & 190 & 475 \\ 1900 & 570 & 190 \\ 950 & 950 & 950 \end{bmatrix}$ 49. $[385]$

Exercises 12.3   Page 587

1. $M_{11} = -2, M_{12} = 3, M_{21} = 0, M_{22} = 4; A_{11} = -2, A_{12} = -3, A_{21} = 0, A_{22} = 4$
3. $M_{11} = 5, M_{12} = 10, M_{13} = 3, M_{21} = -35, M_{22} = 29, M_{23} = -21, M_{31} = -8, M_{32} = -16, M_{33} = 15; A_{11} = 5, A_{12} = -10, A_{13} = 3, A_{21} = 35, A_{22} = 29, A_{23} = 21, A_{31} = -8, A_{32} = 16, A_{33} = 15$
5. -7 7. 12
9. $a^2 + b^2$ 11. 60
13. -61 15. -862
17. $-abcd$ 19. Theorem 12.3.2(ii)
21. Theorem 12.3.2(i)
23. Theorem 12.3.2(v), where -4 times the third row is added to the first row
25. Theorem 12.3.2(iv)

27. 101   **29.** 560
37. $-5, 2$ **39.** $-2, -1, 3$
41. By adding the second row to the third row, we get
$$\begin{vmatrix} 1 & 1 & 1 \\ a & b & c \\ a+b+c & a+b+c & a+b+c \end{vmatrix}$$
$$= (a+b+c)\begin{vmatrix} 1 & 1 & 1 \\ a & b & c \\ 1 & 1 & 1 \end{vmatrix}.$$
The last determinant is 0 from Theorem 12.3.2(iv).
43. -3 **45.** 0
47. -6 **49.** -4

Exercises 12.4   Page 597

3. $\begin{bmatrix} \frac{1}{5} & \frac{1}{5} \\ \frac{2}{5} & -\frac{3}{5} \end{bmatrix}$

5. inverse does not exist

7. $\begin{bmatrix} \frac{1}{2a} & -\frac{1}{2a} \\ \frac{1}{2a} & \frac{1}{2a} \end{bmatrix}$

9. $\begin{bmatrix} 0 & -1 & 1 \\ 0 & 1 & 0 \\ 1 & 1 & -1 \end{bmatrix}$

11. inverse does not exist

13. $\begin{bmatrix} \frac{1}{4} & -\frac{1}{8} & 0 \\ 0 & \frac{1}{2} & 0 \\ 0 & 0 & -\frac{1}{5} \end{bmatrix}$

15. $\begin{bmatrix} -\frac{7}{15} & \frac{4}{15} \\ \frac{2}{15} & \frac{1}{15} \end{bmatrix}$

17. inverse does not exist

19. $\begin{bmatrix} \frac{1}{4} & 0 \\ 0 & \frac{1}{5} \end{bmatrix}$

21. $\begin{bmatrix} -1 & -\frac{1}{3} & \frac{1}{3} \\ \frac{1}{3} & \frac{4}{9} & -\frac{1}{9} \\ -\frac{2}{3} & -\frac{2}{9} & \frac{1}{18} \end{bmatrix}$

23. $\begin{bmatrix} \frac{7}{66} & -\frac{4}{33} & \frac{10}{33} \\ -\frac{2}{33} & \frac{7}{33} & -\frac{1}{33} \\ -\frac{13}{66} & -\frac{2}{33} & \frac{5}{33} \end{bmatrix}$

25. $\begin{bmatrix} -\frac{1}{2} & \frac{1}{4} & -\frac{3}{4} & -\frac{3}{4} \\ 0 & 1 & -2 & -2 \\ 0 & 0 & 0 & 1 \\ 0 & -\frac{1}{2} & \frac{3}{2} & \frac{3}{2} \end{bmatrix}$

27. $A = \begin{bmatrix} -2 & 3 \\ 3 & -4 \end{bmatrix}$

Exercises 12.5   Page 605

1. $\begin{bmatrix} 4 & -6 \\ 1 & 3 \end{bmatrix}, \begin{bmatrix} 4 & -6 & | & 1 \\ 1 & 3 & | & 7 \end{bmatrix}$

3. $\begin{bmatrix} 1 & -1 & 0 \\ 0 & 1 & 1 \\ 2 & 0 & -1 \end{bmatrix}, \begin{bmatrix} 1 & -1 & 0 & | & 1 \\ 0 & 1 & 1 & | & 2 \\ 2 & 0 & -1 & | & 3 \end{bmatrix}$

5. $\left(\frac{1}{2}, \frac{1}{3}\right)$
7. $(3\alpha - 4, \alpha); \alpha$ any real number
9. $(2, 1, -3)$
11. $(-1, \alpha - 4, \alpha); \alpha$ any real number
13. $\left(-\frac{1}{3}\alpha + \frac{8}{3}, \frac{1}{3}\alpha + \frac{1}{3}, \alpha\right); \alpha$ any real number
15. $\left(4, 3, \frac{1}{3}\right)$ **17.** $(3, -1, 0)$
19. no solution   **21.** $(0, 0, 0)$
23. $(10\alpha, 8\alpha, \alpha); \alpha$ any real number
25. $(1, 2, 1, 0)$
27. $(19\alpha + 27, -10\alpha - 10, 2\alpha + 3, \alpha); \alpha$ any real number
29. $\left(-\frac{3}{7}\alpha + \frac{3}{7}, \frac{25}{7}\alpha - \frac{4}{7}, \alpha\right); \alpha$ any real number
31. no solution
33. $2Na + 2H_2O \rightarrow 2NaOH + H_2$
35. $Fe_3O_4 + 4C \rightarrow 3Fe + 4CO$
37. $3Cu + 8NHO_3 \rightarrow 3Cu(NO_3)_2 + 4H_2O + 2NO$
39. $f(x) = -2x^2 + 6x + 4$

41. $x = -2, y = 4, z = 6, w = -5$
43. $i_1 = \frac{18}{25}, i_2 = \frac{6}{25}, i_3 = \frac{12}{25}$ **45.** $x = 0.3, y = -0.12, z = 4.1$
47. $x = 3.76993, y = -1.09071, z = -4.50461, w = -3.12221$

Exercises 12.6   Page 612

1. $x = 2, y = -1$ **3.** $x = \frac{1}{2}, y = -\frac{1}{3}$
5. $x = \frac{13}{22}, y = -\frac{3}{11}$ **7.** $x = 3, y = -3$
9. $x = 2, y = 0, z = 1$ **11.** $x = 3, y = 1, z = 4$
13. $x = -2, y = 1, z = 0$ **15.** $x = \frac{1}{2}, y = \frac{1}{4}, z = 5$
17. $x = 0, y = 1, z = 1, w = 0$ **19.** $x = 22, y = -12$
21. $x = -65, y = 36$ **23.** $y = 0.4x + 0.6$
25. $y = 1.1x - 0.3$ **27.** $y = 1.3571x + 1.9286$
29. $1, 4, 7$

31. (b) $v = \sqrt{\dfrac{d_2^2 - d_1^2}{t_2^2 - t_1^2}}; D = \dfrac{1}{2}\sqrt{\dfrac{t_1^2 d_2^2 - t_2^2 d_1^2}{t_2^2 - t_1^2}}$

(c) 947.9 m; 1531 m/s

Exercises 12.7   Page 617

1. $x = 4, y = -3$ **3.** $x = 2, y = 7$
5. $x = -\frac{5}{3}, y = -\frac{5}{3}$ **7.** $x = 4, y = -4, z = -5$
9. $x = 4, y = 1, z = 2$ **11.** $x = \frac{1}{4}, y = \frac{3}{4}, z = 1$
13. $x = 1, y = 0, z = -1, w = 2$ **15.** $x = 5, y = 3, z = 10$

Exercises 12.8   Page 620

1. (a) $\begin{bmatrix} 35 & 15 & 38 & 36 & 0 \\ 27 & 10 & 26 & 20 & 0 \end{bmatrix}$

3. (a) $\begin{bmatrix} 48 & 64 & 120 & 107 & 40 \\ 32 & 40 & 75 & 67 & 25 \end{bmatrix}$

5. (a) $\begin{bmatrix} 31 & 44 & 15 & 61 & 50 & 49 & 41 \\ 24 & 29 & 15 & 47 & 35 & 31 & 21 \\ 1 & -15 & 15 & 0 & -15 & -5 & -19 \end{bmatrix}$

7. STUDY_HARD   **9.** MATH_IS_IMPORTANT_
11. DAD_I_NEED_MONEY_TODAY

13. (a) $B' = \begin{bmatrix} 15 & 22 & 20 & 8 & 23 & 6 & 22 \\ 10 & 22 & 18 & 23 & 25 & 2 & 25 \\ 3 & 26 & 26 & 14 & 23 & 16 & 12 \end{bmatrix}$

Chapter 12 Review Exercises   Page 622

A. 1. true   **3.** false
5. true   **7.** false
9. true   **11.** false

B. 1. 3×6 **3.** $a_{32} = 10$

5. 7   **7.** $\begin{bmatrix} -4 & 6 \\ 8 & -3 \end{bmatrix}$

9. $\begin{bmatrix} 0 & 1 \\ 1 & 0 \end{bmatrix}$ **11.** 6

C. 1. $x = 1, y = 1, z = 0$
5. $A_{11} = -14, A_{12} = -18, A_{13} = 13, A_{21} = 14, A_{22} = 9,$
$A_{23} = -10, A_{31} = -7, A_{32} = -3, A_{33} = 8$
7. 128   **9.** 2

11. $aeg - cef$ **13.** $\begin{bmatrix} 0 & -\frac{1}{2} \\ \frac{1}{5} & \frac{1}{10} \end{bmatrix}$

15. $\begin{bmatrix} \frac{16}{23} & -\frac{6}{23} \\ \frac{20}{23} & \frac{4}{23} \end{bmatrix}$ **17.** $x = \frac{1}{3}, y = \frac{1}{3}$

19. $x = -1, y = -2, z = -5$ 21. $x = -7, y = 8, z = 9$
23. $x = \frac{75}{13}, y = -\frac{145}{26}$ 25. 68
27. $x = \pm 1, A^{-1} = \frac{1}{x^2 - 1}\begin{bmatrix} x & 1 & -1 \\ 0 & x^2 - 1 & 0 \\ -1 & -x & x \end{bmatrix}$
29. $[2 \ -10 \ 8]$ 31. $[-32]$
33. $[-11 \ -2]$ 35. $\begin{bmatrix} 78 & 54 & 99 \\ 104 & 72 & 132 \\ -26 & -18 & -33 \end{bmatrix}$
37. $\begin{bmatrix} 0 \\ 8 \\ 7 \end{bmatrix}$ 39. $\begin{bmatrix} -13 & \frac{78}{5} & -\frac{91}{5} \\ 18 & -\frac{108}{5} & \frac{126}{5} \\ -2 & \frac{12}{5} & -\frac{14}{5} \end{bmatrix}$

Exercises 13.1 Page 633

1. $-1, 1, -1, 1, -1, \ldots$
3. $1, 3, 6, 10, 15, \ldots$
5. $\frac{1}{2}, \frac{1}{5}, \frac{1}{10}, \frac{1}{17}, \frac{1}{26}, \ldots$
7. $-1, 2, -3, 4, -5, \ldots$
9. $0, \frac{1}{3}, \frac{2}{13}, \frac{5}{17}, \frac{4}{21}, \ldots$
11. $-2, 4, -8, 16, -32, 36, \ldots$
13. $\frac{1}{49}, 8, \frac{1}{81}$
15. $3, \frac{1}{3}, -3, -\frac{1}{3}, 3, \ldots$
17. $0, 2, 8, 26, 80, \ldots$
19. $1, \frac{1}{2}, \frac{1}{6}, \frac{1}{24}, \frac{1}{120}, \ldots$
21. $7, 9, 11, 13, 15, \ldots$
23. $d = -5; a_n = 4 - 5(n-1); a_{n+1} = a_n - 5, a_1 = 4$
25. $r = -\frac{3}{4}; a_n = 4\left(-\frac{3}{4}\right)^{n-1}; a_{n+1} = -\frac{3}{4}a_n, a_1 = 4$
27. $d = -11; a_n = 2 - 11(n-1); a_{n+1} = a_n - 11, a_1 = 2$
29. $r = 0.1y; a_n = 0.1(0.1y)^{n-1}; a_{n+1} = (0.1y)a_n, a_1 = 0.1$
31. $r = -\frac{2}{3}; a_n = \frac{3}{8}\left(-\frac{2}{3}\right)^{n-1}; a_{n+1} = -\frac{2}{3}a_n, a_1 = \frac{3}{8}$
33. 113
35. $\frac{1}{2}$
37. $\frac{1}{8}$
39. 255
41. $4, 7, 10, 13, \ldots$
43. $3870
45. 6.6%
47. $145
49. 57,665
51. 32
53. $1, 1, 2, 3, 5, 8, 13, \ldots$

Exercises 13.2 Page 639

1. 14
3. -40
5. $\frac{23}{2}$
7. $1 + \sqrt{2} + \sqrt{3} + 2 + \sqrt{5}$
9. $1 - 1 + 1 - 1$
11. $\sum_{k=1}^{5}(2k + 1)$
13. $\sum_{k=0}^{5}\frac{(-1)^k}{3 \cdot 2^k}$
15. $\sum_{k=1}^{5}\frac{2^k + 1}{2^k}$
17. 35
19. 564
21. -748
23. $\frac{40}{81}$
25. 66.666
27. $\frac{85}{128}$
29. $a_1 = \frac{9}{2}, a_{10} = \frac{45}{2}$
31. 80
33. 12
35. $\frac{x^{15} + y^{15}}{x^{13}y(x + y)}$
37. $n^2 + n$
39. 500, 500
41. $13,500
43. 72 m
47. approximately 69.73 ft
49. approximately 55.6 mg

Exercises 13.3 Page 648

1. converges to 0
3. converges to 0
5. converges to $\frac{1}{2}$
7. diverges
9. diverges
11. converges to $\sqrt{2}$
13. diverges
15. converges to $\frac{5}{6}$
17. converges to 5
19. converges to 0
21. $\frac{2}{9}$
23. $\frac{61}{99}$
25. $\frac{1313}{999}$
27. 4
29. $\frac{2}{5}$
31. $\frac{81}{7}$
33. diverges
35. $-\frac{3}{10}$
37. (a) $S_n = 1 - \frac{1}{n+1}$ (b) 1
39. 75 ft
41. $\sqrt{3}$

Exercises 13.4 Page 654

1. (i) $2 = (1)^2 + 1$, is true. (ii) Assume that $S(k)$,
$2 + 4 + \cdots + 2k = k^2 + k$, is true. Then

$$2 + 4 + \cdots + 2k + 2(k + 1) = k^2 + k + 2(k + 1)$$
$$= k^2 + k + 2k + 2$$
$$= (k^2 + 2k + 1) + (k + 1)$$
$$= (k + 1)^2 + (k + 1).$$

Thus $S(k + 1)$ is true. By (i) and (ii) the proof is complete.

3. (i) $1^2 = \frac{1}{6}1(1 + 1)[2(1) + 1] = \frac{1}{6}2 \cdot 3$, is true. (ii) Assume that $S(k)$, $1^2 + 2^2 + \cdots + k^2 = \frac{1}{6}k(k + 1)(2k + 1)$, is true. Then

$$1^2 + 2^2 + \cdots + k^2 + (k + 1)^2 = \frac{1}{6}k(k + 1)(2k + 1) + (k + 1)^2$$
$$= \frac{k(k + 1)(2k + 1) + 6(k + 1)^2}{6}$$
$$= \frac{(k + 1)[k(2k + 1) + 6(k + 1)]}{6}$$
$$= \frac{(k + 1)(2k^2 + 7k + 6)}{6}$$
$$= \frac{(k + 1)[(k + 2)(2k + 3)]}{6}$$
$$= \frac{1}{6}(k + 1)[(k + 1) + 1][2(k + 1) + 1].$$

Thus $S(k + 1)$ is true. By (i) and (ii) the proof is complete.

5. (i) $\frac{1}{2} + \frac{1}{2} = 1$, is true. (ii) Assume that $S(k)$,

$$\left(\frac{1}{2} + \frac{1}{2^2} + \cdots + \frac{1}{2^k}\right) + \frac{1}{2^k} = 1,$$

is true. Then

$$\left(\frac{1}{2} + \frac{1}{2^2} + \cdots + \frac{1}{2^k} + \frac{1}{2^{k+1}}\right) + \frac{1}{2^{k+1}} = \left(\frac{1}{2} + \frac{1}{2^2} + \cdots + \frac{1}{2^k}\right) + \frac{2}{2^{k+1}}$$
$$= \left(1 - \frac{1}{2^k}\right) + \frac{2}{2^{k+1}}$$
$$= 1 - \frac{1}{2^k} + \frac{1}{2^k} = 1.$$

Thus $S(k + 1)$ is true. By (i) and (ii) the proof is complete.

7. (i) $\frac{1}{1 \cdot (1 + 1)} = \frac{1}{1 + 1}$, is true. (ii) Assume that $S(k)$,

$$\frac{1}{1 \cdot 2} + \frac{1}{2 \cdot 3} + \cdots + \frac{1}{k(k + 1)} = \frac{k}{k + 1}, \text{ is true. Then}$$

$$\frac{1}{1 \cdot 2} + \frac{1}{2 \cdot 3} + \cdots + \frac{1}{k(k + 1)} + \frac{1}{(k + 1)[(k + 1) + 1]}$$

$$= \frac{k}{k+1} + \frac{1}{(k+1)[(k+1)+1]}$$
$$= \frac{k(k+2)+1}{(k+1)(k+2)}$$
$$= \frac{k^2+2k+1}{(k+1)(k+2)}$$
$$= \frac{(k+1)^2}{(k+1)(k+2)}$$
$$= \frac{k+1}{(k+1)+1}.$$

Thus $S(k+1)$ is true. By (i) and (ii) the proof is complete.

9. (i) $1 = \frac{1}{3}(4-1)$ is true. (ii) Assume that $S(k)$, $1 + 4 + 4^2 + \cdots + 4^{k-1} = \frac{1}{3}(4^k - 1)$, is true. Then
$$1 + 4 + 4^2 + \cdots + 4^{k-1} + 4^k = \frac{1}{3}(4^k - 1) + 4^k$$
$$= \frac{1}{3}4^k + 4^k - \frac{1}{3}$$
$$= \frac{4}{3}4^k - \frac{1}{3}$$
$$= \frac{1}{3}(4^{k+1} - 1).$$

Thus $S(k+1)$ is true. By (i) and (ii) the proof is complete.

11. (i) The statement $(1)^3 + 2(1)$ is divisible by 3, is true.
(ii) Assume that $S(k)$, $k^3 + 2k$ is divisible by 3, is true; in other words, $k^3 + 2k = 3x$ for some integer x. Then
$$(k+1)^3 + 2(k+1) = k^3 + 3k^2 + 3k + 1 + 2k + 2$$
$$= k^3 + 2k + 3k^2 + 3k + 3$$
$$= (k^3 + 2k) + 3(k^2 + k + 1)$$
$$= 3x + 3(k^2 + k + 1)$$
$$= 3(x + k^2 + k + 1),$$

is divisible by 3. Thus $S(k+1)$ is true. By (i) and (ii) the proof is complete.

13. (i) The statement, 4 is a factor of $5 - 1$, is true. (ii) Assume that $S(k)$, 4 is a factor of $5^k - 1$, is true. Then
$$5^{k+1} - 1 = 5^k \cdot 5 - 1$$
$$= 5^k \cdot 5 - 5 + 4$$
$$= 5(5^k - 1) + 4.$$

Since 4 is a factor of $5^k - 1$ and of 4, it follows that 4 is a factor of $5^{k+1} - 1$. Thus $S(k+1)$ is true. By (i) and (ii) the proof is complete.

15. (i) The statement, 7 is a factor of $3^2 - 2^1 = 9 - 2$, is true.
(ii) Assume that $S(k)$, 7 is a factor of $3^{2k} - 2^k$, is true. Then
$$3^{2(k+1)} - 2^{k+1} = 3^{2k} \cdot 3^2 - 2^k \cdot 2$$
$$= 3^{2k} \cdot 9 - 2^k \cdot 2$$
$$= 3^{2k} \cdot (2 + 7) - 2^k \cdot 2$$
$$= 2(3^{2k} - 2^k) + 7 \cdot 3^{2k}.$$

Since 7 is a factor of $3^{2k} - 2^k$ and of $7 \cdot 3^{2k}$, it follows that 7 is a factor of $3^{2k+2} - 2^{k+1}$. Thus $S(k+1)$ is true. By (i) and (ii) the proof is complete.

17. (i) The statement, $(1+a)^1 \geq 1 + (1)a$ for $a \geq -1$, is true.
(ii) Assume that $S(k)$, $(1+a)^k \geq 1 + ka$ for $a \geq -1$, is true. Then, for $a \geq -1$,
$$(1+a)^{k+1} = (1+a)^k(1+a)$$
$$\geq (1+ka)(1+a)$$
$$= 1 + ka^2 + ka + a$$
$$\geq 1 + ka + a$$
$$= 1 + (k+1)a.$$

Thus $S(k+1)$ is true. By (i) and (ii) the proof is complete.

19. (i) Since $r > 1$, the statement $r^1 > 1$ is true. (ii) Assume that $S(k)$,

If $r > 1$, then $r^k > 1$,

is true. Then, for $r > 1$,
$$r^{k+1} = r^k \cdot r > r^k \cdot 1 > 1 \cdot 1 = 1.$$

Thus $S(k+1)$ is true. By (i) and (ii) the proof is complete.

Exercises 13.5 Page 659

1. 6 3. $\frac{1}{60}$ 5. 144 7. 10
9. 7 11. 4 13. n
15. $\dfrac{1}{(n+1)(n+2)^2(n+3)}$ 17. 5! 19. 100!
21. 4!5! 23. $\dfrac{4!}{2!}$ 25. $\dfrac{n!}{(n-2)!}$ 27. true
29. false 31. true 33. $x^4 - 10x^2y^4 + 25y^8$
35. $x^6 - 3x^4y^2 + 3x^2y^4 - y^6$
37. $x^2 + 4x^{3/2}y^{1/2} + 6xy + 4x^{1/2}y^{3/2} + y^2$
39. $x^{10} + 5x^8y^2 + 10x^6y^4 + 10x^4y^6 + 5x^2y^8 + y^{10}$
41. $a^3 - 3a^2b + 3ab^2 - b^3 - 3a^2c + 6abc - 3b^2c + 3ac^2 - 3bc^2 - c^3$
43. $n = 6$: 1 6 15 20 15 6 1; $n = 7$: 1 7 21 35 35 21 7 1
45. $6ab^5$ 47. $-20x^6y^6$
49. $2240x^4$ 51. $2002x^5y^9$
53. $-144y^7$ 55. 252
57. 0.95099

Exercises 13.6 Page 666

1. $abc, acb, bac, bca, cab, cba$
3. Here (x, y) represents the number x on the red die and the number y on the black die:
(1, 1), (1, 2), (1, 3), (1, 4), (1, 5), (1, 6), (2, 1), (2, 2), (2, 3),
(2, 4), (2, 5), (2, 6), (3, 1), (3, 2), (3, 3), (3, 4), (3, 5), (3, 6),
(4, 1), (4, 2), (4, 3), (4, 4), (4, 5), (4, 6), (5, 1), (5, 2), (5, 3),
(5, 4), (5, 5), (5, 6), (6, 1), (6, 2), (6, 3), (6, 4), (6, 5), (6, 6)
5. 576 7. 800
9. 120 11. 6
13. 9900 15. 20, 160
17. 6 19. 1225
21. 78 23. 1
25. 24 27. (a) 5040 (b) 2520
29. 66 31. 252
33. 90 35. 10
37. 56 39. (a) 360 (b) 1296 (c) 2401
41. $C(5, 3) \cdot C(3, 2) \cdot 5! = 3600$ 43. (a) 40 (b) 80

Exercises 13.7 Page 674

1. {HH, HT, TH, TT}
3. {1H, 2H, 3H, 4H, 6H, 1T, 2T, 3T, 4T, 6T}
5. $\frac{3}{13}$ 7. $\frac{1}{6}$
9. $\frac{1}{36}$ 11. $\frac{1}{8}$
13. $\dfrac{13 \cdot 48}{C(52, 5)} \approx 0.00024$ 15. $\dfrac{4 \cdot C(13, 5)}{C(52, 5)} \approx 0.002$
17. $1 - \dfrac{C(39, 5)}{C(52, 5)} \approx 0.78$ 19. $1 - \dfrac{1}{2^{10}} \approx 0.999$
21. $\frac{15}{16}$ 23. (a) $\frac{28}{143}$ (b) $\frac{134}{143}$

25. $\frac{1}{1024}$
27. $\frac{6}{36} + \frac{2}{36} = \frac{2}{9}$
29. $\frac{10}{100} + \frac{5}{100} = \frac{3}{20}$
31. $\frac{1}{6}$
33. $\frac{3}{5}$ or 60%
35. $\frac{2}{3}$

Chapter 13 Review Exercises   Page 677

A. 1. false
   3. true
   5. true
   7. true
   9. true
   11. true

B. 1. $2x + 7, 2x + 10, 2x + 13, \ldots$
   3. $1 + \frac{1}{2} + \frac{1}{3} + \frac{1}{4} + \frac{1}{5}$
   5. $\frac{2}{5}$
   7. 2700
   9. $\frac{9}{4}$
   11. $\frac{8}{15}$

C. 1. $6, 3, 0, -3, -6, \ldots$
   3. $-1, 2, -3, 4, -5, \ldots$
   5. $1, 3, 14, 72, 434, \ldots$
   7. $-\frac{1}{8}$
   9. $\frac{341}{256}$
   11. C
15. (i) $1^2(1 + 1)^2 = 4$ is divisible by 4. (ii) Assume $S(k)$,
    $k^2(k + 1)^2$ is divisible by 4, is true. Then
    $$(k + 1)^2(k + 2)^2 = (k + 1)^2(k^2 + 4k + 4)$$
    $$= k^2(k + 1)^2 + 4(k + 1)^3$$
    is divisible by 4 since each term is divisible by 4. Thus $S(k + 1)$ is true. By (i) and (ii) the proof is complete.
17. (i) $1(1!) = 2! - 1 = 1$, is true. (ii) Assume $S(k)$,
    $$1(1!) + 2(2!) + \cdots + k(k!) = (k + 1)! - 1,$$
    is true. Then
    $$1(1!) + 2(2!) + \cdots + k(k!) + (k + 1)(k + 1)!$$
    $$= (k + 1)! - 1 + (k + 1)(k + 1)!$$
    $$= (k + 1)!(1 + k + 1) - 1$$
    $$= (k + 1)!(k + 2) - 1$$
    $$= (k + 2)! - 1.$$
    Thus $S(k + 1)$ is true. By (i) and (ii) the proof is complete.

19. (i) $\left(1 + \frac{1}{1}\right) = 1 + 1$, is true. (ii) Assume $S(k)$,
    $$\left(1 + \frac{1}{1}\right)\left(1 + \frac{1}{2}\right)\left(1 + \frac{1}{3}\right)\cdots\left(1 + \frac{1}{k}\right) = k + 1,$$
    is true. Then
    $$\left(1 + \frac{1}{1}\right)\left(1 + \frac{1}{2}\right)\left(1 + \frac{1}{3}\right)\cdots\left(1 + \frac{1}{k}\right)\left(1 + \frac{1}{k + 1}\right)$$
    $$= (k + 1)\left(1 + \frac{1}{k + 1}\right)$$
    $$= (k + 1 + 1)$$
    $$= k + 2.$$
    Thus $S(k + 1)$ is true. By (i) and (ii) the proof is complete.
21. 40
23. 28
25. $(n + 1)(n + 2)(n + 3)$
27. $a^4 + 16a^3b + 96a^2b^2 + 256ab^3 + 256b^4$
29. $x^{10} - 5x^8y + 10x^6y^2 - 10x^4y^3 + 5x^2y^4 - y^5$
31. $-175{,}000a^5b^9$
33. $210x^6y^{12}z^{12}$
35. 37th
37. $\frac{1}{5}$
39. converges to 0
41. HHH, HHT, HTH, HTT, THH, THT, TTH, TTT
43. (a) 496   (b) 328
45. 120
47. 720
49. (a) $2 \cdot 10!12! \approx 3.48 \times 10^{15}$ (b) $22! \approx 1.124 \times 10^{21}$
51. $\{(1, 1), (1, 2), (1, 3), (1, 4), (1, 5), (2, 1), (2, 2), (2, 3), (2, 4),$
    $(2, 5), (3, 1), (3, 2), (3, 3), (3, 4), (3, 5), (4, 1), (4, 2), (4, 3),$
    $(4, 4), (4, 5), (5, 1), (5, 2), (5, 3), (5, 4), (5, 5)\}$
53. $\frac{25}{102}$
55. (iii)
57. (a) $\frac{14}{33}$ (b) $\frac{1}{11}$ (c) $\frac{17}{33}$
59. (a) 24   (b) $\dfrac{1}{[C(15, 5)]^4 C(15, 4)} \approx \dfrac{1}{11.1 \times 10^{17}} \approx 9 \times 10^{-18}$

multiplicative inverse of, 96
polar form of, 432
powers of, 436–437
product of, 95, 434
properties of, 95, 97
pure imaginary, 94
quotient of, 434
real part of, 94
roots of, 438
standard form of, 94
subtraction of, 95
sum of, 95
trigonometric form of, 432
Complex plane:
definition of, 431–432
imaginary axis, 432
real axis, 432
Complex zero of function, 242
Component form a vector, 509
Component of a vector on another vector, 519
Components of a vector:
definition of, 509
horizontal, 512
vertical, 512
Composition of functions:
definition of, 194
domain of, 195
Compound interest, 304, 632
Conditional equation, 66
Conic sections:
definitions of, 448, 455, 461
in rectangular coordinates, 449–450, 455, 457, 462, 465
degenerate, 470, 475
eccentricity, 448, 456, 502
in polar coordinates, 502
Conjugate axis of a hyperbola, 463
Conjugate of a complex number, 96
Conjugate hyperbolas, 470
Conjugate solutions of a quadratic equation, 98
Conjugate zeros theorem, 247
Consistent system of equations, 529–530
Constant function, 226
Constant of proportionality, 146
Constraint, 554
Continuous compounding of interest, 304–305
Continuous function, 187
Convergent infinite series, 645
Convergent sequence, 643
Conversion:
from rectangular to polar coordinates, 490
from polar to rectangular coordinates, 489
Convex limaçon, 496
Convex set in the plane:
definition of, 554
vertex of, 555
Coordinate(s):
of a point in the Cartesian coordinate system, 124
of a point on the real number line, 11
polar, 488
x-coordinate, 124
y-coordinate, 124
Coordinate plane, 124

Coordinate system:
Cartesian, 124
polar, 488
rectangular, 124
Copernicus, Nicolaus, 319
Correspondence, rule of, 156
Cosecant:
definition of, 372
graph of cosecant function, 374
inverse of, 395, 396
period of cosecant function, 373
signs of, 372
Cosine(s):
amplitude of cosine function, 365, 366
cycle, 361–362, 364
definition of, 329, 340, 355
difference formula, 380
domain of cosine function, 363
double-angle formula, 384
graph of cosine function, 362
half-angle formula, 384
hyperbolic, 313
inverse of, 392
law of, 422–423
period of cosine function, 357, 364–365
range of cosine function, 363
shifted graph of cosine function, 365–366
signs of, 356
sum formula, 380
Cotangent:
definition of, 372
graph of cotangent function, 373–374
inverse of, 395, 396
period of cotangent function, 373, 375
signs of, 372
Coterminal angles, 320
Counting, principles of, 661
Cramer, Gabriel, 616
Cramer's rule, 616
Cryptography, 618
Cube of a binomial, 39
Cube root, 25
Cubic function, 226
Curve, Gompertz, 318
Curie, Pierre and Marie, 301
Cycle of a graph, 361, 374, 375
Cycloid, 479

D

Dantzig, George B., 527
Dead Sea scrolls, 303
Decay:
carbon dating, 303
constant, 302
half-life, 302
radioactive, 301–302
Decimal form of a real number, 4
Decimals:
nonrepeating, 4
repeating, 4
terminating, 4
Decoding matrix, 618
Decreasing function, 180
Degenerate conic sections, 470, 475

Degree(s):
    as an angular measure, 320
    conversion to radians, 323–324
    fraction of, 321
    of a polynomial, 226
DeMoivre, Abraham, 437
DeMoivre's theorem, 437
Denominator:
    of a fraction, 6
    least common, 53
    rationalization of, 27
Dependent equations in a linear system, 529, 530
Dependent variable, 156
Depression, angle of, 413
Descartes, René, 124
Determinant(s):
    definition of, 581
    expansion by cofactors of first row, 584
    expansion theorem for, 583
    of a square matrix, 581
    minor, 582
    of order n, 581
    properties of, 585
    solving systems of equations by, 615, 616
Diagonal, main, of a matrix, 567
Diameter of the Moon, 328
Difference:
    common, 630
    of complex numbers, 95
    of functions, 193
    of matrices, 572
    of two cubes, 42
    of two squares, 42
    of vectors, 510, 512
Difference quotient, 196
Difference trigonometric formulas:
    for cosine, 380
    for sine, 382
    for tangent, 383
Difference of two squares, 40
Difference of two cubes, 42
Dimension of a matrix, 566
Dimpled limaçon, 496
Diophantus, 65, 71
Direction angle of a vector, 509
Direct proportion, 146–147
Directrix of a parabola, 448
Direct variation, 146
Discontinuous function, 264
Discriminant, 85–86, 245, 474
Disjoint sets, 2, 672
Displacement vector, 508
Distance:
    between points on the number line, 14
    formula, 126
Distributive law, 5
Divergent to infinity, 643
Divergent infinite series, 645
Divergent sequence, 642
Dividend, 236
Divides evenly, 236
Divisible, 236
Division:
    of polynomials, 235–236
    synthetic, 238

Division algorithm for polynomials, 236
Divisor, 236
Domain:
    of a function, 156
    implicit, 157
    of a sequence, 628
    of a variable, 36
Dot product:
    and the angle between two vectors, 518
    definition of, 516, 517
    and orthogonal vectors, 518
    physical interpretation of, 520
    properties of, 516
Double-angle formula:
    for cosine, 384
    for sine, 384
    for tangent, 389
Double-napped cone, 448

E

e:
    as base, 283, 289
    definition of, 283
Eccentricity, 458
Echelon form of a matrix, 600, 603
Echo sounding, 613
Effective half-life, 310
Elementary row operations on an augmented matrix, 594, 600
Element of a set, 2
Elevation, angle of, 412–413
Elimination, method of, 531
Elimination operations symbols, 531
Ellipse:
    applications of, 458–459
    center of, 455
    with center (h, k), 457
    definition of, 455
    eccentricity of, 458
    equations of, 455, 457
    focal width of, 461
    foci of, 455
    major axis of, 456
    minor axis of, 456
    in polar coordinates, 503
    in rectangular coordinates, 449, 450
    standard form for, 455, 457
    vertices of, 456
Ellipsis, 3
Empty set, 2
Encoding matrix, 618
End behavior of a function, 158, 228, 264
Entry of a matrix, 566
Equality:
    of complex numbers, 94
    of matrices, 567
    of vectors, 510
Equation(s):
    absolute value, 105–106
    of a circle, 131
    conditional, 66
    dependent, 529, 530
    of an ellipse, 455, 457
    equivalent, 67, 130

exponential, 294
extraneous solution of, 68, 296
first degree, 528
graph of an, 130
identity, 66
intercepts of an, 134
of a hyperbola, 462, 465
independent, 529, 530
of a line with slope, 141, 142
of a horizontal line, 142
of a vertical line, 142
linear, 67
logarithmic, 294
in one variable, 66
of a parabola, 449, 450
quadratic, 81
root of, 66
satisfy, 66, 130
simultaneous, 528
solution set of, 66, 130
solving, 66, 528
system(s) of, 528
trigonometric, 398
in two variables, 130
Equivalent equations, 67, 130
Equivalent inequalities, 99
Equivalent matrices, row, 599
Equivalent systems of equations, 528
Eratosthenes, 327
Even function, 166
Even-odd trigonometric identities, 357, 374
Event:
    certain, 671
    definition of, 669
    impossible, 671
    mutually exclusive, 671
Expansion theorem for determinants, 583
Exponent(s):
    integer, 17
    irrational, 280
    laws of, 19
    rational, 31
Exponential equations, 294
Exponential function:
    applications of, 300–305
    asymptote for, 282
    base of, 280
    definition of, 280
    domain of, 280
    graph of, 281
    natural, 283–284
    properties of, 282
Extraneous solution, 68, 296, 401

F

Factor completely, 49
Factor a polynomial, 243
Factorial function, 164
Factorial notation, 656
Factoring:
    FOIL method, 40
    polynomials, 45
    quadratic polynomials, 46
Factorization formulas, 48

Factor theorem, 243
Family of lines, 145
Feasible solution, 554
Ferrari, Lodovico, 225
Fibonacci, 634
Finite geometric series, 638
Finite sequence, 628
First-degree equation, 528
Floor function, 187
Focal chord of a parabola, 450
Focal width:
    of an ellipse, 461
    of a hyperbola, 469
    of a parabola, 454
Focus (foci):
    of an ellipse, 455, 457
    of a hyperbola, 461, 462, 465
    of a parabola, 448, 449
Fontana, Gregorio, 487
Fontana, Niccolò (Tartaglia), 225
Force, resultant, 513
Fraction(s):
    denominator of, 6
    numerator of, 6
    partial, 543
    proper, 543
    improper, 547
Fractional expression, 55
Function(s):
    absolute value, 188
    arccosecant, 395, 396
    arccosine, 392
    arccotangent, 395, 396
    arcsecant, 395, 396
    arcsine, 390
    arctangent, 393
    asymptotes for, 263, 264, 265
    ceiling, 192
    circular, 355
    combining, 192–193
    composition of, 194
    constant, 226
    continuous, 187
    cosecant, 329, 340, 355
    cosine, 329, 340, 355
    cotangent, 329, 340, 355
    cubic, 226
    decreasing, 180
    definition of, 156
    difference of, 193
    domain of, 156
    discontinuous, 187
    end behavior of, 158, 228
    even, 166
    exponential, 280
    factorial, 164
    floor, 187
    global behavior of, 158
    graph of, 158–159
    greatest integer, 187
    horizontal-line test, 210
    hyperbolic, 313
    image of, 156
    implicit domain of, 157
    increasing, 180

Function(s) (*Continued*)
   intercepts of, 160
   inverse of, 202
   inverse hyperbolic sine, 314
   inverse trigonometric, 389
   linear, 175, 226
   local behavior of, 229
   logarithmic, 287
   logistic, 301
   natural exponential, 284
   natural logarithmic, 289
   notation, 156
   objective, 554
   odd, 166
   one-to-one, 210
   periodic, 356
   piecewise-defined, 185
   polynomial, 175, 226
   postage stamp, 186
   power, 165, 227
   product of, 193
   quadratic, 175
   quartic, 226
   quintic, 226
   quotient of, 193
   range of, 156
   rational, 261
   reflected graphs of, 169
   secant, 329, 340, 355
   shifted graphs of, 168
   sine, 329, 340, 355
   square root, 157
   squaring, 156
   sum, 164
   sum of, 193
   symmetry of graphs of, 165–167
   tangent, 329, 355
   trigonometric, 355
   of two variables, 554
   value of, 156
   vertical-line test for, 159
   zero, 226
   zeros of, 230, 363
Fundamental counting principle, 661
Fundamental theorem of algebra, 244
Future value, 305

G

Galileo, 447
Gauss, Carl Friedrich, 225, 244
Gaussian elimination, 599
Gauss-Jordan elimination, 599, 603
General angles, trigonometric functions of, 339–340
General term of a sequence, 628
Generation time, 301
Geocentric parallax, 417
Geometric sequence, 631
Geometric series:
   definition of, 638, 645
   finite, 638
   infinite, 646
   sum of, 639, 646
Germain, Marie-Sophie, 409

Global behavior of a function, 158
Golden ratio, 542
Gompertz curve, 316
Gospel of Judas, 303
Graph(s):
   end behavior of, 158, 228–229
   of an equation, 130
   of the exponential function, 281
   of a function, 158
   intercepts of, 134, 160, 230
   of the inverse trigonometric functions, 390, 392, 393
   of linear inequalities in two variables, 549
   of the logarithmic function, 287–288
   points of intersection, 498, 529, 530
   polar, 492–499
   of power functions, 166, 227
   reflected, 169
   shifted, 168
   symmetry of, 166–167
   tangent to an axis, 160
   of the trigonometric functions, 362, 373, 374
Greater than, 12
Greater than or equal to, 12
Greatest integer function, 187
Grouping, factoring by, 45
Growth, population, 303

H

Half-angle formula:
   for cosine, 384
   for sine, 384
   for tangent, 389
Half-life:
   definition of, 302
   effective, 310
   of radium, 302
Half-open interval, 102
Half-plane, 549
Harmonic motion, 429
Heron, 427
Heron's area formula, 427
Hildago, asteroid, 459
Hipparchus, 319, 487
Hohmann transfer, 507
Hole in a graph, 270
Homogeneous system of equations, 534
Hookes law, 146
Horizontal asymptote:
   for exponential functions, 282
   definition of, 264
   for rational functions, 264, 265–266
Horizontal component of a vector, 512
Horizontal compression of a trigonometric graph, 364
Horizontal line, equation of, 142
Horizontal shift of a graph, 168
Horizontal-line test for a one-to-one function, 201
Hypatia, 448
Hyperbola(s):
   applications, 467
   asymptotes of, 463, 465
   auxiliary rectangle of, 463
   branches of, 462
   center of, 461
   conjugate, 470

conjugate axis of, 463
definition of, 461
eccentricity of, 466
equations of, 462, 465
focal width of, 469
foci of, 461
in polar coordinates, 503
in rectangular coordinates, 462, 465
rectangular, 470
standard form for, 462, 465
transverse axis of, 462
vertices of, 462
with center (h, k), 465
Hyperbolic functions:
definitions of, 313
graphs of, 313–314

I

i (imaginary unit), 93
Iceman (Ötzi), 309
Identities, trigonometric:
even-odd, 357
cofunction, 332
Pythagorean, 333, 342, 355, 379
quotient, 330
reciprocal, 330
Identity, 66
Identity element:
for addition, 4, 571
for multiplication, 4, 575
Identity matrix, 575
Image of a function, 156
Imaginary axis, 432
Imaginary number, 94
Imaginary part of a complex
number, 94
Imaginary unit, 93
Implicit domain of a function, 157
Impossible event, 671
Improper fraction, 236, 547
Inconsistent system of equations, 529–530
Increasing function, 180
Independent equations in a linear
system, 529, 530
Independent variable, 156
Index of a radical, 25
Index of summation, 635
Induction, mathematical, 652
Inequality (inequalities):
absolute values and, 106
definition of, 12
equivalent, 99
linear, 100
nonlinear, 551
nonstrict inequality, 12
polynomial, 110
rational, 110
simultaneous, 102, 106
solution of, 99
solving, 100
solving simultaneous, 102
systems of, 550
strict inequality, 12
transitivity property of, 12
triangle, 14
involving two variables, 549
Infinite geometric series, 646
Infinite sequence, 628
Infinity symbols, 101
Initial point:
of a parameterized curve, 477
of a vector, 508
Initial side of an angle, 320
Inner product, 516, 576
Integer exponents, 17, 19
Integers, 3
Intensity level of sound, 311
Intercept(s):
of the graph of an equation, 134
of the graph of a function, 160
x-intercept, 134, 160
y-intercept, 134, 160
Interest:
compound, 304
continuous compounding of, 305
simple, 73
Intermediate value theorem, 257
Intersection of sets, 2
Interval:
closed, 101
endpoints of, 101
graphs of, 102
half-open, 102
notation, 101
open, 101
unbounded, 102
Inverse of a function:
definition of, 202
finding the, 203
properties of, 202, 389
Inverse hyperbolic sine, 314
Inverse of a matrix:
definition of, 590–591
methods for finding an, 591–592, 593, 594
Inverse proportion, 147
Inverse trigonometric functions:
sine, 390
cosine, 392
graphs of, 390, 392, 393
properties of, 394
tangent, 393
Inverse variation, 147
Invertible matrix, 591
Investment problems, 73
Irrational exponents, 280
Irrational number, 3
Irreducible quadratic polynomial, 543

J

Joint variation, 148

K

Kepler, Johannes, 447
Kepler's first law, 447, 459
Kepler's third law, 150

L

Lacroix, Sylvestre François, 487
Lascaux cave drawings, 308
Lauchen, Georg Joachim von, 319
Law(s):
    associative, of addition, 4, 571
    associative, of multiplication, 4
    commutative, of addition, 4, 571
    commutative, of multiplication, 4
    of cosines, 423
    distributive, 5, 575
    of integer exponents, 19
    of logarithms, 290
    of radicals, 25
    of rational exponents, 32
    of real exponents, 280–281
    of sines, 418
Leading coefficient of a polynomial function, 226
Least common denominator, 53
Least squares line, 217, 610
Legs of a triangle, 328
Leibniz, Gottfried Wilhelm, 155
Lemniscate, 498
Less than, definition of, 11
Less than or equal to, 12
Libby, Willard, 303
Liber Abacci, 635
Limaçon, 496
Line(s):
    of best fit, 217, 610
    equations of, 141, 142
    horizontal, 142
    least squares, 217, 610
    parallel, 143
    perpendicular, 143
    point-slope equation of, 141
    segment, midpoint of, 127
    segment in the plane, midpoint of, 127
    slope, 139
    slope-intercept equation of, 141
    vertical, 142
Linear equation:
    in two variables, 142, 528
    in n variables, 528
Linear function:
    of one variable, 174, 226
    of two variables, 554
Linear inequality in one variable, 100
Linear inequality in two variables:
    definition of, 549
    graphing of, 549–550
    half-plane, 549
    solution of, 549
    system of, 550
Linear programming:
    constraints, 554
    definition of, 554
    objective function, 554
Linear speed, 327
Linear system:
    consistent, 529, 530
    definition of, 528
    inconsistent, 529, 530

Local behavior of a function, 229
Local extremum, 230
Logarithm(s):
    applications of, 395
    base e, 289
    base 10, 289
    change of base, 297–298
    definition of, 287
    laws of, 290
Logarithmic differentiation, 291
Logarithmic equations, 294
Logarithmic function(s):
    applications of, 305–306
    asymptote for, 288
    base of, 287
    common, 289
    definition of, 287
    domain of, 287
    graph of, 287–288
    natural, 289
    properties of, 288
Logistic function, 301
LORAN, 467

M

Magnitude of a complex number, 432
Magnitude of a vector, 509
Main diagonal of a matrix, 567
Major axis of an ellipse:
    definition of, 456
    length of, 456
Malthus, Thomas, R., 301
Mathematical model, 146, 181
Mathematical induction, 651
Matrix (matrices):
    addition of, 570
    additive identity, 571
    additive inverse, 572
    adjoint, 593
    associative law of, 571, 575
    augmented, 599
    coefficient, 599, 608
    column (vector), 567
    commutative law of, 571
    decoding, 618
    definition of, 566
    determinant of, 581
    difference of, 572
    dimension of, 566
    distributive law of, 575
    elementary row operations on, 594, 600
    encoding, 618
    entry of, 566
    equality of, 567
    identity, 575
    inner product of, 576
    inverse, 590–591
    invertible, 591
    main diagonal of, 567
    multiplication of, 573–574
    multiplicative identity, 575
    multiplicative inverse, 590–591
    noninvertible, 591
    nonsingular, 591

of order n, 566
product, 573–574
rectangular, 566
reduced row echelon form, 603
row echelon form, 599–600
row equivalent, 599
row operations, 599
row reduction, 599
row (vector), 567
scalar multiple of, 571
singular, 591
square, 566
subtraction of, 572
sum of, 570
transpose of, 568
zero, 571

Maximum (minimum) of a quadratic function, 176–177
Method of elimination, 531
Method of factoring for solving quadratic equations, 82
Method of substitution, 529
Midpoint of a line segment:
    in the plane, 127
    on the real number line, 14
Minor, 582
Minor axis of an ellipse, 456
Minutes, angle measure, 321
Mixture problems, 75
Modulus of a complex number, 432
Monomial, 37
Multiplication:
    of complex numbers, 95
    of matrices, 573–574
Multiplicative identity:
    for $m \times n$ matrices, 575
    for real numbers, 4
Multiplicative inverse:
    for complex numbers, 96
    for $m \times n$ matrices, 590–591
    for real numbers, 4
Multiplicity of a zero, 230, 243
Mutually exclusive events, 672

N

Napier, John, 279
Nappes of a cone, 448
Natural exponential function, 283–284
Natural logarithmic function, 289
Natural numbers, 2
Navigation, 409, 424
Negative integer power of x, 17
Negative real numbers, 3
Negative of a vector, 510
Newton, Isaac, 123, 304, 459
Newton's law of cooling/warming, 303
Noninvertible matrix, 591
Nonlinear equation, 537
Nonlinear system, 537
Nonnegative real numbers, 3, 12
Nonpositive real numbers, 12
Nonrigid transformation of a graph, 170–171
Nonsingular matrix, 591
Nonstrict inequality, 12
Normalization of a vector, 512

Notation:
    factorial, 656
    function, 156
    interval, 101–102
    scientific, 20
    summation, 217
NSAIDS, half-life of, 303
nth-order determinant, 581
nth root, 438
Number(s):
    complex, 94
    imaginary, 94
    integer, 3
    irrational, 3
    natural, 2
    negative, 3
    nonnegative, 3
    positive, 3
    rational, 2
    real, 3
Number line, real, 11
Number system:
    complex, 93
    real, 2
Numerator of a fraction, 6

O

Objective function, 554
Oblique asymptote, 265
Oblique triangle, 425
Obtuse angle, 324
Odd-even trigonometric identities, 357, 374
Odd function, 166
One-to-one function:
    definition of, 201
    horizontal-line test for, 201
Open interval, 101
Order of a matrix, 566
Order relations:
    greater than, 11
    greater than or equal to, 12
    less than, 11
    less than or equal to, 12
Ordered n-tuple, 528
Ordered triple, 528
Orientation of a parameterized curve, 477
Origin:
    in the rectangular coordinate system, 124
    in the polar coordinate system, 488
    on the real line, 11
Orthogonal vectors, 518
Outcomes, 669
Overdetermined system of equations, 217

P

Parabola:
    applications of, 451
    axis of, 448
    definition of, 448
    directrix of, 448
    equations of, 449, 450
    focal chord of, 450
    focal width of, 454

Parabola (*Continued*)
   focus of, 448
   in polar coordinates, 503
   in rectangular coordinates, 449, 450
   standard form for, 449, 450
   vertex of, 448, 450
Paraboloid, 451
Parallax, geocentric, 417
Parallel lines:
   definition of, 143
   slope of, 143
Parameter, 476
Parameterization, 476
Parametric equations, 476
Partial fraction decomposition, 543–546
Partial fractions, 543
Pascal, Blaise, 656
Pascal's triangle, 656
Peacock, George, 487
Perfect square, 48
Perigee, 505
Perihelion, 459
Period:
   of motion, 430
   of the trigonometric functions, 357, 373
Periodic function, 356
Permutation, 662
Perpendicular lines:
   definition of, 143
   slope of, 143
Petals of a rose curve, 497
pH of a solution, 306
Phase shift, 366
Piecewise-defined function, 185
Pisano, Leonardo (Fibonacci), 634
Plane:
   complex, 431–432
   coordinate, 124
   xy-plane, 124
Plane curve, 480
Plotting points, 125
Point(s):
   coordinate of, on the real number line, 11
   distance between, on the number line, 14
   initial point of a vector, 508
   plotting, 125
   polar coordinates of a, 488
   rectangular coordinates of a, 124–125
   terminal point of a vector, 508
   turning, 181, 229
Point-slope equation of a line, 141
Pole, 488
Polar axis, 488
Polar coordinate system:
   conic sections in, 502–503
   conventions in, 488
   conversion of polar coordinates to rectangular, 489
   conversion of rectangular coordinates to polar, 490
   definition of, 488
   polar axis, 488
   polar coordinates, 488
   pole, 488
   tests for symmetry in, 494
Polar equations of conics, 503

Polar form of a complex number, 432
Polar graphs:
   cardioid, 495
   circle, 493, 497
   lemniscate, 498
   limaçon, 496
   line through origin, 493
   rose curve, 497
   spiral of Archimedes, 494
Polygon, 116
Polynomial(s):
   algebra of, 38
   binomial, 37
   complete linear factorization of, 244
   constant, 226
   constant term of, 37, 226
   cubic, 226
   definition of, 37
   degree of, 37, 226
   division algorithm for, 236
   end behavior of, 228–229
   equation, 81
   factoring of, 45, 243, 244, 245, 250
   factor theorem for, 243
   function, 175, 226
   graph of, 227
   leading coefficient of, 37, 226
   linear, 175
   local behavior of, 229
   monomial, 37, 227
   inequalities, 110
   in one variable, 37
   product of, 38–39
   quadratic, 175
   quartic, 226
   quintic, 226
   rational zeros of, 251–252
   relative extremum of, 230
   remainder theorem for, 237
   single term, 227
   standard form of, 37
   synthetic division of, 238
   term of, 37
   in three variables, 41
   trinomial, 37
   turning point of, 229
   in two variables, 41
   zero of, 37, 230, 243
Polynomial equation:
   completing the square, 84
   of degree n, 81
   factoring, 82
   linear, 67
   quadratic, 81
   quadratic formula, 85
   rational roots of, 251–252
   square root method, 83
Polynomial inequalities:
   definition of, 110
   guidelines for solving, 111
Population growth, 300
Position vector, 509
Positive integers, 2
Positive integer power of x, 17
Positive real numbers, 3, 12

Postage stamp function, 186
Power function:
    definition of, 165, 227
    domain of, 165
Powers of a complex number, 436–437
Present value, 310
Principle of mathematical induction, 652
Principle nth root of a real number, 24
Principal square root, 93
Probability of an event, 669
Product:
    of complex numbers, 95, 434
    dot, 516, 517
    of functions, 193
    inner, 516, 576
    of matrices, 573–574
    of polynomials, 38–39
Product-to-sum formulas, 388
Programming, linear:
    basic problem of, 554
    constraints, 554
    objective function, 554
Projection of a vector onto another vector, 520
Proper fraction, 236, 543
Proper rational expression, 543
Properties of determinants, 585
Proportion:
    direct, 146–147
    inverse, 147–148
    jointly, 148
Proportionality, constant of, 146
Psalms scroll, 303
Pure imaginary number, 94
Pythagoras, 87
Pythagorean theorem:
    generalization of, 422–423
    statement of, 87, 422
Pythagorean trigonometric identities, 333, 342, 355, 379

Q

Quadrantal angle, 324
Quadrants, 124
Quadratic:
    equation, 85
    formula, 85, 245
    function, 175, 226
    maximum (minimum) of, 177
    irreducible, 543
    polynomial, 175, 226
Quartic polynomial function, 226
Quintic polynomial function, 226
Quotient:
    difference, 196
    of functions, 193
Quotient trigonometric identities, 330

R

Radian measure of an angle:
    conversion to degrees, 323–324
    definition of, 322
Radical:
    definition of, 24–25
    index of, 25
    laws of, 25
    radicand of, 25
    rationalizing, 27
Radicand of a radical, 25
Radioactive decay:
    carbon dating, 303
    definition of, 301–302
    half-life, 302
Radius of a circle, 131
Range:
    of a function, 156
    of a sequence, 628
Rate problems, 74
Ratio:
    common, 631
    golden, 542
Rational exponents, laws of, 32
Rational expression:
    definition of, 51
    domain of, 51
Rational function:
    asymptotes of, 263
    definition of, 261
Rational inequalities, 110
Rational numbers, 3
Rational zero of a polynomial, 251–252
Rational Zeros Theorem, 252
Rationalizing radicals, 27
Real axis, 432
Real number(s):
    absolute value of, 12
    decimal form of, 4
    integers, 3
    irrational, 3
    negative, 3
    nonnegative, 3
    order relations for, 11–12
    positive, 3
    properties of, 4–8
    rational, 3
    root of, 24
    set R of, 3, 11
Real number line:
    coordinate of a point on, 11
    description of, 11
    midpoint of a line segment on, 14
    origin on, 11
Real number system, 4
Real part of a complex number, 94
Reciprocal, 4
Reciprocal trigonometric identities, 330
Reciprocal trigonometric functions, 330, 372
Rectangular coordinate system:
    definition of, 124
    symmetry of a graph in, 135
Rectangular hyperbola, 470
Rectangular matrices, 566
Recursion formula, 629
Reduced row echelon form, 603
Reference angle:
    definition of, 344, 358–359
    property of, 345
Reflections of a graph:
    in the x-axis, 169
    in the y-axis, 169

Refraction of light, 404
Relations, order, 11–12
Relative extrema:
　definition of, 230
　maximum, 230
　minimum, 230
Remainder theorem, 237
Repeated zero, 230
Repeating decimals, 4
Resultant force, 513–514
Rheticus, Georg Joachim, 319
Rhind, Alexander Henry, 10
Rhind papyrus, 10
Richter scale, 305
Right angle, 324
Right triangle:
　definition of, 324
　solving a, 410
　trigonometry, 328
Rigid transformation of a graph, 168, 169
Rise, 139
Robotic arm, 426
Root(s):
　of complex numbers, 438
　cube, 25
　of an equation, 66, 160
　extraneous, 68
　of multiplicity m, 243
　of a polynomial equation, 82
　principal nth, 24
　principal square, 93
　rational roots (zeros) of a polynomial, 252
　of real numbers, 24–25
Rotation:
　equations, 471
　of polar graphs, 499, 505
　of rectangular axes, 470
Row echelon form, 600
Row matrix, 567, 576
Row vector, 576
Rule of correspondence, 156
Run, 139

S

Saint-Vincent, Grégoire de, 487
Sample space, 669
Scalar, 508
Scalar multiple of a vector, 510, 512
Scientific notation, 20
Secant:
　definition of, 372
　graph of secant function, 374
　inverse of, 395, 396
　period of secant function, 374
　signs of, 372
Secant line, 196
Second, angle measure, 321
Sector of a circle, 327
Semicircle, 133, 159
Sequence:
　arithmetic, 629
　convergent, 643
　definition of, 628
　divergent, 643
　finite, 629
　general term of, 628
　geometric, 631
　infinite, 628
　nth term of, 628
　of partial sums, 645
　recursively defined, 629
Series:
　arithmetic, 637
　　sum of, 639, 646
　geometric, 638–639, 646
　infinite, 645
　partial sums of, 645
　sum of, 637, 639, 645, 646
Set builder notation, 2
Set(s):
　convex, 554
　definition of, 2
　disjoint, 2
　element of, 2
　empty, 2
　intersection of, 2
　of real numbers, 2–3
　subset of, 2
　union of, 2
Shift, phase, 366
Shifted graphs:
　of cosine function, 365–367
　of a function, 168–169
　of sine function, 365–367
Shroud of Turin, 308
Side adjacent, 328
Side opposite, 328
Sigma notation, 217–218
Sign chart method, 110–111
Sign properties:
　of a product, 110
　of a quotient, 113
Significant digits, 20
Simple closed curve, 477
Simple harmonic motion, 429
Simple interest, 73
Simple zero, 230, 243
Simultaneous equations, 528
Simultaneous inequalities, 102
Sine(s):
　amplitude of sine function, 365, 366
　cycle, 361–362, 364
　definition of, 329, 340, 355
　difference formula, 382
　domain of sine function, 363
　double-angle formula, 384
　graph of sine function, 362
　half-angle formula, 384
　hyperbolic, 313
　inverse of, 390
　law of, 418–419
　period of sine function, 357, 364–365
　range of sine function, 363
　shifted graph of sine function, 365–366
　signs of, 356
　sum formula, 382
Single-term polynomial, 227
Singular matrix, 591
Slant asymptote, 265

Slope(s):
    of a line, 139
    of parallel lines, 143
    of perpendicular lines, 143
    rise, 139
    run, 139
    undefined, 140
Slope-intercept equation of a line, 141
Snell's law, 404
Solution:
    of an equation, 66
    extraneous, 68, 296
    feasible, 554
    of inequalities, 100, 102, 106
    of a system of inequalities, 548–549, 551–552
    pH of, 306
    of a triangle, 410
Solution set:
    of an equation, 66, 130
    of an inequality, 99
    of a system of equations, 528
    of a system of inequalities, 550
Solving:
    equations, 66
    simultaneous inequalities, 102
    systems of equations by determinants, 615, 616
Sørensøn, Søren, 306
Sound, intensity level of, 311
Sounding, echo, 613
Space, sample, 669
Space curve, 480
Special angles, 335
Speed:
    angular, 327
    linear, 327
    as the magnitude of velocity, 513
Spiral of Archimedes, 494
Spring constant, 429
Square of a binomial, 39
Square error, 217
Square matrix, 566
Square root, 24–25
Square root, principal, 25
Square root method, 83
Square of a sum, 48
Standard basis vectors, 512
Standard forms for:
    circles, 131
    complex numbers, 94
    ellipses, 455, 457
    hyperbolas, 462, 465
    parabolas, 176, 449, 450
    polynomials, 37
Standard position of an angle, 320
Statuary hall, 459
Straight angle, 324
Strict inequality, 12
Subset of a set, 2
Substitution:
    back, 529–530
    method of, 529
Subtraction:
    of complex numbers, 95
    of matrices, 572
    of vectors, 510

Sum:
    of complex numbers, 95
    of functions, 193
    of a geometric series, 638–639
    of matrices, 570
    of square errors, 217
    of two cubes, 42
    of vectors, 510, 512
Sum trigonometric formulas:
    for cosine, 380
    for sine, 382
    for tangent, 383
Summation index, 217
Summation notation:
    definition, 217
    index of, 217
    properties of, 636
Supplementary angle, 324
Symmetry of a graph:
    of a function, 165
    in polar coordinates, 494
    in rectangular coordinates, 135
    tests for, 136, 166, 494
Sylvester, James Joseph, 565
Synthetic division, 238–239
System of equations:
    consistent, 528–529, 530
    elimination operations, 531
    equivalent, 528
    homogeneous, 534
    inconsistent, 529, 530
    linear, 528
    method of elimination, 531
    method of substitution, 529
    nonlinear, 537
    solving, by determinants, 615, 616
    solving, by matrices, 599–603, 608–609
Systems of inequalities:
    linear, 550
    nonlinear, 551–552

T

Tangent:
    cycle, 373
    definition of, 329, 340, 355
    difference formula, 383
    domain of tangent function, 372
    double-angle formula, 389
    graph of tangent function, 373, 374
    half-angle formula, 389
    inverse of, 393
    period of tangent function, 373, 375
    range of tangent function, 372
    signs of, 372
    sum formula, 383
Tangent to an axis, 160
Tangent line to a circle, 145
Tangent line to an ellipse, 542
Tautochrone, 480
Telescoping series, 649
Term(s):
    of a polynomial, 37
    of a sequence, 628

Terminal point:
    of a parameterized curve, 477
    of a vector, 508
Terminal side of an angle, 320
Terminating decimals, 4
Test point, 549–550
Tests for symmetry of a graph, 136, 167, 494
Train and fly problem, 650–651
Transitivity property, 12
Transpose of a matrix, 568
Transverse axis of a hyperbola:
    definition of, 462
    length of, 462
Tree diagram, 661
Triangle(s):
    oblique, 425
    Pascal's, 656
    right, 324
    solution of, 410
Triangle inequality, 14
Trichotomy law, 12
Trigonometric equations, 398
Trigonometric form:
    of a complex number, 432
    of a product of complex numbers, 434
    of a quotient of complex numbers, 434
    of a vector, 513
Trigonometric function(s):
    of acute angles, 329
    circular, 355
    cycle of, 361–362, 364, 373, 374
    of a general angle, 339–340
    graphs of, 361–367, 373–374
    inverses of, 390, 392, 393, 395, 396
    period of, 357, 365, 366–367, 373, 375
    phase shift for, 366–367
    of a real number, 354–355
Trigonometric identities:
    double-angle, 384
    difference, 380, 382, 383
    half-angle, 384
    Pythagorean, 333, 342, 355, 379
    quotient, 330
    reciprocal, 330
    sum, 380, 382, 383
Trinomial, 37
Trivial solution, 534
Turning point of a graph, 181, 229
2-space, 124

U

Union of sets, 2
Unit, imaginary, 93, 242
Unit circle, 354
Unit vector, 512
Universal gravitation constant, 149

V

Value:
    absolute, 12
    of a function, 156
    future, 305
    present, 310

Variable:
    definition of, 36
    dependent, 156
    domain of, 36
    independent, 156
Variation:
    combined, 148
    direct, 146, 147
    inverse, 147
    joint, 148
Vector(s):
    angle between, 518
    component of a vector on another, 519
    component form of a vector, 509
    components of, 509
    difference of, 510, 512
    definition of, 508
    direction angle of, 509
    displacement, 508
    dot product of, 516
    equality of, 509
    force, 513–514
    geometric interpretations of, 511
    horizontal component of, 512
    i, 512
    initial point of, 508
    inner product of, 526
    j, 512
    linear combination of, 512
    magnitude of, 509
    negative of, 510
    normalization, 512
    operations on, 510
    orthogonal, 518
    perpendicular, 518
    position, 509
    projection of a vector onto another, 520
    properties of, 510
    scalar multiple of, 510, 512
    standard basis, 512
    sum of, 510, 512
    terminal point of, 508
    trigonometric form of, 513
    unit, 512
    vertical component of, 512
    zero, 509
Verhulst, P.F., 301
Vertex (vertices):
    of an angle, 320
    of a convex set, 555
    of an ellipse, 456
    of a hyperbola, 462
    of a parabola, 176, 448
Vertical asymptotes:
    definition of, 263
    for logarithmic functions, 288
    for rational functions, 264, 265
    for trigonometric functions, 373–374
Vertical component of a vector, 512
Vertical compression of a graph, 171
Vertical line, equation of, 142
Vertical shift of a graph, 168
Vertical stretch of a graph, 171
Vertical-line test for a function, 159
Viète, François, 1

W

Whispering gallery, 459
Width, focal:
    of an ellipse, 461
    of a hyperbola, 469
    of a parabola, 454
Witch of Agnesi, 483
Word problems:
    age, 73
    investment, 73
    mixture, 75
    rate, 74
    work, 76
Work as a dot product, 520–521

X

x-axis:
    definition of, 124
    tangent to, 160
x-coordinate, 124
x-intercept, 134
xy-plane, 124

Y

y-axis, 124
y-coordinate, 123
y-intercept, 134, 177

Z

Zero(s):
    bisection method for approximating, 258
    complex, 242
    of a function, 160, 242
    of multiplicity m, 230, 243
    of a polynomial, 111, 243
    real, 250
    rational, 251–252
    repeated, 230
    simple, 230, 243
Zero factor property of real numbers, 6, 82
Zero matrix, 571
Zero polynomial, 226, 261
Zero solution, 509
Zero vector, 509
z-plane, 432

Credits

Title and Half title page © William James Warren/Science Faction/Corbis

Contents
page vii (top) © Doug Pahael/ShutterStock, Inc.; **page vii** (bottom) Courtesy of Joe Ross; **page viii** (top) © Dainis Derics/ShutterStock, Inc.; **page viii** (second from top) © olly/ShutterStock, Inc.; **page viii** (second from bottom) © TebNad/ShutterStock, Inc.; **page viii** (bottom) © Tad Denson/ShutterStock, Inc.; **page ix** (top) © Sergey Shandin/ShutterStock, Inc.; **page ix** (middle) © Gabi Moisa/ShutterStock, Inc.; **page ix** (bottom) © Mikhail Khusid/ShutterStock, Inc.; **page x** (top) Courtesy of NASA/JPL; **page x** (second from top) © Michael Monahan/ShutterStock, Inc.; **page x** (second from bottom) © Andresr/ShutterStock, Inc.; **page x** (bottom) © Skocko/ShutterStock, Inc.; **page xi** © Andrea Danti/ShutterStock, Inc.; **page xii** © William James Warren/Science Faction/Corbis

Review chapter
page 1 © Doug Pahael/ShutterStock, Inc.; **page 4** © Toponium/ShutterStock, Inc.; **page 8** © Inmagine/Alamy Images; **page 10** © The Art Gallery Collection/Alamy Images; **page 21** © Lisa F. Young/ShutterStock, Inc.; **page 23** © Stephen Girimont/ShutterStock, Inc.; **page 28** © Luca Oleastri/Dreamstime.com; **pages 29, 43** © Yuri Arcurs/ShutterStock, Inc.

Chapter 1
page 65 Courtesy of Joe Ross; **page 78** © lighpoet/ShutterStock, Inc.; **page 87** Courtesy of the National Library of Medicine; **page 89** © Tatiana Belova/ShutterStock, Inc.; **page 114** © michaeljung/ShutterStock, Inc.; **page 121** Courtesy of Joe Ross; **page 122** © jokerpro/ShutterStock, Inc.

Chapter 2
page 123 © Dainis Derics/ShutterStock, Inc.; 143, © PhotoSmart/ShutterStock, Inc.; **page 150** (top) © S.Borisov/ShutterStock, Inc.; **page 150** (bottom) © Michael Onisiforou/ShutterStock, Inc.

Chapter 3
page 155, olly/ShutterStock, Inc.; **page 156** © Noah Strycker/ShutterStock, Inc.; **page 161** © ZTS/ShutterStock, Inc.; **page 169** Courtesy of Joanna Lee; **page 184** © John Lund/age fotostock; **page 196** Courtesy of the U.S. Navy; **page 212** © Lisa F. Young/ShutterStock, Inc.; **page 215** © Lev Mel/ShutterStock, Inc.

Chapter 4
page 225 © TebNad/ShutterStock, Inc.; **page 251** Courtesy of Library of Congress, Prints & Photographs Division [reproduction number cph.3c00653]; **page 271** © Pindyurin Vasily/ShutterStock, Inc.

Chapter 5
page 279 © Tad Denson/ShutterStock, Inc.; **page 291** © Stock4B GmbH/Alamy Images; **page 301** (top) Courtesy of Janice Haney Carr/CDC; **page 301** (middle, bottom) © National Library of Medicine; **page 303** (top) © Jones & Bartlett Learning. Photographed by Kimberly Potvin; **page 303** (middle) Jules le Baron, courtesy AIP Emilio Segre Visual Archives; **page 303** (bottom) Courtesy of the Israel Antiquities Authority; **page 304** © Johanna Goodyear/ShutterStock, Inc.; **page 305** Courtesy of the Archives, California Institute of Technology/Photo by James McClanahan; **page 306** Courtesy of The Carlsberg Group; **page 308** (left) © Robert Harding Picture Library Ltd./Alamy Images; **page 308** (right) © Targa/age fotostock; **page 309** (top) Courtesy of the South Tyrol Museum of Archaeology—www.iceman.it; **page 309** (middle) © Ron Hilton/ShutterStock, Inc.;

page 309 (bottom) © Thomas Weißenfels/ShutterStock, Inc.; **page 310** Courtesy of USGS; **page 311** © Edi Engeler, Keystone/AP Photos; **page 313** (top) © Tad Denson/ShutterStock, Inc.; **page 313** (bottom) © Caleb969/Dreamstime.com

Chapter 6
page 319 © Sergey Shandin/ShutterStock, Inc.; **page 327** Courtesy of Mariner 10, Astrogeology Team, and USGS; **page 328** © Monkey Business Images/ShutterStock, Inc.; **page 347**, © Ron Buskirk/Alamy Images; **page 349** © Doug James/ShutterStock, Inc.

Chapter 7
page 353 © Gabi Moisa/ShutterStock, Inc.; **page 385** © Fernando Batista/ShutterStock, Inc., **page 397** Courtesy of National Park Service;

Chapter 8
page 409 © Mikhail Khusid/ShutterStock, Inc.; **page 416** (top) © Paul Fries/ShutterStock, Inc.; **page 416** (bottom) © Peter Elvidge/ShutterStock, Inc.; **page 417** © Mary Lane/ShutterStock, Inc.; **page 425** © Roman Sigaev/ShutterStock, Inc.

Chapter 9
pages 447, 448 Courtesy of NASA/JPL; **page 451** (top) © Christos Georghiou/ShutterStock, Inc.; **page 451** (bottom) Courtesy of NASA/JPL; **page 452** (top) © dragon_fang/ShutterStock, Inc.; **page 452** (middle) © where-@tiscali.it/ShutterStock, Inc.; **page 452** (bottom) © Corbis; **page 459** © Brand X Pictures/Alamy Images; **page 480** © Corbis; **page 481** Courtesy of Robert Guy/National Cancer Institute; **page 486** © David Page/Alamy Images

Chapter 10
page 487 © Michael Monahan/ShutterStock, Inc., **page 494** From Wilson Bentley, The Snowflake Man, Annual Summary of the Monthly Weather Review, 1902. Photo courtesy of National Weather Service Collection/NOAA; **page 499** © Dmitry Nikolaev/ShutterStock, Inc.; **page 505** Courtesy of Mariner 10, Astrogeology Team, and USGS; **page 507** © Datacraft/age fotostock; **page 514** © Elnur/ShutterStock, Inc.

Chapter 11
page 527 © Andresr/ShutterStock, Inc.; **page 547** © Corbis; **page 560** (top) © Robert Kneschke/ShutterStock, Inc.; **page 560** (bottom) © Steve Bower/ShutterStock, Inc.

Chapter 12
page 565 © Skocko/ShutterStock, Inc.; **page 576** © Africa Studio/ShutterStock, Inc.; **page 612** © Matty Symons/ShutterStock, Inc.

Chapter 13
page 627 © Andrea Danti/ShutterStock, Inc.; **page 634** Courtesy of Liam Quin; **page 641** (top) © Melvyn Longhurst/Alamy Images; **page 641** (bottom) © Fesus Robert/ShutterStock, Inc.; **page 648** © Sofos Design/ShutterStock, Inc.; **page 652** © Monkey Business Images/ShutterStock, Inc.; **page 664** © Sergielev/ShutterStock, Inc.; **page 667** (top) © juan carlos tinjaca/ShutterStock, Inc.; **page 667** (bottom) © Tony Rolls/Alamy Images; **page 668** Courtesy of Pressman Toy Corporation; **page 669** (top) © James Steidl/ShutterStock, Inc.; **page 669** (bottom) © Gjermund Alsos/ShutterStock, Inc.; **page 670** © Sebastian Kaulitzki/ShutterStock, Inc.; **page 674** © xiver/ShutterStock, Inc.; **page 677** © Monkey Business Images/ShutterStock, Inc.

Unless otherwise indicated, all photographs and illustrations are under copyright of Jones & Bartlett Learning.